World Geomorphological Landscapes

Series Editor

Piotr Migoń, Institute of Geography and Regional Development, University of Wrocław, Wrocław, Poland

Geomorphology – 'the Science of Scenery' – is a part of Earth Sciences that focuses on the scientific study of landforms, their assemblages, and surface and subsurface processes that moulded them in the past and that change them today. Shapes of landforms and regularities of their spatial distribution, their origin, evolution, and ages are the subject of geomorphology. Geomorphology is also a science of considerable practical importance since many geomorphic processes occur so suddenly and unexpectedly, and with such a force, that they pose significant hazards to human populations. Landforms and landscapes vary enormously across the Earth, from high mountains to endless plains. At a smaller scale, Nature often surprises us creating shapes which look improbable. Many geomorphological landscapes are so immensely beautiful that they received the highest possible recognition – they hold the status of World Heritage properties. Apart from often being immensely scenic, landscapes tell stories which not uncommonly can be traced back in time for millions of years and include unique events. This international book series will be a scientific library of monographs that present and explain physical landscapes across the globe, focusing on both representative and uniquely spectacular examples. Each book contains details on geomorphology of a particular country (i.e. The Geomorphological Landscapes of France, The Geomorphological Landscapes of Italy, The Geomorphological Landscapes of India) or a geographically coherent region. The content is divided into two parts. Part one contains the necessary background about geology and tectonic framework, past and present climate, geographical regions, and long-term geomorphological history. The core of each book is however succinct presentation of key geomorphological localities (landscapes) and it is envisaged that the number of such studies will generally vary from 20 to 30. There is additional scope for discussing issues of geomorphological heritage and suggesting itineraries to visit the most important sites. The series provides a unique reference source not only for geomorphologists, but all Earth scientists, geographers, and conservationists. It complements the existing reference books in geomorphology which focus on specific themes rather than regions or localities and fills a growing gap between poorly accessible regional studies, often in national languages, and papers in international journals which put major emphasis on understanding processes rather than particular landscapes. The World Geomorphological Landscapes series is a peer-reviewed series which contains single and multi-authored books as well as edited volumes.

World Geomorphological Landscapes – now indexed in Scopus® !

Piotr Migoń · Kacper Jancewicz
Editors

Landscapes and Landforms of Poland

Springer

Editors
Piotr Migoń
Institute of Geography and Regional Development
University of Wrocław
Wrocław, Poland

Kacper Jancewicz
Institute of Geography and Regional Development
University of Wrocław
Wrocław, Poland

ISSN 2213-2090 ISSN 2213-2104 (electronic)
World Geomorphological Landscapes
ISBN 978-3-031-45761-6 ISBN 978-3-031-45762-3 (eBook)
https://doi.org/10.1007/978-3-031-45762-3

This Springer imprint is published by the registered company Springer Nature Switzerland AG
The registered company address is: Gewerbestrasse 11, 6330 Cham, Switzerland

Paper in this product is recyclable.

Series Editor's Preface

Landforms and landscapes vary enormously across the Earth, from high mountains to endless plains. At a smaller scale, Nature often surprises us by creating shapes which look improbable. Many physical landscapes are so immensely beautiful that they have received the highest possible recognition—they hold the status of World Heritage properties. Apart from often being immensely scenic, landscapes tell stories which not uncommonly can be traced back in time for tens of million years and include unique events. In addition, many landscapes owe their appearance and harmony not solely to natural forces. For centuries, or even millennia, they have been shaped by humans who modified hillslopes, river courses and coastlines, and erected structures which often blend with the natural landforms to form inseparable entities.

These landscapes are studied by Geomorphology—'the Science of Scenery'—a part of Earth Sciences that focuses on landforms, their assemblages, the surface and subsurface processes that moulded them in the past and that change them today. The shapes of landforms and the regularities of their spatial distribution, their origin, evolution and ages are the subject of research. Geomorphology is also a science of considerable practical importance since many geomorphic processes occur so suddenly and unexpectedly, and with such a force, that they pose significant hazards to human populations and not uncommonly result in considerable damage or even casualties.

To show the importance of geomorphology in understanding the landscape, and to present the beauty and diversity of the geomorphological sceneries across the world, we have launched a new book series *World Geomorphological Landscapes*. It aims to be a scientific library of monographs that present and explain physical landscapes, focusing on both representative and uniquely spectacular examples. Each book will contain details on geomorphology of a particular country or a geographically coherent region. This volume covers Poland—the home country of the series editor. Usually seen through the lens of its vast lowlands, it is in fact far more diverse and includes the whole range of genetically different geomorphological landscapes, from high mountains in the south to the coast in the north. Thus, beside landforms produced by continental ice sheets in the Pleistocene, one will find various examples of rock-controlled denudational relief, impressive landslides, karst on different types of bedrock, loess plateaus, legacy of permafrost and periglacial processes, inland and coastal dunes, cliffs and sandy beaches, and last but not least, impressive evidence of human impact on landforms. More than 35 regional studies contained in this volume offer insights into the geomorphological diversity of Poland.

The *World Geomorphological Landscapes* series is produced under the scientific patronage of the International Association of Geomorphologists—a society that brings together geomorphologists from all around the world. The IAG was established in 1989 and is an independent scientific association affiliated with the International Geographical Union and the International Union of Geological Sciences. Among its main aims are to promote geomorphology and to foster dissemination of geomorphological knowledge. I believe that this lavishly illustrated series, which sticks to the scientific rigour, is the most appropriate means to fulfil these aims and to serve the geoscientific community.

Finally, I want to express my thanks to the co-editor of this volume, Dr. Kacper Jancewicz, for his expert assessment and advice regarding maps and diagrams included into separate chapters, and last-minute help with some of the graphics. I am grateful to all individual contributors (73 in total!) who accepted invitation to contribute and agreed to add the task of writing chapters to their busy agendas and delivered such interesting stories to read.

Wrocław, Poland

Series Editor
Piotr Migoń

Acknowledgements

While editing academic books, ensuring quality is among major concerns. Since the World Geomorphological Landscapes series is intended to serve global geomorphological community and anyone interested in landform diversity, additional presentation challenges emerge. Therefore, to offer the best possible final product, each chapter was reviewed by two persons (only in a few case one reviewer was involved) and, independently, by the editors. Here we want to express our gratefulness to the many reviewers, some of whom are also contributors to this volume, but other are not, for providing most helpful reviews and comments. These are, in alphabetical order: Anita Bernatek-Jakiel, Mirosław Błaszkiewicz, Maciej Dąbski, Radosław Dobrowolski, Filip Duszyński, Marek Ewertowski, Jacek Forysiak, Piotr Gębica, Zdzisław Jary, Andrzej Kacprzak, Marek Kasprzak, Piotr Kittel, Piotr Kłapyta, Kazimierz Krzemień, Agnieszka Latocha-Wites, Tomasz Łabuz, Adam Łajczak, Ireneusz Malik, Włodzimierz Margielewski, Małgorzata Mazurek, Anna Orłowska, Tomáš Pánek, Paweł Prokop, Zofia Rączkowska, Joanna Rotnicka, Artur Sobczyk, Andrzej Traczyk, Andrzej Tyc, Wojciech Tylmann, Jan Urban, Piotr Weckwerth, Grzegorz Wierzbicki, Barbara Woronko, Wojciech Wysota, Joanna Zawiejska, Wojciech Zgłobicki and Tomasz Zieliński.

Introduction

Poland is a country of remarkable geomorphological diversity, even if the distant look at the hypsometric map that shows predominant lowland terrain may suggest otherwise, and has a long history of geomorphological research. Its territory stretches from high mountains in the south through medium–high mountain ranges, uplands, plains and hilly lakelands to the Baltic Sea coast in the north, offering opportunities to investigate very diverse aspects of geomorphology. More than 100 years ago, a Polish scientist, Walery Łoziński, introduced the term 'periglacial' to emphasize specific lines of landform evolution in cold, but non-glaciated regions around the peripheries of large ice sheets, which was then widely adopted and has become a part of international geomorphological vocabulary. In the second half of the twentieth century, despite political and economic barriers, Polish geomorphologists made substantial contributions to global geomorphology, particularly in the fields of detailed geomorphological mapping (Mieczysław Klimaszewski), quantification of slope processes (Alfred Jahn), landform change due to extreme events and valley floor evolution in the temperate climatic zone (Leszek Starkel). Studies of high-mountain morphology on the example of the Tatra Mountains, glacial morphogenesis connected with continental ice sheet advances and decays, periglacial structures and landforms, and inland dunes were distinctive hallmarks of Polish geomorphology of that time. Contemporary geomorphological processes were the subject of considerable interest and numerous field research stations were established to support long-term measurements. A distinctive feature of Polish geomorphology, logically emerging from the history of morphological evolution of the country, was its almost inseparable link with Quaternary geology and palaeoecology. Many geomorphologists were also Quaternary geologists and vice versa. Strongly represented were studies of glacial and aeolian deposits. Among the latter, loess, widespread in the mountain forelands and uplands, has been the traditional subject of research. Polish geomorphologists were also active abroad, particularly in the Arctic (Spitsbergen, Iceland), but also in Antarctica (King George Island), India, Mongolia and occasionally in other European and non-European countries.

Nowadays, when most of the old obstacles vanished, geomorphology in Poland continues to develop, involving new generations of researchers, utilizing all modern tools and techniques available and taking all advantages of international collaboration. High-resolution LiDAR data are now freely available for the entire country and are invaluable for geomorphologists, helping to both recognize landform patterns previously overlooked and to revisit former concepts using objective, precise and quantitative information about land configuration. Thus, traditional themes have not been abandoned, even though some are less exposed than before (e.g. long-term landform evolution over the Cenozoic timescale), but the portfolio of Polish geomorphology has substantially increased. Among these new trends are applied geomorphological studies, especially in the fields of fluvial geomorphology and landslide mapping, geomorphometric approach to landforms and landscapes, use of cosmogenic isotopes to constrain landform evolution, and studies focused on direct and indirect human impact on landform evolution. Geomorphology has also become more transdisciplinary, benefitting from collaboration with structural geologists, sedimentologists, soil scientists, ecologists, archaeologists and engineers. Last but not least, heritage values of landforms are increasingly recognized, and more and more Polish geomorphologists are involved in geoheritage studies, geoconservation, and popularization of geomorphology for the general public.

From an organizational point of view, the Association of Polish Geomorphologists (SGP) represents the geomorphological community and includes members with both geographical and geological background. The SGP was founded in 1991, with Stefan Kozarski as the first President, just two years after the International Association of Geomorphologists (IAG) was formally established and has been its strong national scientific member ever since. Currently, it has around 200 members, many of whom are early career scientists. Major geomorphological conferences of the SGP are held every three years, with many thematic meetings and workshops organized in the meantime. *Landform Analysis* is an international journal published by the SGP since 1991.

These various characteristics of Polish geomorphology are captured in this volume, part of the World Geomorphological Landscapes series. Following the general format adopted for the series, it consists of two parts. In Part I, six chapters provide an overview and set the necessary context for site-specific studies contained in the second part. As the foundations of morphology reside in geology, the opening Chap. 1 by Stanisław Mazur outlines the complex geological evolution of the territory of Poland, whose roots lie in the Precambrian. This is followed by an overview of Quaternary history and palaeogeography of Poland (Zdzisław Jary and Leszek Marks)—a period decisive for the contemporary morphology of most of the country due to almost complete coverage by inland ice sheets in the Middle Pleistocene. In Chap. 3, Marek Błaś and Hanna Ojrzyńska present the main characteristics of the contemporary climatic conditions, particularly emphasizing those relevant for geomorphic processes. With the stage set, in Chap. 4 Piotr Migoń and Kacper Jancewicz introduce geomorphological diversity of Poland, with the framework provided by latitudinal arrangement of distinctive morphogenetic belts. This framework is then adopted in part II of the volume and guides the structure of regional studies. A special chapter (Chap. 5) by Leon Andrzejewski et al. is devoted to rivers and valley floors, the key components of denudational systems. Part I is closed by a look at landscapes and landforms of Poland from the perspective of geoheritage and geoconservation (Piotr Migoń).

Part II contains nearly 40 regional studies, in which either selected, geomorphologically distinctive regions are presented, or the focus is on specific landforms distributed over wider areas (Fig. 1). We begin in the south-western part of Poland, in the medium-high mountains of the Sudetes, which belong to the Variscan system of Central Europe, and their immediate foreland. Six chapters present diverse rock-controlled landscapes from this region: predominantly granitic massif of the Karkonosze, the highest in the Sudetes, and the adjacent Jelenia Góra Basin (Chap. 7, by Marek Kasprzak), the 'Land of Extinct Volcanoes'—an aspiring UNESCO Global Geopark and an area dominated by volcanic rocks of various ages (Chap. 8, by Piotr Migoń), landslide terrain of the Kamienne Mountains (Chap. 9, by Piotr Migoń), unique sandstone tableland of the Stołowe Mountains with its spectacular rock labyrinths (Chap. 10, by Filip Duszyński), the Polish part of the East Sudetes, where the focus is on various anthropic signatures (Chap. 11, by Agnieszka Latocha-Wites), and the solitary gabbro inselberg of Mt. Ślęża in the Sudetic Foreland, with its enigmatic blockfields and blockstreams (Chap. 12, by Andrzej Traczyk).

The next group of chapters shows geomorphological diversity of the Carpathians. The highest part of the Outer (Flysch) Carpathians, the Babia Góra massif, is presented by Adam Łajczak in Chap. 13. Chapters 14 and 15 focus on the only high-mountain terrain in Poland, the Tatra Mountains. Whereas Piotr Kłapyta and Jerzy Zasadni (Chap. 14) provide a general overview and look more closely at glacial landforms and history, Jacek Szczygieł (Chap. 15) introduces karst, particularly caves, which are the longest and deepest in Poland. To the north of the Tatra Mountains extends a large intramontane basin, whose complex morphology is presented by Joanna Zawiejska et al. (Chap. 16). In turn, in Chap. 17 Włodzimierz Margielewski and Leszek Jankowski provide insights into the origin of the Dunajec River Gorge across the Pieniny Mountains—one of the most recognized tourist attractions of

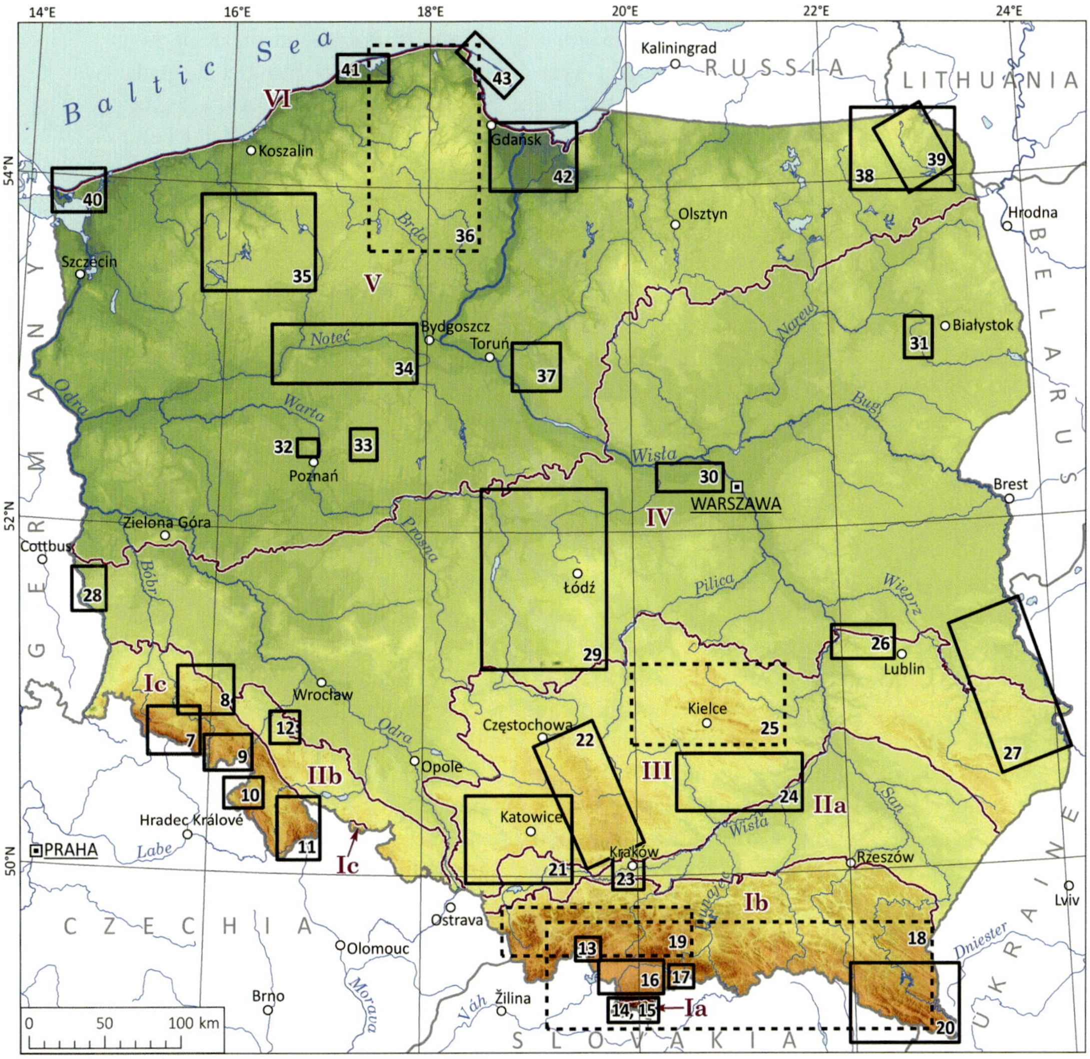

Fig. 1 Location of areas covered by regional studies, in relation to morphogenetic (landscape) belts of Poland (I–VI). Numbers inside/next to black rectangles refer to chapter numbers. Broken lines indicate that the respective chapters are not comprehensive presentations of a region, but focus on specific landforms. Mountains: Ia—high mountains (Tatra Mts.); Ib—medium-high mountains of the Alpine system (Beskidy Mountains); Ic—medium-high mountains of the Variscan system (Sudetes). Fore-mountain areas: IIa—Sub-Carpathian basins; IIb—Sudetic Foreland. III—uplands. IV—'old glacial relief' of central Poland. V—'young glacial relief' of northern Poland. VI—coastal zone

southern Poland. Chapters 18 and 19 adopt a different approach and focus on themes rather than regions, but still within the Outer (Flysch) Carpathians. In the former, Włodzimierz Margielewski and Leszek Jankowski use a number of examples to show how diverse in terms of age and overarching controls are landslides—perhaps the most distinctive medium-size landforms in the region. In the latter, Włodzimierz Margielewski and Jan Urban present several non-karstic caves, emphasizing their close connection with gravity-induced deformations of flysch slopes. Finally, Anita Bernatek-Jakiel (Chap. 20) presents the Bieszczady Mountains—the easternmost part of the Polish Carpathians.

Seven chapters illustrate various geomorphological aspects of the upland belt, going from west to east. The westernmost part of this morphogenetic belt is most altered by human activities, related mostly to coal mining, resulting in specific landscapes presented by Renata Dulias (Chap. 21). Human impact is also the main theme in Chap. 23, focused on urban geomorphology of the city of Kraków (by Adam Łajczak and Roksana Zarychta). The upland belt hosts various karst landscapes, developed upon different lithologies, and these are subject to three chapters. The most extensive limestone karst area in Poland, in the Kraków-Częstochowa Upland, is presented by Andrzej Tyc (Chap. 22), whereas Jan Urban et al. (Chap. 24) provide an overview of gypsum karst in the Nida Basin. Finally, chalk karst of easternmost Poland, extending over the upland and the adjacent part of the lowland belt, is shown by Radosław Dobrowolski in Chap. 27. The last two chapters in this part focus on sandstone crags, which add to the geodiversity of the Kielce Upland (Chap. 25, by Jan Urban), and loess gullies (Chap. 26, by Jan Rodzik et al.).

The central lowland belt is geomorphologically least diverse and hence, despite large areal coverage, is illustrated by four studies only. Jacek Koźma and Piotr Migoń (Chap. 28) present the unique landscape of the Muskau Arch push moraine, considerably re-modelled by protracted mining activity. Jan Goździk and Jacek Forysiak (Chap. 29) reveal the late Quaternary history of the central Poland around the city of Łódź and demonstrate how long-term periglacial morphogenesis turned a glacial landscape into a much subdued landscape of today. The periglacial theme is continued by Maciej Dłużewski and Barbara Woronko (Chap. 30), who focus on one of the largest complexes of inland dunes in Poland, in the Kampinos National Park next to Warsaw. A study by Tomasz Zieliński (Chap. 31) from eastern Poland provides insights into a unique fluvial landscape of an anastomosing Narew River—one of very few river reaches left in natural conditions.

The group of chapters presenting the hilly 'young glacial' relief of northern Poland begins with a study of a truly unique feature—a cluster of mid-Holocene impact craters set into a morainic landscape of the last glaciation near the city of Poznań (Chap. 32, by Witold Szczuciński and Wojciech Włodarski). The surroundings of Poznań were the cradle of the Polish statehood more than 1000 years ago and abound in the respective archaeological evidence, addressed in the geoarchaeological story by Lidia Żuk et al. (Chap. 33). Several types of large landforms genetically connected with glacial morphogenesis are characteristic for northern Poland and are presented in a few separate stories. These are ice-marginal valleys (pradolinas) (Chap. 34, by Leszek Kasprzak et al.), lake-filled tunnel valleys moulded by subglacial discharges (Chap. 36, by Mirosław Błaszkiewicz and Karol Tylmann), intriguing glacial curvilineations, previously interpreted as drumlins (Chap. 37, by Wojciech Wysota et al.) and landform assemblages produced by cataclysmic outburst floods (Chap. 39, by Piotr Weckwerth and Wojciech Wysota). In turn, Chap. 35 by Małgorzata Mazurek et al. and Chap. 38 by Ewa Smolska et al. provide more regional overviews of landform diversity, not only of glacial origin, from two representative lakelands in northern Poland, in the west and east, respectively.

The volume concludes with chapters presenting the coastal belt. Zbigniew Zwoliński et al. (Chap. 40) show the diversity of Wolin Island in the extreme west, typified by the highest coastal cliffs in Poland. However, the majority of the Polish coastline is made of sandy beaches backed by coastal dunes, and the arguably most famous section, in the Słowiński National Park, is presented by Joanna Rotnicka and Krzysztof R. Borówka (Chap. 41). In Chap. 42, Damian Moskalewicz and Piotr Paweł Woźniak reveal the fascinating story behind the landscape of the Wisła (Vistula) River delta, part of which lies below the sea level. Last by not least, Radosław Wróblewski et al. (Chap. 43) explain how the most distinctive littoral landform in Poland, the Hel Peninsula, evolved alongside the Holocene history of the Baltic Sea.

Piotr Migoń
Kacper Jancewicz

Contents

Editors and Contributors

About the Editors

Piotr Migoń is Professor of Geography at the University of Wrocław and former Director of the Institute of Geography and Regional Development. His main research interests are rock control in geomorphology, especially in sandstone and granite areas, mass movements in mountain terrain, geoheritage and geotourism. The Sudetes Mountains in Central Europe is his main research area, but he was also involved in projects in various other European countries (Czechia, Germany, Great Britain, Sweden, Portugal, Spain), in the USA, Mexico, Brazil, China, Jordan and Namibia. He used to be Secretary of the International Association of Geomorphologists (1997–2001), its Vice-President (2009–2013) and member of Executive Committee (2013–2022); he was also President of the Association of Polish Geomorphologists (2005–2008). He is the series editor of 'World Geomorphological Landscapes' series and member of editorial board of *Geomorphology* journal.

Kacper Jancewicz holds Ph.D. degree in Earth Sciences and works as a senior specialist in the Department of Geomorphology, Institute of Geography and Regional Development, University of Wrocław, Poland. His main research interests focus on various LiDAR DEM applications in geomorphological studies, geomorphometric analyses and digital geomorphological mapping in GIS environment. His publication record includes papers in *Geomorphology*, *Catena*, *Transactions in GIS* and *Journal of Maps*.

Contributors

Aleksander Adamczyk State Water Holding Polish Waters, Regional Water Management Authority in Bydgoszcz, Bydgoszcz, Poland

Leon Andrzejewski Department of Geomorphology and Quaternary Paleogeography, Nicolaus Copernicus University, Toruń, Poland

Anita Bernatek-Jakiel Institute of Geography and Spatial Management, Jagiellonian University, Kraków, Poland

Marek Błaś Institute of Geography and Regional Development, University of Wrocław, Wrocław, Poland

Mirosław Błaszkiewicz Department of Environmental Resources and Geohazards, Institute of Geography and Spatial Organization, Polish Academy of Sciences, Toruń, Poland

Krzysztof R. Borówka Institute of Marine and Environmental Sciences, University of Szczecin, Szczecin, Poland

Dorota Chmielowska-Michalak Department of Physical Geography, University of National Education Commission, Krakow, Kraków, Poland

Anna Chwalik-Borowiec The Świętokrzyskie and Nadnidziańskie Landscape Parks Complex, Kije, Poland

Maciej Dłużewski Faculty of Geography and Regional Studies, University of Warsaw, Warsaw, Poland

Radosław Dobrowolski Institute of Earth and Environmental Sciences, Maria Curie-Skłodowska University, Lublin, Poland

Renata Dulias Institute of Earth Sciences, University of Silesia in Katowice, Sosnowiec, Poland

Filip Duszyński Institute of Geography and Regional Development, University of Wrocław, Wrocław, Poland

Marek W. Ewertowski Faculty of Geographical and Geological Sciences, Adam Mickiewicz University, Poznań, Poland

Jacek Forysiak Department of Geology and Geomorphology, University of Łódź, Lodz, Poland

Grzegorz Gajek Institute of Earth and Environmental Sciences, Maria Curie-Skłodowska University, Lublin, Poland

Jan Goździk Department of Geology and Geomorphology, University of Łódź, Lodz, Poland

Kacper Jancewicz Institute of Geography and Regional Development, University of Wrocław, Wrocław, Poland

Leszek Jankowski Polish Geological Institute—National Research Institute, Kraków, Poland

Zdzisław Jary Institute of Geography and Regional Development, University of Wrocław, Wrocław, Poland

Jakub Kalita Faculty of Geographical and Geological Sciences, Adam Mickiewicz University, Poznań, Poland

Leszek Kasprzak Faculty of Geographical and Geological Sciences, Adam Mickiewicz University, Poznań, Poland

Marek Kasprzak Institute of Geography and Regional Development, University of Wrocław, Wrocław, Poland

Andrzej Kasza Świętokrzyski Speleoclub, Kielce, Poland

Piotr Kłapyta Institute of Geography and Spatial Management, Jagiellonian University, Kraków, Poland

Artur Komorowski Beskidy Caving Club, Połaniec, Poland

Andrzej Kostrzewski Institute of Geoecology and Geoinformation, Adam Mickiewicz University, Poznań, Poland

Andrzej Kowalczyk Muzeum Pierwszych Piastów na Lednicy, Lednogóra, Poland

Jacek Koźma Stowarzyszenie "Geopark Łuk Mużakowa", Łęknica, Poland

Kazimierz Krzemień Institute of Geography and Spatial Management, Jagiellonian University, Kraków, Poland

Józef Kukulak Department of Physical Geography, University of National Education Commission, Krakow, Kraków, Poland

Adam Łajczak Association of Polish Geomorphologists, Poznań, Poland

Agnieszka Latocha-Wites Institute of Geography and Regional Development, University of Wrocław, Wrocław, Poland

Włodzimierz Margielewski Institute of Nature Conservation, Polish Academy of Sciences, Kraków, Poland

Leszek Marks Faculty of Geology, University of Warsaw, Warsaw, Poland

Stanisław Mazur Institute of Geological Sciences, Polish Academy of Sciences, Kraków, Poland

Małgorzata Mazurek Institute of Geoecology and Geoinformation, Adam Mickiewicz University, Poznań, Poland

Piotr Migoń Institute of Geography and Regional Development, University of Wrocław, Wrocław, Poland

Damian Moskalewicz Department of Geomorphology and Quaternary Geology, University of Gdańsk, Gdańsk, Poland

Hanna Ojrzyńska Institute of Geography and Regional Development, University of Wrocław, Wrocław, Poland

Renata Paluszkiewicz Institute of Geoecology and Geoinformation, Adam Mickiewicz University, Poznań, Poland

Jan A. Piotrowski Department of Geoscience, Aarhus University, Aarhus C, Denmark

Katarzyna Pochocka-Szwarc Polish Geological Institute-National Research Institute, Warszawa, Poland

Jan Rodzik Institute of Earth and Environmental Sciences, Maria Curie-Skłodowska University, Lublin, Poland

Joanna Rotnicka Institute of Geology, Adam Mickiewicz University in Poznań, Poznań, Poland

Maria Rucińska Department of Geophysics, University of Gdańsk, Gdynia, Poland

Stanisław Rudowski Maritime Institute, Gdynia Maritime University, Gdańsk, Poland

Patryk Sitkiewicz Department of Geomorphology and Quaternary Geology, University of Gdańsk, Gdańsk, Poland

Ewa Smolska Faculty of Faculty of Geography and Regional Studies, University of Warsaw, Warszawa, Poland

Witold Szczuciński Geohazards Research Unit, Institute of Geology, Adam Mickiewicz University, Poznań, Poland

Jacek Szczygieł Institute of Earth Sciences, University of Silesia, Sosnowiec, Poland

Izabela Szuman Faculty of Geographical and Geological Sciences, Adam Mickiewicz University, Poznań, Poland

Piotr Szwarczewski Faculty of Faculty of Geography and Regional Studies, University of Warsaw, Warszawa, Poland

Aleksandra M. Tomczyk Faculty of Geographical and Geological Sciences, Adam Mickiewicz University, Poznań, Poland

Andrzej Traczyk Institute of Geography and Regional Development, University of Wrocław, Wrocław, Poland

Andrzej Tyc Institute of Earth Sciences, University of Silesia, Sosnowiec, Poland

Karol Tylmann Department of Geomorphology and Quaternary Geology, Faculty of Oceanography and Geography, University of Gdańsk, Gdansk, Poland

Jan Urban Institute of Nature Conservation, Polish Academy of Sciences, Kraków, Poland

Piotr Weckwerth Faculty of Earth Sciences and Geospatial Management, Nicolaus Copernicus University in Toruń, Toruń, Poland

Marcin Winowski Institute of Geoecology and Geoinformation, Adam Mickiewicz University, Poznań, Poland

Wojciech Włodarski Department of Palaeoenvironmental Research, Institute of Geology, Adam Mickiewicz University, Poznań, Poland

Barbara Woronko Faculty of Geology, University of Warsaw, Warsaw, Poland

Piotr Paweł Woźniak Department of Geomorphology and Quaternary Geology, University of Gdańsk, Gdańsk, Poland

Radosław Wróblewski Department of Geophysics, University of Gdańsk, Gdynia, Poland

Wojciech Wysota Faculty of Earth Sciences and Geospatial Management, Nicolaus Copernicus University in Toruń, Toruń, Poland

Bartłomiej Wyżga Department of Physical Geography, University of National Education Commission, Krakow, Kraków, Poland

Roksana Zarychta Department of Geodesy, Cartography and Spatial Management, Institute of Law, Economics and Administration, University of National Education Commission, Krakow, Kraków, Poland

Jerzy Zasadni Faculty of Geology, Geophysics and Environmental Protection, AGH University of Science and Technology, Kraków, Poland

Joanna Zawiejska Department of Physical Geography, University of National Education Commission, Krakow, Kraków, Poland

Wojciech Zgłobicki Institute of Earth and Environmental Sciences, Maria Curie-Skłodowska University, Lublin, Poland

Artur Zieliński Institute of Geography and Environmental Sciences, Jan Kochanowski University of Kielce, Kielce, Poland

Tomasz Zieliński Institute of Geology, Adam Mickiewicz University, Poznań, Poland

Lidia Żuk Faculty of Archaeology, Adam Mickiewicz University, Poznań, Poland

Zbigniew Zwoliński Institute of Geoecology and Geoinformation, Adam Mickiewicz University, Poznań, Poland

Part I

Physical Environment

1 An Outline of the Geology of Poland

Stanisław Mazur

Abstract

The complexity of Poland's geology is due to its location at the junction of three major geological provinces of Europe: (1) the East European Precambrian Platform, (2) the Palaeozoic Platform, and (3) the Carpathians. The East European Platform corresponds to an old and stable Precambrian craton, extending over the eastern half of the continent and mostly covered by weakly deformed Proterozoic to Phanerozoic sediments. Although the Precambrian Platform is approximately limited from the southwest by the Teisseyre-Tornquist Zone, the attenuated East European cratonic crust continues in southwesterly direction, underneath the Palaeozoic Platform of central and western Poland. The latter includes Ediacaran-Palaeozoic, pre-Permian sediments, partly folded and thrust faulted during the Caledonian and Variscan cycles. The Palaeozoic Platform also comprises accreted Variscan and Cadomian crustal domains of the Sudetes and the Upper Silesian Block, respectively. Soon after cessation of the Variscan orogeny, the Palaeozoic Platform was affected by early Permian continental rifting and following thermal subsidence in the Mesozoic. The Permian-Mesozoic basin was inverted in the Late Cretaceous-early Paleogene as a far-field effect of the collision between Northern Europe and the Alpine-Carpathian orogen. Both the East European Craton and the Palaeozoic Platform were overridden in the Paleogene-Miocene by the outer Western Carpathians fold-and-thrust belt. Further south, the Pieniny Klippen Belt represents a tectonic suture separating the East European Craton and the Palaeozoic Platform from tectonic elements derived from the southerly Adria Plate.

Keywords

East European Craton · Palaeozoic Platform · Carpathians · Sudetes · Upper Silesian Block · Polish Basin · Pieniny Klippen Belt

1.1 Introduction

Poland is located at the junction of three major geological domains of Europe: (1) the East European Precambrian Platform, (2) the Palaeozoic Platform, and (3) the Carpathians. Each of these domains is characterised by different age, geological evolution, and present-day structure. The East European Platform corresponds to old and stable Precambrian Eastern Europe, extending over the eastern half of the continent. The Palaeozoic Platform belongs to younger and mobile Western Europe and includes the crust accreted during the Caledonian and Variscan orogenies. The Carpathians represent a piece of the Alpine-Carpathian mountain belt formed due to collision between Eurasia and northward drifting terranes. The transition from the Precambrian to Palaeozoic Europe is entirely masked by younger sediments and, thus, it is revealed only by borehole and geophysical data. In contrast, the Carpathian belt, adjoining both the Precambrian and Palaeozoic Platforms from the south (Fig. 1.1), is well exposed at the surface.

Since the publication of Nowak's "Outline of the Tectonics of Poland" (1927), considered to be the first comprehensive and modern view of the geological structure of Poland, almost a century has passed. Since then, many concepts of geological structure and evolution have been proposed (e.g. Pożaryski 1956, 1974; Książkiewicz et al. 1965; Oberc 1967; Stupnicka 1989; Dadlez et al. 1994, 2000; Znosko 1998; Mizerski 2009; Żelaźniewicz et al. 2011; Nawrocki and Becker 2017; Narkiewicz 2021), the discussion of which is beyond the scope of this chapter. In general, these ideas, regardless of specific preferences of individual researchers, reflect a continuing evolution in

S. Mazur (✉)
Institute of Geological Sciences, Polish Academy of Sciences, Senacka 1, 31-002 Kraków, Poland
e-mail: ndmazur@cyf-kr.edu.pl

P. Migoń and K. Jancewicz (eds.), *Landscapes and Landforms of Poland*, World Geomorphological Landscapes, https://doi.org/10.1007/978-3-031-45762-3_1

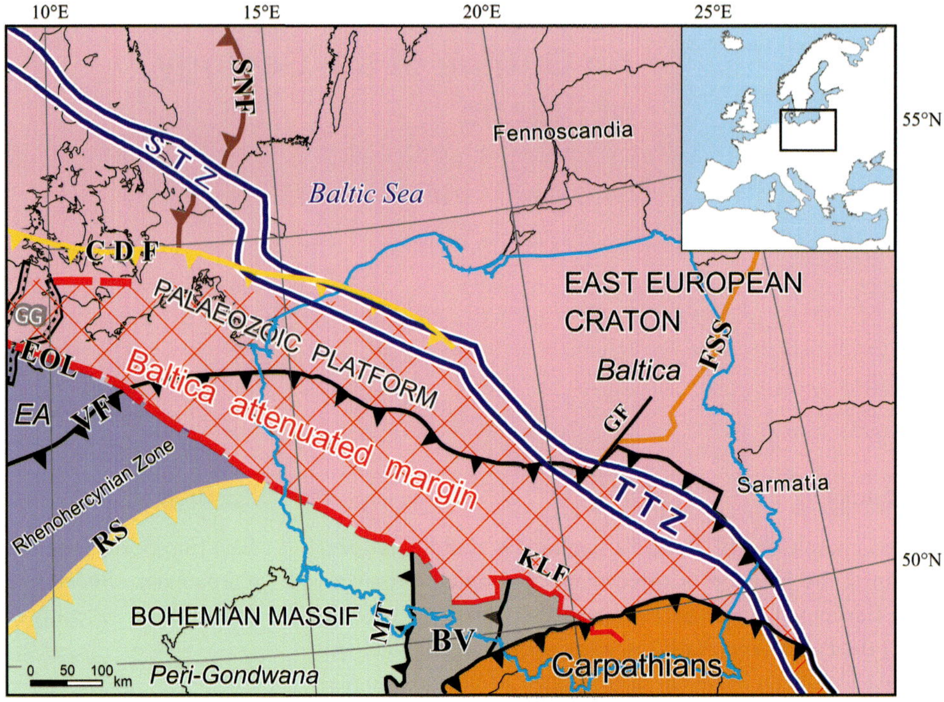

Fig. 1.1 Crustal domains at the transition from the Precambrian East European Platform to the Palaeozoic Platform of Western Europe and Western Carpathians with major structural elements overlaid (modified from Mazur et al. 2021). Blue outline shows a boundary of Poland. Inset map indicates geographic position of the depicted area within Europe. BV—Brunovistulicum; CDF—Caledonian deformation front; EA—East Avalonia; EOL—Elbe–Odra Lineament; FSS—Fennoscandia-Sarmatia Suture; GF—Grójec Fault; GG—Glückstadt Graben; KLF—Kraków–Lubliniec Fault; MT—Moldanubian Thrust; RS—Rheic Suture; SNF—Sveconorwegian front; STZ—Sorgenfrei–Tornquist Zone; TTZ—Teisseyre–Tornquist Zone; VDF—Variscan deformation front

understanding the geological structure of Poland and gradual progress in Earth sciences. The present synopsis mostly relies on previous findings although it also presents some new interpretations based on the synthesis of recently published data. Many aspects of the Poland's geological structure are generalised here to make their overview consistent and comprehensive. Provided references may guide an interested reader to more detailed information.

1.2 Synopsis of Geological Evolution

Most of the Polish territory is underlain by the crystalline crust of the East European Craton. In north-eastern Poland, the Precambrian Platform comprises undeformed to weakly deformed Neoproterozoic-Phanerozoic sediments resting upon a thick cratonic crystalline crust. The Precambrian Platform is approximately limited from the southwest by the Teisseyre-Tornquist Zone (TTZ; Fig. 1.1), a lithospheric necking zone produced during the Ediacaran break-up of Rodinia (Mikołajczak et al. 2019; Mazur et al. 2021) and separation of Amazonia from Baltica (e.g. Johansson 2009). The attenuated East European cratonic crust continues in south-westerly direction, beyond the TTZ and underneath the Palaeozoic Platform of central and western Poland, except for the north-eastern margin of the Bohemian Massif (the Sudetes) and the Upper Silesian Block in south-western Poland and the Western Carpathians in the south (Fig. 1.1). However, the crystalline basement of the East European Craton is overlain within the Palaeozoic Platform by Ediacaran-Palaeozoic, pre-Permian sediments, partly folded and thrust faulted during the Caledonian and Variscan cycles (e.g. Dadlez et al. 1994). During the latter, the Variscan internides of south-western Poland (the Sudetes) were assembled in the Late Devonian-Carboniferous and accreted to the edge of the Palaeozoic Platform (Fig. 1.1). The nature of the contact, thrusting versus strike-slip, remains disputable due to a thick sedimentary cover (Mazur et al. 2020). Soon after cessation of the Variscan orogeny, the Palaeozoic Platform and the Variscan Sudetes were affected by early Permian continental rifting (Dadlez et al. 1995; Mazur et al. 2021). The TTZ acted at that time as a mechanical buffer that prevented penetration of extensional stresses farther to the NE (Mazur et al. 2021). Therefore, the main rift depocenter, the Mid-Polish Trough, was localised directly to the southwest of the TTZ and the East European Craton margin underlying the Palaeozoic Platform was further attenuated (Mazur et al. 2021). Gradual thermal subsidence from the late Permian to the Late Cretaceous caused widening of the Mid-Polish Trough to form the Polish Basin, extending over the entire Palaeozoic Platform and the south-western slope of the East European Craton. The growth of the basin was punctuated by three major pulses of accelerated subsidence in late Permian-Early Triassic, Late Jurassic, and middle Cretaceous (Dadlez et al. 1995). The Polish Basin was inverted in the Late Cretaceous-early Paleogene as a far-field effect of the collision between Northern Europe and the Alpine-Carpathian orogen (Krzywiec 2002). At that time, the Mid-Polish Trough was inverted into the

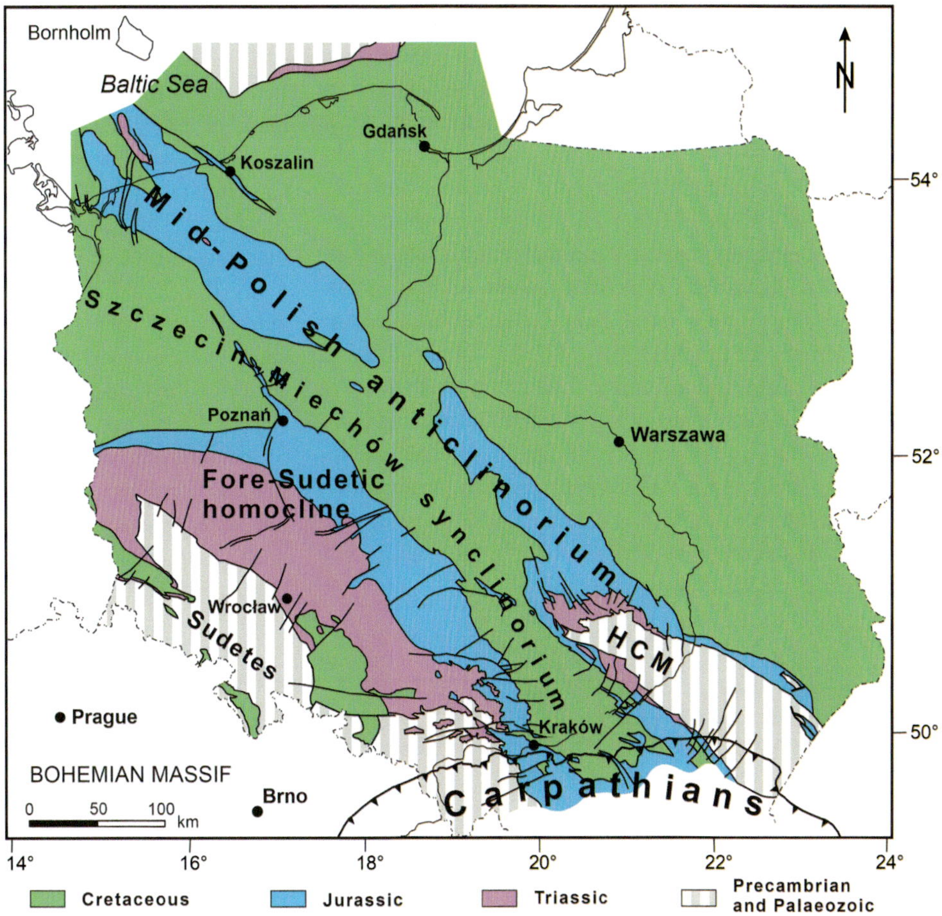

Fig. 1.2 Pre-Cenozoic geology of Poland showing the structure of the Permian-Mesozoic Polish basin developed during Late Cretaceous inversion (modified after Dadlez et al. 2000). HCM—Holy Cross Mountains

Mid-Polish Anticlinorium adjoining the TTZ from the southwest (Fig. 1.2; Krzywiec 2002). The south-western part of the Polish basin was tilted at that time to become the Fore-Sudetic Homocline and its south-eastern extension–the Kraków-Silesian Homocline, with strata gently dipping northeast towards the Szczecin-Miechów Synclinorium (Fig. 1.2). Both the East European Platform and the Palaeozoic Platform, jointly with the Permian-Mesozoic sedimentary cover, were overridden in the Paleogene-Miocene by the outer Western Carpathians fold-and-thrust belt resting upon their margins (Fig. 1.2). Further south, the Pieniny Klippen Belt represents a tectonic suture separating Northern Europe from tectonic elements derived from the southerly Adria Plate.

1.3 East European Platform

The Precambrian Platform of eastern and north-eastern Poland is a stable part of the crust, with little relief. The platform is composed of Precambrian basement overlain by 0.4–8 km of relatively undeformed sedimentary rocks. The latter range in age from Neoproterozoic (mostly Ediacaran) to Cenozoic and reach a maximum thickness along the south-western margin of the platform. The Precambrian basement, jointly with the Baltic and Ukrainian Shields, is collectively referred to as the East European Craton that is an isostatically positive portion of the continent and tectonically stable relative to adjacent orogens. The seismic structure of the craton, revealed by refraction surveys, is characterised by three nearly horizontal and uniform layers, lower, middle, and upper crust, respectively, typical of old cratons worldwide (Grad et al. 2002; Guterch and Grad 2006). The depth to the Moho ranges from 42 to over 50 km beneath north-eastern Poland and Lithuania (Guterch and Grad 2006). The crystalline basement in the Polish sector of the East European Craton is composed of various medium- to high-grade metamorphic rocks of Paleoproterozoic age (c. 2.0–1.8 Ga) that are intruded by the Middle Proterozoic (c. 1.5 Ga) AMGC (Anorthosite-Mangerite-Charnockite-Granite) plutons (Krzemińska et al. 2017). These rocks form NE–SW trending belts, truncated by the NW–SE oriented craton margin (Bogdanova et al. 2008).

The sedimentary cover of the Precambrian Platform in Poland comprises three successions with Ediacaran-early Palaeozoic, Devonian-Carboniferous, and Permian-Mesozoic ages. The uppermost Neoproterozoic-lower Palaeozoic sequence of significant thickness and lateral extent consists of Ediacaran syn-rift sedimentary and volcanic rocks succeeded by Cambrian and Ordovician sediments of the Baltica paleocontinent passive margin and deposits of the rapidly subsiding Silurian Caledonian foredeep (Poprawa et al. 1999; Mazur et al. 2018). This early Palaeozoic basin is almost completely buried beneath younger sediments, with Permian-Mesozoic, Carboniferous, and Devonian strata overlying the top of the Silurian in north-western, central and south-eastern Poland, respectively. The top of the Silurian represents a major unconformity over entire north-western and central Poland, corresponding to a late Caledonian compressional event (e.g. Poprawa et al. 2019). Another unconformity is associated with the base of Permian, representing early Carboniferous uplift in Pomerania and Podlasie and shortening of the external Variscan belt in the Lublin area (e.g. Mazur et al. 2020; Botor et al. 2021).

1.4 Palaeozoic Platform

In contrast to the East European Platform, the basement of the Palaeozoic Platform of Western Europe consists of rocks that were accreted and deformed during the Caledonian (early Palaeozoic) or Variscan (late Palaeozoic) orogenies. They are covered by little deformed sediments of Permian-Mesozoic-Cenozoic ages, with the thickness up to 8 km (Fig. 1.2). The Palaeozoic basement may include metamorphic and igneous rocks as well as pre- to syn-orogenic sediments involved in fold-and-thrust belts. In the Polish sector of the Palaeozoic Platform, the Variscan basement of Carboniferous age largely prevails (Fig. 1.3). It is composed of metamorphic and igneous rocks in south-western Poland (the Sudetes) and deformed pre-Permian sediments in central and western Poland. The Caledonian basement has been recognised by two dozen of deep boreholes in central Pomerania within the Koszalin-Chojnice Zone (Fig. 1.3),

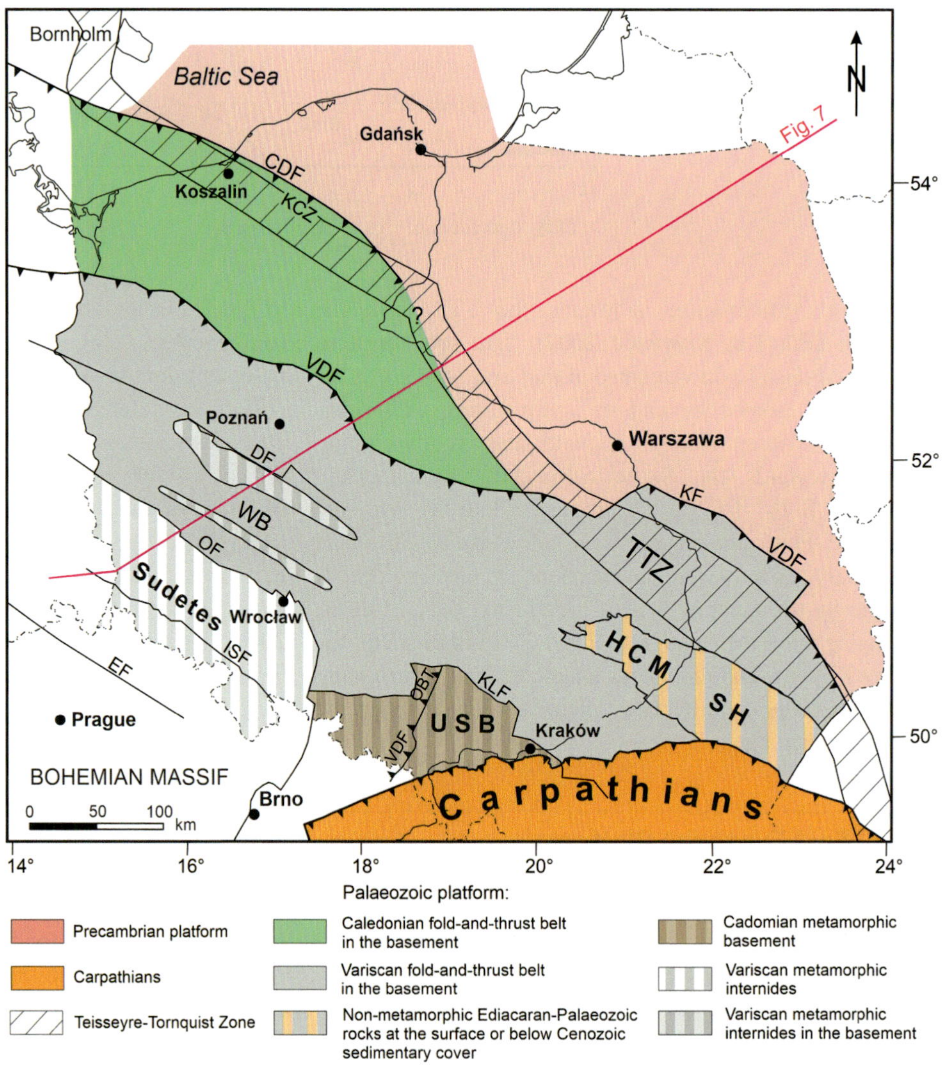

Fig. 1.3 Tectonic provinces of Poland without Permian-Cenozoic platform cover (modified from Aleksandrowski and Mazur 2017). CDF—Caledonian deformation front; DF—Dolsk Fault; EF—Elbe Fault; HCM—Holy Cross Mountains; ISF—Intra-Sudetic Fault; KCZ—Koszalin-Chojnice Zone; KF—Kock Fault; KLF—Kraków-Lubliniec Fault; OBT—Orlová-Boguszowice Thrust; OF—Odra Fault; SH—San Horst; TTZ—Teisseyre-Tornquist Zone; USB—Upper Silesia Block; VDF—Variscan deformation front; WB—Wolsztyn Block

where it is represented by tightly folded Ordovician and Silurian sediments. The similar type of basement may extend farther west, underneath western Pomerania, up to the Polish-German border (Mazur et al. 2018).

Summing up, the Palaeozoic Platform of Poland can be divided into four sub-domains (Fig. 1.3): (1) the Sudetes (south-western Poland), with the Variscan crystalline basement exposed at the surface, (2) central and western Poland with Variscan non-metamorphic fold-and-thrust belt in the basement–the external Variscides, (3) Upper Silesian Block comprising Cadomian basement partly affected by Variscan deformation, and (4) north-western Poland underlain by an early Palaeozoic Caledonian fold-and-thrust belt–the Pomeranian Caledonides. The deep substratum of the Palaeozoic Platform is characterized by the thick upper crust with low velocities of seismic waves (< 6.2 km/s) down to ~ 20 km and thinner lower crust with a high seismic velocity of 6.5–7.2 km/s (Grad et al. 2002; Guterch and Grad 2006). The depth to the Moho ranges between 38 and 31 km beneath central and south-western Poland, respectively (Guterch and Grad 2006).

1.4.1 The Sudetes

The Sudetes are geologically located on the north-eastern margin of the Bohemian Massif, straddling the state boundary between south-western Poland and north-eastern Czechia (Fig. 1.4). They represent the north-easternmost portion of the Variscan basement exposed in Europe. In the geological sense, the Sudetes combine the hilly to mountainous region of the Sudetes Mts. proper together with their foothills (jointly: the geographical Sudetes) and the south-eastern part of the Silesian-Lusatian Plain adjacent to the north (Kondracki 2013). The former two geographical domains are crosscut by the topographically prominent Sudetic Boundary Fault (Fig. 1.4), a Cenozoic feature throwing to the northeast, with a vertical displacement in order of several hundred metres (e.g. Badura et al. 2007; Danišík et al. 2012). Thus, the present-day Sudetes comprises the fault-bounded block of the Sudetic Mountains in the southwest and the Fore-Sudetic Block in the northeast (Fig. 1.4). The entire Sudetic area extends between the WNW–ESE oriented Middle Odra Fault in the northeast and the parallel Upper Elbe Fault in the southwest (Fig. 1.5). The Sudetic basement plunges below the Miocene Carpathian foreland basin to the southeast, and it merges with the Neoproterozoic Lusatian Massif to the northwest. The Sudetes comprises a mosaic of geologically distinct (Cloos 1922; Teisseyre 1957), fault-bounded pre-Permian units affected by mostly Devonian to Carboniferous deformation and metamorphism (Fig. 1.5).

The Carboniferous-early Permian granitoid magmatism and bimodal volcanism affected the Sudetes in a wide time span of about 50 million years, from c. 340 to 290 Ma, and produced a number of igneous suites currently exposed at the surface (for overview see Mazur et al. 2007; Turniak et al. 2014; Oberc-Dziedzic et al. 2015). Early Carboniferous granitoids include a number of variably sized intrusions dispersed across the Sudetes and in the vicinity of the Odra Fault (Fig. 1.5). Late Carboniferous granitoids are the most voluminous intrusions in the Sudetes. They include the shallow, mostly undeformed Karkonosze, Strzegom-Sobótka, and Strzelin plutons with wide contact aureoles (Fig. 1.4). Widespread volcanic and subvolcanic igneous complexes are interstratified in the Carboniferous-Permian sedimentary successions of the intramontane Intra-Sudetic and North-Sudetic Basins (Fig. 1.4). The latter depressions are major late- to post-orogenic intramontane basins, confined between major fault zones. They are both filled with Carboniferous to Upper Cretaceous sediments, with a major hiatus spanning the Upper Triassic, Jurassic, and Lower Cretaceous. The lower part of the basin fills (upper Carboniferous-lower Permian) is a continental, volcano-sedimentary succession of mainly siliciclastic deposits. The upper part (Zechstein, Lower Triassic, Upper Cretaceous) comprises various siliciclastic and carbonate deposits as well as evaporites of shallow marine, epicontinental and, partly, continental origin. A thick succession of the Upper Cretaceous sandstones, related to transgression of a Cenomanian Sea, features a characteristic landscape in the form of large crags or tabular mountains.

The Sudetes belongs to a wide belt of exposures of Variscan crystalline basement in between the Alpine orogen in the south and the Permo-Mesozoic basin in the north. The eastern extremity of this European belt (*Mittelgebirge*) is represented by the Bohemian Massif, including the Sudetic area (Fig. 1.5). Post-Variscan thermal evolution of basement rocks reflects successive phases of late Palaeozoic uplift, Mesozoic subsidence, and the Late Cretaceous/Paleogene inversion, which affected the northern Alpine foreland due to the final phases of the Variscan orogeny, the break-up of Pangea, and the Alpine collision, respectively (Ziegler 1987). Low-temperature thermochronological studies, based on apatite fission track data, suggest that the Sudetes experienced a complex post-orogenic thermal evolution that was influenced by burial under Mesozoic sediments and Late Cretaceous inversion-related exhumation (e.g. Aramowicz et al. 2006; Sobczyk et al. 2015). Moreover, significant sedimentary burial may have caused resetting of zircon (U-Th)/He ages in the central part of the Sudetes during the Late Cretaceous (Danišík et al. 2012; Sobczyk et al. 2015).

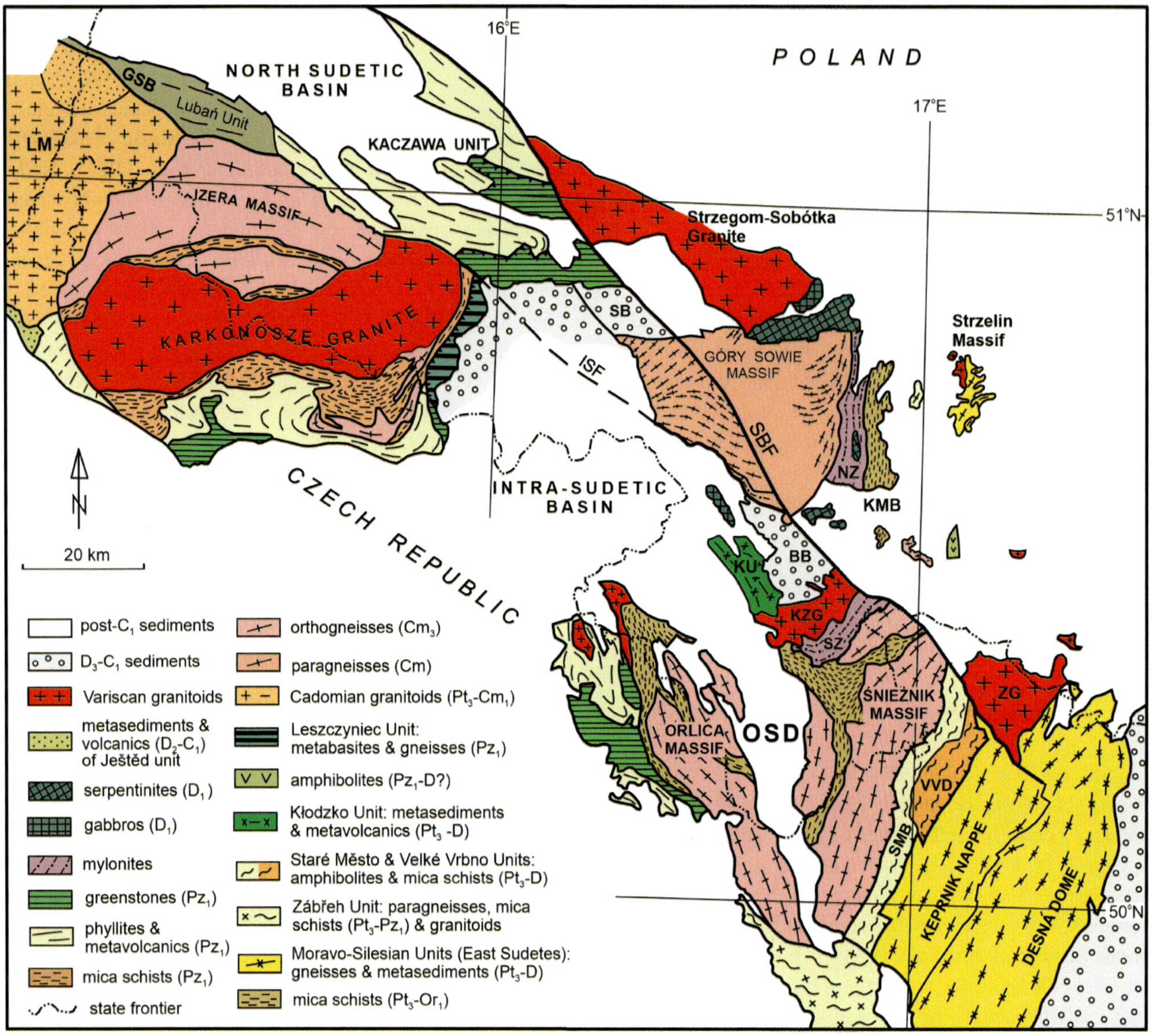

Fig. 1.4 Geological sketch map of exposed pre-Permian basement in the Sudetes. Abbreviations: BB—Bardo Basin; ISF—Intra-Sudetic Fault; GSB—Görlitz Slate Belt; KU—Kłodzko Metamorphic Unit; KMB—Kamieniec Metamorphic Belt; KZG—Kłodzko–Złoty Stok Granite; LM—Lusatian Massif; NZ—Niemcza Shear Zone; OSD—Orlica–Śnieżnik Dome; SB—Świebodzice Basin; SBF—Sudetic Boundary Fault; SMB—Staré Město Belt; ZG—Žulová Granite. Age assignments: Pt—Proterozoic; Pz—Palaeozoic; Cm—Cambrian; Or—Ordovician; D—Devonian; C—Carboniferous; 1—Early; 2—Middle; 3—Late

However, the stratigraphic record suggests less than 1 km of Mesozoic sedimentary cover (e.g. Milewicz 1997). The latter inspired a paleogeographic model of islands emerging from a shallow Late Cretaceous sea (e.g. Milewicz 1997). Nevertheless, recent studies suggest that the Late Cretaceous-Paleogene cooling phase was related to tectonic uplift, which occurred after deposition of a significant thickness of Cenomanian–Turonian sediments in the range of ~4 km (Botor et al. 2019; Sobczyk et al. 2020).

In the Cenozoic, the Sudetes was subjected to intense weathering and erosion, resulting in the development of a peneplain locally overlain by thin Late Eocene to Oligocene fluvial and lacustrine sediments (Migoń and Lidmar-Bergström 2001). At the same time, the widespread Oligocene–Miocene and, in the East Sudetes, Pliocene basaltic volcanism occurred (Birkenmajer et al. 2004), including a variety of forms and structures: stratocones (usually considerably re-shaped by erosion and denudation), lavas and tuffs, plugs, less frequently dykes and sills (Birkenmajer et al. 2011). The Cenozoic basalts of the Sudetes belong to the Central European intraplate volcanic province.

1.4.2 External Variscides of Central and Western Poland

The Variscan external fold-and-thrust belt, forming the basement of the Palaeozoic Platform in central and western Poland (Fig. 1.3), is entirely buried below the Permian-Mesozoic to Cenozoic infill of the Polish Basin. The fold-and-thrust belt comprises a Carboniferous syn- to post-orogenic clastic sedimentary succession of a minimum thickness exceeding 2.5 km, which was deposited in an extensive foreland basin. This basin formed part of the Variscan foreland basinal system extending from England to Poland along the northern margin of the Variscan orogenic belt. The Polish part of the Variscan foreland basin, covering the areas of Wielkopolska, Kujavia and northern Małopolska, developed on a passive margin of Laurussia that was presumably covered by lower Palaeozoic-Devonian sediments. However, the latter rocks were not confirmed by boreholes that did not reach the base of the Carboniferous succession. In contrast, low-grade metamorphic slates were drilled by several boreholes within two WNW–ESE-trending crystalline highs in the Wolsztyn Block directly north of the Odra Fault (Fig. 1.3). While the peak metamorphic age of these low-grade rocks is constrained at 358.6 ± 1.8 Ma (latest Devonian; Mazur et al. 2006), their protolith age remains unknown. Carboniferous clastic rocks occur in a transgressive position on top of the metamorphic basement, marking a backstepping of the foreland basin onto the hinterland. This suggests a connection between the basement of the Wolsztyn Block and the Variscan internides of the Bohemian Massif (Fig. 1.5).

Carboniferous flysch has been drilled region-wide below the base-Permian unconformity. Moderate to steep (30°–80°) dips predominate across the Carboniferous subcrop, with no clear steepening/shallowing trend (Fig. 1.6; Mazur et al. 2020). The predominance of moderate to steep dips may be an expression of thin-skinned folding and/or imbricate stacking of the Carboniferous series (Fig. 1.7; Mazur et al. 2020). The borehole data leave a broad margin of uncertainty for the position of the Variscan deformation front since a big gap exists in borehole coverage across the depocenter of the Polish Basin (Fig. 1.7). Nevertheless, statistical analysis of structural dips allows for delimiting the subsurface position of the Variscan deformation front that separates the external fold-and-thrust belt and mostly undeformed section of the Variscan foreland basin (Mazur et al.

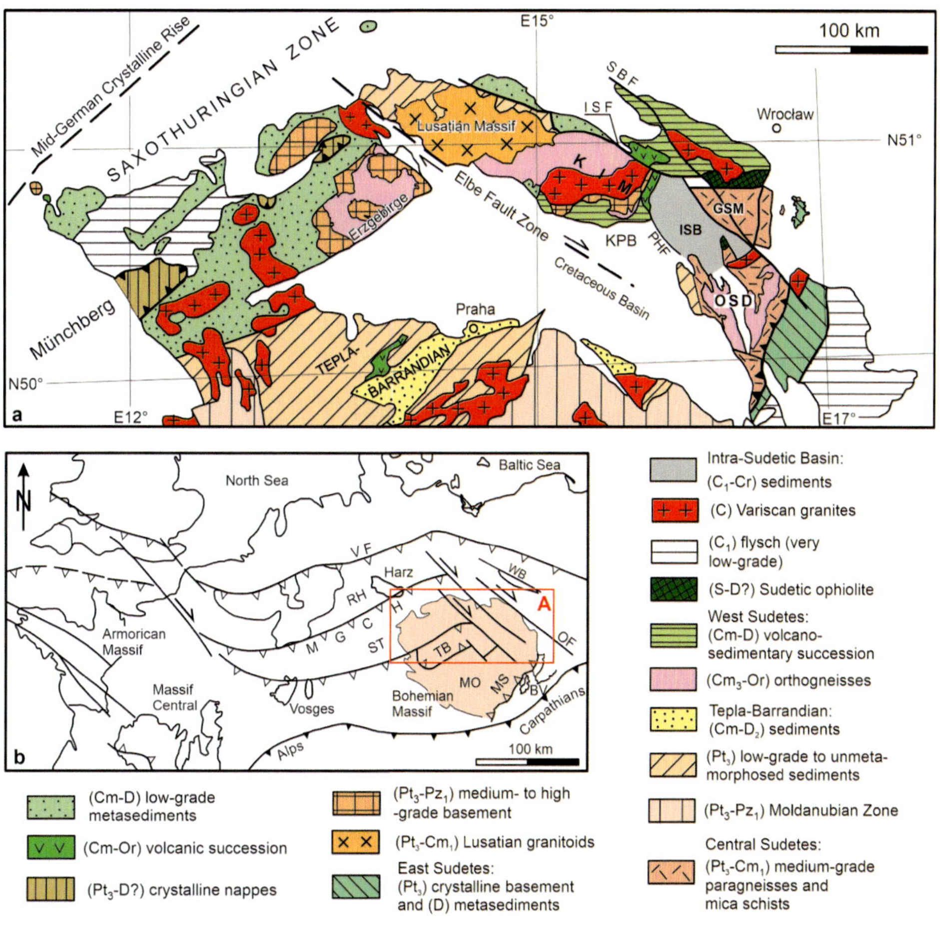

Fig. 1.5 Tectonic setting of the Sudetes within the Variscan belt of Europe (after Botor et al. 2019). Only pre-Premian basement is shown. **a** Geological sketch of the northern part of the Bohemian Massif. **b** Main tectonic units of the Variscides. BV—Brunovistulicum; GSM—Góry Sowie Massif; ISB—Intra-Sudetic Basin; ISF—Intra-Sudetic Fault; KIM—Karkonosze-Izera Massif; KPB—Krkonoše Piedmont Basin; MGCH—Mid-German Crystalline High; MO—Moldanubian Zone; MS—Moravo-Silesian Zone; OF—Odra Fault; OSD—Orlica-Śnieżnik Dome; PHF—Pořičí-Hronov Fault; RH—Rhenohercynian Zone; SBF—Sudetic Boundary Fault; ST—Saxothuringian Zone; TB—Teplá-Barrandien Zone; VF—Variscan deformation front; WB—Wolsztyn Block. Age assignments: Pt_3—late Proterozoic; Cm—Cambrian; Or—Ordovician; S—Silurian; Pz_1—lower Palaeozoic; D—Devonian; C—Carboniferous; Cr—Cretaceous

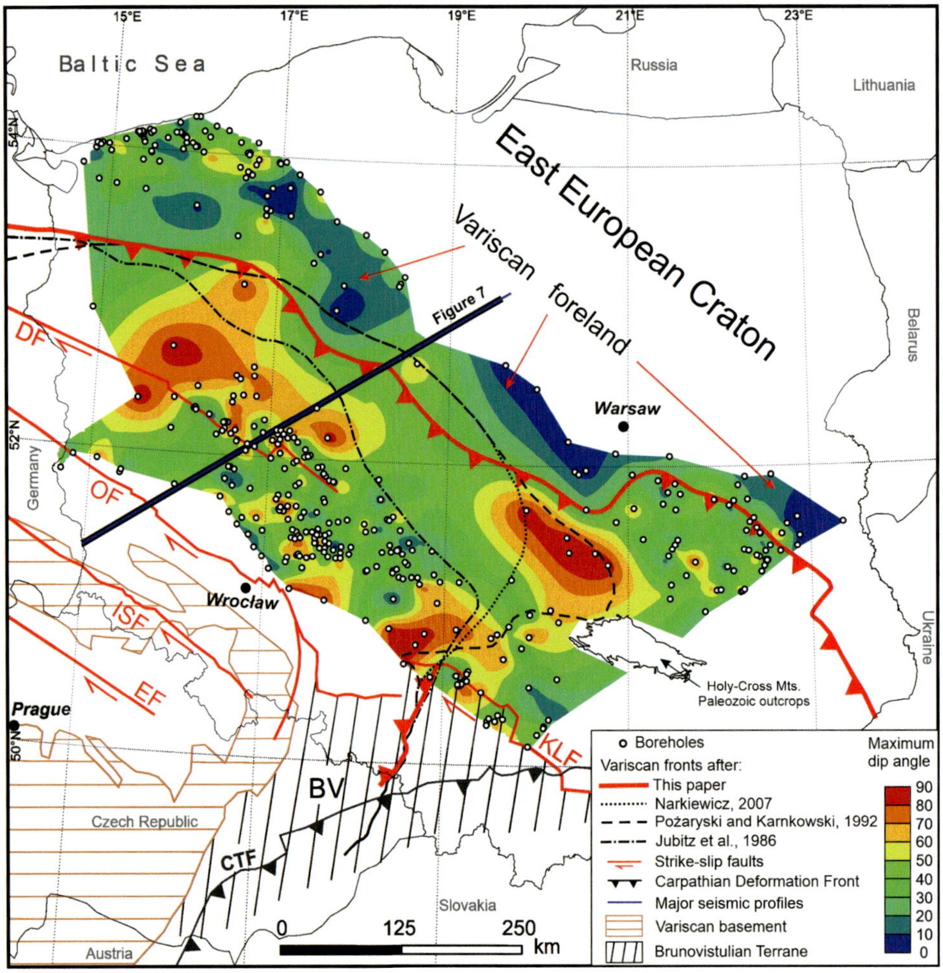

Fig. 1.6 Distribution of maximum dip angles in the Devonian-Carboniferous succession of the external zone of the Variscan orogen in Poland (after Mazur et al. 2020). VDF—admissible trace of the Variscan deformation front in W Poland. Other acronyms: BV—Brunovistulicum; CTF—Carpathian thrust front; DF—Dolsk Fault; EF—Elbe Fault; ISF—Intra-Sudetic Fault; KLF—Kraków-Lubliniec Fault; OF—Odra Fault

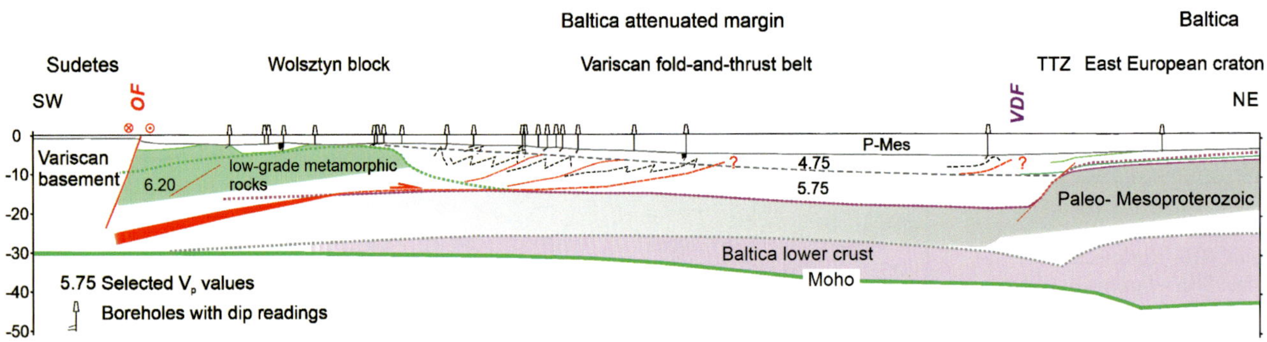

Fig. 1.7 Regional cross-section through the SW margin of the East European craton and the eastern part of the Variscan belt in Poland along seismic refraction line P4 (after Mazur et al. 2020), based on a two-dimensional P-wave velocity model (Grad et al. 2003). OF—Odra Fault; P-Mes—Permian-Mesozoic; TTZ—Teisseyre-Tornquist Zone; VDF—Variscan deformation front

2020). The orientation of the Variscan deformation front beneath the Permian-Mesozoic Polish Basin is roughly WNW–ESE, except for a NNE–SSW-oriented side branch in Moravia (Mazur et al. 2020; Fig. 1.6).

1.4.3 Upper Silesian Block

The Upper Silesian Block, located in Lower and Upper Silesia and western Małopolska in southern Poland, is part of a larger tectonic unit known as Brunovistulicum (Fig. 1.1). The latter represents a fragment of the Precambrian continental crust located between the Variscan orogen in the west, the Carpathians to the southeast, and the attenuated margin of the East European Craton to the north (Dudek 1980; Mazur et al. 2015, 2018). In contrast to the Precambrian Platform of eastern Poland, the majority of Brunovistulian basement was deformed, metamorphosed, and intruded by magmatic bodies in the latest Precambrian during the Cadomian orogeny (650–550 Ma, Buła et al. 2015; Żelaźniewicz et al. 2020). Precambrian rocks are unconformably covered by an extensive lower to middle Cambrian clastic succession and, locally, by Upper Ordovician sediments. These are discordantly overlain, with a large stratigraphic gap, by widespread Lower-Middle Devonian platform-type continental clastic sediments and Upper Devonian-lowest Carboniferous carbonates.

The western part of Brunovistulicum was incorporated in the Variscan orogenic belt as it was reworked within a 50 km wide and 300 km long Moravo–Silesian Zone represented by imbricated and metamorphosed nappes (Fig. 1.5; Suess 1912). This zone is considered a main tectonic boundary between the Variscan internides of the Bohemian Massif and the Brunovistulian promontory farther east (e.g. Collett et al. 2021). The northern section of the Moravo-Silesian Zone is traditionally included into the Sudetes (East Sudetes; Figs. 1.4 and 1.5). Thrusting-related exhumation of the Variscan nappes over the Brunovistulian basement resulted in the development of a 7.5 km thick Carboniferous foreland basin unconformably covering the Precambrian basement and its sedimentary cover. The lower Carboniferous turbiditic succession grades eastward into paralic and continental sediments of the Upper Silesian Coal Basin. Lower Carboniferous sediments of the Variscan foreland basin (Culm facies) and pre-orogenic Devonian sedimentary successions are involved into the NNE–SSW-trending, eastward verging Moravo-Silesian fold-and-thrust belt (Bábek et al. 2006). Deformation gradually declines towards the east, with the Orlová-Boguszowice Thrust conventionally interpreted to delimit the position of the Variscan deformation front in Upper Silesia (Fig. 1.3). This N–S oriented branch of the Variscan front in Poland is terminated against the Kraków-Lubliniec Fault at the northern margin of the Upper Silesian Block and does not link with the NW–SE oriented front of external Variscides in central and western Poland (Figs. 1.3 and 1.6).

Although only the western part of Brunovistulicum was largely affected by Variscan deformation, the weakly deformed eastern part is covered by an extensive Carboniferous clastic succession of the foreland basin. For this reason, the entire Brunovistulicum is included in the Palaeozoic Platform, even though the platform cover is relatively thin and discontinuous, including Permian, Triassic, Jurassic, and Miocene rocks. In the southeast, the Brunovistulian section of the Palaeozoic Platform is overthrust by the Cretaceous–Paleogene nappes of the Outer Western Carpathians and onlapped by their Miocene foredeep (Dudek 1980).

1.4.4 Pomeranian Caledonides

The Pomeranian Caledonides are a belt of deformed Upper Ordovician and Silurian sedimentary rocks that runs NW–SE onshore and offshore north-western Poland (Dadlez et al. 1994). The fold belt is entirely concealed beneath a 1–4 km thick cover of largely undeformed upper Palaeozoic and Permian-Mesozoic sediments. Therefore, the Pomeranian Caledonides were discovered only in the 1960s and 1970s when a deep drilling programme commenced in the area. More than 20 deep boreholes encountered subcrops of deformed lower Palaeozoic rocks forming a narrow and elongated uplift named the Koszalin-Chojnice Zone (Fig. 1.3). Towards the east, the fold belt is in contact with the undeformed lower Palaeozoic sediments of the Baltic Basin belonging to a sedimentary cover of the East European Platform. Towards the west, the lower Palaeozoic basement deepens and has not been penetrated by onshore wells. However, one offshore well reached similar lower Palaeozoic basement almost halfway to the Island of Rügen, a type locality of the North German Caledonides. Therefore, a reasonable assumption is that the Caledonian basement extends underneath the entire western Pomerania from the vicinity of Koszalin to the German border.

The top of crystalline basement beneath the Pomeranian Caledonides, as imaged by seismic data and geophysical modelling, shows smooth geometry, and it is not involved in the early Palaeozoic fold belt (Mazur et al. 2016). Hence, the latter represents a thin-skinned fold-and-thrust belt involving early Palaeozoic sediments of the Caledonian foreland basin (Mazur et al. 2016). This is supported by general similarity of the deformed lower Palaeozoic sediments within the Koszalin-Chojnice Zone and their undeformed equivalents in the Baltic Basin. Also, the provenance study by Poprawa et al. (2006) for the lower Palaeozoic sediments of the Koszalin-Chojnice Zone indicates their source area within the East European Craton.

The Pomeranian Caledonides cannot be traced beyond the town of Chojnice. Farther southeast, the lower Palaeozoic basement is deeply concealed beneath Carboniferous and Permian-Mesozoic sediments and has not been penetrated by wells or imaged by seismic data. The Caledonian orogen must have once extended along the SW margin of the East European Craton as evidenced by the undeformed Caledonian foreland basin onlapping the slope of the Precambrian Platform across entire Poland up to the Ukrainian border (Mazur et al. 2018). However, Caledonian deformations in the Holy Cross Mountains, the only outcrop of non-metamorphosed lower Palaeozoic rocks within the Palaeozoic Platform, are mild compared to the Pomeranian Caledonides (Mazur et al. 2018) and much weaker than Variscan folding of the Holy Cross Mountains (e.g. Mizerski 2004).

1.4.5 Permian-Mesozoic Polish Basin

The Polish Lowlands, between the mountainous area of the Sudetes to the south and the Baltic Sea to the north, are underlain by the extensive Permian-Mesozoic Polish Basin, unconformably covered by an up to 350-m thick Cenozoic sedimentary succession. The basin extends over the parts of Palaeozoic Platform having the Caledonian and Variscan fold-and-thrust belts in their basement. Its margins onlap the Precambrian Platform, the Upper Silesian Block and the Sudetes. The early Permian (Rotliegendes) to Late Cretaceous (Maastrichtian) (Lamarche and Scheck-Wenderoth 2005; Mazur et al. 2005) strata attain the largest present-day thickness of 8 km within the Mid-Polish Trough (Mid-Polish Anticlinorium; Fig. 1.2), an easternmost NW—SE elongated depocenter that adjoins the margin of the Precambrian Platform from the southwest (Dadlez et al. 1995). North of the Sudetes, the present-day thickness of the Permian-Mesozoic strata consistently increases northward, towards the axis of the Mid-Polish Trough, from c. 1 km near to the Odra Fault to more than 4 km north of Poznań. The area of the Polish basin is now fairly flat due to Cenozoic erosion and sedimentation, with more evident large-scale morphological features created by glacial processes.

1.5 Transition from the East European Platform to the Palaeozoic Platform

The boundary between the Precambrian and Palaeozoic Platforms is conventionally defined by the TTZ that is the longest European tectonic and geophysical lineament, extending from the Baltic Sea in the northwest to the Black Sea in the southeast (Fig. 1.1; Pharaoh 1999). This tectonic feature delineates the transition between the thick crust of the East European Craton in the northeast and the thinner crust underlying the Palaeozoic Platform in the southwest (Guterch and Grad 2006). The zone is entirely buried beneath Ediacaran and Phanerozoic sediments and thus, only documented by geophysical data. The TTZ is an up to 50 km wide (Fig. 1.3), corresponding to a change in a Moho depth by 6–8 km and a slope of the Precambrian basement, descending south-eastward c. 10–12 km underneath the extensive Palaeozoic and Mesozoic sedimentary successions (Fig. 1.7; e.g. Mikołajczak et al. 2019). Consequently, the thickness of the crystalline crust is reduced across the TTZ by 16–20 km (Mikołajczak et al. 2019). Jointly, the entire lithosphere changes its thickness from 150 to 200 km under the craton to 80–120 km under the Palaeozoic Platform (e.g. Wilde-Piórko et al. 2010). Although the TTZ has been variously interpreted over the past decades, recent results suggest that it represents a crustal necking zone created by Ediacaran rifting and then, exaggerated by early Permian continental rifting (Mikołajczak et al. 2019). Consequently, the Palaeozoic Platform is underlain by the attenuated crust of the East European Craton that extends in the substratum of the Palaeozoic Platform up to the boundary of the Upper Silesia Block (Brunovistulicum) and the Sudetes (Mazur et al. 2015, 2020).

Although in the past the TTZ was considered the boundary between the Precambrian and Palaeozoic Platforms, this is not exactly the case. Even though the Caledonian deformation front in Pomerania is indeed aligned with the TTZ due to a buttressing effect of the crystalline basement ramp (Mazur et al. 2016), the Variscan deformation front encroaches the area traditionally regarded as the Precambrian Platform in south-eastern Poland (Lublin area; Krzywiec et al. 2017). Consequently, the boundaries of the Palaeozoic Platform in the Lublin area are defined by the Grójec and Kock Faults (Fig. 1.3; Aleksandrowski and Mazur 2017).

1.6 The Carpathians

The current architecture of the Western Carpathians is divisible into three mega-units referred to as the Inner, Central, and Outer Western Carpathians (Froitzheim and Schuster 2008). The latter two crop out in southern Poland. The Central Western Carpathians are composed of thick and thin-skinned thrust sheets involving (from the bottom to top of the nappe pile) the Tatric, Veporic, and Gemeric Variscan basement nappes accompanied by the Fatric, Hronic, and locally Silicic cover nappes. The Polish section of the Tatra Mountains (Fig. 1.8) exposes the Tatricum basement nappe jointly with its Mesozoic autochthonous sedimentary cover and two Mesozoic High-Tatric nappes (Wierchowe nappes). These are overthrust by the Fatric (local names: Krížna, Reglowa dolna) and Hornic (local names: Choč, Reglowa górna) Sub-Tatric cover nappes (Kotański 1963; Jurewicz 2005). The Tatric basement was only weakly overprinted by Alpine metamorphism at sub-greenschist to only locally lower greenschist conditions. The Central Western Carpathians nappe stack was assembled during the middle Cretaceous (c. 120–90 Ma) shortening associated with the southward subduction of the Alpine Tethys Ocean (Schmid et al. 2008).

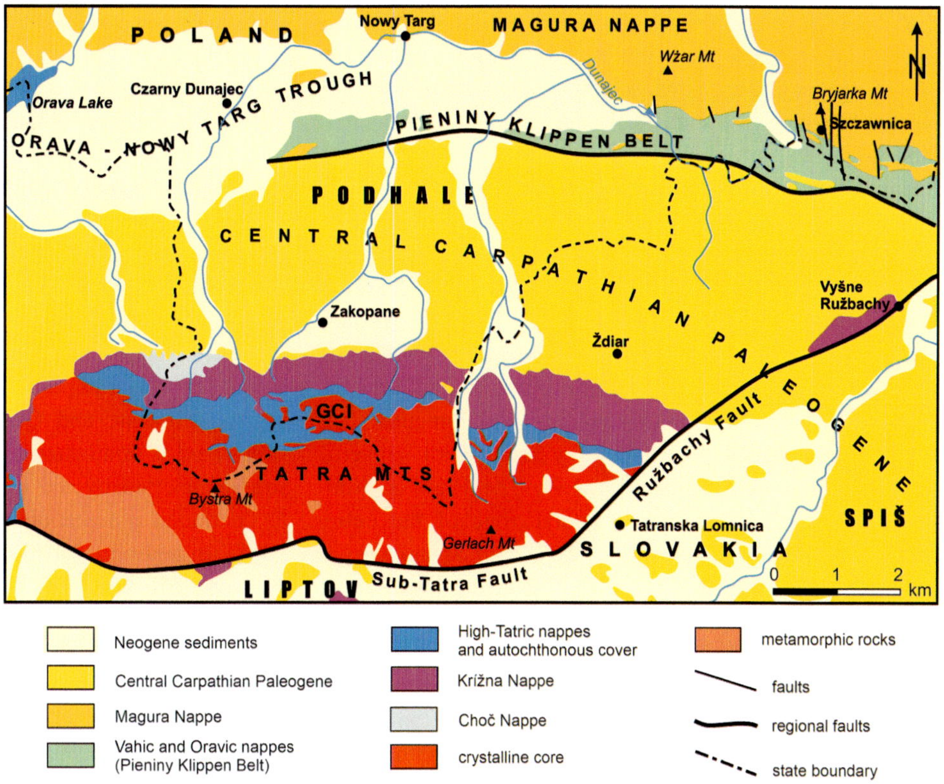

Fig. 1.8 Schematic geological map of the Tatra Mts and Pieniny Klippen Belt after Jurewicz (2005) and Felisiak (2008). GCI—Goryczkowa Crystalline Island

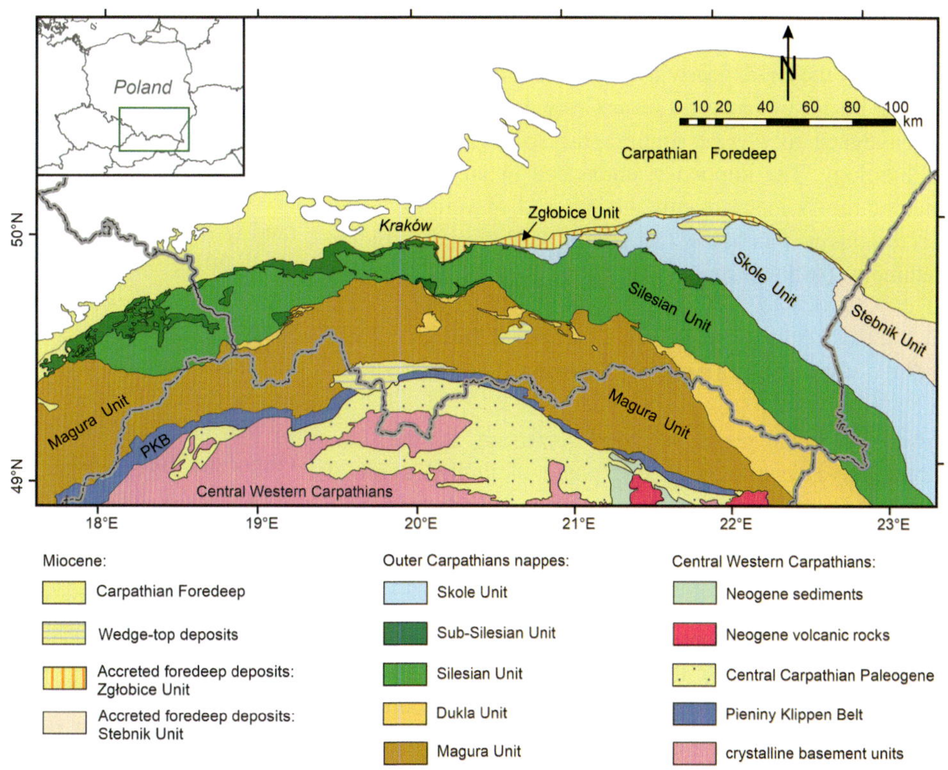

Fig. 1.9 Main structural units of the Polish sector of the Outer Carpathians. From north to south, they comprise structurally higher and older parts of the accretionary wedge. PKB—Pieniny Klippen Belt. Modified after Żytko et al. (1989), Mikołajczak et al. (2021)

The Outer Western Carpathians comprise the Late Cretaceous to Miocene accretionary wedge, including sedimentary units of the Alpine Tethys Ocean and the North European passive margin (Fig. 1.9). The Alpine Tethys in the Western Carpathians is inferred to have been composed of two oceanic domains, Vahic and Magura, with an enigmatic and currently non-existent continental ribbon, the Czorsztyn (Oravic) Ridge, in between (Birkenmajer 1977;

Plašienka 2018). Sediments deposited on the floor of the Magura Ocean are represented by a thick flysch succession of the Magura nappe, that is juxtaposed against the currently lowermost Silesian, Skole, and Dukla flysch nappes (and some less prominent units) derived from the North European passive margin. All these units were juxtaposed due to the earliest Neogene (c. 20 Ma) shortening accompanying the closure of the Carpathian branch of the Alpine Tethys Ocean.

The Alpine Tethys suture zone in the Western Carpathians is manifested by the occurrence of rock formations belonging to the Pieniny Klippen Belt (PKB), a few kilometres wide and hundreds of kilometres long zone of predominantly sedimentary rocks with subvertical orientation of bedding (Fig. 1.9). The PKB is probably the most conspicuous structural zone of the entire Carpathian arc. Picturesque landscape and abundance of Jurassic fossils especially have been drawing attention of geologists for more than 150 years. The belt is positioned at the backstop of the Paleogene accretionary wedge (Outer Carpathians), and in front of the Cretaceous nappe stack (Central Carpathians). The PKB involves predominantly Jurassic, Cretaceous, and Paleogene sediments with exceptionally variable lithology and intricate internal structure (Birkenmajer 1977). In general, two types of rock complexes are distinguished in the PKB. The "klippen" is formed by the rigid, nearly isometric, lenticular, or lozenge-shaped blocks of Jurassic-Lower Cretaceous limestones that are assigned to several stratigraphic successions differing in lithology. The klippen is embedded in an incompetent matrix, the "klippen mantle", composed of various Upper Cretaceous to Paleogene marlstones, shales, and turbiditic sediments (Birkenmajer 1977).

1.7 Cenozoic Tectonic Events

After the main phase of basin inversion in the Late Cretaceous-Paleocene, a second phase of limited uplift of the Mid-Polish Anticlinorium (Swell) occurred during the middle-late Eocene. End-Eocene and early Oligocene subsidence of narrow grabens on the Fore-Sudetic Homocline was coeval with the development of the Central European Rift System. At the same time, the subsidence and erosion patterns were rearranged in the Carpathian foreland from being dominated by the NW–SE trending Mid-Polish Anticlinorium to that influenced by the growth of the W–E trending Meta-Carpathian Swell (Jarosiński et al. 2009). The development of this feature is thought to reflect long-wavelength buckling of the lithosphere in response to the build-up of north-directed compressional intraplate stresses transferred from the Carpathian collision zone and tectonic loading by the outer Western Carpathian nappes. At the end of the Oligocene, the Fore-Sudetic graben system propagated across the Palaeozoic Platform (Fig. 1.10), while the Sudetes remained still active during the Miocene and Pliocene (Różycka and Migoń 2017). This was synchronous with the development of the Carpathian flexural foredeep basin and forebulge (Fig. 1.10). North-directed nappe transport in the Outer Western Carpathian was terminated during the Sarmatian, at c. 12 Ma. This was followed by eastward migration of the subsidence axis of the Carpathian Foredeep basin and the termination of tectonic activity in the grabens of the Polish Lowlands.

After a period of post-orogenic relaxation, built-up of the present-day compressional stress regime continued during the Pliocene and Quaternary (Jarosiński 2005). This stress field is dominated by collision-related compressional stresses exerted by the Carpathians on their foreland due to continued indentation of the Adriatic plate (Jarosiński 2005). In addition, intensified ridge push forces exerted on the North Atlantic passive margins contribute to the present-day stress pattern that is dominated by an approximately N–S direction of maximum compressional stress.

1.8 Concluding Remarks

Despite a long and complicated geological evolution, the relief of Poland is fully related to the Alpine orogenic cycle from the Late Cretaceous to the Cenozoic. Thermochronological studies show that the initial uplift in the Sudetes and Holy Cross Mountains was simultaneous with the inversion of the Permian-Mesozoic basin due to compression exerted by the Alpine orogen on its foreland (Aramowicz et al. 2006; Danišík et al. 2012; Sobczyk et al. 2015, 2020; Botor et al. 2017, 2018, 2019). This Late Cretaceous-Paleocene event led to exhumation of the Variscan crystalline basement and Palaeozoic sedimentary succession in the Sudetes and Holy Cross Mountains, respectively. However, any relief that had been produced by that time was then peneplenised during the Eocene and Oligocene (e.g. Migoń and Lidmar-Bergström 2001). Subsequently, the landscape of the Bohemian Massif, including the Sudetes, was rejuvenated starting from the early Miocene because of renewed compressional stresses emanating from the Alps (Jarosiński et al. 2009). This event was synchronous with the still ongoing extension in the Carpathian foreland as a result of diachronic collision propagating from the west to east.

The formation of the Polish Uplands, extending from the Upper Silesia through the Holy Cross Mountains to the Lublin Upland, was related to the growth of the Carpathian flexural forebulge that replaced the Meta-Carpathian Swell from the Miocene onward (Jarosiński et al. 2009). The forebulge developed in response to tectonic loading of the foreland plate by the Carpathian orogen. Upon the late Miocene

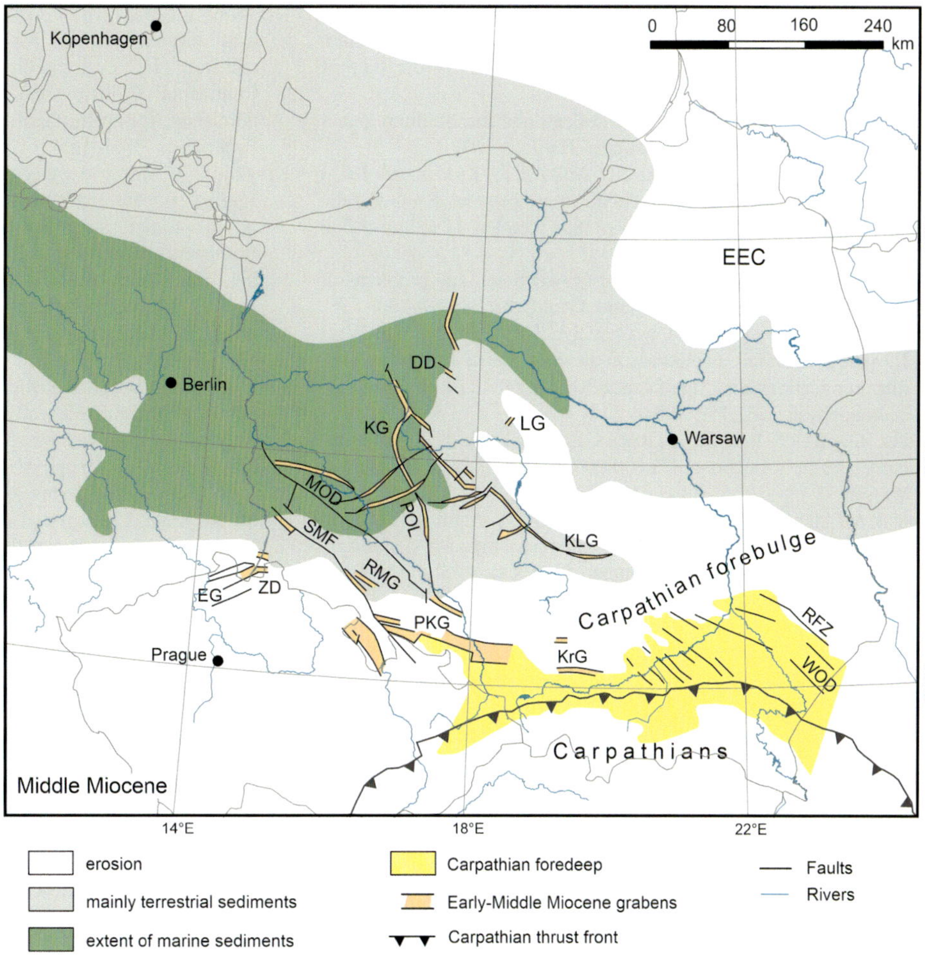

Fig. 1.10 Miocene tectonic features and deposition pattern of the Carpathian Foreland (modified from Jarosiński et al. 2009). Subsidence of the narrow Carpathian Foredeep basin was coeval with uplift of the forebulge and presence of a residual basin in the Polish Lowlands. DD—Damasławek Depression; EG—Eger Graben; KG—Krzywin Graben; KIG—Kleszczów Graben; KrG—Krzeszowice Graben; LG—Lubstów Graben; MOD—Middle Odra Fault Zone; PKG—Paczków-Kędzierzyn Graben; POL—Poznań-Oleśnica Lineament; RFZ—Roztocze Fault Zone; RMG—Roztoka-Mokrzeszów Graben; SMF—Sudetic Marginal Fault; WOD—Wielkie Oczy dislocation; ZD—Zittau Depression

detachment of the subducted North European slab beneath the Western Carpathians, their isostatic uplift commenced and is still active. The contemporary compressional stress regime in Poland results from continued indentation of the Adriatic plate into the Alpine-Carpathian orogen interacting with the far-field North Atlantic ridge push forces.

References

Aleksandrowski P, Mazur S (2017) O nowych rozwiązaniach tektonicznych w „Atlasie geologicznym Polski". Przegl Geol 65:1499–1510

Aramowicz A, Anczkiewicz AA, Mazur S (2006) Fission track dating of apatites from the Góry Sowie Massif, Polish Sudetes, NE Bohemian Massif: implications of post-Variscan denudation and uplift. N Jhrb Miner Abh 182:221–229

Bábek O, Tomek C, Melichar R, Kalvoda J, Otava J (2006) Structure of unmetamorphosed Variscan tectonic units of the southern Moravo-Silesian Massif: a review. N Jhrb Geol Paläont Abh 239:37–75

Badura J, Zuchiewicz W, Stepancikova P, Przybylski B, Kontny B, Cacoń S (2007) The Sudetic marginal fault: a young morphophotectonic feature at the NE margin of the Bohemian Massif, Central Europe. Acta Geodyn Geomater 4(4):7–29

Birkenmajer K (1977) Jurassic and Cretaceous lithostratigraphic units of the Pieniny Klippen Belt, Carpathians, Poland. Studia Geol Polon 45:1–158

Birkenmajer K, Pécskay Z, Grabowski J, Lorenc MW, Zagożdżon PP (2004) Radiometric dating of the Tertiary volcanics in Lower Silesia, Poland. IV. Further K-Ar and palaeomagnetic data from Late Oligocene to Early Miocene basaltic rocks of the Fore-Sudetic Block. Ann Soc Geol Polon 74:1–19

Birkenmajer K, Pécskay Z, Grabowski J, Lorenc MW Zagożdżon PP (2011) Radiometric dating of the Tertiary volcanics in Lower Silesia, Poland. VI. K–Ar palaeomagnetic data from basaltic rocks of the West Sudety Mountains and their northern foreland. Ann Soc Geol Polon 81:115–131

Bogdanova SV, Bingen B, Gorbatschev R, Kheraskova TN, Kozlov VI, Puchkov VN, Volozh YA (2008) The East European Craton (Baltica) before and during the assembly of Rodinia. Precambrian Res 160:23–45

Botor D, Dunkl I, Anczkiewicz A, Mazur S (2017) Post-Variscan thermal history of the Moravo-Silesian lower Carboniferous Culm Basin (NE Czech Republic-SW Poland). Tectonophysics 712:643–662

Botor D, Anczkiewicz AA, Dunkl I, Golonka J, Paszkowski M, Mazur S (2018) Tectonothermal history of the Holy Cross Mountains (Poland) in the light of low-temperature thermochronology. Terra Nova 30:270–278

Botor D, Anczkiewicz AA, Mazur S, Siwecki T (2019) Post-Variscan thermal history of the Intra-Sudetic Basin (Sudetes, Bohemian Massif) based on apatite fission track analysis. Intern J Earth Sci 108:2561–2576

Botor D, Mazur S, Anczkiewicz AA, Dunkl I, Golonka J (2021) Thermal history of the East European Platform margin in Poland based on apatite and zircon low-temperature thermochronology. Solid Earth 12:1899–1930

Buła Z, Habryn R, Jachowicz-Zdanowska M, Żaba J (2015) The Precambrian and Lower Paleozoic of the Brunovistulicum (eastern part of the Upper Silesian Block, southern Poland)—the state of the art. Geol Quart 59:123–134
Cloos H (1922) Der Gebirgsbau Schlesiens und die Stellung seiner Bodenschätze. Gebrüder Borntraeger, Berlin, p 107
Collett S, Štípská P, Schulmann K, Míková J, Kröner A (2021) Tectonic significance of the Variscan suture between Brunovistulia and the Bohemian Massif. J Geol Soc London 178. https://doi.org/10.1144/jgs2020-176
Dadlez R, Kowalczewski Z, Znosko J (1994) Some key problems of the pre-Permian tectonics of Poland. Geol Quart 38:169–190
Dadlez R, Narkiewicz M, Stephenson RA, Visser MTM, Van Wees JD (1995) Tectonic evolution of the Mid-Polish trough: modelling implications and significance for central European geology. Tectonophysics 252:179–195
Dadlez R, Marek S, Pokorski J (2000) Geological map of Poland without Cenozoic formations, 1:1,000,000. Polish Geological Institute, Warsaw
Danišík M, Štěpančíková P, Evans NJ (2012) Constraining long-term denudation and faulting history in intraplate regions by multisystem thermochronology: an example of the Sudetic Marginal Fault (Bohemian Massif, central Europe). Tectonics 31:TC2003
Dudek A (1980) The crystalline basement block of the Outer Carpathians in Moravia: Bruno-Vistulicum. Rozpravy Česko-Slovenské Akademie Věd, Řada Matematicko-Přírodních Věd 90:1–85
Felisiak I (2008) The Tatras—nappes and landscapes. Geotourism/geoturystyka 13:75–75
Froitzheim N, Schuster R (2008) Alpine tectonics of the Alps and Western Carpathians: the Alps. In: McCann T (ed) The geology of Central Europe, vol 2. Geological Society of London, pp 1141–1181
Grad M, Guterch A, Mazur S (2002) Seismic refraction evidence for crustal structure in the central part of the Trans-European Suture Zone in Poland. Geol Soc London Spec Publ 201:295–309
Grad M, Jensen SL, Keller GR, Guterch A, Thybo H, Janik T, Tiira T, Yliniemi J, Luosto U, Motuza G, Nasedkin V, Czuba W, Gaczyński E, Środa P, Miller KC, Wilde-Piórko M, Komminaho K, Jacyna J, Korabliova L (2003) Crustal structure of the Trans-European suture zone region along POLONAISE'97 seismic profile P4. J Geophys Res Solid Earth 108(B11):2541
Guterch A, Grad M (2006) Lithospheric structure of the TESZ in Poland based on modern seismic experiments. Geol Quart 50:23–32
Jarosiński M (2005) Ongoing tectonic reactivation of the Outer Carpathians and its impact on the foreland: results of borehole breakout measurements in Poland. Tectonophysics 410:189–216
Jarosiński M, Poprawa P, Ziegler PA (2009) Cenozoic dynamic evolution of the Polish Platform. Geol Quart 53:3–26
Johansson Å (2009) Baltica, Amazonia and the SAMBA connection—1000 million years of neighbourhood during the Proterozoic? Precambrian Res 175:221–234
Jurewicz E (2005) Geodynamic evolution of the Tatra Mts. and the Pieniny Klippen Belt (Western Carpathians): problems and comments. Acta Geol Polon 3:295–338
Kondracki J (2013) Geografia regionalna Polski. PWN, Warszawa, p 440
Kotański Z (1963) O charakterze mezozoicznej geosynkliny Karpat Zachodnich i o epimiogeosynklinie podhalańskiej. Acta Geol Polon 13:13–25
Krzemińska E, Krzemiński L, Petecki Z, Wiszniewska J, Salwa S, Żaba J, Gaidzik K, Williams IS, Rosowiecka O, Taran L, Johansson Å, Pécskay Z, Demaiffe D, Grabowski J, Zieliński G (2017) Geological map of Crystalline Basement in the Polish part of the East European Platform 1:1 000 000. Państwowy Instytut Geologiczny, Warszawa
Krzywiec P (2002) Mid-Polish trough inversion—Seismic examples, main mechanisms and its relationship to the Alpine-Carpathian collision. In: Bertotti G, Schulmann K, Cloetingh S (eds) Continental Collision and the Tectonosedimentary evolution of Forelands, European Geosciences Union. Stephan Mueller Spec Publ Series 1:151–165
Krzywiec P, Mazur S, Gągała Ł, Kufrasa M, Lewandowski M, Malinowski M, Buffenmyer V (2017) Late Carboniferous thin-skinned compressional deformation above the SW edge of the East European craton as revealed by seismic reflection and potential field data—correlations with the Variscides and the Appalachians. In: Law RD, Thigpen JR, Merschat AJ, Stowell HH (eds) Linkages and feedbacks in orogenic systems. Geological Society of America Memoir, 213, pp 353–372
Książkiewicz M, Samsonowicz J, Rühle E (1965) Zarys geologii Polski. Wyd Geol, Warszawa, p 380
Lamarche J, Scheck-Wenderoth M (2005) 3D structural model of the Polish Basin. Tectonophysics 397:73–91
Mazur S, Scheck-Wenderoth M, Krzywiec P (2005) Different modes of the Late Cretaceous-Early Tertiary inversion in the North German and Polish basins. Intern J Earth Sci 94:782–798
Mazur S, Dunlap WJ, Turniak K, Oberc-Dziedzic T (2006) Age constraints for the thermal evolution and erosional history of the central European Variscan belt: new data from the sediments and basement of the Carboniferous foreland basin in western Poland. J Geol Soc London 163:1011–1024
Mazur S, Aleksandrowski P, Turniak K, Awdankiewicz M (2007) Geology, tectonic evolution and Late Palaeozoic magmatism of Sudetes—an overview. In: Kozłowski A, Wiszniewska J (eds) Granitoids in Poland. Archivum Mineralogiae Monograph 1, pp 59–87
Mazur S, Mikolajczak M, Krzywiec P, Malinowski M, Buffenmyer V, Lewandowski M (2015) Is the Teisseyre-Tornquist Zone an ancient plate boundary of Baltica? Tectonics 34:2465–2477
Mazur S, Mikołajczak M, Krzywiec P, Malinowski M, Lewandowski M, Buffenmyer V (2016) Pomeranian Caledonides, NW Poland—a collisional suture or thin-skinned fold-and-thrust belt? Tectonophysics 692:29–43
Mazur S, Porębski SJ, Kędzior A, Paszkowski M, Podhalańska T, Poprawa P (2018) Refined timing and kinematics for Baltica-Avalonia convergence based on the sedimentary record of a foreland basin. Terra Nova 30:8–16
Mazur S, Aleksandrowski P, Gągała Ł, Krzywiec P, Żaba J, Gaidzik K, Sikora R (2020) Late Palaeozoic strike-slip tectonics versus oroclinal bending at the SW outskirts of Baltica: case of the Variscan belt's eastern end in Poland. Intern J Earth Sci 109:1133–1160
Mazur S, Malinowski M, Maystrenko YP, Gągała Ł (2021) Pre-existing lithospheric weak zone and its impact on continental rifting—the Mid-Polish Trough, Central European Basin System. Glob Planet Change 198:103417
Migoń P, Lidmar-Bergström K (2001) Weathering mantles and their significance for geomorphological evolution of central and northern Europe since the Mesozoic. Earth Sci Rev 56:285–324
Mikołajczak M, Mazur S, Gągała Ł (2019) Depth-to-basement for the East European Craton and Teisseyre-Tornquist Zone in Poland based on potential field data. Intern J Earth Sci 108:547–567
Mikolajczak M, Barmuta J, Ponikowska M, Mazur S, Starzec K (2021) Depth-to-basement study for the western Polish Outer Carpathians from three-dimensional joint inversion of gravity and magnetic data. J Geosci 66:15–36
Milewicz J (1997) Górna kreda depresji północnosudeckiej (lito- i biostratygrafia, paleogeografia, tektonika oraz uwagi o surowcach). Acta Uniw Wratisl 1971. Prace Geol-Miner 61:1–59
Mizerski W (2004) Holy Cross Mountains in the Caledonian, Variscan and Alpine cycles—major problems, open questions. Przegl Geol 52:774–779

Mizerski W (2009) Geologia Polski, 3rd edn. PWN, Warszawa, p 316
Narkiewicz M (2021) Geologiczna historia Polski. Wyd Uniw Warszawskiego, Warszawa, p 280
Nawrocki J, Becker A (ed) (2017) Geological atlas of Poland. Państw Institute of Geology, Warszawa
Nowak J (1927) Zarys tektoniki Polski. II Zjazd Słowiańskich Geografów i Etnografów, Kraków, 160 p
Oberc J (1967) Podział Geologiczny Polski. Kwart Geol 11:389–406
Oberc-Dziedzic T, Kryza R, Pin C (2015) Variscan granitoids related to shear zones and faults: examples from the Central Sudetes (Bohemian Massif) and the Middle Odra Fault Zone. Intern J Earth Sci 104:1139–1166
Pharaoh TC (1999) Palaeozoic terranes and their lithospheric boundaries within the Trans-European Suture Zone (TESZ): a review. Tectonophysics 314:17–41
Plašienka D (2018) Continuity and episodicity in the early Alpine tectonic evolution of the Western Carpathians: how large-scale processes are expressed by the orogenic architecture and rock record data. Tectonics 37:2029–2079
Poprawa P, Šliaupa S, Stephenson R, Lazauskien J (1999) Late Vendian–early Palaeozoic tectonic evolution of the Baltic Basin: regional tectonic implications from subsidence analysis. Tectonophysics 314:219–239
Poprawa P, Paszkowski M, Fanning MC, Pécskay Z, Nawrocki J, Sikorska M (2006) Charakterystyka geochronologiczna obszarów źródłowych dla dolnopaleozoicznych utworów z NW kratonu wschodnioeuropejskiego oraz strefy Koszalin-Chojnice; datowania detrytycznych łyszczyków (K/Ar) i cyrkonów (U/Pb SHRIMP). Prace Państw Inst Geol 186:149–164
Poprawa P (2019) Geological setting and Ediacaran-Palaeozoic evolution of the western slope of the East European Craton and adjacent regions. Ann Soc Geol Polon 89:347–380
Pożaryski W (1956) Podział strukturalno-geologiczny Polski jako podstawa badań. Przegl Geol 6:237–241
Pożaryski W (1974) Podział obszaru Polski na jednostki tektoniczne. In: Pożaryski W (ed) Budowa Geologiczna Polski, T. IV. Tektonika, Cz. 1. Niż Polski. Wyd Geol, Warszawa, pp 24–34
Różycka M, Migoń P (2017) Tectonic geomorphology of the Sudetes (Central Europe)—a review and re-appraisal. Ann Soc Geol Polon 87:275–300
Schmid SM, Bernoulli D, Fügenschuh B, Maţenco L, Schefer S, Schuster R, Tischler M, Ustaszewski K (2008) The Alpine-Carpathian-Dinaridic orogenic system: correlation and evolution of tectonic units. Swiss J Geosci 101:139–183
Sobczyk A, Danišík M, Aleksandrowski P, Anczkiewicz A (2015) Post-Variscan cooling history of the central Western Sudetes (NE Bohemian Massif, Poland) constrained by apatite fission-track and zircon (U-Th)/He thermochronology. Tectonophysics 649:47–57
Sobczyk A, Sobel ER, Georgieva V (2020) Meso-Cenozoic cooling and exhumation history of the Orlica-Śnieżnik Dome (Sudetes, NE Bohemian Massif, Central Europe): Insights from apatite fission-track thermochronometry. Terra Nova 32:122–133
Stupnicka E (1989) Geologia regionalna Polski. Wyd Geol, Warszawa, p 286
Suess FE (1912) Die moravische Fenster und ihre Beziehung zum Grundgebirge des Hohen Gesenkes. Denkschriften Österr Akad Wissen, Math-Naturwiss Klasse 88:541–631
Teisseyre H (ed) (1957) Geologia regionalna Polski, t. III, Sudety, cz. 1: Utwory przedtrzeciorzędowe. Pol Tow Geol, Kraków, p 300
Turniak K, Mazur S, Domańska-Siuda J, Szuszkiewicz A (2014) SHRIMP U-Pb zircon dating for granitoids from the Strzegom-Sobótka Massif, SW Poland: constraints on the initial time of Permo-Mesozoic lithosphere thinning beneath Central Europe. Lithos 208:415–429
Wilde-Piórko M, Świeczak M, Grad M, Majdański M (2010) Integrated seismic model of the crust and upper mantle of the Trans-European Suture zone between the Precambrian craton and Phanerozoic terranes in Central Europe. Tectonophysics 481:108–115
Żelaźniewicz A, Aleksandrowski P, Buła Z, Karnkowski PH, Konon A, Oszczypko N, Ślączka A, Żaba J, Żytko K (2011) Regionalizacja tektoniczna Polski. Komitet Nauk Geol PAN, Wrocław, p 60
Żelaźniewicz A, Oberc-Dziedzic T, Slama J (2020) Baltica and the Cadomian orogen in the Ediacaran–Cambrian: a perspective from SE Poland. Intern J Earth Sci 109:1503–1528
Ziegler PA (1987) Late Cretaceous and Cenozoic intraplate compressional deformations in the Alpine foreland—a geodynamic model. Tectonophysics 137:389–420
Znosko J (ed) (1998) Geological map of Poland 1:500 000. Państwowy Instytut Geologiczny, Warszawa
Żytko K, Gucik S, Ryłko W, Oszczypko N, Zając R, Garlicka I, Nemčok J, Eliáš M, Menčík E, Dvořák J, Straník Z, Rakús M, Matejovska O (1989) Geological map of the Western Outer Carpathians and their foreland without Quaternary formations. In: Poprawa D, Nemčok J (eds) Geological Atlas of the Western Outer Carpathians and their Foreland, 1:500,000. Polish Geological Institute, Warszawa

Stanisław Mazur is Professor at the Institute of Geological Sciences, Polish Academy of Sciences, and currently serves as director. His main research interests are tectonics, structural geology, basin analysis and applied geophysics. He earned his Ph.D. in structural geology from the University of Wroclaw. Then, he attended structural geology training at University College Dublin (1994) and was a postdoctoral fellow at the GeoforschungsZentrum Potsdam (2003–2004). Prior to joining the Institute of Geological Sciences, he worked for British Geophysical Consulting Companies (Getech, ARKeX) for 11 years, where he was responsible for technical supervision of multi-client projects. He was also a visiting research fellow at the University of Leeds in 2008–2011 and 2014–2020. He has been a member of scientific expeditions to Greenland, Spitsbergen, British Columbia and North Labrador. Since 2018, he leads the Depositional Systems Research Group (DEPOS) in the Research Centre in Kraków.

Quaternary History

2

Zdzisław Jary and Leszek Marks

Abstract

Studies of Quaternary deposits in Poland accelerated during the last 70 years when extensive geological mapping programme was initiated. The older Fennoscandian glaciations were generally more extensive, reaching the Carpathians and the Sudetes in the south, whereas the younger ones were less widespread. Deposits of older glaciations are scarce in southern Poland and due to long-lasting fluvial erosion, denudation and loess accumulation, they are only slightly reflected in the present-day landscape. The modern river pattern in Poland inherits large latitudinal reaches of ice-marginal spillways that were the main run-off trains of meltwaters during the glaciations. Repeated climate changes resulted in rebuilding of the landscape and transformation of sedimentary environments. In cold periods, clastic sediments, released by melting of ice sheets, were deposited in different environments, whereas the occurrence of permafrost controlled the periglacial processes. During climate ameliorations, fluvial accumulation and soil formation processes occurred. Human-induced transformations of the landscape, vegetation and habitats in Poland were initiated in the Middle Holocene.

Keywords

Quaternary · Glaciations · Interglacials · River valleys · Ice-marginal spillways · Loess · European Sand Belt

Z. Jary (✉)
Institute of Geography and Regional Development, University of Wrocław, pl. Uniwersytecki 1, 50–137 Wrocław, Poland
e-mail: zdzislaw.jary@uwr.edu.pl

L. Marks
Faculty of Geology, University of Warsaw, Żwirki i Wigury 93, 02–089 Warsaw, Poland
e-mail: leszek.marks@uw.edu.pl

2.1 Introduction

The present territory of Poland is covered in about 70% by Quaternary deposits. This sedimentary cover is considerably thicker in the northern and particularly, in the north-eastern part of the country, generally reaching more than 200 m, but even up to 460 m in areas with glaciotectonic deformations. The Quaternary succession is gradually thinner southwards, where older rocks crop out. In the Carpathians and the Sudetes (cf. Solon et al. 2018), the Quaternary occurrences are limited to river valleys and isolated patches of loess, glacial and hillslope deposits. In the South-Polish Uplands, the Quaternary cover is more compact and dominated by glacial and fluvial deposits, whereas loess is common in the southern part of the region. In general, the Quaternary sedimentary sequence is dominated in Poland by glacial deposits of Pleistocene inland glaciations. They form quasi-latitudinal zones, connected with the extension of successive Fennoscandian ice sheets that advanced several times into the territory of Poland, reaching the least elevated parts of the Carpathians (to ~ 380 m asl) and the Sudetes (to ~ 550 m asl). The ice sheets left mainly vast tills and glaciofluvial sand and gravel series, locally also glaciolacustrine (ice-dammed) fine-grained sediments.

Scientific studies of Quaternary features in Poland were initiated in the first half of the nineteenth century by Staszic (1805; Fig. 2.1a), Pusch (1829–1830, 1831–1836, 1836; Fig. 2.1b). They were the first to notice the presence of Scandinavian erratic boulders in Poland and their accumulation in moraine ridges. Both correctly identified the bones of extinct animal species such as mammoths and rhinos (Staszic 1815; Pusch 1831–1836, 1836). Pusch (1831–1836, 1836) recognized loess in Poland, determined the ranges of erratic boulder accumulations and indicated areas with a continuous cover of Quaternary sediments.

Siemiradzki (1882, 1890; Fig. 2.1c) began a new stage of research on the Quaternary, postulating the occurrence of two continental glaciations within the territory of

P. Migoń and K. Jancewicz (eds.), *Landscapes and Landforms of Poland*,
World Geomorphological Landscapes, https://doi.org/10.1007/978-3-031-45762-3_2

Fig. 2.1 Pioneers of Quaternary studies in Poland. **a** Stanisław Staszic (1755–1826) (source: https://en.wikipedia.org/wiki/Stanisław_Staszic). **b** Georg Gottlieb Pusch, in Polish: Jerzy Bogumił Pusz (1790–1846) (source: http://geoportal.pgi.gov.pl/portal/page/portal/muzeum/historia/poczet/pusch). **c** Józef Siemiradzki (1858–1933) (source: https://rcin.org.pl/dlibra/publication/174199/jozef-siemiradzki?language=pl). **d** Władysław Szafer (1886–1970) (source: Archives of the Institute of Botany, Polish Academy of Sciences)

Poland. Sawicki (1922) added the third, youngest glaciation and Szafer (1928; Fig. 2.1d), based on an analysis of fossil flora, distinguished two separating interglacials and suggested a possible occurrence of the fourth, oldest glaciation. In the twentieth century, glaciations distinguished in Poland were correlated with the ones in the Alps (Penck and Brückner 1909).

Since 1950s, extensive geological mapping carried out in Poland supplied new data and resulted in substantial changes in a number of glaciations and interglacials (Szafer 1953; Rühle 1973; Różycki 1961, 1980; Lindner 1984, 1988; Mojski, 1985), induced by global climatic forcing and correlated with the ones distinguished in the neighbouring countries. A general review of the Quaternary with a full description of glacial and non-glacial phenomena was provided, amongst others, by Różycki (1972), Mojski (1985, 1993, 2005), Lindner (1992). Regional geomorphology was presented by Galon (1972), Klimaszewski (1972, 1988). Figure 2.2 shows global chronostratigraphy and magnetostratigraphy of the last 3 million years, their correlation with marine isotope stages (MIS) and with Quaternary climato- and lithostratigraphy in Western Europe and Poland.

2.2 General Characteristics and Stratigraphy of the Quaternary in Poland

The Quaternary is a short geological period (2.58 million years) that was dominated by alternating coolings and warmings, which resulted in drastic transformations of the environment and changing palaeogeography. During the cool intervals, when the mean annual temperature was 10–15 °C lower than at present, Poland was occupied by the Fennoscandian ice sheet, but glaciers could also develop in the higher parts of the Carpathians and the Sudetes. Atmospheric highs formed above the ice sheet, whereas weaker oceanic circulation and extensive sea ice cover favoured aggradation of permafrost, a southward shift of polar and vegetation zones, and migration of fauna. Huge amounts of water stored in the glacial ice affected drop of the sea level, even to about 130 m (Hughes et al. 2020), which in turn intensified fluvial erosion in terrestrial areas. Lower precipitation and scanty vegetation cover stimulated periglacial processes in extraglacial areas. Growing and melting of ice bodies activated glacioisostatic movements, involving subsidence of the glaciated area and its subsequent uplift during deglaciation. During warmings, when mean annual temperature was occasionally even 2–3 °C higher than at present, reduction of glacial ice released huge volume of water, resulting in a sudden rise of sea level, up to several metres higher than at present, during the last two interglacials (Holsteinian and Eemian).

In spite of cyclic coolings and warmings, mean temperatures have been gradually decreasing in the Quaternary, and this trend was accompanied by increasing temperature amplitudes between warmings and coolings. Simultaneously, participation of thermophilous (exotic) plants was significantly reduced. At the onset of the Pleistocene, open vegetation with taxa typical for a cool steppe predominated (Popescu et al. 2010). During the subsequent warmings, the area was occupied by deciduous and mixed forest, whereas coolings were typified by boreal forest and open, also tundra vegetation.

The first Fennoscandian ice sheet passed across the present-day Baltic Sea basin around 900 ka during the Nidanian Glaciation and occupied central and eastern Poland as far south as the Moravian Gate at the Polish–Czech border (Marks 2012). The extents of successive glaciations are reflected in the present landscape. The ice sheet of the Sanian 1 Glaciation (670–630 ka) was the most extensive, occupied

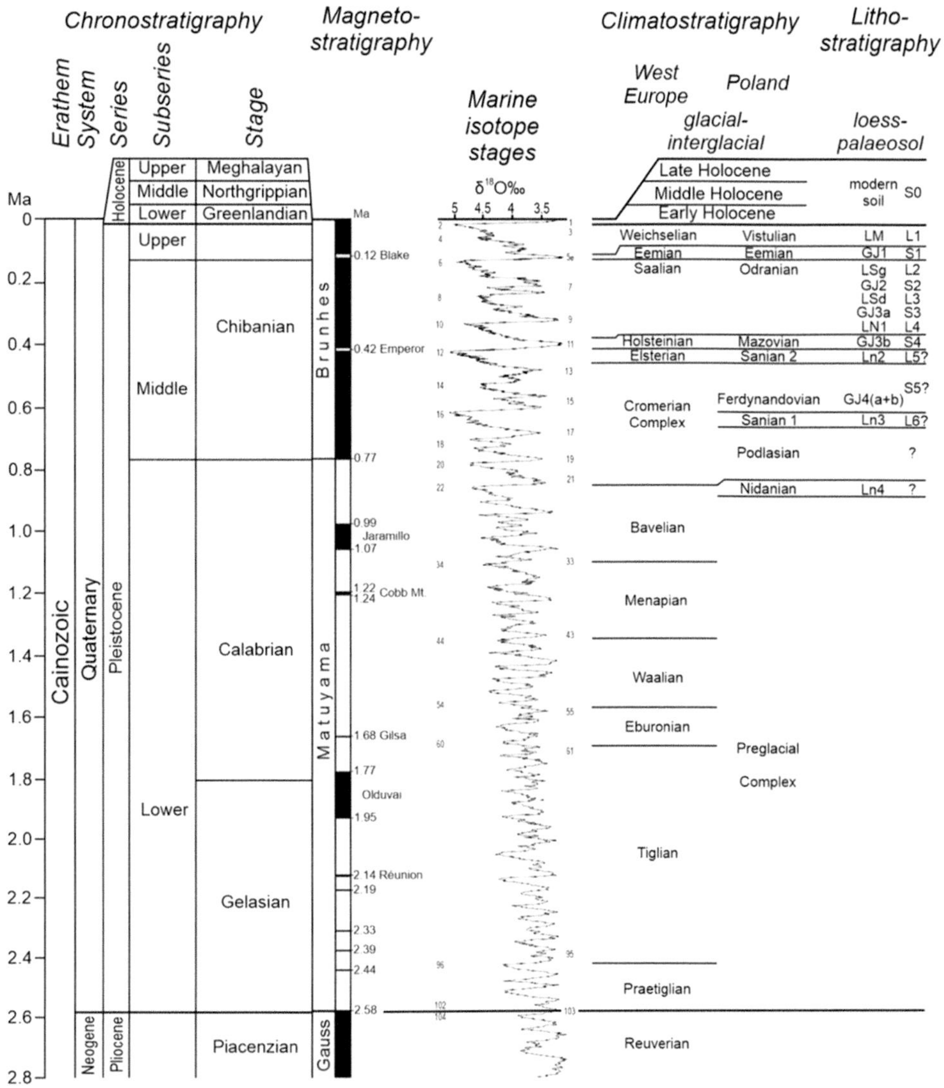

Fig. 2.2 Global chronostratigraphy and magnetostratigraphy of the last 3.0 million years (after Laj and Channel 2007; Head et al. 2021), their correlation with marine isotope stages (after Lisiecki and Raymo 2005) and with the Quaternary climatostratigraphy in Poland and Western Europe after Gibbard et al. (2005), Marks et al. (2016a), modified. Loess–palaeosol lithostratigraphy according to Maruszczak (2001), correlated with unified stratigraphic model of Marković et al. (2015) where loess units are denoted by the letter "L" and soil units by the letter "S"

the Sandomierz Basin and reached the Carpathians (Fig. 2.3). The successive glaciations were more extensive in western Poland, with the younger ones generally less widespread (Marks et al. 2016a). Ice sheets of the Sanian 2 (460–430 ka) and the Odranian (160–130 ka) glaciations had similar extents, reaching the Sudetes and the Holy Cross Mts. The Vistulian (30–15 ka) ice sheet occupied the lakelands of Northern Poland and the western part of central Poland.

The fresh postglacial landscapes at the beginning of the interglacials consisted mostly of lakelands, but they became gradually overgrown and transformed into marshes or were drained by streams. The interglacial vegetation cover in Poland was similar to the modern one (Fig. 2.4). In each of the older interglacials (Podlasian and Ferdynandovian, Fig. 2.2), there were 2–3 warm optima, separated by distinct but non-glacial coolings, whereas each younger interglacial (Mazovian, Eemian and Holocene) had only a single climatic optimum (Lindner et al. 2013).

In the extraglacial zone of the Pleistocene ice sheets, favourable conditions occurred for periglacial processes (Łoziński 1909) and loess accumulation (Tutkovsky 1899). Sequences of loess and intervening interglacial fossil soils provide an excellent archive of Pleistocene climate changes and stratigraphy, independent on the glacial record. Maruszczak (1991a, 2001) distinguished several loess beds in Poland (Fig. 2.2), composed of the younger (LM) through older (LS) to the oldest loess (LN); this scheme was then supplemented by Dolecki (1995). The oldest loess–soil stratigraphic complexes (indicated by question marks in Fig. 2.2) were identified mainly in drilling cores, thus requiring further research and verification in terms of both their genesis and age.

Fig. 2.3 The maximum extents of the main Pleistocene glaciations in Poland and the most important ice-marginal spillways: S1—Sanian 1, S2—Sanian 2, O—Odranian, W—Vistulian

2.3 The River System and Spillways (Ice-Marginal Valleys)

A principal framework of the contemporary river pattern in Poland developed as early as the Early Pleistocene, when the rivers flowed north-westwards to the North Sea or were tributaries of the Baltic river in the north (cf. Gibbard 1988). In the Carpathian foreland, the rivers flowed eastwards to the Dnister valley and the Black Sea. The Pleistocene glaciations considerably transformed the earlier river network, especially through adding large latitudinal reaches of ice-marginal spillways (Fig. 2.3) as important elements inherited up to the modern times (Mojski 1993, 2005). The river pattern in Poland during the Mazovian Interglacial resembled the modern one but the water divide between the Baltic Sea and the Black Sea was farther to the west in south-eastern Poland than at present (Fig. 2.5; Marks and Pavlovskaya 2003).

During the most extensive glaciations (Nidanian, Sanian 1, Sanian 2), the Subcarpathian ice-marginal spillway was the main corridor for meltwater discharge to the Black Sea through the Dnister valley (Lindner and Marks 2015). Meltwaters of the Odranian ice sheet carved the Wrocław–Magdeburg spillway in the west (Keilhack 1913; Woldstedt 1955), the Krzna–Pripyat spillway in the east, whereas in the south they flowed to the Danube drainage basin via the Moravian Gate at the Polish–Czech border. In the successive glacial phases during the Vistulian, meltwaters were collected in the Baruth–Głogów, Warsaw–Berlin and Toruń–Eberswalde spillways (Fig. 2.3). The modern river system in Poland belongs almost entirely to the Baltic Sea drainage basin and is composed mainly of latitudinal and longitudinal reaches. The former are inherited from the Pleistocene ice-marginal spillways which developed during the Pleistocene glaciations outside the glaciated area and collected meltwaters from the ice sheet as well as waters flowing north from the highlands and mountains of southern Poland. The quasi-longitudinal river reaches use either the previous Pleistocene outwash valleys, glacial tunnel valleys and gorges causally connected with ruptures of ice dams of proglacial lakes (Galon 1961; Kozarski 1988; Marks and Pavlovskaya 2003). Final transformations of the present fluvial pattern in Poland occurred at the termination of the last glaciations and in the Holocene.

The rivers were also influenced by glacioisostatic subsidence provoked by successive loading of the area by advancing ice sheets and its uplift during deglaciation. It resulted in repeated rejuvenation of older deep-seated (bedrock) faults and had significant impact on the landscape in Poland (Baraniecka 1971, 1975; Ber 2000), especially on ice sheet extents and the development of river valleys (Marks 1988; Marks et al. 2018). The Sudetes and the Carpathians were generally uplifted during the Pleistocene, and the Carpathian foreland was depressed (Zuchiewicz 1980, 1984a, b, 1989, 1991, 1995). The river network is highly dependent on the location of faults in the mountains and accumulation of vast alluvial cones in their forelands (Starkel, 1985; Krzyszkowski et al. 1995). Higher river terraces in the mountains were formed during the Pleistocene, but mainly during the last glaciation in the lowland area, whereas floodplains developed in the Holocene. In the mountain foreland, vast alluvial fans developed at the outlets of mountain rivers.

2.4 Quaternary Palaeogeography of Poland

The older glaciations (Nidanian, Sanian 1 and Sanian 2) were the most widespread but due to later fluvial erosion, denudation and loess accumulation, their deposits are scarce and their impact is only slightly reflected in the present-day landscape of southern Poland (Fig. 2.3). Deposits and landforms of the Odranian Glaciation are widely distributed in south-western and eastern part of central Poland. The ice sheet of the last Pleistocene glaciation (Vistulian) modelled the landscape in northern and mid-western Poland, at present dominated by glacial landforms.

The last glacial stage in Poland is informally subdivided into Early Vistulian, Lower Plenivistulian, Interplenivistulian or Middle Plenivistulian, and Upper Plenivistulian (Fig. 2.6), and its final part is referred commonly as the Late Glacial. The climate showed considerable variability, with successive warm and cool intervals (Fig. 2.6). The Early Vistulian was characterized by higher continentality in eastern Poland, especially during warm

Fig. 2.4 Pollen diagrams of the interglacials: **a** Ferdynandovian (after Pidek and Małek 2010, modified). **b** Mazovian (after Hrynowiecka and Pidek 2017, modified). **c** Eemian (after Zalat et al. 2021, modified)

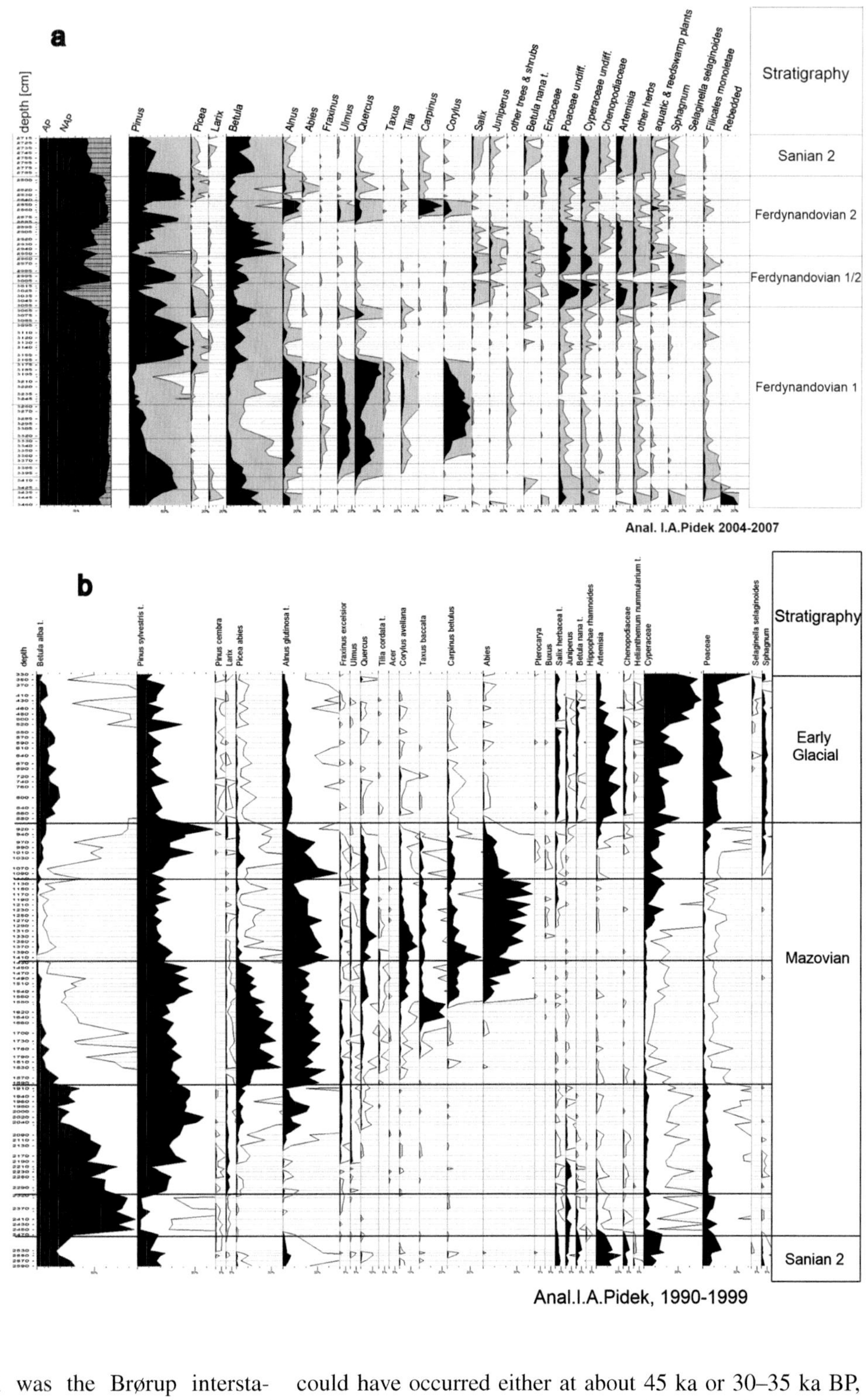

interstadials (most continental was the Brørup interstadial). A limited palaeobotanical record of the Pleniglacial is likely a result of lower mean annual temperatures and/or lower mean annual precipitation, as indicated by desiccation of lakes and replacement of lake sediments by peat or mineral deposits (Marks et al. 2016b).

During the last Vistulian Glaciation, there were at least two major ice sheet advances. The first Vistulian advance could have occurred either at about 45 ka or 30–35 ka BP, or both, and the ice sheet occupied mid-northern Poland (Wysota 2002; Marks 2012). Landforms and sediments of this glacial event are entirely covered by younger glacial deposits. The main ice sheet advance occurred during the Upper Plenivistulian, and the maximum extent of inland ice was asynchronous (Fig. 2.7): at 24 ka BP in western Poland (Leszno Phase) and 19–20 ka BP in central and

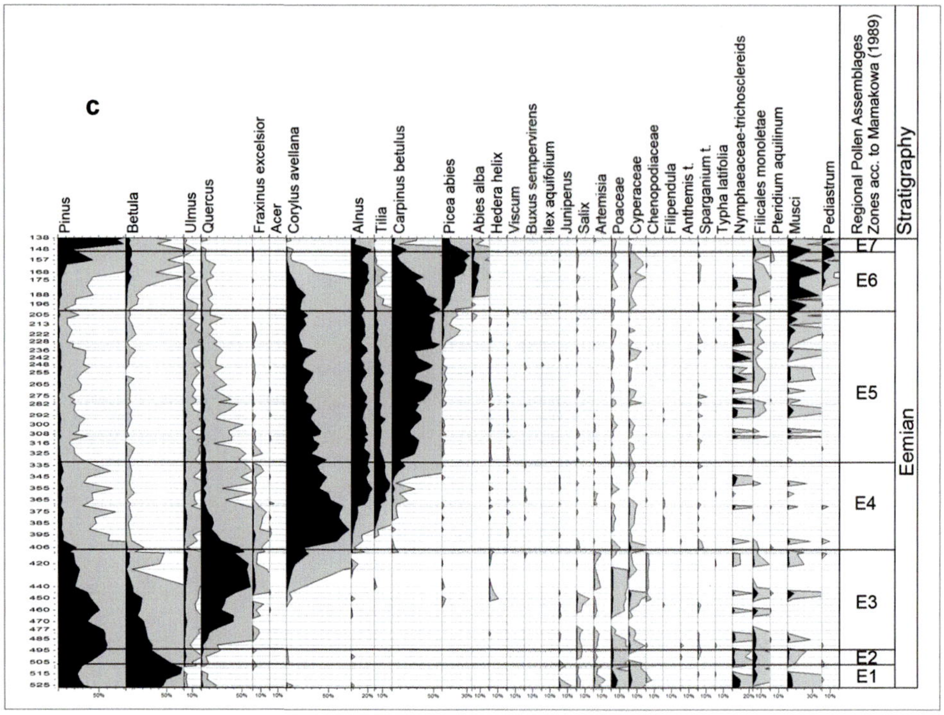

Fig. 2.4 (continued)

Anal. A. Bober, I.A.Pidek

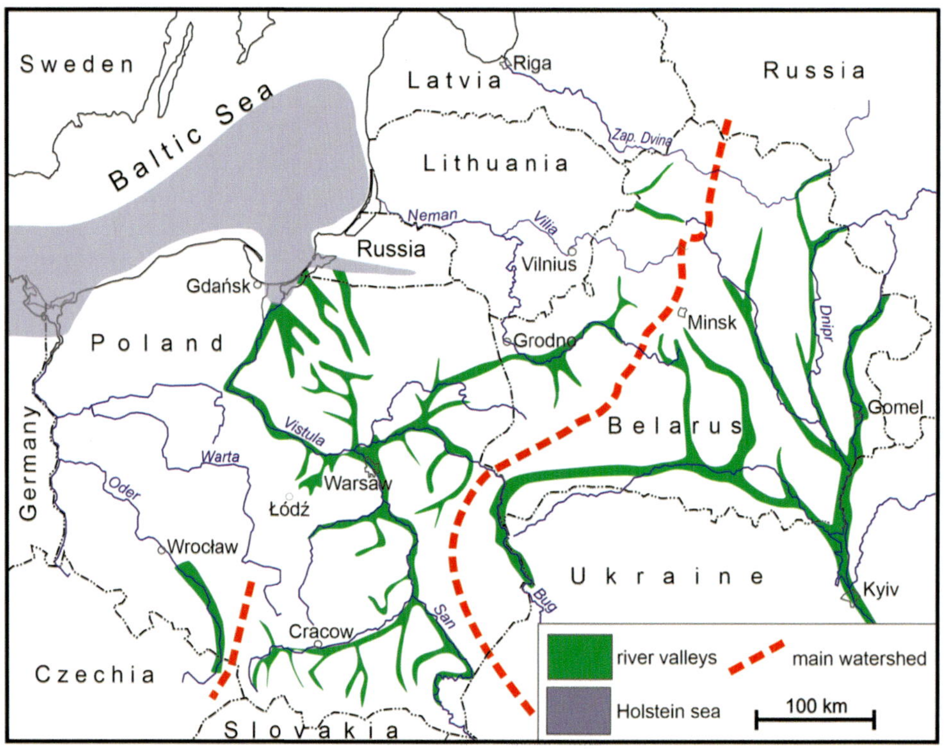

Fig. 2.5 River valleys in the Mazovian Interglacial in mid-eastern Poland and in Belarus (after Marks and Pavlovskaya 2003, modified)

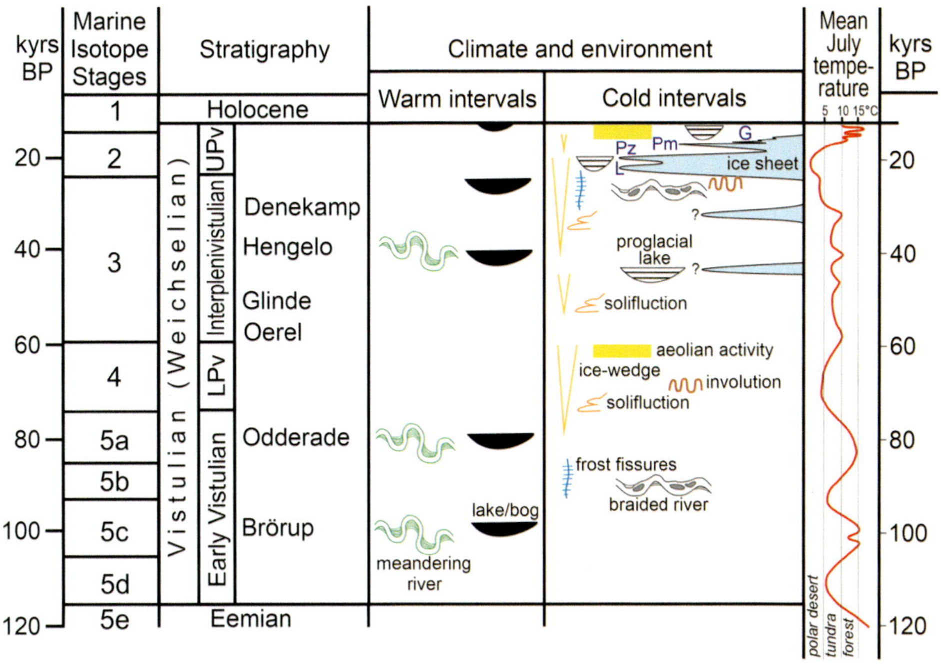

Fig. 2.6 Vistulian stratigraphy, climate and environmental changes in northern and central Poland correlated with Western European stratigraphy, temperature curve after Roman et al. (2014). LPv—Lower Plenivistulian, UPv—Upper Plenivistulian; glacial phases: L—Leszno (Brandenburg), Pz—Poznań (Frankfurt), Pm—Pomeranian, G—Gardno. After Marks et al. (2016b), modified

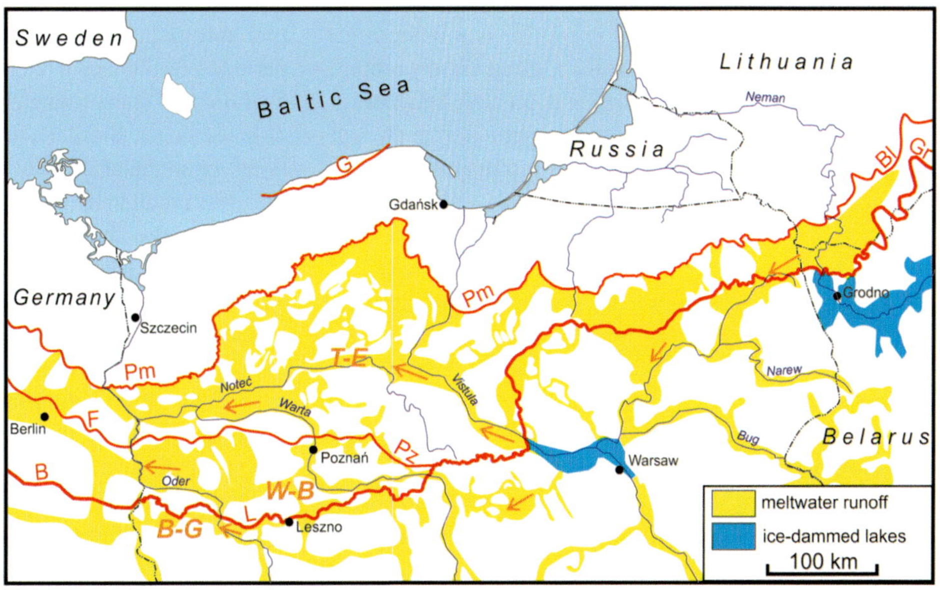

Fig. 2.7 Late Vistulian glacial limits in Poland and neighbouring areas after Marks (2012), modified. Glacial phases: L—Leszno (B—Brandenburg), Pz—Poznań (F—Frankfurt, Gr—Grūda), Pm—Pomeranian (Bl—Baltija), G—Gardno, SB—Słupsk Bank, SMB—Southern Middle Bank; ice-marginal spillways: B-G—Baruth–Głogów, T-E—Toruń–Eberswalde, W-B—Warsaw–Berlin, arrows indicate meltwater run-off routes

eastern Poland (Poznań Phase). They were followed by the less extensive standstill of the ice margin at 16–17 ka BP (Pomeranian Phase).

The northern edge of the Carpathians was reached by ice sheets of the Nidanian and Sanian 1 glaciations, whereas ice sheets of the Sanian 2 and Odranian glaciations advanced to the Sudetes. The highest parts of the both these mountain ranges (Tatra, Karkonosze) were occupied by local glaciers, at least during the last two glaciations.

2.5 Climate Changes Recorded in the Last Glacial Loess–Palaeosol Sequences

The Polish part of the northern European loess belt was further shaped in cold climates of the foreland zone of Pleistocene continental glaciations. Unlike the loess of southern Europe, loess–palaeosol sequences in Poland were strongly influenced by periglacial processes and environmental transformations.

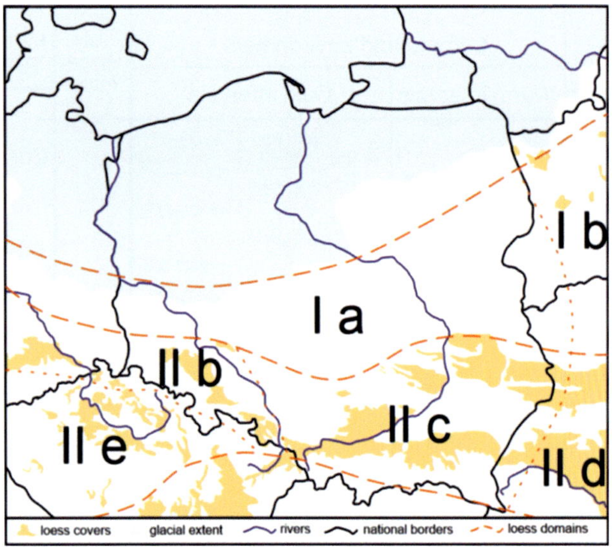

Fig. 2.8 Upper Plenivistulian domains (roman numerals) and subdomains (lower-case letters) of loess and loess derivatives in Poland (after Lehmkuhl et al. 2021, modified)

Lehmkuhl et al. (2021) distinguished two loess domains and three subdomains in the territory of Poland (Fig. 2.8). The western protogenetic subdomain (Ia) is a western part of the domain called the Vistulian marginal or protogenetic zone (I). This subdomain was located between the Vistulian ice sheet limit and the northern European loess belt and is interpreted as a protogenetic transport and accumulation zone with patches of sandy loess and coversands (Kozarski and Nowaczyk 1991). In the Late Vistulian, this area has been transformed into a part of the European Sand Belt. There are two subdomains of the northern European loess belt (II) in Poland: the Western European continental subdomain (IIb) in south-western Poland and the Central European continental subdomain (IIc) in central–eastern Poland. This domain contains loess that was mainly deposited during the Last Glacial. Geochemical and mineralogical analyses show that the majority of loess material which has its origin in northern Europe was delivered by the Fennoscandian ice sheets and contains repeatedly recycled material (Van Loon 2006; Nawrocki et al. 2019; Skurzyński et al. 2020; Baykal et al. 2021; Lehmkuhl et al. 2021). Loess in south-western Poland is distributed in isolated patches differing in sediment thickness, stratigraphy and physical properties (Jary 2010; Jary and Ciszek 2013). The loess material was derived and deposited within a rather narrow corridor between the Vistulian ice sheet and the Sudetes Mountains and then often redistributed by fluvial processes (Smalley et al. 2009; Badura et al. 2013) and blown outside the river valleys. The thickness, continuity and stratigraphic differentiation of loess–palaeosol sequences increase towards the east where loess from several glacial cycles forms thick sequences, locally up to 40 m (Cegła 1972; Jersak 1973; Maruszczak 1991b; Jary and Ciszek 2013).

The Vistulian loess–palaeosol sequences are the most suitable for reliable palaeoclimatic reconstruction. The last glacial loess in Poland was deposited in the dynamically changing environment of the extraglacial zone of the Vistulian ice sheet. Most likely, aeolian deposition in this area was discontinuous (Moska et al. 2015, 2017, 2019), and a characteristic feature of the loess–palaeosol sequences is considerable diversity of soil–sedimentary records, a common presence of erosional unconformities and periglacial features, which usually form well-developed horizons (Jary 2009). Despite regional differentiation, the periglacial record indicates the occurrence of four main cold stages during the last glacial period (Fig. 2.9).

The two earliest cold stages are represented by two generations of cryogenic frost wedges with primary mineral infilling and are related to the S1 pedocomplex development. These sudden, short cold periods of severe, continental climate are correlated with the Early Vistulian Herning (MIS 5d) and Rederstall (MIS 5b) stadials. Opposite to some opinions, such cryogenic wedges cannot be interpreted as a convincing evidence of former permafrost. They probably indicate a deep seasonal frost action through the periods of cold and dry continental climate, with a significant drop of temperature during the coldest month and lack of snow cover. Similarly to the present-day climatic conditions, continental climate in the eastern part of the study area was much more severe (Jary 2009).

During the third cold stage, two periglacial horizons were formed. The first (cryoturbation/gelifluction level in the top of S1 pedocomplex) possibly corresponds to rapid climate shift between MIS 5 and MIS 4, which caused permafrost expansion (Fig. 2.9). Further climate deterioration led to ice wedge formation. The distribution and morphological features of this generation of ice wedge pseudomorphs indicate their development in-between continuous and discontinuous permafrost (Jary 2009, 2010). Mitigation of climatic conditions between MIS 4 and MIS 3 caused permafrost decay.

The first phase of the periglacial stage IV is signified by cryoturbation/gelifluction level in the top of the interplenivistulian soil complex L1SS1. It probably corresponds to a rapid climate change which caused the next permafrost expansion between MIS 3 and MIS 2. Further climate deterioration led to ice wedge formation of younger generation. Despite opinions of their epigenetic origin, the syngenetic and anti-syngenetic origin of ice wedges should be taken into consideration (Mackay 1990). Climate improvement between MIS 2 and MIS 1 caused ultimate permafrost decay.

The common occurrence and considerable size of ice wedge pseudomorphoses of the youngest generation

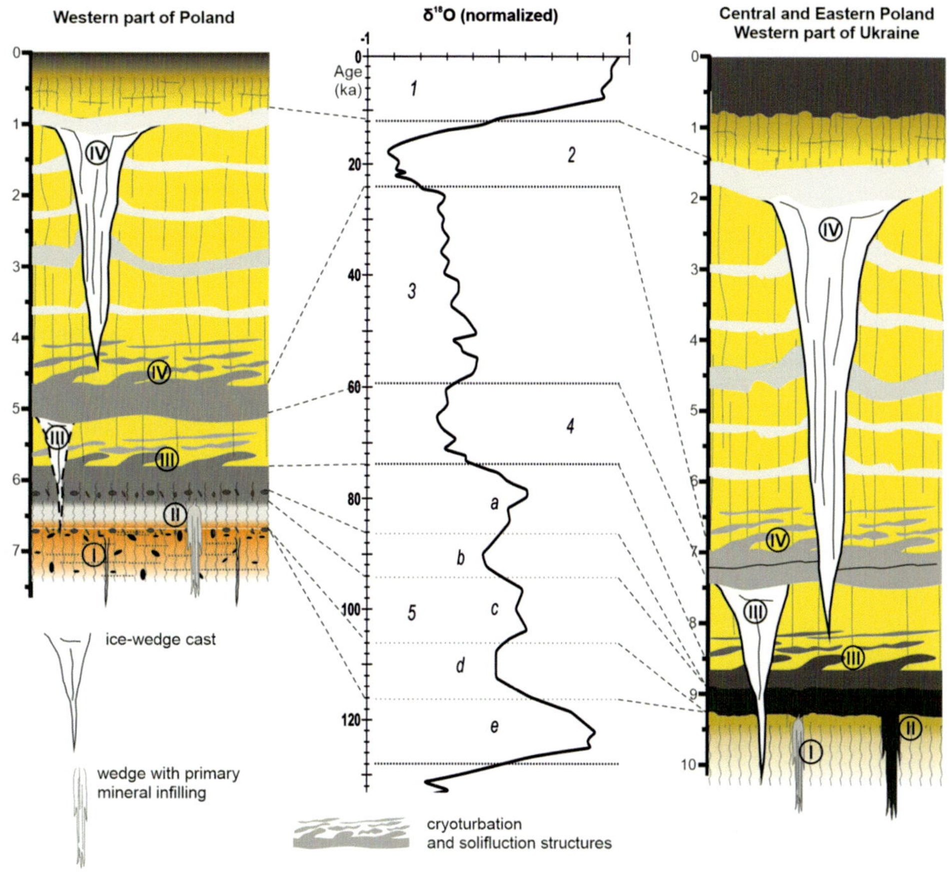

Fig. 2.9 Stages of periglacial activity (I–IV) within the Late Pleistocene loess–palaeosol sequences in the central part of the northern European loess belt and their correlation with the deep-sea record (after Jary 2009, modified)

indicate the occurrence of continuous permafrost in central and eastern part of Poland. However, in south-western Poland, these structures probably developed in the presence of both continuous (Trzebnica Hills) and discontinuous permafrost (Głubczyce Plateau). This record shows clear climatic differences between the western and eastern parts of Poland, as well as a strong meridional climate gradient during the Last Glacial.

2.6 Late Glacial and the European Sand Belt

Climate change at the end of the Late Plenivistulian and in the Late Glacial (14.7–11.7 ka) caused fundamental rebuilding of the landscape and sedimentary environments in all morphogenetic zones in Poland (Kozarski 1993; Weckwerth 2018). A huge amount of clastic sediments was released by melting of the ice sheet and then deposited in different environments by processes modelling ice-marginal valleys, outwash plains, terminal moraines and moraine plateaus. Intensive fluvial processes and rapid permafrost degradation initiated aeolian activity and mobilization of huge amounts of dry sandy deposits. A significant part of the territory of Poland was covered by the European Sand Belt (Zeeberg 1998; Zieliński 2016). Intensive aeolian processes took place in the cold periods of the Late Glacial and were interrupted by climate ameliorations of Bølling and Allerød interstadials, when soil formation processes prevailed (Nowaczyk 1986; Manikowska 1995; Goździk 2007).

Moska et al. (2022) established the chronology of aeolian and soil forming processes from the Late Glacial to the present time in extraglacial, glaciomarginal and glacial zones (Fig. 2.10), indicating distinct differences in aeolian activity phases between them. In the extraglacial zone, aeolian processes started in the Oldest Dryas (coversand deposition) and continued until the beginning of the Holocene (dune deposition). Aeolian phases were terminated by pedogenic processes during climate ameliorations (Bølling and Allerød interstadials) and in the beginning of the Holocene. Holocene phases of reactivation of aeolian processes were induced mainly by human impact and are of local origin (Fig. 2.10).

Remarkable human transformations of the landscape, vegetation and habitats in Poland were initiated around 7.5 ka BP, reflected by an increasing number of the Neolithic settlements and intensified farming activities. Such gradually intensifying human impact continued during the Bronze Age, Iron Age, Roman Period and Medieval Period, but it was interrupted by short withdrawals of people during less favourable climatic conditions. During the last 200

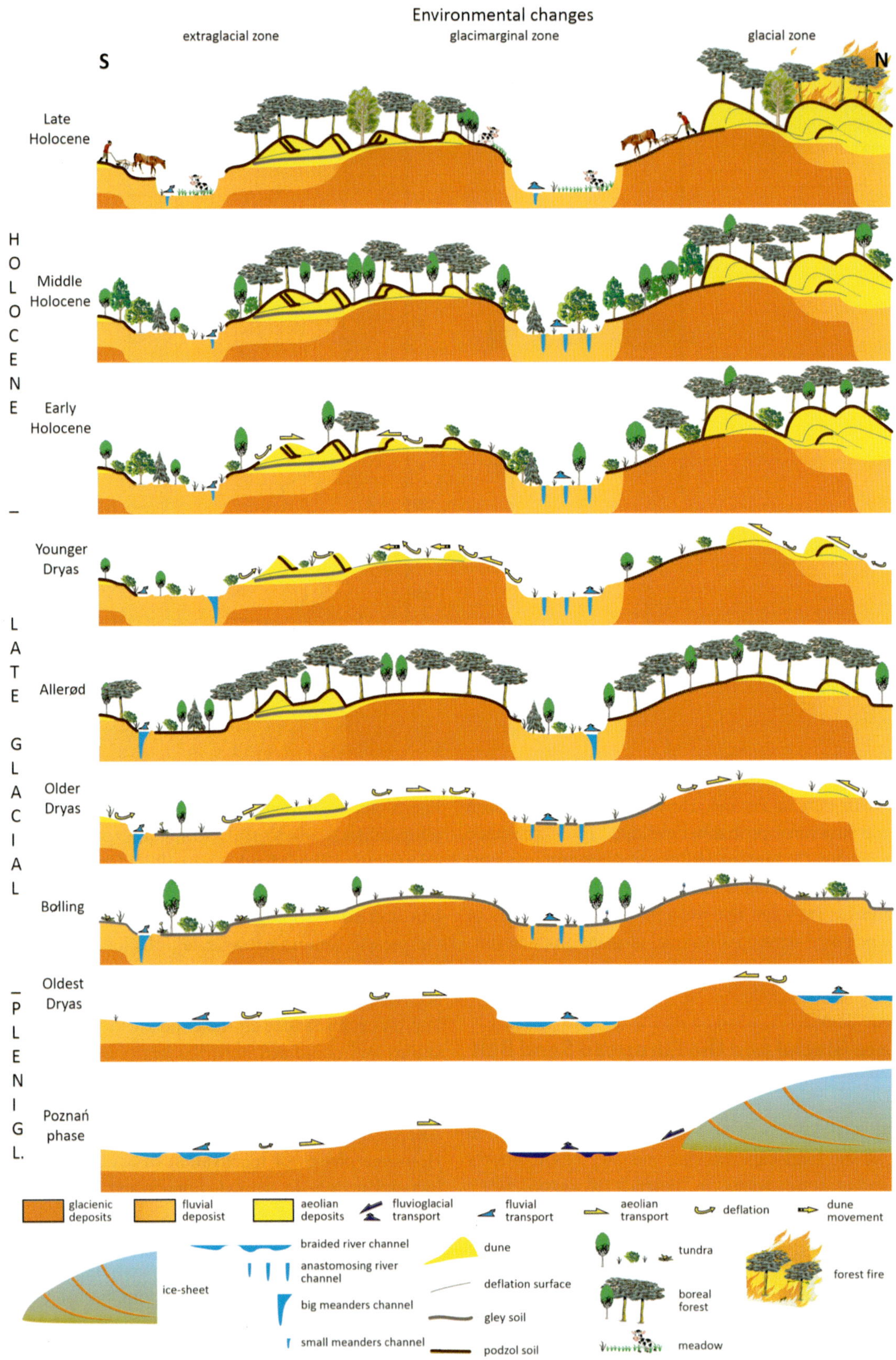

Fig. 2.10 Landscape development in the Pleniglacial, Late Glacial and Holocene, including relief transformation, vegetation change (not to scale) and human impact (after Moska et al. 2022, modified)

years, abrupt expansion of farming, industry and settlements was accompanied by extensive transformation of the landscape and hydrographic network.

2.7 Summary and Conclusions

The first studies of Quaternary deposits in Poland were initiated as early as the beginning of the nineteenth century, but they were considerably supplemented and updated during the last 70 years, with new data supplied by the extensive geological mapping programme. The first Fennoscandian ice sheet passed across the present-day Baltic Sea basin around 900 ka and occupied the territory of Poland. The older glaciations were generally more extensive, reaching the Carpathians and the Sudetes in the south, whereas the younger ones were in principle less widespread. Due to long-lasting fluvial erosion, denudation and loess accumulation in southern Poland, deposits of older glaciations are scarce and only slightly reflected in the contemporary landscape. Lakes of postglacial lakeland landscapes were gradually overgrown during interglacials and transformed into marshes or were drained by streams. The modern river system in Poland developed already in the Early Pleistocene, but the Pleistocene glaciations considerably transformed the earlier river network, adding large latitudinal reaches of the ice-marginal spillways that were the main draining routes of meltwaters during the glaciations. Favourable conditions for periglacial processes and loess accumulation to occur prevailed in the extraglacial zone of Pleistocene ice sheets. Repeated climate changes in the Quaternary resulted in fundamental rebuilding of the landscape and sedimentary environments in every morphogenetic zone in Poland. In cold periods, clastic sediments were released by melting of ice sheets and then deposited in different environments, accompanied by intensive fluvial processes and aggradation of permafrost. During climate amelioration, fluvial accumulation and soil formation processes occurred. Human transformations of the landscape, vegetation and habitats were initiated at around 7.5 ka BP.

References

Badura J, Jary Z, Smalley I (2013) Sources of loess material for deposits in Poland and parts of Central Europe: the lost big river. Quat Intern 296:15–22

Baraniecka MD (1971) Staroczwartorzędowe rowy tektoniczne i ich osady. Kwart Geol 15:358–372

Baraniecka MD (1975) Zależności wykształcenia osadów czwartorzędowych od struktur i dynamiki podłoża w środkowej części Niżu Polskiego. Biul Inst Geol 288:5–97

Baykal Y, Stevens T, Engström-Johansson A, Skurzyński J, Zhang H, He J, Lu H, Adamiec G, Költringer C, Jary Z (2021) Detrital zircon U-Pb age analysis of last glacial loess sources and proglacial sediment dynamics in the Northern European Plain. Quat Sci Rev 274:107265

Ber A (2000) Plejstocen Polski północno-wschodniej w nawiązaniu do głębszego podłoża i obszarów sąsiednich. Prace Państw Inst Geol 170:1–89

Cegła J (1972) Sedymentacja lessów Polski. Acta Univ Wratisl 168, Studia Geogr 17, p 72

Dolecki L (1995) Litologia i stratygrafia mezoplejstoceńskicjh utworów lessowych południowo-wschodniej części Wyżyny Lubelskiej. Wyd UMCS, Lublin, p 169

Galon R (1961) Morphology of the Noteć-Warta (or Toruń-Eberswalde) ice marginal streamway. Prace Geogr Inst Geogr PAN 29:1–129

Galon R (ed) (1972) Geomorfologia Polski, 2: Niż Polski. PWN, Warszawa, p 373

Gibbard PL (1988) The history of the great northwest European rivers during the past three million years. Phil Trans Royal Soc B 318:559–602

Gibbard PL, Boreham S, Cohen KM, Moscariello A (2005) Global chronostratigraphical correlation table for the last 2.7 million years. Boreas 34:(unpaginated)

Goździk J (2007) The Vistulian aeolian succession in central Poland. Sediment Geol 193:211–220

Head MJ, Pillans B, Zalasiewicz JA, Alloway B, Beu AG, Cohen KM, Gibbard PL, Knudsen KL, Kolfschoten T, van Lirer F, Litt T, Liu J, Marks L, McManus J, Piotrowski JA, Räsänen M, Rasmussen SO, Saito Y, Tesakov A, Turner C, Walker M, Zazo-Cardeña C (2021) Formal ratification of subseries for the Pleistocene series of the quaternary system. Episodes 44:241–247

Hrynowiecka A, Pidek IA (2017) Older and Younger Holsteinian climate oscillations in the palaeobotanical record of the Brus profile (SE Poland). Geol Quart 61:723–737

Hughes PD, Gibbard PL, Ehlers J (2020) The "missing glaciations" of the Middle Pleistocene. Quat Res 96:161–183

Jary Z (2009) Periglacial markers within the Late Pleistocene loess-palaeosol sequences in Poland and western part of Ukraine. Quat Intern 198:124–135

Jary Z (2010) Loess-soil sequences as a source of climatic proxies: an example from SW Poland. Geologija 52, 1–4(69–72):40–45

Jary Z, Ciszek D (2013) Late Pleistocene loess-palaeosol sequences in Poland and western Ukraine. Quat Intern 296:37–50

Jersak J (1973) Litologia i stratygrafia lessu wyżyn południowej Polski. Acta Geogr Lodz 32:139

Keilhack K (1913) Die geologischen Verhältnisse des Niederlausitzer Braunkohlengebietes mit besonderer Berücksichtigung der Felder der Ilse B.-A. in Grube Ilse. Selbstverlag der Grube Ilse, p 41

Klimaszewski M (ed) (1972) Geomorfologia Polski, 1: Polska południowa: góry i wyżyny. PWN, Warszawa, p 384

Klimaszewski M (1988) Rzeźba Tatr Polskich. PWN, Warszawa, p 707

Kozarski S (1988) Origin of pradolinas: a discussion of mistakes ideas. Z Gletscherkunde Glazialgeol 24:75–92

Kozarski S (1993) Late Plenivistulian deglaciation and the expansion of the periglacial zone in NW Poland. Geol Mijnb 72:143–157

Kozarski S, Nowaczyk B (1991) Lithofacies variation and chronostratigraphy of Late Vistulian and Holocene aeolian phenomena in northwestern Poland. Z Geomorph NF, Suppl-Bd 90:107–122

Krzyszkowski D, Migoń P, Sroka W (1995) Neotectonic Quaternary history of the Sudetic Marginal Fault, SW Poland. Folia Quat 66:73–98

Laj C, Channel JET (2007) Geomagnetic excursions. Treatise Geophys 5:373–416

Lehmkuhl F, Nett JJ, Pötter S, Schulte P, Sprafke T, Jary Z, Antoine P, Wacha L, Wolf D, Zerboni A, Hošek J, Marković SB, Obreht I, Sümegi P, Veres D, Zeeden C, Boemke B, Schaubert V, Viehweger J, Hambach U (2021) Loess landscapes of Europe—mapping, geomorphology, and zonal differentiation. Earth Sci Rev 215:103496

Lindner L (1984) An outline of Pleistocene chronostratigraphy in Poland. Acta Geol Polon 34:27–49
Lindner L (1988) Stratigraphy and extents of Pleistocene continental glaciations in Europe. Acta Geol Polon 38:63–83
Lindner L (ed) (1992) Czwartorzęd: osady, metody badań, stratygrafia. Wyd PAE, Warszawa, p 683
Lindner L, Marks L (2015) Early and Middle Pleistocene fluvial series in northern foreland of the Carpathians (Poland and Ukraine) and their relation to Dnister River terraces. Quat Intern 357:22–32
Lindner L, Marks L, Nita M (2013) Climatostratigraphy of interglacials in Poland: Middle and Upper Pleistocene lower boundaries from a Polish perspective. Quat Intern 292:113–123
Lisiecki LE, Raymo ME (2005) A Pliocene-Pleistocene stack of 57 globally distributed benthic δ18O records. Palaeoceanography 20:PA1003
Łoziński W (1909) Über die mechanische Verwitterung der Sandsteine im gemässigten klima. Bull Intern L'acad Sci Cracovie Sciences Math Nat 1:1–25
Mackay JR (1990) Some observations on the growth and deformation of epigenetic, syngenetic and anti-syngenetic ice wedges. Permafrost Perigl Proc 1:15–29
Manikowska B (1995) Aeolian activity differentiation in the area of Poland during the period 20–8 ka BP. Biul Perygl 34:125–166
Marković SB, Stevens T, Kukla GJ, Hambach U, Fitzsimmons KE, Gibbard Ph, Buggle B, Zech M, Guo Z, Hao Q, Wu H, O'Hara-Dhand K, Smalley IJ, Ujvari G, Sumegi P, Timar-Gabor A, Veres D, Sirocko F, Vasiljević DA, Jary Z, Svensson A, Jović V, Lehmkuhl F, Kovacs J, Svircev Z (2015) Danube loess stratigraphy—towards a pan-European loess stratigraphic model. Earth Sci Rev 148:228–258
Marks L (1988) Relation of substrate to the quaternary paleorelief and sediments, western Mazury and Warmia (northern Poland). Kwart AGH Geologia 14(1):1–76
Marks L (2012) Timing of the Late Vistulian (Weichselian) glacial phases in Poland. Quat Sci Rev 44:81–88
Marks L, Dzierżek J, Janiszewski R, Kaczorowski J, Lindner L, Majecka A, Makos M, Szymanek M, Tołoczko-Pasek A, Woronko B (2016a) Quaternary stratigraphy and palaeogeography of Poland. Acta Geol Polon 66:403–427
Marks L, Gałązka D, Woronko B (2016b) Climate, environment and stratigraphy of the Pleistocene last glacial stage in Poland. Quat Intern 420:259–271
Marks L, Karabanov A, Nitychoruk J, Bahdasarau M, Krzywicki T, Majecka A, Pochocka-Szwarc K, Rychel J, Woronko B, Zbucki Ł, Hradunova A, Hrychanik M, Mamchyk S, Rylova T, Nowacki Ł, Pielach M (2018) Revised limit of the Saalian ice sheet in central Europe. Quat Intern 478:59–74
Marks L, Pavlovskaya IE (2003) The Holsteinian Interglacial river network of mid-eastern Poland and western Belarus. Boreas 32:337–346
Maruszczak H (1991a) Zróżnicowanie stratygraficzne lessów polskich. In: Maruszczak H (ed) Podstawowe profile lessów w Polsce. Wyd UMCS, Lublin, A, 13–35
Maruszczak H (1991b) Ogólna charakterystyka lessów w Polsce. In: Maruszczak H (ed) Podstawowe profile lessów w Polsce. Wyd UMCS, Lublin, A, 1–12
Maruszczak H (2001) Schemat stratygrafii lessów i gleb śródlessowych w Polsce. In: Maruszczak H (ed) Podstawowe profile lessów w Polsce II. Wyd UMCS, Lublin, 17–29
Mojski JE (1985) Geology of Poland I, Stratigraphy 3b, Cainozoic, Quaternary. Wydawn Geol, Warszawa, p 244
Mojski JE (1993) Europa w plejstocenie, ewolucja środowiska przyrodniczego. Wyd PAE, Warszawa, p 333
Mojski JE (2005) Ziemie polskie w czwartorzędzie, zarys morfogenezy. Państw Institute of Geology, Warszawa, p 404
Moska P, Adamiec G, Jary Z, Bluszcz A (2017) OSL chronostratigraphy for loess deposits from Tyszowce, Poland. Geochronometria 44:307–318
Moska P, Jary Z, Adamiec G, Bluszcz A (2015) OSL chronostratigraphy of a loess-palaeosol sequence in Złota using quartz and polymineral fine grains. Radiat Meas 81:23–31
Moska P, Jary Z, Adamiec G, Bluszcz A (2019) Chronostratigraphy of a loess-palaeosol sequence in Biały Kościół, Poland using OSL and radiocarbon dating. Quat Intern 502, A:4–17
Moska P, Sokołowski R, Jary Z, Zieliński P, Raczyk J, Szymak A, Krawczyk M, Skurzyński J, Poręba G, Łopuch M, Tudyka K (2022) Stratigraphy of the Late Glacial and Holocene aeolian series in different sedimentary zones related to the Last Glacial maximum in Poland. Quat Intern 630:65–83
Nawrocki J, Bogucki AB, Gozhik P, Łanczont M, Pańczyk M, Standzikowski K, Komar M, Rosowiecka O, Tomeniuk O (2019) Fluctuations of the Fennoscandian Ice Sheet recorded in the anisotropy of magnetic susceptibility of periglacial loess from Ukraine. Boreas 48:940–952
Nowaczyk B (1986) Wiek wydm, ich cechy granulometryczne i strukturalne a schemat cyrkulacji atmosferycznej w Polsce w poźnym vistulianie i holocenie. Wyd Nauk UAM, Poznań, p 245
Penck A, Brückner E (1909) Die Alpen im Eiszeitalter. Tauchnitz, Leipzig, p 1199
Pidek IA, Małek M (2010) A bi-partite Ferdynandovian succession from Łuków, Eastern Poland: a new palynostratigraphic approach. Geol Quart 54:69–84
Popescu S-M, Biltekin D, Winter H, Suc J-P, Melinte-Dobrinescu MC, Klotz S, Rabineau M, Combourieu-Nebout N, Clauzon G, Deaconu F (2010) Pliocene and Lower Pleistocene vegetation and climate changes at the European scale: long pollen records and climatostratigraphy. Quat Intern 219:152–167
Pusch GG (1829–1830) Geognostyczny opis Polski. Sławianin 25:392–396
Pusch GG (1831–1836) Geognostische Beschreibung von Polen so wie der übrigen Nordkarpathen-Länder, 1–2. J.G. Cotta'schen Buchhandlung, Stuttgart und Tübingen, p 338+695
Pusch GG (1836) Geognostyscher Atlas von Polen. J.G. Cotta'schen Buchhandlung in Stuttgart, p 10
Roman M, Dzieduszyńska D, Petera-Zganiacz J (2014) Łódź Region and its northern vicinity under Vistulian Glaciation conditions. Quaest Geogr 33:155–163
Różycki SZ (1961) Guide-book of excursion from the Baltic to the Tatras 2: Middle Poland. PWN, Warszawa, p 116
Różycki SZ (1972) Plejstocen Polski Środkowej na tle przeszłości w górnym trzeciorzędzie, 2nd edn. PWN, Warszawa, p 315
Różycki SZ (1980) Principles of stratigraphic subdivision of Quaternary of Poland. Quat Studies Poland 2:99–106
Rühle E (1973) Metodyka badań osadów czwartorzędowych. Wydawn Geol, Warszawa, p 687
Sawicki L (1922) Wiadomość o środkowopolskiej morenie czołowej. Rozpr Wydz Mat-Przyr III 21A:26–35
Siemiradzki J (1882) Nasze Głazy Narzutowe. Pam Fizjogr 2:87–122
Siemiradzki J (1890) O morenach czołowych bałtyckiego lodowca dyluwialnego. Kosmos 15:506–508
Skurzyński J, Jary Z, Kenis P, Kubik R, Moska P, Raczyk J, Seul C (2020) Geochemistry and mineralogy of the Late Pleistocene loess-palaeosol sequence in Złota (near Sandomierz, Poland): Implications for weathering, sedimentary recycling and provenance. Geoderma 375:114459
Smalley IJ, O'Hara-Dhand K, Wint J, Machalett B, Jary Z, Jefferson IF (2009) Rivers and loess: the significance of long river transportation in the complex event-sequence approach to loess deposit formation. Quat Intern 198:7–18

Solon J, Borzyszkowski J, Bidłasik M, Richling A, Badora K, Balon J, Brzezińska-Wójcik M, Chabudziński Ł, Dobrowolski R, Grzegorczyk I, Jodłowski M, Kistowski M, Kot R, Krąż P, Lechnio J, Macias A, Majchrowska A, Malinowska E, Migoń P, Myga-Piątek U, Nita J, Papińska E, Rodzik J, Strzyż M, Terpiłowski S, Ziaja W (2018) Physico-geographical mesoregions of Poland: verification and adjustment of boundaries on the basis of contemporary spatial data. Geogr Polon 91:143–170
Starkel L (1985) Controversial opinions on the role of tectonic movements and climatic changes in the Quaternary evolution of the Polish Carpathians. Studia Geomorph Carp-Balc 19:45–60
Staszic S (1805) O ziemiorodztwie gór dawnej Sarmacji a później Polski. Pierwsza Rozprawa Towarzystwa Naukowego Warszawskiego, Warszawa, p 129
Staszic S (1815) O ziemiorodztwie Karpatów i innych gór i równin Polski. W Drukarni Rządowej, Warszawa, p 390
Szafer W (1928) Zarys stratygrafii polskiego dyluwium na podstawie florystycznej. Roczn Pol Tow Geol 5:21–33
Szafer W (1953) Stratygrafia plejstocenu w Polsce na podstawie florystycznej. Roczn Pol Tow Geol 22:1–99
Tutkovsky PA (1899) K woprosu o sposobje obrazovanija lossa. Zemlevedenie 1–2:213–311
Van Loon AJ (2006) Lost Loesses. Earth Sci Rev 74:309–316
Weckwerth P (2018) Fluvial responses to the Weichselian ice sheet advances and retreats: implications for understanding river paleohydrology and pattern changes in Central Poland. Int J Earth Sci 107:1407–1429
Woldstedt P (1955) Norddeutschland und angrenzende Gebiete im Eiszeitalter. Koehler Verlag, Stuttgart, K.F, p 467
Wysota W (2002) Stratigraphy and Sedimentary Environment of the Weichselian Glaciation in the Southern Part of the Lower Vistula Region. Wyd UMK, Toruń, p 144
Zalat AA, Bober A, Pidek IA, Żarski M (2021) Environmental and climate change during the Late Saalian-Eemian Interglacial at the Struga site (Central Poland): a diatom record against the background of palynostratigraphy. Rev Palaeobot Palynol 288:104386
Zeeberg JJ (1998) The European sand belt in eastern Europe—and comparison of Late Glacial dune orientation with GCM simulation results. Boreas 27:127–139
Zieliński P (2016) Regionalne i lokalne uwarunkowania późnovistuliańskiej depozycji eolicznej w środkowej części europejskiego pasa piaszczystego. Wyd UMCS, Lublin, p 236
Zuchiewicz W (1980) The tectonic interpretation of longitudinal profiles of the Carpathian rivers. Ann Soc Geol Polon 50:311–328
Zuchiewicz W (1984a) Neotectonics of the Polish Carpathians: facts and doubts. Studia Geomorph Carp-Balc 17:29–44
Zuchiewicz W (1984b) Neotectonic movements in the Carpathians. Tectonophysics 104:195–204
Zuchiewicz W (1989) Neotectonics versus gravity and crustal thickness: a case study of the Polish Flysch Carpathians. Tectonophysics 163:277–284
Zuchiewicz W (1991) On different approaches to neotectonics a Polish Carpathians example. Episodes 14:116–124
Zuchiewicz W (1995) Selected aspects of neotectonics of the Polish Carpathians. Folia Quat 66:145–204

Zdzisław Jary is Professor of Geography at the University of Wrocław, head of the Physical Geography Department and former Dean of the Faculty of Earth Sciences and Environmental Management. His main research interests are focused on varied aspects of loess-soil sequences, periglacial and aeolian geomorphology and geoarchaeology. Central Europe is his main research area, in particular loess regions of Poland and Ukraine. He is currently the President of the Committee for Quaternary Research of the Polish Academy of Sciences and the President of the Loess and Pedostratigraphy Working Group of the International Union For Quaternary Research (2023–2027). He is a founder member and the first president of the Polish Quaternary Union (POLQUA).

Leszek Marks is Professor of Earth Sciences at the Polish Geological Institute—National Research Institute in Warsaw, its former Director and Head of the Mapping Department. He is also Professor Emeritus at the University of Warsaw, former Deputy Dean of the Faculty of Geology and Head of the Climate Geology Department. His main research interests are Quaternary geology, climate geology, geological mapping, glacial geology, palaeolimnology and geoarchaeology. Central Europe was his main research area and he was involved in projects in Belarus, Croatia, Finland, Germany, Norway, Poland, Ukraine, United Kingdom, United States and Egypt. He used to be the Vice-President of EuroGeoSurveys (2005), the International Union of Quaternary Research (1999–2003), and a member of the executive committees of its several subordinate bodies. He is the editorial board member of *Boreas* journal.

The Climate of Poland

3

Marek Błaś and Hanna Ojrzyńska

Abstract

Poland is located on the western periphery of the Eurasian continent, which, together with the predominance of western circulation of the temperate zone, translates into a greater influence of the Atlantic. Only in the eastern part of the country, the influence of the continent is more pronounced. The border identifying the prevalence of one of these areas is hard to draw, which classifies Poland's climate as 'transitional'. Due to the positive thermal anomaly in this part of the temperate zone, Poland's climate can be defined as 'warm-temperate', in accordance with the classic classification by Köppen. Only the climate of north-eastern and eastern Poland should be classified as a cooler, snow forest type of temperate climates. In terms of precipitation regime, compared to the conditions on land in a similar geographical zone, Poland's climate belongs to the group of humid climates.

Keywords

Climate · Atmospheric circulation · Air temperature · Precipitation · Climate change

M. Błaś (✉) · H. Ojrzyńska
Institute of Geography and Regional Development, University of Wrocław, pl. Uniwersytecki 1, 50–137 Wrocław, Poland
e-mail: marek.blas@uwr.edu.pl

H. Ojrzyńska
e-mail: hanna.ojrzynska@uwr.edu.pl

3.1 Introduction

This chapter presents general characteristics of climate elements which significantly affect the occurrence, course and the effects of geomorphic processes observed in the area of Poland. In this context, precipitation, temperature and wind speed are the most prominent. Their impact may be long term; hence, the relevant geomorphic processes (e.g. soil creep, sediment transport in rivers and sedimentation in lake reservoirs) are characterized by low intensity and their effects depend on the annual and multi-annual variability of precipitation and temperature (Niedźwiedź and Starkel 2008). The same climatic factors may also show rare (episodic) extreme departures from the long-term means, when certain critical values of precipitation intensity, temperature variation, or wind speed are exceeded. Extreme, widespread and convective torrential precipitation leads to remodelling of slope relief, river beds, or floodplains (Starkel 2008). It triggers surface runoff, mass movements, accounts for flash floods and intensive bed erosion. The emerging cold waves contribute to profound ground freezing, which impedes infiltration of water and conditions surface water runoff during thaw. Frost heave leads to the formation of needle ice, soil crushing and its drying up. The greatest hazard associated with deflation processes occurs during heat waves, which are often accompanied by a significant lack of precipitation. Strong winds, in addition to aeolian erosion and coastal abrasion, additionally contribute to the occurrence of tree uprooting, which, especially in mountainous regions, leads to the transformation of slope covers. Considerable attention has been paid to contemporary climate change and atmosphere dynamics. Trends observed in recent decades indicate that violent storms accompanied by heavy rain and strong wind as well as periods of intensive rainfall resulting in floods are going to alternate with periods of severe drought.

P. Migoń and K. Jancewicz (eds.), *Landscapes and Landforms of Poland*,
World Geomorphological Landscapes, https://doi.org/10.1007/978-3-031-45762-3_3

3.2 Atmospheric Circulation Over Poland—Regional and Continental Context

3.2.1 The Position of Poland in the Context of General Atmospheric Circulation

Poland is largely a lowland country and part of a vast North European Plain. Its topography is band-like, starting from the north with the coastal belt, undulating lake districts, lowlands, uplands, foothills, and ending in the south with parallel running ranges of the Sudetes (exceeding 1000 m asl) and the Western Carpathians (exceeding 2000 m asl, Fig. 3.1). The lowlands of northern and central Poland, which are not protected by any elevations on their western, northern, and eastern sides, offer a potentially easy access to maritime as well as continental air masses (Kottek et al. 2006). For this reason, in the cold half of the year, when the Atlantic becomes a significant source of heat for the western parts of the continent, the temperature over Poland rises by 0.3 °C per one degree of longitude westwards. In the warm half of the year, when the continent becomes warmer than the ocean, the temperature over Poland rises towards the east, but only by 0.1 °C per one degree of longitude (Boryczka 1977; Kożuchowski 2011).

Circulation conditions in the temperate latitudes zone, understood as the formation of low-pressure systems, are closely related to westerly airflow in the middle and upper troposphere and various types of disturbances to which polar jet stream is subjected. These disturbances, that is, deviations from the latitudinal direction of movement of air masses are also the result of the influence of the substrate of the atmosphere, that is, the presence of the continents, oceans with warm and cold currents, and ice sheets present in some sectors of the temperate zone, as well as at high latitudes. Meanders and waves contribute to the intensification of the meridional circulation; hence, the northward movement translates into warm advection towards higher latitudes, while the southward one is associated with cold advection. The formation of areas of low pressure (in the warmer air advection zone) and high pressure (in the cooler air advection zone) is favoured beneath developing long waves (Rossby's waves) and the overlapping short waves. In the cooler part of the year, cold and dense air accumulates over the cooling continent, pressure rises and the

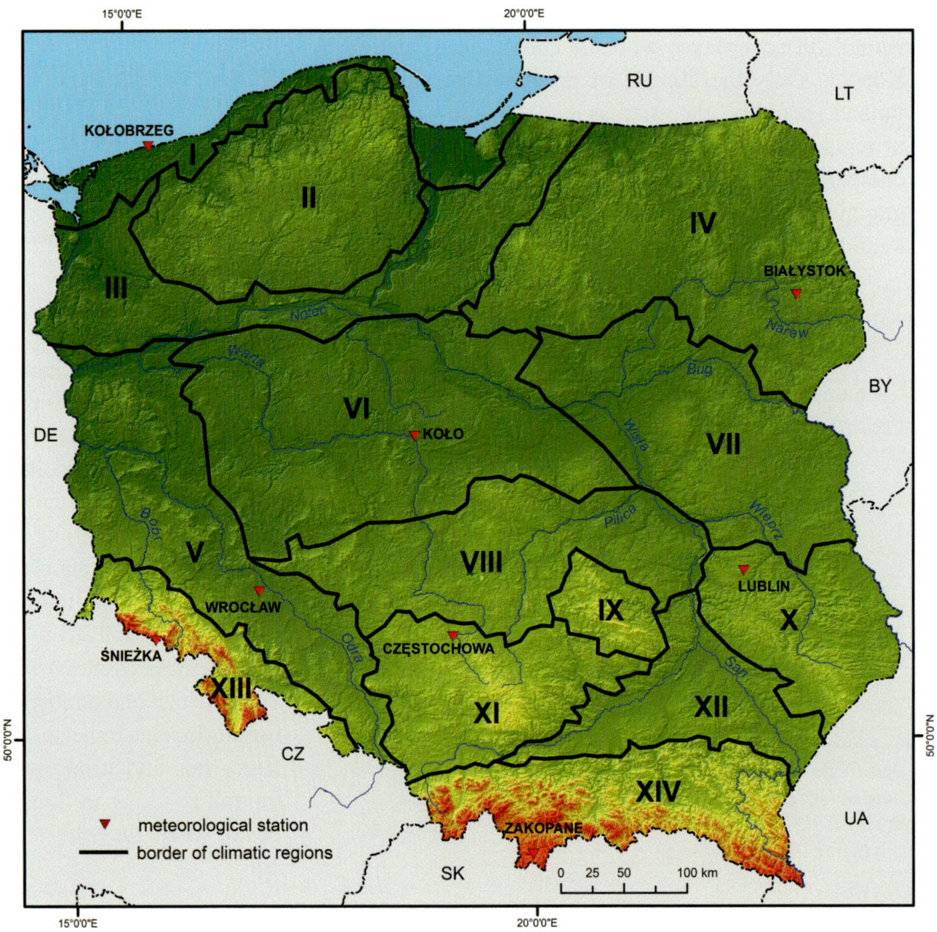

Fig. 3.1 Topography and climate domains of Poland with the position of weather station used for climatic characterisation (see also Fig. 3.6)

middle troposphere is dominated by the subsidence of air masses. In the summer, the pressure over the heated land lowers, and the air tends to ascend.

3.2.2 Atmospheric Pressure Centers

In terms of mean distribution of atmospheric pressure, Poland is under the influence of two year-round pressure systems: the Azores High and the Icelandic Low. Occasionally, there is an influence of the summer low and the winter high from above Asia, as well as cyclonic systems connected with Mediterranean cyclogenesis. In some months, the mean distribution of pressure over central Europe frequently resembles a col. This favours the inflow of air masses from various source areas: the Atlantic, the continent, and also from the Arctic, and even North Africa. The areas defined by the range of advection lasting no more than three days (which ensures not a very advanced transformation of air masses) cover a vast area stretching from Greenland to the Urals and from Spitsbergen in the Arctic to Tunisia and Algeria in Africa (Kożuchowski 2011). Poland is located in the eastern part, which indicates that western influences dominate over the eastern ones, that is, polar–maritime air masses occur more frequently than polar-continental ones. This is manifested in the high concentration of routes of cyclonic systems moving mostly from the west, and also from the south (Fig. 3.2). According to the conventional classification of cyclonic routes developed by van Bebber (Van Bebber 1891; Hann 1906), routes III and Vb cross over north-eastern Poland. Over Denmark, route IV crosses route III, and over Baltic Sea, route II meets route IV. Low-pressure areas migrating along the above-mentioned routes directly cover the territory of Poland, shaping the cyclonic weather-cloudy and humid, bringing thaw in winter and cool conditions in summer. The atmospheric fronts related to them are responsible for weather variability and precipitation. A higher concentration of fronts occurs over the southern Baltic Sea, Denmark and the North Sea (Woś 2010).

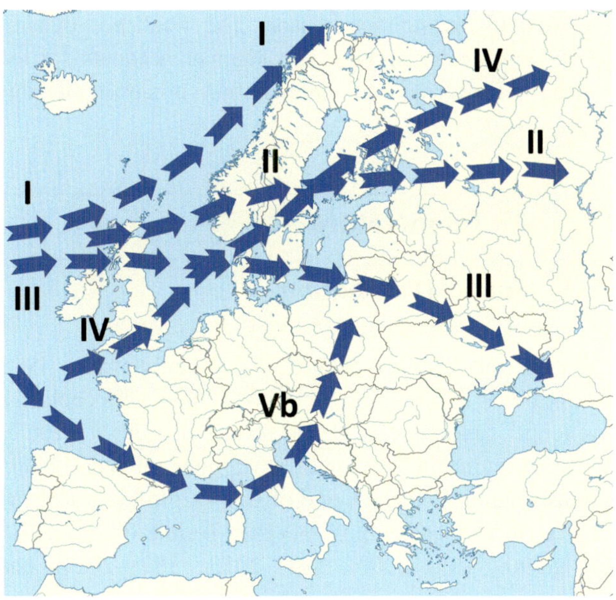

Fig. 3.2 Main trajectories of depressions movement over Europe (adapted from Van Bebber 1891)

The greatest impact on climatic conditions in Poland during the winter season have the lows passing over the Baltic Sea. On the one hand, they cause warming winter thaws, and on the other one, they are followed by intrusions of cold air from the north caused by the cold front. Definitely less frequent are Mediterranean lows over Poland, which transfer warm air rich in water vapour from the eastern and north-eastern sectors (Kożuchowski 2011). They are accompanied by heavy precipitation, which quite frequently poses a threat of flooding. In the warm part of the year, low-pressure areas, which affect weather conditions in Poland, form mostly around Iceland, and secondly over central Scandinavia and the northern Baltic Sea. Occasionally, Genoa lows, forming over the northern part of Italy and the Adriatic Sea, may arrive in Poland. They are often responsible for intense and heavy precipitation. The distribution of lows over northern Europe and the North Atlantic contributes to zonal circulation, and Poland is covered by their southern, warm periphery. The meridional circulation takes various forms depending on Poland's location within the range of a ridge or a longwave trough. According to Girs (1981), the so-called 'C form' of the meridional circulation means that Poland is situated on the eastern side of the upper trough and advection of air masses from the south. 'Form E', in turn, means location in the central or eastern part of the upper trough, which activates advection from the south. Consequently, a given circulation type controls the advection of particular air masses over Poland. It is characterised by the dominance of the polar-maritime air masses (70%), which flow with the cyclonic circulation from the NW, W, and the SW sector (Kożuchowski 2011). They dominate throughout the seasons with the strongest presence in summer. Continental air masses (16%) occur mostly in winter and spring with anticyclonic circulation from the NE, E, and SE directions. The smallest share is represented by the arctic (13%) and tropical (2%) air masses, arriving mostly in spring and autumn. The arctic air flows over Poland with anticyclonic circulation from the NW and NE sectors, and with the cyclonic one from the NE sector. Tropical air masses result from cyclonic circulation from the SE sector and the anticyclonic ones-from the SE and E sectors.

3.2.3 North Atlantic Oscillation Index

The North Atlantic Oscillation (NAO) is the teleconnection indicator of circulation over Poland, which is related to the pressure gradient prevailing far beyond its area. The NAO indicator values are a measure of the intensity of the zonal flow of air masses (Hurrell 1995). As the pressure over the Azores increases and the Icelandic Low deepens, the increase of NAO index is observed. This is the so-called positive phase, characterized by an intensive transfer of heat and humidity from the ocean's surface into the atmosphere and a dynamic flow of polar-maritime air over Europe, including Poland (Marsz 2008). Thus, the positive phase contributes to the amplification of oceanic features of thermal climate of Poland. Żmudzka (2002) showed that fluctuations of the NAO indicator within ± two standard deviations lead to changes in the date of temperature exceeding + 5 °C from about 20 days in the SE of Poland to 50 days in the NW extremities. The closest relationship between temperature in Poland and the NAO index is observed mainly in the winter months, when its value exceeds 1.5 (Marsz 2008). In the warm season, the correlation between temperature and the NAO index is clearly weakened and tends to be negative, as the advection from the west is responsible for cooling. While precipitation amounts in winter and spring are correlated with the intensity of westerly circulation, from June to November the positive phase of the NAO index has a negative effect on precipitation amounts (more frequent, but less intensive precipitation).

The negative phase of the NAO index means that the Azores High is weakening and pressure over Iceland is increasing. Thus, the westerly circulation weakens and blocking pressure systems are developing over Europe. These systems trigger meridional circulation conditioning the inflow of polar-continental, arctic, or tropical air (Kożuchowski 2011). The negative phase is characteristic of transitional seasons (April–May, as well as September–October), when the relationships between temperature and the NAO index become ambiguous. This is because thermal conditions in Poland are determined by the circulation responsible for the advection of heat from the south, or cold air masses from the north.

3.2.4 Moisture Transport

Wypych et al. (2018) used multiyear values of precipitable water content (PW) to identify and characterize humidity regions in Europe. According to this classification, the area of Poland belongs to the Mid-Continental region, characterized by the largest moisture content in summer and moisture deficit in winter. Moisture content at lower levels (850 and 950 hPa) largely depends on the properties of the boundary layer (evaporation, temperature and its vertical structure), whereas in the upper troposphere (500 hPa) it reflects the influence of atmospheric circulation on the continental scale. Degirmendžić and Kożuchowski (2017) showed that extreme values of PW in Poland in the years 1958–2008 ranged from 1.9 to 41.0 mm, with an average of 15.9 mm. Maximum rainfall amount events in Poland occur under relatively high PW, but the height of these maxima is poorly correlated with PW. Thus, the amount of maximum precipitation is primarily determined by the efficiency of precipitation processes (Bordi et al. 2016). The maximum daily rainfall occurred at PW above 30 mm (Kożuchowski 2016).

PW is also used to identify atmospheric rivers, that are long, narrow, and transient corridors of strong horizontal water vapour transport that are typically associated with low-level jet stream ahead of the cold front of an extratropical cyclone (Browning 1990; Neiman et al. 2008). Atmospheric rivers can lead to heavy precipitation whenever these systems are forced upward, either by mountains or by an ascent in the warm conveyor belt. The integrated water vapour transport depends on both the amount of moisture and the strength of the flow, and calculating it requires the profiles of both moisture and wind to be known (Ralph et al. 2004; Neiman et al. 2008; Dettinger et al. 2011). Water vapour transport over Poland changes from about 60–70 kg/m/s in January to over 100 kg/m/s in July. Thus, seasonal changes in the stream of water vapour are considerably lower than changes in the water vapour content. This indicates considerable dynamics of air movement and the intensity of moisture transfer in the winter season compared to the summer season. The weakened dynamics of moisture transport in the summer accounts for the reduced effectiveness of precipitation processes in this period (Kożuchowski 2011).

3.3 Air Temperature

3.3.1 Spatial Distribution

The aerial mean air temperature for Poland between 1991 and 2020 was 8.4 °C. However, this value is not conclusive in a longer timeframe as in the former decades it steadily increased, indicating progressive warming of the climate. For instance, in Wrocław, one of the warmest cities of the country, in the consecutive 30-year periods (1951–1980, 1961–1990, 1971–2000, 1981–2010, 1991–2020) the average temperature consistently increased, being respectively: 8.2 °C; 8.3 °C; 8.7 °C; 9.0 °C and 9.7 °C. On the other hand, in Białystok, one of the coolest cities of the country, it was 6.7 °C; 6.7 °C; 6.9 °C; 7.2 °C and 7.6 °C, respectively.

The spatial distribution of the mean annual air temperature in Poland shows a consistent increase from the north-east, the coolest part of the country, towards the warmest region located in the south-west (Lower Silesia). This is due to the interaction between two factors: the latitude and the distance from the Atlantic. In Fig. 3.3, two warmest areas stand out, with an average annual temperature exceeding 9.0 °C. These are mainly located in less elevated terrains along the valleys of Poland's two largest rivers, which are not deep and secluded enough to become frost hollows. The first area extends along the valley of the Odra River and its major tributaries, from Upper Silesia, via Wrocław, as far as to Szczecin in the north. The other area includes the Wisła and the San valleys in the fore-mountain zone and a narrow belt running to the north along the Wisła River, reaching Warszawa and beyond. The highest average annual air temperatures (> 9.0 °C) are recorded among others in Wrocław (9.7 °C). The cooler areas (except for the mountains), with an average below 8.5 °C, are mainly the Silesian Upland in the south and the slightly elevated (100–250 m asl) lake districts in northern Poland. An average annual temperature lower than 8.0 °C is observed among others in Białystok. Including mountain regions, the lowest average annual temperatures are recorded at the highest weather stations, that is, at Mt. Kasprowy Wierch in the Tatra Mountains (0.1 °C; 1987 m asl) and at Mt. Śnieżka in the Sudetes (1.4 °C; 1602 m asl).

The distributions of the mean minimum temperature of the coolest month (January, TminI) and the mean maximum of the warmest month (July, TmaxVII) are different. TminI consistently decreases from the belt along the western border and the western part of the coast (− 1.0 °C) towards the eastern border of Poland (− 6.0 °C, Fig. 3.4). Thus, the isotherms of the coolest month have a meridional course. Similar is the frequency distribution of days with Tmin < 0 °C, from about 90 days in the northwesternmost part of the coastal belt to 110 days in SE part of the country (Kożuchowski 2011). This results from circulation conditions, that is, advection of warm, polar-atlantic air masses from the W and SW directions, or cold, polar-continental air from the E and NE. Moreover, this translates into a high year-to-year variability in the mean temperature of the coolest month, which increases towards the north-east. Only in the coastal area, the thermal conditions are more stable. The

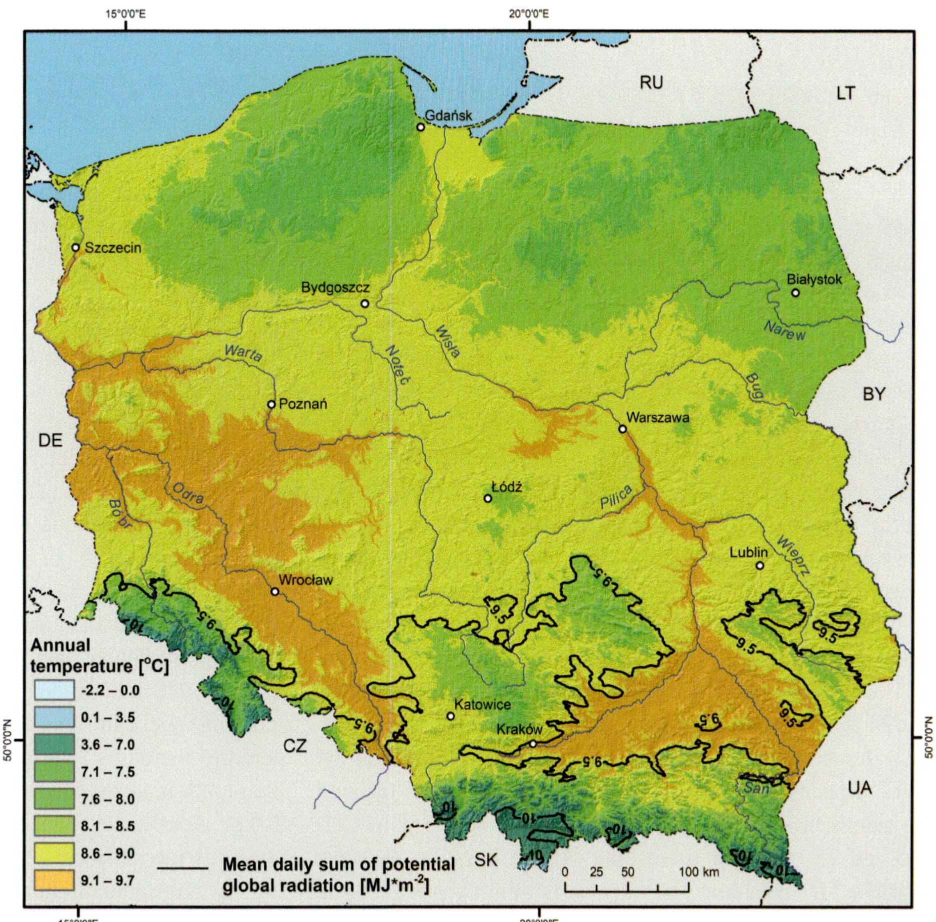

Fig. 3.3 Mean annual temperature across Poland and mean daily sum of potential global radiation for the period of 1990–2020

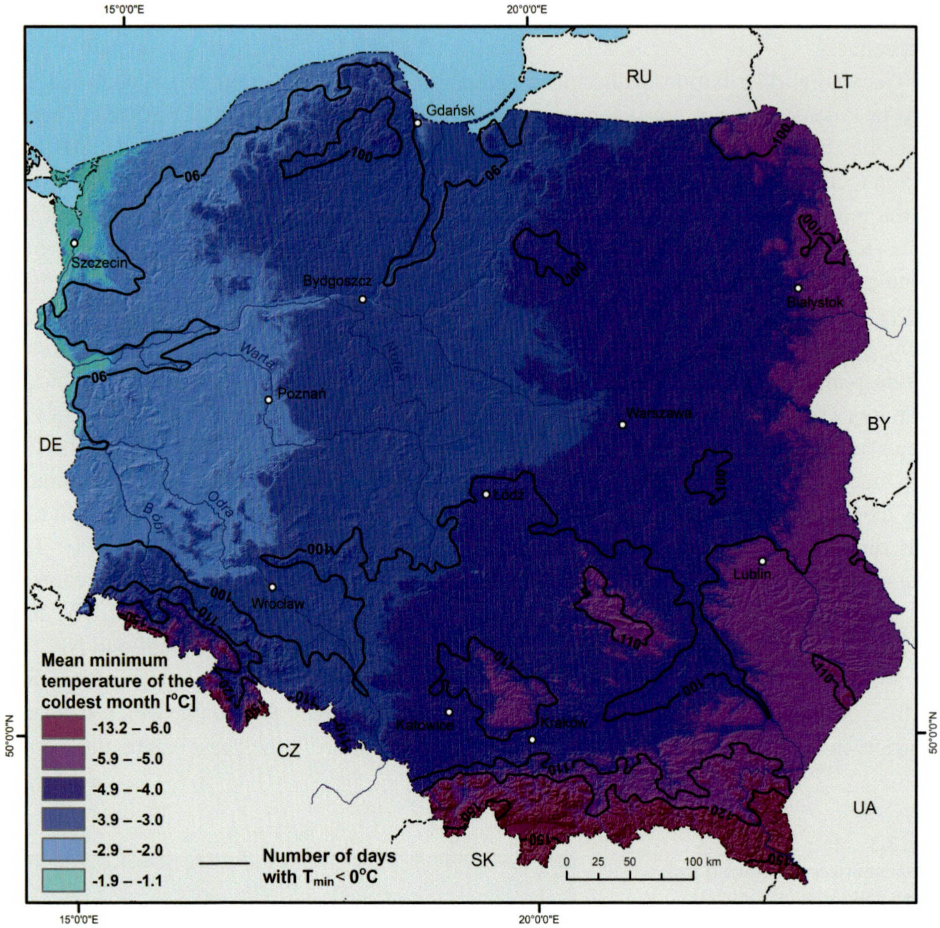

Fig. 3.4 Mean minimum temperature in January (TminI) and annual number of days with Tmin < 0 °C (for the period of 1990–2020)

lowest recorded temperature is observed in the highs of air pressure, with an influx of frosty continental air from the E and NE directions. The influence of absolute altitude, irrespective of the season, is constant (temperature decreases with altitude); hence, the coolest are mountain tops (< – 10.0 °C) and intramontane basins, where additionally the cold air tends to stagnate and forms frost hollows.

In the case of TmaxVII (Fig. 3.5) the isotherms are distributed latitudinally, shaped by the radiative balance and geographical factors linked with the influence of the Baltic Sea in the north and latitudinally arranged mountain ranges in the south. The highest maximum temperatures in Poland are recorded on days with high air pressure without clear advection, or with the flow of tropical air masses from the southern sector. Basically, the entire belt of Polish Lowlands is characterized by TmaxVII exceeding 24 °C and over 40 hot days on average (with maximum daily temperatures exceeding 25 °C). The warmest region is the Silesian Lowland and the Sandomierz Basin (around the confluence of Wisła and San rivers), where TmaxVII exceeds 25 °C, with more than 50 hot days. In the lake districts, uplands, and foothills, it is slightly cooler, with TmaxVII ranging from 21 to 24 °C, and fewer than 35 hot days per annum. In the highest parts of the Sudetes and Carpathians the number of hot days is almost reduced to zero.

3.3.2 Diurnal and Day-to-Day Temperature Changes

One of the features of the thermal regime is variability of the average day-to-day air temperature. In 11% of cases, the absolute magnitude of changes in the mean daily temperature exceeds 4 °C in Poland (Kożuchowski 2011). In extreme situations, temperature drops from day to day by even 20 °C, and increases as much as 15 °C. In this case, the most thermally unstable is winter, while the most thermally stable is autumn. The maximum temperature drops (cooling) occur in May, while the maximum of sudden temperature increases (warmings) is in January. The most stable month in terms of temperature is October (Kożuchowski 2011). Another cyclic component of air temperature variability is the daily range, which increases in spring, reaching its maximum in June (11.1 °C). Then, it decreases in autumn to reach its lowest average value in December

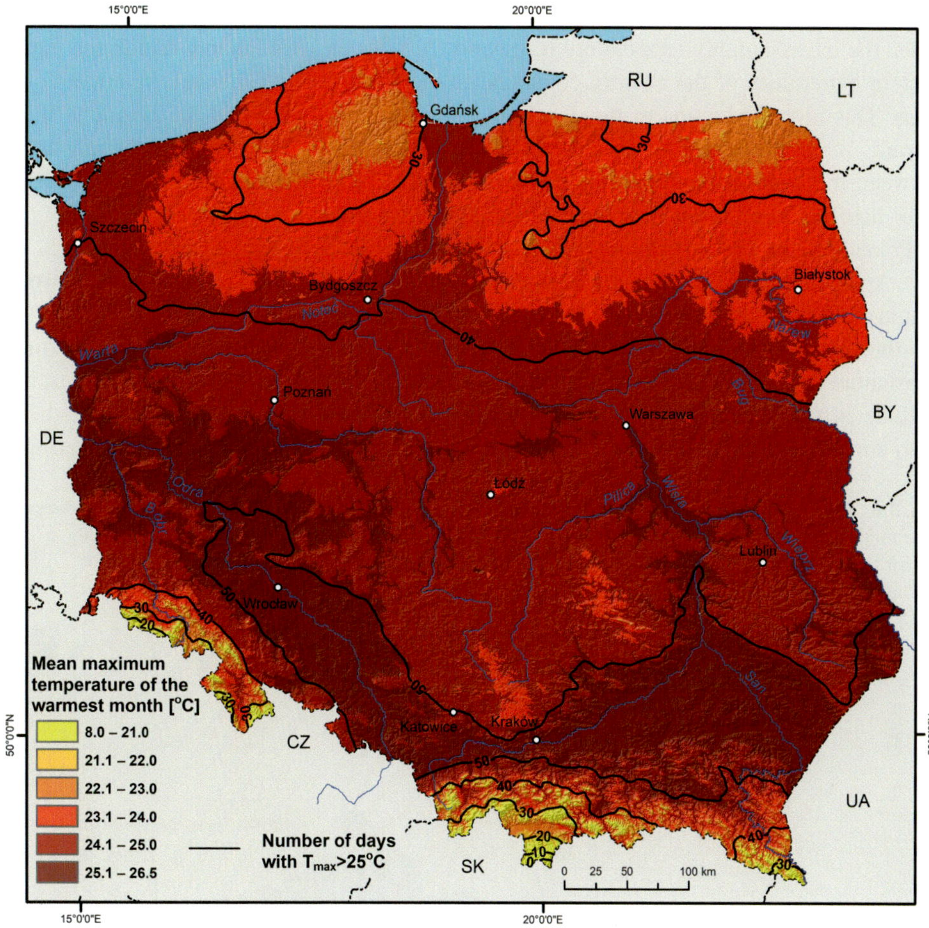

Fig. 3.5 Mean maximum temperature in July (TmaxVII) and annual number of days with Tmax > 25 °C Poland (for the period of 1990–2020)

(4.9 °C). Significantly reduced daily temperature range is recorded in the coastal area, as exemplified by the city of Gdynia (June 7.2 °C and December 3.8 °C) or the Hel Peninsula (June 9.2 °C and December 3.4 °C). As altitude increases, particularly in convex landforms, daily temperature changes decrease, and their annual pattern disappears. This is due to the fact that a significant amount of heat from the insolation of the active surface is dispersed in the atmosphere through intensive air exchange. This is further favoured by more cloudy conditions recorded in mountainous areas. At Mt. Śnieżka, the average daily temperature range reaches its minimum in February and November (4.8 °C) and the maximum daily temperature range in June reaches only 5.7 °C. Maximum daily temperature ranges are observed in the floors of concave landforms, which is explained by the formation of cold air stagnation areas and significant temperature drops at night. In this respect, an exceptional place is Hala Izerska, which is situated in the NW part of the Sudetes, with the surface of more than 10 km^2, and its bottom at 800–840 m asl. Here, daily temperature range frequently exceeds 25 °C. The record 31.2 °C, was measured on 24 February 2003 (Błaś and Sobik 2005). Moreover, this place is characterized by a high risk of frost even in summer months.

3.3.3 Atmospheric Circulation and Continentalism

On the national scale, the degree of continentalism, that is, the impact of the continent on the annual course of climate elements generally increases from the west towards the east, together with the distance from the Atlantic, and also (especially in the northern part of the country) from the north towards the south, that is, with the distance from the Baltic coast. These changes result from the frequency and degree of the transformations of air masses originating from source areas. The annual range of temperature in Poland is about 20 °C and is a few degrees lower compared to the mean zonal range. However, the difference in the annual range between the western and the eastern parts of Poland is about 4 °C (Stopa-Boryczko and Boryczko 1998). A greater gradient of the degree of continentalism is marked in winter, when the mean monthly temperature is lower by 5 °C in the eastern extremities of the country compared to the western ones. This is the reflection of rapidly decreasing insolation towards the north as well as an increasing thermal contrast between the continent and the ocean (Kożuchowski 2011). The distribution of mean temperatures in summer months in Poland is less varied (extreme values differ by about 1.5 °C). Apart from that, summer is warmer in the south of the country than in its more continental, north-eastern part. North–south extent (approximately 6-degree) is not very substantial in this respect. However, while the variation with latitude in the total radiation is hardly noticeable in summer, zonal contrasts are intensifying in winter. Longer days in summer allow for more solar energy to reach the ground in Poland than in the equatorial zone (on 22 June it is up to 20% more). In autumn and winter the length of a day shortens and the height of the sun above the horizon decreases. In December, the amount of solar radiation reaches its minimum, that is, merely 10% of the June value.

3.3.4 Thermal Climates

Figure 3.6 shows thermal conditions for a few selected weather stations representing extreme types of 'thermal climates' found in Poland. The city of Wrocław represents the warmest region covering the SW part of Poland, running

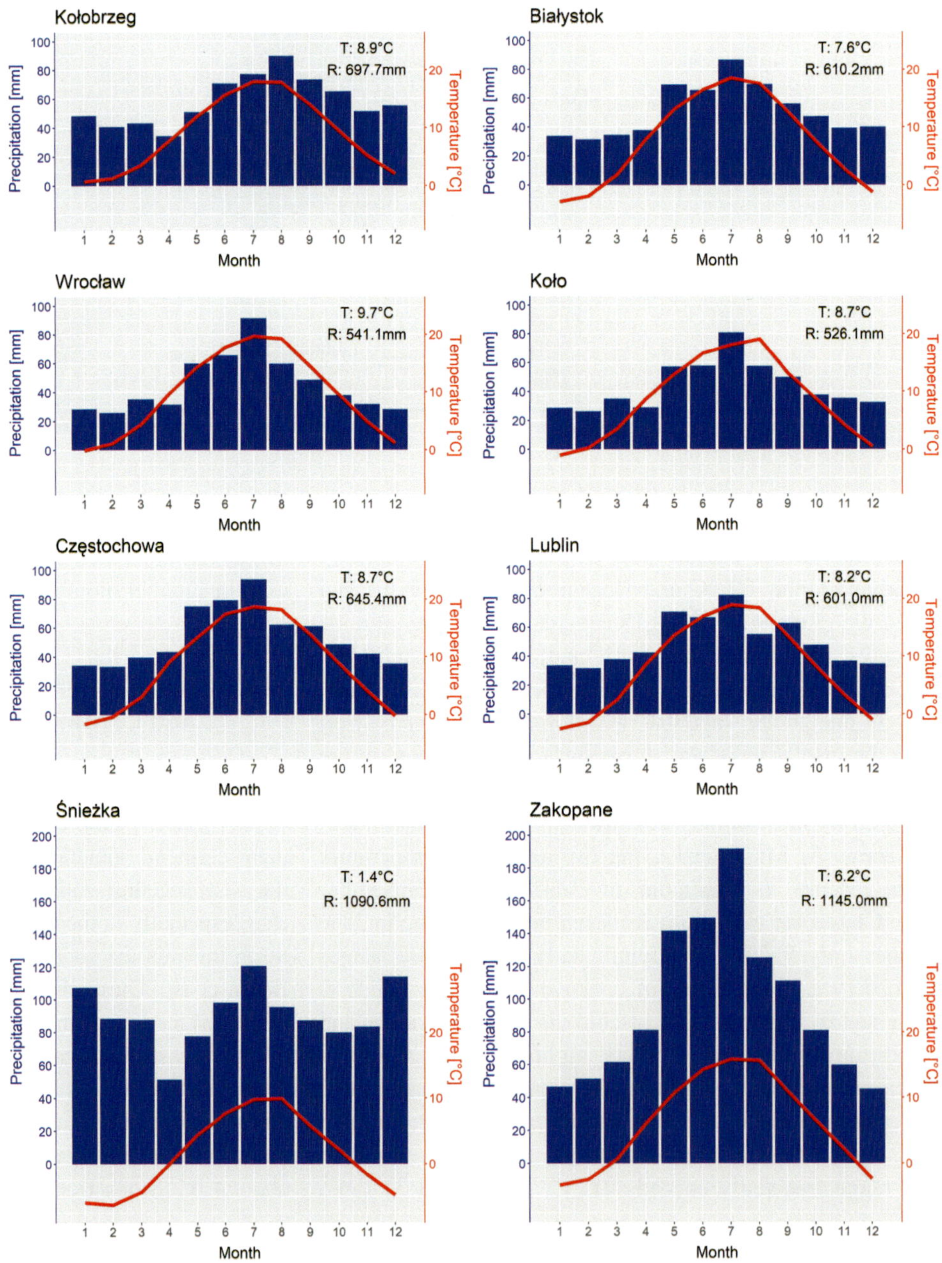

Fig. 3.6 Air temperature and precipitation regimes for selected stations representing climate domains of Poland (for the period of 1990–2020)

along the Odra river valley (Fig. 3.1). It is characterized by mild winters, warm summers and the highest average annual temperature (9.7 °C). The mean monthly temperature drops below zero only in January (− 0.1 °C). The town of Kołobrzeg represents the area with the most evident maritime climate in Poland. It is distinguished by the mildest thermal conditions in winter (the average for January is 0.7 °C), as well as rather cool spring and summer. Autumn is distinctly warmer here compared to spring. The annual temperature range is only 18.6 °C. The city of Białystok represents the NE part of Poland, considered the 'cold pole', where (with the exception of the mountains) the lowest January average (− 2.8 °C) and the lowest annual mean temperature (7.6 °C) are recorded. Thermal continentalism is evident here, which is associated with severe winters, a fairly high rate of temperature change in spring and autumn, and the annual temperature range of 21.2 °C. Mt. Śnieżka, the highest summit of the Sudetes (1602 m asl), represents a severe mountain climate, cool in summer and consistently frosty in the cold part of the year, without extremely low temperatures in winter. It is characterized by temperatures below 0 °C for 6 months, with the annual mean reaching 1.4 °C. The mean for the warmest month, August (9.9 °C), is comparable to the means for April in the lowland areas of Poland. Temperature differences between Mt. Śnieżka and the lowlands increase to 10 °C in summer, and in January they drop to 5.9 °C. The annual temperature range reaches only 16.1 °C, which is indicative of the super-oceanic climate of the mountains. The absolute maximum temperature here reaches 25 °C, while the absolute minimum (− 34 °C) is at the level recorded in the warmest regions of lowland Poland (Kożuchowski 2011).

3.4 Precipitation

3.4.1 Spatial Distribution

The average area precipitation measured in Poland between 1990 and 2020 was 630 mm. Approximately, 63.5% of the country exhibits annual precipitation lower than the area average. This does not apply to mountains, uplands in the south, and elevated parts of lake districts in the northern part. February displays the lowest monthly total, which is merely 5.4% of the average annual precipitation. The most frequently recorded precipitation, expressed in the modal value, reaches 31 mm in February. In contrast, precipitation in July reaches 14% of the annual total, with the modal value of 89 mm. The culmination of precipitation in the summer months is the effect of the development of thermal convection, but also of meridional atmospheric circulation (Mediterranean lows) as well as zonal circulation (cyclogenesis associated with Icelandic Low). Therefore, the annual precipitation rhythm in Poland is of a mixed oceanic-continental character (Kożuchowski 2011).

In the spatial distribution of annual average precipitation, several latitudinal belts can be distinguished (Fig. 3.7). The maximum totals, imposed by orography, are recorded in the southernmost part of the country. The centre of Poland, in turn, features a zone of low precipitation that increases again in the north. The lowest annual precipitation (< 600 mm) occurs along the narrow coastal strip as well as in the lowlands of central Poland, reaching the uplands in the south. The influence of lower ground running along the valleys of the Wisła river (Warsaw, 549 mm), the Odra river (Wrocław, 541 mm), and the Warta river (Koło, 526 mm), is quite evident. Precipitation as low as 500 mm or less is recorded between the regions of Wielkopolska and Kujawy and results in a negative water balance. According to Kożuchowski (2011), the reduction in precipitation, observed mainly in the cool part of the year, is related to a decrease in ground roughness (agricultural areas compared to forest areas) and the resulting increase in atmospheric dynamics. This, in turn, contributes to horizontal divergence, which means greater tendency towards air subsidence. This finally leads to decreased water content of clouds and precipitation intensity. In these 'dry' areas, summer precipitation does not differ from the precipitation recorded in other stations. This justifies the theory of circulation genesis of precipitation in the cold half of the year, which in summer gives way to the influence of convection generating a rather even precipitation field.

Uplands and lake districts are characterized by a slight increase in the annual precipitation (up to 700–750 mm). In the north of the country, it is partly caused by greater frequency of passing atmospheric fronts and the relief of the lake districts. In the mountainous areas in the south of Poland precipitation increases with altitude. This is a consequence of the forced ascent of incoming masses of humid air along the slope, leading to the formation of orographic clouds and an increase in their liquid water content ('feeder' clouds). Their droplets are too small to grow into raindrop-sized particles. However, 'seeder' rain falling from high altitudes (frontal clouds) is sweeping out the 'feeder' cap cloud particles and they grow by accreting 'feeder' cloud drops. Finally, the 'seeder-feeder' effect observed in the mountains, if compared with the surrounding lowlands, was typically accompanied by doubling of rainfall enhancement (Choularton and Perry 1986; Dore and Choularton 1992). In the course of this process, windward slopes of mountain ranges have a particularly important role, where the forced ascent of air masses is the most intense. In the case of the Sudetes, this role take the Izerskie Mountains and the Karkonosze (situated in their NW part), while in the

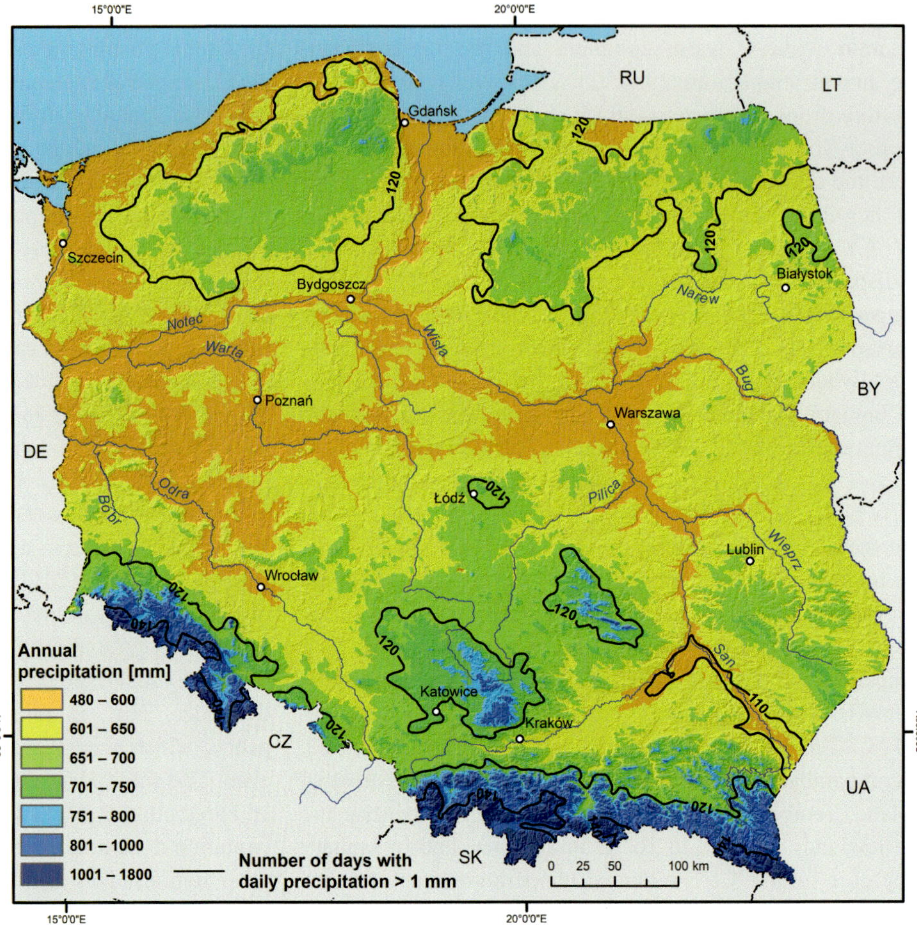

Fig. 3.7 Mean annual precipitation and number of days with daily precipitation > 1 mm (for the period of 1990–2020)

Carpathians it is the Silesian Beskid and Żywiecki Beskid ranges (Sobik and Błaś 2010). In these places, exceptionally high annual precipitation is recorded, often exceeding 1500 mm (Dore et al. 1999). Interestingly, in the Izerskie Mountains an even distribution of precipitation is observed annually. In all the months, precipitation exceeding 100 mm is recorded with double annual maximum observed in January and July (Błaś and Sobik 2005). Reduced precipitation in relation to altitude is recorded on leeward slopes, in the floors of intramontane basins and wide valleys, and on isolated hills. This is due to the presence of precipitation shadow and the mutual shielding of mountain ranges (Sobik et al. 2001).

3.4.2 Annual Variability and Number of Rainy Days

In the north-western part of Poland, precipitation of the cool half-year constitutes over 60% of the precipitation of the warm half-year, and thus, oceanic influences are visible (Kożuchowski 2011), whereas in the southern and south-eastern Poland the whole cool half-year has a 40–45% share in the annual precipitation. In the lake district, the influence of the Baltic Sea is noticeable, which shifts more rainfall from spring to autumn (the influence of relatively cooler Baltic waters in the spring and warmer in autumn). Moving further towards the centre of Poland, towards the south-east, summer precipitation increases over the winter one. The number of rainy days is characterised by a fairly small spatial variation. It ranges from 150 to 160 days in the lowlands, 170–180 in the lake districts and the seaside, to over 200–240 in the mountainous areas. In the course of a year, a slight increase in the number of days with precipitation is observed in autumn–winter months, while a decrease is recorded in spring and at the onset of autumn. In the northern extremities of Poland, in the coastal area, slightly higher precipitation is recorded due to their higher frequency. On the other hand, in the mountains in the south of Poland, an increased frequency of precipitation is coupled with its increased intensity. In the coastal area as well as in central Poland even 40% of rainy days do not exceed a daily total of 1 mm. In the mountains this share is usually below 30%. High daily precipitation totals (> 10 mm), in turn, have less than 10% share in the annual precipitation in the lowlands, while in the mountains it exceeds 20%. The maximum frequency of days with high precipitation across the country occurs in summer months, which confirms the significance of convection processes. In the mountains such cases occur also in autumn–winter months.

3.4.3 Extreme Precipitation

Extreme precipitation (daily sum>50 mm, Fig. 3.8) in Poland can be divided into two main categories: (a) widespread rainfall caused by intense cyclogenesis and (b) convective torrential rainfall (Sobik and Błaś 2010). Widespread precipitation, typically covering a considerable part of the country, does not exhibit particularly high intensity, but due to its persistence translates into large precipitation sums. They occur mainly in summer and cause the greatest hydrological and geomorphological consequences. During a single event, daily precipitation can reach from a few tens to more than one hundred millimetres, exceeding monthly climatological norms by a few times (Table 3.1).

A particularly extreme precipitation event in Central Europe was the widespread rainfall that caused flooding in July 1997 (Müller et al. 2009), which extended over the upper catchments of Odra, Morava and Wisła rivers. High precipitation values in the consecutive days indicate stagnation and partial regeneration of the convergence zone. In the headwater part of the Biała Lądecka river in the East Sudetes the following precipitation totals were recorded on 5th, 6th, 7th 8th and 9th of July 1997: 10, 122, 180, 154 and 16 mm (482 mm in total). These were caused by macro-scale convergence of air masses in cyclonal systems (Storebo 1976). This phenomenon consists of two air streams with high temperature difference meeting in the lower and middle troposphere. On the eastern side of low air pressure area warm and humid air masses (usually tropical-marine air) flow from the southern sector (Bader and Roach 1977). On the western side of the low air pressure area cool polar-maritime air arrives from the north. In the convergence zone, there is a forced upward sliding of large volumes of relatively warm air over the cold air. If during intensive summer convergence the air stagnates or the movement of the frontal zone slows down, then the daily precipitation, even in the lowlands, may reach 50–100 mm. The highest daily precipitation exceeding 200 mm, and in extreme cases even 300 mm, occurs in mountain areas (Table 3.1). This is partly caused by a fast-flowing stream of cold and vapour-saturated air from the north, which is deformed as it crosses mountain ridges, which, in turn, increases the intensity of the aforementioned 'seeder-feeder' mechanism. Such events often occur in connection

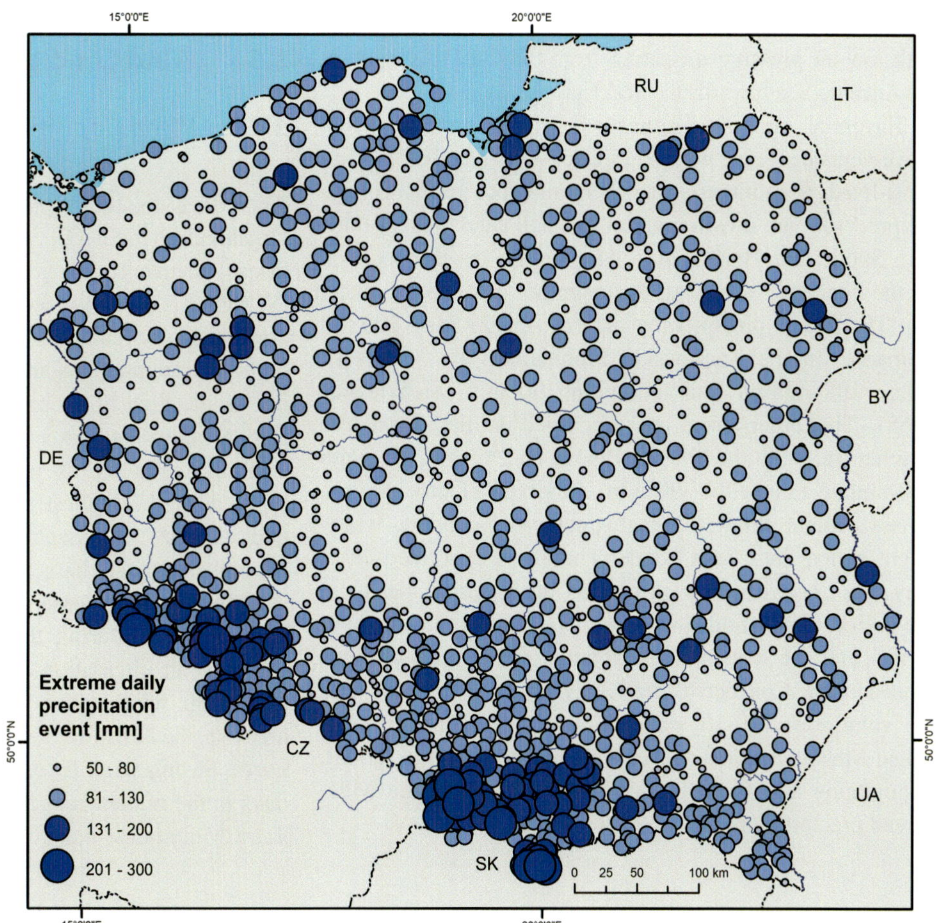

Fig. 3.8 Extreme daily rainfall (> 50 mm) at selected stations of IMGW-PIB in 1951–2020

Table 3.1 Selected episodes with the most intensive precipitation, distinguishing between widespread as well as convective torrential rainfall

	Date	Station name (region)	*H* [m asl]	Precipitation [mm]
Widespread	2 August 1888	Świeradów Zdrój (Sudetes)	550	215 (day)
	29 July 1897	Nová Louka (Sudetes)	805	345 (day)
	2 July 1909	Śnieżne Kotły (Sudetes)	1490	220 (day)
	16-18 July 1934	Witów (Western Carpathians)	815	383 (3 days)
	25 August 1938	Międzylesie (Sudetes)	453	197 (day)
	10 August 1964	Ostroszowice (Sudetes Foreland)	375	192 (day)
	30 June 1973	Hala Gąsienicowa (Western Carpathians)	1521	300 (day)
	21-25 July 1980	Teleśnica (Carpathians Foreland)	441	343 (5 days)
	5-9 July 1997	Šance (Western Carpathians)	509	617 (5 days)
	5-7 July 1997	Hala pod Śnieżnikiem (Sudetes)	1218	557 (3 days)
	13 August 2002	Knajpa (Sudetes)	967	278 (day)
Convective torrential	23 July 1914	Walentynowo (Polish Lowland)	108	91 (30 min)
	2 August 1951	Komorów (Sudetes)	275	193 (150 min)
	23 August 1953	Lubaczów (Carpathians Foreland)	215	63 (10 min)
	23 July 1956	Skrzyczne (Western Carpathians)	1257	220 (90 min)
	29 May 1968	Komarno (Sudetes)	490	182 (90 min)
	6 June 1988	Dopiewiec (Polish Lowland)	86	220 (180 min)
	18 May 1996	Sułoszowa (Świętokrzyskie Mts.)	443	180 (120 min)
	19 July 2002	Nowosielce (Carpathians Foreland)	306	53 (40 min)
	8 September 2022	Gorzów Wielkopolski (Polish Lowland)	72	130 (190 min)

with low air pressure migrating from the Gulf of Genoa or the Adriatic Sea towards Central Europe.

Torrential rainfall is connected with strong ascending movements of the air within large convective cells. They are short-lived, very intense, and their influence on hydrological processes are mostly limited to small catchments (Table 3.1; Sobik and Błaś 2010). Among the cases described in the literature and well-documented, two events stand out. The first is precipitation recorded on 29 May 1968 in Komarno, in the Kaczawskie Mountains (West Sudetes), where 182 mm of rain fell within 90 min (Sienkiewicz 1968). The other event was recorded at the Skrzyczne weather station in the Silesian Beskid on 23 July 1956, with 220 mm of rain within 90 min. This type of precipitation causes violent flows in local small watercourses, which might evolve into flash floods. The above-described cases may occur between late spring and late summer, that is, in the period of maximum water vapour storage in the troposphere and the development of intensive upward currents (convection). Convective cells may be large in size, with the volume of 10^3–10^5 cubic kilometres, forming the so-called Mesoscale Convective Systems (Ashley et al. 2003). Topography affects their development less than the widespread precipitation.

3.5 Wind Conditions

3.5.1 Wind Conditions at a National Scale

According to the report of the World Meteorological Organization (WMO), wind is the second element (after floods) causing the greatest material losses in Europe (WMO 2022). Its speed and direction are derived from atmospheric circulation shaped by baric systems (low and high). In mountainous areas, the force of the wind increases due to the flow of air over mountain ridges or its channelling in valleys and passes.

Winter months, as well as November and March, are typified by the highest wind speeds in Poland, while July and August are associated with the lowest velocities (Lorenc 1996). A characteristic feature of the annual course is a significant increase in wind speed observed at the turn of November. During this period, there is a transition to dynamic air circulation, typical for the winter season. The maximum wind speeds in Poland occur in mountainous areas in the south of the country and coastal areas in the north. Strong wind (exceeding 10 m/s) is recorded for 292 days in the highest range of the Sudetes, for 185 days in the Tatra Mountains, and for about 50–70 days in the coastal belt. In the lowlands, the frequency of strong winds ranges from 10 to 20 days in low-lying areas to 40 days a year in more elevated places (Kożuchowski 2011). The average annual wind speed in Poland varies from 12 m/s in the highest point of the Sudetes to 2.5 m/s in secluded valleys and mountain valleys. Outside the mountains, in the lowlands, the average annual wind speed ranges from 3.0 to 4.0 m/s. It also increases locally to 4.5–5.0 m/s in the windiest, that is, central part of the coast. Due to the varied spatial scale of atmospheric processes that are responsible for the occurrence of high-speed winds, large-scale and local events are distinguished.

3.5.2 Large-Scale Events

In the large-scale approach (synoptic scale), wind speed and direction result from the general or regional atmospheric circulation, shaped by baric systems. It is characterized by intense cyclonic activity in temperate latitudes (Sobik and Błaś 2010). Significant wind speeds in Poland are most often recorded in the case of low-pressure systems moving along routes III and IV designated by Van Bebber (1891; Fig. 3.2). In the case of route III, the deepening low-pressure system moves quickly over Poland from the area of the Norwegian Sea or northern Scandinavia and is called a 'diving low'. Active lows also quite often move from the Atlantic through the North Sea towards the southern part of Scandinavia and the Baltic Sea (route IV). High wind speed may persist continuously for several consecutive days. The best examples are cyclones (hurricanes) Lothar (December 26, 1999), Cyril (January 18–19, 2007) and Emma (March 1, 2008). If such a deep and active low is effectively blocked by an extensive high located on the eastern side of the continent, then the isobars run in the SW–NE pattern (Kwiatkowski and Hołdys 1985; Sobik and Błaś 2010). The wind reaches the highest speeds in the Sudetes and the Carpathians and in their forelands due to the overlapping effect of deformation of air flow over the mountain ridges (foehn wind). This is well illustrated by examples from July 13, 1961; May 6, 1968; February 11, 1978; December 19, 1989; January 6, 1994 and January 17 and 30, 2022, when wind gusts on Mt. Śnieżka in the Sudetes exceeded even 60 m/s (Żurański and Jaśpińska 1996).

3.5.3 Local Events

Extremely high wind speeds can also occur in a relatively small area during intense storms in the form of squalls or whirlwinds. A squall is formed during a thunderstorm as a descending column of cooler air comes into contact with the Earth's surface. It can cause significant material losses due to wind speed of 30–40 m/s. The zone of influence of a squall usually covers a strip up to several kilometres wide and up to several dozen kilometres long (Sobik and Błaś 2010). One such event took place in the north-eastern part of Poland. On July 4, 2002, a squall reached gusts of up to 45 m/s. It caused significant forest damage on a total area of 45.4 thousand ha. In turn, the average frequency of whirlwinds in Poland is about 5–6 events per year, although their number may change significantly from year to year (Taszarek and Brooks 2015). A greater probability of violent storms and accompanying whirlwinds occurs in the southern, upland part of Poland. In the current century, on August 15, 2008, an exceptional event took place. On that day, there were violent storms over Poland combined with strong gusts of wind, extreme precipitation and whirlwinds. One of them covered the route of about 60 km and caused huge material losses. The wind speed in the vortex could have reached even 70–80 m/s (Sobik and Błaś 2010).

3.6 Climate Change

3.6.1 Climate Evolution Since the Last Glacial Cycle

Analysing climate changes occurring on the European continent in the last millennium, several clear phases can be distinguished, two of which are exceptional. The eleventh century was characterized by temperatures higher than the multi-century average and together with the twelfth and thirteenth centuries it is called the Medieval Warm Period. Apart from the increase in temperature currently observed in Europe (in the second decade of the twenty-first century), it was the warmest period in measurement history (Przybylak et al. 2005; Woś 2010). Then, in the middle of the fourteenth century, the air temperature dropped and a period of significantly cooler climate began, which lasted at least until the middle of the fifteenth century. It was followed by a short warming lasting only until the middle of the sixteenth century. Since then, another, even more pronounced decrease in air temperature (much below the average for the entire millennium) called the Little Ice Age began. Brázdil et al. (2010) assumed that it lasted from 1570 to 1860. In the Polish lands, the lowest temperatures occurred in the first half of the seventeenth century. The Little Ice Age began relatively suddenly and had a significant impact on (among other things) the vegetation. It was the last and youngest period of growth of the surface covered with glaciers and ice sheets. As such, it was not homogeneous, as it was characterized by the occurrence of short periods of particularly cold conditions, separated by series of warmer years. There were single years or their series, during which significant changes in the regime of the current weather conditions and a greater frequency of catastrophic events were recorded. In the eighteenth century the climate warmed up again. The air temperature in Poland was close to the multi-century average. The turn of the nineteenth century was the beginning of another colder period, and the lowest temperatures were recorded in the years 1800–1820 (Brázdil et al. 2010).

When looking for the cause of the described temperature changes, feedback effects should be given great importance. For example, the relationships between the circulation of the atmosphere and the transport of water in the north-eastern part of the Atlantic Ocean are significant (García-Ibáñez et al. 2018). In this case, it is of great importance to maintain the thermohaline circulation, whose heating effect on the North Atlantic region may weaken, which results in, among other things, increasing ice cover and extent of drift ice in the ocean (Broecker 1994, 2000; Lohmann and Gerdes 1998). An important role is also played by the level of salinity of surface waters in the Atlantic Ocean, associated with changes in the extent of the ice cover (Aagaard and Carmack 1994; Nørgaard-Pedersen et al. 2003). Another factor are changes in the solar constant and the interactions between the solar wind, cosmic rays and atmospheric processes (Dorman 2009; Mörner 2010). Hence, the Little Ice Age coincided with the lowest solar activity known as the Maunder minimum (1645–1715) and the Dalton minimum (1820–1850) (Rind 2004; Owen et al. 2017). The importance of great volcanic eruptions followed by several years of thermal and precipitation anomalies should not be overlooked either. Exceptionally cool summers were observed in Europe and Poland after the eruptions of Tambora (1815), Krakatau (1883) and Santa Maria (1902) volcanoes (Schaller et al. 2009).

3.6.2 Contemporary Climate Changes

The beginning of the first phase of warming in Poland in the twentieth century occurred in the first decade. At that time, winters became milder, while summers were mostly cool. This phase reached its maximum in the 1930s due to exceptionally warm summers. The mean annual air temperature for the whole decade of 1931–1940 in Warsaw exceeded 8.0 °C. In the 1940s warming stopped and a series of severe winters followed. At the same time, the winter of 1946/47 was the last one when the entire Baltic Sea froze (Kożuchowski 2011). The next cooling wave with frosty winters occurred in Poland in the 1960s and 1970s. On the other hand, the next considerable warming took place at the end of 1980s. The mean annual air temperature in Poland at that time more often exceeded 9.0 °C. The next three consecutive winters of 1987/1988, 1988/1989 and 1989/1990 as well as the summers of 1992, 1994 and 1995 were exceptionally warm. Further increase of the upward air temperature trend occurred after 2014. The calculated area temperature averages indicate that throughout the history of temperature measurements the warmest year in Poland was 2019, with the temperature reaching 10.2 °C, followed by 2020, 2018, 2015 and 2014 (all with the mean temperature > 9.5 °C). A characteristic feature of the currently observed warming is that the biggest changes concern the average winter temperatures, while less pronounced and less regular changes are observed in summer. Although quasi-cyclical fluctuations are still observable, i.e., abnormally warm summers alternate with the cool ones, the latter are less and less frequent.

Poland is among the areas highly sensitive to circulation conditioning, which is responsible for the occurrence of successive phases of warming and cooling. If the zonal (westerly) flow of air masses intensifies, the average annual temperature might rise even by 1.5–2.0 °C (Biernacik et al. 2010). However, changes in the circulation conditions do not fully account for the magnitude of warming observed in recent decades. This means that the entire atmosphere has become warmer and as a result, the same but warmer air masses are arriving over Poland. However, it is also worth noting that temperature fluctuations are not too tightly correlated with changes in the mean hemispheric temperature. Thus, temperature variability in Poland has a considerably individual character and not all observed fluctuations are derivatives of global fluctuations. Marsz (2007) stressed that appropriate distribution of the ocean surface temperature, including atmospheric circulation, may ensure increased atmospheric heat input and its accumulation from year to year. He went on to state that contemporary circulation mechanisms set in motion a series of previously non-existing processes, whose diverse actions force air temperature to rise on the hemispheric scale. A different view on the cause of the warming are formulated in IPCC reports (2021). According to these, changes in radiative balance and global temperature are mainly a result of an increasing accumulation of greenhouse gasses. This is caused by burning fossil fuels and changes in the way the ground is being utilized (anthropogenic factor).

Degirmendžić et al. (2004) described a significantly increasing trend of areal temperature calculated as mean temperature from 51 meteorological stations for the period 1951–2000. The trend in annual means reached 0.2 °C per decade. For monthly means, a significant tendency was observed only in March (0.6 °C per decade) and May (0.3 °C per decade). Degirmendžić et al. (2004) underlined that temperature changes in the first months of the year are mainly responsible for the observed 50-year warming trend. They also showed that the pace of warming increased rapidly during the second half of the analyzed period and reached 0.6

°C per decade for the period of 1976–2000. There are a few longer homogeneous air temperature series in Poland, from Warsaw (since 1779), Wrocław (since 1791), Cracow (since 1792), Poznań (since 1848), Gdańsk (since 1851), Puławy (since 1871) or Mt. Śnieżka (since 1880). They also confirmed an increasing tendency of temperature means, especially significant in the last decades (Górski and Marciniak 1992; Trepińska and Kowanetz 1997; Miętus 1998; Lorenc 2000; Bryś and Bryś 2010; Migała et al. 2016; Kolendowicz et al. 2019). This warming is visible also in a longer period in a reconstructed series of temperature prepared with the use of standardized tree-ring width chronology of the Scots pine (*Pinus sylvestris* L.) (Przybylak et al. 2005). Figure 3.9 presents the time series of annual temperature in the last 120 years (1901–2020) in the city of Wrocław (lowlands) and Mt. Śnieżka (mountains). An increasing temperature trend for both weather stations is significant and reaches 0.14 °C per decade in Wroclaw and 0.13 °C per decade at Mt. Śnieżka. The highest increase is observed in the last 50 years, 0.6 and 0.4 °C per decade, respectively.

Long-term changes in temperature are observed also in the extremes. As reported by Brázdil et al. (1996) for Central and Southern Europe for the years 1951–1990, and later by Wibig and Głowicki (2002) directly for the area of Poland for the years 1951–1998, the mean minimum and mean maximum daily temperatures increased, whereas the daily temperature range decreased. Wibig and Głowicki (2002) showed statistically significant tendencies for most of the analyzed weather stations for mean minimum temperature (0.14–0.31 °C per decade). They noted a decreased rate of daily temperature range from 0.03 to 0.18 °C per decade and a significant trend for 60% of the stations. They also reported a significant decrease in the annual frequency of days with the minimum temperature below 0 °C (i.e., cold days) with a rate from 0.4 to 6.0 days per decade. These tendencies were significant mainly for winter and spring months. Based on a longer period-from 1951 to 2015, Wibig (2017) also showed a significant increasing tendency of summer heat waves with daily maximum temperatures above 30 °C. The results took into account also the last decade of temperature measurements (Tomczyk et. al. 2022) and showed an intensification of the increase in the mean maximum temperature and also in the number of hot days in summer. The reported increasing rate for mean

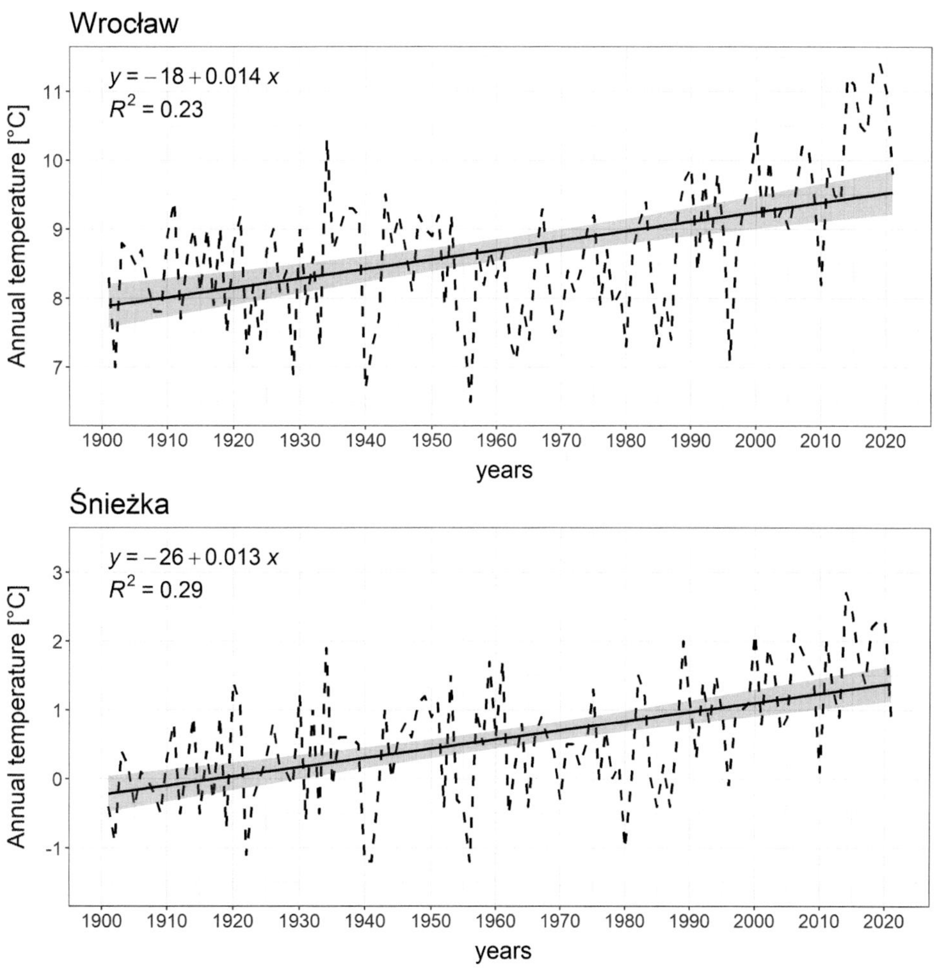

Fig. 3.9 Time series of annual temperature of Wrocław and Śnieżka (dashed lines) and trend lines (solid) with confidence regions (in grey). *Source* Compiled on the basis of data provided by the authors of the data series homogenization: Tadeusz and Krystyna Bryś (Wrocław) and Krzysztof Migała, Grzegorz Urban and Karol Tomczyński (Śnieżka)

maximum temperature was between 0.35 and 0.75 °C per decade, whereas the number of hot days increased between 0.4 and 2.8 days per decade.

Changes in air temperature do not simply translate into trends in precipitation. Degirmendžić et al. (2004) analyzed trends in precipitation totals in the years 1951–2000 and showed a weak increasing trend in monthly values but, excluding March, the tendency was insignificant. They noted, however, a decreased contribution of summer precipitation sums to the annual total. Trends from a longer period (1951–2013) examined by Szwed (2019) confirmed their earlier results. A significantly more distinct increase of monthly precipitation totals were observed for the northern part of Poland. Szwed (2019) also underlined spatial variability of an increasing tendency rate of winter precipitation totals in Poland. Tomczyk et al. (2021), taking into account the increasing maximum and minimum temperature of winter seasons, showed a significant decreasing tendency of snow cover depth. Substantial changes were also reported for extreme precipitation sums. Łupikasza (2010) demonstrated seasonal and spatial variability of decreasing tendency in extreme precipitation sums in the years 1951–2006. Complex changes of extreme precipitation described by Pińskwar (2021) confirmed earlier results of Łupikasza, but only for precipitation above the values of the 95th percentile. For the highest ones, above 99th percentile, the precipitation amount increased.

3.7 Implications for Geomorphological Processes

Among significant features of Poland's climate affecting the occurrence and intensity of geomorphological processes one should distinguish: (1) the annual rhythm of the variability of climate elements; (2) its instability observed from year to year with regard to the duration and structure of the weather types contribution; and (3) the occurrence of extreme meteorological phenomena. Weather variability in Poland is most often considered in terms of four seasons. Ultimately, the described annual climatic cycle, along with the vegetation and hydrological cycles, shape the seasonal character of geomorphological processes.

During winter months, there are several cold waves with snow cover, which are separated by thaw episodes. This frequently results in recorded air temperatures falling below 0 °C. Maximum annual wind speeds are also typical. The occurrence of seasonal frozen ground slows down hillslope processes, but at the same time facilitates surface runoff (Dobiński 1997). In snowless periods of the year, combined with cold and windy weather, aeolian processes intensify (Rodzik et al. 2008). The spring period, due to the prevalence of longitudinal exchange of air masses, is characterized by a wide range of thermal weather types (from short winter episodes with snow cover and frost to daily average temperatures typical of summer). Due to melting of the snow cover and thawing of the ground, the most intensive surface runoff is observed in this period (Migoń 2008). Snow-melt floods increase the dynamics of fluvial processes (Rodzik et al. 2007). Vegetation-free soil creates favourable conditions for the intensification of aeolian processes during drought episodes and for sheet flow during intensive precipitation (Kotarba 2005). Summer is characterised by longer periods of precipitation separated by heat waves, intensive evaporation, and the highest frequency of extreme events such as storms, heavy rain and strong wind. On the one hand, the most developed vegetation cover, which protects against splash erosion, is important for slope stability. On the other hand, however, the occurrence of extreme precipitation results in linear erosion, with catastrophic consequences (Czerwiński and Żurawek 1999; Migoń et al. 2002; Zieliński 2003). Autumn is the most stable season of the year, with a predominance of periods of insufficient precipitation. With the overlapping decline phase of vegetation season, an uptrend is observed in the significance of wind erosion.

Apart from temporal variability of climate elements, which control the intensity of geomorphological processes, their spatial variability is also important. The relief of the Polish Lowland is remodelled by chemical denudation, deflation and aeolian accumulation. On gentle slopes rain-induced erosion dominates, while gully erosion and mass movements are of marginal importance. Due to insignificant slope inclinations there is little erosive activity in the lowlands and even flash floods do not lead to important transformations of floodplains (Falkowski 1971). Overbank flows in rivers are likely to occur in the spring during thaw episodes as well as during spring–summer heavy rainfall (Starkel 1976). Intensive rain erosion occurs during rainfall of 10 mm per day and in the absence of vegetation (Gradziński et al. 2003; Kostrzewski et al. 2008). Frequent thawing and freezing of the ground during oceanic winter leads to the destruction of soil aggregates and promotes the occurrence of rain-induced erosion of the frozen ground layer (Hagen 1991; Twardy 1990). Wind activity is one of the more important morphogenetic processes in the lowland belt (Podsiadłowski and Rząsa 1994; Stach and Podsiadłowski 2002). Wojtanowicz (1999) estimated the intensity of the processes of deflation, aeolian transport and sediment accumulation at 50–300 t/km^2/a. The most intensive deflation occurs on non-forested, open surfaces of sandy accumulation (Galon 1958). Taking into account spatial variability, the greatest danger of deflation occurs in central Poland due to shortages of precipitation recorded there (Kasprzyk 2004).

In upland and mountainous areas, the contemporary morphogenesis is significantly influenced by extreme precipitation triggering diverse slope processes and impacting on the geomorphological effectiveness of water flows in river channels. Rain-induced areal and linear erosion play an important role on slopes used for agriculture (Koćmit 1998). Debris flows, which are the most hazardous and the most destructive mass movements, are found mostly in the highest parts of the Carpathians (Tatra Mountains Kotarba et al. 1987, 2013; Rączkowska 2021) and the Sudetes (Czerwiński and Żurawek 1999; Migoń et al. 2002; Migoń 2008). The Flysch Carpathians, on the other hand, are dominated by landslides, which is due to the occurrence of flysch series prone to periodic changes in strength, structural deformations and, as a result, gravitational mass movements. (Gil and Kotarba 1977; Ziętara 1999; Starkel 2006; Rączkowska et al. 2012). Due to high frequency of foehn wind, areas to the north of the Sudetes and the Carpathians experience a particularly intensive soil drying, which intensifies the role of aeolian erosion (Gerlach and Koszarski 1968; Jahn 1969; Kwiatkowski and Hołdys 1985; Kida and Jary 2002).

Among the factors most intensely affecting shorelines and the seabed in the coastal areas of non-tidal seas, such as the Baltic Sea, is wind (Basiński et al. 1993). Strong wind recorded in the coastal zone brings about storms, which, in turn, cause abrasion of coastal cliffs, the pace of which is several metres per year on average (Subotowicz 1982; Kostrzewski and Zwoliński 1988). At wind speeds exceeding 5–7 m/s under favourable conditions, deflation process of fine particles of desiccated soil and activation of sand on coastal dunes may occur (Wojtanowicz 1999; Łabuz 2005).

References

Aagaard K, Carmack EC (1994) The Arctic Ocean and climate: a perspective, in the Polar Oceans and their role in shaping the global environment. Geophys Monogr 85:5–20

Ashley WS, Mote TL, Dixon G, Trotter SL, Powell EJ, Durkee JD, Grundstein AJ (2003) Distribution of mesoscale convective complex rainfall in the United States. Monthly Weather Rev 131(12):3003–3017

Bader MJ, Roach WT (1977) Orographic rainfall in warm sectors of depressions. Quart J Roy Meteorol Soc 103:269–280

Basiński T, Pruszak Z, Tarnowska M, Zeidler R (1993) Ochrona brzegów morskich. Bibl Nauk Hydrotechnika 17, IBW PAN, Gdańsk, p 536

Biernacik D, Filipiak J, Miętus M, Wójcik R (2010) Zmienność warunków termicznych w Polsce po roku 1951. In: Rezultaty projektu KLIMAT, pp 9–21

Błaś M, Sobik M (2005) Osobliwości klimatu Karkonoszy i Gór Izerskich. In: Krzemień K, Trepińska J, Bokwa A (eds) Rola stacji terenowych w badaniach geograficznych. Uniwersytet Jagielloński, Kraków, pp 109–121

Bordi I, Zhu X, Fraedrich K (2016) Precipitable water vapour and its relationship with the standardized precipitation index: ground based GPS measurements and reanalysis data. Theor Appl Climatol 123:263–275

Boryczka J (1977) Empiryczne równania klimatu Polski. Rozprawy UW (86), Wyd Uniw Warsz, Warszawa, p 288

Brázdil R, Budíková M, Auer I, Böhm R, Cegnar T, Faško P, Lapin M, Gajić-Čapka M, Zaninović K, Koleva E, Niedźwiedź T, Ustrnul Z, Szalai S, Weber RO (1996) Trends of maximum and minimum daily temperatures in central and southeastern Europe. Intern J Climatol 16:765–782

Brázdil R, Dobrovolny P, Luterbacher J, Moberg A, Pfister C, Wheeler D, Zorita E (2010) European climate of the past 500 years: new challenges for historical climatology. Clim Change 101:7–40

Broecker WS (1994) Massive iceberg discharges as triggers for global climate change. Nature 372:421–424

Broecker WS (2000) Was a change in thermohaline circulation responsible for the little ice age? Proc Nat Acad Sci 97(4):1339–1342

Browning KA (1990) Organization of clouds and precipitation in extratropical cyclones. In: Newton CW, Holopainen E (eds) Extratropical cyclones: the erik palmén memorial volume. American Meteorlogical Society, pp 129–153

Bryś K, Bryś T (2010) Reconstruction of the 217-year (1791–2007) Wrocław air temperature and precipitation series. Bull Geogr Phys Geogr Ser 3(1):121–171

Choularton TW, Perry SJ (1986) A model of the orographic enhancement of snowfall by the seeder-feeder mechanism. Quart J Met Soc 112:335–345

Czerwiński J, Żurawek R (1999) The geomorphological effects of heavy rainfalls and flooding in the Polish Sudetes in July 1997. Studia Geomorph Carp-Balc 33:27–43

Degirmendžić J, Kożuchowski K, Żmudzka E (2004) Changes of air temperature and precipitation in Poland in the period 1951–2000 and their relationship to atmospheric circulation. Intern J Climatol 24:291–310

Degirmendžić J, Kożuchowski K (2017) Mediterranean cyclones, the atmospheric moisture content and precipitation in Poland. Geogr Polon 90:5–20

Dettinger M, Ralph FM, Das T, Neiman PJ, Cayan DR (2011) Atmospheric rivers, floods and the water resources of California. Water 3:445–478

Dobiński W (1997) Distribution of mountain permafrost in the High Tatras based on freezing and thawing indices. Biul Perygl 36:29–37

Dore AJ, Choularton TW (1992) A three-dimensional model of airflow and orographic rainfall enhancement. Quart J Roy Meteorol Soc 118:1041–1056

Dore AJ, Sobik M, Migała K (1999) Patterns of precipitation and pollutant deposition by rain and snow in the western Sudete Mountains, Poland. Atmosph Environ 33:3301–3312

Dorman L (2009) Solar wind, cosmic rays and space weather effects on global climate change. Ann Geophys 30:9–19

Falkowski E (1971) Historia i prognoza rozwoju układu koryta wybranych odcinków rzek nizinnych Polski. Biul Geol Uniw Warsz 12:5–45

Galon R (1958) Z problematyki wydm śródlądowych w Polsce. In: Galon R (ed) Wydmy śródlądowe Polski. PWN, Warszawa, pp 13–31

García-Ibáñez M, Pérez F, Lherminier P, Zunino P, Mercier H, Tréguer P (2018) Water mass distributions and transports for the 2014 GEOVIDE cruise in the North Atlantic. Biogeosci 15:2075–2090

Gerlach T, Koszarski L (1968) Współczesna rola morfogenetyczna wiatru na przedpolu Beskidu Niskiego. Studia Geomorph Carp-Balc 2:85–114

Gil E, Kotarba A (1977) Model of slide slope evolution in flysch mountains (an example drawn from the Polish Carpathians). CATENA 4:233–248

Girs AA (1981) K woprosu o formach atmosfernoj cyrkulacji. Trudy AANII 374:4–13

Górski T, Marciniak K (1992) Air temperature at Puławy over the years 1871–1990. Mean monthly temperature. Pamiętnik Puławski 100:7–26

Gradziński R, Baryła J, Doktor M, Gmur D, Gradziński M, Kędzior A, Paszkowski M, Soja R, Soja R, Zieliński T, Żurek S (2003) Vegetation controlled modern anastomosing system of the upper Narew River (NE Poland) and its sediments. Sediment Geol 3–4:253–276

Hagen LJ (1991) Wind erosion mechanics: abrasion of aggregated soil. Trans Am Soc Agric Eng 34(4):831–834

Hann J (1906) Lehrbuch der Meteorologie. Chr Herm Tauchnitz, Lepizig, p 642

Hurrell JW (1995) Decadal trends in the North Atlantic Oscillation. Science, New York, NY, p 269

IPCC (2021) Climate change 2021: the physical science basis. Contribution of Working Group I to the sixth assessment report of the intergovernmental panel on climate change. In: Masson-Delmotte V, Zhai P, Pirani A, Connors LS, Péan C, Berger S, Caud N, Chen Y, Goldfarb L, Gomis MI, Huang M, Leitzell K, Lonnoy E, Matthews JBR, Maycock TK, Waterfield T, Yelekçi O, Yu R, Zhou B (eds) Cambridge University Press. United Kingdom and New York, NY, USA, Cambridge, p 2391

Jahn A (1969) Niveo-eoliczne procesy w Sudetach i ich działanie na glebę. Probl Zagosp Ziem Górskich 5:53–92

Kasprzyk A (2004) Występowanie susz hydrologicznych w Europie środkowej. Prace IG Akad Świętokrz w Kielcach 13:181–201

Kida J, Jary Z (2002) Procesy niveo-eoliczne na przedpolu Gór Bardzkich w grudniu 2001 r. Zesz Probl Post Nauk Roln 487:87–95

Koćmit A (1998) Charakterystyka zmian w morfologii i właściwościach gleb uprawnych spowodowanych erozją wodną na obszarach młodoglacjalnych Pomorza. Zesz Probl Post Nauk Roln 460:531–557

Kolendowicz L, Czernecki B, Półrolniczak M, Taszarek M, Tomczyk AM, Szyga-Pluga K (2019) Homogenization of air temperature and its long-term trends in Poznań (Poland) for the period 1848–2016. Theor Appl Climatol 136:1357–1370

Kostrzewski A, Zwoliński Z (1988) Morphodynamics of the cliffed coast, Wolin Island. Geogr Polon 55:69–81

Kostrzewski A, Zwoliński Z, Andrzejewski L, Florek W, Mazurek M, Niewiarowski W, Podgórski Z, Rachlewicz G, Smolska E, Stach A, Szmańda J, Szpikowski J (2008) Współczesna ewolucja rzeźby młodoglacjalnej Niżu Polskiego. In: Starkel L, Kostrzewski A, Kotarba A, Krzemień K (eds) Współczesne przemiany rzeźby Polski. Uniwersytet Jagiellońksi, pp 135–165

Kotarba A (2005) Geomorphic processes and vegetation pattern changes. Case study in the Zelene Pleso Valley, High Tatra, Slovakia. Studia Geomorph Carp-Balc 39:39–48

Kotarba A, Kaszowski L, Krzemień K (1987) High-mountain denudational system in the Polish Tatra Mountains. Geogr Stud IGiPZ PAN, Spec Iss 3:1–106

Kotarba A, Rączkowska Z, Długosz M, Boltižiar M (2013) Recent debris flow in the Tatra Mountains. In: Loczy D (ed) Geomorphological impact of extreme weather (case studies from Central and Eastern Europe). Springer, Dordrecht, pp 221–236

Kottek M, Grieser J, Beck C, Rudolf B, Rubel F (2006) World Map of the Köppen-Geiger climate classification updated. Meteorol Z 15:259–263

Kwiatkowski J, Hołdys T (1985) Klimat. In: Jahn A (ed) Karkonosze Polskie. Ossolineum, Wrocław, pp 87–116

Kożuchowski K (2011) Klimat Polski – nowe spojrzenie. PWN, Warszawa, p 293

Kożuchowski K (2016) Zawartość wody opadowej w atmosferze i opady w Polsce. Przegl Geofiz 61:151–169

Lohmann G, Gerdes R (1998) Sea ice effects on the sensitivity of the thermohaline circulation. J Climate 11:2789–2803

Lorenc H (1996) Struktura i zasoby energetyczne wiatru w Polsce. Materiały Badawcze IMGW: Meteorologia 25:155

Lorenc H (2000) Studies on 218 years of air temperature data set for Warsaw (1779–1997) and estimation of multiyear tendencies. Materialy Badawcze IMGW: Meteorologia 31:104

Łabuz TA (2005) Brzegi wydmowe polskiego wybrzeża Bałtyku. Czas Geogr 76:19–47

Łupikasza E (2010) Spatial and temporal variability of extreme precipitation in Poland in the period 1951–2006. Intern J Climatol 30:991–1007

Marsz AA (2007) Czy w latach 1976–1978 nastąpiła zmiana reżimu cyrkulacyjnego i klimatycznego półkuli północnej? Przegl Geofiz 52:7–25

Marsz AA (2008) W sprawie genezy Oscylacji Północnego Atlantyku. Przegl Geofiz 53:3–26

Miętus M (1998) The reconstruction and homogenization of long-term series of monthly mean temperature from Gdansk-Wrzeszcz station, 1851–1995. Wiadomości IMGW 21(2):41–64

Migała K, Urban G, Tomczyński K (2016) Long-term air temperature variation in the Karkonosze mountains according to atmospheric circulation. Theor Appl Climatol 125:337–351

Migoń P, Hrádek M, Parzóch K (2002) Extreme events in the Sudetes Mountains. Their long-term geomorphic impact and possible controlling factors. Studia Geomorph Carp-Balc 36:29–49

Migoń P (2008) Współczesna ewolucja rzeźby Sudetów i ich Przedgórza. In: Kostrzewski A, Kotarba A, Krzemień K (eds) Starkel L. Współczesne przemiany rzeźby Polski, Uniwersytet Jagielloński, pp 135–165

Mörner NA (2010) Solar Minima, Earth's rotation and little ice ages in the past and in the future: the North Atlantic-European case. Global Planet Change 72:282–293

Müller M, Kašpar M, Matschullat J (2009) Heavy rains and extreme rainfall-runoff events in Central Europe. Nat Hazards Earth Syst Sci 9:441–450

Neiman PJ, Ralph FM, Wick GA, Lundquist JD, Dettinger MD (2008) Meteorological characteristics and overland precipitation impacts of atmospheric rivers affecting the West Coast of North America based on eight years of SSM/I satellite observations. J Hydrometeorol 9:22–47

Niedźwiedź T, Starkel L (2008) Klimatyczne tło współczesnych procesów morfogenetycznych na obszarze Polski. In: Kostrzewski A, Kotarba A, Krzemień K (eds) Starkel L. Współczesne przemiany rzeźby Polski, Uniwersytet Jagieloński, pp 21–35

Nørgaard-Pedersen N, Spielhagen RF, Erlenkeuser H, Grootes PM, Heinemeier J, Knies J (2003) Arctic Ocean during the last glacial maximum: atlantic and polar domains of surface water mass distribution and ice cover. Paleoceanography 18:1063

Owens MJ, Lockwood M, Hawkins E, Usoskin I, Jones GS, Barnard L, Schurer A, Fasullo J (2017) The maunder minimum and the little ice age: an update from recent reconstructions and climate simulations. J Space Weather Space Clim 7:A33

Pińskwar I (2021) Complex changes of extreme precipitation in the warming climate of Poland. Intern J Climatol 42:817–833

Podsiadłowski S, Rząsa S (1994) The intensity of wind erosion as influenced by agriculture activity. Roczn Akad Roln w Poznaniu 260:95–99

Przybylak R, Majorowicz J, Wójcik G, Zielski A, Chorążyczewski W, Marciniak K, Nowosad W, Oliński P, Syta K (2005) Temperature changes in Poland from the 16th to the 20th centuries. Internat J Climatol 25:773–791

Ralph FM, Neiman PJ, Wick GA (2004) Satellite and CALJET aircraft observations of atmospheric rivers over the eastern North Pacific Ocean during the winter of 1997/98. Monthly Weather Rev 132:1721–1745

Rączkowska Z (2021) Współczesna ewolucja rzeźby Tatr. In: Kostrzewski A, Krzemień K, Migoń P, Starkel L, Winowski M,

Zwoliński Z (eds) Współczesne przemiany rzeźby Polski. Bogucki Wyd Nauk, Poznań, pp 61–94
Rączkowska Z, Łajczak A, Margielewski W, Święchowicz J (2012) Recent landform evolution in the Polish Carpathians. In: Lóczy D, Stankoviansky M, Kotarba A (eds) Recent landform evolution: the Carpatho-Balkan-Dinaric region. Springer, Dordrecht, pp 47–102
Rind D, Shindell D, Perlwitz J, Lerner J, Lonergan P, Lean J, Mclinden C (2004) The relative importance of solar and anthropogenic forcing of climate change between the maunder minimum and the present. J Climate 17:906–929
Rodzik J, Furtak T, Zgłobicki W (2007) Influence of snowmelts and heavy rainfalls on water and sediment yield from the loess gully catchment (Lublin Upland—Poland). In: Casali J, Gimenez R (eds) Progress in gully erosion research. Univ Publ de Navarra, Pamplona, pp 104–105
Rodzik J, Ciupa T, Janicki G, Kociuba W, Tyc A, Zgłobicki W (2008) Współczesne przemiany rzeźby Wyżyn Polskich. In: Kostrzewski A, Kotarba A, Krzemień K (eds) Starkel L. Współczesne przemiany rzeźby Polski, Uniwersytet Jagieloński, pp 385–395
Schaller N, Griesser T, Fischer A, Stickler A, Brönnimann S (2009) Climate effects of the 1883 Krakatoa eruption: Historical and present perspectives. Vierteljahrsschrift Naturforsch Gesell Zürich 154:31–40
Sienkiewicz R (1968) Przyczyny i skutki wystąpienia opadu nawalnego w Komarnie w dniu 29 maja 1968 r. Gazeta Obserwatora PIHM 11:10–12
Sobik M, Netzel P, Quiel F (2001) Zastosowanie modelu rastrowego do określenia pola rocznej sumy opadów atmosferycznych na Dolnym Śląsku. Uniw Gdański, Rocz Fizycznogeogr 6:27–34
Sobik M, Błaś M (2010) Wyjątkowe zdarzenia meteorologiczne. In: Migoń P (eds) Wyjątkowe zdarzenia przyrodnicze na Dolnym Śląsku i ich skutki. Rozpr Nauk Inst Geogr i Rozwoju Reg Uniw Wrocł 14:35–80
Stach A, Podsiadłowski S (2002) Pulverizing and wind erosion as influenced by spatial variability of soil texture. Quaest Geogr 22:67–78
Starkel L (1976) The role of extreme (catastrophic) meteorological events in the contemporaneous evolution of slopes. In: Derbyshire E (ed) Geomorphology and climate. Wiley, London, pp 203–246
Starkel L (2006) Geomorphic hazards in the Polish Flysch Carpathians. Studia Geomorph Carp-Balc 40:7–19
Starkel L (2008) Typy i kierunki współczesnych przekształceń rzeźby Polski. In: Kostrzewski A, Kotarba A, Krzemień K (eds) Starkel L. Współczesne przemiany rzeźby Polski, Uniwersytet Jagieloński, pp 385–395
Stopa-Boryczka M, Boryczka J (1998) Cechy kontynentalne klimatu Polski południowo-wschodniej. In: Nowosad M (eds) Problemy współczesnej klimatologii i agrometeorologii regionu lubelskiego. Wyd UMCS, Lublin, pp 111–114
Storebo P (1976) Small-scale topographical influences on precipitation. Tellus 28:45–59
Subotowicz W (1982) Litodynamika brzegów klifowych. PAN Ossolineum, Gdańsk, p 154
Szwed M (2019) Variability of precipitation in Poland under climate change. Theor Appl Climatol 135:1003–1015
Taszarek M, Brooks HE (2015) Tornado climatology of Poland. Monthly Weather Rev 143:702–717
Tomczyk AM, Bednorz E, Szyga-Pluta K (2021) Changes in air temperature and snow cover in winter in Poland. Atmosphere 12:68
Tomczyk AM, Piniewski M, Eini MR, Bednorz E (2022) Projections of changes in maximum air temperature and hot days in Poland. Internat J Climat 42:5242–5254
Trepińska J, Kowanetz L (1997) Long-term course of monthly values of air temperature in Cracow (1792–1995). In: Trepińska J (eds) Climate variability in Cracow (1792–1995). Institute of Geography UJ, Kraków, pp 99–130
Twardy J (1990) Przebieg spłukiwania w okolicach Bogini koło Łodzi, w cyklu rocznym. Acta Univ Lodz, Folia Geogr 12:15–49
Van Bebber WJ (1891) Die Zugstrassen der barimetrischen Minima. Meteor Zeit, p 361
Wibig J, Głowicki B (2002) Trends of minimum and maximum temperature in Poland. Climate Res 20:123–133
Wibig J (2017) Heat waves in Poland in the period 1951–2015: trends, patterns and driving factors. Meteorol Hydrol Water Manage Res Oper Appl 6(1):37–45
Wojtanowicz J (1999) Procesy eoliczne. In: Starkel L (eds) Geografia Polski. Środowisko Przyrodnicze, PWN, Warszawa, pp 410–413
Woś A (2010) Klimat Polski w drugiej połowie XX wieku. Wyd Nauk UAM, Poznań, p 489
Wypych A (2018) Tropospheric moisture content over Europe. Uniw Jagielloński, Kraków, p 197
WMO (2022) State of the climate in Europe 2021. World Meteorological Organization, Geneva, Switzerland, WMO, No. 1304, p 52
Zieliński T (2003) Catastrophic flood effects in alpine/foothill fluvial system (a case study from the Sudetes Mts, SW Poland). Geomorphology 54:293–306
Ziętara T (1999) The role of mud and debris flows modelling of the Flysch Carpathians relief, Poland. Studia Geomorph Carp-Balc 33:81–100
Żmudzka E (2002) Zmienność terminu wiosennego przejścia temperatury powietrza przez próg + 5 °C w Polsce oraz jego związki ze zmiennością wskaźnika NAO. In: Marsz AA, Styszyńska A (eds) Oscylacja Północnego Atlantyku i jej rola w kształtowaniu zmienności warunków klimatycznych i hydrologicznych Polski. Wyd Uczelniane Akad Morskiej, Gdynia, pp 129–139
Żurański JA, Jaśpińska B (1996) Directional analysis of extreme wind speeds in Poland. J Wind Eng 65:13–20

Marek Błaś is Associate Professor in the Department of Climatology and Atmosphere Protection in the Institute of Geography and Regional Development, University of Wrocław. His main professional interests are mountain meteorology and climatology, boundary layer meteorology, air pollution transport and deposition processes, forest-climate interactions, undergraduate and graduate education. In 1995–2014 he was the head of the mountain Meteorological Observatory at Mt. Szrenica in the Sudetes Mts. Manager or main contractor in over 25 research projects financed, among others, by the National Science Centre in Poland, Swedish Institute, Norwegian funds of European Economic Area, Department of the Environment, Fisheries and Rural Affairs (UK). He managed a project aimed at the inventory of pollutant emissions in Poland, where the main goal was to create a module to manage emission limits in Poland.

Hanna Ojrzyńska is a doctor of geography and Assistant Professor at the University of Wrocław. Her main research interests are mountain climatology, GIS and spatial planning. She was involved in projects focused on climatological conditions of mountain areas (the Sudetes and the Bieszczady Mountains) and air quality over Lower Silesia. In her research she uses GIS methods to recognize the influence of land morphology and type of land cover on the spatial diversity of climate. She is an author of classification of air circulation types for the Sudetes Mountains.

Geomorphological Diversity of Poland—Major Controls and Main Geomorphological Regions

4

Piotr Migoń and Kacper Jancewicz

Abstract

The chapter provides an overview of morphological diversity of Poland. Although predominantly a lowland country, Poland is remarkably diverse in terms of origin and characteristics of relief, which reflects different timescales of relief development in different parts of the country, various geological controls, climatic variability and variable intensity of human impact. At the large scale, its territory can be divided into six main geomorphological belts stretching from west to east. Going from the south, these are mountains, fore-mountain basins and hilly lands, uplands, subdued lowlands of 'old glacial relief', hilly lakelands of 'young glacial relief' and the coastal belt along the Baltic Sea. Mountains belong to the Variscan system in the west (the Sudetes) and the Alpine system in the east (the Carpathians), whereas uplands are mainly supported by sedimentary formations ranging in age from the Palaeozoic to the Miocene. Within the Carpathians the only high-mountain area in Poland, the Tatra Mountains, is located. In the Pleistocene nearly the entire area of Poland was covered by the Scandinavian ice sheet, but the extents of ice-sheet advances varied during different cold phases. Glacial morphology is most evident in the northern part of the country (lakelands), where glaciation was as recent as 23–15 ka. The Polish coast is cut in poorly consolidated deposits and typified by long sandy beaches and, locally, low cliffs. Human impact is ubiquitous and morphology of some areas, especially of mining districts, bears very strong anthropic signatures.

P. Migoń (✉) · K. Jancewicz
Institute of Geography and Regional Development, University of Wrocław, pl. Uniwersytecki 1, 50–137 Wrocław, Poland
e-mail: piotr.migon@uwr.edu.pl

K. Jancewicz
e-mail: kacper.jancewicz@uwr.edu.pl

Keywords

Long-term landform evolution · Geological history · Mountains · Uplands · Glacial landforms · Human impact

4.1 Introduction

Poland is predominantly a lowland country, with only 8.6% of the total area above 300 m asl and a mere 0.18% above 1000 m asl. Most of its territory is underlain by Quaternary deposits, which account for subdued relief, especially if analysed at large spatial scales. This distant view is, however, deceptive and while it is true that the ruggedness of the country is low in comparison with the southern neighbours of Czechia and Slovakia, the territory of Poland comprises an almost complete catalogue of different types of geomorphological landscapes, from the dynamic coastline to high mountains, formed over different timespans, depending on the location within the country. Having said that Quaternary sediments dominate at the surface, it is equally important to emphasize that the southern part of Poland is mountainous, with different types of mountain topography present, whereas an upland belt separates the mountains and the lowlands, except in the south-west. The lowland part, in turn, is not homogeneous either and it would be a considerable oversimplification if all this area is simply labelled as 'drift-covered plains'. First, different parts of Poland were glaciated for the last time in different intervals of the Pleistocene, during different glacial periods, which means that timescales of landscape evolution after the decay of a respective ice sheet were different too. This has clear implications for the preservation of glacially-produced topography, which remains fairly clear in northern Poland, but has been significantly transformed further to the south. Accordingly, the topography in the north is

P. Migoń and K. Jancewicz (eds.), *Landscapes and Landforms of Poland*,
World Geomorphological Landscapes, https://doi.org/10.1007/978-3-031-45762-3_4

much more diverse than in the central part of the country and the southern limit of lake basins roughly indicates the boundary between the two. To emphasize these differences and the main responsible factor, which is time, terms such as 'young glacial relief' and 'old glacial relief' have long been used in physical geography of Poland and are firmly embedded. Second, it would be incorrect to equate the Polish lowlands with glacial inheritance alone. They also host large rivers with their varied fluvial landscapes and considerable areas occupied by inland dunes, now largely stabilized, whereas the 'old glacial relief' has been significantly transformed by periglacial processes related to the presence of permafrost. Last but not least, relief of nearly all parts of Poland was modified by human activity and geomorphic consequences are profound, even if not necessarily realized at the first sight.

The geomorphological diversity of the country is thus reflected in the presence of several morphogenetic (or landscape) belts, stretching east–west and being roughly parallel to one another (Figs. 4.1 and 4.2). These are, from south to north:

- mountains, subdivided into three distinctive types of relief, each within a clearly defined geographical space: (a) high mountains, represented by the Tatra Mountains; (b) medium–high mountains of the Outer Carpathians and their foothills, which are part of the Alpine chain, dominated by flysch successions; (c) medium–high mountains of the Sudetes, which belong to the Variscan system of central Europe.
- fore-mountain terrains, which are represented by different types of relief in the west and east, respectively.

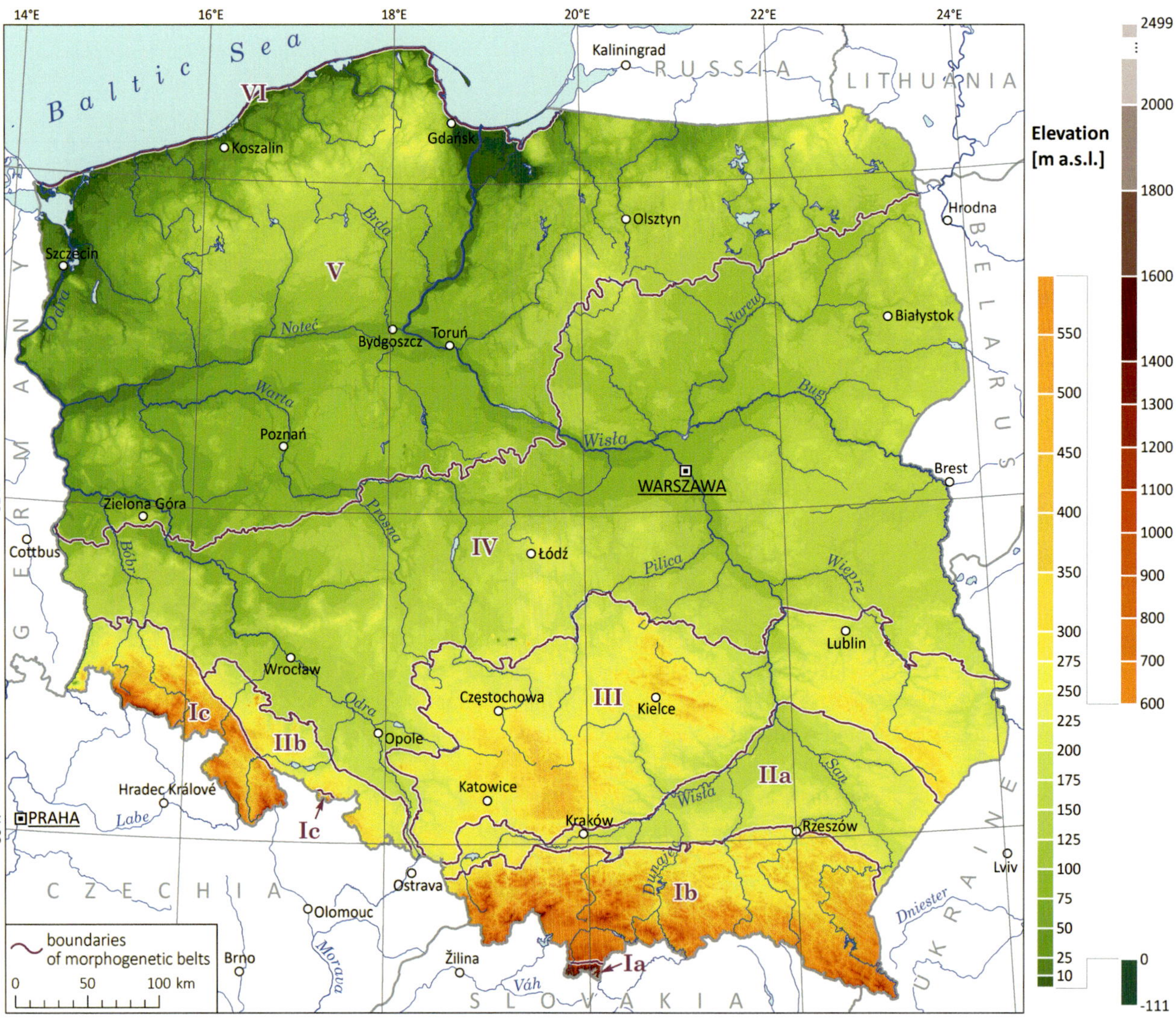

Fig. 4.1 Altitudinal diversity of Poland with superimposed boundaries of morphogenetic belts. Mountains: Ia—high-mountains (Tatra Mts.); Ib—medium-high mountains of the Alpine system (Beskidy Mountains); Ic—medium-high mountains of the Variscan system (Sudetes). Fore-mountain areas: IIa—Sub-Carpathian basins; IIb—Sudetic Foreland. III—uplands. IV—'old glacial relief' of central Poland. V—'young glacial relief' of northern Poland. VI—coastal zone

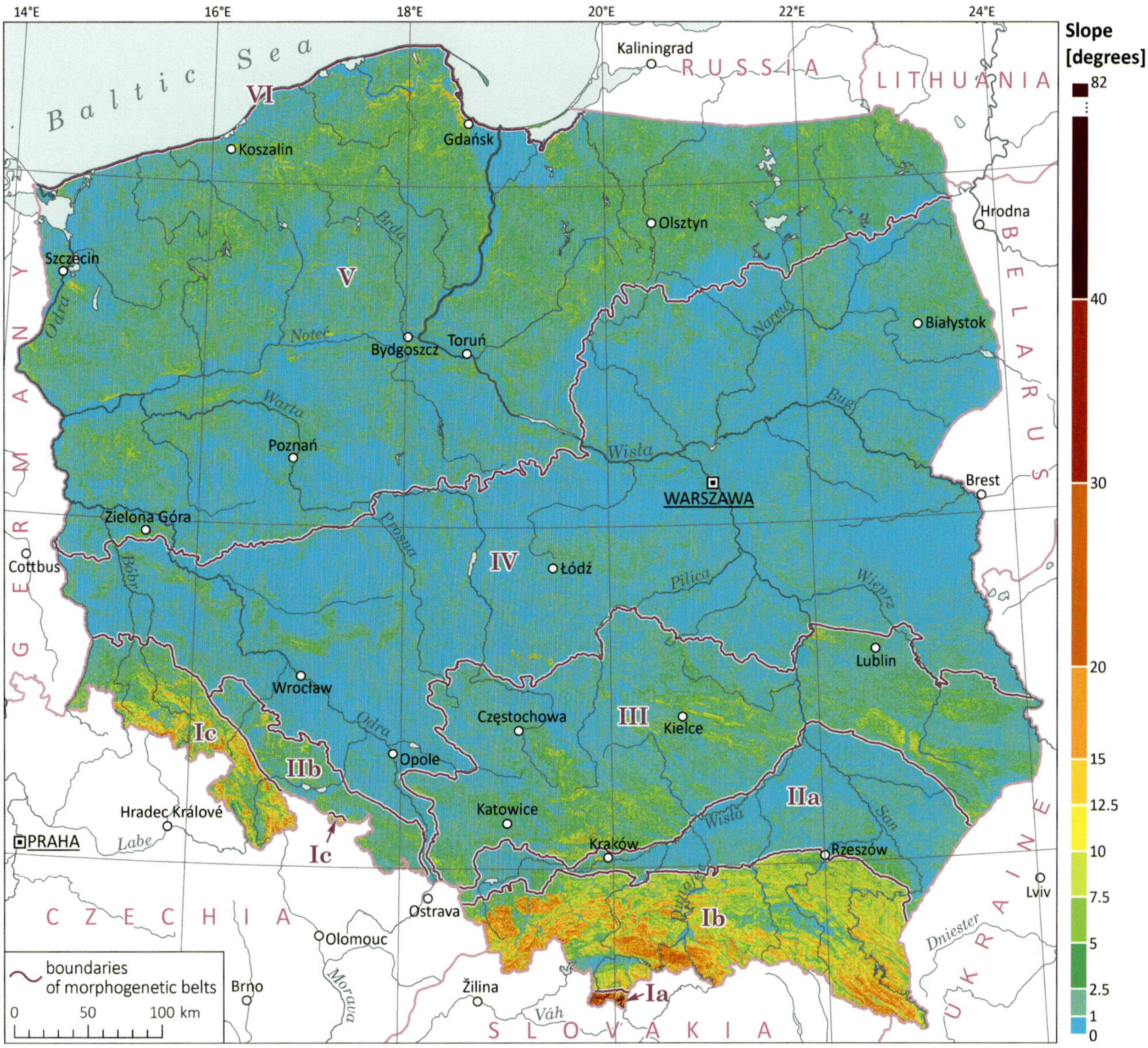

Fig. 4.2 Slope map of Poland. Explanation of morphogenetic belts (I–VI) as on Fig. 4.1

- uplands, extending from the Odra river in the west to the Ukrainian border in the east (and beyond), developed mostly upon sedimentary rocks ranging in age from the Carboniferous to the Neogene, but locally of older ages. Thick loess deposits are characteristic for the upland belt.
- subdued lowland glacial landscapes of central Poland ('old glacial relief'), with the dominance of rather monotonous plains, but crossed by river valleys of variable size and superimposed by dune fields in areas where sand supply was abundant.
- diverse glacial landscapes of northern Poland ('lakelands'), with moraine hills locally exceeding 300 m asl and numerous lake basins of different origin and size.
- the coast, with extensive sandy beaches backed by either dune belts or cliffs in poorly consolidated Pleistocene deposits.

These morphogenetic belts will be presented in more detail below, but prior to that, the main factors controlling the geomorphic evolution of the territory of Poland are briefly discussed. As the descriptions of individual regions are rather general and aim to highlight key geomorphic characteristics, extensive referencing is not considered feasible, unless specific views with clear attribution are quoted. On the other hand, cross-referencing to other chapters in this volume, which explore specific themes or areas in more detail, is implemented to facilitate further reading.

Fig. 4.3 Location of maps presenting representative examples of geomorphological diversity of Poland, in relation to boundaries of morphogenetic belts (explanation as on Fig. 4.1). Numbers refer to specific figures included in this chapter

Different types of relief are illustrated by pairs of maps (18 in total; Fig. 4.3): one showing shaded relief enhanced with hypsometric colours, another one presenting slopes—both based on the 20 m × 20 m DTM, which was derived from the high-resolution (1 m × 1 m) airborne LiDAR dataset. A few characteristic landscapes, where topography is highlighted by land use patterns, are also illustrated using Google Earth imagery.

4.2 Relief-Controlling Factors

4.2.1 Timescales

It is useful to begin the discussion with the brief consideration of timescales involved, as they are vastly different for different parts of Poland. The geomorphological division of the country into morphogenetic belts reflects this variability quite clearly. In the most generalized approach, the timescales relevant for the regions of Poland built of Quaternary deposits, thus for the entire central and northern part, as well as a substantial part of the south-west, are within the limits of the Quaternary. More specifically, the decay of the most recent ice sheet within a given territory is an onset of landform evolution that has continued until the present-day. However, since the oldest among these most recent glaciations is Elsterian (corresponding to the Marine Oxygene Isotope Stage 12), the actual timescales cover the Middle and Late Pleistocene. Consequently, various parts of the 'old glacial relief' have developed since the Elsterian (MOIS 12, 450,000–400,000 years ago) or the Saalian (MOIS 6, 200,000–150,000 years ago), whereas the relevant timescale for shaping the 'young glacial relief' is only 25,000–15,000 years. This does not exclude the possibility that older elements of relief may be locally incorporated into younger landscapes. For example, the long chain of moraine hills in south-western Poland known as the Trzebnica Rampart occurs within the area last glaciated in the Saalian and marks the southernmost extent of the younger Saalian ice sheet (Warthanian stage), but it is interpreted as a glacitectonic structure originated during the Elsterian, large enough not to be obliterated by the subsequent ice-sheet advances (Brodzikowski 1995; Ber and Krzyszkowski 2004). The Polish littoral belt is even younger and began to form after the final disappearance of the ice sheet in the latest Pleistocene. However, despite the short timescale the evolution of the Baltic Sea involved several stages and the present-day coastline, notwithstanding ongoing changes, was established only some 6000 years ago (Tomczak 1995).

The situation in the upland belt is more complicated, since elements of pre-glacial topography coexist with glacial landforms and sediments. The patterns of ice-sheet advances were adjusted to the preglacial rock-cut morphology rather than complete re-modelling of early Pleistocene landscape occurred. In specific areas, weathering residues and their re-deposited equivalents demonstrate that certain features of relief may be traced back to the Miocene, or even the Palaeogene (Liszkowski 1996). On the other hand, large areas have clearly very short histories, such as densely gullied loess plateaus, where the near-surface loess layers are of late Pleistocene age. This part of Poland is therefore best described as a mosaic of geomorphological landscapes, each having evolved within its own timescale, from tens of million years to several tens of thousands of years, but the regional canvas were set in the older Cenozoic. The same is true for the Sudetic Foreland—an area in front of the Sudetes in south-west Poland, last glaciated during the Saalian and re-moulded in subsequent periglacial periods, but containing many elements of bedrock morphology inherited from the Palaeogene (Migoń 1999).

The timescale for the Carpathians is constrained by the timing of tectonic events leading to the origin of this mountain chain. Since the final episodes of overthrusting of Carpathian nappes occurred in the Miocene, it was long believed that the relevant timescale is the late Neogene and Quaternary (e.g. Starkel 1987). However, recent reappraisal of long-term landscape evolution in the Outer Carpathians shows that many gross landforms have originated during the mountain building stage and record both an earlier compressional phase and the subsequent period of gravitational collapse of the orogen (Margielewski and Jankowski 2021). Thus, the entire Neogene should be considered as relevant. Subsequent geomorphic evolution occurred with different intensities in different parts of the Carpathians. The most significant re-shaping of pre-Quaternary topography took place in the most elevated Tatra Mountains, caused by extensive and repetitive mountain glaciation. Consequently, the contemporary relief is largely of Quaternary age (Kłapyta and Zasadni 2023).

The longest timescale applies to the Sudetes and the onset of geomorphic evolution is conventionally set at the Cretaceous/Palaeogene boundary, broadly coeval with the final withdrawal of marine conditions and widespread inversion tectonics. However, it does not mean that any landforms from that distant period survived, which is rather unlikely given the results of thermochronological studies carried out in various parts of the Sudetes (Danišík et al. 2010, 2012). Nevertheless, the area of the present-day Sudetes has remained terrestrial since that time, save some peripheral areas, and was subject to long-term morphogenesis in changing environmental conditions, with an increasing role of tectonics in the late Neogene (Migoń 2011).

4.2.2 Geology

Geology of Poland is very complex (Mazur 2023), even though largely concealed by Cenozoic, mostly Quaternary deposits. The location of Poland at the junction of three continental-scale geological domains, the East European platform, the Palaeozoic platform of western Europe that incorporates areas shaped by the Variscan orogeny, and the Alpine orogenic belt, explains some of the reasons behind the gross geomorphological division of the country. Thus, the Carpathians in the south-east are part of the Alpine belt and the style of tectonics coupled with the magnitude of uplift or, locally, subsidence, are reflected in second-order features of their topography (Margielewski and Jankowski 2021). Areas subject to strongest uplift remain topographically high, especially if built of resistant bedrock such as granites in the High Tatra Mts. (up to 2499 m) or thick packages of Magura Sandstone in the Babia Góra Massif (1725 m) in the Outer Carpathians. In a similar way, outcrops of resistant Jurassic limestones support the rugged terrain of the Pieniny Mts., at the border between the Outer and Central Carpathians, rising high above the adjacent areas. Many topographic escarpments coincide with the frontal parts of large thrusts, typical for the Outer Carpathians. Rock-relief relationships are evident at smaller scale too, reflected in the presence of sandstone crags, gorges, waterfalls (Alexandrowicz 1978, 1994), or the development of karst in the western part of the Tatra Mts. (Szczygieł 2023). Geological relationships are also a significant control of certain geomorphological processes, especially landsliding that is ubiquitous in the flysch Carpathians. Even though direct triggers are usually heavy rainfall events, the pattern of sliding and resultant landforms reflect the internal architecture of the rock massif.

The Sudetes show a different geological structure than the Carpathians, with the dominance of crystalline basement rocks, little deformation of post-Variscan sedimentary cover, and a significant role played by vertical displacements along normal faults, active mostly in the Neogene. Consequently, fault-generated mountain fronts are common and many individual mountain massifs are fault-bounded horsts, including the highest Karkonosze Mountains (1603 m) (Różycka and Migoń 2017). However, tectonic imprints are apparently superimposed on the rock-controlled relief that has developed over protracted intervals of the Cenozoic (Migoń 2011). Hence, there is generally a good correspondence between altitude, local relief and rock resistance, with rocks of ancient volcanic origin playing a particularly important role, ubiquitously supporting elevations. Rock control is also evident in the origin of sedimentary tablelands, granite tors, limestone karst, or the distribution of river gorges (Migoń and Placek 2014).

In the upland belt, tectonics and lithology are also important factors, although rock-controlled relief is less evident than in the mountains. This appears to be due to two interrelated reasons. First, the magnitude of Cenozoic surface uplift was limited, which did not create enough local relief to fully exploit bedrock differences. Second, low elevation allowed the Scandinavian ice sheets in the Middle Pleistocene to override the uplands, erode the higher ground, and leave sediment packages of variable thickness in less elevated areas, concealing the rock-cut topography partially or completely. Nevertheless, the upland belt hosts fine examples of limestone karst (Kraków-Częstochowa Upland), gypsum karst (Nida Basin), and chalk karst (Lublin Polesie region), all covered in specific chapters in this book (Dobrowolski 2023; Tyc 2023; Urban et al. 2023). The upland belt is also known for the cuesta landscape in the western part and rock-controlled ridges in the Świętokrzyskie (Holy Cross) Mountains, built of highly resistant quartzitic sandstones.

In the lowland part of Poland the influence of pre-Quaternary geology may seem negligible and indeed, outcrops of bedrock of that age are extremely rare. However, in the broader perspective, it is the long-term subsidence of the Mid-Polish basin that is responsible for creating an accommodation space for Oligocene and Neogene deposits and low elevation at the onset of the Quaternary. These sub-continental topographic conditions allowed the Scandinavian ice sheets to expand far southward and cover the contemporary territory of Poland, until they met the mountainous relief in the extreme south.

4.2.3 Climate and Climatic History

4.2.3.1 Contemporary Climate and the Role of Extreme Atmospheric Events

The generally subdued topography of Poland corresponds with rather modest contemporary geomorphological dynamics, even in the mountainous areas of the south-west (the Sudetes). However, low long-term rates of geomorphic processes are as much the outcome of low relative relief, as they are the derivatives of climatic conditions, which favour dense vegetation growth and sustain agricultural activity, both limiting the presence of bare ground susceptible to erosion, and do not abound in extreme atmospheric events that would trigger surface processes at a large scale. The caveat in this position is that it is highly generalized and does not take into account site-specific situations, in terms of relief, lithology, and weather-related events.

In reality, extreme rainfalls with daily amounts exceeding 100 mm and hourly intensities of more than 20 mm occur on a regular basis (Niedźwiedź and Starkel 2008; Błaś and Ojrzyńska 2023), especially in the mountains, and these are capable of triggering a range of hillslope processes bringing substantial, even though local change in relief. These include debris flows in the high mountains and those parts of the medium–high mountains, which are located above the timberline; landslides, particularly ubiquitous in the flysch-dominated Outer Carpathians; sheet wash on cultivated slopes in the foothills, uplands, and steeper parts of moraine hills; and gully erosion, especially in loess terrains (Kotarba et al. 1987, 2013; Kotarba 1992; Migoń et al. 2002; Rodzik et al. 2009; Rączowska et al. 2012; Rączkowska 2021). Each of these processes results in the origin of a suite of erosional and depositional landforms, whose lifetimes (visibility in the landscape) depend on the magnitude of the event, time elapsed since the event, and possible human interventions aimed at land rehabilitation. High rainfall, whether of very high short-term intensity or longer duration, contributes to the generation of floods, whose geomorphic impact varies depending on the setting. Floods on mountainous rivers are typically more devastating, causing considerable bed scour, bank erosion, channel change, and overbank accumulation (Czerwiński and Żurawek 1999; Starkel 2006; Rączowska et al. 2012). However, widespread river regulation and training limit the efficacy of floods as geomorphic agents these days and flood-related changes in floodplain morphology are quickly obliterated during remedial operations. Only a few rivers in Poland are in now in natural conditions (perhaps even these should be recognized as semi-natural, because of human impacts in the catchments and the effects of past river regulations) and their channels evolve by constant adjustment to discharge conditions. Depending on gradient, discharge and sediment load, meandering, anastomosing and locally braided patterns may be observed (Andrzejewski et al. 2023).

Extreme events are also important to understand the evolution of the Polish coastline (Zwoliński et al. 2023). In the short-term, summer periods of stability alternate with late autumn to spring periods of geomorphic change, caused by storm events related to low-pressure systems arriving from the north-west. Thus, beaches become depleted or locally completely eroded away, especially if there was an earlier human interference with the natural beach-dune/cliff system, and dune ridges and sea cliffs are undercut and retreat. Cliff retreat may be gradual, but occasionally larger cliff portions fail through landsliding, causing threat to cliff-top infrastructure.

The role of other climate- and weather-related events is rather minor, although should not be ignored. There is a growing appreciation of the role of wind, evident in sustaining dune formation and migration, deflation and dust production on silt-rich soils, typical for central Poland, and causing tree uprooting, which could affect large surfaces. The latter process is particularly important in the mountains, where pit-and-mound microrelief is created and water erosion processes significantly accelerate, taking advantage of high slope and temporarily bare ground (Pawlik et al. 2016; Strzyżowski et al. 2016). Snow avalanches occasionally have geomorphological consequences, if mineral ground is involved, but their activity is restricted to the highest parts of the mountains. The most elevated parts of the Sudetes and the Carpathians are at the lower thermal limit of periglacial processes, but the efficacy of frost sorting is low and most of periglacial landforms documented there are inherited from the Pleistocene.

4.2.3.2 Climatic Inheritance

The logical consequence of the above evaluation of the role of contemporary climatic conditions is an assumption that climates of the past are far more significant for understanding the contemporary relief of Poland. It is however beyond the scope of this chapter to include a comprehensive review of the climatic history of the Cenozoic, assembled from a multitude of direct and proxy sources and subject to various interpretations (and Quaternary is covered elsewhere in the book, see Jary and Marks 2023), so that only a generalized overview is provided. Relating to the timescales of relief evolution presented before (Sect. 4.2.1), climates of the Palaeogene, Neogene and Quaternary are relevant.

In the Palaeogene the territory of Poland was much closer to the Equator than it is now and enjoyed warm subtropical climate, alternating between more and less humid, with the thermal optimum in the Middle Miocene and modest cooling in the Oligocene (Mosbrugger et al. 2005). It used to be believed that a regionally extensive 'Palaeogene planation surface' formed at that time, whose remnants were sought in the high-elevation planar surfaces in the Sudetes, partly buried inselberg landscapes of the Sudetic Foreland, and karst topography of the Kraków-Częstochowa Upland (Klimaszewski 1958). Although it is hardly possible to convincingly demonstrate the existence of such a surface, and many of its alleged elements were reinterpreted to reflect rock control rather than tropical morphogenesis, thick weathering mantles derived from granites, gneisses, schists and serpentinites (locally > 50 m thick), recognized beneath Oligocene and Miocene sediments, show the efficacy of deep weathering at that time (Migoń and Lidmar-Bergström 2001). Weathering residues in karst pockets in the Kraków-Częstochowa Upland are another testament of morphogenesis under warm and humid conditions (Liszkowski 1996).

In the Neogene temperatures initially increased, resulting in the secondary climatic optimum of the Cenozoic in the Early Miocene (ca. 22 Ma), followed by another warmer interval in the Middle Miocene (ca. 15–14 Ma). However, even considering fluctuations, the climate remained warm, with the reconstructed mean annual temperature above 15 °C. The Late Miocene was already cooler and the cooling trend accelerated and intensified in the Pliocene (Mosbrugger et al. 2005). Thus, subtropical conditions may be inferred to last until 5 Ma, whereas the Pliocene was a period of temperate climate, gradually becoming similar to that experienced by the region today. It is apparently more difficult to reconstruct palaeo-precipitation, but according to Mosbrugger et al. (2005), annual rainfall largely remained above 1000 mm and no distinct periods of aridity can be inferred. This is an important finding given that in many previous attempts to recreate the Tertiary morphogenesis intervals of decreasing rainfall were inferred, even with suggestions of short-term spells of semi-desert conditions in the Late Miocene (Jahn 1980; Głazek and Szynkiewicz 1987), with corresponding search for inherited 'arid landforms' in the contemporary relief of Poland.

The Quaternary was typified by climatic fluctuations (cycles) of different durations, with the cooling trend amplified in the Middle and Late Pleistocene (Jary and Marks 2023). Environmental conditions alternated between temperate climate during interglacials, with some being warmer than the Holocene, cool temperate climate during waxing phases of glacial stages and interstadials, to glacial/periglacial conditions during peaks of cold stages. These alternations have had obvious implications for morphogenesis, particularly for the development of permafrost and related periglacial processes during cold stages. Cold-climate processes were not only re-modelling landforms inherited from the Neogene, giving them subtle, but nevertheless distinctive 'make-up' (Jahn 1980; also Starkel 1987), but were very efficient agents to modify glacial landforms produced by repetitive inland glaciations (Kozarski 1995; Goździk and Forysiak 2023). The brief interval at the Late Pleistocene/Holocene transition, with associated ultimate permafrost decay, is considered very important for shaping both lowland Poland, where major changes in fluvial patterns occurred, and the mountains, where acceleration of hillslope and fluvial processes is inferred (Starkel 1991; Starkel and Gębica 1995). Further, although comparatively minor climate alterations occurred in the Holocene, reflected in changes in temperature and humidity. Again, these are reflected in the geomorphic record of fluvial and hillslope processes, especially in phases of enhanced landslide activity in the Carpathians associated with more humid intervals of the Holocene (Margielewski 1998, 2018).

4.2.4 Human Impact

The role of humans as geomorphic agents, in both direct and indirect way, receives an increasing recognition worldwide and Poland is not different. However, in contrast to southern Europe this impact for long had remained rather modest, whereas considerable, large-scale alterations of geomorphological scenery have only occurred in the past several centuries. Before that, even though land-use changes may have been substantial and deforestation was ongoing since the Neolithic, low relief precluded massive geomorphic change due to accelerated, human-induced erosion.

Types of human activity that imparted the most evident impact on landforms are mining and agriculture. Nonetheless, these impacts are difficult to compare for two main reasons. First, mining has been carried out in specific parts of Poland only, mainly in the Sudetes, their foreland, and in the upland belt, whereas agriculture was practised throughout the country, even in the mountains up to 900 m asl or so. Second, geomorphic consequences of mining are primarily manifested in intentionally-made landforms such as quarries, open-cast pits, waste heaps, altered river channels, but non-intentional subsidence hollows are also very widespread in some regions (Wójcik 2013; Jancewicz et al. 2021; Dulias 2023). By contrast, agriculture-related change is more subtle, reflected mainly in soil erosion over time, but slope terracing implemented in mountainous and upland terrains makes it more visible (Latocha-Wites 2023). Landform change due to mining activity can be traced back to the late medieval times, but it has become really overwhelming along with industrial revolution and the advent of new means to dewater the mines, thus in the nineteenth and twentieth centuries. Significant agriculture-related relief modifications occurred over a similar timescale and are more cumulative, in contrast to mining which locally was very short-lived.

Other important human activities were modifications of river channels to suit various purposes, mainly transportation and energy production, land reclamation along some rivers and at their mouths to the Baltic Sea, building construction activities involving land levelling and filling, construction of embankments and excavation of trenches to build roads and railway lines, and introduction of coastal protection measures to prevent erosion. All these undertakings are of relatively recent date, a few centuries old at most, contributing to an increasing anthropic signature in the contemporary morphology of Poland, but clearly not dominating over geomorphic features produced by the long-term natural morphogenesis.

4.3 Main Types of Relief

4.3.1 Mountains

4.3.1.1 High Mountains

The Tatra Mountains in southern Poland are the only high-mountain terrain in the country (Kłapyta and Zasadni 2023), but even this is extremely limited in extent. Less than 200 km^2 of the Tatra Mountains lie with the territory of Poland (Fig. 4.4) and not all of this area shows morphology usually associated with high mountains. Considering geological structure, altitudes and relief, the mountains are further subdivided into the High Tatra and the Western Tatra, both with the crystalline core, with the marginal belt of forested ridges and valleys in sedimentary rocks in the north (Tatry Reglowe in Polish). In the High Tatra altitudes along the ridges exceed 2000 m asl and elevation differences between crests and cirque/valley floors exceed 1000 m. The highest spot in Poland, Mt. Rysy (2499 m), is located in the High Tatra. Bedrock is granite of Carboniferous age. The Western Tatra are lower (2176 m at maximum in Poland) and less rugged in the most elevated parts built of various

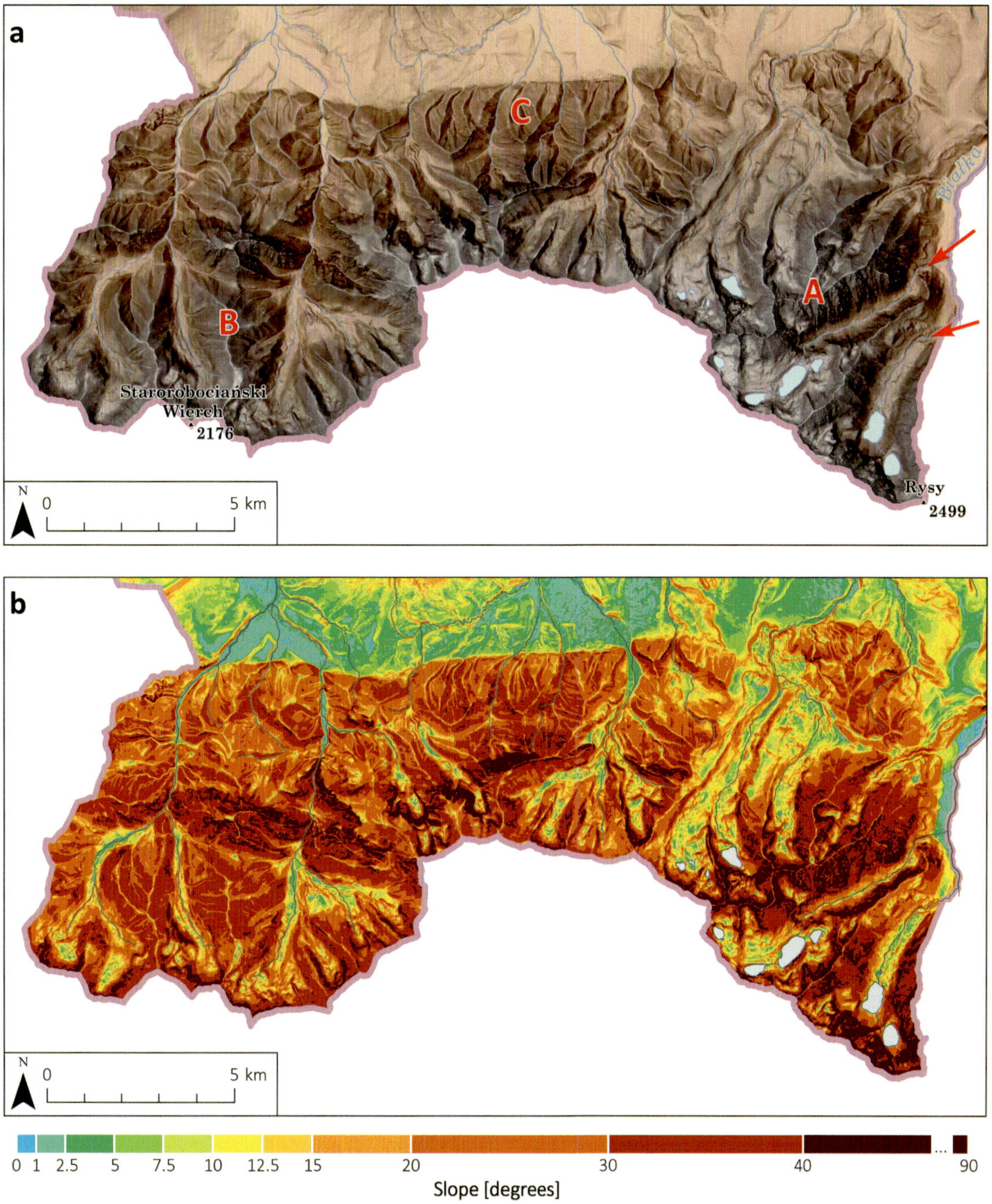

Fig. 4.4 The Tatra Mountains. **a** shaded relief, **b** slope map. Three distinct parts can be distinguished considering relief and geological structure: High Tatra (A), Western Tatra (B), Tatry Reglowe (C). Note the straight fault-related northern margin of the Tatra and more evident glacial imprint in the High Tatra, evidenced by numerous lakes in glacial cirques and troughs and the presence of wide U-shaped valleys open towards the foreland. By contrast, lower valley reaches in the Western Tatra were never glaciated and show typical erosional morphology, with V-shaped cross-sections and very narrow floors. Arrows show exits of tributary valleys hanging above the trunk Białka Valley—another testimony of intensive glacial re-modelling

pre-Variscan granites, gneisses and schists, but locally extremely steep in the northern section, where Mesozoic sedimentary rocks occur. The contemporary relief is mainly the outcome of glacial erosional processes and should be considered as the cumulative effect of several successive glaciations, although most glacial landforms, especially moraines, are of Late Pleistocene age. The longest glaciers within the north-facing slopes of the High Tatra were more than 10 km long and had their distal parts in the foreland, whereas those in the Western Tatra were confined to the headwater reaches of the valleys. Steep relief provides necessary conditions for high-energy mass movements of short duration, not occurring anywhere else in Poland with such a high frequency and magnitude of individual events.

At the scale of individual landforms, three genetic assemblages make the Tatra Mountains so distinctive in Poland. First, these are glacial landforms and their rich inventory includes cirques and cirque staircases, rock bars closing cirques on the downstream side and allowing for lakes to form, roche mountonnées of variable dimensions, U-shaped trough valleys, hanging valleys, lateral and frontal moraines, boulder-rich lag deposits within the valley floors, and rock glaciers, derived mainly from decaying 'proper' glaciers (Kłapyta and Zasadni 2023) (Fig. 4.4). The second group consists of landforms produced by mass movements, both contemporary and dated to earlier intervals of the Holocene and the Late Glacial. These include extensive taluses from major prehistoric rockfalls, scree cones, typically coalescing along the valley shoulders, debris flow tracks and torrential fans (Kotarba et al. 1987; Rączkowska 2021). Rainfall intensities around 40 mm per hour appear sufficient to trigger debris flows, which mainly affect loose regolith accumulated in bedrock ravines and material of scree cones (Kotarba 1992). Finally, the Tatra Mountains are distinctive for the development of karst in the limestone part of the massif (Szczygieł 2023). Surface karst landforms are not as impressive as in the Alps, although dolines, karren fields and large springs do occur, whereas the subterranean karst is very widespread and includes the deepest and longest cave in Poland—Wielka Śnieżna Cave, 824 m deep and more than 23.5 km long.

Besides the Tatra Mountains, elements of high-mountain relief occur in two highest massifs within the medium–high mountains, in the Babia Góra Massif in the Outer Carpathians (Łajczak 2023) and in the Karkonosze Mountains in the Sudetes (Kasprzak 2023). In both cases, they are products of local glaciation and show as rock-walled cirques. Steep relief allows also for the occurrence of debris flows. However, the spatial extent of these 'high-mountain enclaves' is very small.

4.3.1.2 Medium–High Mountains, Intramontane Basins and Foothills of the Alpine System

The Outer Carpathians in Poland represent medium–high mountains and despite considerable length (more than 300 km) and width (up to 80 km) show remarkable similarity in gross relief, arising from common geology. They are collectively called the Beskidy Mountains and are typified by long forested ranges, densely dissected by fluvial valleys and significantly affected by landslides (Fig. 4.5). They are built of several flysch series of Late Jurassic to Miocene age, arranged in west–east striking nappes, thrust upon one another in northerly direction. These flysch series comprise mainly sandstones and shales, with the former being much more resistant than the latter. Consequently, thick sandstone packages support most distinctive relief, with isolated mountain massifs and homoclinal ridges (e.g. Bernatek-Jakiel 2023; Łajczak 2023), whereas at the local scale they crop out in the form of picturesque crest tors and mid-slope crags, as well as support waterfalls and cascades in stream channels. Shales, in turn, preferentially occur in the lower slopes and underlie the valleys. In addition to rather orderly flysch series, chaotic complexes (melanges) produced at the mountain building stage are scattered throughout the Beskidy, and these were preferentially eroded to form topographic basins of different dimensions and ground plans (Jankowski and Margielewski 2021).

The Beskidy Mountains are subdivided into more than a dozen of geomorphic units of lower rank, based on altitude and the presence of major south-to-north flowing rivers separating individual ranges. The southern ranges are higher, reaching altitudes of 1257 m (Beskid Śląski), 1557 m (Pilsko Massif), 1725 m (Mt. Babia Góra), 1311 m (Gorce range), 1266 m (Sądecki Beskid), and 1343 m (Bieszczady), moving from the west to the east. Among them, the part known as the Beskid Wyspowy ("Island Beskid") stands out through its unusual relief consisting of separate massifs surrounded by wide depressions (Fig. 4.6). The northern part of the Beskidy Mountains is at lower elevation, up to 600 m asl, and forms a belt of foothills of variable width, reduced to a few kilometres in the west, where the margin of the Carpathians is very pronounced, to several tens of kilometres in the east. The foothills too are internally arranged into the series of parallel ridges and valleys, perpendicularly crossed by major rivers. They gradually lose height in the northerly direction, so that the northern front of the Carpathians is generally rather subdued. Altitude differences between the southern and northern part of the Outer Carpathians correspond to the tectonic history of the range—the southern part is built of several nappes stacked in an earlier, compressional phase of mountain building (Jankowski and Margielewski 2021).

A distinctive feature of geomorphology of the Outer Carpathians is the sheer scale of slope re-modelling by mass movements, favoured by geological structure and available relief. Large-scale displacements occurred in the post-compressional stage of evolution of the mountain range in the Miocene, accounting for the separation of adjacent ridges and isolation of large massifs (e.g., Beskid Wyspowy unit—Fig. 4.6), whereas the Pleistocene–Holocene history involved the origin and re-activation of numerous landslides, scalloping the slopes and producing thick colluvial packages in the middle and lower slopes (Margielewski and Jankowski 2023b). A large number of non-karstic caves occurs in sandstones of the Outer Carpathians and most of these are genetically related to

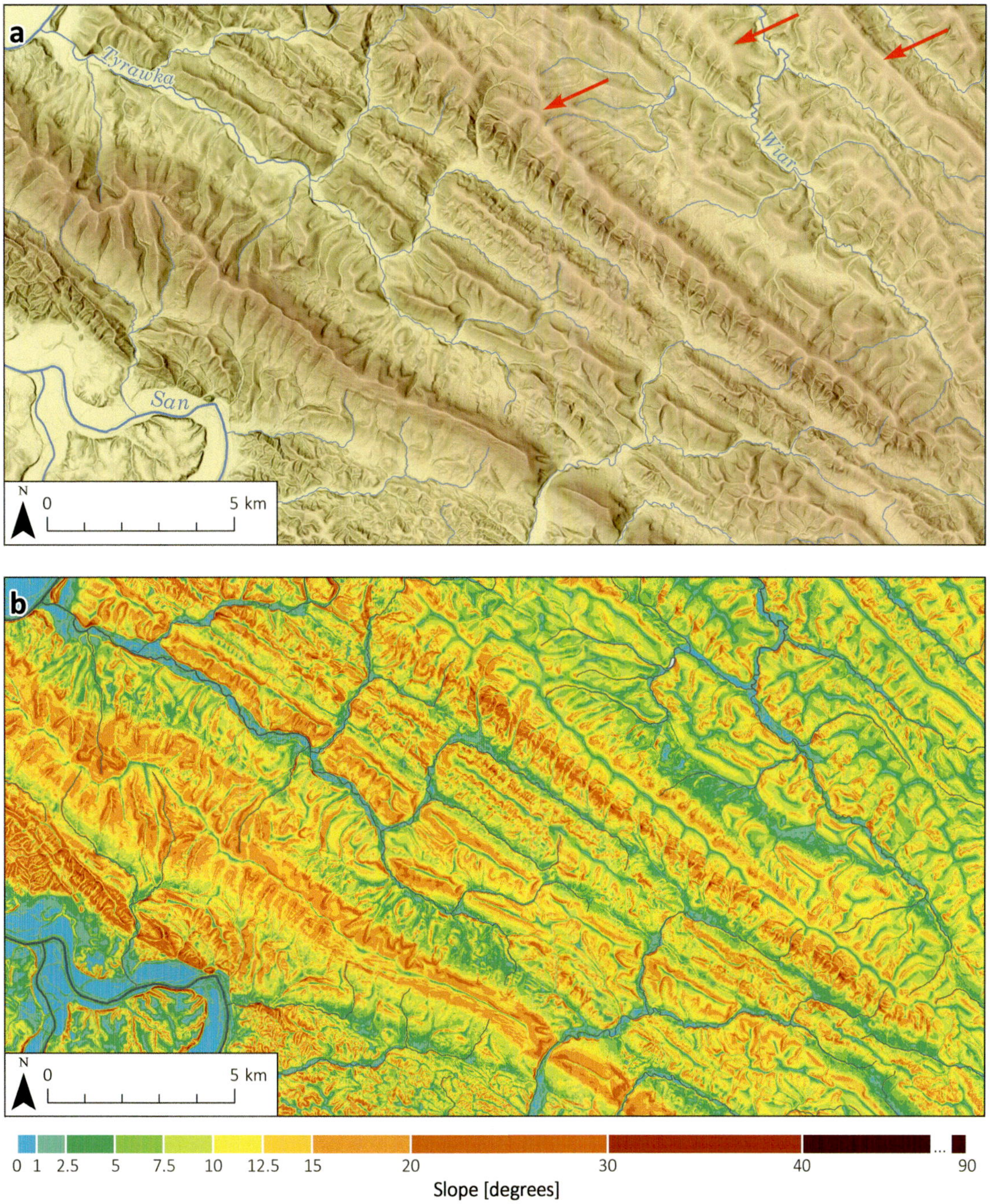

Fig. 4.5 Typical morphology of the Flysch Carpathians on the example of Low Bieszczady Mountains, with ridge-and-valley topography conditioned by alternating occurrence of rock series of variable resistance. **a** shaded relief, **b** slope map. Relief differences are up to 300 m. Note numerous structural water gaps in the drainage basin of the Tyrawka River and flattened water divides in the Wiar drainage basins (arrows)

rock-mass disintegration (Margielewski and Urban 2023). Landslides are being produced every year, usually following heavy or persistent rainfall, being the most hazardous geomorphic process in Poland, repeatedly affecting houses and infrastructure.

In between the flysch-built Outer Carpathians and the high-mountain relief of the Tatra Mountains two geomorphic units are sandwiched, small in terms of area but distinctive in terms of geomorphology and unlike any of the adjacent larger regions. One is the Pieniny Mountains (1050 m), built of impressive limestone crags (Klippen) of Jurassic age and crossed by the picturesque gorge of the Dunajec River that follows a highly sinuous course in a series of entrenched meanders. The origin of the gorge and, more general, of the entire Pieniny Klippen Belt as a geological structure, have remained a matter of considerable

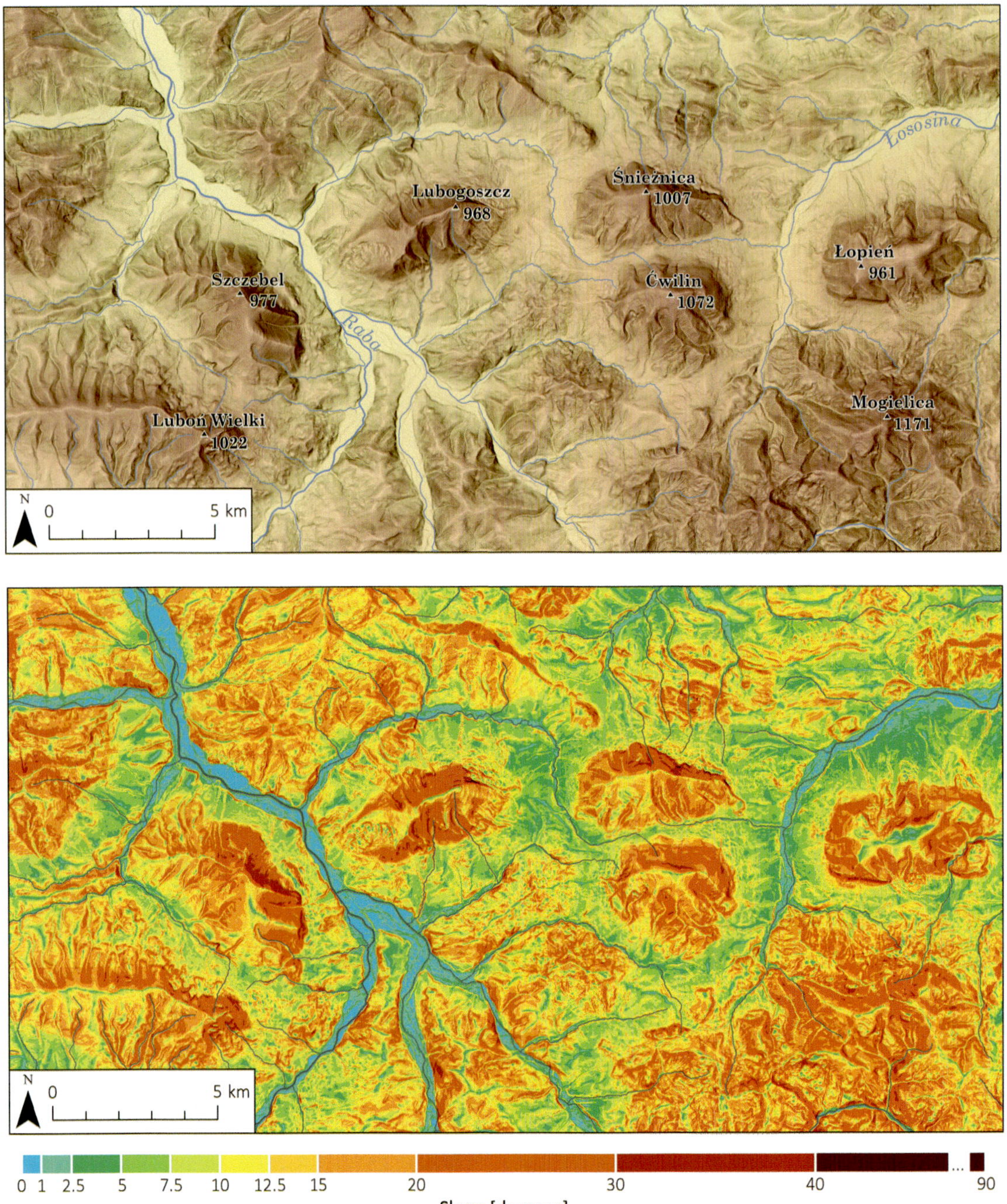

Fig. 4.6 Peculiar topography of the Beskid Wyspowy, with steep-sided isolated elevations of relative height up to 500 m, is the combined product of large-scale gravitational displacements during mountain building stage and rock resistance. Scalloped hillslopes provide evidence of considerable landslide re-modelling. **a** shaded relief, **b** slope map

debate (Margielewski and Jankowski 2023a). The second, much larger unit is the Nowy Targ—Orawa Basin, with flysch-built ridges in the south and flat depositional surfaces of Quaternary age in the north, also known for its widespread peat bogs (Zawiejska et al. 2023).

4.3.1.3 Medium-High Mountains of the Variscan System

The Variscan system, very widespread in the adjacent Czechia and Germany, is represented in Poland only by the Sudetes Mountains in the south-west. This mountain system straddles the Polish/Czech boundary and only a part is located in Poland. The Polish part is nearly 200 km long and up to 50 km wide, with the highest elevations in the Karkonosze Mountains in the west (Mt. Śnieżka 1603 m), followed by the Śnieżnik Massif (1424 m) in the east, both at the state boundary. Otherwise, elevations only locally exceed 1000 m asl and altitudes between 700 and 900 m are most common, with relative relief typically in the order of 300–500 m. The timberline runs at ca. 1200 m asl, so that the majority of the Sudetes is naturally under forest, although large parts were deforested as a result of human impact (Latocha-Wites 2023). Geologically, the Sudetes are extremely diverse and include pre-Variscan crystalline basement (mainly gneiss and schist complexes, but also greenschists, amphibolites, phyliites, and quartzites), Variscan granite intrusions, volcanic rocks from different geological periods, from the Cambrian to the Neogene, and little deformed sedimentary rocks spanning the wide interval from the Carboniferous to the Cretaceous. This lithological variability and complexity account for a multitude of rock- and structure-controlled landforms, such as tablelands, cuesta ridges, domes and ridges built of resistant volcanic rocks, granite inselbergs and tors, or rock-controlled water gaps (Migoń 2023a, b) (Fig. 4.7). However, the regional pattern of highlands, uplands and basins seems primarily controlled by differential uplift and relative subsidence in the Neogene, continuing into the Quaternary. Thus, many second-order mountain groups and ranges in the Sudetes are horsts, not uncommonly asymmetric in cross-section, bound by distinct fault-generated escarpments. The height of individual mountain fronts reaches 500–600 m, whereas the total displacement could have been in excess of 1 km, for instance along the junction of the Karkonosze Mountains and the adjacent Jelenia Góra Basin (Kasprzak 2023). The most distinctive mountain front delimits the Sudetes from the north-east, running straight for more than 150 km in NW–SE direction, although changing the height along the strike.

Two areas within the Sudetes stand out, representing unique types of relief in Poland. One is the Karkonosze Mountains in the West Sudetes (Kasprzak 2023), with high-altitude planar surfaces above the timberline, undercut by several glacial cirques, including two that host lakes. Extensive moraine terrains occur in front of the cirques. The mountains are also famous for various geomorphic features produced under periglacial conditions, including widespread blockfields and various types of patterned ground. It is worth emphasizing that the Karkonosze is the highest mountain group within the entire Central European belt of Variscan ranges, of highest geodiversity. The second area is the sedimentary tableland of the Stołowe Mountains, whose multi-storey topography reflects alternating occurrence of resistant sandstones and weak mudstones and marls in the vertical sequence (Duszyński 2023) (Fig. 4.8). Extensive ruiniform relief, rock labyrinths and hoodoo rocks have made the region famous since the end of the eighteenth century and heavily visited.

4.3.2 Fore-Mountain Terrains

4.3.2.1 Fore-Carpathian Basins

The Outer Carpathians give way to low-lying, but nevertheless geomorphologically complex forelands developed upon sediments of Miocene age, deposited within the Carpathian Foredeep during the closing stages of mountain building. As the forelands are limited from the north by the upland belt, they are termed basins in Polish geomorphological and physico-geographical regionalization. In the west, the Racibórz-Oświęcim Basin separates the Carpathians and the Silesian Upland, but includes a water divide between the drainage basins of Wisła River in the east and the Odra River in the west. Altitudes reach 300 m asl and the generally monotonous relief consists of low plateaus with Quaternary cover overlying Miocene substratum, denudational plains on glacigenic sediments, and fluvial terraces, locally with inactive dune fields. Natural landscapes in some parts of the region have been considerably altered by mining and industrial activities, both directly and indirectly (Dulias 2023).

The much larger Sandomierz Basin is a triangular topographic depression between the subdued front of the Carpathians and the upland belt, reaching as far west as the city of Kraków. Its north-western boundary is drawn along the Wisła river valley, beyond which is the Nida Basin, whereas the San River flows from south-east to north-west, to join the Wisła near the apex of triangle. Altitudes gradually decrease northward, from more than 250 m asl in the south to 140 m asl at the confluence of the major rivers. Similarly to the western part, the topographic surface is cut across Quaternary deposits, mainly glacial, fluvioglacial and fluvial, with the Miocene bedrock at greater depth. Aeolian sediments are widespread, both loess—common along the front of the Carpathians, where thicknesses up to 10 m are recorded, and sands, which give rise to extensive dune fields, with the height of individual linear and transverse dunes up to 25 m. Valley floor terraces are extremely wide, in places more than 10 km, with several generations of palaeomeanders (Fig. 4.9).

4.3.2.2 Sudetic Foreland

The topography in front of the Sudetes is very different from the one in front of the Carpathians. It is dominated by residual bedrock elevations built of crystalline basement rocks, which take the form of low ridges, larger uplands or isolated hills (inselbergs) (Fig. 4.10). Their height is typically 50–100

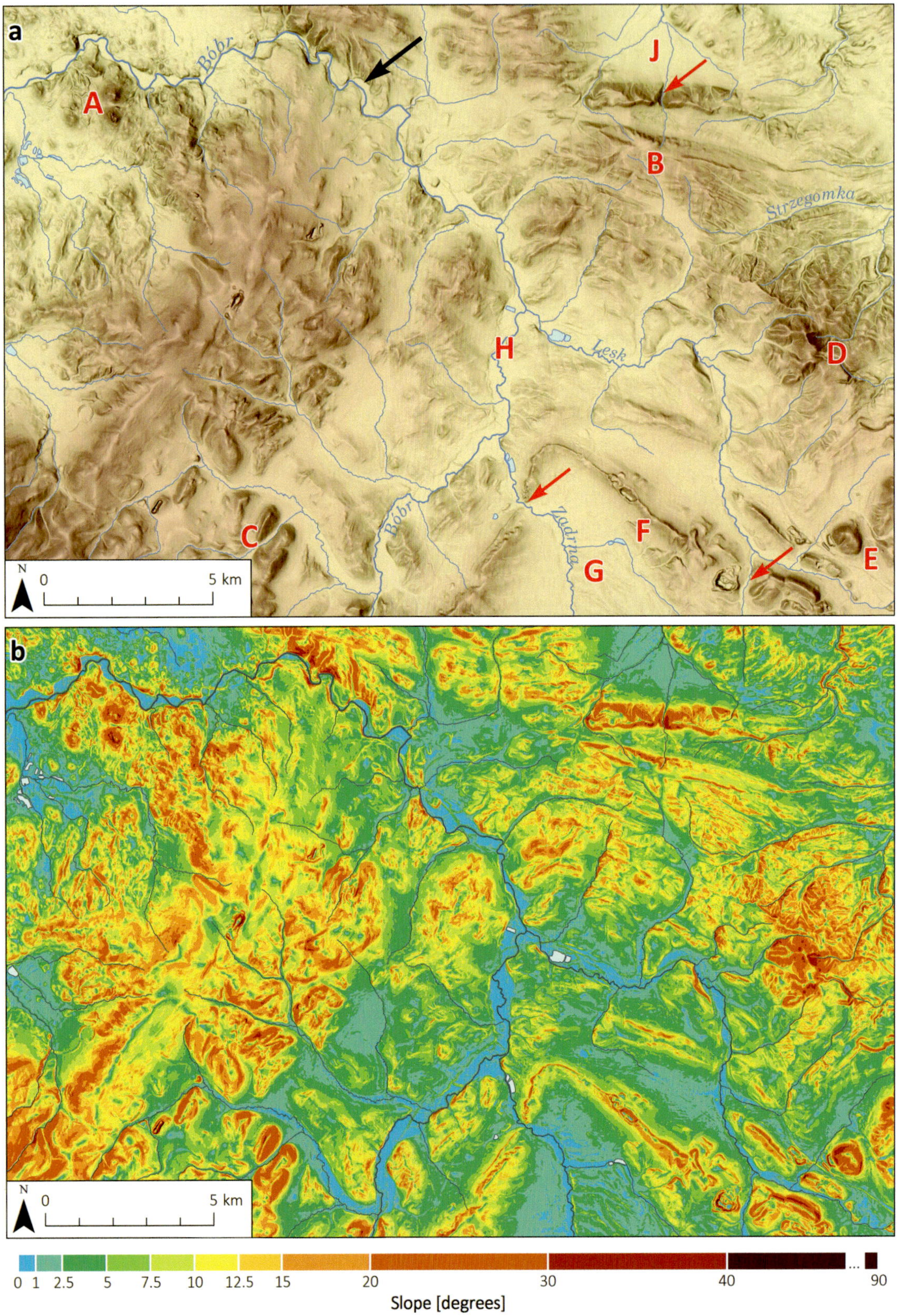

Fig. 4.7 Complex topography of the border area between the Middle and West Sudetes shows primarily the effects of variable rock resistance. **a** shaded relief, **b** slope map. Characteristic landscapes include: isolated granite hills (inselbergs) in the Jelenia Góra Basin (A), parallel ridges and troughs in the thick succession of Lower Carboniferous clastic rocks, alternating between conglomerates, sandstones and mudstones (B), elevations built of Carboniferous (C–E) and Permian volcanic rocks (F), and basins excavated in weak sedimentary rocks (G, H, J). Note also water gaps through outcrops of resistant rocks (red arrows). The long gorge of the Bóbr River with entrenched meanders (black arrow) is probably antecedent

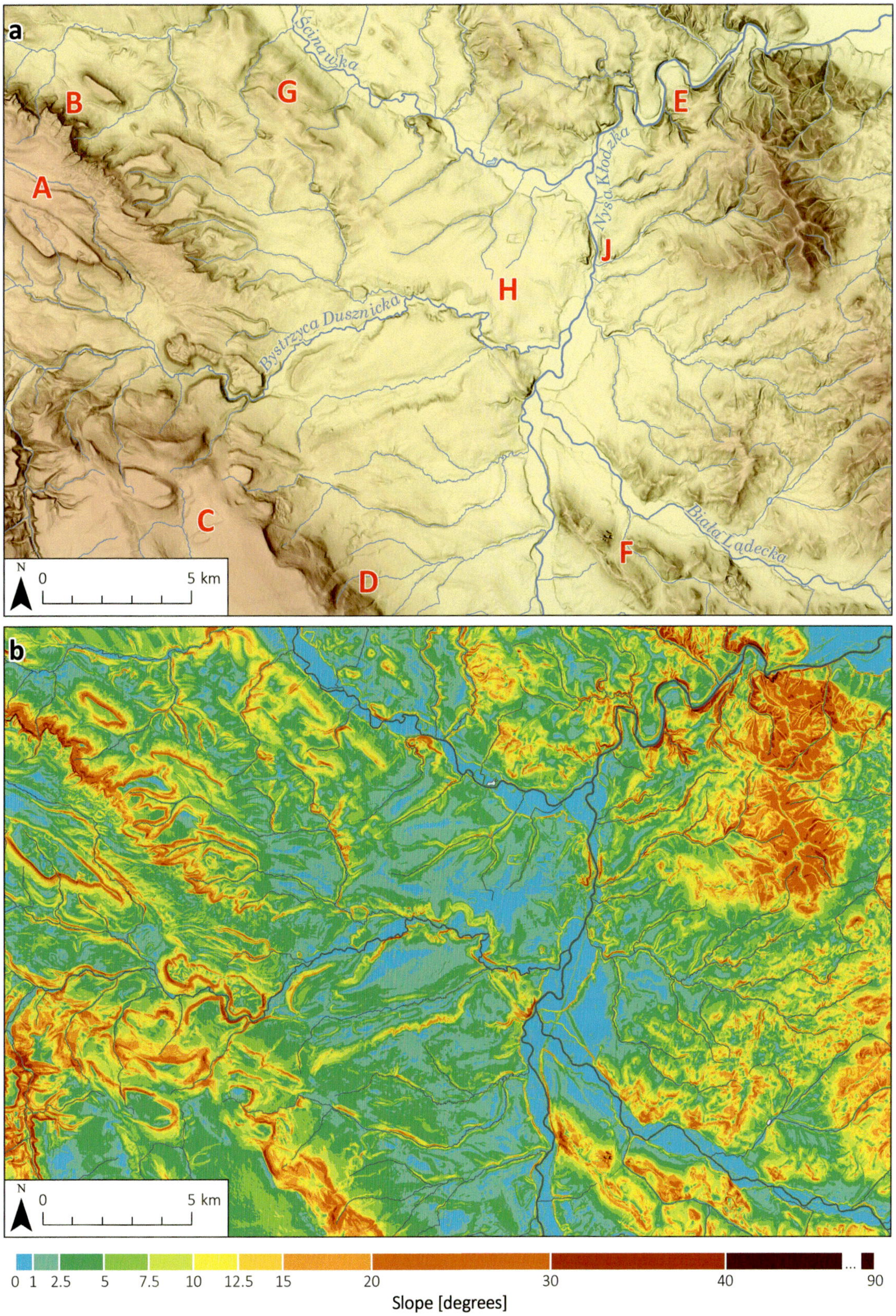

Fig. 4.8 Complex topography of eastern part of the Middle Sudetes, with the intramontane Kłodzko Basin in the centre. **a** shaded relief, **b** slope map. The most distinctive landscape is the stepped tableland of the Stołowe Mountains (A), with its prominent north-eastern escarpment undercut by large amphitheatres (B). The tableland of the Bystrzyckie Mountains (C) terminates against a fault-generated escarpment in the east (D). Entrenched meanders of the Nysa Kłodzka River form the most famous antecedent water gap in the Sudetes (E). Lithological control is seen in elongated ridges built of Proterozoic dolomitic limestones (F) and Permian volcanic rocks (G). The floor of the Kłodzko Basin (H) is underlain by Quaternary glacigenic sediments left by the Scandinavian ice sheet, mantled by younger loess deposits. The narrow section of the Nysa Kłodzka River in the town of Kłodzko (J) results from drainage superimposition during deglaciation

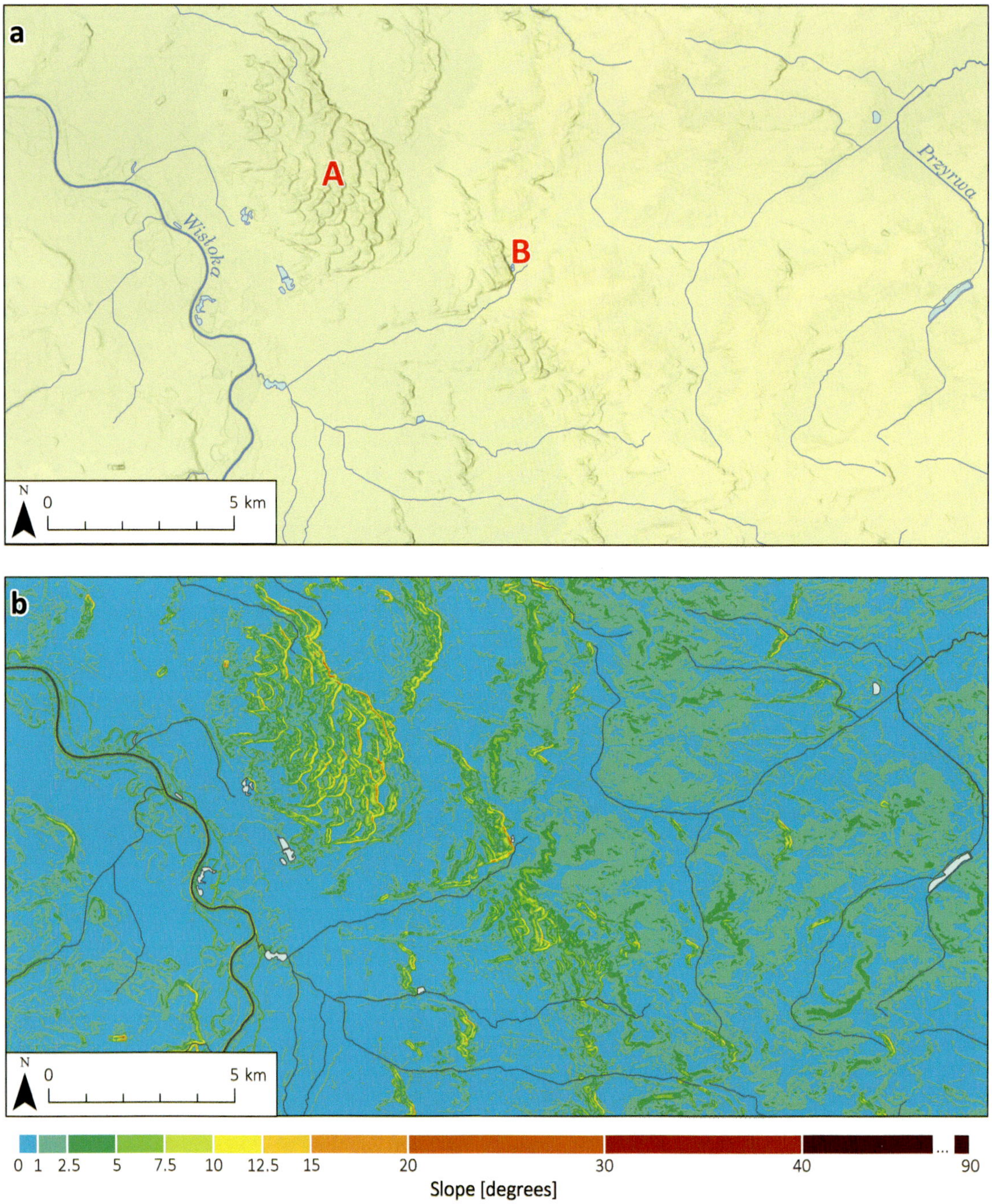

Fig. 4.9 Topography of the western part of the Sandomierz Basin. **a** shaded relief, **b** slope map. Characteristic elements are wide valley floors with generations of paleomeanders, as along the Wisłoka River, and dune fields, now under forest and inactive (A, B)

m, but a few are higher than this, including the Ślęża Massif (718 m), which rises above the adjacent plains by as much as 500 m (Traczyk 2023). The reasons for this anomalously high elevation are yet to be deciphered. Bedrock topography of the Sudetic Foreland is among the oldest landscapes in Poland and acquired its main traits in the Palaeogene, as testified by its incomplete burial by Neogene and Quaternary deposits and the widespread presence of thick weathering mantles pointing to advanced rock decay, apparently connected with warm and humid environmental conditions. The Quaternary record includes glacigenic deposits associated with several advances of Scandinavian ice sheets and the mantle of loess, which locally reaches 10 m thick and has allowed for the development of gully networks. The region also bears a heavy anthropic imprint, mainly in the form of numerous quarries, both disused and still in operation (Fig. 4.10).

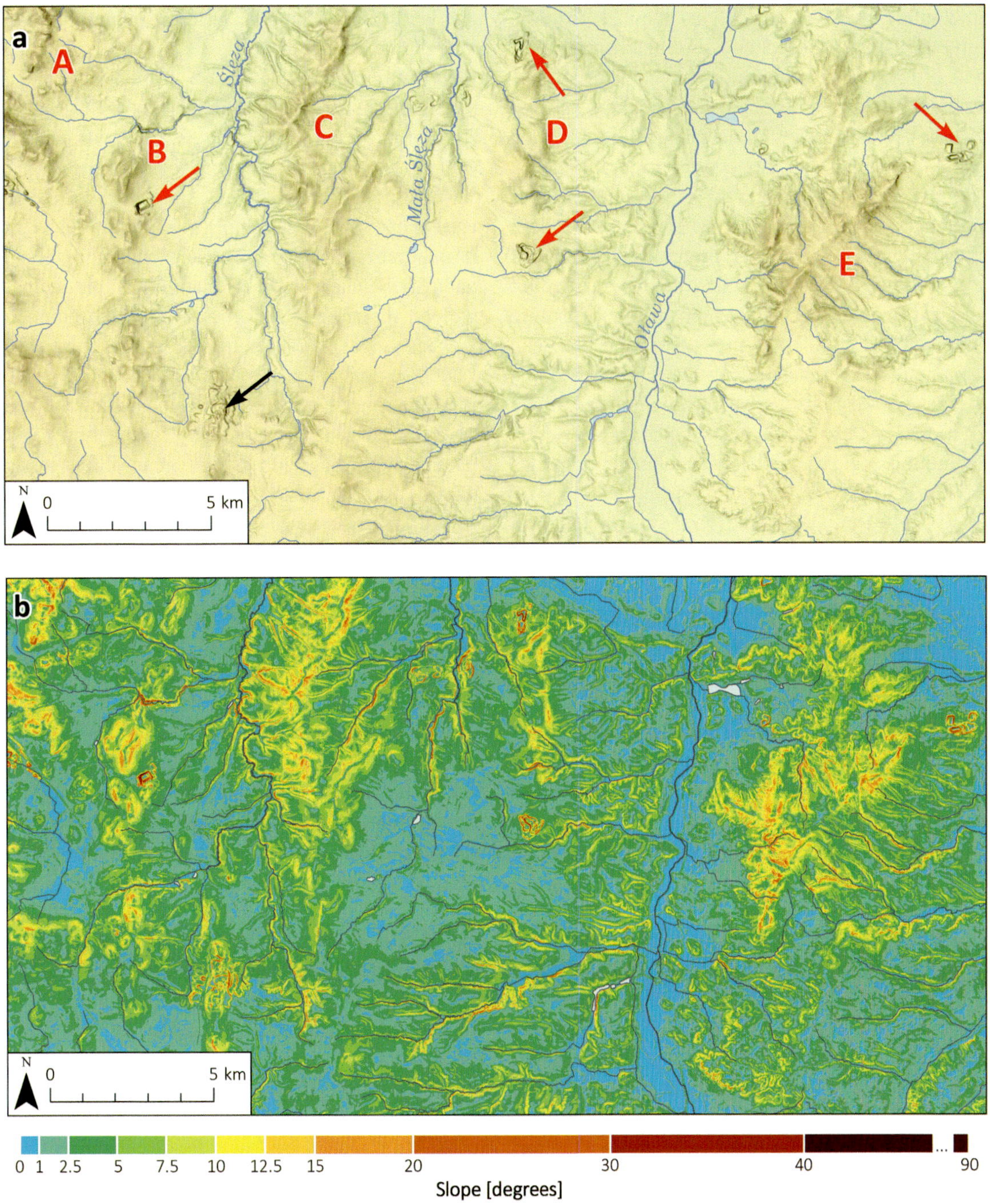

Fig. 4.10 Complex topography of the Sudetic Foreland, with elevations supported by outcrops of various Palaeozoic basement rocks (A–E). **a** shaded relief, **b** slope map. The intervening depressions are filled by Neogene and Quaternary deposits, the latter mainly of glacial and aeolian origin. Geological control at a large scale is evident in the parallel arrangement of elevations and depressions, running north-to-south. Note numerous quarries (red arrows). In the village of Szklary thick nickel-bearing weathering mantle on serpentinite was mined—an important witness of warm and humid Palaeogene environment and evidence of landscape inheritance (black arrow)

4.3.3 Uplands

The upland belt extends from the Odra River valley eastwards and is approximately 400 km long and over 100 km across in the widest sections. It is distinguished within the country mainly on the basis of altitude, ranging from 300 to 500 m asl (612 m in the Świętokrzyskie Mountains), whereas in fact it integrates various types of rock-controlled relief, developed upon various types of bedrock, overprinted in the Pleistocene by the effects of glacial erosion and accumulation processes, as well as loess deposition. Accordingly, the belt is divided into several main regions, each with its distinctive geomorphological characteristics.

The westernmost part is the Silesian Upland, built of Late Palaeozoic up to Triassic sedimentary rocks, concealed by Miocene and younger deposits over considerable areas, especially in the west. The topography of the southern part is mainly of horst-and-graben type, whereas the northern part has cuesta-type relief, with the prominent role played by Triassic dolomites. However, the contemporary relief bears significant, locally spectacular human imprint arising from centuries-long traditions of mining of zinc and lead ores, coal, stone resources, and aggregates. These anthropic signatures can be very distinctive (large quarries, huge waste rock heaps) or more subtle (subsidence areas), nevertheless making this region most altered by human activities in Poland (Dulias 2023).

Further to the east extends the Kraków-Częstochowa Upland, which is the largest limestone karst area in Poland (Tyc 2023). General tilt of Mesozoic strata to the east causes elevation asymmetry, with more than 100 m high cuesta front facing to the south-west. However, its distinctiveness varies along the strike and essentially disappears under Quaternary deposits to the north of the city of Częstochowa (Wieluń Upland). The upland surface above the cuesta rises to 400–450 m asl, with the highest spots exceeding 500 m asl. Unique landforms in the upland are numerous limestone crags, some as high as 30 m, typically forming clusters crowning the local divides (Fig. 4.11). In the southernmost part of the upland morphological effects of Neogene faulting are evident, resulting in horst-and-graben topography, which extends onto the territory occupied by the city of Kraków (Łajczak and Zarychta 2023). Fault-generated escarpments are dissected by limestone gorges. Karst phenomena are widely present in the area and include more than one thousand caves of different origin, including many hypogenic ones, suffosion dolines, dry valleys, blind valleys, and karst springs.

Going eastward, the Kraków-Częstochowa Upland gives way to the large Nida Basin, named after the trunk river of Nida, an important tributary of the Wisła River. Upper Cretaceous marls and limestones occur in the western part of the basin, but most widespread are sedimentary rocks of Miocene age, including gypsum deposits. Gypsum strata support karst and the Nida Basin is the only region in Poland, where gypsum karst occurs (Urban et al. 2023). Gypsum cuestas, dolines, dry valleys and caves are typical landforms.

The Kielce Upland is the largest region in the upland belt, and also very heterogeneous in terms of relief. It extends between the Pilica River in the west and the Wisła River in the east, being 150 km long and up to 60–70 km wide. The northern and southern borders are not evident and landscape change is gradual rather than sharp. The core of the region is made of rocks of older Palaeozoic ages (Cambrian to Devonian), which underlie the highest elevations, up to 612 m asl in the Łysogóry Range. These in turn, are part of the Świętokrzyskie (Holy Cross) Mountains, consisting of numerous parallel ridges of W–E strike, although in some areas the pattern of elevations and depressions is more complicated (Fig. 4.12). Blockfields of quartzitic sandstone of Cambrian age are distinctive landmarks, whereas other features of particular geomorphological interest are rock-controlled homoclinal and hogback ridges, karst landforms (caves, karren fields), and water gaps of not fully deciphered origin. The outer parts of the region are built of Mesozoic sedimentary rocks and although the general topography is rather subdued, numerous sandstone crags add to the geomorphic diversity (Urban 2023).

The eastern part of the upland belt in Poland, located to the east of the Wisła River valley, is usually divided into three major regions. These are the Roztocze elevation in the south, the Lublin Upland in the centre and north, and the Wołyń-Podole Upland in the east, which continues in Ukraine. The Roztocze is a NW–SE-trending zone of higher ground, quite distinctively rising above the Sandomierz Basin in the south-west (Fig. 4.13), but gradually sloping to the north-east. Altitudes exceed 300 m asl and the highest spot, close to the Ukrainian border, is at 391 m asl. This regional elevation is interpreted as an uplifted and tilted block in front of the Carpathians, cut by numerous faults, but with a limited throw. Bedrock is predominantly made of chalk, siliceous-carbonate rocks (opoka), and detrital siliceous deposits (gaize) of Late Cretaceous age, locally superimposed by Miocene limestones, clays and sands. Loess covers are locally widespread and associated with characteristics landscapes of gully erosion (Fig. 4.13).

The Lublin Upland and the Wołyń-Podole Upland are also underlain by predominantly Upper Cretaceous rocks, but altitudes are lower (hardly above 300 m asl) and topography more subdued. Loess is widely present, with thicknesses up to 40 m. A characteristic type of geomorphological landscape for this part of Poland are loess plateaus, densely dissected by gully erosion (Rodzik et al. 2023). In the west, the Lublin Upland terminates against the Wisła River Gap, locally in the form of an impressive escarpment 40–80 m high.

4.3.4 Lowland Landscapes of Central Poland

The central belt, located to the north of the uplands and to the south from the lakelands and morainic hills of northern Poland, is arguably least diverse geomorphologically within

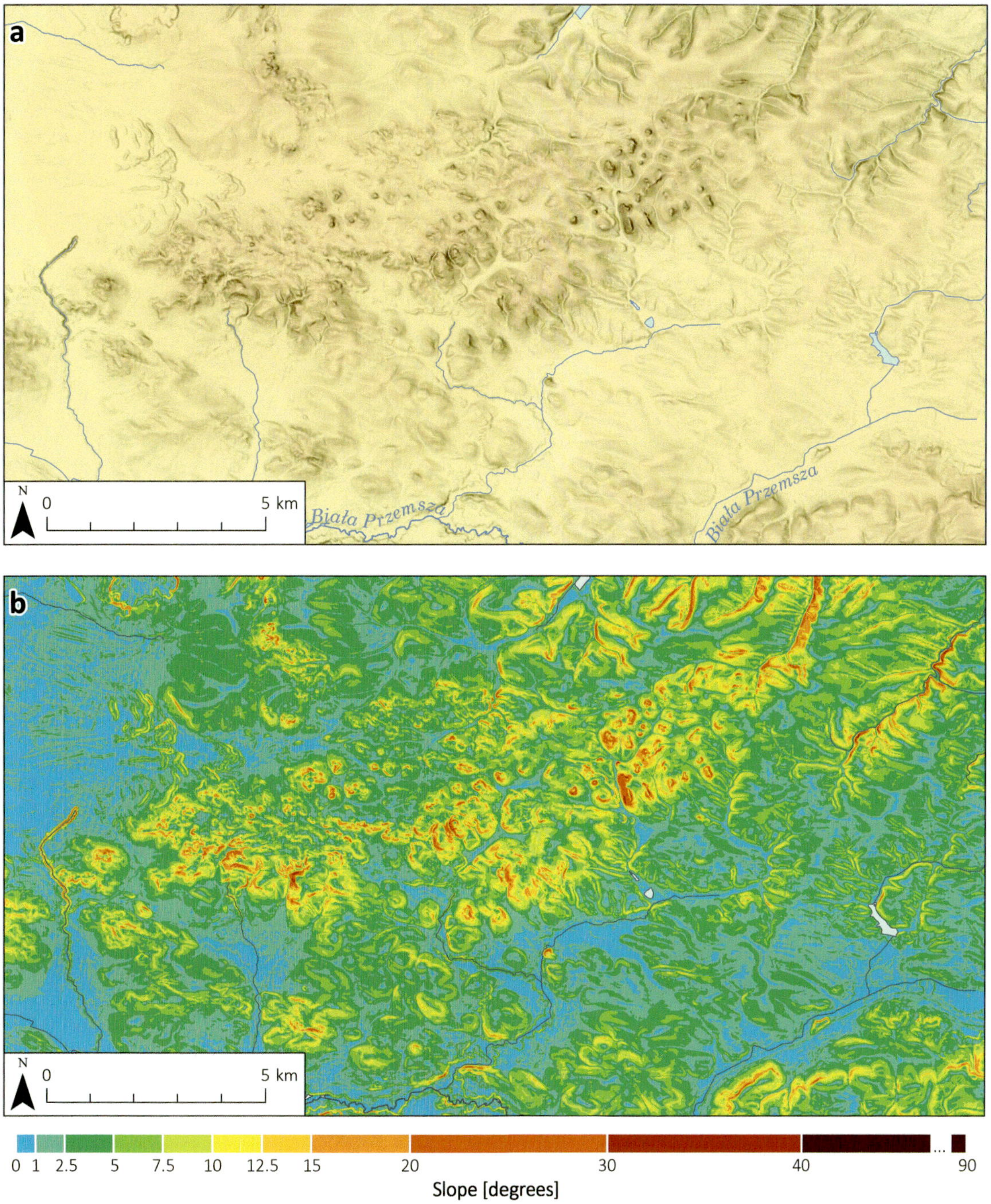

Fig. 4.11 Karstic landscape of the central part of Kraków-Częstochowa Upland, where clusters of rocky limestone hills hosting numerous caves are the most characteristic landform assemblages. **a** shaded relief, **b** slope map

the country. Altitudes typically vary within the 100–200 m asl range, generally increasing to the east, but the most extensive higher ground is located close to the centre of the country, around the city of Łódź, and hence occasionally named as the Łódź Upland. However, it lacks widespread outcrops of pre-Quaternary bedrock, typical for the upland belt proper. The highest spots, in turn, are within the Trzebnica Rampart in the south-west, where altitude of 284 m asl is reached.

This latitudinal belt was repeatedly glaciated, with stacked sequences of glacigenic deposits, but long time that has elapsed since the withdrawal of the last ice sheet allowed the periglacial processes to smooth any more impressive glacial relief that may have existed (Goździk

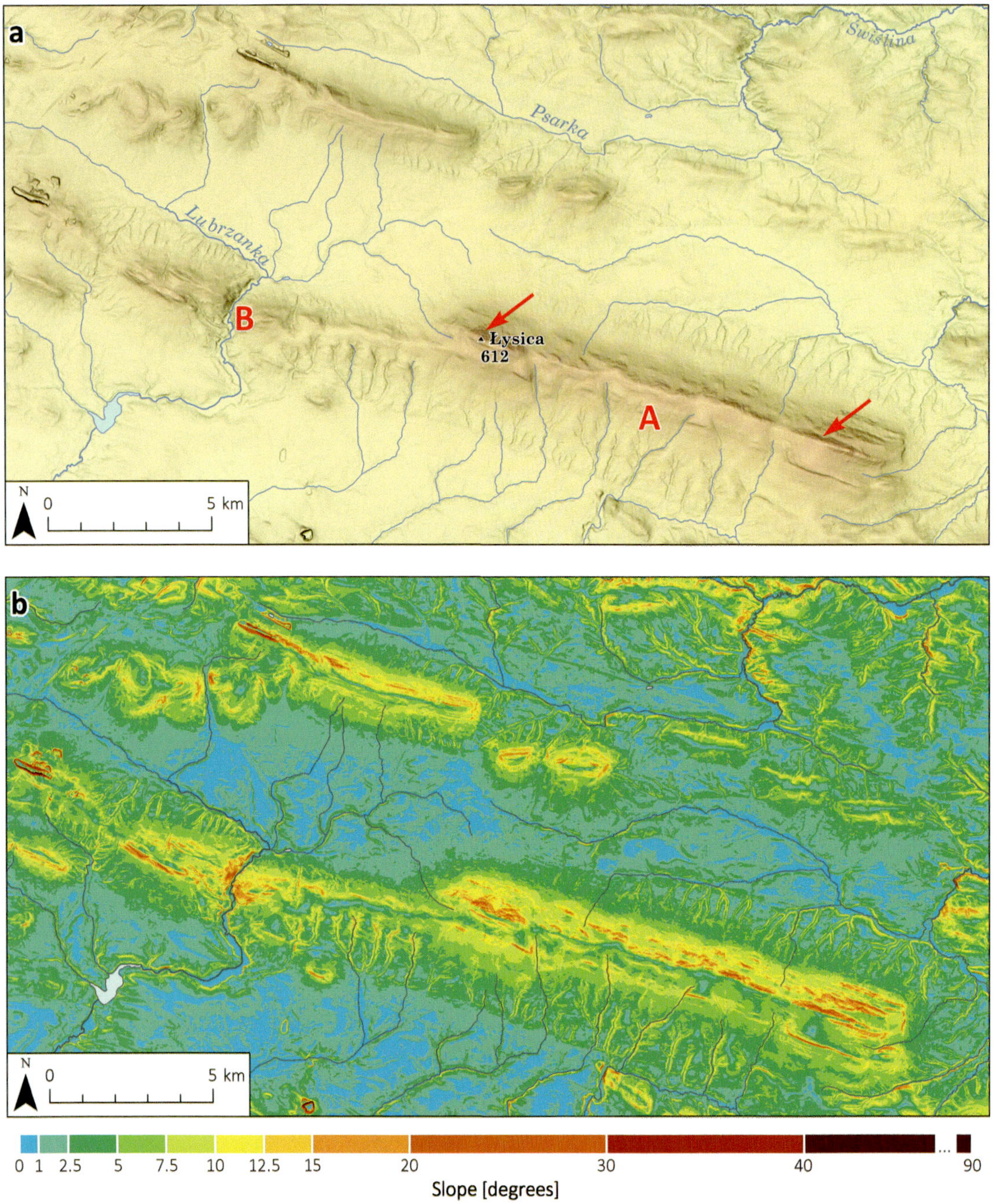

Fig. 4.12 Central part of the Kielce Upland—the Świętokrzyskie Mountains. **a** shaded relief, **b** slope map. They consist of parallel ridges built of Palaeozoic rocks, including resistant quarzitic sandstones which support the highest ridge of Łysogóry (A). Note the structurally-controlled asymmetry of slope inclinations along the ridge, evident on the slope map (**b**). Further to the west a prominent water gap of Lubrzanka River occurs (B). The region is famous for periglacial blockfields (localities shown by red arrows)

and Forysiak 2023). Consequently, extensive flatlands with occasional subdued elevations typify the area. The latter are considerably degraded end moraines, kame plateaus, and glacitectonically-deformed hills. Nevertheless, it would be incorrect to assume that the central belt of Poland is everywhere nearly featureless and lacks landforms of interest.

The most diverse relief occurs in the Trzebnica Rampart, which is a 150 km long and up to 20 km wide belt of hilly terrain, locally elevated up to 150 m above the adjacent plains and river valleys (Fig. 4.14). Its origin is connected with extensive glacitectonic deformations of soft Neogene sediments during the Elsterian (Sanian) glaciation (Koźma

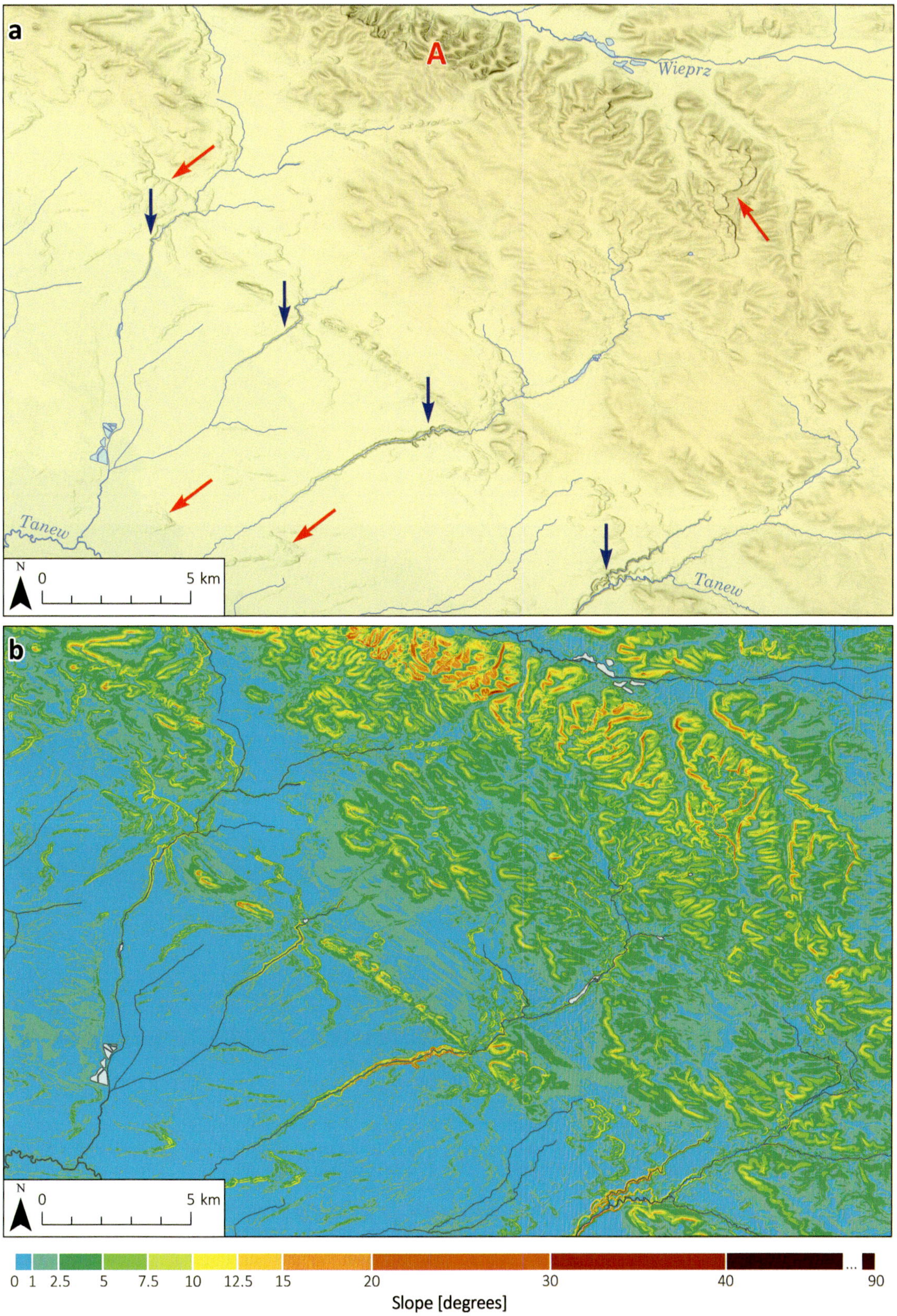

Fig. 4.13 The border zone between the upland of Roztocze in the north-east and the flat floor of the Sandomierz Basin in the south-west. **a** shaded relief, **b** slope map. Note considerable dissection of the upland, with networks of gullies incised in loess (A). Rivers flowing towards the south-west form water gaps with short reaches of bedrock channels (blue arrows), where they cross a belt of resistant Miocene limestones or Cretaceous gaize. Dune fields are typical for the Sandomierz Basin and also occur on the plateau of Roztocze (red arrows)

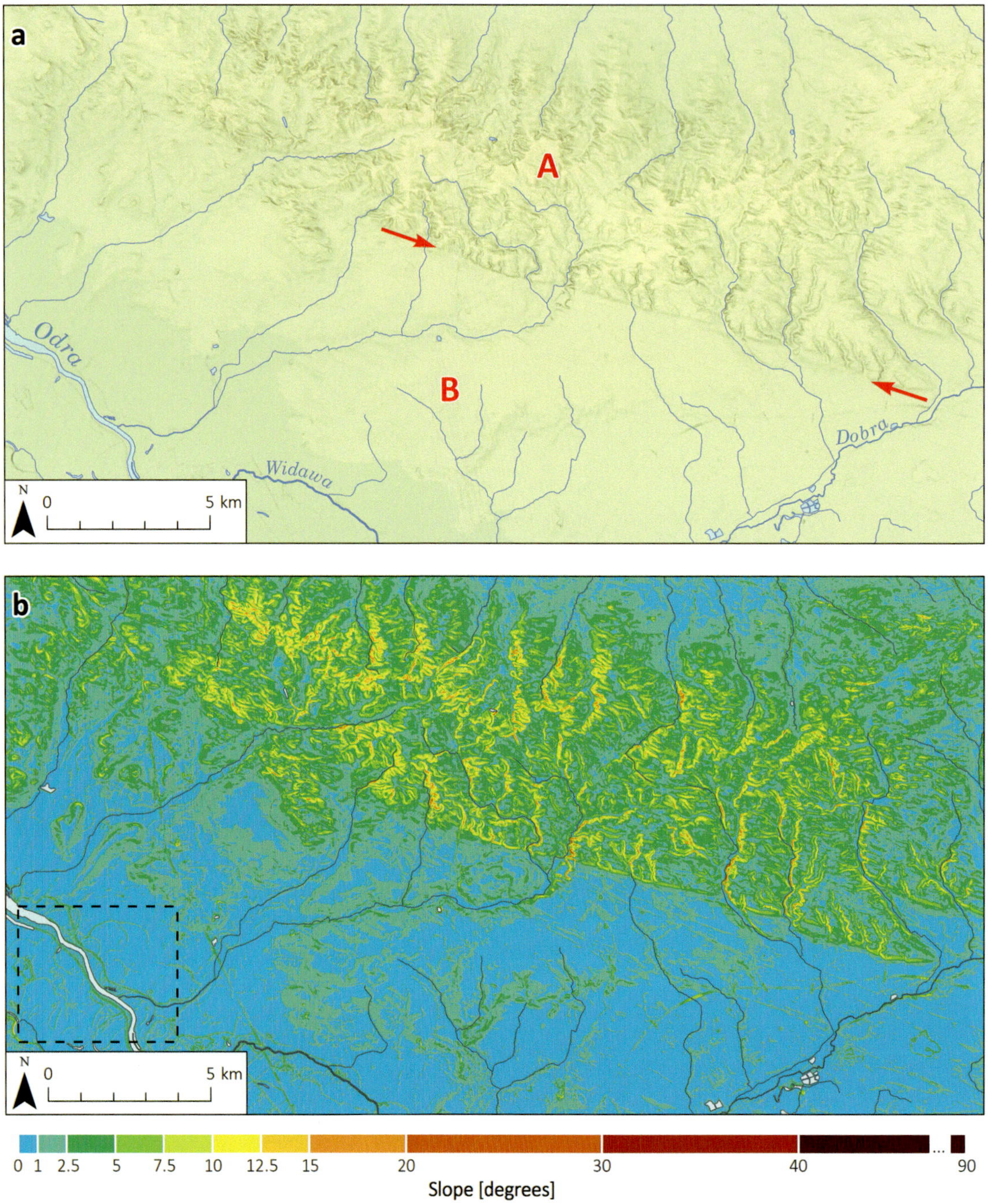

Fig. 4.14 The Trzebnica Rampart (A) and its sharp boundary with the nearly flat Silesian Lowland (B) to the north of the city of Wrocław (arrowed). **a** shaded relief, **b** slope map. Note considerable dissection of the rampart, mainly by dry valleys cut into loess and glacigenic deposits. Slope map (**b**) also reveals several generations of arcuate palaeomeanders along the Odra River (black rectangle)

and Migoń 2023). Given protracted timespan since the Elsterian available for periglacial denudation and subsequent overriding by the Saalian ice sheet, it must have been an impressive structure when newly formed c. 400,000 years ago. A geologically recent addition to landform diversity was gully erosion in the Late Pleistocene loess, deposited against the slopes of the rampart.

More diverse relief also occurs in the vicinity of major river valleys of Poland, such as the Odra, Warta, Wisła, Narew and Bug rivers, which cross the central belt (Fig. 4.15). In certain reaches the rivers are incised by a few tens of metres, forming river-parallel escarpments up to 40 m high, for example in the water gap of Bug River next to the Polish/Belarussian border. Finally, in many localities within

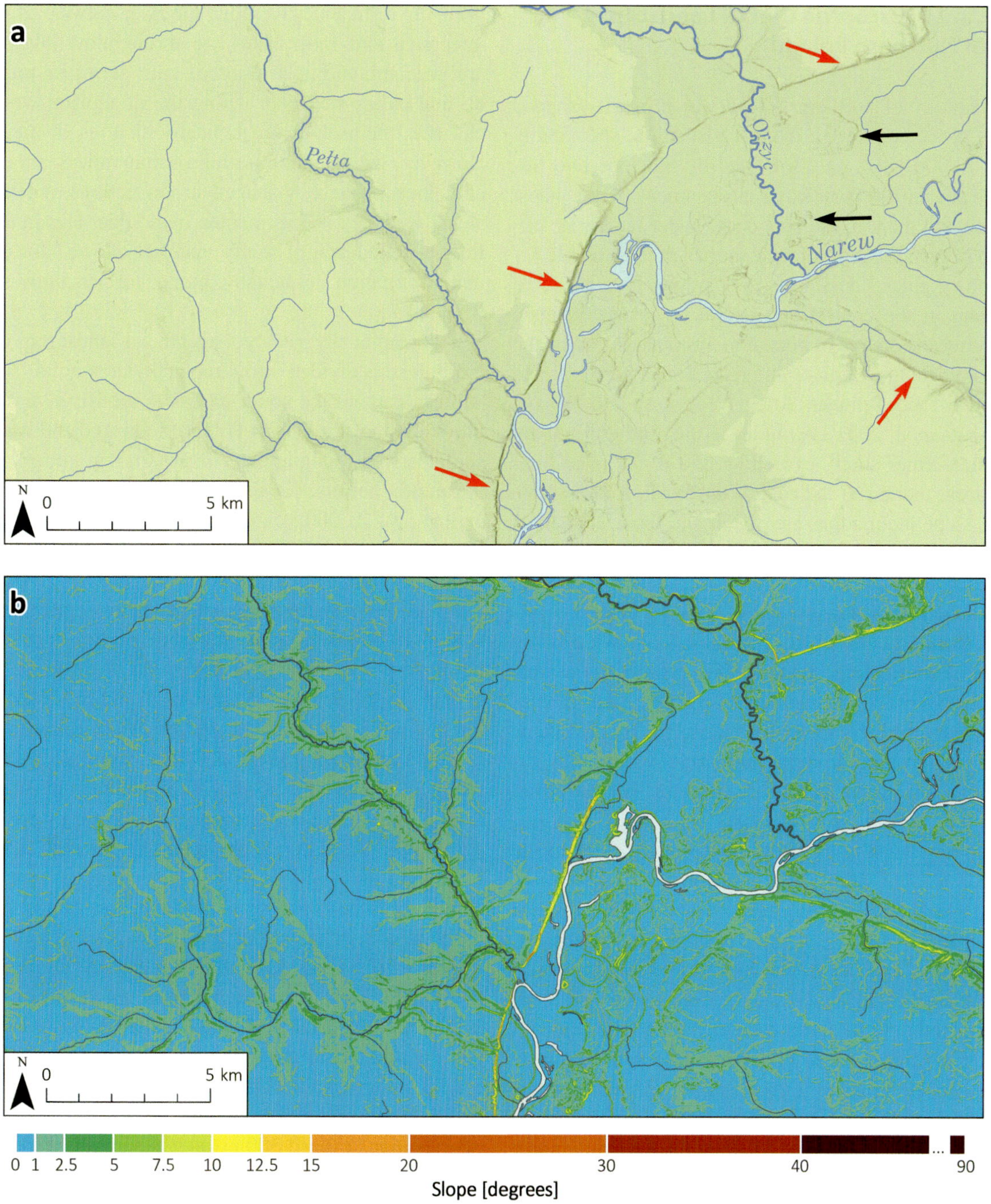

Fig. 4.15 Almost featureless, flat Northern Masovian Lowland is incised by the wide valley of the Narew River, lined by distinct escarpments (red arrows). Dune fields occur on higher fluvial terraces (black arrows), whereas the slope map also reveals the occurrence of many generations of palaeomeanders. **a** shaded relief, **b** slope map

the central belt there was considerable aeolian reworking of older fluvial and fluvioglacial sandy deposits, leading to the origin of inland dunes. These are mainly parabolic dunes, reaching the height of 20–30 m, often coalescing and forming large dune complexes. The most notable dune fields are located in the Opole Plain, next to the middle reach of the Warta river (Goździk and Forysiak 2023), and in the Kampinos Forest, just to the west of the Polish capital of Warsaw (Dłużewski and Woronko 2023). A unique landscape of chalk karst has developed in the eastern extremity of the central lowlands, in the Polesie region next to the Ukrainian border. Its most intriguing surface features are more than sixty lakes, whose basins are causally related to chalk dissolution and permafrost degradation (Dobrowolski 2023).

4.3.5 Late Pleistocene Glacial Landscapes of Northern Poland

The morphology of northern Poland was primarily shaped by glacial processes during the last glaciation, specifically since ca. 23,000 years ago, followed by hillslope and fluvial processes in the periglacial environment of the latest Pleistocene and then, in the Holocene, under temperate climatic conditions. Owing to much shorter time available for denudation of glacial landforms, topography is much more diverse than in the central belt and it is generally much easier to infer the course of glacial morphogenesis since the sedimentological and geomorphic records can be combined (Błaszkiewicz and Tylmann 2023; Weckwerth and Wysota 2023; Wysota et al. 2023). Altitudes in the northern belt are higher than further south, exceeding 250 m asl in a number of places, mainly to the east of the Wisła River valley. Likewise, slope inclinations are higher and those above 10° are not uncommon (Fig. 4.16). Among the most impressive landforms of the Late Pleistocene glacial landsystems are end moraines, eskers, tunnel valleys, parallel glacial curvilineations (formerly described as drumlins), and wide valley tracts used to convey huge amounts of meltwater to the west (pradolinas; Germ. *Urstromtäler*; Kasprzak et al. 2023). The region also features extensive outwash (sandur) plains, till plateaus, kame and drumlin fields, kettles and holes left after melting of buried ice, whereas minor, but nevertheless important geodiversity elements are erratic boulders, locally forming dense packs (Mazurek et al. 2023; Smolska et al. 2023). At the large scale, a characteristic feature of the young glacial relief of northern Poland is the presence of lake basins (Figs. 4.17, 4.18). These are of various origins, dimensions and depth, and include the largest lake in Poland—Śniardwy (113 km^2) and the deepest one—Hańcza (105.5 m), both in the north-eastern part of the country. Although many have already disappeared due to sediment fill and others progressively diminish in size and become shallower, the total number of lakes in northern Poland is still well over 6000, with most of them in the Pomeranian Lakeland in the north-west. However, the largest lake basins occur in the Masurian Lakeland in the north-east.

Even though the main features of relief of northern Poland are genetically linked with glacial processes, other morphogenetic agents should not be ignored. As elsewhere, hillslope processes began to re-shape glacial landforms immediately after ice-sheet decay, with the efficacy dependent on local relief and erodibility of substratum (Mazurek et al. 2023). Thus, steep shoulders of many tunnel valleys are considerably gullied, spring sapping undercuts the slopes of morainic hills, minor terrain depressions are filled with sediment, and landslides occasionally occur on particularly steep slopes. Fluvial processes must have been very efficient, given that the river network is now well integrated and most lakes are throughflow lakes. Valleys are predominantly flat-floored, with series of terrace levels and palaeomeanders within the floodplain, whereas the channels themselves are typically sinuous or meandering. However, within morainic hills some valleys are considerably incised, have V-shaped cross-sections and their erosional tendency is documented by boulder lags in the valley floors. Abundance of sandy material allowed for extensive dune formation, especially along former outwash tracts (Fig. 4.17).

Geographically, the 'young glacial landscape' of northern Poland is divided into three main large units. The south-western part, to the south of the Noteć River, is known as the Wielkopolskie (Great Poland) Lakeland and here the ice sheet-derived topography is least striking among the three, apparently because the ice sheet withdrew from this area at earliest, ca. 22,000 years ago, leaving the longest time for non-glacial re-modelling. Nonetheless, lakes are still abundant and tunnel valleys easy to recognize (Fig. 4.17), especially in the westernmost part of the region, the Lubuskie Lakeland. A geomorphic curiosity of this region are the best preserved impact landforms in Poland, at Morasko in the city of Poznań (Szczuciński and Włodarski 2023). Further north and bounded by the Wisła River in the east is the Pomeranian Lakeland, where the main water divide roughly indicates the position of the Scandinavian ice sheet during the Pomeranian phase, ca. 17,000–16,000 years ago. South of the divide are extensive outwash plains, now largely forested, with numerous kettle holes after melting of buried ice. The most varied relief occurs in the north-east of the region, in the Kashubian Lakeland (Błaszkiewicz and Tylmann 2023) and along the littoral of the Gulf of Gdańsk (Fig. 4.19). Elevations reach 328 m asl. The north-eastern part is the Masurian Lakeland, consisting of a large number of minor geomorphological regions with distinctive relief characteristics. The largest lakes occur in this part of the 'young glacial landscape' (Fig. 4.18), and the highest elevations exceed 300 m asl. Of particular interest is the Suwałki Lakeland in the extreme north-east, as it is here that the ice sheet-related glacial landforms are best preserved and most evident (Smolska et al. 2023; Weckwerth and Wysota 2023). As the Pomeranian Lakeland in the west, the Masurian Lakeland also gives way to extensive outwash plains in the southerly direction.

4.3.6 The Coast

The Polish coastline is more than 500 km long and the major part of it is a rather straight section between the triple mouth of the Odra River in the west and the apex of the Hel

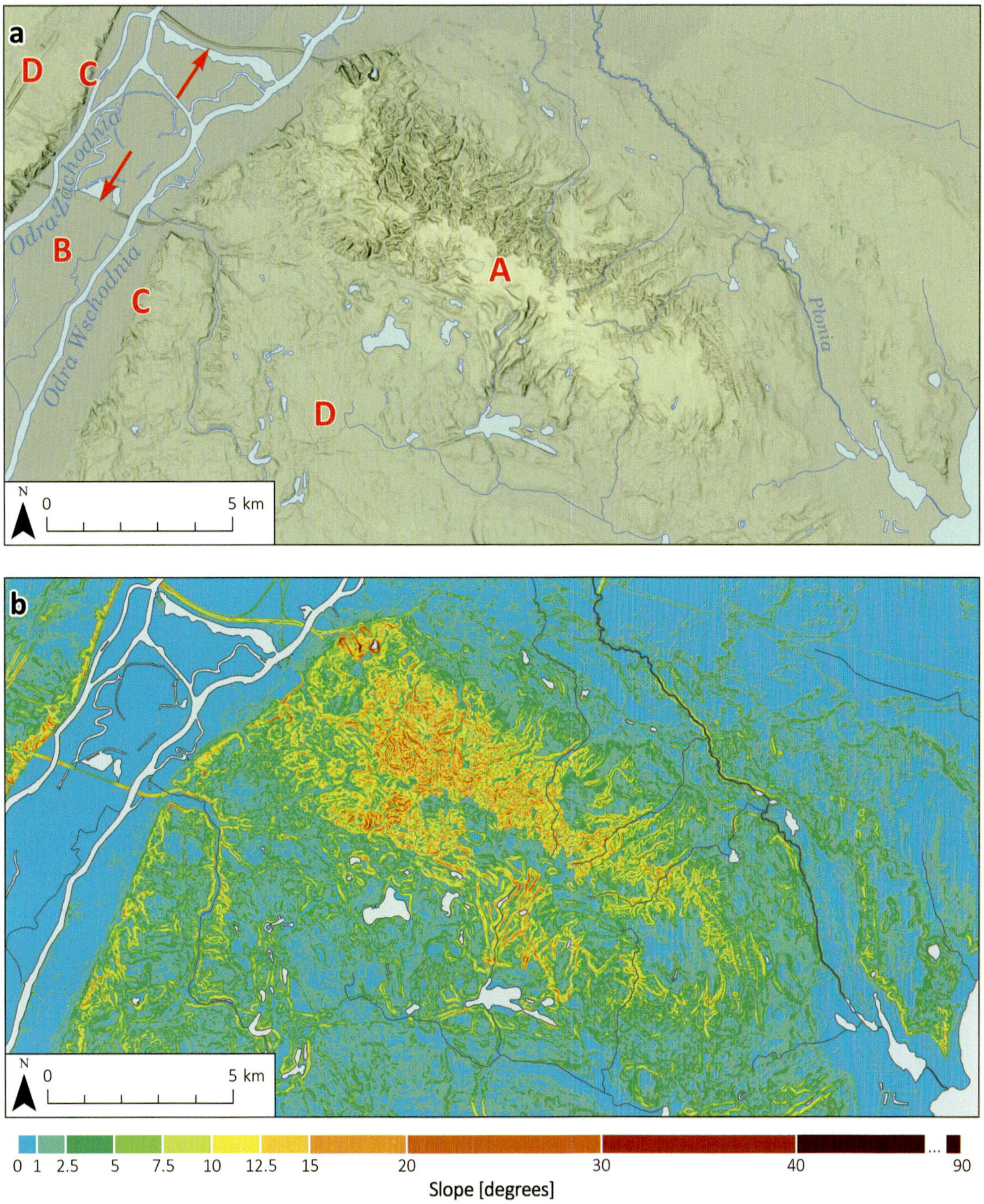

Fig. 4.16 The Bukowe Hills near the city of Szczecin are genetically an end-moraine rampart affected by glacitectonic deformations, one of the most prominent in north-west Poland (A). **a** shaded relief, **b** slope map. Its northern slopes were considerably dissected and turned into a complicated gully network. To the west one can notice the wide floor of the Lower Odra Valley (B), bordered by steep escarpments (C) linking to the undulating moraine plateaus (D). An original anastomosing channel pattern was partly anthropogenically altered to facilitate water transportation. Embankments across the valley (arrowed) are also anthropic features, built as part of road and railway construction

Peninsula in the east. Sandy beaches are almost continuous, although relief on the landward side varies from undercut moraine plateaus to coastline-parallel strips of dunes (Fig. 4.20). The highest cliffs occur in the north-east (Cap Rozewie), where they attain 30 m and show a tendency for fast retreat, counteracted by various preventive engineering solutions. The most renowned dune field, in turn, occurs next to the coastal resort of Łeba (Rotnicka and Borówka 2023). It is famous for its activity and ongoing movement to the east, probably initiated by forest clearance in historical times. A peculiar feature of the coast is the presence of several lakes separated from the sea by narrow sand bars

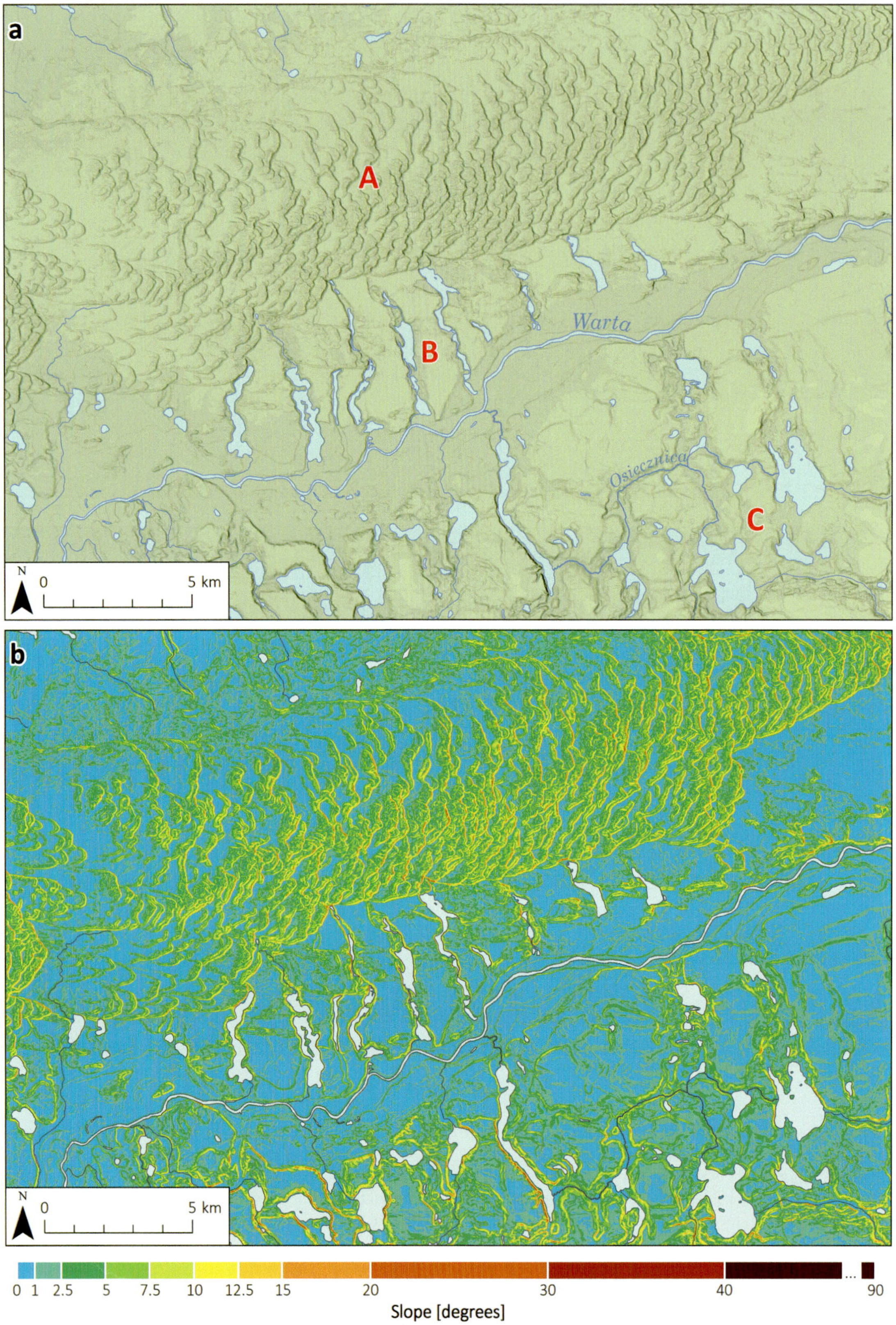

Fig. 4.17 Glacial and postglacial landscapes along the lower reach of Warta River, to the north-west of the city of Poznań. **a** shaded relief, **b** slope map. The floor of the former ice-marginal valley (Toruń-Eberswalde Pradolina) is occupied by large dune fields, with mainly parabolic dunes (A). Further south the relief was shaped by glacial processes in the Late Pleistocene and is outstanding by many lakes. Both elongated (trough) lakes in former tunnel valleys (B) and lakes within moraine plateaus (C) occur

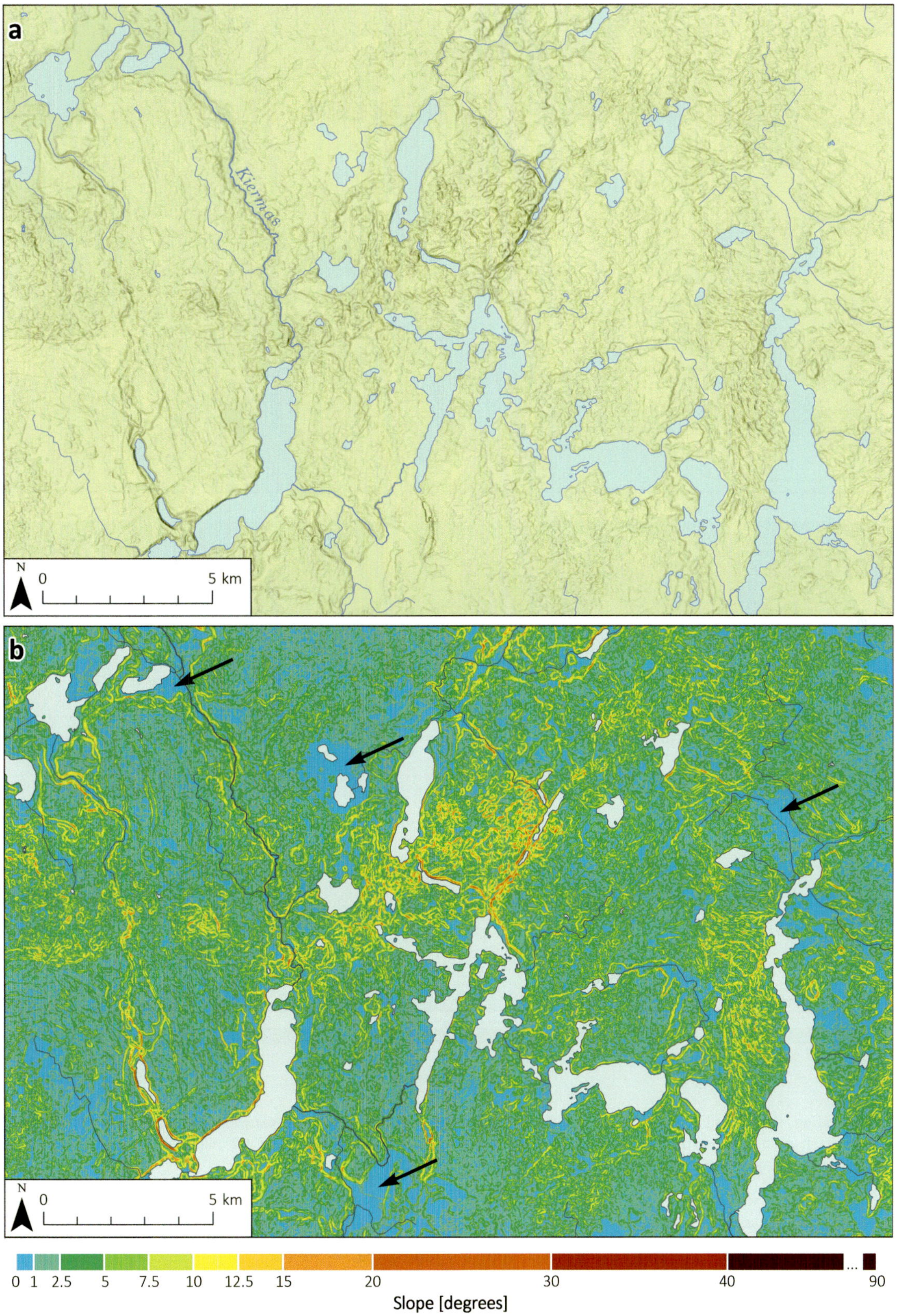

Fig. 4.18 In the Masurian Lakeland lakes of different shapes and dimensions, inset into undulating to hilly glacial relief, are abundant. **a** shaded relief, **b** slope map. On the slope map (**b**) note flat-floored basins (arrowed) with either no lakes or disconnected small lakes. They testify to the ongoing process of filling lake basins and lake disappearance

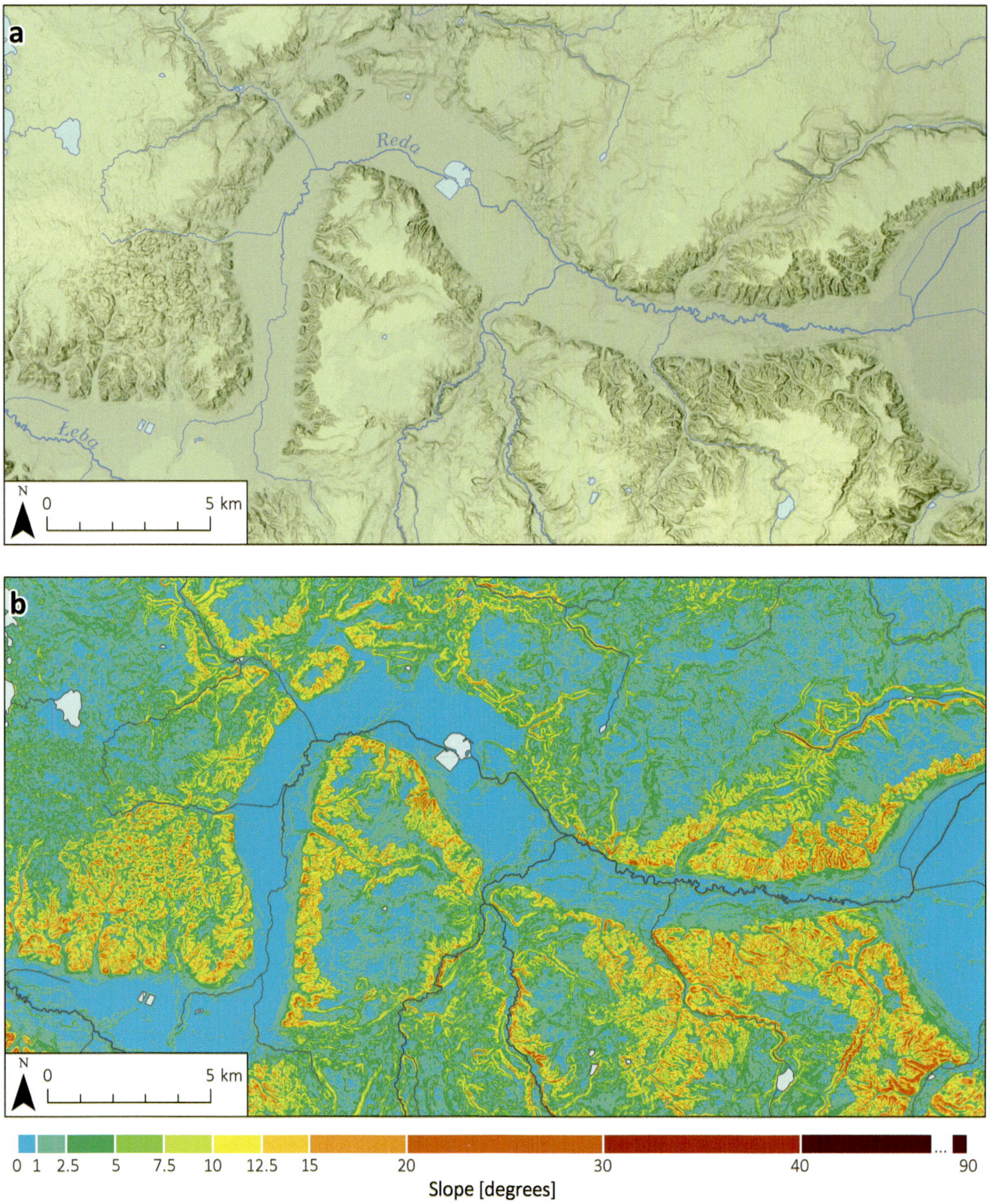

Fig. 4.19 The winding, flat-floored Reda-Łeba ice-marginal valley, incised between moraine plateaus, constitutes the northern border of the Kashubian Lakeland. It is the most northerly located and hence, the youngest landform of this kind in Poland. Note considerable postglacial erosional dissection of steep slopes of the valley. **a** shaded relief, **b** slope map

(Fig. 4.20), among which the Łebsko Lake is the largest (71.4 km^2), although fairly shallow (6.3 m maximum). The origin of coastal lakes was long explained by littoral drift and progressive closure of marine embayments, but this view cannot be taken as a universal explanation. Some coastal lakes along the Polish coast should be seen as an effect of rising sea level in Baltic Sea and flooding of various topographic depressions within the postglacial landscape.

More diverse is coastal morphology in the extreme west of the country, where the Odra River debouches into the Baltic Sea. It does so in three arms, among which two are located in Poland. However, they all begin in the large Szczecin Lagoon. The eastern and middle arm delimit the

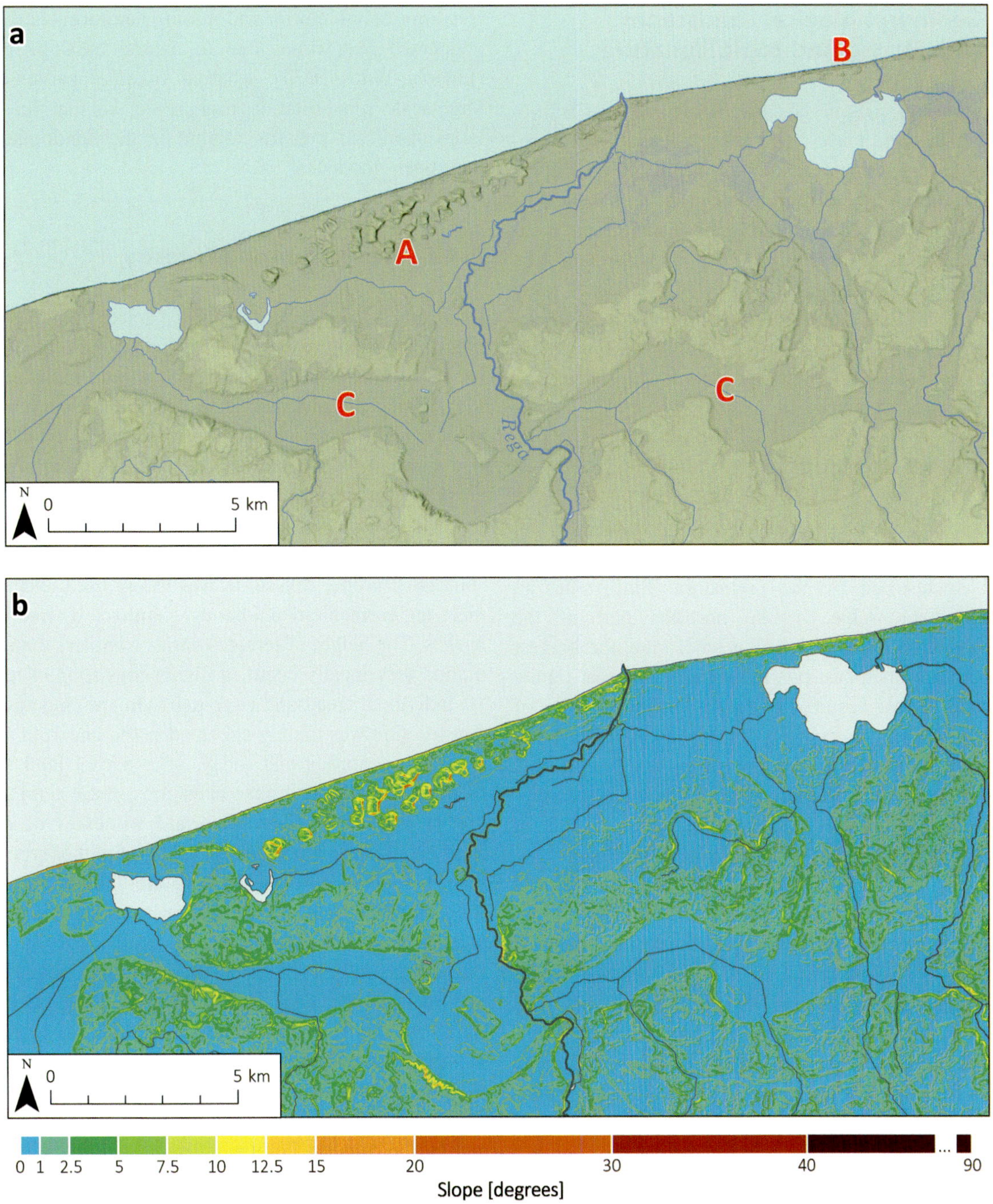

Fig. 4.20 Representative morphology of the western part of Polish coast. **a** shaded relief, **b** slope map. The coastline is fairly straight. Lagoons are separated from the sea by coastal barriers and dune belts of variable width (A—large dune field, B—narrow dune strip) occur along the coast. Inland, underfit valleys (C) are former glacial meltwater tracts

island of Wolin, with its arcuate moraine ridge overlooking the sea and forming the highest cliffs on the Polish coast, which are 93 m high (Zwoliński et al. 2023). In the extreme east, in turn, coastal morphology is dominated by the Hel Peninsula—a narrow spit that is 35 km long and as narrow as 200 m is some sections (Wróblewski et al. 2023), and the Wisła Barrier, which encloses the deltaic plain of the Wisła River and the Wisła Lagoon (Moskalewicz and Woźniak 2023). The coastal section in between, within Gulf of Gdańsk, consists of alternating cliffs and flatlands, the latter used to develop two largest Polish coastal cities, Gdańsk and Gdynia.

4.4 Geomorphological Landscapes with Strong Anthropic Signatures

Superimposed on natural landscapes are various anthropogenic impacts and these range from insignificant to profound, large-scale transformations, which give specific identity to certain regions or areas. The human impact should not be considered purely in terms of intentional modification of natural landforms. Equally important may be landforms developed as a consequence of intentional relief alterations, as well as changes in patterns and rates of processes due to anthropogenic forcings, up to the appearance of processes which otherwise would probably be absent.

4.4.1 Mining and Industry

Some parts of Poland are rich in mineral resources and these have been mined since long ago, even since the Neolithic as testified by the prehistoric flint mines in Krzemionki Opatowskie, in the northern part of the Kielce Upland. In terms of singular landforms, the biggest among them are open-cast pits excavated to extract lignite (brown coal) if this occurred at rather shallow depths, of the order of 100 m or so. Thus, the open-cast excavation in Bełchatów in central Poland is currently 8.5 km long, more than 3 km wide and more than 250 m deep, occupying nearly 24 km^2 (Fig. 4.21). The huge open-cast pit in Turoszów in the West Sudetes has comparable dimensions, whereas an artificial mound (waste rock heap) next to it is 6.5 km × 4.5 km, covers nearly 20 km^2, and is 200 m high. Worth particular attention is the post-mining landscape of the Mużaków Rampart, next to the German border, where numerous excavation pits and subsidence troughs were filled with water, giving rise to a unique anthropogenic lakeland (Koźma and Migoń 2023). Considerable changes took place in areas where quarrying was the main industry, such as in the Strzegom Hills in the Sudetic Foreland, where essentially each granite hill is undercut by a quarry, some being very large and deep (Fig. 4.22). Quarries are common in the Sudetes, in the eastern part of the Silesian Upland and in the Kielce Upland.

However, the most profound alterations of the natural landscape occurred in the Upper Silesian region, where coal has been mined for the last two centuries and the district is still the main supplier of coal as an energy resource (see Sect. 4.3.3; Dulias 2023). No less significant are anthropic changes imparted in the natural landscape of the Wałbrzych Basin in the Middle Sudetes, typified by numerous spoil heaps, subsidence troughs and basins, and large expanses of levelled ground (Jancewicz et al. 2021). Other types of mining, such as ore prospecting in the Sudetes, have left less impressive imprint and mining landforms have become attenuated over times due to natural hillslope processes (Latocha-Wites 2023), whereas in other parts of Poland large-scale rehabilitation took place, so that for instance large open-cast pits for sulphur in the Sandomierz Basin have been flooded.

4.4.2 Agriculture and Forestry

Landform modifications causally related to the development of agriculture are rather minor if the size of individual landforms is considered, but become more significant if the slope or catchment scale is taken into account. The most evident impact is slope terracing, attempted to facilitate agrarian activities on steeper slopes and to prevent soil erosion. Agricultural terraces were built in many parts of the Sudetes and Beskidy Mountains, and they also occur in some areas within the upland belt. However, unlike in southern Europe, dry stone was rarely used and most terraces are earthen embankments, stabilized by tree lines and bushes. The height of terraces varies from less than 1 to 2–3 m, and they locally occur in flights consisting of more than 10 individual tread-and-riser pairs. In the Sudetes, terrace construction was mainly executed in the late 19th and early twentieth centuries and today, after partial land abandonment, especially at higher elevations, these terraces occur under forest, becoming important witnesses of changing local economy (Latocha-Wites 2023). Another type of landform associated with agriculture are sunken lanes (holloways, road gullies), which are former access roads to field plots located at higher elevations, eroded over time due to increasing traffic and poor maintenance, especially lack of floor strengthening (Zgłobicki et al. 2021; Rodzik et al. 2023). In erodible substratum, such as loess or any other silt-rich deposit, the depth of sunken lanes may exceed 5 m. Sunken lanes are common in the upland belt, especially where loess cover is thick, in the flysch Outer Carpathians, and in the Sudetic Foreland, again on loess. Thinner and coarser regolith in the Sudetes did not facilitate considerable deepening of former field access roads. Sunken lanes are also produced by forestry operations due to use of heavy machinery, whereas similar landforms, although with V-shaped cross-profiles, originate through downslope dragging of tree logs. High-resolution DTMs reveal that modification of forested slopes due to road gully evolution can be substantial in specific settings, dictated by relief and bedrock properties.

Associated with land transformation to facilitate agricultural activities is land reclamation, focused on wetlands and frequently flooded terrains along major rivers. Large-scale attempts to gain fertile land were mainly undertaken by

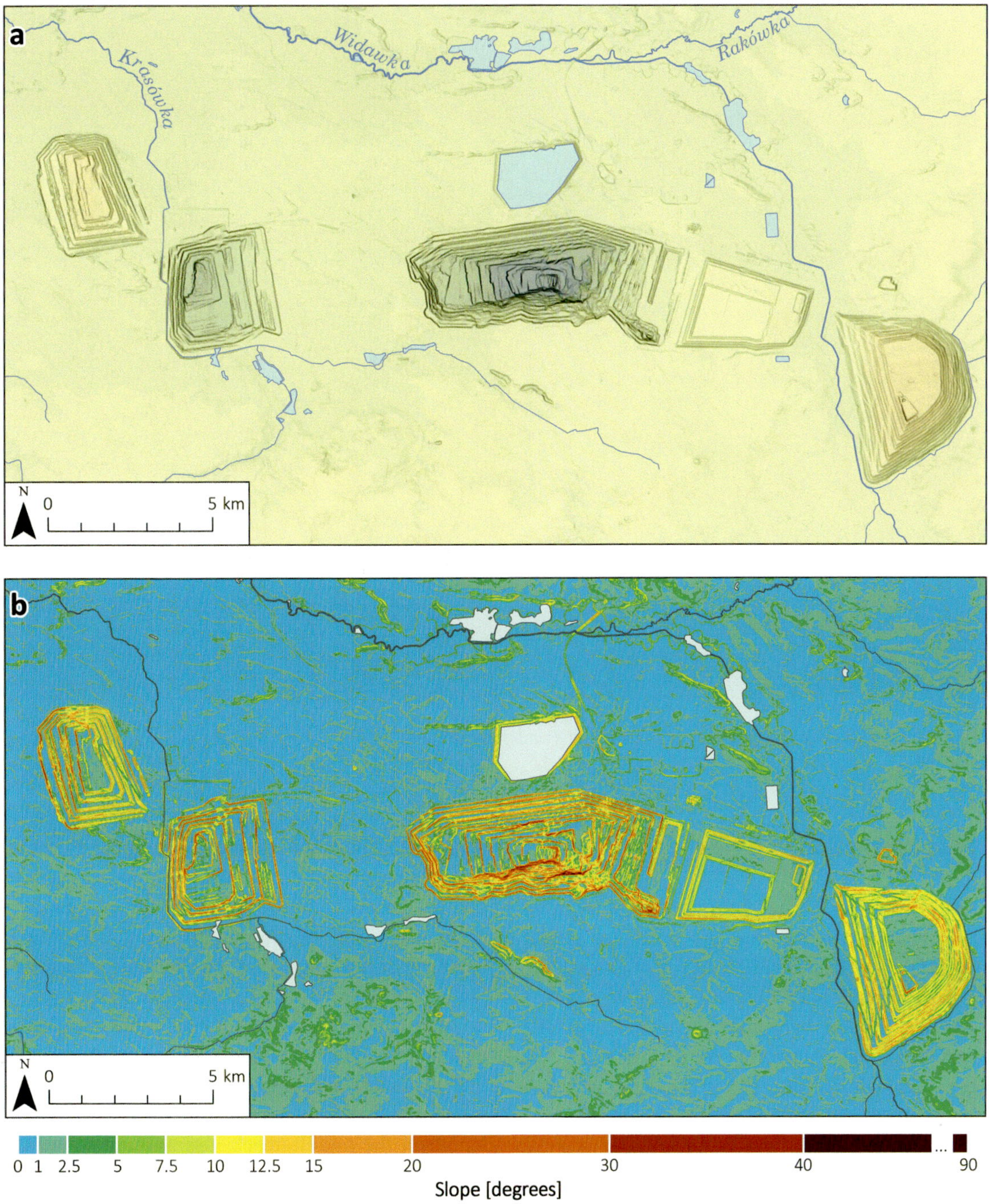

Fig. 4.21 The largest anthropic landforms in Poland—open-cast lignite pits and terraced waste heaps near the town of Bełchatów in the central-south part of the country. **a** shaded relief, **b** slope map

Dutch settlers in what is now western and northern Poland. One example is provided by the Żuławy Wiślane region in the Wisła deltaic plain (Moskalewicz and Woźniak 2023), whereas another such transformation was implemented along the lower Warta River, close to its junction with the Odra River (Fig. 4.23). Among unwanted consequences of land drainage was ground subsidence due to peat oxidation and shrinking, which in turn necessitated the construction of embankments (dykes) along the river channels and their heightening to prevent flooding.

Fig. 4.22 Heavy impact of stone quarrying on the hilly granite landscape near the town of Strzegom in the western part of Sudetic Foreland. Source: Google Earth

4.4.3 River Regulation

Very few larger rivers in Poland are in natural conditions, so that their channels can migrate and transform reflecting only non-anthropogenic controls (e.g. Zieliński 2023). Most rivers were heavily altered, with controls imposed on both channel patterns and hydrological regimes, the latter via dam operations and, indirectly, via land use and land cover changes. Human interventions in river channels have been various, with other measures implemented in mountainous catchments and others applied to lowland rivers. In the mountains, where the principal purpose was to reduce bedload transport during extreme events and to prevent flooding in general, common solutions were building of dry reservoirs to trap sediment, bank and bed strengthening and armouring, and construction of cascading reaches to slow down the flow (e.g., Kasprzak 2023). Typical interventions also included flow diversions into canals to supply water mills, although many of these are now abandoned. Equally important are non-intentional alterations, associated mainly with gravel mining from the alluvial fills, common in the gravel-bed rivers of the Carpathians (Zawiejska et al. 2023).

In the lowlands, hydro-engineering works served two main, interrelated purposes: to make the channels navigable and not to allow the channels to migrate freely across the floodplain, besides more obvious flood prevention. These goals were accomplished mainly through channel straightening, diversion of the current away from the banks by means of series of groynes, cutting off the meanders, building of locks to control the gradient and by-pass channels in some cities (Fig. 4.24). Basins were excavated to serve as temporary harbours for the barges, with some being former meanders adapted to this purpose. The Odra River in western Poland was most affected by these changes and its length in the middle reach was reduced by approximately 20% due to extensive hydro-engineering works in the late 19th and early twentieth centuries.

4.4.4 Urban Development

Spatial expansion of cities was another factor contributing to landform change. Although urbanization processes accelerated in the last 150 years or so, significant alterations of urban space occurred already in medieval times, as comprehensively shown for the city of Kraków—the former capital of Poland (Łajczak and Zarychta 2023). Changes of earlier date were rather minor (Żuk et al. 2023). Considering that

Fig. 4.23 Effects of land reclamation and agricultural use in the lowermost reach of the Warta River. Human intervention also included straightening of the river channel, but the former anastomosing pattern is still visible within the wide floodplain (light stripes). Source: Google Earth

most large cities in Poland are located in lowlands, alterations were not particularly profound and effects of surface levelling are not conspicuous. Valley floors were modified most, especially after insular setting of medieval strongholds was no longer regarded as an advantage, but rather an obstacle hindering further development. Consequently, secondary river channels were filled in and disappeared, floodplains were drained and built up, and flood-protective embankments were constructed. A good example of considerable change of this kind is provided by the city of Wrocław on the Odra River (Kasprzak and Traczyk 2014). The river used to have an anastomosing pattern within this reach, conditioned by slow subsidence, very low gradient, and largely suspended load. The medieval settlement was located on islands and multiple channels were important communication and defence lines. Rapid development of the city since the early nineteenth century resulted in almost complete disappearance of this unique channel pattern, only echoed by a few short branch canals of the Odra that exist today. However, new channels by-passing the city were excavated to serve transportation and flood control, further adding to valley floor modification. Similar works, albeit at a smaller scale, were undertaken in the cities of Kraków, Poznań, Opole and several others.

4.5 Geomorphology of Poland in a Broader European Perspective

None of large geomorphological regions (types of relief) presented above is endemic to Poland. Owing to the location of Poland at the junction of western and eastern Europe, with no clear morphological barriers of meridional strike on either side, the morphogenetic belts recognized in Poland continue both westward and eastward, whereas the mountains also extend further to the south. In fact, there are many specific situations where landform complexes continue beyond the border and they can be properly understood only if transnational research is involved. One example is the push moraine arch of the Mużaków Rampart, cut in two parts by the Polish/German border, which essentially mirror each other (Koźma and Migoń 2023).

Thus, the high mountains of the Tatra have their counterpart on the Slovak side (Boltižiar 2022), in fact much larger in terms of area than the Polish side. Likewise, the limestone crags of the Pieniny Klippen Belt continue in Slovakia (Plašienka and Novotný 2022), whereas the Polish part of the Outer Carpathians (Beskidy Mountains) is but a section

Fig. 4.24 Regulated channel of the Odra River in its middle reach, upstream of the town of Głogów. Note regularly spaced groynes along the banks and remnants of the former channels, some completely disconnected from the contemporary one. Source: Google Earth

of an extended flysch belt that begins in south-west Slovakia and runs across Slovakia, partly in eastern Czechia (Pánek and Lenart 2016), and then onwards to Ukraine (Kovalchuk et al. 2012), with a similar suite of landforms and processes. In a similar way, the Fore-Carpathian basins continue into Czechia and then Austria in the west, and into Ukraine in the east, reflecting the existence of the Carpathian Foredeep. The Sudetes, in turn, are part of the Variscan Bohemian Massif that covers a significant part of Czechia and extends into Germany and Austria, and most types of relief comprised in the Sudetes can also be found elsewhere within this large geographical and geological unit (e.g., Mentlík 2016; Mertlík and Adamovíč 2016; Pilous 2016; Raška and Cajz 2016; Vítek 2016). In an even broader perspective, the Sudetes are part of the belt of Variscan medium–high mountain ranges of central Europe that begins in eastern France and terminates against the Carpathians. The Sudetic Foreland can be compared with denudational landscapes of eastern Germany, north of the Erzgebirge and Harz, where bedrock outcrops co-exist with thick Cenozoic deposits, including those of glacial origin (Wagenbreth and Steiner 1982; Böse et al. 2022). The southern Polish upland belt continues in Ukraine, where for example spectacular examples of gypsum karst occur (Klimchouk 1996). Loess landscapes and chalk karst are also present.

The broader context of morphogenetic belts is even more evident in lowland Poland. Both the subdued 'old glacial landscape' and the 'young glacial landscape' continue into Germany in the west (Böse et al. 2022) and Belarus and Lithuania in the east (Baltrunas et al. 2007). The role of the southern limit of the last ice sheet as a demarcating line between these two types of post-glacial landscapes is also clear, and the lakelands of northern Poland have their equivalents in the eastern part of Germany, towards Berlin and Mecklemburg (Mecklemburgische Seenplatte), as well as in Lithuania, towards the city of Vilnius. Finally, the coastal morphogenetic systems are open and include littoral zones of the neighbouring countries, which is most evident in Gulf of Gdańsk, where the perfect curve of the Wisła Barrier continues beyond the border of the Królewiec (Kaliningrad) area, now part of the Russian Federation.

References

Alexandrowicz Z (1978) Skałki piaskowcowe zachodnich Karpat fliszowych. Prace Geol PAN 113:87

Alexandrowicz Z (1994) Geologically controlled waterfall types in the Outer Carpathians. Geomorphology 9:155–165

Andrzejewski L, Krzemień K, Zwoliński Z (2023) Riverine landscapes. In: Migoń P, Jancewicz K (eds) Landscapes and landforms of Poland. Springer, Cham

Baltrunas V, Švedas K, Pukelytė V (2007) Palaeogeography of South Lithuania during the last ice age. Sediment Geol 193:221–231

Ber A, Krzyszkowski D (eds) (2004) Glacitektonika wybranych regionów Polski. Biul Państw Inst Geol 408:73–125

Bernatek-Jakiel A (2023) Bieszczady Mts.—structural relief with human imprint in the wilderness corner of Poland. In: Migoń P, Jancewicz K (eds) Landscapes and landforms of Poland. Springer, Cham

Błaszkiewicz M, Tylmann K (2023) Tunnel valleys of the Tuchola Forest and Kashubian Lake District. In: Migoń P, Jancewicz K (eds) Landscapes and landforms of Poland. Springer, Cham

Błaś M, Ojrzyńska H (2023) The climate of Poland. In: Migoń P, Jancewicz K (eds) Landscapes and landforms of Poland. Springer, Cham

Boltižiar M (2022) Unique glacial landscape on the roof of the Carpathians—Tatras Mts. In: Lehotský M, Boltižiar M (eds) Landscapes and landforms of Slovakia. Springer, Cham, pp 63–97

Böse M, Ehlers J, Lehmkuhl F (2022) Deutschlands Norden von Erdaltertum zur Gegenwart. Springer, Heidelberg, p 249

Brodzikowski K (1995) Pre-Vistulian glaciotectonic features in south-western Poland. In: Ehlers J, Kozarski S, Gibbard PL (eds) Glacial deposits in North-East Europe. Balkema, Rotterdam, pp 339–359

Czerwiński J, Żurawek R (1999) The geomorphological effects of heavy rainfalls and flooding in the Polish Sudetes in July 1997. Studia Geomorph Carp-Balc 33:27–43

Danišík M, Migoń P, Kuhlemann J, Evans N, Dunkl I, Frisch W (2010) Thermochronological constraints on long-term erosional history of the Karkonosze massif, Central Europe. Geomorphology 117:89–98

Danišík M, Štěpančíková P, Evans NJ (2012) Constraining long-term denudation and faulting history in intraplate regions by multisystem thermochronology: An example of the Sudetic Marginal Fault (Bohemian Massif, central Europe). Tectonics 31:TC2003

Dłużewski M, Woronko B (2023) Inland dunes of the Kampinos National Park. In: Migoń P, Jancewicz K (eds) Landscapes and landforms of Poland. Springer, Cham

Dobrowolski R (2023) Chalk karst in the Lublin region. In: Migoń P, Jancewicz K (eds) Landscapes and landforms of Poland. Springer, Cham

Dulias R (2023) Upper Silesian Region—an example of large-scale transformation of relief by mining. In: Migoń P, Jancewicz K (eds) Landscapes and landforms of Poland. Springer, Cham

Duszyński F (2023) Stołowe Mountains—a unique sandstone tableland. In: Migoń P, Jancewicz K (eds) Landscapes and landforms of Poland. Springer, Cham

Głazek J, Szynkiewicz A (1987) Stratygrafia młodotrzeciorzędowych i staroczwartorzędowych osadów krasowych oraz ich znaczenie. In: Problemy młodszego neogenu i eoplejstocenu w Polsce, Ossolineum, Wrocław, pp 113–130

Goździk J, Forysiak J (2023) Polygenetic lowland relief around the city of Łódź—glacial inheritance and periglacial transformation. In: Migoń P, Jancewicz K (eds) Landscapes and landforms of Poland. Springer, Cham

Jahn A (1980) Main features of the Tertiary relief of the Sudetes Mountains. Geogr Polon 43:5–23

Jancewicz K, Traczyk A, Migoń P (2021) Landform modifications within an intramontane urban landscape due to industrial activity, Wałbrzych, SW Poland. J Maps 17:194–201

Jankowski L, Margielewski W (2021) Geological control of young orogenic mountain morphology: from geomorphological analysis to reinterpretation of geology of the Outer Western Carpathians. Geomorphology 386:107749

Jary Z, Marks L (2023) Quaternary history. In: Migoń P, Jancewicz K (eds) Landscapes and landforms of Poland. Springer, Cham

Kasprzak L, Ewertowski M, Szuman I, Kalita J, Tomczyk A (2023) The Middle Noteć Valley—the landscape of an ice-marginal valley and its surroundings. In: Migoń P, Jancewicz K (eds) Landscapes and landforms of Poland. Springer, Cham

Kasprzak M (2023) Karkonosze Mountains and Jelenia Góra Basin—unique variety of granite landforms. In: Migoń P, Jancewicz K (eds) Landscapes and landforms of Poland. Springer, Cham

Kasprzak M, Traczyk A (2014) LiDAR and 2D electrical resistivity tomography as a supplement of geomorphological investigations in urban areas: a case study from the city of Wrocław (SW Poland). Pure Appl Geophys 171:835–855

Klimaszewski M (1958) Rozwój terytorium Polski w okresie przedczwartorzędowym. Przegl Geogr 30:3–43

Klimchouk A (1996) Gypsum karst in the Western Ukraine. Intern J Speleol 25:263–278

Kłapyta J, Zasadni J (2023) The Tatra Mountains—the only high-mountain landscape in Poland. In: Migoń P, Jancewicz K (eds) Landscapes and landforms of Poland. Springer, Cham

Kotarba A (1992) Denudacja mechaniczna Tatr Wysokich pod wpływem opadów ulewnych. Prace Geogr IGiPZ PAN 155:191–208

Kotarba A, Kaszowski L, Krzemień K (1987) High-mountain denudational system in the Polish Tatra Mountains. Geogr Studies IGiPZ PAN 3:106

Kotarba A, Rączkowska Z, Długosz M, Boltižiar M (2013) Recent debris flow in the Tatra Mountains. In: Loczy D (ed) Geomorphological impact of extreme weather (case studies from Central and Eastern Europe). Springer, Dordrecht, pp 221–236

Kovalchuk I, Kravchuk Y, Mykhnovych A, Pylypovych O (2012) Recent landform evolution in the Ukrainian Carpathians. In: Lóczy D, Stankoviansky M, Kotarba A (eds) Recent landform evolution: the Carpatho-Balkan-Dinaric region. Springer, Dordrecht, pp 177–204

Kozarski S (1995) The periglacial impact on the deglaciated area of northern Poland after 20 kyr BP. Biul Perygl 34:73–102

Koźma J, Migoń P (2023) Mużaków Rampart (Muskau Arch)—the legacy of glacial processes and mining in the UNESCO Global Geopark. In: Migoń P, Jancewicz K (eds) Landscapes and landforms of Poland. Springer, Cham

Latocha-Wites A (2023) Cultural geomorphological landscapes of the East Sudetes. In: Migoń P, Jancewicz K (eds) Landscapes and landforms of Poland. Springer, Cham

Liszkowski J (1996) Poligeneza rzeźby obszarów wyżynnych Polski w świetle analizy reliktowych pokryw wietrzeniowych. Acta Geogr Lodz 71:165–179

Łajczak A (2023) Mt. Babia Góra—the highest flysch ridge in the Western Carpathians. In: Migoń P, Jancewicz K (eds) Landscapes and landforms of Poland. Springer, Cham

Łajczak A, Zarychta R (2023) Kraków—anthropogenic changes in the relief of a large city. In: Migoń P, Jancewicz K (eds) Landscapes and landforms of Poland. Springer, Cham

Margielewski W (1998) Landslide phases in the Polish Outer Carpathians and their relation to climatic changes in the Late Glacial and the Holocene. Quat Studies in Poland 15:37–53

Margielewski W (2018) Landslides fens as a sensitive indicator of the palaeoenvironmental changes since the Late Glacial; Polish Western Carpathians case study. Radiocarbon 60:1199–1213

Margielewski W, Jankowski L (2023a) Dunajec River Gorge through the Pieniny Mountains. In: Migoń P, Jancewicz K (eds) Landscapes and landforms of Poland. Springer, Cham

Margielewski W, Jankowski L (2023b) Impact of mass movements on landscape and landform transformation in the Beskidy Mountains—from gravitational collapse to recent times. In: Migoń P, Jancewicz K (eds) Landscapes and landforms of Poland. Springer, Cham

Margielewski W, Urban J (2023) Gravitationally-induced non-karst caves in the Polish Outer Carpathians. In: Migoń P, Jancewicz K (eds) Landscapes and landforms of Poland. Springer, Cham

Mazur S (2023) An outline of the geology of Poland. In: Migoń P, Jancewicz K (eds) Landscapes and landforms of Poland. Springer, Cham

Mazurek M, Paluszkiewicz R, Zwoliński Z (2023) Glacial and postglacial landforms of the Drawsko Lakeland. In: Migoń P, Jancewicz K (eds) Landscapes and landforms of Poland. Springer, Cham

Mentlík P (2016) Bohemian Forest: landscape and people on the frontier. In: Pánek T, Hradecký J (eds) Landscapes and landforms of the Czech Republic. Springer, Switzerland, pp 87–100

Mertlík J, Adamovíč J (2016) Bohemian Paradise: sandstone landscape in the foreland of a major fault. In: Pánek T, Hradecký J (eds) Landscapes and landforms of the Czech Republic. Springer, Switzerland, pp 195–208

Migoń P (1999) Inherited landscapes of the Sudetic Foreland (SW Poland) and implications for reconstructing uplift and erosional histories of upland terrains in Central Europe. In: Smith BJ, Whalley WB, Warke PA (eds) Uplift, erosion and stability: perspectives on long-term landscape development. Geol Soc London Spec Publ 162:93–107

Migoń P (2011) Geomorphic diversity of the Sudetes—effects of global change and structure superimposed. Geogr Polon 84 2:93–105

Migoń P (2023a) The Kamienne Mountains—Late Palaeozoic volcanic terrain shaped by Late Quaternary landslides. In: Migoń P, Jancewicz K (eds) Landscapes and landforms of Poland. Springer, Cham

Migoń P (2023b) Land of Extinct Volcanoes—rock-controlled landforms, postglacial gorges and faulted margin of the Sudetes. In: Migoń P, Jancewicz K (eds) Landscapes and landforms of Poland. Springer, Cham

Migoń P, Lidmar-Bergström K (2001) Weathering mantles and their significance for geomorphological evolution of central and northern Europe since the Mesozoic. Earth Sci Rev 56:285–324

Migoń P, Placek A (2014) Litologiczno-strukturalne uwarunkowania rzeźby Sudetów. Przegl Geol 62:36–43

Migoń P, Hrádek M, Parzóch K (2002) Extreme events in the Sudetes Mountains. Their long-term geomorphic impact and possible controlling factors. Studia Geomorph Carp-Balc 36:29–49

Mosbrugger V, Utescher T, Dilcher DL (2005) Cenozoic continental climatic evolution of Central Europe. Proc Nat Acad Sci 102:14964–14969

Moskalewicz D, Woźniak PP (2023) Vistula River delta-plain—a region of fluvial, coastal, and land reclamation impact on landscape development. In: Migoń P, Jancewicz K (eds) Landscapes and landforms of Poland. Springer, Cham

Niedźwiedź T, Starkel L (2008) Klimatyczne tło współczesnych procesów morfogenetycznych na obszarze Polski. In: Kostrzewski A, Kotarba A, Krzemień K (eds) Starkel L. Współczesne przemiany rzeźby Polski, Uniwersytet Jagieloński, pp 21–35

Pánek T, Lenart J (2016) Landslide landscape of the Moravskoslezské Beskydy Mountains and their surroundings. In: Pánek T, Hradecký J (eds) Landscapes and landforms of the Czech Republic. Springer, Switzerland, pp 347–360

Pawlik Ł, Migoń P, Szymanowski M (2016) Local- and regional-scale biomorphodynamics due to tree uprooting in semi-natural and managed montane forests of the Sudetes Mountains, Central Europe. Earth Surf Proc Landf 41:1250–1265

Pilous V (2016) Krkonoše Mountains: a case study of polygenetic relief. In: Pánek T, Hradecký J (eds) Landscapes and landforms of the Czech Republic. Springer, Switzerland, pp 177–193

Plašienka D, Novotný J (2022) Limestone Klippen Belt—atypical landforms in flysch uplands. In: Lehotský M, Boltižiar M (eds) Landscapes and landforms of Slovakia. Springer, Cham, pp 189–205

Raška P, Cajz V (2016) Neovolcanic terrain of the České Středohoří Mountains. In: Pánek T, Hradecký J (eds) Landscapes and landforms of the Czech Republic. Springer, Switzerland, pp 139–152

Rączkowska Z (2021) Współczesna ewolucja rzeźby Tatr. In: Kostrzewski A, Krzemień K, Migoń P, Starkel L, Winowski M, Zwoliński Z (eds) Bogucki Wyd Nauk, Poznań, pp 61–94

Rączkowska Z, Łajczak A, Margielewski W, Święchowicz J (2012) Recent landform evolution in the Polish Carpathians. In: Lóczy D, Stankoviansky M, Kotarba A (eds) Recent landform evolution: the Carpatho-Balkan-Dinaric region. Springer, Dordrecht, pp 47–102

Rodzik J, Furtak T, Zgłobicki W (2009) The impact of snowmelt and heavy rainfall runoff on erosion rates in a gully system, Lublin Upland, Poland. Earth Surf Proc Landf 34:1938–1950

Rodzik J, Gajek G, Zgłobicki W (2023) Loess gullies of the Nałęczów Plateau—an outcome of human-environment interactions. In: Migoń P, Jancewicz K (eds) Landscapes and landforms of Poland. Springer, Cham

Rotnicka J, Borówka KR (2023) Sandy coast landforms of the Łeba Barrier, Słowiński National Park. In: Migoń P, Jancewicz K (eds) Landscapes and landforms of Poland. Springer, Cham

Różycka M, Migoń P (2017) Tectonic geomorphology of the Sudetes—review and re-appraisal. Ann Soc Geol Polon 87:275–300

Smolska E, Pochocka-Szwarc K, Szwarczewski P (2023) Suwałki Lakeland—the best example of lowland glacial geomorphology in Poland. In: Migoń P, Jancewicz K (eds) Landscapes and landforms of Poland. Springer, Cham

Starkel L (1987) The role of the inherited forms in the present-day relief of the Polish Carpathians. In: Gardiner V (ed) International geomorphology 1986, Part II. Wiley, Chichester, pp 1033–1045

Starkel L (1991) Environmental changes at the Younger Dryas-Preboreal transition and during the early Holocene: some distinctive aspects in central Europe. Holocene 1:234–242

Starkel L (2006) Geomorphic hazards in the Polish Flysch Carpathians. Studia Geomorph Carp-Balc 40:7–19

Starkel L, Gębica P (1995) Evolution of river valleys in Southern Poland during the Pleistocene-Holocene transition. Biul Perygl 34:177–190

Strzyżowski D, Fidelus J, Żelazny M (2016) Geomorphological changes within a hillslope caused by a windthrow event in the Tatra Mountains, southern Poland. Geogr Ann 98A:347–360

Szczuciński W, Włodarski W (2023) Morasko—meteorite impact craters in the glacial landscape. In: Migoń P, Jancewicz K (eds) Landscapes and landforms of Poland. Springer, Cham

Szczygieł J (2015) Cave development in an uplifting fold-and-thrust belt : case study of the Tatra Mountains, Poland. Intern J Speleol 44:341–359

Szczygieł J (2023) The Tatra Mountains—host of the deepest caves in the Carpathians. In: Migoń P, Jancewicz K (eds) Landscapes and landforms of Poland. Springer, Cham

Tomczak A (1995) Geological structure and Holocene evolution of the Polish coastal zone. In: Rotnicki K (ed) Polish coast: past, present and future; J Coast Res 22:15–31

Traczyk A (2023) Mount Ślęża—the highest inselberg in the Sudetic Foreland and its glacial-periglacial morphogenesis. In: Migoń P, Jancewicz K (eds) Landscapes and landforms of Poland. Springer, Cham

Tyc A (2023) Kraków-Częstochowa Upland—monadnocks and relic caves in the karst landscape. In: Migoń P, Jancewicz K (eds) Landscapes and landforms of Poland. Springer, Cham

Urban J (2023) Sandstone crags and blockfields in the Świętokrzyskie (Holy Cross) Mountains. In: Migoń P, Jancewicz K (eds) Landscapes and landforms of Poland. Springer, Cham

Urban J, Chwalik-Borowiec A, Kasza A, Zieliński A, Komorowski A (2023) The south-eastern Nida Basin (Ponidzie)—karst in gypsum rocks. In: Migoń P, Jancewicz K (eds) Landscapes and landforms of Poland. Springer, Cham

Vítek J (2016) Adršpach-Teplice Rocks and Broumov Cliffs—large sandstone rock cities in the Central Europe. In: Pánek T, Hradecký J (eds) Landscapes and landforms of the Czech Republic. Springer, Switzerland, pp 209–220

Wagenbreth O, Steiner W (1982) Geologische Streifzüge. Landschaft und Erdgeschichte zwischen Kap Arkona und Fichtelberg. VEB Deutscher Verlag für Grundstoffindustrie, Leipzig, p 204

Weckwerth P, Wysota W (2023) Unique landscape originated by cataclysmic glacial floods at the Weichselian glaciation decline in north-eastern Poland. In: Migoń P, Jancewicz K (eds) Landscapes and landforms of Poland. Springer, Cham

Wójcik J (2013) Mining changes on the example of the Wałbrzych Basin relief (The Sudetes, Poland). Z Geomorph NF 57:187–205

Wróblewski R, Rucińska M, Sitkiewicz P, Rudowski S (2023) Hel Peninsula—a distinctive spit on the Polish Baltic Coast. In: Migoń P, Jancewicz K (eds) Landscapes and landforms of Poland. Springer, Cham

Wysota W, Adamczyk L, Piotrowski JA, Weckwerth P (2023) Glacial curvilineations of the Dobrzyń Lakeland—spectacular landscape carved by subglacial meltwater erosion. In: Migoń P, Jancewicz K (eds) Landscapes and landforms of Poland. Springer, Cham

Zawiejska J, Wyżga B, Kukulak J, Chmielowska-Michalak D (2023) Hills, gravelly floodplains and peat bogs of the Podhale—a unique landscape mosaic between the Tatra, the Beskidy and the Pieniny Mountains. In: Migoń P, Jancewicz K (eds) Landscapes and landforms of Poland. Springer, Cham

Zgłobicki W, Poesen J, De Geeter S, Boardman J, Gawrysiak L, Golosov V, Ionita I, Niacsu L, Rodzik J, Stankoviansky M, Stolz C (2021) Sunken lanes—development and functions in landscapes. Earth Sci Rev 221:103757

Zieliński T (2023) The valley of anastomosing Narew River (NE Poland)—a unique fluvial landform of the European lowlands. In: Migoń P, Jancewicz K (eds) Landscapes and landforms of Poland. Springer, Cham

Zwoliński Z, Kostrzewski A, Winowski M, Mazurek M (2023) Wolin Island—outstanding geodiversity on the Polish coast. In: Migoń P, Jancewicz K (eds) Landscapes and landforms of Poland. Springer, Cham

Żuk L, Latocha-Wites A, Kowalczyk A (2023) Lednica Landscape Park—a geoarchaeological landscape in Central Wielkopolska. In: Migoń P, Jancewicz K (eds) Landscapes and landforms of Poland. Springer, Cham

Piotr Migoń is Professor of Geography at the University of Wrocław and former Director of the Institute of Geography and Regional Development. His main research interests are rock control in geomorphology, especially in sandstone and granite areas, mass movements in mountain terrain, geoheritage and geotourism. The Sudetes Mountains in Central Europe is his main research area, but he was also involved in projects in various other European countries (Czechia, Germany, Great Britain, Sweden, Portugal, Spain), in the United States, Mexico, Brazil, China, Jordan and Namibia. He used to be the Secretary of the International Association of Geomorphologists (1997–2001), its Vice-President (2009–2013) and member of Executive Committee (2013–2022); he was also the President of the Association of Polish Geomorphologists (2005–2008). He is the series editor of 'World Geomorphological Landscapes' series and member of editorial board of *Geomorphology* journal.

Kacper Jancewicz holds PhD degree in Earth Sciences and works as a senior specialist in the Department of Geomorphology, Institute of Geography and Regional Development, University of Wrocław, Poland. His main research interests focus on various LiDAR DEM applications in geomorphological studies, geomorphometric analyses and digital geomorphological mapping in GIS environment. His publication record includes papers in *Geomorphology*, *Catena*, *Transactions in GIS* and *Journal of Maps*.

Riverine Landscapes

5

Leon Andrzejewski, Kazimierz Krzemień and Zbigniew Zwoliński

Abstract

The chapter reviews the state of current knowledge and discussion on the evolution and typology of valley and river channel systems in Poland. Contemporary riverine landscapes reflect the overlap of multiple natural and anthropogenic factors. Amongst them, climate changes and regional morphotectonic and morphogenetic determinants are of crucial importance. Over the last decades, these landscapes have undergone substantial transformations, amongst others as a result of climate change and as a consequence of changing hydrological conditions, land-use change, as well as a result of increasing direct human interference with river channel and valley floors. This complexity is reflected in an attempt to classify different evolutionary types and subtypes of river valleys and different types of contemporary river channel patterns and floodplains.

Keywords

Fluvial system · Fluvial processes · Valley types · Channel patterns · Climate change · Human impact

L. Andrzejewski
Department of Geomorphology and Quaternary Paleogeography, Nicolaus Copernicus University, Lwowska 1, 87–100 Toruń, Poland
e-mail: leon@umk.pl

K. Krzemień
Institute of Geography and Spatial Management, Jagiellonian University, Gronostajowa 7, 30–387 Kraków, Poland
e-mail: kazimierz.krzemien@uj.edu.pl

Z. Zwoliński (✉)
Institute of Geoecology and Geoinformation, Adam Mickiewicz University, ul. Krygowskiego 10, 61–680 Poznań, Poland
e-mail: zbzw@amu.edu.pl

5.1 Introduction

Contemporary valley and channel systems in Poland reflect overlapping of multiple natural and anthropogenic conditions. In the differentiation of these systems along the longitudinal profile of country from mountain areas to coastlands, a vital role is played by events that reflected fluvial processes in various geological and historical periods (Andrzejewski and Starkel 2017). As a result, the main valleys in Poland function in morphogenetically diverse zones (Gilewska 1986; Andrzejewski and Starkel 2017). All elements of a fluvial system are described by the universal model by Schumm (1977), which includes three zones, characterised by production (supply), transfer (transport), and sediment deposition, respectively (Fig. 5.1). In the first one, the matter is supplied to the system by surface water and groundwater and includes soluble compounds and clastic and organic material. The dynamics of the supply and further transport of water and sediments was defined by Zwoliński (1986) as the river regime, which means treating the hydrological and sedimentation regimes in river channels as complementary. Intensive evolution and growth of the fluvial system usually occur in the production zone (Mazurek 2010). The second zone is dominated by fluvial transport (transfer, transit) in various types of river channels. The activity, spatial diversity and temporal variability of river channels in this zone are primarily controlled by the contemporary intensity of hydrological and slope–fluvial processes in the first (upstream) zone and, to a lesser extent, by inheritance from earlier landscape development stages such as from glacial periods. However, human impact on a local scale, which interferes with the course of fluvial processes, e.g. through the construction of hydrotechnical structures, cannot be ignored. On the other hand, in the third zone, the deposition (accumulation) of matter delivered from the first two zones, and in fact from the entire basin, dominates, forming alluvial fans or deltas. Spatial succession of these three zones determines the river

P. Migoń and K. Jancewicz (eds.), *Landscapes and Landforms of Poland*,
World Geomorphological Landscapes, https://doi.org/10.1007/978-3-031-45762-3_5

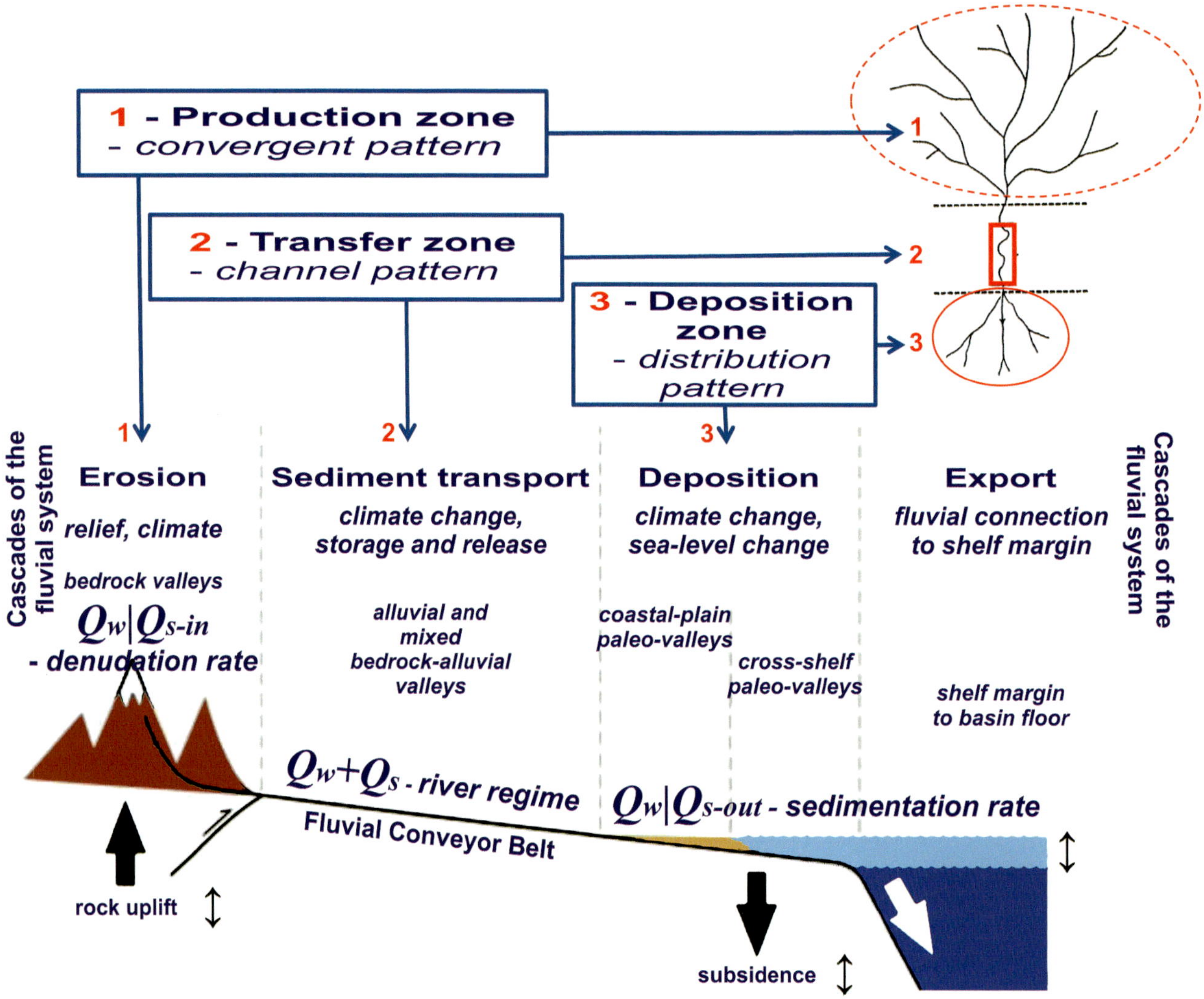

Fig. 5.1 Model of fluvial system in the concepts of Schumm (1977), Zwoliński (1986, 1989, 2007, 2016), Castelltort and Van Den Driessche (2003), Blum and Womack (2009), Tofelde et al. (2019)

system as a cascade system, i.e. a chain of subsystems, each of which has its own spatial dimension and is dynamically interconnected through the circulation of energy and matter (Zwoliński 2016).

The circulation of energy and matter within contemporary fluvial systems and subsystems is individualised and representative of Poland's morphogenetic zones, i.e. mountains, intramontane basins, uplands, old-glacial lowlands, and young-glacial lowlands. The openness of these systems means that they penetrate each other at various levels, creating a kind of continuum of energy flow and matter circulation. The present-day pattern of valley and channel systems (Fig. 5.2), regardless of the conditions resulting from the complex geological past, is to a large extent an expression of changing climatic conditions along longitudinal and latitudinal transects (Jania and Zwoliński 2011; Niedźwiedź and Starkel 2021), determining the directions and trends of changes in river regimes (hydrological and sedimentation). The consequence of such an approach to the functioning of contemporary valley systems is their inter-zone uniqueness, dependent on local and regional environmental conditions and anthropogenic transformations. On the other hand, their universalism results from the supremacy of some independent variables, such as geological structure, lithology, as well as latitudinal and altitudinal zonation of climatic conditions and their effects, e.g. in the form of varied soil and plant cover, physiographic conditions, and others.

The problems of valley systems and river channel systems lie at the interface between geology, geomorphology, sedimentology and hydrology, which is the reason for the diversity of research approaches (Fig. 5.3). The morphostatic, morphodynamic, sedimentological and hydrological approaches correspond, respectively, to the four essential elements that make up the river channel environment:

Fig. 5.2 The pattern of the river network in Poland in relation to the morphogenetic belts. Mountains: Ia—high mountains (Tatra Mts.); Ib—medium-high mountains of the Alpine system (Beskidy Mountains); Ic—medium-high mountains of the Variscan system (Sudetes). Fore-mountain areas: IIa—Sub-Carpathian basins; IIb—Sudetic Foreland. III—uplands. IV—'old-glacial relief' of central Poland. V—'young-glacial relief' of northern Poland. VI—coastal zone

forms, processes, sediment, and water. All features of the channel environment, which characterise the above four components, can be conceptualised structurally or functionally (Kaszowski and Krzemień 1999). The physiographic approach relates to the environmental conditions of the channel system, i.e. to the geographical environment of the river basin. Applying all research approaches simultaneously enables a systemic approach that most fully characterises any channel or valley system in qualitative, quantitative, spatial, or temporal terms. In relation to the research procedures, various approaches to the classification of river channels can be distinguished (Fig. 5.3). The planar and hydraulic geometries of river channels express several natural and anthropogenic conditions throughout the basin. River channels can be treated as synthetic indicators of the dynamics of the basin (Froehlich 1975, 1982; Kaszowski and Krzemień 1977; Zwoliński 1989; Krzemień 2012). They are open geomorphological systems which, along with their development trends and due to their spatial and functional structure, can be studied with various

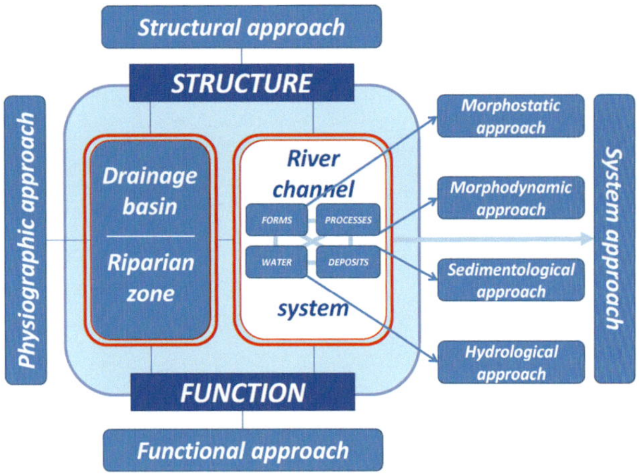

Fig. 5.3 Morphostatic, morphodynamic, sedimentological, hydrological, and system approaches to the fluvial systems (after Kaszowski and Krzemień 1999, modified)

methodological approaches and methods. Therefore, the pattern of a river channel, as well as the processes, forms, and sediments within it, is determined by geological and lithological factors, such as the geological structure of the valley bottom, the type, structure, and dynamics of the load transported by the river; hydrological factors, such as water level and velocity fluctuations, discharge volume, and intensity; and biotic factors which include aquatic vegetation, bank vegetation or wood detritus and logs transported in the river channel. Therefore, the contemporary spatial–temporal differentiation of fluvial systems results from the long-term overlapping of natural and anthropogenic processes in entire river basins and their various parts (Starkel 1991a; Andrzejewski and Krzemień 2017; Andrzejewski and Starkel 2017; Andrzejewski et al. 2018; Kostrzewski et al. 2021).

5.2 Palaeogeographic Setting

Landforms and geological structure of most river valleys in Poland usually express complex evolution resulting from overlapping endogenous and exogenous factors, including anthropogenic ones. In the morphogenetic zones of Poland, from the Carpathians and Sudetes Mts. in the south to the young-glacial landscapes in the north, these relations have undergone significant changes, especially in the Quaternary. Episodes of accelerated erosion caused by climate change occurred many times during successive continental deglaciations (Starkel 1995; Vandenberghe 1995). Many years of geological, geomorphological, and hydrological research carried out within river valleys allowed us for relatively good recognition of their evolution in this period (Kozarski and Rotnicki 1977; Kozarski et al. 1988; Florek 1997; Starkel 2001; Andrzejewski and Starkel 2017). Longitudinal profiles of valleys of such large rivers as the Vistula, Odra, Bug, Warta, or Narew, spanning various morphogenetic zones, are polygenetic and polychronic. Along the course of these rivers, in the Quaternary, the significance of changes in the erosion–accumulation base related to successive advances and recessions of Scandinavian ice sheets and their glacial lobes increased, later also influences by sea-level changes in the Baltic Sea. Simultaneously with the changes in the erosion–accumulation base, climate changes and alternations of successive drier and wetter phases were recorded with corresponding sequences of erosional dissections and accumulative insertions resulting from changing trends in fluvial processes.

Alluvial insertions and the corresponding palaeochannel generations of various ages are particularly characteristic for the mountain foreland, where they record a series of climatic changes consisting of successive wet and dry periods. These periods are recognised based on dating the levels of bogwood in the Carpathian foreland: 8.5–8.0 ka BP, 6.6–6.0 ka BP, 5.5–4.9 ka BP, 4.5–4.1 ka BP, 3.5–2.9 ka BP, 2.2 BC–200 AD, 425–625 AD, 1550–1850 AD (Kalicki 1991; Starkel 1991a; Kalicki et al. 1996). Progressive warming in the Late Weichselian and Holocene changed the water cycle and thus the hydrological regime of rivers. Progressive degradation of permafrost, deep infiltration, and encroaching forests played an important role. As a result, wash-out and deflation processes were suppressed, which was expressed, amongst others, in the reduction of river load and changes of its structure. The decreasing contribution of bedload was reflected in the gradual change of channel patterns from braided to meandering. This process was active in the warm phases of the Late Glacial and at the beginning of the Holocene (Szumański 1972; Falkowski 1975; Kozarski and Rotnicki 1977; Kozarski 1981, 1991; Starkel 1983, 2002; Antczak 1986; Turkowska 1988). Generations of palaeomeanders with different parameters of planar and hydraulic channel geometry found in many valley bottoms testify to changing discharges at that time (Rotnicki 1983; Antczak 1986; Gonera 1986; Rotnicki and Młynarczyk 1989; Andrzejewski 1994a; Starkel et al. 1996; Starkel 2002). However, the transformation from braided to meandering channels and changes in the size of palaeomeanders were not synchronous across the country (Fig. 5.4), as they occurred gradually from the Younger Dryas, through the pre-Boreal to the sub-Boreal (Starkel 2002). It is noteworthy that changing hydrological parameters during the Late Glacial and the Holocene were expressed by changing trends in the course of fluvial processes, including the tendency for increasingly intense flood-induced scouring of channel bottoms. This resulted in the lowering the erosion base level, and the tendency to increase the thickness of deposits of channel and overbank facies, following a

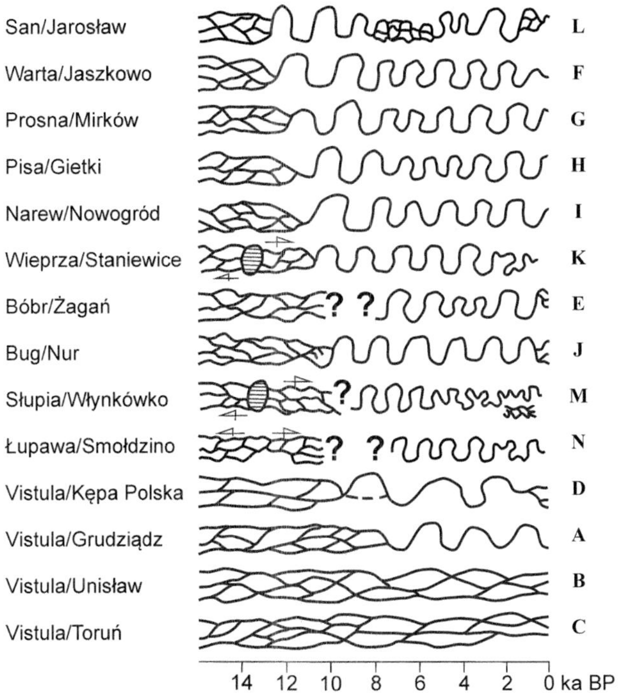

Fig. 5.4 Changes in the river patterns of selected Polish rivers during the last 16,000 years BC (after Florek 1991, modified); rivers are described with their name and the nearest town; the letters A-M indicate the location of the rivers shown on map in Fig. 5.5

change in the structure of river load and an increased tendency to lateral and scouring erosion at increasingly lower discharges during this period. This phenomenon took place not only in the larger valley bottoms, such as the Prosna (Rotnicki 1987; Rotnicki and Młynarczyk 1989), Obra rivers (Słowik 2017), but also in smaller tributaries of the Vistula, e.g. the Wetlina River (Kukulak 2015) or in the lower reaches of the Zgłowiączka, Mień, Tążyna, and Wda rivers (Andrzejewski 1994b). The narrowing of the riverbed migration belt was also significant, and thus reducing possibility of rivers meandering in their valley bottoms.

In the young-glacial landscape zone, apart from climate change and permafrost degradation, a vital role in the formation of discharge was played by intensive melting of blocks of dead glacial ice deposited in numerous terrain depressions and deep subglacial tunnels (Piasecki 1982). Most of them melted down during the Alleröd, but in some depressions, this process did not end until the pre-Boreal (Niewiarowski 1986; Błaszkiewicz 1998). Subglacial tunnels and kettle depressions were subject to gradual fluvial transformation and incorporation into valley systems of young-glacial areas (Niewiarowski 1986; Andrzejewski 1994b; Mazurek 1998). Over time, these postglacial forms acquired typical fluvial features. This trend continues in many reaches of small valleys to the present day. The phenomenon is well illustrated by some fragments of the youngest river valleys, including those of the Wierzyca, Wda, Słupia, Parsęta rivers, and other valleys of in the near-coastal region (Przymorze), as well as smaller tributaries of the Vistula, such as the Zgłowiączka, Mień, Radunia rivers, or the tributary of the Drwęca, the Wel River (Rachocki 1974; Koutaniemi and Rachocki 1981; Zwoliński 1989; Florek 1997; Andrzejewski 1994b; Błaszkiewicz 1998). In some of them, only their lower sections incised into the terraces of the Vistula River have fully fluvial features (Andrzejewski 1994b).

Other valleys, which developed on the outwash routes, including those of the Skrwa, Wkra, Drwęca, Brda, Wda, Gwda rivers, have complex staircases of erosive terraces resulting from climate change and lowering of the erosion–accumulation base as a reaction to the recession of the last Scandinavian ice sheet and the low level of the Baltic Sea in the Late Glacial and at the beginning of the Holocene (Galon 1968; Niewiarowski 1987; Weckwerth 2014). Therefore, from the decline of the upper pleni-Weichselian, through the entire Late Weichselian, increased erosion continued in the lower Vistula and Odra valleys. It can be assumed that this process was most intensive in the earlier part of the Late Glacial. During that time, lasting about 2500 years, the lower sections of the above valleys and their main tributaries were deepened by approximately 40–50 m. Their bottoms in the form of floodplains and river terraces were shaped already in the Late Glacial. In this process, the role of glacitectonics modifying the main valley slopes, and vertical tectonic movements affecting the shape of longitudinal profiles of the valleys and hence, the intensity of erosion and deposition processes within various parts of the valleys, should also be taken into account (Krygowski 1962; Galon 1968; Mojski and Starkel 1990).

Similarly, in the middle courses of main river valleys, climate cycles were expressed by several alluvial levels and palaeochannel generations with diverse geometric parameters. As a result, different evolutionary types are recorded in different regions and thus in different places along the valley longitudinal profiles (Falkowski 1975; Kozarski and Rotnicki 1977; Florek 1997; Starkel 1991a, b, 2001; Kozarski 1983; Antczak 1986; Gonera 1986; Kozarski et al. 1988; Wyżga 1993; Andrzejewski 1994b; Gębica 2013; Krupa 2013; Michno 2013).

A specific element of the lowland valley systems is the latitudinal sections of wide ice-marginal valleys inherited from the Pleistocene with flat and non-rejuvenated bottoms (Kasprzak et al. 2023), which in many parts show a constant tendency to biogenic aggradation (Florek 1991; Andrzejewski 1994b; Kordowski et al. 2014; Żurek 1975). The progressive, gradual recession of the last Scandinavian ice sheet is expressed in water gap sections of lowland river valleys following the meridional direction, evident along rivers such as the Vistula, Odra, and Warta (Kozarski 1965;

Galon 1968). Amongst the inherited ice-marginal valleys, the anastomosing section of the Narew River is distinctive. Due to the deficiency of bed and suspended load, it has been built up over the last millennia by biogenic sedimentation (Gradziński et al. 2000). The record of river valley formation in the Holocene can also be read from the sediments in the fossil valleys of the Przymorze rivers, now to be found in the littoral of the Baltic Sea.

Transit river valleys, flowing from mountainous areas through uplands and lowlands to coastal areas, show a specific evolutionary type (Łajczak et al. 2006, 2008; Ciszewski and Czajka 2021). Within large valleys with a clear development of diverse alluvial facies, there are transit river valleys whose hydrological regime is primarily shaped in the mountain or upland areas. This applies particularly to the Vistula and Odra valleys, and to lesser extent to the Warta or Bug rivers. Visible erosional dissections and alluvial insertions record wetter and drier phases and corresponding palaeochannels in their upper and lower sections. In some of these sections, overbank deposits of a considerable thickness up to 3–5 m occur (Czajka 2007). On the other hand, in the valley basins, e.g. in the Oświęcim Basin, the Płock Basin, Toruń Basin, or the Unisław Basin, valley floors are dissected by deep, relatively narrow channels with patterns similar to anastomosing rivers (Florek et al. 1987; Niewiarowski 1987; Tomczak 1987; Wiśniewski 1987). Amongst the valleys shaped mainly by the lowland hydrological regime, there are examples such as the lower reaches of the Vistula, Warta, Prosna, Bug, and Narew rivers, and the lower reaches of the Bzura and Pilica. Dating of dissections and insertions in the valleys indicate their synchronicity with the valleys in the mountain forelands in southern Poland (Falkowski 1975; Florek et al. 1987; Turkowska 1988; Rotnicki and Młynarczyk 1989; Kozarski 1991).

These natural changes in river valleys are superimposed on regulatory transformations caused by human activity, not always positive (Babiński 2002; Gorczyca 2016). Nowadays, in the last dozen or so years, many of the regulated river sections are subject to renaturalisation, both natural and human-controlled (Żelazo and Popek 2002; Gorczyca 2016; Gorczyca et al. 2017).

5.3 Regional Typology of River Channel Systems

Contemporary types of river channel systems in Poland reflect trans-regional (e.g. Vistula, Odra), regional (e.g. Bug, Warta, Bzura, Pilica, Narew), or local (tributaries of the Vistula and Odra, the Przymorze rivers) environmental conditions and their variability over time. As a result of high diversity of the natural environment and protracted human activity in Poland, twelve types of river channels can be distinguished within nine macroregions, and seven types for transit rivers, flowing across all morphogenetic zones (Fig. 5.5). This typology uses a physiographic approach (Kaszowski and Krzemień 1999), and the following types are distinguished (Andrzejewski and Krzemień 2017; Andrzejewski et al. 2018):

- channels in mountain and upland areas (five types): 1—high-mountain channels, 2—mid-altitude mountain channels (eastern Beskids, western Beskids, Sudetes), 3—channels in intramontane basins, 4—channels in foothill basins, 5—upland channels (semi-natural, regulated);
- channels in the old-glacial area (two types of winding or meandering channels): 1—autochthonous channels (anastomosing channels, marginal and proglacial valley channels), 2—allochthonous channels (excluding transit ones);
- channels in the young-glacial area (five types, most often winding or meandering channels): 1—channels of the southern slope of the Pomeranian phase moraine ridge (channels with flow-through lakes and across vast alluvial plains), 2—channels of the northern slope of the Pomeranian phase moraine ridge (channels of water gap valleys and in the bottoms of kettle depressions), 3—channels within latitudinal marginal valleys or ice-marginal valleys, 4—channels in water gap valleys with a meridional or similar system, 5—tributary channels of large transit rivers (channels within the terraces of transit river valleys, plateaus, outwash plains and water gap valleys, and glacial depressions);
- channels of large transit rivers (Vistula, Odra) (seven types): A. Vistula River: 1—reach to the mouth of the San River, 2—reach across the upland belt, 3—the Masovian section, 4—the Kujawsko-Pomorskie section (the lower Vistula River); B. Odra River: 1—reach to the mouth of the Nysa Kłodzka River, 2—middle reach to Słubice, 3—the lower Odra River.

5.4 Contemporary Facets of River Channel Systems

5.4.1 Channels in Mountainous Areas

The mountain river channels are highly diversified and occur within the mountain ranges of the Carpathians and the Sudetes (Figs. 5.6 and 5.7). In these areas, aggradation in the bottoms of large river valleys was replaced about 100 years ago (Starkel and Łajczak 2008), and in the Sudetes even 150 years ago (Latocha 2007), by gradual incision and narrowing of the channels (Gorajska 2014). During this

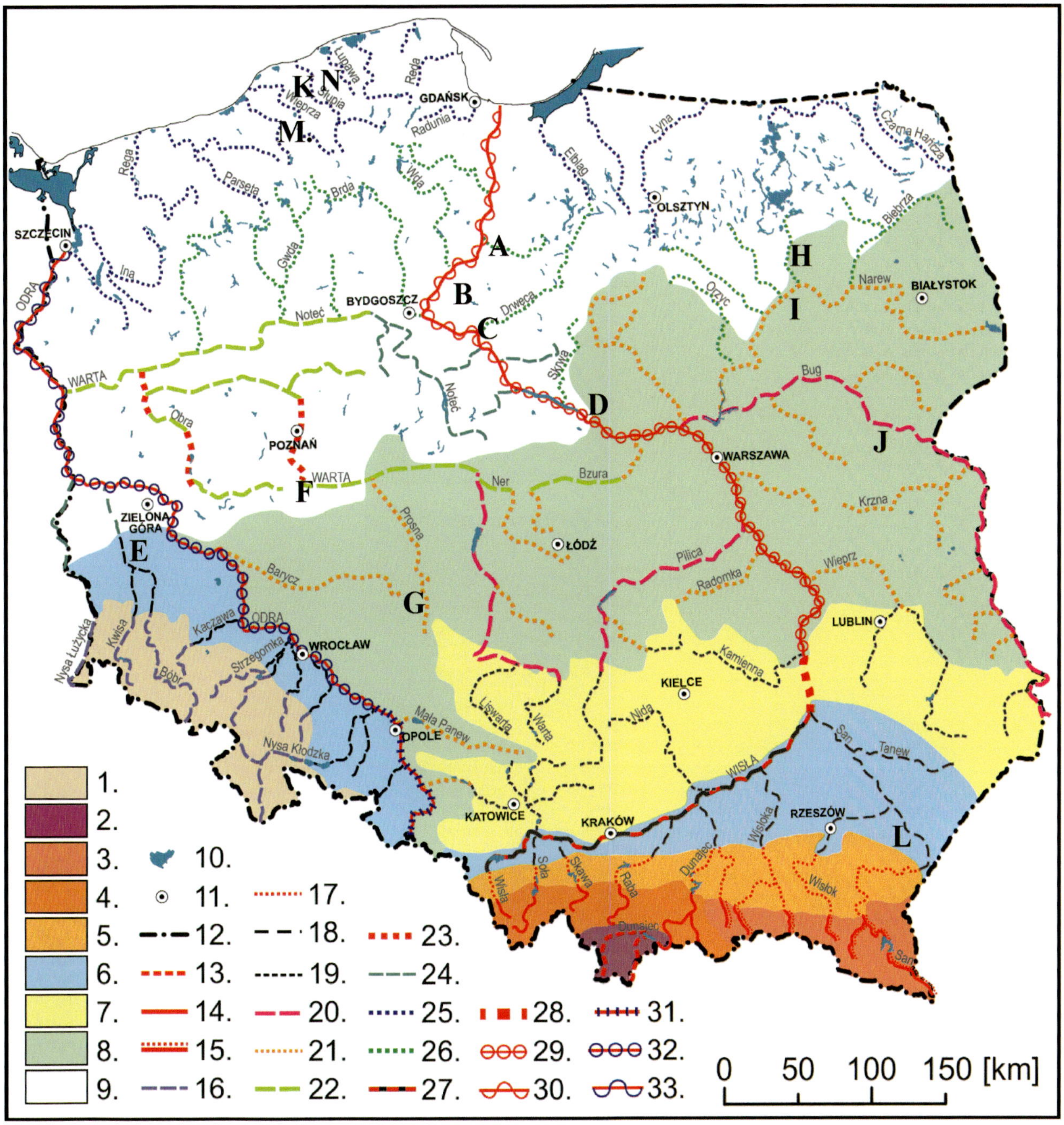

Fig. 5.5 Spatial distribution of river channels in Poland (after Andrzejewski and Krzemień 2017; Andrzejewski et al. 2018, modified); the letters A-M indicate the location of the rivers shown in Fig. 5.4. 1—the Sudetes Mountains and Sudetic Foreland, 2—the Tatras with Podhale, 3—eastern area of the Beskids, 4—western area of the Beskids, 5—the Carpathian Foothills, 6—fore-mountain basins and foreland areas, 7—uplands, 8—old-glacial area, 9—young-glacial area, 10—lakes and reservoirs, 11—main cities, 12—state border, 13—high-mountain river channels, 14—mid-altitude mountain channels in the Western Beskids, 15—mid-altitude mountain channels in the Eastern Beskids, 16—mid-altitude mountain channels in Sudetes, 17—foothill river channels (Carpathian Foothills), 18—intramontane basins channels, 19—upland channels, 20—allochthonous winding or meandering river channels (excluding transit ones), 21—autochthonous winding or meandering river channels, 22—channels formed within latitudinal ice-marginal valleys (mainly in young-glacial areas and partly in old-glacial areas), 23—channels in water gap valleys with meridional or similar orientation, 24—channels of tributaries of large transit rivers (the Vistula and the Odra), 25—channels of the northern slope of the Pomeranian phase moraine ridge, 26—channels of the southern slope of the Pomeranian phase moraine ridge, 27—the upper Vistula River channel to the mouth of the San River, 28—the Vistula River channel in the upland belt, 29—the Vistula River channel in the Mazovian section, 30—the lower Vistula River channel in the Kujawsko-Pomorskie section, 31—the upper Odra River channel to the mouth of the Nysa Kłodzka River, 32—the middle Odra River channel to the mouth of the Warta River, 33—the lower Odra River channel

Fig. 5.6 High-mountain and foreland rivers. **a** Channel of Starorobociański Potok in the Western Tatras incised into moraines of mountain glaciation, with coarse boulder bed. **b** Alluvial channel of Czarny Dunajec in the foreground of the Western Tatras after summer flood. **c** Bedrock channel of Czarny Dunajec in the area of the Gubałowskie Foothills developed from a former alluvial channel. Incision occurred after river regulation and excessive exploitation of in-channel gravels and sands (photographs by K. Krzemień)

Fig. 5.7 Examples of fluvial morphology in medium-altitude mountains. **a** Bedrock channel of the San River near Zatwarnica in the Bieszczady Mountains with longitudinal rock steps, cut in flysch rocks. **b** Alluvial channel of the Wisłoka River near Świątkowa in the Low Beskids. The right bank of the channel is undercut, thanks to which the central bar is enlarged. **c** Alluvial channel of the Białka River in the gorge through the Pieniny Rocks. Below the Białka gorge, channel migration caused inundation of a village. **d** Bedrock channel of the Konina Stream in Gorce Mts. The channel is cut along a fissure in flysch. **e** Bedrock channel of the Skawa River near Maków Podhalański in the Makowski Beskids. The channel is located in a section previously occupied by an alluvial channel regulated in the 1970s. **f** The Szklarka Waterfall in Karkonosze Mountains (photographs **a**, **b**, **d**, **e** by K. Krzemień; **c** by B. Pawłowski; **f** by Z. Zwoliński)

period, intense erosion occurred along the main rivers and then in their tributaries, mainly due to regulatory works, construction of reservoirs of various sizes, sediment exploitation, and reduction of sediment transport due to land-use changes. The emergence of bedrock exposures in many previously alluvial channels in the last 20–30 years indicates a return to the pre-deforestation trend (Starkel 1960). It results from uncontrolled exploitation of riverbed gravel, channel regulation, and changes in land use (Gorczyca et al. 2017). This erosional process was large scale, mainly in the Carpathians (Łajczak et al. 2014). In effect, in the lower and middle reaches of the Carpathian tributaries of the Vistula River, the channels were deepened up to 2–4 m (Klimek 1987; Krzemień 2003; Starkel and Łajczak 2008; Zawiejska and Wyżga 2010; Gorczyca 2016; Gorczyca et al. 2017). Channel deepening is also evident in the tributary valleys due to headward erosion. Such a situation leads to unfavourable changes in the natural environment in the valley bottoms, related to excessive desiccation of many areas. As a result of these changes, in many sections, gravel-bed channels have turned into rocky channels and braided patterns were replaced by wandering rivers, cut into solid bedrock (Starkel and Łajczak 2008; Gorczyca et al. 2017). Under the conditions of decreasing load, the rivers in the Sudetes also tend to incise their former floors and develop a meandering pattern (Kostrzewski 1970; Teisseyre 1977; Klimek and Latocha 2007).

Nowadays, three types of dynamic longitudinal profiles of rivers and floodplains can be observed in the Carpathians (Łajczak et al. 2006; Starkel and Łajczak 2008; Andrzejewski and Krzemień 2017). The first type begins in high mountains (Fig. 5.6) and medium-altitude mountain massifs (Fig. 5.7). Until recently, a tendency to braiding and a significant contribution of bed load have been characteristic, but now many of these rivers are partitioned by retention reservoirs. The second type occurs in low mountains and foothills (Fig. 5.8). Channels tend to deepen over most of the profile length and show a dominant share of suspension load, related to limited sediment supply from the catchment due to changes in land use. The third type concerns foothill rivers (Fig. 5.9), with frequent floods (in spring and summer) and ongoing accumulation on floodplains (Klimek and Starkel 1974; Łajczak 1995, 1999).

In the Sudetes, apart from natural conditions, hydrotechnical structures have a significant impact on the functioning of fluvial systems (Witek 2012). In this area, bed

Fig. 5.8 Mixed bedrock-alluvial channel of the Ropa River below Wysowa in the Low Beskids (photograph by K. Krzemień)

Fig. 5.9 Regulated channels. **a** Raba River near Gdów (Wielickie Foothills). **b** Gravelly, regulated channel of the Vistula River west of Kraków in the area of Czernichów (photographs by K. Krzemień)

load has a significant share in the total river transport during extreme events, with corresponding systems of erosive and accumulative flood forms (Kostrzewski 1970; Teisseyre 1977; Zieliński 2003; Łach 2007; Kasprzak 2008; Pawlik 2010). Remedial actions taken immediately after floods, consisting in blocking the entrances to the flood chute channels, backfilling of flood channels, and strengthening of the undercut banks, have aimed to restore the state from before the flood, with one stable winding channel.

In the historical period, meandering channel patterns developed in the valleys of the Sudetes in their gorge sections, whilst in the intramontane basins, anastomosing patterns prevailed (Teisseyre 1991). These natural channel patterns were largely destroyed as a result of river regulation, mainly at the beginning of the twentieth century. So far, only traces of the previous channels have been preserved. Short sections of anastomosing channels have survived only in the foreland of the Sudetes (the lower course of the Oława River and the Odra River in the Wrocław reach).

Within the valleys and foothills of the Carpathians, in natural conditions, meandering systems typified by low slope and usually low width in relation to depth had developed, but now these channels are mostly regulated. On the one hand, such a topographic situation favours deepening of river channels, but on the other hand, during floods, it favours intensive processes of floodplain build-up (Łajczak 1995) or overbank sedimentation between the embankments (Łajczak 1999; Czajka 2007).

5.4.2 Channels in Upland Areas

Natural, semi-natural, and artificial river channels (or their reaches) can be distinguished in the upland belt (Fig. 5.10). Artificial channels differ in the degree of anthropogenic transformation, from a straight course as a result of shortening of bends, through the construction of artificial and straight sections of the channels, to the formation of winding channels with complex geometric parameters. Bottom aggradation dominates in wide alluvial valleys, shaped by unconstrained (natural) lateral migration (Michno 2013). Locally, rivers tend to develop meandering patterns (Biała Nida, Biała Przemsza, Wieprz rivers) or form multi-channel systems (Harasimiuk 1991). Within narrowings and across bedrock-controlled structural thresholds in the valley bottoms, lateral migration of channels is limited and bedrock structure often controls the shape of meanders. In these sections, well-formed overflow channels (flood and chute channels) are typical. When floods exceed bankfull discharges, the entire floodplain becomes inundated. Along small rivers in the central and western part of the upland belt, partitioning of river channels and valley bottoms by damming structures such as weirs, dykes, and dammed reservoirs with energy production occurs. This can often contribute to the emergence of anastomosing river patterns (Ludwikowska-Kędzia 2000; Łajczak 2004). Channels regulated in the eighteenth and nineteenth centuries, reinforced with fascine, in which no hydrotechnical works have been carried out recently, are subject to renaturalisation and gradually return to their natural geometric parameters as semi-natural channels. Channels in urban areas, whose geometric parameters have been fixed by paving the bottom and banks by concrete, remain stable today (Ciupa 2001).

5.4.3 Channels in the Old-Glacial Belt

The old-glacial zone stretches between the boundary of the last glacial maximum extent (LGM) of the Weichselian glaciation in the north and the Sudetes and the upland belt in the south, forming the southern part of the Polish Lowland. Its relief was shaped during older ice ages, i.e. the Odranian and Wartanian, from about 255–260,000 BP. During the last ice age, this area was subject to intense periglacial transformations, manifested, amongst others, by changes in the pattern of river network (Turkowska 1988). This zone is dominated by autochthonous rivers in naturally homogeneous regions (Klimek 2008), such as drainage basins of the Barycz, Bzura, Narew, Prosna rivers (Fig. 5.11). Next to them, there occur allochthonous rivers, such as the Bug, Pilica, Warta, and Wieprz rivers, with the upper reaches in the neighbouring upland regions (Fig. 5.12). Only the left-bank tributaries of the Odra (including the Nysa Łużycka, Bóbr, Kaczawa, Nysa Kłodzka, and Osłoboga rivers) are classified as foothill rivers.

The dynamics of contemporary fluvial processes in this zone are expressed by the channel patterns, their horizontal stability, and the magnitude of bed and suspended loads (Kaniecki 1976; Młynarczyk and Rotnicki 1989). Winding and meandering channels dominate (Forysiak 2005), whilst in wide basins, anastomosing reaches may be present (Gradziński et al. 2000). Braided channels occur mainly in the transit valley of the middle Vistula River (Falkowski 2006) and in the lower reaches of some of its tributaries. Many river reaches, especially in urban and residential areas or in intensively managed zones, have been embanked, and their natural course has been significantly changed.

5.4.4 Channels in the Young-Glacial Belt

The river network in the young-glacial area is genetically related to the geomorphological and hydrogeological conditions dating back to the LGM period. The polygenetic nature of most valleys (sections of former glacial tunnels, ice-marginal valleys, water gaps, outwash trails, sections linking former kettle forms and within moraine plateaus) determine the geodiversity of contemporary river channels, also within individual rivers (Rachocki 1974; Piasecki 1982; Zwoliński 1989; Smolska 1996; Florek 1997; Mazurek 1998). The modern river channels and floodplains are inherited mainly from the Late Weichselian, but they also record Holocene transformation of the fluvial

Fig. 5.10 Artificially modified upland rivers. **a** Seminatural, straight section of the Pilica River below Pukarzów (Google, Inc.). **b** Artificial channel of the Przemsza River in Będzin (Google, Inc.)

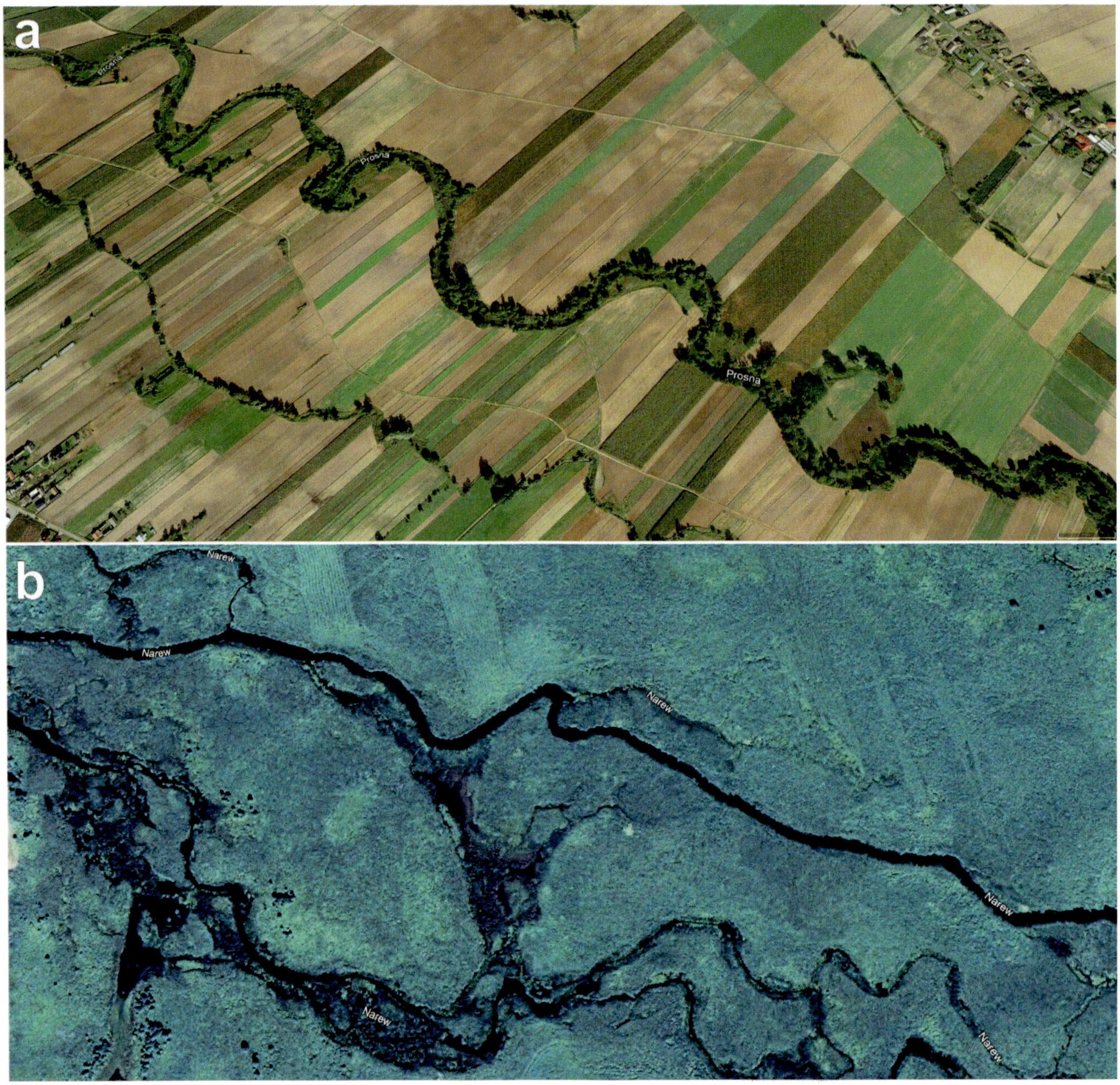

Fig. 5.11 Autochthonous rivers in the old-glacial belt. **a** Prosna River near Wielowieś (Google, Inc.). **b** Anastomosing Narew River near Waniewo (Google, Inc.)

system (Sylwestrzak 1978; Koutaniemi and Rachocki 1981; Piasecki 1982). It should be emphasised that the complex process of merging postglacial depressions and basins into valley networks and the formation of river networks through the development of channel heads have not been completed in the young-glacial areas (Mazurek 2010). Contemporary fluvial processes in river channels and within floodplains, including erosion, transport, and accumulation, are diverse and show different types and rates of river channel transformations, in response to the frequency of above-average events, mainly spring and early summer floods (Zwoliński 1985; Kostrzewski et al. 2021). The present-day river network of this zone includes various outflow directions resulting mainly from the morphogenesis of the relief of the Polish Lowlands (Kostrzewski et al. 2021). As a result, five main types of channels can be distinguished: (a, b) within the southern and northern slope of the Pomeranian phase moraine ridge; (c) straight or winding channels within latitudinal marginal valleys or ice-marginal valleys; (d) gap channels in the meridional or similar valleys; as well as (e) tributary channels of large transit rivers (Fig. 5.5).

The river channels of the southern slope of the Pomeranian phase moraine ridge (Drawa, Gwda, Brda,

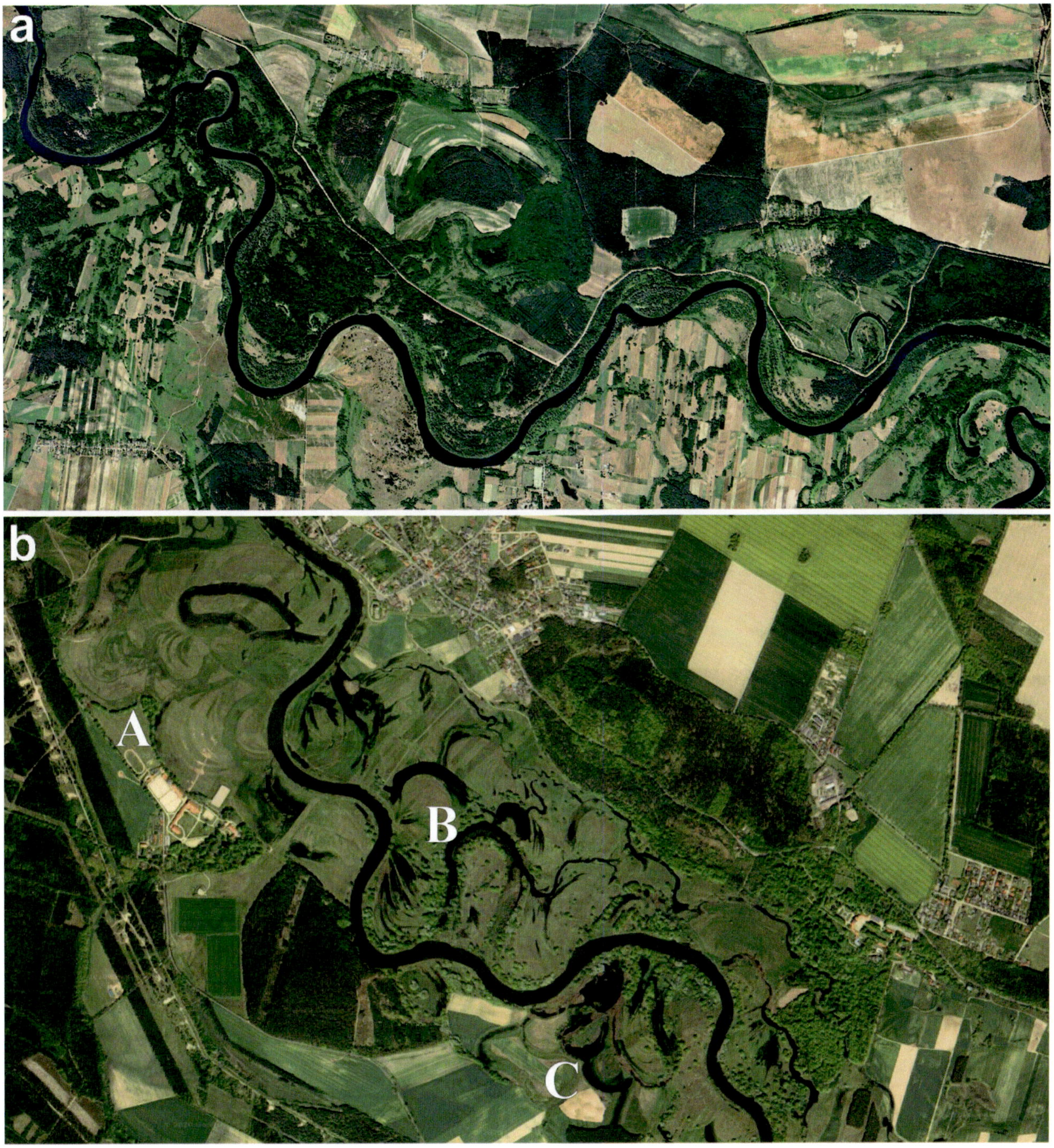

Fig. 5.12 Allochthonous rivers in the Polish lowlands. **a** Bug River within a flat old-glacial area near Zaczopki (Google, Inc.). **b** Meandering Warta River near Poznań within the Warsaw-Berlin Pradolina in a young-glacial area. The image shows three generations of meanders with different geometries and ages from Preboreal (A) to Subatlantic (C) (Google, Inc.)

Wda, Wierzyca, Drwęca, Skrwa, Biebrza rivers; Fig. 5.13) flow approximately longitudinally in the valleys located most often on outwash routes, usually meandering within floodplains, locally including a chain of flow-through lakes. The channels of the northern slope of the Pomeranian phase moraine ridge (Ina, Rega, Parsęta, Grabowa, Wieprza, Słupia, Łupawa, Reda, Radunia, Pasłęka, Łyna rivers; Fig. 5.14) have uneven longitudinal profiles (in the upper reaches slopes of up to 6.3% are found), most often meandering and using (a) gap and kettle melt-out basin sections (roughly meridional), and (b) fragments of indistinct marginal valleys or ice-marginal valleys (approximately latitudinal) (Kostrzewski et al. 2021). The channels of these rivers have evolved in response to changes in the erosional

Fig. 5.13 The bottom of the meandering Drwęca valley in its middle section within the southern slope of the Pomeranian phase moraine ridge (photograph by B. Pawłowski)

base level of the Baltic Sea (Florek 1997). The tributary channels (Ina, Wierzyca, Drwęca rivers; Fig. 5.15) of large transit rivers (e.g. the Vistula and Odra; Fig. 5.16) show a pattern that is often different in the upstream reaches, where they are winding, from the fragments incised into the terraces of the main valleys, where rivers meander within vast floodplains (Andrzejewski 1994b; Błaszkiewicz 1998).

5.5 Selected Impacts on River Channel Systems

The observed increase in the frequency of extreme meteorological, hydrological, and geomorphological events such as torrential precipitation, rapid snowmelt, and high floods causes the shaping of valley bottoms in all morphogenetic zones to be more and more intense today (Jania and Zwoliński 2011). It includes not only changes in the planar patterns of river channels caused by flood discharges, as observed in the middle Vistula reaches (Łajczak et al. 2006) or many mountain and upland rivers, but mainly vertical changes:

- incision into bedrock, as mainly in the case of mountain rivers; or
- intensive reworking of in-channel alluvia of upland and lowland rivers; or
- accretion on floodplain terraces, sometimes also due to breaching of flood embankments.

During these events, the threshold values to trigger some fluvial processes are exceeded and those with a high morphogenetic capacity are initiated (Święchowicz et al. 2021). Frequent crossings of the threshold values of these processes within a short time lead to the change in the direction of evolution of fluvial systems in different regions.

Quickly changing erosion and accumulation processes on floodplains contribute to frequent morphological changes, sometimes from – 2 to + 1 m. This makes the forecasting of flood hazard and risk much more difficult, especially in urbanised or developed areas, e.g. within polders (Warachowska and Zwoliński 2023). Deposition of overbank sandy and silty sediments in terrace depressions (flood basins) frequently dominates amongst fluvial processes in the valley bottoms of upland and lowland rivers, which,

Fig. 5.14 Parsęta River in the young-glacial zone. **a** The river near Dębczyno on the northern slope of the Pomeranian phase moraine ridge. **b** Bank erosion of muddy terrace deposits near Dębczyno. **c** Effects of bank erosion in the bend of the Parsęta River near Bardy (2022). In the middle part of the channel, evidence of previous position of erosional bank in 1970s can be seen (photographs **a**, **c** by M. Winowski; **b** by Z. Zwoliński)

with increasing human impact on the valley bottoms, contributes to more and more frequent changes in the types of land cover and land use on floodplains (Zwoliński 1992; Szmańda 2011).

Many sections of Polish river channels are regulated using hydrotechnical structures located longitudinally and/or transversely to the channel, and the channels of large rivers are additionally lined by embankments. As a result of human interference in the last 50–70 years, bed incision in the rivers of the Carpathians and their forelands has been significantly enhanced. Clastic material derived from these channels is nowadays deposited between the embankments

Fig. 5.15 Palaeomeanders within the vast floodplain of the Drwęca River in its lower section (photograph by B. Pawłowski)

of the Vistula River, especially in the Sandomierz Basin near the town of Sandomierz (Gębica 2004). Significant changes in river channel patterns, and thus changes in the morphological and sedimentological expression of valley bottoms, result from the construction of water mills, various types of barrages and dams, communication junctions and highways, and well as concreting and embanking of river channels. In the case of water mills, the occurrence of which shaped the bottoms of river valleys, it should be noted that the regulatory works related to their existence have lasted for a very long time, practically from the medieval times to the twentieth century (Kaniecki 1999; Podgórski 2004; Korpak 2010; Krupa 2013; Gorczyca 2016; Zwoliński et al. 2018). The impact of barrages and dams on the morphology of valley bottoms, illustrated primarily by erosion processes below these structures, has been often documented in geomorphological studies (Babiński 1992; Korpak 2010; Kaczmarek 2018), but reservoir siltation and the development of channel systems are also an essential issue in these valley stretches (Froehlich and Klimek 1979; Gierszewski 2018).

In many sections of natural and previously regulated channels, spontaneous renaturalisation of rivers takes place due to the lack or decreasing human impact, e.g. the disappearance of water mills or settlement withdrawal, abandonment of arable fields, succession of forest communities in valleys. This process is also accelerated thanks to the activity of beavers (Kobojek 2013; Szpikowski and Szpikowska 2018) and a growing understanding of the need to protect riverine landscapes and ecosystems.

5.6 Final Remarks

The diversity of valley and channel systems in Poland is determined by regional and local factors, expressed in the alternation of morphotectonic and morphogenetic zones, as well as by climatic and hydrological interactions. Regardless of the evolutionary differentiation of individual valley sections, fluvial systems on the river basin scale function as complex, interconnected structures of water and sediment cycles. River channels in the Polish territory depend in the relief inherited from the past. Their primary differentiation relates to the country's latitudinal hypsometric and morphogenetic pattern.

Rivers in mountain and upland areas are now subject to strong impact of extreme natural phenomena and intense human activity in some areas, predominantly in urban settings. By contrast, they are subject to relatively rapid renaturalisation in other areas, mainly in areas abandoned by people. Such strong anthropogenic impulses on the slope-river systems cause evident changes in the functioning of

Fig. 5.16 The largest rivers of Poland. **a** Braided reach with moving alluvial bars in the middle part of the transit Vistula River valley near Kępa Polska (photograph by B. Pawłowski). **b** Multi-channel pattern of the lower Odra River Valley with palaeomeanders and Yazoo type channel within the vast floodplain near the city of Szczecin (Google Inc.)

these systems. Increased erosion, transport, and accumulation in river valleys contribute to changes in the river channel patterns, including changes in their geometry and lateral migration rates, and consequently influence changes in the management strategies in valley areas.

In the old-glacial zone, tectonic stability and lithological diversity of Quaternary deposits, as well as relatively long timescale of fluvial system development, made the longitudinal profiles of the valley bottoms adapted to local and regional erosional base levels. The rivers of this zone usually incise loose Quaternary formations, which means that contemporary alluvial channels are mainly of sand-bed or, less frequently, gravel-bed type, with both aggradation and degradation tendencies. Comparing the diversity of the current channel systems in this zone to the young-glacial zone and the zone of mountains and uplands, it can be stated that it is least pronounced, in spite of their varied patterns that include winding, meandering or even locally anastomosing and braided systems.

In the valleys of the young-glacial zone, as in the mountains and upland areas, overlapping of two opposing tendencies is observed, namely the increasingly deeper reworking of channel alluvia, mainly during high floods, and the build-up of floodplains. In the valley bottoms with low slope (ice-marginal valleys), sedimentation processes induced by floods, including sedentation of peat deposits, occur. In the channels of rivers flowing through biogenic plains, mainly peat ones, filling various depressions (glacial tunnels, kettle depressions), biogenic deposits are gradually replaced with alluvial deposits.

The knowledge about complex natural and anthropogenic conditions shaping ancient and contemporary fluvial systems is of fundamental importance for forecasting their changes in the future. These issues take on a unique role in context of the observed climate changes and increasing human pressure. These changes in the Polish fluvial systems indicate how sensitive river channels are to global, regional and local climate changes and changes in land cover and land use. A systemic approach to valley evolution and recent changes in river channels should be the basis for proper flood risk management, and, on the other hand, for the increasingly widely implemented concept of renaturalisation and revitalisation of waterways.

References

Andrzejewski L (1994a) Paleohydrologiczna interpretacja zmian geometrii paleokoryt wybranych dopływów dolnej Wisły. Acta Univ Nicolai Copernici Geogr 27(98):121–129

Andrzejewski L (1994b) Ewolucja system fluwialnego doliny dolnej Wisły w późnym vistulianie i holocenie na podstawie wybranych dolin jej dopływów. Rozprawy UMK, Toruń, p 112

Andrzejewski L, Krzemień K (2017) Doliny i koryta rzek. In: Jokiel P, Pociask-Karteczka J, Marszelewski W (eds) Hydrologia Polski. PWN, Warszawa, pp 121–127

Andrzejewski L, Krzemień K, Zwoliński Z (2018) Zarys przyrodniczych i antropogenicznych uwarunkowań rozwoju systemów dolinnych i korytowych w Polsce. Landf Anal 37:17–51

Andrzejewski L, Starkel L (2017) Ewolucja systemów dolinnych i zmiany w obiegu wody od ustąpienia ostatniego zlodowacenia. In: Jokiel P, Pociask-Karteczka J, Marszelewski W (eds) Hydrologia Polski. PWN, Warszawa, pp 111–116

Antczak B (1986) Transformacja układu koryta i zanik bifurkacji Warty w Pradolinie Warszawsko-Berlińskiej i południowej części przełomu poznańskiego podczas vistulianu. Wyd Nauk UAM, Poznań, Geografia 35:111

Babiński Z (1992) Współczesne procesy korytowe dolnej Wisły. Prace Geogr IGiPZ PAN 157:171

Babiński Z (2002) Wpływ zapór na procesy korytowe rzek aluwialnych ze szczególnym uwzględnieniem stopnia wodnego „Włocławek". Wyd Akad Bydg, Bydgoszcz, p 185

Błaszkiewicz M (1998) Dolina Wierzycy, jej geneza oraz rozwój w późnym plejstocenie i wczesnym holocenie. Dokument Geogr IGiPZ PAN 10:116

Blum MD, Womack JH (2009) Climate change, sea-level change, and fluvial sediment supply to deepwater systems. In: Kneller B, Martinsen OJ, McCaffrey B (eds) External controls on deep water depositional systems: climate, sea-level, and sediment flux, vol 92. SEPM Special Publication, pp 15–39

Castelltort S, Van Den Driessche J (2003) How plausible are highfrequency sediment supply-driven cycles in the stratigraphic record? Sediment Geol 157:3–13

Ciszewski D, Czajka A (2021) Współczesne przemiany koryta i równiny zalewowej Odry. In: Kostrzewski A, Krzemień K, Migoń P, Starkel L, Winowski M, Zwoliński Z (eds) Współczesne przemiany rzeźby Polski, 2nd edn. Bogucki Wyd Nauk, Poznań, pp 683–716

Ciupa T (2001) Funkcjonowanie koryt rzecznych Silnicy i Sufragańca w strefie miejskiej Kielc. In: Karczewski A, Zwoliński Z (eds) Funkcjonowanie geoekosystemów w zróżnicowanych warunkach morfoklimatycznych. Monitoring, ochrona, edukacja. Stowarzyszenie Geomorfologów Polskich, Bogucki Wyd Nauk, Poznań, pp 103–113

Czajka A (2007) Środowisko sedymentacji osadów przykorytowych rzek uregulowanych na przykładzie górnej Odry i górnej Wisły. Prace Nauk Uniw Śląskiego, Katowice 2534:99

Falkowski E (1975) Variability of channel processes of lowland rivers In Poland and changes of the valley floors during the Holocene. Biul Geol UW 19:45–78

Falkowski T (2006) Naturalne czynniki stabilizujące wybrane odcinki strefy korytowej Wisły środkowej. Rozpr Nauk i Monografie SGGW, Warszawa, p 128

Florek E, Florek W, Mycielska-Dowgiałło E (1987) Morphogenesis of the Vistula Valley between Kępa Polska and Plock in the Late Glacial and Holocene. In: Starkel L (ed) Evolution of the Vistula river valley during the last 15,000 years, part II. Geographical Studies, SI 4, pp 189–205

Florek W (1991) Postglacjalny rozwój dolin rzek środkowej części północnego skłonu Pomorza. Prace WSP, Słupsk 174:238

Florek W (1997) Climatic and anthropogenic impulses in the Late Vistulian and Holocene development of the river channels and valleys of the Baltic Coastal Region and Pomerania. Landf Anal 1:41–50

Forysiak J (2005) Rozwój doliny Warty między Burzeninem i Dobrowem po zlodowaceniu Warty. Acta Geogr Lodz 90:116

Froehlich W (1975) Dynamika transportu fluwialnego Kamienicy Nawojowskiej. Prace Geogr IGiPZ PAN 114:122

Froehlich W (1982) Mechanizm transportu fluwialnego i dostawy zwietrzelin do koryta w górskiej zlewni fliszowej. Prace Geogr IGiPZ PAN 143:144

Froehlich W, Klimek K (1979) Debris sedimentation in overflowing mountain streamlets upstream drop dams. Excursion Guide-book, field meeting of the IGU Commission on field experiments in geomorphology, Poland, pp 135–141

Galon R (1968) Ewolucja sieci rzecznej na przedpolu zanikającego lądolodu. Prace Geogr IG PAN 74:101–120

Gębica P (2004) Przebieg akumulacji rzecznej w górnym vistulianie w Kotlinie Sandomierskiej. Prace Geogr IGiPZ PAN 193:229

Gębica P (2013) Geomorphological records of human activity reflected in fluvial sediments in the Carpathians and their foreland. Landf Anal 22:21–31

Gierszewski P (2018) Hydromorfologiczne uwarunkowania funkcjonowania geoekosystemu Zbiornika Włocławskiego. Prace Geogr IGiPZ PAN 268:220

Gilewska S (1986) Podział Polski na jednostki geomorfologiczne. Przegl Geogr 58:15–40

Gonera P (1986) Zmiany geometrii koryt meandrowych Warty na tle wahań klimatycznych w późnym vistulianie i holocenie. Wyd Nauk UAM, Poznań, Geografia 33:84

Gorajska J (2014) Wpływ regulacji technicznej na równowagę pionową koryta Wisły na przedpolu Karpat. Landf Anal 26:3–9

Gorczyca E (2016) Rozwój górskich żwirodennych koryt rzecznych w warunkach antropopresji. IGiGP UJ, Kraków, p 240

Gorczyca E, Krzemień K, Liro M, Sobucki M (2017) Changes of mountain river channels and their environmental effects. In: Radecki-Pawlik A, Hradecký J, Pagliara S, Hendrickson E (eds) Open channel hydraulics, river hydraulics structures and fluvial geomorphology: for engineers, geomorphologists and physical geographers. CRC Press, Taylor & Francis, pp 303–321

Gradziński R, Baryła J, Dynowski W, Doktor M, Gmur D, Gradziński M, Kędzior A, Paszkowski M, Soja R, Zieliński T, Żurek S (2000) Anastomosing system of the Upper Narew River, NE Poland. Ann Soc Geol Polon 70:217–229

Harasimiuk M (1991) Vistulian glacial cycle of the fluvial processes development in the valley of the middle Wieprz river (SE Poland). Ann UMCS B 46(5):81–109

Jania J, Zwoliński Z (2011) Ekstremalne zdarzenia meteorologiczne, hydrologiczne i geomorfologiczne w Polsce. Landf Anal 15:51–64

Kaczmarek H (2018) Ewolucja strefy brzegowej nizinnych zbiorników zaporowych w warunkach dużych wahań poziomu wody na przykładzie Zbiornika Jeziorsko na Warcie. Prace Geogr IGiPZ PAN 265:140

Kalicki T (1991) The evolution of the Vistula river valley between Cracow and Niepołomice in Late Vistulian and Holocene times. In: Starkel L (ed) Evolution of the Vistula river valley during the last 15,000 years, part IV. Geographical Studies, SI 6, pp 11–37

Kalicki T, Starkel L, Sala J, Soja R, Zernickaya VP (1996) Subboreal paleochannel system in the Vistula valley near Zabierzów Bocheński (Sandomierz Basin). In: Starkel L (ed) Evolution of the Vistula river valley during the last 15 000 years, part VI. Geographical Studies, SI 9, pp 129–158

Kaniecki A (1976) Dynamika rzeki w świetle osadów trzech wybranych odcinków Prosny. Prace Kom Geogr-Geol PTPN 17, Poznań, p 100

Kaniecki A (1999) Młyny wodne w dawnym Poznaniu i ich wpływ na przeobrażenie stosunków wodnych. Acta Univ Nicolai Copernici, Geografia 29:337–346

Kasprzak L, Ewertowski M, Szuman I, Kalita J, Tomczyk A (2023) The Middle Noteć Valley (central section of Toruń-Eberswalde Pradolina) and its surroundings—a characteristic element of postglacial landscape in the European Lowlands. In: Migoń P, Jancewicz K (eds) Landscapes and landforms of Poland (this volume)

Kasprzak M (2008) Strefy erozji i akumulacji podczas fluwialnych zdarzeń ekstremalnych w Sudetach, przykład Wilczej Poręby w Karpaczu. Landf Anal 8:36–40

Kaszowski L, Krzemień K (1977) Structure of mountain channel systems as exemplified by chosen Carpathians streams. Studia Geomorph Carp-Balc 11:111–125

Kaszowski L, Krzemień K (1999) Classification systems of mountain river channels. In: Krzemień K (ed) River channels pattern, structure and dynamics. Prace Geographical Institute of Geography UJ 104, pp 27–40

Klimek K (1987) Mans's impact on fluvial processes in the Polish Western Carpathians. Geogr Ann 69A:221–226

Klimek K (2008) Współczesne procesy rzeczne w strefie staroglacjalnej. In: Starkel L, Kostrzewski A, Kotarba A, Krzemień K (eds) Współczesne przemiany rzeźby Polski. Wyd IGiGP UJ, Kraków, pp 236–239

Klimek K, Latocha A (2007) Response of small mid-mountain rivers to human impact with particular reference to the last 200 years; Eastern Sudetes, Central Europe. Geomorphology 92:147–165

Klimek K, Starkel L (1974) History and actual tendency of flood-plain development at the border of the Polish Carpathians. Abh Akad Wissen Göttingen 29:185–196

Kobojek E (2013) Wpływ działalności bobrów na lokalne procesy fluwialne w wybranych rzekach Równiny Łowicko-Błońskiej. Acta Univ Lodz, Folia Geogr Phys 12:17–32

Kordowski J, Gamrat W, Gierszewski P, Kubiak-Wójcicka K, Szmańda JB, Tyszkowski S, Solarczyk A (2014) Zapis procesów sedymentacji fluwialnej i biogenicznej w osadach dna doliny dolnej Wisły. Landf Anal 25:77–93

Korpak J (2010) Geomorphologic effects of river engineering structures in Carpathian fluvial systems. Landf Anal 14:34–44

Kostrzewski A (1970) Uziarnienie i obróbka współczesnych aluwiów bobru jako wyraz dynamiki rzecznego środowiska sedymentacyjnego. Prace Kom Geogr-Geol PTPN 8(4):171

Kostrzewski A, Zwoliński Z, Andrzejewski L, Florek W, Mazurek M, Niewiarowski W, Podgórski Z, Rachlewicz G, Smolska E, Stach A, Szmańda J, Szpikowski J (2021) Współczesna ewolucja rzeźby młodoglacjalnej Niżu Polskiego. In: Kostrzewski A, Krzemień K, Migoń P, Starkel L, Winowski M, Zwoliński Z (eds) Współczesne przemiany rzeźby Polski, 2nd edn. Bogucki Wyd Nauk, Poznań, pp 483–571

Koutaniemi L, Rachocki A (1981) Palaeohydrology and landscape development in the middle course of Radunia basin, North Poland. Fennia 159:335–342

Kozarski S (1965) Zagadnienie drogi odpływu wód pradolinnych z zachodniej części pradoliny Noteci-Warty. Prace Kom Geogr-Geol PTPN 5(1):97

Kozarski S (1981) River channel changes in the Warta valley. In: Guide-book of excursion symposium Paleohydrology of the temperate zone. Poznań–Poland, pp 6–23

Kozarski S (1983) River channel changes in the middle reach of the Warta valley, Great Poland Lowland. Quat Studies Poland 4:159–169

Kozarski S (1991) Warta—a case study of a lowland river. In: Starkel L, Gregory KJ, Thornes JB (eds) Temperate Palaeohydrology. Wiley, Chichester, pp 189–215

Kozarski S, Gonera P, Antczak B (1988) Valley floor development and paleohydrological changes: the Late Vistulian and Holocene history of the Warta River (Poland). In: Lang G, Schluchter C (eds) Lake, mire and river environments during the last 15,000 years. Balkema, Rotterdam, pp 185–203

Kozarski S, Rotnicki K (1977) Valley floors and changes of river channel pattern in the North Polish Plain during the Late-Wurm and Holocene. Quaest Geogr 4:51–93

Krupa J (2013) Naturalne i antropogeniczne procesy kształtujące dno doliny Czarnej Nidy w późnym vistulianie i holocenie. Folia Quat 81:5–174

Krygowski B (1962) Uwagi o niektórych typach zaburzeń glacitektonicznych niżowej części Polski Zachodniej. Bad Fizj Nad Polską Zach 9:61–85

Krzemień K (2003) The Czarny Dunajec River, Poland, as an example of human-induced development tendencies in a mountain river channel. Landf Anal 4:57–64

Krzemień K (ed) (2012) Struktura koryt rzek i potoków (studium metodyczne). IGiGP UJ, Kraków, p 144

Kukulak J (2015) Terasy Wetliny w Bieszczadach (Karpaty Wschodnie) – próba porównania ich hipsometrii, budowy i wieku z doliną górnego Sanu. Landf Anal 28:29–44

Latocha A (2007) Przemiany środowiska przyrodniczego w Sudetach Wschodnich w warunkach antropopresji. Studia Geogr 81, Wyd Uniw Wrocł, p 216

Ludwikowska-Kędzia M (2000) Ewolucja środkowego odcinka doliny rzeki Belnianki w późnym glacjale i holocenie. Dialog, Warszawa, p 182

Łach J (2007) Wpływ zabudowy koryt rzecznych na przekształcanie den dolinnych rzek sudeckich w czasie ekstremalnych powodzi. In: Kostrzewski A, Szpikowski J (eds) Procesy ekstremalne w środowisku geograficznym. Bogucki Wyd Nauk, Poznań, pp 57–67

Łajczak A (1995) Potential rates of the present-day overbank sedimentation in the Vistula Valley at the Carpathian Foreland, Southern Poland. Quaest Geogr 17(18):41–53

Łajczak A (1999) Współczesny transport i sedymentacja materiału unoszonego w Wiśle i głównych dopływach. Monografie Kom Gosp Wodnej PAN 15:215

Łajczak A (2004) Negative consequences of regulation of meandering sandy river and proposals tending to diminish flood hazard. Case study of the Nida river, Poland. In: Proceedings of 9th international symposium on river sedimentation (IRTCES 2004). Yichang, China, III, pp 1773–1783

Łajczak A, Margielewski W, Rączkowska Z, Święchowicz J (2014) Contemporary geomorphic processes in the Polish Carpathians under changing human impact. Episodes 37:21–32

Łajczak A, Plit J, Soja R, Starkel L, Warowna L (2006) Changes of the Vistula River channel and floodplain in the last 200 Years. Geogr Polon 79(2):65–87

Łajczak A, Plit J, Soja R, Starkel L, Warowna L (2008) Współczesne przemiany koryta i równiny zalewowej Wisły. In: Starkel L, Kostrzewski A, Kotarba A, Krzemień K (eds) Współczesne przemiany rzeźby Polski. Wyd IGiGP UJ, Kraków, pp 349–369

Mazurek M (1998) Dostawa materiału klastycznego do transportu fluwialnego na obszarach młodoglacjalnych (zlewnia Kłudy, Pomorze Zachodnie). Bad Fizj Nad Polską Zach 49:127–146

Mazurek M (2010) Hydrogeomorfologia obszarów źródliskowych (dorzecze Parsęty, Polska NW). Wyd Nauk UAM, Geografia 92, Poznań, p 304

Michno A (2013) Rozwój dolin w ujściowych odcinkach rzek na wyżynach lessowych w Polsce. IGiGP UJ, Kraków, p 228

Młynarczyk Z, Rotnicki K (1989) Flood and vortex scour of the channel bed of the Prosna River and their depth range. Earth Surf Proc Landf 14:365–373

Mojski JE, Starkel L (1990) The physical setting, geological structure and morphology. In: Starkel L (ed) Evolution of the Vistula river valley during the last 15,000 years, part III. Geographical studies, SI 5, pp 34–41

Niedźwiedź T, Starkel L (2021) Klimatyczne tło współczesnych procesów morfogenetycznych na obszarze Polski. In: Kostrzewski A, Krzemień K, Migoń P, Starkel L, Winowski M, Zwoliński Z (eds) Współczesne przemiany rzeźby Polski, 2nd edn. Bogucki Wyd Nauk, Poznań, pp 35–58

Niewiarowski W (1986) The phases of transformation of subglacial channel into River Valley: a case study of the Lower Vistula Region. Acta Univ Nicolai Copernici, Geografia 21:61–72

Niewiarowski W (1987) Evolution of the lower Vistula valley in the Unislaw Basin and at the river gap to the north of Bydgoszcz-Fordon. In: Starkel L (ed) Evolution of the Lower Vistula River valley during the last 15,000 years, part II. Geographical studies, SI 4, pp 233–252

Pawlik Ł (2010) Mapa zagrożeń geomorfologicznych wywołanych wezbraniem rzeki górskiej (na przykładzie dolnej części doliny Ścinawki, Sudety Środkowe). Przegl Geogr 82:367–387

Piasecki D (1982) Ewolucja Dolin Rzek Przymorza. Przegl Geogr 54:49–68

Podgórski Z (2004) Wpływ budowy i funkcjonowania młynów wodnych na rzeźbę terenu i wody powierzchniowe Pojezierza Chełmińskiego i przyległych części dolin Wisły i Drwęcy. Wyd UMK, Toruń, p 200

Rachocki A (1974) Przebieg i natężenie współczesnych procesów rzecznych w korycie Raduni. Dokument Geogr IG PAN 4:121

Rotnicki K (1983) Modeling past discharges of meandering rivers. In: Gregory KJ (ed) Background to palaeohydrology. Wiley, Chichester, pp 321–354

Rotnicki K (1987) Main phases of erosion and accumulation in the Prosna Valley in the last glacial-interglacial cycle. Geogr Polon 53:54–65

Rotnicki K, Młynarczyk Z (1989) Późnovistulianskie i holoceńskie formy i osady korytowe w dolinie środkowej Prosny i ich paleohydrologiczna interpretacja. Wyd Nauk UAM, Geografia 43, Poznań, p 76

Schumm SA (1977) The fluvial system. Wiley, New York, p 338

Słowik M (2017) The formation of an anabranching planform in a sandy floodplain by increased flows and sediment load. Earth Surf Proc Landf 43:623–638

Smolska E (1996) Funkcjonowanie systemu korytowego w obszarze młodoglacjalnym na przykładzie górnej Szeszupy (Pojezierze Suwalskie). WGiSR UW, Warszawa, p 123

Starkel L (1960) Rozwój rzeźby Karpat fliszowych w holocenie. Prace Geogr IG PAN 22:239

Starkel L (1983) The reflection of hydrologic changes in the fluvial environment of the temperate zone during the last 15,000 years. In: Gregory KJ (ed) Background to palaeohydrology. Wiley, Chichester, pp 213–235

Starkel L (1991a) Long-distance correlation of fluvial events in the temperate zone. In: Starkel L, Gregory KJ, Thornes JT (eds) Temperate paleohydrology. Wiley, Chichester, pp 473–495

Starkel L (1991b) The Vistula river valley: a case study for Central Europe. In: Starkel L, Gregory KJ, Thornes JT (eds) Temperate paleohydrology. Wiley, Chichester, pp 171–188

Starkel L (1995) Palaeohydrology of the temperate zone. In: Gregory KJ, Starkel L, Baker VR (eds) Global continental palaeohydrology. Wiley, Chichester, pp 233–257

Starkel L (2001) Historia Doliny Wisły od ostatniego zlodowacenia do dziś. Monografie IGiPZ PAN 2:263

Starkel L (2002) Change in the frequency of extreme events as the indicator of climatic change in the Holocene (in fluvial systems). Quat Intern 91:25–32

Starkel L, Kalicki T, Krąpiec M, Soja R, Gębica P, Czyżowska E (1996) Hydrological changes of valley floors in upper Vistula basin during the last 15,000 years. In: Starkel L (ed) Evolution of the Vistula river valley during the last 15,000 years, part VI. Geographical studies, SI 9, pp 7–128

Starkel L, Łajczak A (2008) Kształtowanie rzeźby den dolin w Karpatach (koryt i równin zalewowych). In: Starkel L, Kostrzewski A, Kotarba A, Krzemień K (eds) Współczesne przemiany rzeźby Polski. Wyd IGiGP UJ, Kraków, pp 95–108

Sylwestrzak J (1978) Rozwój sieci dolinnej na Pomorzu pod koniec plejstocenu. Gdańskie Tow Nauk, Gdańsk, p 168
Szpikowski J, Szpikowska G (2018) Akumulacja osadów w rozlewiskach bobrowych w świetle oceny wielkości denudacji mechanicznej zlewni młodoglacjalnej (Kłuda, Pomorze Zachodnie). In: Kostrzewski A, Stach A, Majewski M (eds) Geneza, litologia i stratygrafia utworów czwartorzędowych, t. VII. Instytut Geoekologii i Geoinformacji UAM, Poznań, pp 197–204
Szmańda J (2011) Zapis warunków depozycji w uziarnieniu aluwiów pozakorytowych. Landf Anal 18:3–97
Szumański A (1972) The valley of lower San river in the Sandomierz Basin. In: Excursion Guide-Book of Holocene Symposium 2, Warszawa, pp 55–68
Święchowicz J, Margielewski W, Starkel L, Łajczak A, Pietrzak M, Krzemień K, Gorczyca E, Bucała-Hrabia A (2021) Współczesna ewolucja rzeźby Karpat Zewnętrznych i Podhala. In: Kostrzewski A, Krzemień K, Migoń P, Starkel L, Winowski M, Zwoliński Z (eds) Współczesne przemiany rzeźby Polski, 2nd edn. Bogucki Wyd Nauk, Poznań, pp 97–222
Teisseyre AK (1977) Współczesne procesy rzeczne w dorzeczu górnego Bobru I Strzegomki (Sudety Środkowe). Geol Sudetica 12(2):93–109
Teisseyre AK (1991) Bank crevassing and channel anastomosis in the upper River Bóbr valley (Central Sudetes, SW Poland). Acta Univ. Wratisl. 1286, Prace Geol-Miner 21, p 109
Tofelde S, Savi S, Wickert AD, Bufe A, Schildgen TF (2019) Alluvial channel response to environmental perturbations: fill terrace formation and sediment-signal disruption. Earth Surf Dyn 7:609–631
Tomczak A (1987) The evolution of the Vistula Valley in the Toruń Basin in the Late Glacial and Holocene. In: Starkel L (ed) Evolution of the Vistula river valley during the last 15,000 years, part II. Geographical studies, SI 4, pp 207–231
Turkowska K (1988) Rozwój dolin rzecznych na Wyżynie Łódzkiej w czwartorzędzie. Acta Geogr Lodz 57:157
Vandenberghe J (1995) Postglacial river activity and climate: state-of-the-art and future prospects. In: Frenzel B, Vandenberghe J, Kasse C, Bohncke S, Gläser B (eds) European river activity and climatic change during the Lateglacial and early Holocene. Paläoklimaforschung 14, pp 1–9
Warachowska W, Zwoliński Z (2023) Gospodarka polderowa w Polsce – wyzwania naturalnej retencji powodziowej. Landf Anal 42:3–23
Weckwerth P (2014) Transformacja form korytowych na przykładzie aluwiów piaskodennych rzek roztokowych środkowego vistulianu w Kotlinie Toruńskiej. Landf Anal 25:169–177
Wiśniewski E (1987) The evolution of the Vistula river valley between Warsaw and Plock Basin during the last 15,000 years. In: Starkel L (ed) Evolution of the Vistula river valley during the last 15,000 years, part II. Geographical studies, SI 4, pp 171–187
Witek M (2012) Wpływ zabudowy hydrotechnicznej na współczesne kształtowanie rzeźby koryt rzek ziemi kłodzkiej. Landf Anal 19:91–102
Wyżga B (1993) Funkcjonowanie systemu rzecznego środkowej i dolnej Raby w ostatnich 200 latach. Dokument Geogr IGiPZ PAN 6:92
Zawiejska J, Wyżga B (2010) Twentieth-century channel change on the Dunajec River, southern Poland: patterns, causes and controls. Geomorphology 117:234–246
Zieliński T (2003) Catastrophic flood effects in alpine/foothill fluvial system (a case study from the Sudetes Mts, SW Poland). Geomorphology 54:293–306
Zwoliński Z (1985) Sedymentacja osadów przyrostu pionowego na terasie zalewowej Parsęty. Bad Fizj Nad Polską Zach 35:205–238
Zwoliński Z (1986) Kooperacja przepływowa jako miernik zmian reżimu rzecznego. In: Kaniecki A (ed) Ogólnopolska Konferencja Hydrograficzna „Hydrologia regionalna i procesy hydrologiczne w zlewniach". UAM, Poznań, pp 221–225
Zwoliński Z (1989) Geomorficzne dostosowywanie się koryta Parsęty do aktualnego reżimu rzecznego. Dokument Geogr IGiPZ PAN 3–4:144
Zwoliński Z (1992) Sedimentology and geomorphology of overbank flows on meandering river floodplains. Geomorphology 4:367–379
Zwoliński Z (2007) Mobilność materii mineralnej na obszarach paraglacjalnych, Wyspa Króla Jerzego, Antarktyka Zachodnia. Wyd Nauk UAM, Geografia 74:266
Zwoliński Z (2016) Solute and solid cascade system in the Antarctic oases. In: Beylich AA, Dixon JC, Zwoliński Z (eds) Source-to-sink fluxes in undisturbed cold environments. Cambridge University Press, Cambridge, pp 183–198
Zwoliński Z, Hildebrandt-Radke I, Mazurek M, Makohonienko M (2018) Anthropogeomorphological metamorphosis of an urban area in the postglacial landscape: a case study of Poznań City. In: Thornbush MJ, Allen CD (eds) Urban geomorphology: landforms and processes in cities. Elsevier, Amsterdam, pp 55–77
Żelazo J, Popek Z (2002) Podstawy renaturyzacji rzek. Wyd SGGW, Warszawa, p 319
Żurek S (1975) Geneza zabagnienia Pradoliny Biebrzy. Prace Geographical IG PAN, p 110

Leon Andrzejewski is Emeritus Professor at the Faculty of Earth Sciences and Spatial Management, Nicolaus Copernicus University (NCU) in Toruń, Poland. For many years he was Director of the Institute of Geography and Head of Geomorphology Department at NCU. His main research interests include morphogenesis and functioning of valley and channel systems in different geological and climatic settings, and Pleistocene and contemporary evolution of postglacial landscapes. He was involved in studies in both Poland and abroad (Iceland, Spitsbergen, Mongolia) and took part in scientific expeditions to, amongst others, China, Nepal, India, south-east Asia, northern, central and southern Africa, South America, and Scandinavia. For many years he was participated in activities of scientific commissions of the Polish Academy of Sciences, scientific societies and organisations. He is an Honorary Member of the Association of Polish Geomorphologists.

Kazimierz Krzemień is Professor at the Institute of Geography and Spatial Management at Jagiellonian University. He specialises in physical geography and geomorphology. His interests include fluvial geomorphology of mountain areas, dynamics of contemporary morphogenetic processes, relief change in mountain areas, and arid zones. He has conducted studies mostly in the Carpathians. His research outside Poland has included sites in Great Britain, Italy, France, Ukraine, and Morocco. He is an Honorary Member of the Association of Polish Geomorphologists.

Zbigniew Zwoliński is geographer, geomorphologist, Professor at the Institute of Geoecology and Geoinformation, Faculty of Geographical and Geological Sciences, Adam Mickiewicz University in Poznań. The most important research issues investigated recently include studies of denudation and environmental changes in different morphoclimatic zones, development of the concept and implementation of geomorphological map of Poland in the scale of 100k and 500k, studies on morphology and sediments in urban areas, development of the concept of geodiversity and methods of its assessment, changes in land cover and land use in selected landscapes of Poland, uncertainty and sensitivity analysis in multi-criteria modelling, creation of geo-questionnaries and geo-discussion applications for spatial planning purposes. He is author of over 300 publications, editor-in-chief of *Landform Analysis*, managing editor of *Quaestiones Geographicae*. Member of the Association of Polish Geomorphologists and its past President (2008–2011), Chairman of the Working Groups on Geodiversity and Global Denudation of the International Association of Geomorphologists. Awarded the Denys Brunsden Medal by the International Association of Geomorphologists.

Geomorphological Heritage of Poland—Diversity, Conservation and Promotion

6

Piotr Migoń

Abstract

Geomorphological heritage of Poland includes landforms and landscapes of various origin and age, but also of varying clarity of expression. This variability arises from various geological and climatic controls, as well as from dissimilar timescales of relief evolution in different parts of the country. In the south, where mountains and uplands are located, significant landforms are those reflecting lithological and structural control, including karst, mass movements and fluvial processes. In the most elevated parts of the mountains, an impressive legacy of glacial and periglacial conditions occurs. In the lowland part of Poland, geomorphological heritage is mostly associated with the history of continental glaciation, with subsequent modifications by periglacial, erosional and aeolian processes. The legal framework for the protection of elements of abiotic nature is defined in the Nature Conservation Law. Protection categories most applicable to geomorphological heritage include national parks, landscape parks, nature reserves, nature monuments and sites of scientific interest. Two UNESCO Global Geoparks exist in Poland, with another one aspiring for this designation. A national geopark network, however, is yet to be built and receive solid organizational foundation.

Keywords

Geoheritage · Geosites · Protected areas · Geoparks

P. Migoń (✉)
Institute of Geography and Regional Development,
University of Wrocław, pl. Uniwersytecki 1, 50–137 Wrocław, Poland
e-mail: piotr.migon@uwr.edu.pl

6.1 Introduction

Poland has rich and diverse geoheritage, with a long history of legal protection and use for education and tourism. The wealth of geoheritage reflects two main factors: (a) the protracted geological evolution of the country, with the evidence of geological processes recorded in rocks that may be as old as the end of the Precambrian, and (b) dissimilar pathways of this evolution in different parts of Poland, in terms of both processes and timescales involved. Hence, geoheritage in various regions of Poland differs from one another and, in particular, the north provides insights into a very different interval of the Earth history than the south, whereas the south is very diverse in itself, so that the south–west and south–east are two different worlds in respect to rock and landform record.

In fact, substantial part of geoheritage of Poland is geomorphological heritage, meaning that information about the past, especially the past few million years, is conveyed by landforms and landscapes rather than by rocks, geological structures or fossils. This is particularly true for northern Poland, whose topography has mainly been shaped by ice sheets, glacially conditioned (e.g. meltwater action) and periglacial processes, which created complex landform assemblages but have left a rather inconspicuous sedimentary record consisting of 'ordinary' gravel, sands and clays, available for viewing mostly in artificial and short-lasting exposures. In the mountains and the upland belt, in turn, a large part of geoheritage is made of landforms of tectonic and erosional origin. However, at the same time, these geomorphological sites may show outcrops of very old rocks and structures, and many specific localities have multiple values from the geoheritage perspective (Alexandrowicz 2006).

Geomorphological heritage of Poland consists of localities, whose scenic attributes are very variable. Consequently, their potential for geotourism and geo-education varies, too. On the one hand, it includes spectacular

P. Migoń and K. Jancewicz (eds.), *Landscapes and Landforms of Poland*,
World Geomorphological Landscapes, https://doi.org/10.1007/978-3-031-45762-3_6

mountain sceneries dominated by deeply incised glaciated valleys, cirques and jagged peaks, fluvial gorges, or extensive tracts of ruiniform relief. On the other hand, scientifically important landform complexes may be difficult to appreciate from the ground perspective due to their subdued topography, large spatial extent and thick vegetation. These circumstances pose various challenges for conservation and promotion of geomorphological heritage, including the development of geotourism.

The legal framework of protection of geoheritage of Poland, including geomorphological heritage, is generally adequate and offers various conservation options (Alexandrowicz 2008). However, they are not necessarily used in proper way and the awareness of the significance of abiotic nature still lags behind the much better appreciation of biotic components. This is particularly true for sites of local significance, where municipalities have a key role to play and are legally empowered to decide about protection measures. Scenic localities are clearly better protected, promoted, and looked after than sites with rather unimpressive evidence of Quaternary glacial, aeolian, fluvial or hillslope processes, even if the latter have long been recognized by scientists as important.

This chapter provides a brief overview of the diversity of geomorphological heritage of Poland,[1] presents various aspects of this geoheritage in the context of nature and landscape conservation, and outlines how the idea of geoparks develops in the country.

6.2 Diversity of Geomorphological Heritage of Poland

Geomorphological heritage of Poland includes landforms and landscapes, along with the associated rock and sedimentary record, which reflect the operation of genetically different processes, occurring at different rates and in different time intervals. They can be divided into nine main groups using the origin/process as a criterion, although such an approach is better understood as based on the recognition of the dominant theme at a particular place. In fact, many localities of geoheritage value, especially the large ones (landform complexes), record the occurrence of various processes.

[1] Geoheritage values of specific localities in Poland emerge from a wide range of papers, both contributing fundamental knowledge of geomorphology and discussing geo-conservation issues and relevance for geotourism. Since size limitations preclude to provide a complete bibliography, a different approach was adopted by the author. Extensive cross-referencing to other chapters in this volume is applied, whereas other papers are referred to if the theme is not covered in regional chapters or they are of special significance.

6.2.1 Weathering and Rock-Controlled Landforms

Rock-controlled landforms reflect the dominant role of passive geological factors, related to lithology and structure of rock complexes. The former is understood as characteristics of particular rock types, such as mineralogy, texture, chemical composition and physical properties. Structure, in turn, is considered at a larger spatial scale and emphasizes co-existence of magmatic bodies of different composition, strike and dip of strata in sedimentary successions and their change within an area, style of folding, the occurrence of thrusts, variations in jointing and bedding density, and various instances of juxtaposition of different kinds of rocks. Consequently, lithology tends to explain the origin of small- and medium-size landforms, whereas controls imposed by structure are usually seen in the regional patterns of landforms, although discontinuities may also play a fundamental role in controlling the evolution of minor landforms. Both lithological and structural controls are revealed by differential weathering, followed by removal of weathering products to leave residual rock-controlled landforms. Although weathering processes continue today, landforms of geoheritage value are generally inherited, having been shaped since the Neogene (major landforms) or the Pleistocene (medium-size and minor landforms).

In Poland, the occurrence of rock-controlled landforms is limited to the mountains and uplands in the south, since solid bedrock does not occur further north. They seem to be more diverse in the Sudetes than anywhere else because of very complex lithology and structure, as well as due to negligible to no role of glacial processes in shaping the hillslopes and the divides. The most typical landforms of this kind are crags, whose occurrence usually indicates higher strength and resistance of the supporting rock. Some rise above the surroundings along all sides and are bedrock outcrops a few metres high, even more than 20 m high, and these are termed tors (Alexandrowicz 1990). In the Sudetes, crags and tors are common in granites (Kasprzak 2023), where they reach up to 25 m high, and in sandstones, where they locally cluster to form larger groups of ruiniform relief or form picturesque hoodoo rocks (Duszyński 2023). Crags are also common in gneisses (Sobczyk 2005), greenschists (Michniewicz 2016), gabbro (Traczyk, 2023) and basalt. In the Outer Carpathians, built of thick flysch successions, thickly bedded sandstones usually give rise to crags emerging from the slopes or crowning the ridges (Alexandrowicz 1990). Among famous and well-popularized examples are the group of impressive sandstone tors near the town of Krosno, known as'Prządki' (The Spinners) (Welc and Miśkiewicz 2020), and 'Skamieniałe Miasto' (The Petrified City) near the town of Ciężkowice—a large cluster of sandstone outcrops in the foothill belt (Alexandrowicz 1970) (Fig. 6.1).

Crags are also present in the upland belt, but not as commonly as in the mountains. The most impressive ones are limestone towers, cones and pinnacles in the Kraków-Częstochowa Upland (Tyc 2023), and sandstone crags in the Świętokrzyskie Mountains (Urban 2016, 2023).

Larger residual landforms reflecting lithological control are hills and ridges built of very resistant rocks, surrounded by topographic depressions (valleys, basins, plains) underlain by much weaker rocks. They are particularly common in the Sudetes and mainly associated with ancient volcanic rocks. Thus, rhyolites and trachybasalts of Carboniferous and early Permian age almost invariably support steep-sided terrain elevations, some up to 400 m high (Migoń 2023a), whereas Cenozoic basalts occur as cones or domes evolved from the former volcanic conduits (Migoń 2023b). A special type of residual landforms is represented by granite hills, once likened to inselbergs rising from plains, although planar surfaces are nowhere particularly extensive. These are usually associated with more massive, poorly jointed parts of granite intrusions and may host numerous crags, rock labyrinths, boulder fields and small non-karstic caves (Migoń 1997; Migoń and Latocha 2008; Kasprzak 2023). Another highly distinctive rock-controlled relief is represented by the Pieniny Mountains in the Carpathians, where large masses of limestone have been exposed as high cliffs and mega-crags (Klippen) due to weathering and erosion of soft flysch formations, into which they were originally embedded (Margielewski and Jankowski 2023a). Weathering has also produced diverse microrelief on exposed surfaces of crags and tors, including weathering pits typical for granites in the West Sudetes (Migoń 1996; Kasprzak 2023) and various manifestations of cavernous weathering on sandstones (arcades, honeycombs, tafoni) (Alexandrowicz 1970; Urban and Górnik 2017; Duszyński 2023).

Large-scale structure-controlled landforms show some relationship to available relief, which in turn reflects the magnitude of Cenozoic uplift and the progress of differential erosion. In the Sudetes, the prime example of such topography is the stepped tableland of the Stołowe Mountains, with escarpments and flat-topped mesas supported by thick sandstone beds of negligible dip (Duszyński 2023). Its total elevation above the surrounding rolling terrain is more than 400 m. However, where the dip increases, the tableland gives way to a cuesta landscape, as in the foothills of the West Sudetes (Migoń 2023b). Cuesta landscape also typifies the eastern part of the Silesian Upland (Dulias 2023), the Kraków-Częstochowa Upland (Tyc 2023), and the Nida Basin (Urban et al. 2023), with limestones or gypsum playing the role of resistant caprock. However, with a few exceptions, cuesta landscapes are not particularly conspicuous and the reasons may include moderate rather than very high rock resistance, as well as truncation of older relief by glacial erosion effected by the Scandinavian ice sheets. In the Świętokrzyskie Mountains, characteristic landforms within this group are hogback ridges, developed upon sedimentary strata inclined at 40° or more, up to vertical. Good examples are provided by the limestone ridges around the town of Chęciny, which together form opposite limbs of a dissected anticline, and the highest ridge of the Świętokrzyskie Mountains, the Łysogóry ridge, built of quartzitic sandstones of Cambrian age (Urban 2023). In the Outer Carpathians, in turn, homoclinal ridges supported by thick sandstone beds present within individual thrust sheets are common. The more resistant and thicker are the sandstone packages, the steeper are the frontal parts of the ridges, as in the Babia Góra Massif (Łajczak 2023). Finally, examples of structure-controlled landforms can be found in the Tatra Mountains, and these are typically built of Mesozoic limestones. Both homoclinal and hogback ridges occur in this region.

6.2.2 Karst Landforms

Even though the total area occupied by soluble rock types is not very large in Poland, different types of karst occur and karst landforms constitute an important part of national geomorphological heritage. Limestone karst of the Kraków-Częstochowa Upland stands out, not only because the surface area of carbonate outcrops is large, but the inventory of landforms is particularly rich and the age of some karst landforms (palaeo-dolines) goes back to the Palaeogene (Tyc 2023). Limestone residual hills, dry valleys, canyons and dolines typify the region, which also hosts more than one thousand caves of both hypogenic and epigenic origin. Limestone karst is also represented in the Tatra Mountains, where the longest and deepest caves in Poland, developed concomitantly with fluvial incision in response to regional uplift during the Quaternary, occur (Szczygieł 2023). They are, however, not readily accessible to the general public. Surface karst is less striking, although limestone gorges, solution dolines, karren fields, and voluminous springs are all present. Limestone outcrop areas in the Sudetes are small and disconnected, and surface karst is poorly developed. Nonetheless, the Niedźwiedzia (Bear) Cave in the East Sudetes is more than 5 km long, with cave chambers up to 100 m long and 20 m high, and impressive sinter decorations. Some sections of the cave are available for the public. Karst features in the Świętokrzyskie Mountains, within the upland belt, developed in widespread Devonian and Jurassic limestones, mainly during the Palaeogene and the Neogene. Therefore, they currently comprise surface karst forms (palaeo-dolines) and underground karst

Fig. 6.1 Sandstone crags in the Outer Carpathians. **a** Prządki group near the town of Krosno. **b** Fragment of Skamieniałe Miasto crag cluster near the town of Ciężkowice in the Carpathian Foothills (photographs by P. Migoń)

systems partly or totally filled by sediments. Approximately 100 caves, which are fragments of the relict karst systems cleaned of the sediments, have been explored, with the longest Chelosiowa Jama-Jaskinia Jaworznicka system ca. 4 km long.

The Nida Basin in the upland belt is the only region in Poland, where gypsum rocks occur extensively, giving rise to gypsum karst. In contrast to the Świętokrzyskie Mountains, it is much younger and has developed mainly recently, in the Late Pleistocene and Holocene, and is currently evolving within the water table zone that is accessible in many caves (Urban et al. 2023). Caves, dolines, dry and blind valleys are the key components of the gypsum karst relief. Karst in the Lublin Upland, developed in rather weak chalk deposits, is less spectacular, lacking residual hills or deep karstic depressions. However, it is unique in Poland and represented by a range of closed depressions and palaeokarst features exposed in former and existing quarries (Dobrowolski 2023). Solution processes, acting in concert with permafrost aggradation and thawing, are responsible for the emergence of a unique natural lakeland (Polesie region), located far to the south from the geographical limit of the last Scandinavian ice sheet.

6.2.3 Other Denudational Landforms

This genetic group includes landforms produced by mass movements and water erosion in settings other than permanent channels (these are covered in Sect. 6.2.4). Mass movements, especially landslides, are usually considered as significant geomorphological hazards and 'troublemakers' rather than processes of any value for geoheritage and geotourism, but this view is changing and there is an increasing recognition of their scientific and educational significance (Alexandrowicz and Margielewski 2010; Morino et al. 2022). Dated landslide deposits, including peat bogs developed over landslide terrain, are carriers of important palaeoenvironmental information, including that about climate fluctuations in the Holocene. Moreover, landslides may be associated with peculiar minor geomorphic features generating public interest, such as ridge-top and mid-slope trenches, rock cliffs, ruiniform relief, crevice caves, boulder fields and water-filled hollows, as well as be sites of biological significance. In Poland, the Outer Carpathians are the main landslide-prone area and thousands of active, dormant and fossil landslides have been recognized and documented (Łajczak 2023; Margielewski and Jankowski 2023b). Most of the caves in the Outer Carpathians, among which are ones as long as 1–2 km and several tens of metres deep, have developed within landslide bodies (Margielewski and Urban 2023). However, landslides are mainly located under forest and they are difficult to appreciate in the full extent.

Landslides also occur in the Sudetes, although the dominance of crystalline rocks and limited thickness of weathering covers make the area much less favourable to large-scale slope deformations. Nevertheless, the Kamienne Mountains are a regional 'landslide hot spot', with several tens of individual localities, representing various mechanisms of displacement and diverse resultant landforms (Migoń 2023a). The best and easiest to access example, though, is the historical landslide in the town of Bardo, which occurred in 1598 and caused diversion of the channel of the Nysa Kłodzka River. The internal structure of the depositional part, undercut by the 1997 flood, with huge, chaotically distributed blocks of greywacke, can still be observed (Fig. 6.2). Landslides also modify high bluffs of certain river valleys, e.g. of the Vistula valley downstream of the city of Toruń, and the coastal cliffs (see Sect. 6.2.8). The occurrence of other types of mass movements, more rapid than landslides, is confined to the most elevated parts of the Tatra Mountains (Kłapyta and Zasadni 2023), Karkonosze Mountains (Kasprzak 2023) and Babia Góra Massif (Łajczak 2023). They include debris flows and rock falls, the latter contributing to the build-up of scree cones, particularly characteristic for the high-mountain landscape of the Tatra Mountains. Until now, ongoing mass movements are regarded as part of active geomorphological systems rather than components of geoheritage, although boulder masses from huge prehistoric rock-slope collapses in the Tatra Mountains (e.g. Wantule) are certainly worth attention from geoheritage perspective.

Runoff-related erosion shares its image with mass movements, as a process that creates environmental problems. But here a distinction can also be made between recent processes damaging crops and infrastructure, whose geomorphic effects are being quickly rehabilitated, and large gullies resultant from cumulative action of many erosional episodes in the past, usually now under forest. Gullies occur in various parts of Poland, especially in the upland belt, where the loess cover is thick, and they may connect into extensive networks of dry valleys of different length, depth, and steepness (Fig. 6.3). The north-western part of the Lublin Upland is one area well known for loess gullies (Rodzik et al. 2023), but they also occur in the Roztocze upland, around the historical town of Sandomierz (Sandomierz Upland), in the Proszowice Plateau and Miechów Upland next to Kraków, in the western part of the Silesian Upland (Góra Świętej Anny), and in the Sudetic Foreland.

6.2.4 Fluvial Landforms

Research in fluvial geomorphology has been very popular in Poland in the past few decades and river systems are

Fig. 6.2 Internal structure of the landslide body in Bardo (Sudetes), exposed thanks to toe undercutting during the flood in 1997. Note huge greywacke boulders inside the colluvial mass and exposed on the landslide surface (upper left) (photograph by P. Migoń)

reasonably well understood (Andrzejewski et al. 2023). Both contemporary fluvial processes and their evolution since the Late Glacial were investigated in different parts of Poland, on rivers varying in length, discharge and geomorphic setting. Longer timescales were considered in the mountains, where the history of drainage network can be traced back to the Neogene, and possibly to the Palaeogene in the Sudetes. However, among the outcomes of these studies is also the recognition of a considerable scale of human impact on alluvial systems, manifested in either direct intervention into channel patterns and valley floor morphology, or alteration of discharge and load characteristics in response to catchment-wide land use and land cover change. As a result of these research activities, various elements of fluvial geomorphological heritage may be recognized. The nature of this heritage in the mountains, where rivers are incised into bedrock or have gravel beds, is different from that in the lowlands, where channel gradients are much lower and valley floors are generally much wider.

Elements of fluvial geoheritage in the mountains can be identified at the scale of a valley and of a channel. The former include gorges and water gaps, present in both the Sudetes and the Carpathians. These higher-gradient and steep-sided valley sections are of different origins, including antecedence in response to the late Neogene uplift (e.g. the Nysa Kłodzka water gap in the Sudetes), superimposition from now eroded sedimentary cover, glacial epigenesis in the peripheral parts of the mountains, reached by the Scandinavian ice sheet, or they simply reflect the presence of more resistant rocks (structural water gaps). The water gap of the Dunajec River across the Pieniny Mountains is considered the most picturesque among large-scale fluvial landforms of this kind in Poland, with not yet fully deciphered origin (Margielewski and Jankowski 2023a). At the channel scale, valuable geoheritage elements include waterfalls (e.g. Wielka Siklawa in the Tatra Mountains—70 m high, Kamieńczyk Waterfall in the Karkonosze Mountains—27 m high), associated slot canyons indicating the progress of waterfall retreat, minor bedrock steps and rapids, chutes, and potholes. Beyond the mountains, a notable fluvial geoheritage site is the series of minor, but picturesque bedrock steps in the Tanew River in the Roztocze Upland, in south-eastern Poland (Fig. 6.4). Rare semi-natural gravel-bed reaches with braiding patterns, present along some rivers in the Carpathians, are also worth attention (Zawiejska et al. 2023).

Fig. 6.3 Deep gullies cut in loess and related deposits in the Niemcza Hills, Sudetic Foreland (photograph by P. Migoń)

Most lowland rivers have been considerably altered by human interventions and their contemporary channel patterns are no longer natural (Andrzejewski et al. 2023). The more valuable are river reaches, where human impact was negligible (or recovery has taken place) and the channels are free to evolve in response to natural factors. Thus, the Vistula River has several braided sections, with large in-channel and lateral sand bars, migrating across the channel, whereas the Narew River in eastern Poland is a unique anastomosing system, conditioned by topographic setting, very low gradient, and dominance of suspended load (Zieliński 2023). By contrast, some minor rivers in the northern lakelands are deeply incised into morainic hills and plateaus, and their channels contain boulder lag deposits, resembling a mountainous rather than a lowland stream (Mazurek et al. 2023). At the larger spatial scale, some rivers flow in between or along high bluffs, locally more than 50 m high and occasionally offering wide panoramic views (Fig. 6.5). The most continuous reaches of this kind are associated with the former ice-marginal valleys, especially the Toruń–Eberwalde Spillway, now used by the Noteć and Warta rivers (Kasprzak et al. 2023).

6.2.5 Glacial Landforms

Glacial geomorphological heritage of Poland is of two kinds, each typified by a distinctive assemblage of landforms. First, these are landforms testifying to mountain glaciation. They are most evident in the Tatra Mountains (Kłapyta and Zasadni 2023), which were extensively and repeatedly glaciated in the Pleistocene, including two ice lobes that were not confined to the valleys, but spread over the hilly foreland of the massif. Inventories of glacial landforms in particular valleys consist of cirques, which are overdeepened or not, in some valleys arranged in a staircase, knock-and-lochan relief, roches moutonnées, U-shaped troughs, hanging tributary valleys, lateral moraines, kettle holes and terminal moraines. The glacial landsystems continue in the foreland in the form of nested outwash surfaces. Glaciation in the Karkonosze Mountains in the Sudetes was much less extensive, but it also left distinct geomorphic evidence that includes six glacial cirques (two twin cirques) and boulder-rich moraines, with lateral and terminal moraine ramparts in some valleys (Kasprzak 2023). Least conspicuous is glacial morphology of Mt.

Fig. 6.4 Bedrock steps in the channel of Tanew River, Roztocze Upland in south-eastern Poland (photograph by P. Migoń)

Fig. 6.5 High bluffs of the Wisła River valley next to the town of Świecie, northern Poland (photograph by P. Migoń)

Fig. 6.6 Examples of erratic boulders of Scandinavian origin in northern Poland. **a** St. Wojciech Boulder near the village of Budziejewko, Gniezno Lakeland, central Poland. **b** Tempelburg boulder near the town of Czaplinek, Pomeranian Lakeland (photographs by P. Migoń)

Babia Góra in the Outer Carpathians, where glacial and landslide-related landforms coexist, but notwithstanding old controversies there is now no doubt that the north-facing side of the massif hosted several glaciers (Łajczak 2023).

A different suite of glacial landforms occurs in the part of Poland, which was within the extent of Scandinavian ice sheet (Jary and Marks 2023). They are less visible in general due to sheer spatial scale, considerable periglacial denudation in southern and most of central Poland since the wastage of the last ice sheet in these areas, vegetation cover, and much lower local relief. Thus, the geomorphic legacy of glaciation in the lowlands is better appreciated on digital terrain models than in the field. Nevertheless, the northern Poland in particular has a number of localities with very distinctive glacial landform assemblages such as tunnel valleys (e.g. Błaszkiewicz and Tylmann 2023), intriguing ridge-and-trough complexes (glacial curvilineations), especially in the vicinity of the city of Toruń (Wysota et al. 2023), eskers, kame plateaus and kame terraces, moraine ridges, boulder fields, with several localities in the Masurian Lakeland (Smolska et al. 2023; Weckwerth and Wysota 2023), outwash (sandur) plains dotted by kettle holes after melting of buried ice, and several extremely wide ice-marginal valleys that carried glacial meltwater in the westerly direction (e.g. Kasprzak et al. 2023). Another testimony of continental glaciation is provided by erratic boulders, which are particularly common in northern Poland, but occur in the southern part, too (Fig. 6.6).

6.2.6 Periglacial Landforms

Given the geographical location of Poland and its Quaternary history, the significant role of periglacial conditions in shaping the contemporary relief is not surprising. This has been documented widely across the country and Poland used to be one of the leading centres of research focused on Pleistocene periglacial forms and structures (Jahn 1975; Goździk and Forysiak 2023). However, the geomorphic legacy of periglaciation and permafrost is not obvious, especially in lowland terrains, where cold-climate surface processes of solifluction, surface wash, ground ice formation and melting attenuated the relief rather than created evident new landforms. It is possible that some closed circular hollows in northern Poland may be pingo remnants (Smolska et al. 2023) and examination of aerial photographs taken at suitable vegetation and soil moisture conditions revealed polygonal patterns of former ice wedges (Dąbski 2011), but these are impossible to see from the ground perspective.

More evident is periglacial geoheritage in the upland belt and, especially, in the mountains. Among the latter, the Karkonosze Mountains in the Sudetes (Kasprzak 2023) and Mt. Ślęża in the Sudetic Foreland (Traczyk 2023) are outstanding. They have a rich inventory of periglacial bedrock landforms such as block fields and block slopes, block streams, relict patterned ground, solifluction steps and lobes, mid-slope crags and cliffs, probable cryoplanation terraces and rock glaciers, and texturally varied hillslope deposits. Most of these features, albeit in less impressive form and partly vegetated, can be found in other parts of the Sudetes as well. Block fields built of quartzite are a regional landmark of the highest part of the Świętokrzyskie Mountains. Moreover, many crag groups were remodelled by gravity-induced slope processes active in the periglacial environment (Urban 2023). The Tatra Mountains, in turn, is home to several rock glaciers, which signify the ultimate phase of glacier decay in the valleys (Kłapyta and Zasadni 2023).

6.2.7 Aeolian Landforms

Aeolian geomorphological heritage of Poland comprises mainly inactive inland dunes, which date to the latest Pleistocene and waning periglacial environment. Alluvial plains, terraces, outwash plains and floors of ice-marginal valleys provided an abundant source of sand, transported by westerly winds across these depositional alluvial surfaces and onto the adjacent higher grounds. Climate amelioration and vegetation development in the early Holocene immobilized the dunes, although reactivation of aeolian processes occurred in many places due to vegetation removal by people. Parabolic dunes are a common variety, but there is in fact a considerable diversity of dune types, with crescentic, transverse, linear and other variants. Unfortunately, nearly all dune fields are under forest and hence, impossible to appreciate in their entire extent. One distinctive dune area is the Kampinos Forest close to Warsaw, with dunes as high as 25–30 m (Dłużewski and Woronko 2023), but similar dune fields occur widely in central and southern Poland, adding to the diversity of otherwise rather monotonous plains.

An aeolian deposit of considerable scientific significance is loess and there are several well-documented localities in south-western and south-eastern Poland, with thick, complex loess-paleosol sequences (Solarska et al. 2013; Jary and Marks 2023). However, they are not part of geomorphological heritage sensu stricto, and typical landforms and landscapes in loess terrains are of erosional nature, associated with runoff erosion (see Sect. 6.2.3).

6.2.8 Littoral Landforms

The Polish coastline is shaped entirely in soft sediments. Therefore, it lacks impressive erosional landforms that are common along rocky coasts (rock cliffs, stacks, sea arches, abrasion platforms, etc.), and is also fairly dynamic, which means constant change of coastal scenery and coastline configuration. This in turn is tried to be counterbalanced by various engineering works, which interrupt the natural functioning of littoral systems, negatively impacting dynamic coastal geoheritage. Sites of particular geoscientific interest and value along the Polish coast are associated with cliffs, usually eroded in till/outwash successions of morainic uplands. They are from c. 10 m high to nearly 100 m high in the island of Wolin (Zwoliński et al. 2023), constantly re-shaped by landslides and minor falls. Among various cliff sites is the iconic locality in the village of Trzęsacz in western Poland, where the centuries-long progress of cliff retreat caused undermining and then a series of collapses affecting a fifteenth century gothic church, whose one remaining wall is now perched on the cliff top (Fig. 6.7). Thus, the site is of both geoheritage and cultural significance.

Long stretches of the Polish coastline are made of sandy beaches, backed by dune strips of variable width. Except the foredunes, dunes are vegetated and generally stable, again not very easy to appreciate, even from vantage points at numerous lighthouses scattered along the coast. However, ancient deforestation of the dunes in the Łeba Barrier caused widespread reactivation of aeolian activity and nowadays the active dunes of the Słowiński National Park are considered among the most valuable landscapes of the Polish coastal belt, of considerable scientific and tourist importance (Rotnicka and Borówka 2023). The inventory of coastal landforms is completed by singular spits (Moskalewicz and Woźniak 2023; Wróblewski et al. 2023), coastal lagoons and the regressive delta of the Odra River (Zwoliński et al. 2023).

6.2.9 Impact Landforms

A unique site with landforms of impact origin in Poland is located at the peripheries of the city of Poznań (Szczuciński and Włodarski 2023). The Morasko crater field consists of seven circular closed hollows, with three of them filled with water. The largest one is approximately 90 m across. The meteorite fall likely occurred at 5.5–5 ka, thus in the mid-Holocene. The site is easily accessible and protected as a nature reserve.

6.3 Geomorphological Heritage and System of Nature Conservation in Poland

6.3.1 Legal Framework

The system of nature conservation in Poland, which is also relevant and applicable to geoheritage, is based on the Nature Conservation Law, adopted in 2004 (with subsequent minor changes). This legal act defines ten categories of nature conservation (translations into English by the author): (1) national parks, (2) nature reserves, (3) landscape parks, (4) areas of protected landscape, (5) Natura 2000 areas, (6) nature monuments, (7) sites of scientific interest (also translated as documentary sites), (8) minor sites of ecological significance, (9) nature-landscape complexes, (10) individual protection of plant, animal and fungi species. Among them, decisions regarding national parks, Natura 2000 areas and individual protection of species are undertaken at the central level, those regarding nature reserves, landscape parks and areas of protected landscape at the regional (provincial) level, whereas the remaining

Fig. 6.7 Cliff-top ruins of a medieval church in Trzęsacz, north-western Poland, now protected using various types of coastal defences (photograph by P. Migoń)

categories of conservation are established at the local (municipality) level.

Geoheritage elements may be included in any of the large protected areas (national parks, landscape parks, areas of protected landscape), and an overarching rationale is that holistic approach to nature conservation applies. Nature reserves have more specifically defined purpose and object of conservation, being classified accordingly. Thus, relevant types of nature reserves are those focussed on abiotic nature (geological reserves) and those protecting the value of the landscape, with its various components (landscape reserves). However, some floristic and forest reserves also include valuable geoheritage. The notion of nature monument implies some scenic and aesthetic value, rarity, and/or distinctiveness. Elements of both biotic (singular trees and their groups) and abiotic nature (crags, caves, waterfalls, erratic boulders) may be declared as nature monuments. Sites of scientific interest were specifically introduced into the Nature Conservation Law to provide a formal basis of recognition and protection of localities that are important for geosciences but do not fit the common perception of a 'nature monument', such as inconspicuous rock and sediment outcrops, mineral or fossil occurrences. Finally, the presence of geoheritage elements may be among reasons to establish a nature-landscape complex, which implies mainly aesthetic value, often associated with some cultural heritage value.

6.3.2 National Parks

As of 2023, there are 23 national parks in Poland, distributed across the country (Fig. 6.8). Among them, six are located in the Carpathians, two in the Sudetes, three in the upland belt, five are within the 'old glacial relief', another five in the northern lakelands, within the limits of the last Scandinavian ice sheet, and two include fragments of the Polish coastal zone (Table 6.1). The prominence of geoheritage in each of the national parks varies, with some including landforms and landscapes of international significance, fundamental to the value of the protected area, whereas others do not have any particularly distinctive geomorphology and were established to protect mainly extensive forest or grassland complexes, with their endangered fauna.

Most diverse and spectacular is geomorphological heritage in four mountainous national parks, all covered in detail elsewhere in this volume (see Table 6.1). These are the following:

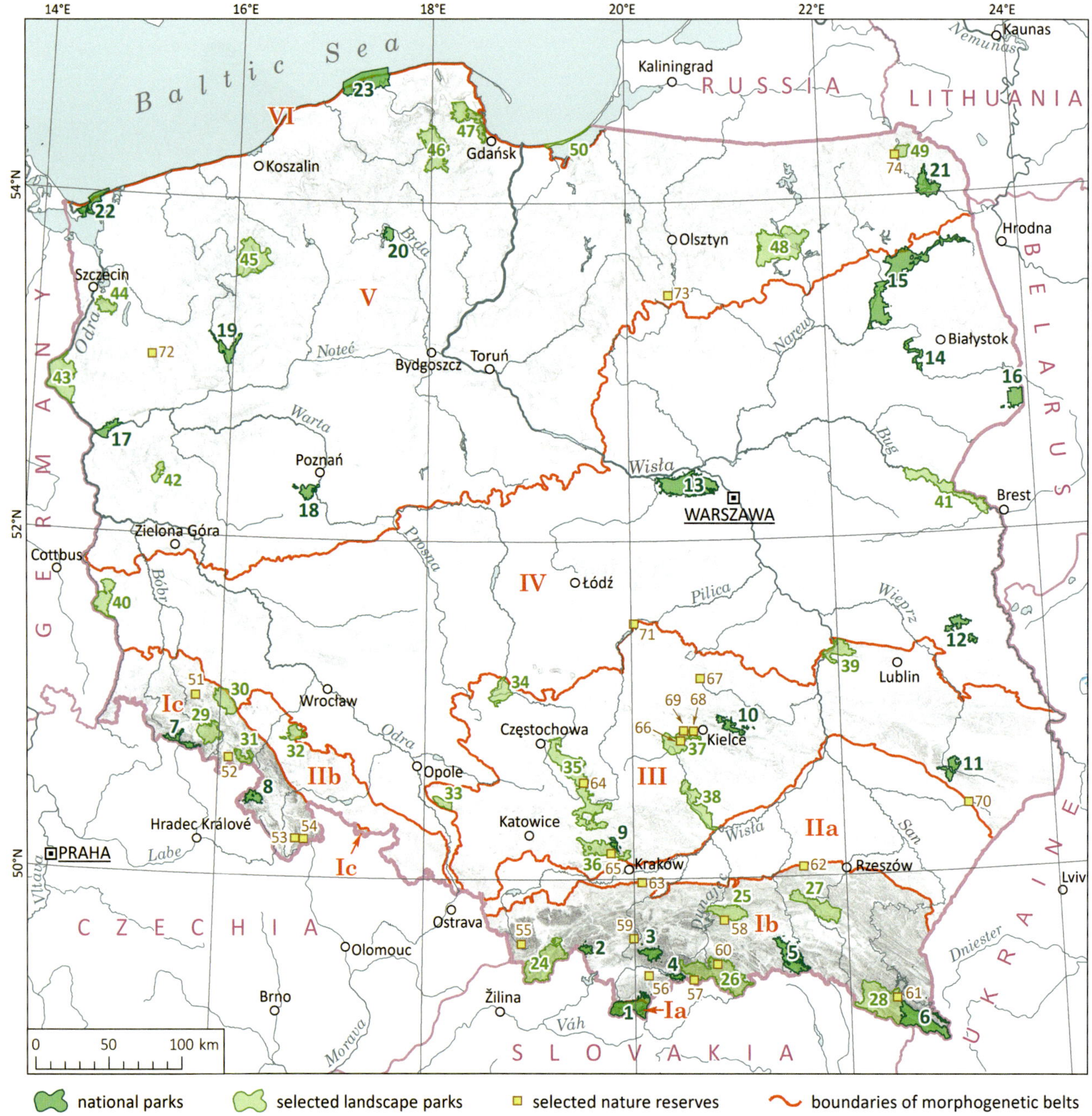

Fig. 6.8 Selected elements of nature conservation system in Poland, with a strong geomorphological component and geoheritage of significant value. Numbers refer to names of protected areas and are explained in the text. Morphogenetic belts: Mountains: Ia—high mountains (Tatra Mts.); Ib—medium–high mountains of the Alpine system (Beskidy Mountains); Ic—medium–high mountains of the Variscan system (Sudetes). Fore-mountain areas: IIa—Sub-Carpathian basins; IIb—Sudetic Foreland. III—uplands. IV—'old glacial relief' of central Poland. V—'young glacial relief' of northern Poland. VI—coastal zone (map courtesy of Kacper Jancewicz)

- Tatra Mountains National Park (Tatrzański Park Narodowy), the only one in Poland representing high-mountain topography, with evidence of extensive glaciation, widespread structural relief, well-developed karst, and considerable contemporary geomorphological dynamics.
- Karkonosze National Park (Karkonoski Park Narodowy), which protects the most elevated part of the Variscan range of the Sudetes, with a range of scenic granite landforms (tors, boulder fields, weathering pits, waterfalls), glacial geomorphic legacy, relict periglacial phenomena and occasional debris flows.

Table 6.1 Summary of geomorphological features of national parks in Poland, according to morphogenetic belts. Bold font highlights national parks with the most conspicuous landforms

Name (numbers refer to Fig. 6.8)	Area [km^2]	Key geomorphological features	Remarks
High mountains (Alpine system)			
Tatrzański [1]	211.8	Glacial relief (cirques, trough valleys, moraines), rock slopes and screes, structure-controlled relief, caves	Chapters 14 and 15
Medium-altitude mountains (Alpine system)			
Babiogórski [2]	33.9	Structural relief on flysch, large landslides, periglacial blockfields	Chapter 13
Gorczański [3]	70.3	Denudational relief on flysch	
Pieniński [4]	23.5	Limestone crags, fluvial gorge (water gap)	Chapter 17
Magurski [5]	123.3	Structural relief on flysch, sandstone crags	
Bieszczadzki [6]	292	Structural relief on flysch, sandstone crags, trellis drainage network	Chapter 20
Medium-altitude mountains (Variscan system)			
Karkonoski [7]	59.5	Granite landforms, cirques and moraines, periglacial blockfields and patterned ground	Chapter 7 Certificate of national geopark
Gór Stołowych [8]	63.4	Sandstone escarpments, rock cities, hoodoo rocks	Chapter 8
Uplands			
Ojcowski [9]	21.5	Limestone plateau, canyons, caves	Chapter 22
Świętokrzyski [10]	76.3	Quartzite ridges and crags, periglacial blockfields	Chapter 25
Roztoczański [11]	84.8	Subdued denudational landscapes on marls and calcareous-siliceous sediments, inland dunes	
Old glacial relief			
Poleski [12]	97.6	Lake basins of mixed karst-periglacial origin	Chapter 27
Kampinoski [13]	385.4	Inland dunes	Chapter 32
Narwiański [14]	73.5	Anastomosing river system	Chapter 33
Biebrzański [15]	592.2	Meandering river and floodplain	
Białowieski [16]	105.2	Moraine plateau, outwash plains, alluvial valley floors	World Heritage for biodiversity values
Young glacial relief			
Ujście Warty [17]	79.6	Anastomosing river system	
Wielkopolski [18]	76.2	Moraine hills, lake basins	
Drawieński [19]	113.4	Outwash plains	
Bory Tucholskie [20]	48	Tunnel valleys, outwash plains	Chapter 36
Wigierski [21]	150.8	Moraine hills, lake basins	Chapter 38
Young glacial relief/coastal			
Woliński [22]	109.4	Coastal cliffs, moraine hills	Chapter 40
Słowiński [23]	327.4	Active coastal dunes	Chapter 41

- Stołowe Mountains National Park (Park Narodowy Gór Stołowych), the only one established specifically to protect geoheritage, whose main values reside in the structure-controlled tableland morphology, with impressive sandstone escarpments, tabular hills, rock labyrinths and hoodoo rocks.
- Pieniny Mountains National Park (Pieniński Park Narodowy), which includes the gorge of the Dunajec River with its entrenched meanders and limestone mega-crags.

Significant landforms and landscapes of different origin (rock-controlled, karst, aeolian, littoral) occur in several further national parks located in the mountains, uplands and coastal Poland (Table 6.1), whereas the contribution of geomorphology to the significance of the remaining national parks is rather modest and less exposed. Nonetheless, the underpinning role of geomorphological diversity for soils, plant communities and habitats is evident. Among the latter parks is Białowieski National Park, which has the status of a UNESCO World Heritage property. However, it was inscribed solely in recognition of its biodiversity values (criteria (ix) and (x)). Likewise, four localities in the Bieszczady Mountains included into the serial transboundary property of 'Ancient and Primeval Beech Forests of the Carpathians and Other Regions of Europe' do not have specific geomorphological values.

6.3.3 Landscape Parks

As of 2023, there are 125 landscape parks in Poland, also broadly distributed across the country, from the Carpathians and the Sudetes in the south to the coastline in the north. They vary in surface area, from nearly 900 km^2 to as little as 17 km^2. Among them, 22 have been established in the mountains (Carpathians—13, Sudetes with the foreland—9), 25 are located in the upland belt, and two border the coast. The remaining 76 are in different parts of the lowlands, both within 'old' and 'young' glacial relief.

From the geomorphological perspective, some landscape parks contain landforms of at least national significance, but these are in minority. In most cases, especially in the lowland part of Poland, relief is rather subdued and the values of particular areas reside in their biodiversity attributes, the occurrence of expansive forest complexes and wetlands, and aesthetic qualities of the landscape associated with traditional land use and limited human impact. Table 6.2 lists selected landscape parks, whose geomorphology is particularly distinctive within Poland and is emphasized as one of the main assets of these protected areas (Fig. 6.8).

6.3.4 Nature Reserves

More than 1500 nature reserves have been established in Poland, but only a small fraction (less than 80) is explicitly focused on the protection of abiotic elements of nature, be they rock outcrops, surface landforms, caves, or soil sequences. The number of landscape-focused reserves is 108, whereas nearly 200 peat bogs, whose existence is crucially underpinned by geomorphology, are also under protection as nature reserves. However, even though some nature reserves may be formally classified as floristic, forest or faunistic, they can still include important geoheritage values. The thematic range of geomorphological features within nature reserves is highly variable, as demonstrated by the selection of areas (numbers after the reserve name refer to Fig. 6.8, where locations are indicated) briefly described below:

- Ostrzyca Proboszczowicka (Sudetes; no. 51)—distinctive Neogene basaltic cone, 150 m high, with numerous crags and periglacial scree slopes.
- Głazy Krasnoludków (Dwarfs' Boulders; Sudetes; no. 52)—minor sandstone cuesta, with a 700 m long line of rock outcrops sculpted into rock walls, spurs and hoodoo rocks, with intervening clefts and roofed fissures.
- Wodospad Wilczki (Wilczka Waterfall; Sudetes; no. 53)—one of the highest waterfalls in the Sudetes (22 m high), with associated slot canyon marking the progress of headward erosion.
- Jaskinia Niedźwiedzia (Bear Cave; Sudetes; no. 54)—the longest, multi-level cave in the Sudetes, with rich sinter decorations and palaeontological findings; surface karst phenomena in the vicinity (ponors, springs, dry valleys).
- Wisła (Vistula River; Carpathians; no. 55)—headwater reaches of the longest Polish river, with numerous waterfall steps and bedrock channels.
- Przełom Białki (Białka Water Gap; Carpathians; no. 56)—breakthrough of the Białka River between two limestone mega-crags, which are parts of the Pieniny Klippen Belt.
- Wąwóz Homole (Homole Gorge; Carpathians; no. 57)—limestone gorge, with numerous crags and caves along valley sides.
- Diable Skały (Devil's Rocks; Carpathians; no. 58)—cluster of sandstone rock outcrops of different shapes, indicating various pathways of selective weathering, gravity-induced caves and weathering caverns.
- Luboń Wielki (Carpathians; no. 59)—landslide system with large rock-cut main scarp, gravity-induced trenches, caves and blockfields.
- Barnowiec (Carpathians; no. 60)—landslide system with large, partly rock-cut main scarp, an assemblage of gravity-induced trenches, and peat bogs.
- Sine Wiry (Carpathians; no. 61)—water gap of the Wetlinka River through structure-controlled flysch ridge, with bedrock channel sections and minor in-channel rock steps.
- Szwajcaria Ropczycka (Carpathians; no. 62)—minor loess gully within the loess-covered front of the Carpathians.
- Groty Kryształowe (Crystal Grottos, Carpathians; no. 63)—unique natural underground caverns with walls covered with halite crystals (accessible from galleries of the Wieliczka Mine).
- Góra Zborów (Zborów Hill, Kraków-Częstochowa Upland; no. 64)—large cluster of limestone crags, forming a rock city, with caves and various surface karst phenomena.

Table 6.2 Selected landscape parks in Poland, with strong geoheritage component, according to morphogenetic belts

Name (numbers refer to Fig. 6.8)	Area [km^2]	Key geomorphological features	Remarks
Medium-altitude mountains (Alpine system)			
Żywiecki [24]	359	Denudational relief on flysch, landslides, waterfalls	
Ciężkowicko-Rożnowski [25]	182	Sandstone tors and crags, non-karstic caves	
Popradzki [26]	534	Denudational relief on flysch, landslides, Poprad river water gap	
Czarnorzecko-Strzyżowski [27]	257	Denudational relief on flysch, sandstone tors	
Ciśniańsko-Wetliński [28]	515	Denudational relief on flysch, structure-controlled ridges, landslides, transverse drainage	Chapter 20
Medium-altitude mountains (Variscan system)			
Rudawski [29]	157	Granite tors and block fields, rock-controlled residual hills, mining landforms	
Chełmy [30]	160	Basaltic hills, sandstone cuesta, fluvial gorges due to meltwater erosion	Chapter 8 Aspiring UNESCO Global Geopark
Sudetów Wałbrzyskich [31]	65	Landslides, rock-controlled relief on volcanic/sedimentary rocks	Chapter 9
Fore-mountain terrains			
Ślężański [32]	82	Gabbro inselberg, periglacial blockfields, crags	Chapter 12
Uplands			
Góra Świętej Anny [33]	51	Karst, loess gullies, remnants of Neogene volcanism	Certificate of national geopark
Załęczański [34]	147	Karst and caves, water gap of Warta river	
Orlich Gniazd [35]	608	Cuesta landscape, residual relief on massive limestone, karst and caves	Chapter 22
Dolinki Krakowskie [36]	207	Limestone gorges, karst and caves, fault-generated escarpments	Chapter 22
Chęcińsko-Kielecki [37]	198	Structure-controlled ridges, karst, mining landforms	UNESCO Global Geopark
Nadnidziański [38]	229	Gypsum karst	Chapter 24
Kazimierski [39]	150	Water gap of Wisła river, loess gullies	Chapter 26
Old glacial relief			
Łuk Mużakowa [40]	187	Push moraine, mining landforms	Chapter 28 UNESCO Global Geopark
Podlaski Przełom Bugu [41]	309	High bluffs along Bug river, meandering river	
Young glacial relief			
Łagowsko-Sulęciński [42]	54	Moraine hills, tunnel valleys, lake basins	
Cedyński [43]	308	Morainic upland, high bluffs along Odra river, floodplain	
Szczeciński [44]	91	Morainic upland, erosional relief, mining landforms	
Drawski [45]	384	Morainic upland, tunnel valleys, erratic boulders	Chapter 35
Kaszubski [46]	332	Morainic upland, tunnel valleys, deeply incised fluvial valleys	Chapter 36
Trójmiejski [47]	199	Morainic upland	
Mazurski [48]	537	Diverse glacial topography, lake basins, boulder fields	
Suwalski [49]	63	Diverse glacial topography, glacial curvilineations, eskers, tunnel valleys, dead ice basins, boulder fields	Chapters 38 and 39
Coastal			
Mierzeja Wiślana [50]	44	Spit, dunes	Chapter 42

Fig. 6.9 Groty Mechowskie (Mechowo Grottos) in the Gdańsk Coastland—the longest cave in the Polish lowland (61 m), eroded in Pleistocene sands under the cemented sandstone cap and partly anthropogenically enlarged, has status of a nature monument. **a** Outside view. **b** Inside the cave (photographs by P. Migoń)

- Wąwóz Bolechowicki (Bolechowice Gorge; Kraków-Częstochowa Upland; no. 65)—limestone gorge incised into the fault-generated escarpment, with tall rock cliffs, scree and minor caves.
- Zelejowa (Kielce Upland; no. 66)—limestone ridge of hogback type, structure-controlled flank of the anticline, karren and other surface karst landforms.
- Skałki Piekło nad Niekłaniem ('Hell' Crags next to Niekłań; Kielce Upland; no. 67)—large group of sandstone tors, with minor non-karstic caverns, various weathering features and evidence of mass movement.
- Kadzielnia (Kielce Upland; no. 68)—caves and palaeokarst phenomena exposed in a disused limestone quarry, in the centre of the city of Kielce.
- Chelosiowa Jama (Kielce Upland; no. 69)—palaeokarst phenomena of late Palaeozoic to Cenozoic age, numerous caves with a maze system of Chelosiowa Jama-Jaskinia Jaworznicka, ca. 4 km long.
- Nad Tanwią (At the Tanew River; Roztocze Upland; no. 70)—minor bedrock steps and rapids in the channel of the Tanew River, in the marginal part of the upland.
- Niebieskie Źródła (Blue Springs; Pilica valley, central Poland; no. 71)—group of large karst springs.
- Skalisty Jar Libberta (Libbert's Rock Gully; Pomeranian Lakeland; no. 72)—gully system incised into morainic upland, with outcrops of unique Quaternary sandstones and conglomerates that form crags and rock pillars a few metres high.
- Źródła Rzeki Łyny (Łyna River Springs; Masurian Lakeland; no. 73)—sapping amphitheatres and headwater parts of the Łyna River within moraine hills.
- Głazowisko Bachanowo (Bachanowo Boulder Field; Suwałki Lakeland; no. 74)—boulder lag deposits left after the wastage of the last Scandinavian ice sheet.

6.3.5 Site Protection

Among many nature monuments, some were designated to protect distinctive geomorphological features. Most represented are bedrock crags and tors in the south of the country, as well as erratic boulders in the central and northern part of Poland. Other geomorphological sites include waterfalls, caves (Fig. 6.9), gullies, scree slopes and springs. Another group are former quarries, which now expose rocks and structures. Although these are not natural landforms, they may be important for geomorphology, helping to decipher geological controls of particular landscape features. For example, abandoned basalt quarries in the West Sudetes, some of which are protected as nature monuments, usually reveal the geometric pattern of columnar jointing, otherwise difficult to discern. These patterns bear on the distribution and shapes of summit and midslope crags, as well as inform about the depth of erosion of former volcanic edifices, showing that the hills are not volcano remnants but rather exposed volcanic vents (necks) (Fig. 6.10).

Sites of scientific interest were introduced as a category of nature conservation in 1991 and are mainly used to protect geological outcrops. However, some of these outcrops are also of some geomorphological significance, illustrating wider relationships between rocks, structures and topography. In addition, some important geomorphological features are located within nature-landscape complexes, whose total number in Poland approaches 350. For instance, this category of protection was selected for the historical 1598 landslide in the town of Bardo in the Sudetes. Other representative sites protected in this way are natural stream channels and valley floors, dune fields, gullied hillslopes, rock cliffs crowned with historical monuments (castle ruins,

Fig. 6.10 Geological outcrop of basaltic columnar jointingJointingcolumnar at Perkun's Cone (West Sudetes) has geomorphological significance, revealing the inner structure of a hill. The site also illustrates problems with maintaining visibility of geological and geomorphological features in humid areas, with expansive vegetation (photograph taken in summer) (photograph by P. Migoń)

churches etc.), as well as surface remnants of historical mining.

As indicated at the beginning of this section, according to the Nature Conservation Law, site protection is now decided at the local level, implying bottom-up approach (even though informed by scientific community) and participatory management of natural resources. In effect, implementation of these means of conservation varies between administrative regions and the number of protected sites does not necessarily reflect the actual value of geoheritage within a given area.

6.3.6 Other Non-Legal Means of Geosite Recognition

Geosites can be defined, after Reynard (2004), as 'portions of the geosphere that present a particular importance for the comprehension of Earth history'. They can illustrate various aspects of geological and geomorphological evolution, and if there is a clear focus on the origin and characteristics of landforms, such geosites are distinguished as geomorphosites (Reynard 2009). Following this understanding, various localities within national parks, landscape parks, nature reserves, as well as nature monuments and sites of scientific interest can be considered as geosites. Some may be of mainly scientific value, with resultant limitations imposed on accessibility, whereas other may combine scientific significance with the suitability to develop geo-education and geotourism. However, a national inventory of geosites, which would also inform conservation efforts at regional and local level, does not exist in Poland. The Polish Geological Survey maintains the Central Register of Geosites (https://cbdgportal.pgi.gov.pl/geostanowiska/; access date 2022-06-08), but the regional coverage is very uneven, many regional and thematic gaps exist, and many site descriptions need update after they were originally provided 10 or so years ago. Moreover, the Register mainly includes entries relevant to 'classic' geological sites such as natural rock outcrops, quarries (existing or abandoned), or mines. Geomorphology is clearly under-represented, with many important sites missing, whereas others, even though included, are only described in terms of rock record. The Central Register of Geosites includes both localities, which are legally protected using one of the means discussed above, and those lacking any formal protection.

For specific regions, gaps may be filled by local inventories prepared as part of various conservation-oriented projects, such as management plans for protected areas, evaluations of nature resources for various administrative entities, or geosite inventories for possible future geoparks. However, these are not always easily accessible, remaining as unpublished studies. Geosite inventories have also been produced as components of research projects, later published in scientific papers (e.g. Chrobak 2016; Różycka and Migoń 2018; Jamorska et al. 2020; Szadkowska et al. 2022), and these can be accessed without difficulties. These inventories may play an important role at the regional/local level and if the results are communicated to the local stakeholders, they help to raise awareness of geoheritage values, ultimately also promoting conservation efforts, better maintenance and development of on-site interpretation (e.g. Pijet-Migoń and Migoń 2019).

6.4 Geoparks

Geoparks are among the most effective means to raise awareness of geoheritage by focusing on geo-education and promotion of geotourism. However, this concept is yet to be fully implemented in Poland. As of 2023, only two UNESCO Global Geoparks exist in Poland and one more is officially aspiring to the Global Geopark Network (GGN). A number of projects were executed some 10–15 years ago, with the support from the Ministry of Environment and the Polish Geological Survey, focused on geosite inventories in areas suggested to become geoparks. However, that work largely followed top-down approach and there was little follow-up on the management side, whereas local interest and participation were often marginal. Consequently, most of these projects were eventually suspended or abandoned (Alexandrowicz and Miśkiewicz 2016). However, in the last decade, several truly bottom-up initiatives emerged and these raise hopes that both a national geopark network will be eventually launched and further applications for membership in the GGN will be submitted.

The first Polish member of the Global Geopark Network is the transboundary Muskau Arch UNESCO Global Geopark, approved in 2011. The key components of its geoheritage of international significance are the sedimentary record from the Neogene to the Holocene, illustrating the transition from warm climate of the Miocene to the Ice Ages, glacial transformation of pre-existing relief, involving large-scale glacitectonic deformations and the origin of a complex push moraine rampart, and varied history of coal, sand and clay mining that left impressive anthropogenic landforms. Geomorphology and geomorphological history of the Muskau Arch Geopark are presented in a separate chapter in this volume (Koźma and Migoń 2023). In 2021, the Holy Cross Mountains Geopark was approved. It includes the urban area of Kielce and the area to the west and south of the city, covering 526 km^2 in total. Despite rather small area, geology is very diverse and all geological periods of the Phanerozoic, except the Palaeogene, are represented by various sedimentary rocks. The diversified lithology, combined with fold-type Variscan and Alpine tectonics, is responsible for specific structure-controlled relief, characterized by parallel ridges separated by wide denudational depressions, not related to fluvial erosion. Among typical morphological features of this area are steep limestone hogback ridges (Fig. 6.11a, b) and crags, caves up to 4 km long, among which are two showcaves (Fig. 6.11d), various palaeokarst and paleo-surface structures exposed in old quarries (Fig. 6.11c), as well as geomorphological legacy of ancient copper mining.

In 2009–2010, an attempt was undertaken to build a network of national geoparks by means of issuing special certificates by the Ministry of Environment. Besides the Muskau Arch Geopark, which then joined the Global Geopark Network, two other areas were awarded in this way. These were the Mt. Góra Świętej Anny in the Silesian Upland (see Table 6.2 for location) and the Karkonosze National Park along with its buffer zone, presented in detail in this volume by Kasprzak (2023). However, this project lacked legal foundation and was soon abandoned. In addition, it was focused on geoheritage alone, missing the vital components of sustainable local development, priority for slow nature-based tourism and active involvement of local communities. An interest in geoparks did not disappear, though, and several regions in Poland consider the geopark formula as an adequate model of local development, with the key role of geotourism based on geo-resources, both geological and geomorphological. As for 2023, the most advanced is the Land of Extinct Volcanoes Geopark in the West Sudetes and the adjacent part of the foreland, which has a status of an Aspiring Global Geopark. Its primarily post-volcanic geomorphology, but not limited to this theme, is presented by Migoń (2023b) elsewhere in this volume. Various activities aimed at promotion of geoheritage (although with focus on rocks rather than landforms) are being undertaken in the Sudetic Foreland Geopark (Szadkowska et al. 2022), which includes Mt. Ślęża (Traczyk 2023). Slowly developing are geopark projects in the western part of the Pomeranian Lakeland (Moryń Geopark and Dębnica–Drawa Geopark, Mazurek et al. 2023), as well as in several other parts of Poland, but their future is at the moment rather uncertain.

Fig. 6.11 Selected elements of geomorphological heritage in the Holy Cross Mountains UNESCO Global Geopark. **a** Limestone hogback ridge of Mt. Miedzianka. **b** Ruins of the royal castle of Chęciny occupy a crest part of another limestone hogback ridge. **c** Paleokarstic features (palaeo-dolines indicated by arrows) in the Kadzielnia nature reserve in the city of Kielce. **d** Inside the cave in the Kadzielnia urban park (photograph **a** by J. Urban; **b**–**d** by P. Migoń)

6.5 Conclusions

Geomorphological heritage of Poland is very diverse, reflecting very complex geomorphic evolution of the country area, over variable timescales, under changing environmental conditions, and on different geological structures. Although the most scenic landforms and landscapes occur in the mountainous south and are related to erosional-denudational processes developed in response to the Neogene–Quaternary uplift of the Sudetes and the Carpathians, the relief of the predominant part of Poland is the outcome of glacial morphogenesis, with subsequent periglacial, fluvial and aeolian modifications. Individual landforms and their assemblages are not as evident as those in the mountains, but they are very important scientifically and include unique assemblages of glacial curvilineations, meltwater channels, glacitectonically deformed moraine hills, lake basins and extensive fields of inactive inland dunes.

Some of these landforms and landscapes are protected using different designations specified in the Nature Conservation Law, such as national parks, landscape parks, nature reserves, nature monuments, sites of scientific significance and nature-landscape complexes. However, various national, regional and local inventories of geosites show that the number of localities of scientific and educational significance is larger and the network of protected areas cannot be considered as complete. Geoheritage of Poland is generally easily accessible for the general public and opportunities to develop geotourism and geo-education are ample. Paradoxically, however, the concept of geoparks is rather slowly implemented in Poland, with only two UNESCO Global Geoparks at present and one Aspiring Global Geopark. Nonetheless, various local initiatives aimed at promotion of geomorphological heritage emerge, arising from good collaboration between scientists and local stakeholders.

Acknowledgements The author is grateful to Jan Urban for careful review of the draft of this chapter, corrections and helpful suggestions, as well as for kindly lending the photo included in Fig. 6.11a.

References

Alexandrowicz Z (1970) Skałki piaskowcowe w okolicy Ciężkowic nad Białą. Ochrona Przyr 35:281–335

Alexandrowicz Z (1990) The optimum system of tors protection in Poland. Ochrona Przyr 47:277–308

Alexandrowicz Z (2006) Framework of European geosites in Poland. Nature Cons 62:63–87

Alexandrowicz Z (2008) Geoconservation in Poland for progresses of long-lasting development. Przegl Geol 56(8/1):579–583

Alexandrowicz Z, Margielewski W (2010) Impact of mass movements on geo- and biodiversity in the Polish Outer (Flysch) Carpathians. Geomorphology 123:290–304

Alexandrowicz Z, Miśkiewicz K (2016) Geopark – od idei do realizacji, ze szczególnym uwzględnieniem Polski. Chrońmy Przyr Ojcz 72:243–253

Andrzejewski L, Krzemień K, Zwoliński Z (2023) Riverine landscapes. In: Migoń P, Jancewicz K (eds) Landscapes and landforms of Poland. Springer, Cham

Błaszkiewicz M, Tylmann K (2023) Tunnel valleys of the Tuchola Forest and Kashubian Lake District. In: Migoń P, Jancewicz K (eds) Landscapes and landforms of Poland. Springer, Cham

Chrobak A (2016) Valorisation and categorisation of the geosites in the Podtatrze area (Southern Poland). Geotourism 46–47:3–26

Dąbski M (2011) Grunty strukturalne w Polsce. Przegl Geogr 83:307–321

Dłużewski M, Woronko B (2023) Inland dunes of the Kampinos National Park. In: Migoń P, Jancewicz K (eds) Landscapes and landforms of Poland. Springer, Cham

Dobrowolski R (2023) Chalk karst in the Lublin region. In: Migoń P, Jancewicz K (eds) Landscapes and landforms of Poland. Springer, Cham

Dulias R (2023) Upper Silesian Region—an example of large-scale transformation of relief by mining. In: Migoń P, Jancewicz K (eds) Landscapes and landforms of Poland. Springer, Cham

Duszyński F (2023) Stołowe Mountains—a unique sandstone tableland. In: Migoń P, Jancewicz K (eds) Landscapes and landforms of Poland. Springer, Cham

Goździk J, Forysiak J (2023) Polygenetic lowland relief around the city of Łódź—glacial inheritance and periglacial transformation. In: Migoń P, Jancewicz K (eds) Landscapes and landforms of Poland. Springer, Cham

Jahn A (1975) Problems of the periglacial zone. PWN, Warszawa, p 240

Jamorska I, Sobiech M, Karasiewicz T, Tylmann K (2020) Geoheritage of postglacial areas in northern Poland—prospects for geotourism. Geoheritage 12:12

Jary Z, Marks L (2023) Quaternary history. In: Migoń P, Jancewicz K (eds) Landscapes and landforms of Poland. Springer, Cham

Kasprzak L, Ewertowski M, Szuman I, Kalita J, Tomczyk A (2023) The Middle Noteć Valley—the landscape of an ice-marginal valley and its surroundings. In: Migoń P, Jancewicz K (eds) Landscapes and landforms of Poland. Springer, Cham

Kasprzak M (2023) Karkonosze Mountains and Jelenia Góra Basin—unique variety of granite landforms. In: Migoń P, Jancewicz K (eds) Landscapes and landforms of Poland. Springer, Cham

Kłapyta P, Zasadni J (2023) Tatra Mountains—the only high-mountain landscape in Poland. In: Migoń P, Jancewicz K (eds) Landscapes and landforms of Poland. Springer, Cham

Koźma J, Migoń P (2023) Mużaków Rampart (Muskau Arch)—the legacy of glacial processes and mining in the UNESCO Global Geopark. In: Migoń P, Jancewicz K (eds) Landscapes and landforms of Poland. Springer, Cham

Łajczak A (2023) Mt. Babia Góra—the highest flysch ridge in the Western Carpathians. In: Migoń P, Jancewicz K (eds) Landscapes and landforms of Poland. Springer, Cham

Margielewski W, Jankowski L (2023a) Dunajec River Gorge through the Pieniny Mountains. In: Migoń P, Jancewicz K (eds) Landscapes and landforms of Poland. Springer, Cham

Margielewski W, Jankowski L (2023b) Impact of mass movements on landscape and landform transformation in the Beskidy Mountains—from gravitational collapse to recent times. In: Migoń P, Jancewicz K (eds) Landscapes and landforms of Poland. Springer, Cham

Margielewski W, Urban J (2023) Gravitationally-induced non-karst caves in the Polish Outer Carpathians. In: Migoń P, Jancewicz K (eds) Landscapes and landforms of Poland. Springer, Cham

Mazurek M, Paluszkiewicz R, Zwoliński Z (2023) Glacial and postglacial landforms of the Drawsko Lakeland. In: Migoń P, Jancewicz K (eds) Landscapes and landforms of Poland. Springer, Cham

Michniewicz A (2016) Skałki zieleńcowe grzbietu Okola w Górach Kaczawskich (Sudety Zachodnie). Chrońmy Przyr Ojcz 72:206–218

Migoń P (1996) Granite landscapes of the Sudetes Mountains—some problems of interpretation: a review. Proc Geol Assoc 107:25–38

Migoń P (1997) The geologic control, origin and significance of inselbergs in the Sudetes, NE Bohemian Massif, Central Europe. Z Geomorph NF 41:45–66

Migoń P (2023a) The Kamienne Mountains—Late Palaeozoic volcanic terrain shaped by Late Quaternary landslides. In: Migoń P, Jancewicz K (eds) Landscapes and landforms of Poland. Springer, Cham

Migoń P (2023b) Land of Extinct Volcanoes—rock-controlled landforms, postglacial gorges and faulted margin of the Sudetes. In: Migoń P, Jancewicz K (eds) Landscapes and landforms of Poland. Springer, Cham

Migoń A, Latocha A (2008) Enhancement of cultural landscape by geomorphology. A study of granite parklands in the West Sudetes, SW Poland. Geogr Fis Dinam Quat 31(2):195–203

Morino C, Coratza P, Soldati M (2022) Landslides, a key landform in the global geological heritage. Frontiers Earth Sci 10:864760

Moskalewicz D, Woźniak PP (2023) Vistula River delta-plain—a region of fluvial, coastal, and land reclamation impact on landscape development. In: Migoń P, Jancewicz K (eds) Landscapes and landforms of Poland. Springer, Cham

Pijet-Migoń E, Migoń P (2019) Promoting and interpreting geoheritage at the local level—bottom-up approach in the Land of Extinct Volcanoes. Geoheritage 11:1227–1236

Reynard E (2004) Geosite. In: Goudie AS (ed) Encyclopedia of geomorphology, vol 1. Routledge, London, p 440

Reynard E (2009) Geomorphosites: definitions and characteristics. In: Reynard E, Coratza P, Regolini-Bissig G (eds) Geomorphosites. Pfeil Verlag, München, pp 9–20

Rodzik J, Gajek G, Zgłobicki W (2023) Loess gullies of the Nałęczów Plateau—an outcome of human-environment interactions. In: Migoń P, Jancewicz K (eds) Landscapes and landforms of Poland. Springer, Cham

Rotnicka J, Borówka KR (2023) Sandy coast landforms of the Łeba Barrier, Słowiński National Park. In: Migoń P, Jancewicz K (eds) Landscapes and landforms of Poland. Springer, Cham

Różycka M, Migoń P (2018) Customer-oriented evaluation of geoheritage—on the example of volcanic geosites in the West Sudetes, SW Poland. Geoheritage 10:23–37

Smolska E, Pochocka-Szwarc K, Szwarczewski P (2023) Suwałki Lakeland—the best example of lowland glacial geomorphology in Poland. In: Migoń P, Jancewicz K (eds) Landscapes and landforms of Poland. Springer, Cham
Sobczyk A (2005) Rzeźba skałkowa środkowej części Gór Złotych. Przyroda Sudetów 8:147–162
Solarska A, Hose TA, Vasiljević DA, Mroczek P, Jary Z, Marković SB, Widawski K (2013) Geodiversity of the loess regions in Poland: inventory, geoconservation issues, and geotourism potential. Quat Int 296:68–81
Szadkowska K, Szadkowski M, Tarka R (2022) Inventory and assessment of the geoheritage of the Sudetic Foreland Geopark (South-Western Poland). Geoheritage 14:24
Szczuciński W, Włodarski W (2023) Morasko—meteorite impact craters in the glacial landscape. In: Migoń P, Jancewicz K (eds) Landscapes and landforms of Poland. Springer, Cham
Szczygieł J (2023) The Tatra Mountains—host of the deepest caves in the Carpathians. In: Migoń P, Jancewicz K (eds) Landscapes and landforms of Poland. Springer, Cham
Traczyk A (2023) Mount Ślęża—the highest inselberg in the Sudetic Foreland and its glacial-periglacial morphogenesis. In: Migoń P, Jancewicz K (eds) Landscapes and landforms of Poland. Springer, Cham
Tyc A (2023) Kraków-Częstochowa Upland—Monadnocks and relic caves in the karst landscape. In: Migoń P, Jancewicz K (eds) Landscapes and landforms of Poland. Springer, Cham
Urban J (2016) The geological constraints of the development of sandstone landforms in Central Europe, a case study of the Świętokrzyskie (Holy Cross) Mountains, Poland. Geomorphology 274:31–49
Urban J (2023) Sandstone crags and blockfields in the Świętokrzyskie (Holy Cross) Mountains. In: Migoń P, Jancewicz K (eds) Landscapes and landforms of Poland. Springer, Cham
Urban J, Górnik M (2017) Some aspects of lithological and exogenic control of sandstone morphology, the Świętokrzyskie (Holy Cross) Mts. case study, Poland. Geomorphology 295:773–789
Urban J, Chwalik-Borowiec A, Kasza A, Zieliński A, Komorowski A (2023) The south-eastern Nida Basin (Ponidzie)—Karst in gypsum rocks. In: Migoń P, Jancewicz K (eds) Landscapes and landforms of Poland. Springer, Cham
Weckwerth P, Wysota W (2023) Unique landscape originated by cataclysmic glacial floods at the Weichselian glaciation decline in north-eastern Poland. In: Migoń P, Jancewicz K (eds) Landscapes and landforms of Poland. Springer, Cham
Welc EM, Miśkiewicz K (2020) The concept of the geotourism potential and its practical application: a case study of the Prządki (the Spinners) Nature Reserve in the Carpathians, Poland. Resources 9:145
Wróblewski R, Rucińska M, Sitkiewicz P, Rudowski S (2023) Hel Peninsula—a distinctive spit on the Polish Baltic Coast. In: Migoń P, Jancewicz K (eds) Landscapes and landforms of Poland. Springer, Cham
Wysota W, Adamczyk L, Piotrowski JA, Weckwerth P (2023) Glacial curvilineations of the Dobrzyń Lakeland—spectacular landscape carved by subglacial meltwater erosion. In: Migoń P, Jancewicz K (eds) Landscapes and landforms of Poland. Springer, Cham
Zawiejska J, Wyżga B, Kukulak J, Chmielowska-Michalak D (2023) Hills, gravelly floodplains and peat bogs of the Podhale—a unique landscape mosaic between the Tatra, the Beskidy and the Pieniny Mountains. In: Migoń P, Jancewicz K (eds) Landscapes and landforms of Poland. Springer, Cham
Zieliński T (2023) The valley of anastomosing Narew River (NE Poland)—a unique fluvial landform of the European lowlands. In: Migoń P, Jancewicz K (eds) Landscapes and landforms of Poland. Springer, Cham
Zwoliński Z, Kostrzewski A, Winowski M, Mazurek M (2023) Wolin Island—outstanding geodiversity on the Polish coast. In: Migoń P, Jancewicz K (eds) Landscapes and landforms of Poland. Springer, Cham

Piotr Migoń is Professor of Geography at the University of Wrocław and former Director of the Institute of Geography and Regional Development. His main research interests are rock control in geomorphology, especially in sandstone and granite areas, mass movements in mountain terrain, geoheritage and geotourism. The Sudetes Mountains in Central Europe is his main research area, but he was also involved in projects in various other European countries (Czechia, Germany, Great Britain, Sweden, Portugal, Spain), in the United States, Mexico, Brazil, China, Jordan and Namibia. He used to be the Secretary of the International Association of Geomorphologists (1997–2001), its Vice-President (2009–2013) and member of Executive Committee (2013–2022); he was also the President of the Association of Polish Geomorphologists (2005–2008). He is the series editor of 'World Geomorphological Landscapes' series and member of editorial board of *Geomorphology* journal.

Part II

Regional Studies

7 Karkonosze Mountains and Jelenia Góra Basin—Unique Variety of Granite Landforms

Marek Kasprzak

Abstract

This chapter presents the essential geomorphological characteristics of the Karkonosze Mts. and Jelenia Góra Basin in southwestern Poland. These two adjacent mesoregions were mostly formed on bedrock of one plutonic body. Together, they present a unique example of a granite landscape differentiated by tectonic and denudational factors in Poland. The Karkonosze Mts., which also constitute a border area between Poland and the Czech Republic, are the highest mountain massif within the Sudetes and the entire Variscan belt of Central Europe. They are distinguished from other massifs in the Polish Sudetes by high local elevation, the presence of tree line, clear morphological evidence of Pleistocene mountain glaciation, and periglacial landforms. Harsh environmental conditions of the ridge parts are often compared to the areas of northern Fennoscandia, especially extensive summit planation surfaces, located at an altitude of 1200–1400 m asl and partially occupied by blanket mires. The morphology and origin of the Jelenia Góra Basin are no less intriguing. Its surface resembles an exposed weathering front. Numerous inselbergs rise above the basin floor, reflecting specific fracture patterns in granites and petrographic diversity. The whole area abounds in tors with a rich inventory of microforms due to granite weathering. In the past, the region was subjected to mountain floods at an exceptional scale. They have been well documented, as are mass movements on the mountain slopes, mainly debris flows. The area is also interesting because of the centuries-old human impact on geomorphology.

M. Kasprzak (✉)
Institute of Geography and Regional Development, University of Wrocław, pl. Uniwersytecki 1, 50-137 Wrocław, Poland
e-mail: marek.kasprzak@uwr.edu.pl

Keywords

Granite landforms · Tors · Inselbergs · Planation surfaces · Neotectonics · Glaciations · Periglacial landforms · Debris flows · Floods · Human impact

7.1 Introduction

The Karkonosze Mountains are a massif in the western part of the Sudetes in southwestern Poland. It straddles the border with the Czech Republic, where its name is similar—*Krkonoše*—and comes from words meaning 'dwarf pine' and 'wear'. In German toponymy, the name is different and sounds *Riesengebirge*, which means the *Giant Mountains*. Although the Karkonosze with its highest peak, Śnieżka (1603 m asl), are the most elevated mountain massif among the Variscan ranges of Central Europe, they are classified as medium-elevation mountains (Germ. *Mittelgebirge*). Thus, the name of *Riesengebirge* emphasised a distinctive, steep rise of the Karkonosze above the floor of the adjacent Jelenia Góra Basin, with a total elevation difference of c 1000 m.

The Karkonosze and the Jelenia Góra Basin are linked by a common lithology—granite—and landforms related to its selective weathering—tors (Migoń 1996). The Karkonosze, their foothills, and the Jelenia Góra Basin are often called the Karkonosze region, and this area will be the subject of this chapter. In addition, the Czech part of the Karkonosze will be considered when necessary, although this area was described separately, similar in scope to this one, by Pilous (2016).

The Karkonosze have distinct, altitude- and vegetation-related morphodynamic zones, including those at the footslope, in largely deforested areas, on forested higher slopes, and above the tree line (Treml et al. 2008). The landscape above 1000 m asl hosts numerous landforms related to mountain glaciations (Nývlt et al. 2011; Engel et al. 2014), giving parts of the mountains specific features of an alpine

P. Migoń and K. Jancewicz (eds.), *Landscapes and Landforms of Poland*, World Geomorphological Landscapes, https://doi.org/10.1007/978-3-031-45762-3_7

Table 7.1 Climatic conditions of the Karkonosze Mts and Jelenia Góra Basin (after: Sobik et al. 2019, supplemented)

	Jelenia Góra (342 m asl)	Karpacz (650 m asl)	Szklarska Poręba (643 m asl)	Mt Szrenica (1332 m asl)	Mt Śnieżka (1602 m asl)
Average annual temperature [°C]	7.9	6.7	6.5	1.9	0.7
Average January temperature [°C]	−2.4	−2.0	−2.1	−6.8	−7.0
Average July temperature [°C]	17.1	15.4	15.2	10.0	8.5
Average annual precipitation [mm]	699	997	1141	1429	1090.6
Annual average number of days with fog	68	61	34	274	296

environment (Fig. 7.1). In the highest altitude zones, landforms characteristic of the periglacial zone occur, including blockfields, cryoplanation terraces, solifluction lobes, and even relicts of rock glaciers. A distinctive feature of the Main Ridge of Karkonosze is the presence of extensive planation surfaces at the elevation of 1200–1400 m asl, partly occupied by blanket mires (Skrzypek et al. 2009). All this makes the Karkonosze more reminiscent of the *fells* of northern Fennoscandia, such as Pallas-Yllästunturi in Finland, than other European medium-altitude mountain massifs. These similarities also apply to climatic conditions (Table 7.1). The Karkonosze, together with the Izerskie Mts. to the west (Migoń 2016), are the significant orographic barriers in the path of Atlantic air masses travelling to the east. This results in the deformation of the airflow field and, consequently, in the occurrence of strong winds, frequent fog (Błaś et al. 2002), and significant moisture input from the atmosphere, including frost and rime (Migała et al. 2002). Rainfall during summer sometimes exceeds 250 mm/day, which is peculiar to this climatic zone and results in debris flows and dynamic fluvial processes in river valleys. Winter snow conditions are conducive to avalanches. In terms of geomorphology, the Jelenia Góra Basin is no less unique (Fig. 7.2). Its undulating to hilly

Fig. 7.1 View of the ridge part of Karkonosze Mts. In the foreground are the twin glacial cirques of Śnieżne Kotły undercutting the summit planation surface, periglacial granite blockslopes in the foreground and left of the cirques. The highest peak of Mt. Śnieżka in the background, to the left (photograph courtesy of B. Schutty, facebook.com/schuttyphotography)

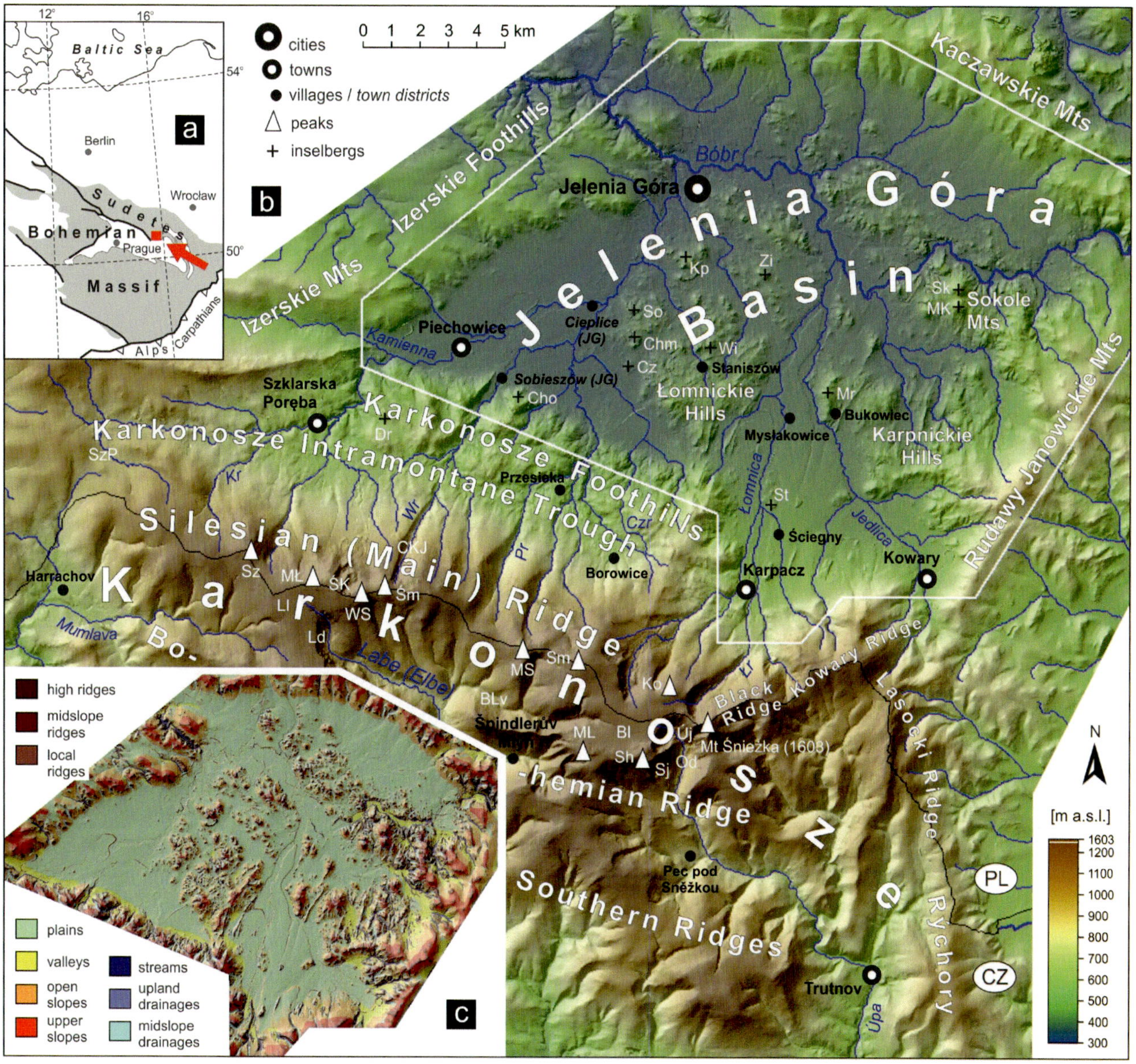

Bl – Bílá louka
BLv – Bílé Labe valley
Chm – Chmielnik
Cho – Chojnik
CKJ – Czarny Kocioł Jagniątkowski
Cz – Czubek
Czr – Czerwonka river
Dr – Drewniak
Ko – Kopa
Kp – Kopki
Kr – Kamieńczyk river
Ld – Labský důl
Ll – Labska louka
Łr – Łomniczka river
MK – Mt Krzyżna
ML – Mt Luční
MŁ – Mt Łabski
Mr – Mrowiec
MS – Mały Szyszak
Od – Obří důl
Pr – Podgórna river
Sj – Studniční jamy
Sk – Sokolik
Sm – Smogornia
So – Sołtysia
St – Straconka
Sz – Szrenica
Szp – Szklarska pass
ŚK – Śnieżne Kotły
Śm – Śmielec
Új – Úpská jáma
Wi – Witosza
Wr – Wrzosówka river
WS – Wielki Szyszak
Zi – Ziębiniec

Fig. 7.2 Topographic situation of the Karkonosze Mts. and Jelenia Góra Basin. **a** Location within the Bohemian Massif (red rectangle). **b** Hypsometric map with hillshade, names of objects mentioned in the chapter are given, white frame marks the range of the map in figure **c**. **c** Inselbergs inside the Jelenia Góra Basin. Landforms were extracted automatically based on geomorphometric modelling with the use of topographic position index (Wilson and Gallant 2000)

relief, with inselbergs, tors, and boulders, is compared to an exposed weathering front and one of the model types of granite morphology (Migoń 1997), referred to as multiconcave relief. It has no equivalent anywhere else in Poland.

Exploration of the Karkonosze massif started with an intensive search for gold and gemstones between the eleventh and thirteenth centuries and for metal ores (silver, tin, lead, iron, copper, and others), particularly intensive since the twelfth century. Since the Middle Ages, the Karkonosze were deforested for the local glass industry and used for cattle grazing. In the seventeenth century, settlement pressure led to the establishment of numerous villages, even at altitudes of 900 m and above (Szmytkie and Kasprzak 2016). Roads, embankments, and other man-made landforms were created.

At the end of the nineteenth century, the need to protect the natural values of the Silesian part (north side) of the Karkonosze region began to be recognised, primarily because of the impoverishment of the habitats of rare plant and animal species. The first protected areas were designated in 1929, and in 1933, the Karkonosze were declared a plant protection area and the first reserves under state protection were established. It was also the time when impressive granite tors started to be treated as nature monuments. After World War II, the new Polish administration led to the creation of the Karkonosze National Park in 1959 (now with an area of 5951.42 ha and 13,093 ha of buffer zone), which together with the Czech Krkonošský národní park established in 1963, since 1992 forms a UNESCO Biosphere Reserve (MaB) with a total area of over 60 000 ha.

7.2 Geological Setting

The Karkonosze are located in the northeastern part of the Bohemian Massif and were uplifted during the late stages of the Alpine orogenesis. The activity of tectonic processes was most significant from the middle Miocene to the beginning of the Quaternary, i.e. roughly between 16 and 2.6 Ma (Ziegler and Dèzes 2007; Jarosiński et al. 2008, 2009). However, geologically, they are part of an old Variscan orogen, formed in the late Devonian and Carboniferous. They represent its deeper part, exposed due to erosive removal of much of the cover rocks. Its central part is occupied by a granitoid pluton formed in the late Carboniferous, within only a few hundred thousand years, between 313 and 312 Ma (Žák et al. 2013; Kryza et al. 2014; Kusiak et al. 2014), at a depth of 7–10 km. The outcrop of the pluton is elongated in the ENE–ESE direction (Fig. 7.3), which is about 60 km long and 10–25 km wide. The marginal parts of the pluton are covered by a metamorphic complex represented by tectonically deformed rocks.

The plutonic rocks of the Karkonosze are formed by several varieties of granite of different grain sizes and mineral compositions (Borkowska 1966). The most common variety is the porphyritic granite with large feldspar crystals in the medium-grained mineral mass and common biotite schlieren. It occurs mainly in the Jelenia Góra Basin and on the lower slopes of the Karkonosze. The mid-slopes are composed of porphyritic granite or granite with occasional large feldspars; however, this boundary is challenging to capture cartographically (Mierzejewski 2005). The Main Ridge is formed by granites with singular large feldspars and equigranular granites. In the region, there also occurs aplite granite with frequent feldspar–quartz intergrowths. During the solidification of the granitic body, veins of microgranite, quartz, aplite, lamprophyre, and pegmatite were formed from remnant magma. Granitic rocks are characterised by a quasi-regular pattern of fractures resulting from horizontal extension of the rock mass (Aleksandrowski et al. 2019).

Minor intrusions of Cenozoic age occur within granites. Due to their location and size, basaltoids are the best known. The age of this rock was determined to be 26 ± 1.2 Ma (Pécskay et al. 2004), and based on detailed chemical analyses, it is classified as basanite (Kozłowska-Koch 1987). The intrusion, regarded as subvolcanic (Zagożdżon and Zagożdżon 2006), is exposed in the wall of the glacial cirque of the Mały Śnieżny Kocioł, at approx. 1400 m asl.

7.3 Main Landscape Features

7.3.1 Karkonosze Mountains

The length of the Karkonosze range is about 40 km, and its width varies from 8 to 20 km. The area of the mountains is about 650 km^2, of which 185 km^2 (28.5%) belongs to Poland. The Karkonosze consist of the latitudinally running Silesian Ridge, also called the Main Ridge, Black Ridge, and Kowarski Ridge, from which shorter ridges perpendicular to them branch off (Fig. 7.2). On the Czech side, one can distinguish the Bohemian Ridge and the Southern Ridges. In the east, the nearly meridional Lasocki Ridge connects with the Rudawy Janowickie Mts. to the northeast, whereas its extension on the Czech side is known as the Rýchory. The lower parts of the Karkonosze on the northern side of the massif are called the Karkonosze Foothills. They are separated from the Main Ridge by the range-parallel Karkonosze Intramontane Trough.

The Karkonosze form an asymmetric horst, with the steeper slope descending towards the north. This asymmetry results from the nonuniform tectonic uplift of the massif, more substantial in its granite part with lower rock density in relation to the metamorphic complex in the south (Migoń 1991; Pilous and Migoń 2007). The upper part of the Main Ridge is morphologically diversified and contains extensive summit surfaces of low relief. The largest, located at the altitude of 1350–1450 m asl, are the flattenings of Równia pod Śnieżką, Bílá louka, and Labska louka (Fig. 7.4). There are no other such extensive, high-elevation planar surfaces in the whole Sudetes. The slopes of the Karkonosze are generally long and uniformly inclined, rarely incised by river valleys. These valleys are deep, especially on the southern side of the mountains. The valley bottoms of the Labe and Úpa rivers are 500–600 m below the water divides.

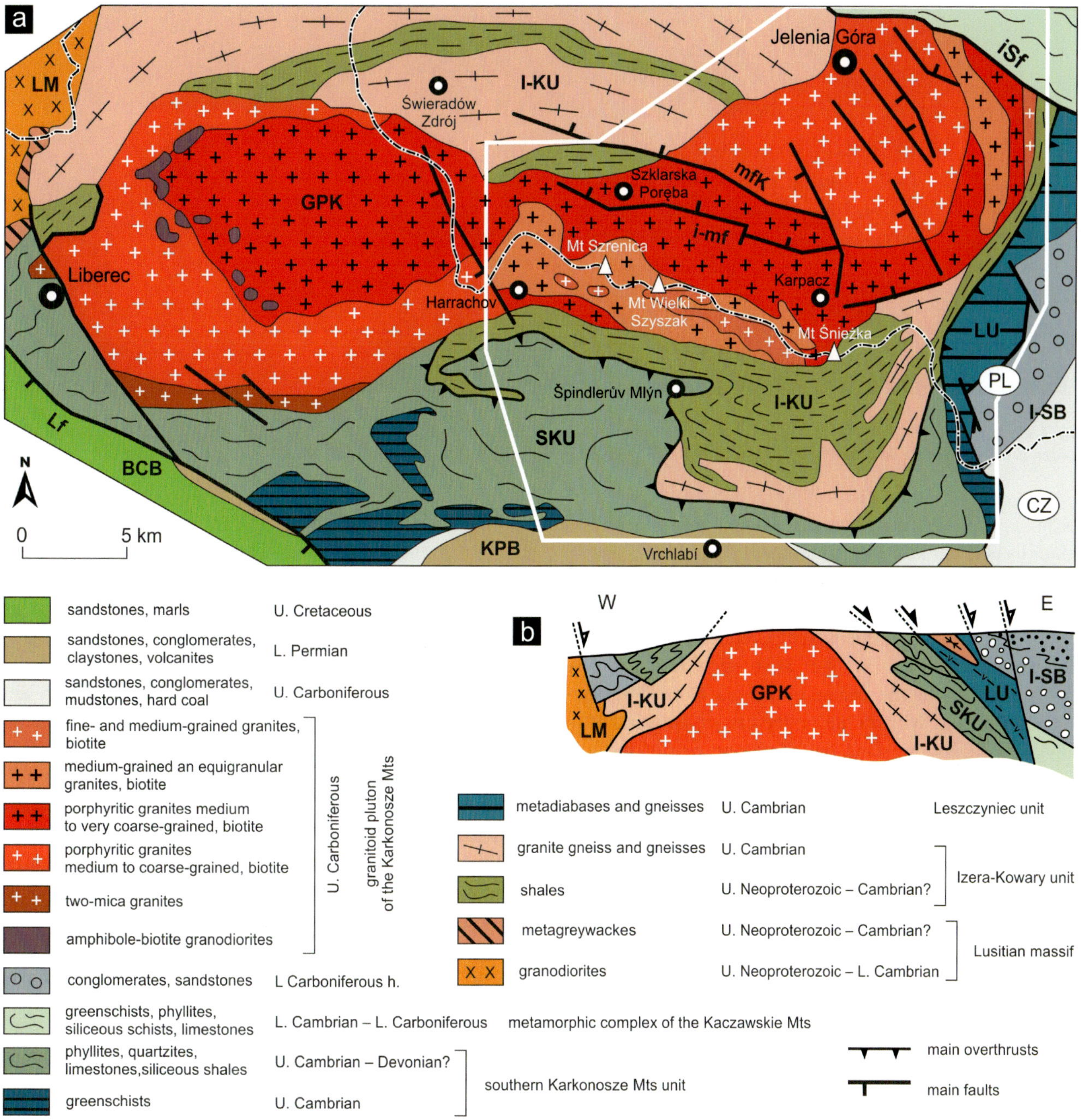

Fig. 7.3 Geology of the Karkonosze Mts. and Jelenia Góra Basin. **a** Simplified geological map. **b** Synthetic cross-section (without scale). Explanations: I-KU—Izera-Kowary unit, LU—Leszczyniec unit, SKU—southern Karkonosze unit, LM—Lusatian massif, I-SB—Intra-Sudetic Basin, KPB—Karkonosze Piedmont Basin, BCB—Bohemian Cretaceous Basin, GPK—granitoid pluton of the Karkonosze; faults: mfK—marginal fault of the Karkonosze, Lf—Lusatian Fault, i-mf—intra-mountain fault, iSf—Intra-Sudetic Fault. The white frame delimits the area shown in Fig. 7.2. Compiled from Žák and Klomínský (2007) and Aleksandrowski et al. (2019), modified by the author

7.3.2 Jelenia Góra Basin

The Jelenia Góra Basin has an area of about 270 km^2. Its floor has a rhomboidal shape and is located at 350–400 m asl. Isolated hills or groups of hills rise above the bottom of the basin (Fig. 7.2c). In its central part, they reach a height of 450–500 m asl (Łomnickie Hills), and in the eastern part, they even rise to 550–650 m asl (Sokole Mts.). The highest point of the Jelenia Góra Basin is Mt. Krzyżna (654 m asl), and the lowest one is in the riverbed of the Bóbr River west of the town of Jelenia Góra (324 m asl). The basin is surrounded by mountain ranges rising 250–1000 m above

Fig. 7.4 View from the slope of Mt. Luční on the vast summit planation surface at an altitude of c. 1500 m. Its part, on the Polish side of the border, is called Równia pod Śnieżką (photograph by M. Kasprzak)

its floor. Only the north-western border with the Izerskie Foothills is lower and does not exceed 150 m high.

The occurrence of inselbergs determines the extraordinary landscape of the Jelenia Góra Basin (Migoń 1993, 1997). The hills in the vicinity of Cieplice (hills: Sołtysia, Chmielnik, Czubek), in the area of Czarne (Kopki, Ziębiniec), east of the centre of Jelenia Góra (Paulinum, Kamienista), and Witosza hill on the eastern outskirts of the village of Staniszów stand out. However, one of the most characteristic and best-recognised views in the Sudetes is the panorama of the Sokole Mts. against the background of the Karkonosze (Fig. 7.5). The Sokole Mts. are situated in the north-eastern part of Jelenia Góra Basin and consist of six hills. The two highest, Mt. Krzyżna (654 m asl) and Sokolik (623 m asl), rise about 200 m above their surroundings.

7.4 Long-Term Landscape Evolution

The long-term denudation history of the Karkonosze region has been inferred from several complementary sources of information, mainly based on thermochronological data (Danišík et al. 2010; Sobczyk et al. 2015), examination of sedimentary records in the adjacent areas, and landform analysis (Migoń and Danišík 2012). The unroofing of the granite pluton is dated to pre-Middle Triassic times, around 230 Ma (Aleksandrowski et al. 2019), or even earlier, to the Early Permian, around 300–270 Ma (Teisseyre 1957). Three phases of increased denudation were distinguished: the early Permian, the early Triassic, and especially the late Cretaceous. Erosion rates in the latter period may have reached up to 300 m/Ma. Since the turn of the Upper Cretaceous, rocks with a thickness of 3.6 to 6 km have been removed, most of them between 100 and 75 Ma ago. No more than 1.2 km of rock was lost after this period.

The most pronounced fault-generated slopes form the margin of the Karkonosze Foothills on the border with the Jelenia Góra Basin. They stretch from the town of Piechowice in the west to the town of Kowary in the east (Fig. 7.2). The escarpment is most pronounced between Piechowice and the village of Miłków, reaching a relative height of 130–150 m near the village of Podgórzyn and 250 m above Piechowice. The effects of block movements are well visible in the valley morphology (Migoń 1992), also in the area of the Karkonosze Intramontane Trough, which is most probably a graben or half-graben, although its structural preconditioning was indicated before (Cloos 1925; Jahn 1954). In the area of the Karkonosze Foothills, indicators of non-uniform uplift include wind gaps within water-divide ridges and deep, heavily indented gorge sections formed by the largest watercourses.

The Jelenia Góra Basin is polygenetic, and particularly, its boundaries have different genesis (Migoń 1993). The northern and western border with the Kaczawskie Mts. and Izerskie Foothills were formed under protracted etchplanation, lasting through the whole Palaeogene to the Late Miocene. The southern (described above) and eastern borders of the basin are escarpments of tectonic origin, younger than the other borders. During their formation (Pliocene?), the previously formed northern edge of the Basin was additionally uplifted along the Intra-Sudetic Fault (Fig. 7.3). The floor of the Basin represents the effect of selective weathering of granite, most substantial within more densely fractured and plagioclase- and biotite-rich parts of the pluton, and less efficient in the massive dome-shaped compartments, equigranular granites and quartz-rich granites (Migoń 1997; Kajdas et al. 2017). Therefore, the latter are marked by morphological domes and ridges. In places subject to stronger weathering, troughs and secondary basins have developed. This resulted in a hilly relief with rocky hills (inselbergs).

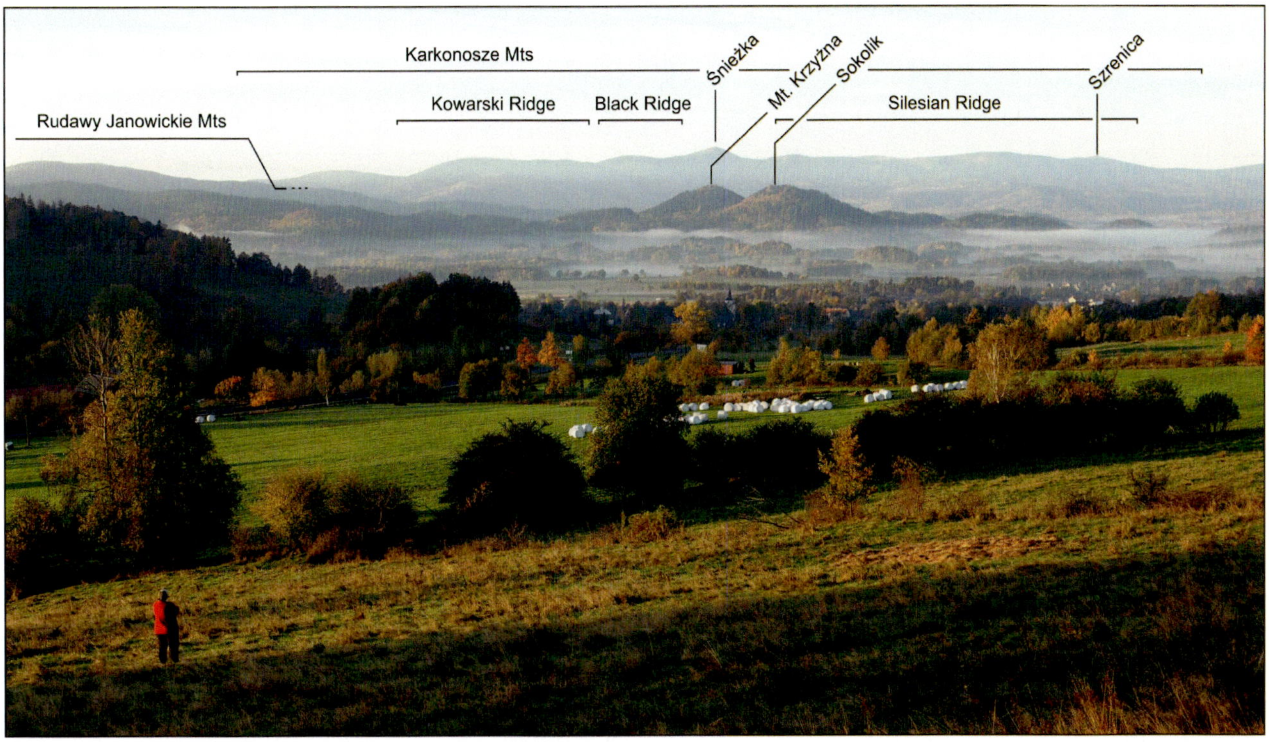

Fig. 7.5 Eastern part of the Jelenia Góra Basin and Karkonosze Mts., a view from the northeast (Radomierska Pass). Note the inselbergs of the Sokole Mts. (Mt. Krzyżna and Sokolik) in the centre of the plan (photograph by M. Kasprzak)

7.5 Specific Landforms

7.5.1 Tors

Tors of various sizes and with numerous peculiar microforms of selective weathering on their surfaces are characteristic denudational landforms of the Karkonosze region (Fig. 7.6). The Karkonosze granite is very heterogeneous and characterised by a great variety of textures. Additionally, it is not uniformly fractured. Local occurrence of particularly fine-grained variants is expressed by the presence of rocky, cone-shaped hills (Mt. Łabski, Mały Szyszak, Szrenica). The tors built of this granite variant are mainly angular in cross-section and consist of regular cuboid blocks forming walls and rock castles. Porphyritic granite, on the other hand, is much more varied, both in terms of its mineralogical composition and orientation of fractures. The weathering is therefore very selective and its efficacy varies from place to place. These tors are usually irregular in shape, formed by chaotically overlapping boulders with varying degree of roundness. Numerous large, loose blocks of granite occur alone or in clusters.

The tors are found in all altitude zones, including the main ridge of the Karkonosze. Their highest location is at 1491 m asl (Kasprzak and Traczyk 2019). They are most numerous at altitudes of 500–700 m asl, on the peaks of the Karkonosze Foothills, and in the 1000–1100 m asl belt, on the second-order ridges branching from the Main Ridge. They occur as singular forms or large clusters.

In the Jelenia Góra Basin, tors accompany inselbergs. One specific locality is Mt. Witosza (484 m asl) near the village of Staniszów. Its attractiveness is mainly due to magnificent rock walls up to 20 m high, widespread talus of massive blocks and boulders, narrow clefts, weathering pits of various sizes, rock shelters, and non-karstic caves (Migoń 1997). These rock formations were created where the regular, rectangular joint patterns in the granite bedrock play a subordinate role in respect to the concentric (dome-like) fracture system. The latter consists of curved joint surfaces, convex upwards, of probable stress-release origin (Migoń 1993). Similar dome forms occur in the Karkonosze Foothills, an example being Mt. Chojnik (627 m asl).

The Sokole Mts. have the highest density of exceptionally large, tower-like tors. The tallest ones can be found at the highest elevations. The reason for the evolution of these forms invariably lies in the properties of granite. Mt. Krzyżna and Mt. Sokolik are built of aplitic granite (aplogranite), a variety of granite with reduced content of sodium-calcium feldspar (plagioclase) and biotite. The predominance of potassium feldspars and quartz makes the rock more resistant to weathering.

Fig. 7.6 Tors of the Karkonosze Mts. and Jelenia Góra Basin. **a** Pielgrzymy tor group. **b** Weathering pits as a characteristic example of microforms due to surface weathering (Pielgrzymy tor). **c** Selective weathering of granite, with core stones embedded in sandy-gravelly mass, exposed in the former quarry on the Straconka hill, Jelenia Góra Basin. **d** Balancing rock with a movable upper boulder. **e** Medieval Chojnik castle integrated with granite outcrops in the surroundings (photographs by M. Kasprzak)

Among peculiarities of the tors in the Karkonosze region are balancing rocks—loose, locally fairly large boulders of granite resting on the base with a minimal surface of contact (Fig. 7.6d). As a result, it is possible to change their position even by pushing them by hand or loading their edges. The best-known balancing rock occurs in the town of Szklarska Poręba. Others can be found in the Karkonosze Foothills and include the Waloński Kamień tor in the village of Przesieka and two balancing rocks on Drewniak hill above Piechowice.

7.5.2 Glacial Landforms—Evidence of Pleistocene Glaciations

The Karkonosze are characterised by the best-developed Pleistocene glacial landforms in the Bohemian Massif, which were the focus of pioneering research conducted at the turn of the twentieth century (Partsch 1882, 1894; Berg 1915; Fig. 7.7). At least 12 snow accumulation zones and glaciers up to 5 km long functioned here simultaneously, and the existence of further ones remains a matter of speculation (Fig. 7.8). Ice tongues were supplied by well-developed cirques, including six on the Polish side of the mountains (twin cirques of Śnieżne Kotły, Czarny Kocioł Jagniątkowski, twin cirques of Mały and Wielki Staw, Łomniczka cirque). Glaciers flowed mainly towards the northern and eastern side of the massif, leaving the most spectacular glacial cirques inset into the edges of summit planation surfaces. Only the double cirque of Kotelní jamy on the Czech side faces south. There are no distinct landforms of this type exposed to the west. This asymmetry was explained by Jeník (1961), who drew attention to the existence of a particular anemo-orographic system. The deep valleys of the Mumlava and the Bílé Labe (Elbe), with their latitudinal course, channelise a local, western circulation of air masses, thanks to which the snow accumulated on the summit flats is then blown to the east, to the lee-side.

A typical example of glacial erosion is presented by the Śnieżne Kotły cirques (Fig. 7.1). They are twin glacial cirques with headwalls about 100 m high and slopes of 65–70°. The most distant terminal moraines are 20–30 m high, while at the mouth of the cirques, there are larger forms with a height of 40–60 m, perhaps hiding, however, the protuberances of the granite bed (Jahn 1960; Traczyk 2009). The height of the intermediate terminal and lateral moraines is, on average, 10–15 m. The glacier tongue reached a maximum distance of 2.4 km from the back wall of the cirque and descended to an altitude of 920–950 m. The glacial landscape is different in the headwater parts of the Łomnica Valley, where two lakes were created: Mały Staw (Small Lake) and Wielki Staw (Big Lake). The latter is the largest (8.32 ha) natural reservoir in the Sudetes (Kasprzak 2021).

The timing of glaciations in the Karkonosze was long debated. The distinctiveness of local moraines allows us to conclude that there were at least three main, separate phases of glacier development (Traczyk 2009). Initially, based on the now contested TL dating, Chmal and Traczyk (1999) indicated that moraines in front of the Śnieżne Kotły cirques were formed in the interval from about 95 to 10 ka BP. More recent studies based on ^{10}Be exposure ages of morainic blocks at the outlet of Śnieżne Kotły area and in the middle part of the Úpa Valley (Obří důl) suggest that the oldest preserved moraines represent early phases of the Last Glacial Maximum (LGM), 24–21 ka (Engel

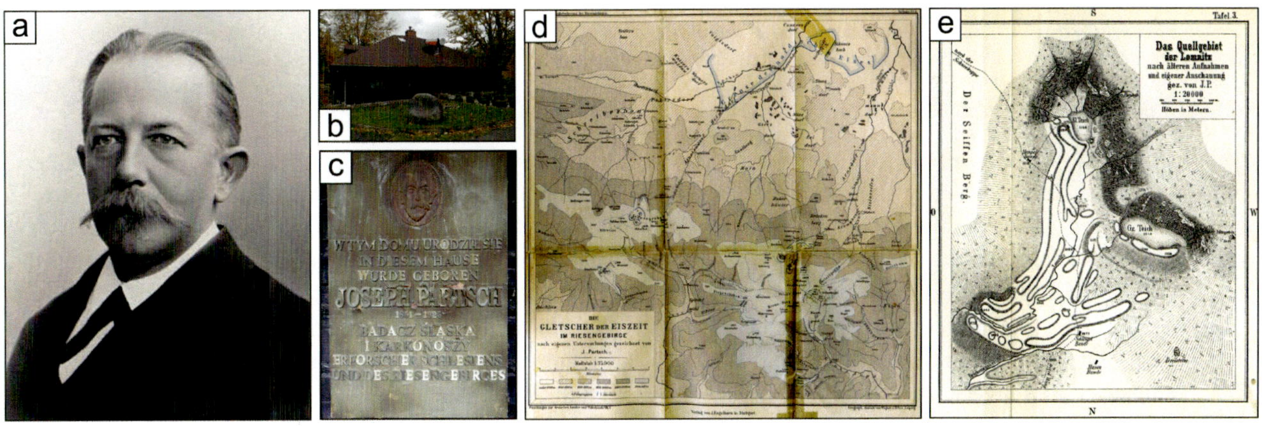

Fig. 7.7 Legacy of Joseph Partsch (1851–1925), pioneer of glacial research in the Karkonosze Mts. **a** Partsch, born in Szklarska Poręba, later rector of the University in Breslau (today University of Wrocław), is considered one of the most prominent geographers of the late nineteenth and early twentieth centuries. **b, c** Plaque commemorating Partsch can be found on the building of the former Josephine glassworks in Szklarska Poręba. **d** Map with the extent of the Scandinavian ice sheet in the Jelenia Góra Basin and the ice cap over the Karkonosze (Partsch 1882), according to contemporary knowledge mountain glaciers occupied only some valleys of this massif. **e** Map of glacial landforms in the Łomniczka Valley (Partsch 1882). Photographs: collection of the University of Wrocław, public domain (**a**), M. Kasprzak (**b, c**)

Fig. 7.8 Examples of landforms and sediments formed in glacial and periglacial conditions. **a** Nivation hollow of Biały Jar and glacial cirques filled with lakes, Mały Staw (volume 86.9 m^3, 4.6 m depth) and Wielki Staw (741.5 m^3, 23.5 m). **b** Moraines from at least three glaciations in the Łomnica Valley, the glacier crossed the watershed and occupied a part of the Złoty Potok Valley (cf. Fig. 7.4e, DTM source: CODGiK, geoportal.gov.pl, PD), there was a third lake in the Łomnica Valley, today filled by mineral and organic sediments (peat bog near Domek Myśliwski hut). **c** One of giant moraine boulders in the Złoty Potok Valley. **d** Patterned ground on the Black Ridge. **e** Pseudomorphosis of an ice wedge on the main ridge. **f** Coarse-grained slope cover transformed by solifluction in Karpacz Wilcza Poręba. Photographs courtesy of B. Schutty, facebook.com/schuttyphotography (**a**) and by M. Kasprzak (**c**–**f**)

et al. 2014). ^{10}Be dating of stable morainic boulders and rock samples from cirque headwalls in the Łomnica and Łomniczka Valleys indicated an even younger age (Engel et al. 2011). These results show that the lowest moraine ridges in the Łomnica Valley were formed around 17–16 ka BP. Moraines deposited in cirques were formed at 14–12 ka BP, and remnants of cirque glaciers existed as late as 9–8 ka BP. However, exposure ages from the lowermost zone refer to the termination of the post-depositional moraine surface transformation rather than the deposition period. Particularly interesting information was obtained from the lake sediments in the Labsky důl Valley on the Czech side of the massif (Engel et al. 2010). The retrieved core documents changes in the environment of the Karkonosze over the last 30 ka. The sedimentary record suggests that the bottom of the cirque was ice-free at the end of MIS 3 and that the glaciation re-developed during MIS 2. In conclusion, the local glaciation in the Karkonosze was possibly more extensive during the LGM than in earlier phases of the Weichselian glacial (Engel et al. 2014), contrary to the views presented before (Traczyk 1989). The preserved landforms and settlements do not allow unambiguous interpretations of the landscape evolution during the earlier glaciations.

A separate and unfinished scientific discussion concerns the maximum extent of the Scandinavian ice sheet and its presence in the Jelenia Góra Basin (Hall and Migoń 2010). Currently, contrary to earlier, long-standing opinions, it is claimed that the ice sheet intruded into the intramontane basins of the Sudetes only during the Elsterian glaciation (Badura and Przybylski 1998; Michniewicz 1998). It probably did not reach much beyond the Bóbr river valley in the northern part of the Basin, as Berg (1927, 1940a, b) implied.

7.5.3 Periglacial Conditions and Frost-Induced Landforms

The high-elevation parts of the Karkonosze are the arena of frost processes (Büdel 1937; Traczyk and Migoń 2000; Křížek 2007; Kasprzak et al. 2021). Well visible legacy of the Pleistocene periglacial environment includes frost cliffs and cryoplanation terraces. Well-developed complexes of such forms are found on the Black Ridge, Mt. Smogornia, and Mt. Wielki Szyszak. On the Czech side of the border, such forms are best developed on the northern slopes of Mt. Luční hora.

Coarse-grained slope covers are common. On granite bedrock they may contain large boulders, as it is the case on the slopes of Mt. Łabski, Wielki Szyszak and Mały Szyszak, Śmielec, Szrenica, and Smogornia peaks. On the outcrops of metamorphic rocks (schists and gneisses), they consist of somewhat smaller fractions (e.g., the slopes of Mt. Śnieżka and the Black Ridge). Blockfields occupy the upper parts of the Main Ridge above 1200–1250 m asl, forming extensive surfaces lacking vegetation cover. Smaller, non-forested blockfields are also found in the central parts of the slopes (e.g., in the Łomniczka Valley, on the northern slopes of Mt. Kopa). They have areas ranging from 0.004 to 77 ha (Kasprzak and Traczyk 2019). They also occur at lower altitudes, but are covered by forest (e.g., lower sections of the slopes of Mt. Szrenica or Mt. Łabski). Surface morphology of blockfields located above the upper forest boundary shows the presence of lobes, boulder steps, and closed depressions. During the cold periods of the Pleistocene, in places cemented by ground ice, boulders moved independently from the rest of the blockfields in a manner similar to rock glaciers (Chmal and Traczyk 1993; Żurawek 1999).

Forms of frost sorting are associated with blockfields on the summit plateau, including fields of patterned ground on the Black Ridge, on the summit plain of Równia pod Śnieżką, slopes of Mt. Luční hora and Wielki Szyszak (Fig. 7.8). Patterned ground is generally in the form of stone circles with a diameter of 1–3 m, where debris surrounds an earthy island (Křížek and Uxa 2013). The debris-loamy slope covers common on the slopes were transformed by solifluction (Büdel 1937).

The periglacial-like conditions persisted in some areas of the mountains despite climate warming in the Holocene. During the Little Ice Age, i.e. as recently as 100 years ago, the average annual temperature of the highest parts of the Karkonosze did not exceed 0 °C (Migała et al. 2016). Although the number of days with negative temperatures is lower than in the High Arctic or the Alps, there are more days with the temperature passing through 0 °C than in the aforementioned areas (Troll 1944). Therefore, under specific terrain conditions, especially on moist, clayey ground, further development of patterned ground still occur (Traczyk 1992; Soukopová et al. 1995).

7.5.4 Fluvial Geomorphology as a Response to Tectonics and Climate Changes

The Labe (Elbe), one of Europe's longest rivers, has its source in the Karkonosze, south of Mt. Łabski, at the altitude of 1387 m. On the northern side of the massif, all streams join in the floor of Jelenia Góra Basin, forming two most significant rivers—Kamienna and Łomnica. They flow into the Bóbr River, one of the main tributaries of the Odra. Many streams in the Karkonosze did not form clear valleys and flow within shallow hillslope hollows and troughs. Larger rivers, however, have greater erosive power and have eroded numerous ravine sections, such as the Kamieńczyk River. The section of its valley below the Kamieńczyk Waterfall has the form of a deep (ca. 25 m) rocky gorge. Bedrock channel reaches are common. During the largest floods, the water flow increases by 100–160 more than an average discharge (Kasprzak 2010) and streams have the capacity to transport the largest bed load elements (Walicka et al. 2019).

Waterfalls are numerous in Karkonosze (Fig. 7.9). They are mainly associated with fault-generated escarpments, as in the case of Kamieńczyk Waterfall, the highest in Polish Sudetes (27 m). Even higher waterfalls descend from the rock walls of glacial cirques on the southern side of the massif, e.g., the Pančavský Waterfall (39 m), as a part of the 162 m high cascade in the Labský důl Valley (Pilous 1989).

The formation of a distinct faulted escarpment in the northern part of the Karkonosze significantly affected the local morphology of the valleys. The largest streams such as Kamienna, Wrzosówka, and Podgórna rivers carried enough water for downcutting to keep pace with the tectonic uplift, which resulted in the creation of notched valleys with steep slopes and bedrock channels. In smaller valleys, the intensity of incision was lower; hence, they took on a bottle shape (like Czerwonka Valley) or their floors are suspended above the footline of the escarpment (Sroka 1991; Migoń 1993; Chmal and Kasprzak 2009). In the case of the Łomniczka Valley, the Pleistocene-age uplift of alluvial sediments is about 15 m. The Pleistocene uplifted terraces with coarse-grained sediments of the Wrzosówka River in Sobieszów and the Jedlica River in Kowary are also truncated by scarps of similar height. Continuous tectonic activity of the Karkonosze area is evidenced by reports of seismic phenomena. Several earthquakes were recorded here in the recent past. One of the strongest tremors took place in Kowary in 1903 (Dyjor and Oberc 1983).

The extension of the terrace system of the Karkonosze rivers are alluvial fans in the Jelenia Góra Basin. The most extensive is the Ściegny fan (Kasprzak 2009), reaching as far as the Bóbr Valley in the north (Fig. 7.9c). Its surface is incised, and younger terraces were inserted into the incisions. The complete terrace system consists of two levels, including the fan surface, regarded as Pleistocene (c. 12–20 m and 3–5 m a.r.l.) and two lower ones from the Holocene (c. 1.5–2.5 m and contemporary). In the mountainous parts, fluvial terraces are preserved only fragmentarily. In addition to the Holocene levels, isolated shelves of higher Pleistocene terraces, characterised by coarse-grained sediments, are common.

7.6 Extreme Events

In Silesia, continuous meteorological measurements have been conducted since 1810, followed by regular hydrological observations since the turn of the twentieth century. Historical literature is also abundant, so the knowledge of local natural extreme events is relatively complete. High-energy mass movements occur in the highest parts of the Karkonosze and include rock falls, snow avalanches, landslides, and debris-mudflows (Fig. 7.10). The latter are characteristic for the Karkonosze and are referred to in older literature as *Muren* (Germ.). They occur mainly in ravines dissecting headwalls of glacial cirques, but also on steep slopes of spring hollows and valleys. The largest cluster of debris-flow landforms is located in the upper part of the Łomniczka Valley, where 27 tracks have been identified. Their formation is favoured by steep slopes and thick layer of fine-grained weathering cover which, saturated with water, can move by gravity, sliding on solid rock beneath. In the whole Karkonosze Mts., traces of more than 250 debris flows of various ages have been recorded so far,

Fig. 7.9 Fluvial landforms in the Karkonosze Mts. and Jelenia Góra Basin. **a** Fault scarps are accompanied by numerous waterfalls and cascades, as in the case of the Piszczak stream, a tributary of the Jedlica River (photograph by M. Kasprzak). **b** Extensive alluvial fans provide evidence of intensive denudation and drainage of the Karkonosze during the cold periods of the Pleistocene, as in the case of the Łomniczka River fan in Karpacz Wilcza Poręba (photograph by M. Kasprzak). **c** Geological-geomorphological sketch of the Ściegny fan stretching from the Karkonosze Foothills to the Bóbr Valley in the southern part of the Jelenia Góra Basin. 1—gravels of river terraces (Holocene), 2—gravels and boulders of river terraces (Late Pleistocene), 3—gravels and boulders of river terraces (alluvial fan, Early/Middle Pleistocene), 4—glacial and glaciolacustrine deposits, 5—boulders of moraines of mountain glaciers, 6—granite bedrock, 7—gneisses and metamorphic schists, 8—glacial cirques, 9—distinct rock bases of river terraces, 10—maximum extent of the Scandinavian ice sheet, JG—Jelenia Góra, Ka—Karpacz, Ś—Ściegny, Ko—Kowary, black frames show tectonically uplifted and incised alluvial fans of the Łomniczka and Jedlica rivers (after Kasprzak 2007). **d** Scheme of river channelisation made in the early twentieth century and common in the region (after Sommer 1913—author of the photograph unknown)

including more than 70 on the Polish side of the mountains (Pilous 1973, 2016; Migoń and Parzóch 2008). The debris-flow furrows visible in the terrain morphology are between 30 and 500 m long and several or so metres wide (Parzóch et al. 2007).

The history of the Karkonosze Mts. and Jelenia Góra Basin contains many accounts of catastrophic floods (Kasprzak and Migoń 2015). The largest were those in 1608, 1736, 1778, 1796, 1813, 1858, 1882, 1897, and 1926. Large floods also occurred in 1958, 1977, and 1997. The flood in 2006 was also peculiar, when intense rainfall from 7 to 8 August was preceded by a four-week long drought. The most severe flood damage in the Karkonosze occurred in the seventeenth and nineteenth centuries. It is worth noting that many valleys experienced catastrophic floods as a result of highly localised storm precipitation at completely different times.

The largest regional flood happened at the end of July 1897 (Fig. 7.11). It was triggered by prolonged rainfall, which on 30 July exceeded the previous records of daily intensity in the Western Sudetes. In the Łomnica drainage basin alone, 20 million m^3 of rain fell daily (average 171 mm, Czerwiński 1991). On the Bóbr River in Jelenia Góra from noon of 29th July to morning of 30th July, 5 m of water was added. The discharge for Pilchowice gauging station to the north of Jelenia Góra was calculated as 1187 $m^3 s^{-1}$ at mean annual flow (MAF) of 11.6 $m^3 s^{-1}$. The discharge of the Łomnica River was 400 $m^3 s^{-1}$ (MAF 2.16 $m^3 s^{-1}$) and that of the Kamienna River 500 $m^3 s^{-1}$ (MAF 3.19 $m^3 s^{-1}$) (Partsch 1911). In Jelenia Góra, the waters inundated a wide floodplain and the water level reached the roofs of the lowest houses. The villages on the Karkonosze Foothills suffered most. The town of Kowary, where the Jedlica river formed a new riverbed, was severely affected (29 houses completely destroyed, 14 partly damaged). Similar destructive processes affected the Łomniczka River. In Karpacz, five houses and a mill were damaged, and in the village of Łomnica, nine houses perished. Losses were also recorded in many other villages. Most mountain roads were destroyed. On the Silesian (northern) side of the Karkonosze the flood claimed four victims, on the Czech side of the mountains, depending on the source, from 120 to 135 victims.

Snow avalanches may also have serious consequences in the Karkonosze. On 20 March 1968 the most tragic snow avalanche in Poland took place here, killing 19 people (Mazurski 1969). It happened in the Biały Jar Valley. The amphitheatrical valley head promotes snow deposition and is the source of frequent avalanches.

7.7 Human Impact

In the Karkonosze Mts., one of the reasons for changes in surface morphology was mining activity. Among other things, cobalt ore, pyrites, and pegmatites were extracted. Numerous heaps and sinkholes marking the location of small shafts and adits, and in some places also, fragments of underground mines, as in the vicinity of Szklarska Poręba, have survived from this period (Borzęcki et al. 2018). There are larger post-mining landforms from iron ore extraction in the Kowary area. Exploitation initiated in the fourteenth century was carried out in small open pits and shallow shafts, and in the sixteenth century reached a depth of 130 m. Mining works connected with uranium ore exploration in the 1950s were even more extensive (Borzęcki and Wójcik 2017). These resulted in a total of 5.5 km of underground mine workings. The remnants of this episode are heaps, on which a bound of 28 thousand m^3 of rock spoil was accumulated.

With intensive deforestation for metallurgy in the Middle Ages, including smelters moving up the valleys to find sources of fuel, e.g. from Piechowice through Szklarska Poręba to Jakuszyce and Orle in the Izerskie Mts., the slopes were adapted for agriculture and pastoralism. Agricultural terraces were formed, and larger stones were removed from the surface and collected along field borders in the form of stone ramparts or stone walls. The remains of such walls can be found in the village of Borowice, as well as in other areas of the Western Sudetes (Duma et al. 2020). In villages, mill canals were created, and the land was planted for weaving and linen-making.

Considerable changes in the landscape were caused by hydro-engineering works (Fig. 7.9d), undertaken after the floods in 1888 and 1897 (Budych and Majewicz 1999). In the first two decades of the twentieth century, a project of regulation of the Odra's mountain tributaries was implemented. It included building of retention reservoirs (Karpacz), dry flood control reservoirs (Mysłakowice, Sobieszów), systems of check dams, and the transformation of riverbeds into artificial cascades. In 2001, a drinking water reservoir next to the village of Sosnówka was put into operation.

In the seventeenth century, the Karkonosze region became a tourist destination, and over time, it developed into one of the most visited tourist regions in Central Europe (Fig. 7.12). Ski and sledge infrastructure was created on the slopes of Karkonosze. Such a complex was created, among others, in Szklarska Poręba, unsuccessfully competing with the Bavarian town of Garmisch-Partenkirchen for the organisation of the Winter Olympic

Fig. 7.10 Relief transformation by debris flows in the Karkonosze Mts. **a** Slopes of the Łomniczka glacial valley between Mt. Śnieżka and Mt. Kopa, arrows show fresh debris-flow scars, dated to 1964 and 1994. **b** Debris-flow tracks on the slope of Mt. Śnieżka. **c** Attempt to stop subsequent erosion within a debris-flow channel on the slope of Mt. Kopa (photographs by M. Kasprzak)

Games in 1936. From the end of the nineteenth century, following the trend of that time, natural areas were artificially made more apparently attractive for tourists. For example, water reservoirs were created above waterfalls to temporarily increase the discharge of rivers, and exotic, high-mountain plant species, such as edelweiss, the symbol of mountain hikers, were introduced into the environment.

Also in the nineteenth century, the Jelenia Góra Basin belonged to the most fashionable holiday and tourist areas. It was visited by Caspar David Friedrich, John Quincy Adams, Frederic Chopin, Izabela Czartoryska, and Johann Wolfgang Goethe. They were attracted by the health resort in Bad Warmbrunn (after 1945 Cieplice Śląskie-Zdrój, from 1976 within Jelenia Góra), castles and palaces, which were the seats of Silesian dukes, the Prussian royal family (e.g., kings of Prussia: Wilhelm III and Wilhelm IV), aristocracy and nobility of Polish, German, Czech, and Austrian families such as Radziwiłł, Czartoryski, Schaffgotsch, or Hohenzollern. In

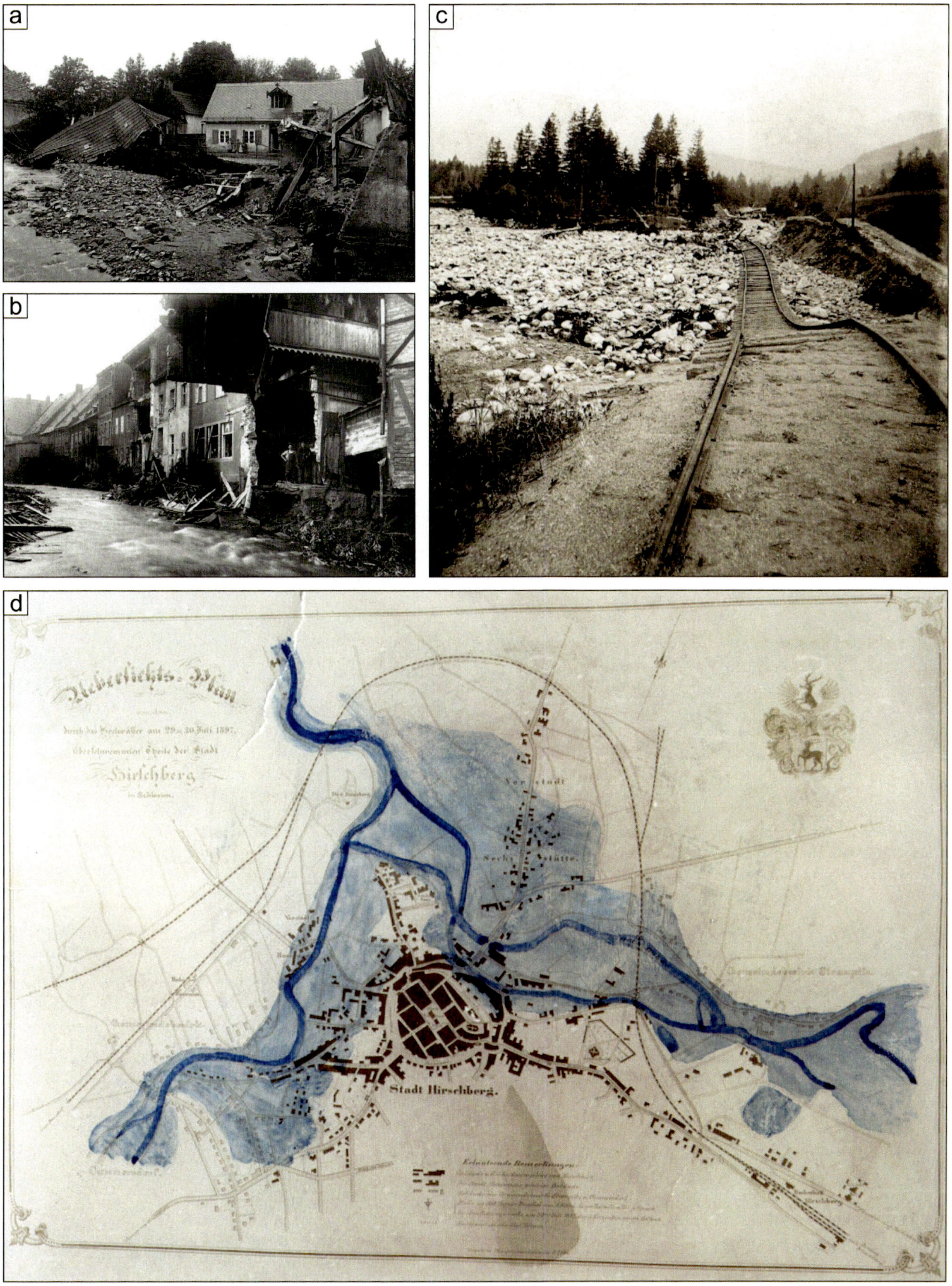

Fig. 7.11 Results of the largest known flood in the Karkonosze region on 29–30 July 1897. **a, b** Destruction of buildings due to lateral erosion in the Jedlica riverbed in Kowary. **c** Damaged railway track over the Łomnica River in Miłków. **d** Extent of flood inundation in Jelenia Góra. Sources: *Die Hochwasser-Katastrophe…* 1897 (**a–c**), State Archive in Wrocław, Jelenia Góra Branch, public domain (**d**)

Fig. 7.12 One of the components of centuries-old human impact on the landscape of the Karkonosze Mts. is mass tourism experienced already in the nineteenth century. **a** Man dressed as Rübezahl (mountain ghost) on the pass at Śnieżka. **b** Tourists at the source of Elbe. **c** Elbe waterfall with artificially increased discharge. **d** Project of a sport complex intended for the Winter Olympics in 1936 (now non-existent). **e** Kamieńczyk gorge open to the public, white arrow shows the ski-jump's landing place visible in picture **d**. Source: Old postcards, authors unknown (**a**, **d**, **e**); Seifert (1906) (**b**, **c**)

many cases, palace and palace gardens were uniquely incorporated into agricultural and forest landscapes. The park establishment in the village of Bukowiec is worth mentioning as an excellent example of a romantic landscape park (Migoń and Latocha 2008). It covers an area of over 20 ha of morphologically diversified terrain with a central depression, several breeding ponds, granite hills on its eastern side, and an elongated ridge on a long microgranite vein on its western side. Further to the west rises Mt. Mrowiec (513 m asl), once a developed and highly regarded lookout point. The landscape of granite rocks has been skilfully integrated with older defensive castles: Sokolec on Mt. Krzyżna and Chojnik in the Karkonosze Foothills. Most of these lost their splendour after World War II and have been restored in the last three decades.

Contemporary geomorphological processes, responsible for denudation (Cielińska 1961; Bieroński et al. 1992; Katrycz 1998), were briefly intensified during the so-called 'environmental catastrophe' of the 1980s. At that time, West Sudetic spruce monoculture forests were dying off en masse due to the acidification of the atmosphere by the nearby lignite-fired power plants (Mazurski 1986). Today, erosion processes on the hiking trails in the Karkonosze are intensified by huge tourist traffic, and haphazard urbanisation becomes a threat to the aesthetic values of the geomorphological and cultural landscape of the Jelenia Góra Basin.

7.8 Conclusions

The Karkonosze Mountains and the adjacent Jelenia Góra Basin constitute an exceptional area on the geomorphological map of Poland. They stand out not only against the background of the Sudetes but also the European medium-altitude mountains. They form a unique landscape, strictly dependent on structural features of granite and its internal differentiation. The Karkonosze are distinguished by strong late Cenozoic uplift, numerous tors, clear traces of Pleistocene glaciation, and periglacial cryogenic processes. The Jelenia Góra Basin may be regarded as the only example of exposed granite weathering front in Poland. In spite of exceptionally long tradition of geological and geomorphological studies of this region, not all issues of relief development have been satisfactorily explained. Some of the older views also need revision given the advance in knowledge and the development of research techniques, including absolute dating of sediments and rocks. Observations of contemporary geomorphological processes await their continuation in the face of climate change and increasing human impact.

References

Aleksandrowski P, Słaby E, Szuszkiewicz A, Galbarczyk-Gąsiorowska L, Knapik R, Madej S, Sobczyk A, Szełęg E (2019) Budowa geologiczna Karkonoszy. In: Knapik R, Migoń P, Raj A (eds) Przyroda Karkonoskiego Parku Narodowego. KPN, Jelenia Góra, pp 7–46

Badura J, Przybylski B (1998) Zasięg lądolodów plejstoceńskich i deglacjacja obszaru między Sudetami a Wałem Śląskim. Biul PIG 385:9–28

Berg G (1915) Die Vergletscherung an den Teichen des Riesengebirges. Z Dt Geol Gesell 67:63–82

Berg G (1927) Zur Morphologie des Riesengebirges. Z Geomorph 2:1–20

Berg G (1940a) Geologische Karte des Deutschen Reiches. Bl. Bad Warmbrunn 1:25 000, Preuss Geol L-A, Berlin

Berg G (1940b) Geologische Karte des Deutschen Reiches. Bl. Krummhübel (u. Erläuterungen) 1:25 000, Preuss Geol L-A, Berlin

Bieroński J, Chmal H, Czerwiński J, Klementowski J, Traczyk A (1992) Współczesna denudacja w górskich zlewniach Karkonoszy. Prace Geogr IGiPZ PAN 155:151–169

Błaś M, Sobik M, Quiel F, Netzel P (2002) Temporal and spatial variations of fog in the Western Sudety Mts., Poland. Atmosph Res 64:19–28

Borkowska M (1966) Petrografia Granitu Karkonoszy. Geol Sudetica 2:7–119

Borzęcki R, Wójcik D (2017) Pozostałości górnictwa rud uranu i żelaza w rejonie Kowar. Część I. Hereditas Minariorum 4:81–108

Borzęcki R, Wójcik D, Kalisz M (2018) Pozostałości górnictwa rud uranu, toru i kruszców w rejonie Jeleniej Góry i Szklarskiej Poręby. Hereditas Minariorum 5:85–121

Büdel J (1937) Eiszeitliche und rezente Verwitterung und Abtragung im ehemals nicht vereisten Teil Mitteleuropas. Petermanns Geogr Mitt Erg 229:5–71

Budych I, Majewicz R (1999) Strategia ochrony przeciwpowodziowej Bobru w ujęciu historycznym, Rocznik Jeleniogórski 31:45–54

Chmal H, Kasprzak M (2009) Stożki Piszczaka – problem wieku teras rzecznych i utworów lodowcowych w Kotlinie Jeleniogórskiej. In: Kostrzewski A, Paluszkiewicz R (eds) Geneza, litologia i stratygrafia utworów czwartorzędowych 5, Ser. Geogr. 88, Wyd. Nauk. UAM, Poznań, pp 31–43

Chmal H, Traczyk A (1993) Plejstoceńskie lodowce gruzowe w Karkonoszach. Czas Geogr 64:253–263

Chmal H, Traczyk A (1999) Die Vergletscherung des Riesengebirges. Z Geomorph NF, Suppl 113:11–17

CielińskA M (1961) Comment on quantitative investigations of some processes of modern denudation in the Karkonosze Mountains. Zeszyty Nauk UWr B8. Nauka o Ziemi 5:57–69

Cloos H (1925) Einführung in die tektonische Behandlung magmatischer Erscheinungen (Granittektonik). Teil I. Das Riesengebirge in Schlesien. Berlin, p 194

Czerwiński J (1991) Powodzie w rejonie Karkonoszy od XV w. do czasów współczesnych. Acta Univ Wratisl 1237. Prace Inst Geogr A6:85–104

Danišík M, Migoń P, Kuhlemann J, Evans NJ, István D, Frisch W (2010) Thermochronological constraints on the long-term erosional history of the Karkonosze Mts., Central Europe. Geomorphology 117:78–89

Die Hochwasser-Katastrophe im Riesengebirge am 29 bis 30 Juli 1897. Hirschberg

Duma P, Larocha A, Łuczak A, Piekalski J (2020) Stone walls as a characteristic feature of the cultural landscape of the Izera Mountains, southwestern Poland. Int J Hist Archaeol 24:22–43

Dyjor S, Oberc J. (1983) Współczesne ruchy skorupy ziemskiej w Polsce SW i wynikające z nich możliwości zagrożeń dla obiektów górniczych i inżynierskich. In: Współczesne i neotektoniczne ruchy skorupy ziemskiej w Polsce, vol 4, Wrocław, pp 7–23

Engel Z, Braucher R, Traczyk A, Laetita L, Aster Team (2014) ^{10}Be exposure age chronology of the last glaciation in the Krkonoše Mountains, Central Europe. Geomorphology 206:107–121

Engel Z, Nývlt D, Křížek M, Treml V, Jankovská V, Lisá L (2010) Sedimentary evidence of landscape and climate history since the end of MIS 3 in the Krkonoše Mountains, Czech Republic. Quat Sci Rev 29:913–927

Engel Z, Traczyk A, Braucher R, Woronko B, Křížek M (2011) Use of ^{10}Be exposure ages and Schmidt hammer data for correlation of moraines in the Krkonoše Mountains, Poland/Czech Republic. Z Geomorph NF 55:175–196

Hall AM, Migoń P (2010) The first stages of erosion by ice sheets: evidence from central Europe. Geomorphology 123:349–363

Jahn A (1954) Karkonosze – rys morfologiczny. Czas Geogr 23(24):107–121

Jahn A (1960) Czwartorzęd Sudetów. In: Teisseyre H. (ed) Regionalna Geologia Polski, t. III, Sudety, z. 2, Utwory trzeciorzędowe i czwartorzędowe oraz pogląd na rozwój budowy geologicznej Sudetów. Pol Tow Geol, Kraków, pp 358–418

Jarosiński M, Poprawa P, Ziegler P (2009) Cenozoic dynamic evolution of the Polish Platform. Geol Quart 53:3–26

Jarosiński M, Zuchiewicz W, Poprawa P, Badura J (2008) Cenozoic geodynamic evolution of the Carpathian Foreland. In: McCann T (ed) The geology of Central Europe, vol 2. Mesozoic and Cenozoic. Geological Society, London, pp 1258–1270

Jeník J (1961) Alpinská vegetace Krkonoš, Králického Sněžníku a Hrubého Jeseníku: teorie anemoorografických systemu. Nakl. ČSAV Praha, p 409

Kajdas B, Michalik MJ, Migoń P (2017) Mechanisms of granite alteration into grus, Karkonosze granite, SW Poland. CATENA 150:230–245

Kasprzak M (2009) Geomorfologia stożka Wilczej Poręby w Karkonoszach. Opera Corcontica 46:19–39

Kasprzak M (2010) Wezbrania i powodzie na rzekach Dolnego Śląska. In: Migoń P (ed) Wyjątkowe zdarzenia przyrodnicze na Dolnym Śląsku i ich skutki. Rozprawy Naukowe Instytutu Geografii i Rozwoju Regionalnego, vol 14. Uniwersytet Wrocławski, Wrocław, pp 81–140

Kasprzak M (2021) Wody Karkonoszy / Vodstvo v Krkonoších. In: Spallek W (ed) Atlas Karkonoszy / Atlas Krkonoš, Karkonoski Park Narodowy, Správa Krkonošského národního parku, Jelenia Góra—Vrchlabí, pp 70–73

Kasprzak M, Migoń P (2015) Historical and recent floods in the West Sudetes—geomorphological dimension. Z Geomorph NF 59(Suppl 3):73–97
Kasprzak M, Traczyk A (2019) Ukształtowanie powierzchni Karkonoszy. In: Knapik R, Migoń P, Raj A (eds) Przyroda Karkonoskiego Parku Narodowego. KPN, Jelenia Góra, pp 47–90
Kasprzak M, Traczyk A, Migała K (2021) Seasonally frozen ground. Reference module in earth systems and environmental sciences.https://doi.org/10.1016/B978-0-12-818234-5.00141-3
Katrycz M (1998) Transport rumowiska wleczonego w zlewni Wilczego Potoku – Karkonosze Wschodnie. Acta Universitatis Wratislaviensis 2061. Prace Inst Geogr A9:37–58
Kozłowska-Koch M (1987) Klasyfikacja i nomenklatura trzeciorzędowych wulkanitów Dolnego Śląska i Śląska Opolskiego. Arch Miner 42:43–91
Křížek M (2007) Periglacial landforms above alpine timberline in the High Sudetes. In: Goudie AS, Kalvoda J (eds) Geomorphological variations. P3K, 1, Praha, pp 313–337
Křížek M, Uxa T (2013) Morphology, sorting and microclimates of relict sorted polygons, Krkonoše Mountains, Czech Republic. Permafrost Perigl Proc 24:313–321
Kryza R, Pin Ch, Oberc-Dziedzic T, Crowley QG, Larionov A (2014) Deciphering the geochronology of a large granitoid pluton (Karkonosze Granite, SW Poland): an assessment of U-Pb zircon SIMS and Rb-Sr whole-rock dates relative to U-Pb zircon CA-ID-TIMS. Int Geol Rev 56:756–782
Kusiak MA, Williams I, Dunkley DJ, Konečný P, Słaby E, Martin H (2014) Monazite to the rescue: U-Th-Pb dating of the intrusive history of the composite Karkonosze pluton, Bohemian Massif. Chem Geol 364:76–92
Mazurski K (1969) Wielka lawina w Karkonoszach i jej morfologiczne skutki. Czas Geogr 40:375–379
Mazurski K (1986) The destruction of forests in the Polish Sudetes Mountains by industrial emissions. Forest Ecol Manag 17:303–315
Michniewicz M (1998) The pre-Elsterian valley system in the Western Sudetes, southwestern Poland, and its later transformation. Geol Sudetica 31(2):317–328
MierzeJewski MP (2005) Karkonosze – ewolucja masywu granitowego. In: Mierzejewski MP (ed) Karkonosze. Przyroda nieożywiona i człowiek. Wydawnictwo Uniwersytetu Wrocławskiego, Wrocław, pp 83–132
Migała K, Liebersbach J, Sobik M (2002) Rime in the Giant Mts. (The Sudetes, Poland). Atmosph Res 64:63–73
Migała K, Urban G, Tomczyński K (2016) Long-term air temperature variation in the Karkonosze mountains according to atmospheric circulation. Theor Appl Climatol 125:337–351
Migoń P (1991) The origin of stepped topography in the northern part of the Karkonosze granite massif. Bull Polish Acad Sci 39:267–276
Migoń P (1992) Tektoniczne formy rzeźby na północnym stoku Karkonoszy. Opera Corcontica 29:5–24
Migoń P (1993) Geneza Kotliny Jeleniogórskiej. Opera Corcontica 30:85–115
Migoń P (1996) Evolution of granite landscapes in the Sudetes (Central Europe): some problems of interpretation. Proc Geol Assoc 107:25–37
Migoń P (1997) The geologic control, origin and significance of inselbergs in the Sudetes, NE Bohemian Massif, Central Europe. Z Geomorph NF 41:45–66
Migoń P (2016) Jizerské Hory—an interplay of rock control, faulting and inland glaciation in the evolution of a granite terrain. In: Pánek T, Hradecký J (eds) Landscapes and landforms of the Czech Republic. Springer, Switzerland, pp 165–175
Migoń P, Danišík M (2012) Erosional history of the Karkonosze Granite Massif—constraints from adjacent sedimentary basins and thermochronology. Geol Quart 56:440–454
Migoń P, Latocha A (2008) Enhancement of cultural landscape by geomorphology. A study of granite parklands in the West Sudetes, SW Poland. Geogr Fis Dinam Quat 31:195–203
Migoń P, Parzóch K (2008) Spływy gruzowe w polskich Karkonoszach – przyczyny, skutki i zagrożenia. Przegl Geogr 80:385–401
Nývlt D, Engel Z, Tyráček J (2011) Pleistocene glaciations of Czechia. Developments in quaternary sciences, vol 15. Elsevier, Amsterdam, pp 37–46
Partsch J (1882) Die Gletscher der Vorzeit in den Karpathen und den Mittelgebirgen Deutschlands. W. Koebner, Breslau, p 198
Partsch J (1894) Die Vergletscherung des Riesengebirges zur Eiszeit. Forsch Dt Landes- Volksk 8(2):103–194
Partsch J (1911) Schlesien. Eine Landeskunde für das deutsche Volk, auf wissenschaftlicher Grundlage, II. Teil. Ferdinand Hirt, Breslau, p 690
Parzóch K, Migoń P, Szymanowski R, Gąsiorek M (2007) Spływy gruzowe w północnej części Karkonoszy. Opera Corcontica 44(1):81–88
Pécskay Z, Lorenc M, BirkenmaJer K, Zagożdżon PP (2004) Recent geochronological studies on Tertiary volcanics in Lower Silesia. In: International workshop 'Basalt 2004', pp 24–25
Pilous V (1973) Strukturní mury v Krkonoších – I. část. Opera Corcontica 10:15–69
Pilous V (2016) Krkonoše Mountains: a case study of polygenetic relief. In: Pánek T, Hradecký J (eds) Landscapes and landforms of the Czech Republic. Springer, Switzerland, pp 177–193
Pilous V, Migoń P (2007) Geomorfologie. In: Flousek J, Hartmanová O, Štursa J. Potocki J (eds) Krkonoše. Příroda, historie, život. Baset, Praha, pp 103–124
Pilous V (1989) Geomorfologie vodopádů Labského a Obřího dolu v Krkonoších. Opera Corcontica 26:5–50
Seifert J (1906) Riesengebirgs-Ansichten. Hirschberg, Schlesien
Skrzypek G, Baranowska-Kącka A, Keller-Sikora A, Jędrysek M-O (2009) Analogous trends in pollen percentages and carbon stable isotope composition of Holocene peat–Possible interpretation for palaeoclimate studies. Rev Palaeobot Palynol 156:507–518
Sobczyk A, Danišik M, Aleksandrowski A, Anczkiewicz A (2015) Post-Variscan cooling history in the central Western Sudetes (NE Bohemian Massif) and its implications for topographic evolution: insights from apatite fission-track and zircon (U-Th)/He thermochronology. Tectonophysics 649:47–57
Sobik M, Błaś M, Migała K, Godek M, Nasiółkowski T (2019) Klimat. In: Knapik R, Migoń P, Raj A (eds) Przyroda Karkonoskiego Parku Narodowego. KPN, Jelenia Góra, pp 147–186
Sommer F (1913) Schlesien. Eine Landeskunde als Grundlage für den Unterricht. Vierte, durchgesehene Auflage mit 91 Bildern und Karten im Text, p 63
Soukupová L, Jeník J, Kociánová M, Sekyra J (1995) Arctic-alpine tundra in the Krkonoše, the Sudetes. Opera Corcontica 32:5–88
Sroka W (1991) Tektoniczny charakter krawędzi Pogórza Karkonoskiego. Acta Univ Wratisl 1375. Prace Geol-Miner 29:239–249
Szmytkie R, Kasprzak M (2016) Rozwój osadnictwa na północnym skłonie Karkonoszy. Opera Corcontica 53:61–72
Teisseyre H (1957) Regionalna Geologia Polski. T. I II. Sudety. Z. 1. Utwory przedtrzeciorzędowe. Pol Tow Geol, Kraków, p 300
Traczyk A (1989) Zlodowacenie doliny Łomnicy w Karkonoszach oraz poglądy na ilość zlodowaceń plejstoceńskich w średnich górach Europy. Czas Geogr 60:267–286
Traczyk A (1992) Formy współczesnego sortowania mrozowego w Karkonoszach i klimatyczne uwarunkowania ich rozwoju. Czas Geogr 63:351–359
Traczyk A (2009) Zlodowacenie Śnieżnych Kotłów w Karkonoszach Zachodnich w świetle analizy morfometrycznej oraz GIS. Opera Corcontica 46:41–56

Traczyk A, Migoń P (2000) Cold-climate landform patterns in the Sudetes. Effects of lithology, relief and glacial history. Acta Univ Carolinae Geogr 35(Suppl):185–210

Treml V, Jankovská V, Petr L (2008) Holocene dynamics of the alpine timberline in the High Sudetes. Biologia 63:73–80

Troll C (1944) Strukturböden, Solifluktion und Frostklimate der Erde. Geol Rund 34:545–694

Walicka A, Jóźków G, Kasprzak M, Borkowski A (2019) Terrestrial laser scanning for the detection of coarse grain size movement in a mountain riverbed. Water 11:2199

Wilson JP, Gallant JC (2000) Primary topographic attributes. In: Wilson JP, Gallant JC (eds) Terrain analysis: principles and applications. Wiley, Chichester, pp 51–85

Zagożdżon PP, Zagożdżon K (2006) Charakterystyka wystąpienia oligoceńskiego bazaltoidu w Małym Kotle Śnieżnym (Karkonosze). Przegl Geol 54:496–500

Žák J, Klomínský J (2007) Magmatic structure in the Krkonoše-Jizera Plutonic Complex, Bohemian Massif: evidence for localized multiphase flow and small-scale thermal–mechanical instabilities in a granitic magma chamber. J Volc Geotherm Res 164:254–267

Ziegler PA, Dèzes P (2007) Cenozoic uplift of Variscan Massifs in the Alpine foreland: timing and controlling mechanisms. Glob Planet Change 58:237–269

Żurawek R (1999) Relict rock glaciers in the Central European Mid-Mountains. State-of-the art. Biul Perygl 38:163–192

Marek Kasprzak is Associate Professor at the Institute of Geography and Regional Development, University of Wrocław, Poland. He authored numerous articles, maps, popular science studies, and environmental assessments on Lower Silesia region. He is interested in periglacial zone and the evolution of mountain and polar areas, participated in many scientific expeditions to the Arctic and Antarctic. He specialises in processing and interpretation of LiDAR- and SfM-based digital terrain models and applied geophysics.

Land of Extinct Volcanoes—Rock-Controlled Landforms, Postglacial Gorges, and Faulted Margin of the Sudetes

8

Piotr Migoń

Abstract

The "Land of Extinct Volcanoes" is an informal name of a part of south-western Poland that includes the Kaczawskie Mountains and the Kaczawskie Foothills in the West Sudetes as well as the adjacent fore-mountain area. The legacy of several periods of volcanic activity in the Paleozoic and Cenozoic is the most distinctive component of regional geoheritage and finds its geomorphic expression in the occurrence of hills and ridges built of more resistant volcanic rocks. Among them, basaltic cones and domes are particularly clear and serve as regional landmarks. Further valuable elements of geomorphological diversity include tors and crags, periglacial blockfields and stepped hillslopes, sandstone cuestas with associated rock towers, boulder fields and non-karstic rock shelters, karstic features, gorges cut by glacial meltwaters during the decay of the Scandinavian ice sheet, and the fault-generated marginal escarpment of the Sudetes. The pending application for the status of a UNESCO Global Geopark reflects both the international dimension of regional geoheritage and the considerable involvement of local communities to promote tourism based on geoscience-related values.

Keywords

Volcanism · Structural geomorphology · Periglacial · Karst · Glaciation · Geopark · Sudetes

8.1 Introduction

The "Land of Extinct Volcanoes" is an informal name of an area in south-western Poland that straddles the low-mountain and upland terrain in the south-west and hilly to lowland terrain in the north-east. Geographically, it includes the regions of the Kaczawskie Mountains, Kaczawskie Foothills (both being parts of the Sudetes mountain range), Silesian Lowland, and in small part the Sudetic Foreland (Solon et al. 2018) (Fig. 8.1). The name "Land of Extinct Volcanoes" is primarily used as a brand name in regional marketing aimed at the promotion of geotourism in the region (Pijet-Migoń and Migoń 2019). It highlights the most distinctive aspect of regional geological history, which is the occurrence of rocks of volcanic origin, widespread across this part of the country. These rocks are of different ages and testify to three different periods of volcanic activity, separated from one another by vast expanses of geological time. The most distant was submarine volcanism in the early Palaeozoic, followed by terrestrial volcanism in the Early Permian and, finally, by volcanic eruptions and extrusions in the late Cenozoic. Each of these phases produced specific rock varieties, which together represent the highest diversity of volcanic materials on the territory of Poland. This eventful volcanic history provided much of the foundation for geomorphic landscape development in the region. As a consequence, distinctive landforms built of volcanic rocks occur widely, with many easily recognizable from afar by their characteristic silhouettes. This is particularly true for the Cenozoic volcanism, the testament of which are numerous basaltic cones, domes, and crags, including Mt. Ostrzyca—the landmark of the "Land of Extinct Volcanoes" (Fig. 8.2).

However, the geodiversity of the region includes much more than remnants of ancient volcanism (Migoń 2021). The rock record not only spans the interval from the Cambrian to the Quaternary, with only the Jurassic being absent, but is represented by an enormous variety of rock

P. Migoń (✉)
Institute of Geography and Regional Development, University of Wrocław, pl. Uniwersytecki 1, 50-137 Wrocław, Poland
e-mail: piotr.migon@uwr.edu.pl

P. Migoń and K. Jancewicz (eds.), *Landscapes and Landforms of Poland*,
World Geomorphological Landscapes, https://doi.org/10.1007/978-3-031-45762-3_8

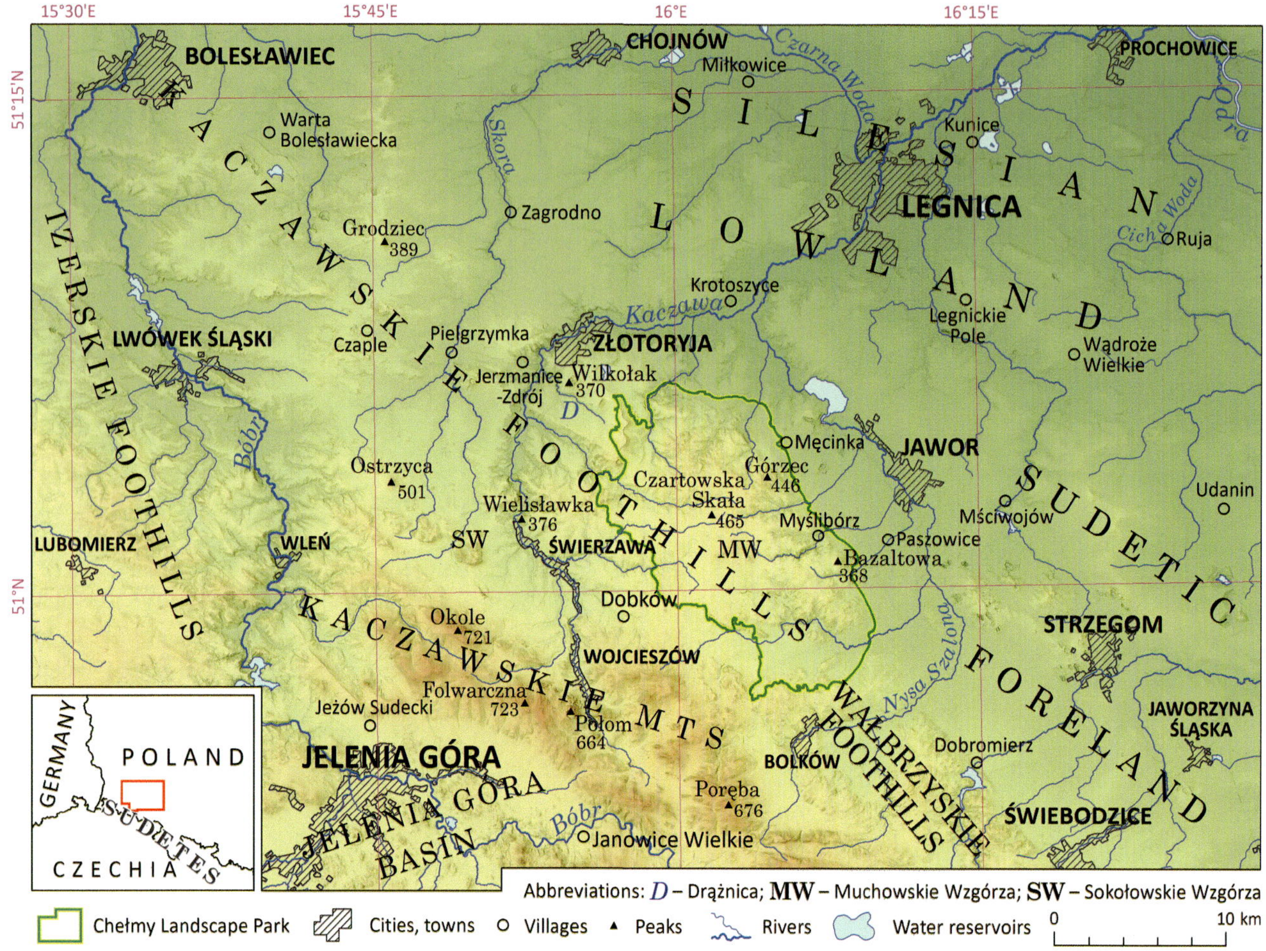

Fig. 8.1 Location map of the "Land of Extinct Volcanoes" at the boundary between the West Sudetes and the Silesian Lowland (map courtesy of Kacper Jancewicz)

Fig. 8.2 Basaltic cone of Mt. Ostrzyca is the landmark of the "Land of Extinct Volcanoes" (photograph by P. Migoń)

types originated in contrasting sedimentary settings, from deep ocean to shallow marine and terrestrial, and includes also plutonic and metamorphic rocks (Baranowski et al. 1990). Consequently, regional topography is also diverse, showing various relationships between rock properties and landform patterns, even though the landforms themselves are rather subdued and the altitude differences within the region do not exceed 500 m. This rock-controlled morphology was partially re-shaped in the Pleistocene, when the Scandinavian ice sheet invaded the area, leaving glacial sediments, outwash landforms, and striking erosional gorges. In the past millennium, rich and varied reserves of mineral resources have been prospected and mined, resulting in evident anthropogenic impact on geomorphology, but also contributing to the regional cultural heritage.

The growing awareness of these different values of the natural and cultural environment prompted local stakeholders to identify geodiversity and geoheritage as the key regional assets and viable foundation of sustainable nature-based tourism, particularly geotourism (Pijet-Migoń and Migoń 2019). After many years of organic growth and steady development of infrastructure and regional identity, an application for the status of UNESCO Global Geopark for the "Land of Extinct Volcanoes" was submitted in 2019 and is currently under evaluation. Consequently, this chapter reviews both geomorphic landscapes and selected landforms of the region, as well as the history of making regional geoheritage more accessible to the general public.

8.2 Geological Background

8.2.1 Pre-Quaternary Rock Record

At the most general level, the geological structure of the "Land of Extinct Volcanoes" can be resolved into two tiers, consistent with the most general division applicable for the entire Sudetes (Żelaźniewicz 2015). The older one includes rocks which originated prior to the final stages of the Variscan orogeny in the Carboniferous and cover the interval from the Late Proterozoic to the Lower Carboniferous (Baranowski et al. 1990; Kryza and Muszyński 1992). Within this tier, most of the region belongs to the so-called Kaczawa Fold Belt, except the eastern extremity of the region. The outcome of the Variscan orogeny was the consolidation of the basement through amalgamation of different terranes, each with a complicated pre-Variscan history, that docked against one another to form a solid foundation for subsequent geological evolution (Aleksandrowski and Mazur 2002). The younger tier includes rocks originated in the post-Variscan period, from the Late Carboniferous to the Quaternary (Fig. 8.3). They are mainly sedimentary formations, deposited in both terrestrial and marine settings, the latter including both very shallow (shelf) and moderately deep environments (Śliwiński et al. 2003; Chrząstek and Wojewoda 2011). In addition, volcanism occurred in the Permian and Cenozoic.

In the rock record, the lower structural tier is represented by various plutonic, volcanic, and sedimentary rocks, which were all subject to metamorphism, often multiple and mostly low-grade (Baranowski et al. 1990; Kryza and Muszyński 1992; Kryza et al. 2007). The dominant rock series are greenschists and phyllites, which are products of metamorphism of primary ocean-floor basaltic lavas and fine-grained clastic sediments, respectively. The greenschists, which are important geomorphologically, occur in several variants, from massive through platy to ones typified by the survival of original pillow-lava textures. Their ages vary from Cambrian to Silurian. In the Cambrian, calcareous deposits of a substantial thicknesses originated in the vicinity of some volcanic fissures, later transformed into crystalline limestones (Lorenc 1983). Other meta-sedimentary rocks include schists, meta-sandstones, and black cherts (lydites). Apart from basaltic extrusions, siliceous volcanism occurred too, testified by isolated outcrops of trachytes and rhyolites in the south-eastern part of the Kaczawskie Mountains (Kryza and Muszyński 1992). In the easternmost part of the region, Late Proterozoic/early Palaeozoic gneisses and mica schists are present, as are granites emplaced much later, by the end of the Carboniferous.

The stratigraphic record of the upper structural tier begins with Upper Carboniferous clastic rocks (mainly sandstones and conglomerates), which are correlative deposits to the erosion of the Variscan mountain range. These are followed by Lower Permian sedimentary series, which are lithologically similar and also include conglomerates, sandstones, mudstones, and shales, but were deposited under different environmental conditions, much drier and with only seasonal or episodic rainfall. They originated as deposits of braided rivers, alluvial fans, debris flows, and ephemeral lakes. Both siliceous and alkaline volcanism occurred at that time, producing rhyolites, trachyandesites, and trachybasalts (Awdankiewicz et al. 2014). Terrestrial sedimentation in the Early Permian was replaced by marine sedimentation in the Late Permian, with limestones, dolomites, sandstones, and shales (Biernacka et al. 2005). Some of these sediments were found to be copper-bearing and gave rise to copper mining industry. The alternation between marine and terrestrial conditions continued in the Early and Middle Triassic, resulting in fluvial sandstones and marine limestones. The period from the Middle Triassic to the beginning of the Late Cretaceous is not represented in the rock record, and regional evidence suggests that

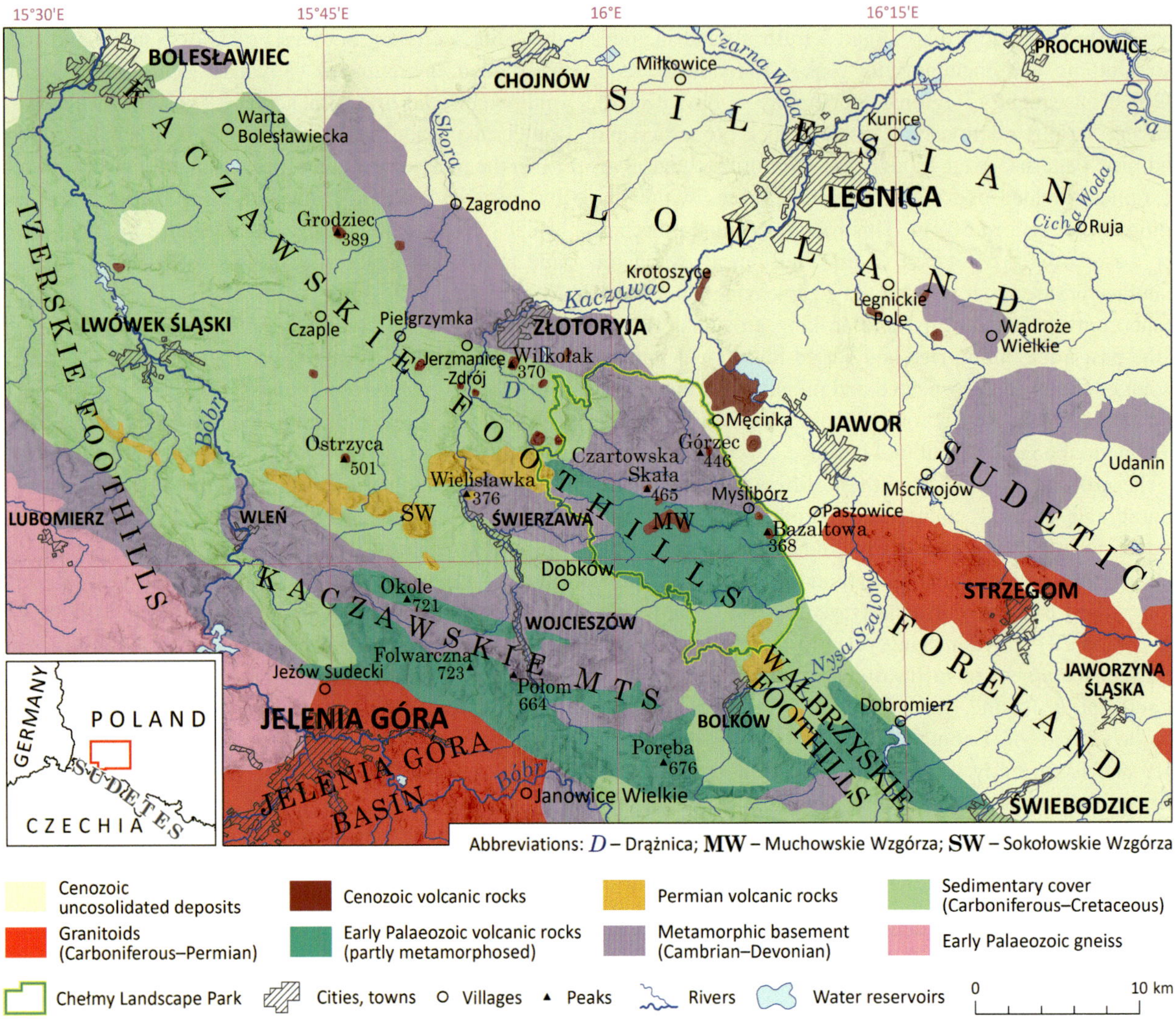

Fig. 8.3 Geology of the "Land of Extinct Volcanoes" (map courtesy of Kacper Jancewicz)

terrestrial conditions prevailed, with efficient deep weathering and the origin of kaolinitic saprolites in the neighbouring areas. Sedimentation resumed with the onset of a marine transgression in the Cenomanian (c. 99 Ma) and continued into the Santonian (c. 85 Ma), resulting in the succession of alternating quartz-rich sandstones, calcareous sandstones, mudstones, and marls (Milewicz 1997; Leszczyński and Nemec 2019). Relating to geological subdivision of the Sudetes, all these sedimentary formations were deposited within the North-Sudetic Synclinorium.

The Cenozoic was the period of dominant denudation and Pliocene/early Quaternary sands and gravels are the only deposits from this protracted period in the Sudetic part of the "Land of Extinct Volcanoes" that preceded Quaternary glaciations (Grodzicki 1972). However, the sedimentary record is much more extensive in the foremountain part, especially in the deep, fault-bounded graben parallel to the margin of the Sudetes. The thickness of the Cenozoic sedimentary series, consisting of unconsolidated gravels, sands, and clays from the latest Oligocene to the early Quaternary, reaches 200 m (Kowalski 1977). It is less thick beyond the graben, a few tens of metres, but this was enough to seal an earlier mid-Cenozoic topography and bury the Palaeogene saprolitic mantles on granites and gneisses (Badura et al. 2004).

The younger part of the Cenozoic was also a period of volcanism, the youngest among the three in the region. According to various radiometric datings, the oldest Cenozoic volcanic rocks originated in the latest Eocene (c. 42 Ma), whereas the youngest ones are from the Middle Miocene (c. 15 Ma), with the majority of dates pointing to the latest Oligocene and the Early Miocene (Badura et al. 2005; Birkenmajer et al. 2007). For simplicity, volcanic products from this broad period are usually called basalts, even though in fact they represent a wider spectrum of mineralogical and chemical compositions, so various specific names apply. Considerable denudation since the period of activity has led to the degradation of primary volcanic landforms, and the contemporary outcrops are mostly exposed volcanic conduits and plugs, veins (probably many are subvolcanic), and only rarely remnants of lava flows and sheets (Birkenmajer 1967).

8.2.2 Cenozoic Faulting, Uplift and Subsidence

Rock formations in the "Land of Extinct Volcanoes" record multi-phase tectonic deformations of different ages and large-scale displacements, especially within the lower structural tier. However, for the contemporary landscapes and landforms, it is the Cenozoic tectonics which is the most important. Within it, two main phases can be distinguished.

At the turn of the Cenozoic, Central Europe was subject to widespread inversion tectonics, with the formation of reverse faults, locally evolving into low-angle thrusts. In the "Land of Extinct Volcanoes", faults of this kind affected the former Permian and Mesozoic sedimentary basins, causing thrusting of metamorphic basement onto the Permian and Cretaceous rocks, deforming the latter in the vicinity of faults, e.g. in the vicinity of the village of Jerzmanice-Zdrój, where Cretaceous sandstones are turned to near-vertical next to the Jerzmanice Fault (Solecki 2008, 2011). They also brought rock series of different strength and resistance into high-angle contact with one another, setting the stage for the subsequent origin of fault-related escarpments. Further consequences included minor regional tilting of sedimentary successions, which later allowed for the development of cuesta landscapes, and—at the site scale—the origin of siliceous bands within sandstones, marking shearing and recrystallization.

In the Neogene and the Quaternary, tectonics was mainly extensional, leading to the development of normal faults and associated fault-generated escarpments (Zuchiewicz et al. 2007). The magnitude of relief differentiation in the "Land of Extinct Volcanoes" was much smaller than in the inner, more elevated parts of the Sudetes (e.g., Kasprzak 2023), but nevertheless it amounted to ~ 300–400 m, as can be inferred from altitude variability. Among the tectonic structures from this period, the most distinctive is the Sudetic Marginal Fault, responsible for the dividing line between the Kaczawskie Foothills in the west and the foreland of the Sudetes in the east (Fig. 8.1). The present-day elevation of the main ridge of the Kaczawskie Mountains is also likely a product of up-faulting, although subsequent erosion and denudation have considerably re-shaped the tectonic relief.

8.3 Regional Landscapes

8.3.1 Kaczawskie Mountains

The Kaczawskie Mountains are a good example of low-altitude mountains that typify the less elevated and peripheral parts of the Central European Variscan belt. The highest peaks only slightly exceed 700 m asl (Mt. Folwarczna—723 m, Mt. Okole—721 m), the height difference between the water divides and the valley floors is 200–300 m, and typical slope inclinations are in the 10–15° range. They are entirely included within the forest belt, although centuries of human impact have resulted in considerable deforestation and replacement of forest communities by arable lands, pastures, and meadows.

Valleys and passes divide the Kaczawskie Mountains into several minor geomorphic units, with the north-trending Kaczawa River valley as the main line separating the WNW–ESE striking ridges of the western part from the less regular pattern of ridges in the eastern part. The western ridges follow the dominant strike of the inherited Variscan structures of the Kaczawa Fold Belt, with more resistant greenschists and meta-sandstones supporting elevations, whereas valley tracts tend to be excavated in phyllites and schists. How these two common rock types, greenschists and phyllites, control relief at a smaller scale, can be observed in the eastern part of the mountain, around Mt. Poręba (676 m), where an alternating sequence of low ridges on the former and shallow trough valleys on the latter occurs (Fig. 8.4). Both the northern and southern margin of the Kaczawskie Mountains are fairly distinct and are most probably degraded fault-generated escarpments. The one facing south, towards the Jelenia Góra Basin, is particularly clear and associated with the Main Intra-Sudetic Fault, which is one of the most important Variscan structures in the Sudetes, very likely to have been reactivated during the Cenozoic. In the central and eastern part of the range, outcrops of crystalline limestones occur in a W–E belt and these support distinctive elevations, with Mt. Połom (664 m) rising nearly 300 m above the adjacent valley floor of the Kaczawa River. These limestones are hosts to karst phenomena, presented in more detail below (see Sect. 8.4.4).

Among minor landforms, natural mid-slope and summit bedrock outcrops add to the geodiversity of the area and are worth particular attention. Some may be called tors, and these rise above the regolith-mantled surfaces in the form of walls, towers, and more complex castellated ruins. Heights above 15 m are recorded. More common are asymmetric outcrops—cliffs, pulpits, and ridge-top crags passing smoothly into debris-covered slopes. Various lithologies support bedrock outcrops, but greenschists, limestones, and trachytes are most represented.

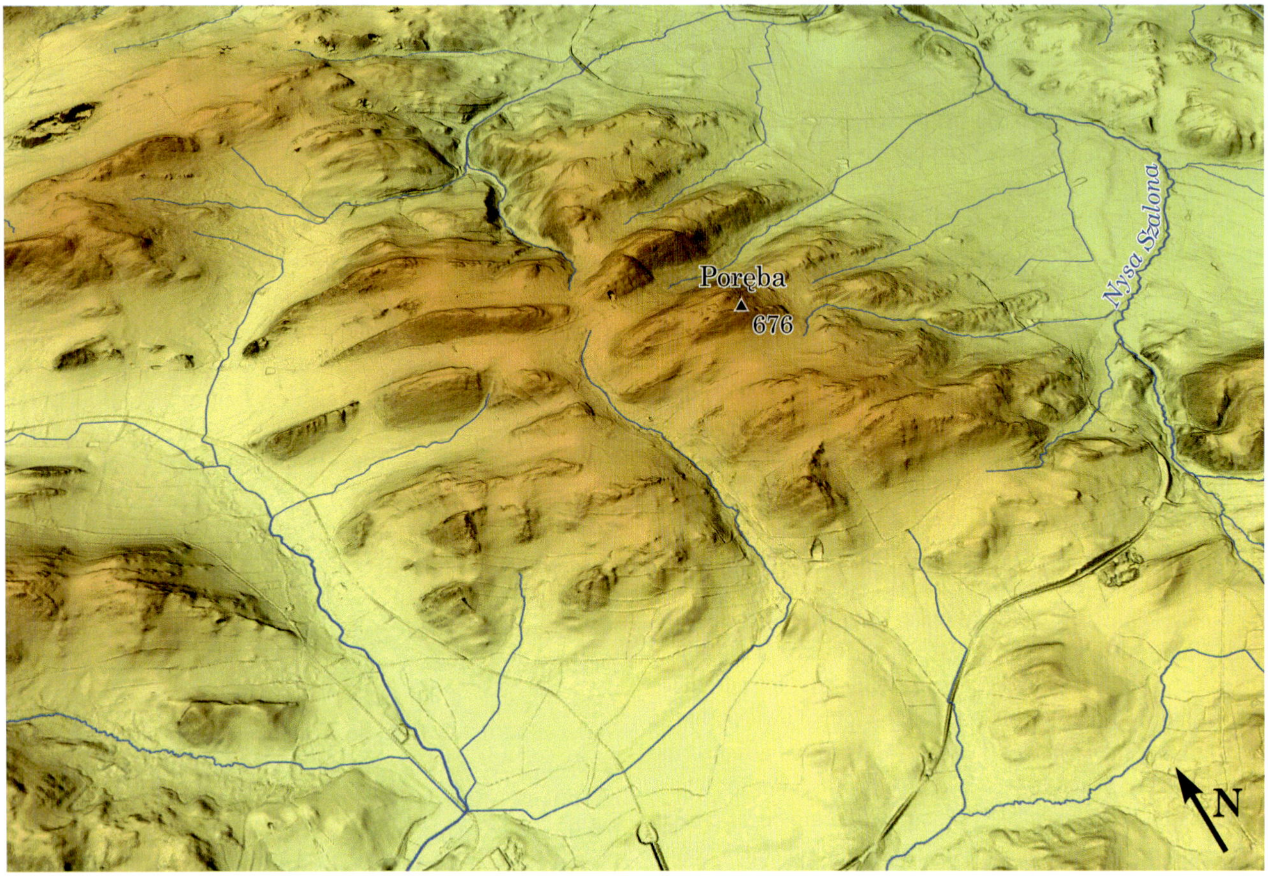

Fig. 8.4 Parallel ridges and troughs around Mt. Poręba (676 m) in the eastern part of the Kaczawskie Mountains illustrate direct rock control on topography. The ridges are built of greenschists, whereas depressions are eroded in phyllites. Beyond the massif of Poręba, along the Nysa Szalona River, a small intramontane basin occurs excavated in weak clastic sedimentary rocks of Permian age. Note also numerous agrarian terraces on slopes, especially in front of the image (visualization courtesy of K. Jancewicz)

8.3.2 Kaczawskie Foothills

To the north of the Kaczawskie Mountains extends a hilly relief of the Kaczawskie Foothills, sharply limited in the east by the faulted margin of the Sudetes, but lacking an evident northern boundary. Altitudes reach 501 m asl (Mt. Ostrzyca), whereas they are typically around 400 m asl, with the eastern part of the sub-region located at higher elevation than the western one. Relative relief does not exceed 200 m, and slopes are generally less than 10°, except numerous but otherwise scattered elevations built of more resistant volcanic rocks from the Permian and Cenozoic volcanic periods.

The Kaczawa valley continues to be a major line separating areas of different topography, related to differences in bedrock. The eastern part is mostly within the metamorphic complex and is dominated by greenschists and phyllites, but the corresponding change in relief is very subtle and the overall relief is rather planar. It becomes more complex in the vicinity of the margin of the Sudetes, where valleys are deeper and locally gorges occur (see Sect. 8.4.5). However, in the north-west, close to the town of Złotoryja, sedimentary rocks occur widely and these have given rise to a distinctive cuesta (see Sect. 8.4.3). The western part is almost exclusively underlain by sedimentary and, subordinately, volcanic rocks and shows low subdued relief, with bedrock extensively concealed by Quaternary deposits of glacial and fluvioglacial origin. However, two prominent basaltic hills (Mt. Ostrzyca, Mt. Grodziec) overlook this rather monotonous landscape.

8.3.3 Mountain Foreland

The fore-mountain part is geomorphologically complex, even though this is not at all evident at the first sight. Immediately before the mountain front of the Sudetes a tectonic graben is located, with the bedrock floor at the depth of more than 200 m. However, the transition to the plains located further to the north is smooth due to the ubiquitous cover of glacial deposits from several inland glaciations. Ice sheets reached the region some 450 ka ago (MIS 12, Elsterian (Sanian) glaciation) and 200–150 ka ago (MIS 6, Saalian (Odranian) glaciation). Rock outcrops built of gneiss, quartzite, and mica schists are few and not very distinctive, although they were more prominent in the Neogene. Topographic lows between the hills are filled with Neogene and Quaternary deposits, which indicates a higher relative relief in the past. In the south-east, elevations are associated with granite of Carboniferous age and these reach 100 m high. However, in contrast to the granite areas inside the Sudetes (e.g., the Karkonosze—see Kasprzak 2023), they almost lack tors and their slopes are very smooth. This may be an effect of repeated glacial erosion in the Pleistocene, probably capable of erasing upstanding, but otherwise rather delicate, bedrock superstructures.

Remnants of the Cenozoic volcanism are not as frequent in the foreland as they are in the mountains. Among them, two lava plateaus are worth mentioning. One is located to the west of the town of Jawor and forms a low plateau, incised by the valley of the Nysa Szalona River. The other one lies further to the north-west and is largely buried under younger sediments. It crops out along the valley side of the Nysa Szalona River and is distinctive by its high degree of in situ saprolitic weathering, as much as 20 m thick. This saprolite owes its survival to the protective sedimentary cover and informs us that any more extensive lava flows in the mountainous part of the region may have been quickly degraded following efficient deep weathering in the warm and humid climate of the Miocene.

8.4 Characteristic Landforms

8.4.1 Geomorphic Expression of Ancient Volcanism

Each of the three periods of past volcanism is associated with specific landforms, but the most evident are those built by volcanic rocks produced during the most recent, Cenozoic phase (Wocke 1927; Birkenmajer 1967). They almost invariably build hills, whose height above the base reaches from 20–25 m to more than 150 m, depending on the area of basaltic outcrop and the difference in strength between the basalt and the host rock (Fig. 8.5a, b). If the latter is weak, as in the case of most sedimentary rocks, the basaltic elevations are more pronounced (Placek 2007). Columnar jointing exposed in numerous natural outcrops and quarries, along with the general absence of pyroclastic materials, shows that most volcanic hills are exposed deeper parts of volcanic conduits and, hence, are best described as necks. Mt. Ostrzyca is the most characteristic among them, reaching 160 m high in an almost perfect cone shape (Fig. 8.2). It has never been quarried, offering a rare opportunity to see the natural slope topography of a volcanic elevation, with extensive scree covers (blockfields) (Fig. 8.5c). It also nicely shows the transition from low-angle footslopes cut in the surrounding Permian sedimentary rocks through moderately inclined mid-slope segments to steep upper slopes dotted by numerous crags, which supply debris to build the voluminous scree below. Other eye-catching elevations are Mt. Grodziec, with its flattened top used to build a castle in the medieval times, Mt. Wilkołak, and Mt. Czartowska Skała, although the latter two have been considerably reshaped by quarrying activity and their sharp silhouettes are deceptive.

Other basaltic elevations are less conspicuous, but no less interesting. The large massif of Muchowskie Wzgórza (Muchów Hills) is of uncertain origin, and both a large composite neck and a fragment of a lava flow have been hypothesized. This massif has been very little quarried too and has retained a fine suite of periglacial landforms and deposits on their slopes (see Sect. 8.4.2). The flat top of Mt. Bazaltowa has developed upon a colonnade of thick basaltic columns, and it is possible that it represents a former lava lake within a crater.

Similar to basalts, the volcanic rocks of Permian age are mechanically strong and resistant to weathering and erosion, rising above the surrounding flat and rolling lands in sedimentary formations. The highest elevations exceed 100 m high (Sokołowskie Wzgórza). By contrast to basalts, in turn, conical forms are absent and typical landforms are dome-shaped hills with steep slopes (locally as much as 40°) and levelled summit surfaces. This shape is represented by Mt. Wielisławka that overlooks the Kaczawa River. It is interpreted as a dome, possibly a laccolith, and has been made famous because of a large outcrop of jointed rhyolite in a former quarry (Fig. 8.5d).

The geomorphic legacy of the Early Palaeozoic submarine volcanism is unlike the other two because of a considerable alteration after lava emplacement, including subsequent erosion and regional metamorphism, and different form of emplacement itself. Therefore, it is practically impossible to link the contemporary occurrence of greenschists and the form of ridges with the primary morphology of lava flows. However, at a small scale, morphological details of various greenschist crags reflect the presence of stacked pillows of lava (Michniewicz 2016) that have apparently survived subsequent deformations or have been only slightly reshaped.

8.4.2 Periglacial Legacy

During the cold periods of the Pleistocene (glacial stages), as the rest of Central Europe, the "Land of Extinct Volcanoes" was repeatedly located in the periglacial zone, where permafrost and frost-related processes decisively

Fig. 8.5 Geomorphic diversity of hills built of volcanic rocks. **a** Low cone of Mt. Czartowska Skała (ca. 30 m high), partly reshaped by quarrying, rises above a denudational surface cut across Palaeozoic metamorphic rocks. **b** Dome-like elevation of Mt. Grodziec is more than 100 m high and the rocks around the hill are sedimentary. **c** Angular blockfield on Mt. Ostrzyca. **d** Columnar jointing in Permian rhyolites at Mt. Wielisławka, exposed in a former quarry (photographs by P. Migoń)

contributed to landform evolution. The geomorphic legacy of periglacial environments is diverse, but most evident are bedrock crags and associated blockfields fields, solifluction landforms and deposits, as well as loess covers (Fig. 8.6).

Causal relationships between crag morphology and periglacial conditions were first explored by Martini (1969), who pointed out that the setting and shapes of greenschist crags in the eastern part of the Kaczawskie Mountains, as well as the characteristics of regolith in their immediate vicinity, are not compatible with the two-stage model of tor evolution through deep weathering and stripping, argued for granite tors in the nearby Karkonosze Mountains (Jahn 1962). Rather, they indicate the dominant role of mechanical weathering, frost wedging, and disintegration into angular fragments, which then accumulated around. Likewise, Michniewicz (2016) explained the origin of greenschist crags on the Okole ridge in terms of periglacial hillslope development, although she also provided evidence of strong structural control on the distribution of crags. They are preferentially associated with south-facing slopes because of the dip of foliation and schistosity planes into the slope. Since mass wasting processes in periglacial environments are more efficient on south-facing slopes (on the northern hemisphere), structural and climatic factors converged to provide an optimal setting for the origin of crags. The height of greenschist tors attains 12 m and the most common shapes are cliffs and mid-slope spurs (Michniewicz et al. 2020) (Fig. 8.6a). Piles and sheets of angular, typically platy debris are ubiquitous on the downslope side of crags, but in contemporary conditions, they are largely overgrown.

Mid-slope crags occur also on those basaltic hillslopes, which have not been altered by quarrying, such as on Mt. Ostrzyca (Migoń et al. 2002; Migoń and Pijet-Migoń 2016). They have shapes and dimensions similar to those formed in greenschists, typically a few metres high and up to 20 m across the slope. Their morphological details clearly reflect the pattern of columnar jointing, and the places, where joints dip into the slope, are preferential locations for basaltic spurs. The crags on Mt. Ostrzyca have supplied debris for blockfields below, which rest at an angle of 15–30° and are only conditionally stable. On the south-facing slopes, they are bare and subject to slow talus creep, as attested

Fig. 8.6 Periglacial inheritance. **a** Greeenschist mid-slope tors on Mt. Okole in the Kaczawskie Mountains. **b** Blockfield in rhyolite in the water gap of the Kaczawa River. **c** Rock cliffs, talus, and possible cryoplanation benches at the Muchowskie Hills. **d** 4-m-thick cover of loess in the marginal zone of the Kaczawskie Foothills, near the town of Złotoryja (photographs by P. Migoń)

by partly buried trees and tree-ring reactions (Remisz and Bijak 2011). However, the blockfields on the north-facing slopes are almost completely covered by trees and rich undergrowth. Blockfields may also occur independently of crags, as in the water gap of the Kaczawa River to the north of Świerzawa, where they exist on rhyolites (Fig. 8.6b).

A different suite of cold-climate hillslope landforms has recently been documented on the low-angle slopes of the Muchowskie Wzgórza massif, in the eastern part of the Kaczawskie Foothills (Migoń et al. 2020). Flattened summit parts terminate against rock cliffs, which are 2–10 m high and supported by thick (up to 2 m in diameter) vertically arranged basalt columns. The lower parts of the cliffs are buried under talus, and there are lateral passages from cliffs to blocky accumulations and vice versa (Fig. 8.6c). Immediately below occur mid-slope benches, with a decreasing density of boulders down the slope. Altogether, this association of forms and deposits is consistent with model examples of frost-riven cliffs and cryoplanation terraces described from various periglacial environments. Further down the slope extend thick solifluction mantles, perhaps up to a few metres thick as inferred from geophysical surveys, with a variable density of basalt blocks at the surface. In some slope sectors, their surfaces are smooth, but in others, high-resolution DTMs derived from LiDAR data revealed steps and risers perpendicular to the general slope. These risers are interpreted as the toe parts of overlapping solifluction sheets.

Loess is present in the northern part of the region, especially around the town of Złotoryja. It is up to 6 m thick and covers the uplands and valley sides (Fig. 8.6d). Being a deposit prone to water erosion, it is associated with gullies, which locally form complex dendritic patterns and may be up to 30 m deep.

8.4.3 Sandstone Cuestas

Cuesta landscape has developed in the northern part of the "Land of Extinct Volcanoes", within the North-Sudetic Synclinorium (Adam 2004). Tilting of strata due to regional tectonics was the necessary prerequisite, whereas differences in rock strength and resistance against weathering and erosion allowed for the emergence of cuestas. These are defined as elongated, asymmetric terrain elevations, adjusted to the strike and dip of strata in such a way that the steep frontal slope is opposite to the dip, the low-angle backslope is generally following the dip, and the lateral extension of the ridge reflects the strike of the strata. Most cuesta landforms in the region have a rather subdued topography, possibly attenuated by ice-sheet erosion during the Pleistocene, but two are very clear. Both are supported by Cretaceous quartz sandstones.

To the south of the town of Złotoryja, one can follow a 15 km long cuesta ridge, cut into two sectors by the meridional Kaczawa valley (Fig. 8.7). The western sector is much lower, up to 35 m high, whereas the eastern one reaches a height of 120 m. Over most of its length, the cuesta follows the WNW–ESE direction, but turns to N–S in the eastern part. The cuesta face in its upper part is built by sandstones of Cenomanian age. They are not exposed as natural outcrops (rock cliffs), but in some sections, sandstone boulders are common, mantling the upper and middle slopes. Some of these boulders are more than 5 m long and occur as a dense pack or are even piled up (Duszyński et al. 2017). The converging backslopes of the cuesta are dissected in the central part by a valley of the Drążnica Stream. Close to the confluence with the Kaczawa River, the stream has significantly undercut its right-side slope, exposing younger sandstones of the Turonian age. Non-karstic caverns a few metres deep have evolved within the valley side, becoming a local geomorphological curiosity.

The other cuesta occurs in the westernmost part of the region, between the valleys of the Skora and the Bóbr rivers. Sandstone of Coniacian age is the main cuesta builder, and the height of the cuesta frontal slope is locally up to 80 m. The backslope is poorly expressed and largely covered by Quaternary sediments. Notable geomorphic features of interest are free-standing rock towers emerging from the cuesta face, with the highest one being 14 m tall, minor rock cliffs, and crevice caves next to the Bóbr Valley (Migoń and Pijet-Migoń 2020) (Fig. 8.8).

Fig. 8.7 Cuesta supported by Upper Cretaceous (Cenomanian) sandstones, with the course of the rim highlighted by dashed red line, is the most distinctive denudational landform in the northern part of the Kaczawskie Foothills. The most important volcanic necks, largely destroyed by quarries, are labelled by their names. 1—section of cuesta front slope with widespread blockfields, 2—location of non-karstic caverns and other rock forms in the Drążnica valley (visualization courtesy of K. Jancewicz)

Fig. 8.8 Sandstone landforms. **a** Isolated rock tower in front of a cuesta. **b** Rock cliffs and non-karstic caves (entrance to one of these shown by arrow) in the Bóbr River valley, north of the town of Lwówek Śląski (photographs by P. Migoń)

8.4.4 Karst

Karst landforms have developed in crystalline limestones of Cambrian age in the Kaczawskie Mountains, but since the limestones occur as separate bodies, apparently without any connection at depth, a large integrated karstic system could not have evolved (Pulina 1977). Instead, each limestone massif hosts its own assemblage of karstic features. In general, local karst is not spectacular, and this is because surface landforms are very poorly expressed. Typical karst mesoforms such as dolines, dry valleys, blind valleys, or karren fields are essentially absent. On the other hand, efficient karstic springs occur in several places, indicating the existence of well-developed subterranean karst. Indeed, caves have long been reported from the area, particularly from Mt. Połom above the town of Wojcieszów, which is the highest limestone hill, once interpreted in terms of tropical karst inheritance and likened to a tropical mogote of the Palaeogene. This hypothesis, however, lacks solid evidence, and the limestone hills are rather rock-controlled forms, which owe their emergence to high rock-mass strength. A few tens of caves have been documented in Mt. Połom, including several deeper than 100 m (Szczelina Wojcieszowska, Jasna Cave) and the longest Gwiaździsta Cave (> 500 m long) (Rogala 2003). Sadly, none of these caves exists, having fallen victims to the ongoing operation in large limestone quarries, although a few others have survived, but are not accessible being located within the mining area. Abandoned quarries within other outcrops, including those at the adjacent Mt. Miłek, host several caves, but these are much shorter, not exceeding 30 m, and quite tight.

8.4.5 Gorges—How Ice Sheets Have Contributed to the Present-Day Relief

Peculiar geomorphic features in the eastern part of the Kaczawskie Foothills are gorge sections of river valleys, otherwise wide open and lacking very steep valley sides. They are characterized by narrow valley floors, locally less than 20 m, very steep and mostly rocky slopes, with bedrock outcrops as high as 20 m, and occasional bedrock steps in the channels (Fig. 8.9). As such, they belong to the most picturesque traits of local landscape and are also localities with considerable biodiversity values, with four being protected as nature reserves. The depth of incision varies from 20–25 m to 70 m. In each case, incision occurred within greenschists, but there is no difference in bedrock between the gorges themselves and more upstream and downstream reaches. Gorges vary in length, from less than 0.5 km to more than 2.5 km in the Jawornik Valley.

The origin of the gorges was proposed to be causally related to the decay of the last ice sheet that covered the Kaczawskie Foothills (Migoń 1999, 2021). The scenario holds that the spatial pattern of preglacial valleys was different from the contemporary one, and the valleys themselves were fairly wide open. During deglaciation, they were for longer filled by ice, including blocks of dead ice, whereas the divides were already ice-free. Under these conditions, meltwaters were incising elevations and margins of the upland in new places, leading to local reorganization of drainage. After all remnant ice melted away, a new drainage network was established, integrating "old" and "new" elements, the latter being deeply incised, rocky gorges (Fig. 8.10). At the same time, abandoned sections of preglacial valleys remained filled by glacigenic sediments and disconnected from the present-day network.

Unfortunately, it is not known with certainty when this could have occurred, whether during the MIS 12 or the MIS 6 glaciation. In the past, it was assumed that these two ice sheets had almost identical extent in the Sudetes, and the currently observed landforms and deposits are the legacy of the younger, MIS 6 glaciation. However, a shift of opinions occurred in the last 25 years or so, and it is now argued that the younger ice sheet reached the marginal parts of the Sudetes, at some 200 m asl, but was unable to transgress into the mountainous terrain. Thus, all glacial elements inside the Sudetes would be associated with the MIS 12 glaciation. Kowalski et al. (2018) have recently documented glacitectonic deformations affecting fluvial and glacial deposits in a gravel pit in the north-western part of the Kaczawskie Foothills (Czaple locality) and assumed that the sedimentary package and the deformation took place during the MIS 6. This locality is at 250–270 m asl, and one can only speculate if this ice sheet could have advanced further to the south, to cover the upland that rises to more than 400 m asl. The clarity of the gorges is not inconsistent with the proposal that it could have indeed been a MIS 6 ice-sheet advance.

8.4.6 Fault-Generated Mountain Front

The eastern morphological boundary of the Kaczawskie Foothills is constituted by a distinctive escarpment (Fig. 8.11), even though it is rather low in terms of relative height (less than 150 m, mostly around 100 m). This escarpment is part of a much larger landform of regional extent, which is the fault-generated mountain front of the Sudetes, related to the Sudetic Marginal Fault. The activity of this fault zone in the Cenozoic is responsible for the separation

Fig. 8.9 Meltwater gorges in the eastern part of the Kaczawskie Foothills. **a** Greenschist cliffs in the longest Myślibórz Gorge. **b** Rockfall-derived greenschist boulders in the bottom of the Siedmica Gorge (photographs by P. Migoń)

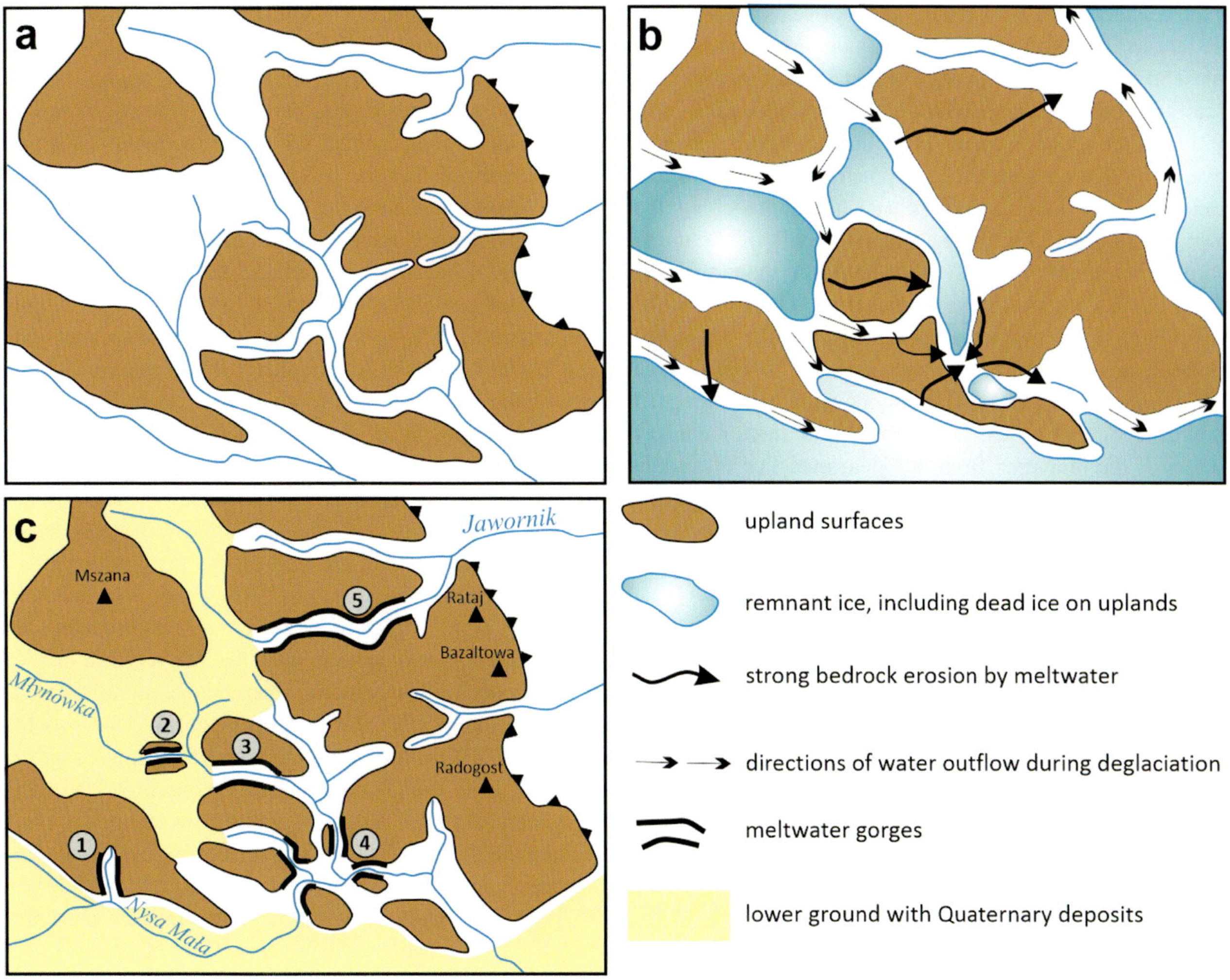

Fig. 8.10 Origin of fluvial gorges in the eastern part of the Kaczawskie Foothills. **a** Probable valley and river network prior to the glaciation. **b** Deglaciation phase: the presence of dead-ice masses in terrain depressions forced meltwater to flow around them and, locally, to cut through bedrock elevations. **c** Contemporary valley and river network. Gorges and water gaps: 1—Lipa, 2—Nowowiejski, 3—Siedmica, 4—gorges near the village of Grobla, 5—Myślibórz

Fig. 8.11 Fault-generated marginal escarpment of the Sudetes is rather subdued along the Kaczawskie Foothills, but is nevertheless a distinctive landform of regional significance (photograph by P. Migoń)

of the Sudetes from its foreland and its rise as a mountainous terrain. The entire length of the mountain front is more than 150 km and its cumulative height in the most uplifted sectors exceeds 500 m (Badura et al. 2007; Różycka et al. 2021). In the north-western part, that is within the area covered by this chapter, the escarpment is much lower and gradually loses height, until it vanishes approaching the town of Złotoryja. The footwall block is built of the metamorphic basement of the Kaczawa Fold Belt, punctuated in places by basaltic outcrops. Wherever the latter occur, the escarpment is steeper, apparently reflecting the resistance of basalt against weathering and erosion. At Mt. Górzec (446 m), the escarpment is partially covered by a vegetated blockfield built of basalt. In contrast to the more southerly sectors of the mountain front, no clear evidence of middle and late Quaternary displacements along the fault in the form of truncated fluvial terrace levels has been recognized (Migoń and Łach 1998), which seems consistent with the subdued nature of the escarpment.

8.5 Towards a UNESCO Global Geopark

Despite the considerable geodiversity and the presence of numerous sites of a very high geoheritage value, the area encompassed by the "Land of Extinct Volcanoes" used to be generally neglected by tourists and rarely visited. Clearly, a mutual negative feedback existed between the poorly developed tourist infrastructure and a limited interest in the region, which was additionally overshadowed by the nearby Karkonosze Mountains—the highest and arguably the most scenic mountain terrain within the Sudetes (see Kasprzak 2023). However, this neglect from the general public contrasted with the popularity of the area among professionals and students of geology and geography, regarding both research (although much more intense in the field of geology than geomorphology) and academic education. Thus, scientific and educational values of the region were realized but poorly communicated outside academia.

The first attempts to increase the visibility of regional geoheritage date back to the 1980s and were focused on the legacy of volcanism (Pijet-Migoń and Migoń 2019). Several basaltic hills and outcrops were connected by a 85-km-long hiking trail named as "The Trail of Extinct Volcanoes". In 1992, the eastern part of the Kaczawskie Foothills became formally protected as a Landscape Park, and an educational centre was opened in the village of Myślibórz. Several publications introducing geology to the general public were launched, and the network of waymarked trails expanded. Parallel to that, various aspects of local mining heritage began to generate interest, including the traditions of gold panning, iron smelting, and lime burning, recreated during annual events. Retrospectively, a milestone development was the establishment of the Local Action Group (LAG) "Kaczawa Partnership" in 2005, which connected more than a dozen municipalities located in the region, both in the Sudetic and fore-Sudetic part. Promotion of sustainable tourism was among the major goals of the LAG, and it was quickly realized that the diverse volcanic heritage, unique within Poland, is among the key assets of the region. In the following 15 years, many new projects aimed at increasing awareness of regional geoheritage were implemented, in the form of publications addressed to different users, trails, interpretation panels, and media promotional campaigns. Last but not least, in August 2015, a geoheritage-oriented educational and information centre was opened in the village of Dobków, in an old Franconian farmstead, specially purchased and adapted to serve these purposes. In 2017, after several years of presenting the idea to various stakeholders, local leaders (municipality mayors) signed a letter of intent confirming the willingness to work towards a UNESCO Global Geopark. In 2018, a complex inventory and valorization of geosites, 130 in total, was prepared, and it confirmed, in a systematic way, that the regional geoheritage is indeed of outstanding international significance, as expected from a UNESCO Global Geopark. This was followed in 2019 by formal submission of an application for membership to the Global Geopark Network of UNESCO. If successful, the "Land of Extinct Volcanoes" would become the third UNESCO Global Geopark in Poland, after the transboundary Muskau Arch (see Koźma and Migoń 2023) and the Holy Cross Mountains Geopark in central Poland. In the meantime, new geosites are being developed for the general public. In 2023, an impressive former basalt quarry at Mt. Wilkołak (Fig. 8.12; see Fig. 8.7 for location) was open to visitors, a few years after basalt excavation came to an end. Its nearly 100-m-high wall and deep pit provide a unique insight into the structure of a volcanic plug, including mega-blocks of thermally-altered Cretaceous sandstone detached and incorporated into the lava.

8.6 Conclusions

The "Land of Extinct Volcanoes" in south-western Poland is both a representative example of less elevated parts of the Central European Variscan ranges (low-altitude mountains and uplands) and an area of outstanding geoheritage in respect mainly to past volcanism that occurred in the Early Palaeozoic, Permian, and Cenozoic. Although the contemporary landscape cannot be considered as volcanic in the strict sense, outcrops of rocks of volcanic and sub-volcanic origin are evident in relief due to their resistance against weathering and erosion. Therefore, they build characteristic landmarks of the region, including conical basaltic hills and rhyolite domes. Numerous quarries, including many long

Fig. 8.12 Basalt quarry at Mt. Wilkołak, itself a volcanic neck, is now developed as a geosite, offering a unique insight into the structure of a lava plug (photograph by P. Migoń)

disused and now open to the public without restrictions, provide insights into the internal structures of volcanic conduits and plugs and reveal spectacular patterns of columnar jointing, which can be linked with the morphological evolution of the hills. However, the regional geoheritage is not limited to the remnants of ancient volcanism. Further localities of geomorphological interest include numerous tors and crags, especially in the more mountainous part of the region, limestone hills, periglacial blockfields on basalt and rhyolite, gullies in loess, and fluvial gorges cut by glacial meltwaters. Thus, even though the subdued landscape of the "Land of Extinct Volcanoes" may look inconspicuous at the first sight, it actually belongs to one of the most diverse in south-western Poland, with its geoheritage being complemented by high-class botanical values and rich cultural heritage that includes, among others, a UNESCO World Heritage listed, seventeenth century Church of Peace in the town of Jawor. The area is therefore a good example of how geodiversity and geoheritage underpin other values and fulfils the ABC approach (abiotic–biotic–cultural) envisaged as optimal for UNESCO Global Geoparks, which is the recognition the "Land of Extinct Volcanoes" is currently seeking.

Acknowledgements I am grateful to Kacper Jancewicz for preparing maps and terrain visualizations for this chapter, as well as to Andrzej Kacprzak for useful comments on the draft version. I am also thankful for constant hospitality offered by Ewelina and Krzysztof Rozpędowscy during field work in the "Land of Extinct Volcanoes".

References

Adam A (2004) Rzeźba strukturalna Pogórza Kaczawskiego i południowo-wschodniej części Pogórza Izerskiego. Przyroda Sudetów 7:175–190

Aleksandrowski P, Mazur S (2002) Collage tectonics in the northeasternmost part of the Variscan belt: the Sudetes, Bohemian Massif. In: Winchester JA, Pharaoh TC, Verniers J (eds) Palaeozoic amalgamation of Central Europe. Geol Soc London Spec Publ 201:237–277

Awdankiewicz M, Kryza R, Szczepara N (2014) Timing of post-collisional volcanism in the eastern part of the Variscan Belt: constraints from SHRIMP zircon dating of Permian rhyolites in the North-Sudetic Basin (SW Poland). Geol Mag 151:611–628

Badura J, Przybylski B, Zuchiewicz W (2004) Cainozoic evolution of Lower Silesia, SW Poland: a new interpretation in the light of sub-Cainozoic and sub-Quaternary topography. Acta Geodyn Geomater 1, 3(135):7–29

Badura J, Pécskay Z, Koszowska E, Wolska A, Zuchiewicz W, Przybylski B (2005) New age and petrological constraints on

Lower Silesian basaltoids, SW Poland. Acta Geodyn Geomater 2, 3(139):7–15

Badura J, Zuchiewicz W, Štěpančíková P, Przybylski B, Kontny B, Cacoń S (2007) The Sudetic Marginal Fault: a young morphotectonic, feature at the NE margin of the Bohemian Massif, Central Europe. Acta Geodyn Geomater 4(4):7–29

Baranowski Z, Haydukiewicz A, Kryza R, Lorenc S, Muszyński A, Solecki A, Urbanek Z (1990) Outline of the geology of the Góry Kaczawskie (Sudetes, Poland). N Jhrb Paläont, Abh 179:223–257

Biernacka J, Borysiuk K, Raczyński P (2005) The Zechstein (Ca1) limestone-marl alternations from the North Sudetic Basin, Poland: depositional versus diagenetic rhythm. Geol Quart 49:1–14

Birkenmajer K (1967) Bazalty dolnośląskie jako zabytki przyrody nieożywionej. Ochrona Przyr 32:225–276

Birkenmajer K, Pécskay Z, Grabowski J, Lorenc MW, Zagożdżon P (2007) Radiometric dating of the Tertiary volcanics in Lower Silesia, Poland. V. K-Ar and palaeomagnetic data from Late Oligocene to Early Miocene basaltic rocks of the North-Sudetic Depression. Ann Soc Geol Polon 77:1–16

Chrząstek A, Wojewoda J (2011) Mezozoik południowo-zachodniej Polski (synklinorium północnosudeckie). In: Ciężkowski W, Wojewoda J, Żelaźniewicz A (eds) Mezozoik i kenozoik Dolnego Śląska. WIND, Wrocław, pp 1–10

Duszyński F, Migoń P, Różycka M, Michniewicz A (2017) Rzeźba progu kredowego i pokrywy blokowe koło Wilkowa (Pogórze Kaczawskie). Przyroda Sudetów 20:269–288

Grodzicki A (1972) Petrografia i mineralogia piasków złotonośnych Dolnego Śląska. Geol Sudetica 6:233–291

Jahn A (1962) Geneza skałek granitowych. Czas Geogr 33:19–44

Kasprzak M (2023) Karkonosze Mountains and Jelenia Góra Basin—unique variety of granite landforms. In: Migoń P, Jancewicz K (eds) Landscapes and landforms of Poland. Springer, Cham

Kowalski A, Makoś M, Pitura M (2018) New insights into the glacial history of southwestern Poland based on large-scale glaciotectonic deformations—a case study from the Czaple II gravel pit (western Sudetes). Ann Soc Geol Polon 88:341–359

Kowalski B (1977) Podłoże trzeciorzędu przedpola Sudetów Zachodnich w rejonie Jawora. Przegl Geogr 49:99–118

Koźma J, Migoń P (2023) Mużaków Rampart (Muskau Arch)—the legacy of glacial processes and mining in the UNESCO Global Geopark. In: Migoń P, Jancewicz K (eds) Landscapes and landforms of Poland. Springer, Cham

Kryza R, Muszyński A (1992) Pre-Variscan volcanic-sedimentary succession of the central southern Góry Kaczawskie, SW Poland: outline geology. Ann Soc Geol Polon 62:117–140

Kryza R, Zalasiewicz JA, Mazur S, Aleksandrowski P, Sergeev S, Presnyakov S (2007) Early Palaeozoic initial-rift volcanism in the Central European Variscides (the Kaczawa Mountains, Sudetes, SW Poland): evidence from SIMS dating of zircons. J Geol Soc London 164:1207–1215

Leszczyński S, Nemec W (2019) Sedimentation in a synclinal shallow-marine embayment: Coniacian of the North Sudetic Synclinorium, SW Poland. Deposit Rec 6:144–171

Lorenc S (1983) Petrogeneza wapieni wojcieszowskich. Geol Sudetica 18(1):61–119

Martini A (1969) Sudetic tors formed under periglacial conditions. Biul Perygl 19:351–369

Michniewicz A (2016) Skałki zieleńcowe grzbietu Okola w Górach Kaczawskich (Sudety Zachodnie). Chrońmy Przyr Ojcz 72:206–218

Michniewicz A, Jancewicz K, Migoń P (2020) Large-scale geomorphological mapping of tors—proposal of a key and landform interpretation. Geomorphology 357:107106

Migoń P (1999) Doliny jarowe na Pogórzu Kaczawskim. Chrońmy Przyr Ojcz 55(2):91–99

Migoń P (2021) Wybrane formy rzeźby terenu w sudeckiej części Geoparku Kraina Wygasłych Wulkanów (pd.-zach. Polska). Landf Anal 40:5–33

Migoń P, Łach J (1998) Geomorphological evidence of neotectonics in the Kaczawa sector of the Sudetic Marginal Fault, southwestern Poland. Geol Sudetica 31:307–316

Migoń P, Pijet-Migoń E (2016) Overlooked geomorphological component of volcanic geoheritage—diversity and perspectives for tourism industry, Pogórze Kaczawskie region, SW Poland. Geoheritage 8:333–350

Migoń P, Pijet-Migoń E (2020) Not simply volcanoes—geoheritage of the Cretaceous system in the Land of Extinct Volcanoes Geopark, West Sudetes (SW Poland). Geotourism 17(61–62):3–22

Migoń P, Maciejak K, Zygmunt M (2002) Peryglacjalna rzeźba wzgórz bazaltowych Pogórza Kaczawskiego (Sudety Zachodnie) i jej znaczenie dla paleogeografii plejstocenu. Przegl Geogr 74:491–508

Migoń P, Kasprzak M, Jancewicz K (2020) Inherited periglacial geomorphology of a basalt hill in the Sudetes, Central Europe: insights from LiDAR-aided landform mapping. Permafrost Perigl Proc 31:587–597

Milewicz J (1997) Górna kreda depresji północnosudeckiej (lito- i biostratygrafia, paleogeografia, tektonika oraz uwagi o surowcach). Acta Univ Wratisl 1971, Prace Geol-Miner 61, Wrocław

Pijet-Migoń E, Migoń P (2019) Promoting and interpreting geoheritage at the local level—bottom-up approach in the Land of Extinct Volcanoes. Geoheritage 11:1227–1236

Placek A (2007) Basaltic hills as structural landforms—morphometry versus rock strength (a study from the Kaczawskie upland, SW Poland). Univ Ostravensis Acta Fac Rer Natur 237, Geographia – Geologia 10:111–127

Pulina M (1977) Zjawiska krasowe w Sudetach polskich. Dok Geogr IGiPZ PAN 2–3:1–118

Remisz J, Bijak S (2011) Stoki usypiskowe Ostrzycy i ich aktywność w świetle badań dendrogeomorfologicznych. Przyroda Sudetów 14:197–206

Rogala W (2003) Pionowy układ jaskiń krasowych na górze Połom w Górach Kaczawskich. Przegl Geol 51:238–242

Różycka M, Jancewicz K, Migoń P, Szymanowski M (2021) Tectonic versus rock-controlled mountain fronts—geomorphometric and geostatistical approach (Sowie Mts., Central Europe). Geomorphology 373:107485

Solecki A (2008) The North-Sudetic Synclinorium – geosites of the inverted basin setting. In: Solecki A (ed) Geoeducational Potential of the Sudety Mts. University of Wrocław, Institute of Geological Sciences, Wrocław, pp 77–89

Solecki A (2011) Rozwój strukturalny epiwaryscyjskiej pokrywy platformowej w obszarze synklinorium północnosudeckiego. In: Ciężkowski W, Wojewoda J, Żelaźniewicz A (eds) Mezozoik i kenozoik Dolnego Śląska. WIND, Wrocław, pp 19–36

Solon J, Borzyszkowski J, Bidłasik M, Richling A, Badora K, Balon J, Brzezińska-Wójcik T, Chabudziński Ł, Dobrowolski R, Grzegorczyk I, Jodłowski M, Kistowski M, Kot R, Krąż P, Lechnio J, Macias A, Majchrowska A, Malinowska E, Migoń P, Myga-Piątek U, Nita J, Papińska E, Rodzik J, Strzyż M, Terpiłowski S, Ziaja W (2018) Physico-geographical mesoregions of Poland. Verification and adjustment of boundaries on the basis of contemporary spatial data. Geogr Polon 91(2):143–171

Śliwiński W, Raczyński P, Wojewoda J (2003) Sedymentacja utworów epiwaryscyjskiej pokrywy osadowej w basenie północnosudeckim. In: Ciężkowski W, Wojewoda J, Żelaźniewicz A (eds) Sudety Zachodnie: od wendu do czwartorzędu. WIND, Wrocław, pp 119–126

Wocke MF (1927) Der Basalt in der Schlesischen Landschaft. Veröffent Schles Ges Erdk 5:1–51

Zuchiewicz W, Badura J, Jarosiński M (2007) Neotectonics of Poland: an overview of active faulting. Studia Quat 24:5–20

Żelaźniewicz A (2015) Przeszłość geologiczna. In: Żelaźniewicz A (ed) Przyroda Dolnego Śląska, 2nd edn. Polska Akademia Nauk – Oddział we Wrocławiu, Wrocław, pp 39–99

Piotr Migoń is Professor of Geography at the University of Wrocław and former Director of the Institute of Geography and Regional Development. His main research interests are rock control in geomorphology, especially in sandstone and granite areas, mass movements in mountain terrain, geoheritage, and geotourism. The Sudetes Mountains in Central Europe is his main research area, but he was also involved in projects in various other European countries (Czechia, Germany, Great Britain, Sweden, Portugal, and Spain), in the United States, Mexico, Brazil, China, Jordan, and Namibia. He used to be the Secretary of the International Association of Geomorphologists (1997–2001), its Vice-President (2009–2013), and member of Executive Committee (2013–2022); he was also the President of the Association of Polish Geomorphologists (2005–2008). He is the series editor of "World Geomorphological Landscapes" series and member of editorial board of *Geomorphology* journal.

The Kamienne Mountains—Late Palaeozoic Volcanic Terrain Shaped by Late Quaternary Landslides

9

Piotr Migoń

Abstract

The Kamienne Mountains are an elongated medium-altitude mountain range in the central part of the Sudetes, built mainly of volcanic and subvolcanic rocks of Permian age. In many places they overlie sedimentary formations, creating significant geomechanical contrasts that facilitate rock-slope instability. Around 60 landslides have been identified in the range using field geomorphological mapping and LiDAR DTM interpretation. The largest singular landslide occupies 25 ha, whereas the largest landslide complexes are nearly 50 ha. Combined landform mapping and geophysical surveys demonstrate a variety of displacement types, including rotational and translational slides, flowslides, rock-slope failures and falls, and lateral spreading. In specific localities slope deformation is ongoing, although historical records of large landslides are not available. Good access to many landslide localities provides a good basis to develop the region as a geotourist destination, with special focus on landslide-related geosites.

Keywords

Landslides · Rock control in geomorphology · Geoheritage · Sudetes

9.1 Introduction

It was long believed that landslides played a very minor part in the geomorphological evolution of the Sudetes range. These opinions were based on the assessment of geological structure of the region, dominated by crystalline rocks and hence considered unfavourable for large-scale, deep-seated rock-mass displacements, and supported by extremely low incidence of large mass movements in historical times. This view has recently proved not entirely correct, as high-resolution LiDAR DTM helped to reveal numerous landslides hidden under forest (e.g. Migoń et al. 2013, 2017b; Kowalski 2017, 2021; Kotwicka et al. 2019; Tichavský and Fabiánová 2022) and therefore difficult to recognize using ground landform mapping or aerial imagery. However, there has always been one notable exception, where the role of landslides received proper attention—the Kamienne Mountains in the central part of the Polish Sudetes. The occurrence of large ground displacements in this region was noted as early as the beginning of the twentieth century, when the first detailed geological mapping was executed, whereas the intensification of research in the last decade or so has resulted in a clear identification of this area as a 'mass-movement hotspot' within the Sudetes. Geological structure and local relief are the key factors that underpin the wide presence of landforms indicative of considerable slope remodelling by mass movements and jointly explain the peculiar position of the Kamienne Mountains within the entire range. This chapter provides an overview of the history of research on mass movements, introduces general characteristics of landslides, and presents a few case studies in more detail, demonstrating the diversity and complexity of mass movements. It also discusses a possible role of landslide landforms as geosites and geotouristic opportunities.

P. Migoń (✉)
Institute of Geography and Regional Development,
University of Wrocław, pl. Uniwersytecki 1, 50-137 Wrocław, Poland
e-mail: piotr.migon@uwr.edu.pl

P. Migoń and K. Jancewicz (eds.), *Landscapes and Landforms of Poland*,
World Geomorphological Landscapes, https://doi.org/10.1007/978-3-031-45762-3_9

9.2 Geographical and Geological Setting

The Kamienne Mountains are a highly elongated, second-order range within the Sudetes, shared by Poland and Czechia (Fig. 9.1). Even though the absolute elevation does not exceed 1000 m asl, the range is very distinctive topographically, rising above the adjacent gently rolling terrain of intramontane troughs and basins by 300–400 m and being bounded by very steep slopes, not uncommonly more than 30° (Fig. 9.2). It consists of several segments of variable strike, separated by water gaps. The westernmost segment is known as the Krucze Mountains and has N–S extension, being 18 km long and 3–5 km wide. The highest peak, Mt. Kralovecký Špičák (881 m), is located in the south, on the territory of Czechia, and this part of the segment is also most dissected by fluvial valleys. To the north altitudes and slope inclinations decrease, to fall below 500 m asl in the town of Kamienna Góra. Beyond the water gap of the Zadrna River in Kamienna Góra the range turns east and follows the WNW–ESE direction for another 6 km. This segment bears the name of Czarny Las and is the lowest within the range, with the highest spot at 660 m asl, and also the most subdued in terms of relief. The next part is located between the valleys of Grzędzki Potok and Ścinawka, with the highest peaks of Mt. Lesista Wielka (851 m) and Mt. Dzikowiec (836 m). Deep river valleys with steep slopes are inset between minor ridges, but Mt. Lesista Wielka itself is a distinctive plateau of minimal relief, ca 1 km long and up to 0.5 km wide.

The water gap of Ścinawka River makes the western border of the largest and highest part of the Kamienne Mountains, known as the Suche (Dry) Mountains. The name relates to the paucity of surface waters and the underdeveloped stream network, although the density of valley network is quite high. The most complicated topography occurs in the western part of this segment, which is also the highest (Mt. Waligóra, 936 m; Mt. Suchawa, 928 m). The

Fig. 9.1 Location map (map courtesy of Kacper Jancewicz)

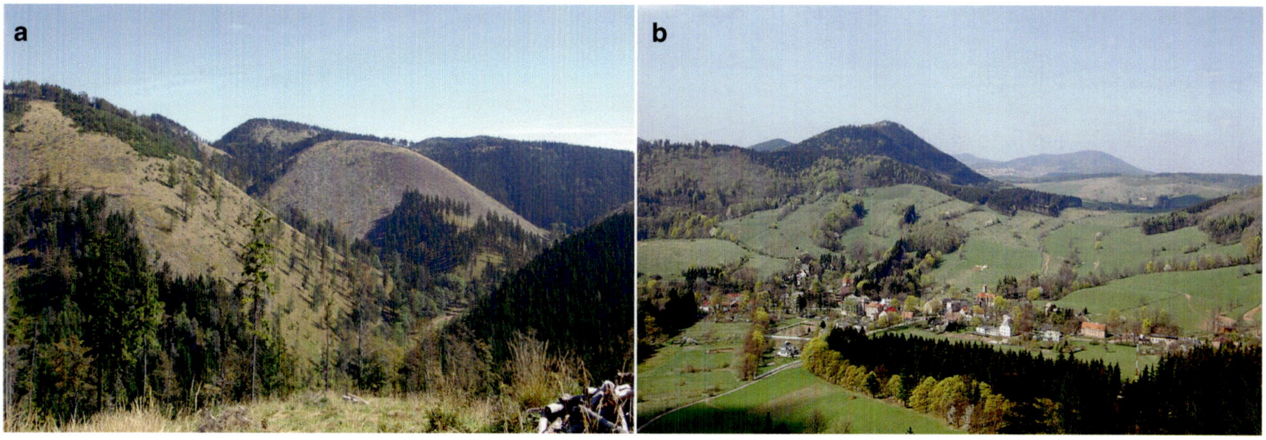

Fig. 9.2 Rock-controlled landscape of the Kamienne Mountains. **a** Steep slopes and narrow valleys are typical for volcanic rocks. **b** A minor intramontane basin occupied by the village of Sokołowsko was eroded in sedimentary rocks and is surrounded by elevations in volcanic rocks. Relative relief exceeds 300 m (photographs by P. Migoń)

local relief attains 400 m within 1 km^2, which is among the highest in the entire Sudetes, and the slopes are commonly steeper than 30°. A small intramontane basin of Sokołowsko stands out (Fig. 9.2b), almost completely encircled by steep-sided hills. Towards the south-east the range becomes narrower, and the altitudes decrease below 750 m asl. This part of the Kamienne Mountains is crossed by the state border, but the Czech part is smaller and apparently less diverse in terms of relief.

Geologically, the Kamienne Mountains are mainly an ancient volcanic terrain that was active in the early Permian (Fig. 9.3a). The history of volcanism was complex, and as a result, several distinctive associations of magmatic rocks are distinguished (Awdankiewicz 1999). The volcanism was bimodal, with both acid and intermediate rocks present. Rhyolites and rhyolitic tuffs are the most widespread, followed by trachyandesites and trachybasalts. They occur in various forms, including large plugs, minor stocks, lava sheets and flows, and sub-horizontal sills. Rhyolitic tuffs formed in the late stage of activity, as an extensive blanket emplaced over older volcanic rocks, and include also welded tuffs (ignimbrites). This volcanism occurred in subaerial conditions, within a vast intramontane trough that originated during the Variscan orogeny (Devonian/Carboniferous). This trough was filled by clastic deposits ranging from coarse gravels transported by high-energy braided rivers and ephemeral debris-mud flows to fine-grained materials laid down in lacustrine setting. Consequently, the sedimentary succession is very complex and consists of alternating conglomerates, sandstones, mudstones, and shales (Bossowski and Ihnatowicz 1994). In respect to the present-day relief, sedimentary rocks often occur beneath those of volcanic origin, in the lower slopes, which results in the 'strong-over-weak' sequence, conducive to rock-mass deformation and slope instability.

During the Pleistocene, the range of Kamienne Mountains was shaped by various periglacial processes, and during MIS12, the time of maximum extent of glaciation, it was beyond the limits of Scandinavian ice sheet, except the water gap of Zadrna river in Kamienna Góra (Traczyk 2004; Pitura and Sobczyk 2017). Direct impact of the ice sheet on hillslope processes can therefore be excluded.

9.3 Landslides—History of Research

The authors of the first detailed geological mapping executed in the region were also the first to recognize that large-scale mass movements must have occurred on steep slopes of volcanic elevations (Berg et al. 1910; Berg 1925). Their reasoning was based on landform interpretation and identification of large allochthonous masses of volcanic rocks in the lower slope positions, apparently detached from their in situ outcrops located in the upper slope. However, these maps did not show the complete extents of landslides and the depletion parts were generally missing. More recent maps (Grocholski 1971; Bossowski et al. 1994) included some landslide terrains in full extent, but at the same time were not entirely consistent regarding cartographic means to show the terrain affected by mass movements. Recent re-appraisal of older cartographic materials, compared with the results of landslide mapping based on LiDAR DTM as a primary source, showed that the actual extent of landslides was severely underestimated (Migoń et al. 2014b). Both the early and late twentieth century maps suggested that landslide-affected terrains occupy 80–90 ha in total in the eastern part of the Kamienne Mountains, whereas LiDAR DTM showed that at least 300 ha were affected. Further research demonstrated that this too was a conservative estimate (e.g. Kasprzak et al. 2019).

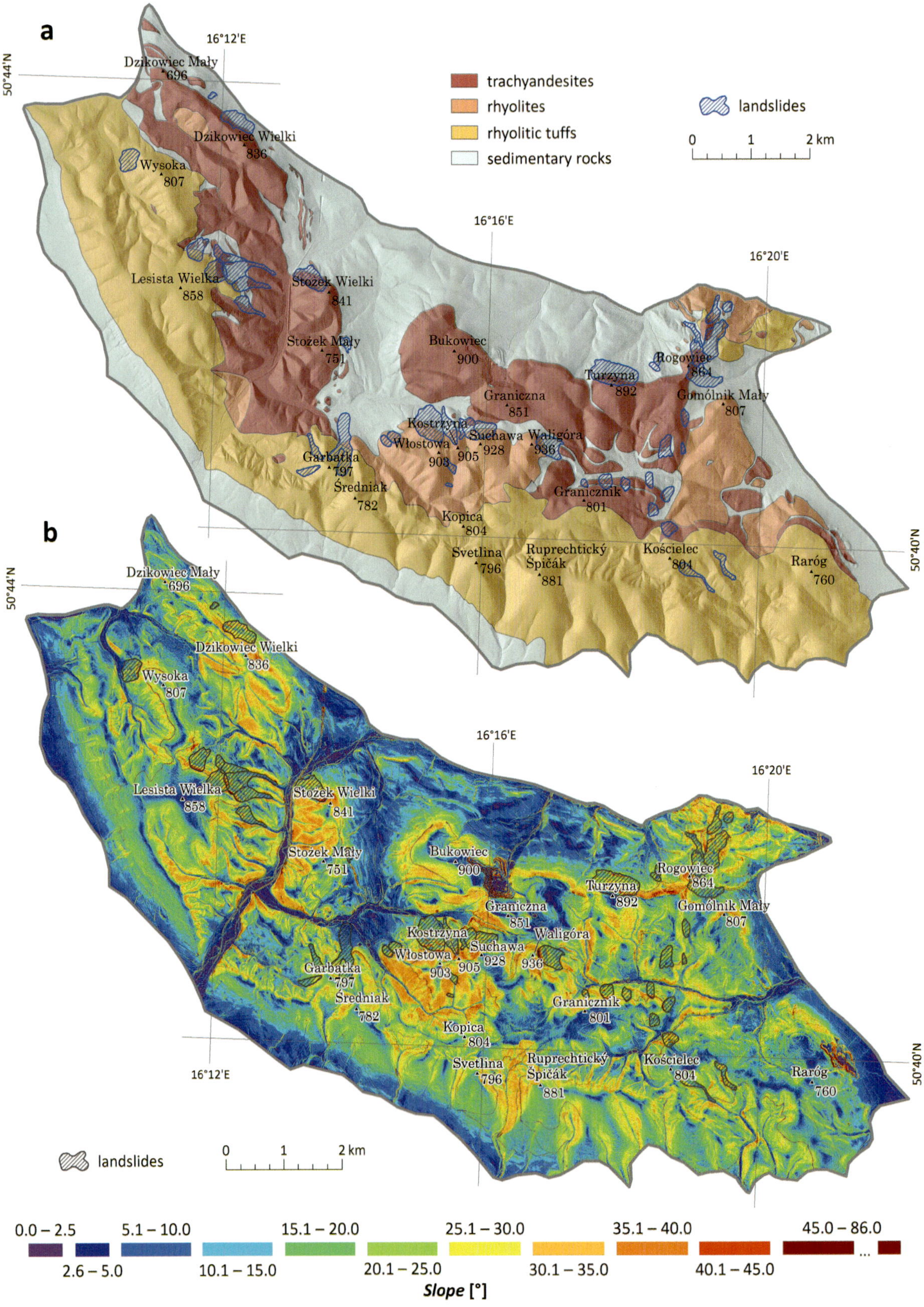

Fig. 9.3 Landslide distribution in the eastern part of the Kamienne Mountains in relation to geology (**a**) and slopes (**b**) (maps courtesy of Kacper Jancewicz)

Despite what geological cartography showed, almost no detailed work was undertaken in the twentieth century, except one brief description of a landslide terrain next to Mt. Rogowiec (Pulinowa 1972), later proved to be but part of a much larger complex of mass movements. Synowiec (2003) was the first to attempt more detailed landform mapping of selected landslides, but the base topographic maps were of limited usefulness, as is also evident from maps included in Migoń et al. (2010). Nevertheless, an intensification of landslide research occurred around the year 2010, with several parallel lines of inquiry. Landslide mapping continued and was considerably boosted in 2013, when the high-resolution LiDAR DTM became available. It was then possible not only to show the extent of landslides and specific landforms within them cartographically, but also to provide quantitative geomorphometric characteristics. This allowed for sound comparative analysis and landslide classification according to quantitative criteria (Migoń et al. 2017b). Several landslide complexes were subject to geophysical survey using electrical resistivity tomography (ERT), particularly useful to detect deep-seated deformations and incipient stages of slope instability (Migoń et al. 2010, 2016, 2017a; Kasprzak et al. 2019). Dendrochronology was applied to detect contemporary movements at several landslides of different types (Migoń et al. 2010, 2014a; Malik et al. 2016). An innovative approach was to look at soil characteristics and maturity of soil profiles within landslides and beyond them, in order to establish relative ages of surfaces. It was shown that within a complex landslide at Mt. Suchawa and Mt. Kostrzyna (Fig. 9.4) soils have different properties despite a uniform parent material, leading to the proposal that landslide activity was multi-phase and the duration of pedogenesis was variable (Kacprzak et al. 2013). Soils could thus become useful tools to differentiate landslides in terms of their relative ages, which is important in the context of paucity of directly datable materials.

Little is still known about the age of landslides in the Kamienne Mountains. Two basal samples from shallow ponds within the Rogowiec landslide complex were dated using radiocarbon, yielding calibrated dates of 4780–4600 BC and 1630–1510 BC (1σ error) (Synowiec 2005). These dates show mid-to-late Holocene onset of sedimentation in the ponds, but the relationship to the chronology of rock-mass movement itself is uncertain. Soil research at several localities (Kacprzak et al. 2013; Migoń et al. 2014a) suggested that some landslides are likely pre-Holocene, whereas others may have originated or been reactivated throughout the Holocene, perhaps up to very recent times. Moreover, the landslides differ in terms of

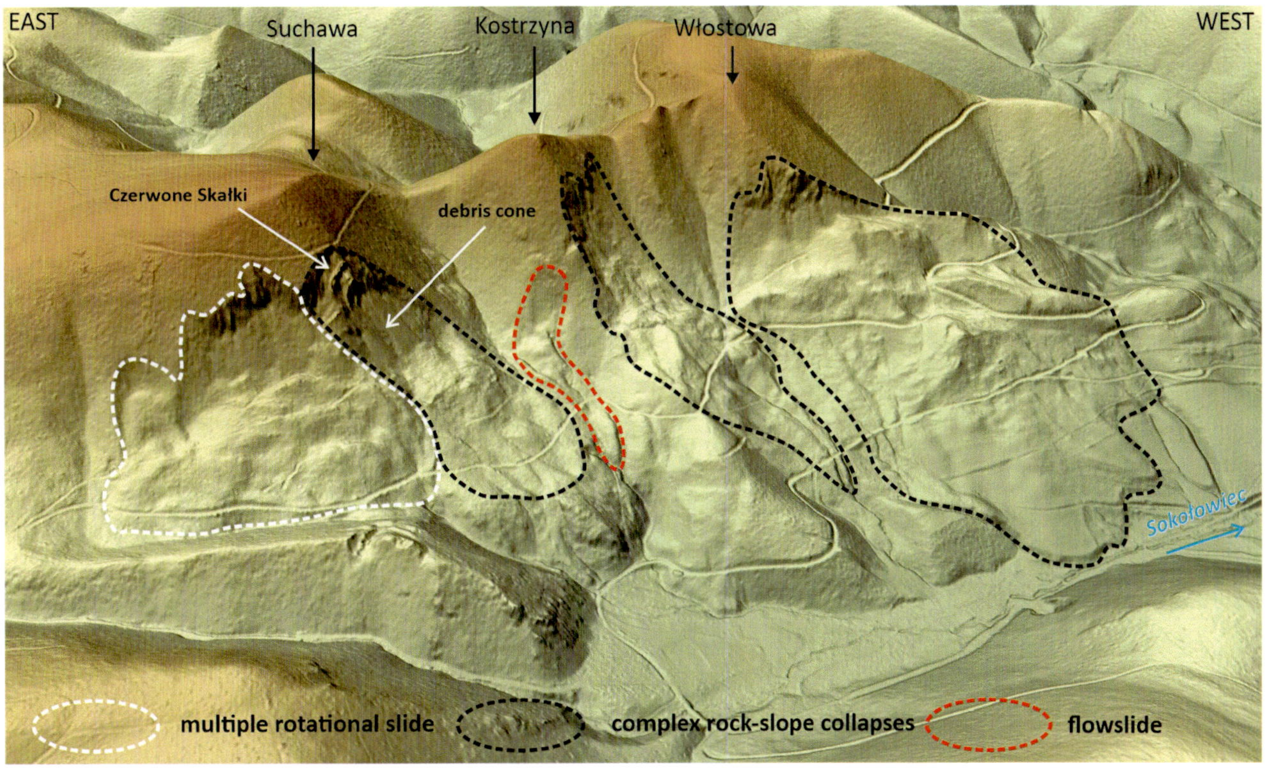

Fig. 9.4 Three-dimensional model of landslide terrain on the northern slopes of Mt. Suchawa–Mt. Kostrzyna–Mt. Włostowa ridge, with the extent of different landslide complexes indicated by broken lines. The image covers an area of approximately 2 km × 1 km (visualization courtesy of Kacper Jancewicz)

geomorphological expression, which also suggests a diversity of ages, but no time constraints can be provided.

9.4 Landslide Typology, Distribution, and Controls

A detailed landslide inventory was prepared for the central-east part of the Kamienne Mountains, where 47 separate landslides occupying the total area of 318 ha (nearly 4% of the range) were recognized, including two beyond the border, in Czechia (Migoń et al. 2017a). However, subsequent research showed (Kasprzak et al. 2019; Migoń et al. 2021) that the actual number is higher, even though the missing landslides at the Schwarzberg locality are not very large (c. 10 ha in total). Finally, no comparable quantitative studies have been carried out on landslides in the Krucze Mountains, where several large landslide areas occur, including one occupying c. 20 ha in the Czech part.

Nevertheless, the part of the area subject to the inventory and cluster analysis may be considered representative for the entire range and appears to have all major types. The following main types can be distinguished:

(a) large rotational or translational slides, laterally extensive in the head scarp part (up to 0.8 km wide) and with singular or multiple rotated/translated blocks below. Hummocky toe parts indicative of significant loss of coherence within the failed mass are absent or not very distinctive. The largest singular landslide in the region, at Mt. Turzyna (25 ha), belongs to this group, although it actually represents a special case, at the boundary between deep-seated gravitational slope deformation (DSGSD) and a proper landslide (Migoń et al. 2017b; Kasprzak et al. 2019).
(b) minor rotational or translational slides, with runout distance less than 300 m long and only one evident displaced block.
(c) flowslides, typified by long depositional parts (even > 1 km) constricted by valley sides. However, their source zones are morphologically diverse and include both distinct slope amphitheatres with bedrock exposed, as well as shallow slope hollows. In a few cases, rigid slid blocks occur just below the head scarps, indicating progressive fragmentation of the displaced mass as the movement developed.
(d) lateral spreads, which affected several ridge-top parts, producing irregular topography with multiple closed depressions. Morphological consequences of spreading are best seen at Mt. Rogowiec locality (see Sect. 9.5.1), but they occur in a few other localities too (Kasprzak et al. 2019; Migoń et al. 2021).

In the real world, these different types may combine into both complex and compound landslides. The former are characterized by a change of the mechanism of movement as it occurred. Typically it was an evolution of a slide into a flow and numerous flowslides apparently evolved from sliding of rigid blocks. Compound landslides are those where adjacent slope sections (sharing the head scarp) were subject to different types of movement. The bipartite Włostowa landslide is an example, with the western part consisting of a series of rigid downthrown blocks below a relatively low (up to 15 m) head scarp and a very high (~ 50 m) head scarp in the eastern part, grading into disintegrated colluvial packages (Fig. 9.4).

In addition, areas of incipient slope deformation were identified, which do not yet have a clear topographic expression and are hardly possible to delimit in space. Indicators include shallow troughs and occasional deeper trenches within plateau/ridge surfaces, lower slope convexities (bulges), and thermal anomalies consisting of winter outflows of warm air in the upper slope. They are most evident at Mt. Lesista Wielka (see Sect. 9.5.3) (Kasprzak et al. 2019).

Landslides in the Kamienne Mountains are clearly associated with lithological contacts between volcanic/subvolcanic and sedimentary rocks and causally related to geomechanical contrasts between these two rock complexes. This was demonstrated using spatial analysis, which indicated that > 80% of the total landslide area is located within 200 m of the contact, on either side of it. Among 47 landslides analysed, 31 occurred in the proximity of contact, and only 2 out of 18 largest landslides (> 7 ha area) were generated in no apparent relation to any lithological boundary. Rigid blocks of heavy volcanic rocks impose load on the underlying sedimentary formations, which cannot withstand the stress and may deform if contain significant amount of clay and silt. This may lead to outward movement (squeezing out) in the lower slope, causing the volcanic overburden to lose vertical support. As a consequence of the resultant tensile stress, cracking occurs, and pre-existing joints may be used to facilitate downslope sliding if the dip is concordant with the topographic slope. In fact, the density of joints in volcanic rocks is quite high, and their rock-mass strength is only moderate. However, in sedimentary rocks new curvilinear sliding planes are likely to develop and accommodate movement, as the original bedding planes dip into the slope, according to the regional geological structure of the post-Variscan intramontane trough (Migoń et al. 2017a).

The second key factor controlling mass movements in the area is relief. The Kamienne Mountains belong to the steepest terrains in the Sudetes (Migoń et al. 2009), with slope inclinations often above 30°. These very steep mid- and upper slope surfaces contrast with low-angle lower slopes in sedimentary formations and flattish topography of minor basins, also excavated in sedimentary rocks. Moreover, the height of slopes and the depth of valleys may reach 300 m or more. As the area was not glaciated, reasons for this rugged topography lie in very efficient erosional removal of weak sedimentary rocks, with erosion in volcanic rocks lagging behind. Consequently, slopes are hypothesized to become too steep for the rock-mass strength and fail, this being a means to restore strength equilibrium.

9.5 Selected Landslides

9.5.1 Rogowiec Landslide Complex

The landslide complex around Mt. Rogowiec (850 m) in the north-eastern part of the Kamienne Mountains (Fig. 9.1) has developed on both sides of a N–S trending ridge, involving also the top part of the ridge itself. The axial part of the ridge is built of volcanic rocks (trachyandesites), resting upon sedimentary formations. It occupies 41 ha in total, which makes it the third largest contiguous landslide area in the range (Migoń et al. 2017b), and consists of several morphologically distinctive units, formed through displacements of various type and, most probably, at different times (Kasprzak et al. 2016; Fig. 9.5). The eastern side of the ridge was affected by large rotational slides, which left nearly rectilinear head scarps and a series of benches perpendicular to the slope below. In a few places minor ponds formed behind or within these benches (Fig. 9.6a), and basal organic deposits retrieved from these minor sediment traps were dated by Synowiec (2005) and yielded mid-to-late Holocene dates (see Sect. 9.3). In the downslope direction displaced blocks apparently disintegrated during motion and nowadays irregular tracts of hummocky terrain with minor erosional incisions occur. The most spectacular

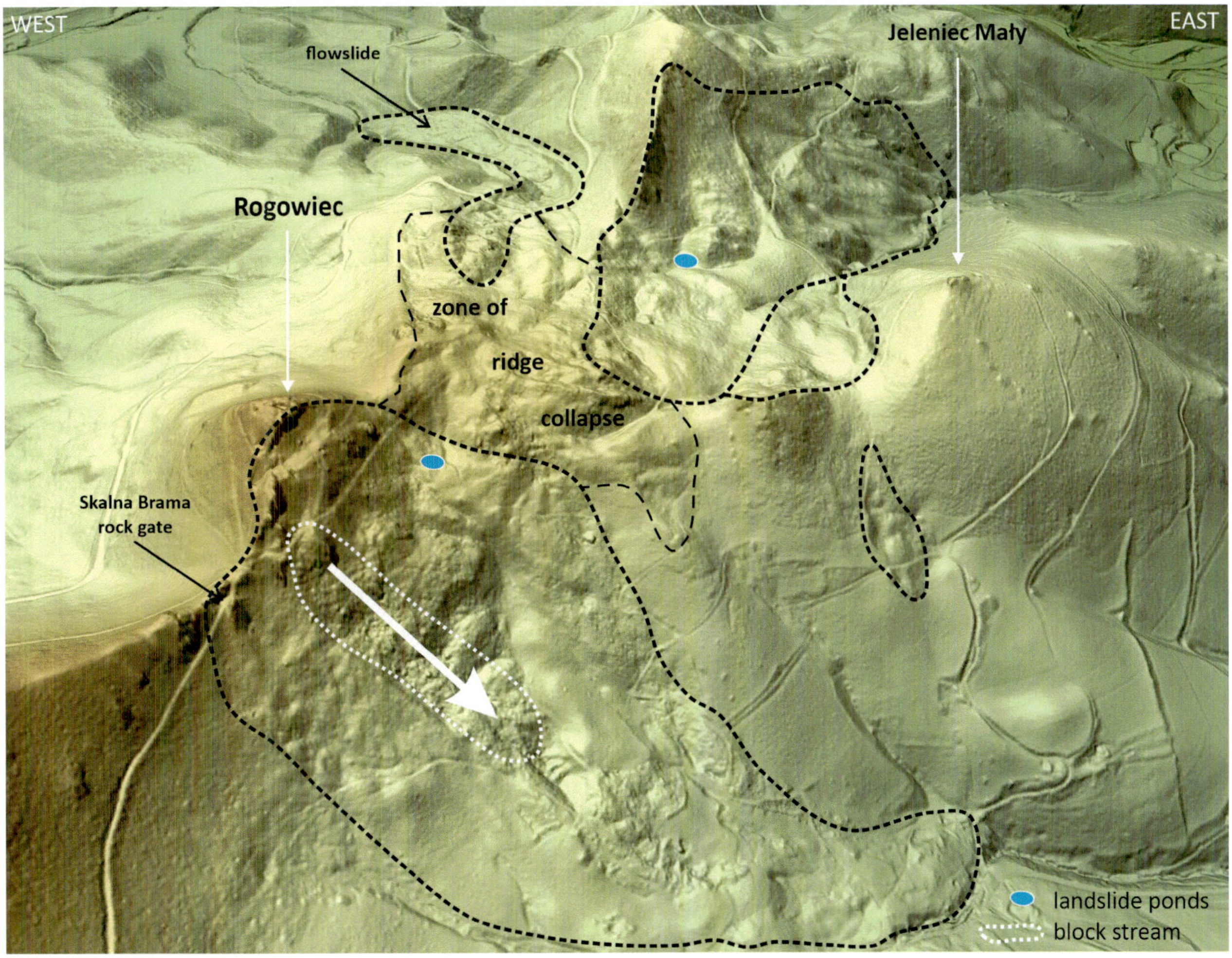

Fig. 9.5 Three-dimensional model of landslide terrain around Mt. Rogowiec, with the extent of different mass movements indicated by broken lines. The image covers an area of approximately 1.5 km × 1.5 km (visualization courtesy of Kacper Jancewicz)

part of the east-facing slopes of Mt. Rogowiec is located in the south, where an impressive block stream, ca 250 m long and up to 80 m wide, begins below the head scarp and terminates in the mid-slope. It is composed of huge angular blocks, piled one upon another, some as big as 10 m, with deep clefts and shafts in between (Fig. 9.6b). It is interpreted as a product of a catastrophic rock-slope collapse from the oversteepened head scarp, subsequent to the main sliding event. Not far away, some 70 m to the south, a large bedrock outcrop (Skalna Brama = '*rock gate*') is detached from the main scarp, displaced 10 m away in downslope direction and tilted, but is still in place (Fig. 9.6c). Its future collapse will produce another, minor block stream.

Of particular interest is the part of the water divide immediately to the north of Mt. Rogowiec. Here, along a length of some 300 m, the relief is no longer of a pointed ridge, typical for the Kamienne Mountains, but it represents an extensive (ca 10 ha) swell-and-hollow topography, with as many as 30 closed depressions of various size (Kasprzak et al. 2019). Some occupy distinct ridge-top trenches (Fig. 9.6d). The convexities are either crowned by trachyandesite crags a few metres high, or by scattered boulders. Concavities, in turn, usually lack bedrock outcrops or boulders. Geophysical (ERT) survey across the water divide revealed high-resistivity blocks, 10–30 m thick, 'floating' over low-resistivity materials, which emerge from beneath them in the mid-slopes. This situation is interpreted as an evidence of lateral spreading and disintegration of the trachyandesite cap (ridge collapse) in the ridge section, where it overlies deformable sedimentary formations. By contrast, if the volcanic cap is rooted in the bedrock, conditions for lateral spreading affecting the entire width of the ridge are not fulfilled, and retrogressive slides occur on slopes (Fig. 9.7). Mass-movement diversity at Mt. Rogowiec is further enhanced by the presence of a flowslide in a valley incised into the NW slopes. An elongated body of allochthonous trachyandesitic materials, including boulders up to 5 m long, fills an originally V-shaped valley along the length of more than 500 m, ending with a distinct toe more than 10 m high.

Fig. 9.6 Landslide morphology around Mt. Rogowiec. **a** Pond impounded by a rotated block. **b** Block stream below the peak of Mt. Rogowiec. **c** Outward tilted block of Skalna Brama (seen on the right) and gravitationally opened cleft. **d** Ridge-top trench as evidence of slow lateral spreading and ridge disintegration (photographs by P. Migoń)

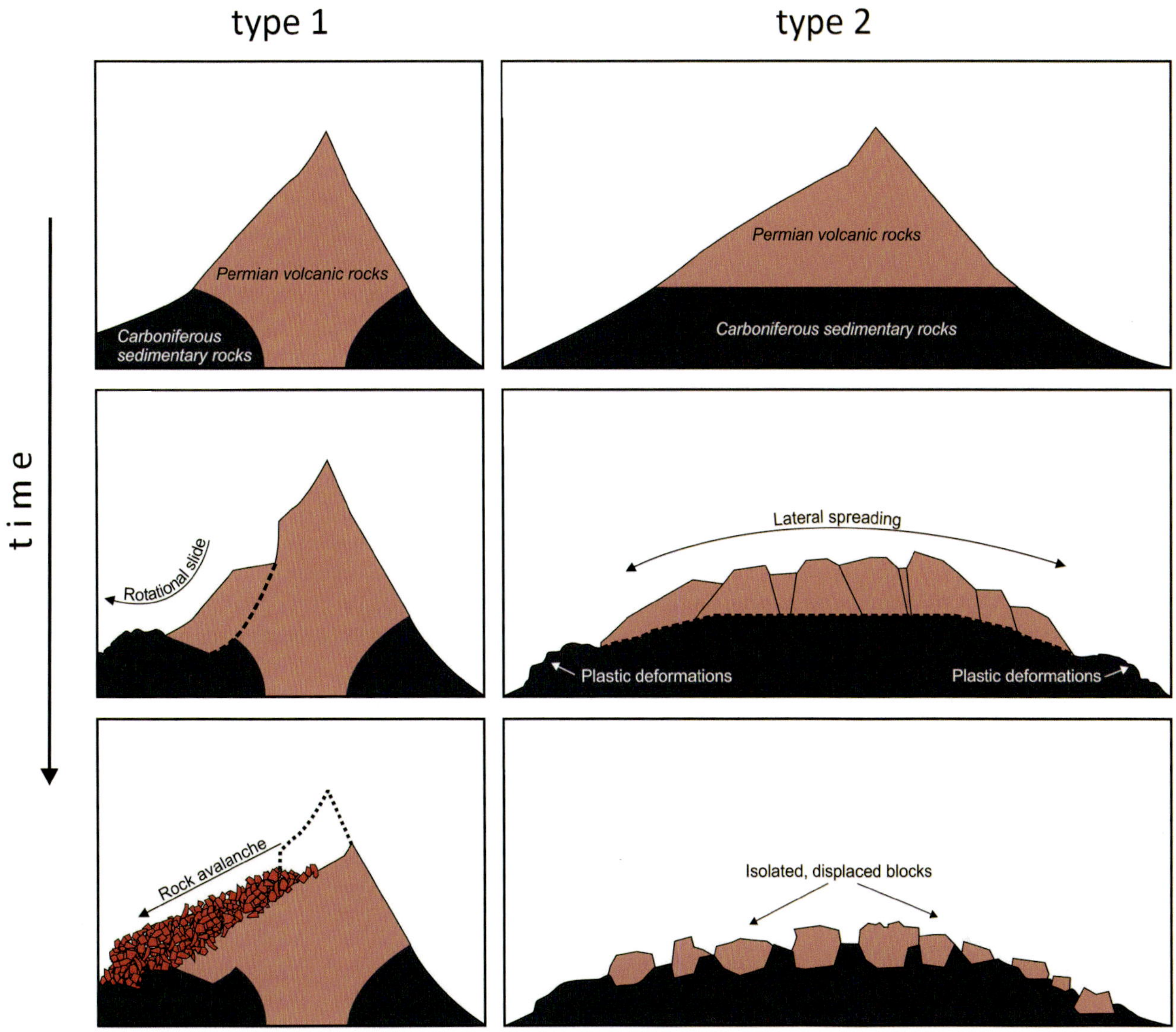

Fig. 9.7 Conceptual model to show two evolutionary pathways of slope deformation, applicable to Mt. Rogowiec, depending on the nature of contact between volcanic and sedimentary rocks. Type 1—volcanic rocks rooted in bedrock; type 2—volcanic cap over sedimentary rocks (after Kasprzak et al. 2016, modified)

9.5.2 Suchawa–Kostrzyna–Włostowa Landslides

Almost the entire north-facing slopes of Mt. Suchawa (928 m), Mt. Kostrzyna (906 m) and Mt. Włostowa (903 m) in the eastern part of the Kamienne Mountains (Fig. 9.1) have been remodelled by landslides, which form a large landslide complex (Fig. 9.4) (Kacprzak et al. 2013; Migoń et al. 2017b). The ridge is built of rhyolites that overlie sedimentary rocks in the form of a thick lava flow fed by the nearby volcanic vent at the present-day Mt. Waligóra. As in most other cases in the region, the combination of heavy load imposed by the volcanic rocks, the presence of poorly permeable sedimentary rocks beneath, steep slopes, and available relief facilitated large-scale rock-mass displacements towards the adjacent valley of Sokołowiec stream.

Landform analysis, based on both field mapping and LiDAR DTM interpretation, led to the recognition of at least five separate landslides (Fig. 9.4) of different types, including a multiple rotational slide in the east, three complex rock-slope collapses in the centre and the west, and one flowslide. Soil mapping provided evidence that the depositional bodies of these landslides differ in terms of soil maturity, which is interpreted to indicate various timings of movement, spanning the Holocene (Kacprzak et al. 2013). The most recent event was the large-scale rock-slope collapse below Mt. Suchawa, which exposed rhyolite cliffs up to 20 m high in the upper slope (known as Czerwone Skałki (= Red Rocks)) and produced a thick pile of thoroughly disintegrated debris covering an older landslide body in the mid-slope. Rhyolite within the head scarp is densely jointed and being subject to ongoing physical and biomechanical weathering, feeds a large scree cone beneath (Remisz and Bijak 2012). The cone is more than 70 m long and 45 m wide, bare except a few scattered trees and enlarging into the forest along its toe (Fig. 9.8). Dendrochronological studies (in Migoń et al. 2010) showed that trees in the vicinity of the cone are under severe environmental stress, showing evidence of scars from both falling or sliding debris and ongoing talus creep. The very fresh appearance of the cone and its ongoing build-up, coupled with extremely poorly developed soil profile on the corresponding debris pile, suggest that the rock-slope collapse may have even been sub-recent (but prior to 1880s as constrained by tree ages). A similar sequence of landforms, from a rocky cliff through talus to overlapping debris piles in the mid-slope, occurs beneath Mt. Kostrzyna, but here the talus is completely overgrown, and hence, the slope collapse is likely to be older.

9.5.3 Mt. Lesista Wielka—Old Landslides and Ongoing Slope Deformations

Mt. Lesista Wielka (850 m) to the west of the Ścinawka river water gap (Fig. 9.1) has long been famous for its 'Wind Clefts'. This name was given to a group of linear hollows at the margin of the summit flat, ranging in appearance from rock-walled trenches to shallow irregular ground surface depressions, and associated with the emergence of warm air in winter. The visual result of this thermal anomaly is either a lack of snow cover or presence of very thin and discontinuous cover around the hollows in winter, whereas its thickness nearby can be 1–2 m. This suggests a deep air circulation in the voids in the rock massif, associated with considerable weakening of the rock-mass coherence due to strong tensile stresses. Similar thermal anomalies typify the secondary ridge of Mt. Lesista Mała (790 m). However, except rather inconsistent landslide symbolization on old geological maps (Berg 1925; Grocholski 1971) no detailed studies on geomorphology of mass movements have been undertaken at this locality.

Fig. 9.8 Debris cone below Mt. Suchawa and rhyolite outcrops of Czerwone Skałki (above) are probably the evidence of the most recent rock-slope failure event in the Kamienne Mountains (photograph by P. Migoń)

Again, the availability of LiDAR DTM helped to appreciate the complicated topography of the eastern slopes of Mt. Lesista Wielka (Fig. 9.9), and the model provided the necessary background for landform mapping, even though the entire pattern is yet to be fully understood (Kasprzak et al. 2019). Moreover, several ERT profiles were executed (Migoń et al. 2010; Kasprzak et al. 2019), and these give further insights into the anatomy of ongoing rock-mass deformations. The 'Wind Clefts', among which the most distinctive form is a trench 6 m long, 3.5 m deep, and 0.5 m wide (Fig. 9.10a), are but parts of a large and complex system (Fig. 9.9) that apparently has a long history. Various sections of the upper slope were affected by deep-seated rotational slides, resulting in arcuate bedrock scarps 100–200 m long (although rock cliffs seldom occur) (Fig. 9.10b) followed in the downslope direction by singular or multiple benches marking the occurrence of rigid slid blocks. The subsequent mobilization of toe parts of the displaced blocks gave rise to flowslides, which moved down the pre-existing valleys and filled them with deposits, possibly up to 20 m thick (Migoń et al. 2010). Flowslide-related landforms, with uneven ground surfaces and distinct toes, have been recognized in two adjacent valleys. Undermining of the plateau caused upslope propagation of instability, manifested by the opening of cracks and the development of deep ventilation of the rock mass. Thus, some sections of the slopes can be considered as subject to deep-seated gravitational slope deformation (DSGSD).

9.6 The Kamienne Mountains as a Geotourist Destination

The occurrence of numerous landslides, generally rare in the Sudetes, makes the Kamienne Mountains a distinctive region within the range and provides a good foundation to promote the area as a geotourist destination. In fact, some attempts to evaluate suitability of the area for geotourism were undertaken in the past (Ihnatowicz et al. 2011), but there has been little follow-up and the assessment was focused on rock record rather than landforms. However, a protected area has existed in the area since 1998 (Sudety Wałbrzyskie Landscape Park), with a few protected geoheritage sites within its borders, and some have interpretation

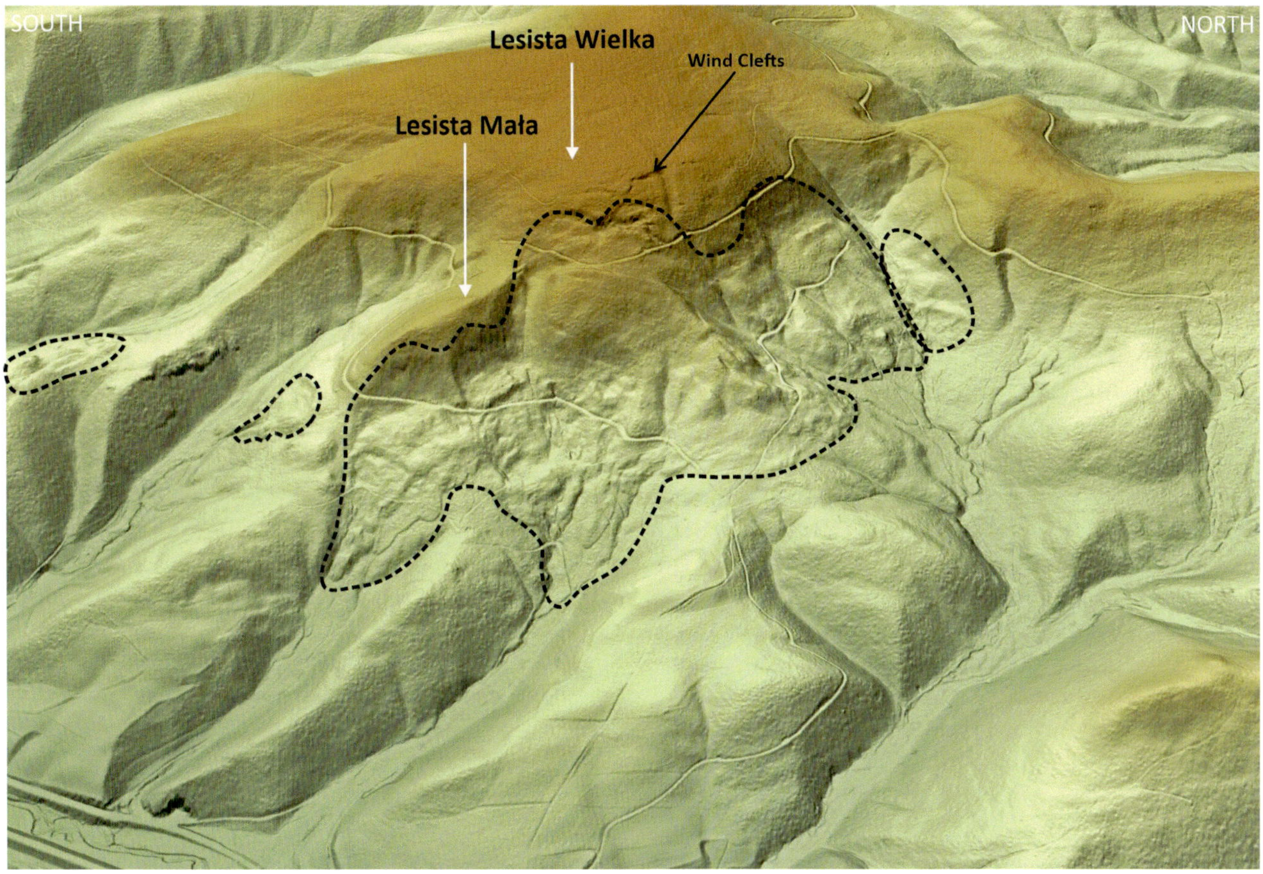

Fig. 9.9 Three-dimensional model of landslide terrain on the eastern slopes of Mt. Lesista Wielka, with the extent of different mass movements and deep-seated slope deformations indicated by broken lines. The image covers an area of approximately 2 km × 2 km (visualization courtesy of Kacper Jancewicz)

Fig. 9.10 Morphological evidence of slope instability at Mt. Lesista Wielka. **a** 'Wind Clefts' with their thermal anomaly indicate ongoing disintegration of rock-mass. **b** Degraded bedrock scarp left by a deep-seated slide (photographs by P. Migoń)

panels (e.g. the Wind Clefts). In the Krucze Mountains in the west a nature reserve 'Kruczy Kamień' to protect rhyolite outcrops near the town of Lubawka has been established. In 2018, a National Geopark Broumovsko was formally established in Czechia, just beyond the border, and it includes parts of the Kamienne Mountains. Several geosites have been identified on the Czech territory (https://geopark.broumovsko.cz/geologicke-lokality-broumovsko; accessed 2021-12-25), but except one, they lack on-site interpretation facilities, and only one addresses geomorphology, whereas the other ones focus on different types of volcanic rocks. In a recent comparative evaluation of geoheritage values of several Central European areas dominated by Permian volcanic rocks Migoń and Pijet-Migoń (2020) argued that landslide-related geomorphology is the main asset of the Kamienne Mountains and adjacent areas, worth promotion, and explanation to the general public.

Even though interpretation of geoheritage is underdeveloped in the Kamienne Mountains, the area in general is easily accessible, and a dense network of waymarked hiking trails connects the most interesting localities. Thus, both the landslide terrains at Mt. Rogowiec and Mt. Lesista Wielka are traversed by marked trails (Mt. Suchawa and Mt. Kostrzyna would require off-trail exploration), and several others can be also partially seen from the existing trails. In the Krucze Mountains, the main site of interest is the 'Kruczy Kamień' nature reserve—a series of impressive rhyolite crags and associated scree cones, attaining the total height of nearly 150 m and towering above the adjacent valley. Several localities have been developed as viewing points, including a few with specially erected viewing towers. Not only they allow the visitors to have panoramic views of the area and more distant parts of the Sudetes, but may be used to offer a commentary regarding relationships between geology and landforms, as these are particularly clear in the Kamienne Mountains. The proximity of the city of Wałbrzych, with its coal mining traditions, geological collection in the local museum, and recently opened interpretation centre in an old coal mine, adds to the potential of the region.

9.7 Conclusions

The Kamienne Mountains in the central part of the Sudetes are geomorphologically very special, even though this is not evident at the first sight. They host the largest assemblage of bedrock landslides in the Polish part of the Sudetes (around 60 in total), which represent a wide diversity of types and sizes. The reasons for this unusual occurrence of so many landslides reside in geological structure facilitating rock-mass displacements due to vertical succession of strong and massive volcanic rocks overlying weak and deformable sedimentary formations, as well as in favourable relief conditions (high frequency of steep slopes). Although unique in the Sudetes, comparable examples of landslide-affected ancient volcanic terrains may be found elsewhere in the Bohemian Massif, for instance in the České středohoří in northern Czechia (Rybář et al. 2000; Raška and Cajz 2016). The wide spectrum of mass-movement phenomena in the Kamienne Mountains includes deep-seated slides, both rotational and translational, flowslides, rock-slope disintegration and fall, shallow regolith slides, and lateral spreading. In addition, evidence of deep-seated gravitational slope deformations is available at several localities, indicating that instability is ongoing. This is consistent with the results of dendrogeomorphological studies. Nonetheless, no historical records of landslide activity are available. The area is crossed by a dense network of marked hiking trails, facilitating access to most localities of geomorphological interest, although only limited on-site interpretation is available so far.

Acknowledgements This chapter builds upon the work carried out over the years by the present author and a large group of collaborators, involved in and responsible for particular aspects of research, be it field mapping, landslide geomorphometry, soil studies, geophysical surveys, or dendrochronology. I am very grateful for the good time we had in the field, productive cooperation towards research papers, and numerous discussions. My thanks particularly go to (in alphabetical order) Kacper Jancewicz, Andrzej Kacprzak, Marek Kasprzak, Ireneusz Malik, and Tomáš Pánek. Participation of Filip Duszyński, Jan Hradecký, Aleksandra Michniewicz, Łukasz Musielok, Piotr Owczarek, Wioleta Porębna (née Kotwicka), Joanna Remisz, Milena Różycka, and Karel Šilhan is also gratefully acknowledged. I am also grateful to Kacper Jancewicz for preparing maps and terrain visualizations for this chapter, as well as to Tomáš Pánek and Andrzej Kacprzak for useful comments on the draft of this chapter.

References

Awdankiewicz M (1999) Volcanism in a late Variscan intramontane trough: Carboniferous and Permian volcanic centres of the Intra-Sudetic Basin, SW Poland. Geol Sudetica 32:13–47

Berg G (1925) Geologische Karte von Preussen 1:25 000. Blatt Waldenburg

Berg G, Dathe E, Zimmermann E (1910) Geologische Karte von Preussen 1:25 000. Blatt Friedland i. Schl

Bossowski A, Ihnatowicz A (1994) Palaeogeography of the uppermost Carboniferous and lowermost Permian deposits in NE part of the Intra-Sudetic Depression. Geol Quart 38:709–726.

Bossowski A, Cymerman Z, Grocholski A, Ihnatowicz A (1994) Szczegółowa Mapa Geologiczna Sudetów 1:25,000, Jedlina Zdrój. Wydawnictwa Geologiczne, Warszawa

Grocholski A (1971) Szczegółowa Mapa Geologiczna Sudetów 1:25,000. Mieroszów. Wydawnictwa Geologiczne, Warszawa

Kacprzak A, Migoń P, Musielok Ł (2013) Using soils as indicators of past slope instability in forested terrain, Kamienne Mts, SW Poland. Geomorphology 194:65–75

Kasprzak M, Duszyński F, Jancewicz K, Michniewicz A, Różycka M, Migoń P (2016) The Rogowiec landslide complex (Central Sudetes, SW Poland)—a case of a collapsed mountain. Geol Quart 60:695–713

Kasprzak M, Jancewicz K, Różycka M, Kotwicka W, Migoń P (2019) Geomorphology and geophysics-based recognition of stages of deep-seated slope deformation (Sudetes, SW Poland). Eng Geol 260:105230

Kotwicka W, Jancewicz K, Migoń P (2019) Rzeźba osuwiskowa południowo-wschodniej części Gór Wałbrzyskich, Sudety Środkowe. Przegl Geol 67:828–836

Kowalski A (2017) Rozmieszczenie i geneza form osuwiskowych w przełomowej dolinie Bobru w okolicach Wlenia (Sudety Zachodnie). Przegl Geol 65:629–641

Kowalski A (2021) Geneza i geologiczne uwarunkowania rozwoju form osuwiskowych w środkowej części Obniżenia Noworudzkiego w Sudetach. Przegl Geol 69:666–679

Malik I, Wistuba M, Migoń P, Fajer M (2016) Activity of slow-moving landslides recorded in eccentric tree rings of Norway spruce trees (*Picea abies* Karst.) – an example from the Kamienne Mts. (Sudetes Mts., Central Europe). Geochronometria 43:24–37

Migoń P, Pijet-Migoń E (2020) Late Palaeozoic volcanism in Central Europe – Geoheritage significance and use in geotourism. Geoheritage 12:43

Migoń P, Placek A, Żyszkowska W (2009) Steep slopes in the Sudetes and their morphotectonic interpretation. Geol Quart 53:219–232

Migoń P, Pánek T, Malik I, Hradecký J, Owczarek P, Šilhán K (2010) Complex landslide terrain in the Kamienne Mountains, Middle Sudetes, SW Poland. Geomorphology 124:200–214

Migoń P, Kacprzak A, Malik I, Kasprzak M, Owczarek P, Wistuba M, Pánek T (2014a) Geomorphological, pedological and dendrochronological signatures of a relict landslide terrain, Mt Garbatka (Kamienne Mts), SW Poland. Geomorphology 219:213–231

Migoń P, Jancewicz K, Kasprzak M (2014b) Zasięg obszarów objętych osuwiskami w Górach Kamiennych (Sudety Środkowe) – porównanie map geologicznych i cyfrowego modelu wysokości z danych LiDAR. Przegl Geol 62:463–471

Migoń P, Duszyński F, Kasprzak M, Różycka M (2017a) Evolving slope instability zone at Mt. Turzyna (Sudetes, SW Poland)—an example of deep-seated gravitational slope deformation. Z Geomorph NF 61:135–148

Migoń P, Jancewicz K, Różycka M, Duszyński F, Kasprzak M (2017b) Large-scale slope remodelling by landslides—geomorphic diversity and geological controls, Kamienne Mts, Central Europe. Geomorphology 289:134–151

Migoń P, Jancewicz K, Porębna W, Różycka M (2021) Nietypowy przykład ruchów masowych w Masywie Lesistej (Góry Kamienne). Przyroda Sudetów 23:189–202

Pitura M, Sobczyk A (2017) Czwartorzęd Kotliny Krzeszowskiej a problematyka zasięgu lądolodu skandynawskiego w Sudetach Środkowych. Przyroda Sudetów 20:315–330

Pulinowa MZ (1972) Procesy osuwiskowe w środowisku naturalnym i sztucznym. Dokument Geogr IG PAN 4:112

Raška P, Cajz V (2016) Neovolcanic terrain of the České středohoří mountains. In: Pánek T, Hradecký J (eds) Landscapes and landforms of the Czech Republic. Springer, Switzerland, pp 139–152

Remisz J, Bijak S (2012) Dendrochronologiczny zapis aktywności stoków usypiskowych Suchawy i Kruczej Skały (Sudety Środkowe). Przyroda Sudetów 15:209–218

Rybář J, Vilímek V, Cílek V (2000) Process analysis of deep seated slope failures in České středohoří neovolcanites. Acta Montana AB 8(115):39–46

Synowiec G (2003) Formy osuwiskowe w Górach Kamiennych. Przegl Geol 51:59–65

Synowiec G (2005) Formy i procesy osuwiskowe w Górach Kamiennych. Unpubl PhD thesis, Instytut Geografii i Rozwoju Regionalnego, Uniwersytet Wrocławski

Tichavský R, Fabiánová A (2022) Contribution of dendrogeomorphology to the dating of secondary processes on dormant rockslides. Quat Geochr 72:101350

Traczyk A (2004) Wpływ rzeźby na przebieg transgresji lądolodu skandynawskiego w Kotlinie Kamiennogórskiej w Sudetach Środkowych. Prace Inst Geogr Akad Świętokrzyskiej w Kielcach 13:151–168

Piotr Migoń is Professor of Geography at the University of Wrocław and former Director of the Institute of Geography and Regional Development. His main research interests are rock control in geomorphology, especially in sandstone and granite areas, mass movements in mountain terrain, geoheritage and geotourism. The Sudetes Mountains in Central Europe is his main research area, but he was also involved in projects in various other European countries (Czechia, Germany, Great Britain, Sweden, Portugal, Spain), in the United States, Mexico, Brazil, China, Jordan and Namibia. He used to be the Secretary of the International Association of Geomorphologists (1997–2001), its Vice-President (2009–2013) and member of Executive Committee (2013–2022); he was also the President of the Association of Polish Geomorphologists (2005–2008). He is the series editor of 'World Geomorphological Landscapes' series at Springer Nature and member of the editorial board of *Geomorphology* journal.

10 The Stołowe Mountains—A Unique Sandstone Tableland

Filip Duszyński

Abstract

The Stołowe Mountains, underlain by alternating sandstones and fine-grained rocks of Upper Cretaceous age, are a classic stepped tableland, rising above the surrounding terrains by some 450 m. Albeit quite small in terms of spatial extent, the mountain range showcases a number of interesting geomorphic features, including prominent mesas and plateaus, rock labyrinths, pedestal rocks, solitary boulders and sandstone canyons. Since the morphology of these landforms is strictly associated with bedrock geology, the tableland represents one of the finest examples of rock control in geomorphology on the territory of Poland. While the time frames of landscape evolution in the Stołowe Mountains remain unknown, the recent decade has brought about a considerable advance in understanding the major geomorphic forces responsible for the retreat of plateau-bounding escarpments. In particular, the role of processes working at depth, which are thought to have primarily contributed to the non-catastrophic evolution of cliff lines and the development of ruiniform relief, has been recognized. The outstanding natural beauty of the area resulted in a long history of tourism development within the time span of more than 200 years.

Keywords

Plateaus · Cliffs · Ruiniform relief · Pedestal rocks · Boulders · Escarpment retreat · Rock control · Sudetes

10.1 Introduction

The Stołowe (= *'Table'*) Mountains in the Central Sudetes are the only example of a stepped tableland in Poland (Duszyński 2018; Migoń and Duszyński 2018) (Figs. 10.1 and 10.2a). Underlain by subhorizontal layers of sandstones and finer-grained rocks, the area abounds with plateaus and mesas crowned by precipitous cliff lines, spectacular rock labyrinths and ruiniform relief, isolated tors and pedestal rocks of bizarre shapes, sandstone canyons—though quite short and relatively shallow—as well as solitary boulders of uncertain origin (Walczak 1963; Pulinowa 1989; Migoń 2008; Migoń et al. 2011). Similar landform assemblages—with a dense network of backslope canyons—occur nearby, in the Czech Broumov Highland, and have been cut in the same strata of the Upper Cretaceous age (Vítek 1979 2016; Mikuláš et al. 2007; Duszyński et al. 2018). There, however, sandstone beds dip to the south-west, giving rise to a cuesta landscape (Fig. 10.2b). Although the Polish part of the tableland is relatively small, extending approximately at a distance of merely 20 km, it offers an excellent insight into rock control in geomorphology and provides some of the most representative examples of sandstone landforms of long-standing scientific interest. It is thus no coincidence that the Stołowe Mountains have attracted attention for 200 years or so, being visited by famous personalities (including the German poet and naturalist Johann Wolfgang Goethe and the future president of the United States of America John Quincy Adams) and a growing number of tourists, among which the mesa of Szczeliniec Wielki became a 'must-see' locality (Migoń and Latocha 2013). The most prominent part of the tableland was designated as the national park in 1993—the only national park in Poland which was established to primarily protect rocks and landforms (Duszyński and Migoń 2022a).

The research conducted in recent years has significantly improved our understanding of the morphogenetic system of the Stołowe Mountains. Thanks to the availability

F. Duszyński (✉)
Institute of Geography and Regional Development,
University of Wrocław, pl. Uniwersytecki 1, 50-137 Wrocław, Poland
e-mail: filip.duszynski@uwr.edu.pl

P. Migoń and K. Jancewicz (eds.), *Landscapes and Landforms of Poland*,
World Geomorphological Landscapes, https://doi.org/10.1007/978-3-031-45762-3_10

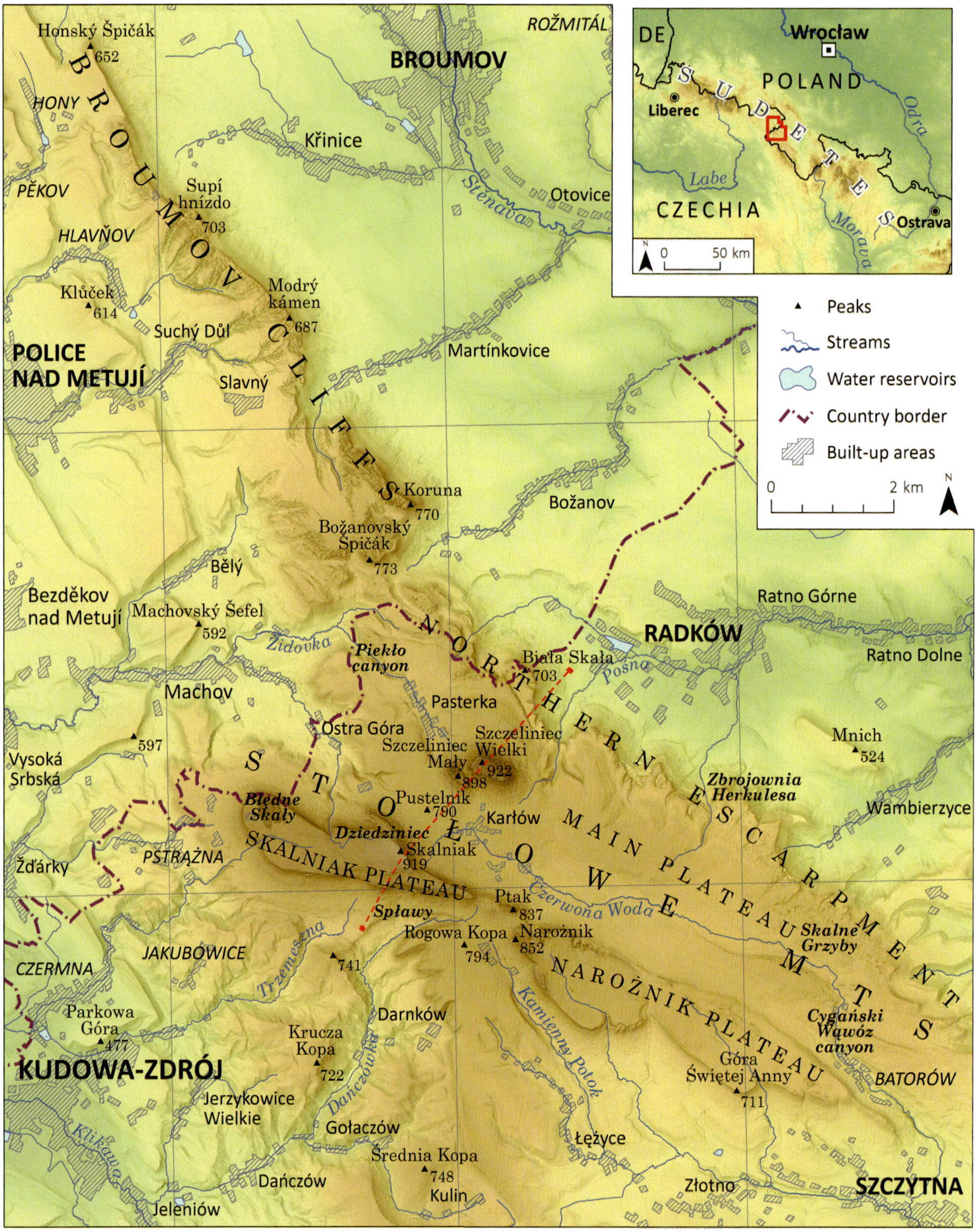

Fig. 10.1 Major geomorphic features of the Stołowe Mountains, with marked locality of specific sites referred to in the text. Dashed line indicates the geological profile presented in Fig. 10.3 (map courtesy of Kacper Jancewicz)

Fig. 10.2 Gross geomorphology of the tableland as visible from two perspectives. **a** Stepped tableland rising above the undulating surface cut in Permian sedimentary rocks (foreground), with a prominent northern escarpment and the mesa of Szczeliniec Wielki above the main plateau. Note a distinctive outlier (Mnich) in front of the escarpment (arrowed). **b** Change of relief from a cuesta landscape (foreground) towards a tableland (background), following change in the dip of sandstone beds, with the twin mesa of Szczeliniec on the skyline (photographs by P. Migoń)

of high-resolution LiDAR-derived DTMs, the morphology of plateau surfaces and escarpments has been studied in greater detail than ever before (Migoń et al. 2013; Kasprzak and Migoń 2015; Migoń and Kasprzak 2015, 2016; Duszyński et al. 2017), and new concepts of escarpment evolution have been developed (Duszyński and Migoń 2015; Duszyński et al. 2016).

This chapter will first provide an overview of the gross geomorphology of the tableland in a wider, geological context, then the diversity of particular groups of sandstone landforms will be presented, with a discussion on their lithological and structural controls. It will then focus on major processes contributing to the geomorphic evolution of the Stołowe Mountains as revealed by the concepts developed so far.

10.2 Geological Context

The Stołowe Mountains, situated in the axial part of a large intramontane depression known as the Intra-Sudetic Synclinorium, are underlain by the Upper Cretaceous strata spanning time interval from the late Cenomanian to late Turonian, possibly early Coniacian, hence c. 6 Ma. In fact, it is part of a much larger sedimentary succession of marine origin (the so-called Bohemian Cretaceous Basin), which extends within the territories of northern Czechia, south-eastern Germany, and south-western Poland. The Upper Cretaceous series of the Stołowe Mountains is nearly 400 m thick and has experienced only minor tectonic deformation. Thus, the strata is characterized by subhorizontal (< 10°) layering, with local and large-scale tilting in the southerly and south-easterly direction, reflected in the general inclination of the plateaus (Migoń and Duszyński 2018). Variations in the depositional environment have resulted in the alternation of strata, which include sandstones and finer-grained rocks (Fig. 10.3). The first ones may be divided into three major groups: the upper quartz sandstone (c. 50 m thick), the middle arkosic sandstone (c. 80 m thick), wedging out in the south-westerly direction and thus cropping out only in the northern part of the tableland, and the lower glauconitic sandstone with conglomerates (c. 15 m thick), present solely in the north as well (Jerzykiewicz 1968; Wojewoda 1997). Minor sandstone insets are locally outcropping, too. The typical structural feature of the Upper Cretaceous sandstones is the presence of an orthogonal jointing pattern showing considerable diversity in joint spacing, from less than 1 m to more than 10 m (Duszyński and Migoń 2015). The sandstone bodies are sandwiched in between the complex of fine-grained rocks, including mudstones, siliceous marls, claystones, calcareous sandstones, spongiolites and gaizes (Rotnicka 1997; Wojewoda 1997). Unlike highly porous sandstones, they are typified by low permeability, which is of crucial importance for the evolution of tableland relief.

The cake-layered geological structure of the Stołowe Mountains is clearly reflected in large-scale morphological patterns, with more resistant sandstones acting as cliff-formers and weaker complexes of fine-grained rocks truncated by less inclined slope sections (Fig. 10.3). Therefore, the tableland is multi-storey and characterized by the presence of remnant plateaus and mesas at different altitudinal levels. The most diversified geomorphologically is the Upper Plateau level, which consists of elevations exceeding the height of 850 m asl (i.e. the mesas of Szczeliniec Wielki and Szczeliniec Mały as well as the Skalniak and Narożnik plateaus; see Fig. 10.1). They are bounded by escarpments, which grade into the most extensive geomorphic entity of the tableland—the Main Plateau, underlain primarily by the complex of fine-grained rocks. Its altitude, reaching c. 750 m in the west, decreases south-easterly by some 200 m, which is related to the general inclination of the strata. To the north, the Main Plateau is truncated by the prominent northern escarpment, rising some 250–300 m above the undulating foreland cut across Permian sandstones and conglomerates (Fig. 10.2).

The morphology of the region changes dramatically to the south-west, where the sedimentary tableland gives way to highly dissected terrain underlain by Carboniferous granites (Fig. 10.1). However, the discussion of geomorphology of this part of the Stołowe Mountains is beyond the scope of this chapter.

10.3 Landform Diversity

10.3.1 Plateaus, Mesas and Escarpments

Textbook examples of flat-topped residual hills form the Upper Plateau level of the Stołowe Mountains (Fig. 10.1). While the plateaus of Skalniak (919 m asl) and Narożnik (852 m asl), extending for 4 and 6 km, respectively, occupy much larger area, the most prominent landmark is the twin mesa of Szczeliniec Wielki (922 m asl) and Szczeliniec Mały (898 m asl), rising some 150 m above the village of Karłów to the south (Fig. 10.4a). Separated by a shallow pass, possibly of tectonic origin, both elevations constitute independent geomorphic entities. The summit surface of Szczeliniec Wielki (c. 630 m × 325 m) is spatially quite limited, yet almost twice as big as the one of Szczeliniec Mały (c. 350 m × 180 m). Much more extensive is the Main Plateau situated below and resembling a giant plinth from which the outliers of the Upper Plateau level rise. It extends at a distance of c. 16 km in the NW–SE direction, but since its undulating surface is dropping considerably towards

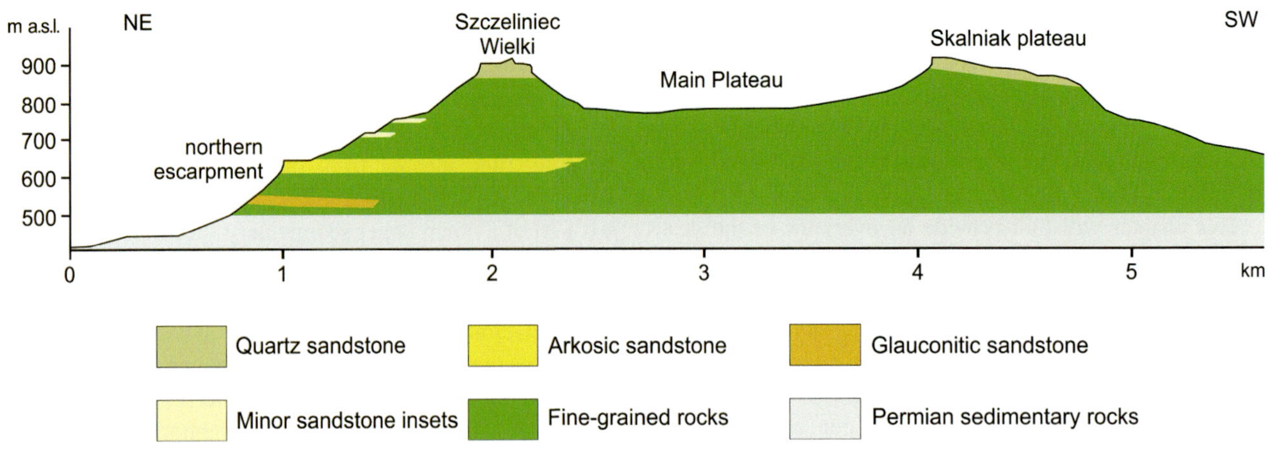

Fig. 10.3 Schematic geological cross section of the Stołowe Mountains (after Duszyński 2018, slightly modified)

Fig. 10.4 Selected geomorphic features of plateaus and mesas of the Stołowe Mountains. **a** Twin mesa of Szczeliniec Wielki (background) and Szczeliniec Mały (foreground). **b** Ideally concave slope profile of south-eastern escarpment of Szczeliniec Wielki. **c** Sandstone cliffs lining the mesa of Szczeliniec Wielki. **d** Large sandstone boulder upon an almost level footslope surface (photograph a by B. Schutty; b, c, d by F. Duszyński)

south-east, it loses its prominence in the vicinity of the spa town of Polanica-Zdrój.

Escarpments bounding the plateaus and mesas of the Stołowe Mountains are generally of bipartite morphology and characterized by concave longitudinal profiles, reflecting strength differences of the underlying rock series (Duszyński 2018) (Fig. 10.4b). In the case of elevations belonging to the Upper Plateau level, the upper Turonian quartz sandstone acts as a cliff-former, giving rise to vertical rock walls. They are usually 10–20 m high but locally rise up to 30–40 m—such impressive cliff lines typify the southern and northern slope sections of the mesa of Szczeliniec Wielki (Fig. 10.4c). Of similar maximum height are the cliffs of the northern escarpment, developed within the middle Turonian arkosic sandstone.

The course of rock walls in detail is dictated by orthogonal jointing, with two dominant vertical joint sets striking 40° and 130°. Thus, cliff lines usually show zig-zag pattern when viewed in groundplan. Joint spacing of sandstone layers is highly varied and seemingly exerts strong control on the cliff form. In general, densely jointed rock wall sections abound with recesses, joint-aligned corridors, and clefts, often filled with boulders (Dumanowski 1961; Duszyński et al. 2016). Possibly, it was the structural factor as well that favoured advanced disintegration of long sections of cliff lines of the Upper Plateau level. Along considerable distances, the cliff lines are absent, having been replaced by minor crags separated by convex slope sections covered by chaotic piles of joint-bounded blocks (Duszyński et al. 2016).

Though geomorphologically less distinctive than the cliff lines, the lower scarp slopes form the longest segments of escarpments. This is due to the fact the complexes of fine-grained rocks they are truncating, together with Permian clastic sediments in case of the northern escarpment, are much thicker than the sandstone caprocks. The inclination of the lower scarp slopes almost continually decreases down the slope, from even > 30° beneath the cliff lines to nearly level footslope surfaces. However, the longitudinal profile is occasionally more complex, with the presence of mid-slope risers and benches, likely originating from shallow landslides (Kasprzak and Migoń 2015; Duszyński et al. 2017).

The escarpment sections below the cliff lines are littered with allochthonous, caprock-derived boulders, often forming continuous aprons extending to the footslope zones (Fig. 10.4d). On the southern slope of the mesa of Szczeliniec Wielki boulders are resting up to some 400 m away from the present cliff lines (Duszyński and Migoń 2015). Even more distant are the dispersed blocks below the Biała Skała cliff segment of the northern escarpment, found 1 km from the rim (Duszyński et al. 2017). The boulders often attain impressive dimensions, with the biggest ones even 15–20 m long. These are particularly common within the northern slope of the mesa of Szczeliniec Wielki, but occur elsewhere as well. A characteristic morphological feature of many boulders in the Stołowe Mountains is their tabular shape, at least partly related to the original configuration of discontinuities in the sandstone caprock.

The escarpments of the Upper Plateau level show a fairly straight course in the groundplan, which might be partly explained by limited surface drainage. As reported by Kasprzak and Migoń (2015), the total of merely eight v-shaped or box valleys as well as two quite shallow troughs occur along the southern rim of the plateaus of Skalniak and Narożnik. Starkly different is the morphology of the northern escarpment, characterized by the presence of alternating spurs and re-entrants, giving the escarpment a highly sinuous course (Fig. 10.5). As many as 14 amphitheatre-like slope hollows may be distinguished, with the most deeply recessed ones in the western part. The cliff lines in these zones are cut by a considerable number of ravines, which testifies to the local significance of fluvial incision (Migoń and Kasprzak 2016).

In most instances, summit surfaces of residual hills in the Stołowe Mountains are regolith-covered and almost lack natural sandstone outcrops. However, some localities in the marginal parts of the plateaus of Skalniak and Narożnik show more diverse sandstone-cut topography, whereas the top surfaces of both Szczeliniec Wielki and Szczeliniec Mały are entirely covered by ruiniform relief (Fig. 10.4a). Of quite different morphology is the western part of the Skalniak plateau, where a distinctive rock labyrinth has developed.

10.3.2 Rock Cities and Ruiniform Relief

Though rock cities and ruiniform assemblages in the Polish part of the Stołowe Mountains are spatially much less extensive and of substantially lower relative heights than their Czech counterparts in the nearby Adršpach-Teplice region (see Vítek 2016), their geomorphic expression might be regarded as truly special for at least two reasons. First, they occupy summit surfaces of distinctively elevated plateaus and mesas, in terms of absolute heights forming the uppermost representatives of this type of morphology in the entire Bohemian Cretaceous Basin (Migoń 2008). Second, the rock labyrinth of the Błędne Skały (= *'Errant Rocks'*) is by far one of the finest examples of joint-controlled sandstone landforms, at least in Central Europe.

The Błędne Skały rock labyrinth, underlain by c. 10 m thick layer of the upper Turonian quartz sandstone, is situated in the westernmost part of the Skalniak plateau (Fig. 10.1), occupying the area of c. 400 m × 200 m. Though relatively small, it consists of a network of

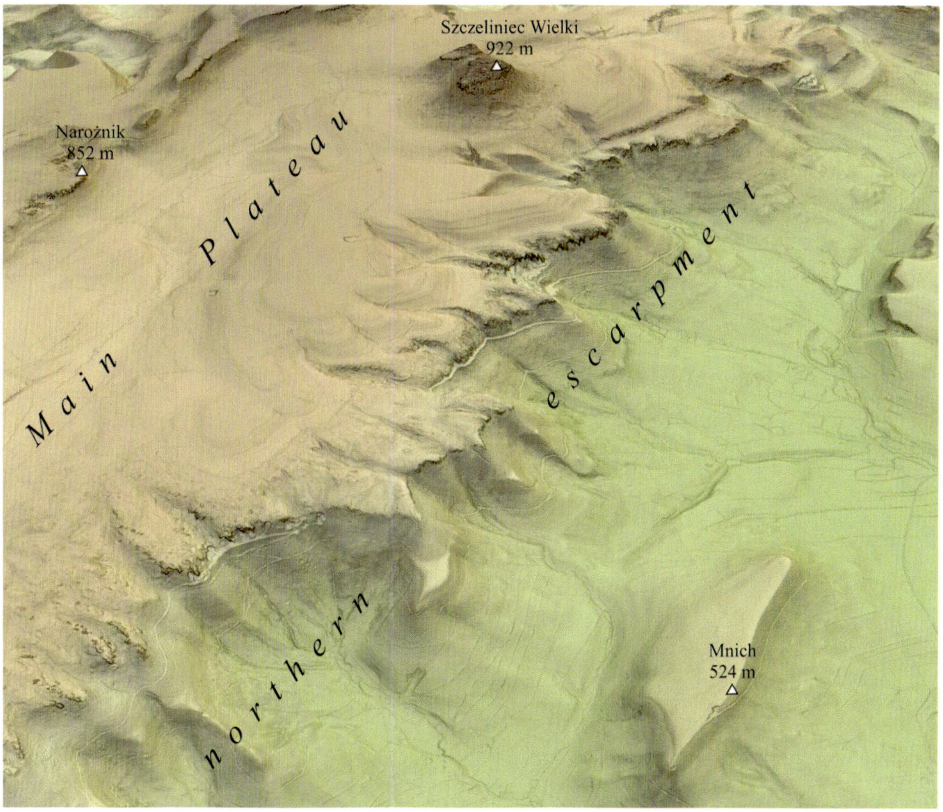

Fig. 10.5 3D image showing highly sinuous course of the northern escarpment, indented by amphitheatres, with the outlier of Mnich left in front (visualization courtesy of Kacper Jancewicz)

corridors and slots, whose cumulative length is estimated to be 20 km (Migoń et al. 2018). The corridors show highly repeatable pattern, clearly following three sets of vertical joints, striking at 15–20°, 85–90°, and 130–135° (Fig. 10.6). With sandstone compartments a few metres high left in between, the rock labyrinth resembles an urban scenery and thus serves as an instructive example of 'rock city' morphology (Migoń et al. 2017). To the south-east, the rock city gradually diminishes, finally giving way to the regolith-covered surface of the main part of the Skalniak plateau.

The pervasive role of structural control in the Błędne Skały rock labyrinth becomes obvious when inspected from two perspectives—the layout of the rock city and the morphology of individual sandstone compartments in the vertical profile. In both cases, the lines of discontinuities have served as the *loci* of enhanced physical and chemical weathering, leading to the preferential disintegration of the sandstone mass. Over a sufficiently long period of time, vertical joints have gradually transformed into corridors, lanes and avenues (Fig. 10.7a). Since selective weathering worked most efficiently at joint intersections, distinctive squares, plazas, and courtyards have developed (Fig. 10.7b). It remains not yet fully understood why the degree of disintegration increases in the south-eastern extremity of the rock city (Fig. 10.7c), but the likely explanation is the growing density of jointing. If this is the case, the lack of sandstone outcrops further to the east becomes a logical consequence. In the vertical profile, the vast majority of sandstone compartments is characterized by bipartite morphology, with the upper part cut in massive sandstone supporting overhangs and the lower one in densely bedded layers, and hence abounding with deeply-recessed basal niches (Fig. 10.7d). Apart from the structural factor, the preferential decay of the lower portion of sandstone appears enhanced by permanently damp conditions in the bottom of the corridors, facilitated by either bedrock impermeability or complete clogging of fissures beneath (Migoń et al. 2017). The widespread presence of collapsed or tilted sandstone blocks resting on the floor of the corridors (Fig. 10.7e) testifies to the concept of the upward propagation of voids following the evacuation of sandy detritus from beneath (see Sect. 10.4.2). It is worth mentioning, however, that along the northern margin of the Skalniak plateau yet another rock labyrinth is present—the Dziedziniec (= *'Courtyard'*). Although much smaller, it shows a fairly similar morphology strictly controlled by structure and illustrates well the processes operating at depth (Duszyński and Migoń 2017).

The most spectacular sandstone-cut scenery of the whole tableland crowns the summit surface of the mesa of Szczeliniec Wielki (Fig. 10.4a). With a number of isolated tors and castellated blocks, often attaining fantastic shapes, as well as criss-crossing corridors and boulder-filled

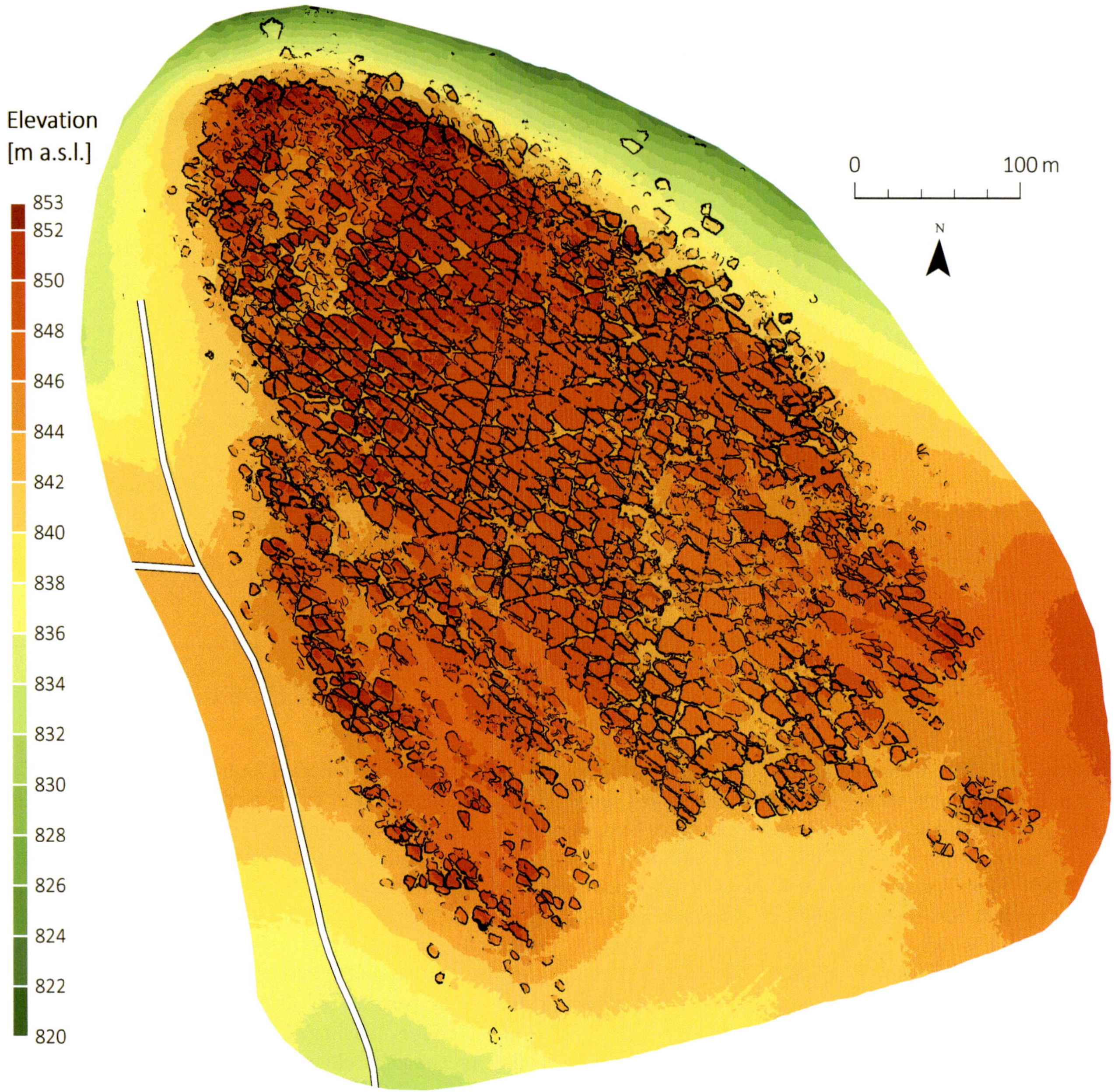

Fig. 10.6 Błędne Skały rock labyrinth showing the pervasive structural control of the spatial patterns of corridors and plazas (map courtesy of Kacper Jancewicz)

depressions, its morphology might be regarded as ruiniform relief (Migoń et al. 2017; Duszyński and Migoń 2018). It is evident that the top surface of Szczeliniec Wielki—just as the Błędne Skały rock labyrinth—reflects strong structural imprint. The tors owe their shapes to selective weathering of the cross-bedded, uppermost portion of the upper Turonian quartz sandstone (Migoń and Kasprzak 2015) (Fig. 10.8a). Due to preferential weathering along discontinuities of variable density, sandstone compartments show a range of micro-weathering features and generally rounded outlines (Fig. 10.8b). The corridors are fracture-guided, too, following two vertical joint sets, striking at 35° and 120–130° (Dumanowski 1961). The latter one controls the course of deep open clefts situated along the northern margin of the mesa, parallel to its rim. While Diabelska Kuchnia (= *'Devil's Kitchen'*) is c. 15 m deep and 40 m long, Piekiełko (= *'Little Hell'*) attains even more impressive dimensions, extending at the distance of c. 100 m and being 17 m deep, with the outer wall tilted from the vertical position by approximately 20° (Fig. 10.8c). To the east, the ruiniform relief of Szczeliniec Wielki changes dramatically into a chaotic boulder accumulation, with deep voids

Fig. 10.7 Błędne Skały rock labyrinth. **a** Joint-aligned corridor. **b** Corridors connected to one another at a right angle, in place of joint intersection. **c** Highly disintegrated eastern part of the labyrinth. **d** Massive sandstone compartment resting upon a densely bedded one. **e** Collapsed block resting on the floor of one of corridors (photographs by F. Duszyński)

Fig. 10.8 Major relief features of Szczeliniec Wielki. **a** Example of a sandstone tor (Wielbłąd = *'Camel'*) on top of the mesa. **b** Microrelief features within the Wielbłąd tor. **c** Open cleft of Piekiełko in the marginal part of the mesa. **d** Substantially lowered eastern part of Szczeliniec Wielki (a) and the rotational slide below (b) (photograph **a** by P. Migoń; **b** and **c** by F. Duszyński; 3D image courtesy of Kacper Jancewicz)

in between, making this part practically impenetrable. This part of the summit surface is lowered by some 15–20 m and separated from the rest of the plateau by a straight terrain step, along which large-scale sagging occurred (Migoń and Kasprzak 2015) (Fig. 10.8d).

10.3.3 Sandstone Tors and Pedestal Rocks

The northernmost part of the Main Plateau abounds with two groups of prominent tors, with many attaining fantastic shapes. One, known as 'Skalne Grzyby' (= *'Rock Mushrooms'*), consists of approximately 100 larger outcrops scattered over an area of c. 2.5 km^2. The second one, situated to the west and spatially much more limited, is called 'Zbrojownia Herkulesa' (= *'the Armoury of Hercules'*) (Fig. 10.1). While both localities are mainly known for their classic pedestal rocks, with a wide cap resting upon a thin stem (Fig. 10.9a), other types of sandstone tors occur too. These involve multi-storey pedestal rocks (Fig. 10.9b), towers, and pinnacles, as well as sandstone walls, extending in accordance with local joint patterns (Fig. 10.9c).

Until the present-day, it remains unclear what controls the distribution of sandstone tors in these two localities. It may be assumed, however, that they are the most resistant remnants of a once continuous sandstone plate (Migoń et al. 2017; Duszyński and Migoń 2022a). In the initial phase, it possibly transformed into a joint-controlled rock city, somehow similar to the Błędne Skały rock labyrinth (see Sect. 10.3.2). In the course of protracted denudation, joint-aligned corridors opened up, which has eventually led to the separation of sandstone compartments to the degree observed today. However, the tors are scattered in such a chaotic manner that the reconstruction of the likely spatial pattern of the previous rock city seems impossible. The only indicator of the structural imprint is the presence of a sequence of dry trough valleys, parallel to each other (20–25° direction), and perpendicular to the rim of the northern escarpment. As pointed out by Walczak (1963), sandstone tors of the Skalne Grzyby locality occur only within the ridges separating the valleys. This implies that the sandstone cap must have been nonuniformly jointed or characterized by some other structural differences that resulted in an uneven rate of denudation.

The discussion over the controls of morphology of individual sandstone tors is less speculative, although it still requires more quantitative data. In general, the bipartite nature of pedestal rocks is explained as the effect of selective weathering (Walczak 1963; Duszyński and Migoń 2022b). The cap, being much more massive and homogenous due to bioturbation phenomenon at the time of sand deposition, is more resistant than the densely bedded stem (Wojewoda et al. 2011). This is because the stem, characterized by high primary porosity, is more prone to various physical and chemical weathering processes. The relationship between the structural patterns (i.e. bedding density) and the morphology of individual sandstone tors in both localities is evident, with a few exceptions where the cap is, for some reasons, thinner than the stem (Migoń 2021). At Zbrojownia Herkulesa, still another factor seems to play the role—the lithological one. Sections of tors built of weaker conglomeratic sandstone (tempestite in terms of origin) are usually distinctively recessed (Fig. 10.9d) (Wojewoda et al. 2011).

The closer to the rim of the northern escarpment, the higher and morphologically more complex are the tors. The reason is the growing depth of trough valleys in this direction, which has led to the exposure of thicker portions of sandstone of variable resistance (Walczak 1963). Thus, in the marginal part of the Main Plateau, tors often appear as compound forms. On the contrary, typical pedestal rocks of bipartite morphology occur where the degree of dissection is lower, i.e. farther away from the escarpment rim.

10.3.4 Solitary Boulders

Among the most intriguing elements of relief of the Stołowe Mountains are residual sandstone boulders scattered across two minor elevations belonging to the Main Plateau, Rogowa Kopa (794 m), and Pustelnik (790 m) (Fig. 10.10). Since both are underlain by the complex of fine-grained rocks, it is clear that the boulders are allochthonous. However, the complete lack of spatial connection between them and block covers on the surrounding escarpments, together with the presence of shallow troughs and valleys separating the elevations from the plateaus nearby, indicate that the boulders are not derived from the escarpments and have not travelled to their contemporary position (Parzóch and Migoń 2015; Migoń and Parzóch 2021).

Almost level (generally $<5°$) top surface of Rogowa Kopa is traditionally termed 'Sawanna Łężycka' (= *'Łężycka Savanna'*). This is due to the fact that the open terrain with a group of isolated boulders resembles an African scenery (Walczak 1968). Nearly 100 boulders occur on the plateau of Rogowa Kopa, with some attaining impressive dimensions up to 11 m long (Migoń and Parzóch 2021). In addition, tens of boulders are dispersed within the valleys undercutting the elevation from the west (Parzóch and Migoń 2015). An analogical situation is observed at Pustelnik. There, the generally featureless plain, sloping at the maximum of 5°, is a host to nearly 20 boulders, with the largest ones up to 10 m long (Migoń and Parzóch 2021). Some boulders are rotated and many show a variety of micro-weathering features, such as arcades, caverns,

Fig. 10.9 Examples of tors developed in the arkosic sandstone. **a** Typical pedestal rock. **b** Multi-storey pedestal rock. **c** Rock wall. **d** One of the tors in the Zbrojownia Herkulesa locality (photographs **a** and **c** by F. Duszyński; **b** and **d** by P. Migoń)

Fig. 10.10 Solitary sandstone boulders scattered within the Main Plateau. **a** Skałki Łężyckie (Sawanna Łężycka) locality. **b** Pustelnik locality (photographs by P. Migoń)

or weathering pits, which indicate a long weathering history. Widespread evidence of past quarrying at both localities, including hollows left after excavation of boulders as well as signs of intentional splitting of boulders that remain in situ, reveals that their original number must have been higher.

Solitary boulders of Rogowa Kopa and Pustelnik are all built of the upper Turonian quartz sandstone, which acts as a caprock of the remnant plateaus and mesas of the Upper Plateau. Yet, they are situated some tens of metres below. Since slope transport from the neighbouring escarpments can be easily excluded, for the reasons presented earlier, it is hypothesized the boulders are remnants of former sandstone-capped outliers that are no longer existent (e.g. Migoń and Parzóch 2021). This scenario fits well the recent ideas on the morphogenesis of the tableland and will be discussed in a separate section (see Sect. 10.4.3).

10.3.5 Canyons

The negligible surface gradient as well as the relatively limited catchment areas are possibly responsible for the presence of only two more prominent sandstone canyons in the Polish part of the Stołowe Mountains. Both occur within the Main Plateau level, incised into the middle Turonian strata (Fig. 10.1). The larger one, named Piekło (= *'Hell'*), is situated to the north-west of the mesa of Szczeliniec Wielki, in the vicinity of the village of Pasterka. The valley is some 30 m deep, 700 m long, and lined with sandstone cliffs of variable height. Interestingly, in the current conditions, the amount of flowing water is minimal, and for long periods of the year, the valley remains almost dry. At the same time, the valley floor is littered with boulders of considerable dimensions. The second canyon, situated in the eastern part of the Main Plateau and cut by the largest stream of the tableland—the Czerwona Woda river—is known as Cygański Wąwóz (= *'Gypsy Gorge'*). Attaining the length of c. 300 m, the canyon is about 30–40 m deep.

10.4 Landform Evolution

10.4.1 General Evolutionary Pathway and Historical Concepts

Undoubtedly, the dominant mode of long-term evolution of the Stołowe Mountains is parallel escarpment backwearing—a pathway that has been recognized in the majority of sedimentary tablelands of the world (Duszyński et al. 2019). Its effects are well exemplified by the outliers of the Upper Plateau level, which once formed a continuous sandstone-capped plateau, then divided into a few separate remnants. Interestingly, unlike in many prominent tablelands of the world (e.g. the Colorado Plateau), where canyon incision clearly dictates the development of an array of plateaus, mesas and buttes, the role of fluvial dissection in the Stołowe Mountains is negligible, especially if considered at a large scale.

However, the contributing role of specific geomorphic processes to the escarpment retreat phenomenon is not yet fully understood and may be site-specific. In this respect, various authors argued for an overarching role of episodic mass movements, although it seems that many of the early claims were based on assumptions rather than empirical evidence. It has thus been explained that the presence of vertical cliff lines, dense boulder covers beneath, and the general analogy of the escarpment morphology to the 'ideal' slope profile of King (1953), with the 'debris slope' in the mid-slope section, must testify to catastrophic rock-slope collapses, such as rockfalls or toppling (Rogaliński and Słowiok 1958; Dumanowski 1961). In reality, only one minor event of this kind was recorded in the tableland a hundred years ago (Duszyński and Migoń 2019) (Fig. 10.11). In the second half of the twentieth century, more attention was paid to the significance of specific hydrogeological conditions in the morphogenesis of the tableland. Since the caprock-forming sandstones are highly porous and jointed, they are permeable as well, which contrasts with the underlying complexes of fine-grained rocks, characterized by much lower permeability. The percolation of meteoric water is thus inhibited at the lithological contact and the subsurface flow continues towards the slope surface. It was suggested that this has several consequences, including the undercutting of rock walls and plastic deformations of saturated marls and mudstones under the rigid sandstone caprock, both resulting in slope instability (Czeppe 1952; Dumanowski 1961). According to Pulinowa (1989), the imprint of subsurface processes is impressively manifested within the northern escarpment, with amphitheatre-like hollows thought to be 'sapping cirques'. Yet, with active sapping processes, one might expect the presence of deep alcoves in the lower portion of cliff lines. These forms, however, are a rarity in the Stołowe Mountains, which does not support the sapping concept—at least in the present-day environmental conditions.

10.4.2 Underground Erosion and the Development of Ruiniform Relief

While groundwater flow is still perceived as the major agent ruling the patterns of cliff-line evolution, it has been realized recently that it primarily contributes to the processes of much lower dynamics. Duszyński et al. (2016) showed that

Fig. 10.11 Old German postcard showing the so-called 'Überhängende Stein' before and after the collapse in May 1921. Image from private collection of Petr Zima

the development of ruiniform relief in the marginal parts of mesas and plateaus, with joint-aligned corridors and slots (Fig. 10.12a), chaotic boulder accumulations as well as isolated crags standing as remnants of once continuous rock walls (Fig. 10.12b), is an effect of 'underground erosion'. The process is most efficient along discontinuities in sandstone caprock, mainly joints, leading to the mechanical removal of sandy detritus by percolating water. In a sufficiently long period, the cliff line loses its coherence and becomes highly disintegrated, giving rise to the blocky chaos and boulder-mantled slopes. The most direct evidence of this phenomenon are sandy cones at the outlets of vertical fissures (Fig. 10.12c). More than 140 forms of this kind were recorded at the foot of cliffs of the Upper Plateau level, with the total volume of sand of nearly 1000 m^3. Since sand is further redeposited onto mid- and lower-slope sections, it forms allochthonous covers of considerable thickness, up to 3 m in the upper slopes. It is estimated that the total volume of sand stored within the escarpments bounding plateaus and mesas of the Upper Plateau level is only one order of magnitude smaller than the total cubature of the remaining sandstone caprock, which testifies to the efficacy of underground erosion (Duszyński et al. 2016). The loss of mass at depth has several consequences visible on the plateau surfaces. First, the development of joint-aligned corridors from the bottom results in the gradual sinking of sandstone compartments situated above. Consequently, some of the voids are roofed by boulders, whereas the floors of the others are covered by chaotic piles of blocks (Fig. 10.12d). Second, with the development of corridors along joints and the removal of sand from below, large sandstone compartments are subject to tilting and subsidence due to the loss of support (Fig. 10.12e). Third, closed depressions are formed and a number of them have been demonstrated by Jancewicz et al. (2019) from the top surface of the mesa of Szczeliniec Wielki.

In general, it seems that the development of rock cities and ruiniform relief in the Stołowe Mountains, particularly in the marginal parts of plateaus, is the effect of underground erosion and sand removal. As pointed out by Duszyński and Migoń (2018), surface weathering processes alone would not be able to produce these landform assemblages. These tend to control the presence of 'architectural detail', i.e. a variety of micro-weathering features within sandstone compartments (Fig. 10.8b). However, the exact weathering mechanism making the sandstone mechanically incoherent and thus subject to physical erosion remains unknown. Since the processes operating at depth show

Fig. 10.12 Evidence and effects of underground erosion and sand removal. **a** Widened corridors. **b** Chaotic accumulations of sandstone blocks in the uppermost section of an escarpment. **c** Fresh sandy cone below an exit of a fissure. **d** Sandstone boulders stacked in the joint-aligned corridor. **e** Tilted block (right) due to the loss of vertical support (photographs by F. Duszyński; except **e** by P. Migoń)

pervasive joint-control and the sandstone of the Upper Plateau level is almost purely quarzitic, it is hypothesized that arenization phenomenon (Wray and Sauro 2017) plays the fundamental role (Duszyński 2018).

10.4.3 Escarpment Retreat and the Origin of Boulders

With a limited role of episodic mass movements, scarce evidence of sapping processes as well as the dominant role of underground erosion, it is now thought that the dominant mode of escarpment retreat in the Stołowe Mountains is the non-catastrophic in situ disintegration of cliff lines (Duszyński and Migoń 2015) (Fig. 10.13). In addition to the evidence presented above, using computer simulations, Duszyński and Migoń (2015) demonstrated that the most distant boulders, often occupying almost level footslope surfaces, are too far away from the cliffs to explain their origin simply by rockfalls and related slope transport. Likewise, the dimensions and the specific, tabular shape of a large number of boulders do not support classic rockfall

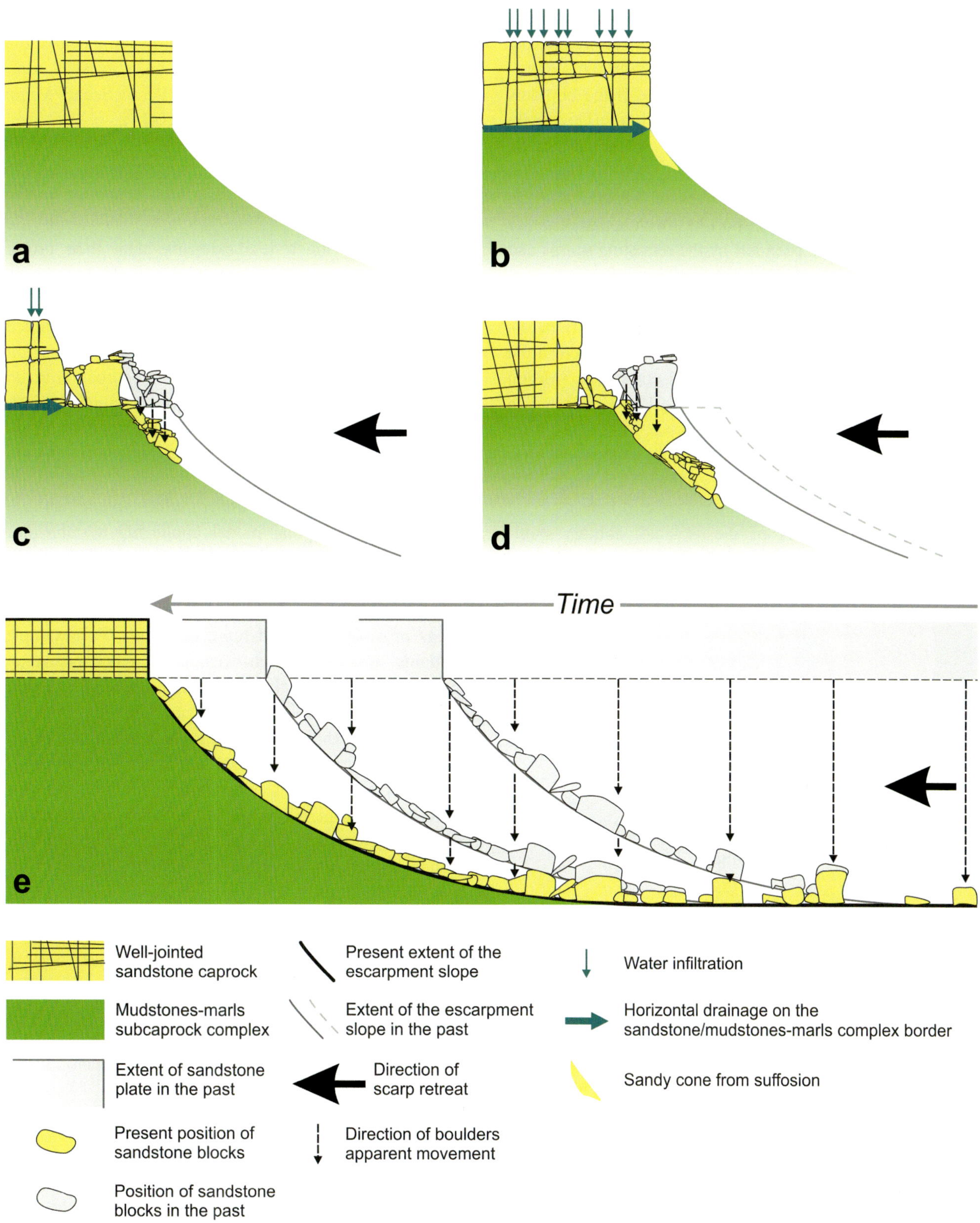

Fig. 10.13 In situ disintegration concept (after Duszyński and Migoń 2015, modified)

scenario or slow ploughing. This is consistent with rock strength tests which revealed that the degree of weathering of boulders is systematically increasing down the slope—such results would not have been achieved if the boulders were products of episodic rock-slope failures (Duszyński and Migoń 2015). Thus, the boulders are thought to be the most resistant remnants of once existent cliff lines that became disintegrated along the lines of weakness in the sandstone caprock. Their transport has been apparent rather than real, as they have been left by the retreating scarps and occupy lower and lower topographic positions.

Locally, the boulders have been subject to passive transport on top of the sliding masses (Duszyński et al. 2017). As many as eight landslide complexes were detected in the tableland, each within the fine-grained complexes (Migoń and Kasprzak 2016; Duszyński et al. 2017). Hence, while the landslides did not involve the caprock itself, they were responsible for further redeposition of allochthonous, caprock-derived boulders, sometimes along very large distances up to 1 km from the contemporary cliff lines.

Although the genesis of the isolated, solitary boulders resting upon the Main Plateau (see Sect. 10.3.4) remains puzzling, it could be also linked with the processes described above. The long-term disintegration of sandstone caprock together with the retreat of escarpments have led to the decay of the once existent outliers to such a degree that only some tens of joint-bounded and most resistant sandstone compartments have been left (Parzóch and Migoń 2015; Migoń and Parzóch 2021) (Fig. 10.14).

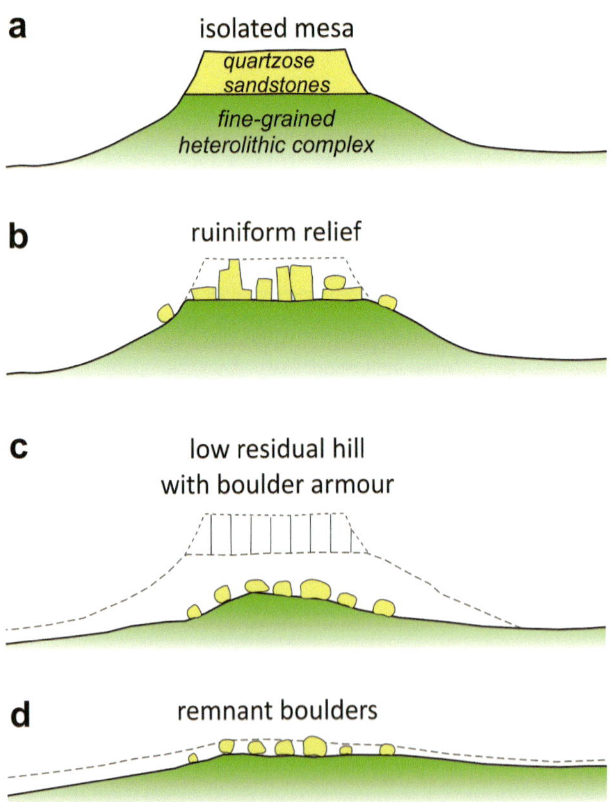

Fig. 10.14 Likely genesis of solitary boulders on the Main Plateau level as proposed by Migoń and Parzóch (2021) (modified)

10.5 Conclusions

Though spatially much less extensive than their prominent analogues from other parts of the world such as the Colorado Plateau, USA, or the nearby Bohemian Paradise in Czechia, the Stołowe Mountains serve as an excellent research polygon to study the principles of landform evolution of the tablelands and the imprint of bedrock structure on the landforms produced. The Stołowe Mountains are the mountain massif where alternative mechanisms contributing to the escarpment retreat have been recognized for the first time. Apparently, non-catastrophic evolution of cliff lines, with the overarching role of underground erosion and sand removal, is a conceptual model applicable to tablelands elsewhere. Examples include the Elbe Sandstone Mountains in Germany (e.g. Migoń et al. 2020) or the Broumov Highland in Czechia (Duszyński et al. 2018). Yet, recent findings provoke more and more questions. For example, it remains unknown which weathering mechanisms are responsible for the detachment of grains of sand and their availability for removal by underground erosion. With the present state of knowledge, it can only be hypothesized that the arenization, i.e. chemical alternation of quartz-rich rocks, is of fundamental importance. Similarly, the time frames of landform evolution in the Stołowe Mountains are virtually unknown. The research projects currently conducted may provide answers in the forthcoming years.

Acknowledgements I express my sincere gratitude to Professor Piotr Migoń, who ignited much of my fascination with the sandstone tablelands. I would also like to thank my colleagues from the Department—Kacper Jancewicz and Marek Kasprzak—who shared their time, knowledge and experience in studying the patterns of escarpment retreat in the Stołowe Mountains. Dorota Duszyńska is gratefully acknowledged for language correction of the manuscript. This study has been financially supported by the National Science Centre, Poland (project no. 2020/39/D/ST10/00861).

References

Czeppe Z (1952) Z Morfologii Gór Stołowych. Ochr Przyr 20:236–252

Dumanowski B (1961) Zagadnienie rozwoju stoku na przykładzie Gór Stołowych. Czas Geogr 32:311–324

Duszyński F (2018) Mechanizmy i uwarunkowania rozwoju progów morfologicznych Gór Stołowych. Przegl Geogr 90:7–33

Duszyński F, Jancewicz K, Kasprzak M, Migoń P (2017) The role of landslides in downslope transport of caprock-derived boulders in sedimentary tablelands, Stołowe Mts, SW Poland. Geomorphology 295:84–101

Duszyński F, Jancewicz K, Migoń P (2018) Evidence for subsurface origin of boulder caves, roofed slots and boulder-filled canyons (Broumov Highland, Czechia). Int J Speleol 47:343–359

Duszyński F, Migoń P (2015) Boulder aprons indicate long-term gradual and non-catastrophic evolution of cliffed escarpments, Stołowe Mts, Poland. Geomorphology 250:63–77

Duszyński F, Migoń P (2017) Zespół skalny Dziedzińca na płaskowyżu Skalniaka w Górach Stołowych. Przyroda Sudetów 20:199–218

Duszyński F, Migoń P (2018) Geneza skalnych miast na płaskowyżach piaskowcowych. Przegl Geogr 90:379–402

Duszyński F, Migoń P (2019) Historyczne przypadki epizodycznych ruchów masowych w Górach Stołowych. Przyroda Sudetów 22:143–160

Duszyński F, Migoń P (2022a) Geomorphological heritage of Cretaceous sandstone terrains in SW Poland—diversity, conservation and interpretation issues. Geoheritage 14:31

Duszyński F, Migoń P (2022b) Skalne grzyby w Polsce i na świecie – terminologia, rozmieszczenie, poglądy na rozwój. Przegl Geogr 94:5–30

Duszyński F, Migoń P, Kasprzak M (2016) Underground erosion and sand removal from a sandstone tableland, Stołowe Mountains, SW Poland. CATENA 147:1–15

Duszyński F, Migoń P, Strzelecki MC (2019) Escarpment retreat in sedimentary tablelands and cuesta landscapes—landforms, mechanisms and patterns. Earth Sci Rev 196:102890

Jancewicz K, Migoń P, Kasprzak M (2019) Connectivity patterns in contrasting types of tableland sandstone relief revealed by Topographic Wetness Index. Sci Total Environ 656:1046–1062

Jerzykiewicz T (1968) Sedymentacja górnych piaskowców ciosowych niecki śródsudeckiej (górna kreda). Geol Sudetica 4:409–462

Kasprzak M, Migoń P (2015) DEM-based analysis of geomorphology of a stepped sandstone plateau, Stołowe Mountains (SW Poland). Z Geomorph NF 59(Suppl 1):247–270

King LC (1953) Canons of landscape evolution. Bull Geol Soc Amer 64:721–752

Migoń P (2008) Rzeźba i rozwój geomorfologiczny Gór Stołowych. In: Witkowski A, Pokryszko BM, Ciężkowski W (eds) Przyroda Parku Narodowego Gór Stołowych. Wydawnictwo Parku Narodowego Gór Stołowych, Kudowa-Zdrój, pp 49–69

Migoń P (2010) Geneza bloków piaskowcowych na Pustelniku (Góry Stołowe). Przyroda Sudetów 13:263–274

Migoń P (2021) Sandstone Geomorphology—recent advances. Geomorphology 373:107484

Migoń P, Duszyński F (2018) Process–form relationship in the Stołowe Mountains tableland (Central Europe)—an example of strong lithostructural control on geomorphic systems of medium-altitude mountains. Studia Geomorph Carp-Balc 51(52):155–178

Migoń P, Duszyński F (2020) Ścieżka Skalnych Grzybów. Przewodnik. Park Narodowy Gór Stołowych, Kudowa-Zdrój, p 69

Migoń P, Duszyński F, Goudie A (2017) Rock cities and ruiniform relief: forms–processes–terminology. Earth Sci Rev 171:78–104

Migoń P, Duszyński F, Jancewicz K, Kotwicka W (2020) Late evolutionary stages of residual hills in tablelands (Elbsandsteingebirge, Germany). Geomorphology 367:107308

Migoń P, Duszyński F, Kasprzak M (2018) Ukształtowanie terenu i procesy rzeźbotwórcze. In: Kabała C (ed) Góry Stołowe – przyroda i ludzie. Park Narodowy Gór Stołowych, Kudowa Zdrój, pp 31–46

Migoń P, Kasprzak M (2015) Analiza rzeźby stoliwa Szczelińca Wielkiego w Górach Stołowych na podstawie numerycznego modelu terenu z danych LiDAR. Przegl Geogr 87:27–52

Migoń P, Kasprzak M (2016) Pathways of geomorphic evolution of sandstone escarpment in the Góry Stołowe tableland (SW Poland)—insights from LiDAR-based high-resolution DEM. Geomorphology 260:51–63

Migoń P, Kasprzak M, Traczyk A (2013) How high-resolution DEM based on airborne LiDAR helped to reinterpret landforms—examples from the Sudetes, SW Poland. Landf Anal 22:89–101

Migoń P, Latocha A (2013) Human interactions with the sandstone landscape of central Sudetes. Appl Geogr 42:206–216

Migoń P, Latocha A, Parzóch K, Kasprzak M, Owczarek P, Witek M, Pawlik Ł (2011) Współczesny system morfogenetyczny Gór Stołowych. In: Chodak T, Kabała C, Kaszubkiewicz J, Migoń P, Wojewoda J (eds) Geoekologiczne warunki środowiska przyrodniczego Parku Narodowego Gór Stołowych, WIND, Wrocław, pp 1–52

Migoń P, Parzóch K (2021) Enigmatic clusters of sandstone boulders on plateaus of the Stołowe Mountains (Sudetes, south-west Poland)—their geoheritage and geotouristic value. Geologos 27:141–155

Mikuláš R, Adamovič J, Hájek A, Spíšek J (2007) Adršpašsko-teplické skály Cliffs and Ostaš Hill (Czech Republic). In: Härtel H, Cilek V, Herben T, Jackson A, Williams R (eds) Sandstone landscapes. Academia, Prague, pp 332–335

Pánek T, Hradecký J (2016) Landscapes and landforms of the Czech Republic. Springer, Switzerland, p 422

Parzóch K, Migoń P (2015) Deciphering the origin of allochthonous sandstone boulder trains within a mudstone escarpment, Stołowe Mountains, SW Poland. Z Geomorph NF 59(Suppl 1):103–122

Pulinowa MZ (1989) Rzeźba Gór Stołowych. Prace Uniwersytetu Śląskiego w Katowicach, p 1008

Rogaliński J, Słowiok G (1958) Rzeźba Gór Stołowych w świetle teorii pedyplanacji. Czas Geogr 29:473–494

Rogowski M (2020) Monitoring system of tourist traffic (MSTT) for tourists monitoring in mid mountain national park, SW Poland. J Mount Sci 17:2035–2047

Rotnicka J (1997) Cykle transgresywne i regresywne w drobnoziarnistych sukcesjach osadowych na przykładzie górnokredowych margli plenerskich (Góry Stołowe, Sudety). Sedimentologica 1(1):18–30

Vítek J (1979) Pseudokrasové tvary v kvádrových pískovcích severovýchodních Čech. Rozpr Čs akad věd, Ř mat přír věd, vol 89, no 4. Academia, Praha, p 58

Vítek J (2016) Adršpach-Teplice rocks and Broumov Cliffs—large sandstone rock cities in the Central Europe. In: Pánek T, Hradecký J (eds) Landscapes and landforms of the Czech Republic. Springer, Switzerland, pp 209–220

Walczak W (1963) Geneza form skalnych na północno-wschodniej krawędzi Gór Stołowych. Acta Univ Wratisl 9 Studia Geogr 1:191–200

Walczak W (1968) Sudety. PWN Warszawa, p 384

Wojewoda J (1997) Upper Cretaceous litoral-to-shelf succession in the Intrasudetic Basin and Nysa Trough, Sudety Mts. In: Wojewoda J (ed) Obszary źródłowe: zapis w osadach. WIND, Wrocław, pp 81–96

Wojewoda J (2011) Geoatrakcje Gór Stołowych – przewodnik geologiczny po Parku Narodowym Gór Stołowych. Park Narodowy Gór Stołowych, Kudowa Zdrój, p 64

Wojewoda J, Białek D, Bucha M, Głuszyński A, Gotowała R, Krawczewski J, Schutty B (2011) Geologia Parku Narodowego

Gór Stołowych – wybrane zagadnienia. In: Chodak T, Kabała C, Kaszubkiewicz J, Migoń P, Wojewoda J (eds) Geoekologiczne Warunki Środowiska Przyrodniczego Parku Narodowego Gór Stołowych. WIND, Wrocław, pp 53–96

Wray RAL, Sauro F (2017) An updated global review of solutional weathering processes and forms in quartz sandstones and quartzites. Earth Sci Rev 171:520–557

Filip Duszyński is Assistant Professor in the Institute of Geography and Regional Development, University of Wrocław. His main research interests include geomorphic evolution of tablelands and scarplands, sandstone geomorphology and mass movements in mountain regions. The majority of his research works has been conducted in the Sudetes Mountains (Central Europe), particularly in the Stołowe Mountains in Poland and Czechia. In recent years, he spent several months as a Visiting Researcher at the University of Oxford, England, and at the University of Zurich (UZH), Switzerland. In 2019, he was granted a prestigious START scholarship for young researchers, funded by the Foundation for Polish Science, whereas in 2021 was awarded a scholarship for outstanding young researchers, funded by the Polish Ministry of Education and Science. He co-authored papers published in international peer-reviewed journals, such as *Earth-Science Reviews*, *Geomorphology*, *Catena* and *Progress in Physical Geography*.

11 Cultural Geomorphological Landscapes of the East Sudetes

Agnieszka Latocha-Wites

Abstract

The chapter focuses on anthropogenic transformation of relief in two subregions of the East Sudetes—the Śnieżnik Massif and the Złote Mountains. They can be considered as representative for the Sudetes, which exemplify a mid-latitude, medium-altitude mountain range distinguished by a high degree of human impact on the geomorphological system due to a great variety of economic activities carried out in the past and now. Human impact on topography is both direct (anthropogenic landforms and features) and indirect (enhancement and then reduction of soil erosion, channel pattern change, and modification of fluvial processes). The recent human-induced landscape changes related to the current economic activities such as forestry and tourism co-occur with relic landforms in the depopulated and abandoned areas, which are the legacy of more intense human impact in the past. The chapter presents the diversity of anthropic landforms of various age according to their origin related to various economic activities such as agriculture, forestry, settlement, mining, quarrying, industry, and water management. The section on indirect human impact includes a discussion on anthropogenic modification of morphological processes and their record in hillslope and alluvial sediments. Finally, the persistence, values, and promotion of cultural geomorphological landscapes are highlighted.

Keywords

Cultural landscape · Human impact · Anthropogenic landforms · Śnieżnik Massif · Złote Mountains

A. Latocha-Wites (✉)
Institute of Geography and Regional Development, University of Wrocław, pl. Uniwersytecki 1, 50–137 Wrocław, Poland
e-mail: agnieszka.latocha@uwr.edu.pl

11.1 Introduction

The Sudetes—in comparison with other *Mittelgebirge* (medium-altitude mountains) ranges in Poland—are distinguished not only by a high degree of anthropogenic transformation of the topography, but above all by a great variety of economic activities carried out here in the past and now. This is mainly due to their great diversity in terms of lithology and related mineral resources and relief. The first factor supported mining and exploitation activity, which had strongly developed in this area since the Middle Ages (thirteenth–fourteenth centuries), while the second factor substantially contributed to the development of tourism. Tourism development is one of the oldest in the present-day Poland—it was carried out in an organized and planned form from the end of the eighteenth century (Mazurski 2012). In addition, mineral and thermal water resources contributed to the creation of numerous health resorts, vast forest areas favoured forest management and timber industry, and arable lands reached a height of up to 1000 m above sea level. A characteristic feature of the development of the Sudetes was also strong industrialization, not limited to towns, but also involving rural areas. Numerous and various industrial and craft plants based on local raw materials were found even in villages located high in the mountains. This is a feature that clearly distinguishes the Sudetes from other mountainous areas in Poland.

All the above-mentioned economic activities have left their tangible evidence in the landscape, both direct, as anthropogenic landforms and features, and indirect, in the form of anthropogenic modification of morphological processes and their records in hillslope and alluvial sediments. The uniqueness of the Sudetes in terms of the possibility of tracing changes in the topography under human impact also results from the changing intensity of anthropogenic activity over time. Especially the period of the last 150 years allows us to trace the landscape effects of first increasing human impact (until 1945), and then its significant weakening, with the cessation of many economic activities, including

P. Migoń and K. Jancewicz (eds.), *Landscapes and Landforms of Poland*, World Geomorphological Landscapes, https://doi.org/10.1007/978-3-031-45762-3_11

land abandonment, especially in the 1980s and 1990s. In the last 20 years, a successive, though generally local, intensification of human intervention in the environment and landscape is observed, leading to new transformations. Fluctuations in the human–environment relationships are directly related to socio-economic changes that have taken place in this area. The most important among them include: (1) depopulation, commonly observed throughout the area from the end of the nineteenth century and lasting until today; (2) the change of the state border after World War II, the displacement of the German population and the influx of new settlers, often from areas with different environmental conditions (lowlands, fertile soils); (3) political and economic changes in Poland after 1989 related to the fall of communism and the development of free market economy; (4) Poland's accession to the European Union (2004) and the Schengen area (2007), which allowed, *inter alia*, for access to funds for new investments and the development of the near-border areas in which the Sudetes are located.

The changing political and socio-economic conditions translate into clear changes in the landscape. Their effect is a unique palimpsest of anthropogenic landforms and, more broadly, of a geomorphological cultural landscape. They represent forms of different ages and origins, and a characteristic feature is the coexistence in close proximity or even overlapping of various anthropogenic landforms: relict and contemporary, newly created and undergoing gradual degradation. For this reason, the Sudetes can be treated as an open-air research laboratory, both for tracing the way in which diversified human activity was recorded in the relief and sediments, and for assessing the degree of persistence of anthropogenic landscape changes in areas where human impact weakened. It also makes it possible to identify processes that contribute to the degradation or consolidation of elements of past cultural landscapes in the relief.

The chapter focuses on the presentation of direct and indirect anthropogenic transformation of the topography in two subregions of the East Sudetes—the Śnieżnik Massif and the Złote Mountains (Fig. 11.1). This choice arises, on the one hand, from the high concentration of various anthropogenic landforms and constructions in this area and, on the other hand, from high dynamics of the observed socio-economic changes, both depopulation and the revival noted in recent years. In addition, it is also dictated by the availability of literature and detailed evaluation of both areas in terms of human impact.

11.2 Setting and Location

The East Sudetes constitute the northern part of the Bohemian Massif and are mostly located in the Czech Republic. The highest culmination is Mt. Praděd on the Czech side (1492 m asl) in the Hrubý Jeseník range. The chapter focuses on two physiographic mesoregions, which form the westernmost part of the East Sudetes: the Śnieżnik Massif and the Złote Mountains, which are both cross-border mountain ranges (Fig. 11.1). Their total area is 471 km^2. The highest culmination in the analysed area is Mt. Śnieżnik (1424 m asl) situated at the state border, and the maximum height difference is 822 m. The European watershed runs through the Śnieżnik Massif. Most of the area is drained to the Baltic Sea—via the Nysa Kłodzka River (left-bank tributary of the Odra River) and its numerous mountain tributaries. The southern end of the region belongs to the North Sea basin, and the eastern slopes of the massif—to the Black Sea basin. Due to the mountainous nature of the watercourses, there is a high hazard of flooding.

The permanent settlement network developed in the thirteenth–fourteenth century. However, human penetration of the region, including trading routes, commenced much earlier (Herzig and Ruchniewicz 2008; Felcman et al. 2012). Due to its borderland location, the statehood of the region often changed (Czech, Austrian-Hungarian, Prussian/German, Polish), and thus numerous wars and diverse cultural and economic influences affected its development. The shift in state borders after the World War II, together with the almost total exchange of population, was one of the most influential transformations in the history of the region, and its socio-economic effects are observed until nowadays.

The analysed area is protected under the Śnieżnik Landscape Park and its buffer zone and is also covered by several Natura 2000 habitat areas. There are also several nature reserves, including the Niedźwiedzia Cave (*Bear Cave*) and Wilczka Waterfall. The areas directly related to the former mining activities are also protected, both as the ecological sites (Rogóżka, Biała Marianna, Orchids' heap) and the documentation site (Adit No. 18 in the Underground Tourist and Educational Route in the Old Uranium Mine in Kletno).

11.3 Geology and Main Features of Relief

Geologically, the Śnieżnik Massif and most of the Złote Mountains are located within the Orlica–Śnieżnik dome, dominated by various types of Lower Palaeozoic gneisses and schists (Don et al. 1990, 2003) (Fig. 11.1). Within them there are, among others, amphibolites and crystalline limestones (marbles). On the other hand, the northern part of the Złote Mountains belongs geologically to the metamorphic Złoty Stok–Skrzynka zone and the Carboniferous granitoid massif of the Kłodzko–Złoty Stok (Mazur et al. 2007). Moreover, in the area of Lądek-Zdrój basalts occur, which are a remnant of late Neogene volcanism (Badura

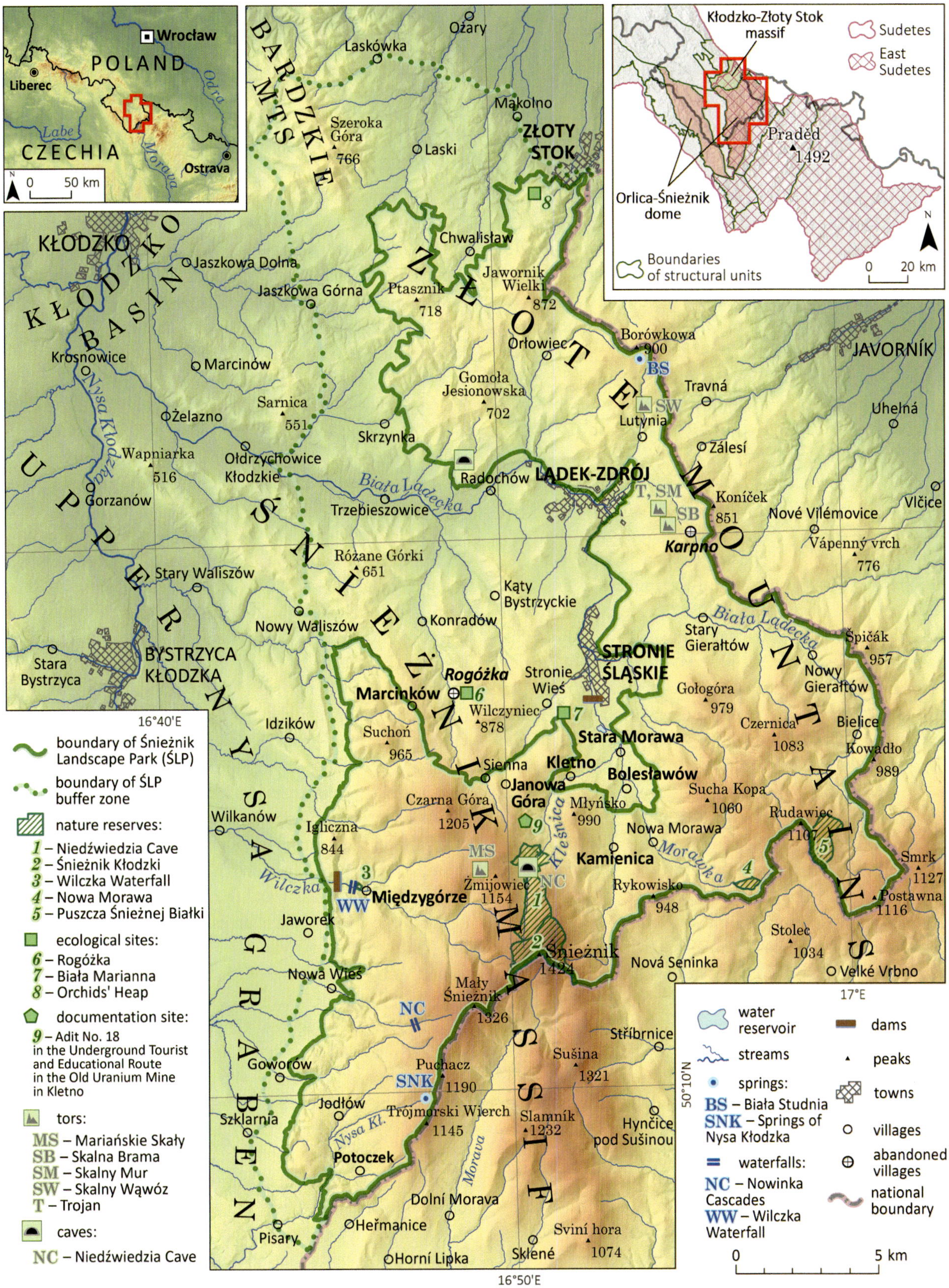

Fig. 11.1 Study area, with localities mentioned in the text and simplified geological background in the inset (map courtesy of Kacper Jancewicz)

et al. 2005). The boundaries of both regions are tectonically conditioned and are legible in the landscape. The mountain ranges represent tectonic horsts associated with the Neogene orogenic movements, which lifted the older denudation surfaces to various heights (Migoń 1996; Sobczyk and Kasprzak 2014).

The highest summit of the Śnieżnik Massif (Mt. Śnieżnik) has the shape of a massive dome raised 100–200 m above the adjacent summit levels, and there are six ridges of various lengths radiating out of the summit, with rounded and gentle summit surfaces interspersed with domed peaks (Figs. 11.1 and 11.2a, b). Characteristic for the western part of the Śnieżnik Massif is the presence of extensive flattenings at an altitude of 700–850 m asl, which create the lower morphological level within the massif (Migoń 1996). The western morphological boundary of the massif is very pronounced due to its tectonic origin (Ranoszek 1999; Badura and Rauch 2014). The mountain ranges are dissected by deep, V-shaped valleys with steep slopes, often following fault lines (Fig. 11.2c). The streams locally form waterfalls (i.e. Wilczka Waterfall and Nowinka cascades). The Wilczka Waterfall is the largest and highest waterfall in the Śnieżnik Massif, and one of the highest in the entire Sudetes. The step of the waterfall, 22 m high, was created on a tectonic fault and at the same time on the lithological boundary of more resistant gneisses above and more rapidly degraded metamorphic schists downstream. In the latter, a deep gorge developed below the waterfall, with almost vertical rock walls reaching up to 25 m in height. Due to the exceptionally picturesque nature of this place, the area around the waterfall was intensively developed for tourists in the mid-nineteenth century, although it was a popular tourist attraction even much earlier (see Sect. 11.4.3).

The Złote Mountains have the form of a long mountain ridge, with varied topography (Fig. 11.2d). The northern and southern parts are more compact, with a few branching ridges, while the central part is a narrow, winding ridge with short arms extending on either side. The sharp north-eastern boundary of the range is tectonic in origin, and numerous tectonically-controlled landforms can be distinguished on the Czech side of the mountains (Štěpančíková et al. 2008; Štěpančíková and Stemberk 2016). The main river of this area is Biała Lądecka, which forms two polygenetic gorges below and above the spa town of Lądek-Zdrój. A characteristic feature of the valley, which in this area is 200–350 m wide, is the presence of several terraces of different ages.

The relief of both regions is also strongly determined by lithology and rock resistance. Therefore, characteristic features of relief are numerous rock outcrops forming tors and mid-slope crags (Fig. 11.2e, f). They developed mainly in gneiss and quartzite and less commonly also in mica schists (with less spectacular sizes). Tors occur in the upper parts of the slopes, in the zones of convex slope breaks, as well as on the plateaus and around the peaks. They take various forms: rock walls, pulpits, rock spurs, and towers. Their shapes follow the pattern of cracks and fissures or cleavage and their common feature is morphological asymmetry (Migoń 1996). In the Śnieżnik Massif, the highest tors are 15 m high (Mariańskie Skały), and the length of the outcrop can reach up to 50–60 m (the rock group at the top of Mt. Młyńsko, the rock wall on Mt. Żmijowiec). In the Złote Mountains, tors are up to 30 m high (Trojan, Skalna Brama/*Rock Gate*) and 100 m long (Skalny Mur/*Rock Wall*). They are most common around the town of Lądek-Zdrój, where they serve as climbing routes and viewpoints (Sobczyk 2005).

Tors within the slopes are found in many deeply incised valleys, where the slope is up to 30°. Rock outcrops are often accompanied by block fields, with single boulders up to 2–3 m long. Blocky-debris and debris-sandy-silty covers are also common, testifying to the intensity of denudation processes in periglacial conditions. Locally, they form tongues and block streams, and the structure of the covers indicates their origin by solifluction processes (Martini 1979; Traczyk 1996). The thickness of slope covers is 1–1.5 m on average, reaching higher values in the lower parts of the slopes, for example, up to 4–6 m in the Kleśnica valley bottom (Traczyk 1996). In the lower slope sections, colluvial sandy loam covers predominate, which may be associated with later anthropogenic activity (see Sect. 11.5). Particular periglacial landforms include frost cliffs and steps several metres high, and cryoplanation terraces up to 50 m long, covered with angular boulders. These forms are closely related to lithology, appearing along the contact of more resistant rocks (gneisses, quartzites) and less resistant schists (Martini 1979). Landforms of periglacial origin within the Śnieżnik dome also include nivation hollows (Martini 1979), patterned grounds, and earth hummocks (Klementowski 1998).

The Śnieżnik Massif and the Złote Mountains combined are also one of two main areas of carbonate rocks occurrence in the Sudetes. Various forms of surface and underground karst have developed, including numerous caves (Pulina 1977). One of the most spectacular—not only at the scale of the Sudetes, but throughout Poland—is the Niedźwiedzia Cave, which is open to the public and characterized by very rich and varied dripstone forms (speleothems) (Jahn et al. 1989). As a result of recent discoveries, the karst system turned out to be much more extensive than originally assumed (Sobczyk et al. 2016; Kasprzak and Sobczyk 2017). The total length of the currently explored corridors is approximately 4.5 km, the height difference is 118 m, and the largest chamber, the Mastodon Hall, is 115 m long, 20 m wide, and 20 m high (Haczek et al. 2014). Furthermore, the best developed examples of surface karst are found in the vicinity of the Niedźwiedzia Cave, in the Kleśnica valley. These are karst springs, ponors, sinkholes, and dry valleys (Pulina 1977; Jahn et al. 1989).

Fig. 11.2 Main features of the relief in the study area. **a, b** Tectonic boundary of the Śnieżnik Massif. **c** V-shaped valleys dissect the Śnieżnik Massif (vicinity of the village of Międzygórze). **d** Undulating relief of the Złote Mountains. **e, f** Gneissic tors in the Złote Mountains (Skalny Wąwóz) (photographs by A. Latocha-Wites)

11.4 Diversity of Anthropogenic Landforms and Features

11.4.1 Landforms Related to Mining and Industry

The topography of the East Sudetes has been modified by mining activities dating back to medieval times, with the peak in the sixteenth century, followed by episodes of local intensification of prospection and extraction in the eighteenth and nineteenth century. The Złote Mountains and the Śnieżnik Massif were important local centres of mining, based mainly on polymetallic ores, including silver, iron, zinc, as well as gold and arsenium in the northern part of the Złote Mountains (i.e. Ciężkowski et al. 1996; Stysz and Mączka 2009). In the 1940s and 1950s, many old adits and mines were re-opened for exploitation of uranium ores (Klementowski 2010). Although it was only a brief episode, intensive works "refreshed" the former mining areas, contributing to the visual enhancement of past mining-related landforms, which can still be recognized in the contemporary topography. The impact of mining and ore prospection is rather spatially limited, but nevertheless it may be locally spectacular. Numerous new landforms result from direct or indirect ground surface transformation, such as open cast pits, adits, shafts, trenches, spoil heaps, heaps of gang stones, disturbed ground, linear surface excavations, trial pits in former mining and prospecting areas, and subsidence hollows. The inventory of mining legacy in the vicinity of the former mining town of Bolesławów (now a village) revealed 40 adits and associated spoil heaps, 1270 exploratory pits, and 8 shafts. And the most recent excavation of underground galleries in this area (47.5 km in total) was related to the exploitation of uranium in 1948–1953 (Ciężkowski et al. 1996). Field mapping and analysis of LiDAR data in three former mining areas in the Śnieżnik Massif (Marcinków, Janowa Góra and Potoczek) revealed ca. 210 stone heaps and ca. 190 mining pits within an area of a mere 0.7 km^2. Hollows related to former adits and shafts are 5–15 m deep and 8–20 m in diameter, trial pits are represented in the contemporary topography by holes 1–2 m deep and 2–5 m in diameter, and prospecting trenches are 1–1.5 m deep. The associated heaps of stones are 1.5–5 m high, with the length up to 30 m (Fig. 11.3) (Migoń and Latocha 2018).

The original topography of the region has also been highly modified by numerous quarries of various age and size. In the study area, the most important and largest quarries were related to the exploitation of marble and basalt, although exploitation of gneiss, schist, and granite was also locally important. In the nineteenth century stone quarrying industry reached its peak, with as many as 118 quarries in the study area (according to the topographic maps Messtischblatt 1:25,000 from the 1880s) with the total surface of ca. 0.3 km^2. Nowadays only two larger quarries are still in operation. In spite of the abandonment of most quarries, usually due to depletion of resources, they provide a long-lasting evidence of past human activities and are

Fig. 11.3 Landforms related to mining in the vicinity of the village of Janowa Góra. **a, b** Stone heaps. **c** Entrance to an adit, now collapsed. **d** Trial pit (photographs by A. Latocha-Wites)

still well visible in the landscape in spite of being usually overgrown by vegetation (Fig. 11.4). Geomorphic legacy of quarrying is diverse and includes both concave (former quarries themselves) and convex features (heaps of unused material), as well as various types of disturbed and man-made ground in the vicinity of former quarries.

Mining and quarrying were often associated with on-site processing of raw materials, and thus, numerous remains of former ironworks, ore processing plants, and lime kilns are still visible. Similarly, glass-making industry, which was among branches very typical for the Sudetes since medieval times and was present in the Śnieżnik Massif up to 2010s (in the town of Stronie Śląskie), has required quarrying of necessary resources (quartz, feldspar). The impact of former industrial sites can also be traced in sedimentological record through the presence of slag or by-product deposits. In addition, many technological processes required the use of timber, and hence, widespread forest clearances followed, enhancing the potential for soil erosion.

11.4.2 Landforms Related to Agriculture, Settlements, and Forestry

Agricultural activity has encroached onto the mountainous areas in the thirteen–fourteenth century, initially following forest clearances for mining and ore processing (Valde-Nowak 1999). However, due to increasing population, which in most locations reached its peak in the nineteenth century, even the less favourable areas were turned into agricultural lands, including steep slopes (more than 15°) up to 1000 m asl. The increased agricultural pressure, along with harsh mountainous climate and shallow and stony soils, resulted in severe soil erosion from arable fields. In order to prevent the undesirable surface wash, new ploughing techniques were used since the nineteenth century (Inglot 1979). They included ploughing parallel to the contour lines and a new type of plough turning soil downslope. As a result, slope profiles were modified from smooth to the stepped ones. The terraced slopes are the most evident manifestation of

Fig. 11.4 Landforms related to quarrying—abandoned marble quarry in the village of Rogóżka (photograph by A. Latocha-Wites)

human impact on the geomorphic slope domain until today, although agricultural activity has long ceased in all more elevated areas in the region (Latocha 2009b). The relic agricultural terraces are currently found both within permanent grasslands and forests, and they are very persistent anthropogenic landforms (Fig. 11.5a, b, f). In some areas, assemblages of several or up to a dozen or so terraces may be found within the slope, with the series of alternating benches and risers. Morphometric characteristics of agricultural terraces vary from site to site, depending on local slope properties. Their length ranges from 50 up to 400 m, and the height of the risers varies from 0.2 to 5.5 m. The usual distance between the risers is 20–100 m, and the inclination of risers ranges from 20° to 60°. Usually, the steeper and the longer is the slope, the higher is the terrace and the steeper is its riser (Latocha 2009a, b). The densities of anthropogenic escarpments, mostly agricultural terraces, calculated for selected catchments in the study area, range from 0.6 to 13.1 km/km^2.

Some terrace risers are reinforced with stone walls or numerous heaps of stones are located along the escarpments. The slope covers in the study area are very abundant in rock debris due to shallow soil and the proximity to crystalline bedrock. Therefore, local farmers were forced to regularly clear their arable lands, gathering and removing the stones, which were constantly appearing on the surface due to ploughing, erosional wash (soil erosion), and frost upheaval. The collected stones were piled up not only along the terraces but also along the field and ownership divisions. They were collected in the form of dry stone walls, or as isolated stone piles within the arable lands (Fig. 11.5c–e). Although ploughing has completely ceased at higher elevations, these landforms are still a very spectacular landmark of many former agricultural areas, even though they are nowadays located in the forest. The height of stone piles can reach up to 4 m, their length up to 15 m, and width up to 10 m. The stone embankments are usually up to 1.5–2 m high and can be several hundred metres long. Also, the density of these landforms is locally impressive. There might be more than 100 stone piles per 1 km^2 in villages located higher up in the mountains. For example, within the areas of two abandoned villages—Karpno (Złote Mountains) and Rogóżka (Śnieżnik Massif)—there are ca. 140 and 130 piles of stones per 1 km^2, respectively.

Slope terracing was also necessary in settlements, where steep slopes had to be levelled for building purposes. Settlement terraces are another very persistent landscape feature, and they last much longer than the buildings themselves, once erected upon them. In some depopulated areas, these terraces are the only traces of the location of former farmsteads. Although most of these relic landforms are now located within forest, they can still be easily identified in the relief as unnatural flats within a slope. Their surface usually reach up to 100 m^2.

Terracing has led not only to changes of topography and slope longitudinal profiles, but it also influenced the dynamics of morphological processes by dividing slopes into separate units with erosion dominating in their upper parts and accumulation in the lower parts, behind the upper edge of each terrace (see Sect. 11.5).

In addition to agriculture, forestry and timber industry were among important economic activities, not only in the past but also nowadays. In the pre-industrial era, timber was used as an important building material and the main source of fuel for most technological processes—in mining, ore processing, glass-making, weaving, and spinning (Fogger 1952). With the advent of coal-based industry at the turn of the nineteenth century, the role of forestry changed but on no account it became less important. The Sudetes became an important supplier of raw timber and processed timber products. Paper industry flourished, with paper mills, manufactures, and factories common across the Sudetes (Szymczyk 2005). The huge demand for timber resulted in widespread clearances of slopes up to the water divides. However, the economic crisis in the nineteen/twentieth century, depopulation, land abandonment, and collapse of handicraft and timber industry resulted in both intentional and spontaneous afforestation. The shifts in forest cover have affected the geomorphic system by alteration of slope-channel coupling and especially by an increase/decrease of sediment wash and transfer. The forest cover is nowadays much more extensive than ca. 150 years ago (the increase in forest area is by ca. 35%, as calculated for the entire Kłodzko region, which includes the analysed area), but the forestry still exerts vital impact on geomorphic systems, mainly causing linear erosion due to logging.

Both in the agricultural lands and in the forested areas, an extensive network of roads has developed over centuries. Most of them are not in use any longer, but nevertheless they can still be recognized in the landscape. On the one hand, roads parallel to the contour lines have substantially changed slope profiles, and in some areas, they form spectacular cuts and escarpments, often reinforced by retaining walls which allow for their better preservation. On the other hand, roads perpendicular to the contour lines have evolved in some places into gullies (sunken lanes) (Fig. 11.6). The gullies within crystalline bedrock (gneiss, schists) usually occur only in the lower or middle parts of slopes, where there is a thick layer of fine-grained weathering mantle or colluvial deposits. Road gullies within agricultural lands may be as deep as 4 m, with an average depth of 1–2 m. Their length does not exceed several hundred metres. Road gullies within forested areas are up to 1.2 m deep, while the depth of logging tracks ranges from 0.1 up to 0.7 m, and their width is 0.3–1.7 m. Most gullies in the study area are relic landforms, with the exception of areas, where intense timber harvesting, involving

Fig. 11.5 Relic features related to agriculture. **a, b** Slopes with agricultural terraces. **c, d** Stone walls. **e** Stone piles, built of materials collected from an agricultural slope. **f** Terrace riser (photographs by A. Latocha-Wites)

the use of heavy machinery, contributes substantially to the deepening of forest roads and logging tracks, transforming them into gullies. They are usually very irregular in both their longitudinal and cross-sectional profiles, which is in opposition to much more regular gully profiles related to agriculture. Within some roads, former cobbling has been preserved, which usually—but not always—prevents further incision.

Fig. 11.6 Old road gullies (sunken lanes). **a** Sunken lane overgrown with vegetation, with evidence of soil creep along the sides. **b** Road gully filled with mineral and organic material (photographs by A. Latocha-Wites)

11.4.3 Landforms and Constructions Related to Water Management

Alterations of channel patterns have a history going back to at least medieval times and were forced by demands of various types (Fig. 11.7). To ensure adequate water supply to mills, parallel channels (leats) maintaining minimal gradient were built to connect with mills and to provide sufficient drop. At the end of the nineteenth century, there were 121 water mills in the study area, as presented in the topographic maps Mestischblatt 1:25,000. Although the leats are no longer in use and most of them are dry, they are still well recognizable in the relief, as are the ruins of the mills. Networks of artificial channels were also built in mineral exploitation areas, to facilitate ore processing or mineral extraction. Widespread dam and reservoir construction started in the beginning of the twentieth century as one among the means to reduce flood hazard. There are two dry reservoirs in the study area, on the Morawka and Wilczka rivers. The latter one is famous by its spectacular stone dam, which is 29 m high and 102 m wide (Fig. 11.7d). Particularly common were engineering works along rivers, which included channel and bank straightening, often using stone armouring, bed grading, building of steps, and check dams to limit bedload transport. They were mainly constructed using local stones as building material.

Two main rivers in the study area, Wilczka and Biała Lądecka, have been channelized along 49% and 26% of their length, respectively (Latocha and Witek 2013). Special engineering works were also necessary to protect road surfaces and disturbed slopes against water erosion. Numerous stone bridges, road culverts, and other underground passages and pipes are the testimony of the extent of the road construction works. The diverse hydrotechnical constructions are very common; they occur even in the upper reaches of the streams, and some of them are very spectacular in size (Fig. 11.7c). In the nineteenth century, special attention was also paid to the springs, many of which became tourist attractions. Water outflows were often enclosed with ornamental stone walls and bas-reliefs, and their surroundings were adopted for tourists needs, with new infrastructure, such as benches, tables, and footpaths. Most of these features have been neglected or destroyed in the post-World War II period; however, some of them have been recently restored, for example, the springs of Nysa Kłodzka in Śnieżnik Massifs and Biała Studnia (*White Well*) in the Złote Mountains. Similarly, unique in-channel features such as waterfalls were also adopted for the tourist needs in the nineteenth century. This is the case of the Wilczka waterfall in the Śnieżnik Massif. Not only a proper tourist infrastructure was built at the site, but also the waterfall itself was modified and made higher through an artificial step, which was destroyed during the flood in 1997 (the pre-WWII height was 27 m and it is 22 m nowadays). Access to the site with improved tourist facilities has been recently restored (Fig. 11.8). Additionally, there is a new recreational water reservoir in the village of Stara Morawa and numerous fishing ponds.

All hydrotechnical constructions, both the old and neglected ones, and the currently used and maintained, have the potential to locally modify fluvial processes. They can influence the intensity and spatial distribution of erosion and accumulation within the channels, which is crucial during extreme hydrometeorological events (Latocha and Parzóch 2010; Kasprzak 2011; Latocha and Witek 2013; Kasprzak and Migoń 2015). Additionally, clogged or destroyed underground drainage systems can substantially alter hydrological properties of soil, which results in the

Fig. 11.7 Diversity of water management constructions; examples. **a** Stone bridge. **b** River bank reinforcement. **c** Check dam (all examples from the upper reach of the Nysa Kłodzka river). **d** Dry flood control reservoir with the dam on the Wilczka river, in the village of Międzygórze (photographs by A. Latocha-Wites)

development of wetlands and marshes within the abandoned settlements.

11.4.4 Other Features of the Cultural Geomorphological Landscape

Anthropogenic landforms related to mining, quarrying, agriculture, forestry, and water management are the most spectacular features of the cultural landscape directly related to geomorphology. However, there are also numerous other elements, which contribute to the richness and diversity of the cultural geomorphological landscape in the study area. First of all, the settlement network and the rural layouts highly depend on the topography, with villages stretching along the valley floors, forming typical mountain settlements (Germ. *Waldhufendorf*). Most of them date back to medieval times and their general layout has been well preserved until the present-day. Local stones were used to construct many buildings, including the representative ones, like churches or manors, as well as smaller architectural features such as commemorative plates and monuments, road-side chapels, crosses, and figures of the saints, which are especially abundant in the study area, forming its specific cultural landscape (Fig. 11.9).

An increasing popularity of tourism since the second half of the nineteenth century has also added to the cultural geomorphological landscape. Local tourist organizations took effort to allow or improve access to various attractive places, such as viewing points, picturesque or unusual rock formations, caves or springs (Dziedzic 2013). Adaptations of natural landforms for tourist purposes were especially widespread around the spa town of Lądek-Zdrój in the Złote Mountains and the mountain resort of Międzygórze in the Śnieżnik Massif. However, minor alterations related to tourist facilities were also common in other areas. They

Fig. 11.8 Tourist facilities at the Wilczka Waterfall (photograph by A. Latocha-Wites)

included the construction of footpaths, rock-hewn staircases, metal or stone barriers, rock-cut viewing platforms or viewing towers, as exemplified by the most recognizable construction on the summit of Mt. Śnieżnik which was built from the local gneiss. Tourism has continued to grow almost without interruptions (save the World War II period), and the infrastructure is being constantly enlarged. In particular, rapid development of ski-resorts (Czarna Góra, Kamienica) and dedicated bike trails (single tracks) has contributed to further anthropogenic transformation of slopes in the past several years.

11.5 Indirect Human Impact on Relief and Sediments

The development of settlements, agriculture, mining, and industry in the historical times resulted in large-scale deforestation of the Sudetes, which triggered intense soil erosion. The severity of past soil erosion emerges from documentary sources, modelled values, specific landforms (stone piles and stone embankments), and slope wash deposits (Latocha et al. 2016; Migoń and Latocha 2018). For the pre-World War II period, there is evidence for episodes of intense rainfalls able to remove the entire soil cover from the slopes, so that the local peasants were forced to carry it back upslope to sustain land productivity (Bac 1950). Calculations of slope material stored within agricultural terraces in three small catchments in the Śnieżnik Massif and the Złote Mountains allowed for an estimation of the annual denudation from arable lands, which was 0.15–2.67 mm year^{-1} depending on topographic location and lithology (Latocha 2009a).

Sediments, which can be linked with past human activities, are widespread both within the slopes and in the valley floors, and they have also modified the local topography. They form a distinct sandy-silty topmost layer in colluvial and alluvial sedimentary successions, which is the result of selective surface wash (Jahn 1968). Fine charcoal particles commonly occur within this layer and thus can be dated

Fig. 11.9 Minor elements of cultural geomorphological landscape, with local stones used in old and modern vernacular architecture. **a** Old lime kiln. **b** New road-side chapel. **c** Ruins of a church in depopulated village. **d** Modern tourist infrastructure (photographs by A. Latocha-Wites)

and related to past soil erosion resulting from forest clearance and agricultural activities. This fine-grained layer differs substantially from the underlying layer of more coarse grained sediments, representing mostly periglacial environments. The human-induced colluvial deposits are found not only at footslopes, but are also common within any flattenings within the slopes higher up. Their thickness on the former arable lands is between 10 and 55 cm, depending on bedrock lithology and topographical position, with the thickest layers developed on schists. The thickness of colluvial sediments increases within agricultural terraces, where the sandy-silty layer is usually ca. 1 m deep, reaching up to 1.6 m in the extreme examples, which proves the effective role of terraces as sediment traps and efficient anti-erosion protection. The thickness of the agricultural sandy-silty layer, which originated from slope wash and ploughing, and accumulated over the terrace riser, increases along with the increase of slope inclination and slope length (Latocha 2009a).

The sandy-silty alluvia with high organic content, including charcoal particles, are also related to surface wash from slopes (e.g. Klimek et al. 2003; Klimek and Latocha 2007). After temporal fluvial transport, they were eventually deposited within the floodplains during the episodes of overbank flow. Sedimentological analysis of alluvial deposits within three small catchments in the study area revealed a high variability of their thickness, ranging from a few centimetres up to over 1.2 m. The variability was evidently dependent on the local variations in the valley morphology, and the alternations of the thickness of alluvial deposits were observed even within short distances (Latocha 2009a). The fine-grained alluvia were found up to 780 m asl, which is much higher than previously reported from other parts of the Sudetes (Teisseyre 1985).

Surface wash from deforested slopes and linear erosion along field access and forest roads, as well as the subsequent deposition of alluvial sediments within the valley floors, have transformed the entire geomorphic system. The floodplains have been built-up, and the original braided-river fluvial system inherited from periglacial environments has evolved into single-channel rivers which are typical for the region until nowadays (Latocha 2009a). The human-induced soil erosion and floodplain accretion were confirmed by the radiocarbon datings of fine charcoal particles found at the base of the sandy-silty colluvial and alluvial deposits. In all investigated catchments, they can be correlated with the advent of human occupation and forest clearances for agriculture or glass industry, which was also confirmed by the historical documents (Latocha 2009a).

Depopulation, abandonment of arable land, decline of industry, and afforestation of upper slopes, which have been observed since the end of the nineteenth century and intensified in the post-War War II period, altered the geomorphic system again. Especially, in many areas, the slope-channel coupling was disrupted. Reduced surface wash and material delivery from the slopes have resulted in an increase of downcutting and lateral erosion of streams and rivers. Additionally, the multi-thread flow during the flood events has been observed in several rivers in the study area, e.g. in the Biała Lądecka and Wilczka channels (Czerwiński and Żurawek 1999; Latocha and Parzóch 2010).

By contrast to the sedimentological record of past agriculture, the indirect legacy of past metallurgy and glass-making is very limited and only site-specific, although the enrichment of local colluvia or alluvia with slag or waste/by-product deposits can be easily detected during field surveys.

11.6 Persistence, Values, and Promotion of Cultural Geomorphological Landscapes

The relic anthropogenic landforms are surprisingly persistent in the landscape, adding to its geodiversity and, indirectly, to the diversity of habitats and wildlife (Latocha et al. 2019). Some former anthropogenic landforms have even been protected as ecological sites due to their unique habitats and species, such as old marble quarries in Rogóżka and Stronie Śląskie (Biała Marianna) and post-mining heap in the town of Złoty Stok overgrown with orchids. The old mines in Złoty Stok and an abandoned basalt quarry near Lądek-Zdrój have also been included into the European Ecological Network Natura 2000 due to their unique habitats. The relic anthropogenic landforms are also able to modify current morphological processes, which is well visible in the fluvial system impacted by diverse water management modifications, especially during the extreme events. Some relic landforms (agricultural terraces, road gullies, prospection pits, embankments, and channel reinforcements) may become obliterated with time due to natural processes such as denudation, erosion, accumulation, impact of vegetation (e.g. tree uprooting), or animals (e.g. burrowing). Observations in former mining areas in the Śnieżnik Massif show that afforestation proceeds quickly and after 20–30 years former exploitation pits, spoil heaps and access roads become completely overgrown (Ciężkowski et al. 1996). Nevertheless, they can still be recognized in the landscape, especially outside the growing season, or on the detailed terrain models based on LiDAR data (Fig. 11.10). Actually, only landforms which were deliberately removed by human action (e.g. stones from buildings or agricultural heaps of stones) have completely disappeared from the landscape.

It is worth to mention that some anthropogenic landforms and features have received new functions nowadays. Especially the former mining and industrial sites have become important tourist attractions. This is the case of Złoty Stok and Kletno where former gold/arsenium and fluorite/uranium mines have been adapted for tourist purposes. There is an art gallery in the old lime kiln in Stara Morawa and a rope park in the former quarry in Złoty Stok. The most known karst cave in the Sudetes, the Niedźwiedzia Cave, was discovered in 1966 due to limestone quarrying. The recent restoration of some springs can be perceived as both the reconstruction of tourist infrastructure reviving the pre-World War II tradition, an enhancement of tourist offer, and promotion of interesting geosites. Nevertheless, there are many other areas, which are not promoted as tourist attractions, but where the diversity and abundance of anthropogenic landforms would be of interest to any

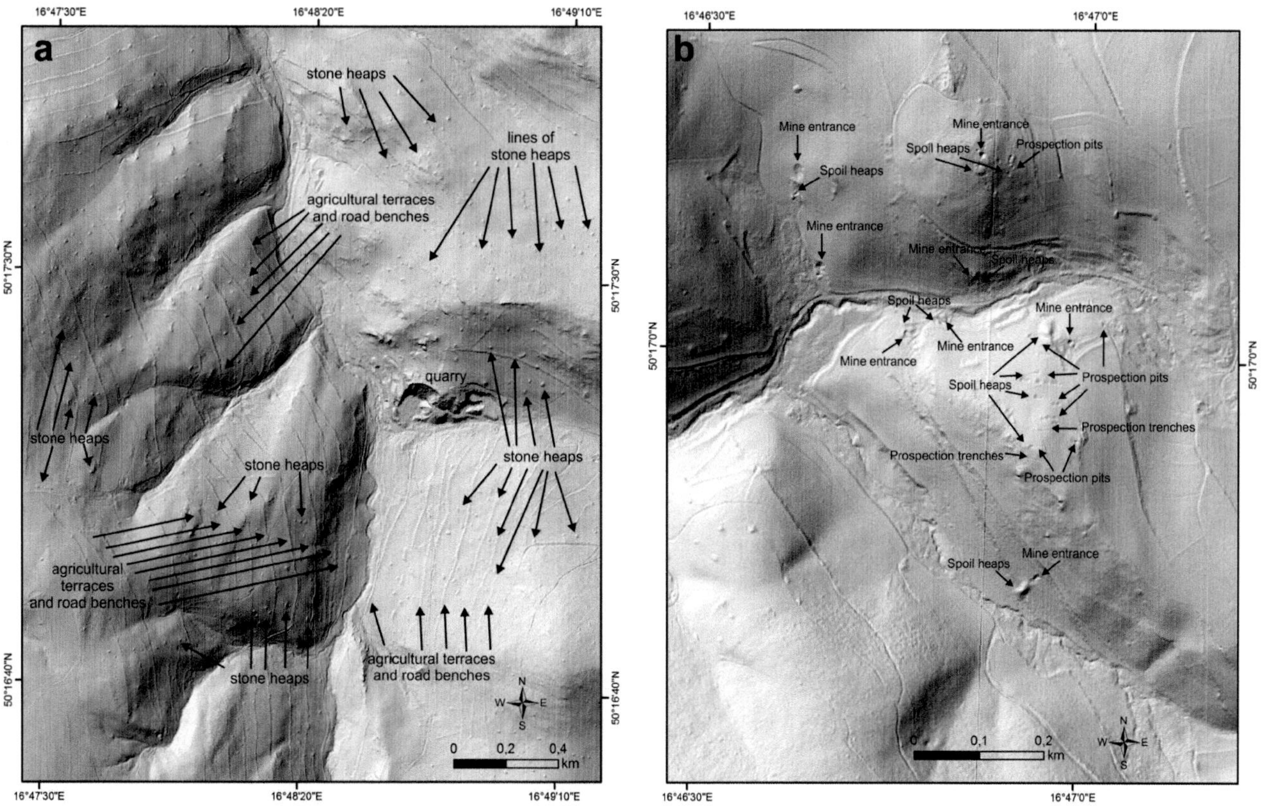

Fig. 11.10 Abundance of relic anthropogenic landforms in the Śnieżnik Massif is still well visible on the LiDAR terrain model. **a** Relic agricultural features. **b** Landforms related to old mining. Source: Migoń and Latocha (2018)

visitors interested in discovering traces of past human activities, which are still detectable in the present-day cultural geomorphic landscape of the East Sudetes.

11.7 Conclusions

The discussed subregions of the Śnieżnik Massif and the Złote Mountains in the East Sudetes can be considered as representative for the Sudetes, which exemplify a mid-latitude, medium-altitude mountain range highly modified by long-term and diverse human activities. Human impact on topography was both direct (anthropogenic landforms) and indirect (enhancement and then reduction of soil erosion, channel pattern change, and modification of fluvial processes). The abundance of anthropogenic landforms of various age and origin is the characteristic feature of the cultural geomorphic landscape of the region. The recent human-induced landscape changes related to current economic activities, such as forestry and tourism, co-occur with relic landforms in the depopulated and abandoned areas, which are "witnesses" of more intense human impact in the past, involving more widespread agriculture, mining, quarrying, and industry.

References

Bac S (1950) Wpływ prac pługa na przemieszczenie gleb. In: Bac S, Ostromęcki J (eds) Badania nad erozją gleb. IUNG, PWRiL, Warszawa, pp 61–80

Badura J, Pécskay Z, Koszowska E, Wolska A, Zuchiewicz W, Przybylski B (2005) New age and petrological constraints on Lower Silesian basaltoids, SW Poland. Acta Geodyn Geomater 2(3/139):7–15

Badura J, Rauch M (2014) Tectonics of the Upper Nysa Kłodzka Graben, the Sudetes. Geol Sudetica 42:137–148

Ciężkowski W, Irmiński W, Kozłowski S, Mikulski S, Przeniosło S, Sylwestrzak H (1996) Zmiany w litosferze wywołane eksploatacją surowców mineralnych. In: Jahn A, Kozłowski S, Pulina M (eds) Masyw Śnieżnika – zmiany w środowisku przyrodniczym. Polska Agencja Ekologiczna, Warszawa, pp 85–119

Czerwiński J, Żurawek R (1999) The geomorphological effects of heavy rainfalls and flooding in the Polish Sudetes in July 1997. Studia Geomorph Carp-Balc 33:27–43

Don J, Dumicz M, Wojciechowska I, Żelaźniewicz A (1990) Lithology and tectonics of the Orlica-Śnieżnik Dome, Sudetes—recent state of knowledge. N Jhrb Geol Paläont, Abh 179:159–188

Don J, Skácel J, Gotowała R (2003) The boundary zone of the East and West Sudetes on the 1:50 000 scale geological map of the Velké Vrbno, Staré Mesto and Śnieżnik metamorphic units. Geol Sudetica 35:25–59

Dziedzic M (2013) Kłodzkie Towarzystwo Górskie 1881–1945. Quaestio, Wrocław, p 509

Felcman O, Gładkiewicz R (eds) (2012) Kladsko. Dějiny regionu. Filozofická fakulta Univerzity Hradec Králové, Polsko-Czeskie Towarzystwo Naukowe, Historický ústav, Powiat Kłodzki, Hradec Králové – Wrocław – Praha – Kłodzko, p 607

Fogger J (1952) Beiträge zur Wirtschaftskunde der Grafschaft Glatz. Zentralstelle der Grafschaft Glatz, Kierspe-Bahnhof i. Westf., p 191

Haczek A, Kostka S, Markowski M (2014) Jaskinia Niedźwiedzia – partie odkryte po roku 2012. In: Materiały 48. Sympozjum Speleologicznego, Sekcja Speleologiczna Polskiego Towarzystwa Przyrodników im. Kopernika, Kraków, pp 44–50

Herzig A, Ruchniewicz M (2008) Dzieje Ziemi Kłodzkiej. Atut, Hamburg, Wrocław, p 568

Inglot S (ed) (1979) Historia chłopów śląskich. Ludowa Spółdzielnia Wydawnicza, Warszawa, p 510

Jahn A (1968) Selektywna erozja gleb i jej znaczenie w badaniach geomorfologicznych. Przegl Geogr 40:419–424

Jahn A, Pulina M, Kozłowski S (eds) (1989) Jaskinia Niedźwiedzia w Kletnie. Ossolineum, Wrocław, p 372

Kasprzak M (2011) The geomorphological effects of extreme flood events in the rivers of Poland's Western Sudetes: historical data, as set against GIS modelling and field observations. Czas Geogr 82:107–135

Kasprzak M, Migoń P (2015) Historical and recent floods in the West Sudetes—geomorphological dimension. Z Geomorph NF 59(Suppl. 3):73–97

Kasprzak M, Sobczyk A (2017) Searching for the void: improving cave detection accuracy by multi-faceted geophysical survey reconciled with LiDAR DTM. Z Geomorph NF 61(Suppl. 2):45–59

Klementowski J (1998) Nowe stanowisko gruntów strukturalnych na Śnieżniku. Czas Geogr 69:73–85

Klementowski R (2010) W cieniu sudeckiego uranu. Kopalnictwo uranu w Polsce w latach 1948–1973. Instytut Pamięci Narodowej, Wrocław, p 376

Klimek K, Malik I, Owczarek P, Zygmunt E (2003) Climatic and human impact on episodic alluviation in small mountain valleys, The Sudetes. Geogr Polon 76(2):55–64

Klimek K, Latocha A (2007) Response of small mid-mountain rivers to human impact with special attention to the last 200 years, Eastern Sudetes, Central Europe. Geomorphology 92:147–165

Latocha A (2009a) Land use changes and longer-term human-environment interactions in a mountain region (Sudetes Mountains, Poland). Geomorphology 108:48–57

Latocha A (2009b) The geomorphological map as a tool for assessing human impact on landforms. J Maps 5:103–107

Latocha A, Parzóch K (2010) Efekty geomorfologiczne powodzi w dolinie Białej Lądeckiej w czerwcu 2009 r. Przyroda Sudetów 13:251–262

Latocha A, Witek M (2013) Dawne obiekty hydrotechniczne na ziemi kłodzkiej – stan zachowania i rola w przebiegu procesów fluwialnych, w tym wezbrań ekstremalnych. In: Kościk E (ed) Gdy nadciągała wielka woda. Klęski powodzi na ziemiach polskich na przestrzeni wieków. Gajt, Wrocław, pp 105–125

Latocha A, Szymanowski M, Jeziorska J, Stec M, Roszczewska M (2016) Effects of land abandonment and climate change on soil erosion. An example from depopulated agricultural lands in the Sudetes Mts., SW Poland. Catena 145:128–141

Latocha A, Reczyńska K, Gradowski T, Świerkosz K (2019) Landscape memory in abandoned areas—physical and ecological perspectives (Central European mountains case study). Landscape Res 44:600–613

Martini A (1979) Peryglacjalny charakter wierzchowiny Masywu Śnieżnika Kłodzkiego. Probl Zagosp Ziem Górskich 20:203–216

Mazur S, Aleksandrowski P, Turniak K, Awdankiewicz M (2007) Geology, tectonic evolution and Late Palaeozoic magmatism of Sudetes—an overview. In: Kozłowski A, Wiszniewska J (eds) Granitoids in Poland. AM Monograph, vol 1, pp 59–87

Mazurski KR (2012) Historia turystyki sudeckiej. Oficyna Wydawnicza "Wierchy", Kraków, p 672

Migoń P (1996) Zarys rozwoju geomorfologicznego Masywu Śnieżnika. In: Jahn A, Kozłowski S, Pulina M (eds) Masyw Śnieżnika – zmiany w środowisku przyrodniczym. Polska Agencja Ekologiczna, Warszawa, pp 35–45

Migoń P, Latocha A (2018) Human impact and geomorphic change through time in the Sudetes, Central Europe. Quat Intern 470(A):194–206

Pulina M (1977) Zjawiska krasowe w Sudetach polskich. Dokumentacja Geograficzna IGiPZ PAN 2–3:118

Ranoszek W (1999) Zastosowanie różnych metod morfometrycznych w analizie morfologii progu tektonicznego na przykładzie zachodniej krawędzi Masywu Śnieżnika. Przegl Geol 47:1027–1031

Sobczyk A (2005) Rzeźba skałkowa środkowej części Gór Złotych. Przyroda Sudetów 8:147–162

Sobczyk A, Kasprzak M (2014) Late Cenozoic tectonic activity of the Śnieżnik Massif area (the Sudetes, SW Poland) in the light of LIDAR DEM morphometric analysis. Studia Geomorph Carp-Balc 48:35–52

Sobczyk A, Kasprzak M, Marciszak A, Stefaniak K (2016) Zjawiska krasowe w skałach metamorficznych w Masywie Śnieżnika (Sudety Wschodnie): aktualny stan badań oraz znaczenie dla poznania ewolucji Sudetów w późnym kenozoiku. Przegl Geol 64:710–718

Štěpančíková P, Stemberk J, Vilímek V, Košťák B (2008) Neotectonic development of drainage networks in the East Sudeten Mountains and monitoring of recent fault displacements (Czech Republic). Geomorphology 102:68–80

Štěpančíková P, Stemberk J (2016) Region of the Rychlebské Hory Mountains—tectonically controlled landforms and unique landscape of granite inselbergs (Sudetic Mountains). In: Pánek T, Hradecký J (eds) Landscapes and landforms of the Czech Republic. Springer, Cham, pp 263–276

Stysz M, Mączka M (2009) Dzieje górnictwa w Marcinkowie. Inwentaryzacja pozostałości robót górniczych dawnych kopalń rud polimetalicznych. In: Zagożdżon P, Madziarz M (eds) Dzieje górnictwa – element europejskiego dziedzictwa kultury, vol 2. Oficyna Wydawnicza Politechniki Wrocławskiej, Wrocław, pp 297–311

Szymczyk M (2005) Zarys dziejów przemysłu papierniczego na Dolnym Śląsku w XIX i XX wieku. Roczniki Biblioteczne 49:35–53

Teisseyre A (1985) Mady rzek sudeckich. Część I: Ogólna charakterystyka środowiskowa (na przykładzie zlewni górnego Bobru). Geologia Sudetica 20(1):113–195

Traczyk A (1996) Formy i osady peryglacjalne w masywie Śnieżnika Kłodzkiego. Acta Univ. Wratisl. 1808, Prace Inst Geogr A8:111–119

Valde-Nowak P (ed) (1999) Początki osadnictwa w Sudetach. Instytut Archeologii i Etnologii PAN, Kraków, p 238

Agnieszka Latocha-Wites is Associate Professor in the Institute of Geography and Regional Development at the University of Wrocław. She is also the coordinator of the Laboratory of Landscape Research. In her research she combines physical and human geography. Her main research interests are long-term human impact on landscape and geomorphic systems, persistence of anthropogenic signatures in depopulated areas, and transformations of cultural landscapes. Her main study area is Lower Silesia, mainly the Sudetes Mountains, but she also conducted her research in Ireland and in the Arctic. She authored numerous articles, chapters and books, including travel books ("Ireland – my paths"; "My Arctic Summer"). She is a member of the Executive Committee of the Polish Geographical Society (2020–2024) and member of editorial board of *Journal of Landscape Ecology*. She was a secretary of the Association of Polish Geomorphologists (2005–2011).

Mount Ślęża—The Highest Inselberg in the Sudetic Foreland and Its Glacial-Periglacial Morphogenesis

12

Andrzej Traczyk

Abstract

Mt. Ślęża is a striking landmark of the Sudetic Foreland in SW Poland, rising as a distinctive inselberg more than 500 m above the surrounding lowlands. It is part of a larger basement massif (Ślęża Massif) characterized by a complex geological structure, which consists of metamorphic rocks (gabbro, serpentinites, amphibolites) included in the Palaeozoic ophiolite structure and late Carboniferous granitoids. Main relief features of the Ślęża Massif were shaped as a result of long-term divergent weathering in the warm climatic conditions of the Mesozoic and most of the Cenozoic. Mt. Ślęża existed in a form similar to the present-day one in the Miocene, as testified by the occurrence of Middle Miocene clays, silts, sands, and lignite on weathered rock-cut surfaces around the massif. Its further morphogenesis in the Pleistocene involved at least two transgressions of the Scandinavian ice sheet and the influence of the periglacial environment. During the maximum extent of the inland glaciation the hilltop of Mt. Ślęża remained as a nunatak, but glacial erosion occurred on the lower slopes. In the younger phase of the Saalian glaciation and during the Weichselian glaciation, the slopes of the Ślęża Massif were modelled by periglacial processes. Crags, frost-riven cliffs, extensive blockfields, solifluction mantles, and congelifluction block tongues, descending to ca. 250 m asl, were formed. Probably soon after the decay of the last ice sheet, as a result of mass movements in a paraglacial environment, large block tongues similar to rock glaciers were formed on the southern and eastern slopes of Mt Ślęża. The diversity of periglacial relief, combined with other natural and cultural values, makes Mt. Ślęża an important element of European heritage.

Keywords

Structural landforms · Inselberg · Trimline · Paraglacial · Periglacial · Crag · Glaciation

12.1 Introduction

In a similar way as Mt. Śnieżka (1603 m) in the Karkonosze Mountains (see Chap. 7) is symbolic for the Sudetes range, Mt. Ślęża (718 m) has become a landmark for the Sudetic Foreland and epitomizes the historical region of Lower Silesia (Fig. 12.1). Rising by more than 500 m above the Silesian Lowland (Figs. 12.2 and 12.3), it is a distinctive element of regional relief, easily identified from afar. It stands in clear isolation and is much higher than other hills of the Sudetic Foreland, which reach 300–450 m asl. The summit of Mt. Ślęża is an excellent vantage point, with a panoramic view of the Silesian Lowland and the distant Sudetes. In the prehistoric times, the forested massif of Ślęża was undoubtedly a place of pagan cult of considerable regional importance, as testified by remnants of stone ramparts surrounding the summit parts of Mt. Ślęża itself and the summits of adjacent Mt. Radunia (573 m) and Mt. Wieżyca (415 m), as well as the presence of megalithic sculptures carved out of local granite. Hence, it is occasionally named as 'Silesian Olympus'. It is also believed that the name of the entire region, Silesia (Polish *Śląsk*), may be derived from the mountain (Wrzesiński 2006). Although historians and archaeologists have debated about the origin, function, and significance of Mt. Ślęża for nearly 200 years, and many aspects of its history remain mysterious, the Ślęża Massif is surely an important part of European heritage (Pawleta 2017).

The significance of the Ślęża Massif is not limited to its geomorphology and history. Despite increasing human impact lasting for more than 2000 years, it also represents valuable natural heritage, being like an island of only moderately altered environment within long deforested plains,

A. Traczyk (✉)
Institute of Geography and Regional Development, University of Wrocław, pl. Uniwersytecki 1, 50–137 Wrocław, Poland
e-mail: andrzej.traczyk@uwr.edu.pl

P. Migoń and K. Jancewicz (eds.), *Landscapes and Landforms of Poland*, World Geomorphological Landscapes, https://doi.org/10.1007/978-3-031-45762-3_12

Fig. 12.1 Ślęża Massif seen form the SE (photograph by K. Jancewicz)

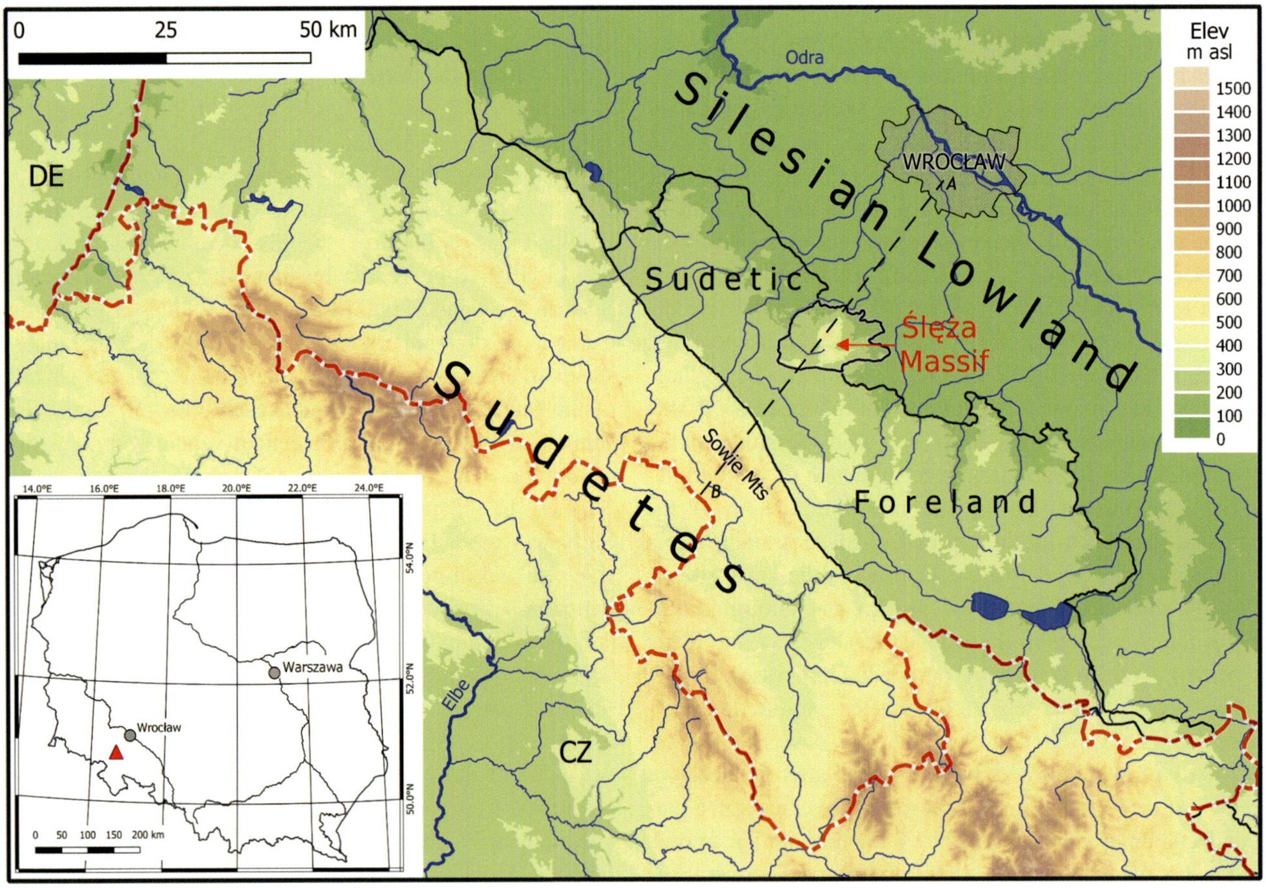

Fig. 12.2 Location of the Sleża Massif within the Sudetic Foreland. Boundary of the macro- and meso-regions after the Solon et al. (2018). Dashed line—morphological cross-section Wrocław—Ślęża Massif—Sowie Mts (see Fig. 12.3)

extensively used for agriculture. Within the slopes of Mt. Ślęża one can find remnants of forest typical for the lower montane forest belt in Central Europe and xerothermic grasslands on serpentinite hills (Żołnierz 2007). To protect these important remnants, three nature reserves and one natural landscape complex were established (Fig. 12.4). Since 1988 the whole area is formally protected as the Ślęża Landscape Park (Ślężański Park Krajobrazowy).

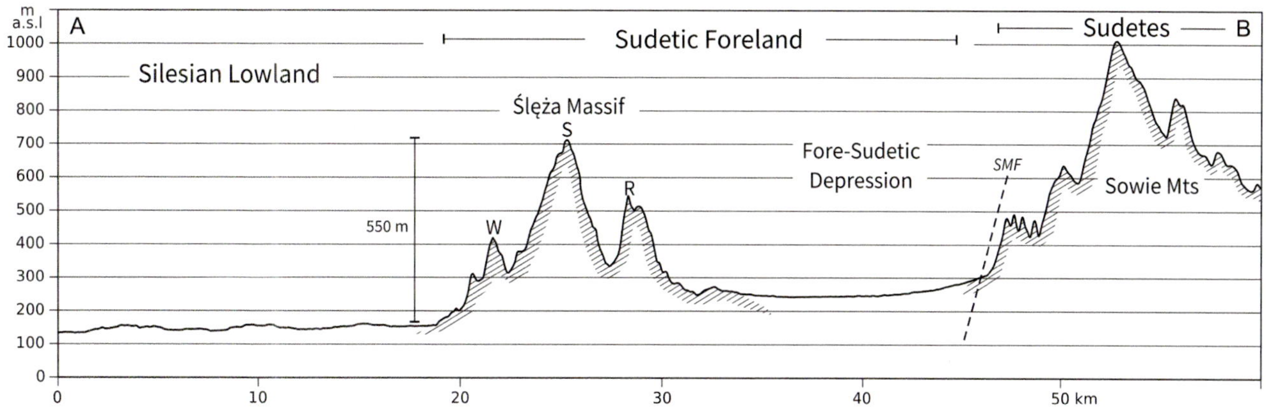

Fig. 12.3 Morphological cross-section Wrocław—Ślęża Massif—Sowie Mts (Sudety Mts) according the DEM (resolution 10 m). Explanations: W—Mt Wieżyca (415 m asl); S—Mt Ślęża (718 m asl); R—Mt Radunia (573 m asl); SMF—Sudety Marginal Fault; hatched—outcrops of the pre-Cenozoic rocks

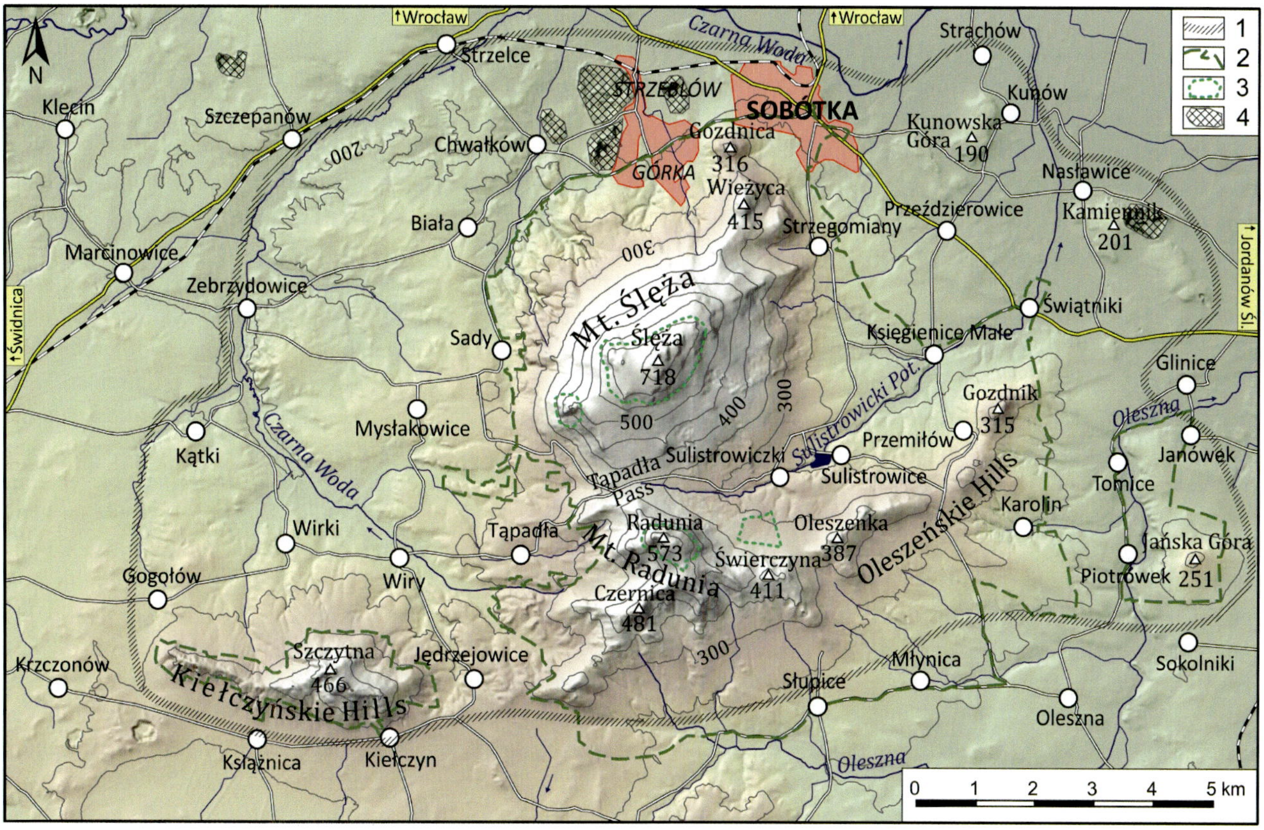

Fig. 12.4 Physiography of the Ślęża Massif. Explanations: 1—boundary of the Śleża Massif (after Solon et al. 2018), 2—boundary of the Śleża Massif Landscape Park, 3—boundaries of protected areas, 4—main quarries. Contour lines every 50 m; shaded relief generated form DEM of 10 m resolution

However, in the context of this volume, geomorphological uniqueness of Mt. Ślęża needs to be emphasized most. Its relative elevation is unusually high, given that other parts of the Sudetic Foreland are built of rocks of apparently similar resistance against weathering and denudation, and yet they are much lower. In the Middle Pleistocene, in the Elsterian and Saalian glaciations (equivalents of Marine Oxygene Isotope Stage 12 and 6, respectively), the Ślęża Massif was at least twice subject to the direct impact of the Scandinavian ice sheet, and at that time, Mt. Ślęża stood out as a nunatak. During intervals of ice-sheet decay and throughout the last cold stage, Weichselian, morphology of the massif was shaped by weathering and hillslope processes typical for periglacial environments. The legacy of periglaciation includes numerous tors, cover deposits, and block streams, some of which resemble rock glaciers. Thus, Mt. Ślęża is a peculiar island of mountainous periglaciation amidst the lowlands.

The island-like nature of the Ślęża Massif determines its special position among other mountain terrains in Poland. In this chapter, the focus will be on three main aspects of this uniqueness: (a) geological controls of the topography, including structure and variable rock resistance; (b) the significance of glacial processes in the Pleistocene; and (c) the role of periglacial processes in the final shaping of morphology.

12.2 Main Relief Features

The Ślęża Massif, surrounded by surfaces of low and subdued relief, is the most distinctive geomorphic region within the Sudetic Foreland. However, it can be further subdivided into four units: Mt. Ślęża, Mt. Radunia, Oleszeńskie Hills, and Kiełczyńskie Hills (Fig. 12.4). The flattened large cone of Mt. Ślęża itself is the dominant feature. It occupies more 30 km^2, with the base extending for 7 km in SW–NE direction and ca. 6 km in NW–SE direction. The slopes are poorly diversified, except a few mid-slope flats and a short ridge descending towards the north-east, from 415 m asl to ca. 170 m asl at the footslope. To the south, separated by the Tąpadła Pass (384 m), the massif of Mt. Radunia (573 m) is located, which continues eastward as a narrow ridge of Oleszeńskie Hills. In between these two elevated units, there is an open basin of Sulistrowicki Potok (creek), drained to east and developed along the zone of major lithological contact between gabbro and serpentinite (see Sect. 12.3). The fourth unit, located at some distance from the previous ones, is the isolated, W–E elongated massif of Kiełczyńskie Hills (Szczytna, 466 m), located in the south-west. Geological affinities suggest that low bedrock hills of Jańska Góra (251 m) and Kamiennik (200 m) in the east and Kunowska Góra (190 m) in the north-east should also be considered as parts of the Ślęża Massif (Solon et al. 2018).

12.3 Geological Background

The main morphological units of the Ślęża Massif directly correspond to its geological structure, the spatial pattern and strike of main structural units of the so-called Ślęża Complex (Sawicki 1995) and its surroundings. This complex, in the central part of the massif, is built of mafic and ultramafic rocks. In the north-west, it is adjoined by the Variscan granite pluton of Strzegom-Sobótka, whereas in the south a dislocation zone separates it from the gneissic block of the Sowie Mountains (Fig. 12.5). The Ślęża Complex is also named as the Ślęża Ophiolite (Cwojdziński 1977; Majerowicz 1981; Floyd et al. 2002). Thus, the Ślęża Ophiolite is composed of serpentinites, meta-gabbro, and amphibolites (Majerowicz 2006). Its constituent rock series are metamorphosed volcanic rocks (meta-volcanics), relatively common among pre-Variscan lithologies in the Sudetes (Krzemińska and Awdankiewicz 2011). Żelaźniewicz (2015) considered the Ślęża Complex as a part of a larger, tectonically dismembered structure defined as the Sudetic Ophiolite. Beside the Ślęża region, its remaining parts are located in various areas in the Sudetic Foreland and in the Middle Sudetes, around the gneissic block of the Sowie Mountains.

From a palaeogeographic perspective, the rocks involved in the Ślęża Ophiolite are an outcome of magmatic processes active some 420–400 million years ago along a divergent plate boundary, in a mid-ocean ridge setting (Krzemińska and Awdankiewicz 2011). In deeper crustal sections, solidification of magma gave rise to gabbro. Closer to the contemporaneous surface subvolcanic rocks of diabase type were formed, whereas the lavas appearing on the ocean floor turned into basalt, possibly with pillow-lava textures. As a consequence of subsequent large-scale tectonic deformations (plate obduction), the primary magmatic rocks became tectonized and altered. Metamorphic alteration occurred in the conditions of greenschist facies in the late Devonian during the Variscan orogeny, at the depth of 5–8 km (Kryza and Pin 2010). Thus, meta-gabbro formed from gabbroic rocks, diabases turned into amphibolites, and basalts were changed into greenschists. Serpentinites, which are also typical components of ophiolitic complexes, formed as a result of metamorphisation of peridotites (Krzemińska and Awdankiewicz 2011). However, the intensity of alteration of the primary rock was non-uniform. Therefore, primary olivine and pyroxene have survived in some places, whereas complete replacement by serpentine may have occurred nearby. The most advanced alteration is hypothesized to have occurred in the border parts of the serpentinite belt and in the eastern part (Szumilas 1963). The highest structural unit of the ophiolitic complex, built of poorly metamorphosed pillow lavas, is missing, but Majerowicz (1984) suggested that they may have already been eroded away. It is assumed that both the ophiolitic massif and the gneissic massif of the Sowie Mountains were exposed at the surface as early as the latest Devonian and have been subject to denudation since that time (Żelaźniewicz 2015).

Among the rock series within the ophiolite the gabbro warrants particular attention. It is usually typified by coarse-grained structure, massive texture, and green colour.

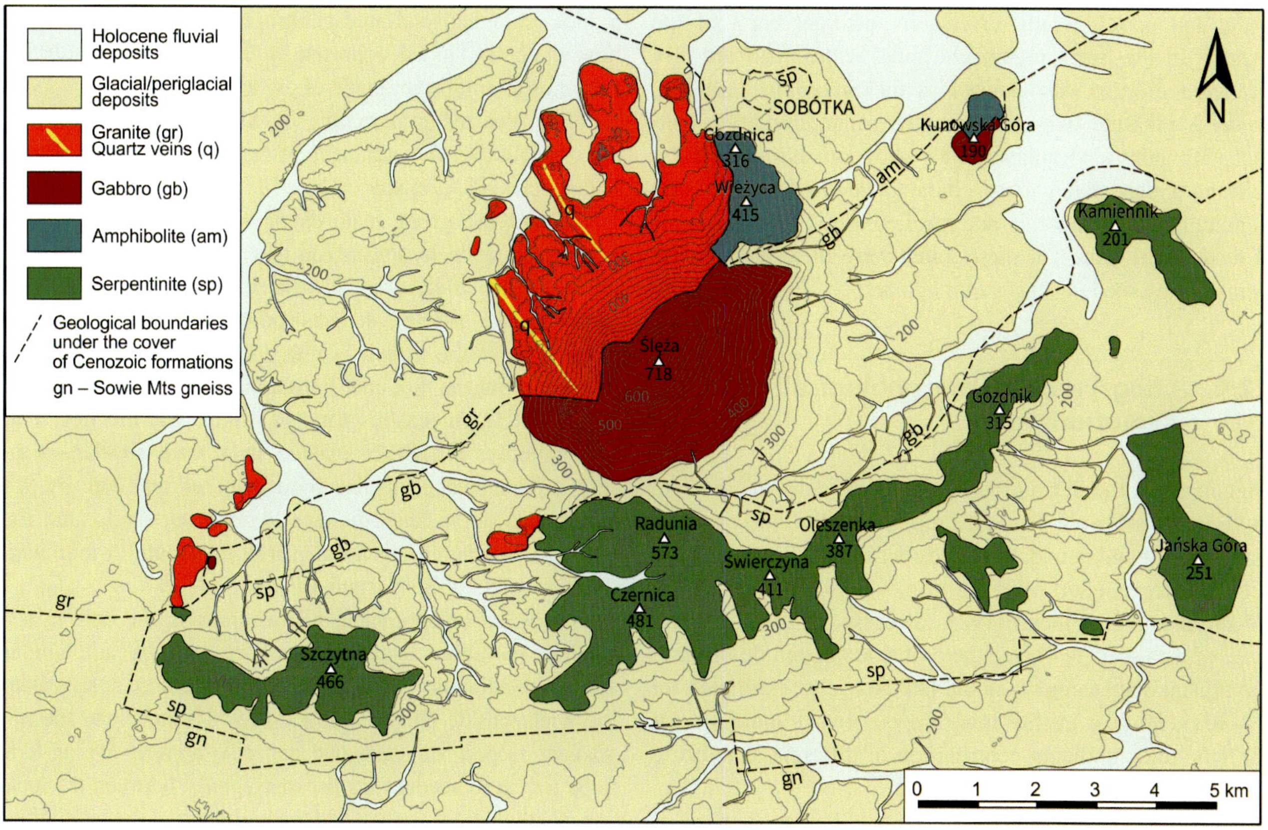

Fig. 12.5 Geological outline of the Ślęża Massif. Sources: Detailed Geological Map of the Sudetes 1:25,0000 and Detailed Geological Map of the Poland 1:50,000—simplified. Additional boundaries of the geological units (dashed lines) according the map by Sawicki (1995)

In some outcrops one can observe laminar arrangement of light and dark minerals (Majerowicz 1984, 2006). The jointing pattern is complicated and non-uniform in space, unlike the adjacent rocks. Three main, steeply-dipping joint sets were distinguished, with the dominant role of NWW–SEE and N–S sets (Buczek 1952) (Fig. 12.6a). Jointing pattern exerts clear influence on how gabbro rocks break down and explains the common presence of large, tetrahedral, or irregular angular blocks (Fig. 12.6b).

The northern and, in part, western slopes of Mt. Ślęża are underlain by granites (granodiorites), which belong to the easternmost part of a late Variscan magmatic intrusive complex of Strzegom-Sobótka massif (Majerowicz 1972). They are mainly medium-grained, locally porphyritic, pale

Fig. 12.6 Examples of gabbro crags, illustrating the nature of jointing. **a** Crag from the NE slope of Mt. Ślęża, with distinct vertical cracks. **b** Joint-controlled, angular outline of a gabbro crag on the top of Mt. Ślęża (photographs by P. Migoń)

grey, and petrologically diverse in different parts of the massif. In the Ślęża region, the most widespread are biotite granodiorites, also known as the Strzeblów Granites, whereas alkaline two-mica granites occur along the contact with gabbro. Granites usually show much more regular jointing pattern than gabbro, characterized by three sets perpendicular to one another and labelled as Q, L, and S joints (Cloos 1922). Some of these joints were filled with quartz veins during the late stages of magmatic activity.

12.4 Long-Term Geomorphological Evolution

The origin of the Ślęża Massif is yet to be fully deciphered. Authors of older publications emphasized the role of structural control and denudation, considering Mt. Ślęża as a relict inselberg (Gellert 1931) or a hill due to rock resistance (Walczak 1970). Finckh (1928) discussed the direct influence of tectonics, but given the probable magnitude of vertical crustal displacements in the late Neogene, estimated for 20–30 m at most in this area (Krzyszkowski and Karanter 2001a), any significant contribution of differential uplift to the shaping of the massif is unlikely.

However, the hypothesis of Mt. Ślęża being simply controlled by rock resistance is problematic too. The main hill is built of two lithologies—gabbro and granite, rather than one, and the contact runs across the western and northern slopes, without being clearly reflected in hillslope topography (Migoń 1997; Żurawek and Migoń 2000). Moreover, the occurrence of gabbro is not restricted to the hill itself, as it also underlies the eastern footslopes and the adjacent basin, at the altitude of c. 200 m asl, that is more than 500 m below the summit. The relationships between geology and landform evolution are thus by no means straightforward.

An overview of various concepts proposed to explain the landform of Mt. Ślęża was comprehensively presented by Żurawek (2002a, b). He confirmed that there is hardly any support for the hypothesis that faulting may have played a significant role. Instead, he suggested that processes of chemical weathering exploiting mineralogical variability within particular lithologies may have been important. In fact, the idea of long-term divergent weathering in the warm climatic conditions of the Mesozoic and most the Cenozoic had been explored before, and Mt. Ślęża was considered as an inselberg, i.e. a part of an old stripped etchplain (Migoń 1999). To account for the current height of Mt. Ślęża, Migoń (1997) inferred its stepwise emergence from the thick weathering mantle, recorded in a staircase of mid-slope flats, which can be identified within the slopes of the main hill. In this context, the finding of a thick, chemically-altered gabbro-derived regolith within a bench at c. 640 m asl, reported by Szczepankiewicz (1958), becomes an important piece of evidence.

Thus, the residual massif of Mt. Ślęża was probably towering above the adjacent lowlands as early as in the Palaeogene (Szczepankiewicz 1958; Jahn 1968; Migoń 1997) and the gross features of relief would be of that age. It was even hypothesized that the morphological evolution of Mt. Ślęża may be traced back to the Mesozoic (Pernarowski 1963a). It was pointed out that the relative height of the massif is much bigger than of any other inselbergs in the Sudetic Foreland, which do not exceed 150 m, although reasons for this uniqueness are not clear. Undoubtedly, Mt. Ślęża existed in a form similar to the contemporary one in the Miocene, as testified by the occurrence of Middle Miocene clays, silts, sands, and lignite on weathered rock-cut surfaces around the massif (Pernarowski 1963b; Migoń 1997).

Geological-topographical relationships in the belt of serpentinite elevations to the south of the main hill are similar. The same lithology supports the low ridges of Oleszeńskie and Kiełczyńskie Hills, the highest Mt. Radunia, and the low gradient slopes, although one can also assume a long-term slope retreat to account for the discrepancy between the location of lithological contact and the perimeter of the hills. Thus, the hypothesis of rock-controlled topography is more credible (Migoń and Kasprzak 2015), although it remains unknown why some sections of the serpentinite belt are associated with higher ground, whereas other coincide with lowlands.

Consequently, it was suggested that variability within one lithology may be actually more important that differences between different rock types. For example, the morphological axis of Mt. Ślęża follows SW–NE direction and is supported by gabbro, although in the vicinity of the gabbro/granite contact. This might suggest an elevated resistance of gabbro in the contact zone due to contact metamorphism (Migoń and Kasprzak 2015). However, such a view is not consistent with the results of rock-strength studies using the Schmidt hammer (Placek 2007). In contrast to previous assumptions, higher resistance of gabbro was not confirmed. The mean value for gabbro in the summit parts of Mt. Ślęża was 52 points (rebound values) on the N-type Schmidt hammer scale, whereas amphibolites of Mt. Wieżyca yielded 58, fresh granites on the western slopes of Mt. Ślęża scored slightly above 55, and the serpentinites of Mt. Radunia were characterized by the mean rebound value of 53. Nonetheless, the strength of gabbro did not prove uniform and minor differences may be responsible for the presence of second-order morphological features of hillslopes. Convex elements of relief (scarps, tors) showed higher intact strength that gabbro outcrops within mid-slope flats (Placek 2007).

12.5 Morphometric Characteristics

Morphometric analysis supplements qualitative characterization of topography and reveals noteworthy differences between different parts of the massif, apparently mainly in relation to lithological diversity. A DTM of 30 m × 30 m resolution was used to capture the main features of relief, with the analysis performed using QGIS (QGIS Development Team 2018) and SAGA (Conrad et al. 2015) software. Several relief properties were subject to analysis: (a) relative elevation calculated from morphological profiles, (b) slope inclination, and (c) convergence index (Table 12.1). The latter was reclassified in such a way that three categories of slopes were returned: convergent (concave), divergent (convex), and straight. These properties were calculated for individual morphological units within the Ślęża Massif and, separately, for the granite and gabbro parts of Mt. Ślęża itself (Table 12.2).

Typical relative elevations within the serpentinite belt are within the 100–140 m range, except the Mt. Radunia, where the mean height exceeds 260 m. This is still much less than for Mt. Ślęża, which rises above the footslopes by 476 m on average (Table 12.1). Slope profiles used to calculate the elevation show an interesting feature of hillslope topography. In the serpentinite part, the slopes are concave (Fig. 12.7). Within Mt. Ślęża slopes are mainly rectilinear, but the inclination increases above 600 m asl. This is at variance with the situation in the Sudetes, where slopes are typically concave in the lower part and convex in the upper part (Jahn 1969, 1980).

The distribution of slope inclinations shows that low-angle slopes (6–12°) dominates in the entire Ślęża Massif, with the exception of the Oleszeńskie Hills, which are even more subdued. However, steeper slopes (12–18°) are also quite common within serpentinites. Within Mt. Ślęża itself the proportions are similar, and slopes of 6–12° inclination are most widespread. A higher percentage of nearly flat terrains (<6°) on gabbro reflects the occurrence of distinctive mid-slope benches at higher elevations (Skalna—c. 520 m asl, Olbrzymki—c. 640 m asl, and unnamed benches at 575 and 625 m asl to the north-east of the summit part, see Fig. 12.8 for location), unknown from the granite part of the massif. Low-angle surfaces in granite tend to occur along local water-divide areas in the footslope areas.

The steepest slopes (>18°) are more widespread in serpentinites than in gabbro or granites. They mainly occur just below the ridge parts, occupying the largest surfaces around the hilltop of Mt. Radunia. Within Mt. Ślęża slopes of this steepness are observed to the west of the summit, framing the mid-slope benches on the downslope side, all cut in gabbro, as well on the west-facing side of Mt. Wieżyca in the amphibolite part. In granite, such slopes are very rare and limited to two settings, namely the contact zone with gabbro below the bench of Olbrzymki and deeply incised valleys within the north-western footslopes.

The last parameter analysed here reveals another distinctive feature of the Ślęża Massif. Within serpentinites, troughs and valleys are generally rare, and more evident ones occur only in the northern foothills of the Kiełczyńskie Hills and the southern foothills of the Oleszeńskie Hills. By contrast, the largest slope hollows are inset into the south- and west-facing slopes of Mt. Radunia. The higher slopes of Mt. Ślęża itself are very poorly dissected, as already highlighted by Żurawek and Migoń (2000). In the gabbro part, slope hollows occur only in the north-eastern part. They are broad and open, originated due to denudation rather than fluvial erosion. No slope hollows are present within the granite slopes at higher elevations either.

Table 12.1 Morphometric characteristic of Ślęża Massif according the GIS analysis (DEM 30 m × 30 m)

Geomorphic unit	Mean (min–max) relative height (m)	Hillslope inclination (%)				Hillslope form (%)		
		<6°	6–12°	12–18°	>18°	Cc	R	Cx
KH	140 (80–182)	30	40	25	5	36	33	32
RM	263 (197–325)	38	36	18	8	35	33	32
OH	101 (84–124)	63	30	6	1	32	32	36
MS	476* (438–518)	26	47	22	5	26	49	25

KH—Kiełczyńskie Hills; *RM*—Radunia Massif; *OH*—Oleszeńskie Hills; *MS*—Mt. Ślęża; *Cc*—concave; *R*—rectilinear; *Cx*—Convex. (*)—without considering the Wieżyca Ridge

Table 12.2 Morphometric characteristics of hillslopes of Mt. Ślęża according to lithology

Lithology	Percentage of slope surfaces within hillslope inclination classes (%)					Hillslope form (%)		
	<6°	6–12°	12–18°	18–24°	>24°	Cc	R	Cx
Gabro	31	46	18	4	1	25	49	26
Granite	23	48	25	4	0	30	53	17

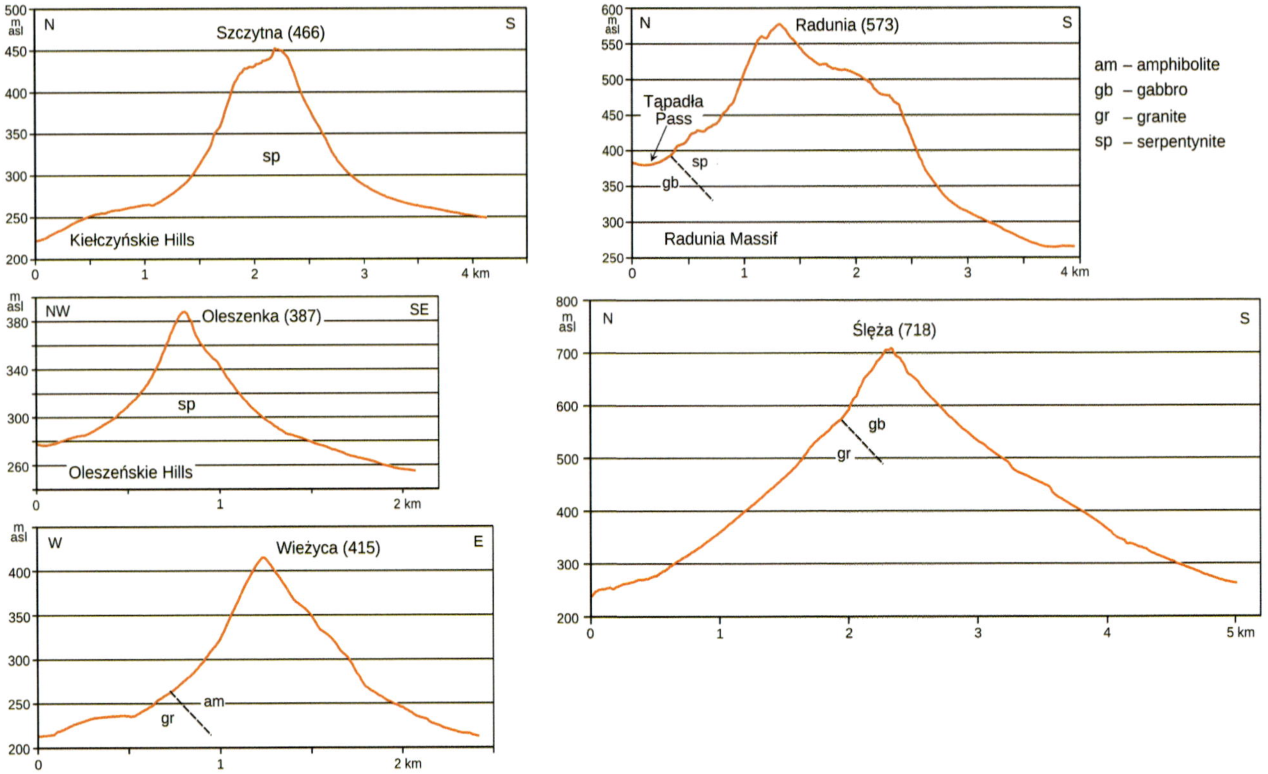

Fig. 12.7 Selected morphological profiles across different parts of the Ślęża Massif

However, the north-western granite footslopes, in the altitude belt of 300–350 m asl, are incised by a network of V-shaped valleys, up to 20–30 m deep, with steep sides and occasional bedrock outcrops.

Morphometric analysis allows one to conclude that the shapes (longitudinal profiles) of slopes in the Ślęża Massif are in clear relation to the lithological diversity. In gabbro and granites slopes tend to be rectilinear, whereas those in serpentinites are predominantly concave, with minor convexities in the top sections. The slope profiles also reveal a concave break at the gabbro/granite contact and even more evidently at the granite/amphibolite contact (Fig. 12.7). Poor development of fluvial network within the massif is another distinctive feature. Larger valley forms are limited to the footslope surfaces in serpentinites and granites and lack in the gabbro part. This is probably related to planar convexity of gabbro slopes, favouring dispersal of surface runoff.

12.6 Ślęża Massif in the Pleistocene

In the synthetic account of geomorphology of the Sudetes, Jahn (1980) argued that elements of glacial and periglacial morphology, as the youngest ones, dominate among meso- and microforms and are responsible for morphological details. However, compared to the 'make-up', they are superimposed on the gross morphology inherited from the preceding, much longer periods of warm climate. This view can be easily applied to the Ślęża Massif, especially to the inselberg of Mt. Ślęża proper, whose slopes in detail contain a diverse assemblage of landforms produced by processes typical for the cold morphogenetic conditions. These are mainly periglacial landforms, but landforms and deposits indicative of inland glaciation occur too.

12.6.1 Inland Glaciation

At least two ice sheets reached the Sudetic Foreland and the mountain front of the Sudetes during the Pleistocene (Elsterian = MOIS 12, Saalian = MOIS 6), but glacial landforms associated with these ice-sheet advances are rather scarce. Deposits of the older glaciation are not exposed at the surface and are only known from boreholes in the adjacent plain. These are typically two–three layers of tills (Elsterian), covered by sands and thin till laid down during the younger glaciation (Saalian) (Krzyszkowski and Karanter 2001a). The thickness of whole glacial package does not exceed 10 m (Jahn 1969), only locally reaching 30 m, in the eastern part of the area (Świątniki) (Krzyszkowski and Karanter 2001b).

In older sources, large accumulations of gabbro boulders on the south-eastern slopes of Mt. Ślęża were interpreted as end-moraines (Finckh 1928), apparently associated with ice-sheet movement or decay, as no one has ever suggested local glaciation of the massif. However, these boulder packs were erroneously considered as of glacial origin and are in fact products of local postglacial hillslope processes, as will be discussed below (see Sect. 6.2.3). Nevertheless, there is no doubt that other remnants of the Saalian ice sheet occur on the slopes of Mt. Ślęża. These include rather infrequent erratic boulders (Finckh 1928), present up to the altitude of 500–550 m asl, and glacial cover deposits. The foot-slopes of the main hill, especially those descending towards the valley of Sulistrowicki Potok, are mantled by a 4–5 m thick till cover up to the altitude of 390 m asl. Its near-surface part is more sandy and shows fine lamination discernible to the depth of 2 m, indicating slope re-deposition (Szczepankiewicz 1958). It is underlain by in situ till, which in turn rests upon chemically-altered gabbro. In the less elevated parts of the Ślęża Massif, the Saalian tills are covered by slopewash and aeolian (loess) deposits. However, isolated till patches occur much higher on the slopes of Mt. Ślęża. Szczepankiewicz (1958, 1969) reported about 6 m thick tills within a mid-slope bench above the rock cliffs of Skalna, at 530 m asl. By contrast, no evidence of glacial overriding was found on the large bench at 640 m asl, and the trench yielded only local hillslope deposits on top of weathered gabbro (Szczepankiewcz 1958; Żurawek 2002a).

In the northern part of Mt. Ślęża, at 380 m asl and close to the minor saddle, a patch of sand and gravel occurs, 8 m thick and covered by flow till and slope deposits. It is interpreted as a remnant of a kame terrace (Szczepankiewicz 1958) that was formed during the decay of the Saalian ice sheet, when the slopes of Mt. Ślęża gradually emerged from beneath the ice.

The distribution pattern of glacial deposits on one hand and the occurrence of chemically-altered bedrock beneath local slope deposits on the other one led to the conclusion that the most recent ice-sheet advance in the region (Saalian glaciation) reached the altitude of 590–600 m asl (Szczepankiewicz 1969). The summit of Mt. Ślęża protruded above the ice surface and remained as a nunatak. Less elevated parts of the massif were covered by ice and may have been subject to glacial erosion during consecutive inland glaciations, which reached either some of the intramontane basins in the Sudetes (Elsterian) or the mountain front (Saalian). The evidence of glacial erosion is provided by gabbro and amphibolite clasts, derived from the Ślęża Massif, in moraine deposits at the front of the Sudetes, in the Sowie Mountains (Dathe 1904). The actual depth of glacial erosion is not known, though. It may be assumed that pre-Quaternary weathering mantles were affected most and their removal exposed solid bedrock. This 'rejuvenation' of morphology of lower slopes of Mt. Ślęża contributed to subsequent increased production of coarse angular debris that makes the extensive periglacial slope covers. The results of Schmidt hammer tests and weathering rind studies are consistent with this hypothesis. Gabbro outcrops in the near-summit parts are more weathered than those located below 450 m asl (Placek 2007) and show finely sculpted microrelief and reddish weathering rinds, which do not occur on boulders found at lower elevations.

12.6.2 Periglacial Legacy

A rich inventory of landforms produced in cold climate but non-glacial conditions is a distinctive feature of Mt. Ślęża, emphasizing its uniqueness among the remaining parts of the Ślęża Massif and other inselbergs of the Sudetic Foreland. It may be linked with the periglacial environment of the youngest phase of the Saalian glaciation (Wartanian) and the last cold period (Weichselian glaciation). Weathering and denudation processes acting in these periods did not erase the main morphological traits of the hill, but have given Mt. Ślęża a unique identity, more typical for the higher parts of the Sudetes rather than for the mountain foreland. This 'periglacial make-up' (Jahn 1980) includes the following landforms: crags, rock cliffs, blockfields built of debris of different size, debris-loamy and debris-silty cover deposits, block streams, and boulder lobes (Fig. 12.8). These landforms are not uniformly distributed within the Ślęża Massif. Whereas minor crags and fine-grained to small debris-dominated slope covers are ubiquitous, coarse-grained deposits, block streams and lobes, and high crags are restricted to the slopes of Mt. Ślęża, where gabbro and granite crop out.

12.6.2.1 Crags and Rock Cliffs

The largest crags have developed in gabbro. They occur both along the axial ridge of Mt. Ślęża (Fig. 12.9a, b) and, less commonly, within the slopes. Their morphology is typically complex, consisting of a series of steps and rock spurs, one above another. The cumulative height of these rock bastions may exceed 50 m, with the height of individual crags up to 8–12 m. Slope crags occur in isolation or in sequences of steps and benches (risers and treads), reminiscent of cryoplanation terraces and frost-riven cliffs, the latter up to 14 m high. In front of the cliffs angular blocky covers are present, with individual blocks up to 2–4 m long. Gabbro crags tend to form associations extended in N–S direction, apparently in relation to the strike of one of main joint sets recognized by Buczek (1952). In turn, joints belonging to the WNW–ESE set control the step-like morphology of rock cliffs and benches, perpendicular to the axis of Mt. Ślęża.

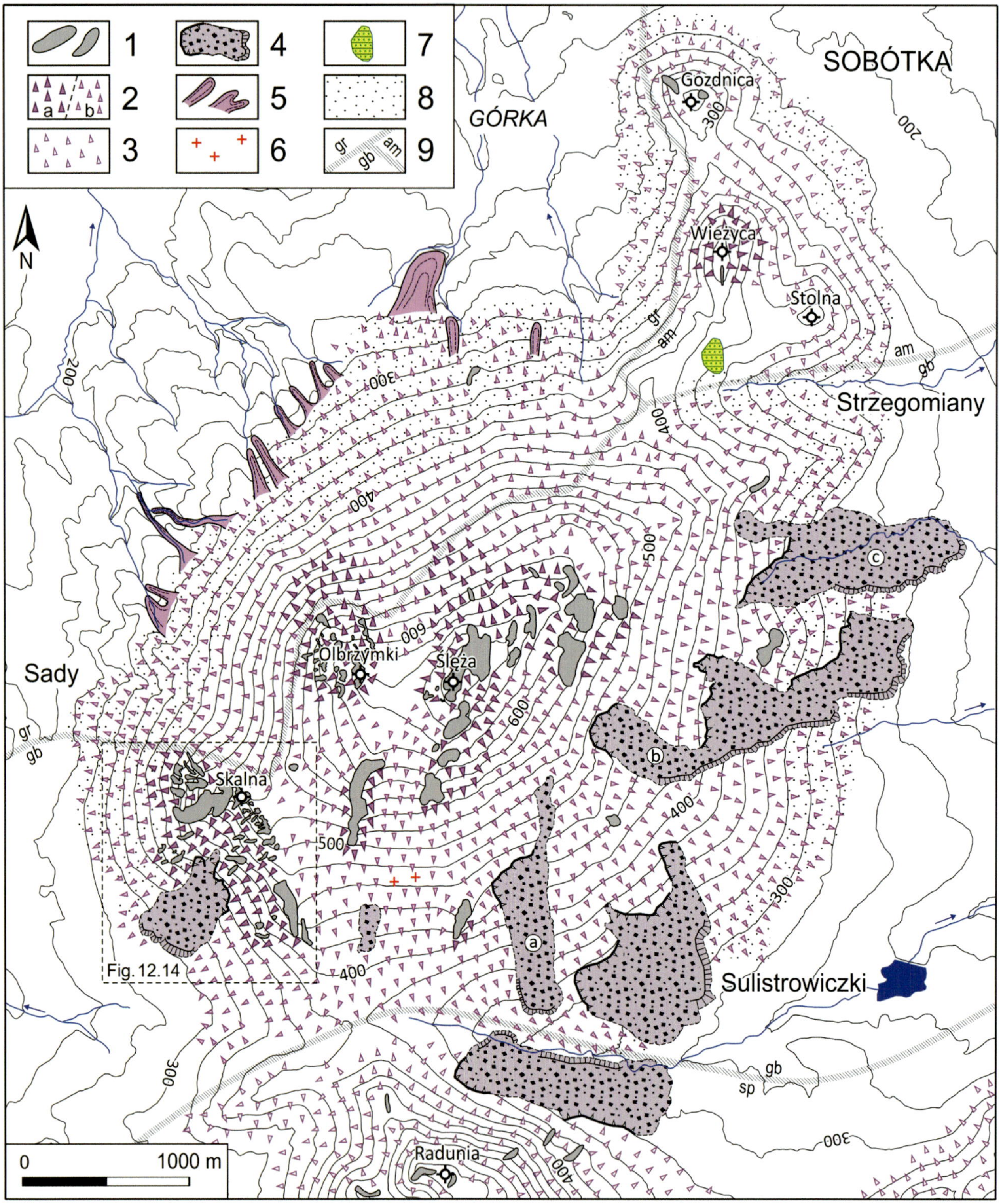

Fig. 12.8 Periglacial and glacial elements of Mt. Ślęża. 1—rock outcrops/crags, 2—coarse periglacial slope covers: **a** bouldery/blocky, **b** bouldery, 3—debris periglacial slope covers, 4—large colluvial tongues (with edges of slope hollows and frontal/lateral risers), 5—large congelifluction boulder-blocky tongues, 6—glacial erratic boulders, 7—remnant of kame (fluvioglacial) terrace, 8—loess deposits within the periglacial slope covers, 9—geological boundaries (gr—granite, gb—gabbro, am—amphibolite, sp—serpentinite); **a**, **b**, **c**—hillslope depositional forms (colluvial tongues) showed in Fig. 12.13

Only a few crags (tors) are present within the granite slopes. One low tor consisting of several tabular blocks occurs close to the contact with gabbro, below the bench of Olbrzymki, whereas more common are rounded granite boulders, some as long as 4 m. They are not rooted in bedrock, but are corestones exposed after loose regolith was eroded away (Żurawek and Migoń 2000). In the footslope parts, granite crops out within sharply incised valleys, forming rock walls 4–8 m high. Genetically, they are linked with fluvial downcutting and separation of less jointed and hence less weathered granite compartments. Above the valley incisions low whaleback elevations with loose granite boulders locally occur, but otherwise no crags are present in this part of the massif. However, a landform in its own is a 1.8 km long bedrock ridge built of a thick quartz vein, with bedrock outcrops more than 10 m high in places. The original morphology is not easy to reconstruct, though, due to protracted mining activity in numerous quarries located along the vein.

In the serpentinite part of the massif crags are restricted to ridge-top settings (Fig. 12.9c, d). A typical form is a 2–4 m high rock cliff, either around the hilltop or forming sequences of morphological steps within the axis of the ridge (Fig. 12.9d). They are most represented on Mt. Radunia and in the highest part of the Kiełczyńskie Hills.

12.6.2.2 Blockfields and Slope Deposits

Bedrock outcrops are associated with coarse slope deposits (Fig. 12.8), whose distribution and properties are conditioned by bedrock characteristics. On serpentinites, these are typically debris covers, with addition of small boulders (<1 m long) below the rock cliffs. Most are vegetated, unless slope angle is higher than 20–22°, facilitating surface wash and debris exposure. Slope covers on amphibolite slopes are similar, as can be inspected on the slopes of Mt. Wieżyca, where inclination up to 25° and slope instability also disfavour the establishment of dense undergrowth. Large clasts of both serpentinite and amphibolite were used in the prehistoric times to build stone rings around the tops of Mt. Radunia and Mt. Wieżyca. Their remains, 3–6 m wide at the base and up to 0.6 m high, can still be seen and are sites of geoarchaeological interest.

Serpentinite- and amphibolite-derived covers pale into insignificance if compared with widely distributed, visually most impressive covers built of gabbro boulders of different

Fig. 12.9 Crags in various lithologies in the Ślęża Massif. **a** Gabbro crag destroyed by periglacial weathering on the eastern slope of Mt Ślęża. **b** Olbrzymki crag gives way to a block cover, western slope of Mt Ślęża. **c** Asymmetric serpentine crag in the upper part of Mt. Radunia hillslope. **d** Degraded serpentine crags and blockfields next to the top of the Mt. Szczytna, Kiełczyńskie Hills (photographs a by A. Traczyk; b-d by P. Migoń)

Fig. 12.10 Blocky-boulder covers derived from weathering of gabbro. **a** Blockfield near Olbrzymki crag at the western hillslope of Mt. Ślęża. **b** Dense gabbro boulder cover ('pavement') in the lower part (c 390 m asl) of the northern slope of Mt. Ślęża (photographs a by P. Migoń; b by A. Traczyk)

dimensions, but generally much larger than the those of serpentinite or amphibolite (Fig. 12.10a). Their occurrence is not limited to gabbro slopes, but owing to gravity-driven displacements, they extend over granite bedrock in the north-western part of Mt. Ślęża, forming dense packs down to the altitude of 230 m asl and the distance of 1.5 km from the gabbro/granite boundary (Fig. 12.10b). However, in the most distal parts both gabbro and granite boulders may occur, the latter occasionally more than 3 m long and clearly distinguishable from the angular gabbro blocks by their rounded shapes.

On the eastern slopes of Mt. Ślęża boulders of gabbro are ubiquitous down to the altitude of 250–270 m asl, although below 400–450 m asl they tend to be rather dispersed, except the immediate vicinity of crags. Densely packed boulder covers, with boulders occupying more than 50% of the surface occur over 18% of the total slope area of Mt. Ślęża (Żurawek et al. 2005). They have an open-work structure, and the distribution pattern of individual boulders does not follow any particular order. Some are in upright position, indicating the role of up-freezing and the past presence of ground ice. In the latest Pleistocene, ground ice melted resulting in non-uniform boulder subsidence and displacements. Dispersed boulder covers are estimated to occur over 70% of the Mt. Ślęża area, and according to statistical analyses, the dominant factor controlling concentration of large boulders is slope angle. Therefore, an important role of selective slope wash as a process leading to the emergence of large rock elements was inferred (Żurawek et al. 2005). Most gabbro boulders are angular, and they show scarce evidence of surface weathering (Fig. 12.11). However, boulders in the summit parts, never covered by inland ice, and those derived from disintegration of crags are more rough, with small cavities and linear furrows on their surfaces, apparently reflecting longer time available for weathering processes to affect the rock.

Fine-grained cover beds, with small gravel-size particles, sand and silt, occur mainly in the least elevated parts of the Ślęża Massif. Only on gabbro slopes the amount of large debris and small boulders is more substantial. Along with periglacial blockfields in the upper slopes and widespread loamy-bouldery beds in the middle slopes, they form a continuum of cover deposits, largely inherited from periglacial conditions. The fine-grained deposits may be subdivided into several layers. The lower part is represented by sandy-silty loam with debris and occasional boulders, whereas the upper part tends to be more silty, with only small debris. An instructive outcrop below Mt. Wieżyca, at 320 m asl (Fig. 12.12a), shows that the lower layer rests upon weathered granite bedrock (1), whereas the constituent larger clasts are amphibolites, indicating downslope movement by solifluction (2). Deformation of the top of granite regolith, with slope-parallel banding grading into downslope bending below (3), provides further evidence of displacement. Notably, none of the layers contains any erratic admixture, which suggests that either glacial erosion rather than deposition was dominant during glaciation or any glacial materials were eroded away long before the currently observed slope cover beds were formed. Erratic materials and insets of glacial deposits can be found at even lower elevations, though. For example, in an outcrop at 215 m asl at the foot of Mt Gozdnica (Fig. 12.12b) in the northernmost part of the massif one can recognize loamy amphibolite debris with occasional clasts of Scandinavian origin in the bottom part (4), overlying jointed bedrock (5). Going upwards, this layer is replaced by silty loam with debris and small boulders, among them frequent erratic quartzites, porphyres and red granites, as well as insets of glacial sands and small debris towards the top (3). The near-surface layer is represented by sandy silts with admixture of small amphibolite debris (2). Thus, the lower and middle layers are interpreted as solifluction deposits, which incorporate elements of the former glacial cover, whereas the upper layer is a slopewash deposit from the waning phase of the last periglacial period.

12.6.2.3 Block Streams and Boulder Lobes

The most spectacular slope deposits in the Ślęża Massif have also clear geomorphological expression. These are block streams and large lobes and boulder tongues. The former occur mainly in the north-western part of Mt. Ślęża, where the undulating footslopes are dissected by fluvial valleys (Fig. 12.9). They are built predominantly of angular boulders of gabbro, although rounded granite boulders may occur too and have moved into the headwater parts of these valleys, bypassing minor elevations of granite bedrock. Typical dimensions are 300–500 m in length, 100–130 m in width, and 4–6 m in height. A few of these block streams continue within the narrow valleys, and these are longer, up to 700 m, but less wide, 30–75 m. Their toe convexities are 2–3 m high (Traczyk and Żurawek 1999). Gabbro-granite block streams represent periglacial, solifluction landforms, whose movement was facilitated by the presence of fine-grained material in the boulder-dominated slope covers in the upper slopes of Mt. Ślęża. Their distribution pattern, in turn, shows the role of pre-existing relief, with minor elevations, slope hollows, and valley incisions. These surface irregularities diverted or channelized the flow and allowed for movement acceleration in terrain concavities due to higher amount of water during melting seasons.

Huge, complex lobes and tongues within the south- and east-facing slopes of Mt. Ślęża form the second group, and these are also the most enigmatic. Depending on delimitation criteria, the number of lobes is from six to eight (Żurawek 1999a). Their relief is very diverse and includes ramparts, steps, closed hollows, semi-circular depressions, and scarps. The internal pattern of these second-order

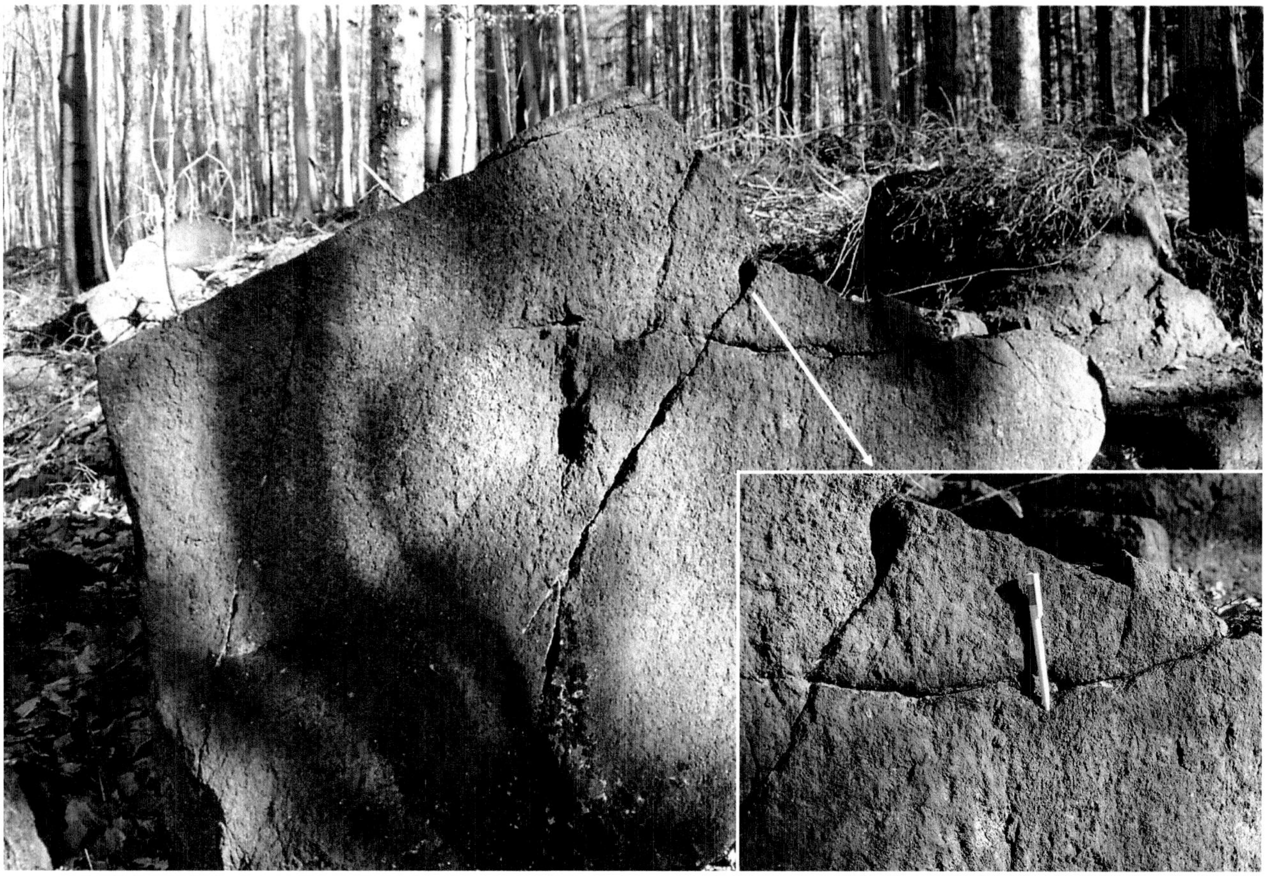

Fig. 12.11 An example of weathering of a gabbro block within the periglacial slope cover (south-western part of hillslope of Mt Ślęża) (photograph by A. Traczyk)

elements of topography can be fully appreciated only on terrain visualizations derived from high-resolution 1 m × 1 m DTM (Fig. 12.13). Table 12.3 provides basic morphometric data, derived from the same source.

The discussion about the origin of these landforms, whose onset may be traced back to Finckh (1928) and Schott (1931), is yet to be concluded. Various interpretations can be found in literature, e.g. as large congelifluction lobes (Szczepankiewicz 1989), periglacial debris-mud flows (Horwath 1981), or rock glaciers (Żurawek 1999a). Evidence in support of the latter view included morphological details and morphometric properties, investigated prior to the availability of LiDAR data (Borowicz and Żurawek 2003), and analysis of sedimentary and deformation structures (Żurawek 2002b, 2003). It was concluded that movement of the lobes involved deformation and creep of frozen ground, but ground ice was not abundant. An attempt was also made to date the lobes using OSL and radiocarbon dating methods (Żurawek 2001). Interpretation of OSL dates is not unequivocal, but the results show that these accumulation landforms were active in several consecutive cold intervals of the Pleistocene, since the Saalian glaciation. The upper temporal boundary of their activity is constrained by the age of organic deposits in the waterlogged closed hollows, determined as 10,270 ± 150 years BP. In comparison with other parts of the medium-altitude mountain ranges of Central Europe, the alleged rock glaciers of the Ślęża Massif are located at very low altitude and do not have an obvious source of material located further upslope, as they originate quite far below the belt of high crags and dense boulder covers.

However, the supply zone is evident above a complex depositional landform that can be recognized in the south-western slope of Mt. Ślęża, below the Skalna mid-slope bench and rock cliffs (Fig. 12.14). The rock slope, up to 50 m high in total and consisting of numerous individual cliffs and spurs, delimits an arcuate slope hollow, downslope of which an extensive boulder accumulation occurs. In its upper part, minor boulder tongues formed at the outlet of slope troughs, whereas the lower part, in the 300–400 m asl belt, is typified by a system of large lobes, steps, and longitudinal ramparts, locally with open-work boulder fields. Further below a slope-perpendicular linear trench occurs, with a small pond, enclosed from the south by narrow, 6–8 m high rampart. This landform complex was investigated using georadar (GPR) and electrical resistivity

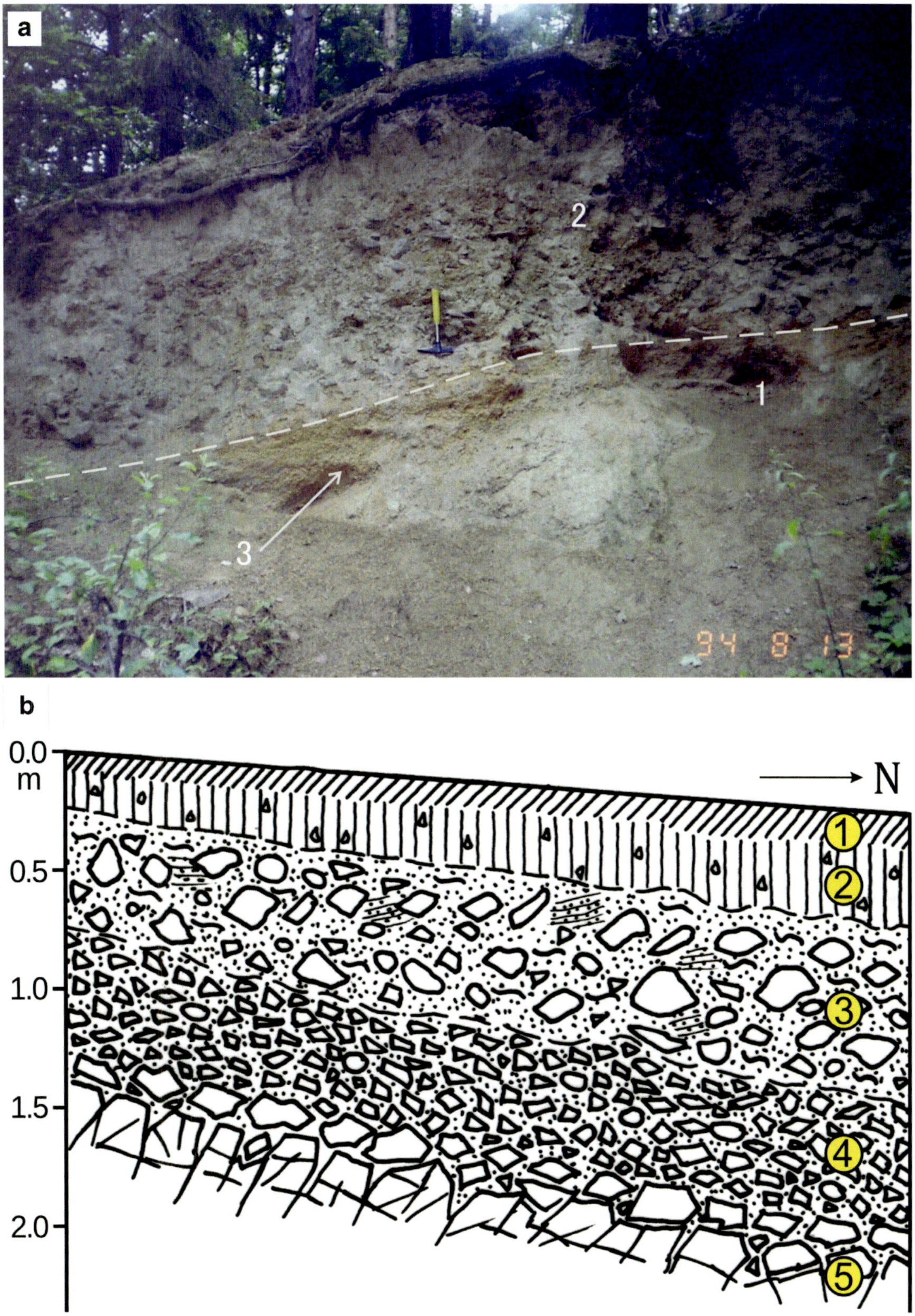

Fig. 12.12 Periglacial hillslope deposits on Mt. Ślęża. **a** Artificial outcrop of the fine-grained slope deposits at the foot of Mt Wieżyca. 1—weathered granitic bedrock; 2—solifluction cover; 3—solifluction deformations within the granite regolith (photograph by A. Traczyk). **b** Profile of colluvial (congelifluctional) and slopewash deposits on the northern footslope of Mt Gozdnica. 1—contemporary soil cover; 2—sandy silts with admixture debris; 3—silty loam with debris and small boulders; 4—loamy debris; 5—jointed amphibolite bedrock

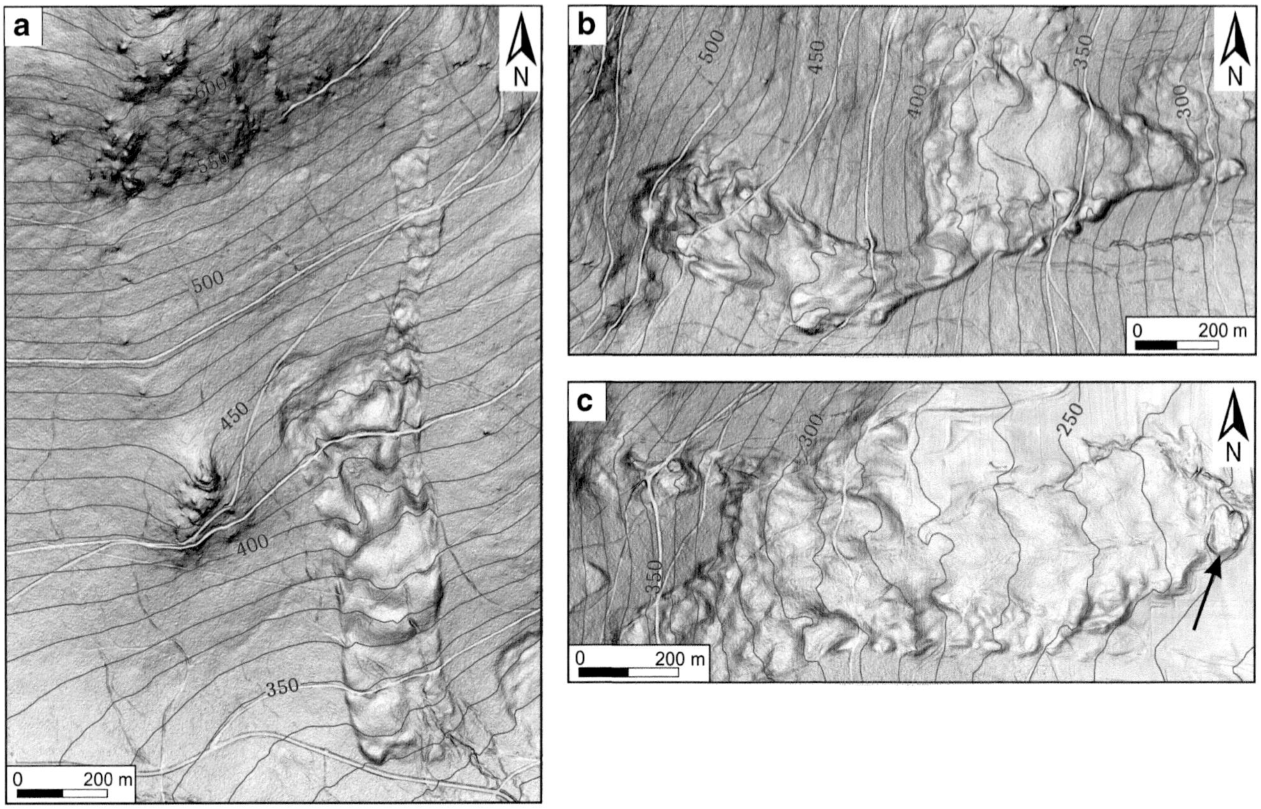

Fig. 12.13 Examples of large blocky tongues (shaded relief generated from 1 m × 1 m DEM) on the eastern slope of the Mt. Ślęża (for location see Fig. 12.8)

Table 12.3 Morphometry of large blocky tongues located on the eastern and southern slopes of the Mt. Ślęża

Statistical parameter	Length (m)	Width (m)	Area (ha)	Front height (m)	Height difference (m)
Min	577	240	17	5	75
Max	1300	695	53	12	157
Mean	905	398	27	8	106

tomography (ERT) geophysical methods. It was revealed that the thickness of regolith varies between the upper and lower part, from 4 to 6 m to locally 20 m, respectively. Moreover, the geometry of regolith/bedrock interface in downslope direction from the large arcuate hollow suggests rotational displacement (Traczyk and Kasprzak 2014).

The data from the Skalna 'rock glacier' allow one to re-evaluate earlier hypotheses regarding the origin of boulder lobes. Combined LiDAR DTM data, geophysical signatures and observations of deposits suggest sliding as the dominant type of downslope movement. Other landforms of this kind also have clear detachment scars in the upper parts and depositional bodies in the lower parts. It is therefore assumed that they formed in effect of mass movements which became active after the decay of Saalian ice sheet, in a paraglacial environment. Important factors facilitating large-scale mass movements included previous glacial buttressing of slopes and forced deposition of weathering-derived debris in the trimline zone, as well as erosional removal of pre-glacial regolith and exposure of new bedrock outcrops subject to efficient mechanical weathering.

12.6.2.4 Magnitude of Glacial and Periglacial Remodelling

The various data presented above allow one to assess the scale of cold-climate remodelling of the slopes of Mt. Ślęża quantitatively. Although the actual depth of glacial erosion remains unconstrained, periglacial reshaping in the presence of permafrost can be crudely assessed. Cold-climate degradation processes may have caused slope surface lowering by as much as 13 m, whereas slope retreat could have locally reached 50 m (Żurawek and Migoń 2000). At the same time, the lower slopes were built up by accumulation of hillslope materials, which are several metres thick over most of the slopes, but may reach 20 m in the case of large boulder lobes ('rock glaciers').

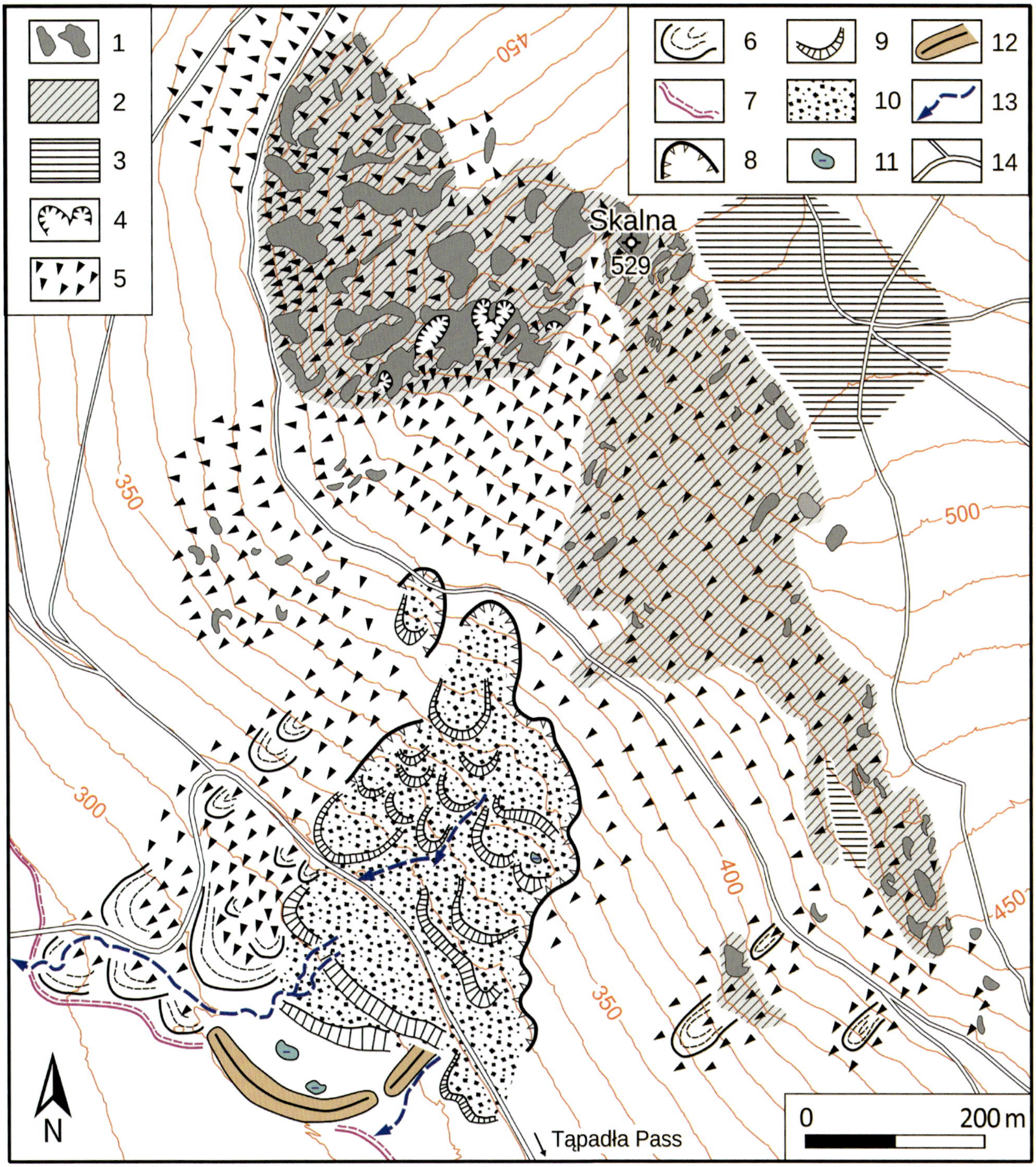

Fig. 12.14 Geomorphological sketch of Skalna area (for location see Fig. 12.8). 1—rock outcrops and crags, 2—hillslopes with low but numerous rock outcrops, 3—mid-slope flats (slope benches), 4—slope hollows, 5—coarse slope covers (blockfields), 6—congelifluction lobes and tongues, 7—lower limit of gabbro boulders, 8—landslide scars, 9—risers within the paraglacial (?) mass wasting bodies, 10—boulder-blocky loam (colluvial) material, 11—closed depressions, 12—slope ramparts of unknown origin (lateral moraine ridges?), 13—channels of periodical watercourses, 14—roads

12.7 Conclusions

The uniqueness of the Ślęża Massif in Poland is not only related to its distant geological history and ophiolitic structure, indeed rare. Most of all, it represents an outstanding landscape and geomorphological unit in the Sudetic Foreland and hence in the entire Silesia. Mt. Ślęża and the adjacent lower hills are examples of inselberg relief, whose main topographic features were shaped in the Palaeogene (possibly even earlier), as an outcome of long-term landform evolution dominated by protracted deep weathering under warm climate.

A distinctive feature of the Ślęża Massif, and Mt. Ślęża itself in particular, is a rich inventory of denudational landforms moulded principally in the periglacial environment of the late Saalian and Weichselian glaciations. Periglacial morphogenesis was longer in the summit part of Mt. Ślęża, since it was certainly a nunatak during the Saalian glaciation, when the ice sheet reached the mountain front of the Sudetes, some 20–30 km further to the south and west. Nowhere else in the Polish mountains one can find such a large number of bedrock outcrops, blockfields and huge boulder lobes formed on gabbro bedrock, all very impressive in terms of visual appearance.

The periglacial legacy of the Ślęża Massif is not specially marked in the field, but the majority of periglacial landforms is easily accessible via a dense network of waymarked trails, supplemented by forestry roads and paths. The Ślęża Landscape Park has set up several nature trails in the massif, but the focus is on biotic components of nature, and periglacial heritage is overlooked. However, recent setting up of a Sudetic Foreland Geopark (Geopark Przedgórze Sudeckie 2021), which includes the Ślęża Massif and aims to promote geoheritage of this part of south-western Poland (Szadkowska et al. 2022) creates an opportunity to redress the balance and bring the fascinating geomorphological history of Mt. Ślęża to the public.

References

Borowicz D, Żurawek R (2003) Topography of a composite relict rock glacier, Ślęża Massif, SW Poland. Geografiska Annaler Ser A 85:31–41

Buczek H (1952) Spękania skalne masywu Sobótki. Ann Soc Geol Pol 22(2):123–161

Closs H (1922) Tektonik und Magma: Untersuchungen zur Geologie der Tiefen 1. Abh der Preussischen Geologischen Landesanstalt 89

Conrad O, Bechtel B, Bock M, Dietrich H, Fischer E, Gerlitz L, Wehberg J, Wichmann V, Böhner J (2015) System for automated geoscientific analyses (SAGA) v. 2.1.4. Geosci Model Dev 8(1991–2007). https://doi.org/10.5194/gmd-8-1991-2015

Cwojdziński S (1977) Główne etapy ewolucji tektogenu waryscyjskiego Polski w świetle teorii tektoniki płyt. Przegl Geol 25:579–582

Dathe E (1904) Erläuterungen zur Geologischen Karte von Preußen und benachbarten Bundesstaaten Blatt Langenbielau. Königlich Preußischen Geologischen Landesanstalt und Bergakademie, Berlin

Finckh L (1928) Erläuterungen zur Geologischen Karte von Preußen und benachbarten deutschen Ländern Blatt Zobten. Preußischen Geologischen Landesanstalt, Berlin

Floyd PA, Kryza R, Crowley QG, Winchester JA, Abdel Wahed M (2002) Ślęża ophiolite: geochemical features and relationship to lower Palaeozoic rift magmatism in the Bohemian Massif. Geol Soc London Spec Publ 201:197–215

Gellert JF (1931) Geomorphologie des mittelschlesischen Inselberglandes. Z Dt Geol Gesell 83:431–447

Geopark Przedgórze Sudeckie. http://geopark.org.pl. Accessed 10 Jul 2021

Horwath BJ (1981) Jęzory gruzowe Ślęży na Dolnym Śląsku. Czas Geogr 52:447–452

Jahn A (1968) Morphological slope evolution by linear and surface degradation. Geogr Polonica 14:9–21

Jahn A (1969) Some problems concerning slope development in the Sudetes. Biul Perygl 22:331–348

Jahn A (1980) Główne cechy i wiek rzeźby Sudetów. Czas Geogr 60:129–153

Korta W (1988) Tajemnice Góry Ślęży. Śląski Inst Nauk, Katowice, p 398

Kryza R, Pin C (2010) The Central-Sudetic ophiolites (SW Poland): petrogenetic issues, geochronology and palaeotectonic implications. Gondwana Res 17:292–305

Krzemińska E, Awdankiewicz M (2011) Historia geologiczna aktywności wulkanicznej na obszarze Polski. Kosmos 60:261–275

Krzyszkowski D, Karanter M (2001a) Glacial sediments and palaeogeography of the northeastern foreland of the Mount Ślęża Massif, Southwestern Poland. In: Krzyszkowski D (ed) Late cainozoic stratigraphy and paleaeogeography of the sudetic Foreland. WIND, Wrocław, pp 109–145

Krzyszkowski D, Karanter M (2001b) Neogene and Pleistocene stratigraphy of the northeastern foreland of the Mount Ślęża Massif, Southwestern Poland, In: Krzyszkowski D (ed) Late cainozoic stratigraphy and paleaeogeography of the Sudetic Foreland. WIND J. Wojewoda, Wrocław, pp 91–108

Majerowicz A (1972) Masyw Granitowy Strzegom-Sobótka. Geol Sudetica 6:7–88

Majerowicz A (1984) Petrografia i geneza skał rodingitowych w serpentynitach ofiolitowego zespołu Ślęży. Geol Sudetica 18(2):109–127

Majerowicz A (1981) Rock series of the Ślęża Mt group in the light of petrologic studies of ophiolite complex. In: Ophiolites and initialites of northern border of the Bohemian Massif, Guidebook of excursion II, Potsdam–Freiberg, pp 172–199

Majerowicz A (2006) Krótki przewodnik terenowy po skałach ofiolitowego zespołu Ślęzy oraz ich petrologicznej i geologicznej historii. Acta Univ Wratisl 2830, Wrocław, p 62

Migoń P, Kasprzak M (2015) Regiony fizycznogeograficzne, In: Żelaźniewicz A (ed) Przyroda Dolnego Śląska. PAN Oddział we Wrocławiu, Wrocław, pp 19–36

Migoń P (1997) Crystalline rock inselbergs in southwestern Poland. Origin and palaeoenvironmental significance. Acta Univ Wratisl 1872 Studia Geogr 66, p 102

Migoń P (1999) Inherited landscapes of the Sudetic Foreland (SW Poland) and implications for reconstructing uplift and erosional histories of upland terrains in Central Europe. In: Smith BJ, Whalley WB, Warke PA (eds) Uplift, Erosion and stability: perspectives on Long-term Landscape Development. Geol Soc London, Spec Publ 162:93–107

Pawleta M (2017) Past in the present: Mount Ślęża according to Pierre Nora's "Sites of Memory" (*Lieux De Mémoire*) concept. Folia Praehist Posnan 22:157–179

Pernarowski L (1963a) Zagadnienie paleogeograficznej predyspozycji północnej krawędzi przedgórza Sudetów Środkowych. Acta Univ. Wratisl. 9 Studia Geogr 1:141–155

Pernarowski L (1963b) Morfogeneza północnej krawędzi Wzgórz Niemczańskich. Acta Univ Wratisl 10 Studia Geogr 2, p 146

Placek A (2007) Rola zróżnicowania wytrzymałości skał w rzeźbie Masywu Ślęży (Przedgórze Sudeckie). Przegl Geol 55:861–869

QGIS Development Team (2018) QGIS Geographic information system. Open Source Geospatial Foundation Project. https://qgis.org/

Sawicki L (1995) Mapa geologiczna regionu dolnośląskiego z przyległymi obszarami Czech i Niemiec (bez utworów czwartorzędowych) 1:100000. PIG, Warszawa

Schott C (1931) Die Blockmeere in den deutschen Mittelgebirgen. Forsch Dt Landes- Volkskunde 29:1–78

Solon J, Borzyszkowski J, Bidłasik M, Richling A, Badora K, Balon J, Brzezińska-Wójcik T, Chabudziński Ł, Dobrowolski R, Grzegorczyk I, Jodłowski M, Kistowski M, Kot R, Krąż P, Lechnio J, Macias A, Majchrowska A, Malinowska E, Migoń P, Myga-Piątek U, Nita J, Papińska E, Rodzik J, Strzyż M, Terpiłowski S, Ziaja W (2018) Physico-geographical mesoregions of Poland: verification and adjustment of boundaries on the basis of contemporary spatial data. Geogr Polon 91:143–170

Szadkowska K, Szadkowski M, Tarka R (2022) Inventory and assessment of the geoheritage of the Sudetic Foreland Geopark (South-Western Poland). Geoheritage 14:1–25

Szczepankiewicz S (1958) Peryglacjalny rozwój stoków Masywu Ślęży. Biul Perygl 6:81–92

Szczepankiewicz S (1969) Sediments and forms of the far extents of scnadinavian glaciations in SW Poland. Geogr Polon 17:149–159

Szczepankiewicz S (1989) Ziemie południowo-zachodniej Polski - morfogeneza i dzieje czwartorzędowe. Acta Univ Wratisl 1029 Studia Geogr 47:136

Szumilas F (1963) Nikiel, kobalt i chrom w serpentynitach okolic Sobótki na Dolnym Śląsku. Arch Miner 24(1):5–101

Traczyk A, Kasprzak M (2014) Plejstoceńska rzeźba stoków masywu Ślęży (Przedgórze Sudeckie)—reinterpretacja na podstawie danych LiDAR oraz pomiarów elektrooporowych (ERT). In: Molewski P, Jaśkiewicz W (eds) Krajobrazy młodoglacjalne ich morfogeneza, teraźniejszość, przyszłość, X Zjazd Geomorfologów Polskich Toruń 16–19 września 2014 r. Streszczenia Wydz Nauk o Ziemi Uniw Mikołaja Kopernika w Toruniu, Stow Geomorfologów Polskich, Toruń, pp 111–112

Traczyk A, Żurawek R (1999) Pleistozäne Schuttdecken und Schuttzungen im nordwestlichen Teil des Śleża-Massivs (Polen) und ihre Entstehung unter den Bedingungen eines Dauerfrostbodens. Peterm Geogr Mitt 143:131–141

Traczyk A (1996) Geneza i znaczenie stratygraficzne rytmicznie warstwowanych osadów stokowych w Sudetach. Acta Univ Wratisl 1808 Prace Inst Geogr UWr A8:93–103

Walczak W (1970) Obszar przedsudecki. PWN, Warszawa, p 416

Wrzesiński W (ed) (2006) Dolny Śląsk. Monografia historyczna. Wyd Uniw Wroc, Wrocław, p 916

Żelaźniewicz A (2015) Przeszłość geologiczna. In: Żelaźniewicz A (ed) Przyroda Dolnego Śląska, PAN Oddział Dolnośląski, Wrocław, pp 39–99

Żołnierz L (2007) Zbiorowiska trawiaste występujące na dolnośląskich serpentynitach—wybrane aspekty ekologii. Zesz Nauk Uniw Przyr we Wrocławiu 555, Rozprawy 247

Żurawek R (1999a) Reliktowe lodowce skalne—nowa interpretacja form akumulacji stokowej na wschodnich stokach Ślęży. Przegl Geogr 71:77–93

Żurawek R (1999b) Relict rock glaciers in the Central European Mid-Mountains. Biul Perygl 38:163–192

Żurawek R (2001) Problem wieku reliktowych lodowców skalnych w Masywie Ślęży w świetle datowań 14C i OSL oraz obserwacji geomorfologicznych. Przegl Geol 49:880–884

Żurawek R (2002a) Geneza głównych rysów rzeźby Masywu Ślęży. Przegl Geogr 74:41–59

Żurawek R (2002b) Internal structure of a relict rock glacier, Ślęża Massif, Southwest Poland. Permafrost Perigl Proc 13:29–42

Żurawek R (2003) The problem of the identification of relict rock glaciers on sedimentological evidence. Landf Anal 4:7–15

Żurawek R, Migoń P (2000) Peryglacjalna morfogeneza Ślęży w kontekście długotrwałej ewolucji rzeźby. Acta Geogr Lodz 76:133–150

Żurawek R, Żyszkowski E, Górecki A (2005) Topographic control of periglacial slope covers, Ślęza Massif, Southwest Poland: a statistical approach. Permafrost Perigl Proc 16:241–248

Andrzej Traczyk is Assistant Professor in the Institute of Geography and Regional Development at the University of Wrocław, where he deals with geomorphology, applications of geographic information systems, and geoarchaeology. He is the author/co-author of publications about glacial and periglacial geomorphology of SW Poland and Svalbard. The main directions of his research concern mountain glaciations (especially the Karkonosze Mts), Scandinavian ice-sheet transgression in the Sudetes (glacial erosion) and periglacial forms and deposits (slope covers, coarse debris accumulation). He also deals with contemporary slope processes (slope erosion, mass movements). He was the main contractor of the Digital Geomorphological Map of the Karkonosze Mountains. His publication record includes papers in *Geomorphology*, *Zeitschrift für Geomorphologie*, and *Journal of Maps*.

Mt. Babia Góra—The Highest Flysch Ridge in the Western Carpathians

13

Adam Łajczak

Abstract

Mt. Babia Góra (1725) is the highest homoclinal ridge in the flysch section of the Western Carpathians, notable for its W–E orientation and structural relief dominated by deep-seated landslides. The most elevated part of the ridge is built of the resistant Magura sandstone, which overlies less resistant older flysch formations. The relief of Mt. Babia Góra is indicative of multi-stage geomorphological development of the ridge from the Late Miocene to the Holocene. The study area has a zonal diversity of relief, with the ridge-top areas above 1400 m asl being the only ones in the Western Flysch Carpathians that penetrate into the subalpine zone. The element that dominates the relief of Mt. Babia Góra is the upper part of the main ridge, which rises above the belt of spring hollows. It is marked by asymmetry in the distribution of landforms, which are different on the precipitous northern slope and the gentle southern slope (penestructural slope). The upper part of the northern slope, where gradients exceed 40° and locally reach 70°, features elements of glacial and nival relief, remodelled landslide forms, periglacial covers, as well as couloirs, torrential fans, and debris flows, which have been preserved to varying degree. The varied relief of the upper part of the northern slope makes this area of Mt. Babia Góra similar to areas with high-mountain relief. The relief of Mt. Babia Góra, which has been only lightly transformed anthropogenically, is one of the elements of nature that were taken into account when the Babia Góra National Park was proposed to be listed as a UNESCO Biosphere Reserve.

Keywords

Homoclinal ridge · Flysch · Structural relief · Landslides · Glaciation · Periglacial

13.1 Introduction

When studying mountain ridges of the flysch section of the Western Carpathian Mountains, which do not rise up to the subalpine zone and are classified as areas with medium-altitude mountain relief, geomorphologists mainly focus on landslide morphology of the slopes, considering lithology and tectonics of the bedrock (Nemčok 1982; Ondrášik 2002; Ziętara 2004; Margielewski 2006; Jankowski and Margielewski 2014, 2021; Mrozek et al. 2014; Pánek et al. 2019; Břežný and Pánek 2017; Jankowski et al. 2019; Kłapyta 2020). There are two exceptions, however, namely two massifs which rise above the timberline—Mt. Pilsko (1557 m) and Mt. Babia Góra (1725 m) where, apart from landslides, attention is drawn to their periglacial, nival, and glacial forms (Łajczak 1998, 2015; Wójcik et al. 2010; Kłapyta 2020).

Mt. Babia Góra and the high-mountain massif of the Tatra Mountains are areas in the Polish part of the Western Carpathians whose topography has attracted the attention of explorers since at least the fifteenth century. This is evidenced by chronicles, which refer to Mt. Babia Góra as "*Mount Baba, a very high massif*" and "*a rocky, inaccessible mountain*", as well as by the fact that Mt. Babia Góra was plotted on sixteenth–eighteenth century maps with a symbol of the same size as that used for the Tatra Mountains (Łajczak 2016). Research on the relief of Mt. Babia Góra began in the end of the nineteenth century and ever since has focused on the description of landslides and on attempts to define the range of glacial and nival forms (Zapałowicz 1880; Rehman 1895; Sawicki 1913; Alexandrowicz 1978,

A. Łajczak (✉)
Association of Polish Geomorphologists,
Bogumił Krygowski 10, 61-680 Poznań, Poland
e-mail: alajczak@o2.pl

P. Migoń and K. Jancewicz (eds.), *Landscapes and Landforms of Poland*,
World Geomorphological Landscapes, https://doi.org/10.1007/978-3-031-45762-3_13

2004; Łajczak 1998, 2014, 2016; Wójcik et al. 2010; Kłapyta 2020; Jankowski and Margielewski 2021). The results of geological surveys in the area of Mt. Babia Góra, which have been under way since the mid-twentieth century (Książkiewicz 1971, 1983), have been useful in analyzing the structural relief of this massif (Łajczak 2013, 2014). In recent years, LiDAR data have been used to assess the location of landforms associated with gravitational displacements of the Mt. Babia Góra ridge, that is deep-seated landslides (Łajczak 2014; Łajczak et al. 2014; Kłapyta 2020). Active debris flows and fossil debris flows' tracks indicate the dynamics of slopes in the Holocene (Matyja 2006; Łajczak and Migoń 2007). There were also studies of cryoplanation terraces and residual crags (Ziętara 1989, 2004), couloirs (Łajczak and Spyt 2016), and the former location of perennial snow patches formed during the Little Ice Age was established (Łajczak 1998, 2016). Much attention has been paid to the extent of periglacial cover deposits (Klimaszewski 1948, 1952; Starkel 1960; Niemirowski 1963; Książkiewicz 1971). Search for stone polygons and permafrost has also been carried out in the vicinity of the summit (Jahn 1958; Dobiński et al. 2016).

To date, studies of the relief of Mt. Babia Góra have been carried out with a degree of selectivity. Some landform types, even those extending over large areas, have been insufficiently surveyed or have been neglected. The purpose of this chapter is to fill this gap and briefly present the general characteristics of the relief of Mt. Babia Góra, taking into account morphometric analysis and vertical zonality, structural pre-conditioning of relief, description of the landforms, development of the relief in the Neogene and Quaternary, and asymmetry of the relief of the massif. The chapter is based on the author's long-lasting research and on an overview of the literature. The relief of Mt. Babia Góra massif has been assessed on the basis of qualitative and quantitative methods, using LiDAR data.

13.2 Study Area

13.2.1 Location

Mt. Babia Góra (1725 m) is the highest mountain massif of the Outer (Flysch) Western Carpathians. It straddles the Polish–Slovak border in the Western Beskidy Mountains (Fig. 13.1) (Starkel 1972; Kondracki 1998), along the main European watershed separating the Baltic Sea and the Black Sea drainage basins, where it forms the headwater area for the Skawica and Polhoranka rivers (belonging to the Vistula and Danube basins, respectively). The 11.3-km-long ridge of Mt. Babia Góra runs along a W–E line and is limited by passes at altitudes of 993 m and 1017 m asl (Fig. 13.1c). The N–S profile of the ridge is asymmetrical, with its northern slope being steeper than the southern one. The massif culminates with two peaks—Diablak or "Devil's Peak" (1725 m) and Cyl (1517 m), separated by a pass at an altitude of 1409 m. The main ridge is joined by lower side ridges that are separated by valleys with a converging pattern in the northern part of Mt. Babia Góra, but with a divergent pattern in the southern part. The surface area of Mt. Babia Góra massif is 75 km^2 (N slope—28 km^2, S slope—47 km^2). Sixty two % of the Mt. Babia Góra massif, including the entire northern slope, lies in Poland, while the rest lies in Slovakia. In total, 40% of Mt. Babia Góra area is protected by the Babia Góra National Park (Fig. 13.1c).

Mt. Babia Góra has different shapes depending on the direction of observation (Fig. 13.2) and rises 200–600 m higher than the neighbouring mountain ridges. Altitude differences within Mt. Babia Góra reach 1095 m on the northern side of the ridge, and 975 m on the southern side. In Poland, greater elevation differences are only found in the Tatra Mountains. The valleys at the foot of the massif are from several dozen metres to over 300 m deep.

13.2.2 Geological Structure

The ridge of Mt. Babia Góra is built of Upper Cretaceous and Palaeogene sandstones and shales of the Carpathian Flysch Belt, which form part of the Magura Nappe. These rock formations were folded and uplifted during the Alpine orogeny in the Miocene (Książkiewicz 1983). The lower parts of the slopes and the footslopes of Mt. Babia Góra are built of sandstones, mudstones, and marly shales, which originated in the Middle Eocene on the northern side of the ridge, and in the Upper Cretaceous, Palaeocene, and the Lower and Middle Eocene on the southern side. On these flysch formations, above 900–1000 m asl, thickly bedded, more resistant Magura sandstones, also known as "ridge-forming" sandstones, occur. These are of Middle and Upper Eocene ages and are over 1000 m thick. The sandstones also occur as small patches on lateral ridges in the northern part of Mt. Babia Góra, below 800 m asl (Książkiewicz 1971, 1983; Geological Atlas 1989; Alexandrowicz 2004). The Magura sandstones make up half of the area of Mt. Babia Góra and occupy larger areas of the neighbouring mountain ridges (Fig. 13.3a). They include medium- and fine-grained sandstones with silica and clay cement and coarse-grained sandstones bound by ferruginous minerals. They have mudstone, clay, and marly shale interbeddings.

The flysch complexes that build Mt. Babia Góra are formed as folds that have dislocated alongside faults oriented NW–SE and locally also N–S (Fig. 13.3a, b). The axes of the folds below 900–1000 m asl run SW–NE and those of the Magura sandstones, which lie above, are trending W–E (Książkiewicz 1971, 1983). The axis of the

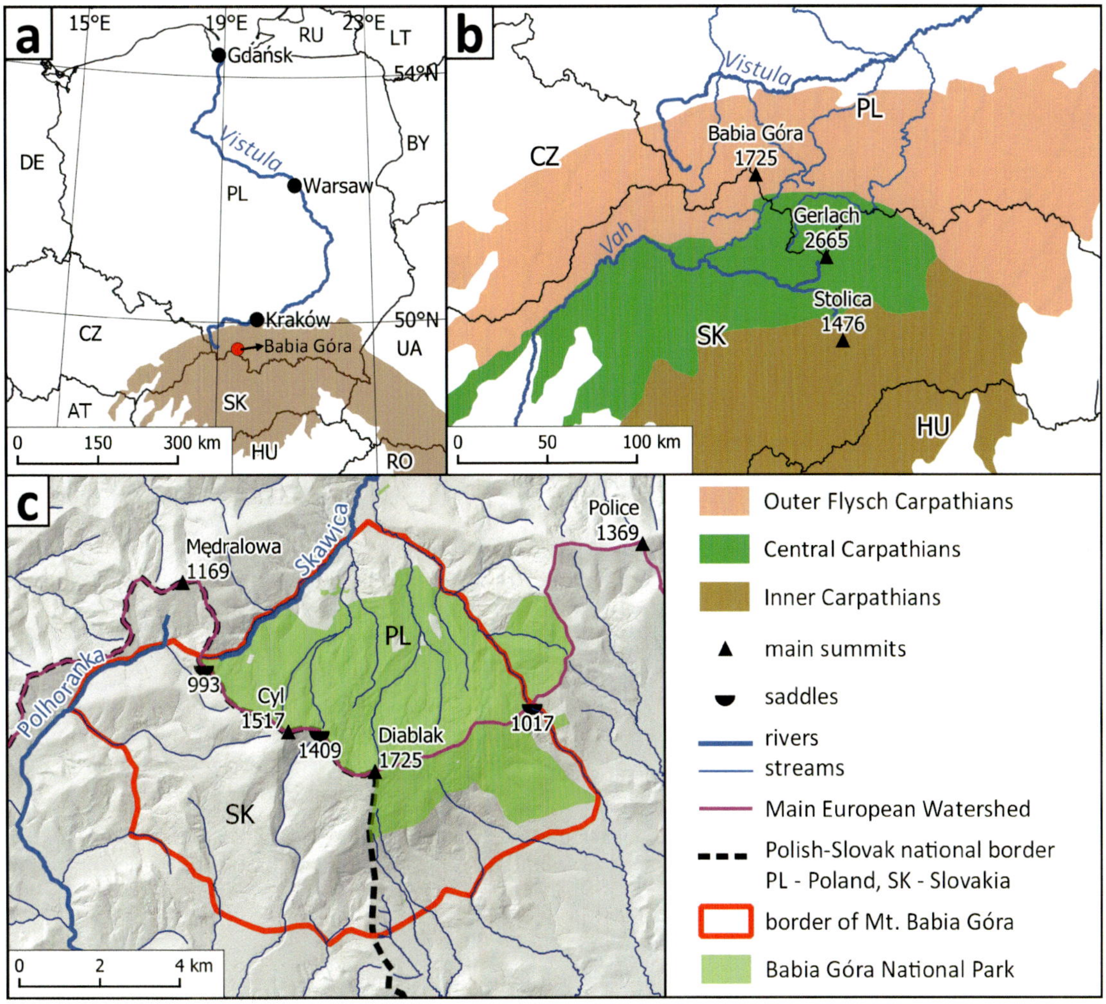

Fig. 13.1 Location of Mt. Babia Góra. **a** In Poland. **b** In the Western Carpathians. **c** Extent of Mt. Babia Góra massif

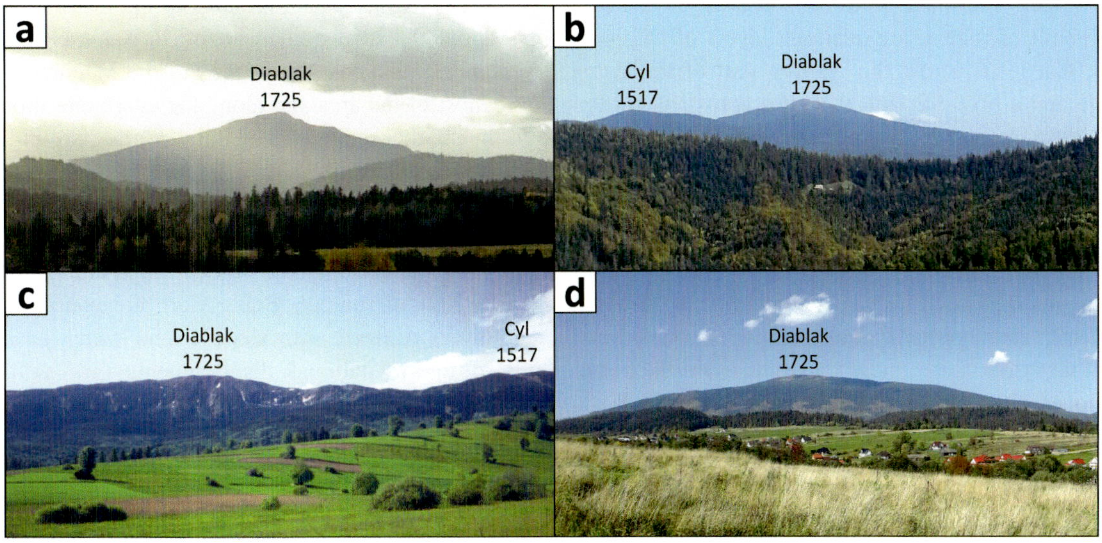

Fig. 13.2 View of Mt. Babia Góra from the east (**a**); west (**b**); north (**c**); south (**d**). From the north and west, the second peak of the massif is also visible (photographs by A. Łajczak)

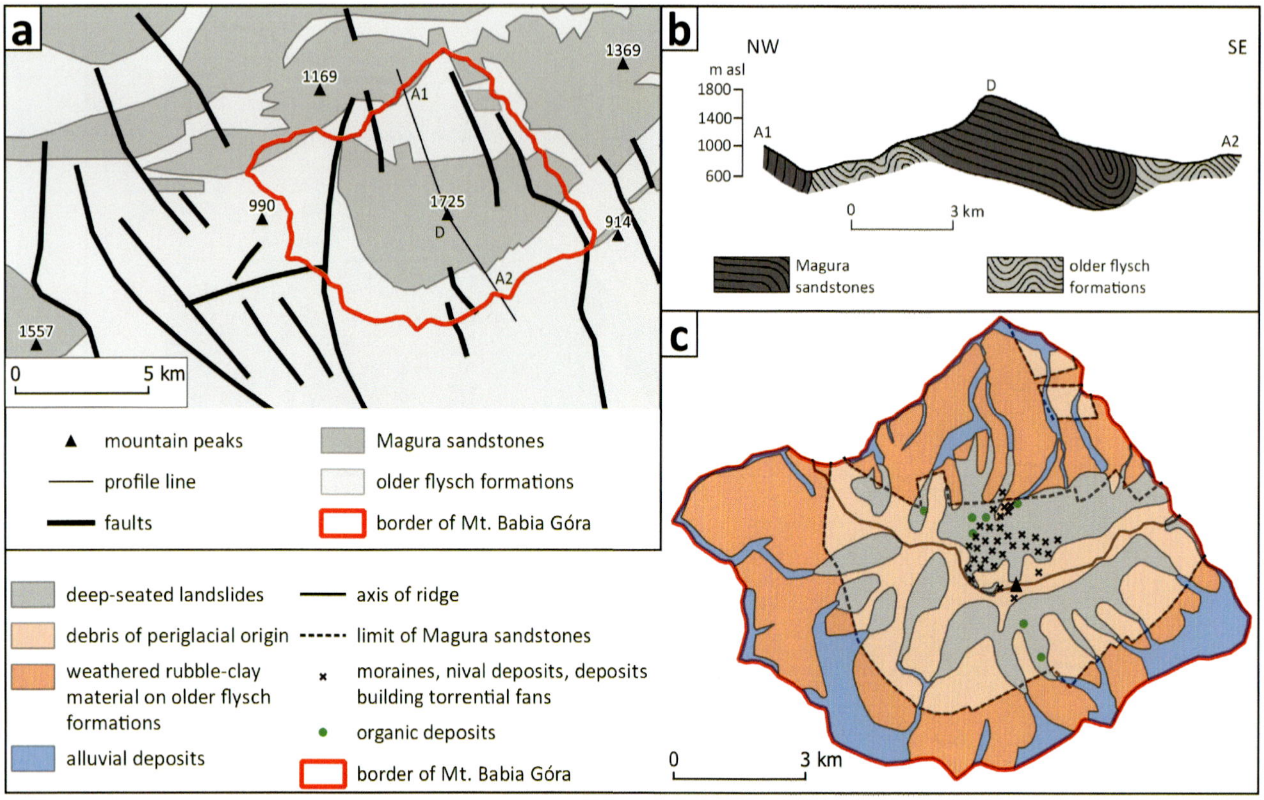

Fig. 13.3 Geological structure of Mt. Babia Góra and its surroundings. **a** Extent of the Magura sandstones and older flysch formations, alignment of faults. **b** Geological profile across the Mt. Babia Góra massif (profile location—see panel a). **c** Extent of slope covers of various origins and of alluvia

syncline within the Magura sandstones runs in the lower part of the southern slope of Mt. Babia Góra; therefore, the more elevated part of the ridge is a cuesta with a steep northern slope and a gentler southern slope, the gradient of which (20°) corresponds to the dip of the rock layers (pene-structural slope). The lower sections of the slopes and side ridges are built of folded, less resistant layers of flysch. A N–S fault that leads across the western part of the Babia Góra ridge and a NW–SE fault that runs across its eastern section delimit the central part of the massif (Fig. 13.3a) that is more uplifted than the nearby mountain ridges. Significant uplift of this vast part of Mt. Babia Góra massif built of Magura sandstones and underlying older flysch formations was initiated by dislocation of large masses of the Magura sandstone in the Middle Miocene towards the southern and northern footslopes (Jankowski and Margielewski 2021). Later, these rock masses on the northern footslopes were transported by large-scale denudation processes, causing exhumation of older flysch formations in this area. As a result, the block of Magura sandstones which forms the upper part of the ridge is surrounded less resistant older flysch formations. The uncovering of these flysch deposits from under the Magura sandstone has led to accelerated headward erosion of the headwater sections of the valleys, which has given rise to a gradual retreat of the slopes of Mt. Babia Góra (Łajczak 2014). In the area located on the western side of Mt. Babia Góra massif, numerous faults striking NW–SE (Fig. 13.3a) (Geological Atlas 1989; Golonka et al. 2009) occur, focusing deep fluvial valleys which reach the axis of the massif main ridge. The mountain ridges adjacent to Mt. Babia Góra from the west, north, and east are formed of the Magura sandstones over most of the area and along the entire elevation profile. On the southern side of Mt. Babia Góra, only older flysch formations are exposed, which are locally overlain by the Magura sandstone packets which have slid down from the southern slope of the ridge (Książkiewicz 1983; Geological Atlas 1989; Jankowski and Margielewski 2021).

The half of the Mt. Babia Góra area built of Magura sandstones is lined with slope covers which have origins related to gravitational displacements (gravitational tectonics) (Fig. 13.3c). These have led to the development of joint fractures in the Magura sandstones reaching down to a depth of at least several dozen metres, with a prevalence of SW–NE and NW–SE directions (Alexandrowicz 1978, 2004; Łajczak 2014). This in turn facilitates block disintegration of the ridge, which includes the formation of deep-seated landslides, the initial stages of which are ridge-top and slope trenches and high rock escarpments along the above directions. Block and debris colluvia with a visible thickness up to 30 m, which form lobes moving to the upper reaches of the valleys, are sliding down the slopes. Over the rest of the ridge, the higher parts of the slopes are covered by debris of periglacial origin, while the slopes made of older flysch formations are covered by weathered rubble-clay material (Klimaszewski 1948, 1952; Niemirowski 1963; Łajczak 2016). In the highest part of Mt. Babia Góra, mainly on the northern slope, one can find moraines and nival sediments as well as debris material of torrential fans. Boulder- and gravel-rich alluvia occur in the bottoms of the valleys. Topographic depressions within landslide bodies are filled with organic sediments (Książkiewicz 1983; Łajczak 1998, 2016, 2018; Alexandrowicz 2004; Obidowicz 2004; Kłapyta 2020).

13.2.3 Morphometry of Mt. Babia Góra

Mt. Babia Góra clearly towers over its surroundings (Fig. 13.4a). The mean height (Hmean) of the massif is 1020 m asl and the median height (Hmed) is 960 m asl. In total, 3.6% of the area of Mt. Babia Góra is situated above

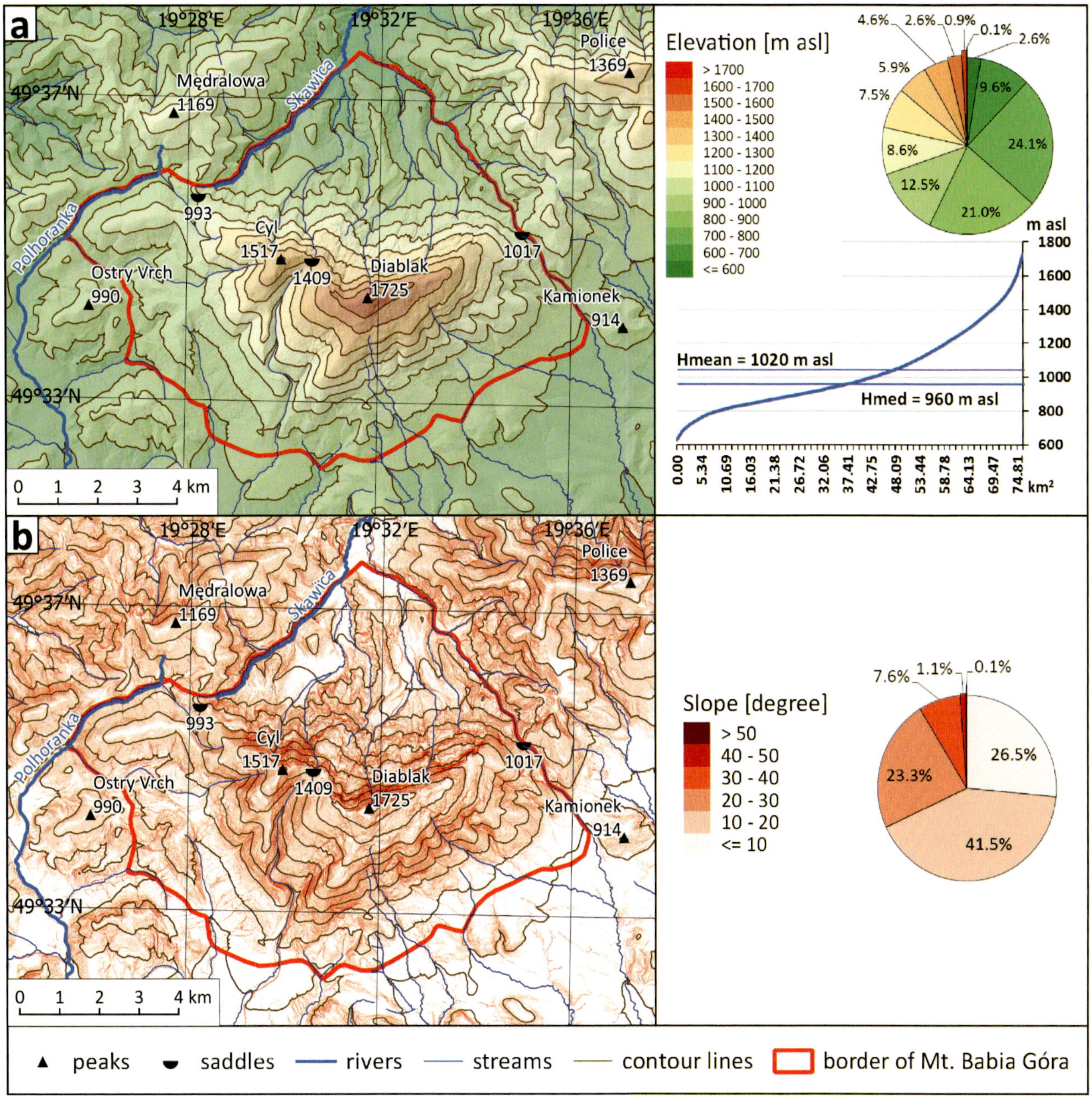

Fig. 13.4 Morphometry of Mt. Babia Góra massif. **a** Hypsometry. **b** Slope

1500 m asl and 42.7% lies above 1000 m asl. It is only the upper part of the northern slope, some valley heads, and upper sections of valleys that have slope gradients greater than 30° (Fig. 13.4b). Areas with gradients > 40° only occur in the precipitous areas within the northern slope. Rock faces in the uppermost parts of the northern slope have gradients > 50° (locally up to 70°). In the upper part of the southern slope, gradients do not exceed 20° and locally are even less than 10°. At the base of "the great escarpment of Mt. Babia Góra", slopes are in the range of 10–30°. Despite the W–E orientation of the main ridge of Mt. Babia Góra, the cumulative area of slopes with N, E, S, and W aspects

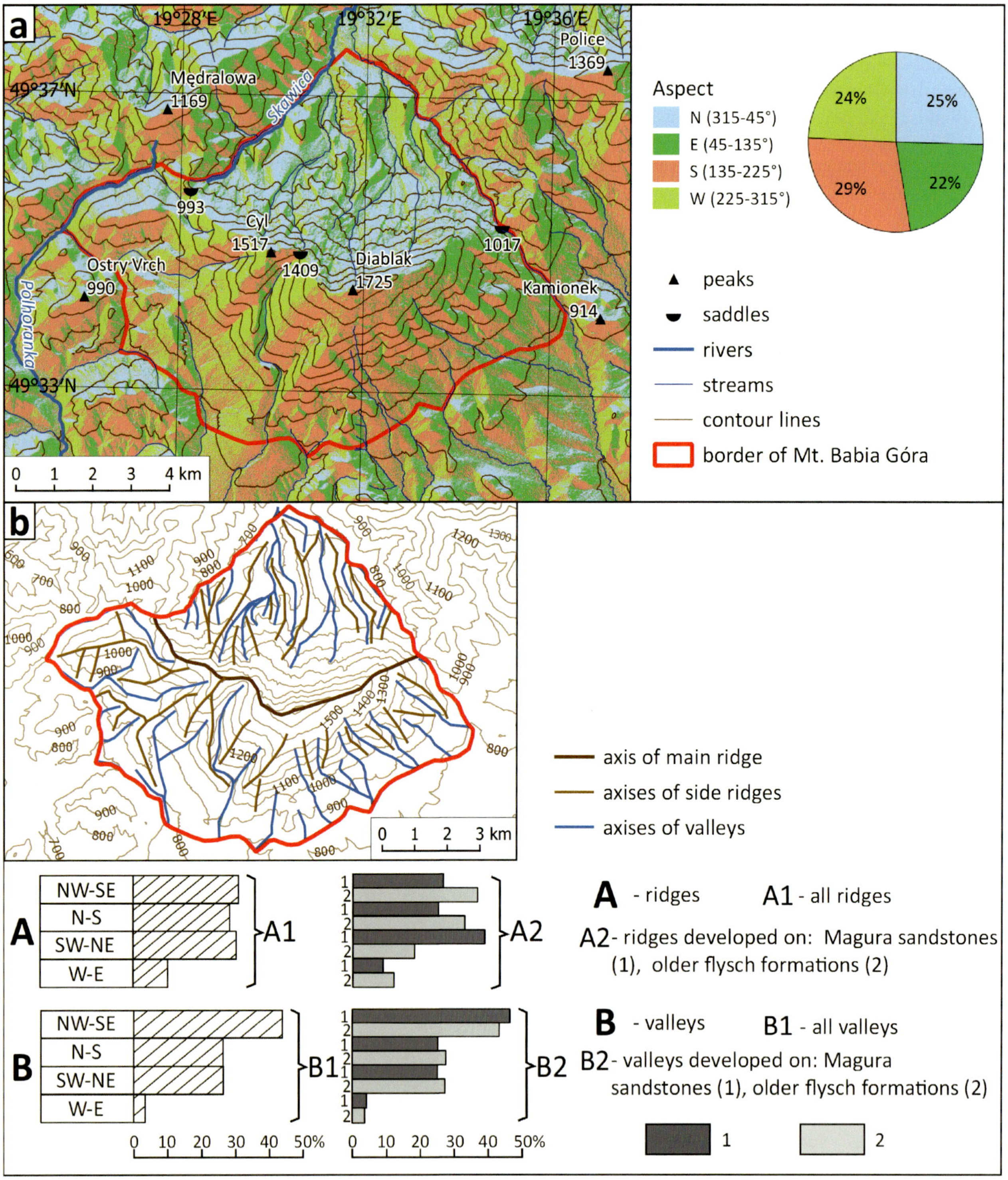

Fig. 13.5 Morphometry of Mt. Babia Góra massif. **a** Aspect. **b** Orientation of ridges and valleys

is similar and ranges between 22 and 29% of the entire area (Fig. 13.5a). In the upper part of the less fragmented northern slope, northern aspect prevails, while in the lower part of this slope, large areas have eastern or western aspects. On the southern slope, regardless of the depth of the headwater hollows and valleys, in addition to areas where the southern aspect predominates, western and eastern aspects are present over large areas.

In the axis of the main ridge and side ridges of Mt. Babia Góra, 1-km-long sections of NW–SE and SW–NE directions predominate (Fig. 13.5b). On bedrock built of the Magura sandstones, the domination of ridge sections of SW–NE direction is even larger. This may be explained by the correlation of orientation of fold axes within the older flysch formations occurring at depth with directions of faults in the overlying Magura sandstones (Książkiewicz 1983; Łajczak 2014). NW–SE directions predominate in valley courses in lower parts of slopes and within footslopes.

13.2.4 Relief-Forming Factors

Modelling of Mt. Babia Góra massif over a protracted timescale mainly depends on lithology and tectonics of the bedrock. This resulted in an asymmetric profile of the main ridge which towers over the surroundings, and also in shallow valleys in the northern, eastern, and southern footslopes. Because of the tectonics affecting the western slope of Mt. Babia Góra and its footslope, the valleys in this area are deeper (Fig. 13.3a, b). High energy of the ridge (1000 m of relief, gradients in the upper part of the northern slope>45°) is conducive to the development of extensive mass movements (Alexandrowicz 1978, 2004), the morphological effect of which is the occurrence of extensive deep-seated landslides (Łajczak 2014; Kłapyta 2020; Jankowski and Margielewski 2021), still activated by seismic shocks (Łajczak 2016). Large inclinations of the slopes foster far-reaching penetration of colluvia into the upper reaches of the valleys (Fig. 13.3c).

Changes in the relief of Mt. Babia Góra are also driven by hydro-climatic factors, the intensity of which is similar to that in high-mountain areas. Precipitation on the northern slope of Mt. Babia Góra reaches 1500 mm, while that on the southern slope is 1200 mm, with the maximum recorded in summer. The highest daily precipitation totals exceed 200 mm (Obrębska-Starkel 2004). In the upper parts of the slopes, especially on the northern side where snow cover can be up to several metres thick, avalanches occur affecting the timberline (Łajczak and Spyt 2016, 2018). The average runoff on the northern slope reaches 35 $dm^3\ s^{-1}\ km^{-2}$, while that on the southern slope is 25 $dm^3\ s^{-1}\ km^{-2}$. Meanwhile, the maximum runoff on the northern slope reaches 2500 $dm^3\ s^{-1}\ km^{-2}$ (Łajczak 2016). Under such conditions, slopes that feature rock debris covers are affected by debris flows, shallow landslides, as well as piping and leaching of slope covers and deeper, jointed bedrock which drives chemical denudation (Matyja 2006; Łajczak and Migoń 2007; Łajczak 2012; Łajczak et al. 2014). At times of sudden rainfall, when runoff exceeds 1000 $dm^3\ s^{-1}\ km^{-2}$ and reaches 2500 $dm^3\ s^{-1}\ km^{-2}$, colluvia are transported from the upper reaches of the valleys, and bedrock channels with small waterfalls are formed further downstream. The hollows on landslide slopes trap the silty-clayey material supplied by suffosion processes within the debris covers. Organic deposits may occur together with these fine-grained materials, and if analyzed jointly, they prove useful in reconstructing the development of the slopes in the Holocene (Łajczak 2018).

During the cold Pleistocene periods, the conditions in the highest part of the ridge were conducive to the development of cryoplanation terraces and residual bedrock outcrops—tors and crags, as well as couloirs (Ziętara 1989, 2004). The morphological and climatic conditions which prevailed at that time led to local glaciation of the upper northern slope, while during the Little Ice Age, they gave rise to the development of perennial snow patches (Łajczak 1998, 2016). Nowadays, above 1650 m asl, the mean annual temperature is below 0 °C (Obrębska-Starkel 2004). Accordingly, a few periglacial structures were found in this area (Jahn 1958; Wójcik et al. 2010).

13.3 Relief of Mt. Babia Góra

13.3.1 Main Elements of Relief and Their Age

Six areas at different altitudes were identified whose relief features were formed at different times and which therefore represent the successive stages of the modelling of the ridge between the Late Miocene and the Late Quaternary (Łajczak 2014). These are, from the summit of the massif to the foothills (Fig. 13.6): A—the highest-elevated area of the southern slope with slight inclination and elevation differences of up to 300 m; B—the precipitous upper part of the northern slope with height differences of up to 400 m ("the great escarpment of Mt. Babia Góra", its central part shaped as a huge amphitheatre); C—more gently sloping foot of this escarpment with height differences of up to 300 m; D—a zone with spring-fed valley heads in the headwater areas on the northern and southern slopes with height differences of up to 350 m; E—lower parts of the slopes situated on the Magura sandstones and locally on older flysch formations. The depth of upper reaches of the valleys on the

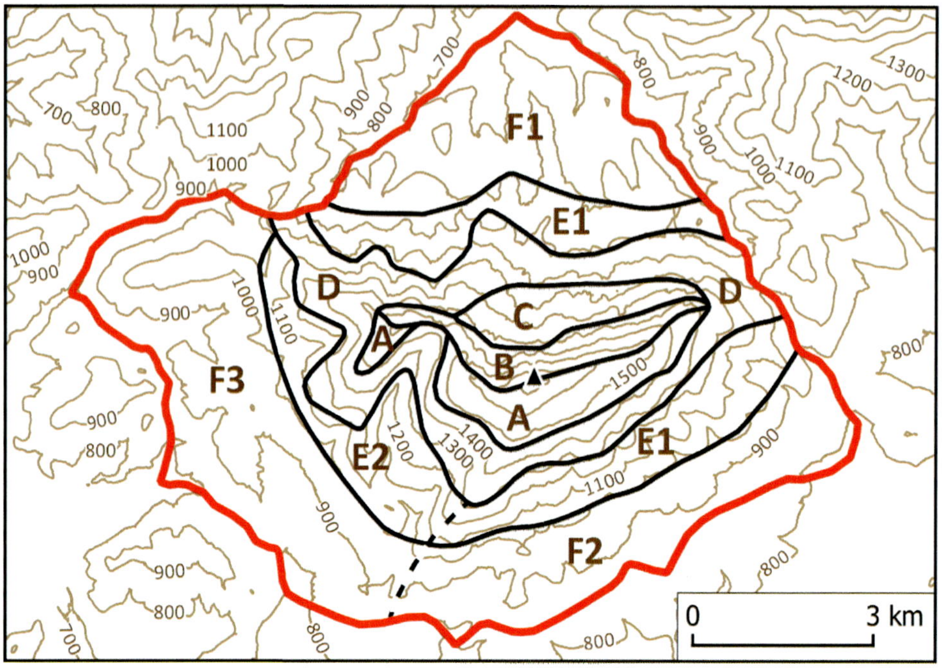

Fig. 13.6 Zonation of relief of Mt. Babia Góra massif. For explanation A–F see the text

northern slope and in the eastern part of the southern slope (E1) does not exceed 150 m, while in the western part of the southern slope (E2), it is greater than 300 m; F—footslope areas of the northern and southern slopes, developed mainly on older flysch formations with valleys of different depths. The depths of the valleys within the northern footslope (F1) reach 100–150 m, do not exceed 50 m in the southeast (F2), and are up to 250 m in the southwest.

The age of these elements of Mt. Babia Góra relief was determined based on the concept of cyclic uplift and tectonic stabilization which produced denudation levels in the Carpathians (see Starkel 1969, 1972, 1988). Recently, however, a new view has appeared (Jankowski and Margielewski 2021) emphasizing foot-wall elevation mechanism, which indicates a permanent uplift of the Flysch Carpathians and does not favour any denudational levels. Nevertheless, the concept by Starkel is followed below since it makes possible to explain the step-wise development of the main features of Mt. Babia Góra relief.

Being a penestructural slope supported by resistant layers of the Magura sandstones, the area A can be considered a relic of the oldest morphological level (Late Miocene). The main features of valley relief in area F were formed in the Late Pliocene, while the valleys and spring hollows in areas E and D were formed at the turn of the Early Quaternary. The key features in areas B and C are considered the youngest and their ages have been attributed to the Early Quaternary; the relief of these areas was locally modified during the Late Quaternary (during the last glaciation). At that time, as a result of the formation of deep-seated landslides, the relief of areas A, B, C, D, and as a result of valley deepening in the Holocene—also that in areas E and F, changed, but without significant large-scale modifications. Generally, the main features of the relief of Mt. Babia Góra were formed in the Neogene and Early Quaternary. Landform changes in the Holocene can be considered negligible. Notably, the most elevated areas that neighbour each other, namely area A (southern slope) and area B (northern slope), have had their key relief features formed at extremely different ages.

13.3.2 Landforms

13.3.2.1 Relief Features of the Main Ridge

The geomorphological landscape of Mt. Babia Góra is dominated by the part of the ridge rising above the valley heads. It represents a homoclinal ridge with an asymmetric N–S profile (Fig. 13.7a, b), comprising areas A, B, and C, which vary in age (Fig. 13.6) and whose present-day relief was formed in the Late Quaternary. The longitudinal section of the ridge axis comprises two summits and 21 structural flats formed at the contact of the thick-bedded Magura sandstones with silica cement and marly slate interbeddings. As a result, the ridge axis has a characteristic step-like profile (Fig. 13.7c), not found on such a scale elsewhere in the flysch Carpathians. The width of the structural flats increases towards the peak up to 150 m (Fig. 13.7d), where periglacial transformation of these forms is recognizable. The following are present in succession below the structural flat surfaces of the steep northern slope (area B):

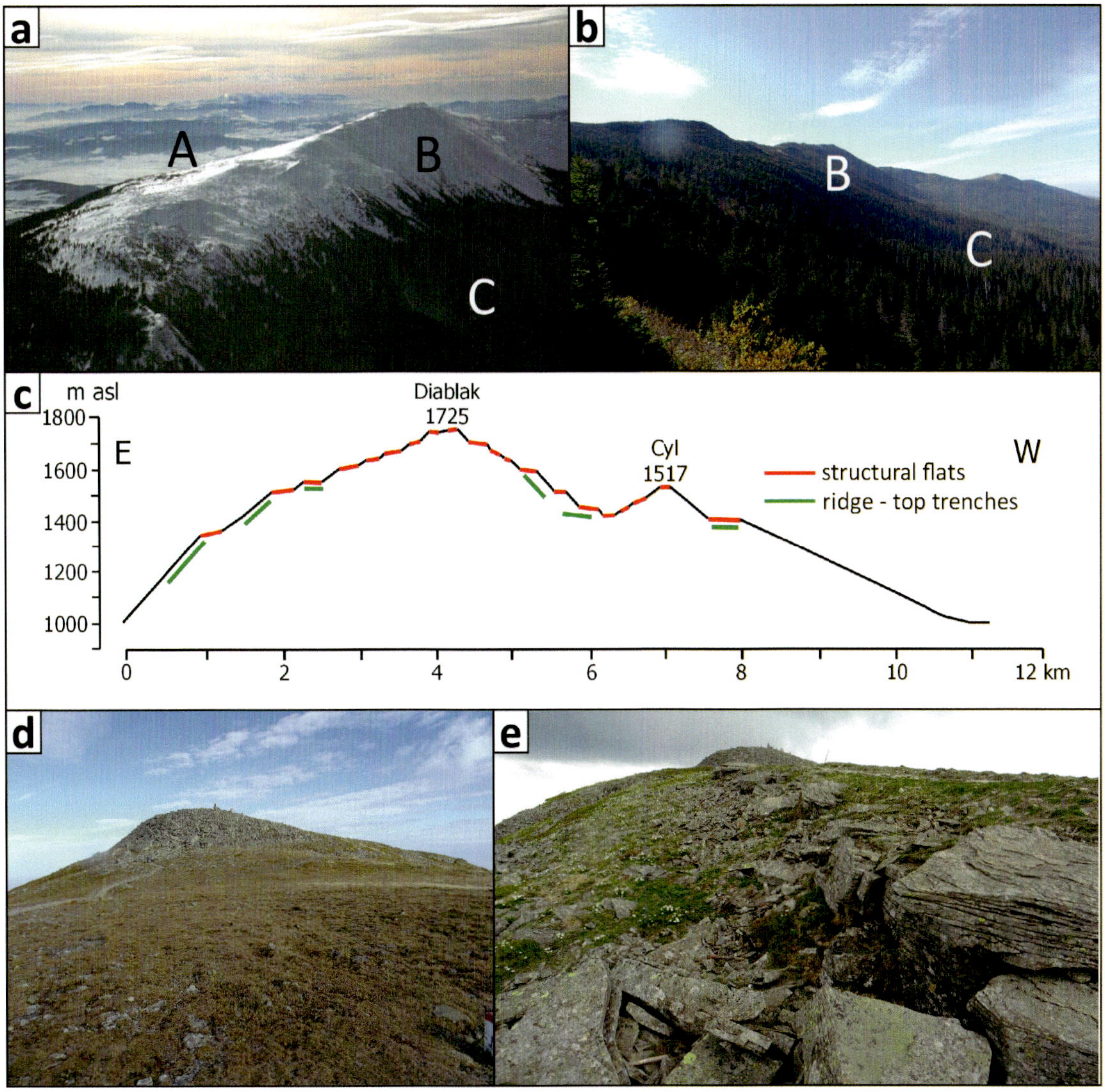

Fig. 13.7 Homoclinal main ridge of Mt. Babia Góra massif. **a** General view. **b** Northern slope. **c** Longitudinal profile of the main ridge. **d** The highest located structural flat transformed by periglacial processes into cryoplanation terrace. **e** Tump (Diablak peak, 1725 m). Relief elements: A—the most elevated part of the gently inclined southern slope; B—precipitous escarpment of the northern slope; C—less steep foot of the escarpment (photographs by A. Łajczak except a, from archives of the Babia Góra National Park)

crags, smoothed slopes covered by stone runs, ridge-top and slope trenches, rock cliffs, landslide hollows, block-fields. The largest landslide hollows have undergone glacial or nival remodelling, and their steep slopes are cut by couloirs below which torrential fans developed. Across a wide range of altitude, there are tracks of recent and past debris flows that reached the more gently sloping foot of the huge escarpment (area C). Shallow landslides, on the other hand, have a limited extent. The southern, gentler slope (area A) is geomorphologically less varied. This area features small stone runs, slope trenches, and extensive landslide hollows giving the slope a stepped profile. The highest located landslide hollows have undergone nival remodelling.

13.3.2.2 Periglacial Landforms

The structural flats in the narrow top area of the ridge above 1600 m asl were likened to cryoplanation terraces (Ziętara 1989, 2004). The width of these forms reaches 150 m, the length is up to 200 m, and their gradient along the ridgeline is 2–5°. They are bordered by precipitous frost-riven scarps (Fig. 13.7d). The summit dome of Mt. Babia Góra is a tump surrounded by a rocky slope, with a flattened top surface (Fig. 13.7e). This is the only such clearly formed tump in the Polish mountains.

Along the northern edge of structural flats, above 1600 m asl, sandstone crags up to 10 m high occur (Fig. 13.8a), which are believed to have been formed under

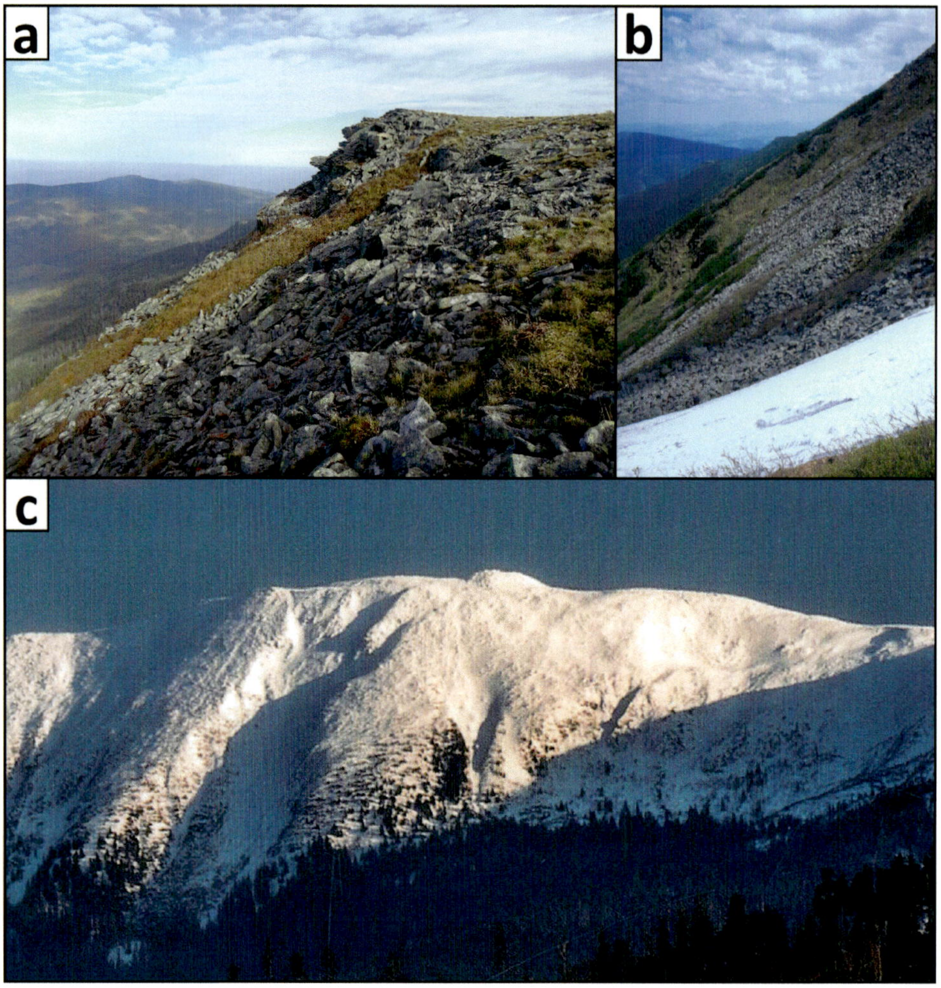

Fig. 13.8 Periglacial landforms. **a** Asymmetric crag up to 10 m high at the northern edge of a ridge-top structural flat. **b** Stone runs. **c** Smoothed areas of the northern slope covered by stone runs above glacially and nivally remodelled landslide hollows (photographs by A. Łajczak)

periglacial conditions towards the end of the last glaciation (Klimaszewski 1948, 1952; Jahn 1958). Sandstone blocks that fall off the rock faces break up into slabs, which feed the stone runs reaching down the northern slope (area B) to the height of 1500 m asl, and locally down to 1280 m asl (Fig. 13.8b). The stone runs are partially overgrown by herbaceous plants, and in the lower areas also by dwarf pine. Within the bare block fields, tongues show the evidence of movement, implied by the "emergent" position of sandstone slabs and the convex profile of the distal parts (Łajczak and Włoch 2001). The several-metres height of sandstone crags can be considered as corresponding to the minimum thickness of the rock material that has been transported down the northern slope by periglacial processes. As a result, the uppermost part of area B is distinguished by a smoothed surface that contrasts with the relief of the lower part, where landslide hollows remodelled by glacial and nival processes occur (Fig. 13.8c).

13.3.2.3 Landslides

In the narrow zone on the border of the northern slope and the ridge-top area in the height range 1150–1600 m asl, six spots with NW–SE or SW–NE-trending ridge-top trenches occur (Fig. 13.9a). Within the areas of structural flats, parallel trenches that run in the NW–SE direction have exposed bedrock faces, are narrow, and are up to 7 m deep, with their bottoms filled with rock blocks, below which are the entrances to crevice-type caves. The SW–NE ridge-top trenches have a similar depth, but they have gentler slopes and are covered with finer-grained weathering material. By contrast, the trenches developed on the steeply sloping sections of the ridge-top area occur as singular forms and are much wider and up to 70 m deep, which stems from the remodelling of these forms by glacial processes (Łajczak 1998). The second place with trenches on the northern slope of Mt. Babia Góra is area C, where the NW–SE or SW–NE directions also prevail. The shape of the trenches and their

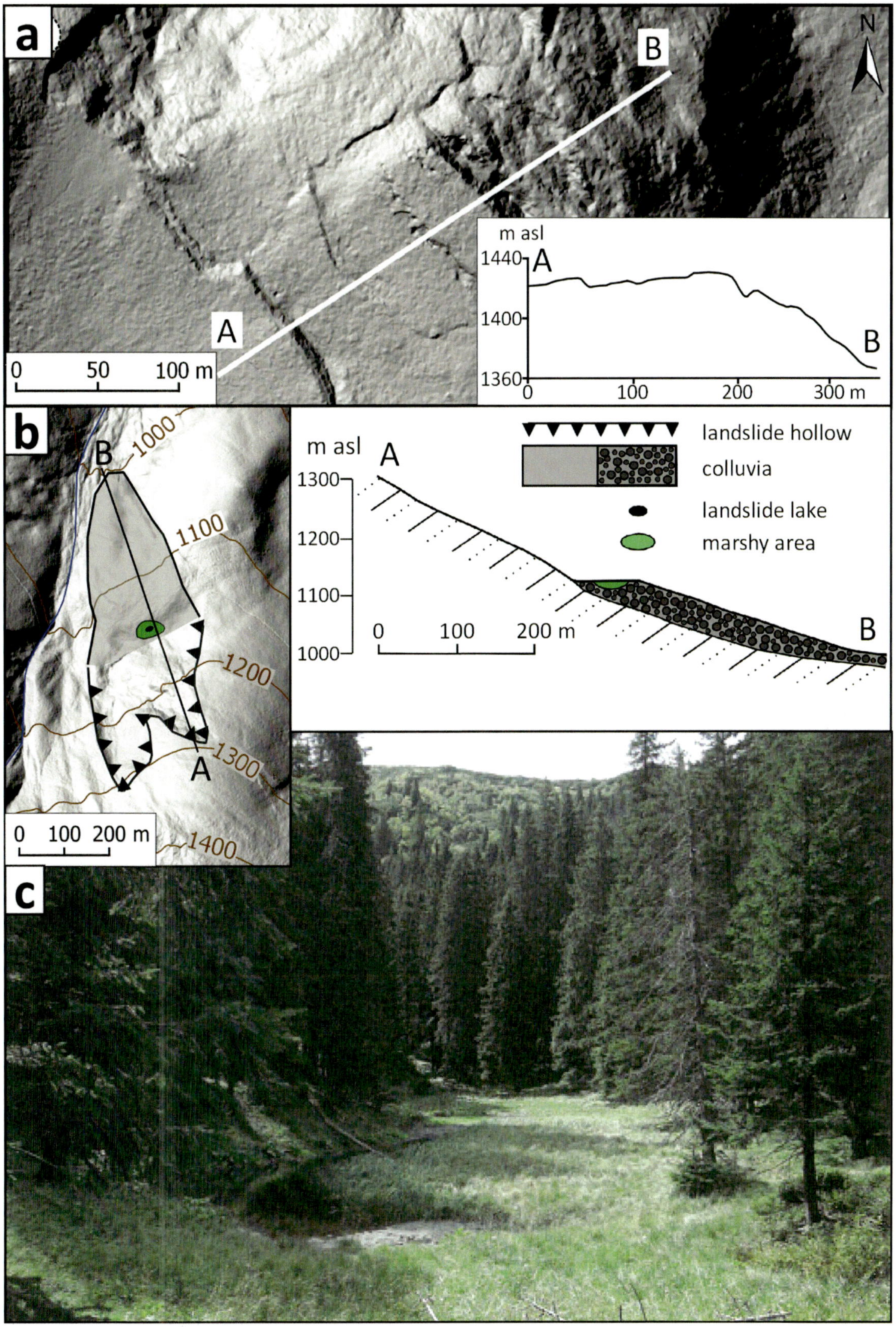

Fig. 13.9 Landslides. **a** System of ridge-top trenches oriented NW–SE and SW–NE. **b** Plan and profile of a deep-seated landslide (hollow, mid-slope flat with local depression occupied by a tiny lake and marshy area). **c** Small lake and neighbouring marshy area within the landslide depositional zone (photographs by A. Łajczak)

parameters in this area are similar to the ridge-top trenches within the structural flats. Slope trenches on Mt. Babia Góra are also found within area A in the vicinity of the ridge-top surface. The depth of the crevice-type caves in these trenches reaches 16 m (Trnka 2005; Szura and Ganszer 2005).

Below the ridge-top and slope trenches, which are initial forms of deep-seated landslides (a detailed statistics of these landforms is available in Kłapyta 2020), there occur stepped rock escarpments, as well as landslide hollows, often with rock faces. Those on the northern slope of Mt. Babia Góra are steeper and deeper than their southern slope counterparts, which, in turn, cover larger areas. Below the hollows, whose bottoms are often step-like, slope flattenings formed on the Magura sandstones occur in places where colluvia are present (Fig. 13.9b). The landslide depressions entrap fine-grained material washed out from the rubble slope covers, with organic material also accumulating in such spots (Fig. 13.9c) (Łajczak 2018). Historical sources and tree age surveys show that two landslides on the northern slope of Mt. Babia Góra were renewed by the 1868 earthquake. Numerous small landslides may be identified on the slopes of the side ridges, which are mostly built of softer flysch material. Colluvia of some of these landslides have blocked the valley bottoms, remodelling their longitudinal profiles.

13.3.2.4 Glacial, Nival, and Other Landforms Connected with Former Glaciation

On the northern slope of Mt. Babia Góra, three largest landslide hollows and a smaller one have been glacially remodelled, while six others (including one on the southern slope) have been reshaped by nival processes (Fig. 13.10a). As a result of the prevailing winds from the W and SW sectors, snow was blown away from the upper, low-angle part of the southern slope and was mainly accumulated, along with avalanche snow, in concave landforms up to 300 m deep within the steep shaded northern slope above 1350 m asl. Because of the winding course of the ridge axis, the supply of drift and avalanche snow to firn fields decreased eastward during the last glacial period, which translated into decreasing glacial remodelling of preglacial landforms in this direction. Glaciers were of the rock apron type, with three of them ending in area C, whereas the longest glacier (2.2 km) terminated at an altitude of about 920 m asl (area D). Łajczak (1981) indicated the limit of glaciers on the northern slope of Mt. Babia Góra in the last glacial period as similar as the one shown in this chapter, but only to the altitude of ca. 1100 m asl. Kłapyta (2020), however, showed that the main glacier reached as low as 900 m asl and was the first to indicate the remnants of a terminal moraine at this altitude.

Undercuts of the northern slope within former firn fields are the most visible glacial features on Mt. Babia Góra. They are parts of glacial cirques with a U-shaped cross profile (Fig. 13.8c, 13.10b, c). A characteristic feature is the occurrence of a convex knickpoint, which separates the higher-lying, smooth, and less inclined slope from the lower, precipitous sections with rock faces dissected by couloirs (Fig. 13.8c). Steep gradients of the northern slope of Mt. Babia Góra did not allow for the formation of classic lateral and terminal moraines. Glacial depositional landforms were subjected to mass movements, were buried by torrential fans, or washed away by streams (Fig. 13.10a). The disappearance of these moraine ramparts is therefore the result of high-energy slope and fluvial processes, pointed out in other areas by Kirkbride and Winkler (2012). Only traces are visible along the route of the longest former glacier. These include, in particular, sub-rounded and rounded Magura sandstone boulders with a diameter up to 5 m, particularly numerous at an altitude of 920–1400 m asl (Fig. 13.10d). Short moraine bulges formed below permanent snow patches (elevation 1520–1650 m asl) slid downhill. The terminal moraine of the longest glacier has been eroded completely, leaving a large concentration of sub-rounded and rounded Magura sandstone boulders at around 920 m asl. The apex of the fluvioglacial fan located at this point indicates the highest location of all alluvial fans at the foot of Mt. Babia Góra.

A spectacular landform in the upper part of the northern slope is a vast boulder field partially dominated by dwarf mountain pine, formed as a bulge with stepped profile at an altitude of 1300–1500 m asl, the axis of which is perpendicular to the contour lines. This landform occurs below a deep, glacially remodelled landslide hollow and its morphological features resemble a non-active (fossil) rock glacier (Fig. 13.10e). This is a novel interpretation of previously indicated rock-avalanche deposits (Alexandrowicz 1978; Kłapyta 2020). It formed during the last deglaciation of Mt. Babia Góra. At present, no permafrost occurs there, whereas Dobiński et al. (2016) showed that no relict permafrost is even present within the culmination of Mt. Babia Góra, located 300 m higher.

13.3.2.5 Couloirs, Torrential Fans, Debris Flows, Shallow Landslides, and Avalanches

Glacial undercuts on the northern slope of Mt. Babia Góra are dissected by couloirs with rock faces that reach a length of 250 m and a depth of 10 m (Fig. 13.11a), below which torrential fans are formed. They host debris flows, whose source areas are located further above, in the area of stone runs. The flows terminate within extensive torrential fans and may reach the altitude of 1275 m asl (fossil debris flows reach 1150 m asl). The youngest debris flows

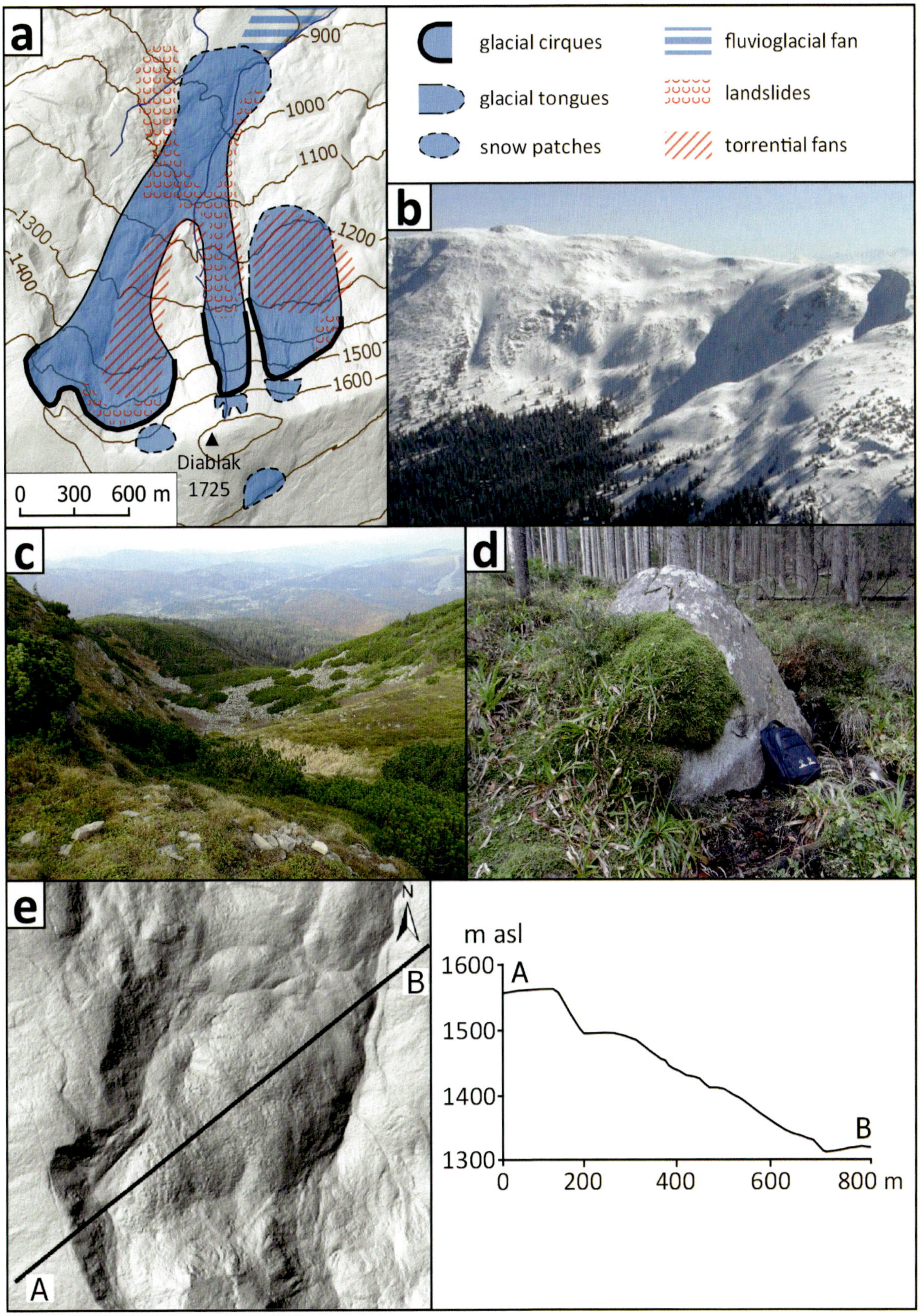

Fig. 13.10 Morphological evidence of former glaciation. **a** Probable maximum limit of rock apron type glaciers on the northern slope of Mt. Babia Góra and distribution of the largest snow patches during the last glaciation. **b** The largest glacial cirque. **c** Smaller glacial cirque, which is a remodelled wide ridge-top trench. **d** Sub-rounded Magura sandstone boulder in a degraded terminal moraine of the longest glacier. **e** LiDAR image of the non-active (fossil) rock glacier below a glacially remodelled deep-seated landslide hollow and the corresponding topographic profile (photographs by A. Łajczak)

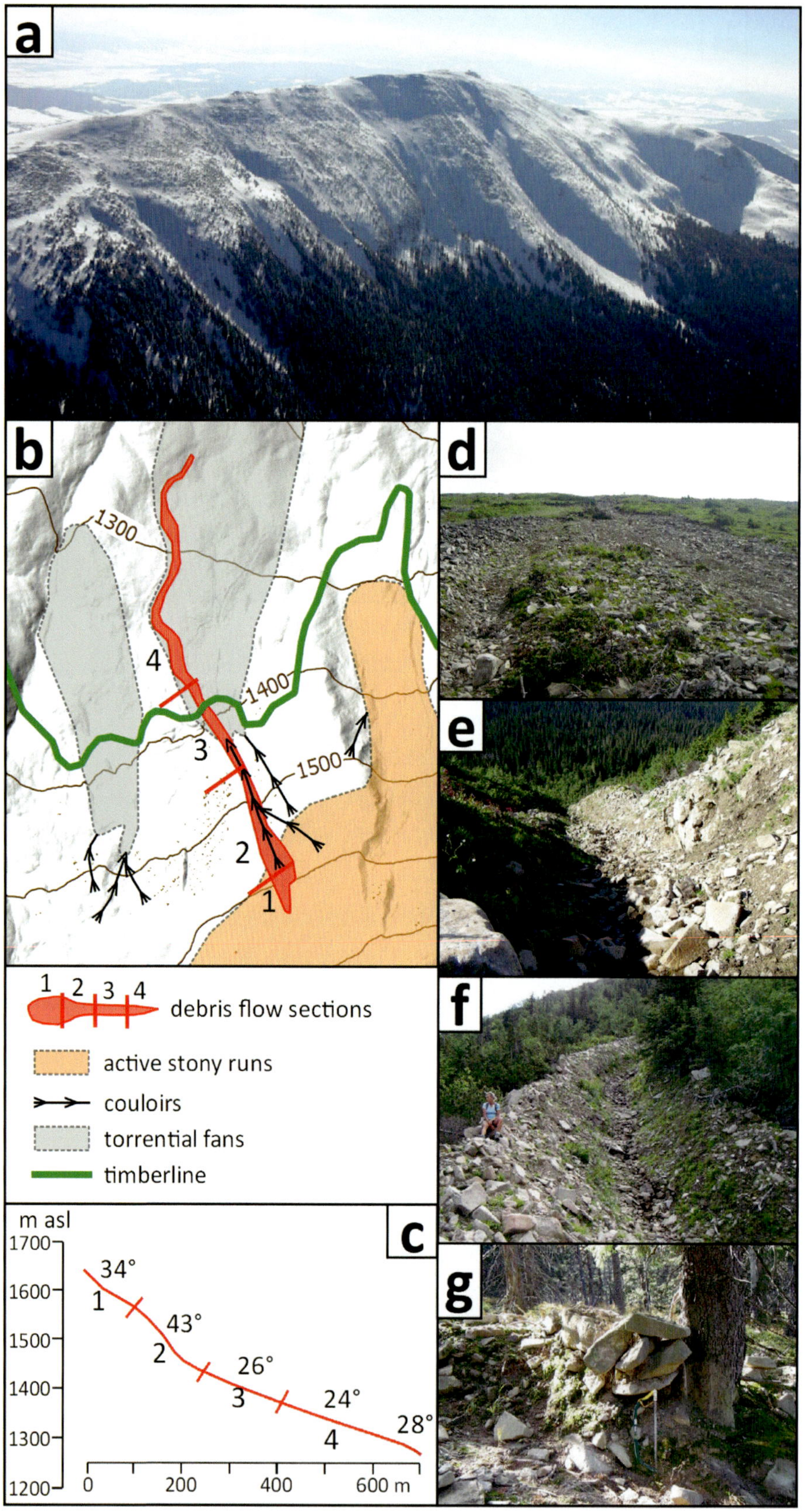

Fig. 13.11 Couloirs and evidence of debris flows. **a** Couloirs dissecting a glacial undercut on the northern slope. The July 2002 debris flow: **b** Plan. **c** Longitudinal profile (sections of the track: 1—head section, 2—transit section, 3—depositional section, 4—terminal section). Views on the debris flow sections: **d** Head section. **e** Transit section. **f** Depositional section. **g** Terminal section (photographs by A. Łajczak except a, from archives of the Babia Góra National Park)

formed after 1990 within the largest glacial cirque. In the longest one (700 m), which dates back to July 2002 and occurs within the height range from 1635 to 1275 m asl, the following sections (Łajczak and Migoń 2007) are distinguished (Fig. 13.11b, c): (1) head section, (2) transit section in the couloir, (3) depositional section on the torrential fan (erosional dissection and levees), (4) terminal section in the forested part of the torrential fan. Fossil levees of debris flows (Matyja 2006) have been recognized on the torrential fans below the timberline, in an old spruce forest more than 200 years old. Some debris flows develop due to transformation of shallow landslides occurring on steep slopes of spring-fed valley heads, covered by thin layer of debris-clay regolith. Such a landform is the Cylowa Zerwa landslide, which has been activated every 40–50 years since 1868 (Łajczak et al. 2023) (Fig. 13.12a, b). On steep slopes, avalanches form above the timberline over a much larger area on the northern side of the ridge than on the southern side. They also move along the couloirs (Łajczak and Spyt 2016, 2018) (Fig. 13.12c–e). Avalanche snow generally accumulates below the timberline, within torrential fans.

13.3.2.6 Fluvial Landforms and Peat-Bogs

In the stream channels at the foot of Mt. Babia Góra, colluvial covers are incised and removed, exposing bedrock. Therefore, bedrock channels and, locally, small waterfalls up to 5 m high occur (Fig. 13.12f, g). Terrace risers' undercut during high water stages represents another source of material transported to stream channels. Incision within short sections of alluvial channels is locally prevented by the exposed large sandstone boulders and wood debris.

On the Pleistocene terraces and alluvial fans, especially on the southern footslope of Mt. Babia Góra, and also on slopes in topographic depressions within landslide bodies, peat-bogs develop (laggs and raised bogs) (Łajczak 2016).

13.3.2.7 Anthropogenic Landforms

The relief of Mt. Babia Góra has been subjected to anthropogenic changes only over a total area of not more than 1%. Anthropogenic landforms include levelled plots of land in built-up areas within area F1, as well as in the vicinity of the tourist shelter on the northern slope at an altitude of 1180–1200 m asl (area D). This category of landform also includes profiled strips of land occupied by municipal roads (areas F, E), forest roads (areas C, D, E), and hiking paths (areas A, B, C, D). Forest roads, especially those running on the less resistant flysch deposits, have turned into 3 m deep sunken lanes which continue to deepen. In their vicinity, one can observe miniature alluvial fans on the slopes and valley bottoms. No erosion is observed along tourist paths, which are reinforced by sandstone slabs. Ever since Mt. Babia Góra has been explored by visitors, which is for at least 250 years, various structures have been erected on the summit dome using local sandstone blocks, only later to be dismantled. As a result, the top of the ridge has been lowered by ca. 2 m. In the nineteenth century, earthen dams were built in the bottoms of some valleys in order to bank up the streams for timber floating purposes, and this continued until the early twentieth century. The build-up of miniature deltas upstreams the dams resulted in the siltation of the dammed lakes. Once the dams were destroyed, terraces were formed at such locations (Łajczak 2006). In summary, the relief of Mt. Babia Góra has its natural features preserved, and the scale of anthropogenic changes should be described as minimal.

13.3.3 Relief Development

In the Middle Miocene, during the uplift of the area currently occupied by the Babia Góra massif and its surroundings (except for the southern footslope), the Magura sandstone was present throughout the area, but its extent shrank over time. In the Pliocene and Quaternary, despite the weakening uplift, the extent of this flysch formation underwent further segmentation and contraction as a result of sliding of large rock packets (Jankowski and Margielewski 2021), erosion of slopes, valley deepening, and fluvial transport of weathered material. The shrinking outcrop of the Magura sandstones translated into an increasing extent of outcrops of the older flysch formations, which now cover a larger area than within the neighbouring, lower mountain ridges. The reduction in the coverage of the Magura sandstone outcrops was associated with the formation of four denudation levels): (1) LM (Late Miocene, this morphological level was later subject to heavy remodelling by tectonic movements), (2) EP (Early Pliocene), (3) LP (Late Pliocene), (4) EQ (Early Quaternary), as distinguished by Starkel (1969, 1972, 1988). He assessed the relative heights of the three younger denudation levels in the Polish Flysch Carpathians as follows: EP—ca. 250–500 m; LP—ca. 150–200 m; EQ—ca. 40–50 m. However, this interpretation assuming the survival of specific denudation levels is not compatible with the newest view concerning large-scale gravitational tectonics of the Carpathians (Jankowski and Margielewski 2014, 2021). It can be assumed that each of the above denudation levels initially functioned as a pediment surrounding the towering ridge of Mt. Babia Góra, to be later incised and fragmented (Łajczak 2014). The uplift of the rock mass and the deepening of the valleys translated resulted in the increase of relative elevation of Mt. Babia Góra, reaching up to almost 1000 min the Early Quaternary. However, its subsequent rise, until the Late Quaternary, was minor (Fig. 13.13a) (Łajczak 2014).

The ridge of Mt. Babia Góra and the neighbouring ridge of Mt. Police are limited to the north by normal faults

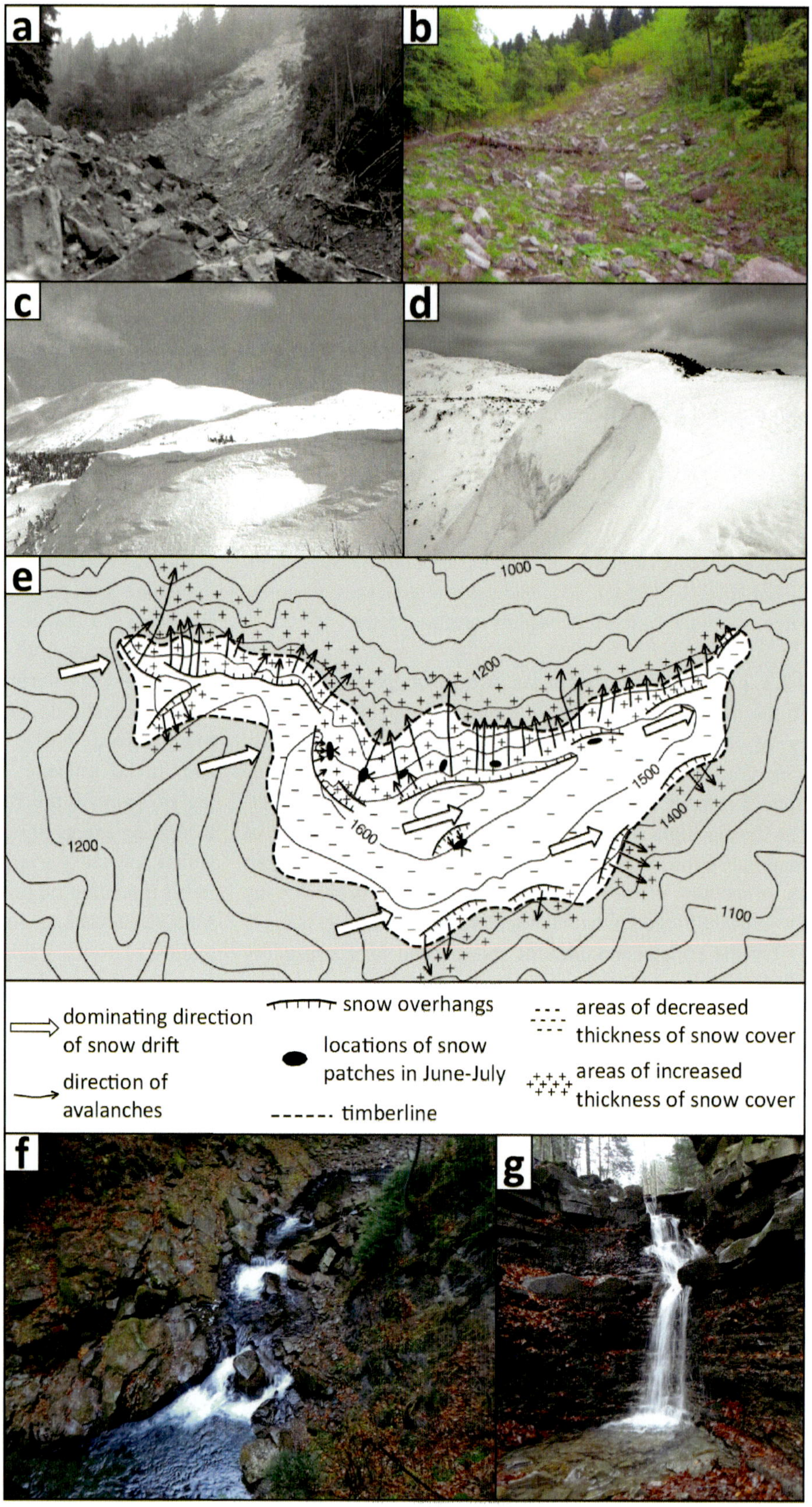

Fig. 13.12 Shallow landslide, avalanches, and bedrock channels. View on the Cylowa Zerwa shallow landslide: **a** In 1962. **b** In 2005. **c–d** Snow overhangs at the northern edge of the ridge-top surface. **e** Avalanche tracks. **f** Bedrock channel. **g** Waterfall (5 m) (photographs by A. Łajczak except a, from archives of the Babia Góra National Park)

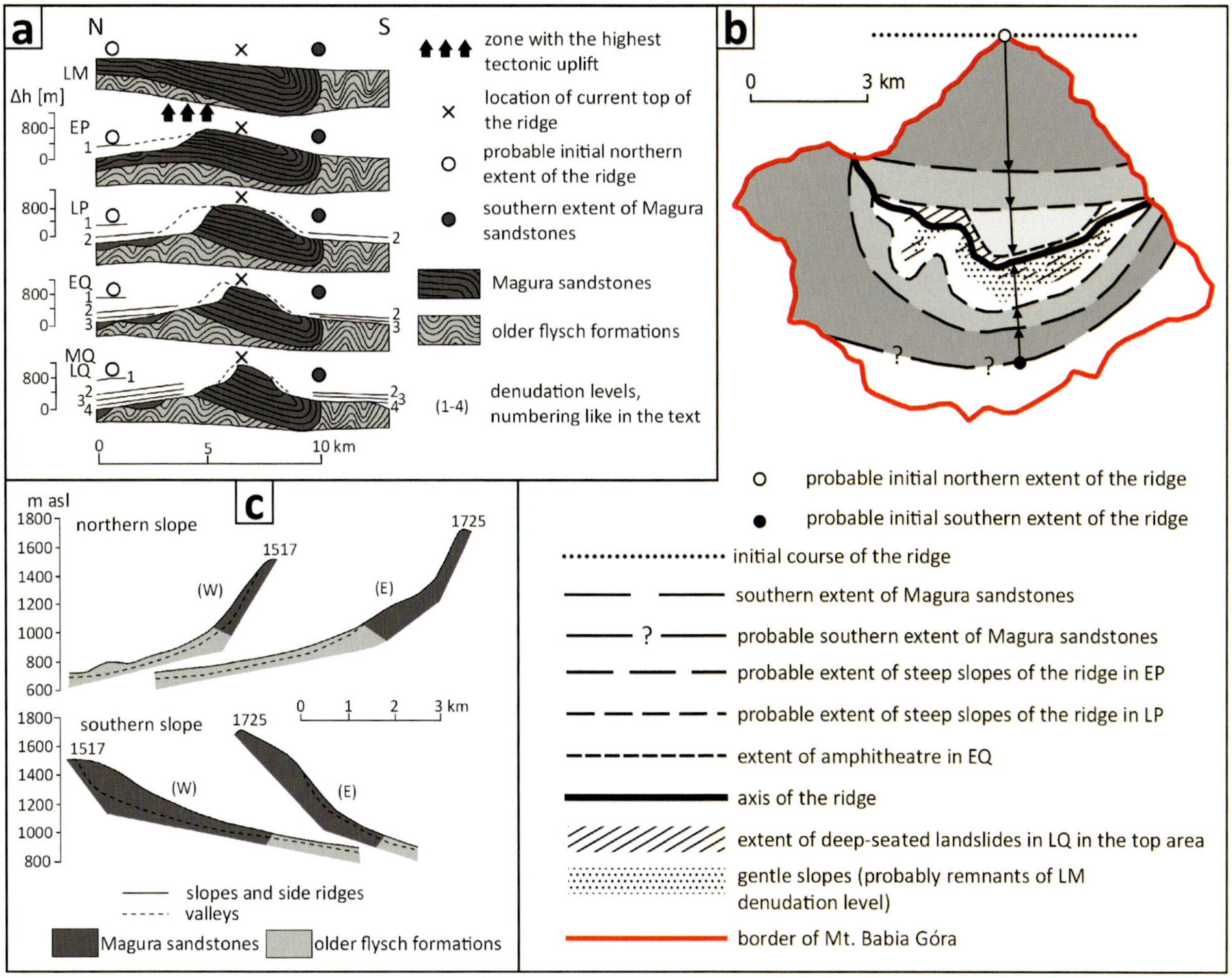

Fig. 13.13 Reconstruction of relief development of Mt. Babia Góra massif. **a** Probable changes in the N–S profile of the area due to uplift, dissection, and slope retreat (changes in relative heights Δh [m] and symbols (LM, EP, LP, EQ, MQ, LQ) are explained in the text). **b** Stages of slope retreat. **c** Profiles of the northern and southern slopes (W—western part, E—eastern part)

exposed in rock formations outcropping in the Skawica River valley (Jankowski et al. 2019). On the southern side, normal faults separate Mt. Babia Góra from its foreland (Pomianowski 2003) and a huge tectonic mélange zone formed as a slip zone occurs here (Jankowski and Margielewski 2014, 2021). The sliding of great rock packets, especially from the northern slopes of Mt. Babia Góra and Mt. Police to the Skawica River valley, is hypothesized to have caused isostatic elevation of both ridges (Jankowski and Margielewski 2014, 2021). The zone with the fastest contemporary isostatic uplift in this part of the flysch Carpathians, presented by Starkel (1972), runs across the northern footslope of Mt. Babia Góra in the W–E direction and across the ridge of Mt. Police. This can explain why, starting from the Middle-Late Miocene—as a result of the largest loss of rock mass in this area—the axis of the ridge of Mt. Babia Góra has retreated approximately 7 km southwards (Alexandrowicz 1978; Łajczak 2014). The current position of the axis of this ridge corresponds to the area where the Magura sandstones within the Babia Góra syncline have the greatest thickness (see Książkiewicz 1983; Fig. 13.3b, 13.13a).

The uplift of Mt. Babia Góra massif and valley deepening went hand in hand with slope retreat. Its rate was likely to be fastest at times of more vigorous rock-mass uplift, when favourable conditions occurred for the development of deep-seated landslides, the retreat of valley heads, and fragmentation of slopes through incision of the upper valley reaches. At times of relative tectonic calm, when the area around the ridge was undergoing planation and pediments' formation, conditions for ridge slope retreat were less favourable. The correlation analysis of traces of former denudation levels, starting from the Late Miocene, allows for the reconstruction of the probable retreat of slopes in Mt. Babia Góra (Łajczak 2014) (Fig. 13.13b). The scale of slope retreat was declining over time from the Late

Miocene to the Early Quaternary, which should be attributed to the increasing relative height of the ridge.

The last phase of slope retreat involved only the central part of the northern slope, when a huge hanging amphitheatre being the part of *"the great escarpment of Mt. Babia Góra"* was formed during the Early Quaternary, evident in the relief of Mt. Babia Góra until today (Alexandrowicz 1978; Łajczak 2014). The origin of this amphitheatre occurred after the formation of general features of relief in valley heads in area D at the turn of the Early Quaternary, and before local glaciation of the northern slope during the Late Pleistocene. The emergence of this large hollow caused the axis of the main ridge of Mt. Babia Góra to retreat by 1–2 km and gave it a sinuous course (Fig. 13.13b). The ridge axis has not shifted much since then, despite the development of deep-seated landslides in the near-surface parts of the northern and southern slopes. Deep landslide hollows have even been formed locally since the Late Pleistocene, but they are of secondary importance for the morphology of the ridge.

During the stepped retreat of the Mt. Babia Góra ridge slopes, the volume of rock material carried away was decreasing. In the final phase of the radical changes in the position of the slopes (until the Early Quaternary), only the upper area of the northern slope (the amphitheatre) had large amounts of rock material removed. Meanwhile, that time interval (and later periods) saw no significant development in the morphology of the upper part of the southern slopes, and therefore, this area is the only one within Mt. Babia Góra massif where remnants of the oldest denudation level dating back to the Late Miocene are extant (Łajczak 2014) (Fig. 13.13b). Late Miocene age of general relief features of the upper part of the southern slope of Mt. Babia Góra is suggested by altitude position of this area (related to the summit morphological level) and also by monotonous relief of this penestructural slope resulting from the lack of typical younger denudational landforms and significant resistance of bedrock. As a result of spatial variations in the retreat pattern of individual areas of the northern and southern slopes of Mt. Babia Góra, the profiles of these slopes have developed differently (Fig. 13.13c). The northern slope has a concave profile, with the western part characterized by mildly sloping incisions of the shallow valleys reaching dar upslope. By contrast, the profile of the central sector, which has undergone most advanced retreat, demonstrates large breaks in inclination and is incised by shallow valleys, which are only present in the lower parts of the slope and at the foot. The southern slope of Mt. Babia Góra has a convex–concave profile. In its western part, valley depth is the greatest within Mt. Babia Góra massif and increases towards the axis of the ridge. Meanwhile, the eastern part of the southern slope is much less fragmented by valleys, which only occur in its lower reach and within the footslope. Therefore, this section of Mt. Babia Góra resembles a "massive ridge" rising above the low-relief land at the foot (Fig. 13.2d).

13.3.4 Spatial Differentiation of Relief—Altitude Zones and Asymmetry

The relief of Mt. Babia Góra reflects relationships with the structure of bedrock and climate changes in the Late Quaternary. As a result of their cumulative impact, these factors have produced relief features that differ depending on the altitudinal zone of the ridge, as well as landforms that make the northern slope different from the southern one, causing an overall relief asymmetry. Three areas are distinguishable in the Mt. Babia Góra elevation profile (Fig. 13.6): (a) lower parts of slopes and footslope areas (areas E, F), where a large area is occupied by fluvial landforms in addition to landslides; (b) an area with valley heads (area D) dominated by landslides; (c) differently shaped uppermost parts of the northern and southern slopes (areas A, B, C), where in addition to landslides with crevice-type caves glacial and nival forms, couloirs with torrential fans, and debris flows tracks occur. Area B satisfies the criteria of an area with high-mountain features as identified by Slaymaker (2004), which makes the ridge of Mt. Babia Góra unique in respect to other high-elevated flysch ridges in the Western Carpathians, with features of medium-altitude mountains (Starkel 1972). According to Kotarba and Starkel (1972), Mt. Babia Góra falls within the range of forest and subalpine morphogenetic zones. The asymmetry in relief between the northern and southern slopes extends from area D towards the top of the ridge and is most noticeable in areas A and B. This contrast is emphasized by the occurrence of deep-seated landslides over a larger area within the northern slope, as well as by the presence of geomorphic features typical of high-mountain relief, which are only found within "the great escarpment of Mt. Babia Góra" (area B).

13.4 Conclusions

Since the nineteenth century, Mt. Babia Góra has been referred to as "the Queen of the Western Beskidy Mountains" on account of its towering height amidst the flysch ridges of the Western Carpathians, its unusual shape, the precipitous northern slope, and its exceptional scenic values. Naturalists, including geomorphologists, consider it a model mountain of the flysch section of the Western Carpathian Mountains, where a rich array of landforms occurs over a relatively small area, depending on the altitude and asymmetry between the northern and southern slopes. On the one hand, the relief of Mt. Babia Góra has features typical of medium-altitude mountains in the flysch section of the Carpathians, and on the other hand, it shows similarity to high-mountain areas. Particularly noteworthy are various forms of deep-seated landslides occurring almost across the whole elevation profile of the massif. They have influenced, inter alia, the distribution of glacial and nival forms, not observed on such a scale elsewhere in the flysch Western Carpathians. In addition to the plant cover (extant in a natural state over large areas), which varies from one elevational zone to another, the relief of Mt. Babia Góra is the element of nature that led the authorities to create the Babia Góra National Park in 1954. It was the first national park in Poland to be incorporated into the network of UNESCO International Biosphere Reserves in 1977. This status of Mt. Babia Góra has led to the rapid growth of tourism and given huge interest in its abiotic nature, also of geotourism.

References

Alexandrowicz SW (1978) The northern slope of Babia Góra Mt. as a huge rock slump. Stud Geomorph Carp-Balc 12:133–148

Alexandrowicz SW (2004) Zarys budowy geologicznej Babiej Góry. In: Wołoszyn W, Jaworski A, Szwagrzyk J (eds) Babiogórski park Narodowy. Monografia przyrodnicza. Babiogórski Park Narodowy, Zawoja, pp 87–107

Břežný M, Pánek T (2017) Deep seated landslides affecting monoclinal flysch morphostructures: evaluation of LiDAR-derived topography of the highest range of the Czech Carpathians. Geomorphology 285:44–57

Dobiński W, Glazer M, Bieta B, Mendecki MJ (2016) Poszukiwania wieloletniej zmarzliny i budowa geologiczna Babiej Góry w świetle wyników obrazowania elektooporowego. Przegl Geogr 88:31–51

Geological Atlas of the Western Outer Carpathians and Their Foreland (1989). Scale 1:500,000, PIG Warsaw

Golonka J, Pietsch K, Marzec P, Stefaniuk M, Waśkowska A, Cieszkowski M (2009) Tectonics of the western part of the Polish Outer Carpathians. Geodin Acta 22:81–97

Jahn A (1958) Mikrorelief peryglacjalny Tatr i Babiej Góry. Biul Perygl 6:57–80

Jankowski L, Margielewski W (2014) Strukturalne uwarunkowania rozwoju rzeźby Karpat zewnętrznych—nowe spojrzenie. Przegl Geol 62:29–35

Jankowski L, Margielewski W, Garecka M, Kowalska S (2019) Stop 6. Skalisty fragment koryta Skawicy w Zawoi, z asocjacją różnowiekowych uskoków. In: Jankowski L (ed) Warsztaty terenowe: Od podnóża Tatr po brzeg Karpat. Współczesne wyzwania kartografii geologicznej. Przewodnik do wycieczki terenowej. PIG-PIB, Kraków, pp 59–65

Jankowski L, Margielewski W (2021) Geological control of young orogenic mountains morphology: from geomorphological analysis to reinterpretation of geology of the Outer Western Carpathians. Geomorphology 386:107749

Kirkbride MP, Winkler S (2012) Correlation of late Quaternary glacier chronologies: impact of climate variability, glacier response, and chronological resolution. Quat Sci Rev 46:1–29

Kłapyta P (2020) Geomorphology of the high-elevated flysch range—Mt. Babia Góra Massif (Western Carpathians). J Maps 16:689–701

Klimaszewski M (1948) Polskie Karpaty Zachodnie w okresie dyluwialnym. Prace Wrocł Tow Nauk 7:233

Klimaszewski M (1952) Zagadnienia Plejstocenu Południowej Polski. Biul PIG 65:137–268

Kondracki J (1998) Geografia regionalna Polski. PWN Warszawa, p 441

Kotarba A, Starkel L (1972) Holocene morphogenetic altitudinal zones in the Carpathians. Stud Geomorph Carp-Balc 6:21–35

Książkiewicz M (1971) Objaśnienia do szczegółowej mapy geologicznej Polski. Arkusz Zawoja. Skala 1:50,000. PIG, Warszawa, p 54

Książkiewicz M (1983) Zarys geologii Babiej Góry. In: Zabierowski K (ed) Park narodowy na Babiej Górze. Przyroda i człowiek. Stud Naturae B 29:25–39

Łajczak A (1981) Formy glacjalne na Babiej Górze. Pr Babiogórskie 3:43–53

Łajczak A (2014) Relief development of the Babia Góra massif, Western Carpathians. Quaest Geogr 33(1):89–106

Łajczak A, Migoń P (2007) The 2002 debris flow in the Babia Góra massif—implications for the interpretation of mountainous geomorphic systems. Stud Geomorph Carp-Balc 41:97–116

Łajczak A, Spyt B (2016) Geomorphological characteristics of avalanche tracks on Babia Góra Mt., Western Carpathians. Stud Geomorph Carp-Balc 50:65–88

Łajczak A, Spyt B (2018) Differentiation of vertical limit of forest at the Babia Góra Mt., the Western Carpathian Mountains. Geogr Polon 91:217–242

Łajczak A, Włoch E (2001) Pokrywy blokowe Babiej Góry. Mat Konf Pokrywy blokowe jako zapis zmian klimatycznych w późnym Vistulianie i holocenie. Publ Uniw Śląski, Sosnowiec, pp 29–37

Łajczak A, Margielewski W, Zielonka T, Pasierbek T, Lamorski T, Kozina P, Izworska K (2023) Cylowa Zerwa landslide—debris flow forms on Mount Babia Góra (1725) and their development over the last ca. 150 years, Western Carpathians. Geogr Polon 96:79–102

Łajczak A, Czajka B, Kaczka RJ (2014) The new features of landslide relief discovered using LiDAR. Case study from Babia Góra massif, Western Carpathian Mountains. Quaest Geogr 33(3):73–84

Łajczak A (1998) Uwagi dotyczące występowania form glacjalnych i niwalnych na Babiej Górze, Zachodnie Karpaty. In: IV Zjazd Geomorfologów Polskich. Głównie kierunki badań geomorfologicznych w Polsce. Stan aktualny i perspektywy, vol. 1. Lublin, pp 349–356

Łajczak A (2006) Deltas in dam-retained lakes in the Carpathian part of the Vistula drainage basin. In: Czałow RS, Kamykowska M, Krzemień K (eds) Channel processes in the rivers of mountains, foothills and plains. Prace Geogr IGiGP UJ 116:99–110

Łajczak A (2012) Water circulation and chemical denudation within the upper Skawica River flysch catchment, Western Carpathian Mountains. Z Geomorph NF 56, Suppl 1:69–86

Łajczak A (2013) Relief development of a highly elevated monoclinal Babia Góra range built by Magura sandstone, Western Carpathian Mts. In: Migoń P, Kasprzak M (eds) Sandstone landscapes. Diversity, ecology and conservation. Proceeding of the 3rd Intern Conf on Sandstone Landscapes, Kudowa-Zdrój (Poland), 25–28 Apr 2012. Inst Geogr i Rozwoju Reg Uniw Wrocł, Wrocław, pp 100–105

Łajczak A (2015) Monografia masywu Pilska, Beskid Żywiecki. Publ Inst Botaniki im W. Szafera Polska Akademia Nauk, Kraków, p 262

Łajczak A (2016) Wody Babiej Góry. Monografie Babiogórskie. Babiogórski Park Narodowy, Zawoja, p 423

Łajczak A (2018) Osady wypełniające zagłębienia osuwiskowe na stokach Babiej Góry (Karpaty Zachodnie) jako archiwum paleogeografii obszaru. In: Kostrzewski A, Stach A, Majewski M (eds) Geneza, litologia i stratygrafia utworów czwartorzędowych, vol. VII. Publ Inst Geoekologii i Geoinformacji Uniw A. Mickiewicza w Poznaniu, pp 109–112

Margielewski W (2006) Structural control and types of movements of rock mass in anisotropic rocks: case studies in the Polish Flysch Carpathians. Geomorphology 77:47–68

Matyja M (2006) The significance of trees and coarse debris in shaping the debris flow accumulation zone (north slope of the Babia Góra Massif, Poland). Geogr Polon 80:83–99

Mrozek T, Kułak M, Grabowski M, Wójcik A (2014) Landslide counteracting system (SOPO) inventory database of landslides in Poland. In: Sassa K, Canuti P, Yin Y (eds) Landslide science for a safer geoenvironment. Springer, Switzerland, pp 815–820

Němčok A (1982) Zosuvy v slovenskych Karpatoch. VEDA, Bratislava, p 320

Niemirowski M (1963) Szkic geograficzny obszaru babiogórskiego. In: Szafer W (ed) Babiogórski Park Narodowy. Publ Zakład Ochrony Przyrody Polska Akademia Nauk, Kraków, No 22, pp 21–43

Obidowicz A (2004) Holoceńska historia roślinności Babiej Góry. In: Wołoszyn W, Jaworski A, Szwagrzyk J (eds) Babiogórski park Narodowy. Monografia przyrodnicza. Babiogórski Park Narodowy, Zawoja, pp 423–428

Obrębska-Starkel B (2004) Klimat masywu Babiej Góry. In: Wołoszyn W, Jaworski A, Szwagrzyk J (eds) Babiogórski Park Narodowy. Monografia przyrodnicza. Babiogórski Park Narodowy, Zawoja, pp 137–151

Ondrášik R (2002) Landslides in the West Carpathians. In: Rybář J, Stemberk J, Wagner P (eds) Landslides. Proceeding first European conference on landslides, Prague, Czech Republic, 24–26 June 2002. Balkena, Lisse, pp 45–57

Pánek T, Břežný M, Kapustová V, Lenart J, Chalupa V (2019) Large landslides and deep-seated gravitational slope deformations in the Czech Flysch Carpathians: new LiDAR-based inventory. Geomorrphology 346:106852

Pomianowski P (2003) Tektonika Kotliny Orawsko-Nowotarskiej—wyniki kompleksowej analizy danych grawimetrycznych i geoelektrycznych. Przegl Geol 51:498–506

Rehman A (1895) Ziemie Polski i sąsiednich krajów słowiańskich opisane pod względem fizyczno-geograficznym. Cz. I – Karpaty opisane pod wględem fizyczno-geograficznym. Publ Akad Umiejętności, Lwów-Kraków, p 683

Sawicki L (1913) Krajobrazy lodowcowe Zachodniego Beskidu. Rozprawy Akad Umiejętności, Wydz Matem-Przyr A 3(13):1–22

Slaymaker O (2004) Mountain geomorphology. In: Goudie AS (ed) Encyclopedia of geomorphology, vol 2. Routledge, London, pp 701–703

Starkel L (1960) Rozwój rzeźby Karpat fliszowych w holocenie. Prace Geogr IG PAN 22:1–239

Starkel L (1969) Reflection of the geological structure in the relief of the Polish Flysch Carpathians. Stud Geomorph Carp-Balc 3:61–71

Starkel L (1972) Karpaty Zewnętrzne. In: Klimaszewski M (ed) Geomorfologia Polski, t. 1—Polska Południowa. PWN Warszawa, pp 52–115

Starkel L (1988) O genezie i wieku zrównań w Polskich Karpatach. Przegl Geogr 60:401–408

Szura C, Ganszer J (2005) Jaskinie Babiej Góry. Zacisk 25:29–30

Trnka R (2005) Przyroda słowackiej części Babiej Góry. In: Ptaszycka-Jackowska D (ed) Światy Babiej Góry. Babiogórski Park Narodowy, Zawoja, pp 121–142

Wójcik A, Rączkowski W, Mrozek T, Nescieruk P, Marciniec P, Zimnal Z (2010) Babiogórski Park Narodowy. Mapa geologiczno-turystyczna. Skala 1:13,000. PIG-PIB, Warszawa

Zapałowicz H (1880) Roślinność Babiej Góry pod względem geograficzno-botanicznym. Sprawozd. Kom Fizjogr Akad Umiejętności Kraków 14:79–250

Ziętara T (1989) Rozwój teras krioplanacyjnych w obrębie wierzchowiny Babiej Góry w Beskidzie Wysokim. Folia Geogr, Geogr-Phys 21:79–92

Ziętara T (2004) Rzeźba Babiej Góry. In: Wołoszyn W, Jaworski A, Szwagrzyk J (eds) Babiogórski Park Narodowy. Monografia przyrodnicza. Babiogórski Park Narodowy, Zawoja, pp 109–135

Adam Łajczak is Professor of Geography at the Pedagogical University of Krakow. His main research interest is geomorphology, hydrology, and geoecology, especially of mountainous areas (structural geomorphology, slope and fluvial processes, water resources, floods, siltation of dam reservoirs, geoheritage, nature protection) and urbanized areas (anthropogenic changes in relief and water circulation). The Carpathian Mountains and the Vistula River valley are his main research areas. He was involved in projects in various European and other countries and participated in geographical expeditions, among others, to the Tibetan Plateau, the Himalayas, the Caucasus, and the Alps. He is a member of European Geoscience Union, DENUCHANGE Working Group of the IAG, and International Association of Hydrological Sciences. He participates in the IGU Commission "Geomorphology and Society".

14 Tatra Mountains—The Only High-Mountain Landscape in Poland

Piotr Kłapyta and Jerzy Zasadni

Abstract

The Tatra Mountains represent the only high-mountain range in Poland, with a typical alpine landscape shaped by Pleistocene glaciers. They stand out as the highest mountain range in Poland as well as in the entire Carpathians, but are not high enough to support present-day glaciers. The landscape of the Tatras is dependent on complex nappe-type alpine geological structure, considerable neotectonic uplift, and impact of multiple glaciations during the Pleistocene. The exceptionally well-preserved glacial landscape includes cirques and deeply incised U-shaped valleys, some with overdeepened bedrock floors, hanging valleys, distinct trimlines, truncated spurs, arêtes, and even palaeonunataks, which are all an effect of strong imprint of glacial erosion. The freshly shaped terminal and recessional moraines of the last glaciation contrast with subdued moraines and till covers from older glaciations which are only preserved locally in the High Tatra Mountains. The Tatra Mountains with patches of discontinuous permafrost and perennial firn-ice patches (glacieretes) are the only area with the present-day periglacial phenomena in Poland. Nevertheless, all rock glaciers are there relict, inherited from the Late Pleistocene cold phases. The Tatra Mountains represent a dynamic landscape that is highly responsive to climate change; however, the assessment of the potential impact of contemporary climate warming on the patterns and dynamics of geomorphologic processes appears to be a key challenge for future geomorphologic studies.

Keywords

Tatra mountains · High-mountain landscape · Quaternary glaciation · Rock-slope failure · Extreme geomorphological events · Human impact

14.1 Introduction

The Tatra Mountains are the highest (Gerlachovský štít, 2654 m asl) mountain range in the Carpathians. They constitute an outstanding topographic culmination, stretching over 50 km in west–east direction and rising 1800–2000 m above the surrounding intramontane depressions in the central part of the Western Carpathians (latitude 49°8′–49°17′ N; longitude 19°36′–20°20′ E). In this chapter, we present the landscape of the Polish part of the Tatra Mountains, which encompasses one-fifth (ca. 175 km^2) of the entire mountain range area (785 km^2) and is located on its northern slopes (Fig. 14.1a). Basic geomorphological information on the Slovak part of the Tatra Mountains is presented in Boltižiar (2022) and Zasadni et al (2022a). The Tatras are often called 'pocket mountains' in comparison to the Alps and the Caucasus (Kotarba et al. 1987). The main crest of the Polish Tatras stretches over a distance of 25 km and culminates in Mt. Rysy at 2499 m asl. In this relatively small area, a complex nappe-type alpine geological structure occurs, which together with considerable neotectonic uplift provides high relief energy and exceptionally high geodiversity. The Polish Tatras also hosted the largest Pleistocene mountain glaciers in the country, which left behind a vast assemblage of freshly shaped glacial landforms, providing an excellent area to examine the impact of former mountain glaciation (Fig. 14.1b). There are also other glaciated massifs in the Western Carpathians (Low Tatras, Babia Góra, Mala Fatra), but the Tatras experienced the strongest glacial transformation evidenced by the abundance of glacial erosional landforms including deeply incised glacial troughs, hanging valleys, overhanging

P. Kłapyta (✉)
Institute of Geography and Spatial Management,
Jagiellonian University, Gronostajowa 7, 30–387 Kraków, Poland
e-mail: piotr.klapyta@uj.edu.pl

J. Zasadni
Faculty of Geology, Geophysics and Environmental Protection,
AGH University of Science and Technology, al. Adama Mickiewicza 30,
30–059 Kraków, Poland
e-mail: zasadni@agh.edu.pl

P. Migoń and K. Jancewicz (eds.), *Landscapes and Landforms of Poland*,
World Geomorphological Landscapes, https://doi.org/10.1007/978-3-031-45762-3_14

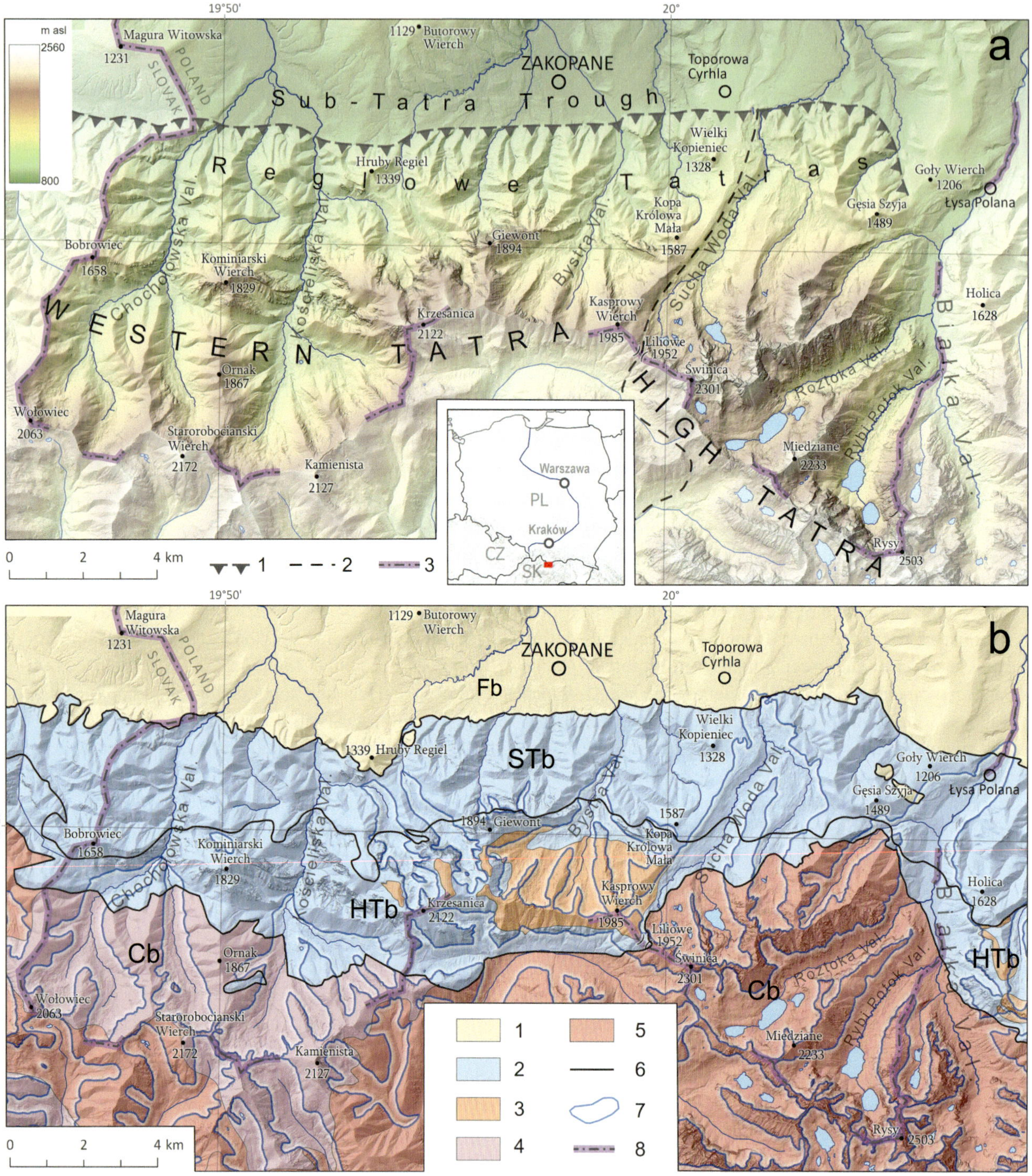

Fig. 14.1 Polish Tatra Mountains. **a** Hypsometry, location, and major topographic subdivision: 1—northern mountain front, 2—border between Western and High Tatras, 3—state border. **b** Lithology (simplified after Piotrowska et al. 2016) and morphotectonic belts: 1—Central Carpathian Palaeogene, sandstones and mudstones (flysch strata); 2—Mesozoic sedimentary cover: limestones, dolostones, marls, sandstones, and quartzites; 3—granitoids incorporated into High-Tatric nappes; 4—metamorphic rocks: gneisses, migmatites, amphibolites, and leukogranites; 5—granitoids; 6—boundaries of morphotectonic belts; 7—extent of LGM glaciers (after Zasadni and Kłapyta 2014); 8—state borders; Cb—crystalline core belt, HTb—High-Tatric belt, STb—Sub-Tatric belt, Fb—Flysch belt

cirque backwalls, aretés, and palaeonunataks. One of the showpieces of this area is an exceptionally large cluster of cirque lakes filling glacially overdeepened basins which demonstrate the vigorous erosion of ice. Among the cirque lakes, the Wielki Staw in the Pięć Stawów Polskich Valley is one of the deepest in Poland (79 m). It is also the only area where perennial snow patches and small glacieretes with associated active periglacial processes occur on the territory of Poland. High relief and the common presence of oversteepened slopes due to glacial erosion provide suitable conditions for the formation of diverse types of rock-slope failures involving the most spectacular catastrophic rockfalls and rock avalanches. Therefore, the Tatras are the only high-mountain landscape in Poland and present one of the most scenic mountain landscapes in Central Europe, which has attracted researchers for years.

A long tradition of geomorphological studies in the Tatras dates back to the pioneering observations of S. Staszic (1815). The history of research on glaciations in the Tatra Mountains began in 1856 when L. Zejszner discovered glacially transported granite boulders on Mesozoic carbonate rocks in the Bystra valley near the town of Zakopane (Zejszner 1856; Kłapyta and Zasadni 2018; Zasadni et al. 2022a). It was one of the earliest glacial discoveries in European mountains, made just after the establishment of the Ice Age theory in the Alps by L. Aggasiz in 1840 (Aggassiz 1840). The geomorphological mapping after World War II (Klimaszewski 1988) and the development of glacial chronology based on radiometric dating of landforms (Baumgart-Kotarba and Kotarba 1997, 2001) brought significant progress in understanding the Late Pleistocene evolution of the Tatra Mts. Currently, research based on the use of cosmogenic nuclides for dating (Dzierżek 2009; Makos et al. 2013, 2018) represents another important step towards comprehensive recognition of the evolution of high-mountain landscape.

14.2 Geology and Structural Relief

The Tatra Mountains form a highly confined and isolated geomorphic unit within the Western Carpathians. Following differences in bedrock lithology and amplitude of relief, two major geological–geomorphological units are distinguished in the Polish Tatra Mountains: the High Tatras and the Western Tatras, separated by the Liliowe Pass (1952 m) located on the main crest (Fig. 14.1a). The High Tatras (Rysy, 2499 m) are built of relatively homogenous Variscan/Carboniferous age granites and are dominated by rocky relief with arêtes, sharp peaks, and oversteepened rock slopes that experienced strong glacial modification (Fig. 14.1b). The Western Tatras (Starorobociański Wierch, 2176 m) are built of metamorphic, crystalline, and sedimentary rocks and are characterised by less pronounced local topography and gentler slope gradients with moderately developed glacial relief (Fig. 14.1b). The High Tatras stand about 400 m higher in respect to the sea level (mean elevation of the crest line ~ 2400 m asl) than the Western Tatras (mean elevation of the crest line ~ 2000 m asl) (Pánek et al. 2020), highlighting the unequal uplift of these two tectonic blocks within the massif. Additionally, the medium-altitude mountain belt of the Reglowe Tatras (Kopa Królowa Mała, 1587 m) is distinguished along the northern rim of the range (Klimaszewski 1988; Balon et al. 2015) (Fig. 14.1a). This area lacks glacial erosional landforms. The sedimentary bedrock built mainly with Middle Triassic dolomites and cuesta-like ridges on monoclinally dipping incompetent beds are characteristic features of the Reglowe Tatras (Klimaszewski 1988; Rączkowski et al. 2015).

The main feature of structural relief of the Tatra Mountains are four parallel morphotectonic belts, which correspond to the main tectonic units and determine the general appearance of the mountain landscape (Fig. 14.1b).

(1) Crystalline belt forms the core of the mountains and takes 41% of the Polish Tatras. Different types of late Carboniferous (360–340 Ma; Kohút and Larinov 2021) intrusive granodiorites built the High Tatras. The most common are porphyritic monzogranites and syenogranites, known as common Tatra-type granites (Gawęda et al. 2015), whereas the upper reaches of the Rybi Potok valley are built with biotitic monzogranites with abundant xenoliths of metamorphic rocks (High Tatra-type granites; Gawęda et al. 2015). In the Western Tatra crystalline core, metamorphic rocks predominate and include mainly migmatites, orthogneisses, paragneisses, and amphibolites together with leucogranite lens-shaped intrusions (Gawęda and Szopa 2011). The most important characteristic of granite bedrock is its jointing. Magmatic-tectonic jointing resulted from cooling of the granitoid pluton, and superimposed tectonic compression is typical for the Tatra Mountains (Jaroszewski 1985). The regular (orthogonal) joint system consists of three sets, perpendicular to one another. Two are nearly vertical and strike NW–SE and NE–SW, and the third set is close to horizontal or gently inclined in the central part of the High Tatras, but inclined to NW at the northern margin (Jaroszewski 1985; Piotrowska 1997), which is well marked in the shape of Mt. Kościelec (2158 m) and Żółta Turnia (2087 m) peaks (Fig. 14.2a). Vertical joint sets are responsible for the formation of smooth monolithic rock plates which are best exposed on the eastern flank of Mt. Mnich (2068; Fig. 14.2b) and the southern flank of the Mt. Zamarła Turnia (2172 m). Extensive tectonization of the crystalline basement resulted in the formation of dislocations and mylonitic/cataclastic shear zones (Piotrowska 1997), reduction of rock mass strength, and predisposition to the formation of deep gullies and cols along the NE–SW and NW–SE running zones (Fig. 14.2c). The crystalline basement in the Western Tatras forms numerous thrust sheets separated by competent fault rocks (tectonic breccias, fault gauges, and phyllonites) developed as the result of brittle shearing during the Alpine orogenesis (Cymerman 2010).

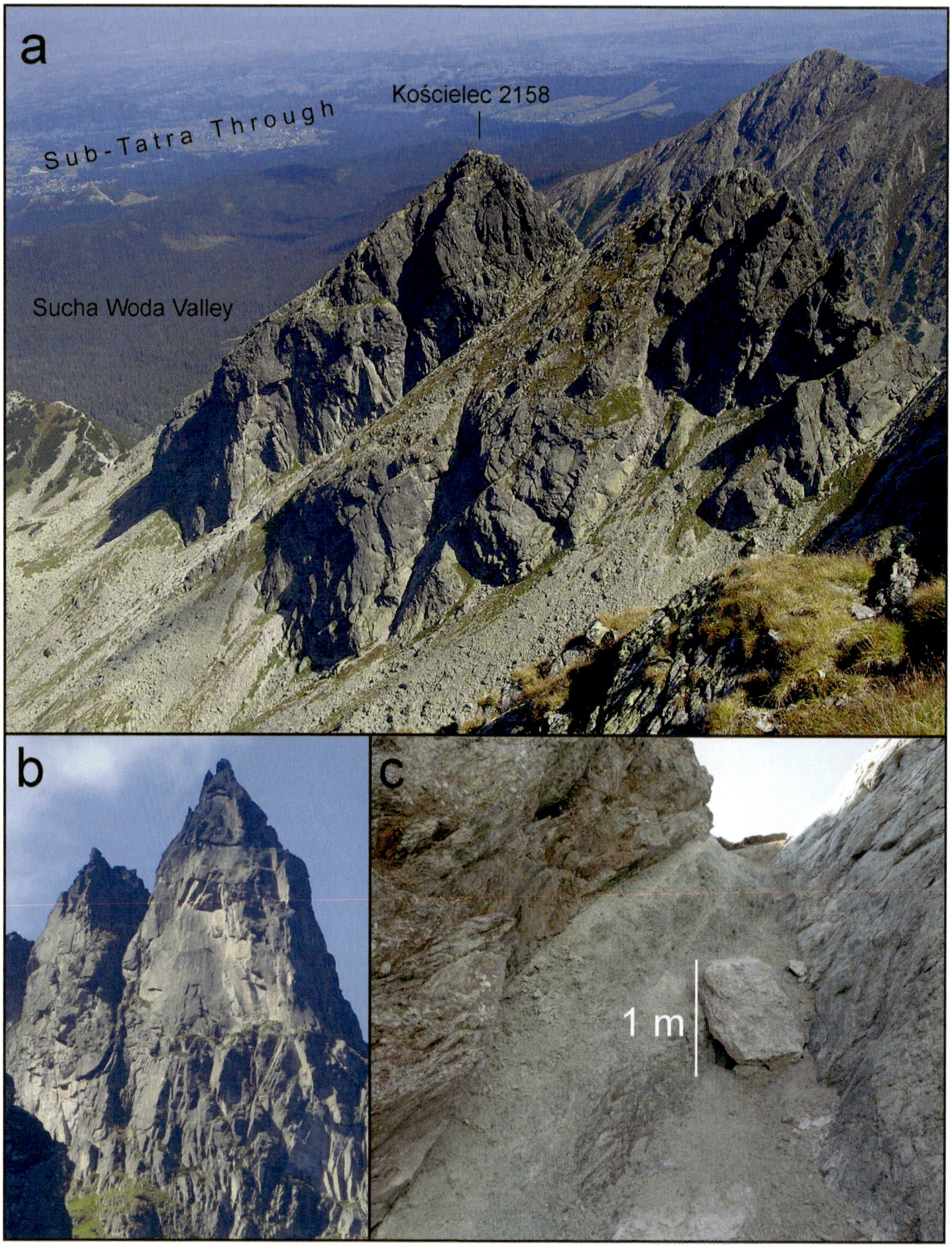

Fig. 14.2 Structural relief in the crystalline core of the High Tatras. **a** NW-inclined set of joints controls the asymmetric shape of Mt. Kościelec. **b** Smooth monolithic rock plates on the eastern flank of Mt. Mnich (2068 m), corresponding to a vertical set of joints; **c** couloir developed along mylonitic zone with a competent chlorite-rich fault-gauge in the tourist route to the Pod Chłopkiem Pass (Rybi Potok Valley)

(2) High-Tatric belt (*pas wierchowy* in Polish) encompasses 19% of the Polish Tatras and includes Mesozoic sedimentary cover which was deposited directly over the crystalline core (autochtonous cover), as well as the Czerwone Wierchy and Giewont tectonic nappes (para-autochtonic cover), which together belong to the Tatric sedimentary realm (Nemčok et al. 1994) (Fig. 14.1b). The High-Tatric belt concentrates most of the high-mountain surface and subsurface karst phenomena, including the longest and deepest caves

in Poland (Szczygieł 2022). Within a thick sequence of High-Tatric sedimentary rocks, Lower Triassic red quartzitic sandstones and Upper Jurassic-Lower Cretaceous limestones are particularly prominent in the morphology (Jach et al. 2014). Due to high strength and resistance, red quartzitic sandstones form extensive block fields (Fig. 14.3b), while massive and thick-bedded limestones support precipitous rock walls (Mt. Giewont, Raptawicka Turnia, northern slopes of the Czerwone Wierchy, Zawrat Kasprowy) (Fig. 14.3a) and narrow sections of valleys locally known as 'gates'. Steeply dipping (above 45°) incompetent limestone beds form narrow hogback-type ridges. Among them, the 3-km-long Mt. Giewont ridge stands out as a prominent geomorphic landmark that rises above the surrounding terrain by 600 m (Fig. 14.3a).

(3) Sub-Tatric belt (*pas reglowy* in Polish) takes 33% of the Polish Tatras and consists of Fatric (Križna nappe) and Hronic tectonic units (Choč nappe; Nemčok et al. 1994) (Fig. 14.1b). Because of the deeper sedimentary environment than that of the High-Tatric rocks, brittle densely fractured dolomites and soft marls and mudstones dominate in the Sub-Tatric belt (Jach et al. 2014). The medium-altitude landscape of the Sub-Tatric belt is formed by numerous thrust sheets that dip monoclinally to the north and form asymmetric ridges, with rock outcrops on the escarpment slopes. Approximately half of the area is dominated by bedded dolomites and dolomite breccia, which, due to selective mechanical breakdown, form many picturesque rocky towers, tors, rocky steps, and ruiniform relief (Fig. 14.3c). The presence of weak mudstones and siltstones promotes the formation of shallow landslides.

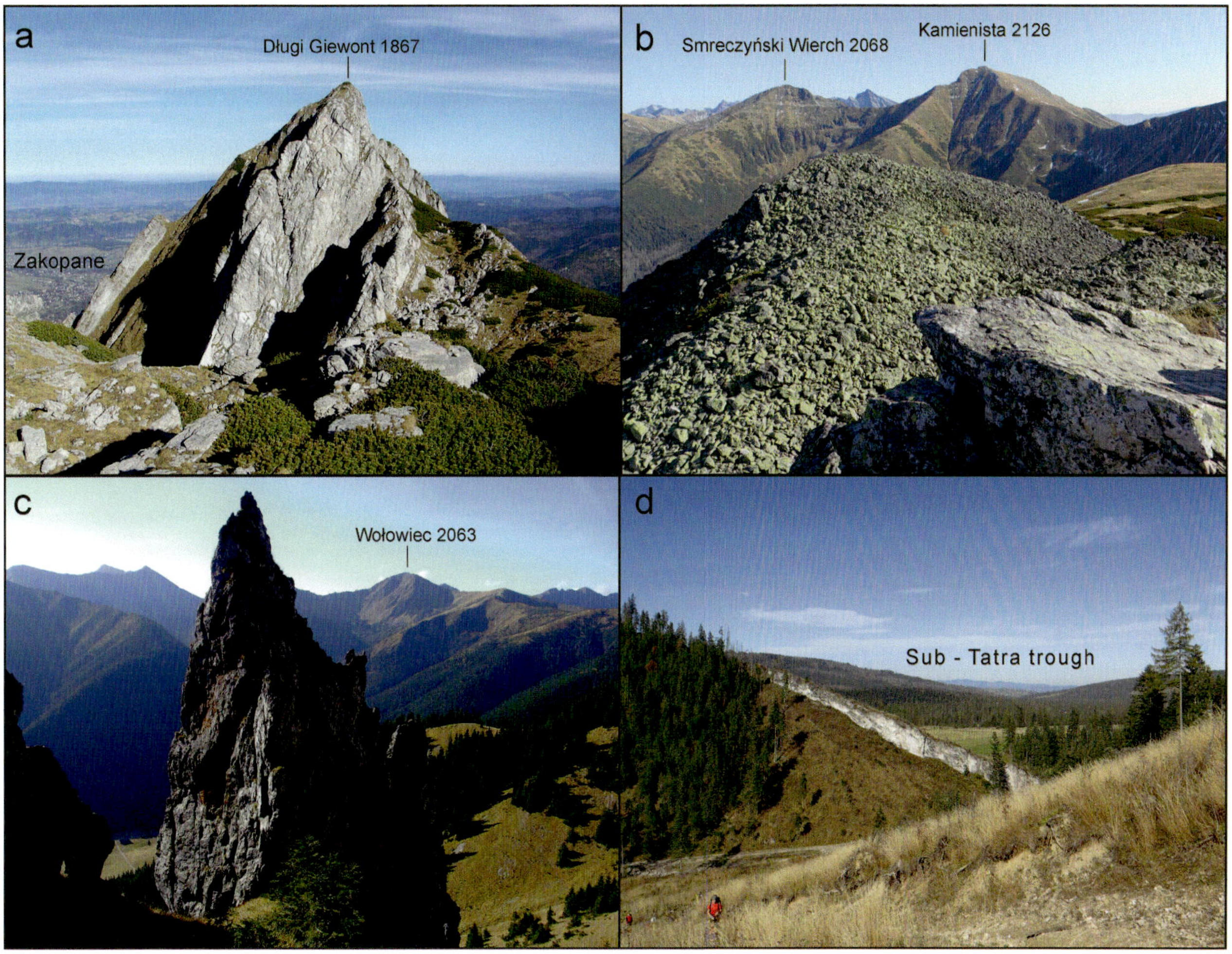

Fig. 14.3 Relief of the High-Tatric and Sub-Tatric belts. **a** Narrow hogback type ridge of Mt. Giewont developed along steeply dipping (80°) Upper Jurrasic-Lower Cretaceous limestones. **b** Mt. Zadni Ornak (1867 m), with block fields developed on very resistant Lower Triassic quartzitic sandstones. The main ridge of the Western Tatras built by metamorphic rocks is visible in the background (crystalline belt). **c** Mnichy Chochołowskie—rock towers built of Middle Triassic dolomites. **d** Rock wall developed along monoclinally dipping nummulitic limestone beds in the mouth of the Lejowa Valley close to the northern mountain front rising above the Sub-Tatra trough (Flysch belt)

(4) The Flysch belt encompasses 7% of the Polish Tatras and is limited to the narrow zone (200–700 m wide) stretching in the northernmost part of the mountains (Fig. 14.1b). The southern limits of this belt correspond to the extent of post-tectonic sedimentary cover of Paleogene age, built of conglomerates, nummulitic limestones, and dolomitic sandstones that are traditionally called "Nummulitic Eocene" (Jach et al. 2014). Only the oldest part of the transgressive succession is preserved on the uplifted margins of the Tatra tectonic block, while the younger, up to 3-km-thick mudstone and turbiditic sandstone successions (so-called Podhale Flysch) are widely distributed in the northern Tatra foreland (Chowaniec et al. 2009) (Fig. 14.3d).

The northern slope of the Tatra Mountains between the Chochołowska and Sucha Woda valleys forms a distinct, almost straight mountain front (Figs. 14.1a and 14.4) descending steeply (25–35°) into the Sub-Tatra trough (900–700 m asl). Traditionally, this mountain front was linked with the lithological difference between high-resistance Mesozoic strata in the Tatras and low-resistance Podhale flysch strata in the foreland. The mountain front is, however, built locally of flysch strata (e.g. Mt. Hruby Regiel) and shows well-developed triangular facets (Fig. 14.4), which suggest its tectonic (faulted) origin (Buczek and Górnik 2020).

14.3 Main Features of Glacial Relief

14.3.1 Erosional Landforms

The Tatra Mountains, situated midway between the former rim of the Scandinavian ice sheet and Alpine ice cap, represent one of the most impressive glaciated landscapes in Central Europe. In the Polish Tatras 13 glaciers existed during the Last Glacial Maximum (LGM), covering an area of 57.25 km^2 (Fig. 14.1b). This was, however, only 20% of

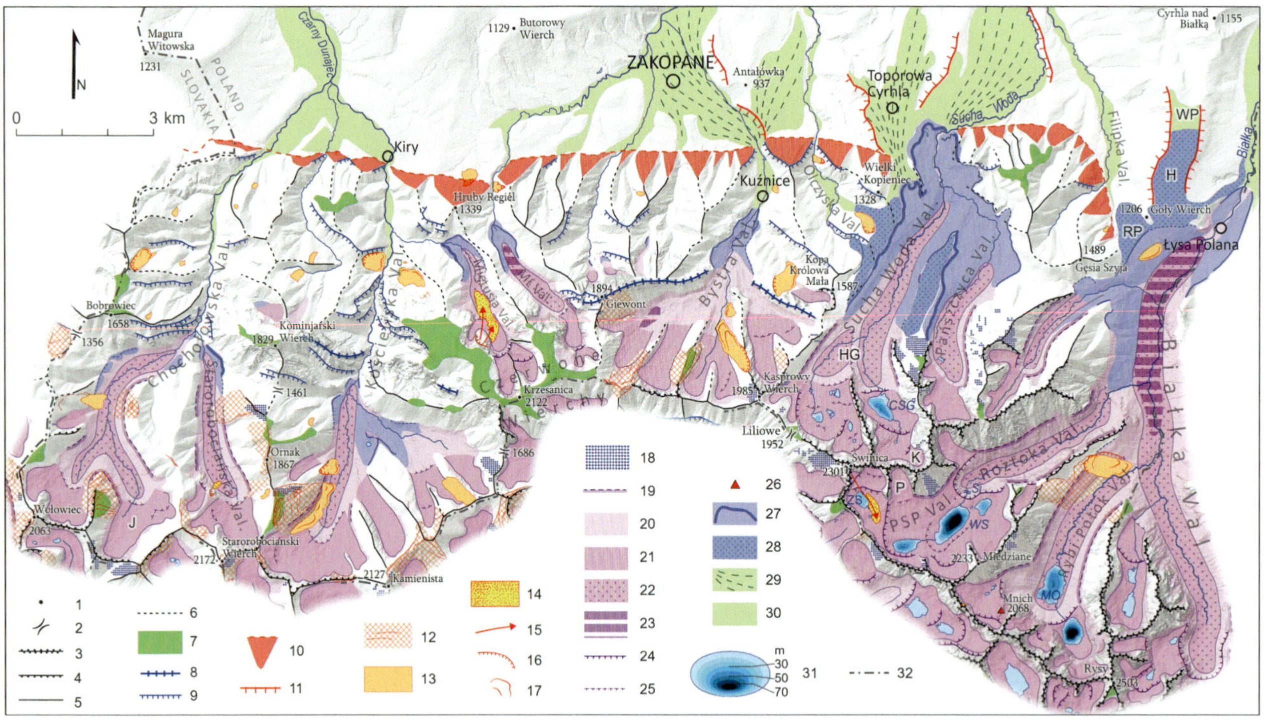

Fig. 14.4 Geomorphological map of the Polish Tatra Mountains. 1—summits, 2—major passes, 3—arêtes, 4—asymmetrical ridges, 5—sharp ridges, 6—flat ridges, 7—mountain-top flat surfaces, 8—vertical (hogback) ridges with rock walls, 9—monoclinal ridges with rock escarpments, 10—triangular facets of mountain front, 11—distinct erosional edges, 12—deep-sated gravitational deformations with gravitational faults and ridge-top trenches, 13—landslides, 14—rock avalanches, 15—travel paths of rock avalanches, 16—landslide scarps, 17—bulges of landslide masses, 18—periglacial block fields, 19—glacial trimlines, 20—glaciated area during the LGM, 21—floors of glacial cirques, 22—floors of glacial troughs (U-shaped valleys), 23—sedimentary fills of overdeepenings in glaciated valleys, 24—main glacial thresholds with rock walls, 25—minor glacial thresholds, 26—palaeonunatak, 27—LGM till cover and latero-frontal moraines, 28—pre-LGM degraded till covers, 29—alluvial/fluvioglacial fans 30—alluvial/fluvioglacial plains, 31—glacial lake bathymetry in metres, 32—state border. J—Jarząbcza Valley, MŁ Val.—Mała Łąka Valley, PSP Val.—Pięć Stawów Polskich Valley, HG—Hala Gąsienicowa, K—Kozia Dolinka Valley, P—Pusta Valley, RP—Rusinowa Polana glade, H—Hurkotne site, WP—Wierch Poroniec site, MO—Morskie Oko Lake, CSG—Czarny Staw Gąsienicowy Lake, WS—Wielki Staw Lake, ZS—Zadni Staw Lake, Si—Siklawa waterfall

the glaciated area in the entire Tatra massif (a total of 55 glaciers covering 280 km^2). The Białka glacier located on the northern slope of the High Tatras, along the present-day Polish-Slovak border, was the largest Tatra's glacier in every respect (area 43.6 km^2, length 13.4 km, max. thickness ca. 400 m, and the lowermost terminus elevation 930 m). However, most glaciers in the Polish Tatras were typical valley glaciers between 4 and 8 km long, terminating inside the Tatra massif in the elevation range of 1050–1100 m asl. Only the glaciers in the Białka and Sucha Woda valleys in the High Tatras reached the mountain front (Zasadni and Kłapyta 2014; Zasadni et al. 2022c) (Figs. 14.1b and14.4).

A striking feature of the Tatra glacial landscape is a well-developed set of glacier erosional landforms inherited from Pleistocene glaciations, consistent with an advanced stage of landscape transformation by glaciers (Zasadni et al. 2021). Due to the high strength of granites and relatively large ice thickness, erosional landforms are especially well-developed and preserved in the High Tatras. Glacial erosion was especially vigorous in the NW corner of the High Tatras which was exposed to the prevailing westerlies which produced orographic-induced precipitation that enhanced high rates of ice accumulation and mass turnover (Zasadni et al. 2021, 2022c). The assemblage of large-scale glacial erosional landforms includes nested (compound) cirques, glacial troughs with hanging valleys, trimlines, arêtes, truncated spurs, and palaeonunataks (Lukniš 1973; Klimaszewski 1988; Makos and Nitychoruk 2011; Křížek and Mida 2013; Zasadni and Kłapyta 2014; Zasadni et al. 2022c) (Fig. 14.4). The strongest glacial modification in the High Tatras is revealed by numerous examples of staircase cirques with overhanging backwalls and large overdeepened basins on its floors, now occupied by lakes. The compound glacial cirque of the Pięć Stawów Polskich valley represents the largest glacial cirque in the Polish Tatras (ca. 4.5 km^2), which is suspended 200 m above the head of glacial trough of the Roztoka Valley (Fig. 14.4).

Cirque overdeepenings in the Białka and Sucha Woda valleys filled by lake waters record the largest (34 ha—Morskie Oko lake, 33 ha—Wielki Staw lake) and deepest lakes in the Tatras (79 m—Wielki Staw lake) (Fig. 14.4). Concentration of cirque lakes is among the largest in Central Europe (Kłapyta et al. 2016). It should be noted that 35% of the cirque lake area of all glaciated massifs of the Carpathians is concentrated within the Polish High Tatras (Zasadni et al. 2021, 2022c). Most cirques in the Western Tatras are simply valley-head or valley-side features, with a backwall engraved in gentler sloping terrain but without overdeepened floors (Fig. 14.4).

In the High Tatras, protracted process of cirque enlargement on two or more sides produced serrated ridgelines known as arêtes (Fig. 14.4). As a result of the advanced stage of rock wall retreat leading to the complete destruction of arêtes in some places in the High Tatras, isolated arête-type summits were formed. During the ice age, such summits acted as nunataks because they were surrounded by ice, but in terms of the present-day relief, they should be named paleonunataks (Rae et al. 2004). Among six examples of palaeonunataks known in the entire Tatras, Mt. Mnich is the only example in the Polish Tatras (Zasadni and Kłapyta 2014; Zasadni et al. 2022c) (Fig. 14.4).

Classic examples of glacial troughs with well-developed U-shaped cross-sections and staircase longitudinal profiles occur in the Białka Valley catchment (Fig. 14.5a). The vertical extent of glacial modification in the formerly glaciated valleys is indicated by glacial trimlines (Fig. 14.5b). They can be observed as steepened valley sides or rock walls carved by lateral erosion of glaciers during its maximal extent, e.g. during the LGM (Makos et al. 2013). Therefore, trimlines can be used as indicators of former ice thickness and ice surface geometry where moraine record is lacking (Zasadni and Kłapyta 2014). Above the trimlines occurs gently sloping, rugged terrain, often with blockfields developed under periglacial conditions (Fig. 14.5b). Trimlines commonly occur in the High Tatras (Makos et al. 2013; Zasadni and Kłapyta 2014; Zasadni et al. 2022c), whereas they are rare in the Western Tatras, with the exception of the Starorobociańska Valley.

Unique glacial erosional features found in the lower sections of the Mała Łąka (Kotarba et al. 1977) and Białka valleys (Baumgart-Kotarba et al. 2008) are terminal overdeepened basins recognized by geophysical methods (Fig. 14.4). Overdeepening of the valley floor up to 100 m is also suspected in the Kondratowa Valley (Bystra Valley catchment) (Tomecka-Suchoń et al. 2017) (Fig. 14.4). During deglaciation, overdeepened basins likely acted as proglacial lakes that were subsequently filled with up to 65–150 m thick alluvial sediments.

The occurrence of glacially polished bedrock, roche moutonnées, and striae is restricted to the crystalline belt and headwater reaches of High Tatras valleys: Sucha Woda, Rybi Potok, and Pięć Stawów Polskich (Fig. 14.4). Due to the medium- to coarse-grained structure of the High Tatra granite, striae are poorly developed and preserved. The best locations to observe these landforms are rims of glacial thresholds, e.g. the Siklawa waterfall threshold between the Pięć Stawów Polskich and Roztoka valleys (Fig. 14.5d, e). Glacial erosion often formed crescentic gauges or p-forms on polished bedrock, but such landforms are rarely found in the Polish Tatra Mts.

Fig. 14.5 Glacial erosional landforms in the Tatra Mountains. **a** Glacial trough with hanging tributary troughs (Rybi Potok and Roztoka valleys). WM—location of the Wodogrzmoty Mickiewicza waterfall. **b** Glacier trimline (dashed line) on the eastern side of the Rybi Potok Valley. **c** Mt. Mnich (2068 m) rock needle is an isolated sharply pointed summit (arête type) in the upper part of the Rybi Potok Valley. It is the only representative of palaeonunatak landform in the Polish Tatras. **d** Glacially polished bedrock (proximal side of roche moutonnée) on the Siklawa waterfall glacial threshold in the Pięć Stawów Polskich Valley. **e** Glacially polished granite bedrock with barely visible striae (Pięć Stawów Polskich Valley). In d and e, white arrows show the direction of ice movement

Fig. 14.6 Moraines and relict rock glaciers in the Tatra Mountains. **a** Boulder-rich till cover of the penultimate glaciation at Rusinowa Polana glade. **b** Sharp-crested LGM lateral moraine in the Bystra Valley close to the Kuźnice settlement. **c** Exposure of till of terminal moraine in the Sucha Woda Valley (LGM). **d** "Palenica boulder"—the largest glacially transported boulder in the Tatras located in the lower section of the Białka Valley on the LGM moraine. **e** Openwork boulder accumulation on the western lateral moraine of the Białka glacier near Rusinowa Polana glade. **f** Dead-ice depressions with a small lake (Toporowy Staw lake) and peat bogs in the Sucha Woda Valley LGM moraine complex, High Tatra Mts. **g** Recessional moraines (dashed lines) and moraine-derived relict rock glacier (solid line) in the upper part of the Jarząbcza Valley, Western Tatra Mts

14.3.2 Depositional Landforms

The fresh-shaped outer moraines in the Tatra Mountains are traditionally attributed to the last glacial cycle/Last Glacial Maximum (Partsch 1923; Romer 1929; Klimaszewski 1988; Baumgart-Kotarba and Kotarba 1997; Zasadni and Kłapyta 2014; Zasadni et al. 2022c) (Fig. 14.6b). In the Polish Tatra Mountains, the spatial extent of LGM glaciers is comparable with previous glacial maxima. Older moraines (older than LGM) are preserved locally and only in the High Tatras. They occur as highly degraded ridges that locally could be traced up to 1 km beyond the limit of the LGM moraine (Fig. 14.4). The best example of such a moraine sequence occurs at the Hurkotne plateau and the Rusinowa Polana in the Białka Valley, where a unique Pleistocene terminal moraine complex with two pre-LGM moraine units was recognized (Zasadni et al. 2021, 2022b). The Hurkotne plateau stands about 220 m above the contemporary valley bottom and is covered with highly degraded tills with completely disintegrated boulders, which are considered as the oldest preserved glacial deposit in the Tatras (MIS 12–18?). The younger pre-LGM till cover with abundant granitic boulders up to 4–5 m long is known from the Rusinowa Polana glade, where local diffluence toward the Filipka Valley occurred during the penultimate glaciation (MIS 6) (Figs. 14.4 and 14.6a) (Baumgart-Kotarba and Kotarba 1997; Zasadni et al. 2021, 2022b). The pre-LGM glacier diffluence also occurred from the Sucha Woda Valley to the Olczyska Valley (Klimaszewski 1988; Fig. 14.4). The Mesozoic carbonate bedrock on the floor of the Olczyska Valley is covered with glacial till and large granitic boulders derived from the headwater parts of the Sucha Woda Valley.

During the LGM, the lower parts of the ablation zones of many glaciers were strongly overloaded by thick supraglacial debris (Zasadni and Kłapyta 2014; Zasadni et al. 2022c) (Fig. 14.6e). A unique example of a morainic amphitheatre formed by a debris-covered glacier was developed in the mouth of the Sucha Woda valley (Fig. 14.4). Distal slopes of massive latero-frontal dump moraines reach here up to 60 m high and bound inner hummocky terrain with numerous dead-ice hollows filled by small lakes and peat bogs (Fig. 14.6f). The formation of debris-covered glaciers could be related to weak sediment removal from the glacier system by meltwater. In this context, the exceptionally well-developed karstic system in the Sucha Woda Valley (Głazek 1964) acted as efficient subglacial meltwater drainage which caused limited sediment evacuation from the Sucha Woda glacier system and promoted increased debris build-up.

A large concentration of open work debris found on LGM moraines indicates passive glacial transport of rock material originating from rockfalls that occurred on the surrounding slopes. The 'Palenica boulder' located on the left-hand side of the Białka Valley is the largest glacial boulder within the moraines in the Tatra Mountains (16.2 m long, 11.3 m wide, and 9.3 m high, 514 m^3 and 1336 t; Zasadni et al. 2021) (Fig. 14.6d). Incidentally, it is similar in size to the Trygław boulder in the Pomeranian Lakeland, considered to be the largest erratic boulder in Poland (Górska-Zabielska 2015).

Climate fluctuations during deglaciation made glaciers oscillate, re-advance, and form 5–10 recessional moraine in the High Tatras (Baumgart-Kotarba and Kotarba 1997, 2001; Zasadni and Kłapyta 2016) and 1–4 moraine/rock glaciers generations in the Western Tatras (Kłapyta 2012). Relict moraine-derived and talus-derived rock glaciers are especially well developed in the Western Tatras (Fig. 14.6g), highlighting the simultaneity of glacial and periglacial processes during deglaciation.

14.4 Mass Movements

The Tatra Mountains are characterized by frequent occurrences of various types of rock-slope failures, including large rockslides (above 0.25 km^2), rock avalanches, and rock-slope deformations (sackungen) (Fig. 14.7). High magnitude and frequency of mass movements were predisposed by the presence of glacially steepened slopes and a dense network of tectonically weakened mylonitic zones. Rock-slope deformations involving tension cracks, ridge-top trenches (Fig. 14.7d, e, f), downhill- and uphill-facing scarps, and bulging toes with diffuse margins and convex cross-sections are the most frequent in the Western Tatras, whereas they are almost lacking in the High Tatras, where rockfalls and rock avalanches predominate (Němčok 1982; Pánek et al. 2017) (Fig. 14.4).

Sackung scarps are concentrated above the glacially steepened slopes where slope extension is enhanced. This led to the opening of joints and short horizontal displacements of bedrock packages. The most extensive sackung deformations in the Polish Tatras (1.2 km^2 area) have affected the ridge and the eastern slopes of Mt. Ornak in the Western Tatras (Figs. 14.4 and 14.7e). The lower portion of the sackung-deformed slope collapsed as a rockslide into the Pyszniańska Valley. Chronological studies on sackungen in the Tatra Mountains (Pánek et al. 2017) indicated that some landforms may have originated soon after ice retreat, but the majority could be the result of long-term stress relaxation within the rock mass that lasts several (>5) millennia or could be triggered by heavy rainfall or seismicity.

Rock avalanches represent the most spectacular types of rock-slope failure resulting from sudden slope collapses

Fig. 14.7 Examples of rock-slope failures in the Polish Tatra Mountains. **a** Rock avalanche tongue in the Pod Kołem cirque (the upper part of Pięć Stawów Polskich valley) with longitudinal and transversal ridges and avalanche source area located within the backwall. **b** Extent of the Wantule rock avalanche (red solid line) in the Miętusia Valley (derived from LiDAR DEM) in relation to the extent of the LGM ice extent (dashed white line). **c** Active rockfall niche below the Dudowe Turnie in the Starorobociańska valley, Western Tatras. **d** Hlińska Pass with double-crested ridge and one of the longest (approximately 1 km) ridge-top trenches in the Tatras. **e** Sackung scarps with sag pond on the ridge on Mt. Ornak. **f** Ridge-top trenches at the Pyszniańska Pass. **g** Large rockslide on the western slope of Mt. Kasprowy Wierch

(Fig. 14.7c). The most impressive examples can be found in the Pod Kołem cirque in the High Tatras, where 660-m-long and 180-m-wide debris lobes built of disintegrated, open-work angular debris occur (Fig. 14.7a). Rock avalanche deposits with extraordinarily large blocks, often exceeding 10 m in length (maximum 32 m), had travelled partly on a glacier surface at the end of the Oldest Dryas (ca. 15 ka) when the Zadni Staw Lake depression was still filled with ice (Zasadni and Kłapyta 2016) (Fig. 14.7a). The Wantule rock avalanche deposit located in the Miętusia Valley

(1200 m long, 0.23 km^2 area) is the best example of catastrophic rock-slope failure in the Western Tatras, formed by limestone rock wall collapse at ca. 11 ka BP (Tołoczko-Pasek et al. 2020) (Fig. 14.7b).

14.5 Landscape Evolution

14.5.1 Neotectonic Uplift

The impressive present-day landscape and geomorphological individualization of the Tatra Mountains resulted from the youngest phase of its morphodynamic evolution related to neotectonic uplift during the Late Miocene, Pliocene, and Pleistocene, at 6.5–1 Ma (Králiková et al. 2014). Accelerated uplift (up to 1 mm/a) followed a period of relatively slow (0.2 mm/a) exhumation during the late Middle to early Late Miocene when no prominent mountain relief existed and the Tatras were still entirely covered with the Podhale flysch (Králiková et al. 2014). Neogene uplift of the massif was asymmetrical and occurred along the 50-km-long Sub-Tatra Fault, which represents a deeply rooted basement structure located on the Slovak side of the mountains, with a total dip-slip offset of ca~9 km (Anczkiewicz et al. 2015). The Tatra Mountains are one of the most neotectonically active areas in the Western Carpathians (Minár et al. 2011; Pánek et al. 2020). Large-scale tilting of the horst to the north (40°) led to the formation of N–S altitudinal asymmetry with north-facing valley heads located ~ 150 m lower than the south-facing ones (Křížek and Mida 2013), less efficient deepening of the northern side valleys and monoclinal dip of the sedimentary cover. The remnants of wide ridges and flattened mountain top surfaces are locally preserved, especially in the Western Tatras (Czerwone Wierchy) (Fig. 14.4). Their previous interpretation as planation surfaces formed during several periods (Miocene–Pliocene) of tectonic stabilisation (Klimaszewski 1988) is inconsistent with thermochronological data which evidence continuous neotectonic uplift of the Tatra block since the 20 Ma, with the phase of accelerated uplift since 7 Ma up to the late Quaternary (Králiková et al. 2014; Pánek et al. 2020).

Geomorphological differentiation between the Western and High Tatras resulted from the higher uplift rate along the eastern segment of the Sub-Tatra Fault, where postglacial fault scarps are by 15% higher along the margin of High Tatras than in the Western Tatras (Pánek et al. 2020). The most recent Pleistocene uplift rate along the Sub-Tatra Fault is up to 0.75 mm/a (Pánek et al. 2020), but much lower (0.4 ± 0.2 mm/a) on the northern side of the mountains (Makowska and Jaroszewski 1987; Makowska 2003). Morphological contrast along the northern Tatra foreland suggests, however, different tectonic regimes in its western and eastern parts. A low-relief (<50 m) morphology with broad-spread alluvial terraces and fans in the Sub-Tatra Trough between the Oravice and Zakopane settlements points to a lack of Pleistocene tectonic uplift. This is in line with cave data from the Bystra, Kościeliska, and Chochołowska valleys, where the oldest speleothems in cave conduits at or below the modern valley floor show no valley incision in Western Tatras after ca. 330 ka (Szczygieł et al. 2020).

In contrast, uneven uplift (~100–220 m) of smaller tectonic blocks (Bystra-Antałówka, Toporowa Cyrhla, Hurkotne-Wierch Poroniec) is suggested for the northern foreland of the High Tatras (Makowska and Jaroszewski 1987; Makowska 2003;). The culmination of uplift (~220 m) occurred at the location of the Hurkotne plateau, whose low relief is interpreted as the remnant of the originally flat (<5°) foreland surface, which after the earliest glacier advance (or advances) (MIS 12–18 ?) was tectonically uplifted and subsequently dissected by the Białka and Filipka valleys (Zasadni et al. 2022b) (Fig. 14.4).

Recent findings of seismic morphology along the Sub-Tatra Fault (Pánek et al. 2020) document postglacial faulting (between 18 ka and the Holocene) and high-magnitude earthquake(s) with possible magnitude Mw>7, which could have been a trigger for many rock-slope failures in the interior of the Tatras. The mechanism of this exceptionally large seismic event(s) with an estimated slip rate between 0.1 and 0.3 mm/a is unclear, taking into account modest recent seismicity in the area, with historical earthquakes reaching only Mw of ~ 4.5–5.5 and limited to the Orava-Nowy Targ and Podhale Basins (Pagaczewski 1972; Pánek et al. 2020). It is suggested therefore that the trigger mechanism of faulting was related to the far-distance glacial isostatic adjustment of Earth crust during the retreat of the Scandinavian ice sheet after the Weichselian glaciation (Pánek et al. 2020).

14.5.2 Glacial Chronology

14.5.2.1 Maximum Glaciations

The scenic alpine-type mountain landscape in the Tatra Mountains is an effect of strong and effective relief transformation during several Pleistocene glacial cycles; however, the exact number of glaciations is unknown. Up to the eight glaciations were proposed by Lindner et al. (2003) based on alluvial terrace morphostratigraphy and TL dating in the Tatra foreland; however, only up to three moraine units were distinguished within the sequences of Pleistocene moraines inside the Tatra Mountains

(Baumgart-Kotarba and Kotarba 1997; Zasadni et al. 2021). Due to poor preservation of the old moraine landscape and effective reworking of glacial deposits on steep valley sides, the chronology of pre-LGM glaciations is poorly constrained. The age of the oldest preserved glacial deposit (Hurkotne plateau) inferred from the rate of tectonic uplift in the Tatras can be approximated as 0.55–0.73 Ma (MIS 12–18), which corresponds to the end of the Mid-Pleistocene Transition (Zasadni et al. 2021, 2022b). The younger pre-LGM moraine unit preserved ca. 1 km beyond LGM moraines may represent the penultimate glaciation (MIS 6). As a consequence of coupled glacial erosion and tectonic uplift, the extent of the Białka glacier during the last glaciation was ca 1 km (7%) shorter, and due to the valley erosion and incision up to 220 m, its front was placed at lower altitude than the terminal moraines from the oldest glaciation (Zasadni et al. 2021, 2022b, c).

During the last glacial cycle, local glaciers achieved their maximum extent during the global Last Glacial Maximum (LGM; 26.5–19 ka; Clark et al. 2009) (Fig. 14.8). In the Polish Tatras, the best-documented LGM chronologies are presented using ^{10}Be datings for the Sucha Woda Valley, where the mean age of terminal moraine stabilization was around 21.5 ± 2.5 ka (Makos et al. 2018). The maximal glacier expansion was caused by a cold and dry climate when the altitude of the glacier equilibrium line (1580 m asl) was lowered about 1000 m below its hypothetical present-day position (Zasadni and Kłapyta 2009, 2016; Makos et al. 2018).

14.5.2.2 Last Deglaciation

The onset of deglaciation after LGM in the Tatra Mountains is dated to ca 20–18 ka, when valley and piedmont glaciers abandoned their outer moraines (Engel et al. 2015; Makos et al. 2018). Contemporarily with the onset of frontal retreat, ice-surface lowering began in the glacier accumulation areas, in the high-elevated trimline positions (1900–2100 m asl) at ca 21.5 ± 1.6 ka (Makos et al. 2013). Further deglaciation formed multi-moraine sequences, which in the High Tatras can be grouped into five major Lateglacial stages (Zasadni and Kłapyta 2016; Zasadni et al. 2023a). The oldest group of recessional moraines mark at least four re-advances of valley glaciers during the Oldest Dryas, dated for 17.3 ka (Dzierżek 2009), 16.6 ± 0.3 ka, 15.7 ± 1.0 ka (Makos et al. 2018), and 14.9 ± 0.4 (Zasadni et al. 2020, 2023a). Recent chronological data (Makos et al. 2018; Zasadni et al. 2020) show advanced deglaciation at the end of the Oldest Dryas (ca 15 ka), with only small cirque glaciers existing in the High Tatras (Fig. 14.9a). This suggests that all glaciers have vanished even in the highest cirques of the High Tatras during the subsequent Bølling/Allerod interphase (14.7–12.9 ka) (Zasadni et al. 2023b). As a consequence, all the cirques in the Polish Tatras were already ice-free at 14–15 ka (Fig. 14.9a). The youngest glacial advance in the Tatras took place during the succeeding Younger Dryas (12–11 ka) climate deterioration (Makos et al. 2018; Zasadni et al. 2020, 2023c), which was the last cold spell of the Pleistocene. During this cold phase, only small, apron-type cirque-niche glaciation was rebuilt in the most elevated and most sheltered cirques. Glaciers were located very close to the headwalls (<500 m), at the elevation range ~1900–2000 m asl (Zasadni and Kłapyta 2016; Zasadni et al. 2020, 2023c) (Fig. 14.9).

The regrowth of glaciers during the Younger Dryas resulted mainly from greater winter cooling (3–4 °C colder than today) and increased seasonality of temperature, which persisted until the Early Holocene (ca. 11–10 ka). Such climatic conditions and marginal glaciation favoured the formation of rock glaciers, which represent a phenomenon of creep of permanently frozen hillslope cover deposits in the periglacial realm (Haeberli 1985). Rock glacier activity prolonged into the beginning of the Holocene (10.4 ka; Zasadni et al. 2020, 2023c) and ceased as the result of increasing summer temperatures (~2.7 °C) and the onset of a pronounced precipitation phase dated at ~10.2 ka in Poland and the Western Carpathians (Starkel et al. 2013; Zasadni et al. 2020, 2023c). Finally, the history of glaciation was completed with the beginning of the Holocene. The present-day snow-ice patches (glacieretes) nested in sheltered slope hollows in the High Tatras (Fig. 14.10) are too small to be considered as evidence of the Late Holocene (Neoglacial) glaciation (Gądek 2014; Kłapyta et al. 2023).

14.5.3 Present-Day Geomorphological Processes

The Tatra Mountains represent the only high-mountain sediment transport system in Poland, which is characterized by high production rates of loose debris and its effective transfer, mainly accomplished by rapid mass movements (rockfalls, debris flows) (Kotarba et al. 1987; Kotarba and Migoń 2010). High-mountain areas are affected by a set of morphogenetic processes determined by local relief and topoclimate (physical weathering, rockfall, debris accumulation, nivation), with debris flows as the most important among the processes. By contrast, in the medium-altitude mountain areas, the most important geomorphic processes are fluvial and slow mass movements (Rączkowska 2008).

Significant landform re-modelling is associated with extreme hydrometeorological events during heavy rainfall episodes that occur especially during summer. For example, on 30 July 1973, the highest ever daily precipitation in the Polish Tatras (and also in Poland) was recorded at Hala Gąsienicowa station—300 mm (Niedźwiedź et al. 2015). Local heavy downpours with the intensity of more than 1 mm/min (such as 2007 and 2018) trigger floods, debris

Fig. 14.8 Tatra Mountains during the Last Glacial Maximum (LGM). **a** Visualization of glaciers (after Zasadni and Kłapyta 2014) based on ESRI City Engine 3D interactive model available online https://bit.ly/3mwgthn. **b, c** and **d** Photorealistic reconstruction of the Białka glacier (Zasadni and Świąder 2015) rendered in Terragen ™ software: (b) 2-km-wide glacier tongue in the vicinity of Rusinowa Polana; (c) Rybi Potok tributary glacier; (d) General view on the glacier. RP—Rybi Potok glacier, PSP—Pięć Stawów Polskich cirque ice-field, P—Pańszczyca glacier, MŁ—Mała Łąka glacier, Mt—Miętusia glacier, K—Kościeliska glacier, St—Starorobociański glacier, Ch—Chochołowski glacier

flows, and mudflows in high-mountain sections, and torrential flows in the less elevated mountain belt (Gorczyca et al. 2014; Krzemień and Kłapyta 2018). Debris movement affecting entire hillslopes requires precipitation magnitudes of 35–40 mm per hour or at least 80–100 mm/day (Kotarba 1992); however, such thresholds vary with lithology and relief (Kotarba et al. 2013). Debris flow tracks occur in almost every glacial cirque. One of the longest debris flows

Fig. 14.9 Youngest moraines and relict rock glaciers in the Tatras. **a** Distribution of Late glacial moraines and rock glaciers the in the Pięć Stawów Polskich Valley. The blue arrows mark the direction of glacier advances during the Younger Dryas (YD) cold phase. RA—rock avalanche; OstD—Oldest Dryas age moraines; 1—^{10}Be ages of moraines (according to Makos et al. 2018); 2—drilled lake sediments with the Oldest Dryas bottommost sediments (Wicik 1979); 3— talus and debris flow deposits; 4—alluvial and fluvioglacial deposits; 5—till cover and moraine ridges; 6—relict rock glaciers. **b** Boulder-rich outer latero-frontal moraine of the Younger Dryas advance in the Pusta Valley. **c** Massive tongue of a relict rock glacier (YD advance) at the Zadni Staw lake. **d** Relict rock glacier body of the Younger Dryas advance in the vicinity of the Przedni Staw lake

Fig. 14.10 Present-day perennial snow-ice patches in the Polish Tatra Mountains. **a** General view on the avalanche-supplied Mięguszowiecki glacieret located in the upper part of the Bańdzioch cirque. **b** Upper part of the Mięguszowiecki glacieret with a marginal crevasse and superficial rockfall-derived debris layer

in Polish Tatra Mountains reaching 1700 m was observed in the Piszczałki chute in the Kościeliska Valley following a 5-h rainstorm event yielding 73.8 mm of rainfall (Krzemień et al. 1995).

According to dendrochoronological data, such extreme events occur every 15 years (Gorczyca et al. 2014). However, in the last 20–30 years, an increased frequency of such extreme events is observed as a result of general increase in temperature (Żmudzka et al. 2015; Gądek et al. 2016) and enhanced circulation types associated with heavy rainfall exceeding 50 mm daily (Niedźwiedź et al. 2015) and a higher frequency of protracted wet periods (Górnik et al. 2017).

When a high-magnitude rainfall follows a period of wet weather, a few-days precipitation totals could excess even 300 mm (e.g. 1934—422 mm, 1973—298 mm, 1997—330 mm, 2001—500 mm, 2010—504 mm) (Niedźwiedź et al. 2015; Pińskwar et al. 2017). Although such episodes of significant re-modelling are rare, they cause devastating floods and widespread mass movements (debris flows, landslides, rock falls) due to liquefaction of regolith covers. A large proportion of debris, including boulders up to 2 m long, may become displaced and reach river channels indicating hillslope–valley floor coupling (Krzemień 1991; Kotarba 2002). Furthermore, an increased frequency of strong winds causing catastrophic windthrows and total damage to the tree canopy is also locally observed in the Tatras (Koreň 2006; Strzyżowski et al. 2016, 2018).

The High Tatras are below the contemporary glaciation threshold, which occurs at an elevation of 2530 m asl in the northern sector of the mountains (Zasadni and Klapyta 2009). Present-day glaciers are absent and there is neither geomorphological nor direct observational evidence of the presence of glaciers during the Little Ice Age (Kędzia and Kotarba 2018). However, some small perennial snowfields or glacieretes supplied by avalanches occupy north-exposed slope hollows below the rock walls (Gądek 2014). The largest firn-ice patch (0.5–0.3 ha area and 22–14 thick) in the Polish Tatras is the Mięguszowiecki glacierete in the Rybi Potok Valley (Fig. 14.10).

The present-day activity of periglacial processes is connected with diurnal and seasonal ground freezing, manifesting in small and currently active landforms such as very small sorted polygons (Rączkowska 2007, 2023). The lower elevation limit for frost-conditioned landforms was established at 1900–2000 m asl (Rączkowska 2007), which roughly corresponds to the lower limit of the potential presence of present-day permafrost (Dobiński 2004). Contemporary active permafrost is present in the form of sporadic and discontinued patches of frozen debris and soil above 1950 m asl in the lower sections of selected north-exposed debris slopes in the High Tatra Mountains. The Kozia Dolinka Valley (Fig. 14.4) represents the most comprehensively studied permafrost polygon in the Polish Tatra Mountains (Mościcki and Kędzia 2001; Lamparski and Kędzia 2007; Gądek and Kędzia 2008, 2009). The development and preservation of permafrost are controlled by local microtopography and are more related to exposure and local circulation of cold air on the surface and low solar radiation than elevation and development of snow cover (Gądek and Kędzia 2008, 2009). The absence of indicative permafrost landforms in the periglacial zone was linked with the high activity of avalanches and gravitational processes, as well as significant thickness of the active permafrost layer (Rączkowska 2007).

14.5.4 Human Impact

The high-mountain environment of the Tatra Mountains is regarded now as being only slightly changed by human activity after the foundation of Tatra National Park in 1954. Direct anthropogenic impact is limited to forest management and tourism with 4 million visitors per year (Rączkowska 2019). The onset of intense human impact in the Tatra Mountains is documented from the fifteenth century by sediment data from the Western Tatra Mountains (Krowi Żleb, Pysznianska glade, Starorobociański cirque; Libelt 1988; Kłapyta 2012; Kłapyta et al. 2016). The period of intense human activity in the Tatras between the fifteenth and mid-twentieth century (with a climax in the eighteenth–nineteenth century) was related to mining, metallurgy, animal grazing, and forestry, which caused forest clearings, lowering of the timberline, as well as changes in the type and rates of activity of morphogenetic processes. The main period of direct anthropogenic influence on the geomorphological landscape coincided with significant worsening of climate conditions during the Little Ice Age (Kędzia and Kotarba 2018), and thus, anthropogenic forcing seems to have compounded the impact of climate variability and contributed to a marked increase in slope erosion rates. The rates of slope processes during the sixteenth–twentieth century were 7–10 times larger than during the whole Holocene and led to the deposition of 1.5–2 m thick layers of gravel-type sediments.

The onset of natural protection in the Tatra Mountains reduced human impact, leading to reforestation and increased hillslope stability supported by secondary plant succession (Wasak and Drewnik 2015). Contemporary human impact in the Tatras is associated mostly with tourism and wood transport following windfall events (Western Tatras) on private lands belonging to the Witowska Community, as well as with skiing in the Mt. Kasprowy Wierch area (Krzemień and Kłapyta 2018). The greatest impact on the present-day high-mountain landscape of the Tatras is anthropogenic erosion along tourist trails (Fidelus 2016; Fidelus-Orzechowska et al. 2017). All tourist trails in the Tatras experience the strongest physical changes during snowmelt in spring as well as in summer during heavy rainfall events (Krzemień and Kłapyta 2018). The effect of tourist erosion is widening and deepening of tourist paths, lowering of slope surface, and formation of various microforms including erosive undercuts, nival niches, and gelideflation terraces. The rate of tourist incision is up to 1.3 cm per year (Fidelus 2016), whereas the total widening of tourist paths can reach up to 20 m (Fidelus-Orzechowska et al. 2017).

14.6 Conclusions

The Tatra Mountains demonstrate a unique diversity of high-mountain landscape, dependent on the complex geological structure, considerable neotectonic uplift, and strong imprint of Pleistocene glacial erosion. This area is typified by high relief energy and exceptionally large landscape geodiversity, providing an excellent area to examine the dynamics of a high-mountain denudational system. It is also the only area in Poland, where relict rock glaciers, perennial snow patches, and small glacieretes with associated active periglacial processes occur. Although there are no active rock glaciers and present-day glaciers in the Tatras, this massif features the best-developed present-day periglacial phenomena in Poland and the distinct assemblage of the freshly shaped landforms of mountain glaciation. The largest Pleistocene mountain glaciers in the country left behind a sequence of moraines that mark the maximal glacial advance during global Last Glacial Maximum (26–21 ka), as well as stages of deglaciation which was almost completed during the Bølling–Allerød interphase. During the succeeding Younger Dryas cold stage (12.9–11.5 ka), small glaciers and rock glaciers have developed anew in the Tatra Mountains, but final permafrost melting and rock glacier stabilization were delayed until the Early Holocene (10.5 ka). The unique alpine landscape together with diversified natural vegetation cover has led to the rapid growth of scientific interest and mountain tourism since the nineteenth century. Currently, the Tatras are the most visited mountain area in Poland, with 4 million visitors per year.

References

Agassiz L (1840) Études sur les glaciers. Jent et Gassmann, Neuchâtel, p 346

Anczkiewicz AA, Danišík M, Środoń J (2015) Multiple low-temperature thermochronology constraintson exhumation of the Tatra Mountains: new implication for the complex evolution of the Western Carpathians in the Cenozoic. Tectonics 34:2296–2317

Balon J, Jodłowski M, Krąż P (2015) The tatra mountains physico-geographical regions. Sheet I.4. In: Dąbrowska K, Guzik M (eds), Atlas Tatr. Przyroda nieożywiona. Zakopane, Tatrzański Park Narodowy

Baumgart-Kotarba M, Kotarba A (1997) Würm glaciation in the Biała Woda Valley, High Tatra Mountains. Stud Geomorph Carp-Balc 31:57–81

Baumgart-Kotarba M, Kotarba A (2001) Deglaciation in the Sucha Woda and Pańszczyca Valleys in the Polish High Tatra. Studia Geomorph Carp-Balc 35:7–38

Baumgart-Kotarba M, Dec J, Kotarba A (2008) Glacial trough and sedimnets infill of the Biała Woda valley (The High Tatra Mountains) using geophysical and geomorphological methods. Stud Geomorph Carp-Balc 42:75–108

Boltižiar M (2022) Unique glacial landscape on the roof of the Carpathians—Tatras Mts. In: Lehotský M, Boltižiar M (eds) Landscapes and landforms of Slovakia. Springer, Cham, pp 63–97

Buczek K, Górnik M (2020) Evaluation of tectonic activity using morphometric indices: case study of the Tatra Mts. (Western Carpathians, Poland). Env Earth Sci 79:176

Chowaniec J, Duliński M, Mochalski P, Najman J, Śliwka I, Zuber A (2009) Znaczniki środowiskowe w wodach termalnych niecki podhalańskiej. Przegl Geol 57:685–693

Clark PU, Dyke AS, Shakun JD, Carlson AE, Clark J, Wohlfarth B, Mitrovica JX, Hostetler SW, McCabe AM (2009) The Last Glacial Maximum. Science 325(5941):710–714

Cymerman Z (2010) Krystaliczne łuski tektoniczne w polskiej części Tatr Zachodnich. In: Kotarba A (ed) Nauka a zarządzanie obszarem Tatr i ich otoczeniem. Materiały IV Konferencji „Przyroda Tatrzańskiego Parku Narodowego a człowiek", Zakopane, 14–16 października 2010, t. 1. TPN, Zakopane, pp 47–53

Dobiński W (2004) Wieloletnia zmarzlina w Tatrach: geneza, cechy, ewolucja. Przegl Geogr 76:327–343

Dzierżek J (2009) Palaeogeography of selected areas of Poland during the last glaciation. Acta Geogr Lodz 95:1–112

Engel Z, Mentlík P, Braucher R, Minár J, Léanni L, Arnold M, Aumaître G, Bourlès D, Keddadouche K, Aster Team (2015) Geomorphological evidence and 10Be exposure ages for the Last Glacial Maximum and deglaciation of the Velká and Malá Studená dolina valleys in the High Tatra Mountains, central Europe. Quat Sci Rev 124:106–123

Górska-Zabielska M (2015) Erratic blocks in NW Poland—geological heritage, conservation and geotourism promotion. In: Geophysical Research Abstracts. EGU General Assembly, Vienna

Fidelus J (2016) Slope transformations within tourist footpaths in the northern and southern parts of the Western Tatra Mountains (Poland, Slovakia). Z Geomorph NF 60:139–162

Fidelus-Orzechowska J, Gorczyca E, Krzemień K (2017) Geomorfologiczne skutki gospodarki turystycznej w Tatrach. Tatrzański Park Narodowy, Instytut Geografii i Gospodarki Przestrzennej UJ, Kraków, p 239

Gądek B (2014) Climatic sensitivity of the non-glaciated mountains cryosphere (Tatra Mts., Poland and Slovakia). Glob Planet Chang 121:1–8

Gądek B, Kędzia S (2008) Winter ground surface temperature regimes in the zone of sporadic discontinuous permafrost, Tatra Mountains (Poland and Slovakia). Permafrost Perigl Proc 19:315–321

Gądek B, Kędzia S (2009) Problem detekcji wieloletniej zmarzliny na podstawie temperatury u spągu zimowej pokrywy śnieżnej na przykładzie Tatr. Przegl Geogr 81:75–91

Gądek B, Grabiec M, Kędzia S, Rączkowska Z (2016) Reflection of climate changes in the structure and morphodynamics of talus slopes (the Tatra Mountains Poland). Geomorphology 263:39–49

Gawęda A, Szopa K (2011) The origin of magmatic layering in the High Tatra granite, Central Western Carpathians—implications for the formation of granitoid plutons. Earth Env Science Trans Royal Soc Edinburgh 102:129–144

Gawęda A, Burda J, Klötzli U, Golonka J, Szopa K (2015) Episodic construction of the Tatra granitoid intrusion (Central Western Carpathians, Poland/Slovakia): consequences for the geodynamics of Variscan collision and Rheic Ocean closure. Intern J Earth Sci 105:1153–1174

Głazek J (1964) Kras podmorenowy doliny Pańszczycy w Tatrach. Kwart Geol 8:161–170

Gorczyca E, Krzemień K, Wrońska-Wałach D, Boniecki M (2014) Significance of extreme hydro-geomorphological events in the transformation of mountain valleys (Northern Slopes of the Western Tatra Range, Carpathian Mountains, Poland). CATENA 121:127–141

Górnik M, Holko L, Pociask-Karteczka J, Bičárová S (2017) Variability of precipitation and runoff in the entire Tatra mountains in the period 1961–2010. Zesz Nauk UJ, Prace Geogr 151:53–74

Haeberli W (1985) Creep of mountain permafrost: internal structure and flow of alpine rock glaciers. Mitteilungen der Versuchsanstalt fur Wasserbau, Hydrologie und Glaziologie an der Eidgenoss. Tech. Hochschule, Zurich 77, p 140

Jach R, Rychliński T, Uchman A (eds) (2014) Skały osadowe Tatr. Wydawnictwa Tatrzańskiego Parku Narodowego, Zakopane, p 278

Jaroszewski W (1985) Magmatic-tectonic jointing in granitoids of the Tatra Mts. Acta Geol Polon 35:173–187

Kędzia S, Kotarba A (2018) The little ice age in the Tatra mountains. Cuad Invest Geogr 44:47–67

Kłapyta P, Zasadni J (2018) Research history on the Tatra Mountains glaciations. Stud Geomorph Carp-Balc 51:43–85

Kłapyta P, Zasadni J, Pociask-Karteczka J, Gajda A, Franczak P (2016) Late glacial and holocene paleoenvironmental records in the Tatra mountains (East-Central Europe) based on lake, peat bog and colluvial sedimentary data: a summary review. Quat Intern 415:126–144

Kłapyta P, Zasadni J, Gądek B (2023) Glacial landscape evolution during the Holocene in the Tatra Mountains. In: Palacios D, Hughes PD, Jomelli V, Tanarro LM, (eds) European glacial landscapes. The Holocene. Elsevier. Amsterdam, pp 315–330

Kłapyta P (2012) Ewolucja rzeźby wysokogórskiej Tatr Zachodnich w późnym glacjale i holocenie. Uniwersytet Jagielloński, Kraków, p 266

Klimaszewski M (1988) Rzeźba Tatr Polskich. PWN, Warszawa, p 707

Kohút M, Larinov AN (2021) From subduction to collision: genesis of the Variscan granitic rocks from the Tatric Superunit (Western Carpathians, Slovakia). Geol Carp 72:96–113

Koreň M (2006) Vetrova kalamita 19. novembra 2004: nove pohl'ady a konsekvencie. Tatry 44:7–28

Kotarba A (1992) Natural environment and landforms dynamics of the Tatra Mountains. Mountain Res Dev 12:105–129

Kotarba A, Kaszowski L, K rzemień K (1987) High-Mountain denudational system of the Polish Tatra Mts. GeoStD IGiPZ PAN, Special Issue 3

Kotarba A, Migoń P (2010) Góry wysokie a góry średnie Europy—spojrzenie geomorfologa. Czas Geogr 81:3–19

Kotarba A, Smolak W, Sroka J (1977) Some remarks on the modelling of glacial valley-floors in the Polish Tatra Mts in the light of geophysical measurements. Stud Geomorph Carp-Balc 11:67–78

Kotarba A, Rączkowska Z, Długosz M, Boltižiar M (2013) Recent debris flow in the Tatra Mountains. In: Loczy D (ed) Geomorphological impact of extreme weather (Case studies from central and Eastern Europe). Springer, Dordrecht, pp 221–236

Kotarba A (2002) Współczesne przemiany przyrody nieożywionej w Tatrzańskim Parku Narodowym. In: Borowiec W, Kotarba A, Kownacki A, Krzan Z, Mirek Z (eds) Przemiany środowiska przyrodniczego Tatr. TPN, Kraków– Zakopane, pp 13–19

Králiková S, Vojtko R, Sliva Ľ, Minár J, Fügenschuh B, Kováč M, Hók J (2014) Cretaceous-quaternary tectonic evolution of the Tatra Mts. (Western Carpathians): constraints from structural, sedimentary, geomorphological and fission track data. Geol Carp 65:307–326

Křížek M, Mida P (2013) The influence of aspect and altitude on the size, shape and spatial distribution of glacial cirques in the High Tatras (Slovakia, Poland). Geomorphology 198:57–68

Krzemień K, Kłapyta P (2018) Current state of knowledge and turning points in geomorphologic studies on the present-day evolution of the Tatra Mountains. Stud Geomorph Carp-Balc 51:107–137

Krzemień K, Libelt P, Mączka T (1995) Geomorphological conditions of the timberline in the Western Tatra Mountains. Zesz Nauk UJ, Prace Geogr 98:155–170

Krzemień K (1991) Dynamika wysokogórskiego systemu fluwialnego na przykładzie Tatr Zachodnich. Rozprawy habilitacyjne UJ 215, Kraków, p 160
Lamparski P, Kędzia S (2007) Permafrost occurence in Kozia Dolinka (High Tatra Mountains) in light of ground penetrating radar investigations. Geomorph Slov Bohem 7:82–88
Libelt P (1988) Warunki i przebieg sedymentacji osadów postglacjalnych cyrkach lodowcowych Tatr Zachodnich na przykładzie Kotła Starorobociańskiego. Stud Geomorph Carp-Balc 22:63–82
Lindner L, Dzierżek J, Marciniak B, Nitychoruk J (2003) Outline of quaternary glaciations in the Tatra Mts.: their development, age and limits. Geol Q 47:269–280
Lukniš M (1973) Relief Vysokych Tatier a ich predpolia. Veda, Bratislava, p 375
Makos M, Nitychoruk J (2011) Last glacial maximum climatic conditions in the Polish part of the High Tatra Mountains (Western Carpathians). Geol Quart 55:253–268
Makos M, Nitychoruk J, Zreda M (2013) Deglaciation chronology and paleoclimate of the Pięciu Stawów Polskich/Roztoki Valley, High Tatra Mountains, Western Carpathians since the Last Glacial Maximum, inferred from 36Cl exposure dating and glacier-climate modeling. Quat Intern 293:63–78
Makos M, Rinterknecht V, Braucher R, Tołoczko-Pasek A, Aster Team (2018) Last glacial maximum and lateglacial in the polish high Tatra mountains—Revised deglaciation chronology based on the 10Be exposure age dating. Quat Sci Rev 187:130–156
Makowska A, Jaroszewski W (1987) O współczesnych ruchach pionowych w Tatrach i na Podhalu. Przegl Geol 37:506–512
Makowska A (2003) Dynamika Tatr wyznaczana metodami geodezyjnymi. Inst Geodezji i Kartografii, Monografie 6, Warszawa, p 201
Minár J, Bielik M, Kováč M, Plašienka D, Barka I, Stankoviansky M, Zeyen H (2011) New morphostructural subdivision of the Western Carpathians: an approach integrating geodynamics into targeted morphometric analysis. Tectonophysics 502:158–174
Mościcki WJ, Kędzia S (2001) Investigation of mountain permafrost in the Kozia Dolinka valley, Tatra Mountains, Poland. Norsk Geogr Tidsskr 55:235–240
Němčok J, Bezák V, Biely A, Gorek A, Gross P, Halouzka R, Janák M, Kotański Z, Lefeld J, Mello J, Reichwalder P, Raczkowski W, Roniewicz P, Ryka W, Wieczorek J, Zelman J (1994) Geological map of the Tatra mountains. Ministerstvo Životného Prostredia Slovenské republiky, Bratislava
Němčok A (1982) Zosuvy v slovenských Karpatoch. Veda, Bratislava, p 306
Niedźwiedź T, Łupikasza E, Pińskwar I, Kundzewicz ZW, Stoffel M, Małarzewski Ł (2015) Variability of high rainfalls and related synoptic situations causing heavy floods at the northern foothills of the Tatra Mountains. Theor Appl Climatol 119:273–284
Pagaczewski J (1972) Catalogue of earthquakes in Poland 1000–1970 years. Publ Inst Geophys Pol Acad Sci 51:3–36
Pánek T, Minár J, Vitovič L, Břežný M (2020) Post-LGM faulting in Central Europe: LiDAR detection of the ~50 km-long Sub-Tatra fault, Western Carpathians. Geomorphology 364:107248
Pánek T, Mentlík P, Engel Z, Braucher R, Zondervan A, Aster Team (2017) Late quaternary sackungen in the highest mountains of the Carpathians. Quat Sci Rev 159:47–62
Partsch J (1923) Die hohe Tatra zur Eiszeit. Ferdinand Hirt & Sohn, Leipzig, p 220
Pińskwar I, Kundzewicz ZW, Choryński A (2017) Projections of changes in heavy precipitation in the northern foothills of the Tatra Mountains. Meteorol Hydrol Water Manag 5(2):21–29
Piotrowska K (1997) Cios, spękania ścięciowe i uskoki w trzonie granitoidowym polskich Tatr Wysokich. Przegl Geol 45:904–907
Piotrowska K, Michalik M, Rączkowski W, Iwanow A, Wójcik A, Derkacz M, Wasiluk R (2016) Szczegółowa mapy geologiczna polski, 1:50 000, Arkusz Tatry Wysokie (1061), Tatry Zachodnie (1060). PIG-PIB, Warszawa
Rączkowska Z (2007) Współczesna rzeźba peryglacjalna wysokich gór Europy. Pr Geograficzne PAN 212:252
Rączkowska Z (2019) Human impact in the Tatra Mountains. Cuad Invest Geogr 45:219–244
Rączkowska Z (2023) The Carpathians. In: Oliva M, Nývlt D, Fernández-Fernández JM (eds) Periglacial landscapes of Europe. Springer, Cham, pp 253–279
Rączkowska Z (2008) Współczesna ewolucja rzeźby Tatr. In: Starkel L, Kotarba A, Kostrzewski A, Krzemień K (eds) Współczesne przemiany rzeźby Polski. IGiGP UJ, Kraków, pp 35–56
Rączkowski W, Boltižiar M, Rączkowska Z (2015) Plate V.1. Relief. Map 1. Relief 1: 100 000. In: Dąbrowska K, Guzik M (eds) Atlas Tatr. Przyroda nieożywiona. Tatrzański Park Narodowy, Zakopane
Rae AC, Harrison S, Mighall T, Dawson AG (2004) Periglacial trimlines and nunataks of the Last Glacial Maximum: the Gap of Dunloe, southwest Ireland. J Quat Sci 19:87–97
Romer E (1929) Tatrzańska epoka lodowa. Pr Geogr (Lwów) 11:3–186
Starkel L, Michczyńska D, Krąpiec M, Margielewski W, Nalepka D, Pazdur A (2013) Progress in the holocene chrono-climatostratigraphy of Polish territory. Geochronometria 40:1–21
Staszic S (1815) O ziemiorodztwie Karpatow. Wydawn Geol, Warszawa (reprint 1955)
Strzyżowski D, Fidelus J, Żelazny M (2016) Geomorphological changes within a hillslope caused by a windthrow event in the Tatra Mountains, Southern Poland. Geogr Ann 98A:347–360
Strzyżowski D, Fidelus J, Żelazny M (2018) Sediment transport by uprooting in the forested part of the Tatra Mountains, southern Poland. CATENA 160:329–338
Szczygieł J (2022) The Tatra Mountains: host for the deepest caves in the Carpathians. In: Migoń P, Jancewicz K (eds) Landscapes and landforms of Poland. Springer, Cham, pp 000–000
Szczygieł J, Hercman H, Hoke G, Gąsiorowski M, Błaszczyk M, Sobczyk A (2020) No valley deepening of the Tatra Mountains (Western Carpathians) during the past 300 ka. Geology 48:1006–1011
Tołoczko-Pasek A, Braucher R, Makos M, Aster Team (2020) Wantule—kiedy to się oberwało ? In: VI Konferencja Przyroda Tatrzańskiego Parku Narodowego a Człowiek Zmiany w Tatrach—zagrożenia istniejące i potencjalne. Wydawnictwo TPN, Zakopane, pp 52–53
Tomecka-Suchoń S, Zogała B, Gołębiowski T, Dzik G, Dzik T, Jochymczyk K (2017) Application of electrical and electromagnetic methods to study sedimentary covers in high mountain areas. Acta Geophys 65:743–755
Wasak K, Drewnik M (2015) Land use effects on soil organic carbon sequestration in calcareous Leptosols in former pastureland–a case study from the Tatra Mountains (Poland). Solid Earth 6:1103–1115
Wicik B (1979) Postglacjalna akumulacja osadów w jeziorach Tatr Wysokich. Przegląd Geol 7:403–404
Zasadni J, Kłapyta P (2009) An attempt to assess the modern and Little Ice Age climatic snowline altitude in the Tatra Mountains. Landf Anal 10:124–133
Zasadni J, Kłapyta P (2014) The Tatra Mountains during the last glacial maximum. J Maps 10:440–456
Zasadni J, Kłapyta P (2016) From valley to marginal glaciation in alpine-type relief: lateglacial glacier advances in the Pięć Stawów Polskich/Roztoka Valley, High Tatra Mountains, Poland. Geomorphology 253:406–424
Zasadni J, Kłapyta P, Broś E, Ivy-Ochs S, Świąder A, Christl M, Balážovičová L (2020) Latest Pleistocene glacier advances and post-Younger Dryas rock glacier stabilization in the Mt. Kriváň group, High Tatra Mountains, Slovakia. Geomorphology 358:107093

Zasadni J, Kałuża P, Kłapyta P, Świąder A (2021) Evolution of the Białka Valley Pleistocene moraine complex in the High Tatra Mountains. CATENA 207:105704

Zasadni J, Makos M, Kłapyta P (2022a) Glacial landscapes of the Tatra Mountains. In: Palacios D, Hughes PD, García-Ruiz JM, Andrés N (eds) European Glacial Landscapes. Maximum Extent of Glaciations. Elsevier, Amsterdam, pp 103–107

Zasadni J, Kłapyta P, Kałuża P, Makos M (2022b) The Tatra Mountains: glacial landforms prior to the Last Glacial Maximum. In: Palacios D, Hughes PD, García-Ruiz JM, Andrés N (eds) European glacial landscapes. Maximum Extent of Glaciations. Elsevier, Amsterdam, pp 271–275

Zasadni J, Kłapyta P, Makos M (2022c) The Tatra mountains: glacial landforms from the last glacial maximum. In: Palacios D, Hughes PD, García-Ruiz JM, Andrés N (eds) European Glacial Landscapes. Maximum Extent of Glaciations. Elsevier, Amsterdam, pp 435–440

Zasadni J, Kłapyta P, Makos M (2023a) The evolution of glacial landforms in the Tatra Mountains during the deglaciation. In: Palacios D, Hughes PD, García-Ruiz JM, Andrés N (eds) European glacial landscapes. The Last Deglaciation. Elsevier, Amsterdam, pp 157–164

Zasadni J, Kłapyta P, Makos M (2023b) The evolution of glacial landforms in the Tatra Mountains during the Bølling Allerød Interstadial. In: Palacios D, Hughes PD, García-Ruiz JM, Andrés N (eds) European glacial landscapes. The Last Deglaciation. Elsevier, Amsterdam, pp 341–346

Zasadni J, Kłapyta P, Tołoczko-Pasek A, Makos M (2023c) The evolution of glacial landforms in the Tatra Mountains during the Younger Dryas. In: Palacios D, Hughes PD, García-Ruiz JM, Andrés N (eds) European glacial landscapes. The Last Deglaciation. Elsevier, Amsterdam, pp 509–515

Zasadni J, Świąder A (2015) Visualisation of the białka glacier. In: K. Dąbrowska, M. Guzik (eds.), Atlas of the TatraMountains, Abiotic Nature. Sheet V.4. Glaciation of the Tatra Mountains. Tatrzański Park Narodowy, Zakopane

Zejszner L (1856) Über eine Längenmoräne im Thale des Biały Dunajec be idem Hochofen von Zakopane in der Tatra. Sitzungsberichte D. K. Akademie Wiss, Math Naturwiss I 21:259–262

Żmudzka E, Nejedlik P, Mikulova K (2015) Temperature, thermal indices. Plate II.2. In: Dąbrowska K, Guzik M (eds) Atlas Tatr. Przyroda nieożywiona. Zakopane, Tatrzański Park Narodowy

Piotr Kłapyta holds Ph.D. degree in Geography and works at the Jagiellonian University in Cracow in the Institute of Geography and Spatial Management. His main research interests are mountain geomorphology, especially in formerly glaciated areas, glacial and periglacial geomorphology and geochronology in mountain terrains, paleogeography, and sedimentology. His main research area are high-mountain areas in the Western Carpathians (Babia Góra massif, Tatra Mountains, Low Tatra Mountains), the Eastern Carpathians in the territory of Ukraine and Romania, and the Southern Carpathians. This research is focused on the reconstruction of glaciers and palaeoclimate during the Last Glacial Maximum.

Jerzy Zasadni holds Ph.D. in Geology from the AGH University of Science and Technology in Kraków and is employed in the Faculty of Geology, Geophysics, and Environmental Protection. His main research interests are Quaternary geology and geomorphology in high mountains, with special focus on glacial and periglacial geomorphology. He used exposure age dating and field mapping to reconstruct glacier chronology and the extent of Pleistocene glaciations in the Carpathians and the Alps.

The Tatra Mountains—Host of the Deepest Caves in the Carpathians

15

Jacek Szczygieł

Abstract

The Tatra Mountains are the highest mountain massif in the Carpathian range, hosting the deepest caves in the belt, reaching a depth of 824 m in the the Wielka Śnieżna Caves System. Cave evolution started at the end of the Miocene and is related to the Tatras' asymmetrical uplift and, during the Pleistocene, to the existence of mountain glaciers. Karst in the Tatras is strongly controlled by fold-and-thrust setting, which influences both underground flow pathways and cave passage morphology. Caves in the Tatras are the only alpine-type karst phenomena in the Western Carpathians and can be defined by containing deep, epigenic multi-level cave systems with proglacial vadose avens, all in the high-relief area shaped by glaciers. Although surface forms are poorly developed, the underground karstic landscape is rich and diverse, with more than 175 km of cave passages. During over a century of research, Polish speleologists have made numerous discoveries, and several of them have become embedded in the global literature, such as the theory of proglacial caves.

Keywords

Cave · Karst · Speleogenesis · Geochronology · Tatra Mts · Carpathians

15.1 Introduction

Caves in the Tatra Mountains (also referred to as "the Tatras") may not seem highly spectacular since speleothems are rare and underground voids are not as roomy as those in other Carpathian caves south of the Tatras. Yet, the uniqueness of the Tatra caves lies in their evolution linked to an uplifting mountain range and mountain glaciers. Interpenetrating these two phenomena on different time scales drove the deepest cave formation in the entire Carpathians. Caves in the Tatras are the only alpine-type karst phenomena in the Western Carpathians. They are deep, epigenic multi-level cave systems with proglacial vadose avens in the high-relief area shaped by glaciers. The only factor distinguishing the karst in the Tatras from those in the Alps or Pyrenees is the scarcity of surface karst phenomena due to the domination of glacial morphology (Kłapyta and Zasadni 2024).

Karst phenomena, especially caves, have been of interest to travellers to the Tatras for centuries. Random visits in Tatra caves began in the early nineteenth century, but the first regular, well-organized exploration was led by the brothers Stefan and Zbigniew Zwoliński from 1913. They explored several caves known by shepherds, mostly horizontal ones, with some short vertical sections. After World War II, caving took on a sporting character, and rope techniques, climbing, diving, and camping in caves started to be popular. The first speleoclub was founded in Kraków in 1950, and later, caving clubs were established in most major Polish cities. Thus, cave mountaineering was born, but since clubs were often located at academic centres, shortly after cave mountaineering, cave research began. In 1960, the seventh deepest cave in the world was discovered, the Śnieżna Cave, with a depth of 563 m, which, in 1968, has been connected to the Nad Kotliny Cave, reaching a depth of 742 m and becoming the sixth deepest cave. Since then, several generations of Polish cavers have been trained in and explored more inaccessible caves. Although the karst area of the Polish part of the Western Tatra only covers about ca. 30 km^2, new caves or new parts in known systems are still being discovered.

In the Polish part of the Tatra Mountains, 860 caves have been mapped to date, with a total length exceeding 175 km; 79 caves are longer than 100 m, and 23 caves are

J. Szczygieł (✉)
Institute of Earth Sciences, University of Silesia, Będzińska 60, 41–200 Sosnowiec, Poland
e-mail: jacek.szczygiel@us.edu.pl

P. Migoń and K. Jancewicz (eds.), *Landscapes and Landforms of Poland*,
World Geomorphological Landscapes, https://doi.org/10.1007/978-3-031-45762-3_15

Table 15.1 The longest and the deepest caves in the Polish part of the Tatra Mountains

	Cave	Depth (m)	Cave	Length (m)
1	Wielka Śnieżna	824	Wielka Śnieżna	23,619
2	Śnieżna Studnia	805	Śnieżna Studnia	14,020
3	Bańdzioch Kominiarski	562	Wysoka—Za Siedmiu Progami	11,700
4	Mała w Mułowej	555	Miętusia	10,780
5	Wysoka—Za Siedmiu Progami	435	Bańdzioch Kominiarski	10,010
6	Kozia	389	Czarna	7247
7	Ptasia Studnia	352	Ptasia Studnia	6283
8	Miętusia	305	Zimna	5480
9	Czarna	303	Mała w Mułowej	3863
10	Siwy Kocioł	295	Kozia	3470

Source Tatra National Park

longer than 1 km. Only 38 caves are deeper than 50 m. The longest system in Poland and the deepest one in the whole Carpathians is the Wielka Śnieżna cave system (Table 15.1), combining five caves (entrances) with the uppermost one, Wielka Litworowa Cave, located at 1906 m above sea level. Currently, all data, including maps, entrance location, morphometric data, and cave descriptions are freely available at the Central Geological Database, hosted by Polish Geological Institute (http://jaskiniepolski.pgi.gov.pl/). In the Tatra National Park, seven caves have been made accessible for tourists, with marked tourist trails. These are Mylna, Raptawicka, Obłazkowa, and Mroźna in the Kościeliska Valley, Smocza Jama in the Kraków Gorge (with a descent to the Kościeliska Valley), and Dziura Cave in the Ku Dziurze Valley.

A detailed literature review of Tatra cave research was presented by Gradziński et al. (2009). Therefore, the aim of this chapter is to highlight both key historical as well as the most recent (post-2009) discoveries relevant to karst and cave development in the Tatra Mountains.

15.2 Geological Setting and Its Influence on Caves

The Tatra Mountains are a small, 57-km-long range on the Polish-Slovak border, extending W-E, located in the northernmost part of the Central Western Carpathians (Fig. 15.1a). They represent a basement-involved fold-and-thrust structure (Jurewicz 2005) composed of a Palaeozoic crystalline basement overlain by Mesozoic sedimentary rocks in the north and west (Nemčok et al. 1994; Fig. 15.1b). The Tatric sedimentary cover is topped with nappes belonging to Tatric (Czerwone Wierchy, Giewont, Široká nappes), Fatric (Križna nappe), and Hronic units (Choč nappe; Nemčok et al. 1994; Fig. 15.1b). Simplified, the Tatric sedimentary succession starts with Lower Triassic terrestrial siliciclastic deposits and shallow-marine carbonates, is followed by Middle Triassic to Lower Cretaceous limestones (with an Upper Triassic-Lower Jurassic hiatus in the nappe successions), and topped with Albian to Cenomanian marls and sandstones (Nemčok et al. 1994). Therefore, successive tectonic units are separated by non-karstic Lower Triassic or Albian strata, forming distinct hydrogeologic units (Fig. 15.1c, d). Limestone thickness does not exceed 600 m, with an average thickness varying around 400 m (Kotański 1961). Yet, a cave with a depth of more than 800 m could develop due to the asymmetrical uplift. The Tatras constitute a horst exhumed from a depth of at least 5 km since the Miocene to present, with rock uplift rates of 100–500 m/My (Králiková et al. 2014; Anczkiewicz et al. 2015). There is a strong west-to-east gradient in the initiation of rock uplift, with the highest rates in the south-eastern corner of the Tatras (Anczkiewicz et al. 2015). Yet, limestone has been exhumed first in the Western Tatras in the latest Miocene (Králiková et al. 2014), which was the beginning of karst system evolution as we know it today. With the highest exhumation rates along the Sub-Tatra Fault, bounding the Tatras from the south, the Tatra block has been tilted northwards by ~ 40° (Jurewicz 2005). This rotation gave rise to conditions where, despite 400–600 m thickness of limestone, elevational differences between the uppermost limestone outcrops in the summit parts, where the karstic system is recharged, and the lowermost limestone outcrops in the valleys, where karstic springs are located, reached ~ 900 m.

Most caves are in the most uplifted Tatric unit, where carbonate successions are folded and cut by faults that determine the course of drainage and morphology of the cave passages (Szczygieł 2015). In contrast to many large alpine cave systems developed in a monoclinal structure, in which passages developed along facies particularly susceptible to dissolution due to physical, lithological, or chemical properties, the so-called "inception horizons" (e.g. Filipponi et al. 2009), cave passages in the Tatra Mountains were mainly guided by tectonically involved steep initial fissures in both phreatic and vadose passages (Szczygieł 2015; Szczygieł et al. 2015a). The fold geometry has also been influential. Fold hinges facilitated hosting of roomy chambers in chevron folds, whereas in the concentric folds, chambers developed in the fold core and deep shafts in the outer part, along steep bedding planes (Fig. 15.2) (Szczygieł 2015). Faults were used by speleogenesis primarily within the massive and thick-bedded limestone of Upper Jurassic-Lower Cretaceous age (Szczygieł 2015; Szczygieł et al. 2015a). In the thin-bedded limestone and

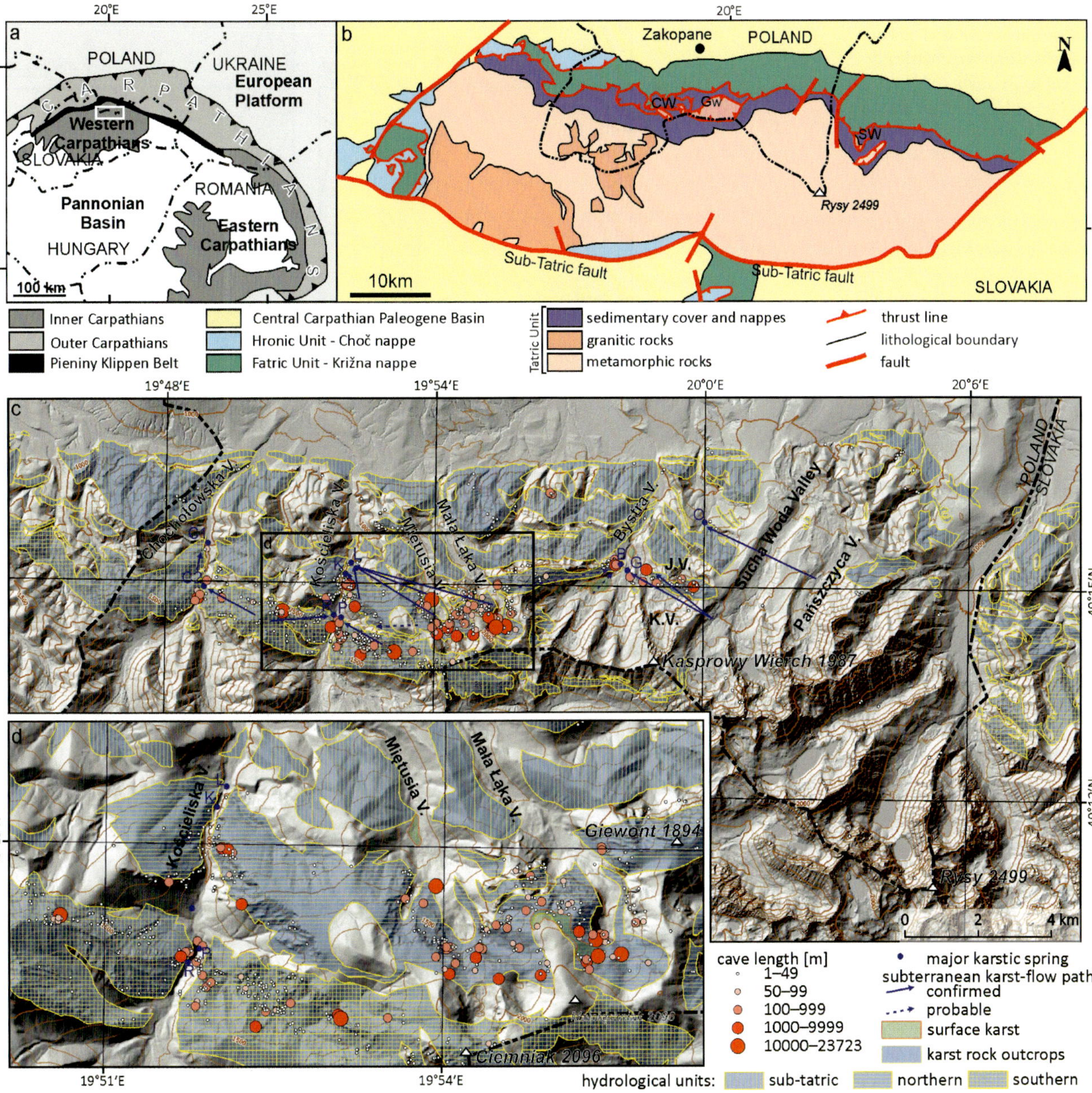

Fig. 15.1 Location of the study area. **a** Simplified geological sketch of the Carpathians with the location of the Tatras marked as the white rectangle. **b** Tectonic map of the Tatra Mountains (after Nemčok et al. 1994, modified), explanation: CW—Czerwone Wierchy nappe, Gw—Giewont nappe, SW—Szeroka Jaworzyńska nappe. **c, d** Outcrops of carbonate rocks and concentrations of surface karst landforms and caves in the Polish part of the Tatra Mountains, overprinted on shaded relief with digital elevation model (based on Szczygieł et al. 2015b); hydrological units after Głazek (1995)

dolomitic limestone of Middle Triassic age, which are usually steeply dipping, the most favourable structures guiding speleogenesis are bedding planes (Szczygieł 2015). The steep bedding resulted in the development of characteristic caves of the Tatra Mountains, which comprise steep vadose passages, the ramps, in Polish called "płytowiec". The floor of such a steep (40–80°) ramp, which is nothing more than a bedding plane, can be as long as ~ 300 m, with some ledges. Not only tectonic structural geometry affects the drainage, but also neotectonic processes, before and during the formation of the caves. Extending fissures to proto-conduit size, the Quaternary faulting set favourable paths for water drainage, often using the same fissures in subsequent phases of speleogenesis (Szczygieł 2015).

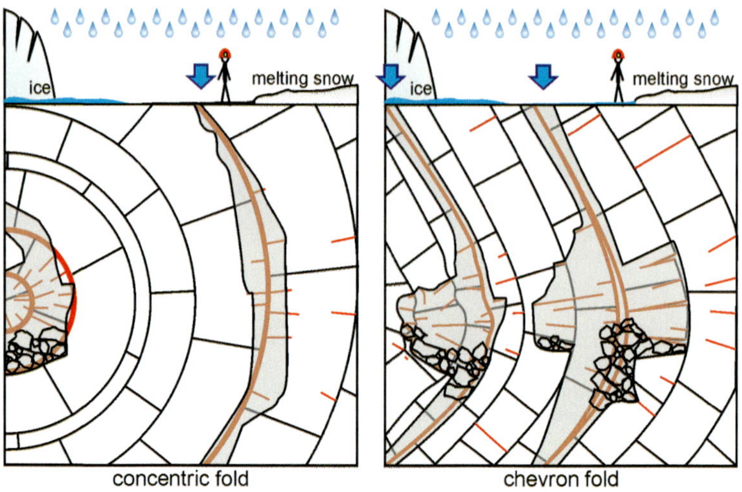

Fig. 15.2 Schematic depiction of the development of chambers and shafts within geometrically different folds; thick red lines—fault; thin red lines—fractures; black lines—bedding planes; grey polygons—cave cross-section. Caver figure is posted to give the definition of a human-accessible passage (a real cave); schematic rain, melting glacier, and snow represents past and present water recharge sources in the Tatra Mts

15.3 Karst Hydrology

Two flow directions dominate in the rocks of the Tatra Mountains: latitudinal flow, consistent with bedding strike and the general structural trend and longitudinal flow, parallel to the surface drainage, i.e. main valley direction. These two directions allow the contact of surface waters with groundwaters of all types (pore, fissure, fissure-karstic groundwater; Małecka 1993). Since the limestone successions are separated from one another by non-karstic rocks, mostly Albian marls, the Tatra karst does not form a single hydrological system but a series of smaller systems isolated from one another. Głazek (1995) distinguished the following hydrological units (Fig. 15.1c, d):

(1) Southern unit, consisting of the Tatric autochthonous sedimentary cover, within which two karstic systems have been developed. The first is the Wypływ spod Pisanej outflow, draining the Kraków Gorge and including the Wysoka—Za Siedmiu Progami Cave; the second one contains the Chochołowskie Spring and Wypływ spod Raptawickiej outflow, draining, among others, Bańdzioch Kominiarski and Szczelina Chochołowska caves. The mean discharge for the Chochołowskie Spring is about 400 dm^3/s (Barczyk 2003).
(2) Northern unit, consisting of the Tatric nappes. It includes the Lodowe (Ice) Spring with a mean minimum discharge of 700 dm^3/s (Barczyk 2003) that drains the Czerwone Wierchy nappe, from the Mała Łąka Valley through the Miętusia Valley to the Kościeliska Valley, and, among many others, the Wielka Śnieżna cave system; the Kraszewski Gate outflow, to which water is supplied, among others, by the Zimna Cave; Bystra and Goryczkowe Springs draining the Giewont nappe, as well as allogenic waters from the upper part of the Sucha Woda Valley (Barczyk 2003), built of crystalline rocks. Spring mean discharges have been estimated at 350 and 800 dm^3/s for Bystra and Goryczkowe Springs, respectively (Barczyk 2003).
(3) Sub-Tatric (*reglowy* in Polish) unit, in which the Olczyskie Spring is located, draining the Pańszczyca Valley, with a mean discharge of 780 dm^3/s (Barczyk 2003). There are no large caves in this unit, and the relatively high discharge is due to allogenic supply from crystalline outcrop areas.

15.4 Surface Karst Landforms

Surface karst forms in the Tatra Mountains are rare and usually isolated or clustered in small areas up to 0.5–2 ha. The only karren field morphologically comparable with the alpine landscape is located in the upper part of the Mała Łąka Valley on the NE slopes of Mt. Małołączniak (Fig. 15.3a). Here, solution dolines and truncated karstic shafts occur within partly bare limestone surfaces with karren, among which rinnenkarren dominate over rundkarren. The most widespread and yet least evident ones are suffosion dolines developed above focused karstic drainage on the top surface of limestone concealed beneath unconsolidated glacial deposits (Fig. 15.3b). Suffosion dolines are commonly found in the lower parts of valleys in the northern part of the Tatras (Sucha Woda Valley, Pańszczyca Valley—Fig. 15.1c), where glacial sediments cover the limestones (Szczygieł et al. 2015b).

Kraków Gorge is a unique karstic landform in the Tatras (Fig. 15.3c) with its upper part called "13-steps couloir" (*pol. Żleb Trzynastu Progów*; Fig. 15.3d). It is a non-glaciated valley, only periodically drained by a surface stream, characterized by vertical or overhanging walls of which the highest one, the Ratusz, is 170 m high. The meandering gorge is locally as narrow as 3 m and is interspersed with a series of steps ranging in height from 2 to 8 m, with plunge pools beneath. Lindner (1985) hypothesized that the Kraków Gorge developed from a collapsed cave. However,

Fig. 15.3 Examples of surface karst morphological features from Tatra caves. **a** Karren in the upper part of the Mała Łąka Valley (photograph by M. Golicz). **b** Suffosion doline in the upper part of the Sucha Woda Valley (photograph by A. Tyc). **c** Kraków Gorge (photograph by J. Nowak). **d** "13-steps couloir" *(pol. Żleb Trzynastu Progów)* in the upper part of the Kraków Gorge (photograph by J. Szczygieł)

this model is neither supported by the gorge morphology, which indicates successive deepening, nor by the chronology that clearly shows that the gorge is older than Lindner (1985) suggested (Szczygieł et al. 2020).

15.5 Cave Morphology

Cave systems in the Tatras have developed simultaneously to mountain uplift and following valley incision, expressed in a vertical pattern where paleo-phreatic cave levels are connected or intersected with vadose vertical sections. Cave levels as an indicator of former positions of valley bottoms have been recognized in the Polish Tatra by Rudnicki (1958), who first distinguished three levels and, later, four (Rudnicki 1967). Wójcik (1960), based on the occurrence of allochthonous gravels within the caves, claimed the existence of eight levels. However, this was not reflected in the vertical distribution of the corridors relative to the valley bottoms. Rudnicki´s observations, based on approximately 20 km of known cave passages at that time, are still valid, even with a much larger dataset established recently. One more level has, however, been added, which was undiscovered at that time, namely the uppermost relict cave level. Currently, five cave levels are distinguished in the Tatras, including the active one (Szczygieł et al. 2020; Fig. 15.4). The vertical span of the cave levels is more extensive in the lower parts of the valleys. Moreover, according to the analysis of paleoflows, some deep paleo-phreatic caves in the Tatras, e.g. the Czarna Cave with the depth exceeding 200 m, represent one cave level, resulting from the formation by deeper circulation in the so-called "phreatic loop" (sensu Ford and Ewers 1978). The above conditions have caused more cave levels to appear in purely morphometric terms than implied by the karst system's evolution, which might explain the excessive number of cave levels designated by Wójcik (1960, 1968) and later by Grodzicki (1991).

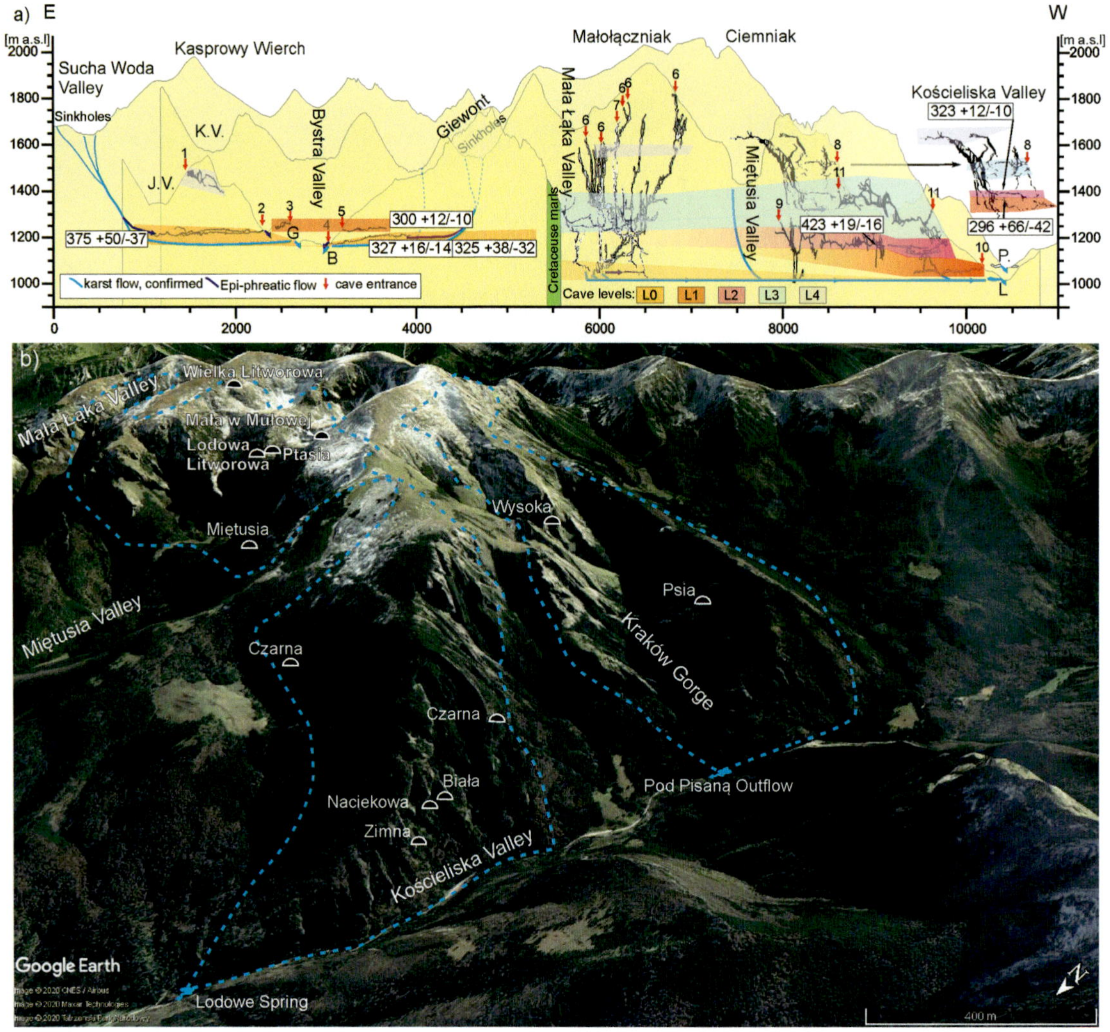

Fig. 15.4 Setting of major cave systems in the Polish part of the Tatra Mountains. **a** Cave systems projected onto superimposed topographic profiles across the Bystra and Kościeliska valleys. Cave levels are colour-coded, from the oldest (L4) to the youngest (L0) (after Szczygieł et al. 2020). K.V.—Kasprowa Valley, J.V.—Jaworzynka Valley; Springs: G—Goryczkowe, B—Bystrej, P—Pod Pisaną, L—Lodowe; Caves: 1—Magurska; 2—Kasprowa Niżna; 3—Goryczkowa; 4—Bystra; 5—Kalacka; 6—Wielka Śnieżna; 7—Śnieżna Studnia; 8—Wysoka; 9—Miętusia; 10—Zimna; 11—Czarna. **b** Panoramic view of the central-eastern Kościeliska Valley with the position of caves mentioned in text and the outlines of catchments (blue dashed lines)

At the passage scale, cave morphology is dominated by two genetic types: vadose and phreatic. Vadose conduits, if horizontal or gently inclined, are high and narrow canyons, often meandering (e.g. Wodociąg passage in the Śnieżna Cave). Sub-horizontal sections are frequently divided by vertical steps. The steeper sections often widen downwards, with overhanging walls due to headward erosion (e.g. Majowy Meander in the Ptasia Studnia Cave; West Gang in the Śnieżna Studnia cave, Fig. 15.5a). The canyons usually have a width of 0.3–2 m (e.g. Lodowa Litworowa Cave). Although height cannot be measured precisely, the most spectacular passages could be as high as 40 m (e.g. Wielki Kanion passage in the Wysoka—Za Siedmiu Progami cave system, and Kanion in the Ptasia Studnia Cave, Fig. 15.5b). Among steep and vertical conduits, besides shafts, the above-mentioned ramps are common (Fig. 15.5c). The deepest, 230-m-deep Wazeliniarzy shaft (Fig. 15.5d), is located in the Śnieżna Studnia Cave. Shafts follow the canyon/meander curvature due to headward erosion or are tube-shaped (e.g. Wielka Studnia shaft in the Śnieżna Cave, see Fig. 15.5e, or the entrance shaft of the Ptasia Studnia Cave). It is assumed that the vertical tubes have been produced by whirlpools resulting from the vast amount of glacial meltwater during deglaciation, hence the term "proglacial cave" (Głazek et al. 1977). The original concept by Głazek et al. (1977), linking the origin of high-elevation vertical caves, as well as steep parts in other caves, with the action of waters from melting glaciers, should be emphasized since the expression "proglacial cave" has subsequently become an established term in the world literature.

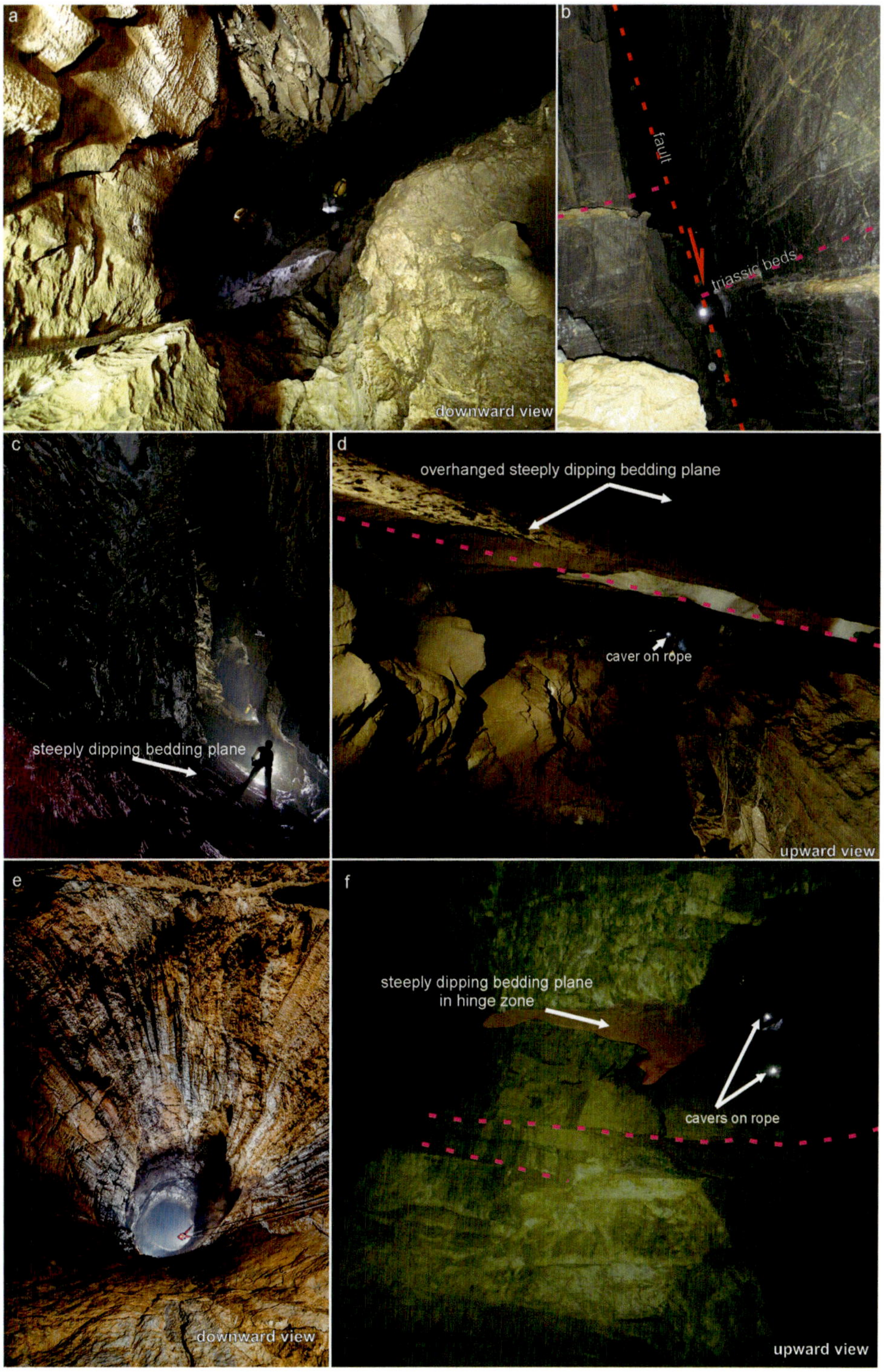

Fig. 15.5 Examples of vadose morphological features from Tatra caves. **a** overhanging step in the Śnieżna Studnia cave. **b** Kanion passage in the Ptasia Studnia cave. **c** Płytowiec in the Wielka Śnieżna cave. **d** Wazeliniarzy Shaft in the Śnieżna Studnia Cave, note the guiding bedding plane. **e** Wielka Studnia in the Wielka Śnieżna Cave, note circular cross-section. **f** Fakro Chamber in the Mała w Mułowej Cave (photographs c and e by J. Kućmierz; a, d, and f by M. Golicz; b by J. Szczygieł)

Along with typical tube-like phreatic conduits (Fig. 15.6a), lens-shaped passages, elongated vertically or diagonally, are also frequent. There are cases in which the ratio of height to width of such lenses exceeds 10, or several enlargements are formed on a single fissure, and the cross-section resembles an elongated figure eight (common in the Szczelina Chochołowska cave—Fig. 15.6b; Zimna and Czarna caves). Phreatic passages are from 0.3 to 2–4 m in diameter (e.g. Wysoka, Miętusia, Szczelina Chochołowska caves). The exceptionally roomy ones are conduits in the Czarna Cave, with an average height of 10–20 m (locally even ~ 30 m) and a width of 5 m (Fig. 15.6d, e). In cases where water was drained through the fault zone in massive limestones, via a network of

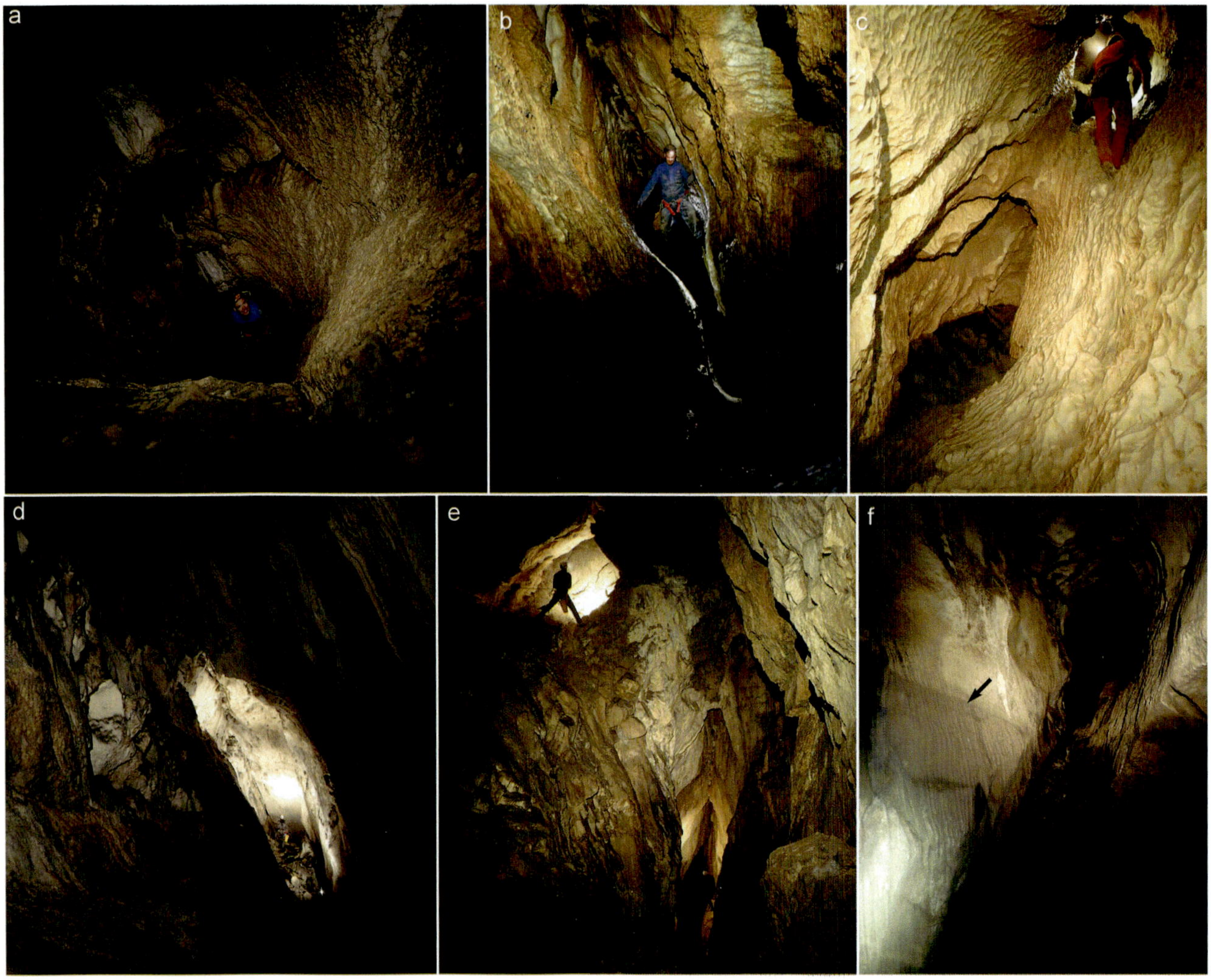

Fig. 15.6 Examples of phreatic morphological features from the Tatra caves. **a** Phreatic tube, the Szczelina Chochołowska Cave, note circular cross-section. **b** Passage with the cross-section of an elongated "8", the Szczelina Chochołowska Cave. **c** Looping conduits in the Kasprowa Niżna Cave. **d, e** Conduits in the Czarna Cave. **f** Keyhole passage leading to the Zwolińskich Sump in the Miętusia Cave, note the line indicating sump-level fluctuation (black arrow) (photographs by J. Szczygieł)

fissures, maze caves with loops and anastomoses have developed (e.g. Goryczkowa and Kasprowa Niżna caves, Fig. 15.6c); those passages are usually narrow (up to 1 m wide). Noteworthy are the conduits going around the siphon filled with deposits, first described in the Zimna Cave by Rudnicki (1960) as a corkscrew (pol. *korkociąg*), later described by Ford (1965) as a bypass. If the paleo-phreatic cave level was intersected with vadose conduits, water used the existing paleo-phreatic conduits by cutting a channel into their bottoms, creating a keyhole cross-section (e.g. Galeriowy passage in the Zimna Cave, or passage leading to Zwolińskich Sump in the Miętusia Cave, Fig. 15.6f).

In both genetic types, paleo-phreatic and vadose, conduits may expand over time due to collapses and/or local geological settings (see Sect. 15.2) to form larger voids customarily referred to as chambers or halls. Although chambers are not precisely defined, and generally, their designation is quite arbitrary, there are several distinctly more "roomy sites" in the Tatra caves. The biggest one is the Fakro Chamber in the Mała w Mułowej Cave, with floor dimensions of 85 m × 45 m and 90 m in height (Fig. 15.5f). It has been formed at the junction of two vadose shafts, in the core part of the concentric and recumbent Organy syncline hinge zone (Szczygieł et al. 2014; Szczygieł 2015). Deeper in the Mała w Mułowej Cave, the Na Luzie Chamber is located. Although not as large, the chamber is genetically unique since it developed along the thrust of the Czerwone Wierchy nappe over the autochthonous sedimentary cover. The Na Luzie Chamber and the passages leading to and from it were formed within the Albian marls due to mechanical fluvial erosion of low-resistant rocks. The Middle Triassic limestones form an overhanging ceiling (Szczygieł et al. 2014), often bearing no signs of karst erosion. One of the roomiest among the caves of the Tatra Mountains is the entrance part of the Magurska Cave, which is situated within the syncline hinge (Hercman 1989), just as the Fakro chamber. These entrance chambers, however, are paleo-phreatic, with a more arch-like ceiling. They are, however, more spacious than the passages leading deeper into the massif, which were closed by sediment-filled siphons until as late as the twentieth century. The disproportionately large volume of the entrance part may be a partial result of frost weathering, a process that has been described in detail in the Alps (Oberender and Plan 2015). Thus, the passage volume, or rather the cross-section area, is not necessarily proportional to the paleoflow as it may have been enlarged after the cave was drained.

A unique cave in the Polish part of the Tatras is the Dziura Cave, as it is hypogenic and not epigenic as the other caves. Bac and Rudnicki (1978) inferred the thermal stage of development of this cave based on the ceiling morphology of this roomy chamber, which was later confirmed by geochemical studies (Gradziński et al. 2007).

15.6 Cave Development in the Tatra Mountains

The cave levels described above demonstrate a multi-stage development of the karst system in the Tatras. Turmoil, with the determination of these levels, indicates that the evolution of caves in the Tatra Mountains is not easy to decipher. In general, dating of cave sediments is crucial in reconstructing karst system evolution. Isotopic dating of sediments from the Tatra caves began early and has long followed global trends (Głazek 1984; Hercman 1991). However, in the twenty-first century, Polish speleothem researchers turned to palaeoclimatic studies, whereas palaeogeographic reconstructions (in terms of dating) became less popular. Consequently, the U–Pb speleothem dating or Be/Al cosmogenic nuclide burial age dating, which have been developed in the last two decades, have not yet been employed. However, this does not mean that data obtained from "classic" U-Th dating of speleothems cannot be used to disentangle the evolutionary history of the caves; quite to the contrary.

The oldest relict karst forms are assumed to have formed in the latest Miocene or Pliocene (Rudnicki 1967; Hercman 1991). Although there is no direct geochronological evidence, this is inferred from their topographic position relative to Miocene exhumation rates (Szczygieł et al. 2020). Hitherto, the oldest ages were yielded by speleothems from the Czarna Cave, cave level L3 (Fig. 15.4a). Although the speleothems' age is out of range of the applied dating method (i.e. 0.6 Ma), based on the U^{234}/U^{238} ratio, their age can be estimated at being older than 1.2 Ma, indicating that the phreatic evolution of the Czarna Cave had ended by that time (Nowicki 2003). The age of speleothems in the Naciekowa Cave (cave level L2; Fig. 15.4a) ranges between 0.6 and 1.2 Ma, pointing to the interval in which the L2 level was dried out (Nowicki 2003); thus, the L1 level must have operated as active drainage at that time. The speleothems from active epiphreatic caves, at or below the valley bottoms, belong to the modern cave level (L0; Fig. 15.4a), with yield ages between 284 and 325 ka. This shows that the modern karst drainage system of the Tatra Mountains was established before the late Middle Pleistocene (Szczygieł et al. 2020). Since then, no significant valley deepening occurred, which would force the water table to lower and result in the development of new phreatic conduits.

From the palaeogeographic point of view, the main directions of the drainage network did not change over time. Already since the formation of the oldest cave levels, the major valleys ran south to north, and groundwater was transported along the W–E structural trend, with minor exceptions representing underground flow of surface streams from S to N. As inferred from scallops—erosional forms on cave walls whose asymmetry is related to the flow direction—the paleoflow directions in caves located between the Mała Łąka and Kościeliska valleys were from east to west (Gradziński and Kicińska 2002; Kicińska 2005; Fryś et al. 2006). The Lodowe Spring system has not experienced direction change but only successive deepening following base-level lowering. Only in some caves, whose entrances were covered by glaciers, the paleoflows from entrances into the cave were recorded. These scallops, however, are overprinted on the original ones and represent the stage of backflooding connected with the rising of the water table during glaciation (Kicińska 2005).

The karst system in the Bystra Valley is fed, in addition to autigenic supply, with allogenic water from the upper part of the Sucha Woda Valley. As inferred from the scallop asymmetry and heavy minerals analysis, the allogenic water was initially transported to the Jaworzynka Valley, which is now dry. Later, this water was captured by the incising Kasprowa Valley (Kicińska et al. 2017), which must have taken place no later than ~ 300 ka, as inferred from the oldest dates obtained from the Kaprowa Niżna Cave (Szczygieł et al. 2020).

15.7 Conclusions

The caves in the Tatra Mountains are an outstanding phenomenon of both scientific and touristic/sport value. This system represents the only high-mountain karst system in the Western Carpathians and the deepest one in the whole Carpathian Mountain range.

Polish speleologists began research in this area in the early twentieth century, but great progress has been made since the

1950s. Jan Rudnicki made many pioneering observations, published in Polish, such as the idea of "bypass passages" (1960). One of the most important contributions to speleology seems to be the concept of proglacial caves, described by Głazek et al. (1977). Geochronological research followed global trends in the twentieth century, slightly lagging behind in the last 20 years. However, nowadays, research using cosmogenic nuclide dating is being carried out, which will certainly help to set another milestone in the research of the Tatra Cave evolution. Although much has been done, there are still many unsolved questions about the timing and rates of the morphogenetic processes in terms of karst system development as well as the Quaternary evolution of the Tatra Mountains.

Acknowledgements I would like to dedicate this work to Dr. Jan Rudnicki and Prof. Jerzy Głazek, whose pioneering studies are still inspiring today. I address my appreciation to cavers, whose exploratory and survey work is always the first step to following research. Michał Gradziński, Helena Hercman, Ditta Kicińska, and Andrzej Tyc are thanked for a decade of discussions on this matter. For constructive review, I thank Dr. Andrzej Tyc, an anonymous reviewer, and the editors Prof. Piotr Migoń and Dr. Kacper Jancewicz.

References

Anczkiewicz AA, Danišík M, Środoń J (2015) Multiple low- temperature thermochronology constraints on exhumation of the Tatra Mountains: new implication for the complex evolution of the Western Carpathians in the Cenozoic. Tectonics 34:1–22

Bac-Moszaszwili M, Rudnicki J (1978) O mozliwości hydrotermalnej genezy jaskini Dziura w Tatrach. Kras i Speleologia 2:84–89

Barczyk G (2003) Karst and vaucluse springs from the Polish Tatra Mts. Results of long-term stationary investigations. Acta Carsologica 32:145–155

Filipponi M, Jeannin PY, Tacher L (2009) Evidence of inception horizons in karst conduit networks. Geomorphology 106:86–99

Ford DC (1965) The origin of limestone caves: a model from the central Mendip Hills, England. Bull Nat Speleol Soc 27(4):107–132

Ford DC, Ewers RO (1978) The development of limestone cave systems in the dimensions of length and depth. Canad J Earth Sci 15:1783–1798

Fryś P, Gradziński M, Kicińska D (2006) Development of Miętusia Cave, Western Tatra Mountains, Poland. Slov Kras 44:55–69

Głazek J (1984) Pierwsze datowania izotopowe nacieków z jaskiń tatrzańskich i ich konsekwencje dla stratygrafii plejstocenu Tatr. Przegl Geol 32:39–42

Głazek J (1995) Hydrografia krasowa Tatr Polskich. In: Grodzicki J (ed) Jaskinie Tatrzańskiego Parku Narodowego, vol 5. Polskie Towarzystwo Przyjaciół Nauk o Ziemi, Warszawa, pp 11–30

Głazek J, Rudnicki J, Szynkiewicz A (1977) Proglacjal caves—a special genetic type of caves in glaciated areas In: Proceedings of 7th international congress speleology, Sheffield, pp 215–217

Gradziński M, Kicińska D (2002) Morphology of Czarna Cave and its significance for the geomorphic evolution of the Kościeliska Valley (Western Tatra Mts.). Ann Soc Geol Polon 72:255–262

Gradziński M, Duliński M, Hercman H, Żywiecki M, Baryła J (2007) Cave development influenced by hydrocarbon oxidation: an example from the Polish Tatra Mts. Aragonit 12:129

Gradziński M, Hercman H, Kicińska D, Barczyk G, Bella P, Holúbek P (2009) Kras tatrzański—rozwój wiedzy w ostatnich trzydziestu latach. Przegl Geol 57:674–684

Grodzicki J (1991) Geneza i ewolucja jaskiń Tatr Zachodnich. In: Grodzicki J (ed) Jaskinie Tatrzańskiego Parku Narodowego, vol 1. Polskie Towarzystwo Przyjaciół Nauk o Ziemi, Warszawa, pp 11–41

Hercman H (1989) Z geologii Jaskini Magurskiej w Tatrach. Kras i Speleologia 6(15):79–84

Hercman H (1991) Rekonstrukcja elementów środowiska geologicznego Tatr Zachodnich na podstawie datowania izotopowego nacieków jaskiniowych. Zesz Nauk Polit Śląskiej, Mat-Fiz, Geochronometria 8:139

Jurewicz E (2005) Geodynamic evolution of the Tatra Mts. and the Pieniny Klippen Belt (Western Carpathians): problems and comments. Acta Geol Polon 55:295–338

Kicińska D (2005) Reconstruction of paleoflows direction in the western part of Lodowe Spring Cave System. Kras i Speleologia 11:107–124

Kicińska D, Hercman H, Najdek K (2017) Evolution of the Bystrej Valley Caves (Tatra Mts, Poland) based on corrosive forms, clastic deposits and U-series speleothem dating. Ann Soc Geol Polon 87:101–119

Kłapyta J, Zasadni J (2023) The Tatra Mountains—the only high-mountain landscape in Poland. In: Migoń P, Jancewicz K (eds) Landscapes and landforms of Poland. Springer, Cham, pp 000–000

Kotański Z (1961) Tektogeneza i rekonstrukcja paleogeografii pasma wierchowego w Tatrach. Acta Geol Polon 11:187–476

Králiková S, Vojtko R, Sliva Ľ, Minár J, Fügenschuh B, Kováč M, Hók J (2014) Cretaceous—quaternary tectonic evolution of the Tatra Mts (Western Carpathians): constraints from structural, sedimentary, geomorphological, and fission track data. Geol Carp 65:307–326

Lindner L (1985) Origin and age of the Cracow Canyon, Western Tatra Mts. Acta Geol Polon 35:189–202

Małecka D (1993) Hydrogeologia krasu tatrzańskiego. In: Grodzicki J (ed) Jaskinie Tatrzańskiego Parku Narodowego, vol 3. Polskie Towarzystwo Przyjaciół Nauk o Ziemi, Warszawa, pp 11–35

Nemčok J, Bezák V, Biely A, Gorek A, Gross P, Halouzka R, Janák M, Kahan Š, Kotański Z, Lefeld J, Mello J, Reich-Walder P, Raczkowski W, Roniewicz P, Ryka W, Wieczorek J, Zelman J (1994) Geological map of the Tatra Mountains. MŽP SR, GÚDŠ, Bratislava

Nowicki T (2003) Ewolucja jaskiń systemu Lodowego Źródła w świetle datowania nacieków jaskiniowych metodą uranowo-torową. Unpublished PhD thesis, ING PAN, Warsaw

Oberender P, Plan L (2015) Cave development by frost weathering. Geomorphology 229:73–84

Rudnicki J (1958) Geneza jaskiń systemu Lodowego Źródła i ich związek z rozwojem Doliny Kościeliskiej. Acta Geol Polon 8:245–270

Rudnicki J (1960) Geneza korkociągów w Jaskini Zimnej. Speleologia 2:51–55

Rudnicki J (1967) Geneza i wiek jaskiń Tatr Zachodnich. Acta Geol Polon 17:521–591

Szczygieł J (2015) Cave development in an uplifting fold-and-thrust belt: case study of the Tatra Mountains, Poland. Intern J Speleol 44:341–359

Szczygieł J, Borowska U, Jaglarz P (2014) Budowa geologiczna masywu Ciemniaka na podstawie danych z Jaskini Małej w Mułowej (Tatry Zachodnie). Przegl Geol 62:349–355

Szczygieł J, Gaidzik K, Kicińska D (2015a) Tectonic control of cave development: a case study of the Bystra Valley in the Tatra Mts, Poland. Ann Soc Geol Polon 85:387–404

Szczygieł J, Gradziński M, Pavlarčík S et al (2015b) Jaskinie i zjawiska krasowe/Jaskyne a krasové javy/Caves and karst phenomena. In: Dąbrowska K, Guzik M (eds) Atlas Tatr. Przyroda nieożywiona. Tatrzański Park Narodowy, Zakopane

Szczygieł J, Hercman H, Hoke G, Gąsiorowski M, Błaszczyk M, Sobczyk A (2020) No valley deepening of the Tatra Mountains (western Carpathians) during the past 300 ka. Geology 48:1006–1011
Wójcik Z (1960) Allochtoniczne żwirowiska jaskiń tatrzańskich. Acta Geol Polon 10:401–443
Wójcik Z (1968) Rozwój geomorfologiczny wapiennych obszarów Tatr i innych masywów krasowych Karpat Zachodnich. Prace Muz Ziemi 13:3–169

Jacek Szczygieł is Associate Professor of Geology at the University of Silesia in Katowice. His PhD thesis focused on tectonic control of cave development in the Tatra Mts. Afterwards, his research focused on neotectonics and paleoseismology based on deformation in caves, but he also studied mountain landscape evolution and its relations with cave development and speleogenesis. The Tatra Mountains is his main research area, but he also researched the Bohemian Massif (Poland), Slovakian Carpathians, and the South China Karst, and he currently works in the Alps. He has been a board member of the Speleological Section of the Polish Society of Naturalists since 2016. From 2021 to 2022, he was a postdoctoral fellow at the University of Vienna.

Hills, Gravelly Floodplains and Peat Bogs of the Podhale—A Unique Landscape Mosaic Between the Tatra, the Beskidy and the Pieniny Mountains

16

Joanna Zawiejska, Bartłomiej Wyżga, Józef Kukulak and Dorota Chmielowska-Michalak

Abstract

The Podhale region is located between the Tatra Mountains, the Pieniny and the Beskidy Mountains in southern Poland. Its northern part—the intramontane Orawa–Nowy Targ Basin—and its immediate surroundings represent an unusual landscape diversity resulting from the geological history of the surrounding regions, a legacy of Pleistocene glaciations in the Tatra Mountains and a relatively recent but pervasive effect of human activity in the region. Wedged into the region is the Pieniny Klippen Belt with isolated carbonate rocky hills, which contrast with flysch hills to the south and the flat basin floor covered with a succession of glaciofluvial–alluvial fan terraces formed by rivers draining the Tatra Mountains. Severe climate conditions together with the occurrence of a poorly permeable layer of silty, loess-like deposits formed in the Pleistocene allowed for the development of once extensive, and now fragmented peatland on the basin floor. Finally, human impacts have led to the transformation of fluvial systems which now span an unusual range of river styles from island- and bar-braided morphologies functioning in relatively natural manner to regulated and incised, bedrock channels. This diverse region also contains an extensive sedimentary record of the evolution of the surrounding mountain ranges and the post-glacial palaeoenvironmental history of the area.

J. Zawiejska (✉) · B. Wyżga · J. Kukulak · D. Chmielowska-Michalak
Department of Physical Geography, University of National Education Commission, Krakow, Podchorążych 2, 30–084 Kraków, Poland
e-mail: joanna.zawiejska@up.krakow.pl

J. Kukulak
e-mail: jozef.kukulak@up.krakow.pl

D. Chmielowska-Michalak
e-mail: dorota.chmielowska-michalak@up.krakow.pl

Keywords

Western Carpathians · Neotectonic movements · Klippen · Fluvioglacial–alluvial fans · Peat bogs · River styles

16.1 Introduction—Geographical and Geological Background

The Podhale region is located in the northern foreland of the high-mountain Tatra massif (Fig. 16.1), in the Polish part of the Western Carpathians (Fig. 16.2), and represents 5.5% of the area of the Polish Carpathians. Situated in the Inner Western Carpathians and the contact zone of the Inner and Outer Western Carpathians (or Central and Outer Western Carpathians according to physiogeographic subdivisions), it encompasses a geologically and morphologically diverse group of subregions between the Tatra Mountains to the south, Pieniny Mountains to the north-east and flysch ranges of the Beskidy Mountains belonging to the Outer Carpathians in the north. Geological contrasts in this contact zone resulted in a unique landscape mosaic of the Podhale region, unmatched in other Polish Carpathian basins that are located in more monotonous, flysch-dominated geological settings.

An oblique uplift of the Tatra Mountains and their northern foreland during the Neogene and Quaternary coupled with subsidence of the area near the contact zone of the Inner and Outer Western Carpathians (Golonka et al. 2005) resulted in a general morphological pattern of the Podhale region (Klimaszewski 1972).

Consequently, the Tatra Mountains with alpine morphology and elevations up to 2655 m asl neighbour to the north of the Gubałówka–Spisz Hills typified by undulate topography and elevations varying between 650 and 1238 m asl, which pass farther north to the intramontane Orawa–Nowy

P. Migoń and K. Jancewicz (eds.), *Landscapes and Landforms of Poland*,
World Geomorphological Landscapes, https://doi.org/10.1007/978-3-031-45762-3_16

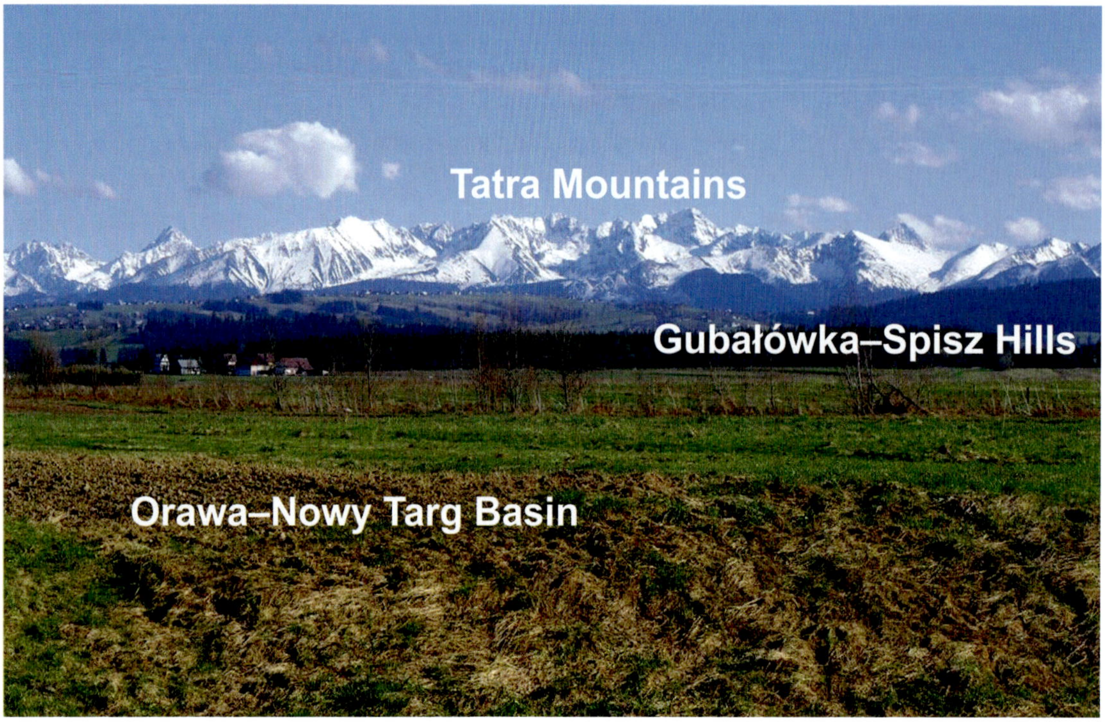

Fig. 16.1 Landscape of the Podhale region, with the Tatra Mountains in the background (photograph by J. Zawiejska)

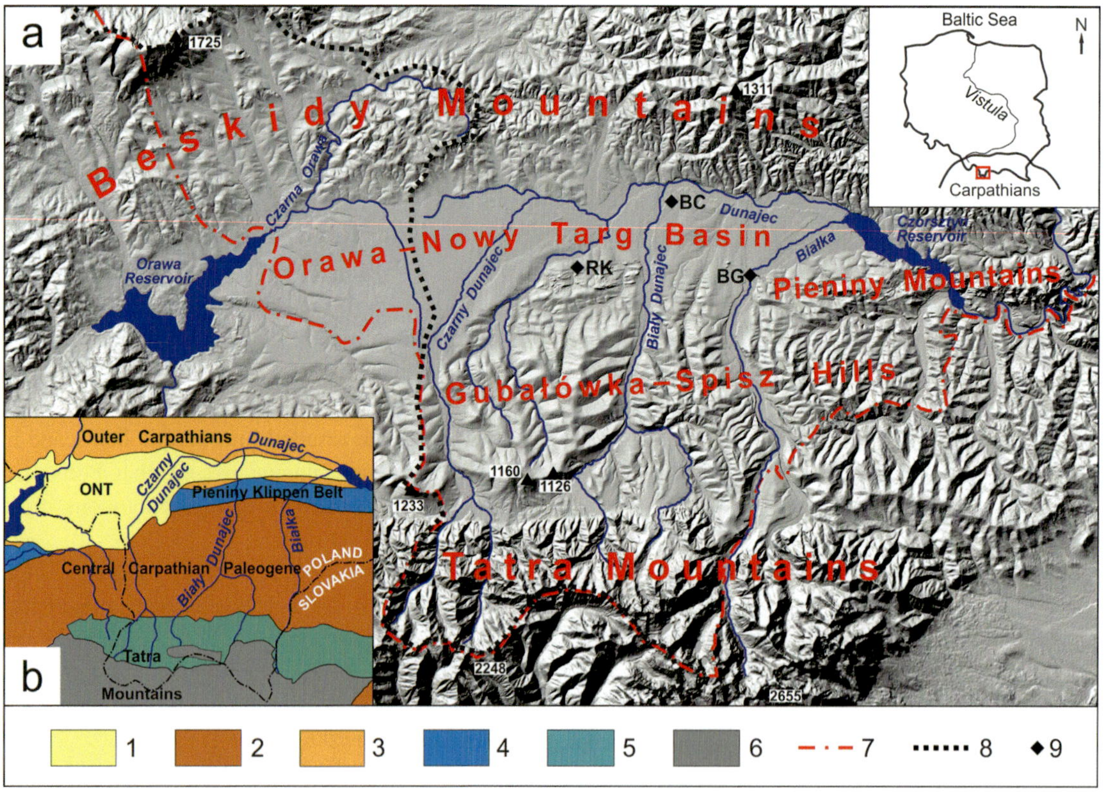

Fig. 16.2 Location **a** and Geological setting **b** of the Podhale region. 1—Neogene-Quaternary sediments of the Orawa and Nowy Targ depressions (ONT); 2—Central Carpathian Paleogene; 3—Cretaceous-Paleogene flysch of the Magura Nappe (Outer Carpathians); 4—Mesozoic sedimentary series of the Pieniny Klippen Belt; 5—Mesozoic sedimentary series of the high-tatric and sub-tatric units; 6—Paleozoic igneous and metamorphic rocks of the crystalline core of the Tatra Massif; 7—State boundary; 8—Main European water divide; 9—Nature reserves: RK—The Rogoźnik Klippe; BG—the Białka Gorge; BC—Bór na Czerwonem raised bog

Targ Basin with smooth topography and elevations ranging from 490 to nearly 800 m asl (Figs. 16.1 and 16.2). Rivers draining the Podhale region and the surrounding mountains combine into the Dunajec River (Fig. 16.2b), which leaves the region through the Pieniny Gorge dissecting the Pieniny Mountains (Margielewski and Jankowski 2022) that border the NE part of the Podhale.

The oldest rocks outcropping in the Podhale region are Jurassic–Cretaceous sedimentary series of the Pieniny Klippen Belt that formed in deep and shallow areas of the Tethys Sea (Birkenmajer 1986); currently, they occur in a few kilometres wide belt in the contact zone between the Inner and Outer Western Carpathians (Fig. 16.2b). Triassic–Cretaceous sedimentary rocks of the high-Tatric and sub-Tatric units, which outcrop in the Tatras (Fig. 16.2b), in the area of the Gubałówka–Spisz Hills are deeply buried under the cover of Paleogene sediments (Lefeld 2009). In the Paleogene, the northern part of the Inner Western Carpathians, including the Tatra Mountains and the Podhale was covered with a marine basin (called Central Carpathian Paleogene Basin), in which thick complexes of flysch sediments were deposited (Fig. 16.2b). In the north, this basin covered the Pieniny Klippen Belt and was connected with the Magura flysch basin of the Outer Western Carpathians existing since the Late Cretaceous. A collision of the ALCAPA micro-plate (encompassing the Eastern Alps, Inner Carpathians and the Pannonian Basin) with the North European Plate in the Late Oligocene and Early Miocene resulted in the formation of flysch nappes of the Outer Western Carpathians and terminated the marine evolution of the Podhale region. Subsequent evolution of the Podhale region under terrestrial and lacustrine conditions has led to the formation of the contemporary relief and landscape of the area described in this chapter.

16.2 Gubałówka–Spisz Hills

North of the Tatra Mountains and an erosional trough at their foothills are the Gubałówka–Spisz Hills (Fig. 16.2) underlain by the Central Carpathian Flysch of Paleogene age. As a result of the Neogene–Quaternary uplift common with the Tatra Mountains and subsequent erosion, the hills reach up to 1233 m asl, with local relief amounting to 200–300 m. Following the orientation of major faults (Wołosiewicz 2016), the valleys of three major rivers flowing from the Tatra Mountains to the north cut through the Gubałówka–Spisz Hills towards the Orawa–Nowy Targ Basin. Relatively wide, the valleys are lined with gravelly terraces reaching 60 m above valley floors and bear evidence to repeated Pleistocene glaciations of the Tatra Mountains. In contrast to the steep valley sides, the long, gentle flat-top hills radiate from the Gubałówka (1126 m) and Butorowy Wierch (1160 m) and only drop steeply to the north at the boundary with the Orawa–Nowy Targ Basin. Wider and higher parts of the hilltops reflect the occurrence of thick complexes of flysch sandstones.

The long and steep slopes of the Gubałówka–Spisz Hills are subject to a variety of mass movements. Shallow and small slides develop in weathered, loamy slope covers, while underlying flysch strata promote formation of deep-seated large landslides (> 1.5 km^2); finally, landslides are common on steep slopes undercut by rivers. Landsliding is additionally enhanced by the occurrence of a dense network of tectonic fracturing of the sandstone–shale complexes, particularly along fault lines (Bober 1971; Rączkowski 2007). The fracturing within sandstone beds controls water circulation and facilitates sliding on the contact surfaces with shale complexes. Most landslides occur in isolation, usually in headwater areas; however, zones of multiple, overlapping landslides occur on the western side of the Czarny Dunajec valley and along fault systems/lines at the boundary of the Gubałówka Hills and the Orawa Basin. Several of these landslide systems represent failures of thick flysch complexes, dislocated on various distances from their original surfaces of rupture. Major landslide zones are usually old but their fragments are occasionally remobilized. Small rockfalls may occur on steep slopes where flysch strata are exposed as a result of fluvial erosion of high river banks or valley sides (Fig. 16.3).

The hills are largely deforested and traces of relatively recent, widespread cultivation of the slopes and hilltops are evident in the landscape, with boundaries of individual ploughland plots and agricultural terraces still visible (Fig. 16.4). With unfavourable topographic conditions, poor shallow soils and challenging climate, especially in the last decades of the Little Ice Age, the region was intensely used for agriculture by poverty-stricken rural communities until around the 1950s. With the emergence of new sources of income and departure from cultivation, the slopes started to be used mostly as meadows and pastures (Fig. 16.4), which must have dramatically reduced the amount of sediment delivered from the slopes to the channels and facilitated rapid incision observed in watercourses of the region.

16.3 Orawa–Nowy Targ Basin

On its northern side, the Gubałówka–Spisz Hills neighbour the radically different Orawa–Nowy Targ Basin. It forms a distinct topographic depression consisting of a wider section in the west (Orawa Basin) and a narrower one in the east (Nowy Targ Basin). In turn, the northern edge of the basin borders on the low- and medium-altitude mountain ranges underlain by Cretaceous–Paleogene flysch of the Magura Nappe, the highest nappe of the Outer Western

Fig. 16.3 Paleogene flysch strata undercut by the Białka River. Small rockfalls may develop on the steep slope of the undercut valley side (photograph by J. Kukulak)

Fig. 16.4 Agricultural terraces visible on the eastern side of the Białka River valley are the legacy of the once intensive agricultural use of slopes within the Gubałówka–Spisz Hills. During the second half of the twentieth century, arable fields on the slopes were abandoned and turned into meadows or pastures (photograph by B. Wyżga)

Carpathians. This intramontane basin is a geomorphological expression of two tectonic depressions bounded by normal and strike-slip faults (Pomianowski 2003) and filled with Neogene–Quaternary sediments (Watycha 1977; Baumgart-Kotarba 1996). However, the horizontal extent of Quaternary sediments and thus the extent of the basin is larger than that of the tectonic depressions (Pomianowski 2003). The depressions are superimposed on the boundary zone between the Inner and Outer Western Carpathians and their origin and development was associated with a change in the regional stress pattern, when the collision of the ALCAPA terrane with the North European Plate finished and was followed by a rotation of the ALCAPA micro-plate (Tokarski et al. 2016; Ludwiniak et al. 2019).

The depressions are filled with two sedimentary complexes of contrasting lithologies: the Neogene complex mostly composed of claystones and siltstones and the Quaternary one dominated by gravels, with the change in the style of sedimentation reflecting acceleration of uplift of the surrounding mountains during the Quaternary (Tokarski et al. 2012). However, considerable differences occurred between the development of the western and eastern parts of the Orawa–Nowy Targ Basin. In the Orawa Basin, the thickness of Neogene–Quaternary sediments reaches 1300 m (Watycha 1977), indicating that subsidence of the basin and its infilling with molasse deposits were the principal processes forming this part of the intramontane depression. In the Nowy Targ Basin, the thickness of analogous sediments does not exceed 200 m (Pomianowski 2003); this, together with 500–700 m of difference between the elevation of the flysch rocks underlying the Neogene–Quaternary sediments in the basin and forming the mountain ridge to the north of it, testifies to an important role of erosional processes in the formation of the eastern part of the intramontane basin (Klimaszewski 1972).

Most of the Orawa–Nowy Targ Basin lies at the elevation of 490–650 m asl, with the exception of the Czarny Dunajec valley and its sides, where its surface reaches up to nearly 800 m asl. The W–E running valley of the Dunajec, the main river draining the Podhale region (Fig. 16.2), occupies the northern part of the Nowy Targ Basin, whereas nearly the entire floor of the Orawa–Nowy Targ Basin is covered with extensive glaciofluvial–alluvial fans of major rivers carrying vast amounts of gravelly material from the Tatra Mountains in the Pleistocene and the Holocene. Successive fan terraces in the basin constitute an important record of Pleistocene glaciations of the Tatra Mountains (Baumgart-Kotarba 1991–1992). Interestingly, in the Orawa Basin, the fan terraces are asymmetrically positioned in relation to the W–E axis of the basin: they are well developed and extensive in its western part with only sections of the terraces located in its eastern part. This asymmetry has been attributed to Quaternary uplift of the SW part of the basin and simultaneous subsidence of its NE part (Baumgart-Kotarba 1991–1992). As a result of the subsidence of the NE section, older parts of the fan were successively covered with the younger ones (Watycha 1977), which now constitute the surface levels in this part of the Orawa Basin. These differential tendencies of neotectonic movements also promoted changes in the surface drainage system. During the Pleistocene, the Czarny Dunajec must have flowed both to the west and the north-east (perhaps periodically shifting its position on the fan and the direction of drainage), whereas the NE direction became established in the Holocene (Baumgart-Kotarba 1991–1992). The line dividing these uplifted and lowered sections of the fan currently constitutes a part of the main European water divide between the basins of the Baltic Sea and the Black Sea.

Abnormal relief and altitudinal position of the European water divide are peculiar features of the Orawa–Nowy Targ Basin. In the Carpathians, this water divide typically runs along mountain ridges, with its elevations exceeding 2000 m asl in the Tatra Mountains. In the Orawa Basin, the water divide descends to an elevation of ~650 m asl and runs through the domes of peat bogs and meadows with a completely flat surface (Fig. 16.5). Here, the water divide separates tributaries of the Czarny Dunajec River to the east from those of the Czarna Orawa River to the west (Fig. 16.2). The Czarny Dunajec River flows towards the Vistula and belongs to the Baltic Sea basin, whereas the Czarna Orawa flows to the Vah and then to the Danube, belonging to the Black Sea basin.

The ongoing tectonic activity in the region is confirmed by seismic events, with recorded recent earthquakes reaching the magnitude of M = 4.4 (Guterch 2009), as well as by the deformation of terrace levels, presence of small faults, fractured clasts within the basin deposits (Fig. 16.6) (Tokarski et al. 2016) and cracked building walls (Wiejacz and Dębski 2009).

16.4 Klippen

To the east of the Podhale region, Jurassic–Cretaceous rocks of the Pieniny Klippen Belt are elevated, forming mountains that reach 1050 m asl in elevation, with numerous klippen (also: crags) and steep rock walls. In the western part of the Podhale region, rock complexes of the Pieniny Klippen Belt are deeply lowered (up to 1.5 km) and occur beneath the Neogeone-Quaternary sediments of the

Fig. 16.5 Section of the European water divide in the northern part of the Orawa Basin. The water divide runs here through the dome of a raised peat bog on the left and the flat meadows in the centre of the photograph, separating areas drained off to the Baltic Sea in the foreground from those drained off to the Black Sea in the background (photograph by B. Wyżga)

Fig. 16.6 Cracking of clasts caused by neotectonic movements visible in Miocene gravels exposed in SE part of the Orawa Basin (photograph by J. Kukulak)

tectonic Orawa depression. To the south of the Nowy Targ Basin, however, these rock complexes are lowered by only 200–300 m in relation to the Pieniny Mountains. Although in physiographic divisions this area is included into the Gubałówka–Spisz Hills, it has specific relief distinguishing it from the surrounding areas to the south and the north. The rock complexes of the Pieniny Klippen Belt appear here on the surface as isolated low hills or klippen reaching up to ~ 770 m asl in elevation that protrude 20–70 m from the Quaternary glaciofluvial–alluvial fans of the Biały Dunajec and Białka rivers (Fig. 16.7). These klippen are composed of Jurassic to Early Cretaceous limestones and radiolarites that were tectonically detached from the source area and embedded in less resistant rocks, mostly marls, shales and flysch (Birkenmajer 1986). After the denudation of the cover of Paleogene flysch rocks during the Miocene, in the Pliocene and the Quaternary the rock complexes of the Pieniny Klippen Belt were subjected to differential erosion. Less resistant rocks were then dissected by the rivers flowing from the Tatras and covered with gravels of the glaciofluvial–alluvial fans, whereas more resistant rocks remained as protruding klippen.

Two groups of klippen in the Podhale region are protected as nature reserves. One of the reserves, the Rogoźnik Klippe situated to the south of Rogoźnik Stream, was established to protect a group of klippen and two abandoned quarries with natural and artificial exposures presenting the geological structure (both lithology and tectonics) of this part of the Pieniny Klippen Belt (Birkenmajer 1962). This place was famous among geologists because of the occurrence of muschelkalk with late Jurassic fossils, especially ammonites and brachiopods, typical of shallow areas of the Tethys Sea. The second reserve, the Białka Gorge, protects the landscape where the Białka River flows between the Obłazowa Klippe to the west and the Kramnica Klippe to the east (Fig. 16.7b). This is a uniquely picturesque location with a braided channel of the river upstream and downstream of the klippen and steep limestone walls of the klippen adjacent to the river channel.

Following the exposure of klippen on the surface, their carbonate rocks were subjected to karstic processes. Despite its small size, the cave in the Obłazowa Klippe,

Fig. 16.7 Klippen belt. **a** Isolated hill of the Cisowa Klippe protruding from the surrounding flat surface. **b** Kramnica Klippe and the Białka River in the nature reserve "The Białka Gorge" (photographs by M. Michalak)

close to the left bank of the Białka River, is famous because of archaeological finds in the sedimentary filling of its bottom parts (Valde-Nowak 2009). Excavations in the Obłazowa Cave revealed ten layers containing artefacts of Palaeolithic cultures, with lower six layers associated with the cave occupation by Neanderthals and the upper four layers recording activities of modern humans. In layer 8, containing numerous artefacts of Upper Palaeolithic reindeer hunters and dated at ~ 30,000 BP, archaeologists excavated a boomerang made from a mammoth tusk, being the oldest confirmed find of a weapon of this type in the world. In the same layer, two finger bones of *Homo sapiens* were found, which represent the oldest skeletal material of modern humans in Poland. The arrangement of the artefacts and particularly the presence of human bones seem to indicate that the cave was a place of some cult or ritual for Palaeolithic people (Valde-Nowak 2009).

16.5 Loess-Like Deposits or 'Orawa Loams'

Pleistocene terraces of glaciofluvial–alluvial fans of the Czarny Dunajec, Biały Dunajec and Białka rivers in the Orawa–Nowy Targ Basin are covered with a layer or patches of fine-grained deposits, masking irregularities of local subsurface topography. The deposits with a thickness of 1–3.5 m have a considerably larger extent in the western part of the basin (Fig. 16.8a) and so were also called 'Orawa loams'. Subject to an ongoing debate, the origin of the deposits has been attributed to long-term weathering of glaciofluvial gravels as well as to aeolian or fluvial processes (Watycha 1977; Baumgart-Kotarba 1991–1992). Recent studies indicate a mostly silty composition and suggest that these are in fact polygenetic, loess-like deposits (Chmielowska and Woronko 2019). Massive structure is typical of these sediments that blanket the surfaces of glaciofluvial fan terraces (Fig. 16.8b), regardless of their age, whereas those with weak horizontal lamination and a major proportion of sandy silt can be found on Pleistocene terraces in the valleys of the Czarny Dunajec and Czarna Orawa rivers. Absolute age of these deposits is unknown, but they are generally considered to be of Pleistocene age, as inferred from their stratigraphic position above glaciofluvial sandy gravels and below Holocene peat deposits. Single TL datings of these deposits from Riss terraces of the Czarny Dunajec indicated the age of 146 ± 40 ka and 110 ± 15 ka (Baumgart-Kotarba 1991–1992).

Studies of heavy mineral content and composition showed that the primary source of these silty deposits were the igneous and metamorphic rocks in the glaciated Tatra Mountains (Chmielowska and Salata 2020). Glacial and then glaciofluvial transport supplied the weathered material to the Tatra foreland. Under cold and dry periglacial conditions, fine particles were easily entrained by wind from extensive outwash plains, transported a short distance and deposited locally, possibly 'trapped' on the wet tundra vegetation in topographically favourable locations. While such deposits might have formed throughout glacial periods, their preservation was only possible in the final parts of the glacials, when the reduction of fluvial sediment flux resulted in the change of river behaviour. The rivers ceased to rework their outwash plains and began incising into their surface, leaving lateral parts of the plains as the locations favouring accumulation of the silty sediments. All these conditions eventually led to the formation of the veneer of the massive loess-like layer that effectively seals the current topography (Chmielowska and Woronko 2019). Given the elevation of the Orawa–Nowy Targ Basin and the location

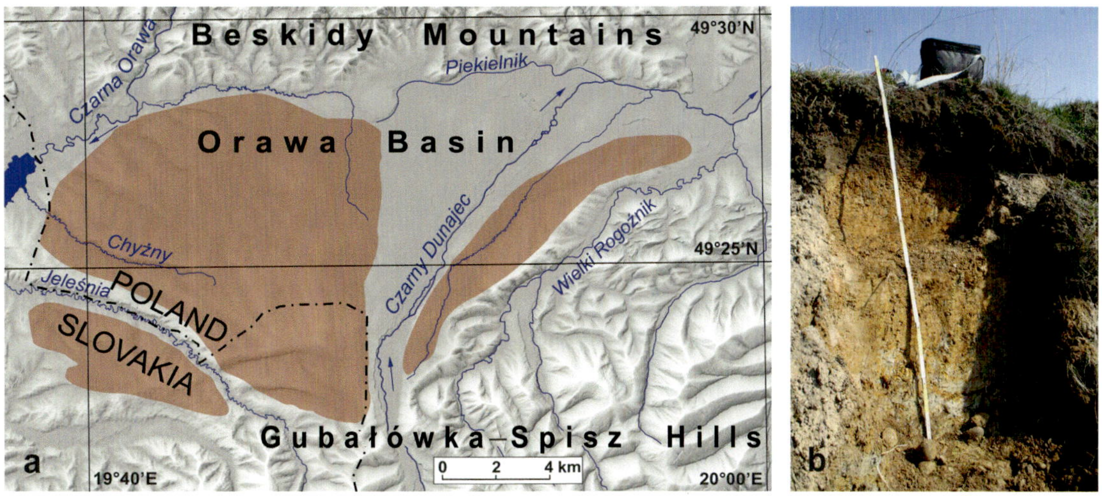

Fig. 16.8 Loess-like deposits in the Orawa Basin. **a** Extent of the Pleistocene loess-like deposits in the Orawa Basin. **b** Massive loess-like deposits covering the Würm terrace of the Czarny Dunajec fan (photograph by D. Chmielowska-Michalak)

Fig. 16.9 Peat bogs in the Orawa–Nowy Targ Basin. **a** Raised peat bog of Puścizna Długopole overgrown with pine forest (Photo by B. Wyżga). **b** Place of a former peat exploitation on the raised bog of Baligówka, with a few metres high wall of undercut peat dome visible in the background and a flat, swampy surface remaining after the exploitation, transformed into a secondary lagg of the bog in the foreground (photographs by J. Zawiejska)

of the deposits on glaciofluvial fans (600–800 m asl), the loess-like deposits are the highest located ones in the Polish Western Carpathians.

16.6 Peat Bogs

In contrast to other intramontane basins in the Polish Carpathians, the Orawa–Nowy Targ Basin supports numerous raised bogs and fens, with the former representing 5% of the total area of raised bogs in Poland. Raised bogs are well-visible in the basin landscape as they form domes that stand a few metres above the surrounding terrain and are overgrown with sparse coniferous forest or dwarf pine (Fig. 16.9a). Currently peat bogs cover 12% of the basin area, but before the beginning of extensive agricultural use of the area in the fifteenth century, they might have covered up to 40% of the basin and up to 70% of its Polish part (Łajczak 2009). Since then peat has been exploited to gain agricultural lands especially in the nineteenth and twentieth centuries, for fuel and fertilizer (Łajczak 2007). Places of relatively recent and incomplete peat exploitation are visible as relatively steep, a few metres high walls of peat bounded by a flat, swampy terrain (Fig. 16.9b). The bogs that remained are now protected under the regulations of Natura 2000 area (Łajczak 2007) and one of raised bogs, Bór na Czerwonem, has been protected as a nature reserve since 1925.

Fens and raised bogs have developed on the terraces of glaciofluvial–alluvial fans of the Czarny Dunajec, Biały Dunajec, Białka and Czarna Orawa rivers formed during the Mindel, Riss and Würm glacials (Łajczak 2009), whereas they are generally lacking on the Holocene floodplains of these rivers. Conditions leading to extensive peat formation and bog development in the basin comprised: (i) the occurrence of a thick layer of loess-like deposits on the surface of the terraces, preventing the infiltration of precipitation water into the underlying gravels; (ii) outflows of groundwater at the toe of the escarpments of higher terraces and the banks of paleochannels as well as in the lowest, distal parts of the fans in the northern part of the basin and (iii) prevalence of flat surfaces inhibiting drainage of precipitation water and groundwater (Łajczak 2009). Throughout the Holocene groundwater outflows in the northern parts of the Czarny Dunajec fan were probably strengthened by the infiltration of river water into fan gravels near the fan apex, but this phenomenon must have ceased with deep river incision into impermeable Miocene bedrock in the twentieth century.

Scarce radiocarbon dates, obtained from the bottom of peat profiles from raised bogs in the basin, indicate variable time of the onset of peat formation, with the oldest dates reaching ~ 10,000 years BP (Obidowicz 1988). Diverse time of the inception of individual bogs can also be inferred from a variable thickness of accumulated peat, with the maximum peat thickness in the domes of raised bogs ranging from 2.3 to 11 m and in fens reaching 3 m. All this evidence indicates that particular raised bogs started to form at different times during the Holocene and that the inception of extensive fens took place later than that of the majority of raised bogs (Łajczak 2009).

Peatlands of the Orawa–Nowy Targ Basin have an exceptional value for preservation of the biodiversity of the area. The water storage capacity of peatlands existing in the basin prior to the onset of intense agricultural use of the area was estimated at 170 million m^3, with 140 million m^3 of water stored in raised bogs, whereas the storage capacity of contemporary peatlands amounts to 60 million m^3 (Łajczak 2017). It is worth to compare these values with 63 million m^3 of the flood capacity of the large Czorsztyn Reservoir impounding the Dunajec River at the eastern end of the basin. Even though only a proportion of the total storage capacity of the peatlands contributes to water retention during rainfalls and is released during droughts, it is clear that the even the remaining peatlands have a great potential for regulating water runoff from the upper Dunajec catchment. The peatlands have also a great scientific value as archives of environmental changes during the Holocene. For instance, dendrochronological and palynological analyses of the Puścizna Wielka raised bog indicated that trees overgrew the bog only during phases of drier and warmer regional climate, whereas phases of cooler and more humid climate were reflected in the treeless conditions of bog development (Krąpiec et al. 2016). All this justifies the demand to protect a greater number of peat bogs in the area as nature reserves (Łajczak 2007).

16.7 River Diversity

Diverse physiographic conditions and spatially variable human impact result in an astounding variety of river styles in the relatively small Podhale region (Fig. 16.10). Small streams autochthonous to the Orawa–Nowy Targ Basin occur only in its western, more extensive part, where they flow on the terraces of glaciofluvial–alluvial fan of the Czarny Dunajec River. They have gravelly beds and banks composed of either peat or mud deposits consisting of the material redeposited from the fine-grained covers of the fan. The streams form irregular meanders and are typified by high sinuosity (Fig. 16.10a), reflecting cohesiveness of the banks and a relatively steady runoff controlled by a high proportion of peat bogs in their catchments.

Streams and rivers originating in the Spisz–Gubałówka Hills flow in deep, V-shaped valleys in their upper courses and in flat-bottomed valleys in the middle courses. With widespread cultivation of the surrounding hillslopes in the nineteenth century, these watercourses formed braided channels in their middle courses, but currently—with a decline

Fig. 16.10 Contemporary diversity of river styles in the Podhale region. **a** Small meandering channel of Borowy Stream draining the Orawa Basin. **b** Rogoźnik Stream originating in the Spisz–Gubałówka Hills forms a meandering channel in the Orawa–Nowy Targ Basin. **c** Braided channel of the Białka River in its lower course. **d** Wandering channel of the Czarny Dunajec River in an unmanaged reach. **e** Regulated channel of the Czarny Dunajec River, with weirs. **f** Bedrock channel of the Białka River in its middle course, with sandstone layers of Paleogene flysch exposed in the bed (photographs a, b, d by P. Mikuś; photos c, e, f by J. Zawiejska)

of hillslopes cultivation in the area—they typically flow in straight, incised channels. In their lower courses within the Orawa–Nowy Targ Basin, these gravel-bed watercourses (e.g. Rogoźnik Stream) as well as those entering the basin from the flysch mountains on the north (e.g. the Czarna Orawa River) developed wide floodplains covered with thick mud deposits. Although these watercourses are typified by relatively rapid runoff reflecting low water storage capacity of the flysch bedrock (Kundzewicz et al. 2014), high cohesiveness of bank sediments conditions the formation of meandering channels in their lower courses (Fig. 16.10b). Severe climate of the basin causes that frost processes—recurrent freezing and thawing of the bank material during winters—considerably contribute to the bank retreat of these watercourses, particularly in the upper parts of the banks composed of fine-grained alluvium (Augustowski and Kukulak 2021).

The largest watercourses of the Podhale region are the Czarny Dunajec, Biały Dunajec and Białka rivers that rise in the Tatra Mountains and run through the Gubałówka–Spisz Hills to the Orawa–Nowy Targ Basin, and the Dunajec River originating from the confluence of the two first rivers and flowing eastwards along the basin (Fig. 16.2b). Despite high annual averages of precipitation in the Tatras (up to 1700 mm), these rivers are typified by moderate irregularity of runoff, which reflects deep, karstic water circulation in the high-mountain areas underlain by carbonate rocks and effective infiltration of water into thick slope mantles in the Tatra massif. However, the flashiness of runoff increases down the rivers with an increasing proportion of flysch areas in their catchments (Kundzewicz et al. 2014). Non-cohesive alluvial plains of these rivers are composed of coarse granitic and quartzitic material supplied in the Tatras and an admixture of sandstone clasts delivered from the flysch parts of the catchments, and have only a thin cover of sandy to muddy overbank sediments.

In the 19th and the first half of the twentieth century, these rivers lacked riparian forest on the banks and flowed in wide, braided channels with extensive gravel bars (Wyżga et al. 2012, 2016), but since then they have experienced remarkable changes as a result of direct human impacts on their channels and environmental changes in the catchments (Wyżga et al. 2016). A decline in agricultural and pastoral use of riparian areas and artificial channel narrowing resulted in widespread development of riparian forest along the rivers (Fig. 16.10c–f), partly on lateral, higher parts of their former channels (Wyżga et al. 2012; Hajdukiewicz and Wyżga 2019). A wide, braided channel has persisted only in the Białka River, especially in its lower course (Fig. 16.10c). In the other rivers, unmanaged reaches are currently scarce, and with a considerable reduction of the sediment flux over the twentieth century, the previous bar-braided channel pattern of these reaches has evolved into a wandering (Fig. 16.10d) or island-braided pattern (Wyżga et al. 2016). Most reaches of the rivers were channelized; their wide, multi-thread channels were replaced with single-thread, narrow channels and some of these channelized reaches were partitioned by weirs (Fig. 16.10e) (Korpak 2007; Zawiejska and Wyżga 2010). In the reaches where the vertical position of the regulated channels was not stabilized by weirs or where intense in-channel gravel mining was conducted in the second half of the twentieth century, rapid channel incision has been induced (Zawiejska et al. 2015), resulting in the transformation of the gravelly channel beds into bedrock beds (Fig. 16.10f) (Hajdukiewicz and Wyżga 2019).

Decreased agricultural activities on hillslopes, an increase in forest cover of the catchments and a change in precipitation pattern after the end of the Little Ice Age caused reduction in flow and sediment dynamics of the rivers in the Podhale region during the twentieth century (Wyżga et al. 2012). As a result, forest developed on many exposed surfaces within these rivers, and currently islands may constitute a large proportion of active river zones in unmanaged reaches (Figs. 16.10d and 16.11a) (Mikuś et al. 2013). Forested river banks and islands became a source of trees and shrubs fallen to river channels, especially during floods that not only intensify bank retreat but also transport and distribute large wood in the rivers. During floods

Fig. 16.11 Riparian forests and large wood—key elements of the contemporary dynamics of rivers in the Podhale region. **a** Island overgrown with mature alder forest in a multi-thread reach of the Czarny Dunajec River in the Orawa–Nowy Targ Basin. **b** Large quantities of driftwood deposited in a wide, unmanaged channel of the Czarny Dunajec River (photographs by B. Wyżga)

large wood is preferentially retained in unmanaged river reaches with a wide and shallow channel (Fig. 16.11b), whereas it can be transported long distances in channelized and incised reaches with a narrow and deep channel (Ruiz-Villanueva et al. 2016; Wyżga et al. 2017). Large wood beneficially influences the functioning of aquatic and riparian ecosystems of mountain rivers; for instance, vegetative regeneration of the living driftwood of willows and alders deposited on gravel bars leads to the inception and growth of islands in the rivers (Mikuś et al. 2013). However, it can also generate flood hazard if deposited at vulnerable sites such as bridge cross-sections (Ruiz-Villanueva et al. 2017).

Apart from introducing a new, major landscape element in the region, the construction of the Czorsztyn Reservoir (Fig. 16.2) in 1997 has initiated changes in the upstream morphology and hydrodynamics of the Dunajec River (Liro 2016; Liro et al. 2020). In the 1.5-km-long reach upstream of the reservoir, periodic inundation by reservoir backwater results in increased water depth and decreased velocity of flood flows in comparison with a channelized river reach not affected by the reservoir backwater. The rise of base level has induced channel slope flattening and fining of the bed material, but increased cohesiveness of river banks, resulting from the deposition of thick overbank sediments, has limited channel widening in this reach (Fig. 16.12). In the immediately upstream reach, which is not subjected to the direct influence of backwater inundation and where only a thin cover of overbank sediments occurs on river banks, the Dunajec has widened substantially and eventually, years after channel regulation, formed large gravel bars (Fig. 16.12).

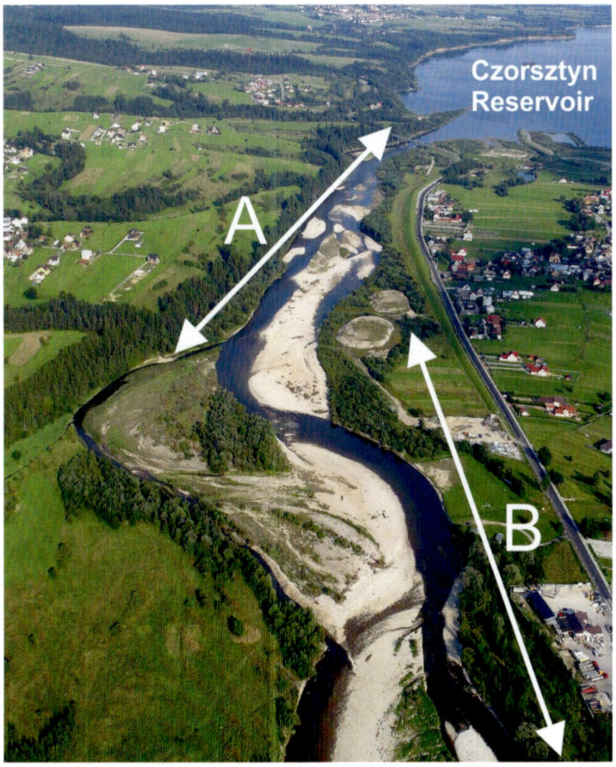

Fig. 16.12 Section of the Dunajec River immediately upstream of the Czorsztyn Reservoir. Before 1997 the Dunajec flowed here in a narrow, regulated channel, but since then marked adjustments in river morphology have been induced by direct (in reach A) and indirect (in reach B) impacts of reservoir backwater (photograph by M. Liro)

16.8 Conclusions

Nested between the spectacular nature and tourism hotspots of the Pieniny and the Tatra Mountains and the somewhat less diverse Beskidy Mountains, the Podhale region has long remained rather unrecognized and its landscape also escaped popular attention. However, a closer look and decades of geological and geomorphological studies confirm this intramontane depression boasts an unexpected landscape diversity resulting from the geological history of the surrounding regions, a legacy of Pleistocene glaciations in the Tatra Mountains, and a relatively recent but pervasive effect of human activity in the region. Complex relief of the region is a record of a multitude of events and processes, starting from plate collision and uplift of the surrounding mountain ranges through the climate shifts in the Pleistocene to the ongoing neotectonic movements and a range of modern human impacts on the environment. This intricate history has resulted in a unique landscape mosaic in this relatively small region. That mosaic is a landscape of contrasts: flysch hills with steep slopes and limestone klippen walls close to gentle peat domes on the flat surface of the basin, extensive gravelly plains reworked by steep, highly dynamic Tatra-born rivers running parallel and in close vicinity to slow, meandering streams, or nearly horizontal European water divide against the backdrop of the high-mountain Tatra massif in the south. It is also a landscape of unexpected diversity and contrast of river styles, semi-natural and those resulting from decades of intense engineering intervention and human transformation of the environment of the basin. From scientific point of view, this diverse region also harbours an extensive sedimentary record of the evolution of the surrounding mountain ranges, while the peatland of the Orawa–Nowy Targ Basin provides an invaluable source of information on the post-glacial palaeoenvironmental history of the area.

References

Augustowski K, Kukulak J (2021) The role of frost processes in the retreat of river banks. Water 13:1812

Baumgart-Kotarba M (1991–1992) Rozwój geomorfologiczny Kotliny Orawskiej w warunkach ruchów neotektonicznych. Stud Geomorph Carp-Balc 25–26:3–28

Baumgart-Kotarba M (1996) On the origin and age of the Orawa Basin, West Carpathians. Stud Geomorph Carp-Balc 30:101–116

Birkenmajer K (1962) Zabytki przyrody nieożywionej pienińskiego pasa skałkowego. Część II. Skałki w Rogoźniku koło Nowego Targu. Ochr Przyr 28:159–185

Birkenmajer K (1986) Stages of structural evolution of the Pieniny Klippen Belt, Carpathians. Stud Geol Polon 88:7–32

Bober L (1971) Structural landslides of the Podhale region (Polish Inner Carpathians). Acta Geol Acad Sci Hung 15:389–403

Chmielowska D, Salata D (2020) Heavy minerals as indicators of the source and stratigraphic position of the loess-like deposits in the Orava Basin (Polish Western Carpathians). Minerals 10:445

Chmielowska D, Woronko B (2019) A source of loess-like deposits and their attendant palaeoenvironment—Orava Basin, Western Carpathian Mountains, S Poland. Aeolian Res 38:60–76

Golonka J, Aleksandrowski P, Aubrecht R, Chowaniec J, Chrustek M, Cieszkowski M, Florek R, Gawęda A, Jarosiński M, Kępińska B, Krobicki M, Lefeld J, Lewandowski M, Marko F, Michalik M, Oszczypko N, Picha F, Potfaj M, Słaby E, Ślączka A, Stefaniuk M, Uchman A, Żelaźniewicz A (2005) The Orava deep drilling project and post-Palaeogene tectonics of the Northern Carpathians. Ann Soc Geol Polon 75:211–248

Guterch B (2009) Sejsmiczność Polski w świetle danych historycznych. Przegl Geol 57:513–520

Hajdukiewicz H, Wyżga B (2019) Aerial photo-based analysis of the hydromorphological changes of a mountain river over the last six decades: the Czarny Dunajec, Polish Carpathians. Sci Total Environ 648:1598–1613

Klimaszewski M (1972) Karpaty Wewnętrzne. In: Klimaszewski M (ed) Geomorfologia Polski, tom 1. PWN, Warszawa, pp 25–52

Korpak J (2007) The influence of river training on mountain channel changes. Geomorphology 92:166–181

Krąpiec M, Margielewski W, Korzeń K, Szychowska-Krąpiec E, Nalepka D, Łajczak A (2016) Late Holocene palaeoclimate variability: the significance of bog pine dendrochronology related to peat stratigraphy. The Puścizna Wielka raised bog case study (Orawa–Nowy Targ Basin, Polish Inner Carpathians). Quat Sci Rev 148:192–208

Kundzewicz ZW, Stoffel M, Kaczka RJ, Wyżga B, Niedźwiedź T, Pińskwar I, Ruiz-Villanueva V, Łupikasza E, Czajka B, Ballesteros-Canovas JA, Małarzewski Ł, Choryński A, Janecka K, Mikuś P (2014) Floods at the northern foothills of the Tatra Mountains—a Polish-Swiss research project. Acta Geophys 62:620–641

Łajczak A (2007) Natura 2000 in Poland, area PLH120016 The Orawsko-Podhalańskie Peatlands. Institute of Botany PAS, Kraków, p 139

Łajczak A (2009) Warunki rozwoju i rozmieszczenie torfowisk w Kotlinie Orawsko-Nowotarskiej. Przegl Geol 57:694–702

Łajczak A (2017) Surface water storage in the Orava—Nowy Targ Basin, Western Carpathians. Geogr Polon 90:383–399

Lefeld J (2009) Alpejskie fazy orogeniczne w Tatrach. Przegl Geol 57:669–673

Liro M (2016) Development of sediment slug upstream from the Czorsztyn Reservoir (southern Poland) and its interaction with river morphology. Geomorphol 253:225–238

Liro M, Ruiz-Villanueva V, Mikuś P, Wyżga B, Bladé-Castellet E (2020) Changes in the hydrodynamics of a mountain river induced by dam reservoir backwater. Sci Total Environ 744:140555

Ludwiniak M, Śmigielski M, Kowalczyk S, Łoziński M, Czarniecka U, Lewińska L (2019) The intramontane Orava Basin—evidence of large-scale Miocene to Quaternary sinistral wrenching in the Alpine-Carpathian-Pannonian area. Acta Geol Polon 69:339–386

Margielewski W, Jankowski L (2022) Dunajec River gorges through the Pieniny Mts. and the Beskid Sądecki Mts. In: Migoń P, Jancewicz K (eds) Landscapes and landforms of Poland. Springer, Switzerland, pp 000–000

Mikuś P, Wyżga B, Kaczka RJ, Walusiak E, Zawiejska J (2013) Islands in a European mountain river: linkages with large wood deposition, flood flows and plant diversity. Geomorphology 202:115–127

Obidowicz A (1988) The Puścizna Rękowiańska raised bog. In: Starkel L, Rutkowski J, Ralska-Jasiewiczowa J (eds) Late Glacial and Holocene environmental changes Vistula basin, Excursion guide book. Wyd. AGH, Cracow, pp 87–90

Pomianowski P (2003) Tektonika Kotliny Orawsko-Nowotarskiej – wyniki kompleksowej analizy danych grawimetrycznych i geoelektrycznych. Przegl Geol 51:498–506

Rączkowski W (2007) Landslide hazard in the Polish flysch Carpathians. Stud Geomorph Carp-Balc 41:61–67

Ruiz-Villanueva V, Wyżga B, Hajdukiewicz H, Stoffel M (2016) Exploring large wood retention and deposition in contrasting river morphologies linking numerical modeling and field observations. Earth Surf Proc Landf 41:446–459

Ruiz-Villanueva V, Wyżga B, Mikuś P, Hajdukiewicz M, Stoffel M (2017) Large wood clogging during floods in a gravel-bed river: the Długopole bridge in the Czarny Dunajec River, Poland. Earth Surf Proc Landf 42:516–530

Tokarski AK, Świerczewska A, Zuchiewicz W, Starek D, Fodor L (2012) Quaternary exhumation of the Carpathians: a record from the Orava-Nowy Targ Basin, Western Carpathians (Poland and Slovakia). Geol Carpath 64:257–266

Tokarski AK, Márton E, Świerczewska A, Fheed A, Zasadni J, Kukulak J (2016) Neotectonic rotations in the Orava-Nowy Targ Intramontane Basin (Western Carpathians): an integrated palaeomagnetic and fractured clasts study. Tectonophysics 685:35–43

Valde-Nowak P (2009) Obłazowa and Hłomcza: two Paleolithic sites in the North Carpathians province of Southern Poland. In: Adams B, Blades BS (eds) Lithic materials and Paleolithic societies. Wiley-Blackwell, Chichester, pp 196–207

Watycha L (1977) Objaśnienia do szczegółowej mapy geologicznej Polski 1:50000, arkusz Czarny Dunajec. Instytut Geologiczny, Warszawa, p 102

Wiejacz P, Dębski W (2009) Podhale, Poland, earthquake of November 30, 2004. Acta Geophys 57:346–366

Wołosiewicz B (2016) Morphotectonic control of the Białka drainage basin (Central Carpathians): insights from DEM and morphometric analysis. Contemp Trends Geosci 5:61–82

Wyżga B, Zawiejska J, Radecki-Pawlik A, Hajdukiewicz H (2012) Environmental change, hydromorphological reference conditions and the restoration of Polish Carpathian rivers. Earth Surf Proc Landf 37:1213–1226

Wyżga B, Zawiejska J, Hajdukiewicz H (2016) Multi-thread rivers in the Polish Carpathians: occurrence, decline and possibilities for restoration. Quat Intern 415:344–356

Wyżga B, Mikuś P, Zawiejska J, Ruiz-Villanueva V, Kaczka RJ, Czech W (2017) Log transport and deposition in incised, channelized and multithread reaches of a wide mountain river: tracking experiment during a 20-year flood. Geomorphology 279:98–111

Zawiejska J, Wyżga B (2010) Twentieth-century channel change on the Dunajec River, southern Poland: patterns, causes and controls. Geomorphology 117:234–246

Zawiejska J, Wyżga B, Radecki-Pawlik A (2015) Variation in surface bed material along a mountain river modified by gravel extraction and channelization, the Czarny Dunajec, Polish Carpathians. Geomorphology 231:353–366

Joanna Zawiejska is an associate professor in geomorphology at the University of National Education Commission, Kraków (Department of Physical Geography). Her research interests focus on the following areas: (i) fluvial geomorphology of mountain regions with particular focus on natural and anthropogenic drivers of change and restoration of ecological quality of rivers, (ii) environmental education and education for sustainable development, (iii) environmental perception and (iv) sustainable tourism. She is also currently the chair of the international scientific network *Science for the Carpathians*.

Bartłomiej Wyżga (deceased) was a Professor at the Institute of Nature Conservation, Polish Academy of Sciences, specializing in interdisciplinary research in the fields of fluvial geomorphology, hydrology, biogeomorphology and ecology. He conducted pioneering research on large wood dynamics and its environmental significance in the mountain rivers of the Polish Carpathians. His research interests included assessment of river ecological state and restoration of mountain rivers, flood hazard and flood management in mountain regions as well as pro-environmental methods of river management. He was involved in several international projects, including pioneering studies on social perception of mountain rivers.

Józef Kululak is a professor in geomorphology at the University of National Education Commission, Krakow, Poland; specializing in geomorphology and Quaternary geology. His research focuses on interactions between geomorphic processes and neotectonics in mountain regions, fluvial and glacifluvial sediment analysis and dating, and chronostratigraphy of fluvial terraces in mountain regions. He conducts his research mainly in the Polish Carpathians and the Peruvian Andes.

Dorota Chmielowska-Michalak is assistant professor in physical geography at the University of National Education Commission, Krakow, Poland, with research interests in geomorphology, paleogeography and methodology of Quaternary sediment analysis. She specializes in automated grain shape image analysis and her research concerns reconstruction of sedimentary environments and their interactions based on the analysis of grain characteristics, particularly of repeatedly redeposited grains. Her expertise comprises both field-based and laboratory research of contemporary fine sediments and covers in mountain regions.

Dunajec River Gorge Through the Pieniny Mountains

17

Włodzimierz Margielewski and Leszek Jankowski

Abstract

New concepts of the development of the Pieniny Mts. (geologically the Pieniny Klippen Belt) as a regional flower structure, as well as a detailed analysis of the Digital Terrain Model (DTM), indicate that the Dunajec River Gorge in the Pieniny Mts. (called also the Pieniny Gorge) is regressive and structural. The Dunajec River, originally flowing along the northern margin of the Pieniny Mts., was intercepted due to river piracy by the Niedziczanka River flowing along the southern margin of the Pieniny Mts. In the formation of the 8 km long Dunajec River Gorge, which cuts through siliceous limestones, an important role was played by the transversal tearing of the flower structure of the klippen belt and the formation of cracks along the reactivated faults, subsequently used by the Dunajec River. Rise and lowering of individual blocks of the flower structure along the faults, as well as footwall elevation phenomenon, had a significant impact on the formation of high walls of the gorge. The Dunajec River Gorge through the Pieniny Mts. is therefore not strictly antecedent, related to the elevation (*en masse*) of the mountain massif transversely cut by the river, but connected with the selective elevation or lowering of the elements (blocks) of the Pieniny flower structure, transverse torn. In addition, the course of the Dunajec River was blocked in the gorge section by huge rock blocks displaced by gravity towards the south along with the massif of Mt. Trzy Korony, which resulted in the formation of a dammed lake and then an overflow through a complicated system of extensional fissures established along faults network associated with the flower structure. Therefore, the most important phenomenon influencing the formation of the Pieniny Gorge was the transverse tearing of the Pieniny flower structure.

Keywords

Dunajec River Gorge · Pieniny flower structure · Gravitational collapse · Pieniny Klippen Belt · Polish Western Carpathians

17.1 Introduction

In the Polish Carpathians, sections of river valleys with narrow floors and high, steep and locally rocky slopes frequently occur (Zuchiewicz 1978, 2010). Until recently, the formation of these river gorges (river gaps) has been considered to be associated with the Late Neogene and the Quaternary tectonic movements (called neotectonic movements) (Zuchiewicz 1978, 1984; Rączkowski et al. 1984). It was commonly assumed that during the neotectonic elevation of the axial parts of mountain ranges (the Beskid Żywiecki Mts., the Beskid Mały Mts., the Beskid Sądecki Mts., the Pieniny Mts. and the Bieszczady Mts.) and simultaneous lowering of intra-mountain depressions (e.g. the Żywiec Basin, the Orawa-Nowy Targ Basin, the Nowy Sącz Basin), rivers cut into the gradually rising blocks, forming antecedent gorges (Fig. 17.1a) (Starkel 1972; Zuchiewicz 1978, 1980, 1987, 2010).

One of the most famous and characteristic river gorges in the Polish Carpathians is the Dunajec River Gorge (called also the Pieniny Gorge) through the Pieniny Mountains, between the village of Czerwony Klasztor and the town of Szczawnica (Figs. 17.1b, 17.2a—site 5)

W. Margielewski (✉)
Institute of Nature Conservation, Polish Academy of Sciences, Adama Mickiewicza Ave. 33, 31-120, Kraków, Poland
e-mail: margielewski@iop.krakow.pl

L. Jankowski
Polish Geological Institute-National Research Institute, Carpathian Branch, Skrzatów 1, 31-560, Kraków, Polande-mail: leszek-jankowski@pgi.gov.pl

P. Migoń and K. Jancewicz (eds.), *Landscapes and Landforms of Poland*,
World Geomorphological Landscapes, https://doi.org/10.1007/978-3-031-45762-3_17

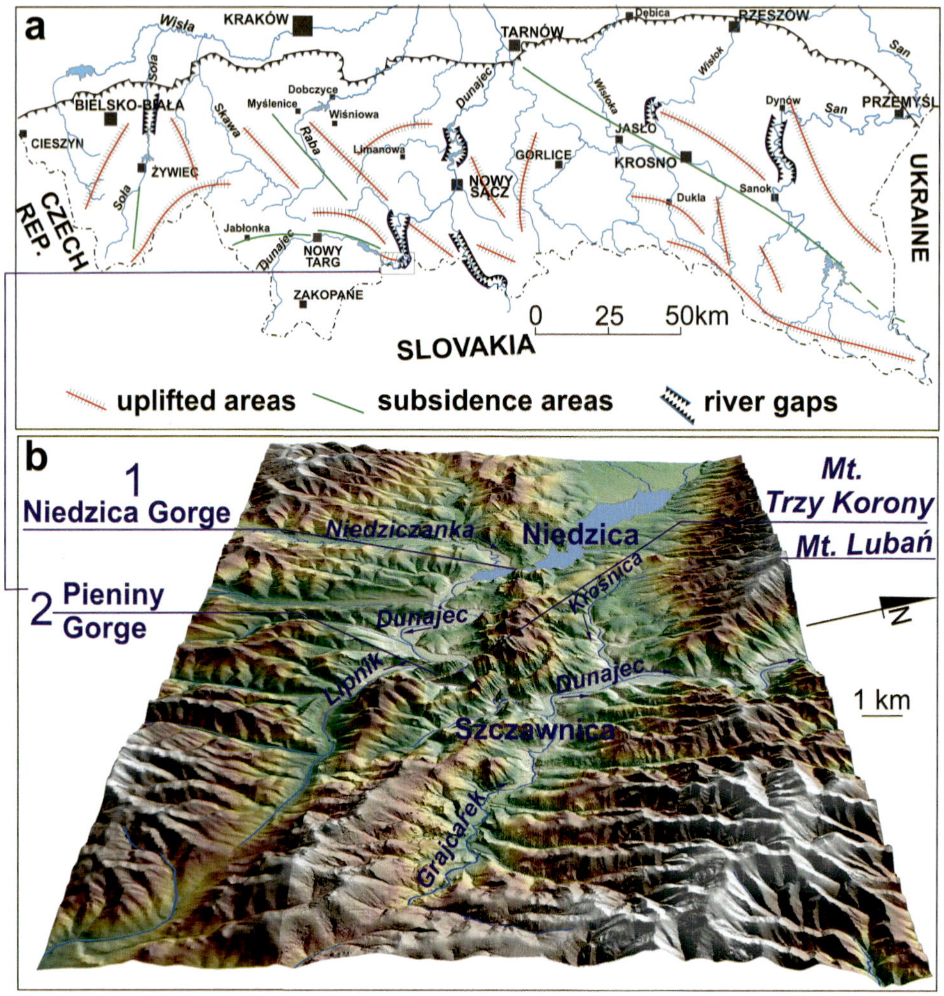

Fig. 17.1 Study area. **a** Neotectonic sketch of the Polish Carpathians, with the position of the Dunajec River Gorge through the Pieniny Mts. (after Zuchiewicz 1984, 2010). **b** SRTM-based Digital Terrain Model of the Dunajec River Gorges through the Pieniny Klippen Belt (1—Niedzica Gorge, 2—Pieniny Gorge)

(Zuchiewicz 1978, 1982; Birkenmajer 2006). In the Pieniny Mts., the Dunajec River, meandering for a distance of 8 km, cuts deeply (up to 400 m) into the rock massifs, creating a narrow valley of gorge-type with very steep (frequently sub-vertical), often rocky slopes (Figs. 17.3a, b, 17.4). In the gorge section, the Dunajec River crosses transversally the Pieniny Klippen Belt (Figs. 17.1b, 17.5a, b). So far, the genesis of this water gap has not been clearly explained, despite various concepts presented in the literature (see Zuchiewicz 1980, 1982, 1987; Birkenmajer 2006).

17.2 Geological Setting of the Pieniny Mts.

The Pieniny Mts. (geologically: the Pieniny Klippen Belt-PKB) represents a long, but very narrow geological unit, only 3–6 km wide, separating the Inner and the Outer Carpathians (Książkiewicz 1972; Żytko et al. 1989). According to the previous concepts, the Pieniny Mts. were characterized by a very complicated tectonic evolution and tectonic structure. Spatial relations between lithostratigraphic formations and tectonic units (nappes) distinguished here are very complex (Birkenmajer 1979, 1985, 2017; Jurewicz 2005, 2018; Oszczypko 2006; Oszczypko et al. 2006; Oszczypko and Oszczypko-Clowes 2014). According to more recent concepts, the Pieniny Klippen Belt was formed as a flower structure of a regional character (Plašienka and Mikuš 2010; Jankowski 2015; Golonka et al. 2018; Plašienka et al. 2020). The flower structure forms a deep-rooted folded structure associated (framed by) with strike-slip faults, where the rock mass is divided into disintegrated rock blocks, which in turn are faulted upward or dropped down (Harding 1985). Huge rock masses of carbonate and siliceous rocks, mainly siliceous limestones, of Mesozoic age are embedded into a tectonic mélange formed during the evolution of the flower structure. They were gravitationally displaced from the Mesozoic carbonate platform to the flysch Jarmuta-Inoceramian basin, and pulled later into the Pieniny flower structure (Plašienka and Mikuš 2010; Golonka et al. 2015; Jankowski 2015). The flower structure of the Pieniny Mts. also includes flysch rocks of the Magura Unit, flysch rock

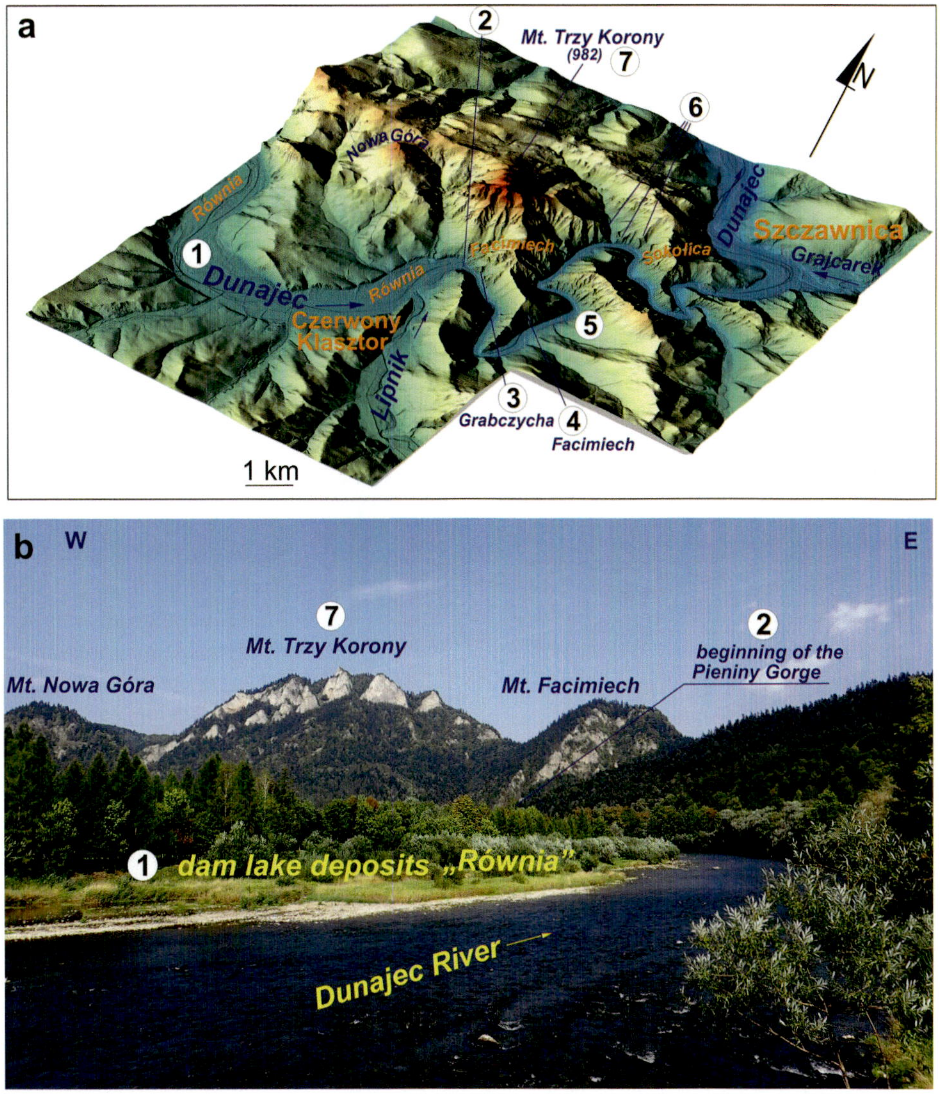

Fig. 17.2 Dunajec River Gorge. **a** Characteristic relief elements discussed in the text and figures as seen on LiDAR DTM: 1—Site of former Dunajec dammed lake (called Równia) with multi-channel (wandering) river; 2—Entrance to the narrow Pieniny Gorge, 3—Triangular facets of the Grabczycha indicating normal faults; 4—Large triangular facets due to reactivation of a normal fault) frame the river meander around Mt. Facimiech; 5—Pieniny Gorge meanders; 6—Triangular facets in the vicinity of Mt. Sokolica; 7—Mt. Trzy Korony. **b** Dam lake deposits of the Dunajec River at the site called "Równia" (no. 1 on this figure part label Fig. 17.2a) and Mt. Trzy Korony (no. 7) (photograph by Jankowski)

of the Podhale Basin and the depositional sequence of the Malcov succession (of Lower Miocene age), which covers this segment of both the Inner and Outer Carpathians (Jankowski 2015; Jankowski and Margielewski 2021). The formation of this flower structure since the Miocene has been associated with the counterclockwise rotation of the Inner Carpathians in relation to the Outer Carpathians by 50° (Márton et al. 2013; Jankowski 2015; Golonka et al. 2018). Within this structure, massifs of carbonate and (less frequently) flysch rocks are stuck in tectonic mélanges (sensu Festa et al. 2019), alternately uplifted or lowered during the tectonic evolution (see model on Fig. 17.5c). Minor andesite intrusions were placed along the northern margin of the Pieniny Klippen Belt during the Miocene (Birkenmajer and Pécskay 2000; Birkenmajer 2017).

17.3 Pieniny Gorge Phenomenon

The Dunajec River, leaving the Orava-Nowy Targ Basin–a flat-floored intermountain depression, suddenly changes direction and twice breaks through the narrow (3–6 km) belt of the Pieniny Mts. In the area of Niedzica village, it turns sharply to the south, forming a short Niedzica Gorge (Fig. 17.1b—site 1) and circumnavigating the Pieniny Mountains from the south, it follows a slightly meandering, but very wide valley (600–700 m wide) up to the village of Czerwony Klasztor (Figs. 17.1b and 17.2a, b—site 1). Between Czerwony Klasztor, located at the foot of Mt. Trzy Korony, and the town of Szczawnica (ca. 2.5 km far in a straight line from each other), the Dunajec River has cut a narrow, strongly meandering gorge through the Pieniny

Fig. 17.3 Triangular facets of active normal fault scarps, framing the Dunajec River channel in the Pieniny Gorge. **a** Grabczycha. **b** huge triangular facet along a reactivated normal fault cutting the Facimiech massif (see Fig. 17.2a for location) (photographs by W. Margielewski)

Mts., transverse to their generally west–east strike. It is about 8 km long and shows height gradient of about 20 m, being framed by steep, often (sub) vertical rock walls with a height somewhere exceeding 300–400m (Birkenmajer 2006) (Figs. 17.3a, b and 17.4). Within the gorge, the river creates very narrow-radius meanders, rather unheard of in mountainous areas (Fig. 17.2a—site 5). The meandering nature of the channel is controlled by a series of faults, reflected today in the form of flatirons (triangular facets) forming steep and high valley sides of the Pieniny Gorge (Figs. 17.2a—sites 3, 4, and 6, 17.3a, b and 17.4). After leaving the gorge, in the town of Szczawnica, the Dunajec River valley widens significantly, the bed is overloaded with alluvial material, and the river is heading north, cutting through the Beskid Sądecki Mts. built of flysch formation. The Dunajec River Gorge was formed mainly in siliceous limestones of Upper Jurassic and Lower Cretaceous ages, resistant to denudation, intercalated with less resistant marly limestones, shales and marls (Birkenmajer 1958, 1979, 2006).

17.4 Pieniny Gorge: From Past to the Recent Concepts

The origin of the picturesque gorge of the Dunajec River with its distinctive meandering course has been so far a subject of numerous studies (Limanowski 1904; Romer 1905; Klimaszewski 1937; Starkel 1972; Zuchiewicz 1978). As already mentioned, the river cut into the Pieniny orogen creating a gorge with steep, rocky walls 300–400 m high, and a strongly meandering narrow channel, which is rather rare in mountainous areas (Figs. 17.2a—sites 2–6, 17.3a, b and 17.4). It is also significant that the river breaks twice through a narrow PKB, first towards the south and then towards the north, forming the Pieniny Gorge proper (Figs. 17.1b and 17.5b). Hence, based on numerous concepts regarding the origin of the (especially) Pieniny Gorge, attempts have made to explain the origin of such a specific natural landform.

17.4.1 The Origin of the Dunajec River Gorge in the Light of Previous Concepts

So far, the Dunajec River Gorge through the Pieniny Mountains has been considered to be antecedent, i.e. formed owing to the gradual uplift of the Pieniny ridge across the river course (Klimaszewski 1937; Starkel 1972; Zuchiewicz 1978, 2010). The process of fluvial incision was to be fostered by the uplift of the Pieniny Mts. along faults cutting the Pieniny Klippen Belt, occurring at a rate of 0.5–1.0 mm/year (Kowalski and Liszkowski 1972; Ząbek et al. 1993), with the simultaneous subsidence of the Orava-Nowy Targ Basin (Makowska 2003; Zuchiewicz 2010). However, this hypothesis has not been fully documented and does not explain all features of the gorge (see Birkenmajer 2006).

In the evolution of ideas on the origin of the Dunajec River Gorge, its epigenetic (superimposed) nature was also taken into account. In this concept, limestone rocks in which the gorge was finally formed are considered to be previously covered by weakly resistant rock formations of Cretaceous-Paleogene age, over which the meandering ancient Dunajec River flowed freely, incising gradually into bedrock (Limanowski 1904; Romer 1905). On

Fig. 17.4 Triangular facets forming the valley sides of the Pieniny Gorge near Mt. Sokolica (see Fig. 17.2a for location) (photograph by W. Margielewski)

the other hand, according to the concept of S. Staszic and W. Pol, the gap of the Dunajec River was considered as an overflow river gorge (cf. Klimaszewski 1961). According to another hypothesis, tectonic structure/conditions of the Pieniny Klippen Belt (Birkenmajer 1958) allowed for the recognition of the Dunajec River Gorge as a structural water gap, in which the river valley used numerous tectonic discontinuities (mainly faults) to break through the ridge (Klimaszewski 1972). In the light of yet another hypothesis, the Dunajec River Gorge formed owing to backward erosion and regressive development of the river valley (Birkenmajer 1979; Zuchiewicz 1982). A river draining the northern slopes of the Pieniny Mts. was supposed to intercept, in the Late Miocene and Early Pliocene, the river systems of the Orava-Nowy Targ Basin (the Biały Dunajec and the Białka rivers), previously directed to the east, towards the Poprad River (Birkenmajer 1979). A significant role in shaping the Dunajec River Gorge was assigned to considerable differences in bedrock resistance. According to Birkenmajer (2006), the occurrence of "patches of soft rocks of the Paleogene flysch strongly folded together with Jurassic-Cretaceous units" conditioned river erosion, which in the soft rocks were exploited to host characteristic meanders "bypassing" the resistant limestone rocks. The Dunajec River, subsequently cutting into the hard Jurassic limestones of the Pieniny Klippen Belt, maintained the original course of the meandering channel, keeping it to the present-day in a form of a winding, very deep gorge (Birkenmajer 2006).

According to Zuchiewicz (1980, 1982), in the Pliocene, the Dunajec River, leaving the Nowy Targ Basin, originally flowed straight towards the east and along the northern margin of the Pieniny Mts., crossing the Snozka Pass and following the present-day Krośnica River valley and further within the present-day Grajcarek River valley towards the Poprad River (Fig. 17.5b). Due to the blockage of the river outflow to the east, the Dunajec River broke through the Pieniny Mts. twice, leading to the origin of the Niedzica Gorge towards the south (Fig. 17.1b—site 1), and then, the Pieniny Gorge towards the north (Fig. 17.1b—site 2). According to Zuchiewicz (1982), the beginning of the formation of the Pieniny Gorge was supposed to take place at some time between the Middle Pliocene and Early Quaternary.

17.4.2 Towards the New Concept

17.4.2.1 Geomorphic Evolution

New concepts of the development of the Pieniny Klippen Belt (Plašienka and Mikuš 2010; Golonka et al. 2018; Plašienka et al. 2020; Jankowski and Margielewski 2021),

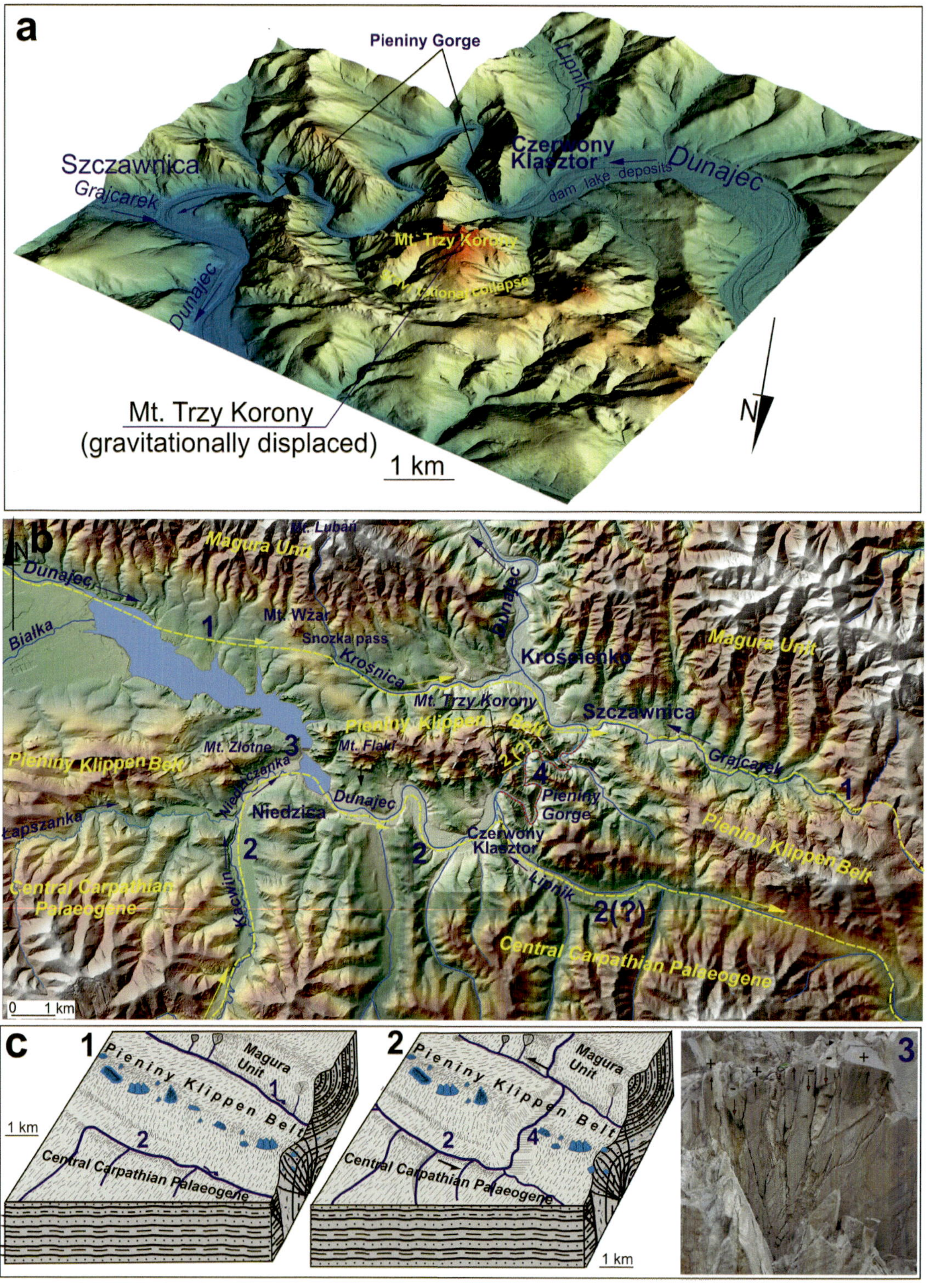

Fig. 17.5 Origin of the Dunajec River Gorge through the Pieniny Mts. **a** LiDAR DTM of the gorge shows the massif of Mt. Trzy Korony gravitationally displaced towards the south (the landslide scarp is marked with a white dashed line). **b** Evolutionary history of water gaps of the Dunajec River: 1—Previous flow of the Dunajec River along the northern margin of the Pieniny Mts (after Zuchiewicz 1982); 2—Wide valley of the Niedziczanka River to the south of the Pieniny Mts; 3—Site of river piracy of the Dunajec by the Niedziczanka River (Niedzica gap); 4—Contemporary Pieniny Gorge (marked by red dashed line). Different colours of fluvial tracts show the history of drainage network evolution: yellow—previous flows of Dunajec (1) and Niedziczanka rivers (2), prior to the formation of the gorge. Note that the Niedziczanka River may have flowed either to the east and south along the contemporary Lipnik River valley, or to the north (as indicated by blue dashed line); red—Contemporary course of the Dunajec River Gorge. Arrows show the direction of gravitationally displaced huge rock mass. **c** Flower structure: casual connection between transversal tear of the Pieniny Klippen Belt and the origin of the Dunajec River gorge (1–2); example of the flower structure (positive *sensu* Harding 1985) to show vertical displacements of rock blocks within this structure (Magura Sandstones, Wierchomla Quarry in the Beskid Sądecki Mts.) (3) (photograph by W. Margielewski)

well as a detailed analysis of the Digital Terrain Model suggest a slightly different genesis of the Dunajec River Gorge in the Pieniny Mts., than considered up to now. It is remarkable that the Dunajec River, originally flowing along the northern margin of the western segment of the Pieniny Mts. (Pieniny Klippen Belt) through the flat bottom of the Orawa-Nowy Targ Basin filled with soft sediments of the Pliocene and Quaternary suddenly turns south and crosses the axial parts of the central Pieniny Mts. in the vicinity of the village of Niedzica (Mt. Złotne and Mt. Flaki) and along a distance of ca. 3 km it forms the so-called Niedzica Gorge (water gap) (Figs. 17.1b—site 1 and 17.5b—site 3) (Birkenmajer 2006). Flowing further east along the southern margin of the Pieniny Mts., from Niedzica to the mouth of the Lipnik Stream in Czerwony Klasztor, the Dunajec River forms a broad valley reach about 11 km long, with the valley floor 300–700 m wide and hydraulic gradient of 3.5‰. The river follows a meandering pattern and runs along the border of the Pieniny Klippen Belt and the Central Carpathian Palaeogene (the Podhale Flysch) (Figs. 17.2a, b—site 1 and 17.5b—site 2).

Detailed analysis of regional morphology indicates that the Dunajec River could indeed have originally flowed through the Snozka Pass and the contemporary Krośnica Valley, bypassing the Pieniny Mountains from the north as it previously suggested by Zuchiewicz (1982) (segment of the Pieniny Mts.—site 1). The level of filling up of the Nowy Targ Basin with Pliocene and Quaternary sediments was higher than today, because the Pliocene sediments have been preserved in higher positions both near the village of Czorsztyn (Szafer 1954) and in the present-day Krośnica valley (Krościenko-Potoczki, Dziadowe Kąty), as well as on the southern slopes of Mt. Lubań (Szafer 1946; Birkenmajer et al. 2010).

At the same time when the Dunajec River flowed along the northern margin of the Pieniny Mts., the Niedziczanka River valley existed in the southern margin of the central segment of the Pieniny Mts. (Fig. 17.5b–site 2). It was formed owing to the junction of the Łapszanka and Kacwin streams, and creating wide meanders, could flow through the valley of present-day Lipnik Stream into the Poprad River catchment or, more likely, directly towards the town of Szczawnica, cutting across the flower structure of the Pieniny Klippen Belt, transversally torn due to the reactivation of strike-slip sinistral faults framing the Pieniny Mts. (Fig. 17.5b—site 4; c—models 1–2). The upper section of the Niedziczanka River valley was formed within the Central Carpathian Palaeogene flysch formations (Podhale Flysch). A significant part of the valley runs along the tectonic mélanges formed in the contact zone of the Pieniny Klippen Belt and the Central Carpathian Palaeogene Flysch. The meandering nature of the river in this section was probably caused by mass movements through which masses of resistant limestones were displaced by gravity from Mt. Złotne or Mt. Flaki, in some places pushing segments of the riverbed towards the south, producing the "meanders" (Figs. 17.1b—site 1, 17.2b—site 1 and 17.5b—site 2). The existence of these large-scale mass movements can be inferred from digital terrain models. Differences in rock resistance, previously suggested by Birkenmajer (2006), could have also played a role in the formation of meanders, but the zones of weakness exploited were tectonic mélanges rather than the soft rocks intercalated with the limestones. An alternating occurrence of wide and narrow reaches of the river valley in this section could have been related to the presence of pull-apart structures in the Carpathians, often associated with strike-slip zones (Jankowski and Margielewski 2014, 2021).

The Dunajec River, originally flowing to the east over the Pliocene deposits of the Nowy Targ Basin and along the northern margin of the Pieniny Mts. (Zuchiewicz 1982), incising into bedrock, ultimately reached the andesite intrusion of Mt. Wżar. Uplift of the central segment of the Pieniny Mts., with the simultaneous lowering of the bottom of Nowy Targ Basin (also observed nowadays–see Makowska 2003), interrupted the river outflow to the east and a dammed lake in the Nowy Targ Basin formed. Subsequently, the Proto-Dunajec River erosionally cut Mt. Złotne, forming the Niedzica Gorge and flowed into the Niedziczanka Stream valley, as an effect of river piracy phenomenon (Figs. 17.1b—site 1 and 17.5b—sites 2–3).

Within the Dunajec River Gorge, the bottom of the Dunajec valley suddenly narrows from 600–700 to 50 m (Fig. 17.2a—sites 1–2). The hydraulic gradient of the river also decreases, from 3.5 to 2.6‰, which seems unusual, considering the tectonics of the gorge section cut by numerous faults (causing elevation or lowering of individual blocks) and the narrow riverbed. A significant part of the rock walls forming steep, often (sub) vertical slopes of the gorge, was formed here along reactivated faults. There are numerous triangular facets (flatirons) of active fault scarps of various strikes, indicating isostatic elevation of footwall parts of normal faults or even splits (Figs. 17.2a—sites 3, 4 and 6, 17.3 and 17.4). The presence of tectoglyphs and slickensides on the rock walls of the Dunajec River Gorge indicates that river meanders have formed along faults, often reactivated within the flower structure. High triangular facets of active fault scarps can indicate strong uplift of valley slopes (and indeed, of the entire rock massifs) caused by footwall elevation, reaching several dozen metres or even more (Figs. 17.3 and 17.4).

17.4.2.2 Evolution of Pieniny Klippen Belt Flower Structure

The Pieniny Klippen Belt is a regional flower structure, the individual elements of which have been uplifted or lowered (see model on Fig. 17.5c—no. 3) (Plašienka and Mikuš 2010; Jankowski 2015; Golonka et al. 2018). As a result of the evolution of this structure due to the activity of sinistral faults framing the Pieniny Klippen Belt from the north and south, the whole Klippen Belt was torn transversely and dissected into minor massifs/blocks, which enabled the formation of the gorge section of the Dunajec River valley, meandering between such individual elevated blocks (Figs. 17.1b—site 2 and 17.5a, b). The evolution of this structure, which is ongoing as indicated by results of geodetic measurements (Makowska 2003), resulted in strong elevation of some elements of PKB and lowering of the others along reactivated faults. In some parts, the river could have used zones of tectonic mélanges which are omnipresent within the flower structure and related to relative displacements of individual elements (blocks) within this structure (see model on Fig. 17.5c—no. 3). This process might explain the formation of very high rock walls of the gorge. Therefore, antecedent nature of the gorge was not associated with the uplift of the mountain massif *en masse*, but with elevation and lowering of individual elements of the flower structure, especially in the area of its transverse tearirng (see model on Fig. 17.5c—no. 1–2). Such selective elevation of rock blocks along reactivated dislocations would explain both the formation of very high walls of the Dunajec River Gorge, often formed along high-angle faults, as well as a narrow river valley, which proves the very fast rate of incision into bedrock.

17.4.2.3 Gravitational Collapse

The digital terrain model indicates that the Dunajec River Gorge is located in the zone of a huge landslide, in which the entire massif of Mt. Trzy Korony was gravitationally displaced to the south, entering the river valley and blocking its outflow (Fig. 17.5a–see Mt. Trzy Korony). It is possible that gravitationally displaced huge fragments of rock massifs (deep seated gravitational slope deformations–type sensu Agliardi et al. 2013; Agliardi and Crippa 2022) could have dammed the Dunajec River valley in the section of Czerwony Klasztor–Szczawnica, forcing the river to break through the displaced rock massifs, cut by a system of cracks formed along tectonic discontinuities, mainly faults and tectonic mélanges associated with them. Some meandering sections of the gorge could have also developed along narrow tectonic mélange belts formed during gravitational displacements of mega-landslides displaced from the Mt. Trzy Korony massif (Fig. 17.5a). Blocking the outflow of the Dunajec River (or Niedziczanka River, which subsequently captured the Dunajec River) with huge fragments of rock masses, mainly limestones, is also suggested by the occurrence of the wide Dunajec valley upstream the Pieniny Gorge, which gives an impression of filling a dammed lake created as a result of a sudden flow stop of the river (Figs. 17.1b—site 1 and 17.2a, b—site 1). Even today, upstream of the Pieniny Gorge, the Dunajec River is overloaded with alluvial material and develops a multi-channel (wandering) pattern or forming oxbow lakes (Figs. 17.1b—site 1 and 17.2a—site 1). As a result of outflow blockage, dammed river waters could overflow through a complicated system of fissures and formed a narrow, meandering river valley along a network of faults associated with the transversal tearing of the flower structure (see Figs. 17.2a—sites 3,4 and 6, 17.5c—models 1–2). Interestingly, the oldest hypotheses concerning the origin of the Dunajec River gap, formulated as early as the nineteenth century, assumed that it can be overflow gorge of the river (Stanisław Staszic, Wincenty Pol, in: Klimaszewski 1961).

17.5 Conclusions

The analysis of the course and relief of the Dunajec River Gorge through the Pieniny Mts as well as of valley sections preceding and following the gorge shows that the Dunajec River Gorge is regressive, formed due to the river piracy by the Niedziczanka Stream and structural. A significant part of its formation is related to the reactivation of the Pieniny flower structure and its transverse tearing, as well as to gravitational displacements of huge rock masses blocking the flow of the river, formed at the stage of a gravitational collapse.

The Dunajec River, flowing initially through the Orawa-Nowy Targ Basin along the northern margin of the western and central segments of the Pieniny Mts., was blocked due to the tectonic elevation of this part of the Pieniny Klippen Belt. Consequently, the Dunajec River cut the PKB, creating the Niedzica Gorge, 3 km long, and joined the Niedziczanka valley that had formed along the border between the PKB and the Central Carpathian Paleogene. The formation of the Dunajec River Gorge was associated with transversal tearing of the Pieniny Klippen Belt, which is a flower structure, formed due to the activity of strike-slip sinistral faults, framing the PKB from the north and south. The rock massifs uplifted and lowered within the reactivated flower structure, as well as the tectonic mélanges attributed to this structure, undoubtedly contributed to the formation of very high, steep walls of the river valley in its section crossing the Pieniny Mts. Consequently, the Dunajec River Gorge is not of a strictly antecedent character and was not the result of transversal cutting of uplifted (*en masse*) mountain range by the river. It was formed owing to the selective uplift and subsidence of individual elements of the Pieniny flower structure. It is very probable that the final shape of the Pieniny Gorge could have been caused by large-scale mass movements related to the gravitational collapse of the Carpathians, which resulted in the gravitational displacement of the Mt. Trzy Korony massif towards the south and its partial disintegration. This caused the filling of the Dunajec River valley in the area of present-day gorge with huge rock blocks, blockage of the outflow, and formation of a dammed lake. This is indicated by the very wide bottom of the river valley upstream the gorge, filled by alluvial deposits. The water of dammed river overflowed through a complicated system of extensional fissures, forming picturesque meandering narrow valley developed along the network of faults and tectonic mélanges associated with the flower structure. The most important phenomenon influencing the formation of the Dunajec Gorge across the Pieniny Mts. was however the transverse tearing of the Pieniny flower structure.

References

Agliardi F, Crippa C (2022) Deep-seated gravitational slope deformations. In: Shroder JF (ed) Treatise on geomorphology, 2nd edn, vol 5, pp 183–199

Agliardi F, Crosta GB, Frattini P, Malusá MG (2013) Giant non-catastrophic landslides and the long-term exhumation of the European Alps. Earth Planet Sci Lett 356:263–274

Birkenmajer K (1958) Przewodnik geologiczny po pienińskim pasie skałkowym, p. I–IV. Wyd Geol, Warszawa, p 350

Birkenmajer K (1979) Przewodnik geologiczny po pienińskim pasie skałkowym, Wyd Geol, Warszawa, p 237

Birkenmajer K (1985) Stages of structural evolution of the Pieniny Klippen Belt, Carpathians. Studia Geol Polon 88:7–32

Birkenmajer K (2006) Przełom Dunajca w Pieninach—fenomen geologiczny. Pieniny Przyroda i Człowiek 9:9–22

Birkenmajer K (2017) Geologia Pienin. Monografie Pienińskie 3:5–66

Birkenmajer K, Pécskay Z (2000) K-Ar dating of the Miocene andesite intrusions, Pieniny Mts. West Carpathians, Poland: a supplement. Studia Geol Polon 117:7–25

Birkenmajer K, Hrynowiecka-Czmielewska A, Stuchlik L (2010) Pollen-bearing Middle Pleistocene deposits at Huba, southern Poland (West Carpathians). Acta Palaeobot 50:89–99

Festa A, Pini GA, Ogata K, Dilek Y (2019) Diagnostic features and field-criteria in recognition of tectonic, sedimentary and diapiric mélanges in orogenic belts and exhumed subduction-accretion complexes. Gondwana Res 74:7–30

Golonka J, Krobicki M, Waśkowska A, Cieszkowski M, Ślączka A (2015) Olistostromes of the Pieniny Klippen Belts, Northern Carpathians. Geol Mag 152:269–286

Golonka J, Krobicki M, Waśkowska A (2018) The Pieniny Klippen Belt in Poland. Geol Geophys Environm 44:111–125

Harding TP (1985) Seismic characteristics and identification of negative flower structures, positive flower structures, and positive structural inversion. AAPG Bull 69:582–600

Jankowski L (2015) Nowe spojrzenie na budowę geologiczną Karpat – ujęcie dyskusyjne. Prace Nauk Inst Nafty i Gazu PIB 202:1–154

Jankowski L, Margielewski W (2014) Strukturalne uwarunkowania rozwoju rzeźby Karpat zewnętrznych – nowe spojrzenie. Przegl Geol 62:29–35

Jankowski L, Margielewski W (2021) Geological control of young orogenic mountain morphology: from geomorphological analysis to reinterpretation of geology of the Outer Western Carpathians. Geomorphology 386:107749

Jurewicz E (2005) Geodynamic evolution of the Tatra Mts. and the Pieniny Klippen Belt (Western Carpathians): problems and comments. Acta Geol Polon 55:295–338

Jurewicz E (2018) The Šariš transitional zone, revealing interaction between Pieniny Klippen Belt, Outer Carpathians and European Platform. Swiss J Geosci 111:245–267

Klimaszewski M (1937) Morfologia i dyluwium doliny Dunajca od Pienin po ujście. Prace Inst Geogr UJ 18:1–54

Klimaszewski M (1961) Geomorfologia ogólna. PWN, Warszawa, p 522

Klimaszewski M (1972) Karpaty Wewnętrzne. In: Klimaszewki M (ed) Geomorfologia Polski, t1. Polska południowa. Góry i Wyżyny. PWN, Warszawa, pp 25–52

Kowalski WC, Liszkowski J (1972) Współczesne pionowe ruchy skorupy ziemskiej w Polsce na tle jej budowy geologicznej. Biul Geol Uniw Warsz 14:5–19

Książkiewicz M (1972) Karpaty budowa geologiczna polski IV, Tektonika 3. Wyd. Geol Warszawa 228

Limanowski M (1904) Wycieczka w Tatry i Pieniny. Pam Tow Tatrzańskiego 25:131–183

Makowska A (2003) Dynamika Tatr wyznaczana metodami geodezyjnymi. Inst Geodezji i Kartografii, Monografie 6, Warszawa, p 201

Márton E, Grabowski J, Plašienka D, Túnyi I, Krobicki M, Haas J, Pethe M (2013) New paleomagnetic results from the upper cretaceous red marls of the Pieniny Klippen Belt, Western Carpathians: evidence for general CCW rotation and implications for the origin of the structural arc formation. Tectonophysics 592:1–13

Oszczypko N (2006) Late Jurassic-Miocene evolution of the Outer Carpathians fold-and-thrust belt and its foredeep basin (Western Carpathians, Poland). Geol Quart 50:169–193

Oszczypko N, Uchman A, Malata E (eds) (2006) Palaeotectonic evolution of the Outer Carpathian and Pieniny Klippen Belt basins. Inst Geol Sci, Jagiellonian Univ, Kraków, p 199

Oszczypko N, Oszczypko-Clowes M (2014) Geological structure and evolution of the Pieniny Klippen belt to the east of the dunajec river—a new approach (Western Outer Carpathians, Poland). Geol Quart 58:737–758

Plašienka D, Mikuš V (2010) Geological setting of the Pieniny and Šariš sectors of the Klippen belt between Litmanová and Drienica villages in the eastern Slovakia. Miner Slov 42:155–178

Plašienka D, Bučova J, Šimonová V (2020) Variable structural styles and tectonic evolution of an ancient backstop boundary: the Pieniny Klippen Belt of the Western Carpathians. Intern J Earth Sci 109:1355–1376

Rączkowski W, Wójcik A, Zuchiewicz W (1984) Late Neogene-Quaternary tectonics of the Polish Carpathians in the light of neotectonic mapping. Tectonophysics 108:51–69

Romer E (1906) Ziemia. Geografia fizyczna ziem polskich. In: Konopnicka M, Romer E, Koneczny F et al Polska. Obrazy i opisy. Macierz Polska, Lwów, pp 1–67

Starkel L (1972) Charakterystyka rzeźby Polskich Karpat i jej znaczenie dla gospodarki ludzkiej. Probl Zagosp Ziem Górskich 10:75–150

Szafer W (1946) Flora plioceńska z Krościenka nad Dunajcem. Rozpr Wydz Mat-Przyr PAU 72(1):1–126

Szafer W (1954) Plioceńska flora okolic Czorsztyna i jej stosunek do plejstocenu. Prace Inst Geol 11:1–238

Ząbek Z, Barlik M, Knap T, Margański S, Pachuta A (1993) Continuation of geodynamic investigations in the Pieniny Klippen Belt, Poland, from 1985 to 1990. Acta Geophys Polon 41:131–150

Zuchiewicz W (1978) Czwartorzędowe ruchy tektoniczne a rzeźba przełomu Dunajca przez Beskid Sądecki. Ann Soc Geol Polon 48:517–531

Zuchiewicz W (1980) Młode ruchy tektoniczne a morfologia Pienin. Ann Soc Geol Polon 50:263–300

Zuchiewicz W (1982) Geneza Przełomu Dunajca Przez Pieniny. Wszechświat 83:169–173

Zuchiewicz W (1984) Late Neogene-Quaternary tectonics of the Polish Carpathians in the light of neotectonic mapping. Tectonophysics 108:51–69
Zuchiewicz W (1987) Ewolucja i strukturalne założenia sieci rzecznej Karpat w późnym neogenie i wczesnym czwartorzędzie. In: Jahn A, Dyjor S (eds) Problemy młodszego neogenu i eoplejstocenu w Polsce. Ossolineum, Wrocław, pp 211–225
Zuchiewicz W (2010) Neotetonika Karpat polskich i zapadliska przedkarpackiego. Wyd AGH, Kraków, p 234
Żytko K, Zając R, Gucik S, Ryłko W, Oszczypko N, Garlicka I, Nemčok J, Eliaš M, Menčik E, Stranik Z (1989) Map of the tectonic elements of the Western Outer Carpathians and their foreland 1: 500 000. In: Poprawa D, Nemčok J (eds) Geological atlas of the Western Outer Carpathians and their Foreland, Państw Inst Geol, Warszawa

Włodzimierz Margielewski is a geologist and geomorphologist, employed as Associate Professor in the Institute of Nature Conservation, Polish Academy of Sciences in Krakow, Poland. His research is focused on geology, geomorphology, and geoconservation in the Carpathians and their Foreland, with particular emphasis to structural geology, geological control of relief, and wide aspects of mass movements. He also works on the reconstruction of palaeoenvironmental changes during the Late Glacial and Holocene in the Polish territory, based on studies of peatland deposits, including landslide fens (a type recognized by him).

Leszek Jankowski is a geologist, tectonic and cartographer. He is Associate Professor at the Polish Geological Institute—National Research Institute in Krakow, Poland. His research is focused on geology, tectonics and sedimentology of the Carpathians and their forelands, as well as on the geological controls behind the morphology of the Carpathians. He is an author or co-author of numerous geological maps of the Carpathians (including also Slovak, Ukrainian and Romanian parts), and scientific articles on the geology of the Carpathians and their Foreland, containing new, original concepts of the development of the Carpathian Orogen.

Impact of Mass Movements on Landscape and Landform Transformation in the Beskidy Mountains—From Gravitational Collapse to Recent Times

18

Włodzimierz Margielewski and Leszek Jankowski

Abstract

Impact of mass movements on morphological development of the Outer (Flysch) Carpathians (Beskidy Mountains) is significant, and their landforms have been transformed by gravitational displacements since the stage of the gravitational collapse of the orogen (Neogene) through the Quaternary stage of relief evolution until the present, partly under the conditions of increasing human impact. At the stage of the gravitational collapse the rock massifs, somewhere of the size of mountain ranges or large hills underwent gravitational displacements. Along with the tectonic mélanges developed during gravitational displacements, river valleys were formed. During the Quaternary stage of relief development, characteristic elements of the gravitationally induced morphology and landscape of the Beskidy Mountains were created, such as concave, convex or convex-concave slopes, rock walls and trenches, block fields, tors, landslide lakes, and fens. A comprehensive analysis of mass movements shows that the size of slopes transformed by gravitational displacements during the Neogene and Quaternary is much larger than it was previously considered. In turn, the present-day landslides, although commonly formed after catastrophic rainfalls, contribute to the transformation of relief only at local scale. Most of them are shallow landslides that are quickly degraded and vanish.

Keywords

Outer (Flysch) Carpathians · Landscape transformation · Mass movements · Gravitational collapse

18.1 Introduction

Mass movements in the western parts of the Outer Carpathians, called the Beskidy Mounatins, are common phenomena. Gravitational displacements have transformed both bedrock and slope covers (e.g. Starkel 1960; Ziętara 1968; Bober 1984; Kotarba 1986; Bober et al. 1997; Margielewski 2006a). Landform analysis shows that mass movements significantly influenced the transformation of slope relief in the Carpathians. In some regions of the Beskidy Mts., they even constitute dominant phenomena in relief transformation, occupying 30–40% of the slope area (Starkel 1960; Ziętara 1968; Bober 1984; Długosz 2011; Chalupa et al. 2018; Pánek et al. 2019). In case of some particular mountain ranges (e.g. Beskid Niski Mts.), "a landslide type of relief" has been specified (Starkel 1960). A significant number of about 60,000 landslides recorded so far in the Polish Outer Carpathians, especially as an effect of the Landslide Counteracting System (LCS: SOPO in Polish) (Mrozek et al. 2014; Marciniec et al. 2019), indicate that in many parts of these mountains, mass movements have played predominant role among geomorphological processes (Starkel 1960; Kotarba 1986; Poprawa and Rączkowski 2003; Łajczak et al. 2014). The characteristic slope features in the Beskidy Mts. landscape (concave, convex slopes, ridge top trenches), as well as specific smaller landforms related to mass movements such as rock walls, tors, debris covers, landslide fens and lakes indicate an important role of mass movements in the transformation of mountain relief during the Quaternary (Starkel 1960; Alexandrowicz 1978a; Alexandrowicz and Alexandrowicz 1988; Alexandrowicz and Margielewski 2010; Pánek et al. 2013; Pánek and Klimeš 2016; Chalupa et al. 2018).

W. Margielewski (✉)
Institute of Nature Conservation, Polish Academy of Sciences, Adama Mickiewicza Ave. 33, 31-120 Kraków, Poland
e-mail: margielewski@iop.krakow.pl

L. Jankowski
Polish Geological Institute—National Research Institute, Carpathian Branch, Skrzatów 1, 31-560 Kraków, Poland
e-mail: leszek-jankowski@pgi.gov.pl

P. Migoń and K. Jancewicz (eds.), *Landscapes and Landforms of Poland*,
World Geomorphological Landscapes, https://doi.org/10.1007/978-3-031-45762-3_18

Recent studies have shown, however, that huge rock massifs, somewhere of a size of mountain ranges, underwent gravitational displacements during the gravitational collapse (sensu Dewey 1987) which has developed since the Middle Miocene (Jankowski and Margielewski 2014, 2021; Jankowski 2021).

Consequently, the morphology of the Beskidy Mts. has been predominantly shaped by gravitational displacements taking place already at the stage of gravitational collapse since Neogene (Jankowski and Margielewski 2014, 2021), as well as later, during the Quaternary stage of the relief evolution, and in present-day period, partly under the conditions of increasing human impact (Starkel 1960; Kotarba 1986; Alexandrowicz and Alexandrowicz 1988; Poprawa and Rączkowski 2003; Rączkowska et al. 2012).

18.2 Geological and Geomorphological Setting

The Carpathians are an arc-shaped mountain range more than 1300 km long (Fig. 18.1a, b). In geological terms, the Polish part of the Carpathians was traditionally divided into the so-called Inner Carpathians (Tatra Mts. and the Podhale flysch), and the Outer (Flysch) Carpathians (Lexa et al. 2000). The boundary between these regions was considered the Pieniny Klippen Belt (PKB) (Birkenmajer 1985; Żytko et al. 1989; Lexa et al. 2000). The Western Outer Carpathians and parts of the Eastern Outer Carpathians (up to the Cheremosh River in Ukraine) are called the Beskidy Mountains (Kondracki 1978).

The Polish parts of the Outer Carpathians (the Polish Beskidy Mts.) are built of siliciclastic turbidites, occasionally of carbonate rocks, formed from the Late Jurassic to the Early Miocene. The folded flysch rocks form several tectonic units: Magura, Dukla, Silesian, Subsilesian, Skole, Stebnik and Zgłobice units (nappes). During the compressional stage of mountain building, they were thrusted northwards over one another and onto the foreland, called the Carpathian Foredeep, built of Middle and Late Miocene sediments (Książkiewicz 1972; Żytko et al. 1989; Lexa et al. 2000; Oszczypko 2006; Oszczypko et al. 2006; Jankowski 2021).

The morphology of the Outer Carpathians is zoned. The southern zone, called the Beskid relief zone, is elevated at about 200 m in relation to the foothills relief zone present in the northern part of the Outer Carpathians (Fig. 18.1c, d) (Starkel 1972). This zonation of relief has tectonic foundations and is caused by rock mass stacking during the compression stage. The tectonic elements incorporated into the orogen during the earliest compressional stage form the elevated Beskidy relief zone, whereas the tectonic units stacked to the orogen in the later phase build the lowered foothills relief zone (Fig. 18.1d) (Jankowski and Margielewski 2014, 2021). The main features of the Outer Carpathians' morphology were formed at the stage of a gravitational collapse lasting since the Middle and Late Miocene until contemporary times, when—after the thrusting (compression) stage—gravitational disintegration of the orogen took place (Jankowski 2015; Jankowski and Margielewski 2021). At this time, the contemporary river network began to form. At the later stage of tectonic evolution of the Carpathians, during isostatic rebound, some ranges of the Beskidy Mts. underwent isostatic elevation (Mt. Babia Góra, Mt. Cergowa) (Jankowski and Margielewski 2014, 2021).

In more detail, four relief types in the Outer Carpathians were distinguished: (a) high mountains (only Mt. Babia Góra—1725 m asl, and Mt. Pilsko—1557 m asl); (b) medium–high mountains (800–1300 m asl); (c) low mountains (in the Beskidy relief zone: 600–800 m asl); and (d) foothills and valley bottoms (mainly in the foothills relief zone) (Fig. 18.1c, d) (Starkel 1972). Medium–high mountains with steep slopes include both compact mountain groups (Beskid Śląski and Żywiecki Mts., Beskid Sądecki Mts.), isolated ridges of the Beskid Wyspowy Mts. and the Bieszczady Mts., with local relief reaching 400–800 m. Low mountains usually occur in the surroundings of medium–high mountains or constitute isolated ridges with steep slopes and a relative height of 200–400 m. A common relief feature in the foothills stretched in the northern foreland of the Beskidy relief zone is the presence of wide humps of various slopes (8–35°), usually convex-concave slopes, and flat valley bottoms (Starkel 1972).

18.3 Gravitational Collapse: Impact of Giant Gravitational Displacements on Morphology and Landscape

The end of the compressional stage in the tectonic evolution of the Carpathian orogen was followed by the gravitational collapse stage, during which gravitational disintegration of the rock massifs occurred (Tokarski et al. 2006; Jankowski and Margielewski 2014, 2021; Jankowski 2015). In the Middle Miocene, within the Carpathians and in their forelands, extensive depressions formed. These included internal (Pannonian Basin), external (Carpathian Foredeep) and smaller intra-mountain depressions such as the Orawa-Nowy Targ Basin, Nowy Sącz Basin (Oszczypko 2006), Iwkowa Basin (Połtowicz and Waśniowska 1977) and the Zakliczyn Basin, all filled with Neogene sediments (Fig. 18.1c, d). These depressions formed also in the tectonic window zone, due to the denudation of strongly tectonically disintegrated rocks (Jankowski 2015). All these depressions constituted accommodation spaces for

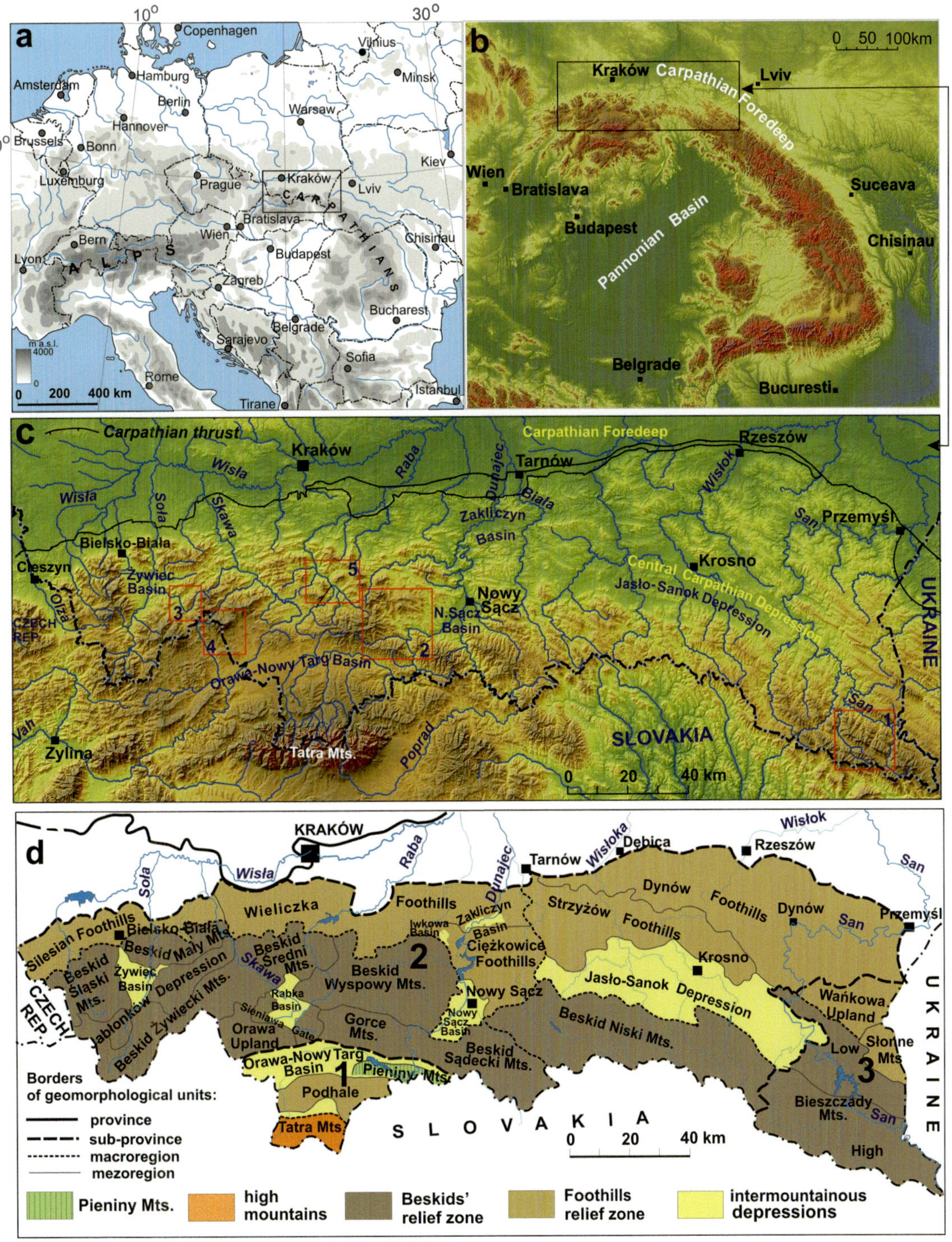

Fig. 18.1 Polish Carpathians. **a** Location in Europe. **b** SRTM model of the Carpathian arch, with the position of the Pannonian Basin and the Carpathian Foredeep. **c** SRTM model of the Polish part of the Outer Carpathians, with location of the study area (1 to 5—position of study areas mentioned in the text: 1—Bieszczady Mts. (see Fig. 18.2a), 2—Gorce Range and Beskid Wyspowy Mts (see Fig. 18.3a, b), 3—Mt. Lasek (see Fig. 18.4a), 4—Mt. Babia Góra (see Fig. 18.5a), 5—Beskid Wyspowy Mts. (see Fig. 18.6f). **d** Geomorphological units and mountain ranges of the Polish Carpathians: 1—Central Western Carpathians, 2—Outer Western Carpathians, 3—Outer Eastern Carpathians (after Starkel 1972, 1991)

gravitational displacements of rock massifs previously piled up during the compressional stage, which could have occurred in various directions (Jankowski 2015; Jankowski and Margielewski 2021).

At the stage of the gravitational collapse in the Neogene, rock massifs, some of a size of a mountain range, underwent gravitational displacements. Large isolated hills—"island massifs", surrounded by river valleys with semi-circular courses—were also formed. Other features of this morphology are characteristic low foothills surrounding the elevated mountain ranges. The contemporary shape of many Carpathian mountain ranges was formed during the gravitational collapse (Jankowski and Margielewski 2014, 2021) (Figs. 18.1c, 18.2, 18.3 and 18.4).

Fig. 18.2 Mountain ridges formed through gravitational displacements of huge fragments of rock massifs: Połoniny Ridge, Graniczne Ridge, and Mt. Wielka Rawka in the Bieszczady Mts. **a** SRTM model of gravitational displacement of mountainous ranges. **b** Gravitational displacements (arrowed) towards the Pannonian Basin (Graniczne Ridge and Mt. Wielka Rawka). **c** Gravitational displacements (arrowed) towards to the Central Carpathians Depression (Mt. Jaworniki and Otryt Ridge). **d** Tectonic mélange in the Wetlinka River valley (photographs by W. Margielewski)

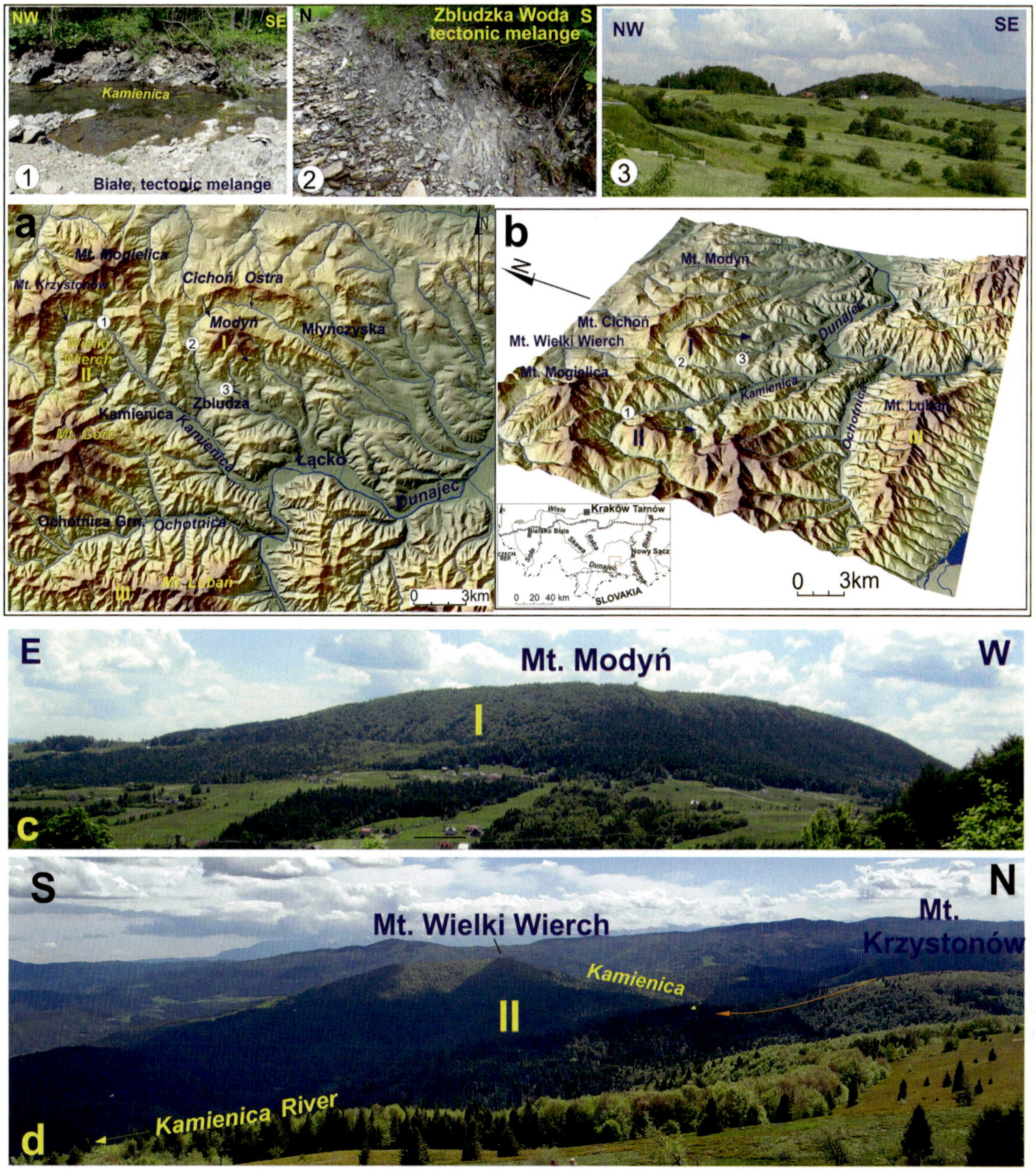

Fig. 18.3 Gravitational collapse in the Carpathians: gravity-induced mountain groups. **a, b** SRTM model presenting the giant-scale gravitational displacement of Mt. Modyń in the Beskid Wyspowy Mts. (I) and Mt. Ostry Wierch in the Gorce Mts. (II), III—Lubań Range. **c** Mt. Modyń (view from Mt. Ostra). **d** Mt. Ostry Wierch: geomorphological consequences of gravitational displacements. Upper photographs: 1, 2—tectonic mélanges formed in slip surfaces/shear zones of giant gravitational movements and exposed in river beds, 3—hills formed due to gravitational displacements of rock blocks in southern part of the Mt. Modyń massif (the location of sites 1–3 is marked in Fig. a and b) (photographs 1–3 and c by W. Margielewski, photograph d by K. Buczek)

18.3.1 Gravity-Induced Mountain Ranges and Their Groups

An example of the formation of mountain ranges as a result of gravitational disintegration of rock massifs is the Połoniny Ridge in the Bieszczady Mts. (Mt. Połonina Wetlińska—1228 m asl, and Mt. Połonica Caryńska—1297 m asl) (Jankowski and Margielewski 2021) (Fig. 18.1c—site 1, 18.2a, b—site 1). Gravitational disintegration of this mountain group in two directions and the creation of several mountain ridges which formed towards SW and NE from the Połoniny Ridge, separated from one another, was possible due to the presence of two extensive accommodation spaces in the foreland of the Połoniny Ridge: the Pannonian

Basin to the SW and the Central Carpathian Depression to the NE (Jankowski and Margielewski 2021). Part of the rock mass overburden of the Połoniny Ridge with a thickness estimated at ca. 3 km (Mazzolli et al. 2010) was gravitationally displaced towards the SW to the Pannonian Basin, creating the so-called Graniczne Range (Fig. 18.2a, b—site 2). In the slip (shearing) zone of the rock massifs, a wide zone of tectonic mélange developed and was subsequently eroded by the Wetlinka River (Fig. 18.2d). Owing to the antithetical displacements, rock masses were detached from the Graniczne Range and slipped towards NE. Today they form Mt. Wielka Rawka elevation (1307 m asl) (Fig. 18.2a, b—site 3). In turn, gravitational displacement of rock masses from the Połoniny Mts. towards the NE, into the Central Carpathian Depression, resulted in the formation on the NE slopes of the Połoniny Mts. of a series of characteristic hills separated by trenches, as e.g. Mt. Jaworniki (887 m asl) (Fig. 18.2a, c—site 4). It is likely that the Otryt Ridge with Mt. Trohaniec (939 m asl), an elongated mountain range separated from the Połoniny Mts. by the San River Valley, also formed as a result of gravitational displacement of huge fragments of the Połoniny mountain group (Fig. 18.2a, c—site 5) (Jankowski and Margielewski 2021). Analyses of tectonic structures (faults) within the disintegrated rock massifs indicate that the gravitational breakdown of the Połoniny mountain group can represent the type of giant-scale deep-seated gravitational slope deformations (DSGSDs) (*sensu* Agliardi et al. 2013).

Giant-scale gravitational displacements connected with gravitational collapse of the orogen occur in other parts of the Beskidy Mts. as well. The Lubań Ridge (1211 m asl) in the Gorce Mts. is another large mountain range (15 km long) which formed due to the gravitational displacement of huge fragments of central part of the Gorce Mts. massif towards the south, into the Orawa-Nowy Targ Basin (Fig. 18.1c—site 2, 18.3a, b—site III).

Another type of landforms formed owing to mass movements associated with the gravitational collapse are huge landslides with typical elements of landslide relief such as head scarp, landslide body and landslide tongue (*see* Dikau et al. 1996). Such a landform (12 km long) developed in the boundary zone between the Gorce Mts. and the Beskid Wyspowy Mts (Fig. 18.1c—site 2). Huge rock masses displaced by gravity formed the mountain group of Mt. Wielki Wierch (1007 m asl) (Fig. 18.3a, b—site II). Giant-scale gravitational displacement of a rock mass detached from the Mt. Krzystonów (1012 m asl) in a form of classic landslide formed a huge amphitheatric scarp with a diameter of 4 km and a height of 350–400 m. The length of this giant landslide reaches ca 12 km (Fig. 18.3a, b—site II, d). The tectonic mélange formed in the slip/shearing zone has been subsequently used by the Kamienica River to erode its valley, which forms a characteristic bend here (Fig. 18.3a, b—site 1). The displacements are confirmed by strike-slip faults framing the range from the west (strike-slip dextral fault) and from the east (strike-slip sinistral fault).

A similar landslide form developed in the Beskid Wyspowy Mts (Fig. 18.1c—site 2). The giant landslide has displaced from the Mt. Cichoń (928 m asl) and Mt. Ostra (928 m asl) areas to the Nowy Sącz Basin and is 12–14 km long (Fig. 18.3a, b—site I). Its main element, gravitationally displaced southwards, is Mt. Modyń (1029 m asl) which is a giant landslide body *sensu* Dikau et al. (1996) (Fig. 18.3c). Tectonic mélange zones, exposed mainly within the edges of this mega-landslide (Fig. 18.3a, b—site 2), are used by the Zbludzka Woda Stream (western edge) and the Młynne Stream (eastern edge) (Fig. 18.3a, b—I).

Characteristic, isolated hills or mountain massifs with insular relief, with huge amphitheatric main scarps, surrounded by river valleys with semi-circular courses formed during the gravitational collapse, are very common in the Beskidy Mountains (Margielewski and Jankowski 2019; Jankowski and Margielewski 2021) (Fig. 18.3c, d). Owing to the gravitational displacement of smaller rock massifs within the landslide of Mt. Modyń, picturesque hills were formed, well-exposed in the landscape (Fig. 18.3a—no. 3 and photo 3). In addition to the huge landslides described above, there are also smaller forms with characteristic relief elements, formed as individual, isolated hills: Mt. Klimas in the Beskid Makowski Mts., Mt. Obocz and Mt. Maślana Góra in the Beskid Niski Mts. (Jankowski and Margielewski 2021).

One of the most picturesque landforms is the Lasek Ridge (868 m asl) in the Beskid Makowski Mts. (Fig. 18.1c—site 3 and 18.4). A cone-shaped mountain massif (with a volume of about 4 km^3) about 5 km in diameter, rises over 350 m above the narrow valley of the Koszarawa River which takes a semi-circular course in this reach (Fig. 18.4a, b). The river valley was entirely developed within the tectonic mélange formed in the slip/shearing zone of rock mass, gravitationally displaced here towards the north (into the Żywiec tectonic window). Tectonic mélanges and strongly tectonized rocks with numerous faults occur along the whole channel of the Koszarawa River that flows around the Lasek Ridge (Fig. 18.4b and photos 1–7) (Margielewski and Jankowski 2019).

18.3.2 Gravitationally-Induced Foothills

Another type of landslide relief formed at the stage of the gravitational collapse had developed in the zone of contact between the middle-high mountain belt and intra-mountain depressions. Characteristic topography in this zone consists

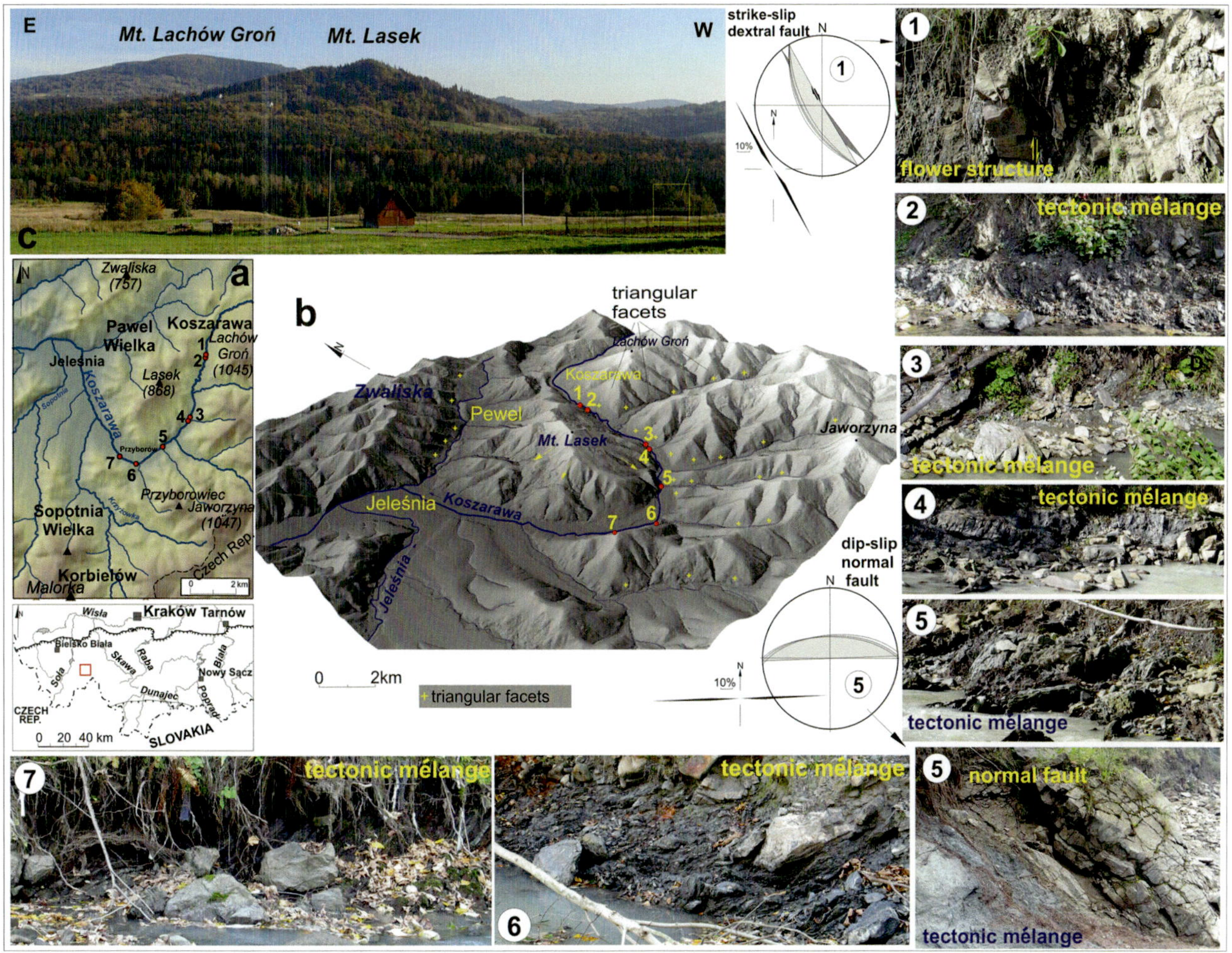

Fig. 18.4 Landforms induced by gravitational collapses. Insular Mt. Lasek in the Beskid Makowski Mts. formed as a giant rock massif gravitationally displaced to the north, with the shear zone used by the Koszarawa River valley (Margielewski and Jankowski 2019). **a** Topographic position of Mt. Lasek. **b** LiDAR DTM of Mt. Lasek, surrounded by the Koszarawa River valley adopting semi-circular course; 1 to 7—sites of tectonic mélanges within the shear zone (with faults), exposed in various parts of the Koszarawa River channel. On the rose diagrams: direction of faults, on the great circle diagram: position of faults surfaces (projection on lower hemisphere). **c** View on Mt. Lasek (photographs by W. Margielewski)

of small hills, often isolated from others or forming clusters, situated in the foreland of elevated mountain ranges (Fig. 18.1c—site 4; 18.5a–d—no. 1 and 3). Such foothill relief is the result of gravitational sliding of large rock blocks detached from the elevated mountain ridges towards the intra-mountainous depressions. Gravitational displacements are accompanied by tectonic mélanges (shearing zones–according to Festa et al. 2019), often exposed in the bottoms of river valleys (Fig. 18.5a—site 2, and c).

Such gravitational-induced foothills occur, among others, in the southern slopes of Mt. Babia Góra. They were created as a result of gravity-driven displacements of rock masses from the southern slopes of the Mt. Babia Góra to the Orawa-Nowy Targ Basin (Fig. 18.5a, b—no. 1). The northern slope of Mt. Babia Góra is cut by a huge scarp with a relative height of 800 m, which was interpreted as an element of huge rock slump (Fig. 18.5a—no. 3) (Alexandrowicz 1978a). Gravitational displacements of rock blocks forming successive foothills were directed towards the tectonic graben of the Skawica River (Fig. 18.5a, d—no. 3). Similar foothills also developed in the northern foreland of the elevated Radziejowa Range in the Beskid Sądecki Mts. The rock masses forming the hills were gravitationally displaced towards the intra-mountain depression of the Nowy Sącz Basin.

The described above large-scale mass movements have much larger spatial range and were formed in the older stage of the morphogenesis of the Carpathians during the Neogene.

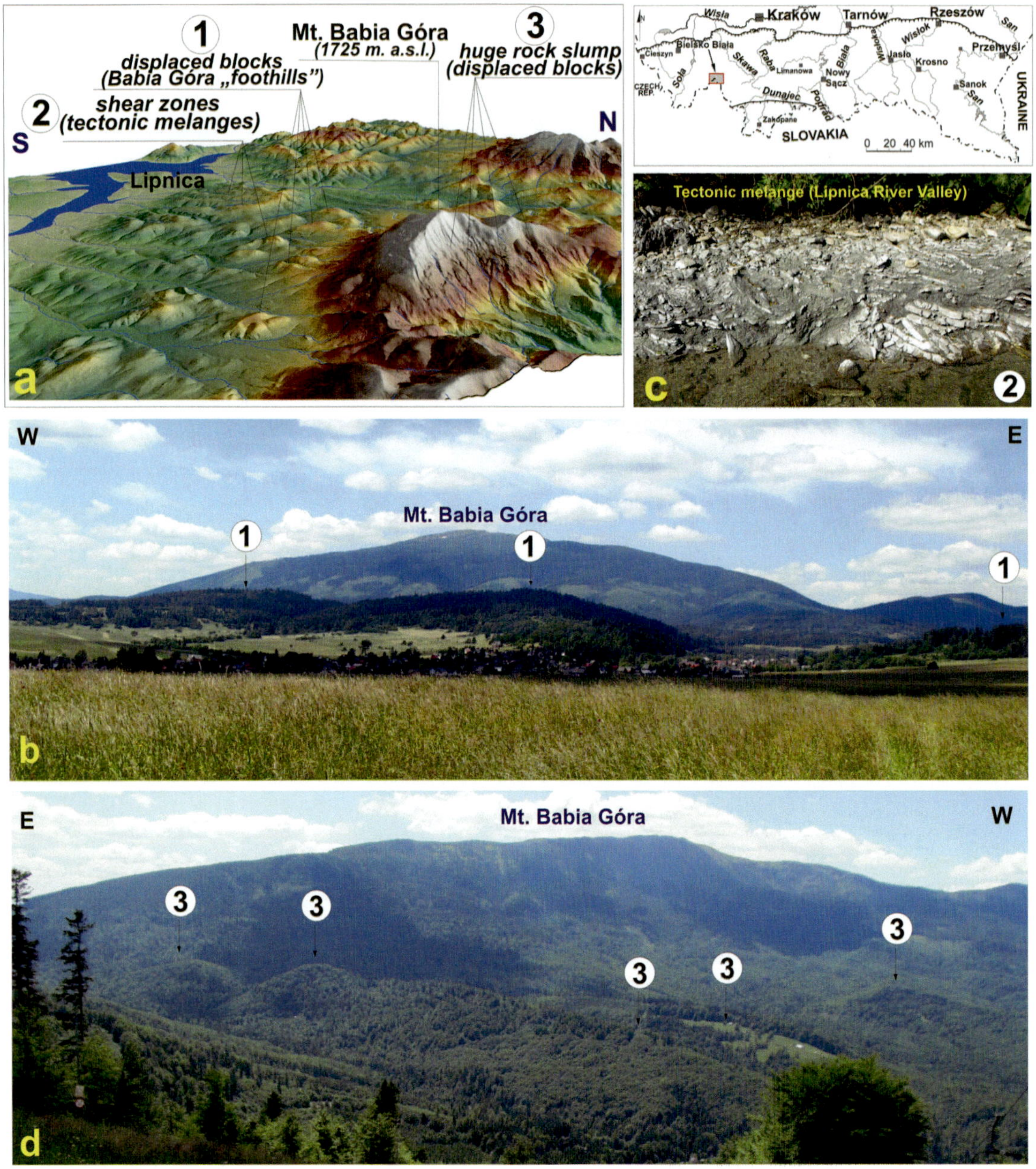

Fig. 18.5 Foothills formation owing to gravitational collapses (partly after Jankowski and Margielewski (2021). **a** Mt. Babia Góra (SRTM model) and their foothills (1 and 3) formed due to gravitationally displaced rock massifs in two directions (mountain sides): towards to the south, into the Orawa Basin (a—1, b—1) with formation of tectonic mélange (a—2, c—2), and towards the north (a—3 and d—3) into the Skawica valley river. **b** View on Babia Góra foothills from the Orawa Basin. **c** Tectonic mélange in the Lipnica River channel. **d** foothills formed in the northern slope of Mt. Babia Góra in (view from Mt. Mosorny Groń) (photographs b and d by W. Margielewski, c by L. Jankowski)

18.4 Quaternary Stage of Landform Evolution

Up to recently, morphological transformation of the Beskidy Mts. by mass movements was considered to be associated with the Quaternary stage of the relief development, mainly with the Late Pleistocene–Holocene interval (Starkel 1960, 1997; Alexandrowicz 1997; Margielewski 2006b; Pánek et al. 2013, 2014). Numerous datings of Carpathian landslides performed using radioisotope methods (^{14}C, ^{10}Be) indicate that extensive landslides were formed mainly during the humid phases of the Late Glacial and the Holocene (Gil et al. 1974; Alexandrowicz 1997; Starkel, 1997; Margielewski 2006b, 2018; Pánek et al. 2013; Pánek 2019; Břežný et al. 2018). Less common are landforms developed earlier during the Last Glacial (Weichselian, Plenivistulian) (Pánek et al. 2014; Jankovská et al. 2018; Wójcik 2019).

18.4.1 Large-Scale Slope Transformations

Specific features of the Beskidy landscape associated with mass movements are characteristic shapes of slopes. The sizes and shapes of the landslides modifying the slopes suggest that they represent several generations of mass movements developed at different times. This is also indicated by the radiocarbon datings of these landslides (Alexandrowicz 1997; Margielewski 2018; Břežný et al. 2018). Among the largest mountain slopes shaped by mass movements in the Polish part of the Beskidy Mts. is the northern slope of the Mt. Babia Góra, with a characteristic concave shape and a height of about 800 m (Fig. 18.5a, and d—no. 3) (Alexandrowicz 1978a), which was formed by numerous generations of mass movements (Kłapyta 2020).

Due to relief transformation by mass movements, there are numerous slopes with convex (Fig. 18.6a), concave (Fig. 18.6b) and concave-convex profiles (Fig. 18.6c) in the Beskidy Mountains (Alexandrowicz and Margielewski 2010). A particularly picturesque landscape element is ridge-top trenches and double ridges transforming the most elevated parts of mountainous topography, caused by gravitational spreading (Fig. 18.6d, e) (Alexandrowicz and Alexandrowicz 1988; Jankowski and Margielewski 2021).

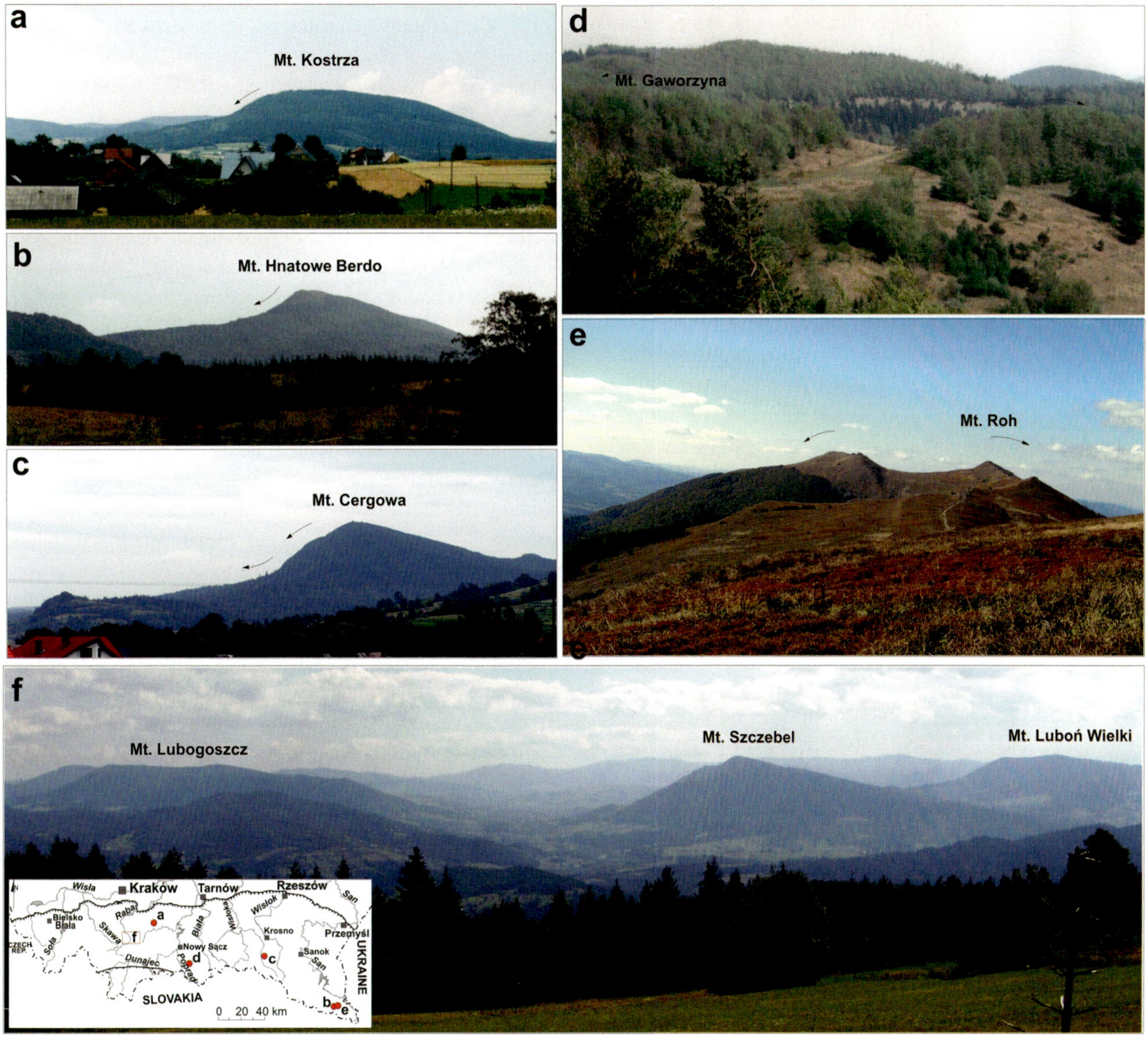

Fig. 18.6 Impact of large-scale landslides (arrows show the direction of movement) on the shape of slopes of the Carpathian Mts. **a** Convex slope of Mt. Kostrza, Beskid Wyspowy Mts. **b** Concave slope of Mt. Hnatowe Berdo, Bieszczady Mts. **c** Concave-convex slope of Mt. Cergowa, Beskid Niski Mts. **d** Ridge-top trench on Mt. Gaworzyna, Beskid Sądecki Mts. **e** Double ridge of Mt. Roh, Bieszczady Mts. (formed as DSGSDs) **f** Landslide morphology of the Beskid Wyspowy Mts., with slopes transformed by mass movements (partly after Alexandrowicz and Margielewski 2010) (photographs by W. Margielewski)

In some places, the slopes of a shape determined by gravitational displacements are visible in the landscape of entire mountain ranges (Fig. 18.1c—site 5 and 18.6f).

18.4.2 Local-Scale Elements of Landslide Morphology

The areas transformed by mass movements are characterized by a specific relief (the so-called landslide relief), in which characteristic landforms are associated with individual elements of a landslide. In case of relatively young mass movements, a complete set of elements of landslide relief can be preserved (Fig. 18.7a–h) (Margielewski and Alexandrowicz 2004; Alexandrowicz and Margielewski 2010). Picturesque rock walls (cliffs) of various sizes, oriented according to the direction of tectonic discontinuities, are the remains of head scarps of landslides (Fig. 18.7b) (Margielewski 2006a). Above these scarps, extensional trenches have often developed (Fig. 18.7a). Locally, as a result of gravitational displacements, only trenches occur and form systems of steps on the slope, as an effect of spreading of the rock mass (Alexandrowicz and Alexandrowicz 1988; Margielewski 2006a). Due to fragmentation of rock cliffs within the head, scarps spectacular rock crags (rock spurs) are formed, reaching significant heights in certain places (Fig. 18.7c). Singular, outlying rock crags (tors or hoodoo rocks) can also be remnants of landslides destroying the top parts of ridges on several sides (Baumgart-Kotarba 1974; Alexandrowicz 1978b; Alexandrowicz and Margielewski 2010). The other type of gravity-induced landforms is caves (non-karstic caves), which are formed both in the initial phase of mass movements and during the development of main deformations within the landslide body (Fig. 18.7d) (Margielewski and Urban 2017). Rock masses disintegrated during gravitational displacements form picturesque block fields (Fig. 18.7e). In depressions present in various parts of landslides water basins (lakes) are formed (Fig. 18.7g), which in the process of paludification turn into fens, usually minerogenic mires (see Fig. 18.7f) (Margielewski 2018). Studies of deposits of these peatlands provide a valuable knowledge about the changes in the palaeogeography and the palaeoenvironments in the Quaternary (Margielewski 2006b, 2018). Locally, where a river valley was blocked by a landslide tongue, landslide dam lakes developed, becoming unique landscape elements in the Beskidy Mountains (Figs. 18.7h and 18.8) (Nowalnicki 1971; Margielewski et al. 2008). Generally, gravitationally induced landforms are characterized by high variability and mosaic character of relief forms, and have a large impact on the geodiversity of the mountain area (Alexandrowicz and Margielewski 2010).

18.5 Contemporary Mass Movements

Contemporary landslides (in a broader sense: mass movements) occur commonly in the Outer Carpathians. Particular intensification of their development is related to catastrophic rainfalls and then their impact on the transformation of relief becomes significant (Fig. 18.9a–e) (Sawicki 1917; Ziętara 1968; Poprawa and Rączkowski 2003; Mrozek et al. 2000; Gorczyca 2004; Rączkowska et al. 2012; Łajczak et al. 2014). Generally, it is difficult to determine universal precipitation thresholds causing the formation of landslides in larger areas, although they can be easier to determine for individual landslides (Wistuba et al. 2021b). However, it is assumed that in the Outer Carpathians, continuous rains delivering 150–400 mm within 2–5 days lead to the formation of landslides within the regolith slope cover (Starkel 1996). On the slopes formed within sandstone and shale complexes, landslides are formed as a result of precipitation of 250–300 mm during 20–45 days, whereas within the thick-bedded sandstone, precipitation thresholds initiating landslides increase to 400–550 mm during 20–40 days (Gil 1997). In the Carpathians, which are a dynamic, young orogen, earthquakes could also play a triggering role (Poprawa and Rączkowski 2003; Wistuba et al. 2018, 2021a).

Apart from a very few landslides with landforms permanently marked in the relief (Fig. 18.8), nowadays usually shallow landslides within regolith (Fig. 18.9b, c), and less often bedrock landslides (Fig. 18.9d and e), are formed. Usually, the relief of shallow landslides becomes blurred quite quickly as a result of denudation processes, thus they only moderately influence relief transformation and landscape change in the Beskidy Mts.

One of a few large landslides having a significant impact on the transformation of slope relief is the landslide at Mt. Kicarz (Beskid Sądecki Mts.). It was formed after catastrophic rainfall in 1813 and dammed the Poprad River. Consequently, the water of landslide-dammed lake destroyed the church in the town of Mníšek (Nowalnicki 1971, 1976). Nowadays, the landslide is still clearly visible in slope morphology, accounting for its concave-convex profile (Fig. 18.8a–c). In turn, the remnants of another recent landslide, which occurred in 1907 in Duszatyn at the western slopes of Mt. Chryszczata (the Bieszczady Mts.), are the picturesque small lakes of Jeziorka Duszatyńskie, formed as a result of damming the Olchowaty stream by landslide colluvium (Schramm 1925). The landslide at the slope of Mt. Połoma (Bieszczady Mts.), which in 1980 dammed the Wetlinka stream and created a dammed lake, has preserved some elements of landslide relief (rock walls and lateral edges, colluvial ramparts) up to now, but they are covered with dense vegetation and do not stand out in

Fig. 18.7 Small-scale landforms and landscape elements developed due to mass movements, presented on the schematic cross-section of hypothetical landslide, with real-life examples (partly after Alexandrowicz and Margielewski 2010). **a** Rock trenches in Mt. Wierch nad Kamieniem, Beskid Sądecki Mts. **b** Rock cliff (landslide scarp) in Mt. Zadnie Góry, Beskid Sądecki Mts. **c** Crags, Devils' Crags, Beskid Sądecki Mts. **d** Gravity-induced cave, Diabla Dziura cave, Ciężkowice Foothills. **e** Rock debris (talus), Mt. Luboń Wielki, Beskid Wyspowy Mts. **f** Landslide fen, Wierchomla Mała, Beskid Sądecki Mts. **g** Landslide lake, Pucołowski Staw, Gorce Mts. **h** Landslide-dammed lake, Szmaragdowe Jezioro, Bieszczady Mts. (photographs by W. Margielewski)

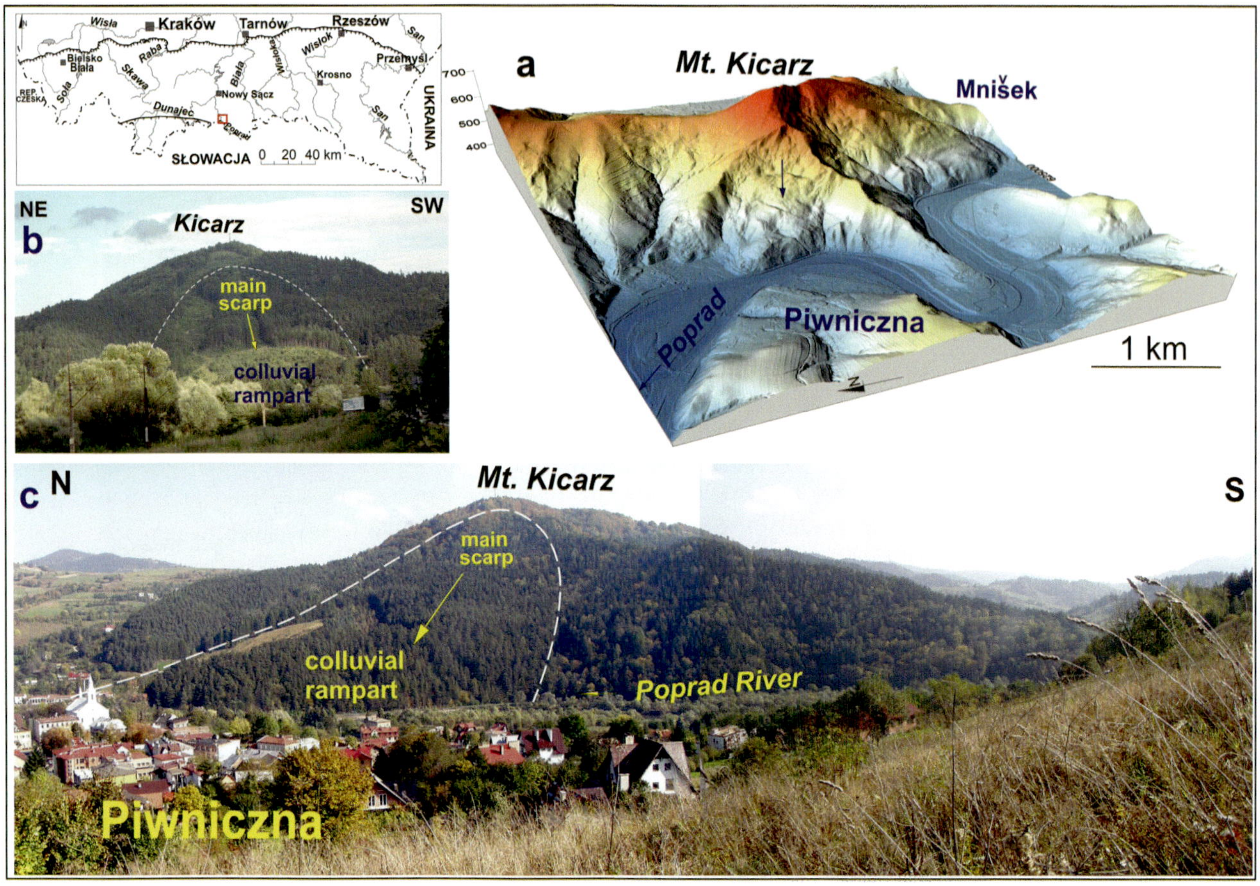

Fig. 18.8 Landslide at Mt. Kicarz (Beskid Sądecki Mts.), formed in 1813. **a** LiDAR DTM (arrow shows the direction of displacement). **b, c** Present-day views of the landslide at Mt. Kicarz (photographs by W. Margielewski)

the landscape. The landslide-dammed lake which was created at that time has been slowly disappearing (Fig. 18.7h) (Margielewski et al. 2008).

Another locality with distinctive landslide relief is the landslide in the village of Lachowice, formed in 2001 (Beskid Makowski Mts.) (Fig. 18.9d—no. 1 and 2). At that time, several buildings in the area affected by mass movements were destroyed. The landslide tongue blocked the Lachówka stream and formed a temporary landslide-dammed lake (Oszczypko et al. 2002). The detachment of rock masses took place along the surface of a strike-slip dextral fault, now exposed in the high scarp of the lateral edge of the landslide (Fig. 18.9d—no. 1 and 4) (Margielewski 2009). In the lower parts of the landslide, the displaced rock material formed characteristic high ramparts, well exposed in the landscape (Fig. 18.9d—no. 3). A further contemporary landslide with distinctive morphological features is the landslide in the village of Kłodne (Beskid Wyspowy Mts.), formed in 2010 (Fig. 18.9e) (Chowaniec and Wójcik 2012). Owing to shallow but very extensive sliding (the slip surface is exposed on the surface) and accumulation of displaced rocks, a vast colluvial rampart 20 m high was formed, still well-marked in the landslide relief (Fig. 18.9e—no. 1 and 3) (Łajczak et al. 2014).

Generally, the relief of contemporary landslides is obliterated quite quickly due to denudational and—on the other hand—vegetation growth. Only forms created in areas previously transformed by mass movements may, in the long term (in effect of landslide succession), contribute to the widening of river valleys, expansion of valley heads or formation of a wavy slope morphological profile (Ziętara 1968; Kotarba 1986; Margielewski et al. 2008).

18.6 Carpathian Relief and Landscape Transformation by Landslides: Towards Geodiversity

The common occurrence of gravitational displacements in the Beskidy Mts. conditioned by the presence of accommodation space, slope inclination, its height, geological structure as well as the initiating factors indicates that apart from the flat surfaces of top parts of mountains and ridges, all slopes (regardless whether they are valley slopes or mountain slopes)

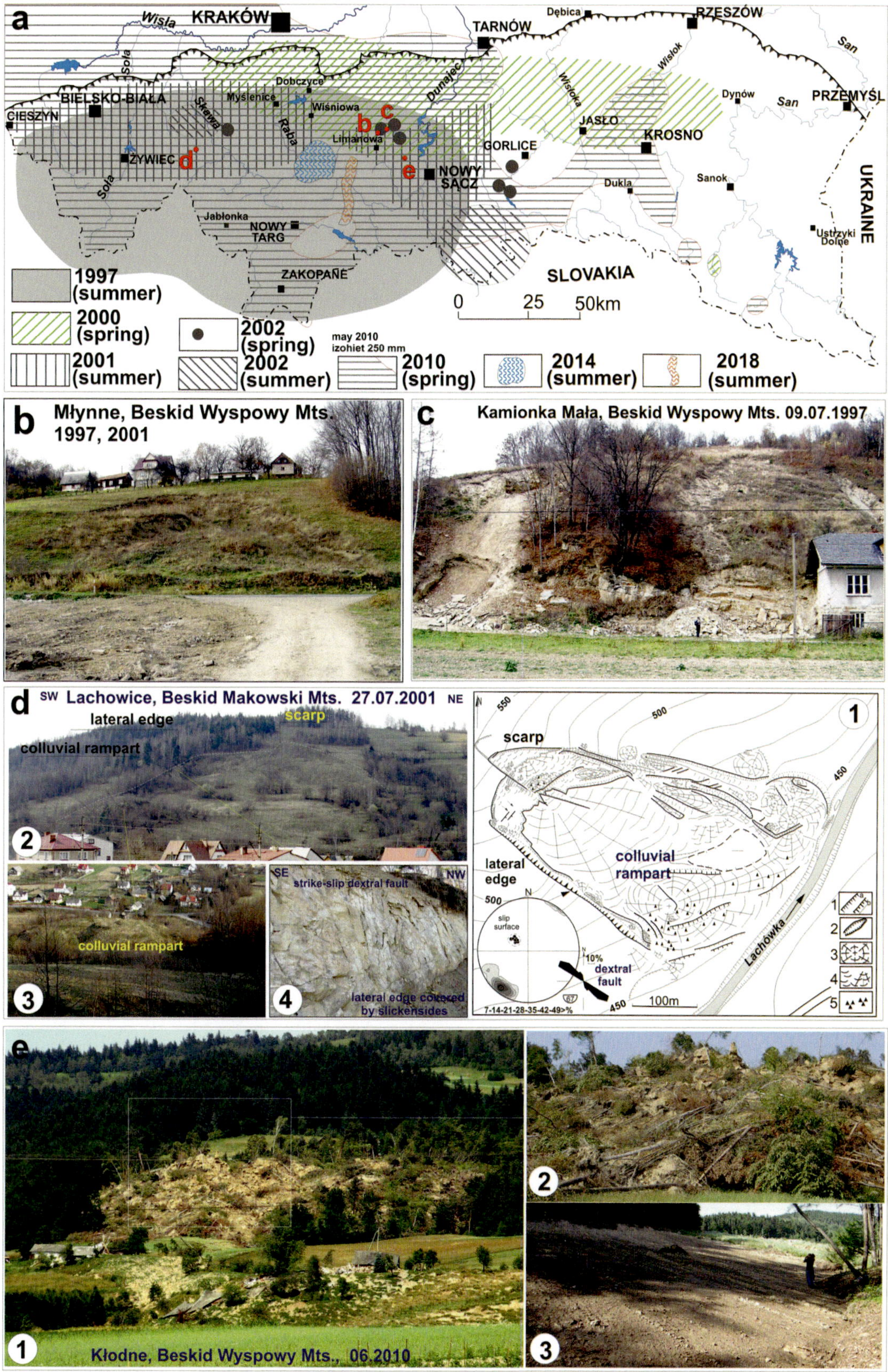

a
Wisła
KRAKÓW
TARNÓW
Dębica
RZESZÓW
San
Soła
Wisłok
Wisłoka
Dunajec
Dobczyce
Myślenice
Wiśniowa
Skawa
Raba
BIELSKO-BIAŁA
CIESZYN
ŻYWIEC
Limanowa
GORLICE
JASŁO
KROSNO
Dynów
San
PRZEMYŚL
NOWY SĄCZ
Dukla
Sanok
UKRAINE
Ustrzyki Dolne
Jabłonka
NOWY TARG
ZAKOPANE
SLOVAKIA
0 25 50km
1997 (summer)
2000 (spring)
2002 (spring)
may 2010 izohiet 250 mm
2001 (summer)
2002 (summer)
2010 (spring)
2014 (summer)
2018 (summer)
b Młynne, Beskid Wyspowy Mts. 1997, 2001
c Kamionka Mała, Beskid Wyspowy Mts. 09.07.1997
d SW Lachowice, Beskid Makowski Mts. 27.07.2001 NE
lateral edge
scarp
colluvial rampart
2
3
colluvial rampart
SE strike-slip dextral fault NW
4
lateral edge covered by slickensides
1
550
500
450
scarp
lateral edge
colluvial rampart
500
N
slip surface
10%
dextral fault
7-14-21-28-35-42-49>%
450
100m
Lachówka
1
2
3
4
5
e
1
Kłodne, Beskid Wyspowy Mts., 06.2010
2
3

Fig. 18.9 Present-day transformation of the Carpathians slopes by mass movements. **a** Map with the distribution of catastrophic precipitations (downpours and continuous rain) that have recently caused mass movements in the Carpathians (partly after Poprawa and Rączkowski 2003; Margielewski et al. 2008; supplemented). **b, c** Shallow landslides formed (or reactivated) owing to catastrophic rains, presented in part a (position of each landslides indicated by respective letters **b**–**e**), which have not caused persistent geomorphological consequences. **d** Lachowice landslide (Beskid Makowski Mts.) formed in 2001 along strike-slip dextral fault: 1—map of the landslide with fault plane position and course presented on contour diagram (equal area plot, projection of poles on lower hemisphere with the number of measurements and contour intervals), and rose diagram, bedding plane in the pole point diagram (legend symbols: 1—scarp (a—rocky, b—creeped), 2—trench, 3—colluvial rampart, 4—creeping, 5—rock debris), 2—view on the landslide, 3—colluvial rampart 4—landslides' lateral edge formed on dextral fault (partly after Margielewski 2009). **e** Landslide in Kłodne (Beskid Wyspowy Mts.) orginated in 2010: 1, 2—vast, high colluvial rampart as a large and significant element of landslide relief; 3—enormous slip surface exposed due to gravitational displacements is, in fact, the surface of a small-angle dip-slip normal fault (photographs by W. Margielewski)

had to be transformed by mass movements. In fact, even ridges may be also transformed by the evolution of ridge-top trenches in some parts (Fig. 18.6d, e). A comprehensive analysis of the transformation of the Carpathian relief by mass movements shows that most slopes of the Beskidy Mountains have been re-modelled by gravitational displacements. The scale, size and extent of transformation of mountain slopes by mass movements is therefore much greater than it has been considered up till now (Starkel 1960; Bober 1984; Bober et al. 1997; Długosz 2011; Pánek et al. 2019). Mass movements undoubtedly play a leading role among the denudation processes that have shaped the morphology and landscape of the Carpathians. The diversity of relief forms and landscape features formed by mass movements, exposure of geological formations and structures in rock outcrops within landslides, and finally the impact of mass movements taking place at the stage of a gravitational collapse on the geology of the Carpathians, all these features make mass movements a very important factor influencing the geodiversity of the Beskidy Mountains and consequently affecting also their biodiversity (Margielewski and Alexandrowicz 2004; Alexandrowicz and Margielewski 2010).

Landslide areas with unique relief and specific plant communities are undoubtedly worthy of protection. However, the problem of their conservation has been so far underestimated. Only a few landslide areas in the Polish Carpathians have been legally protected due to their unique relief. Landslide landforms are only protected in 7 geological nature reserves. In the Luboń Wielki Nature Reserve (Beskid Wyspowy Mts.) unique trenches, rock walls, caves and blockfields of landslide origin are under protection (Alexandrowicz and Alexandrowicz 1988; Alexandrowicz and Margielewski 2010). Similar landforms are protected in the Kuźnie Nature Reserve at Mt. Muronka in the Beskid Śląski Mts. (rock walls, trenches and caves). In some nature reserves focused on protection of plant communities (15 nature reserves), landslide relief is also subject to conservation, but not as a primary value. A perfect example is the Barnowiec Nature Reserve in the Beskid Sądecki Mts., where the main object of conservation is the rare forest community of *Phylitido aceretum*, whereas the reserve also includes unique landslide landforms, including 50 m high rock walls of head scarp, trenches and caves (Alexandrowicz and Margielewski 2010). The same problem applies to NATURA 2000 sites, protecting unique plant habitats, without due regard to their relation with landslide areas within which they arose (Alexandrowicz et al. 2003). Fortunately, beside the nature reserves, there are over 50 nature monuments or documentary sites with selected elements of landslide landforms, mainly caves, crags, rock walls, lakes and fens, protected as unique landslide landforms (Alexandrowicz and Poprawa 2000; Margielewski and Alexandrowicz 2004; Alexandrowicz and Margielewski 2010).

18.7 Conclusions

Mass movements significantly influenced the relief and landscape of the Outer Flysch Carpathians (Beskidy Mountains). At the stage of the tectonic evolution of the Carpathians, during the gravitational collapse in the Neogene, giant-scale mass movements had an essential impact on the spatial arrangement of some mountain ridges and ranges, the distribution of individual mountain massifs (often with insular relief), as well as the course of river valleys developed within tectonic mélanges formed due to gravitational displacements. In turn, in the Quaternary stage of relief evolution, mass movements transformed slopes providing them with convex, concave and convex-concave morphological profiles exposed in the present-day landscape. The diverse landslide relief of individual landslide forms, with abundant rock cliffs, crags, talus slopes, trenches, landslide lakes and fens, significantly influenced the landscape and, therefore, also the geodiversity of the Beskidy Mts. Present–day landslides, although commonly developing during episodes of catastrophic rainfall, have a relatively small impact on the morphology and landscape of the Beskidy Mts. Only occasionally the present-day landslides leave a permanent mark in the landscape of the Carpathians.

The common nature of mass movements that have transformed the morphology of the Carpathians since the gravitational collapse phase strongly suggests that the area and number of the Carpathian slopes that are reshaped by mass movements is much larger than it was previously considered.

References

Agliardi F, Crosta GB, Frattini P, Malusá MG (2013) Giant non-catastrophic landslides and the long-term exhumation of the European Alps. Earth Planet Sci Lett 365:263–274
Alexandrowicz SW (1978a) The northern slope of Babia Góra Mts. as a huge rock slump. Studia Geomorph Carp-Balc 12:133–148
Alexandrowicz Z (1978b) Skałki piaskowcowe zachodnich Karpat fliszowych. Prace Geol PAN 113:87
Alexandrowicz Z, Poprawa D (eds) (2000) Geodiversity conservation of the Polish Carpathians, with Map of protected and proposed for protection areas and objects of inanimate nature 1:400,000. Wyd Kart, Warszawa, p 142
Alexandrowicz SW, Alexandrowicz Z (1988) Ridge top-trenches and rifts in the Polish Outer Carpathians. Ann Soc Geol Polon 58:207–228
Alexandrowicz Z, Margielewski W (2010) Impact of mass movements on geo- and biodiversity in the Polish Outer (Flysch) Carpathians. Geomorphology 123:290–304
Alexandrowicz Z, Margielewski W, Perzanowska J (2003) European ecological network Natura 2000 in relation to landslide areas diversity: a case study in the Polish Carpathians. Ekológia 22(4):404–423
Alexandrowicz SW (1997) Holocene dated landslies in the Polish Carpathians. In: Frenzel B (ed) Rapid mass movements as a source of climatic evidence for the Holocene (Palaeoclim Res 19:75–83)
Baumgart-Kotarba M (1974) Development of the mountain ridges in the Flysch Carpathians. Prace Geogr IG PAN 106:136
Birkenmajer K (1985) Stages of structural evolution of the Pieniny Klippen Belt, Carpathians. Studia Geol Polon 88:7–32
Bober L (1984) Rejony osuwiskowe w polskich Karpatach fliszowych i ich związek z budową geologiczną. Biul PIG 340:115–158
Bober L, Thiel K, Zabuski L (1997) Zjawiska osuwiskowe w polskich Karpatach fliszowych. Geologiczno-inżynierskie właściwości wybranych osuwisk. Wyd. IBW PAN, p 102
Břežný M, Pánek T, Lenart J, Zondervan A, Braucher R (2018) ^{10}Be dating reveals pronounced Mid-to Late Holocene activity of deep-seated landslides in the highest part of the Czech Flysch Carpathians. Quat Sci Rev 195:180–194
Chalupa V, Pánek T, Tábořik P, Klimeš J, Hartvich F, Grygar R (2018) Deep-seated gravitational slope deformations controlled by the structure of flysch nappe outliers: Insight from large-scale electrical resistivity tomography survey and LiDAR mapping. Geomorphology 321:174–187
Chowaniec J, Wójcik A (eds) (2012) Osuwiska w województwie małopolskim. Atlas-Przewodnik. Compass, Kraków, p 143
Dewey JF (1987) Extensional collapse orogens. Tectonics 7:1123–1139
Dikau R, Brunsden D, Schrott L, Ibsen ML (eds) (1996) Landslide recognition. Identification, movement and causes. Wiley, Chichester, p 251
Długosz M (2011) Podatność stoków na osuwanie w polskich Karpatach fliszowych. Prace Geogr IGiPZ PAN 230:112
Festa A, Pini GA, Ogata K, Dilek Y (2019) Diagnostic features and field-criteria in recognition of tectonic, sedimentary and diapiric mélanges in orogenic belts and exhumed subduction-accretion complexes. Gondwana Res 74:7–30
Gil E (1997) Meteorological and hydrological conditions of landslides, Polish Flysch Carpathians. Studia Geomorph Carp-Balc 31:143–157
Gil E, Gilot E, Szczepanek K, Kotarba A, Starkel L (1974) An Early Holocene landslide in the Beskid Niski and its significance for palaeogeographical reconstructions. Studia Geomorph Carp-Balc 8:69–83
Gorczyca E (2004) Przekształcanie stoków fliszowych przez ruchy masowe podczas katastrofalnych opadów (dorzecze Łososiny). Wyd. Uniw Jagiellońskiego, Kraków, p 101
Jankovská V, Baroň I, Nývlt D, Krejčí O, Krejčí V (2018) Last Glacial to Holocene vegetation succession recorded in polyphase slope-failure deposits on the Maleník Ridge, Outer Western Carpathians. Quat Int 470:38–52
Jankowski L (2015) Nowe spojrzenie na budowę geologiczną Karpat—ujęcie dyskusyjne. Prace Nauk Instytutu Nafty i Gazu Państwowego Instytutu Badawczego 202:154
Jankowski L, Margielewski W (2014) Strukturalne uwarunkowania rozwoju rzeźby Karpat zewnętrznych—nowe spojrzenie. Przegl Geol 62:29–35
Jankowski L, Margielewski W (2021) Geological control of young orogenic mountain morphology: from geomorphological analysis to reinterpretation of geology of the Outer Western Carpathians. Geomorphology 386:107749
Jankowski L (2021) Przeglądowa Mapa Geologiczna Karpat w skali 1:500,000. Państwowy Instytut Geologiczny—Państwowy Instytut Badawczy, Warszawa
Kłapyta P (2020) Geomorphology of the high-elevated flysch range—Mt. Babia Góra Massif (Western Carpathians). J Maps 16:689–701
Kondracki J (1978) Karpaty, 2nd edn. WSiP, Warszawa, p 261
Kotarba A (1986) Rola osuwisk w modelowaniu rzeźby Beskidzkiej i pogórskiej. Przegl Geogr 58:119–129
Książkiewicz M (1972) Karpaty. In: Pożaryski W (ed) Budowa Geologiczna Polski, t. IV, Tektonika, cz. 3. Wyd Geol, Warszawa, p 228
Łajczak A, Margielewski W, Rączkowska Z, Święchowicz J (2014) Contemporary geomorphic processes in the Polish Carpathians under changing human impact. Episodes 37:21–32
Lexa J, Bezák V, Elečko M, Mello J, Polák M, Potfaj M, Vozár J (eds) (2000) Geological Map of Western Carpathians and adjacent areas 1:500,000. Geological Survey of Slovak Republic, Bratislava
Marciniec P, Zimnal Z, Wojciechowski T, Perski Z, Rączkowski W, Laskowicz I, Nescieruk P, Grabowski D, Kułak M, Wójcik A (2019) Osuwiska w Polsce—od rejestracji do prognozy, czyli 13 lat projektu SOPO. Przegl Geol 67:291–297
Margielewski W (2006a) Structural control and types of movements of rock mass in anisotropic rocks: case studies in the Polish Flysch Carpathians. Geomorphology 77:47–68
Margielewski W (2006b) Records of the Late Glacial-Holocene palaeoenvironmental changes in landslide forms and deposits of the Beskid Makowski and Beskid Wyspowy Mts. area (Polish Outer Carpathians). Folia Quat 76:149
Margielewski W (2009) Problematyka osuwisk strukturalnych w Karpatach fliszowych w świetle zunifikowanych kryteriów klasyfikacji ruchów masowych—przegląd krytyczny. Przegl Geol 57:905–917
Margielewski W (2018) Landslides fens as a sensitive indicator of the palaeoenvironmental changes since the Late Glacial; Polish Western Carpathians case study. Radiocarbon 60:1199–1213
Margielewski W, Alexandrowicz Z (2004) Diversity of landslide morphology as a part of geoconservation pattern in the Polish Carpathians. Pol Geol Inst Spec Pap 13:65–72
Margielewski W, Urban J (2017) Gravitationally induced non-karst caves: tectonic and morphological constrains, classification and dating; Polish Flysch Carpathians case study. Geomorphology 296:160–181
Margielewski W, Jankowski L (2019) Zakole rzeki Koszarawy utworzone w obrębie melanży tektonicznych występujących w strefie poślizgu masywu skalnego przemieszczonego grawitacyjnie na etapie kolapsu grawitacyjnego Karpat. In: Jankowski L (ed) Warsztaty terenowe: „Od podnóża Tatr po brzeg Karpat. Współczesne wyzwania kartografii geologicznej". Przewodnik do wycieczki terenowej. PIG-PIB, Kraków, pp 65–67
Margielewski W, Świechowicz J, Starkel L, Łajczak A, Pietrzak M (2008) Recent evolution of the Flysch Carpathians relief. In: Starkel L, Kostrzewski A, Kotarba A, Krzemień K (eds) Współczesne

przemiany rzeźby Polski. Instytut Geografii i Gospodarki Przestrzennej Uniwersytetu Jagiellońskiego, Kraków, pp 57–133
Mazzoli S, Jankowski L, Szaniawski R, Zattin M (2010) Low-T termochronometric evidence for post-thrusting (<11 Ma) exhumation in the Western Outer Carpathians, Poland. CR Geosci 342:162–169
Mrozek T, Rączkowski W, Limanówka D (2000) Recent landslides and triggering climatic conditions in Laskowa and Pleśna Regions, Polish Carpathians. Studia Geomorph Carp-Balc 34:89–112
Mrozek T, Kułak M, Grabowski M, Wójcik A (2014) Landslide counteracting system (SOPO): inventory database of landslides in Poland. In: Sassa K, Canuti P, Yin Y (eds) Landslide science for a safer geoenvironment, pp 815–820
Nowalnicki T (1971) Beskidzkie jeziorka zaporowe. Wierchy 40:274–280
Nowalnicki T (1976) Jeziorka osuwiskowe w Beskidzie Sądeckim. Wierchy 45:182–198
Oszczypko N (2006) Late Jurassic-Miocene evolution of the Outer Carpathians fold-and-thrust belt and its foredeep basin (Western Carpathians, Poland). Geol Q 50:169–193
Oszczypko N, Golonka J, Zuchiewicz W (2002) Osuwisko w Lachowicach (Beskidy Zachodnie): skutki powodzi z 2001 r. Przegl Geol 50:893–898
Oszczypko N, Uchman A, Malata E (eds) (2006) Palaeotectonic evolution of the Outer Carpathian and Pieniny Klippen Belt Basins. Institute of Geological Sciences, Jagiellonian University, Kraków, p 199
Pánek T (2019) Landslides and Quaternary climate changes—The state of the art. Earth Sci Rev 196:102871
Pánek T, Klimes J (2016) Temporal behavior of deep-seated gravitational slope deformations: a review. Earth Sci Rev 156:14–38
Pánek T, Smolková V, Hradecký J, Baroň I, Šilhán K (2013) Holocene reactivations of catastrophic complex flow-like landslides in the Flysch Carpathians (Czech Republic/Slovakia). Quat Res 80:33–46
Pánek T, Hartvich F, Jankovská V, Klimeš J, Tabořík P, Bubík M, Smolková V, Hradecký J (2014) Large late Pleistocene landslides from the marginal slope of the Flysch Carpathians. Landslides 11:981–992
Pánek T, Bžežný M, Kapustová V, Lenart J, Chalupa V (2019) Large landslides and deep-seated gravitational slope deformations in the Czech Flysch Carpathians: new LiDAR-based inventory. Geomorphology 346:106852
Połtowicz J, Waśniowska J (1977) Miocen Iwkowej. Rocznik Pol Tow Geol 45:433–445
Poprawa D, Rączkowski W (2003) Osuwiska Karpat. Przegl Geol 51:685–692
Rączkowska Z, Łajczak A, Margielewski W, Święchowicz J (2012) Recent landform evolution in the Polish Carpathians. In: Lóczy D, Stankoviansky M, Kotarba A (eds) Recent landform evolution: The Carpatho–Balkan–Dinaric region. Springer, Dordrecht, pp 47–102
Sawicki L (1917) Osuwiska ziemne w Szymbarku i inne zsuwy powstałe w 1913 r. w Galicji Zachodniej. Rozpr Wydz Mat-Przyr Akademii Umiejętności w Krakowie 56A:227–313
Schramm W (1925) Zsuwiska stoków górskich w Beskidach. Wielkie zsuwisko w lesie wsi Duszatyn Ziemi Sanockiej. Kosmos 50:1355–1374
Starkel L (1960) Rozwój rzeźby Karpat fliszowych w holocenie. Prace Geogr IG PAN 22:239
Starkel L (1972) Charakterystyka rzeźby polskich Karpat (i jej znaczenie dla gospodarki ludzkiej). Problemy Zagosp Ziem Górskich 10:75–150
Starkel L (1991) Rzeźba terenu. In: Dynowska I, Maciejewski M (eds) Dorzecze Górnej Wisły, cz. I. PWN, Warszawa-Kraków, pp 42–54
Starkel L (1996) Geomorphic role of extreme rainfalls in the Polish Carpathians. Studia Geomorph Carp-Balc 30:21–38
Starkel L (1997) Mass movement during the Holocene: Carpathian example and the European perspective. In: Frenzel B (ed) Rapid mass movements as a source of climatic evidence for the Holocene (Palaeoclim Res 19:385–400)
Tokarski A, Świerczewska A, Zuchiewicz W, Márton E, Hurai V, Anczkiewicz A, Michalik M, Szeliga W, Rauch-Włodarska M (2006) Structural development of the Magura Nappe (Outer Carpathians): from subduction to collapse. GeoLines 20:145–153
Wistuba M, Malik I, Krzemień K, Gorczyca E, Sobucki M, Wrońska-Wałach D, Gawior D (2018) Can low-magnitude earthquakes act as a triggering factor for landslide activity? Examples from the Western Carpathian Mts., Poland. Catena 171:359–375
Wistuba M, Gorczyca E, Malik I (2021a) Inferring precipitation thresolds of landslide activity from long-term dendrochronological and precipitation data: Case study on the unstable slope at Karpenciny, Poland. Eng Geol 294:106398
Wistuba M, Malik I, Gorczyca E, Ślęzak A (2021b) Establishing regimes of landslide activity—analysis of landslide triggers over the previous seven decades (Western Carpathians, Poland). Catena 196:104888
Wójcik A (2019) Późnoglacjalny rozwój osuwisk w polskich Karpatach zewnętrznych. Przegl Geol 67:397–404
Ziętara T (1968) Rola gwałtownych ulew i powodzi w modelowaniu rzeźby Beskidów. Prace Geogr IG PAN 60:116
Żytko K, Zając R, Gucik S, Ryłko W, Oszczypko N, Garlicka I, Nemčok J, Eliaš M, Menčik E, Stranik Z (1989) Map of the tectonic elements of the Western Outer Carpathians and their foreland 1: 500 000. In: Poprawa D, Nemčok J (eds) Geological Atlas of the Western Outer Carpathians and their Foreland. PIG, Warszawa

Włodzimierz Margielewski is a geologist and geomorphologist, employed as Associate Professor in the Institute of Nature Conservation, Polish Academy of Sciences in Krakow, Poland. His research is focused on geology, geomorphology, and geoconservation in the Carpathians and their Foreland, with particular emphasis to structural geology, geological control of relief, and wide aspects of mass movements. He also works on the reconstruction of palaeoenvironmental changes during the Late Glacial and Holocene in the Polish territory, based on studies of peatland deposits, including landslide fens (a type recognized by him).

Leszek Jankowski is a geologist, tectonic and cartographer. He is Associate Professor at the Polish Geological Institute–National Research Institute, in Krakow, Poland. His research is focused on geology, tectonics and sedimentology of the Carpathians and their forelands, as well as on the geological controls behind the morphology of the Carpathians. He is an author or co-author of numerous geological maps of the Carpathians (including also Slovak, Ukrainian and Romanian parts), and scientific articles on the geology of the Carpathians and their Foreland, containing new, original concepts of the development of the Carpathian orogen.

Gravitationally Induced Non-karst Caves in the Polish Outer Carpathians

19

Włodzimierz Margielewski and Jan Urban

Abstract

The Outer Carpathians, formed of siliciclastic-clayey rocks, abound in non-karst (pseudokarst) caves. In their Polish part, 1646 such caves with a total length of 27,857 m have been recorded up to July 2021. The longest Jaskinia Wiślańska cave reaches 2275 m, whilst 43 caves are longer than 100 m and 31 caves are 15 m deep or more (up to 60 m). Most of them are genetically related to slope gravitational deformations, but there are also caves—principally only several metres long—formed due to non-uniform weathering of sandstone crags. Amongst the gravitationally induced Carpathian caves, several types can be distinguished. In the morphogenetic classification, considering the succession of processes, three types of caves are recognized: initial caves (preceding the main slope failures), subsequent caves (formed during the main slope deformations) and intermediate caves. Regarding the geomechanic criterion also three principal groups are discriminated: dilational caves (formed due to simple tensional fracturing), dilatancy caves (related to a fissure macrodilatancy) and boulder caves (developing within very strongly deformed slopes). Four examples representing various morphogenetic and geomechanic types of caves are described in this chapter. The gravitationally induced caves have been formed since the Late Glacial, when the destruction of slopes started along with permafrost degradation, but have fully developed and transformed during the Holocene, with several periods of intensification connected with climate changes (moistening).

Keywords

Non-karst caves · Pseudokarst · Gravitational slope deformations · Flysch · Late Glacial · Holocene · Outer Carpathians

19.1 Introduction

Most caves are of karstic origin, but there are regions lacking soluble rocks, which are nevertheless abundant in caves of various non-karstic origin. The Outer Carpathians formed mainly of siliciclastic-clayey rocks are such a region, where 1646 non-karst caves of a total length of 27,857 m were recorded up to July 2021 (Fig. 19.1a). The longest cave, Jaskinia Wiślańska, is 2275 m long (cave position on Fig. 19.1b—4), three other caves are longer than 1 km, and totally 43 caves are longer than 100 m, whilst 31 caves are at least 15 m deep (Klassek and Mleczek 2018; Gądek 2021). The majority of these caves have originated due to gravitational deformations of slopes, whereas a much smaller number has formed due to weathering of sandstone crags and these are principally very short caves, rarely more than 10 m long.

Caves in the Carpathians have been recognized and explored for a long time. The first explorations took place in the nineteenth century, whereas the first descriptions and maps were published in the first half of the twentieth century. A significant increase in the number of explored and documented caves occurred in the 1970s–1980s, when the Bielsko-Biała Speleoclub intensified its activities. Since the beginning of the twenty-first century, exploration was followed by the Beskidy Caving Club (created by amalgamation of the Dębica and Limanowa speleoclubs) as well as the Association for the Cave Conservation "Malinka Group". The number of explored caves increased from several tens in the 1960s to more than six hundred, of the total length higher than 10 km at the beginning of the twenty-first century (Klassek and

W. Margielewski (✉) · J. Urban
Institute of Nature Conservation, Polish Academy of Sciences, Adama Mickiewicza Ave. 33, 31-120 Kraków, Poland
e-mail: margielewski@iop.krakow.pl

J. Urban
e-mail: urban@iop.krakow.pl

P. Migoń and K. Jancewicz (eds.), *Landscapes and Landforms of Poland*, World Geomorphological Landscapes, https://doi.org/10.1007/978-3-031-45762-3_19

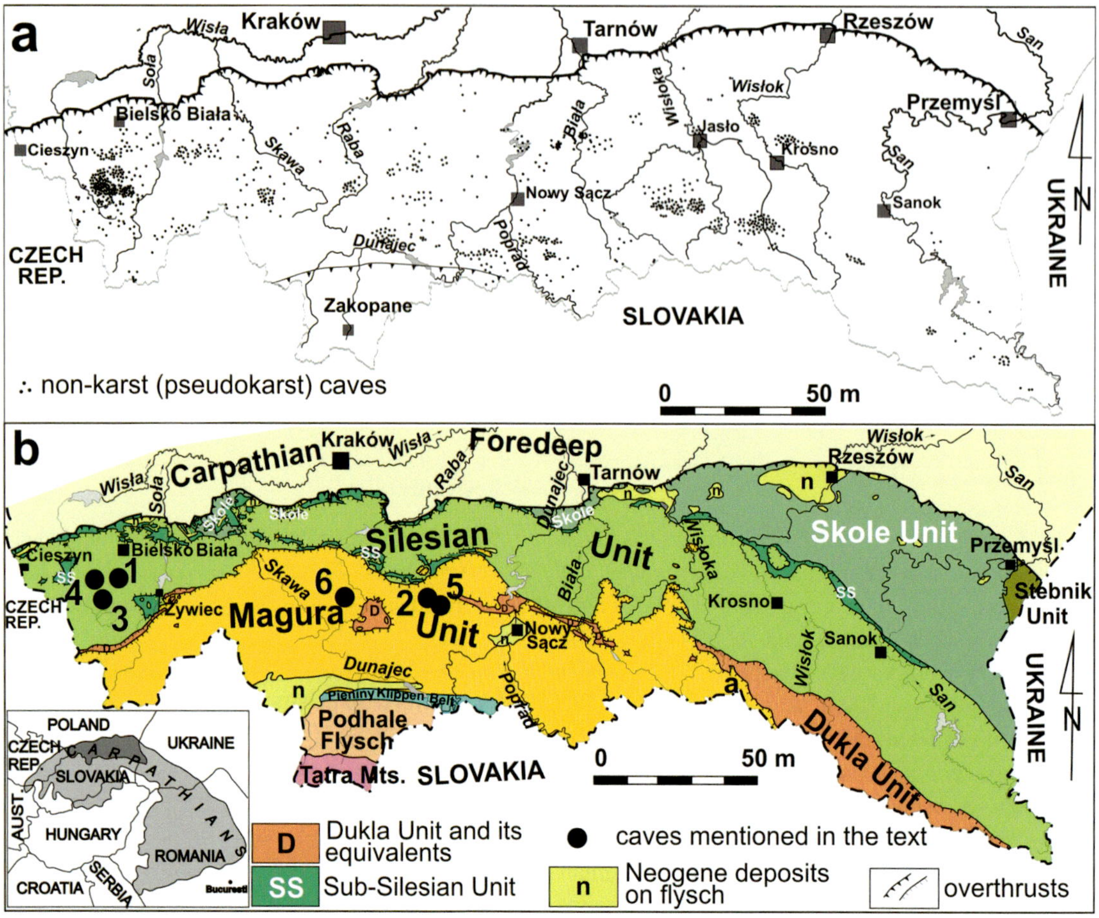

Fig. 19.1 The occurrence of non-karst caves in the Polish Outer Carpathians. **a** Distribution of caves (after Margielewski and Urban 2017). **b** Geological map of the Carpathians (after Żytko et al. 1989), with location of caves mentioned in the text. 1—Jaskinia Malinowska cave, 2—Jaskinia Złotopieńska cave, 3—Jaskinia Miecharska cave, 4—Jaskinia Wiślańska cave, 5—Jaskinia Zbójecka na Łopieniu cave, 6—Zimna Dziura w Szczeblu cave

Mleczek 2008; Urban and Margielewski 2013). At that time also the first scientific studies of these caves were performed and first their genetical descriptions and categorization were proposed (Janiga 1974; Puchejda 1989; Klassek and Mikuszewski 1997). The categorization proposed by Klassek and Mikuszewski (1997), combining many features such as the shape of caves, their spatial situation and selected genetical attributes, was very complex; therefore, the most popular classification of non-karst caves used by Polish speleologists during the first decade of this century was the one proposed by Vítěk (1983) and a similar one by Bella and Gaál 2013.

Executed in the first decade of the twenty-first century, intensive research of non-karst caves in the Polish Carpathians focused on detailed observations of spatial situation of the upslope and downslope cave walls as well as interrelations between subsurface and superficial slope deformations (also using ERT—electrical resistivity tomography), brought about much better understanding of reasons and mechanisms behind the development of these caves. In turn, the analysis of speleothems enabled one to determine the age and environments of cave formation and evolution (Margielewski and Urban 2002, 2003; Margielewski 2006a; Margielewski et al. 2007; Urban et al. 2007, 2015; Pánek et al. 2010). These studies showed that the classification by Vítěk (1983) is not sufficient for unequivocal interpretation of cave types and made possible the elaboration of a new, more adequate cave categorization (Urban and Margielewski 2013; Margielewski and Urban 2017). This chapter aims to describe a very interesting phenomenon of non-karst, gravity-induced caves that commonly occur in mountainous areas formed of siliciclastic-clayey rocks, basing on their recently proposed classification and genetical interpretation.

19.2 Geological and Geomorphological Setting

The Outer Carpathians represent an Alpine orogen formed of the Upper Jurassic–Lower Miocene flysch rocks: sandstone, sandstone–siltstone–claystone, siltstone–claystone and sandstone–conglomerate series, being sediments of deep depositional marine basins. In the Polish Outer Carpathians, such rocks, strongly folded and faulted, form several lithostratigraphic–tectonic units that are thrust upon one another towards the north. Amongst them, the following main units should be mentioned (from the south): Magura Unit, Dukla Unit (and their equivalents), Silesian Unit, Sub-Silesian Unit, Skole Unit and Stebnik Unit (Fig. 19.1b) (Żytko et al. 1989; Lexa et al. 2000).

The Polish Outer Carpathians, in geographical terms comprising the Western Beskidy and Bieszczady (Eastern Beskidy) Mountains and their foothills, are composed of many hill ranges and groups reaching 800–1300 m asl in the mountainous part and standing up to 800 m above river valleys (Starkel 1972). The specific sequence of initially tectonic (compressional–extensional) and then tectonic-gravitational (gravitational collapse) and gravitational processes brought about the present-day morphology of mountain massifs (Jankowski and Margielewski 2021). Since the Late Glacial (uppermost Pleistocene), the steep slopes of the Beskidy and Bieszczady Mountains have been shaped principally by gravitational processes, such as landsliding, toppling and spreading, which have been conditioned by geological structures: tectonic (spatial) positions and lithology of rocks (Margielewski 2006a). These processes have been also responsible for the development of underground discontinuities and cavities (caves being the cavities accessible and recognized by people).

19.3 Classification of Gravity-Induced Caves in the Outer Carpathians

The formation of gravity-induced underground cavities, also voids accessible for people, i.e. caves, consists in opening and widening of discontinuities, principally joints. It is caused by disturbance of slope gravitational equilibrium by loading and relaxation of shear or extensional stresses in rock massifs related to various factors such as rainfall, earthquakes or erosion. Shearing stress is unloaded along potential or existing tectonic discontinuities in rock massifs such as joints and faults, which are opened and widened as extensional cracks (caves) (Fig. 19.2a, b). This process continues until the shearing stress limit is exceeded, and the fragment of a rock massif, for a long time separated by cracks, is gravitationally displaced down the slope within the relatively short time period (Fig. 19.2c). The formation of extensional cracks in the first stage of stress unloading is associated with the phenomenon of dilation (dilatation), i.e. a change in the rock medium volume without changing its shape (Neuendorf et al. 2005; Rustan et al. 2010). During this process, the rock massif gradually increases its volume by the appearance and expansion of fissures (the rock mass "swells"), but there is no destruction (disintegration) of rock masses, e.g. due to shearing processes. There are no brittle deformations transforming the massif in the shearing zone as a result of rock grinding.

During the second stage of gravitational displacements of rock masses along the shearing zone, the other type of voids (caves) is formed, as a result of fissure macrodilatancy (Fig. 19.2a and c, no. 2–4) (Kwaśniewski 1986; Margielewski and Urban 2017). Fissure macrodilatancy comprises an increase in the volume of the rock medium along with the change in its shape, owing to the extensive opening of fractures (Fig. 19.2c, no. 3–4) (Kranz and Scholz 1977; Kwaśniewski 1986, 2007). Dilatancy phenomenon occurs in the zone of rock mass destruction, i.e. in the failure zone formed along the border of a displaced fragment of the massif and bedrock (hence dilation is considered as a "herald" of the destruction of the rock massif—Kwaśniewski 1986). As a result of such disintegration of rock material in the failure zone, its volume increases (the rock mass "swells", also with the participation of fissures/caves—see Fig. 19.2c, no. 3). During this process, the shape of the rock medium is distinctly changed and rocks break down as a result of shearing (Fig. 19.2c, no. 3). Cataclasites, tectonic breccia and tectonic powder (gouge) are usually formed in the failure zone.

Consequently, the categorization of caves originated due to these processes makes sense if it enables for the determination of the interrelations between cave formation and slope evolution, as well as covers the nature and range of rock massif deformation. Therefore, two criteria of classification have been proposed. The first is morphogenetic criterion that determines the relationship between the origin of the cave and the stage of slope evolution. According to this criterion, three cave types are distinguished (Fig. 19.3) (Urban and Margielewski 2013; Margielewski and Urban 2017):

- Initial caves, which develop before the main slope failure, i.e. prior to the typical landslide. They are usually simple passages or bundles of passages formed within slightly deformed slope sections within more advanced slope failures. Most them, but not all, can be identified with crevice-type caves in the classification by Vítek (1983).
- Subsequent caves formed during the principal phase of slope deformation—landslide development. These caves are situated in the slope sections apparently modified by

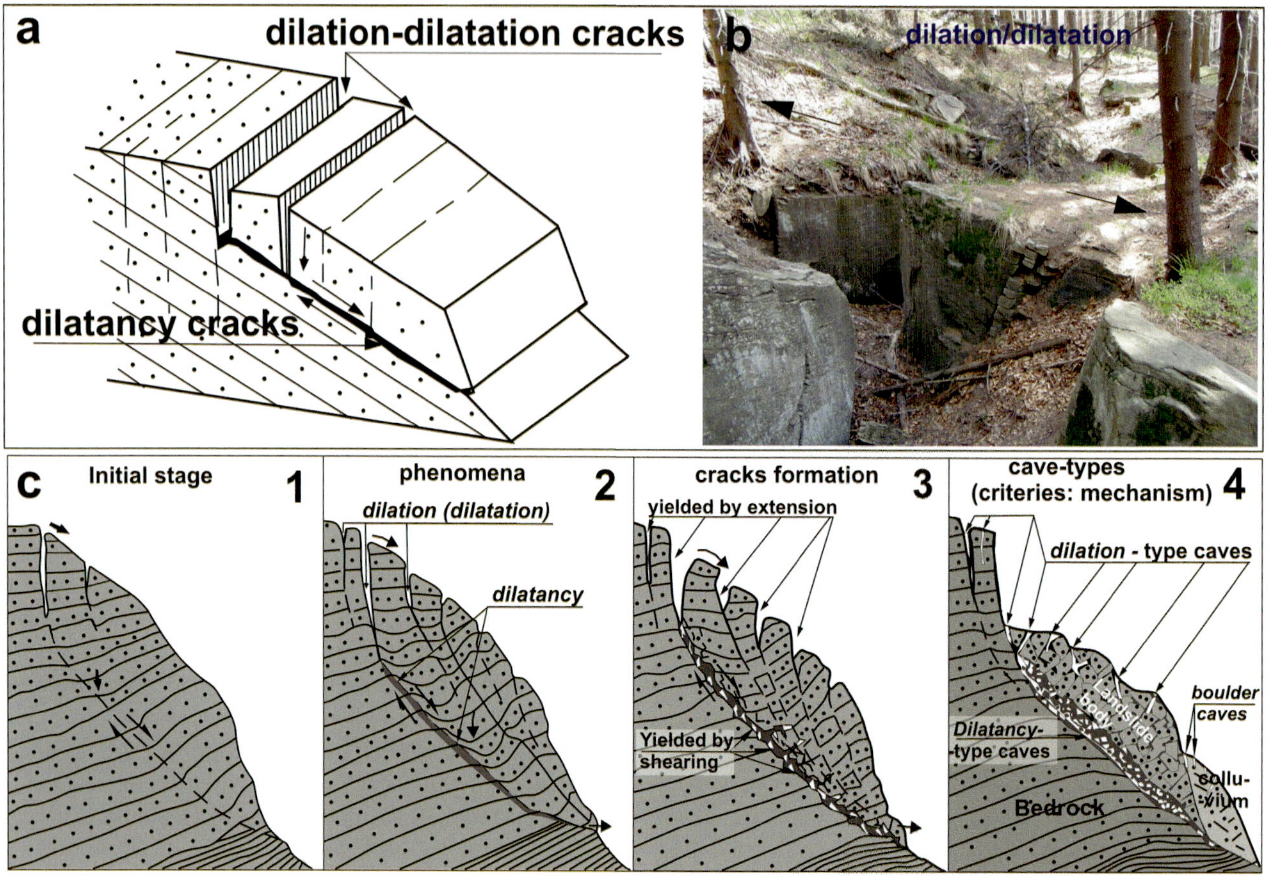

Fig. 19.2 The conceptual model of formation of non-karst caves due to dilation (dilatation) and dilatancy phenomena. **a** Cracks formed owing to dilation and dilatancy phenomena in hypothetical slope. **b** Trenches formed in effect of dilation (sandstones of the Magura Beds, Mt. Wierch nad Kamieniem, Beskid Sądecki Mts.). **c** Stages of cave development due to dilation and dilatancy phenomena (after Margielewski and Urban 2017). Failure zone formation (with fissure macrodilatancy) is preceded by rock flow (Dikau et al. 1996) (see: c 1–2) (photograph b by W. Margielewski)

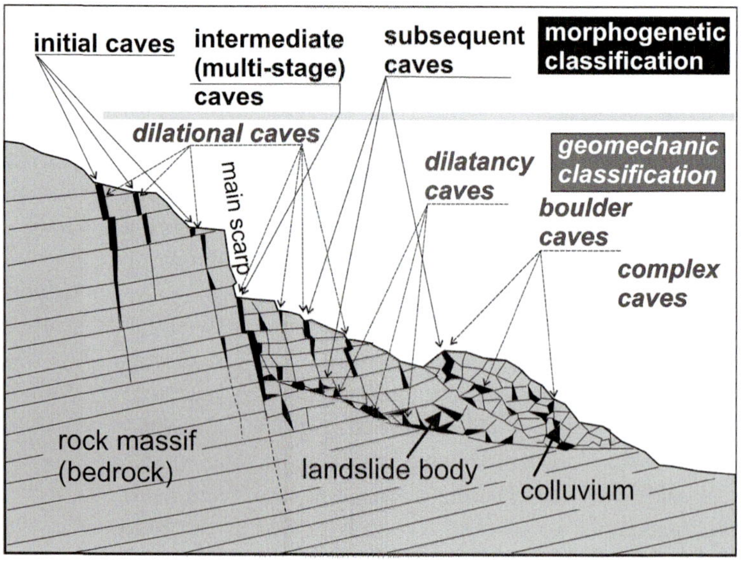

Fig. 19.3 Conceptual model and proposal of classification of non-karst caves (after Margielewski and Urban 2017)

gravitational movements, within the landslide bodies, and represent usually complex systems of passages but not always very long. In respect to the classification by Vítěk (1983), most of them belong to the group of talus caves.

- Intermediate (multi-stage) caves, which occur in a zone of the main landslide scarp, and can form during both initial and final stages of slope gravitational evolution. In this first stage, they are initial caves, which are partly destructed (due to the slope failure) and re-modelled during the main movements. They are not very frequent and usually represent bundles of more or less parallel cavities.

According to the second, geomechanic criterion, describing the nature and range of deformations of the rock massif that are genetically connected with the cave formation, also three types of caves are discriminated:

- Dilational caves, formed owing to the process of rock dilation caused due to simple tensional stresses. Most of these caves, but not all, represent initial caves in the morphogenetic classification, so they are simple passages or benches of parallel passages; often crevice-type caves in the classification by Vítěk (1983).
- Dilatancy caves, developed owing to fissure macrodilatancy (Kwaśniewski 2007) due to more advanced slope failure, produced by shear stress or combination of shear and extensional stresses. They are connected with significant modifications of slope within landslides and often represent networks or maze systems of passages, corresponding in part to caves qualified by Vítěk (1983) as of the talus type.
- Boulder caves, developed owing to chaotic distribution of large boulders in totally disintegrated rock masses forming landslide colluvia. They can be very irregular, short or forming maze systems if they are longer. In other classifications, they are described as talus caves (Vítěk 1983) or boulder caves (Bella and Gaál 2011).

19.4 Case Studies of Representative Caves

19.4.1 Jaskinia Malinowska—Initial and Dilational Cave

The Jaskinia Malinowska cave in the Beskid Śląski Mts. (Fig. 19.1b, site 1), 230 m long and up to 19.5 m deep, is one of the earliest explored caves in the Beskidy Mts. and a legendary hiding place of mountainous robbers in the eighteenth century (Klassek and Mleczek 2008). It has been known for a long time because of its location near the ridge of Mt. Malinów, within its southern slope, and the presence of a large entrance comprising a shaft ca. 10 m deep (Fig. 19.4). It is formed in thick- and very thick-bedded sandstones and conglomerates of the Godula Beds of the Silesian Unit. The strata are slightly inclined towards the south and south-west, thus similarly to the slope inclination (Fig. 19.4c) (Margielewski and Urban 2003).

The main part of the cave comprises a system of passages, which represent opened and widened sections of joints that belong to two crossing joint sets (Fig. 19.4a, b, d). The "zigzag system" of such high, straight and narrow passages is particularly regular in the central and western parts of the cave (Fig. 19.4g). Comparison of structures on both upslope and downslope cave walls, as well as the lack of slickensides on their surfaces, indicates that the main movement producing these passages was simple extension, with participation of slight vertical translation (ca. 10–30 cm). Consequently, dilation was the principal process responsible for cave formation (Margielewski and Urban 2003).

The cave formation took place within the slope not deformed by a landslide. However, the cave shape, including the direction of its central section parallel to the slope contours and downslope directions of its western and eastern sections, resembles a ground plan typically followed by the main scarp of a landslide (Fig. 19.4e, f). Therefore, it is interpreted as an extensional feature developed due to stress relaxation along the structural discontinuities of a rock massif antecedent to the main stage of deformation. Such suggestion is confirmed by geoelectrical (ERT) images, showing various zones of relaxation, also inaccessible one in the vicinity of the cave (Margielewski and Urban 2003; Pánek et al. 2010).

The Jaskinia Malinowska cave, protected as nature monument, is now the only cave in the Beskidy Mts. equipped with specific facilities for touristic exploration, namely ladders in the entrance shaft.

19.4.2 Złotopieńska Dziura—Intermediate, Dilational Cave

The Złotopieńska Dziura cave, 105 m long and 10 m deep, is situated in the Beskid Wyspowy Mts. (Fig. 19.1b, site 2). The cave is located in the northern slope of Mt. Łopień, within the long (ca 400 m) and relatively high (up to 20 m) landslide main scarp, called Czartorysk (Fig. 19.5). Within this scarp zone (also upslope but close to it), 12 caves have been found and explored by cavers from the Limanowa Speleoclub since 1995 (Gubała and Kapturkiewicz 2018). The whole landform assemblage is formed in thick-bedded sandstones of the Magura Beds, Magura Unit. The entrance

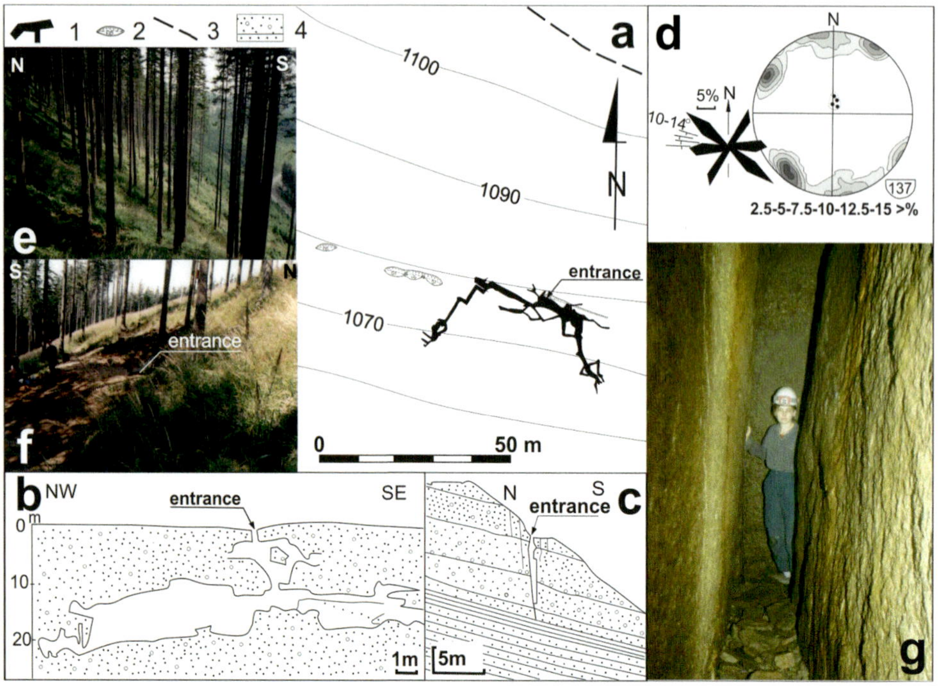

Fig. 19.4 Jaskinia Malinowska cave. **a** Map (cave contours after Rachwaniec and Holek in: Pulina 1997); explanation of symbols: 1—cave galleries, 2—surface depressions, 3—ridge path (touristic trail), 4—sandstones and conglomerates (on cross sections). **b** Longitudinal cross section of the cave (simplified after Ganszer in: Pulina 1997). **c** Conceptual model of transversal cross section of the cave and geological structure (after Margielewski and Urban 2003, modified). **d** Joints and spatial position of strata on the upslope walls in the central and western parts of the cave: joints in rose and contour diagrams—equal area plot, projection of poles in the lower hemisphere, with the number of measurements (137) and contour intervals (5%); bedding plane in the pole point diagram. **e** Slope below the cave entrance. **f** Cave entrance. **g** Passage in the western part of the cave (photographs by W. Margielewski)

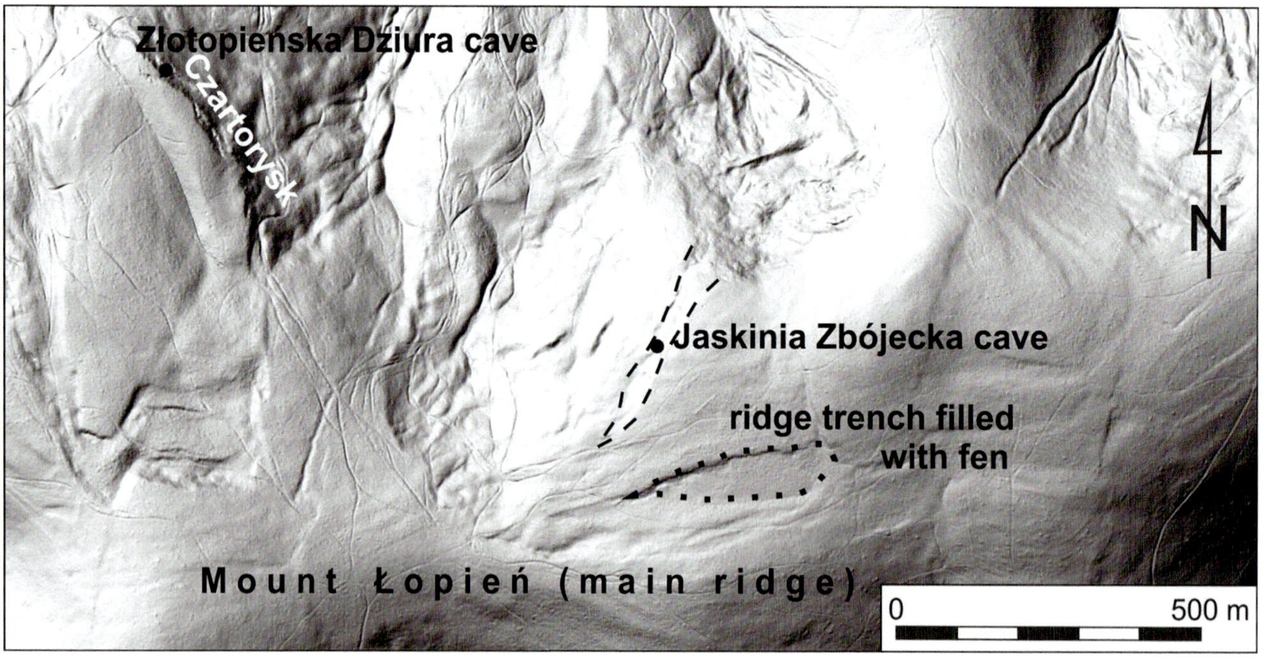

Fig. 19.5 Position of the Jaskinia Złotopieńska cave and the Jaskinia Zbójecka cave in the gravitationally deformed northern slope of Mt. Łopień (on the background of a digital terrain model downloaded from the public domain, www.geoportal.gov.pl)

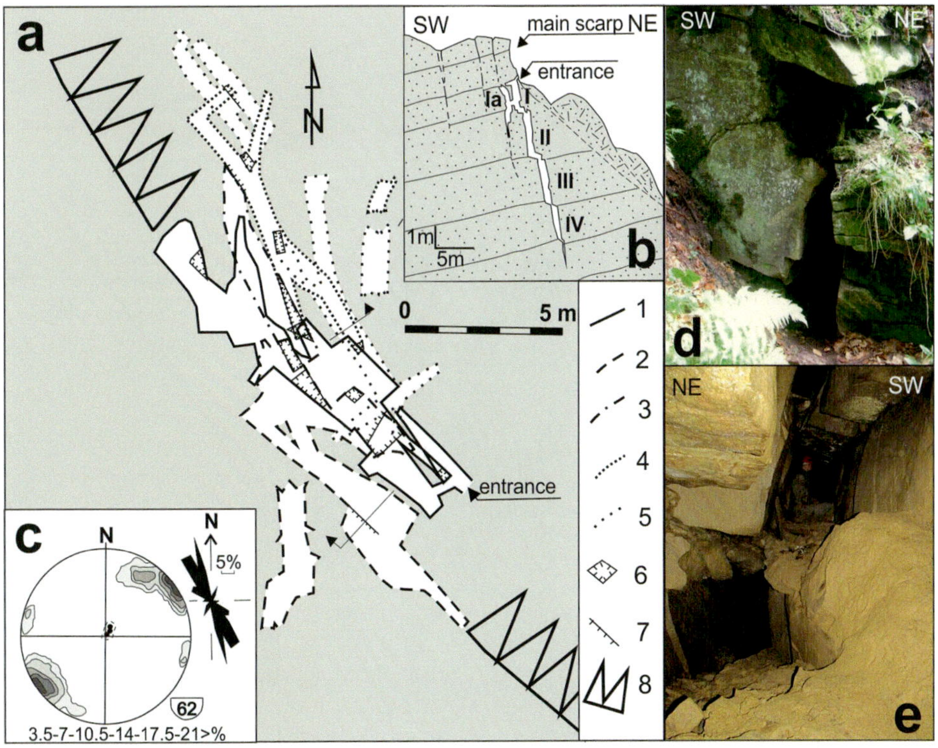

Fig. 19.6 Złotopieńska Dziura cave. **a** Map of cave (after Mleczek in: Pulina 1997); explanation of symbols: 1 to 5—contours of the cave passages (1—level I, 2—level Ia, 3—level II, 4—level III, 5—level IV), 6—cave shaft or chimney, 7—scarp within the cave, 8—landslide main scarp. **b** Conceptual model of transversal cross section of the cave and geological structure (after Margielewski and Urban 2003); **c** Joints and spatial positions of strata of the upslope passage walls in the cave (after Margielewski and Urban 2003); for further explanation see Fig. 19.4. **d** Cave entrance within the main scarp; its shape indicates the share of backward rotation in the cave development. **e** Two levels of passages: lower passage is shifted towards slope dip (photographs by J. Urban)

of the Złotopieńska Dziura cave is situated in the middle part of the vertical scarp, whereas its passages form four levels developed below it, downslope. They are located under one another, but slightly shifted towards the slope dip (Fig. 19.6a, b, d, e). The uppermost level is situated within the more disintegrated part of rock massif, whereas lower levels consist of passages up to 15 m long and 0.4–1.0 m wide, fairly regular in cross section. These passages originated due to the widening of the joints—the same ones which shape principal faces of the main scarp (Fig. 19.6a, c). However, some walls, especially in the deeper levels, are slightly bent (with concave and convex walls), which suggests their formation along joints induced during gravitational and dilational movements (Margielewski and Urban 2003).

The observations of structures on the walls (lack of slickensides, etc.) and their spatial orientation indicate that the main process forming the cave passages was an almost horizontal translation, with backward rotation in the uppermost parts and slight (up to 20 cm) vertical shift downslope. Consequently, it can be postulated that the landslide developed owing to spreading and subsequent massif disintegration into rock packets currently constituting the landslide body and colluvial swell, whereas the Złotopieńska Dziura cave, being part of the main detachment zone and its prolongation into the slope massif, experienced only spreading and hence dilation (Margielewski and Urban 2003).

The whole landslide zone of the Czartorysk is protected within the Special Area of Conservation Nature 2000 called Uroczysko Łopień.

19.4.3 Jaskinia Miecharska Cave—Subsequent and Dilatancy Cave

The Jaskinia Miecharska cave is situated in the Beskid Śląski Mts (Fig. 19.1b, site 3) and formed in the Upper Godula Beds (Upper Cretaceous), Silesian Unit, that are composed of thick-bedded sandstones and conglomerates intercalated with shales and thin-bedded sandstones (Burtan 1973).

The cave, 1838 m long, has formed in the northern slope of the Malinka Stream valley (tributary of the Vistula River). The vertical difference between its highest and lowest parts reaches 55 m (Szura and Juroszek 2016). The cave is composed of a maze system of galleries and chambers, 0.5–10 m high and 0.4–8.0 m wide, formed along joints (Fig. 19.7a, b). The central section of the cave comprises a regular network of passages, whilst other parts are composed of irregular caverns developed along highly disintegrated and displaced rock blocks, often rotated horizontally and vertically (Margielewski and Urban 2017).

The cave was formed within the body of large landslide of complex type, developed in a valley side setting. Therefore, the cave represents a subsequent cave in the morphogenetic classification. The cave system formed along the shear (slip) zone of the landslide consequent to the strata dip, ca. 10–12 m below the ground surface (see Fig. 19.7b), due to the opening of cracks along joints, during gravitational displacement, so the cave is of dilatancy type in the geomechanical classification. In the cave tectonic deformations typical for the shear zone have been

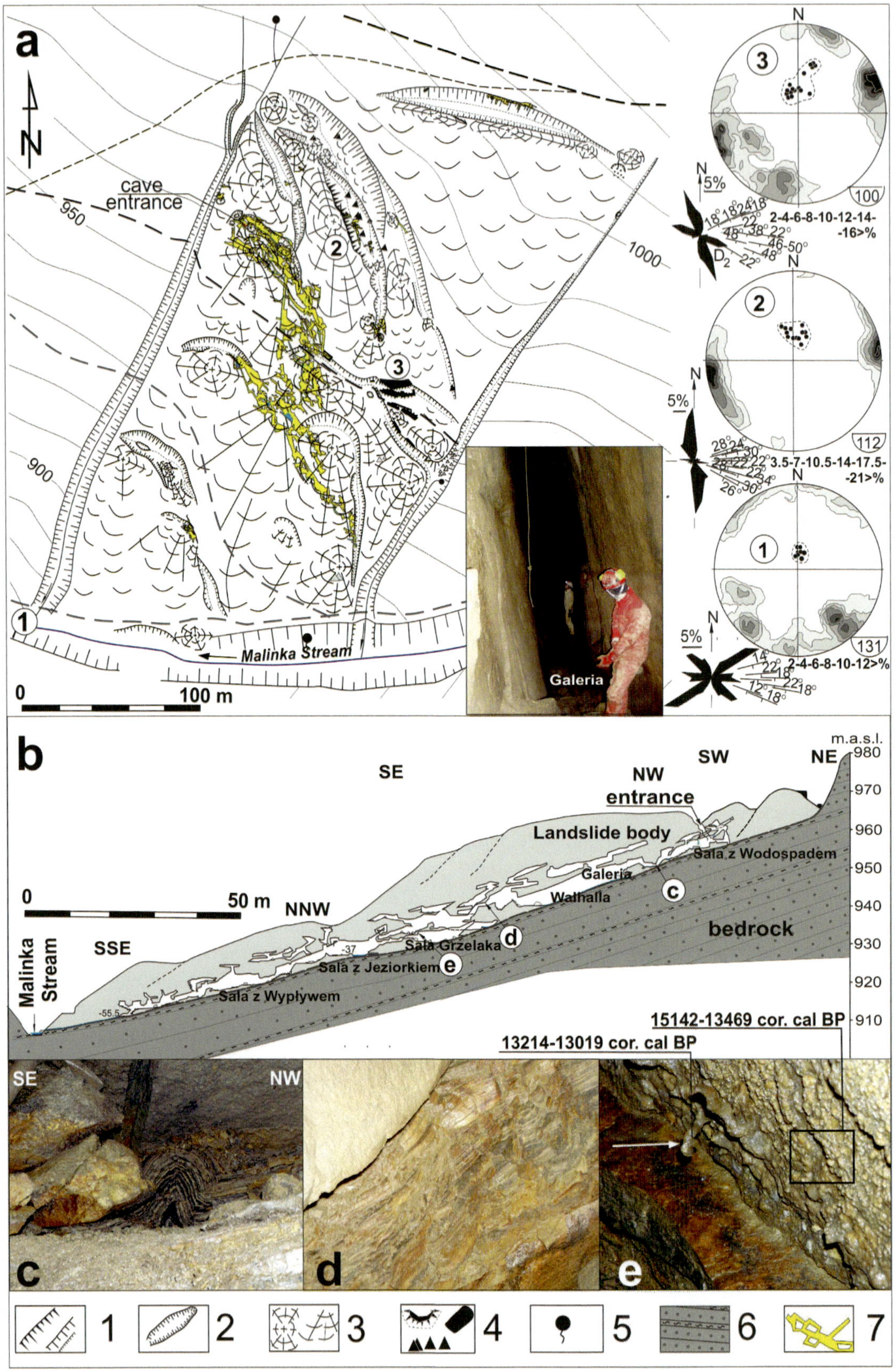
a
cave
entrance
950
1000
900
Malinka Stream
0
100 m
Galeria
5%
2-4-6-8-10-12-14-
-16>%
3.5-7-10.5-14-17.5-
-21>%
2-4-6-8-10-12>%
100
112
131
b
m.a.s.l.
SE
NW
SW
NE
entrance
Landslide body
Sala z Wodospadem
Galeria
Walhalla
bedrock
0
50 m
NNW
SSE
Malinka Stream
Sala Grzelaka
Sala z Jeziorkiem
Sala z Wypływem
15142-13469 cor. cal BP
13214-13019 cor. cal BP
980
970
960
950
940
930
920
910
c
d
e
1
2
3
4
5
6
7

Fig. 19.7 Jaskinia Miecharska cave (after Margielewski and Urban 2017, modified). **a** Map of the landslide with plan view of the cave (marked in yellow colour) and photograph of cave main gallery; joints of rocks outcropping in the landslide (localities 1, 2, 3) area are presented on rose and contour diagrams (for explanation see Fig. 19.4). **b** Longitudinal cross section of the cave and its position along the slip surface of landslide, with location of photographs **c**–**e**. Phenomena related to the cave formation: **c** Shearing fold. **d** Tectonic breccia. **e** Speleothems (helictite) dated by radiocarbon method. Explanation of symbols: 1—scarps, 2—trenches and depressions, 3—landslide body, 4—crags; 5—springs, 6—sandstones and shales on cross section, 7—cave outline (photographs by W. Margielewski)

recognized. They include both plastic (shear folds in claystones) and brittle-type deformations (tectonic breccias) (Fig. 19.7c, d). The geoelectrical survey (ERT) of the landslide recorded other cavities (cave systems) formed due to macro-fissuring process along the shear zone (Pánek et al. 2010), later discovered by cavers.

In the Jaskinia Miecharska cave, various speleothems occur such as carbonate stalactites and unique helictites (up to 4 cm long; Fig. 19.7e), small stalactites (2–4 cm long), and flowstones formed of siliceous–argillaceous mineraloids and organic substance (encrusted by mineraloid material) (Urban et al. 2007, 2015). Radiocarbon datings (regarding a reservoir effect) indicate that the helictites were formed in the Late Glacial (ca. 13.21–13.02 ka corrected cal BP and 15.14–13.47 ka corrected cal BP). The younger carbonate and organic stalactites were dated at ca 8.44–8.37 ka corrected cal BP; 7.51–6.48 ka corrected cal BP; 7.67–7.28 corrected cal. BP) (Margielewski and Urban 2017).

Another example of a dilatancy cave is Jaskinia Wiślańska cave—the longest cave (2275 m) discovered so far in the Polish Beskidy Mts., situated close to the Jaskinia Miecharska cave, in the Beskid Śląski Mts. (Fig. 19.1b, site 4) (Margielewski and Urban 2017).

19.4.4 Jaskinia Zbójecka na Łopieniu—Subsequent, Boulder Cave

The Jaskinia Zbójecka na Łopieniu cave, 433 m long and 20 m deep, is situated in the Beskid Wyspowy Mts (Fig. 19.1b, site 5). It is formed in thick-bedded sandstones of the Magura Beds, Magura Unit, and located within the northern slope of Mt. Łopień, close to its northern culmination (Fig. 19.5). The whole southern part of Mt. Łopień massif is probably significantly deformed by spreading—gravitationally induced extensional movement which broke and detached this part of rock massif from the main ridge. This is evidenced by a ridge depression situated between the southern and northern parts of the massif, which is currently filled by a fen and periodically overflowed (outlined by dotted line on Fig. 19.5). The fen, up to 2.5 m thick, is composed mainly by sedge peat covered by a thin layer of *Sphagnum* peat The datings of organic matter at the bottom of the fen indicate that its accumulation started at ca. 5.8 ka cal BP, at the beginning of the Subboreal phase of the Holocene, so the movements deforming the mountain massif must be of similar or older age than this date (Margielewski 2006b).

The large-scale gravitational deformation of Mt. Łopień rock massif encompassed its highest, relatively flattened and extensive parts, along with the southern culmination. It did not bring about the formation of large and distinct scarps in these parts (which are present in the lower, steeper slopes), but significantly disintegrated the structure of the rock massif, especially in the most deformed zones. This is perfectly visible in the area above the Jaskinia Zbójecka cave and its vicinity. The ground surrounding the main cave entrance is uneven and formed of numerous rock boulders, but it is not very steep and no evident scarp occurs. However, on the digital terrain model, the characteristic zone of SSW–NNE direction can be recognized (outlined by narrow dashed line on Fig. 19.5). It is most likely a tectonic discontinuity significantly deformed due to the gravitational movements. This rock massif failure is evidenced by the nature and shape of the Jaskinia Zbójecka cave, which developed in this zone. The cave is a system composed of several large chambers of very various shapes and sizes, ranging 3–6 m in horizontal plan and 2–10 m in vertical dimensions. These chambers are connected by narrow and very irregular (horizontal, vertical and diagonal) passages that form a very dense, complex labyrinthine network of voids often overlapping one another (Fig. 19.8a–d). This is why the area covering the cave, which is more than 400 m long, is ten times smaller, ranging only 40 m × 15 m. It is not possible to find any predominating or regular direction in the elongation of cavities and their walls related to a regular joint set. However, the elongation of the general cave contours is in accordance with the strike of the tectonic deformation zone visible on the digital terrain model (Figs. 19.5 and 19.8). The formation of such a cave, as a set of large chambers, was possible owing to the formation of very large rock blocks ranging the length of 10 m and more, which were involved in movements of different types, including rotation, translation and their combinations. In this sense, the Jaskinia Zbójecka cave is a boulder-type cave according to the geomechanic criterion, but it is very specific, related to strongly disintegrated landslide body rather than to classic colluvium *sensu* Dikau et al. (1996).

Another interesting geological feature of the Jaskinia Zbójecka cave is non-calcareous stalactites, tooth-like and ribbed sheetings, as well as small calcite coralloids. These last ones were dated by radiocarbon methods at 2.68–2.35 ka corrected cal BP (Margielewski and Urban 2017).

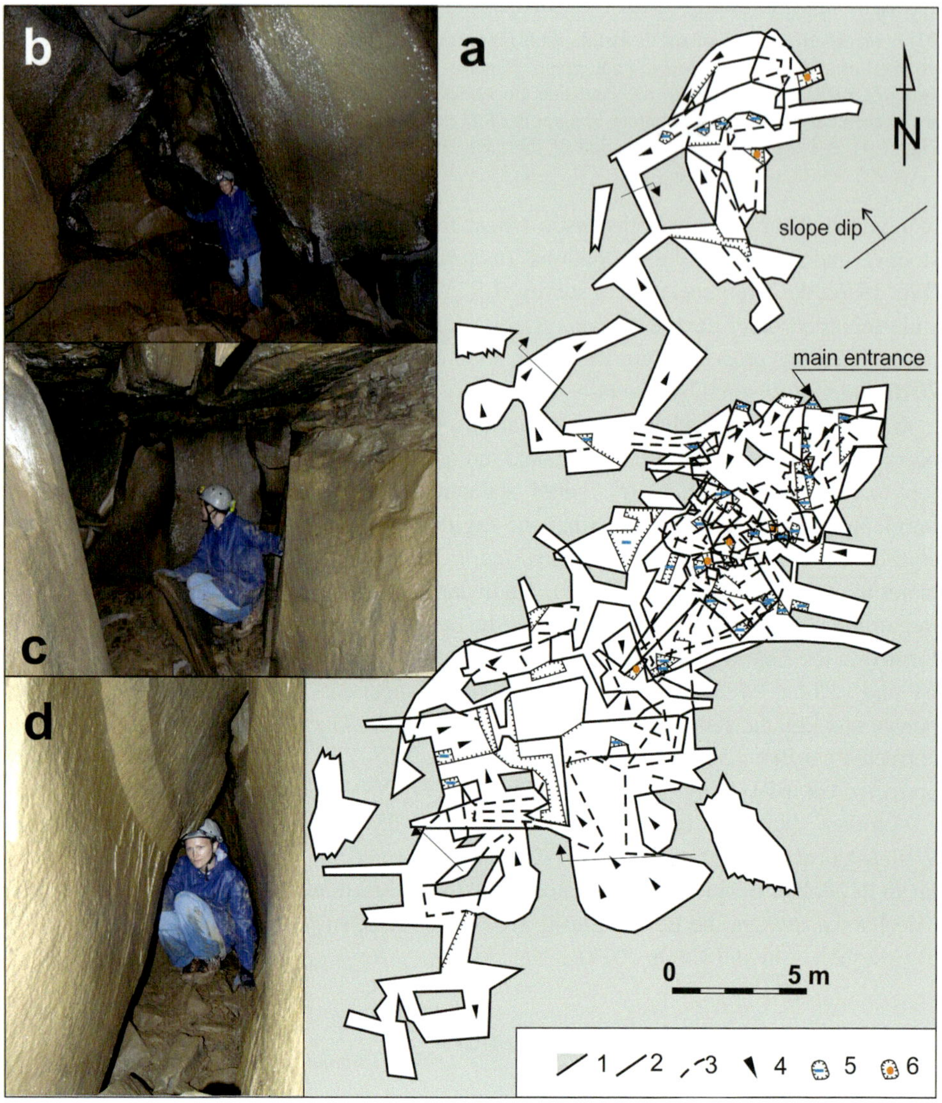

Fig. 19.8 Jaskinia Zbójecka cave. **a** Map (after Mleczek: in Pulina 1997, modified); explanation of symbols: 1—main outline of the cave, 2—contour of passage/chamber of the highest level, 3—contour of passage/chamber overlapped by higher cavities, 4—cave floor inclination, 5—shaft, 6—chimney. **b** Janosik Chamber (4 m long, 3 m wide and up to 10 m high): irregularly situated large boulders (rotated and fallen) form walls and ceiling of this chamber. **c** Rondo Chamber: cavity with relatively regular ceiling related to claystone layer, but with walls formed along irregular, uneven joints induced during the massif disintegration. **d** Łoś Passage: shell-like, concave and convex walls were formed along joints induced during disintegration of the massif, but the hackle marks indicate previous, tectonic stresses recorded in the rock structure (photographs by J. Urban)

The cave has been known for a long time (named also Jaskinia w Łopieniu), which is documented by legends and inscription inside the cave, but its entrance was forgotten in the twentieth century and it was repeatedly explored and documented in 1997–1998 (Gubała and Kapturkiewicz 2018). The cave is the greatest known bat hibernaculum (of mainly *Rhinolophus hipposideros*) amongst the caves of the Polish Outer Carpathians (Gubała and Piksa 2009). It is legally protected as a nature monument and situated within the Special Area of Conservation Nature 2000 called Bagna Łopieńskie.

19.5 The Age and Circumstances of Formation and Evolution of Caves

The development of caves within the mountain slopes of the Outer Carpathians has been connected with the same gravitational rock mass movements which were responsible for slope evolution since the permafrost deterioration in the Late Glacial. Radiocarbon dates of calcareous and organic speleothems indicate that the beginning of formation of some caves took place during the Late Glacial. The subsequent cave development and transformation, marked as the change from concentric to eccentric speleothem growth caused by rotation of rock blocks, occurred in the periods of increasing in climate humidity during the Holocene, particularly in the Preboreal phase, at the transition from the Boreal phase to the Atlantic phase, during the climatic optimum in the decline of the Atlantic phase, at the beginning of the Subboreal phase, and at the beginning of the Subatlantic phase, as well as in the Little Ice Age (Fig. 19.9a) (Urban et al. 2015; Margielewski and Urban 2017). During these phases of more humid climate, an increase in the intensity of mass movements is observed in the Outer Carpathians (Alexandrowicz 1997; Margielewski 2006b, 2018; Pánek et al. 2013; Starkel et al. 2013).

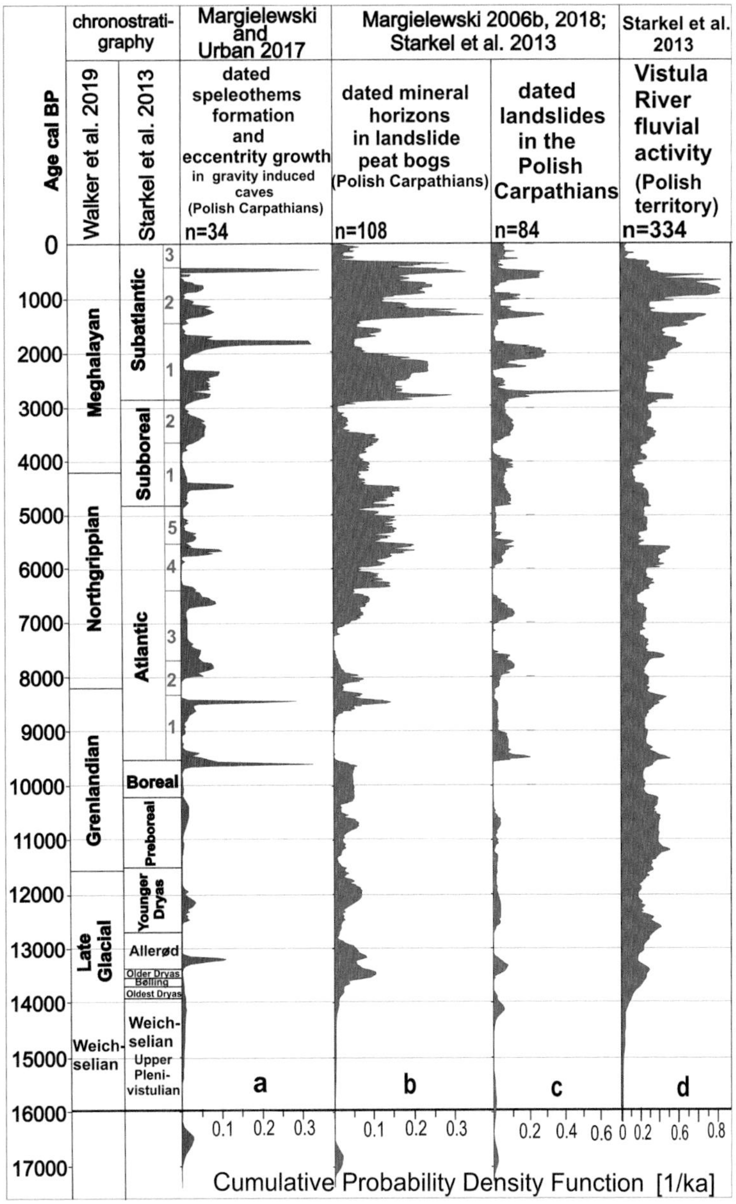

Fig. 19.9 Time range of the formation and transformation of non-karst caves in the Polish Carpathians, on the basis of distribution of radiocarbon dates of non-karst cave speleothems on the cumulative probability density function (column **a**) and correlation with the activity of mass movements (columns **b**, **c**) (after Margielewski and Urban 2017; Margielewski 2018, supplemented), compared with Late Glacial–Holocene phases of climate humidity growth expressed as Upper Vistula River activity (column **d**) (after Starkel et al. 2013). Holocene chronostratigraphy after Starkel et al. (2013) and Walker et al. (2019). Late Glacial chronostratigraphy after Litt et al. (2001). Cumulative probability density function shows the distribution (concentration) of calibrated radiocarbon dates (see Sokal and Rohlf 1981; Michczyńska and Pazdur 2004)

Confirmed by radiocarbon dates, the long-term existence of cracks/caves indicates a long-lasting, continuous evolution of mountain slopes of the flysch Carpathians since the Late Glacial. The system of widened cracks (caves) has existed for thousands of years, which means that the rock massifs, cut by cracks, are in a state of permanent gravitational stress. During more humid phases of the Late Glacial and Holocene, characterized by heavy rainfalls, continuous rains and intensification of river erosion, strong intensification of mass movements in the Carpathians resulted in the finalization of disintegration of rock massifs, which had been long prepared due to crack propagation in response to the gravitational stress. Consequently, the phases of intensification of mass movements (Fig. 19.9b, c) are well correlated with the phases of the increase in fluvial activity of rivers which reflected climate humidity growth (Fig. 19.9d) (Starkel et al. 2013).

19.6 The Value of Caves, Their Protection and Accessibility

The development of gravitationally induced caves has been closely connected with the evolution of slopes in the Outer Carpathians since the Late Glacial. Therefore, scientific studies of these caves, involving observations of their shapes, spatial positions within the slopes/rock massifs, structures outcropped in the caves, ages of cave formations and deformations, provide basic data about the nature of these slopes. Consequently, studies of non-karst caves have been performed in the Outer Carpathians, not only in the Polish segment, for at least two decades (e.g. Margielewski and Urban 2003, 2017; Lenart et al. 2014; Urban et al. 2015; Lenart and Miklin 2016), in order to better understand processes responsible for the morphogenesis of this mountain range. These caves also have significant educational significance as objects illustrating the internal structure of slopes in the Outer Carpathians and processes that have been shaping these slopes. However, the majority of them is hardly accessible for tourists and general public. Firstly, there are natural reasons of this inaccessibility such as difficult exploration of dark, "dirty", and narrow or vertical passages, but the second reason, which recently has limited access to many caves, is closures of entrances to several tens of long caves by steel bars, installed by forestry services in order to protect bat hibernacula and colonies, although the validity and effectiveness of this activity is questioned by some speleologists and naturalists.

The only cave which is intentionally left as accessible for the public and adequately prepared (equipped with ladders) is the Jaskinia Malinowska cave (see Sect. 19.4.1) in the Beskid Śląski Mts. Moreover, descriptions of several other caves situated close to touristic trails and easy to penetrate were published in touristic guidebooks (e.g. Zimna Dziura w Szczeblu—Fig. 19.1b—site 6).

The state of legal protection of caves in Polish part of the Outer Carpathians is not sufficient. Some 20% of these caves, representing about 10% of their total length, are protected in national parks and nature reserves, whereas several per cent of them, but ranging some 25% of their total length, have been declared as nature monuments. Amongst them are four longest caves, exceeding the length of 1 km. Much more caves are situated in landscape parks and special areas of conservation Natura 2000. Nevertheless, the relevance of these caves for abiotic and biotic nature as well as their importance for scientific research has been increasingly realized, and the public attitude to them is changing. This is due to the general increase in environmental awareness in the society related to the development of media, especially the internet, intensively used by groups of naturalists, as well as touristic and caving clubs.

References

Alexandrowicz SW (1997) Holocene dated landslides in the Polish Carpathians. In: Frenzel B (ed) Rapid mass movement as a source of climatic evidences for the Holocene. Palaeoclim Res 19:75–83

Bella P, Gaál L (2011) Terminology and genetic types of boulder caves. Pseudokarst Comm Newsl 22:1–3

Bella P, Gaál L (2013) Genetic types of non-solution caves. In: Filippi M, Bosák P (eds) Proceedings of 16th international congress of speleology, vol 13, Brno, pp 237–242

Burtan J (1973) Szczegółowa mapa geologiczna Polski 1:50,000, arkusz Wisła. Wyd Geol, Warszawa

Dikau R, Brunsden D, Schrott L, Ibsen M-L (eds) (1996) Landslide recognition. Identification, movement and causes. Wiley, Chichester, p 251

Gądek P (2021) Exploration and investigation of the caves in the Polish Flysch Carpathians (August 2020-July 2021). In: Urban J (ed) Materiały 55. Sympozjum Speleologicznego, Bartkowa 11–14.10.2018. Sekcja Speleologiczna PTP im Kopernika, Kraków, pp 47–48

Gubała W, Kapturkiewicz A (2018) History of cave exploration in the Beskid Wyspowy Mountains, Polish Outer Carpathians. Pseudokarst Comm Newsl 28:18–24

Gubała W, Piksa K (2009) The bats of the flysch zone of the Polish Carpathians. Pseudokarst Comm Newsl 19:9–12

Janiga S (1974) Uwagi o genezie jaskiń z okolic Dukli. Czas Geogr 45:355–360

Jankowski L, Margielewski W (2021) Geological control of young orogenic mountain morphology: from geomorphological analysis to reinterpretation of geology of the Outer Western Carpathians. Geomorphology 386:107749

Klassek G, Mikuszewski J (1997) Przyrodnicza charakterystyka jaskiń, warunków ich występowania i rozwoju w obszarze polskich Karpat fliszowych. In: Pulina M (ed) Jaskinie polskich Karpat fliszowych, vol 1. Pol Tow Przyjaciół Nauk o Ziemi, Warszawa, pp 7–18

Klassek G, Mleczek T (2008) Exploration of caves in the Polish Outer Carpathians in the past and nowadays. Zacisk—special issue (9th Intern Symp on Pseudokarst). Speleoclub Bielsko-Biała, pp 4–6

Klassek G, Mleczek T (2018) Exploration and investigation of the caves in the Polish Flysch Carpathians (August 2017-July 2018). In: Czyżewski Ł, Sudoł-Procyk M (eds) Materiały 52. Sympozjum Speleologicznego, Toruń 11–14.10.2018. Sekcja Speleologiczna PTP im. Kopernika, Kraków, pp 48–51

Kranz RL, Scholz CH (1977) Critical dilatant volume of rocks at the onset of tertiary creep. J Geophys Res 82:4893–4898

Kwaśniewski M (1986) Dylatancja jako zwiastun zniszczenia skały. Cz. II. Mechanizm zjawisk poprzedzających zniszczenie. Przegl Górn 42(6):184–196

Kwaśniewski M (2007) Mechanical behavior of rocks under true triaxial compression conditions—volumetric strain and dilatancy. Arch Mining Sci 52:409–435

Lenart J, Miklin J (2016) Pseudokarst caves of the Outer Western Carpathians, Czechia. J Maps 13:37–46

Lenart J, Pánek T, Dušek R (2014) Genesis, types and evolution of crevice-type caves in the Flysch Belt of the Western Carpathians (Czech republic). Geomorphology 204:459–476

Lexa J, Bezák V, Elečko M, Mello J, Polák M, Potfaj M, Vozár J (eds) (2000) Geological map of Western Carpathians and adjacent areas 1:500,000. Geol Survey of Slovak Republic, Bratislava

Litt T, Brauer A, Goslar T, Merkt J, Bałaga K, Muller H, Ralska-Jasiewiczowa M, Stebich M, Negendank JF (2001) Correlation and synchronization of Late Glacial continental sequences in northern central Europe based on annually laminated lacustrine sediments. Quat Sci Rev 20:1233–1249

Margielewski W (2006a) Structural control and types of movements of rock mass in anisotropic rocks: case studies in the Polish Flysch Carpathians. Geomorphology 77:47–68

Margielewski W (2006b) Records of the Late Glacial-Holocene palaeoenvironmental changes in landslide forms and deposits of the Beskid Makowski and Beskid Wyspowy Mts. area (Polish Outer Carpathians). Folia Quaternaria 76:149

Margielewski W (2018) Landslide fens as a sensitive indicator of the palaeoenvironmental changes since the Late Glacial; Polish Western Carpathians case study. Radiocarbon 60:1199–1213

Margielewski W, Urban J (2002) Initiation of mass movement in the Polish Flysch Carpathians studied in the selected crevice type caves. In: Rybář J, Stemberk J, Wagner P (eds) Landslides. AA Balkema Publ, Lisse-Abingdon-Exton-Tokyo, pp 405–409

Margielewski W, Urban J (2003) Crevice-type caves as initial forms of rock landslide development in the Flysch Carpathians. Geomorphology 54:325–338
Margielewski W, Urban J (2017) Gravitationally induced non-karst caves: tectonic and morphological constrains, classification, and dating; Polish Flysch Carpathians case study. Geomorphology 296:160–181
Margielewski W, Urban J, Szura C (2007) Jaskinia Miecharska Cave (Beskid Śląski Mts., Outer Carpathians): case study of the crevice type cave developed on sliding surface. Nat Conserv 63(6):57–68
Michczyńska DJ, Pazdur A (2004) Shape analysis of cumulative probability density function of radiocarbon dades set in the study of climate change in the late glacial and Holocene. Radiocarbon 46:733–744
Neuendorf KKE, Mehl JP Jr, Jackson JA (2005) Glossary of Geology, 5th edn. Amer Geol Inst, Alexandria, Virginia, p 791
Pánek T, Margielewski W, Tábořík P, Urban J, Hradecký J, Szura C (2010) Gravitationally-induced caves and other discontinuities detected by 2D electrical resistivity tomography: case studies from the Polish Flysch Carpathians. Geomorphology 123:165–180
Pánek T, Smolková V, Hradecký J, Baroň I, Šilhán K (2013) Holocene reactivations of catastrophic complex flow-like landslides in the Flysch Carpathians (Czech Republic/Slovakia). Quat Res 80:33–46
Puchejda W (1989) The origin of the Cave in Trzy Kopce in the Silesian Beskidy Mts. Kras i Speleologia 6(XV):66–78
Pulina M (ed) (1997) Jaskinie polskich Karpat fliszowych, vol 1, 2. Polskie Tow. Przyjaciół Nauk o Ziemi, Warszawa, p 250 (vol 1), 228 (vol 2)
Rustan A, Cunningham C, Fourney W, Simha KRY, Spathis AT (eds) (2010) Mining and rock construction technology desk reference. Rock mechanics, drilling and blasting. CRC Press, Leiden, p 454
Sokal RR, Rohlf FJ (1981) Biometry. The principles and practice of statistics in biological research. Freeman & Co, New York, p 859
Starkel L (1972) An outline of the relief of the Polish Carpathians and its importance for human management. Probl Zagosp Ziem Górskich 10:75–150
Starkel L, Michczyńska DJ, Krąpiec M, Margielewski W, Nalepka D, Pazdur A (2013) Progress in the Holocene chrono-climatostratigraphy of Polish territory. Geochronometria 40:1–21
Szura C, Juroszek B (2016) Jaskinia Miecharska. In: Grodzicki J (ed) Jaskinie polskich Karpat fliszowych, Uzupełnienia, vol 1. Państwowy Instytut Geologiczny—Państwowy Instytut Badawczy, Warszawa, pp 212–224
Urban J, Margielewski W (2013) Types of non-karst caves in Polish Outer Carpathians—historical review and perspectives. In: Filippi M, Bosak P (eds) Proceedings of 16th international congress of Speleology, vol 13, Brno, pp 314–319
Urban J, Margielewski W, Schejbal-Chwastek M, Szura C (2007) Speleothems in some caves of the Beskidy Mts., Poland. Nature Conserv 63(6):109–117
Urban J, Margielewski W, Hercman H, Žák K, Zernitska V, Pawlak J, Schelbal-Chwastek M (2015) Dating speleothems in sandstone caves—methodological aspects and practical interpretation, Polish Outer Carpathians case study. Z Geomorph NF 59(Suppl 1):183–208
Vítěk J (1983) Classification of pseudokarst forms in Czechoslovakia. Int J Speleol 13:1–18
Walker M, Gibbard P, Head MJ, Berkelhammer M, Björck S, Cheng H, Cwynar LC, Fisher D, Gknis V, Long A, Lowe J, Newnham R, Rasmussen SO, Weiss H (2019) Formal subdivision of the Holocene series/epoch: a summary. J Geol Soc India 93:135–141
Żytko K, Zając R, Gucik S, Ryłko W, Oszczypko N, Garlicka I, Nemčok J, Eliáš M, Menčik E, Stránik Z (1989) Map of the tectonic elements of the Western Outer Carpathians and their foreland. PIG Warszawa, GUDS Bratislava, UUG Praha

Włodzimierz Margielewski is a geologist and geomorphologist, employed as Associate Professor in the Institute of Nature Conservation, Polish Academy of Sciences in Krakow, Poland. His research is focused on geology, geomorphology and geoconservation in the Carpathians and their foreland, with particular emphasis to structural geology, geological control of relief and wide aspects of mass movements. He also works on the reconstruction of palaeoenvironmental changes during the Late Glacial and Holocene in the Polish territory, based on studies of peatland deposits, including landslide fens (a type recognized by him).

Jan Urban is a geologist and geomorphologist, Associate Professor, employed first in the Świętokrzyskie Mountains Branch of the Polish Geological Institute in Kielce and then in the Institute of Nature Conservation, Polish Academy of Sciences in Krakow, where he recently retired. He deals with the theory and practice of geological heritage protection. The Świętokrzyskie (Holy Cross) Mountains, the Nida Basin (Ponidzie region) and the Outer Carpathians (Beskidy Mountains), situated in the central and southern Poland, are the main areas of his scientific interest. He deals with karst forms in carbonate and gypsum rocks, including caves, as well as palaeokarst forms. He is a co-author of cave inventories in the two first regions. He also studies sandstone crags and non-karst caves in these rocks as well as explores and documents historical mines and open-pits. He is a President of the Pseudokarst Commission of the International Union of Speleology and President of the Speleological Section of the Polish Copernicus Society of Naturalists.

Bieszczady Mts.—Structural Relief with Human Imprint in the Wilderness Corner of Poland

20

Anita Bernatek-Jakiel

Abstract

The Bieszczady Mts. in the south-eastern part of Poland are an excellent area to examine the structural relief of a flysch-built orogen. The main relief features reflect the tectonic evolution of the Carpathians and the lithology of flysch rocks, which consist of resistant sandstones and less resistant shales and mudstones. The general topography of parallel-oriented ridges characterized by the NW–SE direction was formed during the earliest compressional stage in the evolution of the Carpathians. Then, at the extensional stage when gravitational collapses took place and due to the secondary tectonic deformation, the rectangular river pattern was formed. The diversity of slope relief is enhanced by rock crags, blockfields, landslides, and soil piping forms, whereas in bedrock channels, one can find fluvial erosion forms such as bedrock steps, plunge pools, and potholes. The recent history of the Bieszczady Mts. is written in the landscape. The dense population that occupied this region since the fifteenth century was displaced after World War II. Deforested slopes and valleys are now subject to reforestation, but abandoned agricultural terraces and former roads are still well preserved. The structure-controlled landscape with an imprint of the turbulent history of the region makes the Bieszczady Mts. one of the most attractive areas not only in the Polish Carpathians, but in whole Poland.

Keywords

Carpathians · Flysch · Rock crags · Rectangular drainage pattern · Tectonic evolution · Cultural landscape

A. Bernatek-Jakiel (✉)
Institute of Geography and Spatial Management,
Jagiellonian University, Gronostajowa 7, 30-387 Kraków, Poland
e-mail: anita.bernatek@uj.edu.pl

20.1 Introduction

The Bieszczady Mts. are located in the extreme south-east of Poland (Fig. 20.1) and perceived as one of the wildest parts of the country. They represent medium-altitude mountains with a structure-controlled landscape where parallel-oriented ridges interrelate with a grid-like pattern of streams and rivers. Subalpine meadows called polonynas (Pol. *połoniny*) that are stretching above the upper tree line represent a unique landscape in Poland. The upper tree line is created by beech forests, and unlike the other Carpathian ranges (Beskidy Mts.), there is no subalpine spruce forest belt, making the landscape even more attractive. Although the Bieszczady Mts. are now sparsely populated, they were densely inhabited in the past, resulting in picturesque valleys where the villages were located. In many places, rock crags built of resistant sandstones stand out from the forestless ridges. Streams and rivers expose the rock beds. The main European watershed between the Baltic Sea and Black Sea basins runs across the Bieszczady Mts. All these features make this part of the Polish Carpathians fascinating and appealing to people. The most attractive part of the Bieszczady Mts. is protected within the Bieszczady National Park and the East Carpathian Biosphere Reserve, which is a transboundary protected area designated as an area of global importance under UNESCO's Programme "Man and Biosphere". Moreover, in 2021, ancient and primeval beech forests of the Bieszczady Mts. were inscribed on the UNESCO World Heritage List as a part of a serial pan-European property. In addition, in the northern part of the region, in the San River valley, an artificial lake called the Solina Lake was created by dam building and displays another interesting example of human–environmental interactions.

The principal aims of this chapter are to present the structure-controlled landscape of the area and to outline its geomorphic history that interplays with human activity. Over the years, people have been attracted to the unique

P. Migoń and K. Jancewicz (eds.), *Landscapes and Landforms of Poland*,
World Geomorphological Landscapes, https://doi.org/10.1007/978-3-031-45762-3_20

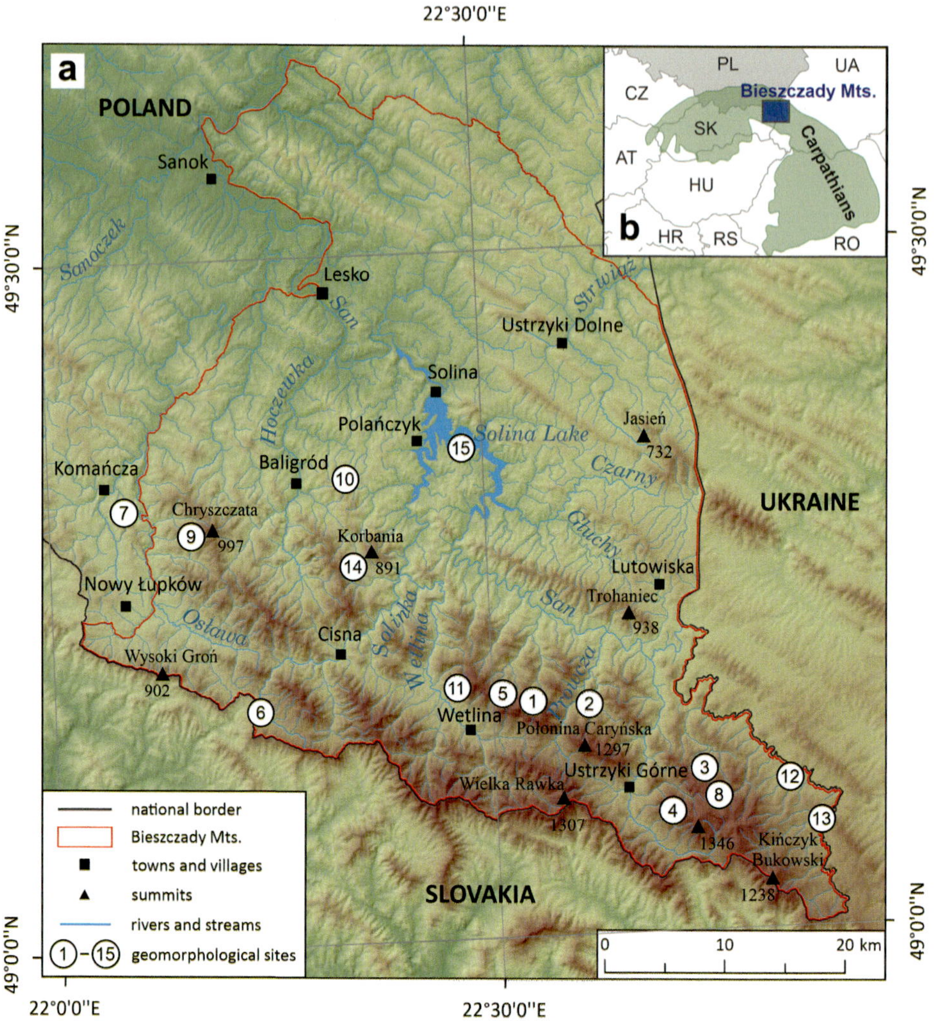

Fig. 20.1 Setting and location of the Bieszczady Mts. **a** General relief. **b** Location within the Carpathians. Numbers 1–15 indicate the location of sites mentioned in the text: 1—Połonina Wetlińska Ridge, 2—Połonina Caryńska Ridge, 3—Mt. Bukowe Berdo, 4—Mt. Szeroki Wierch, 5—Mt. Roh, 6—the place of possible capture of the Upper Solinka River, 7—the Osława River meander with Mt. Łokieć, 8—Mt. Krzemień, 9—Mt. Chryszczata, 10—Bereźnica Wyżna village, 11—the "Siklawa Ostrowskich" waterfall, 12—the former Tarnawa Wyżna village, 13—the former Beniowa village, 14—the former Tyskowa village, 15—Solina Lake. Source: Author's work based on LiDAR data and National Database of Topographic Objects (BDOT10k) obtained from the Head Office of Geodesy and Cartography, Poland (GUGiK)

landscape and wilderness of the Bieszczady Mts. and some of the mysteries hidden in these mountains will be revealed.

20.2 Setting and Location

The Bieszczady Mts. are part of the Eastern Carpathians. They run from the extreme south-east of Poland and north-east of Slovakia to western Ukraine, with the highest peak of Mt. Pikui (1405 m) in Ukraine (Fig. 20.1). They border against (Fig. 20.2):

- the Low Beskids (the Beskid Niski Mts.) in the north-west which constitute a narrow and less elevated part of the Carpathian belt (300–400 m lower than the neighbouring mountain ranges);
- the Mid-Beskidy Foothills (the Carpathian Foothills in general) in the north, which constitute the margin of the Outer Western Carpathians;
- the Polonynian Beskids in the east and south which are a part of the Outer Eastern Carpathians.

This chapter will focus on the Polish part of the Bieszczady Mts., with the highest peak of Mt. Tarnica (1346 m). In the geomorphological division of the Polish Carpathians (Klimaszewski 1972), the Bieszczady Mts. are divided into the lower part (the Low Bieszczady Mts.) representing low mountains and foothill relief with several ridges rising by 200–300 m above the surrounding area, and the higher part (the High Bieszczady Mts.) with several steep-sided ridges with relative altitudes up to 700 m, separated by intramontane depressions. The conventional border between these two units lies along the northern footslopes of Mt. Korbania and the long (approx. 18 km) and straight Otryt ridge, with the highest peak of Mt. Trohaniec (938 m; Figs. 20.1a and 20.2b). The widest and highest part is the Połoniny Range (Fig. 20.2a), with the most prominent Polish polonynas: the Wetlińska and Caryńska Połonina.

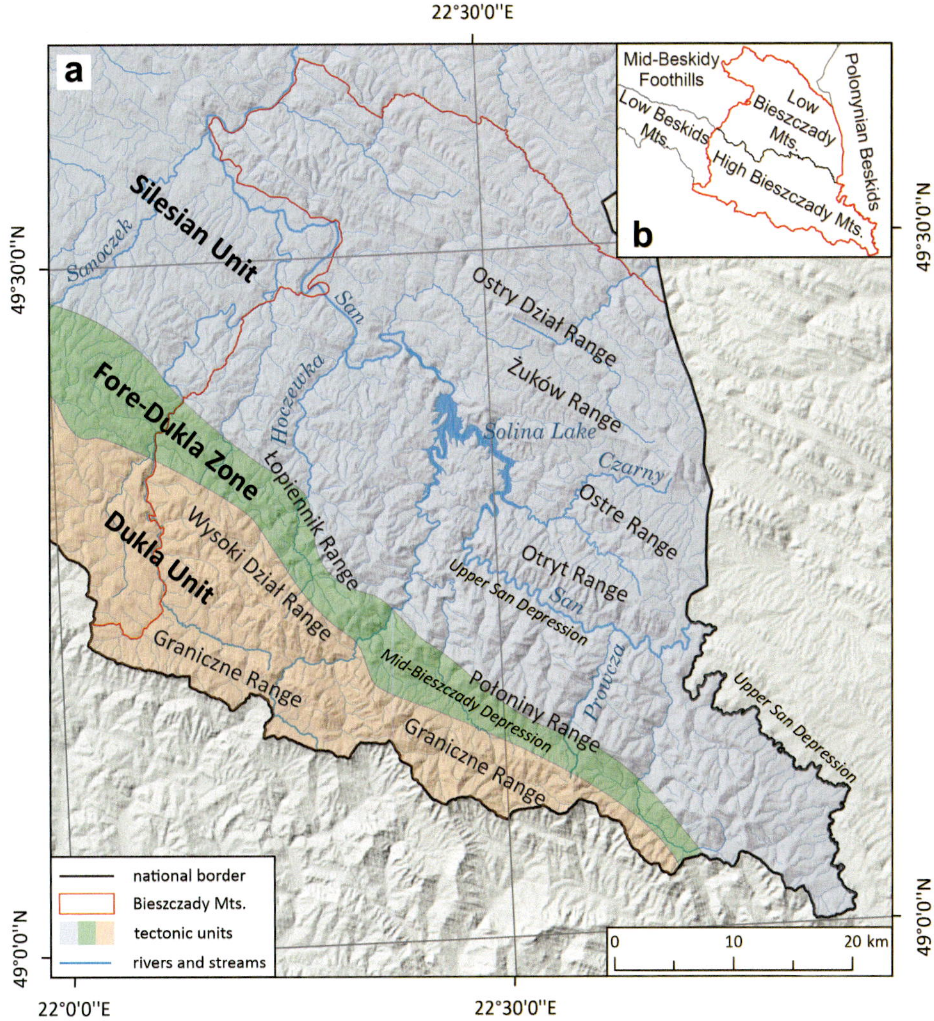

Fig. 20.2 Geological units in the Bieszczady Mts. **a** Parallel-oriented ridges of the Bieszczady Mts. **b** Regions bordering the Bieszczady Mts. in Poland. Source: Author's work based on Haczewski et al. (2007) and Solon et al. (2018)

20.3 Geology

Geologically, the Outer Carpathians, to which the Bieszczady Mts. belong, are the northernmost section of the Alpine mountain belt in Europe (Żytko et al. 1989; Nawrocki and Becker 2017). As a young, active orogen, the relief of the Carpathians generally reflects lithology and tectonics (Jankowski and Margielewski 2014, 2021), and this is especially well visible in the Bieszczady Mts. The landscape of this region bears an imprint of the tectonic evolution of the orogen and variable resistance of the rocks.

20.3.1 Lithological Diversity of Flysch Rocks

This region is built primarily of flysch rocks with occasional intercalations of marls and silicates (Oszczypko et al. 2006; Haczewski et al. 2007; Nawrocki and Becker 2017). The flysch consists of a sequence of thin- to thick-bedded sandstones interbedded with shales and mudstones. The diversity of flysch rocks is manifested both in the vertical profile (variable thickness of shale and sandstone layers) and in the horizontal profile (regional variability). Generally, the Bieszczady Mts. lie in the contact zone of two major tectonic units of the Outer Carpathians, i.e. the Dukla Unit and the Silesian Unit (Haczewski et al. 2007; Fig. 20.2a). The part of the Silesian Unit in front of the Dukla Unit, traditionally called the Fore-Dukla Zone, is an example of chaotic complexes developed as tectonic mélanges (Jankowski and Margielewski 2021).

The Silesian Unit builds the NE part of the region and consists mainly of the Krosno Beds of Oligocene–Lower Miocene age. In the High Bieszczady Mts., the middle section of the Krosno beds comprises of thick beds of the so-called Otryt sandstones, which are not present in the Low Bieszczady Mts. The Otryt sandstones are thick-bedded and represent one of the most resistant rocks in the region, which often build impressive rock crags protruding from the slopes (Haczewski et al. 2007).

The Fore-Dukla Zone extends up to 2 km in width and represents tectonic chaotic complexes built of different series of flysch rocks (Jankowski and Probulski 2011; Jankowski and Margielewski 2014, 2021). The oldest beds, the so-called Hieroglyphic Beds, comprise of thin- and thick-bedded sandstones with characteristic sedimentary structures, as well as feeding traces preserved on the bedding surface. Above them are poorly developed globigerine marls, and the Menilite Beds with black or brown shales intercalated with dark grey, non-calcareous sandstones and mudstones, occasionally with ferruginous dolomites (Haczewski et al. 2007).

In the Dukla Unit, the oldest rocks are flysch, dominated by dark grey, black, marly shales with a high content of calcium carbonate, the so-called Łupków Beds. Above them lie thick packages of thick- and very thick-bedded sandstones, i.e. the Cisna Beds, which are often exposed on the surface, locally building rock crags. The youngest rocks within this succession are the Hieroglyphic Beds (Haczewski et al. 2007).

Folded flysch series are overlain by various Quaternary sediments, which include slope deposits derived from the weathering of flysch rocks and associated with mass movements, fluvial deposits, and peat. The Bieszczady Mts. were not glaciated in the Pleistocene, but were influenced by periglacial conditions. The maximum extent of the Scandinavian ice sheet is recognized along the front of the Carpathian foothills, some 60–80 km to the north. Local glaciation developed in the most elevated parts of the Polish Carpathians (e.g. Tatra Mts. and Babia Góra massif, see Łajczak 2023 and Kłapyta and Zasadni 2023, this volume), but not in the Bieszczady Mts. The slope covers are mainly silty-clayey, which determined the dominant geomorphic processes such as mass movements, gully erosion, soil piping, and tree uprooting (Starkel 1965, 2006; Rączkowska et al. 2012). Steeper slopes are mantled by thin skeletal covers, which become thicker and less skeletal in the less inclined, lower parts. Thick solifluction materials formed in the periglacial environments during the Pleistocene occur in the footslope belt (Starkel 1965). Moreover, at the base of rock walls and rock crags, blockfields occur. Older fluvial deposits are coeval with the Pleistocene glaciations, whereas younger ones are related to the Holocene fluvial processes. Older gravels are overlain, with a sharp boundary, by an extensive cover of sand and silt enriched in wood debris, partially burned. This sediment seems to be a record of a relatively sudden onset of deforestation and cultivation of the valley bottom and slopes in the Upper San basin in the fifteenth and sixteenth centuries (Kukulak 2003). In the valleys of the Upper San River and its tributaries, there occur extensive peatbogs, whose accumulation is exceptional for the Carpathians. The peat cover has developed on fluvial terraces in the San and Wołosaty river valleys and on alluvial fans of their tributaries, as well as within landslide hollows on slopes (Haczewski et al. 2007; Łajczak, 2016).

20.3.2 The Impact of Sedimentary Basin Stage and Tectonic Evolution on the Landscape

So far, the relief development of the Carpathians, including the Bieszczady Mts., has been mainly explained as an interplay of variable rock resistance, tectonic structures (course of folds, overthrusts, and fault zones), and neotectonic activity (Starkel 1965, 1969; Zuchiewicz 2010; Rączkowska et al. 2012). Recent geological studies have suggested that the interpretation of relationships between the morphology of the Carpathians and the geological structure is frequently inconsistent and occasionally even contradicts geological conditions (Jankowski and Margielewski 2021). Therefore, a new concept of geological control in the Carpathians has been presented highlighting consecutive stages of the evolution of sedimentary basins, compression and formation of the orogen, and finally the stage of secondary tectonic deformations, such as gravitational collapses (Jankowski and Margielewski 2021).

The structure of sedimentary basins of the Carpathians was already influenced by orogenic movements that predated the Alpine orogeny. From a geomorphological point of view, an important issue related to the basin stage is the resistance variability of the sediments (Jankowski and Margielewski 2014, 2021), mainly related to the deposition of more resistant sandstones and less resistant shales and mudstones (Jankowski and Margielewski 2021).

When the sedimentary basin was closed and the orogen was formed, the rock complexes formed into tectonic units, then detached, and overthrusted on the foreland and one upon another. During this compression stage, the main geometric pattern of the Carpathian orogen was formed, reflected in the currently observed morphology. The emerging relief zonation was also observed in the Bieszczady Mts. (Fig. 20.1a). The highest structural and morphological position is occupied by tectonic elements that were first incorporated into the orogen (Jankowski and Margielewski 2021). In the Bieszczady Mts., they are represented by the Dukla Unit and the southern part of Silesian Unit (Fig. 20.2a). In turn, the lowest topographical position is occupied by the units that were attached to the orogen as the last ones. In the Bieszczady Mts., one can observe the highest mountain ridges in the SE part of the region, whereas the ridges are getting lower towards NW, and hence, the High and Low Bieszczady Mts. are distinguished (Figs. 20.1a and 20.2b). This suggests that for the general topography of the region, the structural position of orogen-forming units was more important than the diversity of rock resistance within these tectonic units (Jankowski and Margielewski 2021). Nevertheless, lithology has played a significant role in further relief transformation.

The next, extensional stage of the evolution of the mountain chain includes gravitational collapses and is also reflected in the present-day mountainous morphology, especially in the Bieszczady Mts. (Jankowski and Margielewski 2014). During this stage, some mountain ranges were disintegrated, and the river network was formed. In the Bieszczady Mts., the Dukla Unit, which was overthrusted on the Silesian Unit, moved back to the SW. As a result, the Połoniny Range (Mt. Połonina Wetlińska and Mt. Połonina Caryńska, see sites 1 and 2 in Fig. 20.1a) built of the Silesian Unit rocks was tectonically exhumed from the "overburden" of rock formations belonging to the Dukla Unit (Fig. 20.2a). Subsequently, the entire Połoniny Range disintegrated, which in some places resulted in the occurrence of double ridges in the top parts of the range. In the SW, the Graniczne Range was formed as a gravitationally displaced rock massif. At this stage, during the gravitational collapse of the Dukla Unit, the tectonic mélange zone was formed as a rock mass slippage zone, subsequently exploited by deeply incised valleys such as that of the Wetlinka River (Jankowski and Margielewski 2014, 2021).

The gravitational collapse stage has had a clear impact on the formation of river valleys. The rectangular river pattern in the Bieszczady Mts., as in the neighbouring Beskid Niski Mts. (Starkel 1972), seems to directly reflect the intersecting network of normal faults (generally of NW–SE direction) associated with perpendicular faults (Fig. 20.1a). The alternation of narrow and wide reaches of the valleys may be related to the fact that rivers use displacement zones associated with the gravitational collapse of the orogen. This is seen in the course of the Upper San River that mainly drains the Bieszczady Mts. Moreover, some smaller valleys in the Bieszczady Mts. have developed in the tectonic mélange zones associated with the gravitational collapse stage, e.g. the Wetlinka and Prowcza river valleys (Jankowski and Margielewski 2014, 2021).

After the gravitational collapse and subsequent erosion stage, which caused the removal of rocks, isostatic uplift of individual parts of the orogen occurred, influencing the landscape of the Bieszczady Mts. The Połoniny Range, tectonically exhumed from a significant rock overburden during the gravitational collapse stage, was elevated along normal faults. This is indicated by high triangular facets of an active fault-generated escarpment that borders the Połoniny Range from the south (Jankowski and Margielewski 2014, 2021).

20.4 Main Relief Features

20.4.1 Parallel-Oriented Ridges with Structural Slope Relief

The Bieszczady Mts. are characterized by structural relief (Starkel 1960, 1969; Henkiel 1977; Pękala 1997; Haczewski et al. 2007; Gorczyca et al. 2016). The main ridges are relatively narrow and elongated, running from NW towards SE (Figs. 20.1a and 20.2a). They have developed in the resistant Otryt and Cisna sandstones, whereas valleys are cut into less resistant rocks, predominantly shales and mudstones (Henkiel 1982; Haczewski et al. 2007). The height of the ridges increases towards the SE up to 1300 m asl, with the highest peak of Mt. Tarnica (1346 m). Also, the altitude of valley bottoms rises in the same direction. Relative altitudes are up to 700 m, resulting from considerable incision by rivers, facilitated by the intensive uplift of the Bieszczady Mts. (Haczewski et al. 2007; Zuchiewicz 2010).

The diverse structural conditions are clearly reflected in the landscape. In steeply sloping strata, narrow and rocky ridges are formed (e.g. Mt. Bukowe Berdo, see site 3 in Figs. 20.1a and 20.3a), whilst in less inclined beds, flattened and wide ridges occur (Mt. Szeroki Wierch, see site 4 in Figs. 20.1a and 20.3b). In both cases, stepped slope profiles have formed as a result of differential erosion of flysch rocks, highlighting more resistant sandstones against less resistant shales and mudstones (Baumgart-Kotarba 1974; Gorczyca et al. 2016). Locally, discrepancies between morphological and tectonic ridgelines are noted because of the non-uniform incision of streams and retreat of gully heads on both sides of the ridges. In some places, peatlands formed on slopes above the upper tree line (Ralska-Jasiewiczowa et al. 2006; Drewnik et al. 2015; Gorczyca et al. 2016).

One of the most impressive ridge-top landscapes, specific for the Bieszczady Mts., are trenches (so-called "double ridges"). These characteristic landforms, perfectly visible in the landscape, were formed by mass movements after gravitational collapse of the entire Połoniny Range (Alexandrowicz and Margielewski 2010; Jankowski and Margielewski 2021). They can be observed, amongst others, at Mt. Roh, the highest point of the Połonina Wetlińska ridge (see site 5 in Figs. 20.1a and 20.3c).

The shape, gradient, and fragmentation of slopes reflect the strike and dip of rocks and their resistance to weathering and denudational processes (Fig. 20.4). The slopes of main ridges in the Połoniny Range (the High Bieszczady Mts.) are benched (Fig. 20.4a) (Baumgart-Kotarba 1974), whereas slopes in the Graniczne Range and in the Low

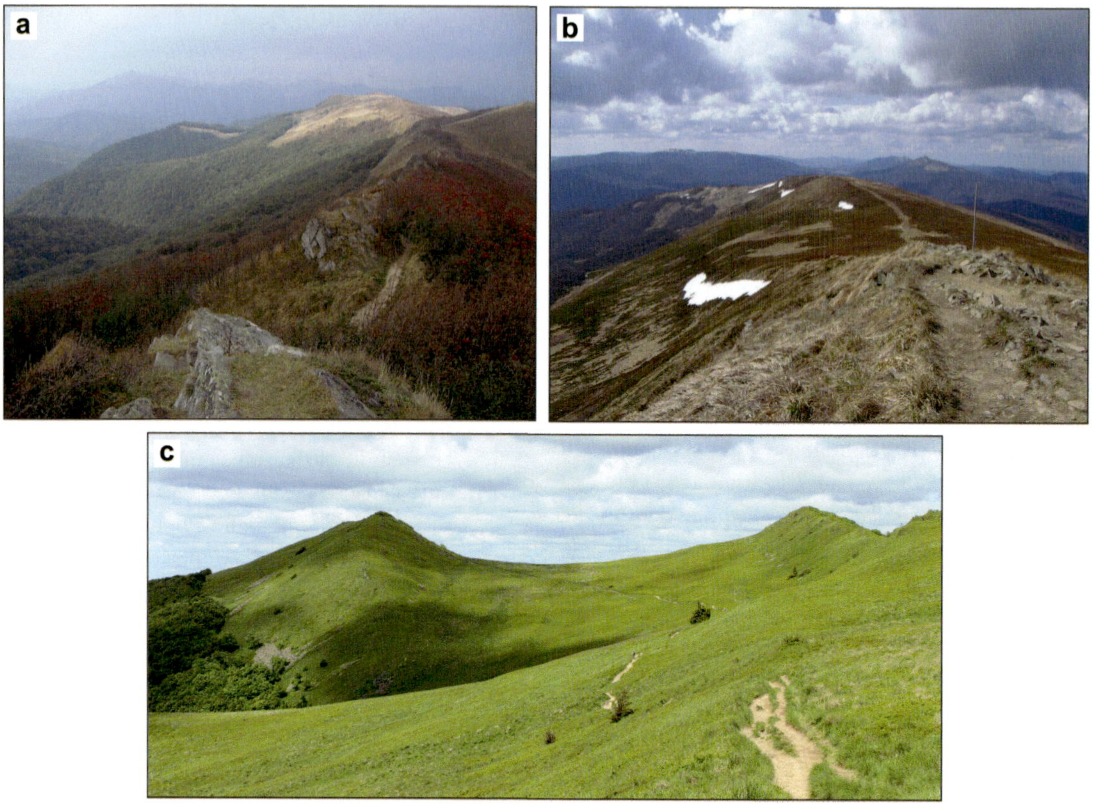

Fig. 20.3 Different types of ridges in the Bieszczady Mts. **a** narrow ridge of Mt. Bukowe Berdo. **b** Wide ridge of Mt. Szeroki Wierch. **c** Double ridge in the Połoniny Range, Mt. Roh (photographs **a**, **b** by A. Bernatek-Jakiel; **c** by M. Stolarczyk)

Fig. 20.4 The slopes of the Bieszczady Mts. **a** Stepped slope of the Połonina Caryńska Ridge. **b** Convex–concave slopes of Mt. Dział. **c** Stepped slope of Mt. Kińczyk Bukowski. Letter "A" indicates geomorphological levels (photographs by A. Bernatek-Jakiel)

Bieszczady Mts. are mainly convex–concave (Fig. 20.4b). On these slopes, a clearly developed edge between convex slope sections formed on more resistant sandstones and concave parts developed in less resistant shales and mudstones typically occurs. In the past, flattened parts of the slopes in the Bieszczady Mts. (here called geomorphological levels), and in the Polish Carpathians in general, were interpreted as relict planation surfaces, although recent geological studies have questioned this theory emphasizing the impact of tectonic evolution on the present-day morphology of the Carpathians instead (Zuchiewicz 2011; Jankowski and Margielewski 2021). Slope relief is further diversified by the occurrence of rock crags, blockfields, landforms connected with mass movements, and soil piping (Pękala 1997; Haczewski et al. 2007; Bernatek 2015), described below.

20.4.2 Grid-Like Pattern of River Valleys

The valley grid reflects the spatial pattern of the ridges, with the main directions of NW–SE and NE–SW (Fig. 20.1a). The streams and rivers flow according to the orientation of ridges along the main tectonic structures (NW–SE, Fig. 20.2a). Outflow towards the north takes place through infrequent gorge-like valley reaches that cut transversely bedrock structures, formed mainly along joints and faults of NE–SW orientation (Henkiel 1982).

Two main depressions in the Bieszczady Mts. associated with the valley network are the Mid-Bieszczady Depression (Pol. *Obniżenie Śródbieszczadzkie*) and the Upper San Depression (Pol. *Obniżenie Górnego Sanu*). The depressions are composed of wider basin-like parts separated by transverse hills and narrower gorge-like valleys. In the larger valleys, such as the San and Wetlinka rivers, meandering channels occur and systems of fluvial terraces can be recognized. The higher ones (20–35 m, 11–15 m, 6–10 m above the current riverbed) are associated with three glacial periods during the Pleistocene, during which Scandinavian ice sheets reached Poland. These were the Elsterian (South Polish, MIS 12), Saalian (Middle Polish, MIS 6), and Weichselian (North Polish, MIS 2) glaciations that correspond to Mindel, Riss, and Würm Glaciations respectively in the glacial chronology established for the Alps (note that the Bieszczady Mts. themselves were under periglacial conditions and never glaciated). The lower terraces (4–6 m, 2–4 m, 0.5–3 m) are connected with climate fluctuations and increased frequency of river floods in the Holocene (Haczewski et al. 2007). The lowest and youngest terraces are also associated with intensive human activity in the past (Kukulak 2004; Haczewski et al. 2007). From the first half of the fifteenth century up to the mid-twentieth century, the region was subjected to agricultural, forestry, and settlement activity, which resulted in intensive erosion on slopes and change of floodplain deposition from gravels to muds. The Pleistocene terraces are preserved in small fragments on valley sides as flattenings with remnants of river gravels, as well as on alluvial fans. The Holocene terraces are located in the valley bottom; locally overlain by peat, both raised (ombrogenic) and fens (minerogenic mires) (Kukulak 2004; Haczewski et al. 2007; Gorczyca et al. 2016). Recently, beaver activity has been observed in the valleys, which has resulted in the construction of woody debris dams up to 2 m high (Gorczyca et al. 2016).

The gorge-like fragments of valleys (mainly in SW–NE direction) are narrow and steep, with exposed bedrock in the riverbed, over which the river flows. In many places, there occur systems of bedrock steps in the riverbeds that are formed on resistant sandstones, often following NW–SE direction. Locally, they are called "*berdo*" (Gorczyca et al. 2016).

The side slopes of the main valleys are dissected by smaller streams characterized by V-shaped valleys. They are generally straight with steep slopes, narrow bedrock-cut floors and high stream gradient, locally with bedrock steps. Downstream, they are getting wider and form the flat-bottomed valleys.

In the future, an interesting geomorphological phenomenon may occur, i.e. stream capture (also called stream piracy), which takes place when a stream or river is diverted from its own bed and flows down towards the bed of a neighbouring stream. This will be even more fascinating, as it will have continental importance. The site of predicted capture is a place where the Upper Solinka River (now in the drainage basin of the Baltic Sea) flows very close to the continental divide between the Baltic Sea and the Black Sea, located along a ridge that rises only a few metres above the Solinka River bed. The Upper Solinka River will be captured southward by the Ondava River, itself a part of the Danube drainage basin that belongs to the Black Sea catchment, with a much lower base level (see site 6 in Figs. 20.1a and 20.5a).

On the western border of the Bieszczady Mts., another interesting example of a fluvial landform occurs—the meander bend of the Osława River near the village of Duszatyn (see site 7 in Figs. 20.1a and 20.5b). This is a unique meander loop, not typical for mountainous areas. It has the shape of an almost closed circle (the smallest distance between the river and the highest point of the bend and lowest one at the neck of the meander is only 120 m) and delimits a denudational spur called Mt. Łokieć (515 m). The characteristic shape of the Osława River meander and its denudational spur has been controlled by differences in bedrock lithology, attitude of the prevailing joint systems, and the occurrence of fault zones that border the area. Moreover, the final shape of the meander has been

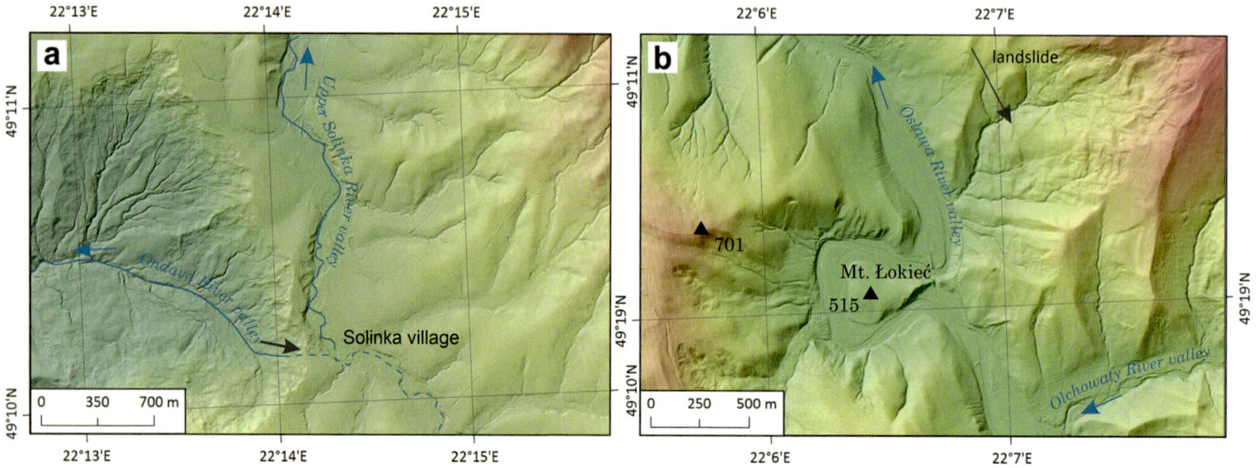

Fig. 20.5 Various geomorphological phenomena in the fluvial systems in the Bieszczady Mts. **a** Site of a possible stream capture of the Upper Solinka River by the Ondava River (arrow). **b** Meander of the Osława River surrounding Mt. Łokieć. Source: Author's work based on LiDAR data obtained from the Head Office of Geodesy and Cartography, Poland (GUGiK)

influenced by landslides, which periodically blocked the valley, thus affecting the fluvial system (Margielewski et al. 2002; Margielewski 2004). The area is protected within the Nature Reserve of the Osława River Meander.

20.5 Medium-Size and Minor Landforms

20.5.1 Rock Crags

Rock crags are characteristic elements of the landscape of the High Bieszczady Mts. They are built of the resistant Otryt sandstones. Their genesis is associated with denudational processes, slope retreat, and selective erosion of rocks characterized by variable resistance (Pękala 1997). Upper surfaces of the crags are inclined to the SW in accordance with the dip of flysch beds. They are often asymmetrical, with a rock cliff on one side and a flat, gently sloping surface on the other side, locally highly weathered and overgrown. Their diversity is expressed by different shapes and height; they may reach up to 15 m high (Fig. 20.6). In some places, they create impressive ridges, like in Mt. Krzemień (see site 8 in Figs. 20.1a and 20.6a). They are found in different topographical positions, i.e. on the ridges (Fig. 20.6a, b), slopes (Fig. 20.6c), and even within the valley sides and gully slopes (Fig. 20.6d; Haczewski et al. 2007; Gorczyca et al. 2016).

20.5.2 Blockfields

The occurrence of blockfields in the Bieszczady Mts. is strictly limited to the thick-bedded sandstones (i.e. Otryt sandstones) that belong to the Krosno Beds. These sandstones are strongly fractured, showing considerable resistance and high porosity. Physical weathering leads to their disintegration into angular debris (Fig. 20.6e, f). These materials form narrow tongues on steep slopes (above 25°) and blockfields on the gentler ones (Fig. 20.6a, e, f). The blockfields gradually pass into loamy-skeletal covers, whereas the block tongues end with the convex fronts manifesting their contemporary development. Their origin is related to periglacial conditions during glacial periods in Poland. Some of them formed due to frost activity and eluviation, others are due to the mass movements at the foot of rock crags, whilst others were formed by solifluction and washing out of fine particles. Often they were formed and then transformed by multiple processes (Pękala 1971).

20.5.3 Landslides

Some of the slopes in the Bieszczady Mts. have been transformed by mass movements. Past displacements are evidenced by old, deep-seated landslides that may cover large areas, e.g. on Mt. Szeroki Wierch (see site 4 in Figs. 20.1a and 20.7). Nowadays, rather only small and shallow slides develop in the lower part of slopes and within gully heads (Haczewski and Kukulak 2004; Haczewski et al. 2007; Gorczyca et al. 2016).

One of the most impressive landslides in the Bieszczady Mts., and the largest landslide in the Outer Polish Carpathians in terms of volume of displaced rock mass, occurs on the slopes of Mt. Chryszczata in the Duszatyn village (see site 9 in Fig. 20.1a). The landslide formed in 1907 after heavy rainfall. As the colluvium blocked the Olchowaty stream, several lakes developed (Pol. *Jeziorka Duszatyńskie*), which are amongst the biggest

Fig. 20.6 Rock crags and blockfields in the Bieszczady Mts. **a** Mt. Krzemień with blockfields formed below ridge-building rock crags. **b** Mt. Kińczyk Bukowski with rock outcrops on the ridge. **c** Forested slope of Mt. Bukowe Berdo with a rock crag. **d** Rock crag within a gullied slope, Mt. Bukowe Berdo. **e, f** Blockfields on the Połonina Caryńska Ridge (arrowed) (photographs by A. Bernatek-Jakiel)

landslide lakes in the Polish Carpathians (Schramm 1925; Kardaszewska 1968). In effect, part of the montane forest has been drowned by lake waters. This area is protected as a nature reserve "Zwiezło" due to its geological and geomorphological values. There are also other examples of large landslides that dammed streams in the Bieszczady Mts., e.g. in the Muczny valley, where the landslide occurred in the early Atlantic stage of the Holocene and caused inundation of the valley by a lake, which was drained afterwards (Haczewski and Kukulak 2004).

20.5.4 Soil Piping Forms and Features

In predominantly fine-grained slope covers, one can find interesting forms and features related to subsurface erosion by soil piping, understood as a water erosion process that leads to the formation of underground channels called pipes (Fig. 20.8). The evidence of this process becomes visible on the surface when a pipe roof collapses (Fig. 20.8a), or a pipe inlet or outlet is recognized (Czeppe 1960; Bernatek 2015). Surface indicators of soil piping can take different

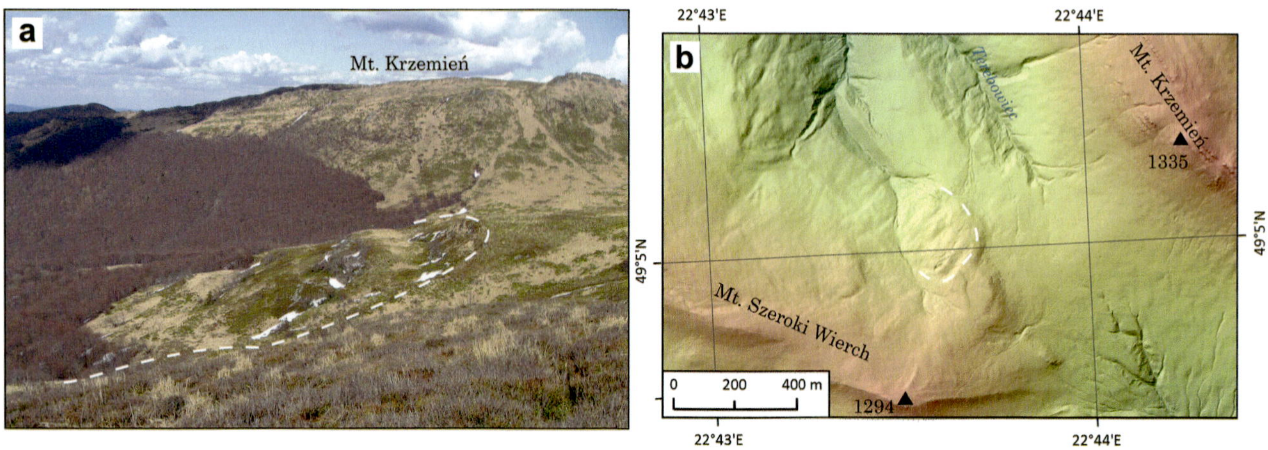

Fig. 20.7 Landslide on Mt. Szeroki Wierch. **a** As seen on the photograph (photograph by A. Bernatek-Jakiel). **b** Identified on the shaded relief model. The outline of the landslide is marked with a dashed line. Source: Author's work based on the LiDAR data obtained from the Head Office of Geodesy and Cartography, Poland (GUGiK)

forms, from shallow sagging depressions, through sinkholes (Fig. 20.8c), up to blind gullies (Fig. 20.8d). In the Low Bieszczady Mts., systems of piping-related features are clearly visible in grasslands and areas used as pastures in the past (Bernatek-Jakiel et al. 2019). There are also some impressive examples of piping features in the High Bieszczady Mts. above the upper tree line, e.g. in Mt. Kińczyk Bukowski and Mt. Krzemień (Haczewski et al. 2007; Bernatek 2015). Generally, soil pipes develop at the soil–bedrock interface at a depth of 0.7–0.8 m, and the diameter of the main pipes is up to 0.3–0.4 m (Bernatek 2015).

In the Bieszczady Mts., soil pipes can mainly be found in relationship to the gully network, i.e. the collapses of pipes lead to the gully formation or pipes develop above gully heads leading to their extension upslope (Fig. 20.8).

Fig. 20.8 Landforms related to soil piping developed in silty-clayey soils in the Bieszczady Mts. **a** Pipe collapses seen above the gully head (within forest), Bereźnica Wyżna village (see site 10 in Fig. 20.1a). **b** Soil pipes. **c** Sinkhole. **d** Blind gully ended with a pipe inlet (photograph a by M. Liro; b, c and d by A. Bernatek-Jakiel)

Soil pipes also occur in gullies subjected to landslides, reflecting landforms response to the new run-off condition (Bernatek 2015; Bernatek-Jakiel et al. 2017).

It seems that soil piping forms and features in the Bieszczady Mts. are more prominent than in other parts of the Polish Carpathians because of geological and pedological conditions, additionally supported by human impact expressed in significant land use changes. The Silesian Unit that dominates in the Bieszczady Mts. is built of the Krosno Beds, which are quite silty compared to other rock formations of the Carpathian flysch (Starkel 1960). Weathering of these materials leads to the formation of silty-clayey slope covers, which are additionally enriched in aeolian sediments (Kacprzak et al. 2015). Moreover, the Bieszczady Mts. were subjected to significant land use changes, i.e. intense human activity from the fifteenth century up to World War II, when local population was abruptly displaced for political reasons. Piping forms have developed in areas where grasslands/pastures have existed for 150 years or where grasslands/pastures existed in the past and were subsequently abandoned (Bernatek-Jakiel and Jakiel 2021). The pedological properties together with intense biological activity on grasslands/pastures favour pipe development in the region.

20.5.5 Erosional Landforms in River Channels

River and stream channels in the Bieszczady Mts. are characterized by long sections with rock beds, reflecting ongoing uplift of the region and steeply dipping flysch rocks. Such conditions lead to the formation of impressive erosional landforms. Resistant sandstones support bedrock steps (Fig. 20.9a) and, locally, waterfalls (Fig. 20.9b). One of the best known and most easily accessible tourist attractions is the "Siklawa Ostrowskich" waterfall located in the Wetlina village (see site 11 in Fig. 20.1a). It developed on the "Słowiński" stream, the tributary of the Wetlinka River, and is 5 m high (Fig. 20.9b). Individual bedrock steps have different shapes, depending on the thickness and structure of sandstone layers, as well as spatial arrangement of joints and faults.

Other interesting erosion landforms that develop in the channels are plunge pools and potholes. Plunge pools are formed below bedrock steps and waterfalls. They achieve depths up to 3 m, and their shape depends on the rock resistance at the bottom. Plunge pools, which developed in solid and resistant sandstones, have regular, rounded shapes and strikingly smooth surfaces, whereas forms in the shales are more irregular. Particularly characteristic erosional landforms in rock beds with dynamic flow are potholes, which develop on bedrock steps and elevated fragments of the rock bed. They are smooth, bowl-shaped, or cylindrical hollows, generally deeper than wide. Typical potholes in the beds of large streams are 10–25 cm in diameter and 15–50 cm deep (Haczewski et al. 2007; Gorczyca et al. 2016).

20.5.6 Peatlands

Peatlands in the Bieszczady Mts. develop in three topographic positions, i.e. on fluvial terraces and alluvial fans in the valley of the Upper San River and its tributaries, in closed depressions within landslide terrains, and on mountain ridges in closed depressions between rock blocks (Haczewski and Kukulak 2004; Ralska-Jasiewiczowa et al. 2006; Haczewski et al. 2007; Drewnik et al. 2015; Łajczak 2016).

Fig. 20.9 Examples of fluvial erosion landforms. **a** Sandstone bedrock steps in the bedrock channel of the Wetlinka River (photograph by K. Krzemień). **b** Waterfall "Siklawa Ostrowskich" in the village of Wetlina (photograph by M. Kisiel)

Fig. 20.10 The landscape of the Bieszczady Mts. with the raised bog in the wide Upper San River valley in the former village of Tarnawa Wyżna (photograph by K. Krzemień)

Most visible in the landscape are raised bogs that are fed mainly by precipitation (ombrotrophic peatlands), occasionally by groundwater, such as raised bogs in the former village of Tarnawa Wyżna (see site 12 in Fig. 20.1a). They have a characteristic shape of a broad dome, with a height of up to 4.5 m (Fig. 20.10). The highest point is located in the central part, where peat has grown at the fastest rate (about 1 mm y^{-1}) and often for the most protracted time span (Haczewski et al. 2007). The marginal parts gradually lose elevation and are typically covered by woodland. The domes are mostly asymmetrical, sloping down towards the river beds, and their heights are proportional to the surface of the peatbogs (Haczewski et al. 2007; Łajczak 2016).

Peatbogs have developed mainly in the Holocene, but some began to develop in the Late Glacial Interstadial. Their formation is controlled by a relatively cold and humid climate (annual precipitation 1000–1200 mm), which results in precipitation exceeding evaporation. However, the most important factors are hydrogeological conditions depending on geological structure and local topography (Łajczak 2016).

20.6 Human Impact and Cultural Geomorphological Landscape

The Bieszczady Mts. were settled relatively late, which was connected with a considerable distance from other human settlements and unfavourable natural conditions for settlement and agriculture. The beginnings of the permanent settlement are dated to the fifteenth century (Kukulak 2004; Wolski 2007), and since then, human impact on the landscape has started. Initially, it was mainly related to deforestation, which had at least two intensive phases, i.e. at the beginning of agricultural colonization (fifteenth–seventeenth centuries) and following the development of timber industry (nineteenth–twentieth centuries). The first phase had a more permanent character and covered mainly valleys and lower sections of slopes, where conditions were more suitable for the development of settlement and agriculture. Simultaneously, the upper tree line was lowered as an effect of cattle and sheep grazing. The second phase was characterized by greater pressure on forest areas (Kukulak 2004). At that time, a dense network of narrow-gauge railways was built to carry timber. Furthermore, in the nineteenth century, on the tributaries of the Upper San River, small water reservoirs were created in order to discharge water to mills, but were later filled with fine-grained sediments. The analysis of these sediments allows for the assessment of changes in the intensity of fluvial processes that occurred after the displacement of the local population after World War II (Kukulak 2004).

After World War II, the majority of population was forcibly displaced; from more than 100 villages existing since the fifteenth and sixteenth centuries approximately 90% of the total number of inhabitants were relocated (Wolski 2016). Abrupt depopulation had substantial effect on the landscape of the Bieszczady Mts. (Wolski 2007; Affek et al. 2021). In the former villages, e.g. in Beniowa and Tyskowa (see sites 13 and 14 in Fig. 20.1a), deforested valley bottoms and lower parts of the slopes have been successively reforested (Fig. 20.11a), although abandoned agricultural terraces and former roads are still well-preserved landforms (Fig. 20.11b, c). All these features create unique landscapes of abandoned villages that testify to several centuries of human economic activity in the region. This once densely populated rural region has become one of the largest "wilderness" areas in Central Europe (Affek et al. 2021).

Despite the resettlement of the region, which began in the second half of the 1950s due to the development of forest management, the population has still not reached the pre-World War II level, and many villages remain uninhabited. Nevertheless, seasonally, this region attracts many tourists, especially during summer and early autumn, which brings a riot of colours. Beech and fir forests on mountain slopes transform into vibrant landscapes with vivid red, orange, yellow, green, and brown colours. The golden colour of alpine meadows adjoins the red beech forests that dominate in the Bieszczady Mts. (Fig. 20.12).

Nowadays, human impact on the landscape is limited mainly to agricultural and forest roads in areas not protected in the Bieszczady National Park, where intense forest management takes place. In the most attractive tourist sites, primarily in the Bieszczady National Park, major human-induced transformations occur on the roads and touristic paths.

Fig. 20.11 Landscapes of abandoned villages. **a** Surroundings of the former village of Beniowa in the High Bieszczady Mts. (photograph by K. Krzemień). **b** Abandoned agricultural terraces (indicated by black arrows) in the former village of Tyskowa in the High Bieszczady Mts. (photograph by A. Bernatek-Jakiel). **c** Hillshaded digital terrain model of slope in the former village of Tyskowa. Black arrows indicate abandoned agricultural terraces shown in the photograph. Source: Author's work based on LiDAR data obtained from the Head Office of Geodesy and Cartography, Poland (GUGiK)

Fig. 20.12 A riot of autumn colours in the Bieszczady Mts. **a** The High Bieszczady Mts. (photograph by K. Mostowik). **b** The Low Bieszczady Mts. (photograph by A. Bernatek-Jakiel)

The landscape of the Low Bieszczady Mts. has been significantly changed in the 1960s when the Solina Dam was constructed on the San River. The concrete Solina Dam is 82 m high and 664 m long, making it the largest dam in Poland. Its construction resulted in the formation of the largest artificial lake in Poland (see site 15 in Fig. 20.1a). It has an area of 22 km^2, stores 472,000,000 m^3 of water, and the length of its coastline reaches 150 km. The lake's great depth, water clarity, and mountainous scenery make it a very popular tourist destination.

20.7 Conclusions

The Bieszczady Mts. represent an excellent example of structural relief in the Polish Carpathians. The topography shows a dominant NW–SE grain where parallel-oriented ridges with structural slope relief interplay with a grid-like river valley pattern. Recent geological studies have revealed the impact of tectonic evolution on the present-day mountain landscape with significant lithological control. Folded and dislocated flysch rocks with strong sandstones alternating with less resistant shales and mudstones respond selectively to fluvial erosion and denudational processes. The most prominent landforms such as rock crags and block-fields on slopes, as well as plunge pools, potholes, bedrock steps in bedrock channels have developed in resistant sandstones.

The onset of human settlement in the Bieszczady Mts. is dated to the fifteenth century, but due to the sudden displacement of local population after World War II, the region has been significantly depopulated. This resulted in a unique landscape with former villages that still carries a clear imprint of several centuries of human economic activity, such as abandoned agricultural terraces and former roads. Nowadays, the Bieszczady Mts. are considered one of the wildest areas in Poland, where nature slowly transforms the landscape of slopes and valleys due to intense reforestation. Fortunately, the highest ridges in the High Bieszczady Mts. seem to have an established upper tree line above which large subalpine meadows stretch, allowing one to have an overview of larger areas. In recent years, this region has become, together with the Tatra Mts., one of the most attractive areas in the Polish Carpathians.

Acknowledgements The author is grateful to Kazimierz Krzemień, Marta Kisiel, Karolina Mostowik, Mateusz Stolarczyk (Jagiellonian University), and Maciej Liro (Polish Academy of Sciences) for providing a few photographs that illustrate this chapter. The work of A. Bernatek-Jakiel is supported by the National Science Centre, Poland, within the SONATINA 1 project (UMO-2017/24/C/ST10/00114).

References

Affek A, Wolski J, Zachwatowicz M, Ostafin K, Radeloff VC (2021) Effects of post-WWII forced displacements on long-term landscape dynamics in the Polish Carpathians. Landsc Urban Plan 214:104164

Alexandrowicz Z, Margielewski W (2010) Impact of mass movements on geo- and biodiversity in the Polish Outer (Flysch) Carpathians. Geomorphology 123:290–304

Baumgart-Kotarba M (1974) Rozwój grzbietów górskich w Karpatach fliszowych. Prace Geogr IG PAN 106:136

Bernatek A (2015) The influence of piping on mid-mountain relief: a case study from the Polish Bieszczady Mts. (Eastern Carpathians). Carp J Earth Env Sci 10:107–120

Bernatek-Jakiel A, Jakiel M (2021) Identification of soil piping-related depressions using an airborne LiDAR DEM: Role of land use changes. Geomorphology 378:107591

Bernatek-Jakiel A, Jakiel M, Krzemień K (2017) Piping dynamics in mid-altitude mountains under a temperate climate: Bieszczady Mts., Eastern Carpathians. Earth Surf Proc Landf 42:1419–1433

Bernatek-Jakiel A, Gutiérrez F, Nadal-Romero E, Jakiel M (2019) Exploring frequency—size relationships of piping-related collapse sinkholes in different morphoclimatic environments. Geomorphology 345:106845

Czeppe Z (1960) Zjawiska sufozyjne w glinach zboczowych górnej części dorzecza Sanu. Biul Inst Geol 150:297–324

Drewnik M, Walas J, Stolarczyk M (2015) Ogólna charakterystyka i właściwości gleb torfowiska stokowego na północno-wschodnim skłonie Szerokiego Wierchu (Bieszczady Zachodnie). Roczn Bieszczadzkie 23:319–333

Gorczyca E, Izmaiłow B, Krzemień K, Łyp M, Wrońska-Wałach D (2016) Rzeźba i jej współczesne przemiany. In: Górecki A, Zemanek B (eds) Bieszczadzki Park Narodowy—40 lat ochrony. Bieszczadzki Park Narodowy, Ustrzyki Górne, pp 51–68

Haczewski G, Kukulak J (2004) Early Holocene landslide-dammed lake in the Bieszczady Mts. (Polish East Carpathians) and its evolution. Studia Geomorph Carp-Balc 38:83–96

Haczewski G, Kukulak J, Bąk K (2007) Budowa geologiczna i rzeźba Bieszczadzkiego Parku Narodowego. Wydawnictwo Naukowe Akademii Pedagogicznej w Krakowie, Kraków, Poland, p 160

Henkiel A (1977) O pochodzeniu rzeźby polskich Karpat fliszowych. Annales UMCS Sec B 32–33:1–35

Henkiel A (1982) O pochodzeniu kratowej sieci rzecznej Beskidów Wschodnich. Czas Geogr 53:19–27

Jankowski L, Margielewski W (2014) Strukturalne uwarunkowania rozwoju rzeźby Karpat zewnętrznych—nowe spojrzenie. Przegl Geol 62:29–35

Jankowski L, Margielewski W (2021) Geological control of young orogenic mountain morphology: from geomorphological analysis to reinterpretation of geology of the Outer Western Carpathians. Geomorphology 38:107749

Jankowski L, Probulski J (2011) Rozwój tektoniczno-basenowy Karpat zewnętrznych na przykładzie budowy geologicznej złóż Grabownica, Strachocina i Łodyna oraz ich otoczenia. Geologia 37:555–583

Kacprzak A, Szymański W, Wójcik-Tabol P (2015) The role of flysch sandstones in forming the properties of cover deposits and soils—examples from the Carpathians. Z Geomorph NF 59(Suppl. 1):227–245

Kardaszewska E (1968) Osuwisko w Duszatynie. Annales UMCS Sec B 21:1–25

Klimaszewski M (1972) Podział geomorfologiczny Polski Południowej. In: Klimaszewski M (ed) Geomorfologia Polski. PWN, Warszawa, Poland, pp 5–17

Kłapyta J, Zasadni J (2023) The Tatra Mountains—the only high-mountain landscape in Poland. In: Migoń P, Jancewicz K (eds) Landscapes and landforms of Poland. Springer, Cham, pp 000–000
Kukulak J (2003) Impact of mediaeval agriculture on the alluvium in the San River headwaters (Polish Eastern Carpathians). CATENA 51:255–266
Kukulak J (2004) Zapis skutków osadnictwa i gospodarki rolnej w osadach rzeki górskiej na przykładzie aluwiów dorzecza górnego Sanu w Bieszczadach Wysokich. Wydawnictwo Naukowe Akademii Pedagogicznej w Krakowie, Kraków, Poland, p 128
Łajczak A (2016) Torfowiska—morfologia i stosunki wodne. In: Górecki A, Zemanek B (eds) Bieszczadzki Park Narodowy—40 lat ochrony. Bieszczadzki Park Narodowy, Ustrzyki Górne, pp 79–90
Łajczak A (2023) Relief of Mt. Babia Góra—the highest flysch ridge in the Western Carpathians. In: Migoń P, Jancewicz K (eds) Landscapes and landforms of Poland. Springer, Cham, pp 000–000
Margielewski W (2004) Geological control on the Osława River meander at Duszatyn, Western Bieszczady Mountains, Polish Outer Carpathians. Ann Soc Geol Polon 74:325–338
Margielewski W, Łajczak A, Szczęsny B (2002) Rezerwat przyrody "Przełom Osławy pod Duszatynem"—wartości przyrodnicze i zagrożenia. Chrońmy Przyrodę Ojczystą 2:39–65
Nawrocki J, Becker A (eds) 2017 Atlas Geologiczny Polski. Państwowy Instytut Geologiczny—Państwowy Instytut Badawczy, Warszawa, p 170
Oszczypko N, Uchman A, Malata E (eds) (2006) Rozwój paleotektoniczny basenów Karpat zewnętrznych i pienińskiego pasa skałkowego. Institute of Geological Sciences, Jagiellonian University, Kraków, p 199
Pękala K (1971) The development of block fields in the Western Bieszczady. Studia Geomorph Carp-Balc 5:191–198
Pękala K (1997) Rzeźba Bieszczadzkiego Parku Narodowego. Roczn Bieszczadzkie 6:19–38
Rączkowska Z, Łajczak A, Margielewski W, Święchowicz J (2012) Recent landform evolution in the Polish Carpathians. In: Lóczy D, Stankoviansky M, Kotarba A (eds) Recent landform evolution: the Carpatho-Balkan-Dinaric Region. Springer, Dordrecht, pp 47–55
Ralska-Jasiewiczowa M, Madeyska E, Mierzeńska M (2006) Vegetational changes in the montane grassland zone of the High Bieszczady mountains (southeast Poland) during the last millennium—Pollen records from deposits in hanging peat-bogs. Veget Hist Archaeobot 15:391–401
Schramm W (1925) Zsuwiska stoków górskich w Beskidach. Wielkie zsuwisko w lesie wsi Duszatyn Ziemi Sanockiej. Kosmos 50:1355–1374
Solon J, Borzyszkowski J, Bidłasik M, Richling A, Badora K, Balon J, Brzezińska-Wójcik T, Chabudziński Ł, Dobrowolski R, Grzegorczyk I, Jodłowski M, Kistowski M, Kot R, Krąż P, Lechnio J, Macias A, Majchrowska A, Malinowska A, Migoń P, Myga-Piątek U, Nita J, Papińska E, Rodzik J, Strzyż M, Terpiłowski S, Ziaja W (2018) Physio-geographical mesoregions of Poland: verification and adjustment of boundaries on the basis of contemporary spatial data. Geogr Polon 91:143–170
Starkel L (1960) Rozwój rzeźby Karpat Fliszowych w holocenie. Prace Geogr IG PAN 22:239
Starkel L (1965) Rozwój rzeźby polskiej części Karpat Wschodnich. Prace Geogr IG PAN 50:160
Starkel L (1969) Odbicie struktury geologicznej w rzeźbie polskich Karpat fliszowych. Studia Geomorph Carp-Balc 3:61–70
Starkel L (1972) Charakterystyka rzeźby polskich Karpat (i jej znaczenie dla gospodarki ludzkiej). Problemy Zagosp Ziem Górskich 10:75–150
Starkel L (2006) Geomorphic hazards in the Polish Flysch Carpathians. Studia Geomorph Carp-Balc 40:7–19
Wolski J (2007) Przekształcenia krajobrazu wiejskiego Bieszczadów Wysokich w ciągu ostatnich 150 lat. Prace Geogr IGiPZ PAN 214:228
Wolski J (2016) The landscape of abandoned villages in the Western Bieszczady: the problem of definition and classification. Geogr Polon 89:371–387
Zuchiewicz W (2010) Neotektonika Karpat polskich i zapadliska przedkarpackiego. Wydawnictwo AGH, Kraków, Poland, p 234
Zuchiewicz W (2011) Planation surfaces in the Polish Carpathians: myth or reality? Geogr Polon 84. Spec Iss 2:155–178
Żytko K, Zając R, Gucik S, Ryłko W, Oszczypko N, Garlicka I, Nemcok J, Elias M, Mencik E, Stranik Z (1989) Map of the tectonic elements of the Western Outer Carpathians and their foreland 1:50,000. In: Poprawa D, Nemcok J (eds) Geological Atlas of the Western Outer Carpathians and their Foreland. Państwowy Instytut Geologiczny, Warszawa

Anita Bernatek-Jakiel is an Assistant Professor at the Jagiellonian University in Kraków (Poland). Her research interests focus on soil erosion and its significance as a morphogenetic process. She mainly explores gully erosion and subsurface soil erosion (i.e. soil piping), as they are the most overlooked soil erosion processes. She does research in the mid-altitude mountains in the temperate zone (Poland), and in semi-arid regions of Spain and Iran. She is the member of Executive Committee of the International Association of Geomorphologists (IAG) responsible for Young Geomorphologists Training Programmes, and she leads the group of Young Geomorphologists in Poland. She is the member of the Association of Polish Geomorphologists, European Geosciences Union (EGU), IAG Badlands Working Group, and the European Soil Observatory (EUSO) Technical Working Group on Soil Erosion. She is a member of the Advisory Board of *Earth Surface Processes and Landforms* journal.

Upper Silesian Region—An Example of Large-Scale Transformation of Relief by Mining

21

Renata Dulias

Abstract

The Upper Silesian Region stands out from other areas of Poland because of its rich natural resources—hard coal, zinc and lead ores, iron ores, stowing sands, dolomites, and limestones. The beginnings of mining date back to the Middle Ages, but it was most intense in the twentieth century. Mining of hard coal in the Upper Silesian Coal Basin, currently the largest coalfield in Europe, was fundamental. As a result of long-term underground and opencast exploitation, in many areas the former agricultural and forest landscape has been transformed into a mining landscape. Its characteristic elements are anthropogenic landforms such as spoil tips, sandpits, quarries, subsidence troughs, and sinkholes occurring in the highly urbanized conurbations of Katowice and Rybnik and on their peripheries. The consequences of mining include significant changes in the course of watersheds, lowering of the base level of erosion, changes in the geometry of riverbeds, and creation of numerous water reservoirs in excavations and subsidence troughs. After 1989, due to the restructuring of the economy, half of the mines were closed. Currently, mining focuses on the southern and central parts of the Upper Silesian Region, while in the northern part many anthropogenic landforms are reclaimed and revitalized. They are transformed into attractive recreational and leisure areas, where the mining heritage has been properly exposed and is often under legal protection. The values of the anthropogenic landscape make the Upper Silesian Region an interesting tourist destination.

Keywords

Mining · Subsidence · Anthropogenic landforms · Landscape · Upper Silesian Region

21.1 Introduction

In Europe, one of the oldest and largest coal mining areas, next to the British coalfields (Northumberland, Durham, North and South Wales, Yorkshire, and others), the Ruhr Area in Germany and Nord-Pas de Calais in France, is the Upper Silesian Coal Basin (USCB), located in southern Poland (Fig. 21.1). As a result of long-term extraction of coal and stowing sands, zinc and lead ores, and building stones, the former agricultural and forest landscape of a significant part of this area has been transformed into a mining/post-mining landscape. The USCB covers various geological and geomorphological units and various historical and economic regions; therefore, it is most often called the Upper Silesian Region.

The Upper Silesian Region's anthropogenic landscape results first of all from mining and metallurgy, industry, urbanization, transport, water economy, and agriculture. The area is rich in deposits of various mineral raw materials, which have been mined since the Middle Ages. Galena and silver mining has almost a thousand-year tradition, and it marks the symbolic beginning of the economic development of the Upper Silesian Region. Deforestation for the needs of ore smelting and construction of drainage adits left its imprint on the landscape of over 30 square kilometers of the so-called Błędów Desert (Rahmonov et al. 2006). At the turn of the twentieth century, this region was one of the leading producers of calamine in the world. However, the primary importance has been given to hard coal mining, developing since the Industrial Revolution from the end of the eighteenth century. After World War II, coal constituted the foundation of the Polish energy sector and steel industry for many years and played an essential role in trade

R. Dulias (✉)
Institute of Earth Sciences, University of Silesia in Katowice, Będzińska 60, 41-200 Sosnowiec, Poland
e-mail: renata.dulias@us.edu.pl

P. Migoń and K. Jancewicz (eds.), *Landscapes and Landforms of Poland*, World Geomorphological Landscapes, https://doi.org/10.1007/978-3-031-45762-3_21

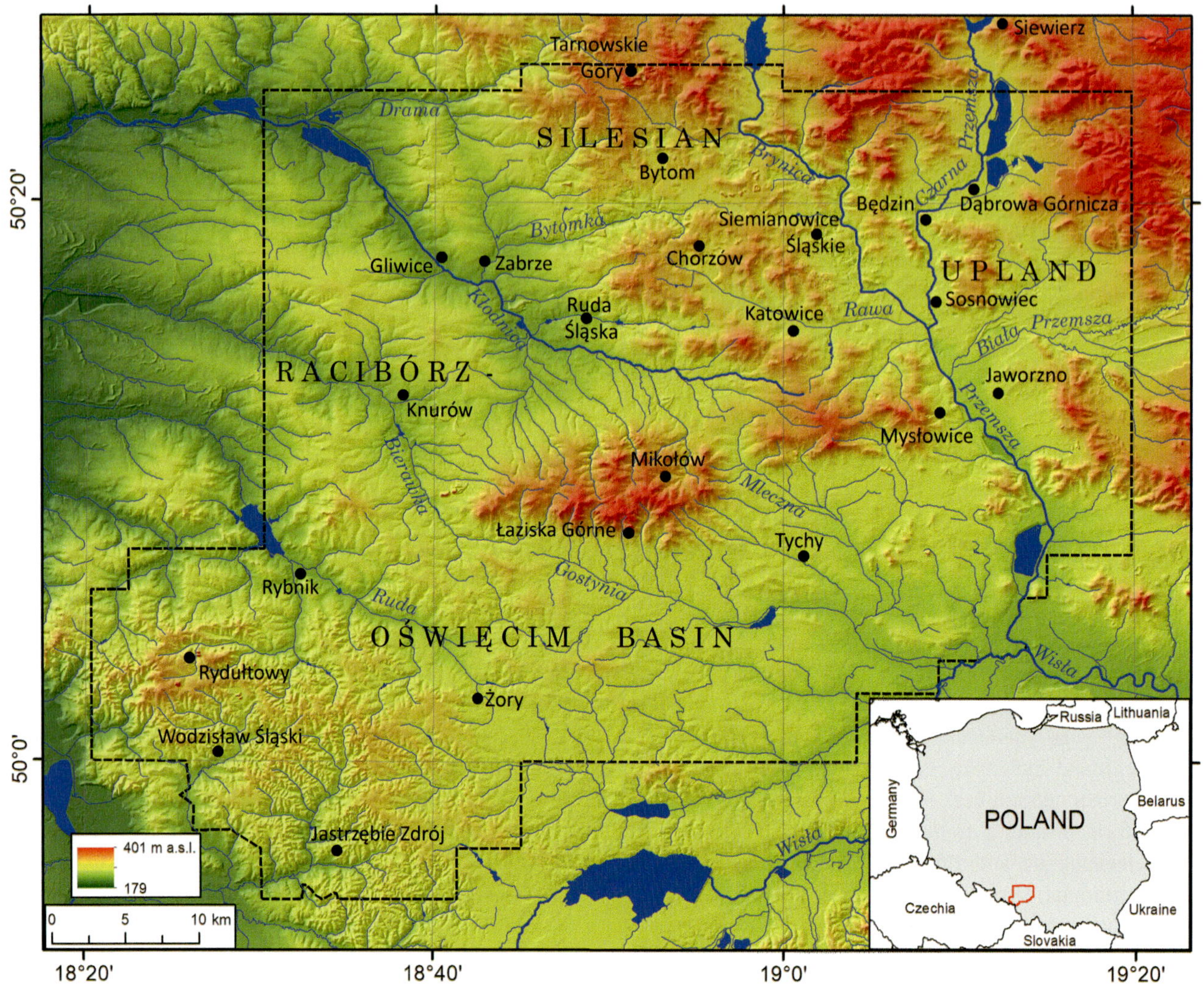

Fig. 21.1 Location of the study area in relation to hypsometry (based on the Numeryczny Model Terenu Polski, DTED-2 2001)

exchange. Among rock resources, exploitation of stowing sands was of principal importance, but dolomites, limestones, marls, and gravel aggregates were also mined.

As a result of long-term mining in the Upper Silesian Region, more than 13 billion tonnes of mineral resources and approximately 2–4 billion tonnes of waste rock were extracted. It resulted in the emergence of various anthropogenic landforms, including excavations, spoil tips, subsidence troughs, and sinkholes. Most mineral resources (98%) were exploited in the twentieth century; therefore, it was the period of the most significant changes in the landscape of the Upper Silesian Region (Dulias 2016; Machowski et al. 2016). After the political transformation in Poland (1989), the regional economy was deeply restructured, and more than half of large plants and mines were closed. The northern part of the region has mainly been "freed" from mining, and the anthropogenic landscape is being reclaimed and revitalized. Currently, coal mining focuses on the southern and central parts of the region (the Rybnik conurbation and the southern part of Katowice conurbation), and anthropogenic relief there is very distinctive. A vast complex of Pszczyna-Ruda forests separates these two highly urbanized conurbations.

21.2 Main Geological and Geomorphological Characteristics

Regarding geological context, the Upper Silesian Region overlaps with the Upper Silesian Foredeep filled with Lower Carboniferous deposits, formed in the Culm or shelly limestone facies, and the Upper Carboniferous ones, which are predominantly sandstones, claystones, mudstones, and hard coal seams (Jureczka et al. 2005). The overburden of Carboniferous strata in the northern and eastern parts of the Upper Silesian Region consists of Triassic rocks belonging to the Silesian-Cracow Monocline. These

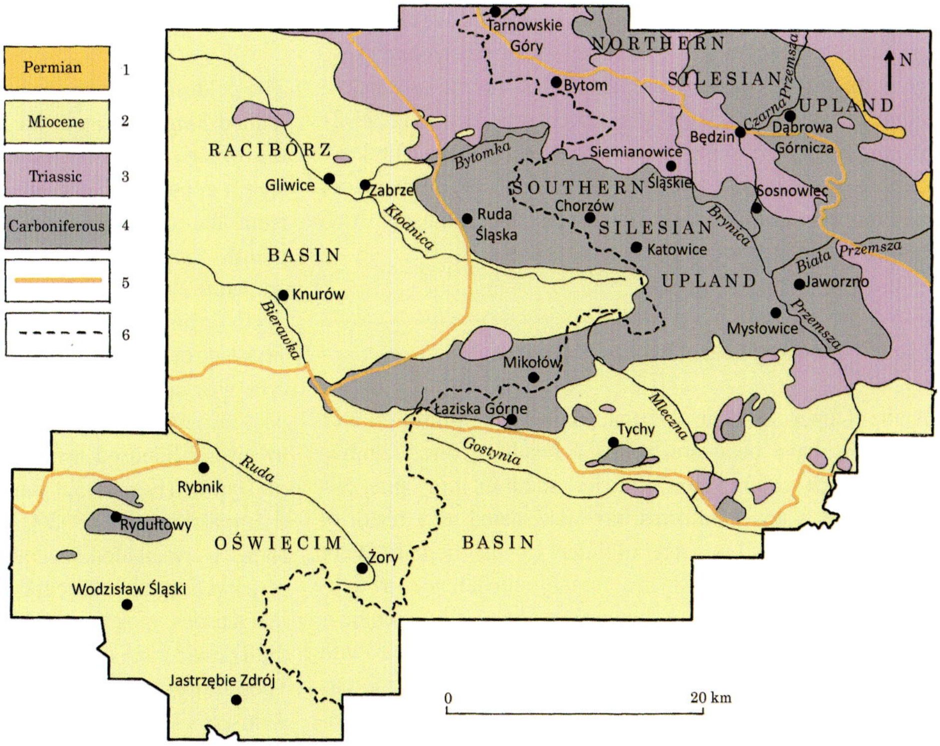

Fig. 21.2 Study area on the background of geological zones. 1—Permian deposits, 2—Miocene zone; 3—Triassic zone, 4—Carboniferous zone, 5—boundaries of geomorphological mesoregions, and 6—the Vistula-Odra watershed

are mainly limestones, dolomites, and marls, while ore-bearing dolomites with zinc and lead ores are particularly important from the mining point of view. A significant part of the Upper Silesian Region in the south and west is covered by Neogene deposits belonging to the Carpathian Foredeep, including clays, mudstones, gypsum, and rock salt. Carboniferous outcrops are located in the central and north-eastern parts of the Upper Silesian Region (Fig. 21.2).

Pleistocene and Holocene deposits are genetically and lithologically varied, and their thickness reaches several tens of meters. They are mainly of glacial, fluvioglacial, luvial, and aeolian origin. Generally, in the western part of the area, boulder clays are present; in the southern belt loess predominates, whereas in the center and the east mostly sandy deposits occur.

The area of the Upper Silesian Region is divided into two geomorphological units (macroregions): Silesian Upland and Racibórz-Oświęcim Basin (Fig. 21.1). The relief of the northern and eastern parts, belonging to the Silesian Upland, was shaped over a long period (about 65 million years) on lithologically diverse Palaeozoic-Mesozoic bedrock (Lewandowski 2015). Their northern parts are characterized by scarpland relief developed within the Silesian-Cracow Monocline, whereas the southern parts have a horst-and-graben relief.

The southern and western parts of the Upper Silesian Region were shaped in a relatively shorter period (over a dozen million years), within generally lithologically homogeneous Carpathian Foredeep deposits. This area belongs to the Racibórz-Oświęcim Basin (Fig. 21.1). Its morphological diversity is associated with the nature of younger Pleistocene sediments. The western part is a high plain with widespread clay deposits, in the south-west there is a loess-covered plateau with numerous gullies, high plains with a loess cover typify the southern part, and the remaining area consists of a vast, predominantly sandy plain.

The Upper Silesian Region can be divided into three zones which, in their general outlines, reflect the presence of older bedrock. These zones, conventionally called as Carboniferous, Triassic, and Miocene, differ in the geological structure, the lithology of bedrock, hydrogeological conditions, mineral resources, and the consequent degree and character of mining-related human impact (Fig. 21.2). The Carboniferous zone covers the central and north-eastern parts of the region, with deeply drained bedrock resulting from intensive coal mining. Outcrops of Carboniferous rocks cover an area of over 500 km^2. Sandstone horsts are noticeable in the highly urbanized landscape, sometimes capped with Triassic rocks, separated by tectonic, erosional, and denudational depressions.

The Triassic zone (about 300 km^2) covers the northern and eastern parts of the Upper Silesian Region (Fig. 21.2). In this area, the Carboniferous overburden consists of carbonate bedrock of Triassic age that accumulates large groundwater reserves and is prone to karst processes. It is an area of escarpments and horst-and-graben relief. Mining activities affected the Carboniferous bedrock, together with the Triassic overburden and Quaternary deposits.

The Miocene zone is the largest in terms of areal coverage (almost 860 km^2) and includes the southern and western parts of the region, where the Carboniferous substrate is found under impermeable Miocene overburden (Fig. 21.2). Groundwaters are found in Quaternary sediments. The landscape is dominated by poorly diversified clay high plains, sandy outwash areas, and loess plateaus, intensely fragmented by gullies and quite heavily urbanized.

21.3 Mining-Related Landforms

In the Upper Silesian Region, numerous mining-related landforms have been created as a result of direct human activity such as spoil tips, quarries, sandpits, clay pits, etc. In addition, some landforms have originated as a result of natural processes but were initiated by the human factor (indirect activity). These are mainly subsidence troughs and sinkholes. The most distinctive and easily recognizable landscape elements are landforms associated with direct mining activities. Many of them have existed for tens or even hundreds of years, which results from the resistance of the material that builds them and vegetation overgrowth. Although they are the most visible landmarks of an anthropogenic landscape, spoil tips and various excavations occupy a relatively small area of the Upper Silesian Region, less than 150 km^2. On the other hand, induced landforms such as subsidence troughs, which occupy an area many times larger than heaps and pits, are generally poorly recognizable in relief due to their similarity to natural landforms, which is also emphasized by frequent occurrence of water bodies within them. Nevertheless, it was mining subsidence, through changes in hypsometric relationships on an area of over a thousand square kilometers that has caused significant changes in the functioning of the geomorphic system in the Upper Silesian Region.

21.3.1 Excavations

The oldest anthropogenic landforms are shallow holes (called *pinga*) created due to former opencast or shaft mining of ores in the Middle Ages and the seventeenth-nineteenth centuries. They are accompanied by mounds formed from the removed overburden and waste rock called, in plural, *warpie*. The dimensions of the holes and mounds are small, usually several meters in diameter and of similar depth or height. They are found on terrain elevations built of Triassic rocks, where shallow mining of silver, zinc, and lead ores was carried out down to the groundwater level. These post-mining holes appear in large clusters at many sites, up to several hundred forms in each, for example, in the cities of Bytom (Fig. 21.3a) and Dąbrowa Górnicza (Fig. 21.3b) (Lamparska-Wieland 2003). Although many of these landforms were created several hundred years ago, they are well preserved in the landscape and covered with forests.

Quarries, mostly of dolomites and limestones, are found in the Triassic zone in the northern and eastern parts of the Upper Silesian Region (Fig. 21.4). They are about 100–150-year-old forms. Six of over one hundred former quarries are still active, with the largest one near the town of Siewierz (about 1.1 km^2) accounting for over 50% of the Polish production of dolomite in 2020 (Nita 2013). Quarries have several to several dozen meters of depth and can be multilevel. Their volume is estimated at 100 millions m^3 (Dulias 2016).

In the Upper Silesian Region, there were many open-pit mines, mainly of hard coal (37) and in fewer numbers also of zinc and lead ores. The oldest coal open-pits were "Reden" (1785) in Dąbrowa Górnicza and "Koszelew" (1825) in the town of Będzin. The "510 seam" (called also *Reden*, *Bolesław Chrobry*), which is the thickest coal seam in the USCB (up to 24 m), was excavated in both mines. The "Reden" open-pit had a length of over 1 km and a width of 200 m. The excavation, despite partial backfilling,

Fig. 21.3 Post-mining landforms after shallow exploitation of zinc and lead ores. **a** Srebrna Góra hill in Bytom (LiDAR image). **b** Bukowa Góra hill in Dąbrowa Górnicza (photograph by R. Dulias)

Fig. 21.4 Gródek Park in a reclaimed dolomite quarry in Jaworzno (photograph by R. Dulias)

is still visible in the relief. In 1956, a large open-pit "Brzozowica" in Będzin was established, and it operated for a dozen years. The total area of the mine was 0.5 km^2, with a depth of 35–90 m.

Large sandpits are a very characteristic element of the landscape of the Upper Silesian Region. There are about a hundred pits where sand used to be exploited, mostly stowing sand, covering a total area of more than 75 km^2, of which over 40% is attributable to the "Szczakowa" mine—the largest in Poland (Dulias 2010). The volume of sandpits exceeds 1 billion m^3. In extensive excavations, intensive aeolian processes occur (Szczypek and Wach 1991). Water reservoirs occur in many sandpits, including the largest ones whose edges are subject to erosive and accumulative shoreline processes (Rzętała 2003).

Clay pits are numerous (over a hundred) but highly dispersed. Most of them are located in the western part of the Upper Silesian Region. Generally, they are small, currently filled with water bodies. Examples of large forms are in the city of Katowice in the districts of Giszowiec and Brynów. The morphological importance of clay pits in the landscape is persistent, but the landforms themselves are rather insignificant.

21.3.2 Spoil Tips

Spoil tips are anthropogenic landforms that are particularly highlighted in the landscape of the Upper Silesian Region. These are mainly heaps of waste rock and tailings associated with mining of hard coal, zinc, and lead ores and, to a lesser extent, stone resources. The mining waste represents about 80% of all waste that was deposited in heaps, with the rest coming from power plants and ironworks. Spoil tips take different shapes. The older forms are most often conical, for example, the characteristic heaps of the "Dębieńsko" mine in the town of Czerwionka-Leszczyny near Knurów. Many spoil tips have the shape of massive mesas (tabular hills), among them is the great heap of Skalny (90 m) by the "Bolesław Śmiały" coal mine in the town of Łaziska Górne. Many spoil tips are irregular in shape, with several culminations. Some have impressive heights, with the highest one being the Szarlota (Charlotte) in the town of Rydułtowy, which is 134 m (Fig. 21.5). The most extensive forms include the heaps of the "Knurów" mine (1.7 km^2) and the "Sośnica" mine (1.5 km^2). Nowadays, the spoil tips remarkably stand out in the landscape of the south-western part of the Upper Silesian Region. The total volume of waste on spoil tips was calculated at 370 million m^3 (Dulias 2016). A similar amount of waste was disposed of by means other than storage in spoil tips. A substantial part of mining waste filled subsidence troughs and former sandpits and also was used for ground leveling for construction purposes. Since 1968, mining waste has been used to build embankments along riverbeds, which are affected by mining subsidence, to prevent flooding of adjacent areas. At the end of the twentieth century spoil tips covered almost 50 km^2, but it has changed dynamically in recent years because the material from heaps has been widely used in road and highway construction.

From the beginning of mining waste disposal in heaps, some of them ignited spontaneously due to the content of

Fig. 21.5 Highest spoil tip in the Upper Silesian Region, called Szarlota (Charlotte) in Rydułtowy. **a** LiDAR image from 2020. **b** Photo of the top of the spoil tip from 2021 (photograph by R. Dulias)

carbonaceous substances. Burning of the heap is a long-lasting process, counted in tens of years. Recently, over a dozen thermally active heaps in the Upper Silesian Region occur (Ciesielczuk et al. 2013).

21.3.3 Ground Subsidence

The underground extraction of raw materials results in the geomechanical transformation of bedrock due to the gravity movements of the overlying rocks into the workings. Displacements affect increasingly higher-lying rock layers, causing their collapse, cracking, and bending. Once the rock-mass displacements reach the terrain surface, they are revealed as ground deformations. The continuity of layers may be retained during the process, leading to the formation of subsidence troughs, or broken, resulting in the origin of sinkholes, fissures, and steps. The size and spatial distribution of bedrock and surface deformations depend on many factors, both natural and mining-engineering.

21.3.3.1 Subsidence Troughs

The continuous deformations result from deep mining and come in the form of extensive depressions of the terrain, generally with gentle slopes, called subsidence troughs. They have oval or elliptic shape and usually are several hundred meters or more in diameter. Subsidence caused by the underground exploitation has nowhere else in Poland reached such large dimensions as in the Upper Silesian Coal Basin. It primarily happened due to the enormous underground mining of hard coal and zinc and lead ores, mainly by the roof-collapsing method of exploitation. Greatest mining subsidence was observed in the 1960s–1980s and refers to three decades of very intensive coal mining. The economic crisis at the end of the last century, the subsequent closure of many mines, and a reduction in coal production, partly limited the scale of subsidence. The extent of continuous deformations in the Upper Silesian Region is at least 1125 km^2 (Perski 2000; Dulias 2016).

In the Carboniferous zone, mining activities were, for a long time, conducted by the oldest mines in the USCB (e.g., "Murcki", "Polska", and "Katowice" mines). The most significant subsidence in this zone was reported, amounting to nearly 1.24 billion m^3. Continuous deformations covered almost 75% of the zone, causing its lowering by an average of 3.2 m. To the greatest extent, mining subsidence affected areas of the cities of Ruda Śląska and Zabrze, where the so-called Saddle Beds were exploited, with particularly thick coal seams. In over 260-year-long mining period, around 1 billion tonnes of coal and waste rock were extracted and caused the lowering of the area by 5.7 m on average. In the southern part of the "Wesoła" mine in the town of Mysłowice, one of the hills subsided by 20–30 m (Dulias 2016).

In the Triassic zone, mining activities were conducted by numerous coal mines: "Siemianowice", "Rozbark", "Centrum", and others. The subsidence volume is over 925 million m^3, resulting in surface lowering by 4.3 m on average. Large-scale subsidence results from the superposition of impacts from two kinds of mining—hard coal and zinc and lead ores. The deepest subsidence trough (about 35 m) was formed in the city of Bytom (Solarski and Pradela 2010) (Fig. 21.6). Significant subsidence was also observed in the town of Siemianowice Śląskie, 5 m on average (Dulias 2016).

In the Miocene zone, mining activities have been conducted by 25 coal mines, of which as many as 13 mines

Fig. 21.6 Deepest subsidence trough in the Upper Silesian Region located in the north part of the city of Bytom (photograph by R. Dulias)

are new, built after the 1960s. In the area within the range of mining impact, the volume of subsidence is nearly 1100 million m^3. Half of the mines are located between the cities and towns of Rybnik, Rydułtowy, Wodzisław Śląski, and Jastrzębie Zdrój. Their mining activities contributed to surface lowering by 2.4 m on average. Subsidence troughs occur in various morphological situations, including hilltops, slopes, and valley floors. The largest ones, with a depth of 20–25 m, were found next to the mines of "Chwałowice" and "Jankowice" (Dulias 2008a).

Mining subsidence decreased after 1989 due to the closure of numerous mines and reduced output. However, in active mining fields, the subsidence is still significant. An example of particularly severe social consequences is the district of Bytom—Karb, where heavy mining damage forced the eviction of hundreds of people in 2011 (Popiołek and Januszek 2016; Solarski et al 2022) (Fig. 21.7).

21.3.3.2 Sinkholes

Discontinuous deformations are classified primarily considering their shape, distinguishing fissures, and terrain steps (linear discontinuities) from sinkholes (surface discontinuities). Linear forms are created under mining activities conducted at different depths and usually accompany continuous deformations. Sinkholes are a result of collapse processes caused by mining activities. Most of them (about

Fig. 21.7 Examples of mining damage caused by mining subsidence. **a** City of Bytom. **b** City of Jastrzębie Zdrój (photographs by R. Dulias)

80%) result from shallow exploitation (up to 80 m), mainly conducted by roof-caving method, and the reactivation of former, shallow workings due to drainage, surface load by buildings, transportation vibrations, or quakes caused by rockburst or landslide movements. Deformations preceding collapses may last from several hours to several decades after the formation of underground working. However, the collapse itself happens suddenly, in a matter of days, hours, or even minutes. Sinkholes contrast with the surrounding area, which is hardly subject to any transformations (Fig. 21.8). Most sinkholes take the shape of a conical funnel with a depth up to several dozen meters and a diameter exceeding even 100 m.

In the Upper Silesian Region, sinkholes and linear deformations occur in areas of shallow exploitation of hard coal and zinc and lead ores, mainly in the north-central part of the USCB. To a lesser extent, they occur in the south-western part of the region. At the end of the twentieth century, about 1000 discontinuous deformations were registered (Goszcz 1996). Still, their actual number is suspected to be much higher. First, there is a lack of information about the number of deformations created in the initial mining phase. Second, most of the mines did not begin to maintain an inventory until the post-war period.

In the Carboniferous zone, sinkholes occur in many areas due to easy access to shallow coal seams. In the southern part of the city of Sosnowiec, nine large sinkholes were formed in the early twentieth century, one of which had a diameter of nearly 100 m. One of the largest sinkholes, with a diameter of about 100 m and volume of 75,000 m^3, formed in 1964 in the the district of Jaworzno—Podłęże (Wilk 2003). Sinkholes still threaten many areas of former shallow exploitation. For example, within the area of closed "Porąbka-Klimontów" mine in Sosnowiec, in the area of 1.3 km^2, 248 underground workings have been found with connections to the surface (Duży et al. 2000; Dulias 2008b).

Fig. 21.8 Sinkholes in the area of shallow coal mining in the 1960s in Katowice (photograph by J. Greszta)

In the Triassic zone, many sinkholes were created in areas, where the impact of hard coal mining and zinc and lead ores overlapped. Until the mid-1970s, within the "Orzeł Biały" Mining and Metallurgy Plant, 177 sinkholes were inventoried in an area of 5.8 km^2. In the next two decades, 91 deformations emerged with a maximum recorded diameter of 28 m. In the vicinity of the town of Tarnowskie Góry, sinkholes frequently appear in the proximity of shafts associated with zinc and lead ore mining (Lamparska-Wieland 2003).

In the Miocene zone, sinkholes are not found, except a small area near Rydułtowy, where the Carboniferous bedrock occurs under the cover of Quaternary deposits. In the 1970s, the area of sinkholes covered more than 2 km^2 (Wrona 1975). 242 sinkholes were recorded, almost all in the eastern part of the "Rydułtowy" mine area (Palki 1982).

21.4 Hydrological and Hydromorphological Consequences of Ground Deformations

21.4.1 Closed Depressions and Shifting Water Divides

The appearance of numerous mining landforms in the relief of the Upper Silesian Region, especially of subsidence troughs, resulted in significant changes in sediment transfer in catchments and closed drainage basins. Underground mining is carried out irrespective of the hypsometric relationships on the surface, and subsidence affects various parts of the catchments, including water divide areas, slopes, and valley bottoms. If mining subsidence occurs at the river mouth, the base level of erosion is lowered; if they include water divide areas, their parts may be excluded from the catchment and transformed into closed drainage basins. Contrary to the pre-mining period, such closed drainage basins are characteristic elements of the contemporary relief. More than 230 such areas without external drainage were identified, covering over 120 km^2 in total (Dulias 2016). Most of them occur in the north-central part of the Upper Silesian Region. An absolutely unique position in this regard is held by the Siemianowice Plateau (the area between the cities of Katowice, Siemianowice Śląskie and Bytom), where closed drainage basins occupy almost 40 km^2, meaning that half of the area of this geomorphological unit is excluded from fluvial drainage.

One of the consequences of mining are changes in the course of water divides (Fig. 21.9). During a hundred years, the Vistula-Odra divide within the Katowice conurbation has changed position along 75% of its length, mainly due to ground subsidence (Czaja 1988). Changes in the areas of

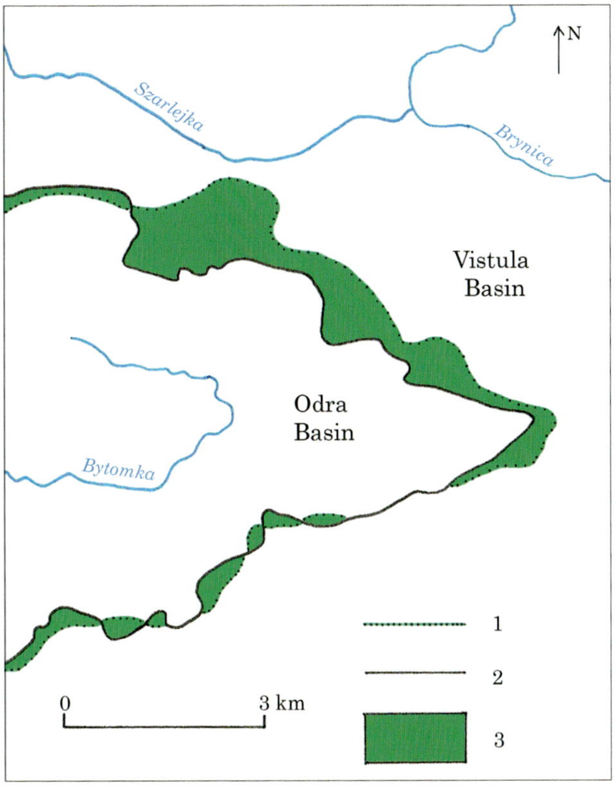

Fig. 21.9 Changes in the position of the first-order water divide between the Vistula and Odra basins caused by mining subsidence in the years 1883–1993. 1—water divide in 1883; 2—water divide in 1993; 3—the area affected by drainage divide changes

some catchments are significant. An extreme example is the Upper Rawa catchment located in the Carboniferous zone, which has almost entirely become a closed drainage basin. In some catchments located in the Triassic zone, not less than 70% of their areas is drained internally, and a specific case is observed in the Michałkowice Ditch catchment (in the town of Siemianowice), where closed drainage basins occupy over 20 km^2. In the Miocene zone, the most significant changes in water divide positions took place in the Dębinka catchment (in the city of Jastrzębie Zdrój), where almost half of its surface is currently excluded from fluvial drainage (Dulias 2008c).

21.4.2 Changes in Stream Longitudinal Gradients

Subsidence in river valleys leads to flooding of surrounding areas. To prevent floods, the riverbeds are embanked, often over long sections and to considerable height. In the Upper Silesian Region, the length of embankments along subsided riverbeds is about 300 km (Fig. 21.10). In areas with poorly diversified relief, even relatively small subsidence necessitates the construction of embankments along riverbeds. An example is the lower sections of the Pszczynka and Korzenica streams in the southern part of the region, where subsidence by about 2.5 m resulted in flooding of a vast area. It forced lateral displacement of more than half-a-kilometer long sections of both rivers and construction of embankments at a length of about 2 km.

Fig. 21.10 Embankments along the Brynica riverbeds and its left tributary, Jaworznik, were created to protect adjacent areas against flooding. Embankments up to 9 m high (264 m above sea level) hold the river water a few meters above the bottom of the large subsidence trough (255 m above sea level) (Source: Google Earth)

At the beginning of the twentieth century, most rivers flowed in natural channels, whereas in the twenty-first century this is the case of only about one third watercourses. The most noticeable anthropogenic changes in riverbeds are straightened courses, bed pavings, channel deepening, and strengthening of river banks. The Brynica River, below the mouth of the Szarlejka tributary, instead of following the natural meandering riverbed flows within an artificial, straight, paved, and embanked channel. The river's length in the analyzed segment decreased by 15 km. No other river in the Upper Silesian Region has been transformed to such an extent (Czaja 1999).

For most rivers in the Upper Silesian Region their base level of erosion was lowered due to mining subsidence, by 3.4 m on average in the Vistula basin and by 4.8 m on average in the Odra basin (Dulias 2016). Lowering of the base level of erosion mostly affected rivers in the western part of the region. The maximum value of 13.2 m was recorded in the Cienka stream bed near the city of Gliwice.

Mining subsidence causes deformations of riverbeds leading to changes in river gradients, which are either increasing or decreasing (Jankowski 1986; Dulias 2011). The passage of the exploitation front under the riverbed may also lead to the reversal of the gradient and the emergence of inundated areas, unless the riverbed is contained within embankments. These areas act as sedimentary basins for material carried by rivers. An increase in river gradient was noted for most rivers in the Upper Silesian Region. For some of them, it was quite significant, for example, the Radlin stream near Rydułtowy (+4.4‰) and the Jordanek stream near Knurów (+2.5‰). A decrease in an average river gradient was primarily observed for rivers belonging to the Vistula basin. An extreme example is the Orzeł Biały Ditch near Siemianowice Śląskie. In the pre-mining period, it had a gradient of 4.1‰, but due to extensive subsidence in the source area (11 m) it decreased to only 0.2‰ (Dulias 2016).

A characteristic feature of many rivers of the Upper Silesian Region is above-norm water pollution by suspended load. The source of contamination is waste from coal flotation and material stored on spoil tips in the vicinity of riverbeds. The increased amount of material carried by rivers accelerates aggradation of floodplains. However, the greatest role in capturing the material transported by the rivers is played by water reservoirs. In the nineteenth century, fish ponds were of major importance in this respect. Today, water reservoirs occur in the overdeepenings of the valleys caused by subsidence. Generally, water bodies in subsidence troughs and post-mining excavations are very characteristic elements of the Silesian landscape. For this reason, the term Upper Silesian Anthropogenic Lakeland was coined (Rzętała 2008).

21.5 Post-Mining Landscape

Today, the landscape of the Upper Silesian Region is changing very dynamically. Over only 100–200 years, some areas have transformed from an agricultural landscape, through a typical mining landscape, to a post-mining landscape (Myga-Piątek 2014). The distinctive feature of the latter is the disappearance of characteristic mining infrastructure (mines, shafts, and railways) from the landscape, but above all, the removal of anthropogenic landforms. Many heaps have been removed, and numerous excavations and subsidence troughs were filled and leveled. The large "Brzozowica" coal open-mine in Będzin, the largest in post-war Europe, was wholly covered with ashes from power plants, and a unique exposure of the thick Reden coal seam was irretrievably lost. A similar fate befell the outcrops of Carboniferous deposits with standing trunks of Carboniferous *Sigillaria* and *Calamites* in the excavation in Katowice-Giszowiec.

On the other hand, many anthropogenic landforms and various brownfields have been transformed into attractive recreational and leisure areas. Part of the mining heritage has been adequately exposed and is protected by law. In 2017, the Tarnowskie Góry Lead-Silver-Zinc Mine and its Underground Water Management System was inscribed on the UNESCO World Heritage List. Apart from the underground mine and shafts, the heap and two fragments of the post-mining landscape have become protected (bearing the names: post-mining landscape from nineteenth century and Srebrna Góra). The heap in the town of Tarnowskie Góry was formed from washing ore-bearing dolomites to separate zinc and lead ores (Fig. 21.11b). For nearly 100 years, about 900 thousand m^3 of rock waste was deposited. The heap has a unique orange color and is partially covered by galmei grasslands. In one of the above-mentioned post-mining landscapes (from the nineteenth century), about 20,000 shafts were dug in search of silver and lead ores in the sixteenth century, and then the area was transformed by the exploitation of zinc and iron ores in the nineteenth century. Currently, it is characterized by very chaotic relief and is covered with galmei grasslands. A more distinct post-mining landscape is found on the Srebrna Góra hill in the Segiet Nature Reserve, where numerous post-mining holes and mounds occur (see Fig. 21.3a). The exploitation of silver and lead ores started in the twelfth century, but the main period of mining coincided with the fifteenth and sixteenth centuries. In the nineteenth century, zinc ores were mined for over 60 years and, to a lesser extent, iron ores.

Among principal geotourist attractions of the region is the Sadowa Góra quarry in the town of Jaworzno, where the GEOsfera ecological and geological education center

Fig. 21.11 Examples of anthropogenic landforms of importance in geotourism. **a** Megaripplemarks exposed during the exploitation of Middle Triassic limestones at the bottom of the Sadowa Góra quarry in the town of Jaworzno. **b** Waste rock heap after washing zinc and lead ore, one of the sites of the Tarnowskie Góry Lead-Silver-Zinc Mine and its Underground Water Management System, inscribed on the World Heritage List in 2017 (photographs by R. Dulias)

was established. The bottom of the quarry is an undulating surface of the Middle Triassic limestones, representing megaripplemarks created by sea waves during a hurricane (Fig. 21.11a). Nearby is the flooded Gródek quarry, where a diving center was set up, and an attractive park with an exposed dolomite wall (Fig. 21.4). Protected as documentation sites are the Carboniferous sandstone profiles in the quarries in Łaziska Górne and Rydułtowy, and the Middle Triassic dolomite and limestone profiles in Blachówka and Srocza Góra quarries (respectively, in Bytom and in Dąbrowa Górnicza).

Many other landforms and post-mining areas have been reclaimed and revitalized. An example is the Silesian Park in the city of Chorzów, the largest city park in Poland established in the area of former heaps, sinkholes, and subsidence basins; a complex of Pogoria I-III water reservoirs in the former sandpits in Dąbrowa Górnicza; heaps of the "Piast" mine, or the nature and landscape complex of Żabie Doły in Bytom and Chorzów on the post-mining area with anthropogenic water reservoirs, heaps, and railway embankments. These examples of landforms and post-mining landscapes prove the great geotourist potential of the Upper Silesian Region (Racki et al. 1999; Jureczka 2001; Nita 2013; Lamparska 2017; Świtała-Trybek and Świtała-Mastalerz 2018). The post-mining landscape of the northern and eastern part of the region is especially appreciated, where mining has ended. However, the landscape of the rest of the area also has large mining heritage and geomorphological values and can be a tourist destination. It is not without significance that the post-mining areas are often characterized by high biodiversity and the occurrence of rare and protected species of plants and plant communities.

21.6 Conclusions

In Poland, no other area was subject to such intense mining activity as the Upper Silesian Region. Although mining origins go back to the Middle Ages, and the related landforms are still visible in the landscape, mining in the last century, mainly of hard coal, was essential for shaping the contemporary landscape. Consequently, agricultural or forestry land was converted into mining and built-up areas. Landscape changes took place with varying intensity and at different times, significantly depending on the geology of the area and the distribution of natural resources.

In the northern and eastern part of the Upper Silesian Region (Triassic zone), opencast or shallow underground mining of silver and lead ores dominated for hundreds of years. From the nineteenth century, zinc and lead ores, iron ores, limestones, and dolomites were recovered. Deep underground mining was carried out in the Carboniferous hard coal deposits. The area has a varied anthropogenic landscape with numerous quarries, excavations, sinkholes, large subsidence troughs, spoil tips, and high embankments along the riverbeds. Anthropogenic landforms are found both in the highly urbanized northern part of the Katowice conurbation and its northern agricultural periphery. Mining has already been terminated, and some anthropogenic landforms and post-mining areas are being reclaimed, revitalized, and subject to various forms of protection.

In the central and north-eastern part of the Upper Silesian Region (Carboniferous zone), in the mid-seventeenth century opencast and shallow underground mining of hard coal began, followed by more and more intensive mining at larger depths. Anthropogenic landforms arose between the chaotically dispersed urban and industrial buildings of the rapidly developing central part of the Katowice conurbation. Spoil tips, extensive subsidence troughs, opencast coal mines, numerous sinkholes, and large sandpits were important. Some of the mines are already closed in this part of the Upper Silesian Region, and the rest are less mined. Numerous post-mining areas have been reclaimed.

Mining has the youngest history in the southern and western part of the Upper Silesian Region (Miocene zone), where the Carboniferous coal-bearing strata occur at great depths under the cover of Miocene deposits. Most of the coal mines were established only several dozen years ago and carry out very intensive underground mining. Anthropogenic landforms are dominated by large spoil tips and subsidence troughs, many small clay pits, and numerous embankments along the riverbeds. The mining landscape occurs within the Rybnik conurbation, but also in formerly typically agricultural areas. Due to ongoing mining activities, the landscape will undergo further transformation.

The mining and post-mining landscape of the Upper Silesian Region is a heritage of its long and complex economic history. Thanks to a well-thought-out concept of its reclamation and revitalization, it can gain new functions (recreational, tourist, and educational) and set new directions for regional development.

References

Ciesielczuk J, Janeczek J, Cebulak J (2013) Przebieg i przyczyny endogenicznego pożaru węgla kamiennego na zrekultywowanym składowisku odpadów komunalnych w Katowicach. Przegl Geol 61:764–772

Czaja S (1988) Zmiany działu wodnego Wisła—Odra w obrębie Górnośląskiego Okręgu Przemysłowego, Geographia. Studia Et Diss 11:95–100

Czaja S (1999) Zmiany stosunków wodnych w warunkach silnej antropopresji (na przykładzie konurbacji katowickiej). Wydawnictwo Uniwersytetu Śląskiego, Katowice, p 189

Dulias R (2008a) Mining subsidence in Oświęcim Basin (Carpathian Foredeep). Geomorph Slov Bohem 8(2):8–13

Dulias R (2008b) Płytka eksploatacja węgla kamiennego na Płaskowzgórzach Dańdówki. Acta Geogr Silesiana 3:13–17

Dulias R (2008c) Wpływ górnictwa węglowego na zmiany krajobrazu w zlewniach Dębinki i Pniówka na Płaskowyżu Rybnickim. Dokument Geogr IG PAN 37:144–149

Dulias R (2010) Landscape planning in areas of sand extraction in the Silesian Upland, Poland. Landsc Urban Plan 95:91–104

Dulias R (2011) Impact of mining subsidence on the relief of the Rybnik Plateau, Poland. Z Geomorph NF 55:25–36

Dulias R (2016) The impact of mining on the landscape. A study of the Upper Silesian Coal Basin in Poland. Springer, Switzerland, p 209

Duży S, Kleta H, Plewa F (2000) Zagrożenie powierzchni ze strony starych zrobów w obszarze likwidowanej kopalni. Zesz Nauk Polit Śląskiej Górnictwo 246:111–117

Goszcz A (1996) Powstawanie zapadlisk i innych deformacji nieciągłych powierzchni na obszarach płytkiej eksploatacji górniczej. In: Szkoła Eksploatacji Podziemnej '96, PAN w Krakowie, Szczyrk, pp 119–137

Jankowski AT (1986) Antropogeniczne zmiany stosunków wodnych na obszarze uprzemysłowionym i zurbanizowanym (na przykładzie ROW). Uniwersytet Śląski, Katowice, p 280

Jureczka J (2001) Odsłonięcia powierzchniowe w koncepcji ochrony georóżnorodności Górnośląskiego Zagłębia Węglowego. Przegl Geol 49:1101–1114

Jureczka J, Dopita M, Gałka M, Krieger W, Kwarciński J, Martinec P (2005). Atlas geologiczno-złożowy polskiej i czeskiej części Górnośląskiego Zagłębia Węglowego. Państwowy Instytut Geologiczny, Warszawa

Lamparska M (2017) Turystyka wśród górniczych szybów. Szlak turystyczny i przewodnik po dawnych i współczesnych kopalniach na obszarze Górnośląskiego Zagłębia Węglowego. „Śląsk" Wyd Nauk, Katowice, p 203

Lamparska-Wieland M (2003) Atlas zmian wybranych elementów krajobrazu terenów górniczych i pogórniczych Płaskowyżu Tarnowickiego. Prace WNoZ Uniw Śląski Sosnowiec 27:4

Lewandowski J (2015) Kenozoik region śląsko-krakowskiego. Wyd Uniw Śląskiego, Katowice, p 200

Machowski R, Rzetala MA, Rzetala M, Solarski M (2016) Geomorphological and hydrological effects of subsidence and land use change in industrial and urban areas. Land Degr Dev 27:1740–1752

Myga-Piątek U (2014) Landscape management on post-exploitation land using the example of the Silesian Region, Poland. Env Socio-Econ Studies 2:1–8

Nita J (2013) Zmiany w krajobrazie powstałe w wyniku działalności górnictwa surowców skalnych na obszarze Wyżyn Środkowopolskich. Uniw Śląski, Katowice, p 184

Numeryczny Model Terenu Polski poziom 2 (DTED-2) (2001) Wydział Projektowania i Wdrożeń, Wojskowy Ośrodek Geodezji i Teledetekcji, Warszawa

Palki J (1982) Zagadnienie zagrożenia powierzchni zapadliskami w Rybnickim Okręgu Węglowym. In: Materiały Konferencji VIII Tydzień Techniki ROW, Jastrzębie Zdrój, pp 102–125

Perski Z (2000) Zastosowanie satelitarnej interferometrii radarowej do określania dynamiki i zasięgu górniczych deformacji terenu na przykładzie wybranych obszarów Górnośląskiego Zagłębia Węglowego. Prace WNoZ Uniw Śląski Sosnowiec 8:9–39

Popiołek K, Januszek M (2016) Miasto w kryzysie. Przypadek Bytomia Czas Psych 22:277–282

Racki G, Bardziński W, Zieliński T (1999) Z kamiennej księgi pradziejów Górnego Śląska. Przewodnik geologiczny. Wyd Uniw Śląskiego, Katowice, p 168

Rahmonov O, Szczypek T, Wach J (2006) The Błędów Desert (Pustynia Błędowska)—a unique phenomenon of the Polish landscape. Ann Geogr 39:34–41

Rzętała M (2008) Funkcjonowanie zbiorników wodnych oraz przebieg procesów limnicznych w warunkach zróżnicowanej antropopresji na przykładzie regionu górnośląskiego. Wyd Uniw Śląskiego, Katowice, p 171

Rzętała MA (2003) Procesy brzegowe i osady denne wybranych zbiorników wodnych w warunkach zróżnicowanej antropopresji (na przykładzie Wyżyny Śląskiej i jej obrzeży). Wyd Uniw Śląskiego, Katowice, p 159

Solarski M, Pradela A (2010) Przemiany wybranych form rzeźby Wyżyny Miechowickiej w latach 1883–1994. Z Badań Nad Wpływem Antropopresji Na Środowisko 11:78–92

Solarski M, Machowski R, Rzetala M, Rzetala MA (2022) Hypsometric changes in urban areas resulting from multiple years of mining activity. Sci Rep 12:2982

Szczypek T, Wach J (1991) Human impact and intensity of aeolian processes XE "aeolian processes" in the Silesian-Cracow Upland (Southern Poland). Z Geomorph NF Suppl 90:171–177
Świtała-Trybek D, Świtała-Mastalerz J (2018) Hałdy poprzemysłowe—ich kulturowy i turystyczny potencjał (na przykładzie wybranych obiektów w województwie śląskim). Zesz Nauk Turystyka i Rekreacja 2(22):125–141
Wilk Z (ed) (2003) Hydrogeologia polskich złóż kopalin i problemy wodne górnictwa. AGH, Kraków, p 611
Wrona A (1975) Problemy degradacji i ochrony powierzchni ziemi w Rybnickim Okręgu Węglowym. Przegl Geogr 47:295–311

Renata Dulias is Associate Professor of Geography at the University of Silesia in Katowice. She is interested in the human impact on the landscape in the past and present, mainly caused by deforestation and mining. The Silesian-Cracow Upland in south Poland is her main research area, but she also conducts research in other countries (Germany, Czechia, Hungary, Portugal, Spain, Greece, and Israel). She was the Vice-President of the Katowice Branch of the Polish Geographical Society (2002–2019). She is a member of the Scientific Council of the Complex of Landscape Parks of the Silesian Voivodeship (Poland). She participates in the work of many expert teams for nature protection in the Upper Silesian region and is involved in regional education of students. She is the author of several books for teachers and students and a juror in geography olympiads. She is the editor-in-chief of *Environmental & Socio-economic Studies* and a member of the editorial board of *Geographia Technica*.

Kraków-Częstochowa Upland—Monadnocks and Relic Caves in the Karst Landscape

22

Andrzej Tyc

Abstract

The Kraków-Częstochowa Upland is the largest carbonate karst area in Poland and belongs to the largest in Central Europe. It is built of limestones with a thickness of over 250 m. They were deposited in a relatively shallow sea as a carbonate ramp on the northern shelf of the Tethys Ocean in the Late Jurassic. Limestone facies differentiation associated with carbonate buildups and related variable resistance to erosion and karstification contributed to the geomorphic diversity within the region. Karst relief is an essential component of the landscape of the Kraków-Częstochowa Upland. Limestone rocky hills, monadnocks, and tors are residual forms that dominate in the surface relief. Numerous cave relics, including hypogenic ones, occur within these landforms. The rocky valleys and gorges, including the picturesque Prądnik Valley, cut the southern part of the Upland. The northern part was the only area of the Upland covered by the Pleistocene ice sheet, and the rocky hills and monadnocks were transformed by glacial erosion into asymmetric roche moutonnées. In the areas mantled by till and loess, covered karst developed. Although the origin and age of karst and caves are still discussed, the Late Cretaceous marine recession and the Holocene mark the general temporal constraints for the development of karst landforms in the Kraków-Częstochowa Upland. The karst landscape attracted humans to settle in this region since the Palaeolithic. Numerous archaeological sites in caves, rock shelters, and ruins of medieval castles testify to the rich cultural heritage of the region, underpinned by geodiversity.

Keywords

Karst landforms · Monadnocks · Tors · Caves · Kraków-Częstochowa Upland

22.1 Introduction

The Kraków-Częstochowa Upland represents a unique geomorphological region in southern Poland, located in the western, highest part of the upland belt in the northern foreland of the Western Carpathians. The area presented in this chapter is the most homogeneous and interesting part of the region and includes the Częstochowa Upland and the Ojców Plateau (Fig. 22.1; Solon et al. 2018). Sawicki (1920) called this region 'Kras' (Karst)—a name reflecting the karst nature and its similarity to karst areas in central and southern Europe. Other names are also used for the uplands located between the cities of Kraków in the south and Częstochowa in the north—the Kraków Upland, the Kraków-Częstochowa Jura, the Polish Jura or just the Jura.

The northern part of the Kraków-Częstochowa Upland is formed by a meridional belt of the Częstochowa Upland (Fig. 22.1). The varied surface of the area rises to an altitude of 300–450 m asl and reaches its maximum height of 515.6 m asl in Podzamcze near the town of Ogrodzieniec. The southern segment of the region, the Ojców Plateau, reaching the mean altitude of 450–480 m asl, with the highest elevation at 512 m asl, has more uniform relief. The plateau passes southward to the area characterized by the presence of numerous faults, horsts and grabens, constituting the Krzeszowice Trench and the Wisła valley. The latitudinal lowering of the Biała Przemsza valley creates the transition zone between two main parts of the Kraków-Częstochowa Upland. Less pronounced latitudinal structures (e.g. Pilica or Biała grabens) dissect the Częstochowa Upland, dividing it into smaller microregions.

Covering an area of almost 2000 km^2, the Kraków-Częstochowa Upland belongs to the most significant

A. Tyc (✉)
Institute of Earth Sciences, University of Silesia, Będzińska 60, 41–200 Sosnowiec, Poland
e-mail: andrzej.tyc@us.edu.pl

P. Migoń and K. Jancewicz (eds.), *Landscapes and Landforms of Poland*,
World Geomorphological Landscapes, https://doi.org/10.1007/978-3-031-45762-3_22

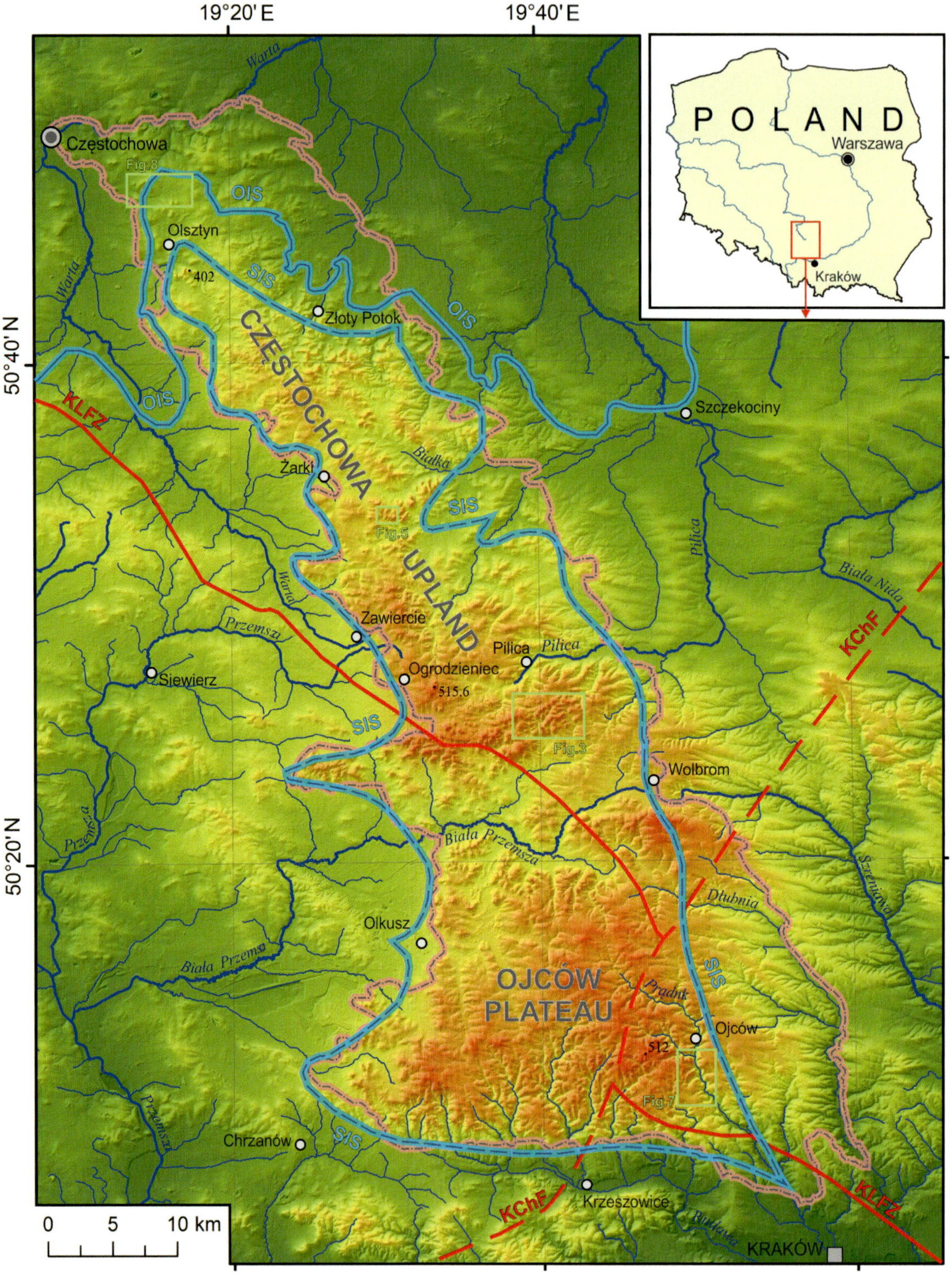

Fig. 22.1 Location of the Kraków-Częstochowa Upland (boundaries as dashed-and-dotted line on pink background; after Solon et al. 2018). KLFZ—Kraków-Lubliniec fault zone and KChF—Krzeszowice-Charsznica fault (after Buła et al. 2002); SIS—the maximal extent of the Sanian I glaciation; and OIS—the maximal extent of the Odranian glaciation (after Lewandowski 2011)

areas of carbonate karst in Central Europe (Tyc 2004). Following views presented by Gilewska (1964), Głazek et al. (1972), Głazek (1989), the region has been regarded in geomorphological and karst literature mainly as an example of fossil or relict karst (palaeokarst) and, more recently, also as a part of an important region (Silesian-Kraków) of hypogene karst and caves (Gradziński and Tyc 2017).

22.2 Geological Setting

Upper Jurassic limestones play a dominant role in the geological structure of the Kraków-Częstochowa Upland. On the surface, they reveal themselves in the form of picturesque rocky hills, monadnocks, and isolated limestone tors, otherwise being present under the cover of younger sediments, mainly of Quaternary age. These limestones occur as a long belt of outcrops stretching from Kraków in the south towards Częstochowa. Structurally, the area belongs to a larger geological region known as the Kraków-Silesian Homocline.

Upper Jurassic limestones form a complex with a thickness of over 250 m, deposited in a relatively shallow sea, on the northern shelf of the Tethys Ocean, as a carbonate ramp (Matyszkiewicz 1997; Matyszkiewicz et al. 2006a; Krajewski and Matyszkiewicz 2009). They mainly represent the Oxfordian, locally early Kimmeridgian stage of the Jurassic. Traditionally, three main facies are distinguished within the limestone succession—massive and nodular limestones (so-called rocky limestone), thick-bedded limestones, usually with cherts, and thin-bedded (platy) limestones. The former are distinctly diversified and represent both small bioherms developed within platy and marly limestones as well as large carbonate buildups that pass laterally to bedded limestones (Dżułyński 1952; Matyszkiewicz 1997; Matyszkiewicz et al. 2006a, b; Krajewski and Matyszkiewicz 2009). Compact and resistant bioherms are associated with lithified but less resistant granular sediments—turbidities, debrites or even debris of submarine landslides (Matyszkiewicz et al. 2006a; Krajewski and Matyszkiewicz 2009). An extensive carbonate platform developed in the vicinity of villages of Smoleń and Strzegowa in the southern part of the Częstochowa Upland, which, punctuated by clusters of rocky hills and monadnocks is still clearly visible in the relief of this area (Matyszkiewicz et al. 2006a). Facies differentiation of the Upper Jurassic limestones and related differential resistance to erosion contributed to the geomorphic diversity within the region.

The tectonic settings of the Palaeozoic basement of the Kraków-Częstochowa Upland control the formation and spatial distribution of Upper Jurassic carbonate buildups and platforms. The prominent Kraków-Lubliniec fault zone (KLFZ) crossing the upland (Fig. 22.1) is responsible, among others, for the occurrence of Palaeozoic magmatic intrusions in its basement, detected by drillings (e.g. Jędrys et al. 2004; Matyszkiewicz et al. 2006b). The major carbonate build-ups and the carbonate platform near Smoleń and Strzegowa are located above these intrusive bodies. Most of the main systems of joints and faults are related to the KLFZ. The latter is of the nature of overpass structures, among them positive flower structures, related to the displacements in the basement along the KLFZ (Żaba 1994, 1999). These structural conditions were favourable for the ascending fluids, and presumably, low-temperature hydrothermal processes were especially present in the northeastern and eastern margin of the KLFZ (Żaba 1999; Pulina et al. 2002, 2005; Żaba and Tyc 2007; Tyc 2009a; Matyszkiewicz et al. 2015; Gradziński and Tyc 2017). Lithological differentiation of the Upper Jurassic limestones and complex tectonic settings of the Kraków-Częstochowa Upland were among the factors affecting the evolution of karst and caves.

The geomorphological landscapes of Poland north of the Sudetes and the Carpathians were, to a large extent, shaped by the Pleistocene glaciations. The Kraków-Częstochowa Upland is one of the exceptions, as in the Pleistocene it was never entirely covered by the ice sheet. This view was formulated by Łoziński (1912) and developed by Różycki (1960), who proposed the term 'Jurassic Glacial Oasis' for the most probable concave nunatak in this region. Różycki (1960) based his statement mainly on the absence of erratics in the sedimentary cover of this region. The central area of the upland, elevated 150–200 m above the surrounding terrains, is considered either not to be covered by inland ice or perhaps only local ice streams penetrated its peripheries during the maximum extent of the Sanian I (Elsterian, MIS 12) glaciation (Lewandowski 2011) (Fig. 22.1). A consequence of the concave nunatak during this glaciation, in the Częstochowa Upland especially, is the presence of a sandy glacifluvial cover, preserved in most of the topographic depressions. Some of these sands were later reworked by aeolian processes. The Odranian (Saalian) glaciation reached only the northern part of the Częstochowa Upland (Fig. 22.1). The subsequent ice-sheet advances during the Warthanian (Late Saalian) and the Weichselian did not reach the upland, but left a record of intense periglacial processes, both in the sedimentary cover and the landscape. It includes debris and loess covers, patterned grounds in the upper layer of thin-bedded limestones, and clusters of rock blocks at the foot of tors resulting from lateral retreat of rock walls due to frost weathering and rock fall (e.g. Tyc 1997, 2009a; Alexandrowicz and Alexandrowicz 2003; Pawelec 2008; Lewandowski 2011).

22.3 Karst Relief

22.3.1 Surface Landforms

The Kraków-Częstochowa Upland has a very diverse landscape, the significant element of which is karst relief. However, 'classical' karst landforms such as solution dolines and karrenfields do not occur in the region. Instead, the characteristic landforms are rocky hills, monadnocks and isolated limestone tors appearing both on the plateaus and valley sides. In the northern part of the upland, within the range of the Pleistocene ice sheet, the shapes of rocky hills and monadnocks were modified by glacial erosion.

Fig. 22.2 Landscape with the rocky hills near Olsztyn town in the north part of the Kraków-Częstochowa Upland (photograph by A. Tyc)

Fig. 22.3 Monadnocks in the highest part of the Ojców Plateau near the village of Jerzmanowice; in the foreground, spongy structures of micro-relief on the top surface of the Łysa Skała Hill (photograph by A. Tyc)

Moreover, in the areas mantled by Pleistocene glacial sediments and loess, covered karst has developed.

22.3.1.1 Rocky Hills, Monadnocks, and Limestone Tors

The surface of the Kraków-Częstochowa Upland is generally flat, becoming undulated in places. Such gentle relief is characteristic of plateaus built mainly of bedded limestones. This image is contrasted with rocky hills and monadnocks built mainly of massive and nodular limestones. The former are large limestone hills that stand out in regional relief. Representative, non-forested hills of this type occur near the town of Olsztyn in the northern part of the Częstochowa Upland (Fig. 22.2). In turn, monadnocks are residual forms of lower rank, built of more resistant massive facies of Upper Jurassic limestone. A separate group of rock relief of the Kraków-Częstochowa Upland are minor, isolated rocks—limestone tors. However, rocky hills and monadnocks are characteristic of the plateaus, so tors occur both in the top parts and slopes of the limestone hills and on slopes of valleys.

These rock landforms are not evenly distributed in the area of the Kraków-Częstochowa Upland. Rocky hills and monadnocks form clusters, the spatial pattern of which refers to the distribution of the above-mentioned carbonate

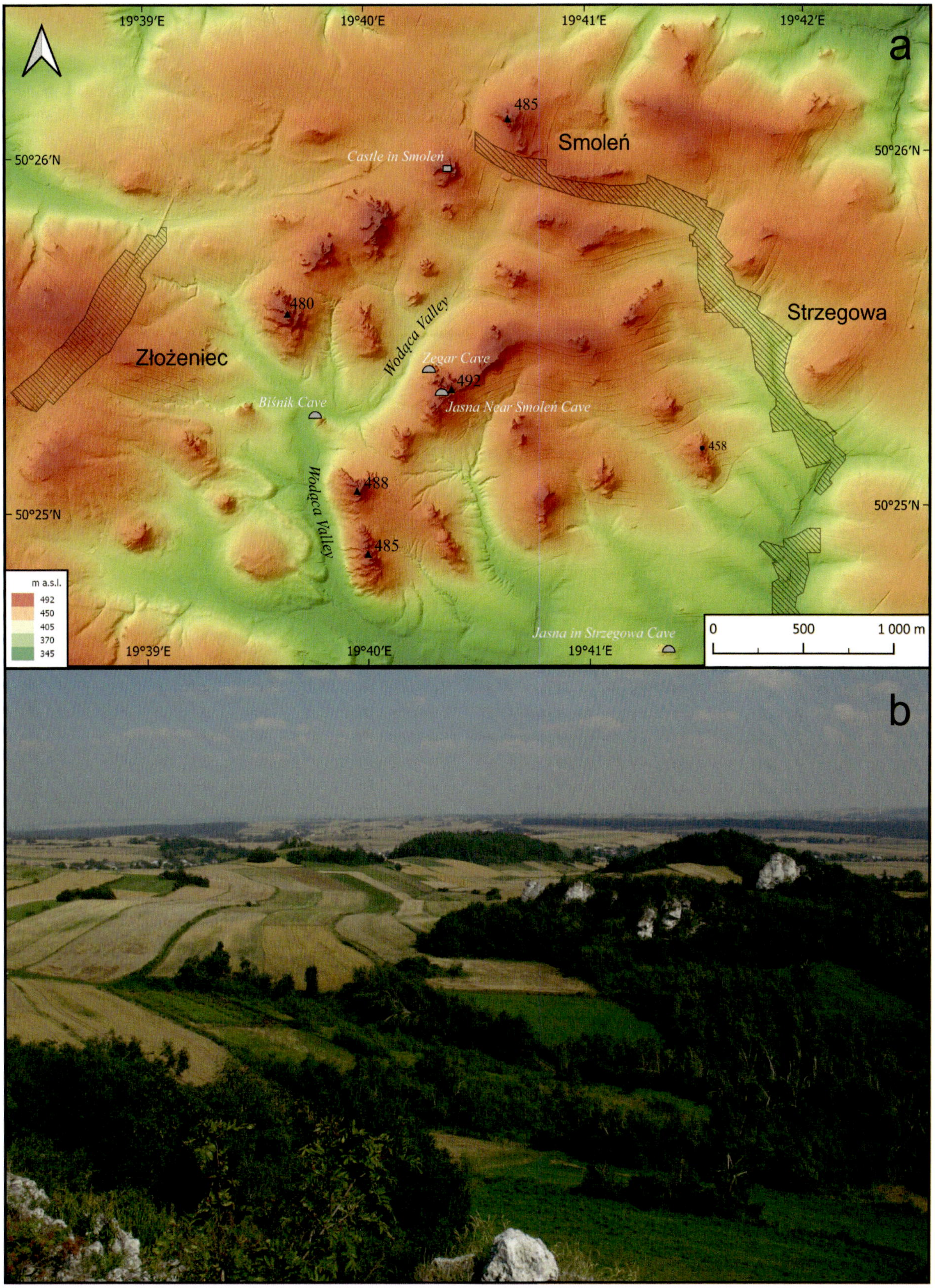

Fig. 22.4 Relief of the carbonate platform near the villages of Smoleń and Strzegowa, with rocky hills and monadnocks, and dry valleys (location on Fig. 22.1). **a** Digital terrain model. **b** Panoramic view of the landscape (photograph by A. Tyc)

Fig. 22.5 Monadnocks and limestone tors of the so-called 'rock city' in Podzamcze village near Ogrodzieniec (photograph by A. Tyc)

buildups (see Matyszkiewicz et al. 2006b). The largest ggregation of such features on the Ojców Plateau lies next to the village of Jerzmanowice, to the west of the Prądnik Valley (Figs. 22.1 and 22.3). The largest monadnocks in this area reach up to 20–30 m high, 100–150 m long, and almost 25 m wide (covering a surface of over 2500 m^2), but most of them are much lower and smaller, up to 50 m long and 20 m wide, covering an area of 700–1000 m^2 (Alexandrowicz and Alexandrowicz 2003). Rocky hills and monadnocks of the Częstochowa Upland are much more numerous and extensive than those on the Ojców Plateau (Figs. 22.4, 22.5 and 22.6). In the village Podzamcze near Ogrodzieniec and on the Zborów Hill next to the village of Podlesice, massive hills are topped with complex rock landforms. They form picturesque landscapes called 'rock cities' or 'rock towns' (Figs. 22.5 and 22.6). These groups include specific minor features such as walls, towers or spires up to 25–30 m high, and the perimeter of the rock groups ranges from several dozen to several hundred metres.

The size of monadnocks depends significantly on the size and shape of carbonate build-ups. Forms developed on extensive and deep-rooted massive limestone bases are more extensive and stable. On the other hand, flat-rooted carbonate build-ups and small bioherms give small rocky forms, which are susceptible to mechanical disintegration (Alexandrowicz and Alexandrowicz 2003).

Relief of rocky hills and monadnocks is determined mainly by the original features of Upper Jurassic limestones and their significant facies differentiation. The presence of joints, primarily vertical and widened by karst or gravitational processes, is essential. Along these joints, individual tors and crags are isolated. The pattern and density of jointing are essential for the shape of rock relief. A crucial structural factor influencing the morphological evolution of monadnocks is the system of fissures and faults developed in several tectonic phases in the Neogene (e.g. Żaba and Tyc 2007). Monadnock morphology is also influenced by differential resistance of the limestones to chemical and mechanical weathering. An essential factor, in this case, is the presence of detrital, grained limestones among the massive and nodular variants. In addition, there are places where discontinuities within massive limestone are observed. They can be interpreted as a record of breaks in bioherm growth.

Moreover, relict caves have been preserved within monadnocks and isolated limestone tors of the Kraków-Częstochowa Upland (Tyc 2005, 2009a, b; Fig. 22.7). The recent relief of monadnocks reflects mainly structural elements of the Upper Jurassic carbonate massif. However, stages of the early development of underground karst (speleogenetic stages) are related to the same structural conditions. Considering the fact that the part of the karst system which developed in bedded limestones became almost destroyed by denudation and erosion, monadnocks and limestone tors are the only elements in the region where relics of early stages of speleogenesis were preserved (Tyc 2005, 2009a, b). In some cases, the walls of monadnocks and tors are relics of cave walls, and they became remodelled by superficial corrosion and mechanical weathering. Relief of the cliffs of many monadnocks shows that their gravitational decay takes place along the walls of these relict caves (Fig. 22.7a).

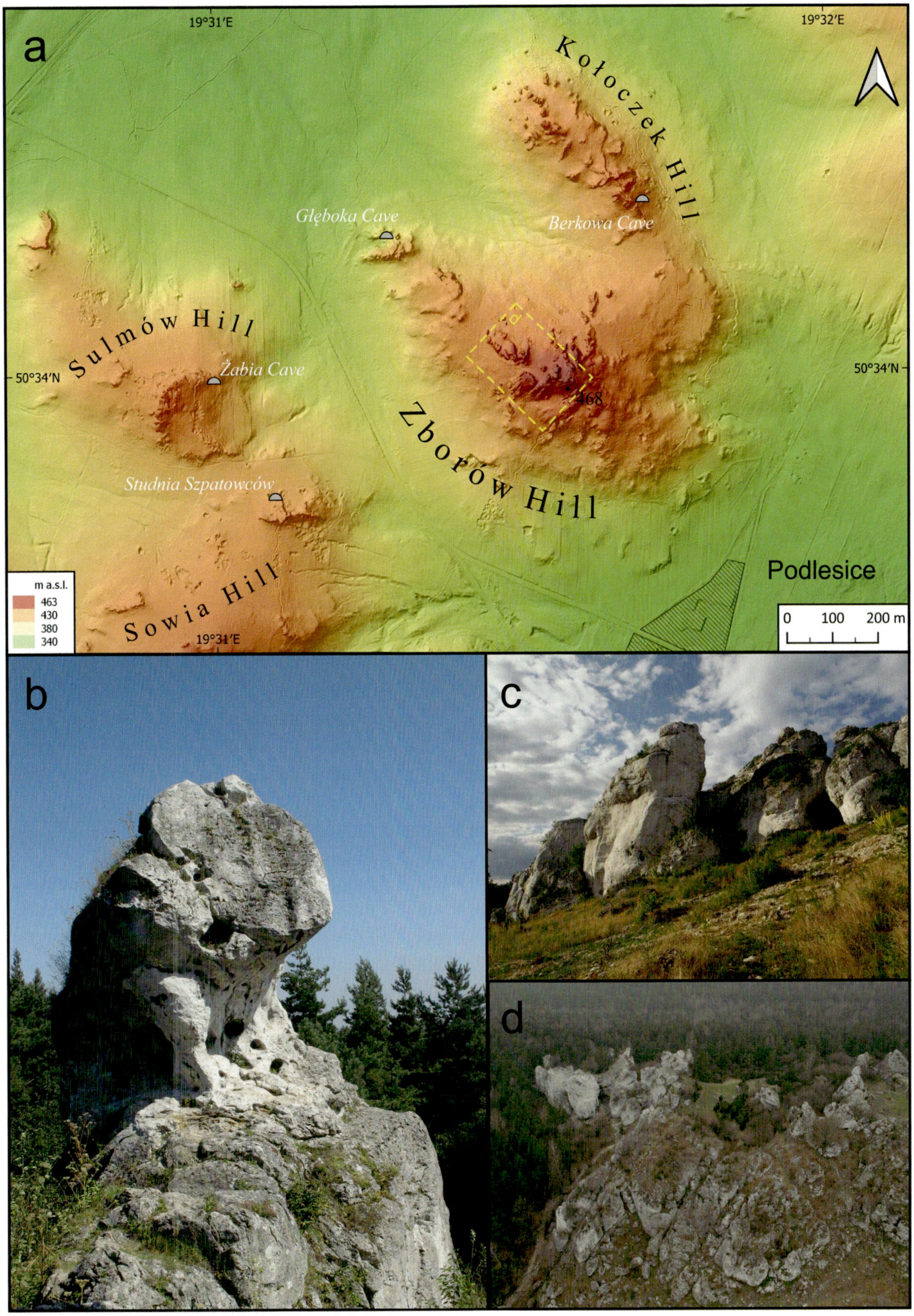

Fig. 22.6 Karst landform with monadnocks, tors, and caves of the rocky hills next to the village of Podlesice (location on Fig. 22.1). **a** Digital terrain model. **b** Limestone tor of Sowa with relics of a cave. **c** Monadnocks on the top of the Zborów Hill with the rock shelter of Kazamar. **d** General view of the highest part of the Zborów Hill (photographs by A. Tyc)

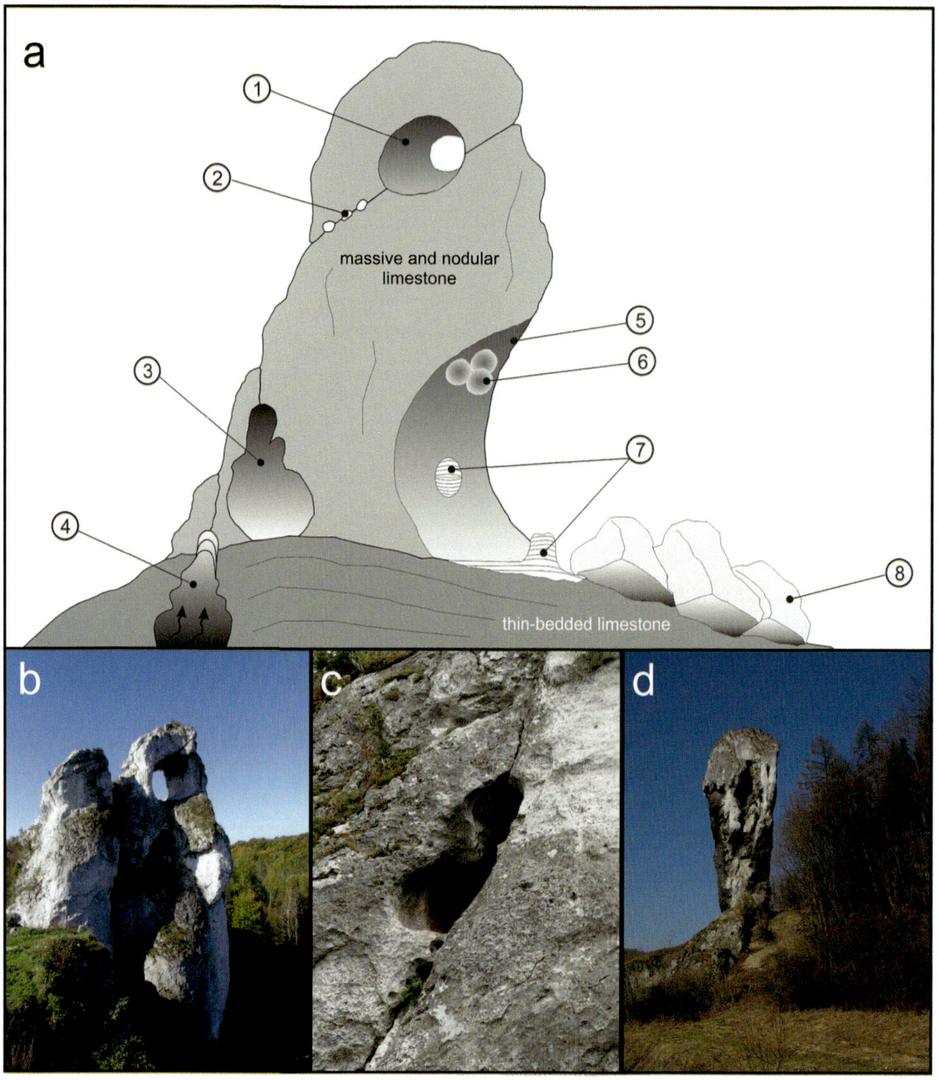

Fig. 22.7 Relics of caves in the relief of monadnocks of the Kraków-Częstochowa Upland. **a** Model of the monadnock with relics of caves (after Tyc 2009b, modified). 1—rock window—relic hemispherical cave chamber, 2—small anastomosis channels related to lithological discontinuity, 3—small relic cave conduit, 4—relic cave with rising half-tubes, 5—overhang being relic cave wall, 6—hemispherical cupolas and niches on the overhanging rock wall, 7—relics of cave clastic sediments and speleothems, 8—rock blocks derived from breakdown of former cave roof. **b** Monadnock with rock window—Okiennik Wielki (Great Window) near the village of Skarżyce. **c** Relics of phreatic tube developed along a joint in massive limestone visible on the wall of a monadnock. **d** Maczuga Herkulesa (Hercules Club)—rocky form (tor) isolated from the slopes of the Prądnik Valley (photographs **b**–**d** by A. Tyc)

To the north and northeast of Olsztyn, rocky hills and monadnocks were reshaped in the Pleistocene by glacial erosion (Fig. 22.1). They were transformed into asymmetric *roche moutonnées*, with gentle proximal (NE) and steep distal (SW) rock slopes. Różycki and Lamparski (1982) linked the remodelling of monadnocks near Olsztyn with the maximum extent of the Odranian glaciation. According to the latest study by Lewandowski (2011), they should be rather associated with the previous Sanian I glaciation.

The rock walls of monadnocks and limestone tors have a very diverse micro-relief. It includes various superficial weathering forms: (1) pock-marked structures—a dense network of shallow, trough-shaped depressions with an oval or round outline, 0.3–3.0 cm in diameter and up to 2.5 cm deep; (2) spongy structures—consisting of irregularly shaped depressions with diameters over 3 cm, short narrow grooves, up to 15 cm long, shallow vertical channels up to 20 cm long, and small nodules; (3) kamenitzas—solution pans, elliptical and isolated depressions, lacking an outlet and with a nearly flat bottom (Otęska-Budzyn 1987). Spongy structures are the most characteristic and frequent features of micro-relief on the monadnocks in the Kraków-Częstochowa Upland. They strictly respond to the structure of limestone building a monadnock (Fig. 22.3). Pock-marked structures occur on inclined wall surfaces. In contrast, spongy structures and kamenitzas are present on more gentle top surfaces of monadnocks and tors. The structurally-conditioned micro-relief of monadnocks and tors of the upland is characterized by the lack of classic hydraulically-controlled karrens such as rills and runnels.

The origin of monadnocks within the landscape of the Kraków-Częstochowa Upland has been discussed since the early 1950s. The karst origin of these rocky forms was formulated by Klimaszewski (1958). According to his concept, the rock landforms standing out from the gentle surface of the Upland are ‘mogotes’. The proposed term refers to the tower karst known from tropical and subtropical karst areas in the Caribbean. Using the term ‘mogote’, Klimaszewski referred to the shape of rock forms and tower karst landscape rather than its origin. The concept of Klimaszewski

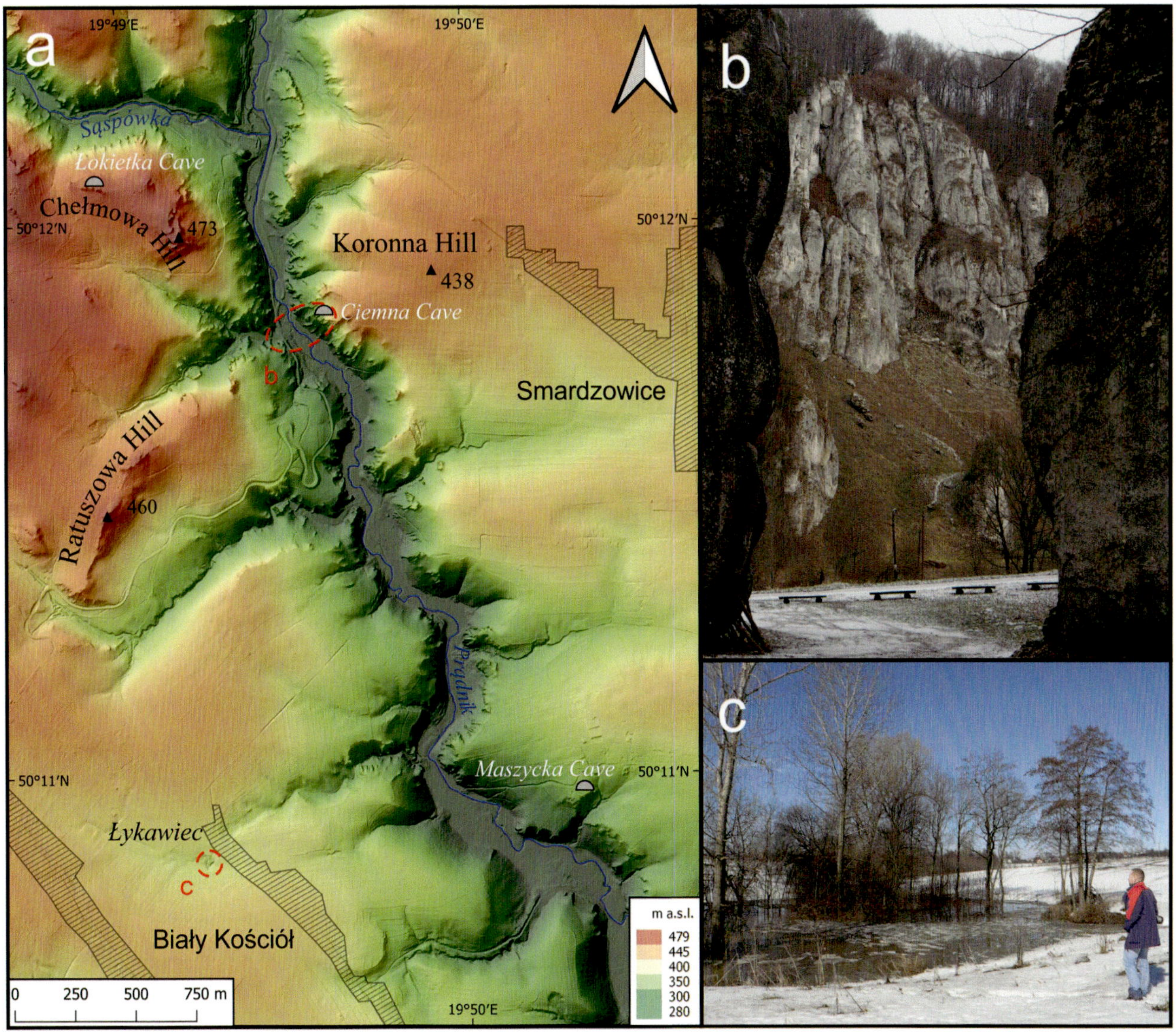

Fig. 22.8 Relief of the central part of the Prądnik Valley (location on Fig. 22.1). **a** Digital terrain model. **b** Brama Krakowska (Kraków Gate) in the foreground and rock walls of the Prądnik Valley—the Rękawica (Glove) Tor in the massif of the Koronna (Crown) Hill in the background (photograph by A. Tyc). **c** Swallow hole Łykawiec during intense snow melting in 2006 (photograph by R. Jach)

(1958) was developed by Polichtówna (1962), Pokorny (1963). The last of the cited authors proposed a schematic profile of the 'mogote' for the area of the Ojców Plateau. This scheme has been reproduced in the literature for decades. However, despite accepting Klimaszewski's view on 'mogotes', Pokorny (1963) envisaged that these are residual forms developed in more resistant rocks than present in the surroundings, and hence, they are monadnocks. The conceptual model of the origin of rocky hills and monadnocks could have been developed after Dżułyński's (1952) publication on the facies differentiation of Upper Jurassic limestones and the origin of rocky limestones (massive and nodular ones). This concept was supported by Szaflarski (1955), who made the first detailed overview and description of monadnocks of the Kraków-Częstochowa Upland. Szaflarski (1955) and Pokorny (1963) considered these forms polycyclic and polygenic. They argued that their relief depends not only on climatic conditions, karst processes and time, but also on their structural features.

22.3.1.2 Rocky Valleys and Gorges

Beside the vast plateaus, punctuated with rocky hills and monadnocks, a characteristic feature of the Kraków-Częstochowa Upland is the presence of deep valleys cutting through the complex of Upper Jurassic limestones. These valleys begin on the plateau as trough-shaped depressions and continue in the form of narrow ravines and then deep rocky gorges with steep slopes, on which limestone tors occur. The upper parts of the valleys are buried by Quaternary sediments.

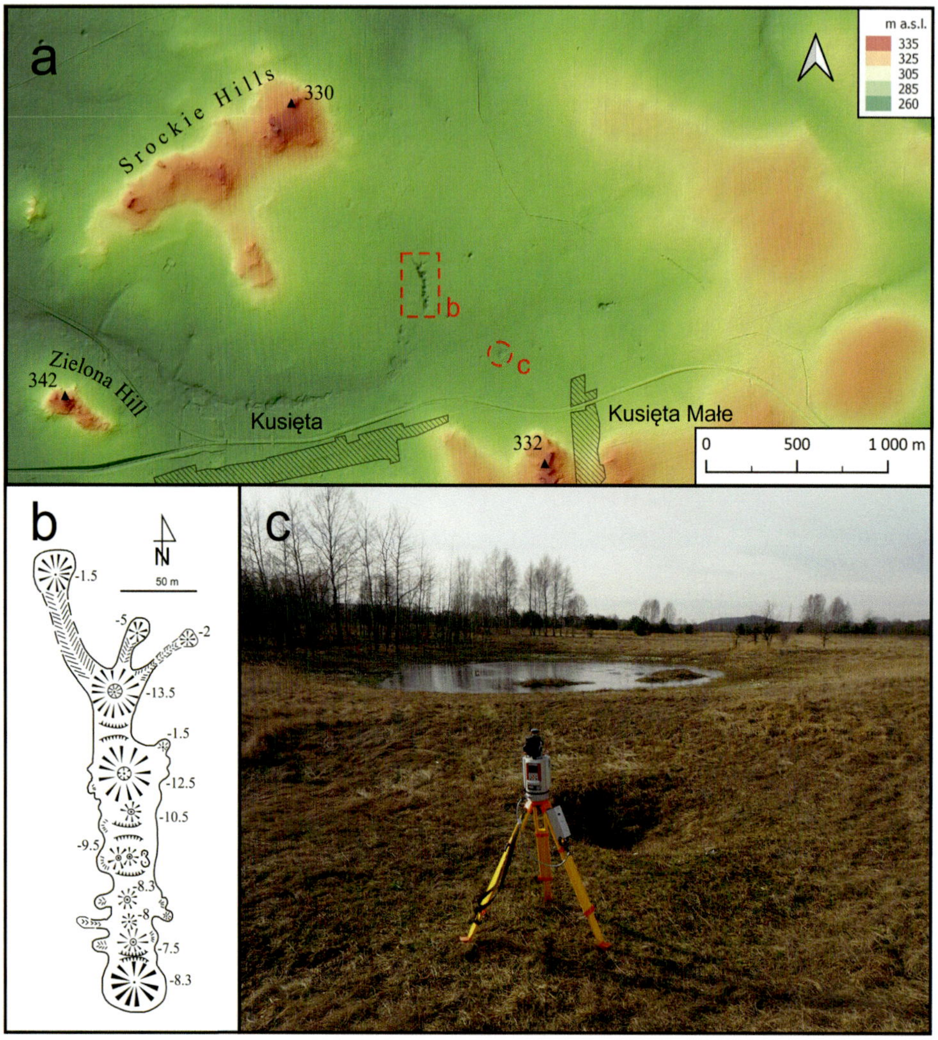

Fig. 22.9 Covered karst to the north of the village of Kusięta (location on Fig. 22.1). **a** Digital terrain model. **b** Complex of suffosion dolines called Koziniec (after Nowak 1993). **c** Shallow suffosion doline filled temporarily with water, small dropout doline in the foreground (photograph by A. Tyc)

The most characteristic and picturesque forms of rocky valleys and gorges are found in the southern part of the Ojców Plateau. The Prądnik Valley is among them, being the most extensive such valley in the entire region (Figs. 22.1 and 22.8a). Tributary valleys of the major river valleys in the plateau are rock gorges without permanent water flow (see tributary gorges of the Prądnik Valley in Fig. 22.8a). Localized, short, and narrow valley reaches (rock gates) are among the characteristic relief elements in the valleys of the Ojców Plateau (Dżułyński et al. 1966; Gradziński et al. 2008). They are often developed in places where a tributary gorge joins the main valley. The most spectacular example is the Brama Krakowska (Kraków Gate) in the Prądnik Valley near Ojców (Fig. 22.8b). Rocky features on the slopes of the Prądnik Valley and side valleys determine the unique landscape of this area. Many rock landforms reach a height of several dozen metres and extend over hundreds of metres, e.g. the rock walls of the Koronna (Crown) Hill (Fig. 22.8b). Maczuga Herkulesa (Hercules's Club), isolated from the slopes, is one of the most spectacular and picturesque rocky forms (Fig. 22.7d). Valleys typical of the Częstochowa Upland are deep and dry, often with rocky upper parts of slopes. They occur in the upper reaches of river valleys draining the upland area. The Wodąca Valley near the villages of Smoleń and Strzegowa is a classic example of dry valleys in this region (Fig. 22.4a).

Rocky valleys and gorges drain the karst massifs and play a crucial geomorphological role as local base levels. Permanent and temporary karst springs are located in these valleys. In the dry valleys of the Częstochowa Upland springs are located in their distal reaches (e.g. at the mouth of the Wodąca Valley, Fig. 22.4a). On the other hand, in the valleys and gorges cutting through the southern part of the Ojców Plateau, springs occur in many places at the bottom of the valley. The karst drainage system, unique at the scale of the whole region, is the underground flow from the semi-blind valley of Łykawiec on the plateau to the spring in the bottom of the Prądnik Valley (Fig. 22.8a, c). Łykawiec is a swallow hole (ponor) developed in loess-like sediments covering this part of the Ojców Plateau, located at a horizontal distance of ca. 1 km from the spring located at the base of rock slopes in the Prądnik Valley. The elevation difference between the swallow hole and spring is ca. 90 m (Gradziński et al. 2008).

22.3.1.3 Covered Karst—Dropout and Suffosion Dolines

Karst areas in Poland that were glaciated in the Pleistocene host characteristic landforms of covered karst. These phenomena are limited to areas with karst in the subsurface zone, and at the same time, where the cover of glacial sediments, clays, and sands (preferably inter-layered) is sufficiently thick, reaching several to several dozen metres. The main manifestations of covered karst are dolines (sinkholes) (Veress 2016; De Waele and Gutiérez 2022). They are mainly dropout dolines developed due to collapses of cover sediments into the cavity developed within karstified bedrock, and suffosion dolines developed due to wash of cover deposits into fissures and voids in a bedrock. In areas with a higher proportion of cohesive sediments (loam and silt layers) above karstified bedrock, dropout dolines develop. On the other hand, in non-cohesive sediments (sands or sandy clay), suffosion processes dominate in the development of shallow depressions. An important factor determining the development of doline forms at the surface is the location of groundwater level below the top layer of karstified rocks.

Favourable conditions for the development of covered karst landforms occur in the northern part of the Kraków-Częstochowa Upland. Discontinuous cover of glacial deposits from at least one glaciation appears to the north and northeast of the town of Olsztyn (Figs. 22.1 and 22.9). Both dropout and suffosion processes dominate the latter and contribute to the development of covered karst relief in this region. The area in the vicinity of the village of Kusięta is particularly interesting (Fig. 22.9a). The groundwater level in the massif built of Upper Jurassic limestones occurs here at a depth of 25–55, 15–30 m below the top layer of limestones. The density of suffosion and dropout dolines in Kusięta is up to 15 forms per 1 km^2 (Nowak 1993; Tyc 2009a). The most interesting example of covered karst in the Kraków-Częstochowa Upland is the complex of 16 suffosion dolines called Koziniec (Fig. 22.9a, b). The complex is 300 m long and 40 m wide. Two dolines are more than 12 m deep, whereas others are shallower. Most dolines have swallow holes at the bottom, active after heavy rains or intense snow melting (Tyc 2009a). Dropout dolines are also developing in the southern and eastern parts of the Kraków-Częstochowa Upland, where karstified limestones are covered with loess or loess-like sediments.

22.3.2 Caves

Karst landforms of the Kraków-Częstochowa Upland include numerous caves. They are distributed unevenly throughout the area. Their occurrence and origin are conditioned mainly by their location in relation to the main structural elements—major faults and the distribution of carbonate build-ups. Despite the presence of Upper Jurassic limestones in the whole upland area, most of ca. 1600 caves and rock shelters in this region lie to the northeast and east of the KLFZ. In the southwestern part of the Kraków-Częstochowa, on the opposite side of the KLFZ, there are only non-karst caves of crevice and fissure types. Moreover, numerous rock shelters within the monadnocks result from frost weathering. They developed actively in the cold and humid conditions of the Pleistocene and continue to enlarge under contemporary environmental conditions.

The karst caves of the Kraków-Częstochowa Upland are generally small. The majority is not more than 40 m long, and most caves have depths up to several metres only. The longest caves are just over 1 km (Nietoperzowa-Zygmunta Cave System—1047 m; and Wierna Cave—1027 m). The deepest one, Studnisko Cave, has a depth of 77.5 m. Their entrances are on the slopes of limestone hills and, in some cases, even in the top parts of the monadnocks. In the case of valleys cutting through the Ojców Plateau or dry valleys of the Częstochowa Upland, cave entrances are located on their slopes. The occurrence of karst caves is related to the massive limestone facies, and their morphology is strongly controlled by the internal lithological diversity and jointing. Their development occurred via gradual widening of pre-existing joints. Most often, the direction of cave passages is related to the directions of such joints. Considering only the karst caves, the majority consists of isolated chambers integrated later into a larger cave along the joint system (Fig. 22.10). The shape of individual chambers is usually hemispherical or elliptical. The largest chamber in the Ciemna Cave (Prądnik Valley in the Ojców Plateau; Fig. 22.8a) is elongated, 88 m long and 23 m wide, with an average height of 8 m (counting from the surface of the bottom sediments).

Most of the karst caves of the Kraków-Częstochowa Upland are relics of the phreatic drainage system within the Upper Jurassic carbonate massifs. The abundance of phreatic-rising morphology of caves in the whole region is one of the manifestations of ascending water circulation in confined conditions, presumably in the presence of elevated temperature (Rudnicki 1978; Głazek and Szynkiewicz 1980; Tyc 2009b; Gradziński et al. 2011; Gradziński and Tyc 2017). One of the most spectacular examples of a cave, the origin of which Rudnicki (1978) linked to the thermal convection process, is the Berkowa Cave in the village of Podlesice (see Fig. 22.6 for location). The cave has an original morphology, with its walls composed of hemispherical niches forming complex grape-like structures. Evidence of convection and rising water in confined conditions can be found in many caves in the region. High cupolas (up to 2–3 m in diameter and up to a few metres high), dead-end chimneys, rising wall channels, and half-tubes are

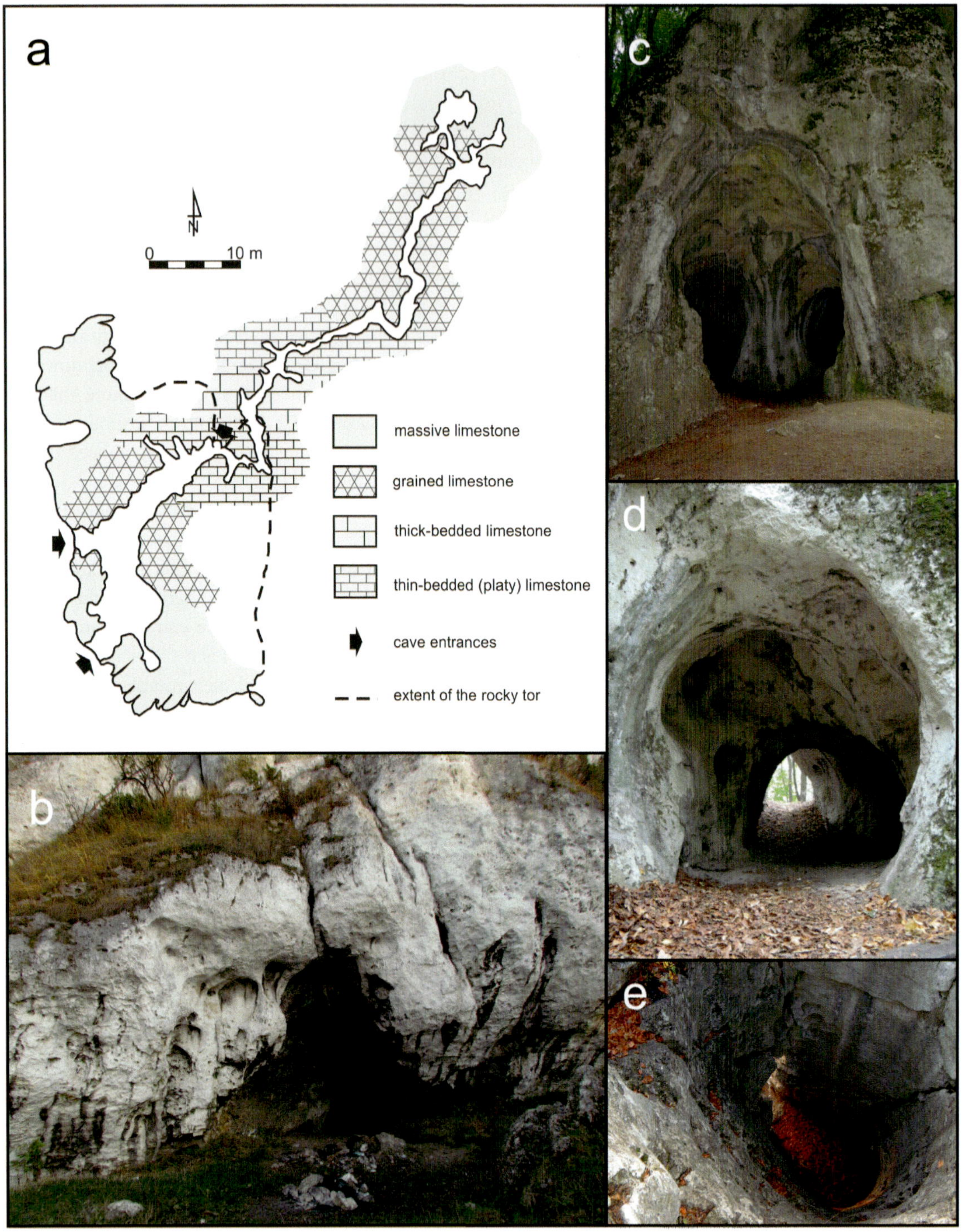

Fig. 22.10 Examples of caves in the Kraków-Częstochowa Upland. **a** Cave morphology versus facies differentiation of Upper Jurassic limestones—W Straszykowej Górze Cave (after Tyc 2009a, modified). **b** Entrance of the Towarna Cave at the foothill of the Duża Towarna Hill near the town of Olsztyn. **c** Entrance chamber of the Ostrężnicka Cave in the village of Złoty Potok—cave rocky relief was modified by mechanical weathering due to frost action. **d** Tunel w Górze Smoleń Cave near the village of Smoleń—cupolas and other cave relief transformed by superficial chemical and mechanical weathering. **e** Studnia w Amfiteatrze in the Sokole Hills near the town of Olsztyn—relic shaft, which was in the past infilled with residual clays and moulding sands (photographs by A. Tyc)

probably traces of the hypogene stage of speleogenesis. Morphological evidence indicates the ascending stage in cave evolution, but is still insufficient to confirm its hypogene nature. Additional evidence of the hypogene stage in the speleogenesis in the Kraków-Częstochowa Upland comes from calcite spar occurrence in caves. Calcite crystalized underwater in phreatic conditions, presumably under slightly elevated temperatures, was found in caves of the Sokola Hill near Olsztyn (Gradziński et al. 2011). Calcite veins and spar were quite common in caves in the northern

Fig. 22.11 Kettle-shaped doline with overhanging sides—an example of an unroofed hypogenic cave (Zborów Hill in the village of Podlesice) (photograph by A. Tyc)

part of the region, but most were mined in the late nineteenth and early twentieth centuries (Tyc 2009a).

Relics of hypogene caves morphology are present among forms preserved on monadnocks and tors. Interesting geomorphological features are kettle-shaped dolines, which can be regarded as unroofed hypogene caves (Fig. 22.11).

It should be emphasized, however, that in the Pleistocene the morphology of caves was modified by other geological and geomorphological processes, e.g. condensation corrosion within cupolas on the cave ceiling, corrosion on the contact with infilling sediments, and gelivation (Fig. 22.10). Almost all karst caves of the Kraków-Częstochowa Upland are filled or partly filled with fine-grained sediments and debris (Madeyska and Cyrek 2002; Krajcarz et al. 2014, 2016).

22.4 Evolution of Karst Landforms

The origin and age of karst of the Kraków-Częstochowa Upland are still a subject of discussion and require further research. The non-uniform Neogene uplift of different parts of this region impedes the interpretation of karst development. The upland forms the western part of the Meta-Carpathian Ridge in the foreland of the overthrust Carpathians (Lewandowski 1993). This uneven uplift, much more significant in the south, means that in the recent relief of the Ojców Plateau there may be relics of different stages of karstification than in the more northerly Częstochowa Upland. The Late Cretaceous marine recession and the Holocene constrain the general time frame for karst development in the Kraków-Częstochowa Upland. The extent of the Cretaceous sedimentary cover and the rate of its subsequent removal were essential for the evolution of karst in different parts of the upland. Hence, several stages can be distinguished in the nearly 70 million years of karst development and landform evolution.

- ***Stage I—Paleogene and Lower Miocene***
  Paleogene to Lower Miocene timespan is the oldest and the most controversial stage of karstification. The Paleogene denudation affected vast areas of Central Europe and took place in subtropical climate conditions. This long terrestrial period has been associated with intensive development of karst, which is recorded in numerous palaeokarst forms (e.g. Gradziński 1962; Gilewska 1964; Głazek and Szynkiewicz 1980; Głazek 1989; Felisiak 1992). The most significant hypogene event recorded in the phreatic stage of cave evolution occurred in the Paleogene, most probably up to Eocene. The ascending groundwater circulation within the Upper Jurassic limestones was active before the removal of the confining Cretaceous marls (Głazek and Szynkiewicz 1980; Gradziński and Tyc 2017). Relics of the hypogene karst of this stage have been preserved in the relief of the Częstochowa Upland (e.g. in the vicinity of Podlesice and Olsztyn).

  After the removal of the Cretaceous sedimentary cover in the Paleogene, intense chemical weathering and early development of residual surface forms (monadnocks) took place (e.g. Pokorny 1963; Gilewska 1964; Felisiak 1992). One of the effects of this stage of denudation are the so-called moulding sands. These are sediments most probably redeposited to karst depressions and shafts from sandy Cretaceous deposits present in the proximity of the deposition area (Gradziński 2004). Relics of these sediments have been preserved in the fossil surface karst forms and caves of the Częstochowa Upland.

- ***Stage II—Miocene***
  The Miocene witnessed the main phase of tectonic development of the Carpathian foreland. Most likely, this stage can be linked with the formation of joints and fissures essential for the development of caves in the subsequent phases. Latitudinal tectonic grabens dividing the Kraków-Częstochowa Upland into smaller segments were also formed at that time, and faults of the KLFZ were reactivated.
- ***Stage III—Pliocene and Early Pleistocene***
  This stage of karst evolution can be associated with the uplift of the Meta-Carpathian Swell. As a result of significant lowering of the erosion base due to the Late Miocene and Pliocene uplift, the pre-existing caves were exhumed and distinct vadose entrenchments developed. Many caves, especially in the Częstochowa Upland, were subsequently filled with sequences of bone-bearing sediments from the Lower Pliocene to Early Pleistocene (e.g. Głazek 1989). Monadnocks were isolated and reshaped in this stage. The layout of the current network of valleys has formed, including the main geomorphic features of rocky valleys and gorges (Alexandrowicz and Alexandrowicz 2003).
- ***Stage IV—Middle and Late Pleistocene***
  Since the Kraków-Częstochowa Upland was only marginally covered by the ice sheet, the Pleistocene stage of karst development involves mainly periglacial processes. Mechanical weathering during multiple Pleistocene glaciations intensified and partial burial of the entire region by glacial, glacifluvial, and periglacial (loess) sediments occurred. Over the remains of red and reddish-brown allochthonous Pliocene and Early Pleistocene sediments, thick series of Middle and Late Pleistocene deposits were deposited in caves (Głazek 1989; Madeyska and Cyrek 2002; Krajcarz et al. 2014). Speleothems were formed in more humid periods. Mechanical weathering and gravitational processes caused the retreat of monadnock walls and accumulation of rock debris and blocks at their base. In the periglacial environment of the Late Pleistocene, a specific epikarst zone with debris mantles and patterned ground developed (Tyc 1997).
- ***Stage V—Holocene***
  The present-day development of karst landform is associated with the exhumation of relief from beneath clastic sediments of Pleistocene age. Within the range of glacial deposits and loess varieties, processes related to covered karst are developing. The walls of monadnocks and tors continue to be modelled by weathering, giving rise to specific micro-relief.

22.5 Geomorphological Landscape and Cultural Heritage

The Kraków-Częstochowa Upland is known for numerous well-studied archaeological sites. Caves and rock shelters with well-exposed and accessible entrances and overhanging rock walls attracted early humans to settle in the upland. Inhabitation of caves in the region from the

Fig. 22.12 Ruins of a medieval castle built on the monadnocks of a vast rock hill in Podzamcze near Ogrodzieniec town (photograph by A. Tyc)

Middle Palaeolithic to historical times has been confirmed (e.g. Cyrek 2009). The earliest relics of human presence have been found in the Biśnik Cave in the Wodąca Valley (Fig. 22.4). The cave and its surroundings were inhabited in the period between the Odranian and Late Weichselian glaciations, i.e. over 250,000 years. At least 18 settlement phases have been distinguished, which can be interpreted as single campsites set up by cave inhabitants (Cyrek et al. 2014). A distinct intensification of Neanderthal settlement, which is characteristic of the whole of Europe, took place in the early Weichselian, with representatives of Middle Palaeolithic—Micoquian, especially Micoquo-Prondnikian, and Mousterian cultures. Caves and rock shelters in the Ojców area in the south (e.g. Koziarnia Cave) can be regarded as more frequently settled basic encampments. At the same time, other ones more to the north were briefly occupied as occasional camps (Cyrek 2009). The second period of intense human settlement in the Kraków-Częstochowa Upland is dated to the Upper Palaeolithic. Aurignacian and Gravetian cultures created by *Homo sapiens* have left their rich testimony in cave settlements of the entire region.

The karst landscape, with rocky hills and monadnocks, was attractive for humans also in the following periods of the Neolithic, Bronze Age, and Middle Ages. In the latter, rocky hills and individual monadnocks were used to construct defensive castles at the former western border of the Polish Kingdom. They formed the entire defensive chain along the main axis of the upland, between Kraków and Olsztyn, the so-called Eagle nests (Fig. 22.12).

The Kraków-Częstochowa Upland has a long tradition in nature conservation in Poland. Protection of karst phenomena dates back to the first half of the twentieth century when projects of nature reserves were created to protect caves and karst landforms in the vicinity of Ojców, Olsztyn, and Złoty Potok. Finally, more than 30 years later, the Ojców National Park was established in 1956, covering the Prądnik Valley. During that time, nature reserves 'Parkowe' next to the village of Złoty Potok, 'Sokole Góry' close to the town of Olsztyn, and 'Góra Zborów' in the vicinity of the village of Podlesice were created. Individual protection as nature monuments and documentary sites was established to protect monadnocks and caves. In the early 1980s, a complex of landscape parks was established, covering substantial areas of the Kraków-Częstochowa Upland.

22.6 Conclusions

The Kraków-Częstochowa Upland has a very diverse landscape with significant karst landforms. The most distinctive landforms of the region are limestone rocky hills, monadnocks, and isolated tors appearing both on the plateaus and within the slopes of hills and valleys. Significant lithological differentiation of the Upper Jurassic limestones and complex tectonic settings have controlled the spatial differentiation and evolution of karst and caves of this region.

Considering the structural settings, different magnitude of Neogene uplift and variable position in respect to the extent of the Pleistocene inland glaciation, karst landforms and landscapes of the Ojców Plateau and the Częstochowa Upland are significantly different. The rocky valleys and gorges, the picturesque Prądnik Valley among them, cut the southern part of the Ojców Plateau. On the other hand, the northern part of the Częstochowa Upland was the only part of the region covered by the Pleistocene glaciation, and the rocky hills and monadnocks were transformed by glacial erosion into asymmetric roche moutonnées. In the areas mantled by glacial deposits and loess covered karst developed.

Although the origin and age of karst and caves are still discussed, the Late Cretaceous marine recession and the Holocene constrain the timespan for the development of karst landforms in the Kraków-Częstochowa Upland. Karst landforms and caves of the Kraków-Częstochowa Upland are polygenetic and multi-stage. Several stages can be distinguished within the nearly 70 million years-long history of karst evolution, with a significant hypogene stage in the early phase of karst development.

References

Alexandrowicz SW, Alexandrowicz Z (2003) Pattern of karst landscape of the Cracow Upland (South Poland). Acta Carsologica 32:39–56

Buła Z, Habryn R, Krieger W, Kurek S, Markowiak M, Woźniak P (2002) Atlas geologiczny paleozoiku bez permu w strefie kontaktu bloków górnośląskiego i małopolskiego. Państwowy Instytut Geologiczny, Warszawa

Cyrek K (2009) Archaeological studies in caves of the Częstochowa Upland. In: Stefaniak K, Tyc A, Socha P (eds) Karst of the Częstochowa Upland and of the Eastern Sudetes: Palaeoenvironments and protection. Studies of the Faculty of Earth Sciences, University of Silesia 56, Sosnowiec–Wrocław, pp 145–160

Cyrek K, Sudoł M, Czyżewski Ł, Osipowicz G, Grelowska M (2014) Middle Palaeolithic cultural levels from Middle and Late Pleistocene sediments of Biśnik Cave, Poland. Quat Intern 326–327:20–63

De Waele J, Gutiérez F (2022) Karst hydrogeology, geomorphology and caves. Wiley Blackwell, 888 p

Dżułyński S (1952) Powstanie wapieni skalistych jury krakowskiej. Rocz Pol Tow Geol 21:125–180

Dżułyński S, Henkiel A, Klimek K, Pokorny J (1966) Rozwój rzeźby dolinnej południowej części Wyżyny Krakowskiej. Rocz Pol Tow Geol 36:329–343

Felisiak I (1992) Oligocene-Early Miocene karst deposits and their importance for recognition of the development of tectonics and relief in the Carpathian Foreland, Kraków Region, Southern Poland. Ann Soc Geol Pol 62(2):173–207

Gilewska S (1964) Fossil Karst in Poland. Erdkunde 18:124–135

Głazek J, Dąbrowski T, Gradziński R (1972) Karst of Poland. In: Herak M, Stringfield VT (eds) Karst—important Karst Regions of the Northern Hemisphere. Elsevier, Amsterdam, pp 327–340

Głazek J, Szynkiewicz A (1980) Kras między Pilicą a Wartą i jego praktyczne znaczenie. In: Przewodnik 52 Zjazdu Pol Tow Geol, Bełchatów. Inst Geol, Warszawa, pp 84–99

Głazek J (1989) Palaeokarst in Poland. In: Bosák P, Ford DC, Głazek J, Horáček I (eds) Palaeokarst. A systematic and regional review. Elsevier and Academia, Amsterdam–Prague pp 77–105

Gradziński M, Gradziński R, Jach R (2008) Geologia, rzeźba I zjawiska krasowe okolic. In: Klasa A, Partyka J (eds) Monografia Ojcowskiego Parku Narodowego. Przyroda. Ojcowski Park Narodowy, Ojców, pp 31–95

Gradziński M, Hercman H, Kicińska D, Pura D, Urban J (2011) Ascending speleogenesis of Sokola Hill: a step towards a speleogenetic model of the Polish Jura. Acta Geol Pol 61:341–365

Gradziński M, Tyc A (2017) Hypogene Speleogenesis in the Karst of Poland—regional review. In: Klimchouk A, Palmer AN, De Waele J, Auler AS, Audra P (eds) Hypogene Karst Regions and caves of the world. Springer, Cham, pp 349–362

Gradziński R (1962) Rozwój podziemnych form krasowych w południowej części Wyżyny Krakowskiej. Rocz Pol Tow Geol 32:429–492

Gradziński R (2004) Piaski formierskie:problemy wieku i genezy. In: Partyka J (ed) Zróżnicowanie i przemiany środowiska przyrodniczo-kulturowego Wyżyny Krakowsko-Częstochowskiej. Tom 1 Przyroda. Ojcowski Park Narodowy, Ojców, pp 43–46

Jędrys J, Grabowska T, Krajewski M, Matyszkiewicz J, Żaba J (2004) Założenia strukturalne górnojurajskich budowli węglanowych na Wyżynie Krakowsko-Wieluńskiej w świetle danych magnetycznych. In: Partyka J (ed) Zróżnicowanie i przemiany środowiska przyrodniczo-kulturowego Wyżyny Krakowsko-Częstochowskiej. Tom 1 Przyroda. Ojcowski Park Narodowy, Ojców, pp 19–26

Klimaszewski M (1958) Nowe poglądy na rozwój rzeźby krasowej. Przegl Geogr 30:421–438

Krajcarz MT, Bosák P, Šlechta S, Pruner P, Komar M, Dresler J, Madeyska T (2014) Sediments of Biśnik Cave (Poland): lithology and stratigraphy of the Middle Palaeolithic site. Quat Intern 326–327:6–19

Krajcarz MT, Cyrek K, Krajcarz M, Mroczek P, Sudoł M, Szymanek M, Tomek T, Madeyska T (2016) Loess in a cave: Lithostratigraphic and correlative value of loess and loess-like layers in caves from the Kraków-Częstochowa Upland (Poland). Quat Intern 399:13–30

Krajewski M, Matyszkiewicz J (2009) Upper Jurassic deposits in the Częstochowa Upland. In: Stefaniak K, Tyc A, Socha P (eds) Karst of the Częstochowa Upland and of the Eastern Sudetes: Palaeoenvironments and protection. Studies of the Faculty of Earth Sciences, University of Silesia 56, Sosnowiec–Wrocław, pp 37–56

Lewandowski J (1993) Rzeźba podczwartorzędowa regionu śląsko-krakowskiego i jej ewolucja morfogenetyczna. Folia Quat 64:101–121

Lewandowski J (2011) 'Jurajska Oaza Śródlodowa' w świetle badań ostatniego półwiecza. Przegl Geol 59:732–738

Łoziński W (1912) Beiträge zur Oberflächengeologie des Krakauer Gebietes. Jb Geol Anst 62:71–86

Madeyska T, Cyrek K (2002) Cave fillings—a chronicle of the past. An outline of the Younger Pleistocene cave sediments study in Poland. Acta Geol Pol 52:75–95

Matyszkiewicz J (1997) Microfacies, sedimentation and some aspects of diagenesis of Upper Jurassic sediments from the elevated part of the Northern peri-Tethyan Shelf: a comparative study on the Lochen area (Schwäbische Alb) and the Kraków area (Kraków-Wieluń Upland, Poland). Berliner Geowiss Abh E21, 111 p

Matyszkiewicz J, Krajewski M, Kędzierski J (2006a) Origin and evolution of an Upper Jurassic complex of carbonate buildups from Zegarowe Rocks (Kraków-Wieluń Upland, Poland). Facies 52:249–263

Matyszkiewicz J, Krajewski M, Żaba J (2006b) Structural control on distribution of Upper Jurassic carbonate buildups in the Kraków-Wieluń Upland (South Poland). N Jahrb Geol Paläont Miner 3:182–192

Matyszkiewicz J, Kochman A, Rzepa G, Gołębiowska B, Krajewski M, Gaidzik K, Żaba J (2015) Epigenetic silicification of the Upper Oxfordian limestones in the Sokole Hills (Kraków-Częstochowa Upland): relationship to facies development and tectonics. Acta Geol Pol 65:181–203

Nowak WA (1993) Skrasowienie podziemne wapieni I jego odzwierciedlenie w rzeźbie Wyżyny Krakowsko-Wieluńskiej w rejonie Częstochowy. Studia Ośrodka Dokumentacji Fizjograficznej 21, 169 p

Otęska-Budzyn J (1987) Wapienne ostańce koło Ogrodzieńca na Wyżynie Częstochowskiej. Ochr Przyr 45:325–345

Pawelec H (2008) Origin of hilltop monadnocks on the Ojców Plateau as reconstructed from slope deposits. Geologos 14:163–176

Pokorny J (1963) The development of mogotes in the southern part of the Cracow Upland. Bull Acad Pol Sc Ser Geol-Geogr 11(3):169–175

Polichtówna J (1962) Ostańce Wyżyny Krakowsko-Częstochowskiej, ich geneza i znaczenie w krajobrazie. Ochr Przyr 28:255–284

Pulina M, Tyc A, Żaba J (2002) The role of endogenic processes in evolution of karst in Central European Platform (example of South Polish Uplands). In: Gabrovsek F (ed) Evolution of Karst: from Prekarst to Cessation. Zbirka Carsologica, Ljubljana–Postojna, pp 425–432

Pulina M, Żaba J, Polonius A (2005) Związek rozwoju form krasowych Pasma Smoleńsko-Niegowonickiego z tektoniczną aktywnością podłoża Wyżyny Krakowsko-Wieluńskiej. Kras i Speleologia 11:39–85

Różycki SZ (1960) Czwartorzęd regionu Jury Częstochowskiej i sąsiadujących z nią obszarów. Przegl Geol 8:421–429

Różycki SZ, Lamparski Z (1982) Exaration phenomena at the terminal ice sheet parts of the Middle-Polish Glaciation along the northern slopes of the Central Polish Uplands. Acta Geol Pol 32:207–215

Rudnicki J (1978) Role of convection in shaping subterranean karst forms. Kras i Speleologia 2:92–101

Sawicki L (1920) Zarys ogólnej geografii ziem polskich – wykłady wygłoszone na Kursie Nauczycielskim w Cieszynie 1919. Cz. 1, Polskie Towarzystwo Pedagogiczne na Śląsku, Cieszyn, 88 p

Solon J, Borzyszkowski J, Bidłasik M, Richling A, Badora K, Balon J, Brzezińska-Woźniak T, Chabudziński Ł, Dobrowolski R, Grzegorczyk I, Jodłowski M, Kistowski M, Kot R, Krąż P, Lechnio J, Macias A, Majchrowska A, Malinowska E, Migoń P, Myga-Piątek U, Nita J, Papińska E, Rodzik J, Strzyż M, Terpiłowski S, Ziaja W (2018) Physico-geographical mesoregions of Poland: verification and adjustment of boundaries on the basis of contemporary spatial data. Geogr Polon 91:143–170

Szaflarski J (1955) Wierzchowinowe formy skalne Wyżyny Krakowsko-Częstochowskiej. Polskie Towarzystwo Turystyczno-Krajoznawcze Oddział w Częstochowie, Muzeum Regionalne w Częstochowie, 29 p

Tyc A (1997) Epikarstic features in zones affected by periglacial processes—example of the Silesian-Cracow Upland (Poland). In: Proceedings of the 12th international congress of speleology, La Chaux-de-Fonds, vol 1, pp 289–292

Tyc A (2004) Europe, Central. In: Gunn J (ed) Encyclopedia of caves and Karst Science. Taylor & Francis, London, New York, pp 684–688

Tyc A (2005) Relikty krasu podziemnego we współczesnej morfologii ostańców Wyżyny Krakowsko-Częstochowskiej. In: Kotarba A, Krzemień K, Święchowicz J (eds) Współczesna ewolucja rzeźby

Polski. VII Zjazd Geomorfologów Polskich, UJ, Kraków, pp 481–486

Tyc A (2009a) Karst and caves of the Częstochowa Upland—morphology and the outline of speleogenesis. In: Stefaniak K, Tyc A, Socha P (eds) Karst of the Częstochowa Upland and of the Eastern Sudetes: palaeoenvironments and protection. Studies of the Faculty of Earth Sciences, University of Silesia 56, Sosnowiec–Wrocław, pp 11–36

Tyc A (2009b) Hypogene ascending speleogenesis in the Kraków-Częstochowa Upland (Poland)—evidence in cave morphology and surface relief. In: Klimchouk AB, Ford DC (eds) Hypogene Speleogenesis and Karst hydrogeology of Artesian Basins. Ukrainian Institute of Speleology and Karstology, Special Paper 1, Simferopol, pp 201–208

Veress M (2016) Covered karsts. Springer, Dordrecht, p 536

Żaba J (1994) Mezoskopowe struktury kwiatowe w dolnopaleozoicznych utworach NE obrzeżenia GZW – rezultat transpresyjnego ścinania w strefie dyslokacyjnej Kraków-Myszków (Hamburg-Kraków). Przegl Geol 42:643–648

Żaba J (1999) Ewolucja strukturalna utworów dolnopaleozoicznych w strefie granicznej bloków górnośląskiego i małopolskiego. Prace Państw Inst Geol 166:1–162

Żaba J, Tyc A (2007) Tectonic and karst phenomena of the Ogrodzieniec area. In: Socha P, Stefaniak K, Tyc A (eds) Karst and Cryokarst—guidebook and abstracts, Sosnowiec–Wrocław, pp 120–121

Andrzej Tyc is a geographer, Assistant Professor at the University of Silesia in Katowice. His main research interests are karst geomorphology, speleology and structural geomorphology. The Kraków-Częstochowa Upland in Poland is his main research area. He also investigates the evolution of relief and karst development in the Dinaric Mountains in Slovenia and in the Central Andes in Peru. He was also involved in research projects in Australia, China, Czechia, Siberia and Cuba. He is the co-chairman of the Structural Geomorphology Commission of the Association of Polish Geomorphologists and a member of the *International Journal of Speleology* editorial board.

Kraków—Anthropogenic Changes in the Relief of a Large City

23

Adam Łajczak and Roksana Zarychta

Abstract

The location of various types of anthropogenic landforms is analysed in the context of major natural landforms within the city of Kraków. While anthropogenic landforms first appeared in the early Middle Ages, the city area is dominated by those that emerged in the nineteenth and twentieth centuries. One notable trend points to the diminishing of the relative elevation differences between the main natural landforms as a result of an equalising trend whereby the lower-lying landforms are being built-up and the higher-located ones are being lowered. On the other hand, the appearance of a number of convex and concave anthropogenic landforms, especially adjacent to each other and measuring more than five metres in either height or depth, increased local elevation differences. Overall, the historic city centre is marked by a dominant process of raising and flattening of the terrain as cultural sediments are deposited, while at a local scale the processes in action lead to the differential lowering and raising of adjacent terrain, thus increasing the undulation of the topographical surface.

Keywords

Human impact · Relief transformation · Anthropogenic landforms · Quarrying · Kraków

A. Łajczak (✉)
Association of Polish Geomorphologists, Bogumił Krygowski 10, 61-680 Poznań, Poland
e-mail: alajczak@o2.pl

R. Zarychta
Department of Geodesy, Cartography and Spatial Management, Institute of Law, Economics and Administration, University of the National Education Commission, Krakow, Podchorążych 2, 30–084 Kraków, Poland
e-mail: roksana.zarychta@uken.krakow.pl

23.1 Introduction

Investigations of the relief of the city of Kraków, which was the capital of Poland from the eleventh to the turn of seventeenth century, started in the nineteenth century and until recently focused on the morphogenesis of this area and the distribution of landforms (Tyczyńska 1968; Tyczyńska and Chmielowiec 1988; Izmaiłow 2013). Within the boundaries of Kraków, which is one of the largest and oldest towns in Poland, there are landforms which document human impact on the natural environment from early medieval times to the modern era. Changes in the relief within the contemporary boundaries of Kraków were caused by similar processes as in other cities, i.e. deposition of anthropogenic sediments, land drainage, river regulation, quarrying, earthworks (ground levelling, creation of convex linear landforms, and mounds of symbolic or cultural importance), and tipping of mineral waste. The assessment of anthropogenic changes of relief in the Kraków area has, until the present, been marginalised, and papers concerning these issues currently concentrate on the reconstruction of the original relief of the city and changes in the morphometry of its historic centre (Zarychta 2019; Łajczak and Zarychta 2020; Zarychta et al. 2020; Łajczak et al. 2021a, b).

Analyses of anthropogenic changes of the relief of urban areas have been undertaken in many geographical settings (Kaniecki 2013; Molewski and Juśkiewicz 2014; Brandolini et al. 2017; Reynard et al. 2017; Pröschel and Lehmkuhl 2019). The magnitude of anthropogenic changes in such areas is influenced by the duration of human impact (age of towns), location in a more or less geomorphologically varied area, and the functions of towns. Therefore, the greatest horizontal and vertical dimensions of anthropogenic landforms (both convex and concave) are observed in towns with a long history, with mining or quarrying industries, with large height differences, or in towns with dynamic development (Campagna and Scardozzi 2016; Dulias 2016; Van et al. 2015; Dall'Aglio et al. 2017; Pica et al. 2017; Luberti 2018; Qiao et al. 2019; Jancewicz et al. 2021).

P. Migoń and K. Jancewicz (eds.), *Landscapes and Landforms of Poland*,
World Geomorphological Landscapes, https://doi.org/10.1007/978-3-031-45762-3_23

In addition to information concerning geomorphology, Quaternary geology (Vogel et al. 2011; Pröschel and Lehmkuhl 2019), geoengineering and archaeology (Setmajer 1973; Radwański 1975; Carver 1983; Mozzi et al. 2018), cartographic and remote sensing materials (Van et al. 2015; Qiao et al. 2019) are also used in research of relief changes in urban settings. Changes in the relief are defined as: (a) overlaying of the ground surface with deposits leading to a flattening of the area or to an increase in local relief and (b) lowering of the ground surface as a result of earthworks, mining, and quarrying or river regulation. In towns, flat areas are often adjacent to areas which were topographically raised or lowered, so that local height differences increased. These changes in the relief of certain areas of a town are usually described in a qualitative way, and only sporadically in a quantitative way, with the application of topographic parameters such as altitude, local relative height, mean slope, and the extent of areas without aspect and with different aspects (Łajczak et al. 2021a, b).

In this chapter, the distribution of different types of anthropogenic landforms is analysed, and these are also classified to distinguish height or depth ranges within the main individual landforms in the Kraków area. On this base anthropogenic changes in the relief of the area within the city boundaries are assessed to show predominant trends in relief development. The analysis of changes in relief was carried out for the whole area of Kraków, with special reference to its historic centre.

23.2 Study Area

23.2.1 Location of Kraków

The city of Kraków is located in southern Poland, on the upper reach of the River Vistula in the borderland of the Polish Uplands, Northern Subcarpathians and Western Carpathians (Fig. 23.1a, b) (Solon et al. 2018). The city area is 327 km^2, and it is located at an altitude of 188–358 m asl; however, one of the anthropogenic landforms reaches 394 m asl. As much as 88% of the territory of Kraków is located within the Northern Subcarpathians (Kraków Gate, Vistula Lowland, Kraków Submontane Region). The Kraków Gate separates the Polish Uplands from the Western Carpathians. The zone of the Kraków Gate is used by the Vistula River, which flows across it in a narrow water gap. The Vistula Lowland represents a plain, whereas the Kraków Submontane Region shows features of the northern borderland of the Western Carpathians (Izmaiłow 2013).

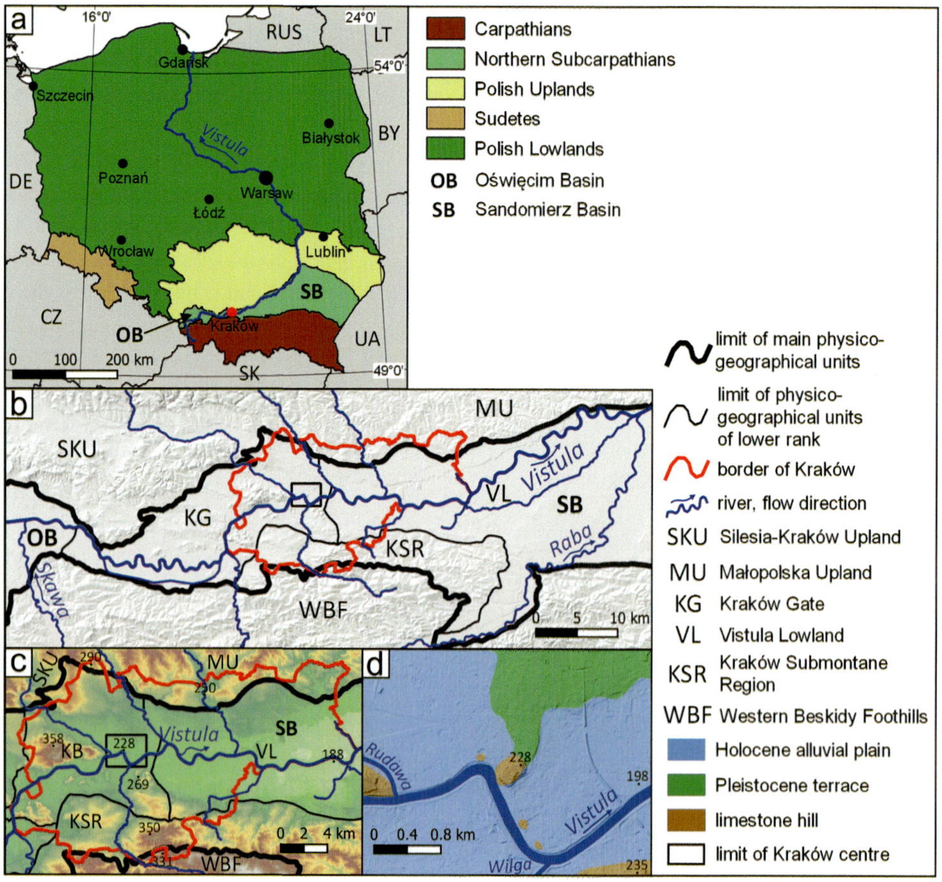

Fig. 23.1 Location of study area. **a** Kraków in Poland (limit of main physico-geographical regions and location of the largest cities are marked). **b** Location of Kraków at the borders of the Polish Uplands, Northern Subcarpathians and Western Carpathians (borders of physico-geographical units of lower rank are marked). **c** Limit of physico-geographical units of different rank in Kraków. Numbers indicate altitudes. **d** Main natural landforms in the historic centre of Kraków

The northern part of Kraków (11% of the city's territory) is located within the borders of the Polish Uplands, its western part belonging to the Silesia-Kraków Upland and the central and eastern parts to the Małopolska Upland. The remaining 1% of the area of Kraków, including the southern peripheries of the city, belongs to the Western Carpathians and forms part of the Western Beskidy Foothills. The historic centre of Kraków covers 2.84% of town's territory and lies in the Kraków Gate at an altitude of 198–235 m asl (Fig. 23.1c, d).

23.2.2 Relief

Because the city of Kraków is located at the junction of three main physico-geographical units of Poland, it shows the most diversified relief among all large Polish cities. Within the boundaries of Kraków, there are eight main landform types, which occupy from 0.9 to 39.4% of the area of the city (Fig. 23.2). They include: (*A*) the Vistula riverbed (1.2% of the city territory). At present, the length of the Vistula within the city is 35 km; however, in the beginning of the nineteenth century, it was 49 km (Łajczak 1995); (*B*) the Holocene alluvial plain, confined to the Vistula valley within the Vistula Lowland and the Kraków Gate and flat-bottomed valleys of the Vistula tributaries (39.4% of the city territory). The meanders, which were intentionally cut off from the river, are filled with fluvial sediments, and—in the city centre—also with anthropogenic deposits (Radwański 1975); (*C*) Pleistocene terraces along the Vistula river and its tributaries (23.2% of the city area). The surface of these terraces includes Pleistocene alluvial fans deposited by the tributaries of the Vistula (young alluvial fans of these tributaries are building up on top of the Holocene alluvial plain); (*D*) hummocks developed on the substratum of Miocene sediments in the western and southern parts of Kraków (Kraków Gate, Kraków Submontane Region) (14.4% of the city area); (*E*) hummocks developed on the substratum of Miocene sediments in the northern part of Kraków (Małopolska Upland) (9.7% of the city area); (*F*) limestone hills partly covered with Cretaceous marls and Quaternary loess deposits (Kraków Gate, Silesia-Kraków Upland) (8.9% of the city area); (*G*) limestone hills covered with Miocene sediments, mainly within the Kraków Gate (2.3% of the city area); (*H*) flysch hills in the part belonging to the Western Beskidy Foothills (0.9% of the city area). The historical centre of Kraków (area of 9.29 km^2) also shows diversified relief. 72.7% of this area is occupied by Holocene alluvial plain, followed by Pleistocene terraces (19.0%), the Vistula riverbed (5.2%), and six limestone hills (3.1%).

The western part of the city shows a more varied relief than the eastern part. Height differences (above the Vistula River) decrease from the west to the east, from 158 to 62 m. Areas next to the northern and southern boundaries of the city rise by 85 m and 150 m above the River Vistula, respectively, and by 65 m and 110 m above local valleys. Height differences along the N–S profile decrease towards the Vistula River; however in the proximity of the river, within the limestone hills, they increase again to 10–58 m near the city centre and to 158 m in the western part of the city (Fig. 23.3).

The distribution of the main natural landforms, their relief, and their morphometric features constitute the base to assess anthropogenic changes in the relief of the city, especially in its historic centre. This is possible to attempt since the limits of main natural landforms and major features of their relief, reflected in their morphometric properties, have generally preserved their primary features despite changes which have occurred as a result of anthropogenic land transformations.

In the most geomorphologically varied part of the city, i.e. within the Kraków Gate and the Silesia-Kraków Upland, Upper Jurassic limestone is exposed, locally overlain by Cretaceous marls and Quaternary loess. This area shows fault-generated block relief (Rutkowski 1992). Tectonic grabens, which separate tectonic horsts, are partly filled with Miocene deposits and show W–E and N–S strike. The Vistula River makes use of the grabens and creates a narrow valley with a winding course which is locally 600 m wide. The exposed parts of the limestone horsts tower over the level of the Vistula River from several to 158 m, and they form hills with steep rocky slopes and flat plateaux (Fig. 23.4). The altitude of these hills in the western part of Kraków exceeds 300 m asl and reaches 358 m asl at maximum (Fig. 23.4a). Towards the east, the altitude decreases to 254 m asl, and in the city centre it is only 210–235 m asl (Fig. 23.5a). On the slopes of limestone hills, there are rocky gorges and caves (Fig. 23.5b, c). In the south-eastern part of the Kraków Gate, tectonic horsts are overlain by Miocene deposits, and limestones and marls are only locally exposed. These are 60 m high and reach an altitude of 269 m asl. On the other hand, in the northern part of the Kraków Gate, limestones and marls do not crop out. Most of the Kraków Gate is covered by Quaternary deposits, mainly of fluvial origin (Izmaiłow 2013). Limestone hills located further north and belonging to the Silesia-Kraków Upland are up to 65 m high and reach an altitude of 290 m asl.

In the northern part of Kraków, which belongs to the Małopolska Upland, hummocks up to 40 m high occur, which reach an altitude of 250 m asl. They are underlain by Miocene clay and slate deposits and overlain by till and loess. The slopes of these hummocks are dissected by dry valleys up to 15 m deep. In the southern part of Kraków, in the area of the Kraków Submontane Region, hummocks

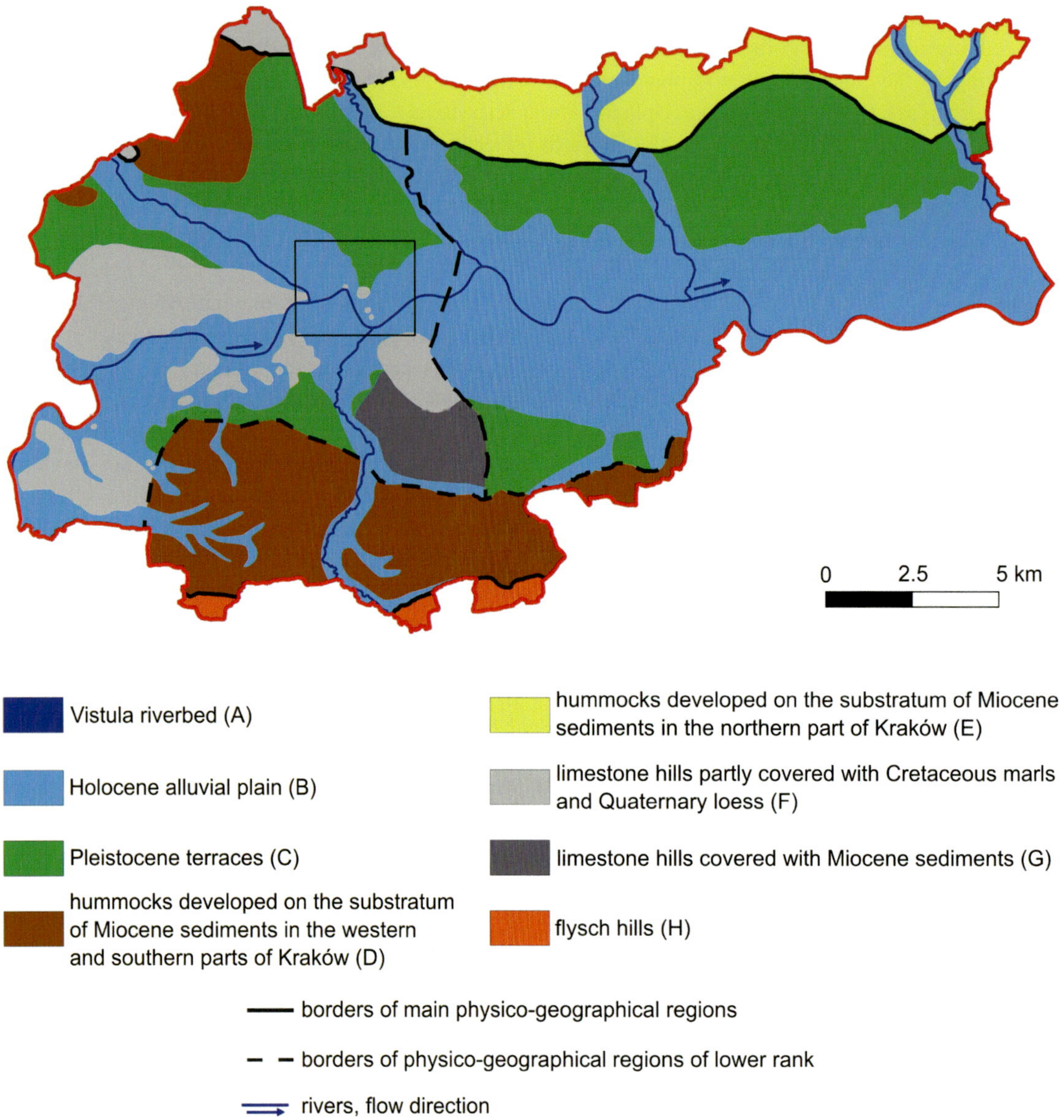

Fig. 23.2 Spatial pattern of main natural landforms in Kraków

up to 55 m high (locally even 140 m) occur, reaching an altitude of 255–350 m asl. They are underlain by Miocene clay deposits. Their top surface is flat, whereas the slopes are dissected by denudation hollows and locally affected by landslides. In the southernmost part of Kraków, which belongs to the Western Beskidy Foothills, hills built of flysch rocks (Cretaceous sandstones, shales, and marls) are present. The altitude of this part of the city reaches 331 m asl, with the elevation differences up to 110 m. The eastern part of Kraków is occupied by the broad, 10-km-wide Vistula Lowland which is eroded in Miocene clay and filled-in by 10–35 m thick Quaternary deposits. Most of this area is occupied by a Holocene alluvial plain, and the remaining part can be subdivided into four Pleistocene terraces separated by 10–15 m high risers (Izmaiłow 2013).

The Vistula River in Kraków is dammed by three low water dams along its whole length which causes doubling of the width of the river water table. Consequently, the natural riverbed is hidden under water (Fig. 23.5d). The cut-off meanders are visible in many places on the alluvial plain. They are filled with deposits to a varying degrees. Such landforms are more abundant in the wide Vistula Lowland in the eastern part of the city than in the extension of the Vistula Valley within the Kraków Gate.

23.2.3 Urban Development

Anthropogenic changes of relief in the contemporary area of Kraków started in the seventh/eighth century and have

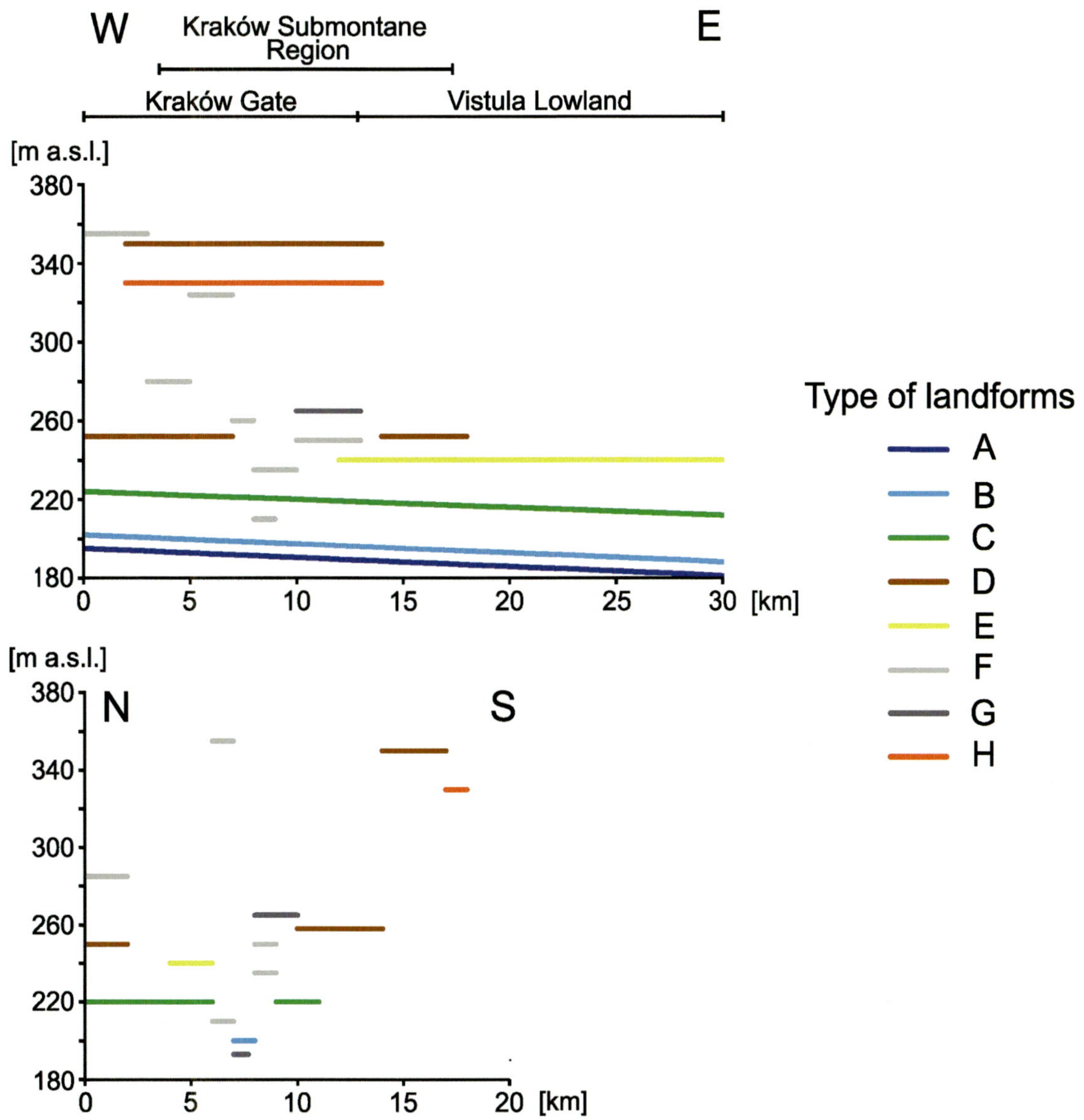

Fig. 23.3 Altitudinal range of landforms within the Kraków area in W–E and N–S profiles. Explanation of letter codes in Fig. 23.2

intensified since the tenth century, and especially in the thirteenth century, when the organisational structure of the town changed (Radwański 1975). In the fourteenth century, other towns were established in the proximity of Kraków (Kazimierz, Kleparz). The eighteenth century marked the foundation of another town named Podgórze, on the southern bank of the river. In the beginning of the nineteenth century all these towns began to merge with Kraków, the boundaries of which by the end of that century mark the extent of Kraków's current historic centre. A very rapid increase of the city's territory occurred in the twentieth century, when its area reached 327 km^2 (Fig. 23.6).

During the historical development of Kraków, the city territory covered not only larger and larger number of natural landforms, but also larger and larger area occupied by these landforms. Until the twelfth century and also after location in the thirteenth century, the contemporaneous Kraków occupied a small area within Pleistocene terrace and the whole area of limestone Wawel Hill. In the next century, the town boundaries extended over part of the Holocene alluvial plain, other small limestone hills, and a further part of Pleistocene terrace, and in the eighteenth century a large part of another limestone hill was included. Since the beginning of the twentieth century the lands which covered a vast area of the Holocene alluvial plain were joined to the town, and on the southern and northern sides of the Vistula River also more elevated areas were incorporated into the city limits, consisting of further fragments of Pleistocene terraces and limestone hills, hummocks developed on the substratum of Miocene sediments,

Fig. 23.4 Selected landforms within Kraków. **a** Plateau of the highest limestone hill (358 m asl) in the western part of the city. **b** Wawel Hill plateau (228 m asl) within the limits of the historical settlement (The Royal Castle, The Cathedral). **c** Rocky slope of the limestone hill in Tyniec undercut by the Vistula River (western part of the city). On the plateau there is Benedictine monastery from the eleventh century (photographs by A. Łajczak)

and flysch hills. The altitude of the latter three groups of landforms, in many place above 300 m asl, significantly exceeded the altitude of limestone hills in historical centre of Kraków. Therefore, Kraków in the contemporary boundaries includes the areas which show much more variability in terms of geomorphology than at the turn of the twentieth century.

The earliest local changes to the relief of the Kraków area introduced over the last 1300 years included historic human-made mounds (seven in total were built between seventh/eighth and twentieth centuries, two of which have been later demolished), a gord established around an early medieval centre of settlement (tenth century), and large-scale earthworks in the historic city centre (intensified since the thirteenth century). In the thirteenth century, efforts to straighten the course of the meandering Vistula River started, which continued until the twentieth century. The cut-off meanders filled up with sediment. In the 1840s, the construction of railways started, involving building of embankments, as did the construction of flood control levees along the Vistula and its tributaries to reduce the impact of catastrophic floods. At that time, operations were intensified to extract limestone, marl, gravel, and loess. In the second part of the nineteenth century, a system of forts was built to protect the city. In the nineteenth–twentieth centuries, the construction and chemical industries expanded, and after 1949, a large integrated steel plant was built in the eastern part of the city called Nowa Huta. Industrialisation of the city caused further expansion of the railway system and the emergence of various types of industrial spoil tips. The

Fig. 23.5 Selected landforms within Kraków—cont. **a** Panoramic view of Kraków (localities of landform types A, B, D, F, and H are shown, view towards the south-east—for explanation of letter codes see Fig. 23.2). **b** Rocky gorge on the northern slope of the highest limestone hill. **c** Rocky slope of the Wawel Hill, with the entrance to the karstic Dragon Cave. **d** Vistula River near the Wawel Hill (visible river stretch is subjected by an effect of backwater of low dam) (photographs by A. Łajczak)

rapid increase of population (currently over 800,000) has resulted in the development of the road system (including on embankments in specific low-elevation locations), the construction of bridges with high heads, and the development of a motorway ring road with associated road junctions. In the nineteenth–twentieth centuries, boulevards were built along the Vistula River in the city centre, and in the beginning of the twenty-first century other recreation and sport structures were added (vantage mounds, a whitewater racecourse on a high embankment next to the River Vistula).

23.3 Methods

Anthropogenic changes in the relief were studied using qualitative and quantitative methods, for identification purposes taking into account all possible anthropogenic landforms with a height/depth greater than 2 m. Although landforms that are lower or shallower than 2 m commonly occur within the city, it was not possible to determine their territorial limits, and therefore, they were omitted. Based on

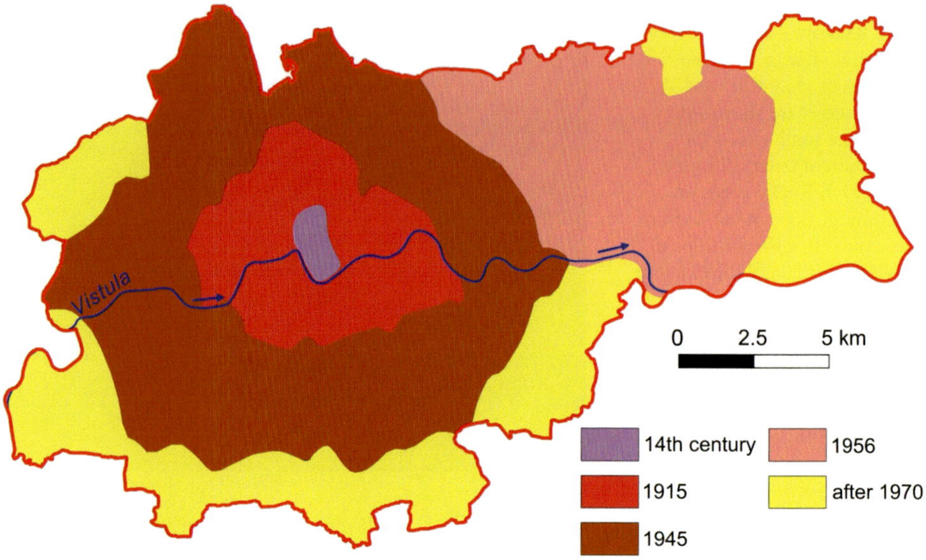

Fig. 23.6 Territorial development of Kraków. In the periods/years cited the area of the city increased. Territorial development of Kraków before the fourteenth century was omitted

field observations, different basic types of anthropogenic landforms were identified in the area of Kraków. Their distribution within the main natural landforms was determined from ALS (airborne laser scanning) data, orthophotomaps, the geomorphological map of Kraków (Izmaiłow 2013), and topographic maps at a scale of 1:10,000. Based on these sources, the horizontal and vertical extents of anthropogenic landforms were determined.

The identified basic types of anthropogenic landforms were subjected to three independent classifications taking into account their shape in horizontal scheme (1), vertical scheme (2), and height range (> 2 m) (3). The following detailed types of anthropogenic landforms were identified: (1) "surface", "linear", and "point" landforms; (2) convex, concave, and complex landforms (the latter include elements of both convex and concave landforms); (3) landforms, the height/depth of which fits the following ranges: 2–5, 5–15, > 15 m. Thus, three attributes (e.g. point/convex/> 15 m) were attributed to each basic anthropogenic landform. In contrast with "surface" landforms, "linear" landforms have their lengths much larger than the widths, whereby the width usually does not exceed 10 m and extremely reaches 70 m. On the other hand, "point" landforms are these, the width of which usually does not exceed 100 m. The next step in quantitative analysis was the determination of the total area covered by the anthropogenic landforms identified and additionally by their individual types within the boundaries of the city and in detail within the individual main natural landforms [km^2, % of area].

In the analysis of anthropogenic changes in the relief of Kraków, the scale of vertical changes of the ground surface (raising or lowering) was considered as a key one within the individual main natural landforms. They indicate the dominant trends in the changes of relief in the Kraków area. More detailed analysis of this type was carried out for the historical centre of Kraków. In this situation, information concerning archaeology, geoengineering, and geomorphology was used, which made it possible to prepare a hypsometric map and geomorphological map of the area, which is now included in the city centre, in a time snapshot of ca. 1000 AD. Based on ALS data, the geomorphological map of Kraków (Izmaiłow 2013), a topographic map and the results of the authors' fieldwork, and analogous maps were prepared, which show the contemporary situation. The comparison of the abovementioned maps makes it possible to assess in detail the changes in relief of the centre of Kraków which have occurred in the last millennium.

23.4 Types of Anthropogenic Landforms

Within the boundaries of Kraków, 22 types of basic anthropogenic landforms were distinguished, each showing elevation differences exceeding 2 m. Table 23.1 contains specification of these landforms together with their areas and additionally with their division into shape in horizontal scheme ("*A*", "*B*", "*C*") and vertical scheme (*a*, *b*, *a*/*b*). The distribution of anthropogenic landforms in Kraków is shown in Fig. 23.7. Surface landforms occupy 3798 ha, linear landforms—1883 ha and point landforms—182 ha (11.61%, 5.76%, and 0.56% of the city territory, respectively). Another division includes anthropogenic landforms for which: (*a*) altitude increased (convex landforms predominate; however, some flat and slightly concave landforms also occur); (*b*) altitude decreased (concave landforms); (*a*/*b*) altitudes in different parts of individual

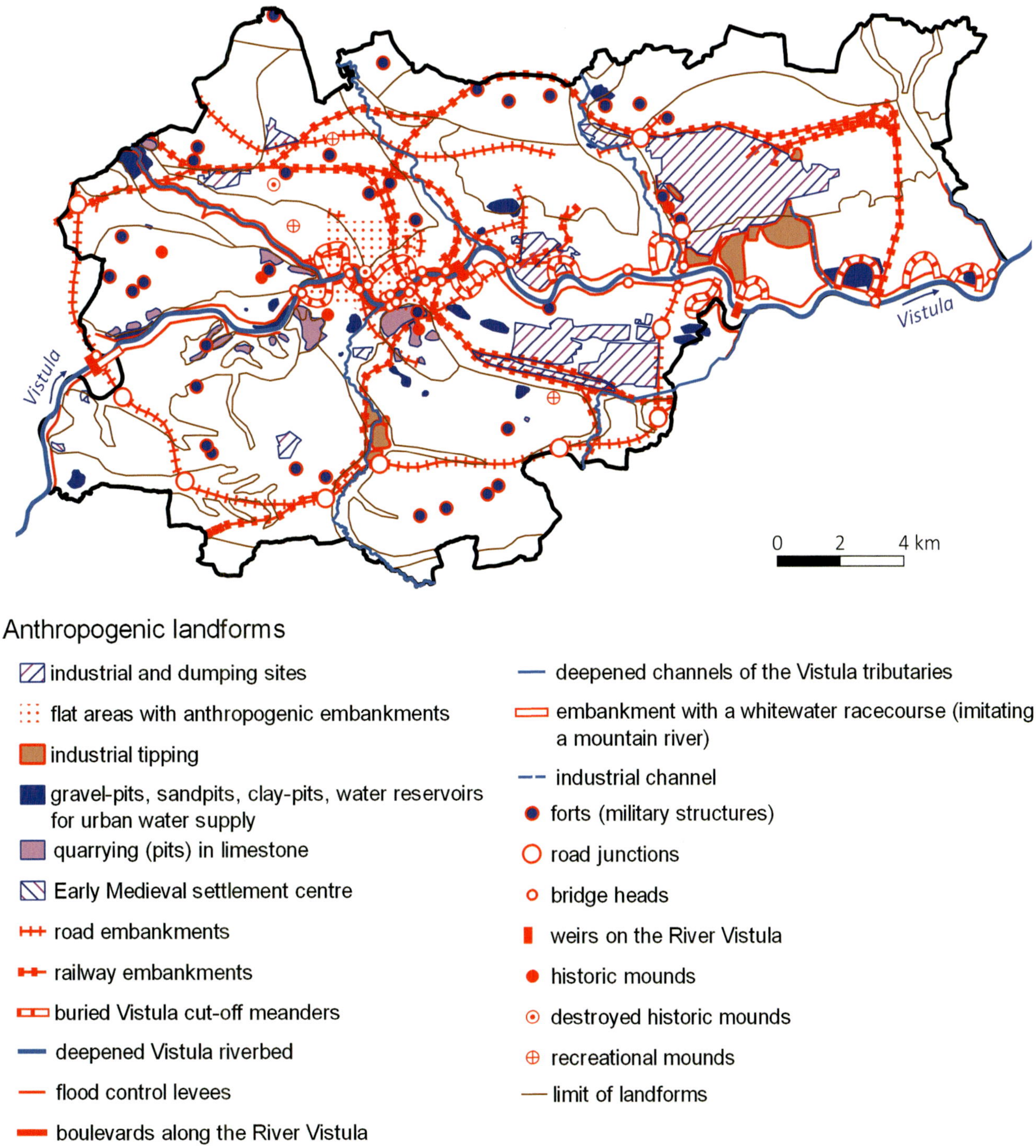

Fig. 23.7 Distribution of anthropogenic landforms in Kraków (compare with Fig. 23.2 for relation to the extent of main natural landforms)

landforms either increased or decreased. Landforms of class (*a*) cover a total area of 2465 ha, landforms of class (*b*) occupy 738 ha, whereas the extent of landforms of class (*a*/*b*) is 2660 ha, which represent 7.54%, 2.26%, and 8.13% of the city area, respectively. The largest area is occupied by anthropogenic landforms of type "*A*" (*a*/*b*) and "*A*" (*a*) (no. 1–3; Table 23.1) which cover 3315 ha (56.5% of the total area of anthropogenic landforms studied and 10.1% of the city area). These landforms mainly occur in the industrial districts of Kraków, and a smaller proportion is also present in the historic city centre. More visible anthropogenic landforms of types "*A*" (*b*), "*B*" (*a*), "*C*" (*a*/*b*), and "*C*" (*a*) (no. 4, 5, 7, 8, 11, 14, 16–18; Table 23.1), including features fulfilling tourist and recreation functions (no. 12, 20, 22; Table 23.1), cover a smaller area (1972 ha, i.e. 33.6% of total area of anthropogenic forms and 6.0% of the city area). The

Table 23.1 Typology and extent of anthropogenic landforms in Kraków (only landforms with a height or depth > 2 m are considered), divided according to spatial pattern ("*A*", "*B*", "*C*") and vertical characteristics (*a*, *b*, *a*/*b*)

No.	Basic types of anthropogenic landforms	Type of landform	Total area [ha]	% of Kraków area	% of Σ
Surface landforms ("A")					
1	Industrial and dumping sites	(*a*/*b*)	2540	7.77	43.32
2	Flat areas with anthropogenic embankments	(*a*)	484	1.48	8.26
3	Industrial tippings	(*a*)	291	0.89	4.96
4	Gravel-pits, sandpits, clay-pits, water reservoirs for urban water supply	(*b*)	240	0.73	4.09
5	Quarrying (hollows) in limestone and marl	(*b*)	233	0.71	3.97
6	Town around an early medieval castle	(*a*/*b*)	10	0.03	0.17
Linear landforms ("B")					
7	Road embankments	(*a*)	746	2.28	12.73
8	Railway embankments	(*a*)	386	1.18	6.58
9	Buried Vistula cut-off meanders	(*a*·)	294	0.90	5.02
10	Deepened Vistula riverbed	(*b*)	234	0.72	4.00
11	Flood control levees	(*a*)	120	0.37	2.05
12	Boulevards along the River Vistula	(*a*)	48	0.15	0.82
13	Deepened channels of the Vistula tributaries	(*b*)	30	0.09	0.51
14	Embankment with a whitewater racecourse	(*a*)	24	0.07	0.41
15	Industrial channel on the surface	(*b*)	1	< 0.01	0.02
Point landforms ("C")					
16	Forts (military structures)	(*a*/*b*)	110	0.34	1.86
17	Main road junctions	(*a*)	36	0.11	0.61
18	Bridge heads	(*a*)	22	0.07	0.38
19	Weirs on the River Vistula	(*a*)	6	0.02	0.10
20	Historic mounds	(*a*)	6	0.02	0.10
21	Historic mounds no longer exist	(*a*)	1	< 0.01	0.02
22	Recreational mounds	(*a*)	1	< 0.01	0.02
1–22			Σ = 5863	17.93	100.0

Landforms: (*a*)—convex, (*b*)—concave, (*a*/*b*)—complex (convex/concave), (*a*·)—initial concave landforms which became shallower due to becoming filled with sediments

remaining anthropogenic landforms cover an area of 576 ha (9.9% and 1.8%, respectively).

The area covered by anthropogenic landforms, in terms of ranges determined for their height or depth, was also taken into consideration (Table 23.2). In terms of surface area, anthropogenic landforms within the height range of 2–5 m predominate (4812 ha, 82% of the total area of all the anthropogenic landforms). Among them the landforms of "*A*" (*a*/*b*) type are the most common (2540 ha). Anthropogenic landforms of 5–15 m height/depth range cover an area of 936 ha (16% of the area of all anthropogenic landforms), with landforms of type (*a*) being dominant. The total area of the highest and deepest anthropogenic landforms (> 15 m) is only 115 ha (2%), with landform type (*b*) being dominant.

Height differences within the area of the most extensive anthropogenic forms, i.e. the areas used for industry or tipping (no. 1, Table 23.2; mainly eastern districts of Kraków), do not exceed ± 5 m. Building up of the initial surface by anthropogenic deposits in the historic centre of Kraków (no. 2; Table 23.2) reaches up to 14 m, but is 2–5 m on average. Local height differences in the next group of anthropogenic forms, i.e. industrial tip heaps in the eastern and southern parts of the city (no. 3; Table 23.2), reach 15 m. The depth of concave landforms in locations of former sand, gravel, clay and loess extraction, and places where water reservoirs were created for urban water supply (which no longer operate) (no. 4; Table 23.2) reach 7 m. The depth of pits in closed limestone quarries (no. 5; Table 23.2) reaches 20–30 m and locally 60 m (Fig. 23.8a–c). The height or depth of linear landforms is included in the range of 3–8 m. Height differences among point landforms reach 7 m in the case of 35 forts (military objects) and 9 m in the case of road junctions. The height of the embankments within the bridge heads on the River Vistula exceeds 7 m, and at the weirs

Table 23.2 Morphometric properties of anthropogenic landforms in Kraków

Number (see Table 23.1) and type of anthropogenic landform	Total area of landforms [ha] classified to groups based on the ranges of height or depth				Max. height (+), max. depth (−) [m]
	2–5 m	5–15 m	>15 m	Σ	
1. "*A*" (*a*/*b*)	2540	–	–	2540	± 5
2. "*A*" (*a*)	365	119	–	484	+ 14
3. "*A*" (*a*)	–	291	–	291	+ 15
4. "*A*" (*b*)	180	60	–	240	− 7
5. "*A*" (*b*)	35	88	110	233	− 60
6. "*A*" (*a*/*b*)	10	–	–	10	± 5
7. "*B*" (*a*)	716	30	–	746	+ 8
8. "*B*" (*a*)	361	25	–	386	+ 7
9. "*B*" (*a*·)	186	108	–	294	+ 7
10. "*B*" (*b*)	234	–	–	234	− 4
11. "*B*" (*a*)	100	20	–	120	+ 6
12. "*B*" (*a*)	48	–	–	48	+ 5
13. "*B*" (*b*)	30	–	–	30	− 3
14. "*B*" (*a*)	–	24	–	24	+ 8
15. "*B*" (*b*)	1	–	–	1	− 3
16. "*C*" (*a*/*b*)	–	110	–	110	± 7
17. "*C*" (*a*)	–	36	–	36	+ 9
18. "*C*" (*a*)	6	16	–	22	+ 7
19. "*C*" (*a*)	–	6	–	6	+ 9
20. "*C*" (*a*)	–	1	5	6	+ 36
21. "*C*" (*a*)	–	1	–	1	+ 10
22. "*C*" (*a*)	–	1	–	1	+ 10
Total	4812 (*a*) 1782 (*b*) 480 (*a*/*b*) 2550	936 (*a*) 678 (*b*) 148 (*a*/*b*) 110	115 (*a*) 5 (*b*) 110 (*a*/*b*) –	5863 (*a*) 2465 (*b*) 738 (*a*/*b*) 2660	+ 36 − 60 ± 7

Explanations as in Table 23.1

they reach 9 m. Historic mounds are from 7 to 36 m high (Fig. 23.8d, e). The heights of recreation mounds which were built in the twentieth–twenty-first centuries are from 5 to 10 m.

23.5 Anthropogenic Changes in the Relief of Kraków

23.5.1 The Whole Area of the City

The largest area occupied by anthropogenic landforms in Kraków occurs on the Holocene alluvial plain (*B*) and on the Pleistocene terraces (*C*) (Fig. 23.7; Table 23.3). Overall, 22.4% of the area (*B*) is occupied by almost all the anthropogenic landforms, among which 2–5 m high landforms predominate (18.1%). The greatest extent is seen in convex landforms (*a*). Anthropogenic changes in the relief of Pleistocene terraces (*C*) affect 28.8% of this area, and they originated as a result of the development of nine types of landforms, with the dominating role of complex landforms (*a*/*b*), mainly industrial and dumping sites (21.3%). Anthropogenic changes in the relief of hummocks which developed on the substratum of Miocene sediments in the western and southern parts of Kraków (*D*) cover only 8.2% of the area of this part of the city. These changes are caused by the creation of only five types of anthropogenic landforms, with the predominating convex landforms covering 7.6% of the area. Changes in the relief of the hummocks, which developed on the substratum of Miocene sediments in the northern part of Krakow (*E*), are minimal (1.5%). The anthropogenically transformed relief of limestone hills in

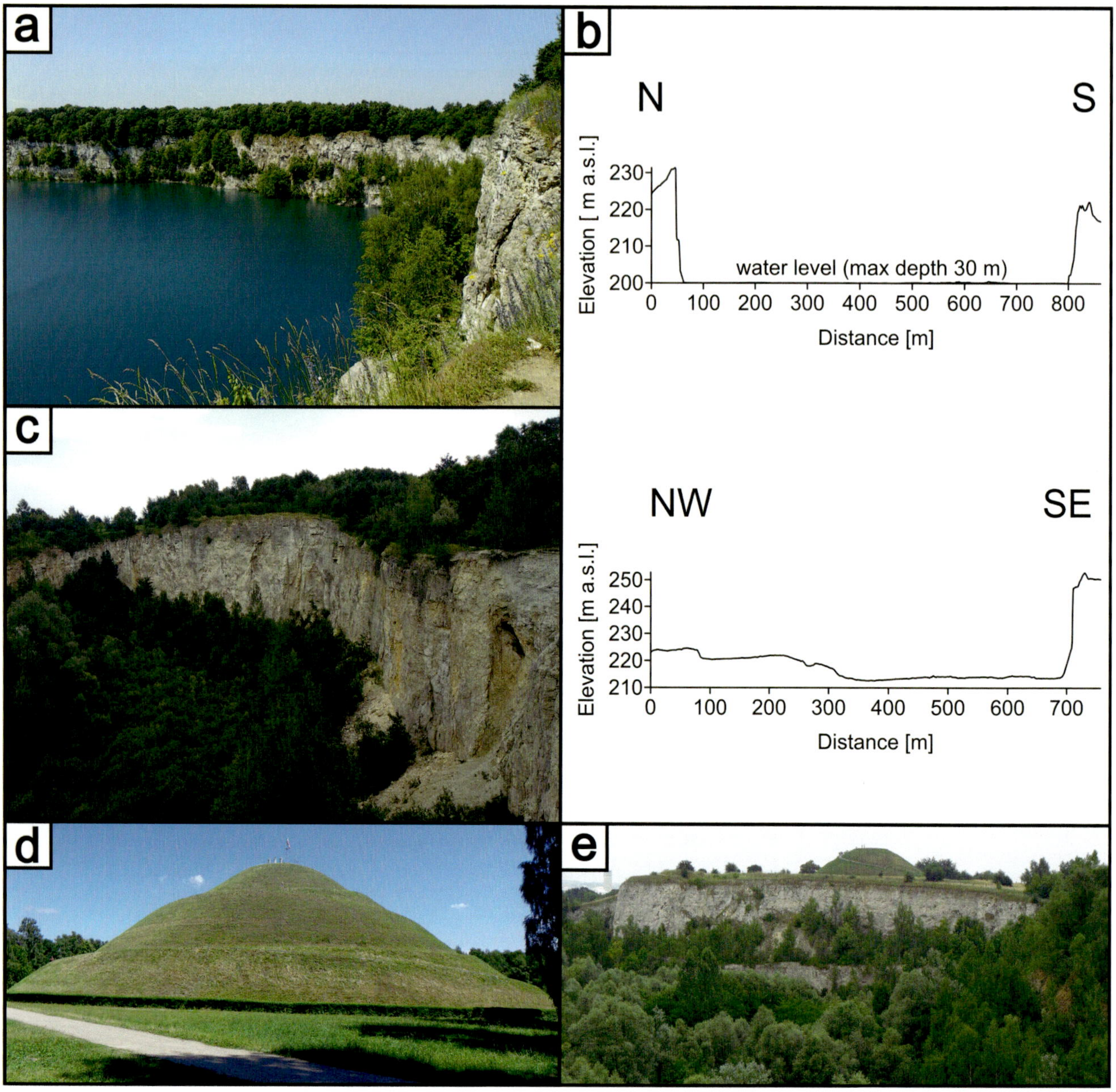

Fig. 23.8 Anthropogenic landforms in Kraków representing the highest increases in local relief. **a** Exemplary headwall of closed limestone quarry (quarry pit is partly filled with water; Zakrzówek quarry). **b** cross-section profiles of two quarries (Zakrzówek and Liban quarries). **c** 40 m high headwall of a closed quarry (rocky slope and debris slope are visible; Liban quarry). **d** The highest anthropogenic, memorial mound (Piłsudski Mound), 36 m high (built in 1934–1938). **e** The proximity of the convex and the concave anthropogenic landforms on a limestone hill (one of the oldest mounds, 16 m high, from seventh/eighth century AD and a closed quarry), total elevation difference is 50 m (Liban quarry and Krak Mound) (photographs **a**, **c**, **e** by R. Zarychta; **d** by A. Łajczak)

the central and western parts of the city (*F*) includes 11.6% of the area of this landform category and developed due to the creation of seven types of anthropogenic landforms which belong to types (*a*), (*b*), and (*a*/*b*). In terms of the area occupied, the pits of closed limestone quarries predominate (11.1% of the area), and they include all the depth ranges (maximum 60 m deep). Convex landforms include the highest historic mounds in Krakow (16–36 m high), which occupy only 0.3% of the area of the limestone hills. Limestone hills covered with Miocene sediments (*G*) have been only slightly transformed by human impact.

The lowest and shallowest (2–5 m) anthropogenic landforms occur within all the main landforms and show the largest extent within the Pleistocene terraces and Holocene alluvial plain (Table 23.3). Anthropogenic landforms of the height/depth of 5–15 m occur within most of the main landforms (*B*, *C*, *D* and *F*) and occupy much smaller areas. The highest and deepest anthropogenic landforms (> 15 m)

Table 23.3 Percentage of the area of main natural landforms (*B–G*) occupied by different types of anthropogenic landforms (*a*, *b*, *a*/*b*), within pre-defined ranges of their height or depth

Main landform	Ranges of height or depth	Types of anthropogenic landforms			
		a	*b*	*a*/*b*	Σ(*a*, *b*, *a*/*b*)
B	*h*1	11.9	3.6	6.9	22.4
	*h*2	7.6	3.6	6.9	18.1
	*h*3	4.3	–	–	4.3
	*h*4	–	–	–	–
C	*h*1	7.2	0.3	21.3	28.8
	*h*2	6.4	0.3	20.9	27.6
	*h*3	0.8	–	0.4	1.2
	*h*4	–	–	–	–
D	*h*1	7.6	0.1	0.5	8.2
	*h*2	6.0	0.1	0.5	6.6
	*h*3	1.6	–	–	1.6
	*h*4	–	–	–	–
E	*h*1	1.4	–	0.1	1.5
	*h*2	1.4	–	0.1	1.5
	*h*3	–	–	–	–
	*h*4	–	–	–	–
F	*h*1	0.3	11.1	0.2	11.6
	*h*2	0.1	6.6	0.1	6.8
	*h*3	0.1	1.8	0.1	2.0
	*h*4	0.1	2.7	–	2.8
G	*h*1	1.9	–	1.9	3.8
	*h*2	1.9	–	1.9	3.8
	*h*3	–	–	–	–
	*h*4	–	–	–	–

Main natural landforms *A* and *H* are omitted. Explanation of (*B–G*) and (*a*, *b*, *a*/*b*) are shown earlier in the text. Ranges of height/depth (in metres): *h*1 (2–> 15); *h*2 (2–5); *h*3 (5–15); *h*4 (> 15). Further explanations of letter codes in Fig. 23.2 and Table 23.2

occur only within the limestone hills and occupy 2.8% of this area.

The largest variations in relief of the Kraków area result from two anthropic activities. These were the excavation of limestone quarries and building of the highest historic mounds within the natural landform types of (*C*) and (*F*). The area of these anthropogenic landforms is 2.8 km^2 and 0.06 km^2, respectively; however, they produce considerable changes in height and range from − 60 m to + 36 m (Fig. 23.8e shows the range from − 34 m to + 16m: Liban quarry and Krak Mound).

The distribution of quarries within the limestone hills is uneven (Fig. 23.7, anthropogenic landform no. 5; Table 23.1); therefore, the most considerable transformations of relief usually are observed in small hills (in terms of area), where large quarries were operating. The depth of pits in the quarries often represents over half of the height of limestone hills, and in some situations, it is even higher. The neighbourhood of the historic mound from the seventh/eighth century and the deep quarry on one of the limestone hills in the centre of Kraków show the most spectacular anthropogenic transformations of relief in the city (Fig. 23.8e), where the total change of height differences within this hill reaches 50 m. Although this change is 10 m less in relation to the deepest pit partly filled with water, within the Zakrzówek quarry on another limestone hill in Kraków (Fig. 23.8a, b), it is visible for the whole height profile.

23.5.2 Historic City Centre

The probable hypsometry and distribution of landforms within the area, which now belongs to the historic centre of Kraków as it was in the tenth century, i.e. in the early stage of its development, is shown in Fig. 23.9a, b. The contemporary situation for the same area is shown in Fig. 23.9c, d. Figure 23.9e presents the hypsometric curve of the area in both variants.

The holocene alluvial plain was laid down in the area now forming the centre of Kraków at an altitude of 198–203 m asl about 1000 years ago and showed variation in relief (Łajczak and Zarychta 2020). The following landforms were present: sandbars locally higher than 3 m (Setmajer 1973), depressions with long-lasting stagnant water, and peatbog plains within backswamps. The alluvial plain was cut in many places by the winding channel of the Vistula River and the channels of its tributaries, which were accompanied by levees. Overbank channels occurred in many places. Within the area, the riverbed of the meandering Vistula was 7 km long, and its sinuosity was 2.1. The mean width of the channel was 130 m, and the mean depth was 4.0 m, with a maximum of about 5 m. Along a section of 2.5 km, two riverbeds of the Vistula were functioning. The Pleistocene terrace occurred at an altitude of 203–213 m asl and was bordered by a high escarpment (Radwański 1975). The terrace gently sloped down towards the Holocene alluvial plain, and in the area studied its extent narrowed towards the south and finally joined the Wawel Hill in the form of a narrow neck. Below the neck of this terrace, the tops of limestone horsts occur at different heights. In historical times the outcrops of limestone or overlying marls were locally exposed at the ground surface, and the oldest churches of early Romanesque style were located in such places. There are four limestone hills in the study area with the heights from several to 30 m (the highest is Wawel Hill—228 m asl), and there are fragments of two higher limestone hills (235 m asl). The highest hills used to have flat plateaux and rocky slopes, whereas the slopes of the other hills were more gently sloping.

As a result of many centuries of development, the Holocene alluvial plain is now located on average 3 m higher than in the historical situation and now lies within the altitudinal range of 199–208 m asl (Fig. 23.9c). The

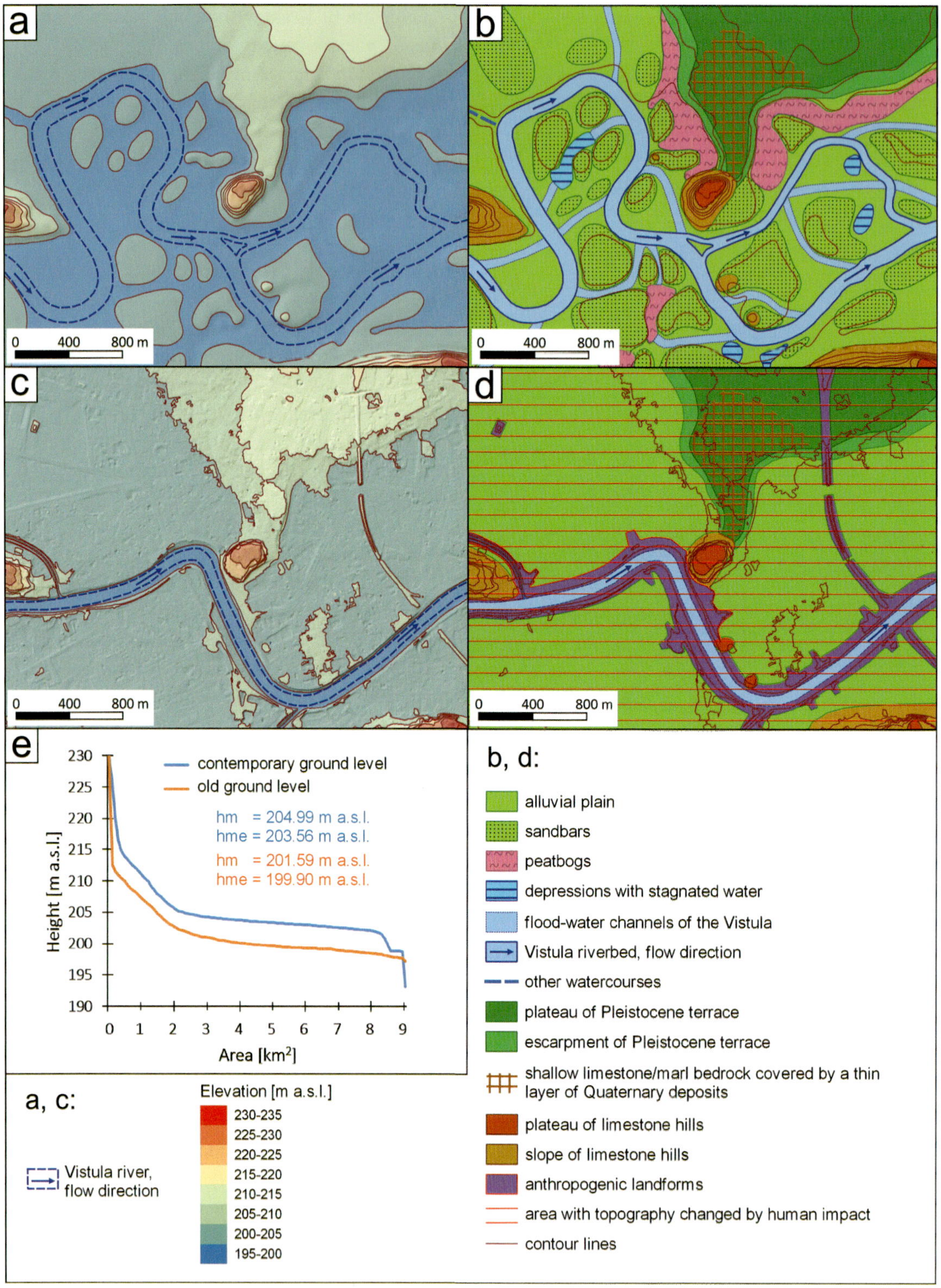

Fig. 23.9 Hypsometry (**a**) and distribution of landforms (**b**) in the centre of Kraków by ca. 1000 AD (probable situation), contrasted with the current hypsometry (**c**) and distribution of landforms (**d**) in the same area. **e** Comparison of hypsographic curves for two time snapshots variants. Based on Łajczak et al. (2021a, b)

increase of altitude of this area was caused by long-lasting formation of anthropogenic ground, road and railway embankments, flood control levees, boulevards, bridge heads, and, at a local scale, historic and vantage mounds. Former convex and concave landforms in the area of the alluvial plain became fossil landforms overlain by anthropogenic ground (Fig. 23.9d). The alluvial plain became flattened, but locally elevation differences within its area increased by over 5 m as a result of the development of convex anthropogenic landforms, mainly linear (flood control levees, boulevards, and railway and road embankments) and also point landforms (bridge heads and mounds). Particularly large changes occurred due to shortening of the Vistula River course, deepening of the river channel as a result of its regulation and filling-in of cut-off meanders. The Vistula channel was shortened by 3 km, mainly due to cutting off of the meander upstream of the Wawel Hill in the thirteenth century, which produced a reduction in sinuosity to 1.2. In the 1870s, the secondary river channel was buried downstream of the Wawel Hill (Łajczak and Zarychta 2020). As a result of regulation works which were carried out in the nineteenth century, the Vistula channel was deepened by 3–4 m (Łajczak 1995), with the result that its width decreased to 65 m. At present the river banks near the Wawel Hill lie at 199 m asl and the river channel at 191 m asl. After construction of a sluice in the 1960s, located 1.7 km downstream of the area studied, the water level in the river was raised by 3.5 m, which caused the width of the river to increase to 110 m.

Table 23.4 Area [ha] of convex (*a*), concave (*b*) and complex (*a*/*b*) anthropogenic landforms within individual main natural landforms (*B*–*G*) and the percentage of the area of Kraków occupied by types of landforms (*a*, *b*, *a*/*b*)

Main landforms	Types of anthropogenic landforms		
	a (↑)	*b* (↓)	*a*/*b* (↑↓)
B 13,276 ha (m)	1484 ha (4.54%) *2–5*	464 ha (1.44%) *2–5*	957 ha (2.93%) *2–5*
C 7586 ha (m)	518 ha (1.58%) *2–5*	18 ha (0.06%) *2–5*	1656 ha (5.06%) *2–5*
D 5199 ha (m)	399 ha (1.22%) *2–5*	1 ha (0.01%) *5–15*	–
E 3172 ha (m)	44 ha (0.13%) *2–5*	–	1 ha (0.01%) *5–15*
F 2518 ha (m)	7 ha (0.02%) *> 15*	255 ha (0.78%) *> 15*	36 ha (0.11%) *2–5*
G 622 ha (m)	13 ha (0.04%) *2–5*	–	10 ha (0.03%) *2–5*
B–*G* 32,373 ha (m)	2,465 ha (7.54%) *2–5*	738 ha (2.26%) *2–5*	2,660 ha (8.13%) *2–5*

Main natural landforms *A* and *H* are omitted. In each situation, the dominant height/depth range of anthropogenic landforms is indicated by *italics*. Further explanations of letter codes in Fig. 23.2 and Table 23.2

The Pleistocene terrace is currently located at an altitude of 208–216 m asl, which is on average 3 m higher than in the historical situation. As a result of anthropogenic deposition in the entire area studied, the cross-profile profile of this terrace became smoothed over and in places more distant from the Vistula, and the former terrace riser became fossilised. Outcrops of limestone and marls that were exposed or occurred at shallow depth within the neck of the terrace in the past became covered by anthropogenic deposits. For instance, within the Market Square in the Old Town, they occur at present at a depth of 2–6 m. Limestone tors and the lowest limestone hills became erased or occupied by sacral buildings. The slopes of the Wawel Hill became overlain by up to 14 m thick soil banks, and its plateau became flattened.

Similarly to other parts of Kraków, the relief of its centre was locally significantly transformed by limestone quarrying. Within the northern slopes of the highest hills, broad pits originated, which are bordered by over 20 m high rocky walls. It should be stressed, however, that in the entire area of the Wawel Hill with monumental buildings (e.g. Cathedral, Royal Castle), no limestone quarrying has ever occurred. The altitude of limestone hills has not changed, except of the smallest hills which were erased. The relative height of limestone hills decreased by 3 m due to the increase of the altitude of the alluvial plain.

23.6 Trends of Changes in the City Relief

Convex (*a*), concave (*b*), and complex—convex/concave (*a*/*b*) anthropogenic landforms only occur in the least elevated parts of the city, within the Holocene alluvial plain and Pleistocene terraces and also within the limestone hills (Table 23.4). Within the remaining main landforms, i.e. within the area of hummocks which developed on the substratum of Miocene sediments and limestone hills covered by Miocene sediments and Vistula riverbed, only two types of anthropogenic landforms occur: (*a*) and (*b*) or (*a*) and (*a*/*b*). Within the area of flysch hills, anthropogenic landforms over 2 m high or deep do not occur.

Trends in vertical changes of the topographic surface of the city are determined by the height range of (*a*) or (*b*) types of anthropogenic landform. The type (*a*/*b*) of anthropogenic landform, despite its extent, is neutral in this respect. Thus, within the main landforms (except the Vistula riverbed and limestone hills) a trend to elevate the surface predominates as the area of convex anthropogenic landforms is larger than the area of concave landforms by 3–400 times or concave landforms do not occur at all. Territorially dominant convex anthropogenic landforms belong to the height range of 2–5 m, and in the case of the Holocene alluvial plain, they occupy 4.5% of the area of Kraków. Only within limestone hills with deep quarries, there is a trend to lower the topographic surface; the area occupied by concave anthropogenic landforms is 36 times larger than convex landforms. The fact that concave anthropogenic landforms fall into the depth range > 15 m also indicates the extremely large surface dissection within limestone hills, where at a local scale altitudes decreased even by as much as 60 m.

The building up of the area as a result of the development of the territorially dominant convex anthropogenic landforms shows increasing trends from the northern and southern parts of the city towards the River Vistula. The highest positive changes of ground surface altitude were recorded within the Holocene alluvial plain, where convex anthropogenic landforms with a dominant height range of 2–5 m (usually closer to 5 m) cover an area larger than the total area of concave and complex landforms. Only within limestone hills located near the Vistula River is this trend disrupted because of the occurrence of deep depressions (quarries).

At the scale of the entire area of the city, there is a trend for height differences to decrease between the individual natural main landforms (except the Vistula riverbed and flysch hills). This trend is due to an increase in the elevation of the terrain within the Holocene alluvial plane, to a lesser extent within the Pleistocene terraces, and to the least extent within the Miocene hummocks and limestone hills covered with Miocene sediments. This was caused by the formation of convex anthropogenic landforms. On the other hand, the average altitude (asl) of the limestone hills with exposed bedrock has been reduced due to the formation of numerous quarry pits. The creation of convex and concave anthropogenic landforms, especially with a height/depth > 5 m (e.g. flood control levees, railway embankments, mounds, forts, and quarry pits caused local increase of relief). The most spectacular changes occurred in places where convex and concave anthropogenic landforms of large vertical range are adjacent to one another (Fig. 23.8e).

Increases in altitude within the Holocene alluvial plain in the area of the centre of Kraków were mainly caused by the century-lasting deposition of anthropogenic deposits and, to a smaller extent, by creation of convex linear anthropogenic landforms with steep slopes. Intentional changes in the river system (infilling of cut-off meanders and of one of the Vistula channels) which have occurred since the thirteenth century caused an increase of the area occupied by the Holocene alluvial plain at the expense of the Vistula riverbed. The deepening of the Vistula channel as a result of river regulation and the creation of flood control levees (with high bridge heads, sluices and other constructions) led to two-directional changes in altitude (up and down), which resulted in the increase of local relief in the belt of terrain including the riverbed and the adjacent zones of the alluvial plain, which locally exceed 10 m.

Anthropogenically-conditioned changes in relief of the lowest-lying places in Kraków, which include both the Holocene alluvial plain and the Vistula riverbed, therefore involve raising the ground surface with a dominant trend to flatten it, and at a secondary scale, to lower the ground surface (riverbed) or raise it (linear or spot landforms) with a trend towards increasing undulation of the land surface. The overlaying of Pleistocene terrace with anthropogenic deposits caused the obliteration of the natural extent of this type of landform, which locally hinders its delimitation.

Limestone tors or the lowest limestone hills were completely removed, and those which were buried by anthropogenic deposits or included in sacral objects, became fossilised. As a result of the increase in altitude of the alluvial plain and high terrace, locally by as much as 8 m, the elevation of higher limestone hills (e.g. the Wawel Hill) decreased. The development of deep quarries with flat bottoms and rocky walls within the highest limestone hills caused an increase of local relief and decrease of mean altitude of these hills. These changes in the relief of limestone hills are consistent with the trend mentioned earlier of the lowering of the main landforms (F) in Kraków.

23.7 Conclusions

The relief of the territory of Kraków, which is located on the border of geomorphologically different physico-geographical regions (Fig. 23.1), should be recognised as the most diverse of all large cities in Poland. The area included in the contemporary boundaries of Kraków is also distinctive in terms of its anthropogenic changes of relief which have occurred under the influence of a large number of factors since early medieval times. However, in terms of territorial range, and locally in terms of vertical scale, anthropogenic landforms which developed in the nineteenth–twentieth centuries are particularly significant. Compared to other cities, where broader landforms developed due to mining/quarrying or the development of other sectors of the economy (pits, heaps, and subsidence basins), with transport infrastructure (motorways and road junctions on high embankments, tunnels, and airports), or with large areas over which there is an increase of thickness of anthropogenic ground (that have buried broad areas of ancient towns) (e.g. Campagna and Scardozzi 2016; Dall'Aglio et al. 2017; Luberti 2018), the territorial extent of anthropogenic landforms within the boundaries of Kraków is not large, and they are assumed to cover ca. 18% of the city area. More striking is the vertical scale of these changes, which reaches from − 60 m (concave landforms) to + 36 m (convex landforms).

The extent of anthropogenic ground in Kraków, which has formed in the city centre for several centuries and contributed to the flattening of the topographic surface, may be compared with vast areas in other towns, e.g. in Warsaw, where large amounts of rubble were deposited during the reconstruction works after damages of World War II. Also in this situation a considerable overbuilding of the land and their flattening occurred, however these changes occurred very rapidly (Ecophysiographical Atlas of Warsaw 2018).

At the local scale of Kraków, the development of convex and concave anthropogenic landforms has led to an increase in relative height. On the other hand, in the area occupied by the Holocene alluvial plain and the Pleistocene terrace in the historic centre of the city, the trend of surface flattening predominates (Łajczak et al. 2021a, b). At the scale of the entire city, the general trend of decreasing height differences between the main natural landforms occurs, which results from larger vertical and horizontal extent of convex anthropogenic landforms within low-lying areas. This trend also results from the large horizontal and, in particular, vertical extent of quarry pits set within the limestone hills. Changes of the vertical range of main natural landforms conditioned by human impact are, except for the limestone hills, relatively small in relation to natural elevation differences within these landforms.

The comparison of horizontal and vertical extents of individual anthropogenic convex and concave landforms indicates that building up the ground is the dominant process in Kraków. This results from the deposition of the material quarried not only within the city boundaries but also beyond the city limits. However, it is not possible at present to assign a value to the trend towards raising the ground surface in Kraków.

The results of the investigation outlined in this chapter show that anthropogenic landforms should be given more consideration in heritage tourism and geotourism in Kraków due to considerable landscape values of some such landforms (mounds and limestone quarries) which are located within attractive limestone hills and close to historic buildings.

References

Brandolini P, Faccini F, Paliaga G, Piana P (2017) Urban geomorphology in coastal environment: man-made morphological changes in a seaside tourist resort (Rapallo, Eastern Liguria, Italy). Quaest Geogr 36:97–110

Campagna L, Scardozzi G (2016) The archaeological map for the reconstruction of the ancient topography of the Greek and Roman City of Taormina. In: LAC2014 proceedings, pp 1–10

Carver M (1983) Forty French towns: an essay on archaeological site evaluation site evaluation and historical aims. Oxford J Archaeol 2:339–378

Dall'Aglio P, de Donatis M, Franceschelli C, Guerra C, Guerra V, Nesci O, Piacentini D, Savelli D (2017) Geomorphological and anthropic control of the development of some Adriatic historical towns (Italy) since the Roman Age. Quaest Geogr 36:111–123

Dulias R (2016) The impact of mining on the landscape. A study of the Upper Silesian Coal Basin in Poland. Springer, Switzerland, p 209

Ecophysiographic Atlas of Warsaw (2018) Warsaw

Izmaiłow B (2013) Rzeźba terenu . In: Degórska B, Baścik M (eds) Środowisko przyrodnicze Krakowa. Zasoby – ochrona – kształtowanie. Urząd Miasta Krakowa, Instytut Geografii i Gospodarki Przestrzennej Uniwersytetu Jagiellońskiego, Wydział Geodezji I Kartografii Politechniki Warszawskiej, pp 21–32

Jancewicz K, Traczyk A, Migoń P (2021) Landform modifications within an intramontane urban landscape due to industrial activity, Wałbrzych, SW Poland. J Maps 17:194–201

Kaniecki A (2013) Environmental changes in the Warta valley within the Poznań area connected with antropopressure. Landf Anal 24:23–34

Luberti G (2018) Computation of modern anthropogenic-deposit thicknesses in urban areas: a case study in Rome, Italy. Anthr Rev 5:2–27

Łajczak A (1995) The impact of river regulation, 1850–1990, on the channel and floodplain of the Upper Vistula River, Southern Poland. In: Hickin EJ (ed) River geomorphology. Wiley, Chichester, pp 209–233

Łajczak A, Zarychta R (2020) Reconstruction of the morphology and hydrography of the centre of Kraków before the mid-13th century. Geogr Polon 93:25–50

Łajczak A, Zarychta R, Wałek G (2021a) Changes in the topography of Kraków city centre, Poland, during the last millennium. J Maps 17:186–193

Łajczak A, Zarychta R, Wałek G (2021b) Quantitative assessment of changes in topography of town caused by human impact, Kraków city centre, Southern Poland. Remote Sens 13:2286

Molewski P, Juśkiewicz W (2014) An attempt to reconstruct the primary relief of the Old Town of Toruń and its close suburbs on the basis of the geological and historical geoinformation. Landf Anal 25:115–124

Mozzi P, Ferrarese F, Zangrando D, Gamba M, Vigoni A, Sainati C, Fontana A, Ninfo A, Piovan S, Rossato S, Veronese F (2018) The modeling of archaeological and geomorphic surfaces in a multistratified urban site in Padua, Italy. Geoarchaeology 33:67–84

Pica A, Luberti G, Vergari F, Fredi P, Del Monte M (2017) Contribution for an urban geomorphoheritage assessment method: Proposal from three geomorphosites in Rome (Italy). Quaest Geogr 36:21–36

Pröschel B, Lehmkuhl F (2019) Paleotopography and anthropogenic deposition thickness of the city of Aachen, Germany. J Maps 15:269–277

Qiao W, Gao J, Guo Y, Ji Q, Wu J, Cao M (2019) Multi-dimensional expansion of urban space through the lens of land use: the case study of Nanjing City, China. J Geogr Sci 29:749–761

Radwański K (1975) Kraków przedlokacyjny: Rozwój przestrzenny. Polskie Towarzystwo Archeologiczne i Numizmatyczne Oddział w Nowej Hucie, Muzeum Archeologiczne w Krakowie, Kraków, p 403

Reynard E, Pica A, Coratza P (2017) Urban geomorphological heritage. An overview. Quaest Geogr 36:7–20

Rutkowski J (1992) Szczegółowa mapa geologiczna Polski 1:50 000, arkusz Kraków (973). Wydawn Geol, Warsaw
Setmajer J (1973) Główne rysy budowy geologicznej oraz pierwotnej topografii Krakowa i strefy przełomowej Wisły. Acta Archaeol Carp 13:139–151
Solon J, Borzyszkowski J, Bidłasik M, Richling A, Badora K, Balon J, Brzezińska-Woźniak T, Chabudziński Ł, Dobrowolski R, Grzegorczyk I, Jodłowski M, Kistowski M, Kot R, Krąż P, Lechnio J, Macias A, Majchrowska A, Malinowska E, Migoń P, Myga-Piątek U, Nita J, Papińska E, Rodzik J, Strzyż M, Terpiłowski S, Ziaja W (2018) Physico-geographical mesoregions of Poland: verification and adjustment of boundaries on the basis of contemporary spatial data. Geogr Polon 91:143–170
Tyczyńska M (1968) Rzeźba i budowa geologiczna terytorium miasta Krakowa. Folia Geogr, Ser Geogr-Phys 1:9–34
Tyczyńska M, Chmielowiec S (1988) Mapa geomorfologiczna 1:50,000 (9). In: Trafas K (ed) Atlas miasta Krakowa. PPWK, Warszawa–Wrocław
Van T, Phuong D, Van P, Bao H (2015) Mapping changes of surface topography under urbanization process in Ho Chi Minh City, Vietnam, using satellite imagery. In: Publication of 1st international electronics conference on remote sensing, 22 June–5 July 2015, pp 1–7
Vogel S, Märker M, Seiler F (2011) Revised modelling of the post-AD 79 volcanic deposits of Somma-Vesuvius to reconstruct the pre-AD 79 topography of the Sarno River plain (Italy). Geol Carp 62:5–16
Zarychta R (2019) Krajobraz poeksploatacyjny kamieniołomu Liban w Krakowie. Przegl Geol 67:1002–1011
Zarychta R, Zarychta A, Bzdęga K (2020) Progress in the reconstruction of terrain relief before extraction of rock materials—the case of Liban quarry, Poland. Remote Sens 12:1548

Adam Łajczak is a Professor of Geography recently retired from the University of the National Education Commission, Krakow. His main research interest are geomorphology, hydrology, and geoecology, especially of mountainous areas (structural geomorphology, slope and fluvial processes, water resources, floods, siltation of dam reservoirs, geoheritage, and nature protection) and urbanised areas (anthropogenic changes in relief and water circulation). The Carpathian Mountains and the Vistula River valley are his main research areas. He was involved in projects in various European and other countries and participated in geographical expeditions, among others, to the Tibetan Plateau, the Himalayas, the Caucasus, and the Alps. He is a member of European Geoscience Union, DENUCHANGE Working Group of the IAG, and International Association of Hydrological Sciences. He participates in the IGU Commission "Geomorphology and Society".

Roksana Zarychta is an assistant professor at the University of the National Education Commission, Krakow. Her main research interests concern the evolution of relief in the conditions of diverse human activity, with particular emphasis on areas left after exploitation of rock raw materials, changes in the topography of historical city centres, and application of advanced geostatistical, GIS, remote sensing, and photogrammetric methods and tools in the reconstruction of relief of anthropogenically-modelled areas.

The South-Eastern Nida Basin (Ponidzie)—Karst in Gypsum Rocks

24

Jan Urban, Anna Chwalik-Borowiec, Andrzej Kasza, Artur Zieliński and Artur Komorowski

Abstract

The south-eastern part of the Nida Basin region is the only area in Poland of gypsum karst development. Karstification, developed in several areas of the region, occurred in a 30–40 m thick gypsum series of Miocene age. The most specific karst area within the Nida Basin is the western part of the Solec Basin, with large karst-denudational depressions, karst blind valleys, ca. 100 caves as well as several clusters of dolines. The large karst-denudational depressions represent senile landforms, whereas karst valleys, most caves and dolines are young morphological features, of Late Pleistocene or Holocene age. They are genetically connected with the present-day water table and are classified as open karst. In the Szaniec Plateau, depressions (dolines, swallow holes) and valleys are the most typical karst landforms, while caves are practically absent. They mainly represent subjacent karst, since the gypsum series is still covered by younger rocks over most of the area. In other areas, gypsum karstification represents intrastratal karst that has developed and currently develops under the impermeable clays and claystones overlying gypsum series. Genetic differences of gypsum karst and their geomorphic consequences decide about scientific importance, whereas the occurrence of open, currently active karst in the Solec Basin has high educational value.

Keywords

Gypsum karst · Blind valleys · Karst depressions · Caves · Open karst · Nida Basin

24.1 Introduction

The Ponidzie hilly upland in the south-eastern part of the Nida Basin (south-central Poland) is characterised by distinctive morphology, which is conditioned by lithology of its substratum and active, ongoing tectonics. Principal morphological elements such as basins and hill chains comprise tectonic grabens and horsts, which have been forming since the beginning of the Cenozoic, whereas secondary landforms, developed during the late Neogene and Quaternary, are elements of structural relief that depends on rock lithology. Gypsum series as well as bioclastic limestones occurring below and above the gypsum series are the most significant lithostratigraphic units responsible for the structural relief of the Ponidzie area, and gypsum karstification has been an important process shaping the relief at the local scale. Lithology and tectonics, as well as diverse water circulation patterns, are reasons of complex morphogenesis, with crucial participation of karstification. Therefore, the geomorphological description of the Ponidzie region is very instructive for a better understanding of interrelations between lithology, tectonics and specific morphogenic processes, such as karstification. Hereafter, the description of areas of gypsum karst in the Ponidzie is given, with special attention focused on the best recognised and most specific karst in the Solec Basin.

J. Urban (✉)
Institute of Nature Conservation, Polish Academy of Sciences, Adama Mickiewicza Ave. 33, 31–120 Kraków, Poland
e-mail: urban@iop.krakow.pl

A. Chwalik-Borowiec
The Świętokrzyskie and Nadnidziańskie Landscape Parks Complex, Umianowice, Parków Krajobrazowych 7, 28-404 Kije, Poland
e-mail: achwalik@pk.kielce.pl

A. Kasza
Świętokrzyski Speleoclub, Żurawia 23, 25-653 Kielce, Poland
e-mail: andrzejka@poczta.onet.pl

A. Zieliński
Institute of Geography and Environmental Sciences, Jan Kochanowski University of Kielce, Uniwersytecka 7, 25–406 Kielce, Poland
e-mail: artur.zielinski@ujk.edu.pl

A. Komorowski
Beskidy Caving Club, Połaniec, Poland
e-mail: akomorowski@op.pl

P. Migoń and K. Jancewicz (eds.), *Landscapes and Landforms of Poland*, World Geomorphological Landscapes, https://doi.org/10.1007/978-3-031-45762-3_24

24.2 Geographical and Geological Setting

The informal Ponidzie region is the south-eastern part of the Nida Basin macroregion (Solon et al. 2018; Richling et al. 2021), within which the following mesoregions are distinguished: Połaniec Basin, Pińczów Hummock, Solec Basin, Wodzisław Hummock and the Nida River Valley (Figs. 24.1 and 24.2). In the geological terms, the direct substratum of the Ponidzie is formed of two geological (tectonic) units: (1) Alpine unit of the Miechów Synclinorium formed at the Cretaceous-Palaeogene transition and built of Upper Cretaceous marls and marly-siliceous rocks with limestone inserts; and (2) young Alpine unit of the Carpathian Foredeep, developed in the Neogene and built

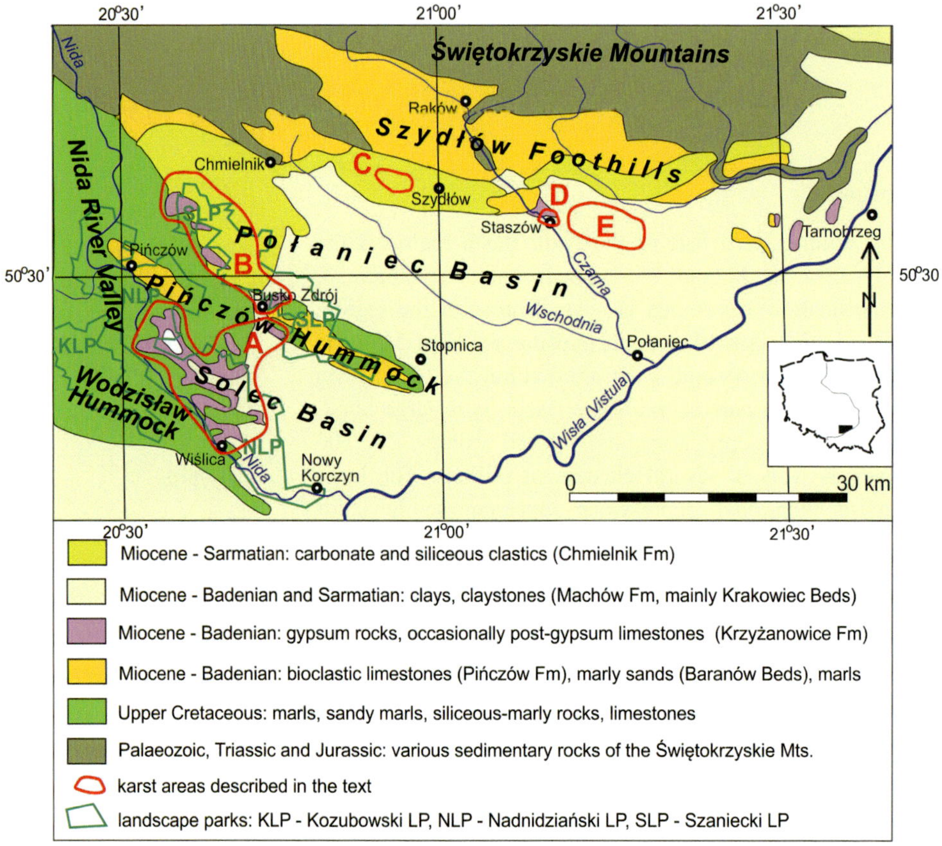

Fig. 24.1 Geological map of the Ponidzie region (SE part of the Nida Basin) with gypsum karst areas: A—Solec Basin, B—Szaniec Plateau, C—Skadla-Grabki Duże area, D—Staszów, E—Lasy Golejowskie; geological map after Jurkiewicz and Woiński (1979, 1981)

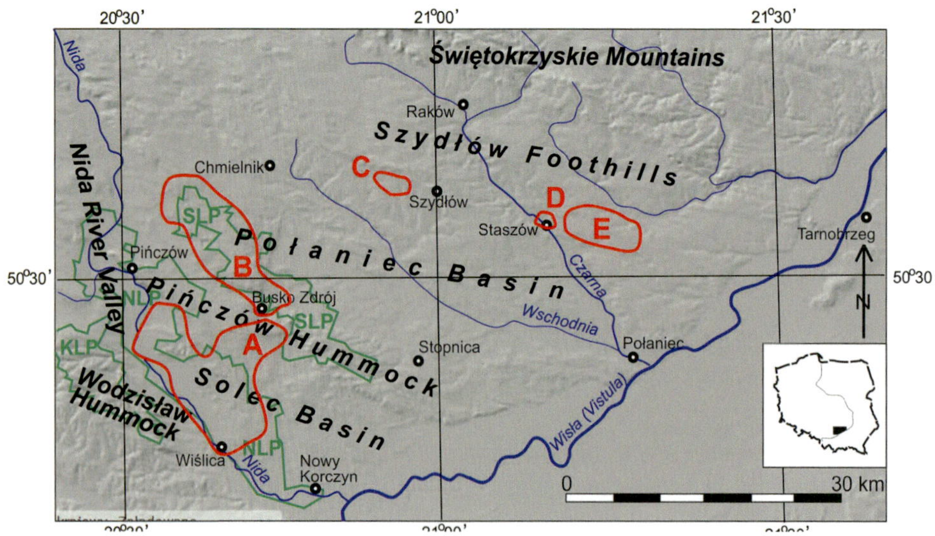

Fig. 24.2 Digital elevation model of the Ponidzie region with the gypsum karst areas described in the text (explanations of letter codes as in Fig. 24.1)

of rocks deposited during the Middle Miocene in various marine environments, from very shallow and littoral to deep and fast deepened ones. The area of gypsum outcrops and its occurrence in the shallow subsurface covers the northern part of the Carpathian Foredeep, composed of thin and lithologically differentiated sedimentary series representing shallow-water facies (Fig. 24.1). Consequently, the gypsum, called the Krzyżanowice Formation (Fm) or Chemical Series and formed during the Lower Badenian (Middle Miocene), overlays in different places the following rocks: (i) Upper Cretaceous marls or marly-siliceous rocks, (ii) Lower Badenian limestones or marls of the Pińczów Fm, or (iii) Lower Badenian marly-sandy sediments of the Baranów Beds. The Krzyżanowice Fm is overlain in the lithostratigraphic sequence by either (i) Upper Badenian and Sarmatian clays and claystones (with siltstone or sandstone inserts) of the Machów Fm, divided into older Pecten Beds and younger Krakowiec Beds, or by (ii) Sarmatian carbonate and/or siliceous sandstones and conglomerates of the Chmielnik Fm, deposited in the coastal, northern part of the basin (Czapowski 2004).

The Krzyżanowice Fm is up to 30–40 m thick and built of various types of gypsum in which primary sedimentary structures are preserved. From the bottom of the formation, the following principal structural types are distinguished (Fig. 24.3): (a) 3–4 m thick layer of columnar, giant gypsum crystals ranging a length up to 3.5 m, and forming a palisade fabric, called glassy gypsum; (b) 2–4 m thick section of fine-grained and crystalline gypsum alternations, called grassy gypsum; (c) several metres to more than 10 m thick, bedded gypsum composed of selenite crystals up to 30 cm long, often elongated and bent, called sabre gypsum; and (d) several metres to more than 10 m thick fine-crystalline (micro-crystalline) and brecciated gypsum, thin-bedded to massive. This structural differentiation results from nonuniform depositional conditions and their changes through time, involving factors such as basin depth, submarine flows, water mineralisation and oxidation as well

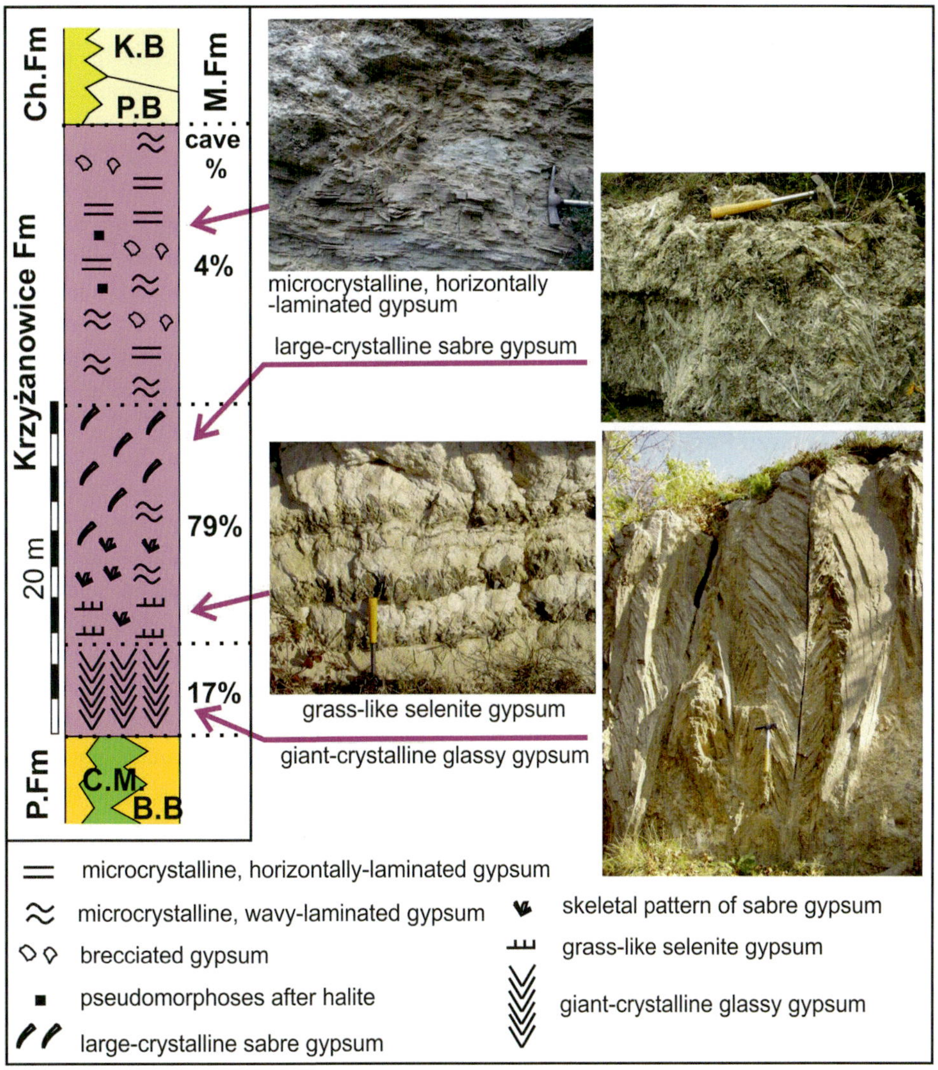

Fig. 24.3 Lithostratigraphical sequence of Miocene formations and their direct substratum in the Ponidzie area. Explanation of the abbreviations: B.B.—Baranów Beds, Ch.Fm—Chmielnik Fm, C.M.—Upper Cretaceous, K.B.—Krakowiec Beds, M.Fm—Machów Fm, P.B.—Pecten Beds, P.Fm—Pińczów Fm; lithostratigraphic column compiled after Bąbel (1999a, b), Kasprzyk (1999). Percentages refer to the number of caves in each gypsum unit

as—especially in the upper section—gravitational processes and halite dissolution (Bąbel 1999a, b; Kasprzyk 1999). Regular jointing system is visible only in the upper, fine-grained section of the formation. Within the Krzyżanowice Fm, in particular in the eastern part of the region, limestone and marl bodies originated due to the metasomatism of gypsum and bearing sulphur deposits occur. They are called post-gypsum or sulphur-bearing limestones (Kubica 1993).

The principal tectonic elements of the Ponidzie area are two horsts elongated in NW–SE direction: the northern Pińczów Horst and the southern Nida Horst. The former separates two grabens: the northern Połaniec Depression and the southern Solec Depression (Krysiak 2000). These tectonic units were formed at a turn of the Palaeogene, but owing to the continuation of tectonic processes during the Cenozoic, the horsts currently form elevated morphological elements of the Pińczów Hummock and the Wodzisław Hummock, while the tectonically depressed areas coincide with the Połaniec Basin and the Solec Basin (Figs. 24.1 and 24.2).

24.3 Karst Areas

In this section, all currently accessible gypsum karst areas within the Ponidzie region are described (omitted are karst sites in sulphur open-cast mines, closed and reclaimed by now). The most extensive chapter is dedicated to karst in the Solec Basin, characterised by the abundance of karst forms and their predominant role in the relief. It is also most comprehensively researched. However, even the least extensive karst area in the town of Staszów is worth a detailed description due to associated engineering problems.

24.3.1 Western Part of the Solec Basin

24.3.1.1 Geological Background

The western part of the Solec Basin (Figs. 24.1 and 24.2, area A) is the only area in the Ponidzie region, where outcrops of gypsum of the Krzyżanowice Fm are common, mainly in the limbs of folds and brachyfolds. In the eastern part of the Solec Basin, gypsum rocks are concealed below the Upper Badenian and Sarmatian clays and claystones. The principal tectonic structure in the Solec Basin is the Skorocice Syncline parallel to the elongation of the basin. It strikes NW–SE, but there are several brachyfolds, mainly brachyanticlines, which developed along the principal tectonic Chmielnik-Busko Zdrój-Wiślica zone crossing the central part of the Solec Basin in N–S direction. The Skorocice Syncline is also accompanied by several secondary synclines and faults (Fig. 24.4). Generally, the inclination of beds within the Krzyżanowice Fm varies from horizontal to 20–30° (Łyczewska 1972a, b, 1975; Krysiak 2000; Urban 2012).

24.3.1.2 Karst Landforms

Gypsum rocks play a crucial role in the morphology of the western part of the Solec Basin, because this series—in particular its lower part, and especially the glassy gypsum layer—represents the hardest lithostratigraphic unit in relation to other rocks emerging at the surface, such as the Cretaceous and Lower Badenian marls and marly-siliceous rocks as well as clays and claystones overlying gypsum. Consequently, gypsum rocks of the Skorocice Syncline limbs form long cuestas as well as monadnocks ranging in height from 10 to 20 m (Figs. 24.4 and 24.5).

Five large karst-denudational depressions surrounded by gypsum hill ranges, with escarpments similar to those mentioned above, have been identified in the area (Figs. 24.2 and 24.6). They represent inverted relief developed in the brachyanticlines built of the Krzyżanowice Fm and the underlying marls. One similar landform is a product of brachysyncline denudation. The areas of these depressions vary from 1.0 to 4.6 km^2, their lengths range from 1.3 to 5.3 km, whereas the depths vary from 14.5 to 46.1 m. These depressions originated due to chemical and mechanical denudation of gypsum, marls and clays or claystones (Chwalik-Borowiec 2013; Urban et al. 2015).

Smaller karst depressions are typical dolines of a diameters ranging from a few metres to 20–30 m. They are concentrated in several sites within the Solec Basin (Fig. 24.4), including karst valleys described below. They are mostly gravitational landforms, i.e. developed due to collapses of karst cavities. The sinkholes are usually dry, formed in "bare" gypsum (not mantled by other rocks), characterised by steep, often rocky slopes and bowl-shaped bottoms (Urban et al. 2015).

Three typical karst blind valleys are situated in the area (Fig. 24.4): Skorocicka Valley, Aleksandrowska Valley and Bronina Valley. The Skorocicka Valley is relatively wide and characterised by uneven bottom, punctuated by numerous hummocks and hollows which are essentially more or less regular sinkholes. It is composed of two parts, separated by a natural rock bar. The upper, blind section is 600 m long and closed by the bar, whereas the lower section, 200 m long, is open towards the south. They are hydrologically connected by the Jaskinia Skorocicka cave. Apart from the cave, numerous karst conduits comprising relict or active underground water channels occur within the hummocks, valley slopes and the rock bar crossing the valley. In total, 35 caves have been explored in this valley (Urban et al. 2015).

The Aleksandrowska Valley is of erosional origin only in its uppermost section. Its middle section represents a typical karst blind valley devoid of a streambed, but framed by

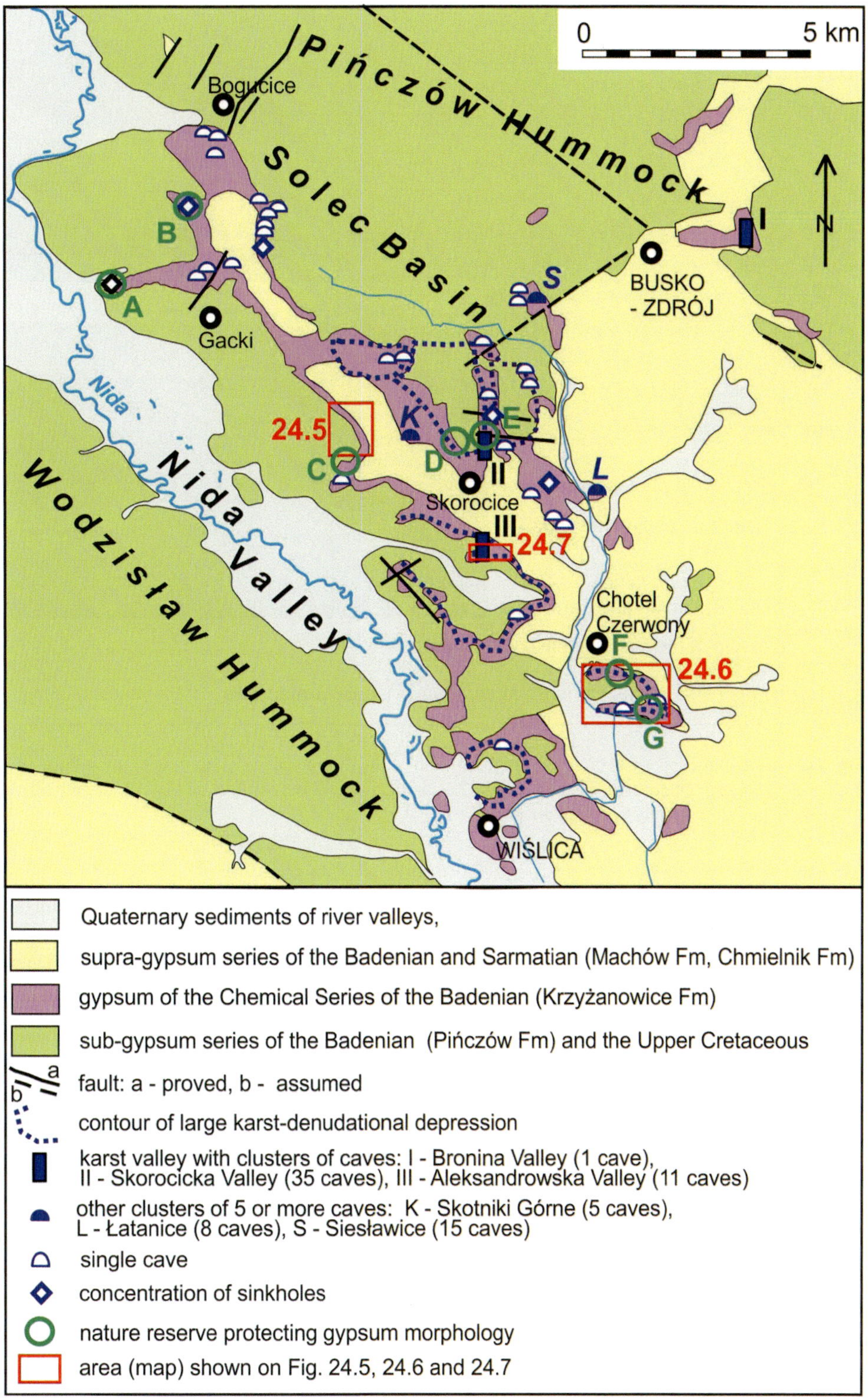

Fig. 24.4 Karst features in the western part of the Solec Basin in relation to the main geological units of this region; geological map after Łyczewska (1972a, b). Nature reserves protecting karst morphology: A—Krzyżanowice, B—Las Grabowiec, C—Skotniki Górne, D—Winiary Zagojskie, E—Skorocice, F—Przęślin, G—Góry Wschodnie

underground channels (carrying water flows periodically) on both sides, and closed by a rock bar, crossed in turn by conduits of the Jaskinia w Ryglu Cave (A on Fig. 24.7). The lowermost part of this hydrological system comprises a sequence of several sinkholes, which are connected by three caves: Jaskinia w Ryglu (A), Jaskinia w Aleksandrowie (B) and Jaskinia Flisa (C). These caves form an underground streambed (Fig. 24.7).

The Bronina Valley seems to be the most advanced in its evolution because it is most extensive and flattened.

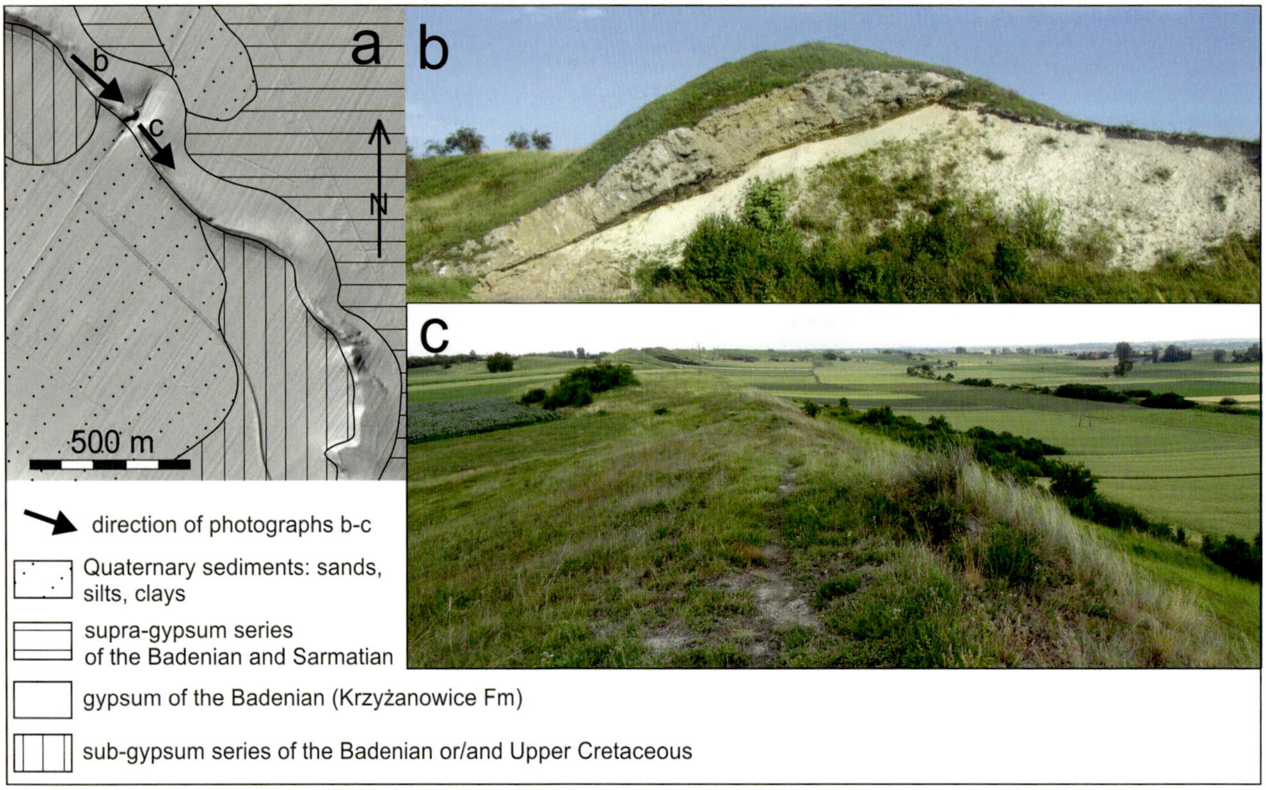

Fig. 24.5 Gypsum-capped cuesta in the southern limb of the Skorocice Syncline next to the village of Wola Zagojska. **a** Digital terrain model and geological map (map after Łyczewska 1972a, modified). **b** Cross-section of the cuesta and lithological units involved in its formation. **c** Cuesta ridge (photographs by J. Urban)

However, this valley is closed by a fairly high (up to 10 m) rock bar crossed with the maze system of Jaskinia Sawickiego Cave (173 m long, Fig. 24.8i), which discharges the seasonal lake formed in this part of the valley.

These three valleys, despite some topographical differences, show many karst and hydrological similarities indicating their development due to the activity of a hydrological system related to the water table. Their bottoms are situated close to the water table which in the cases of the Skorocicka Valley and the Aleksandrowska Valley is stabilised owing to the close occurrence of the bottom of the gypsum series, while in the Bronina Valley the water table exhibits seasonal fluctuations. The first two valleys are characterised by the occurrence of numerous landforms indicating their development due to breakdowns of ceilings of karst galleries with underground streambeds. Both valleys are stretched parallel to the gypsum strata extension, which clearly indicates underground water flow at the water-table level along the bedding planes, as the only conduits enabling water circulation at the beginning of the karstification. The more advanced development of the upper section of the Aleksandrowska Valley can be explained by higher aggressiveness of water flowing in this part of valley that is not yet mineralised. They are partly rain waters and partly allochthonous surface waters (Flis 1954; Urban et al. 2015).

In the Solec Basin, 105 caves have been recorded and are currently accessible. The longest cave is Jaskinia Skorocicka, 352 m long (the longest gypsum cave in Poland), and only a few caves are more than 100 m long. Most caves (56%) are less than 10 mlong. Also the vertical ranges of the caves are usually small (59% less than 2 m). About half of the caves (48%) are concentrated in the karst valleys (Skorocicka Valley—33, Aleksandrowska Valley—11 caves, Fig. 24.4). Most caves are situated very close to the water-table level (Table 24.1), which clearly indicates their origin within the epiphreatic zone. About 40% of the caves are simple, horizontal conduits of branchwork or rudimentary branchwork type (Table 24.1; Figs. 24.8a, d and 24.9), according to Palmer's (1991, 2007) classification. Such conduits develop owing to horizontal flows of underground streams within the water-table zone (Palmer 1991, 2007; Lauritzen and Lundberg 2000). It is illustrated by horizontal concavities preserved in the walls of some caves (e.g. in the Jaskinia Skorocicka—Fig. 24.9) at different levels, which are remnants of the meandering streambed that has gradually descended along with lowering of the water table. The other 40% caves are low but extensive

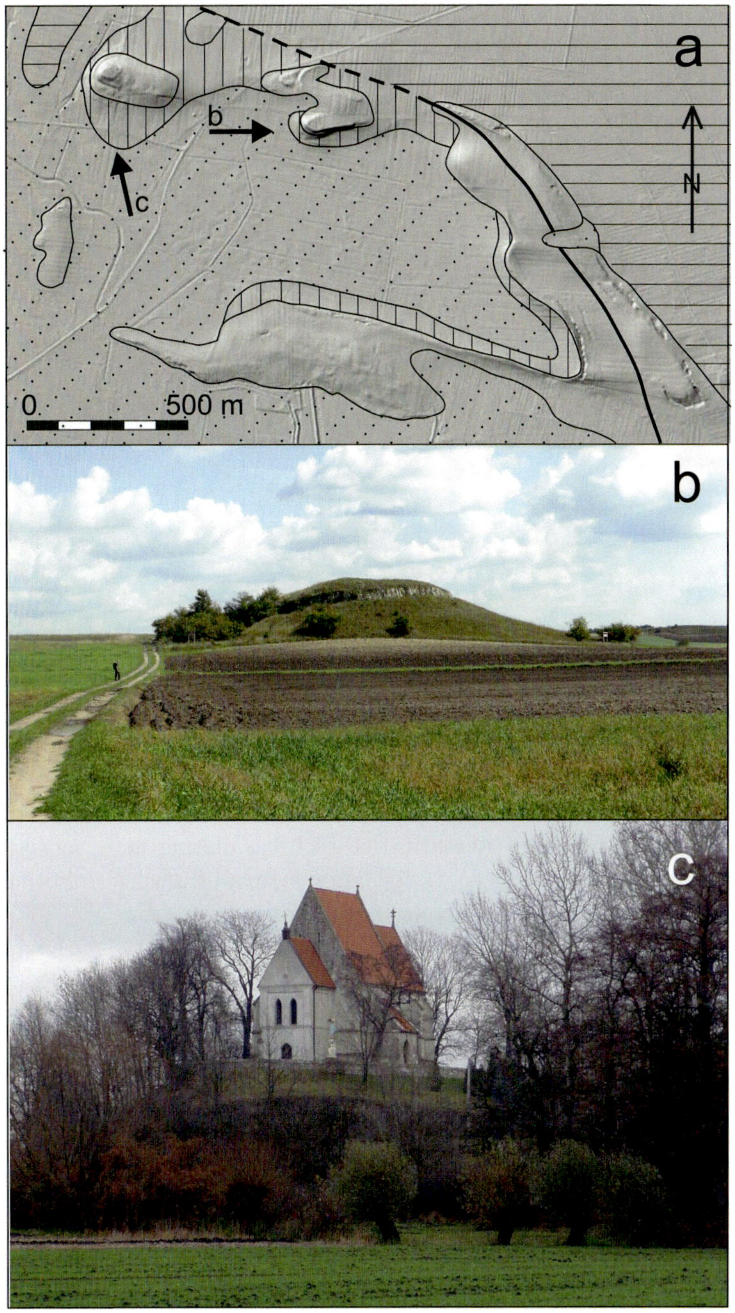

Fig. 24.6 Large karst-denudational depression formed within the brachyanticline in the village of Chotel Czerwony. **a** Digital terrain model and geological map (map after Łyczewska 1972a, significantly modified), for explanation of symbols see Fig. 24.5. **b**, **c** Two "table mountains"—Przęślin (**b**) and the mount with St. Bartholomew Church (**c**)—within the gypsum hill chain surrounding the depression, formed due to higher resistance of the giant-crystalline gypsum layer—see Bąbel et al. 2018 (photographs by J. Urban)

chambers or irregular cavities (Table 24.1; Fig. 24.8b, h), formed also within the water-table zone, but in a stagnant water environment (Urban et al. 2008, 2015).

The occurrence of caves is controlled by the lithology of gypsum, strictly—by its fabric and mechanical properties. The majority of caves (79%) occur in the selenite complex, whereas caves are very rare (4%) in the fine-crystalline and brecciated gypsum complex of similar thickness (Fig. 24.3; Table 24.1). This is due to much higher durability of karst conduits in the large-crystalline gypsum lithologies, whereas cavities in "soft", fine-crystalline gypsum, although easily corroded (dissolved—see Lebedev and Kosorukov 2017) are eroded fast, collapse and get clamped (Urban et al. 2015).

Four caves distinctly differ from the majority of the caves. Two caves, Jaskinia w Krzyżanowicach Górna (Fig. 24.8c) and Jaskinia Chotelecka, are composed of domed chambers characterised by the occurrence of oval and lenticular concavities in the ceilings and walls (Fig. 24.8f, g) (Urban et al. 2008, 2009, 2015). Such karst formations are interpreted as indicators of intrastratal speleogenesis due to very slow and deep water circulation in confined water-table conditions (Palmer 1991, 2007; Lauritzen and Lundberg 2000) often related to ascending

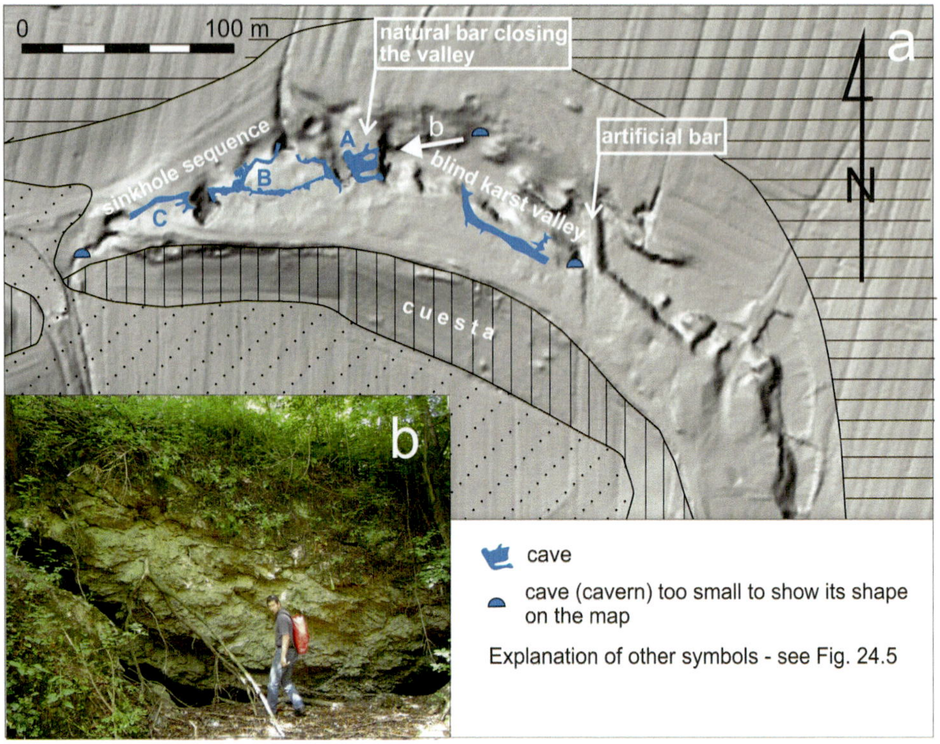

Fig. 24.7 Aleksandrowska Valley and its caves. **a** Digital terrain model with caves at the background of geological map (map after Łyczewska 1972a, significantly modified), for explanation of symbols see Fig. 24.5; caves: A—Jaskina w Ryglu (Cave in the Bar), B—Jaskinia w Aleksandrowie (Cave in Aleksandrów), C—Jaskinia Flisa (Flis Cave). **b** Rock bar closing the middle section of the valley (photograph by J. Urban)

water flow (Klimchouk 2009, 2019). The third cave, Jaskinia Sawickiego, situated in the rock bar closing the Bronina Valley (Fig. 24.8e, i), is a maze system of tubular conduits, which was most probably formed in the epiphreatic zone and related to the seasonal water-table oscillations still observed in the valley (floodwater caves—Lauritzen and Lundberg 2000; Palmer 2001). The fourth cave distinctly different from the others is Schronisko Piecowe (Fig. 24.10), a dome-like cavity, which is most probably a tumulus, i.e. a result of weathering deformation of gypsum strata. Similar tumuli were described in the region by Bąbel (2007), as well as in gypsum areas of Spain and Italy (Calaforra and Pulido-Bosch 1999; Ferrarese et al. 2003).

Secondary chemical formations in the gypsum caves of the Solec Basin are represented mainly by gypsum linear incrustations formed on cave walls (gypsum surfaces) along large (glassy) crystal margins or joints, as well as small crystal aggregates (cauliflower, mushroom-like formations) covering the walls. Gypsum and gypsum-calcite stalactites are very rare and small, reaching a length of a few centimetres. The fourth type of such formations are very rare micro-crystalline gypsum sheets, resembling flowstone, but probably of a different origin (Urban et al. 2008).

Mechanical cave sediments are not very abundant and represented principally by autogenic gypsum blocks and debris, which are ceiling breakdown products, as well as allogenic black gypsum rendzina, which is a kind of soil, washed into the caves after the area was deforested in the Middle Ages. The only two palaeontological-stratigraphical sites recently recorded in the cave sediments represent very young, Holocene (Neo-Holocene) molluscs and bone remains (Urban et al. 2009, 2015).

24.3.1.3 Karst Formation and Evolution

In the Solec Basin, the evidence of different stages of gypsum karstification is well preserved and possible to identify. As Bąbel (1999a) suggested, the oldest, syndepositional karst that developed during the syngenetic stage of gypsum evolution is recorded by gypsum breccias formed after halite dissolution in the upper section of the Krzyżanowice Fm, as well as by uneven top surface of this formation. This karstification phase was caused by changes in water mineralisation in the late and final phase of evolution of the depositional basin.

Two caves composed of dome-like chambers, Jaskinia w Krzyżanowicach Górna and Jaskinia Chotelecka, formed during the mesogenetic stage of gypsum evolution and represent the intrastratal, deep-seated karst. They formed in confined hydrogeological settings, when the gypsum strata were covered by impermeable clays (or claystones) of the Pecten Beds and Krakowiec Beds. Since the Miocene gypsum series itself (not yet karstified by that time) had low permeability, the impermeability of the overburden rock and low permeability of marls and marl-sandy rocks of gypsum substratum were reasons of very weak karst development at this stage, which clearly differentiates karst in the Solec Basin from the gypsum karst of the Carpathian Foredeep in Western Ukraine (Urban et al. 2008). Both

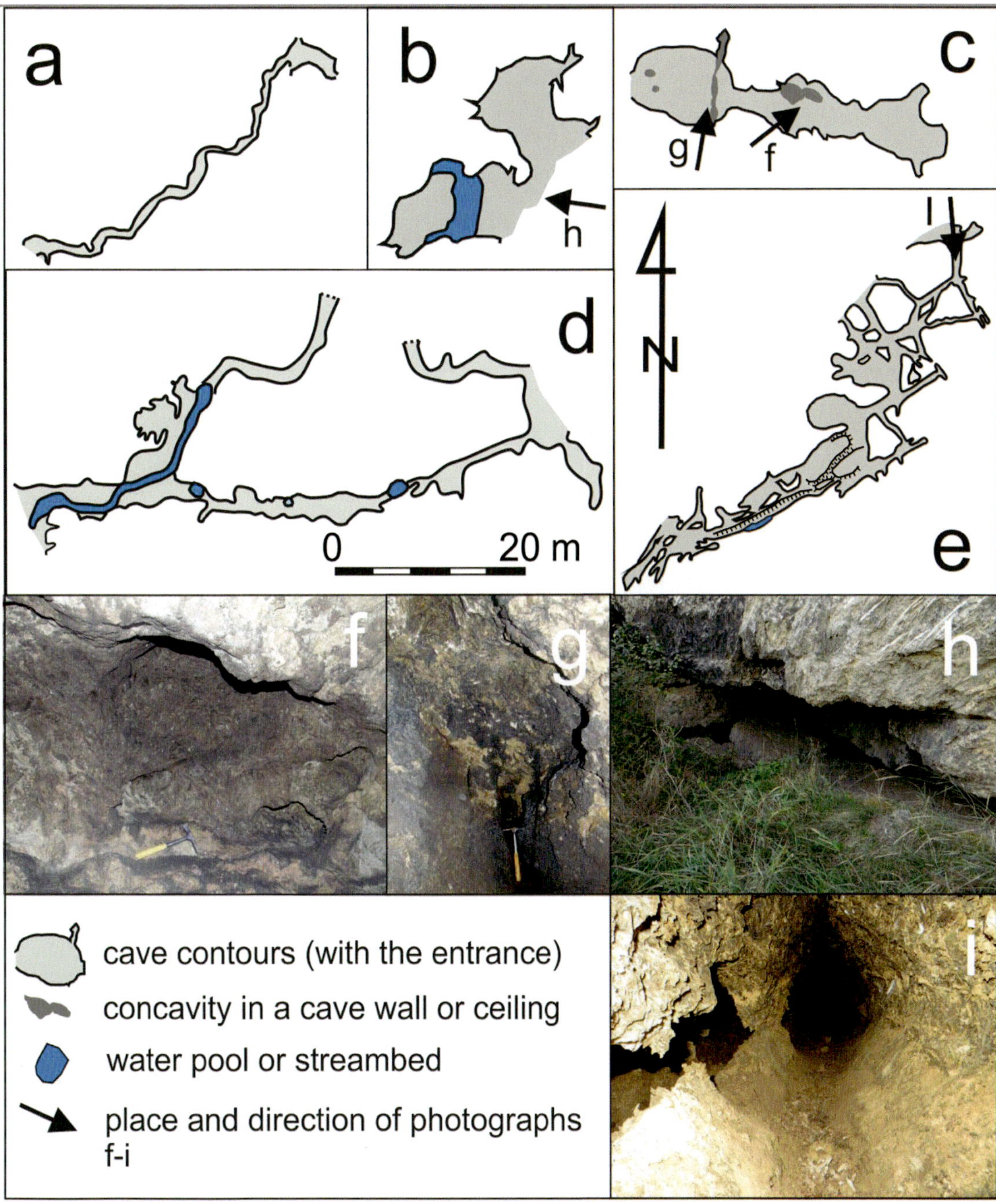

Fig. 24.8 Selected caves in the Solec Basin. Ground plans: **a** Jaskinia w Gackach (Cave in Gacki). **b** Jaskinia Szeroka w Siesławicach (Wide Cave in Siesławice). **c** Jaskinia w Krzyżanowicach Górna (Upper Cave in Krzyżanowice). **d** Jaskinia w Aleksandrowie (Cave in Aleksandrów). **e** Jaskinia Sawickiego (Sawicki Cave); scale bar applies to all cave maps. Details of cave morphology: **f** Karst pothole with smaller cupola on the ceiling in the Jaskinia w Krzyżanowicach Górna. **g** Lenticular concavity in the ceiling, developed along a joint in the Jaskinia w Krzyżanowicach Górna. **h** Entrance to the Jaskinia Szeroka w Siesławicach. **i** Gallery in the Jaskinia Sawickiego (photographs by J. Urban)

Table 24.1 Characteristics of caves in the Solec Basin (after Urban et al. 2015)

Feature	Percentage (in brackets percentage of caves longer than 10 m)		
Cave length	More than 100 m: 7 (15)	10–100 m: 37 (85)	less than 10 m: 56 (–)
Cave depth	More than 5 m: 6 (13)	2–5 m: 35 (61)	less than 2 m: 59 (26)
Relation to higher-rank landforms	Caves concentred in karst valleys: 48 (44)	Caves located within hill and hillslopes: 38 (41)	Caves in depressions and low plains: 14 (15)
Lithology of host rocks	Glassy gypsum: 17 (20)	Grassy and sabre gypsum: 79 (74)	Fine-grained and brecciated gypsum: 4 (6)
Relation to water table	Higher than 5 m: 8 (6)	2–5 m: 25 (11)	Up to 2 m: 67 (83)
Shape of cave system	Horizontal karst channels: 39 (41)	Horizontal chambers and irregular cavities: 37 (50)	Other caves: 24 (9)
Specific morphologies	Caves with horizontal channels: 20 (26)	Caves with karst cupolas: 2 (4)	Other caves: 78 (70)

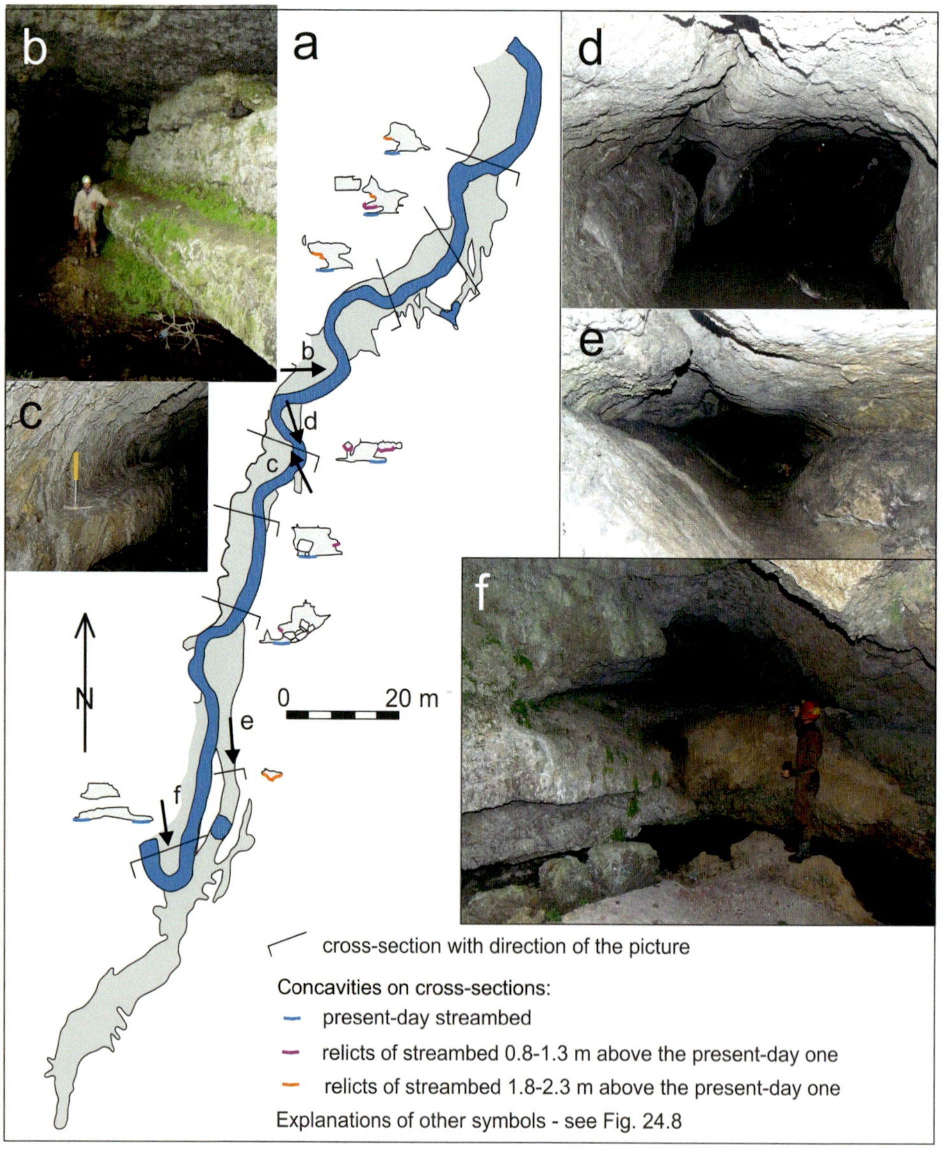

Fig. 24.9 Map, cross-sections and pictures of the Jaskinia Skorocicka that show various stages of lowering of a subsurface meandering streambed. **a** Map and cross-sections. **b** Remnants of streambeds 0.8–1.3 m and 1.8–2.3 m high. **c** Remnants of meandering streambed of the 0.8–1.3 m high level; **d** Present-day streambed and relicts of streambed 0.8–1.3 m high. **e** Conduit comprising streambed at 1.8–2.3 m high level. **f** Present-day streambed and relicts of streambed 0.8–1.3 m high (photographs by J. Urban)

Fig. 24.10 Schronisko Piecowe, a probable tumulus, as suggested by characteristic bending of gypsum beds in its ceiling (photograph by J. Urban)

caves mentioned above are genetically connected with the hydrogeologically specific zones. The Jaskinia w Krzyżanowicach Górna Cave is located close to the fault, while the Jaskinia Chotelecka Cave is situated in the axial zone of a narrow syncline sunk into marls of the substratum and probably densely jointed.

The telogenetic, i.e. denudational stage of the Krzyżanowice Fm evolution, is represented by gypsum morphologies which are now perfectly visible at the surface and underground. Among them, large karst-denudational inversional depressions seem to be the most specific features. Due to their formation within the brachyanticlines, and consequent important role of mechanical weathering and erosion of rocks in their central parts, they cannot be perceived as typical poljes. Nevertheless, they are remnants of a long-lasting karstification that started after the partial removal of cover rocks, when the gypsum was corroded by meteoric water directly infiltrating to the ground. As suggested by the comparison of their original volumes with typical rates of chemical and mechanical denudation assessed for various climatic environments, these depressions must have been developed since the Middle Pleistocene or even the Pliocene (Urban et al. 2008, 2015). However, generally weak development of typical karst landforms in the upper parts of gypsum hills and cuestas could have been caused by their mechanical erosion during cold periods in the Pleistocene, as suggested for the central European gypsum karst by Kempe (1996).

Most typical karst landforms in the area represent open karst, according to its definition given recently by Klimchouk (2019). It means that these forms, including karst valleys, caves and dolines, formed owing to epigenic karstification concentrated in the water-table zone, which has been situated at relatively shallow depth, several metres to slightly more than 10 m (Urban et al. 2008, 2009, 2015). Morphological features of the majority of karst landforms are typical for such morphogenesis (Klimchouk 2019) and regions of similar morphogenetic history (e.g. De Waele et al. 2017; Pisani et al. 2019). As noticed in the Solec Basin long ago by Flis (1954), most karst landforms are elongated concordantly with the strike of gypsum strata stretching, similarly to the epigenic karst in Italy. This indicates the principal role of bedding planes in water circulation within the water-table zone, while joint system in the lower section of the Krzyżanowice Fm is practically absent. The first karst elements produced owing to such circulation are sub-horizontal conduits, where a hydrostatic gradient stimulates adequate flow, or chambers in the water-table zone, where water is relatively stagnant. The intensive gypsum karst development within this zone has been caused by fast gypsum dissolution and subsequent fast solution saturation that has restricted this process to this zone (e.g. Lebedev and Kosorukov 2017; Klimchouk 2019). Along with the enlargement of the cavities, gravitational and erosional processes have gained a significant role in speleogenesis, although their participation is controlled by gypsum lithology. They are almost "powerless" in the giant-crystalline glassy gypsum, fairly efficient in bedded, sabre gypsum due to detachments of individual rock beds, and become the dominant morphogenic factors in fine-grained and brecciated gypsum. Therefore, it is not easy to qualify the caves to the "multi-strata caves" (controlled by lithostratigraphy) or to the "multi-level caves" (controlled by tectonics), following the categories proposed by Calaforra and Gazquez (2017).

Breakdowns of cavities have brought about the formation of dolines, usually grouped in areas of relatively stagnant water, such as flat water-divide areas within morphological elevations (hilltops, plateaus), but also within plains. These processes are also responsible for the development of karst valleys in the zones of intensive water flow, e g. along fold limbs, such as in the Skorocicka Valley and Aleksandrowska Valley. In the Skorocicka Valley, karst conduits occur at various levels above its present-day bottom, but it is not possible to assign relict stages of the karst system and valley development to particular climate changes in the past, as it was possible for the Italian karst (De Waele et al. 2019; Pisani et al. 2017). The lack of clearly defined cave levels, as well as the absence of karst shafts, suggests fast and continuous evolution of karst systems in the Solec Basin.

Although the calculation of chemical denudation rates suggests the Middle Pleistocene as the onset of the Skorocicka Valley formation, most caves, dolines and valleys in the Solec Basin seem to be relatively young landforms, developed during the Late Pleistocene and the Holocene. Many caves comprise still active underground streambeds or lakes situated at the water-table zone and continuous evolution of valleys, and underground karst systems is still very dynamic, as documented by water mineralisation in the region (Urban et al. 2008, 2015; Chwalik-Borowiec 2013).

24.3.2 Szaniec Plateau (Western Part of the Połaniec Basin)

The western part of the Połaniec Basin, called the Szaniec Plateau (Figs. 24.1 and 24.2, area B), differs from the western segment of the Solec Basin in respect to several geological and morphological features. First, outcrops of gypsum rocks are much less abundant in this area. Second, in the lithostratigraphic sequence the gypsum-dominated Krzyżanowice Fm is underlain by limestones of the Pińczów Fm and crowned with discontinuous sheets of clays of the Machów Fm (Pecten Beds and Krakowiec Beds) as well as clastic and permeable rocks of the Chmielnik Fm. Third, tectonics of the Neogene formations is less complex, and the rock strata are generally horizontal.

Such a lithostratigraphic sequence, structure and tectonics are the reasons that the limestones of the Pińczów Fm and conglomerates and sandstones of the Chmielnik Fm are associated with predominant geomorphic elements controlling the landscape, including steep slopes and escarpments (Fig. 24.11a), rather than the gypsum series. The same factors and various hydrogeological conditions, from vadose to phreatic, in places even confined, are responsible for specific karst development, called the "Szaniec type karst" by Liszkowski (1979). It is characterised by the lack of caves, although relatively frequent dolines and several swallow holes indicate that underground karst systems exist here. The most common and characteristic karst features in this area are very diverse dolines (Fig. 24.11). They differ in size, having the diameter from several metres up to about 100 m and depth up to more than 10 m, shape, water occurrence, as well as genesis. Most of them are caprock dolines, formed in gypsum, but reproduced in clays of the Machów Fm or clastic deposits of the Chmielnik Fm. In this first case, they are often filled with water, supporting lakes or wetlands. But there are also bare dolines with gypsum outcrops. Their shape is controlled by lithology: dolines and swallow holes in gypsum are conical if formed in fine-grained rocks, or cylindrical if formed in giant-crystalline gypsum (Fig. 24.11b, c). The dolines reproduced in clays comprise shallow oval depressions with lakes or peatbogs, whereas depressions in clastic materials are dry and usually irregular. Rare solutional dolines/shafts occur too (Fig. 24.11d, e). TL dating of lacustrine sediments indicates the Plenivistulian as the period of their deposition (34.2 ± 5.1 ka) (Urban 2008, 2013; Zieliński 2013a, b).

Several karst valleys that occur in this area are shallow and irregular. Although they display typical karst features such as sequences of closed depressions, and gypsum outcrops are very rare within them, suggesting that they are mostly reproduced in gypsum caprock. In the map shown in Fig. 24.11a, two types of valleys are visible: valleys with gypsum outcrops (Fig. 24.11d, e) and valleys developed within supra-gypsum clays (Fig. 24.11f).

A comparison of the Szaniec Plateau karst with that in the Solec Basin shows that the former represents a less advanced stage of development. According to Klimchouk's (2019) categorisation, it can be classified as a predominantly subjacent karst. Gypsum series in the major part of the area is still covered by younger rocks representing different lithologies, which makes the morphological results more complex. Thus, the relief of the Szaniec Plateau shows a different type of karstification and—consequently—distinct karst evolution history from that in the Solec Basin.

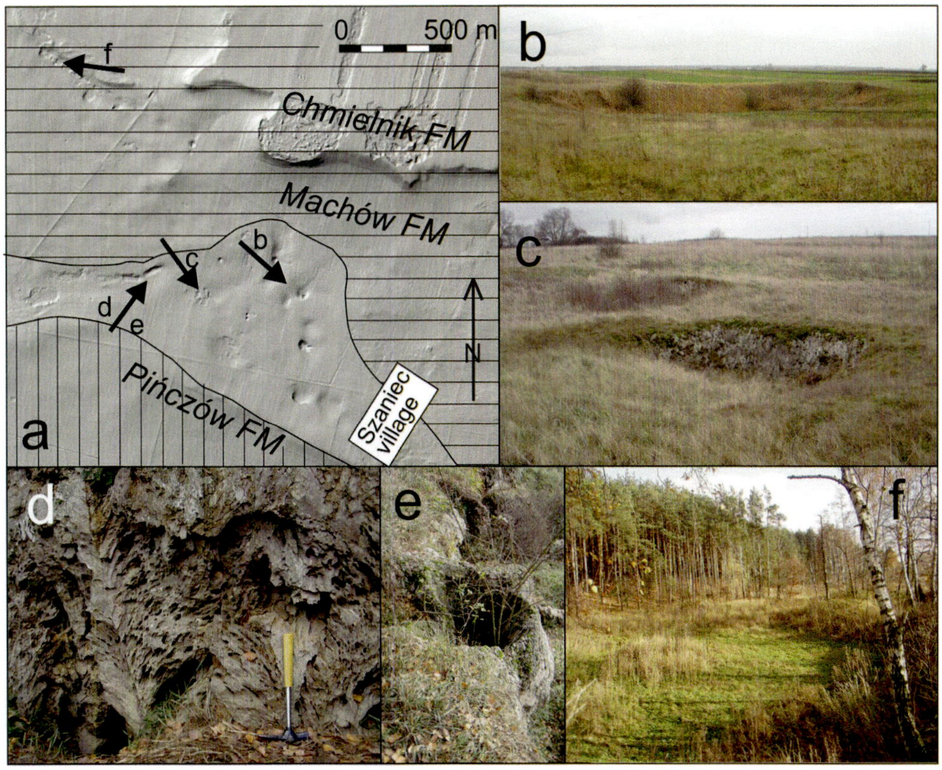

Fig. 24.11 Karst features in the southern part of the Szaniec Plateau. **a** Digital terrain model and geological map (map after Senkowicz, 1958, significantly modified); for explanation of symbols see Fig. 24.5. **b** Conical doline within fine-grained gypsum of the upper part of the Krzyżanowice Fm. **c** Small cylindrical dolines within glassy gypsum. **d** Giant-crystalline gypsum undergoing surface weathering (karstification). **e** Dissolution shaft/doline. **f** Shallow wet depressions within karst valley (photographs by J. Urban)

24.3.3 Skadla-Grabki Duże Area

In the relatively small area between the villages of Skadla and Grabki Duże in the central part of the Połaniec Basin (Figs. 24.1 and 24.2, area C), more than ten evident morphological depressions, usually irregular and of various sizes, up to 200 m in diameter, occur. Small lakes or peatlands occupy the floors of some of these hollows (Fig. 24.12). Although the geological map shows sheets of Quaternary glacigenic clays and sands at the surface, that overlie the Pecten Beds and Krakowiec Beds in this area (Romanek 1982), these depressions are undoubtedly of karst origin, because they are closed and undrained. Radiocarbon dating indicates the transition from the Middle Ages to the Modern Era (1439–1517 cal AD, Kyiv Radiocarb. Lab, Ki-12204) as the age of a peat accumulation in a karst depression.

24.3.4 Staszów Town

The town of Staszów (Figs. 24.1 and 24.2, area D), with a population of some 15 thousand people, has been built on the gypsum-carbonate rocks of the Krzyżanowice Fm and the underlying limestones of the Pińczów Fm, which are in places covered with sheets of Quaternary clays or sands (Walczowski 1965; Urban et al. 2020). In the northern segment of this town, the bedrock comprises various types of gypsum with large bodies of post-gypsum limestones of diversified lithology, including homogenous or cavernous and hard or soft (marly) variants of the total thickness usually greater than 20 m. These rocks are strongly karstified. Therefore, the construction of the Staszów North estate in 1970s has caused severe engineering problems, such as ground surface deformations affecting streets and pavements, as well as cracking of recently constructed buildings (Fig. 24.13). In order to prevent this hazardous situation in last decade of the twentieth century, a dense network of geoelectrical and micro-seismic surveys was set up, as well as more than 200 boreholes up to 20 m deep were drilled in the estate area (ca. 0.3 km^2). These boreholes enabled for geological investigation of the problem, but subsequently 177 boreholes were also used to inject concrete-ash-water mixture into the karst system. The volume injected into the karst cavities varied from less than one cubic metre up to 453 m^3. The quantity of injected mixture into individual boreholes was often not adequate to the expectations inferred on the basis of geophysical and drilling data. The high-resistivity anomalies were interpreted as large empty cavities, but in most cases, this was not confirmed by injected volumes (Urban et al. 2020).

As suggested by borehole and geotechnical data, the underground karst beneath the Staszów North estate is a maze system fragmented into separated and irregular

Fig. 24.12 Floor of a karst depression, marshy in the deepest part, at Captoryje site within the Skadla-Grabki Duże area (photograph by J. Urban)

cavities of diverse shapes and volumes, partly filled with sediments. Such a nature of this system is caused by lithological differentiation of the Krzyżanowice Fm and geological history of its development, from intrastratal karst developed in confined settings to entrenched karst modified by meteoric waters that directly infiltrate into the rock massif. The intrastratal karst developed in gypsum-carbonate, structurally diversified, and thus permeable series, under the impermeable clays or claystones of the Pecten Beds and Krakowiec Beds. An additional factor was the deposition of Pleistocene sands and clays, which partly covered the evaporate series and filled karst depressions. Recent deformations of the ground surface have been produced by gravitational and compaction processes in karst caverns within soft and plastic rocks (fine-grained gypsum and marls) as well as by piping (suffosion) that moves the loose Quaternary material into the voids. Currently, this process has been accelerated by anthropogenic changes which generate concentration of water infiltration and inflows of aggressive water from a failed sewage system. It makes karst-erosion conduits prone to consequent piping-gravitational processes (Urban et al. 2020).

Fig. 24.13 Surface ground deformations evidenced by broken pavements in the Staszów North estate (photographs by A. Komorowski)

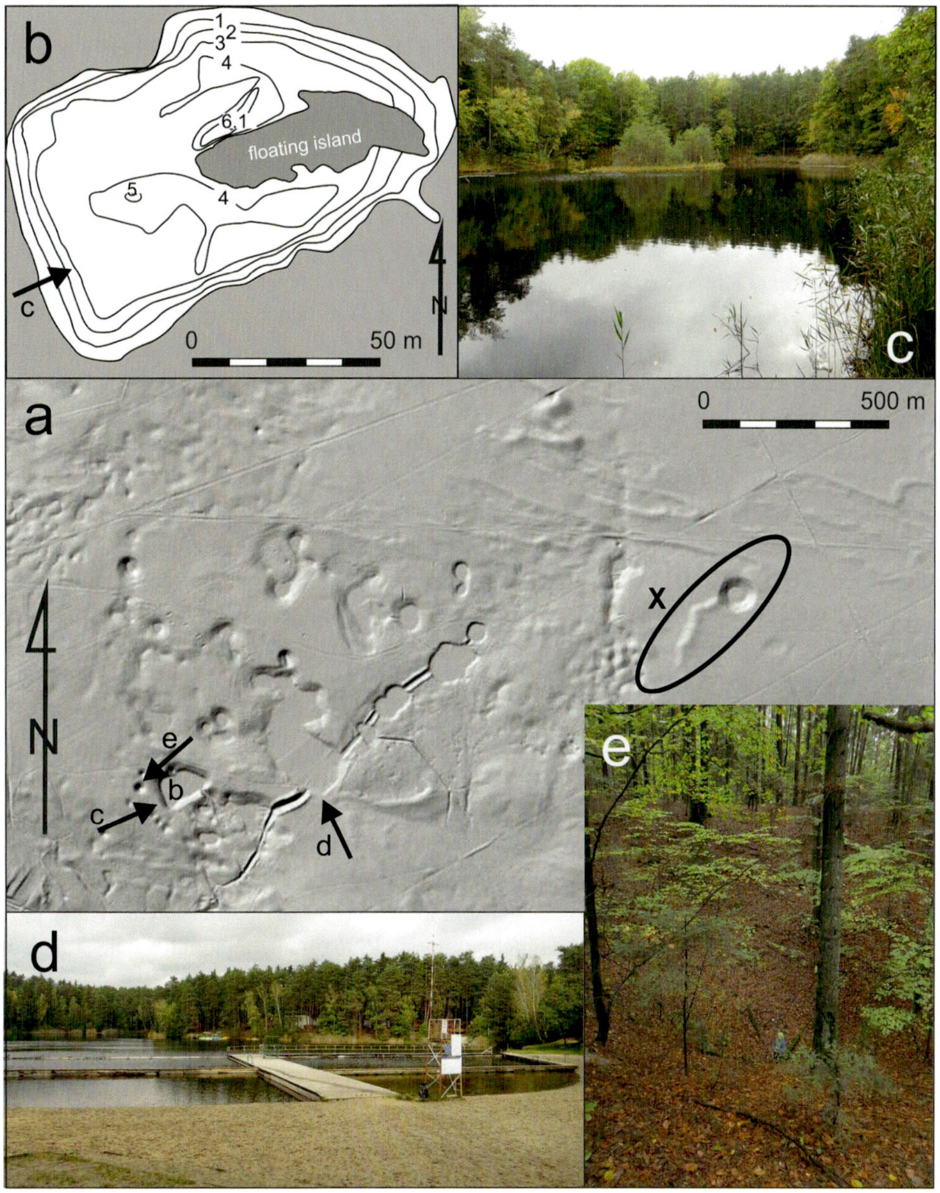

Fig. 24.14 Karst depressions and lakes in the western part of the Lasy Golejowskie. **a** Digital terrain model, x—Late Pleistocene-Early Holocene aeolian dune truncated by a karst depression, which must be therefore younger. **b** Map of the Donica Lake with its depth and floating island. **c** Donica Lake with the floating island. **d** Duży Staw (Large Pond) with its recreation facilities. **e** Wilcze Oczy (Wolf Eyes), one of the deepest dry dolines in the area (photographs by J. Urban). Arrows show the direction of photographs

24.3.5 Lasy Golejowskie (Golejowskie Forests)

In the Lasy Golejowskie (the Golejowskie Forests), situated to the east of Staszów (Figs. 24.1 and 24.2, area E), within an area up to 10 km long, numerous closed and undrained depressions occur (Fig. 24.14). Bedrock in this area is formed of clays and claystones of the Pecten Beds and Krakowiec Beds, Machów Fm, which are in places covered with Late Pleistocene-Early Holocene sand dunes (Walczowski 1965). Although karstic rocks are nowhere present at the surface, the depressions are obviously of karst origin. Karst has developed owing to artesian water inflow from the underlying limestones of the Pińczów Fm and circulation within gypsum and post-gypsum limestones of the Krzyżanowice Fm in a hydrologically confined setting. Such type of karstification, called by Liszkowski (1979) as the "Staszów type karst", can be defined as intrastratal, deep-seated karst according to the commonly accepted karst classification (Klimchouk 2009, 2019). However, the karstification process in this area differs from the intrastratal karst responsible for the formation of large caves in Western Ukraine (see Klimchouk et al. 1996; Klimchouk 2009, 2019), because the rocks overlying the karstified horizon, clays and claystones of the Machów Fm, are not permeable (Urban et al. 2009).

The TL datings of sands from these depressions indicate their deposition during the cold stage of the Plenivistulian (34.2 ± 5.1 – 20.4 ± 3.1 ka), whereas radiocarbon datings of organic sediments from the lakes are younger (10.2 ± 0.1 – 8.1 ± 0.1 ka) (Zieliński 2013a). Palaeobotanical studies suggest the Late Pleistocene and Holocene age of organic matter filling the depressions (Szczepanek 1971). The relation of the depressions to the Late Pleistocene-Early Holocene dunes also suggest younger age of ground subsidence (Fig. 24.14a, area x). Morphological development of these depressions has been complex and has involved, apart from karstification and direct gravitational breakdowns, piping (suffosion) processes, as well as water pressure generated by freezing (Zieliński 2013a).

The depressions are characterised by very various shapes, sizes and depths. The largest depression, Duży Staw (Large Pond), reaches an acreage of 3.63 ha, whereas the other ones usually do not exceed 2 ha. The deepest measured depression is 23 m deep. Some were artificially deepened due to peat extraction, and many have been filled with water forming picturesque lakes within the forest. Therefore, some are currently used as a recreation and touristic places (Fig. 24.14d). One of the most attractive phenomena developing on these lakes are "floating islands" formed due to the formation of gas bubbles under the organic traps (Zieliński 2013a; Stachura et al. 2018).

24.4 Scientific and Educational Values of Gypsum Karst in the Ponidzie

The gypsum karst in the Ponidzie region does not occupy a large area in relation to other regions of gypsum karst in the world (comparing to data given by Klimchouk et al. 1996), but it comprises very diverse types of karst and perfectly illustrates how geological (lithological and tectonic), hydrological and geomorphological conditions control karst evolution. These features can be observed within a relatively small area of a length of ca. 50 km (Figs. 24.1 and 24.2) and in generally homogenous environmental and climatic conditions.

The Solec Basin is an area of typical open karst, with minor participation of "old" intrastratal karst, and the dominant role of gypsum rocks in the morphology. This is conditioned by lithology and tectonics of the Cretaceous and Neogene rocks, as well as by morphological evolution of the area. Some similarities, especially in lithological and tectonic pre-conditioning of gypsum karst development, can be found in the Northern Apennines and Sicily (DeWaele et al. 2017; Pisani et al. 2019), perhaps also in the Harz in Germany (Kempe 1996). In the Szaniec Plateau, different lithologies of lithostratigraphic units underlying and overlying the gypsum horizon, as well as different tectonic patterns, are responsible for different karst features and general morphological evolution. This area represents a much more complex type of karstification, with greater participation of subjacent karst, and diverse karst features, including landforms. The other two areas, Skadla-Grabki Duże and Lasy Golejowskie, comprise intrastratal karst, revealed at the ground surface by various depressions.

Although the same karst type has developed in the Neogene gypsum unit in Western Ukraine (Klimchouk 2009), important geological constraints, such as lithology and permeability of the unit overlying gypsum horizon, and the stage of karst evolution make the Ponidzie area significantly different from that region. In two areas of the Ponidzie mentioned above, the intrastratal karst has been just developing and its underground elements are not accessible for people, whereas in Western Ukraine relicts of this stage of karst development are available for direct observations in the currently evolving entrenched or denuded karst (Urban et al. 2009).

The Staszów area is a specific site where features of intrastratal, entrenched and denuded karst can be recognised. However, the most important problem is the interrelation between karst development and human activity, and hazards caused by gypsum karst, which is a world-wide discussed problem (e.g. Satkunas et al. 2006). Similar hazards manifested by unexpected and instantaneous ground collapses also occur in the Solec Basin, Szaniec Plateau and Lasy Golejowskie, but they have hardly caused real danger due to dispersed human settlements.

The state of scientific evaluation and natural protection of karst phenomena in the Ponidzie region differs between individual areas. Karst in the Solec Basin is situated within the Nadnidziański Landscape Park, whereas most part of the Szaniec Plateau is covered by the Szaniecki Landscape Park (Figs. 24.1 and 24.2). The most interesting morphological and karst elements of the Solec Basin are protected in natural reserves. Among them, the "Skorocice" nature reserve protecting the Skorocicka Valley seems to be the most important from the scientific and educational points of view. The table hills, sections of the gypsum cuesta and escarpments are protected in the "Przęślin", "Krzyżanowice", "Góry Wschodnie", "Winiary Zagojskie" and "Skotniki Górne" nature reserves, while a sinkhole assemblage is located within the "Las Grabowiec" nature reserve (Fig. 24.4). Several nature monuments protecting karst features have been established in the Solec Basin and Szaniec Plateau (Urban 2008, 2012).

Despite adequate legal protection of karst features in the two areas mentioned above, educational and geotouristic use of these features in all karst areas of the Ponidzie is insufficient and needs to be improved. The natural, scientific, educational and aesthetic values of the gypsum karst of the Ponidzie area deserve adequate appreciation and relevant promotion at the European scale.

References

Bąbel M (1999a) History of sedimentation of the Nida gypsum deposits (Middle Miocene, Carpathian Foredeep, Southern Poland). Geol Quart 43:429–447

Bąbel M (1999b) Facies and depositional environments of the Nida gypsum deposits (Middle Miocene, Carpathian Foredeep, Southern Poland). Geol Quart 43:405–428

Bąbel M (2007) Kopuły gipsowe w rzeźbie krasowej Ponidzia. Prace Inst Geogr Akad Świętokrzyskiej w Kielcach 16:71–89

Bąbel M, Urban J, Chwalik-Borowiec A (2018) Stop. 2.5. Chotel Czerwony. In: Głowniak E, Wasiłowska A (eds) IX symposium ProGEO, Poland 2018, field trip guidebook. Faculty of Geology, Univ. of Warsaw, Warsaw, pp 38–45

Calaforra JM, Gazquez F (2017) Gypsum speleogenesis: a hydrogeological classification of gypsum caves. Intern J Speleol 46:251–265

Calaforra JM, Pulido-Bosch A (1999) Genesis and evolution of gypsum tumuli. Earth Surf Proc Landf 24:919–930

Chwalik-Borowiec A (2013) Rzeźba terenu Parków Krajobrazowych Ponidzia. ZŚiNPK w Kielcach, Kielce, p 160

Czapowski J (2004) Otoczenie Gór Świętokrzyskich. In: Peryt T, Piwocki M (eds) Budowa geologiczna Polski t. 1, Stratygrafia, cz. 3a, Kenozoik. Państw Inst Geol, Warszawa, pp 239–246

De Waele J, Piccini L, Columbu A, Madonia G, Vattano M, Caligaris C, D'Angeli IM, Parise M, Chiesi M, Sivelli M, Vigna B, Zini L, Chiarini V, Sauro F, Drysdale R, Forti P (2017) Evaporite karst in Italy: a review. Intern J Speleol 46:137–168

Ferrarese F, Macaluso T, Madonia G, Palmeri A, Sauro U (2002) Solution and recrystalisation processes and associated landforms in gypsum outcrops of Sicily. Geomorphology 49:25–43

Flis J (1954) Kras gipsowy Niecki Nidziańskiej. Prace Geogr Inst Geogr PAN 1:73

Jurkiewicz H, Woiński J (1979) Mapa geologiczna Polski 1:200,000, arkusz Tarnów. Wyd Geol, Warszawa

Jurkiewicz H, Woiński J (1981) Mapa geologiczna Polski 1:200,000, arkusz Mielec. Wyd Geol, Warszawa

Kasprzyk A (1999) Sedimentary evolution of Badenian (Middle Miocene) gypsum deposits in the northern Carpathian Foredeep. Geol Quart 43:449–465

Kempe S (1996) Gypsum caves in Germany. Intern J Speleol 25:209–224

Klimchouk A (2009) Morphogenesis of hypogenic cave. Geomorphology 106:100–117

Klimchouk A (2019) Chapter 58—gypsum caves. In: White WB, Culver DC, Pipan T (eds) Encyclopedia of caves. Elsevier, Amsterdam, pp 485–495

Klimchouk A, Forti P, Cooper A (1996) Gypsum karst in the world: a brief review. Intern J Speleol 25:159–181

Krysiak Z (2000) Tectonic evolution of the Carpathian Foredeep and its influence on Miocene sedimentation. Geol Quart 44:137–156

Kubica B (1994) Metasomatism of Badenian sulphates of the Carpathian Foredeep and its paleogeographic conditions. Geol Quart 38:395–414

Lauritzen SE, Lundberg J (2000) Solutional and erosional morphology. In: Klimchouk A, Ford DC, Palmer AN, Dreybrodt W (eds) Speleogenesis. Evolution of Karst Aquifers. National Speleological Society, Huntsville, pp 408–426

Lebedev AL, Kosorukov VL (2017) Gypsum solubility in water at 25°. Geochem Intern 55:205–210

Liszkowski J (1979) Morphogenetic types and mechanisms of development of surface forms of covered karst in Poland. Biul Geolgiczny Uniw Warsz 23:155–168

Łyczewska J (1972a) Objaśnienia do Szczegółowej mapy geologicznej Polski 1:50,000, arkusz Busko Zdrój. Wyd Geol, Warszawa, p 58

Łyczewska J (1972b) Szczegółowa mapa geologiczna Polski 1:50,000, arkusz Busko Zdrój. Wyd Geol, Warszawa

Łyczewska J (1975) Zarys budowy geologicznej Pasma Wójczo-Pińczowskiego. Biul Inst Geol 283:151–189

Palmer AN (1991) Origin and morphology of limestone caves. Geol Soc Amer Bull 103:1–21

Palmer AN (2001) Dynamics of cave development by allogenic water. Acta Carsologica 30(2):13–32

Palmer N (2007) Cave geology. Cave Books, Dayton, p 454

Pisani L, Antonellini M, De Waele J (2019) Structural control on epigenic gypsum caves: evidences from Messinian evaporites (Northern Apennines, Italy). Geomorphology 332:170–186

Richling A, Solon J, Macias A, Balon J, Borzyszkowski J, Kistowski M (ed) (2021) Regionalna geografia fizyczna Polski. Generalna Dyrekcja Ochrony Środowiska. Poznań, p 608

Romanek A (1982) Szczegółowa mapa geologiczna Polski 1:50,000, arkusz Chmielnik. Wyd Geol, Warszawa

Satkunas J, Taminskas J, Dilys K (2006) Geoindicators of changing landscapes—an example of karst development in North Lithuania. Geol Quart 50:457–464

Senkowicz E (1958) Szczegółowa mapa geologiczna Polski 1:50,000, arkusz Pińczów. Wyd Geol, Warszawa

Solon J, Borzyszkowski J, Bidłasik M, Richling A, Badora K, Balon J, Brzezińska-Wójcik T, Chabudziński Ł, Dobrowolski R, Grzegorczyk I, Jodłowski M, Kistowski M, Kot R, Krąż P, Lechnio J, Macias A, Majchrowska A, Malinowska E, Migoń P, Myga-Piątek U, Nita J, Papińska E, Rodzik J, Strzyż M, Terpiłowski S, Ziaja W (2018) Physico-geographical mesoregions of Poland: verification and adjustment of boundaries on the basis of contemporary spatial data. Geogr Polon 91:143–170

Stachura M, Wieczorek D, Zieliński A (2018) An attempt at a typology of karst lakes in the Połaniec Basin (Małopolska Upland, Poland). Bull Geogr Phys Geogr Ser 15:63–74

Szczepanek K (1971) The Staszów karst in the light of palaeobotanical studies. Acta Palaeobot 12(2):63–120

Urban J (2008) Kras gipsowy w Nadnidziańskim i Szanieckim Parku Krajobrazowym. Zespół Nadnidziańskich i Świętokrzyskich PK, Kielce, p 87

Urban J (2012) Dziedzictwo geologiczne. In: Świercz A (ed) Monografia Nadnidziańskiego Parku Krajobrazowego. Uniw J. Kochanowskiego w Kielcach, Kielce, pp 35–81

Urban J (2013) Stanowisko 2. Szaniec: zagłębienia krasowe w gipsach. In: Fijałkowska-Mader A, Urban J, Zieliński A (eds) Inst. of Geography, Jan Kochanowski Univ. in Kielce, Kielce, pp 46–47

Urban J, Andreychouk V, Gubała J, Kasza A (2008) Caves in gypsum of the Southern Poland and the Western Ukraine—a comparison. Kras i Speleologia 12(21):15–38

Urban J, Andreychouk V, Kasza A (2009) Epigene and hypogene caves in the Neogene gypsum of the Ponidzie area (Niecka Nidziańska region), Poland. In: Klimchouk A, Ford D (eds) Hypogene Speleogenesis and Karst Hydrology of Artesian Basins. Ukrainian Inst. Speleology and Karstology, Simferopol, pp 223–232

Urban J, Chwalik-Borowiec A, Kasza A (2015) The development and age of the karst in gypsum deposits of the Niecka Solecka (Solec Basin) area. Biul Państw Inst Geol 462:125–152

Urban J, Komorowski A, Porzucek S (2020). Development of karst in gypsum in the light of analysis of building hazard in Staszów town, Ponidzie region, Poland. In: Bąbel M, Olszewska-Nejbert D, Nejbert K (eds) The Weathering of Anhydrite and Gypsum Rocks. Wydz Geol, Uniw Warsz, Warszawa, pp 279–294

Walczowski A (1965) Szczegółowa mapa geologiczna Polski 1:50,000, arkusz Staszów. Wyd Geol, Warszawa

Zieliński A (2013a) Rozwój jezior krasowych w Niecce Połanieckiej. Uniw J. Kochanowskiego w Kielcach, Kielce, p 186

Zieliński A (2013b) Stanowisko 3. Jezioro Pleban jako przykład jeziora krasowego na Ponidziu. In: Fijałkowska-Mader A, Urban J, Zieliński A (eds) Inst Geogr, Uniw J Kochanowskiego w Kielcach, Kielce, pp 48–49

Jan Urban is geologist and geomorphologist, Associate Professor employed first in the Świętokrzyskie Mountains Branch of the Polish Geological Institute in Kielce and then in the Institute of Nature Conservation, Polish Academy of Sciences in Kraków, where he recently retired. He deals with the theory and practice of geological heritage protection. The Świętokrzyskie (Holy Cross) Mountains, the Nida Basin (Ponidzie region) and the Outer Carpathians (Beskidy Mountains), situated in the central and southern Poland, are the main areas of his scientific interest. He deals with karst forms in carbonate and gypsum rocks, including caves, as well as palaeokarst forms. He is a co-author of cave inventories in the two first regions. He also studies sandstone crags and non-karst caves in these rocks as well as explores and documents historical mines and open-pits. He is a President of the Pseudokarst Commission of the International Union of Speleology and President of the Speleological Section of the Polish Copernicus Society of Naturalists.

Anna Chwalik-Borowiec holds Ph.D. degree in Earth Sciences and is employed in the Świętokrzyskie and Nadnidziańskie Landscape Parks Complex as the main specialist for protection of landscape as well as historical and cultural values. She has conducted research mainly in the area of the gypsum karst in the Nida Basin. Her Ph.D. thesis concerned the pace of development of karst relief, including physical and chemical properties of waters conditioning this process. She is also an author of a monograph of the relief of the Nadnidziańskie Landscape Parks. Currently, she deals with issues related to nature, culture and landscape protection as well as education in this field.

Andrzej Kasza is geologist and speleologist, a member of the Świętokrzyski Speleoclub association in Kielce and the Regional Council for Nature Conservation in Kielce. He explores, documents and protects caves in the Świętokrzyskie (Holy Cross) Mountains region and the Nida Basin. He was a co-author of cave inventories/monographs for these both regions. He also conducts research and publishes on historical copper, lead, iron and uranium ore mining in the Świętokrzyskie Mountains region.

Artur Zieliński is physical geographer, Associate Professor in the Institute of Geography and Environmental Sciences of the Jan Kochanowski University of Kielce. He started his scientific career in the Institute of Ecology, Polish Academy of Sciences, and in a period of 1988–1990, he participated in the Antarctic Expeditions to the Henryk Arctowski Polar Station on the King George Island. He is a member of the Polish Geographical Society and Holy Cross Society of the Friends of Geological Sciences in Kielce. His main research interests are focused on hydrology, geomorphology and geotourism. Special areas of his studies are the Świętokrzyskie (Holy Cross) Mountains and the Nida Basin. He published a monograph of surface karst (karst lakes) in the eastern part of the Nida Basin.

Artur Komorowski graduated from the AGH University of Science and Technology in Kraków. He is a speleologist, member of the Beskidy Caving Club and the High-Mountain Climbing Club in Kraków. He conducts exploration of karst in the Neogene gypsum and limestones of the Nida Basin (Ponidzie area). The results of his studies were presented during national Speleological Symposia. Collaborator of the Institute of Paleobiology, Polish Academy of Sciences in Warsaw and the Institute of Earth Sciences, Maria Curie-Skłodowska University in Lublin, as a participant of the scientific project on taphonomy of Cretaceous vertebrates in the north-eastern margins of the Świętokrzyskie (Holy Cross) Mountains.

25 Sandstone Crags and Blockfields in the Świętokrzyskie (Holy Cross) Mountains

Jan Urban

Abstract

The majority of ca. 90 sites of sandstone crag groups and individual crags in the Świętokrzyskie (Holy Cross) Mts. region are built of the following crag-forming lithostratigraphic units: Cambrian Wiśniówka Fm, Devonian Barcza Fm and Zagórze Fm, Triassic Zagnańsk Fm and Krynki Beds as well as Jurassic Skłoby Fm and Ostrowiec Fm. Specific features of these rocks are the occurrence of sandstone series up to 20 m thick above plastic, clayey or heterolithic series, high-energy depositional environments, and siliceous composition. The crag-forming sandstones differ in the amount of siliceous cement, from strongly cemented Palaeozoic quartzitic sandstones to porous Mesozoic sandstones with poor cement, which determines diverse mechanical properties. Strongly cemented Palaeozoic rocks display high strength and abrasion resistance, while porous Mesozoic sandstones are characterised by high grain packing due to compaction. Gravitational disintegration of rock massifs under the periglacial conditions in the Pleistocene has played the principal role in crag origin. Other factors influencing crag formation and shaping are the spatial (tectonic) orientation of strata and joints, adequate joint spacing, and bed thickness. The interrelations between specific lithological and structural features as well as tectonics of sandstones that conditioned weathering and gravity-induced processes, resulted in the formation of the crags.

Keywords

Crags · Sandstones · Morphogenesis · Lithology · Tectonics · Periglacial · Świętokrzyskie Mts

25.1 Introduction

Crags are relatively rare landforms in a zone of temperate climate. In the European lowlands and uplands, they are considered to be relict forms developed under the periglacial conditions of the Pleistocene, although conditioned by earlier deep weathering in the Neogene (Palmer and Radley 1961; Tricart and Cailleux 1972; Robinson and Williams 1981; Alexandrowicz 1990; Cìlek et al. 2007; Young et al. 2009; Migoń et al. 2017; Duszyński et al. 2019). However, these relict forms still exist here, and one can ask after Robinson (2007): "why such rocky landforms exists in the green, forested uplands of Central Europe?". Strictly: in which way their lithological, structural and tectonic features stimulated their exposure and shaping in the Pleistocene, as well as which features are responsible for their present-day existence. In the European case, the ongoing discussion centres upon the role of climatic conditions—firstly periglacial and then temperate ones—versus lithology, mechanical properties, structure and tectonic features as factors responsible for the persistence of crags (see e.g., Cìlek et al. 2007; Young et al. 2009).

The Świętokrzyskie (Holy Cross) Mountains is an adequate place to seek answers to these questions, because in this area of upland relief and generally homogenous morphological evolution under uniform climatic condition, occur sandstones of very diverse age, lithology and tectonic setting (Urban and Gągol 2008), which form crags (usually crag groups) and blockfields. Some 90 crag sites have been recorded (Urban 2015, 2016, 2020). Up to the last two decades, these problems had not been practically considered, although some sites were known, described and protected as nature reserves or monuments (Alexandrowicz

J. Urban (✉)
Institute of Nature Conservation, Polish Academy of Sciences, Adama Mickiewicza Ave. 33, 31–120 Kraków, Poland
e-mail: urban@iop.krakow.pl

P. Migoń and K. Jancewicz (eds.), *Landscapes and Landforms of Poland*, World Geomorphological Landscapes, https://doi.org/10.1007/978-3-031-45762-3_25

1990; Alexandrowicz and Urban 2005). The only genetical study concerned the Piekło pod Niekłaniem crags and a significant role of aeolian processes in their emergence and shaping during the coldest period of the Vistulian (Würm) glaciation in the Pleistocene was suggested (Lindner 1972; Chlebowski and Lindner 1989). The subsequent studies of crags in the Świętokrzyskie Mountains, conducted mainly in 2006–2010, aimed to: (a) explain how the lithostratigraphic position of sandstone formations, mineral composition and fabric of sandstones, their mechanical properties, bedding and jointing as well as tectonic situation influenced the development and existence of the crags; (b) reconstruct processes giving rise to the crags; (c) explain the role of internal features (porosity) and external factors (insolation) in the shaping of crag surfaces (Urban 2015, 2016, 2020; Urban et al. 2015; Urban and Górnik 2017).

25.2 Geographical and Geological Setting

In geological terms, the Świętokrzyskie (Holy Cross) Mountains is the most elevated element of the Mid-Polish Anticlinorium Belt, with an outcropped Palaeozoic (Cambrian to Lower Carboniferous) Core tectonically shaped during the Caledonian and Variscan orogenesis. The Palaeozoic Core is surrounded by a Permian-Mesozoic Marginal Zone ('Marginal Zone' further on), which was subject to less intensive tectonic movements, mainly faulting, during the Cenozoic (Fig. 25.1). Both the Palaeozoic Core and the Marginal Zone are formed of sedimentary rocks, among which siliciclastic-argillaceous rocks play a significant role. The geological histories of the southern part of the region, called the Kielce Subunit, and its northern part, the Łysogóry Subunit, that are separated by the regional Świętokrzyski Fault (Fig. 25.1), were different, which resulted in diverse lithostratigraphic sequences, tectonics, as well as morphology of these parts of the region (Konon 2008; Urban and Gągol 2008).

The area comprises an upland with typical structural relief. Hill chains built of hard rocks such as quartzitic sandstones, sandstones, limestones and dolomites stand from several tens of metres to 300 m above the surrounding plains that have developed on outcrops of soft, easily denuded rocks, including clays, claystones, mudstones and marls (Fig. 25.2). The pattern of parallel hill ridges and intervening elongated depressions is controlled by the tectonics, strictly by the shapes of Variscan or Alpine folds constituted of rocks of different mechanical, chemical

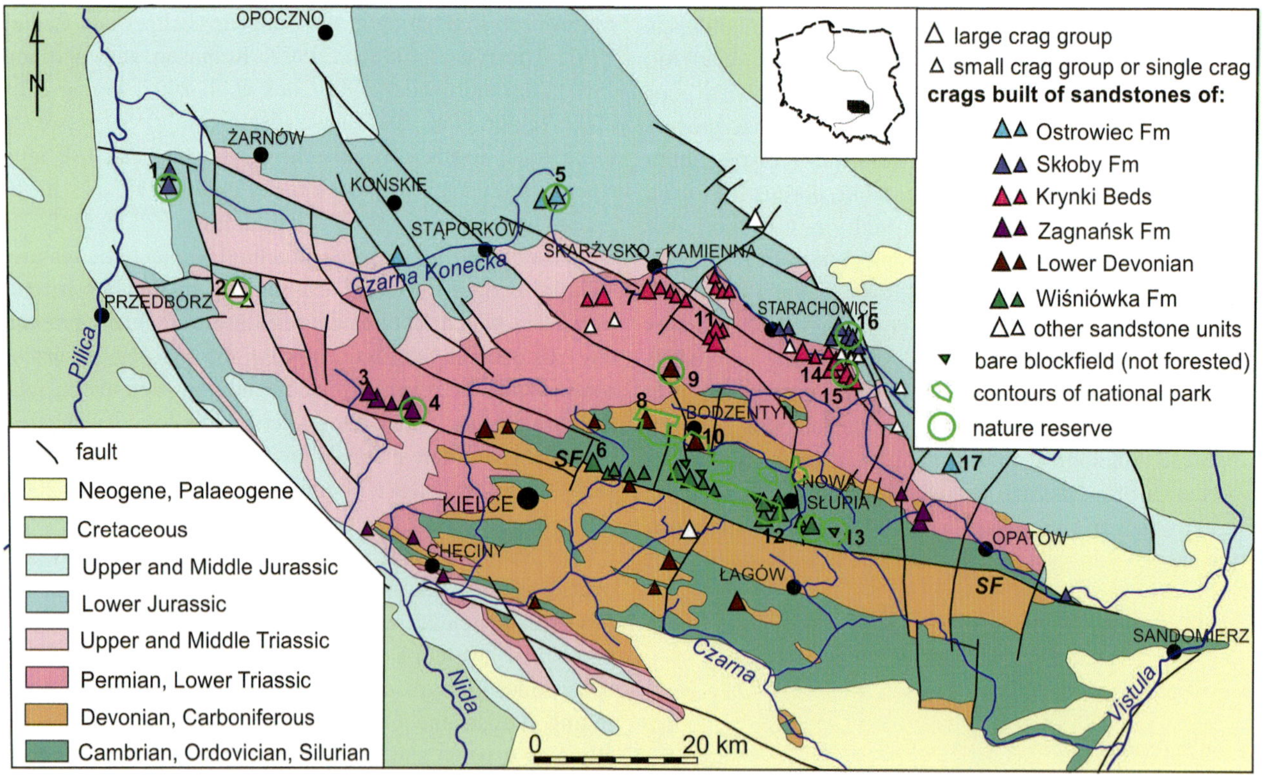

Fig. 25.1 Distribution of crag sites against the generalized geological map (map after Urban 2014; site location after Urban 2020, supplemented). The sites mentioned in the text are numbered: 1—Diabla Góra, 2—Piekiełko Szkuckie, 3—Nowek-Biała Góra, 4—Perzowa Góra, 5—Pieklo pod Niekłaniem, 6—Wielki Kamień, 7—Rejów, 8—Bukowa Góra, 9—Kamień Michniowski, 10—Miejska Góra, 11—Biały Kamień, 12—Góra Jeleniowska, 13—Szczytniak, 14—Ruda, 15—Skały w Krynkach, 16—Skały pod Adamowem, 17—Sfinks; SF—Świętokrzyski Fault

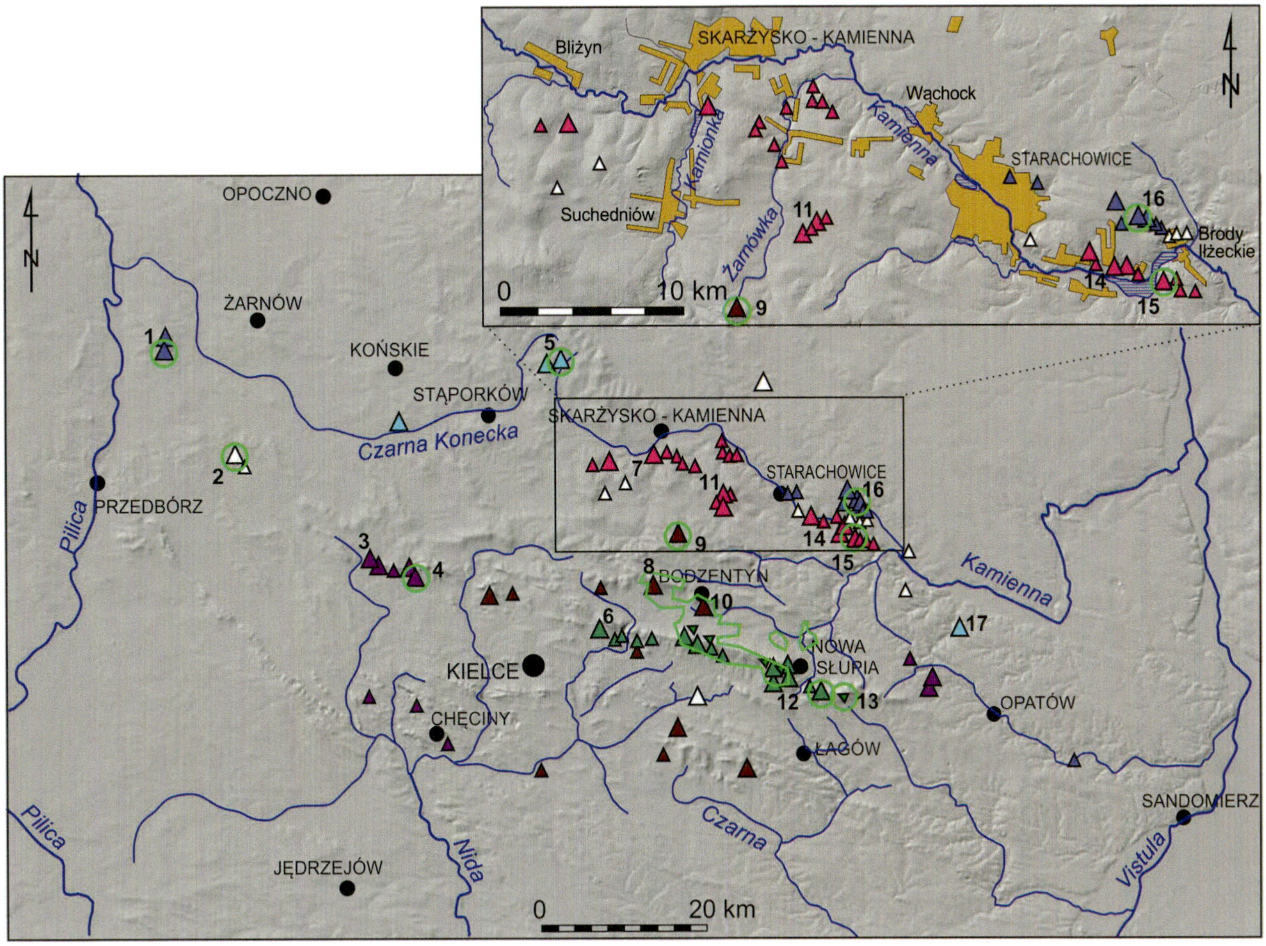

Fig. 25.2 Distributions of crag sites against the digital terrain model; for explanation of all symbols see Fig. 25.1

and structural properties determining rocks hardness. Consequently, the relatively wide depressions are not simple river valleys, although they are used by stream and river networks. Such a relief is an effect of protracted morphogenesis/denudation that occurred in hot and then warm climate of the Palaeogene and Neogene. In the Pleistocene, this morphology was partly modified during the glacial periods, when the ice-sheets stepped in and covered the area or when the area underwent processes related to the periglacial conditions, mainly mechanical weathering and slope processes, related to the cold climate of the glacier foreground. Morphological modifications were less significant during the interglacial periods characterised by moderate climate (Olędzki 1976; Urban 2014).

Within the lithostratigraphic sequence of the Świętokrzyskie Mountains, the following formations have been recognised as crag-forming units (Figs. 25.1, 25.2 and 25.3) (Urban 2016, 2020):

- Quartzitic sandstones of the Upper Cambrian Wiśniówka Sandstone Formation, which form the highest belt of elevations in the region reaching 450–614 m asl, composed of three segments: Masłowskie Range, Łysogórskie Range and Jeleniowskie Range, with the central and the highest section called Łysogóry.
- Sandstones and quartzitic sandstones of the Lower Devonian Barcza Formation and Zagórze Formation in the Łysogóry Subunit that build the Klonowskie Range (450–484 m asl) and other Devonian sandstones in the Kielce Subunit.
- Sandstones of the Lower Triassic (Lower Buntsandstein in the lithostratigraphic division) Zagnańsk Formation in the western and eastern segments of the northern part of the Marginal Zone.
- Sandstones of the Middle Triassic (Upper Buntsandstein) Krynki Beds in the central segment of the northern part of the Marginal Zone (middle section of the Kamienna River catchment).
- Sandstones of the Lower Jurassic Skłoby Formation in the eastern segment of the northern part of the Marginal Zone.
- Sandstones of the Lower Jurassic Ostrowiec Formation in the western segment of the northern part of the Marginal Zone.

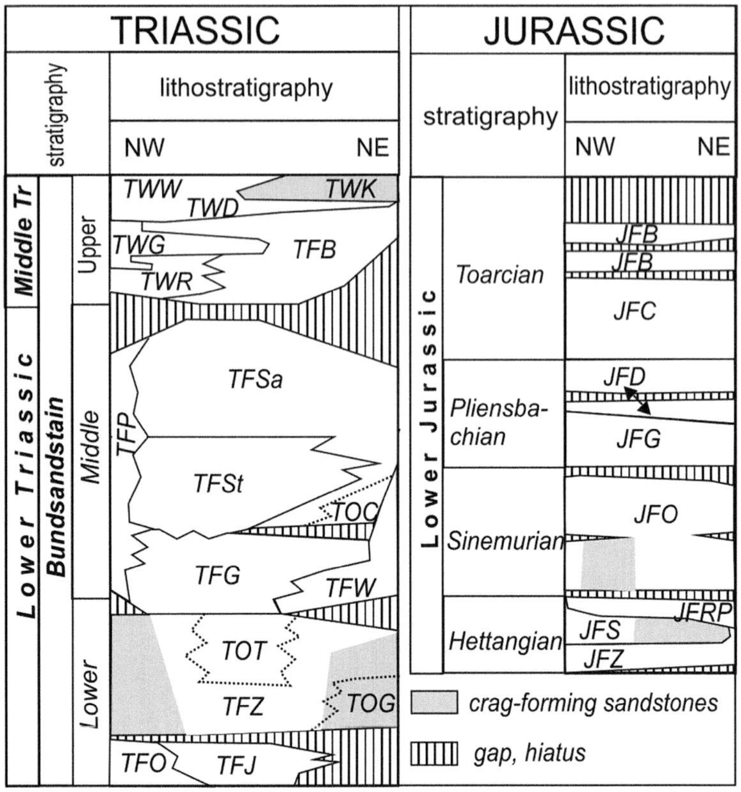

Fig. 25.3 Distribution of crag-forming sandstones in the lithostratigraphic sequences of Lower Triassic and Lower Jurassic (lithostratigraphy after Pieńkowski 2004; Kuleta and Zbroja 2006, modified). Explanation of abbreviations: JFB—Borucice Formation, JFC—Ciechocinek Fm, JFD—Drzewica Fm, JFG—Gielniów Fm, JFO—Ostrowiec Fm, JFRP—Ore-bearing Przysucha Fm, JFS—Skłoby Fm, JFZ—Zagaje Fm, TFB—Baranów Fm, TFG—Goleniawy Fm, TFJ—Jaworzna Fm, TFO—Opoczno Fm, TFP—Piekoszów Fm, TFSa—Samsonów Fm, TFSt—Stachura Fm, TFW—Wióry Fm, TFZ—Zagnańsk Fm, TOC—Cierchy Member, TOG—Czerwona Góra M., TOT—Tumlin M., TWD—Dalejów Beds, TWG—Gypsum Beds, TWK—Krynki Beds, TWR—Radoszyce Beds, TWW—Wilczkowice Beds

25.3 Crags in the Świętokrzyskie Mountains—Examples

The crags in the Świętokrzyskie Mountains, built of sandstones of the abovementioned formations, usually form groups composed of several to several dozens of individual outcrops or boulders, which are not very high, usually ranging 2–6 m, but differ in terms of position within morphological units of higher order (hillslopes, ranges) and shape. These differences are usually conditioned by features and parameters determined by the lithostratigraphical position of sandstones, as exemplified by several examples of crags built of various sandstones provided below.

25.3.1 Upper Cambrian Wiśniówka Fm

Quartzitic sandstones of the Wiśniówka Fm are very fine-grained and very hard, usually, but not everywhere, thin-bedded. Crags (15 sites) form crests or irregular domes crowning the ridges. Such an irregular dome, conditioned in part by thick or indistinct bedding, is Wielki Kamień (Large Rock), 5–6 m high, and more than 10 m in diameter, protected as a nature monument. It crowns the Klonówka Hill ridge in the Masłowskie Range (Figs. 25.1 and 25.4a, c) that rises 150 m above the surrounding plains (Urban 2016). The strata dip reaches 45°, while the strike is consistent with the hill-range direction. The rock is densely jointed in at least three directions (NNW–SSE, ENE–WSW and N–S, Fig. 25.4b), but some fissures are sealed with quartz; average joint spacing for the two joint sets are 1.1 and 2.1 m. Consequently, the crag surface is very irregular having many sharp edges (due to the rock hardness), with an exception of the northern, bedded part, where the bedding planes form more regular surfaces. The northern side of the crag, which is also the uppermost part of the slope, dips very gently, while the southern slope is more inclined (15–20°) and characterised by an occurrence of boulders (Fig. 25.4a, c).

Another very specific and even unique phenomenon formed of the quartzitic sandstones of the Wiśniówka Fm are bare (not forested) blockfields that occur on slopes of the highest parts of the hill chain, the Łysogóry Range (up to 614 m asl) and the Jeleniowskie Range (up to 554 m asl). They are composed of angular blocks reaching the size of several tens of centimetres, usually up to 0.5 m (Figs. 25.1 and 25.5a), and considered to be remnants of periglacial block-debris-silty covers formed during the last Pleistocene glacial period (Vistulian, Würm), maybe also during the earlier glacial period, and subsequently washed and cleaned from the fine fractions at the Pleistocene-Holocene transition (Klatka 1962). Consequently, these blockfields are

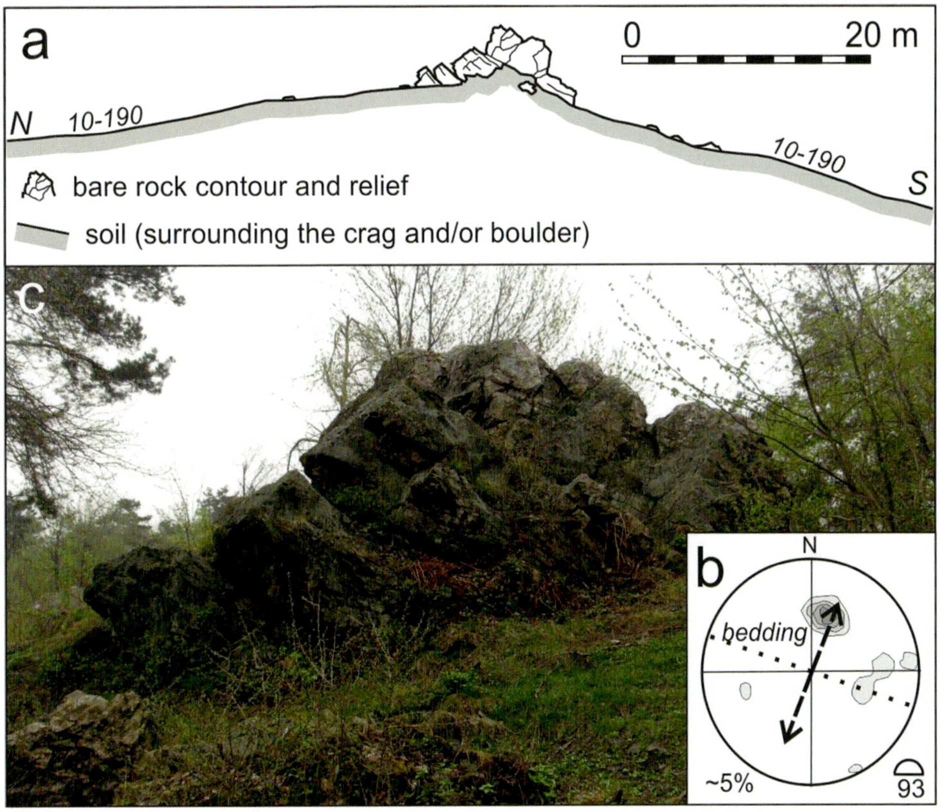

Fig. 25.4 Wielki Kamień crowning the Klonówka Hill. **a** Morphological profile with azimuth (after Urban 2016). **b** Contour diagram of joints and bedding planes (projection of normals to measured planes on the upper hemisphere reflected on horizontal surface) with the number of measurements (93), isoline density (~ 5%), slope dip (arrows) and ridge stretching (dotted line). **c** Crag from the west (WWN) side (photograph by J. Urban)

direct evidence and relicts of periglacial weathering and slope processes, and in such way they were interpreted by the first geomorphologists who identified and studied periglacial phenomena in Poland (Łoziński 1912). As landforms unique for the low mountains of Central Europe, these blockfields have evoked wide scientific interest. Some authors postulated their genetical connection with the Pleistocene and present-day neotectonic movements (Łyczewska 1972; Kowalski 1993), while others denied such a concept (Sedlak 1964).

The best preserved, bare blockfields are on the northern slope of the Łysogóry ridge in zones at the following elevations: 400–425 m asl, ca. 445 m asl, 500–525 m asl and 540–600 m asl. Currently, they reach a length up to ca. 0.8 km and a width of 0.1–0.3 km. Similar blockfield zones are also discernible on the southern slope of the Łysogóry, at the elevations of 480–510 m asl and 510–590 m asl. They are all protected in the Świętokrzyski National Park. The only one situated out of the Łysogóry and the National Park is located on the northern slope of Mt Szczytniak, the highest summit of the Jeleniowskie Range at an elevation of 520–540 m asl, and protected as the Szczytniak nature reserve. The occurrence of blockfield zones is correlated with steeper sections of slope profiles, inclined at 20–40°, and therefore, the zones are considered to reflect outcrops of the sandstone series, while the less inclined sections are regarded to be soft shale outcrops. Kowalski and Jaśkowski (1986) interpreted the blockfields as structurally-conditioned cryoplanation terraces. Recently, these blockfields have been gradually overgrown and ultimately forested (Kobendza 1939; Klatka 1962; Huruk 1986), becoming relict forms. However, Jaśkowski et al. (2002) suggested slight movements and modifications of block covers still in the Holocene.

Similar blockfields, but overgrown by forests, occur on hillslopes built of various Mesozoic sandstones, including units, which do not build crags (Fig. 25.5b).

25.3.2 Lower Devonian Sandstones and Quartzitic Sandstones

Sandstones and quartzitic sandstones of the Lower Devonian are usually fine-grained with coarser and even conglomeratic inserts, medium- and thick-bedded, often displaying diagonal lamination within beds. They usually form individual crags situated in various parts of slopes, such as inclined platforms or (higher) "rock tables" on slopes, near-ridge spurs and irregular cliffs (11 sites). Two large crag groups, Bukowa Góra and Kamień Michniowski (Figs. 25.1 and 25.2), represent this type of sandstones, but one presented here has developed in the uppermost part of Mt Miejska Góra slope and is situated within the Świętokrzyski National Park. It is

Fig. 25.5 Blockfields. **a** Bare blockfield on the northern slope of the Łysogóry ridge. **b** Blocks of Lower Triassic sandstones in the Suchedniowskie Forests—forested remnant of the periglacial slope cover (photographs by J. Urban)

composed of "rock tables" derived from disintegration of a thick sandstone layer (locally two layers) that dips parallel to the slope surface (30–40° towards N). The dense cluster of crags, cut by a system of N–S and E–W joints with average spacing of 3.4 m and 4.0 m, respectively, is surrounded by more detached boulders or lower rock platforms (Fig. 25.6). The geological and morphological setting of this crag group is identical to the Bukowa Góra site, while the shape is partially similar. Three rock shelters formed under or among the crags were registered as caves: Miejski Schron (6 m long), Bodzentyńska Szczelina (6 m—Fig. 25.6f) and Podgórski Schron (5 m—Fig. 25.6d). They formed mainly due to weathering of the lower parts of the sandstone layer, followed by gravitational transport and wash of loose, sandy material. However, typical crevice passages also developed here (Fig. 25.6f) due to gravitational movement of a rock slab (Urban and Kasza 2009).

25.3.3 Sandstones of the Lower Triassic (Lower Buntsandstein) Zagnańsk Formation

Crags built of sandstones of the Lower Triassic Zagnańsk Fm (11 sites) are located in the eastern and western parts of the northern segment of the Permian-Mesozoic Marginal Zone (Figs. 25.1 and 25.2), where this formation is represented by thick or indistinctly bedded, poorly sorted and occasionally conglomeratic fluvial sandstones, while in the central part of this segment, devoid of crags, these sediments are replaced by aeolian, thin-bedded and well-sorted sandstones of the Tumlin Member of the same formation. Most crag sites comprise singular forms or small groups. The exception is Mt. Perzowa Góra site, where 83 individual crags and boulders have been recorded (Czarnocki 1932; Urban 2015). Most of them are protected in the Perzowa Góra nature reserve. The crag-supporting series is at least 8 m thick and formed of thick and indistinctly bedded, poorly sorted sandstones, which are cut by several joint sets. The main ones have S–N and W–E directions and average spacing of 4.0 and 6.0 m, respectively, whereas less frequent joints are running WSW–ENE and WNW–ESE. The series dips slightly (10–20°) to SW, thus—owing to the specific shape of the hill—it crops out in its top part (390–395 m asl), the uppermost part of its northern slope (375–380 m asl) and at similar elevation in the western slope (Fig. 25.7a). Only some 15% of rock forms, mainly cliffs, but also spurs and platforms, are in situ crags. The other rock elements are ex situ boulders of various shapes (domes, tables, oval forms, spurs and platforms), situated downslope from the crags, strewn across the surface dipping usually 10–20° (up to 30°). The distance between in situ crags and boulders varies from several metres to 200–300 m (Fig. 25.7a–c). Most boulders display features indicating their simple sliding and slight forward toppling, but some were rotated backward, rotated around vertical axis or even turned over (Fig. 25.7i–k) (Urban 2015).

The signs of human activity are common at this site. They are represented by the St. Rosaly Chapel located in a partly natural large cavity (Fig. 25.7f, g), as well as numerous remnants of rock extraction such as abandoned small quarries and partly broken cliffs and boulders.

25.3.4 Sandstones of the Middle Triassic (Upper Buntsandstein) Krynki Beds

Sandstones of the Krynki Beds represent high-energy, foreshore or terrestrial sediments, medium- or thick-bedded, but bedding planes are very irregular and discontinuous.

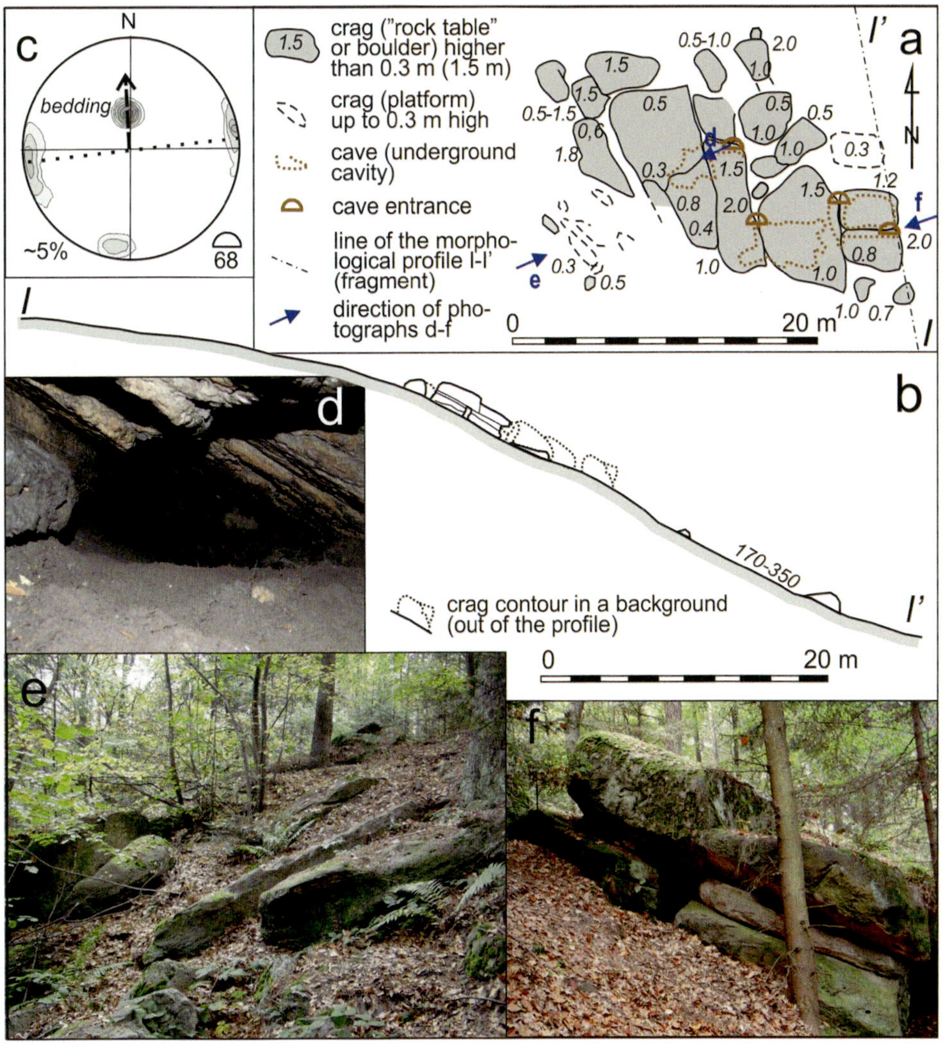

Fig. 25.6 Crags at Mt Miejska Góra site. **a** Map of the cluster. **b** Morphological profile; symbols as on Fig. 25.4. **c** Contour diagram of joints and bedding planes in the whole crag cluster. **d** Passage of Podgorski Schron. **e** Upper part of the group: rock platforms and boulders in the foreground. **f** East side of the "rock table" with an entrance to the crevice-type passage of Szczelina Bodzentyńska cave (photographs by J. Urban)

They are poorly sorted with quartz pebbles, and described by Barczuk (1979) as quartzitic sandstones due to their almost pure siliceous composition. Among 26 crag sites, most comprise groups composed of more than ten individual crags and boulders. Among them is the largest group, Skałki w Krynkach, numbering more than 100 individual outcrops and protected as a nature reserve of the same name (Figs. 25.1 and 25.2). These groups are usually situated in the marginal zones of plateaus or ridges, where in situ crags form more or less continuous cliffs and groups of boulders are located below, on slopes, also far away from the cliffs (Urban 2015, 2016, 2020).

Another example of a crag site supported by these sandstones is a medium-large Ruda group situated on the northern slope of the Kamienna river valley, consisting of 22 individual elements and protected as a nature monument (Fig. 25.8). Sandstone strata are nearly horizontal, cut by rare W–E running joints, with the mean spacing of 8.4 m, and less frequent joints of NW–SE and NNE–SSW directions (Fig. 25.8a, c). The plateau marginal cliff is high in the SE section (7–9 m), but towards NW it disappears and grades into several small steps. The second specific element of this group is the largest boulder—a large rock table of a height of 5–7.5 m. It was translated for ca. 9 m and toppled (10–15°) (Fig. 25.8a, b, d, g, h). There are many other blocks detached from the in situ massif, e.g., the eastern group was shifted 2–3 m towards the south, which created a corridor between it and the cliff (Fig. 25.8e) (Urban 2016). Two small caves, Jaskinia nad Zalewem (8 m long) and Piaskowa Jama (3 m; Fig. 25.8e), formed owing to weathering and erosion along bedding planes.

The other phenomenon observed at this site is a remnant of thick ferruginous (iron oxide) incrustation on a northern face of the large rock table (Fig. 25.9). This is a relic of material that filled the crevice during the stage of deep weathering, probably in the Neogene. Similar remnants were observed at many other sites, especially in the zones of tectonic deformations, e.g., at the Skały w Krynkach site (Urban 2015).

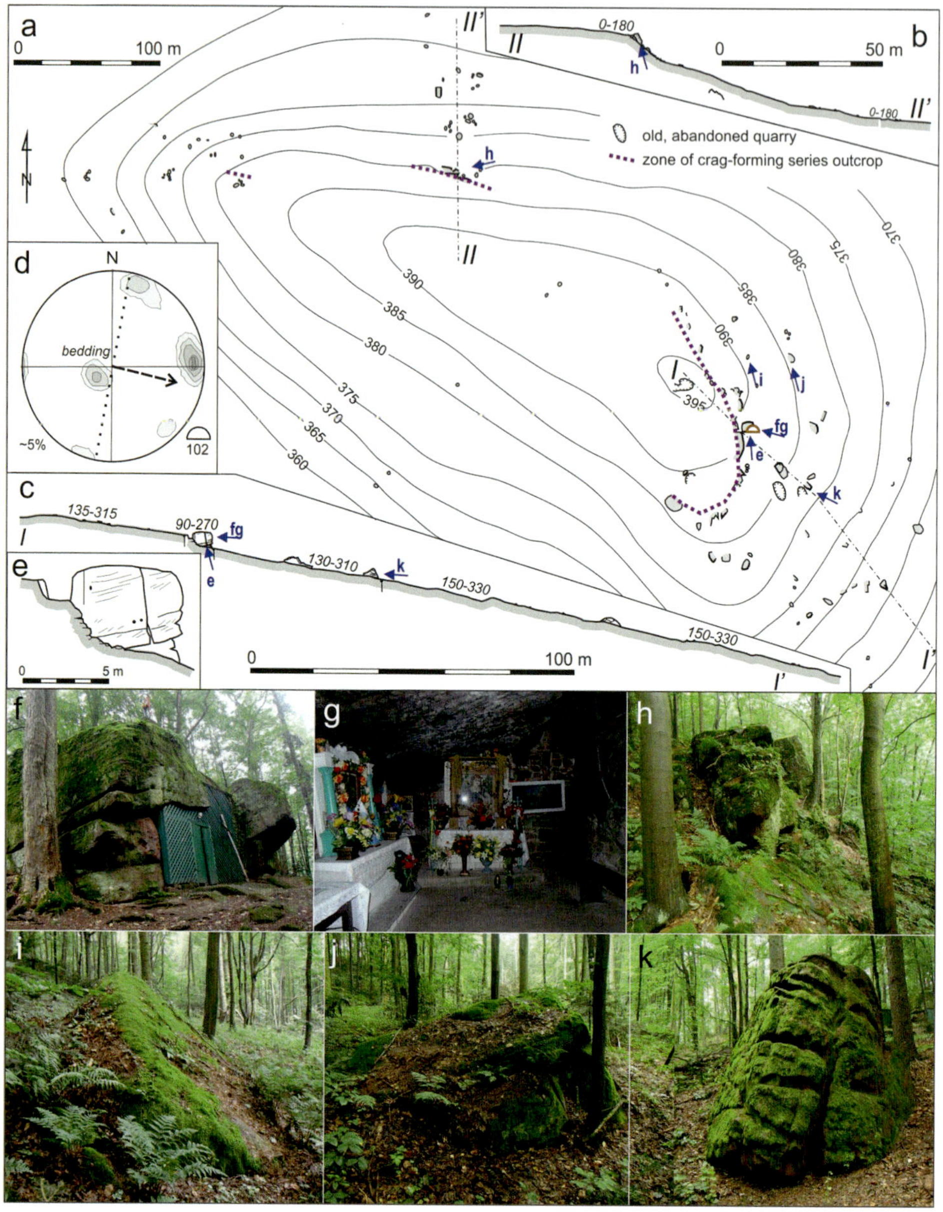

Fig. 25.7 Central part of the crag group at Mt Perzowa Góra site. **a** Map of the group (symbols—see Fig. 25.6). **b–c** Morphological profiles I-I' and II-II' (for symbols see Fig. 25.4). **d** Contour diagram of joints and bedding planes in crags situated next to the summit (for explanations see Fig. 25.4). **e** Details of the largest crag from the south. **f** Largest crag with the St. Rosaly chapel door. **g** Inside the chapel. **h** Rock spurs in the northern sub-group. **i** Boulder slipped downslope and toppled (inclined concordantly to the slope dip). **j** Backward rotated boulder (inclined discordantly to the slope). **k** Boulder fallen and rotated around vertical axis (photographs by J. Urban)

25.3.5 Sandstones of the Lower Jurassic Skłoby Formation and the Ostrowiec Formation

The crag-forming sandstones of the Skłoby Fm and Ostrowiec Fm (not all sandstones within these formations) are generally similar. They are sediments of upper foreshore (very shallow marine to off-shore) depositional systems (Pieńkowski 2004) and include medium to thick and irregularly bedded, poorly sorted, fine- to medium-grained sandstones. Consequently, they form similar crag groups. In the Skłoby Fm, 11 sites occur, whereas four sites are recognized in the Ostrowiec Fm (Figs. 25.1 and 25.2). They resemble the crags built of sandstones of the Krynki Beds of similar depositional origin and lithology. These groups are usually composed of a series of cliffs and spurs situated along the plateau marginal zone, and ex situ boulders situated downslope, locally even 200–300 m far from the cliffs (Urban 2015). Crag groups formed of sandstones of the Skłoby Fm are situated within the relatively short, middle section of the Kamienna River basin.

The largest group, Skałki pod Adamowem, is ca. 1 km long and composed of almost one hundred crags and boulders (Fig. 25.10) (Urban 2015). Its part is legally protected as a nature reserve of the same name. The sandstone strata building the crags dip very gently towards NE, opposite to the slope inclination, so they are often practically horizontal, due to slight gravitational tilting (e.g., Fig. 25.10d, f, g). They are cut by two complementary joint sets, NNE–SSW

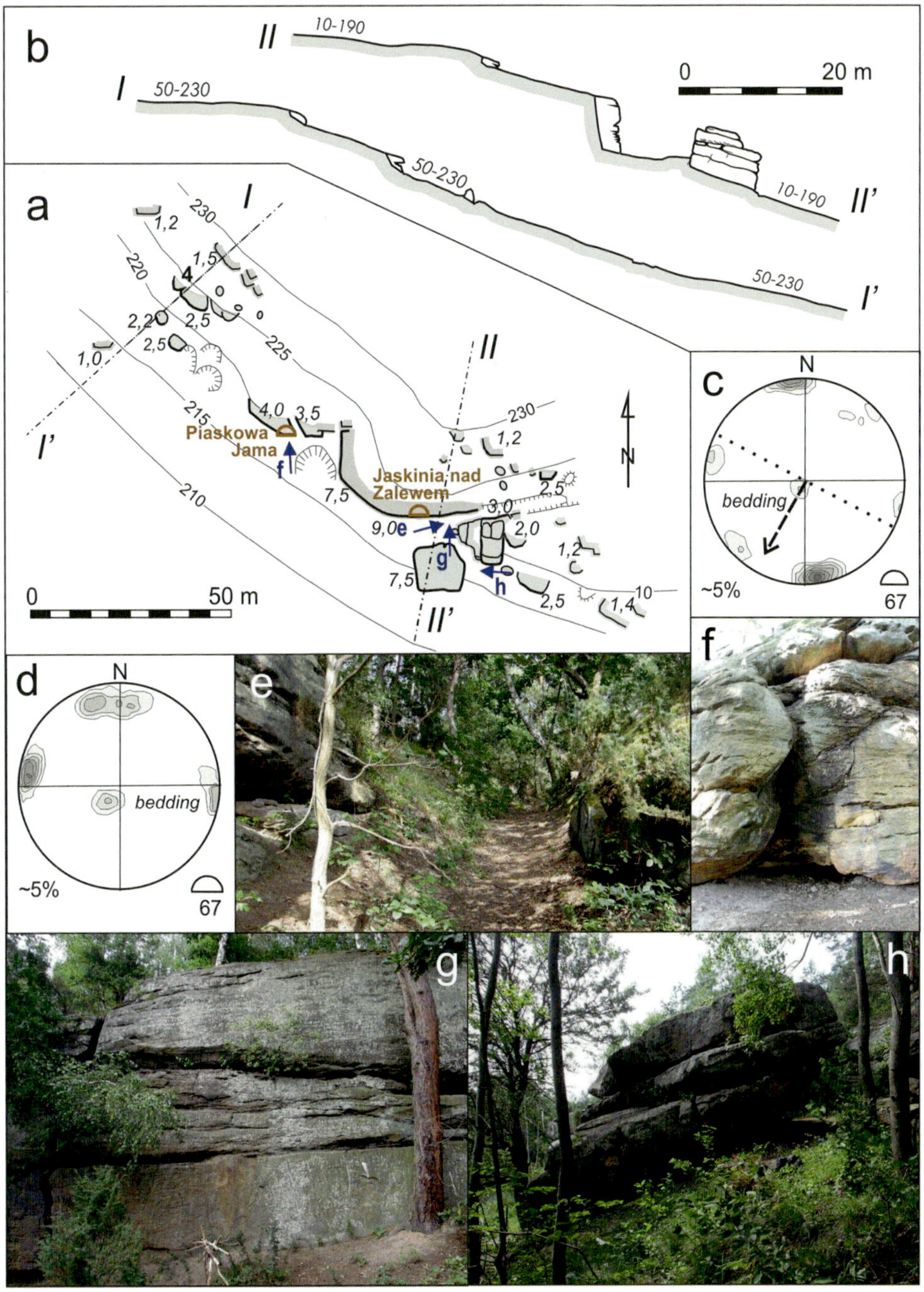

Fig. 25.8 Crags at the Ruda site. **a** Map of the crag cluster. **b** Morphological profiles. **c** Contour diagram of joints and bedding planes measured in the highest cliff. **d** Contour diagram of joints and bedding planes in the large rock table. **e** "Corridor" formed owing to the gravitational shift of the south-east part of the rock massif. **f** Entrance to Piaskowa Jama cave. **g** Eastern part of the highest cliff. **h** Large rock table (photographs by J. Urban); for explanation of symbols—see Figs. 25.4, 25.5 and 25.6

and ESE–WNW (Fig. 25.10f), with average spacing ranging 2.9 m and 6.2 m, respectively. Although the crags along the plateau margin are very evident and continuous in some central sections, composed of cliffs (Fig. 25.10i), spurs (Fig. 25.10c–e, h, l) and even hoodoo rocks (Fig. 25.10k) of a height reaching usually 2–5 m, most bedrock elements are represented by ex situ boulders (Fig. 25.10j, k): rock tables, domes, pyramids and platforms of a very diverse size. They are usually moved several to several tens of metres downslope and situated in the upper section of slope, where the inclination is 15–20°. However, some boulders are situated 150–250 m far from the outcrops, in the lower section of the slope, where inclination is less than 10° (Fig. 25.10b). Although most boulders are tilted forward, some are rotated backward, rotated around vertical axis or even overturned (Fig. 25.10m).

The geophysical electrical resistivity tomography (ERT) surveys performed at this site (e.g., Fig. 25.10g) and other sites (e.g., Skały w Krynkach site mentioned above) proved that the sandstone series (visible on the ERT images as high-resistivity bodies), of a thickness ranging from 10 to 20 m, are disintegrated in the surface zones and form separated "corestones" translated downslope. These features, as well as other observations, such as phenomena observed in

Fig. 25.9 Fragment of several centimetres thick ferruginous incrustation on the northern face of the large rock table at the Ruda site—a remnant of the crevice fill formed in the stage of deep weathering of the rock massif (photograph by J. Urban)

boreholes and relicts of ferruginous incrustations (Fig. 25.9) prove that the gravitational deformations and other exogenic processes (weathering) were active relatively deep and brought about the disintegration of sandstone series long before the formations of crags (Urban et al. 2015).

At this site, five small caves of a length ranging from 2.5 m up to 7.5 m have been explored (Fig. 25.10a, j). They represent various genetical types, not always clearly determined, but more or less closely related to gravity-driven processes, from gravitational shifts of large boulders to sandstone weathering and subsequent removal of loose material by downwashing or gravitational creep (Urban 2015; Urban et al. 2015).

The crags built of sandstones of the Ostrowiec Fm are represented by three sites located in the Gielniów Hillock, with two of them situated next to village of Niekłań. Among them is the famous and the most impressive crag group in the Świętokrzyskie Mts., described already in the first part of the twentieth century and protected in the Skałki Piekło pod Niekłaniem nature reserve since 1958 (Massalski and Kaznowski 1928; Alexandrowicz 1990; Urban et al. 2015). The fourth crag, a large spur called Sfinks (Sphinx), is situated in the eastern part of the northern Permian-Mesozoic Marginal Zone, in the village of Szewna (Massalski 1936). It is a tor up to 7.5 m high, situated within a relatively steep slope. It is famous for anthropogenic relief of its top surface, which results from a local tradition that every newly married couple in the village has to inscribe their initials in a relatively soft sandstones of the crag (Fig. 25.11).

25.4 Lithological and Structural Controls of Crag Formation and Preservation

25.4.1 Lithology and Physico-Mechanical Properties

The characteristic lithological features of the crag-forming sandstones are their occurrence within geological sequences displaying great lithological differentiation and high-energy depositional environments. The crag-forming sandstone series are several metres to ca. 20 m thick, underlain and overlain by claystone, siltstone or heterolithic series (Urban 2016). In the case of Devonian, Triassic and Jurassic sandstones, the high-energy depositional systems are reflected in specific structures such as irregular bedding, diagonal lamination and poor sorting. Characteristic is, for example, the lack of crags in the western section of the Ostrowiec Fm outcrop area, where the sandstones are thin-bedded, very fine-grained and interlayered by siltstone/claystone inserts. Also the thin-bedded aeolian sandstones of the Tumlin Member within the Zagnańsk Fm do not build crags. Although the crag-forming Cambrian sandstones are very fine- and fine-grained, they are considered to be also high-energy sediments (Trela 1998; Pieńkowski 2004; Kuleta and Zbroja 2006; Urban 2016, 2020).

A very characteristic, distinctive feature of the crag-forming sandstones is their almost exclusively siliceous composition, both grain framework and cement. In petrographic terms, these sandstones are usually quartz arenites, occasionally sublithic arenites or subfeldspathic arenites;

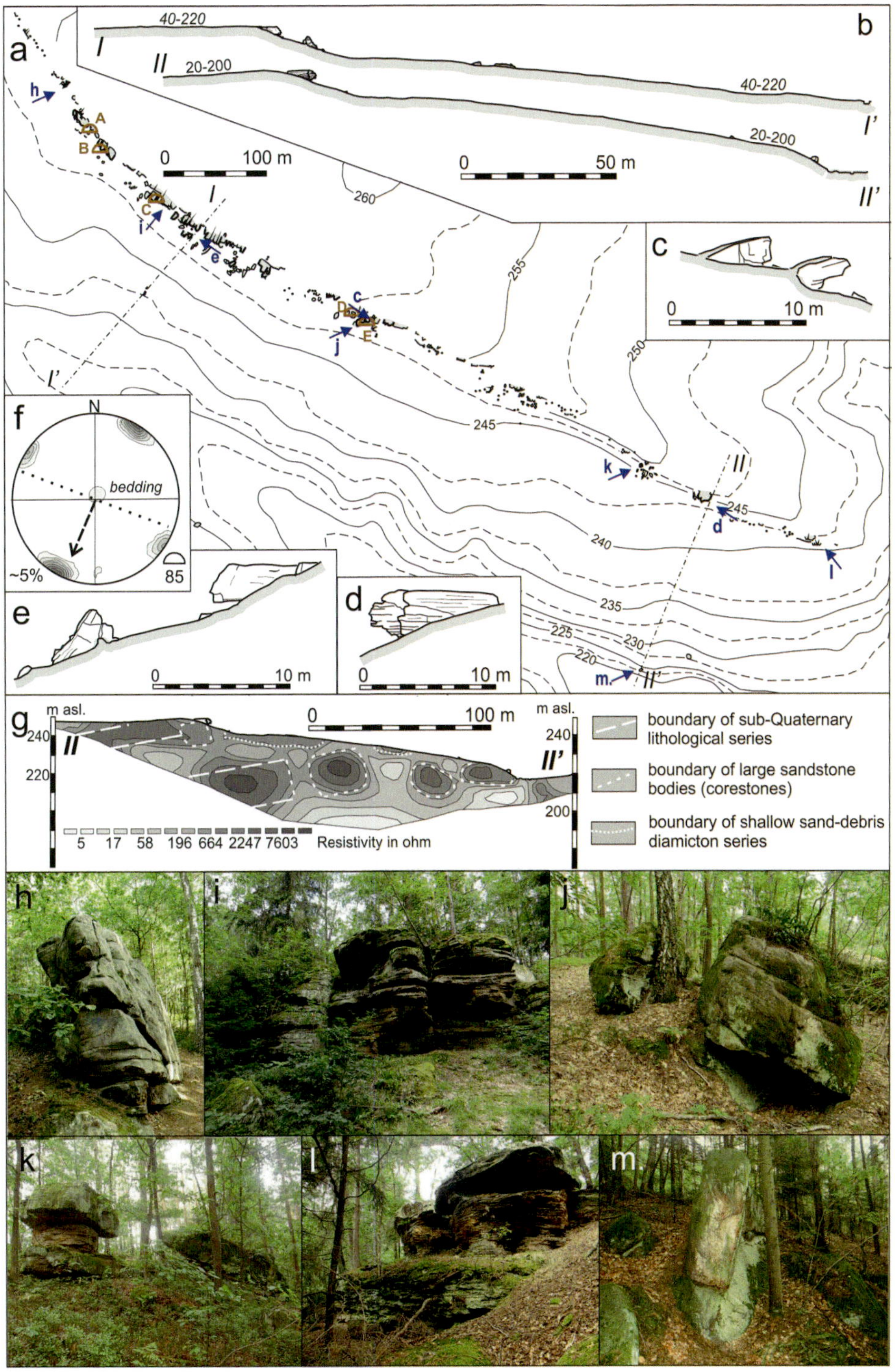

Fig. 25.10 Skałki pod Adamowem site. **a** Map, including location of caves: A—Dziura pod Drutami (4.5 m long), B—Schronisko na Przestrzał (7.5 m), C—Tunelik w Skałkach Adamowskich (4.0 m), D—Schronisko Osobne (2.5 m), E—Stalagmitowy Schron (4.0 m). **b** Morphological profiles. **c** Detailed sketch of two crags, for location see part a. **d** Detailed fragment of the II-II' profile, for location see part a. **e** Detailed fragment of the I-I' profile, for location see part a. **f** Contour diagram of joints and bedding planes. **g** ERT image along the II-II' profile. **h** Rock spur. **i** Rock cliff. **j** Schronisko Osobne cave formed under the boulder that shifted downslope. **k** Hoodoo (mushroom) rock formed owing to thin-bedded insert and a tilted boulder next to it. **l** Two-tiered cliff (spur), with the thin-bedded insert responsible for the recess. **m** Overturned boulder situated very far from the main rock outcrop (photographs by J. Urban); for explanation of symbols see Figs. 25.4, 25.5 and 25.6

only the Zagnańsk Fm sandstones display weaker mineralogical maturity and are classified as sublithic or lithic arenites, occasionally greywackes, but containing principally grains of siliceous rocks (Urban 2016, 2020).

The volume of siliceous cement is very different in various crag-forming rocks. In the sandstones of the Wiśniówka Fm and Devonian sandstones, the cement fills the whole intergranular space and is often difficult to distinguish from the grains, although in the case of the Devonian sandstones the stage of recrystallisation is usually less advanced (Fig. 25.12a, b). These sandstones are thus characterised by very low porosity/water absorption (Peszat 1973; Kowalczewski et al. 1989; Sikorska 2000; Urban 2016, 2020).

The stage of filling of the intergranular space in the sandstones of the Zagnańsk Fm is different in various parts

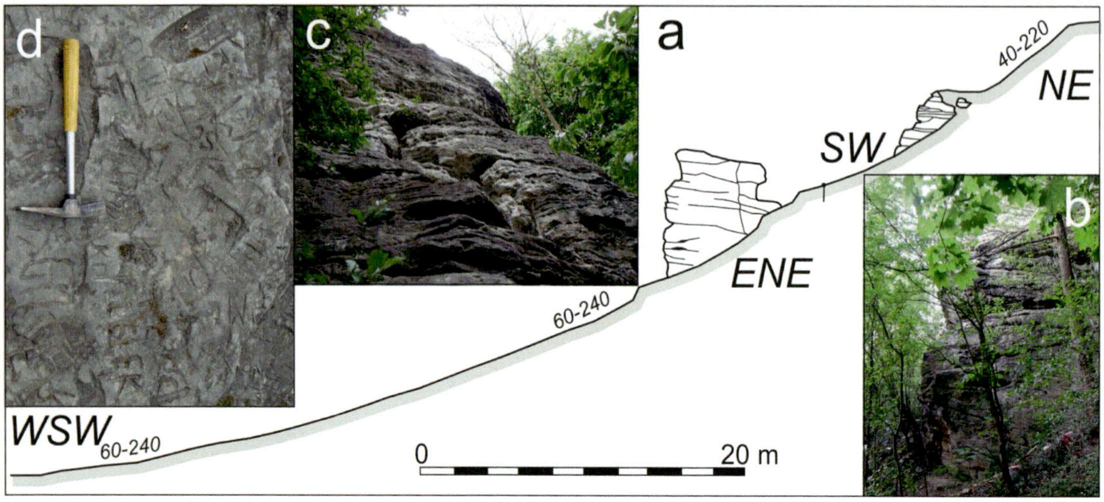

Fig. 25.11 Sfinks site. **a** Morphological profile. **b** Main crag—a tor. **c** Details of relief of the main tor face. **d** Anthropogenic relief of the tor top surface (photographs by J. Urban); for explanation of symbols see Figs. 25.4, 25.5 and 24.6

of the area of their occurrence. It is the most advanced next to the Świętokrzyski Fault, so exactly in the area of crag occurrence (Fig. 25.12c) (Urban 2016, 2020). In turn, in the Triassic sandstones of the Tumlin Member and the Wąchock Sandstones, which do not build crags, iron oxides are significant components filling intergranular space (Urban 2016, 2020).

The Triassic sandstones of the Krynki Beds, called quartzitic sandstones (Barczuk 1979), as well as the Jurassic sandstones of the Skłoby Fm and Ostrowiec Fm are characterised by very poor cement, but exclusively siliceous. Moreover, the grains of these formations are very densely packed, which is illustrated by concave-convex, suture and even healed contacts (Fig. 25.12d–f). The original (depositional) intergranular space of the Skłoby Fm sandstones has been diminished by 30–50% owing to the chemical compaction during their diagenesis (Urban 2016, 2020). This extraordinary compaction of sandstones of the Skłoby Fm resulted in their specific hardness already discerned by Teofilak-Maliszewska (1968), Kobylec (2008).

In the consequence, although the porosity/water absorption of sandstones of the Krynki Beds and the Jurassic sandstones is much higher than that of the Cambrian and Devonian sandstones, the value of this parameter of the crag-forming sandstones is still low in relation to other Mesozoic sandstones, ranging from several per cent to 10% (Kozłowska and Kuberska 2014; Urban 2016, 2020). This seems to be a very specific attribute of crag-forming sandstones (Cìlek et al. 2007). Although sandstones with non-siliceous cement also build crags, these crags occur within steep slopes in areas of much higher topographic gradient, mainly in the high mountains (see e.g., Alexandrowicz 1978; Alexandrowicz 1990; Colbach 2005; Adamovič et al. 2006; Zhu et al. 2010).

The silica content in the rocks determines physio-mechanical parameters of sandstones. The saturation of Palaeozoic rocks with silica and the resultant extremely low porosity contribute to their high compressive strength and abrasion resistance, which are significant features responsible for the formation and existence of crags. In turn, the crag-forming Triassic and Jurassic sandstones are characterised by medium or low compressive strength and medium or low abrasion resistance (Peszat 1973; Kowalczewski et al. 1989; Sikorska 2000; Guzik 2017). However, their very dense grain packing/chemical compaction creates specific internal cohesion (see e.g., Bruthans et al. 2013, 2014; Stück et al. 2013; Young and Wray 2015). Their siliceous composition practically eliminates salt weathering as a potent agent of rock decay, while relatively high porosity does not contribute to efficient frost weathering due to fast water circulation within the rocks (Urban 2016, 2020).

Therefore, when the hardness of the Cambrian and Devonian sandstones, and consequently, durability and stability of rock forms and faces are conditioned by high content of siliceous cement, the feature determining the permanence of rock walls in the Triassic and Jurassic sandstones is internal cohesion and specific resistance for weathering. The sandstones of the Triassic Zagnańsk Fm are in the intermediate position between the Palaeozoic and other Mesozoic sandstones—they have more siliceous cement than other Mesozoic ones and their grains are densely packed (Urban 2016, 2020).

Not very hard crag-forming sandstones of the Świętokrzyskie Mts. are not an exception in Europe, because crags built of rocks of similar physico-mechanical parameters occur in other regions of moderate and warm climate, e.g., in the Cretaceous sandstones of England, Fontainebleau Sandstones in the Paris Basin (France), as

Fig. 25.12 Microscopic images of crag-forming sandstones (after Urban 2020); N II—parallel nicols, N X—crossed nicols; blue colour in Mesozoic sandstone images marks empty intergranular space, red arrows point selected concave-convex, suture and healed contacts. **a** Cambrian sandstone from the Góra Jeleniowska site. **b** Devonian sandstone from the Bukowa Góra site. **c** Triassic sandstone of the Zagnańsk Fm from the Nowek-Biała Glina site. **d** Triassic sandstone of the Krynki Beds from the Biały Kamień site. **e** Jurassic sandstone of the Skłoby Fm from the Skały pod Adamowem site. **f** Jurassic sandstone of the Ostrowiec Fm from the Piekło pod Niekłaniem site

well as in sandstones of the Vosges and Black Forest (Blume and Remmele 1989; Robinson and Williams 2005; Robinson 2007; Thiry 2005; Young et al. 2009; Young and Wray 2015).

25.4.2 Bed Thickness and Joint Spacing

Bed thickness and joint spacing are significant properties that control the formation and shape of crags. This is evident if crag-forming sandstones and ones that do not support crags are compared with one another.

The Cambrian sandstones of the Wiśniówka Fm that build crags in the Łysogórskie Range are thin-bedded (average bed thickness 0.3–0.6 m), while in other crag localities these sandstones are partly thick-bedded. The Devonian sandstones are usually thick-bedded (average bed thickness 1.2 m), but bedding thickness varies greatly (Table 25.1). The Triassic sandstones of the Zagnańsk Fm are also thick-bedded (average bed thickness 1.5 m), but they also display great variability (Table 25.1). Other Triassic and Jurassic crag-forming sandstones are characterised by irregular and discontinuous bedding and, consequently, varying bed thickness. In the case of sandstones of the Krynki Beds, bedding thickness varies from 0.5 to 2.0 m, with the mean value of 0.9 m. Similar values are recorded in the sandstones of the Skłoby Fm and Ostrowiec Fm (Table 25.1) (Urban 2016, 2020).

The Cambrian quartzitic sandstones of the Wiśniówka Fm underwent three stages of tectonic transformations during the Caledonian, Variscan and Alpine orogeny, and are cut by numerous joint sets, with some healed with quartz veins. Consequently, the joint set directions are very diverse, and joint density is relatively high. Joint spacing is close to 1 m; ranging at individual sites from 0.4 to 2.1 m (Table 25.1; Fig. 25.4). The crag-forming Devonian, Triassic and Jurassic sandstones are characterised by the occurrence of two (sub)orthogonal joint sets perpendicular to bedding (with the exclusion of the Bukowa Góra site, where the Devonian rocks are cut by joints of Quadersandstein pattern). In the Devonian sandstones, the average spacing is ca 2.0 m, but varies a lot at individual sites. Analogous values for the younger, Middle Triassic and Lower Jurassic crag-forming units are greater and less differentiated, ranging from 4.2 to 6.5 m. The relevant detailed data are given in Table 25.1 (Urban 2016, 2020).

Bed thickness and joint spacing of other sandstones series in the Triassic and Jurassic sequences of the Świętokrzyskie Mts. that do not build crags significantly differ from the crag-forming sandstones. Most of these series are thin-bedded and joint spacing may be two or three times greater than in the crag-forming rocks (Urban 2016, 2020).

Table 25.1 Typical features of crag-forming sandstones in the Świętokrzyskie Mts. (after Urban 2016)

	Cambrian	Devonian	Triassic		Jurassic	
	Wiśniówka Fm	Barcza Fm, Zagórze Fm	Zagnańsk Fm	Krynki Beds	Skłoby Fm	Ostrowiec Fm
Depositional system (conditioning structural pattern)	High-energy depositional environments					
	Marine: shallow or deep	Shallow marine or fluvial	Fluvial (braided river)	Shallow marine or fluvial	Shallow marine (upper shoreface)	
Mineralogic composition	Quartzose or sublithic arenites, siliceous cement					
Fabric	Very fine and fine grained	Fine grained, occasionally coarser	Poorly sorted, fine pebbles	Poorly sorted	Fine-medium grained	
				High compaction		
Intergranular space [%]	0	0–2	5–10	5–15	10–15	
Water absorption [%]	ca 0.5	0–2	3–5	4–6	4–8	
Compressive strength [MPa]	ca 200	ca 150	30–60	30–80	20–80	30–50
Bedding thickness [m]	0.7 (0.3–1.0)	1.2 (0.7–2.0)	1.3 (0.9–2.1)	0.9 (0.2–2.1)	1.1 (0.9–1.6)	
Joint spacing [m]	ca 1	ca 2	3 (2–6)	6 (3–11)	4 (2–6)	
Strata dip [°]	10–50	10–40	5–20	0–5		

25.4.3 Spatial (Tectonic) Constraints of Crag Formation

Spatial position of sandstone strata (their inclination or horizontal setting) determines the occurrence of crag groups in relation to superior morphological elements such as ridges, slopes and plateaus, as well as the general distribution pattern of rock elements within these groups. The tectonic position of strata in the Świętokrzyskie Mts. is strictly related to their ages (i.e., the number and character of tectonic phases), and therefore the morphological position and shape differ between the Palaeozoic crag-forming sandstones and the Mesozoic ones. Strongly dipping Cambrian and Devonian sandstones, as well as the Triassic sandstones of the Zagnańsk Fm at a few sites, build ridge crests or domes superimposed on the ridges (Fig. 25.4) or near-ridge spurs and rock tables with the upper surfaces dipping parallel to the slope (Fig. 25.5). Other Triassic and Jurassic sandstones, gently dipping or horizontal, form crag groups spatially related to plateau or hill margins that comprise cliffs and spurs, accompanied by numerous boulders situated downslope (Urban 2015, 2016, 2020).

The situation of crags in relation to superior morphological elements is evidently connected with the strike of the strata and/or jointing. The concordance between the strike or one of joint set directions and elongation of a superior landform (hill ridge, slope) has been documented for 75% of crag sites (Urban 2016, 2020).

25.4.4 Evolution of Crags

The formation and shaping of the crags in the Świętokrzyskie Mts. is a result of interaction of selective weathering and gravity-induced processes: (i) deformation of large rock masses within hillslopes; (ii) downslope movements of boulders; (iii) transformation of detrital slope covers (Urban 2016, 2020). Deep weathering, developed since the Neogene, contributed to the disintegration of sandstone series, which is currently marked by remnants of iron oxide veins/covers on crag surfaces (Fig. 25.9) (see Cílek et al. 2007). The subsequent near-surface weathering was very efficient under the periglacial conditions of the Pleistocene (see e.g., Curry and Morris 2004; Cílek and Žák 2007) and produced detrital, mainly sandy sheets covering the hillslopes. Large-scale displacements included detachment, lateral spreading and toppling of fragments of sandstone series. Massive fragments several tens of metres long were subject to these processes, while the range of movement varied from several to several hundred metres. Displacements at such a scale, commonly observed in high mountains (Agliardi et al. 2012), seem to be incredible in the area of very low topographic gradient, but are evident on ERT images at the Skały w Krynkach and Skały pod Adamowem sites, as well as may be inferred from direct observations at the sites of Rejów and Skałki Piekło pod Niekłaniem (Urban et al. 2015). The conditions enabling such movements included the presence of rocks prone to plastic deformations beneath the sandstones, such as claystones and siltstones, and the occurrence of factors stimulating such deformations. The most probable mechanism was cambering—rigid rock strata downslope movement and tilting owing to a permafrost disappearance and consequent thawing of frost-susceptible rocks that underlain series of rigid rocks (sandstones). Fast thawing of these rocks substantially increases their plasticity and lubrication due to an immediate excess of water pore pressure and disintegration of rock fabric (Parks 1991; Harris 2006). Such a process could have occurred many times owing to the permafrost build-up and decay during climatic fluctuations, especially within the decline of the last Glacial (climatic phases of Bölling, Older Dryas, Alleröd, Younger Dryas, and Preboreal). Another factor inducing rock-mass deformation, applicable to the Kamienna River valley, where many crag sites are located, could have been ice-sheet pressure combined with higher topographic gradient in the proglacial valley formed during the Odranian (Saalian) Glacial (Urban et al. 2015; Urban 2020).

Beside large rock-mass displacements, translational and rotational movements of boulders on slopes, such as toppling, reverse rotation, side tilting, rotation around vertical axis, overturning and even uplift, played a significant role in the shaping of crag groups. These movements occurred on slopes of various inclination, often at gently inclined ones (ca. 5°), which was possible owing to the existence of plastic rocks (clay/siltstones, marls) as a substratum. There were two mechanisms of these movements: (i) passive transport caused by plastic deformations, including boulder uplift due to the pressure of overlying sandstone series and extrusion of plastic rocks; (ii) active gravitational downslope creep of boulders due to the high plasticity and lubrication of the ground. The growth of plasticity, caused by cambering, occurred mainly during the Late Pleistocene. Although these movements were not fast, their role seems to be indisputable (Urban 2015, 2020).

The gravity-induced removal of weathering sheets that covered the hillslopes was the principal mechanism of the emergence of crags. Stripping was possible when the pace of weathering of sandstones was slower than the rate of removal of weathered material. These weathering sheets are sands and diamictons containing sandstone debris. The OSL dating indicated the Late Pleniglacial and the Late Pleistocene to early Holocene age of these sheets (Urban 2015). These covers were subject to wash and creep; however, the creep was not driven by typical solifluction because granulation of the material was too coarse for frost-susceptible soils (as defined by Matsuoka 2001). Nevertheless, such type of sheets also undergo creep (Ballantyne 1993; Goździk and French 2004). The role of aeolian process, emphasized by Lindner (1972), is difficult to determine, although the occurrence of aeolian grains in the covers indicate the contribution of this process (Urban 2015).

25.5 Features and Factors Involved in Crag Formation—A Synthesis

The crag formation process is conditioned by a set of interrelated features and factors. The principal petrographic feature determining its feasibility is siliceous composition of sandstones (Table 25.1) and consequential physico-mechanical properties: hardness and low porosity of the Palaeozoic rocks and specific weathering resilience of the Triassic and Jurassic crag-forming sandstones. Another feature stimulating selective weathering and gravity-induced processes that contribute to stripping and formation of crags is the occurrence of sandstone series between series of more plastic claystones, siltstones and/or heterolithes. Such lithostratigraphic position of sandstone series caused not only selective weathering, faster in the case of softer rocks above and below the sandstones and, hence, resulting in the origin of slope breaks (see Olędzki 1976), but contributed to the development of gravitational processes, including boulder movements as well as creep and wash of weathering covers that caused stripping of rock cliffs and formation of crag groups.

The role and range of these processes, and consequently, possibility of crag formation, were conditioned by spatial (tectonic) situation of the sandstones series, as well as the bedding and jointing of sandstones (Table 25.1) (Urban 2015, 2016, 2020). The (sub)horizontal position of the Triassic and Jurassic sandstones did not require large joint spacing for sandstone fragmentation, formation of cliffs and movements of boulders, but just adequate (i.e., sufficiently large) for preserving these cliffs and boulders from their destruction by weathering under periglacial conditions. In the case of the Palaeozoic rocks, usually inclined and harder, both joint spacing and bed thickness could be more diversified, but their adequate values have provided characteristic crag shapes and positions in relation to superior morphological elements. Therefore, it is possible to determine the optimal values of bed thickness and joint spacing, specific for the particular types of crag-forming sandstones, and conditioned by their petrographic and physico-mechanical properties as well as spatial position (Table 25.1) (Urban 2016, 2020).

25.6 Scientific, Educational and Aesthetic Values of Crags

The crags and bare blockfields are considered to be unique, extraordinary landforms in the generally "green" landscape of the Świętokrzyskie Mts. They are characterised by specific aesthetic values, although the general knowledge about their occurrence, origin and characteristics is relatively low. Systematic scientific studies of the crags are also of fairly recent date (Urban 2015, 2016, 2020; Urban et al. 2015; Urban and Górnik 2017).

Despite limited knowledge about the occurrence of crags in the region, they have been recognised as features in the need of legal protection since the first half of the twentieth century (Massalski 1927, 1936, 1951, 1967; Massalski and Kaznowski 1928; Richling 1938; Czarnocki 1932). The most spectacular site, Piekło pod Niekłaniem crags, has been legally protected as a nature reserve since 1958, while famous blockfields and some crags situated in the Łysogóry Range have been protected within the Świętokrzyski National Park since 1950. The Bukowa Góra and Miejska Góra sites were included into the National Park area in 1996. There are also several other nature reserves protecting crag groups, such as Góra Jeleniowska, Kamień Michniowski, Perzowa Góra, Skały w Krynkach, Piekiełko Szkuckie, Skały pod Adamowem and Diabla Góra. The bare blockfields are protected in Świętokrzyski National Park and Szczytniak nature reserve (Figs. 25.1 and 25.2). A total of 32 sandstone crag sites are protected as nature monuments (Alexandrowicz et al. 1975; Alexandrowicz 1990; Alexandrowicz and Urban 2005; Urban and Gągol 2008). Currently, the most picturesque crag sites are accessible by touristic trails, whereas at a few sites, educational trails presenting scientific values of these landforms have been set up. These have been arranged at the Piekło pod Niekłaniem and Skałki w Krynkach sites.

References

Adamovič J, Mikuláš R, Cílek V (2006) Sandstone districts of the Bohemian Paradise: emergence of a romantic landscape. Geolines 21:1–100

Agliardi F, Crosta GB, Frattini P (2012) Slow rock-slope deformation. In: Clague JJ, Stead D (eds) Landslides. Types, mechanics and modelling. Cambridge Univ Press, New York, pp 207–221

Alexandrowicz Z (1978) Skałki piaskowcowe Zachodnich Karpat fliszowych. Prace Geol PAN 113:1–87

Alexandrowicz Z (1990) The optimum system of tors protection in Poland. Ochr Przyr 47:277–308

Alexandrowicz Z, Drzał M, Kozłowski S (1975) Katalog rezerwatów I pomników przyrody nieożywionej w Polsce. Studia Naturae B26:1–298

Alexandrowicz Z, Urban J (2005) Sandstone regions of Poland—geomorphological types, scientific importance and problems of protection. Ferrantia 44:131–136

Ballantyne CK (1993) An Tellach. In: Gordon JE, Sutherland DG (eds) Quaternary of Scotland. Chapman & Hall, London, pp 110–115

Barczuk A (1979) Studium petrograficzne utworów pstrego piaskowca w północno-wschodniej części obrzeżenia Gór Świętokrzyskich. Arch Miner 35:87–155

Blume H, Remmele G (1989) A comparison of Bunter Sandstone scarps in the Black Forest and the Vosges. Catena, Suppl 15:229–242

Bruthans J, Soukup J, Schweigstillová J, Vaculiková J, Smutek D, Mayo AL, Felteisek L (2013) Origin of "rock cities", pillars and cleft conduits in kaolinite-bounded sandstone: new insight from study in sandstone quarry where land form recently evolve. In: Filippi M, Bosák P (eds) Proceedings of 16th international congress speleology, 21–28.07.2013, Brno. Czech Speleology Society, Praha, vol. 3, pp 247–252

Bruthans J, Soukup J, Vaculiková J, Schweigstillová J, Mayo AL, Masin D, Kletetchka G, Rihosek J (2014) Sandstone landforms shaped by negative feedback between stress and erosion. Nature Geosci 7:597–601

Chlebowski R, Lindner L (1989) Weathering processes and the Pleistocene periglacial environment and origin of loess. In: Balasubramaniam KS (ed) Weathering; its products and deposits, vol I. Processes. Theophrastus Publication of Athens, pp 439–456

Cìlek V, Williams R, Osborne A, Migoń P, Mikulaš R (2007) The origin and development of sandstone landscapes. In: Härtel H, Cìlek V, Herben T, Jackson A, Williams R (eds) Sandstone landscapes. Academia, Praha, pp 34–43

Cìlek V, Žak K (2007) Late Glacial and Holocene sedimentation under sandstone rock shelters of Northern Bohemia. In: Härtel H, Cìlek V, Herben T, Jackson A, Williams R (eds) Sandstone landscapes. Academia, Praha, pp 133–138

Colbach R (2005) Overview of the geology of the Luxembourg Sandstone(s). Ferrantia 44:155–160

Curry AM, Morris CJ (2004) Lateglacial and Holocene talus slope development and rock wall retreat on Mynydd Du, UK. Geomorphology 58:85–106

Czarnocki J (1932) Mniej znane zabytki przyrody geologiczne Gór Świętokrzyskich. Ochr Przyr 12:74–81

Duszyński F, Migoń P, Strzelecki MC (2019) Escarpment retreat in sedimentary tablelands and cuesta landscapes—landforms, mechanisms and patterns. Earth Sci Rev 196:102890

Goździk L, French HM (2004) Apparent upfreezing of stones in Late Pleistocene coversand, Bełchatów vicinity, Central Poland. Permafrost Perigl Proc 15:359–366

Guzik K (2017) Możliwości wykorzystania piaskowców jurajskich północno-zachodniego obrzeżenia gór świętokrzyskich jako kamieni architektonicznych w zależności od ich litologii. Instytut Gospodarki Surowcami Mineralnymi i Energią PAN. Studia, Rozprawy, Monografie 205:1–140

Harris C (2006) Cambering and valley bulging. In: Goudie AS (ed) Encyclopedia of geomorphology. Taylor & Francis, London, pp 115–116

Huruk S (1986) Gołoborza Świętokrzyskiego Parku Narodowego. Chrońmy Przyr Ojcz 46(1):14–22

Jaśkowski B, Kowalski B, Sołtysik R (2002) Geneza i wiek pokryw wietrzeniowych na stokach Łysej Góry w Górach Świętokrzyskich. Prace Inst Geogr Akademii Świętokrz w Kielcach 8:107–138

Klatka T (1962) Geneza i wiek gołoborzy Łysogórskich. Acta Geogr Lodz 12:1–124

Kobendza R (1939) Gołoborza i ich stosunek do lasu w Górach Świetokrzyskich. Instytut Badawczy Lasów Państwowych, Warszawa, pp 1–76

Kobylec K (2008) Dolnojurajskie piaskowce z okolic Przysuchy-Opoczna, ich wykształcenie I możliwości wykorzystania. Gospodarka Surowcami Mineralnymi 24:59–72

Konon A (2008) Regionalizacja tektoniczna Polski – Góry Świętokrzyskie I region przyległe. Przegl Geol 56:921–926
Kowalczewski Z, Kowalski B, Janiec J (1989) Wpływ budowy geologicznej na rzeźbę Pasma Klonowskiego. Biul Państw Inst Geol 362:65–95
Kowalski B (1993) Punkt 3B. Święty Krzyż. Morfologiczne konsekwencje neotektonicznej aktywności w obrębie głównego pasma Gór Świętokrzyskich. In: Barcicki M, Mityk J (eds), 42 Zjazd Pol Tow Geogr, Kielce, 27–30.06.1993. Przewodnik terenowy. WSP w Kielcach, Kielce, pp 63–72
Kowalski B, Jaśkowski B (1986) Litologiczno-strukturalne uwarunkowania taras krioplanacyjnych na stokach masywu Łysej Góry w Górach Świętokrzyskich. Przegl Geogr 58:493–514
Kozłowska A, Kuberska M (2014) Diageneza a porowatość piaskowców jury dolnej na Niżu Polskim. Biul Państw Inst Geol 458:39–60
Kuleta M, Zbroja S (2006) Wczesny etap rozwoju pokrywy permskomezozoicznej w Górach Świętokrzyskich. In: Skompski S, Żylińska A (eds) Procesy i zdarzenia w historii geologicznej Gór Świętokrzyskich, Mat 77 Zjazdu Nauk Pol Tow Geol, Ameliówka k Kielc, 28–30.06.2006 r. Warszawa, pp 104–125
Lindner L (1972) Geneza i wiek skałek piaskowcowych góry Piekło oraz ich stosunek do pokrywy lessowej region świętokrzyskiego. Acta Geol Polon 22:168–180
Łoziński W (1912) Die periglaziale Fazies der mechanischen Verwitterung. In: XI international geological congress, Stockholm 1910, Compte rendu 1912, pp 1039–1053
Łyczewska J (1972) Niektóre problem czwartorzędu świętokrzyskiego. Roczn Pol Tow Geol 42:81–92
Massalski E (1927) Bukowa Góra. Ziemia 2:18–19
Massalski E (1936) Skałka Pod Szewną. Ziemia 2–3:85–86
Massalski E (1951) Najcenniejssze pod względem dydaktycznym zabytki skalne Gór Świętokrzyskich. Zab Przyr Nieożyw 1(4):41–43
Massalski E (1967) Góry Świętokrzyskie. Seria: Przyroda Polska. Wiedza Powsz. Warszawa
Massalski E, Kaznowski K (1928) Piaskowcowe skałki góry Piekło koło Niekłania. Ochr Przyr 8:29–33
Matsuoka N (2001) Solifluction rates, processes and landforms: a global review. Earth Sci Rev 55:107–134
Migoń P, Duszyński F, Goudie A (2017) Rock cities and ruiniform relief: forms—processes—terminology. Earth Sci Rev 171:78–104
Olędzki J (1976) Wpływ zróżnicowania budowy geologicznej na Irzeźbę w Górach Świętokrzyskich. Prace i Studia Inst Geogr Uniw Warsz 17. Geogr Fiz 6:1–66
Palmer J, Radley J (1961) Gritstone tors of the English Pennines. Z Geomorph NF 5:37–52
Parks CD (1991) A review of the mechanism of cambering and valley bulging. In: Forster A, Culshlaw MG, Cripps JC, Little JA, Moon CF (eds) Quaternary engineering geology. Geological Society, London, Spec Publ 7, pp 373–380
Peszat C (1973) Własności techniczne piaskowców Gór Świętokrzyskich. Zesz Nauk AGH 378. Geologia 18:1–59
Pieńkowski G (2004) The epicontinental Lower Jurassic of Poland. Pol Geol Inst Spec Papers 12:1–154
Pulinowa M (1989) Rzeźba Gór Stołowych. Prace Uniw Śląskiego w Katowicach 1008:218
Richling W (1938) Brama Piekło pod Suchedniowem. Wiad Muz Ziemi 1:37–38
Robinson DA (2007) Geomorphology of the inland sandstone cliffs of Southeast England. In: Härtel H, Cìlek V, Herben T, Jackson A, Williams R (eds) Sandstone landscapes. Academia, Praha, pp 44–51
Robinson DA, Williams RBG (1981) Sandstone cliffs on the high weald landscape. Geogr Mag 55:587–592
Robinson DA, Williams RBG (2005) Comparative morphology and weathering characteristic of sandstone outcrops in England, UK. Ferrantia 44:41–46
Sedlak W (1964) Teoria łysogórskich gołoborzy. Roczniki Filozoficzne KUL 12(3):45–67
Sikorska M (2000) Historia sylifikacji w piaskowcach kambru z rejonu Wiśniówki w Górach Świętokrzyskich. Przegl Geol 48:251–258
Stück H, Koch R, Siegesmund S (2013) Petrographical and petrophysical properties of sandstones: statistical analysis as an approach to predict material behaviour and construction suitability. Environ Earth Sci 69:1299–1332
Teofilak-Maliszewska A (1968) Petrografia osadów liasu w północnym obrzeżeniu Gór Świętokrzyskich. Biul Inst Geol 16:107–192
Thiry M (2005) Weathering morphologies of the Fontainebleau Sandstones and related silica mobility. Ferrantia 44:47–51
Trela W (1998) Środowisko sedymentacji piaskowców "warstw z Krynek" w rejonie Nietuliska (NE obrzeżenie Gór Świętokrzyskich). Przegl Geol 46:67–70
Tricart J, Cailleux A (1972) Introduction to climatic geomorphology. Longmans, London, p 295
Urban J (2014) Cechy rzeźby strukturalnej Gór Świętokrzyskich oraz południowo-wschodniej częsci Niecki Nidziańskiej. Przegl Geol 62:44–50
Urban J (2015) The role of gravitational processes in shaping sandstone rock landforms in low mountains: Świętokrzyskie (Holy Cross) Mountains, central Poland. Z Geomorph 59(Suppl 1):35–79
Urban J (2016) The geological constraints of the development of sandstone landforms in Central Europe, a case study of the Świętokrzyskie (Holy Cross) Mountains, Poland. Geomorphology 274:31–49
Urban J (2020) Strukturalne, litologiczne oraz tektoniczne uwarunkowania rozwoju I ewolucji świętokrzyskich skałek piaskowcowych. Przegl Geol 68:112–124
Urban J, Gągol J (2008) Geological heritage of the Świętokrzyskie (Holy Cross) Mountains (Central Poland). Przegl Geol 56:618–628
Urban J, Górnik M (2017) Some aspects of lithological and exogenic control of sandstone morphology, the Świętokrzyskie (Holy Cross) Mts. case study, Poland. Geomorphology 295:773–789
Urban J, Kasza A (2009) Genetical types of the caves in sandstones of the Swietokrzyskie (Holy Cross) Mountains, Central Poland. In: Proceedings of 10th international symposium on Pseudokarst, 29.04–2.05.2008, Gorizia, pp 43–52
Urban J, Pánek T, Hradecký J, Táboŕík P (2015) Deep structures of slopes connected with sandstone crags in the upland area of the Świętokrzyskie (Holy Cross) Mountains, Central Poland. Geomorphology 246:519–530
Young RW, Wray RAL (2015) Rock control in sandstone geomorphology: a tribute to Eiju Yatsu with some Australian examples. Z Geomorph 59(Suppl 1):3–17
Young RW, Wray RAL, Young ARM (2009) Sandstone landforms. Cambridge Univ Press, Cambridge, p 304
Zhu C, Peng H, Quyang J, Hu Z, Li L (2010) Rock resistance and the development of horizontal grooves on Danxia slopes. Geomorphology 123:84–96

Jan Urban is geologist and geomorphologist, Associate Professor employed first in the Świętokrzyskie Mountains Branch of the Polish Geological Institute in Kielce and then in the Institute of Nature Conservation, Polish Academy of Sciences in Kraków, where he recently retired. He deals with the theory and practice of geological heritage protection. The Świętokrzyskie (Holy Cross) Mountains, the Nida Basin (Ponidzie region) and the Outer Carpathians (Beskidy Mountains), situated in the central and southern Poland, are the main areas of his scientific interest. He deals with karst forms in carbonate and gypsum rocks, including caves, as well as palaeokarst forms. He is a co-author of cave inventories in the two first regions. He also studies sandstone crags and non-karst caves in these rocks as well as explores and documents historical mines and open-pits. He is President of the Pseudokarst Commission of the International Union of Speleology and President of the Speleological Section of the Polish Copernicus Society of Naturalists.

Loess Gullies of the Nałęczów Plateau—An Outcome of Human–Environment Interactions

26

Jan Rodzik, Grzegorz Gajek and Wojciech Zgłobicki

Abstract

The western part of the Nałęczów Plateau (E Poland) features numerous permanent gullies whose concentration locally exceeds 10 km/km^2. Their formation is linked to agricultural activity and natural factors conducive to intensive gully erosion such as the presence of thick loess covers, diverse relief, or frequent occurrence of torrential rainfall. Gullies form multi-branched systems usually dissecting the floors and slopes of dry valleys. The main forms usually reach the depth of 10–20 m and, due to their high density, they play an important environmental role and have an impact on hydrology, biodiversity, and microclimate. At present, most gullies do not show any significant activity but, under certain conditions, agricultural practices in their catchments are conducive to more intensive geomorphological processes.

Keywords

Loess cover · Gully erosion · Human impact · Land use

26.1 Introduction

Under the conditions of Poland's temperate transitional climate, *Tilio-Carpinetum* forest communities predominated on soils formed from silt and clay formations. Exploitation of forests, especially their clearing and conversion of land into farming, intensified surface runoff, the concentration of which, in an area of diverse relief, resulted in gully erosion, leading to the development of permanent gullies (Schmitt et al. 2006; Dotterweich et al. 2012). Gully development was taking place under favourable natural conditions, including the susceptibility of the substratum, large elevation differences and steep slopes, and the presence of both oceanic and continental climate features such as relatively high precipitation resulting in increased water content in the substratum, spring and summer rainstorms, occasional cold and snowy winters, and intense snowmelt (Rodzik et al. 2009). An important role was played by the properties of the silty substratum—the loess, malleable and susceptible to erosion processes when wet, but able to maintain steep walls when dry, which influenced the relative durability of the gullies (Frankowski and Grabowski 2006). The environmental determinants mentioned above occur in the western part of the Nałęczów Plateau, where the density of the gullies is exceptionally high (Maruszczak 1973; Gawrysiak and Harasimiuk, 2012).

The Nałęczów Plateau lies along the north-western edge of the Lublin Upland, between the valleys of the Vistula between the towns of Kazimierz Dolny and Puławy, and the Bystrzyca river valley in the city of Lublin. Its shape resembles a rectangle, measuring 40–45 and 10–12 km, extending from WNW to ESE (Fig. 26.1). It is distinguished by a nearly continuous loess cover with an undulating top, at 190–230 m asl, bounded by escarpments of varying height and origin. The northern escarpment, 25–30 m high, of complex tectonic, denudation and aggradation origin, is also the boundary of the Lublin Upland. The southern escarpment, up to a dozen or so metres high, results exclusively from the deposition of the loess cover. The sides of river valleys, i.e. the right-hand side of the Vistula valley (70–90 m high) and the left-hand side of the Bystrzyca valley (25–30 m), form the western and eastern boundaries of the Plateau, respectively. The floors of these valleys constitute the base level of erosion for the Plateau. In the western part, it is the valley floor of the Vistula River and its

J. Rodzik · G. Gajek · W. Zgłobicki (✉)
Institute of Earth and Environmental Sciences,
Maria Curie-Skłodowska University, Kraśnicka 2d, 20-718
Lublin, Poland
e-mail: wojciech.zglobicki@mail.umcs.pl

J. Rodzik
e-mail: jan.rodzik@mail.umcs.pl

G. Gajek
e-mail: grzegorz.gajek@mail.umcs.pl

P. Migoń and K. Jancewicz (eds.), *Landscapes and Landforms of Poland*,
World Geomorphological Landscapes, https://doi.org/10.1007/978-3-031-45762-3_26

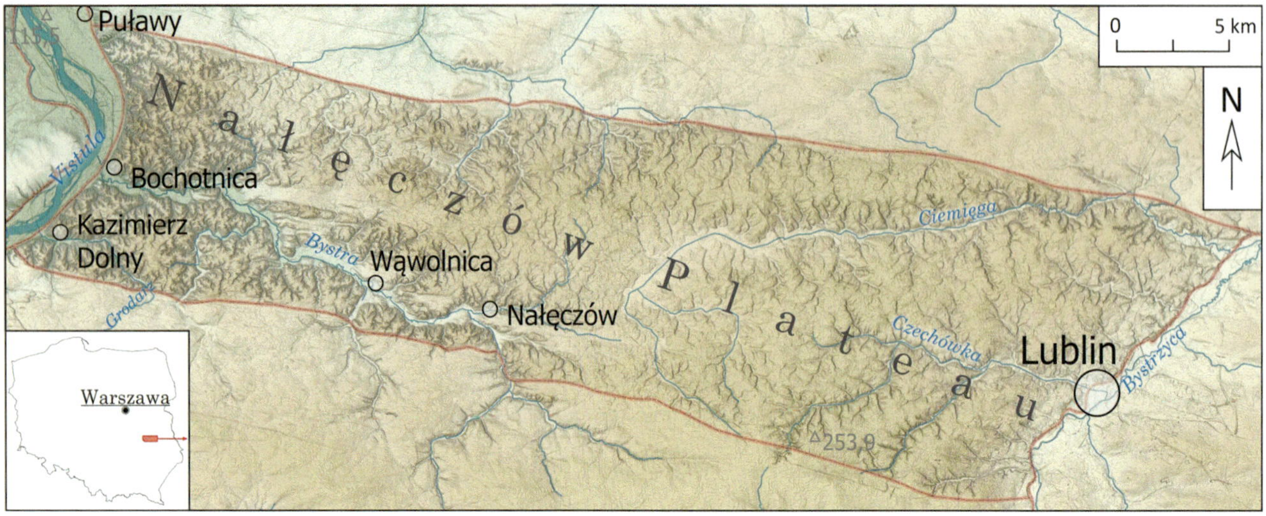

Fig. 26.1 Location and morphology of Nałęczów Plateau. Note contrasts in the intensity of erosional dissection with the adjacent areas

tributaries, the Bystra stream and the Grodarz streams (at about 120 m asl), while in the eastern part, it is the valley floor of the Bystrzyca and its tributaries, the Ciemięga River and the Czechówka River (Maruszczak 1972; Harasimiuk and Henkiel 1976), located nearly 50 m higher (Figs. 26.1, 26.2a).

Loess relief is distinguished by considerable fragmentation of slopes, which are convex in their upper parts and concave in the lower parts. It is internally differentiated by elevation differences, slope gradients, and the degree of dissection with young dry valley forms. Therefore, three subtypes of loess relief are distinguished in the Nałęczów Plateau: denudational, erosional–denudational, and erosional (Rodzik et al. 2021). These subtypes occur in areas with specific elevation differences between the plateau top and the floors of river valleys: small (up to 30 m), medium (30–60 m), and large (above 60 m), respectively (Maruszczak 1963).

The denudational type predominates in the watershed area of the central part of the region, where the loess cover is not more than 10 m thick (Fig. 26.2b). It is characterized by small elevation differences and low density of the valley network, whereas the top surface of the loess plateau features extensive shallow denudation troughs and numerous closed depressions. The erosional–denudational subtype, prevalent in the eastern part of the region, is characterized by medium elevation differences and a high density of multi-branched dry valleys (about 2.5 km/km^2) with 5° to 10° slopes (Maruszczak 1963, 1973; Rodzoś et al. 2006). Dry valleys are locally incised by gullies, mostly formed by traffic-induced erosion. The erosional subtype occurs in the western part of the Nałęczów Plateau and is characterized by large elevation differences, while landforms have developed in the loess cover that is up to 30 m thick and dissected by a network of dry valleys with slopes from 5° to 15°. This subtype is distinguished by a dense system of gullies, mostly dissecting the floors and slopes of valleys. Their mean density is as high as 2.5 km/km^2, whereas in the Nałęczów–Kazimierz Dolny–Puławy triangle it reaches 5 km/km^2, locally even exceeding 10 km/km^2 (Maruszczak 1973; Gawrysiak and Harasimiuk 2012). The base of the elongated triangle is formed by the Middle Vistula Gap, while the valley of its tributary, the Bystra, is the axis of the triangle (Fig. 26.2a, b). Such a strong dissection was determined by the concurrence of anthropogenic factors and natural conditions facilitating erosion.

26.2 Natural Conditions of the Western Part of the Nałęczów Plateau

The bedrock of the Nałęczów Plateau is a Cretaceous–Palaeocene complex of marine carbonate–silica sediments. Cretaceous and Palaeocene rocks are overlain by Pleistocene deposits. Glacial deposits are represented by a bipartite series of till that formed during the Elsterian 2 (MIS 12) glaciation and Saalian (MIS 6) glaciation (Pożaryski et al. 1994). Till occurs in patches which, in the northern part, form elevations beneath the loess. Glaciofluvial deposits, including sand, sand with gravel, and locally gravel, occur on the slopes of the Bystra valley (Harasimiuk and Henkiel 1976).

The younger Pleistocene is represented by loess that was deposited mostly in the Pleniglacial of the Weichselian (Vistulian) glaciation (MIS 2), at 30,000–15,000 BP, in the form of patches of varying size and thickness (Maruszczak 1995). The thickest loess, up to 30 m, occurs along the Vistula valley. Locally the loess cover descends on the

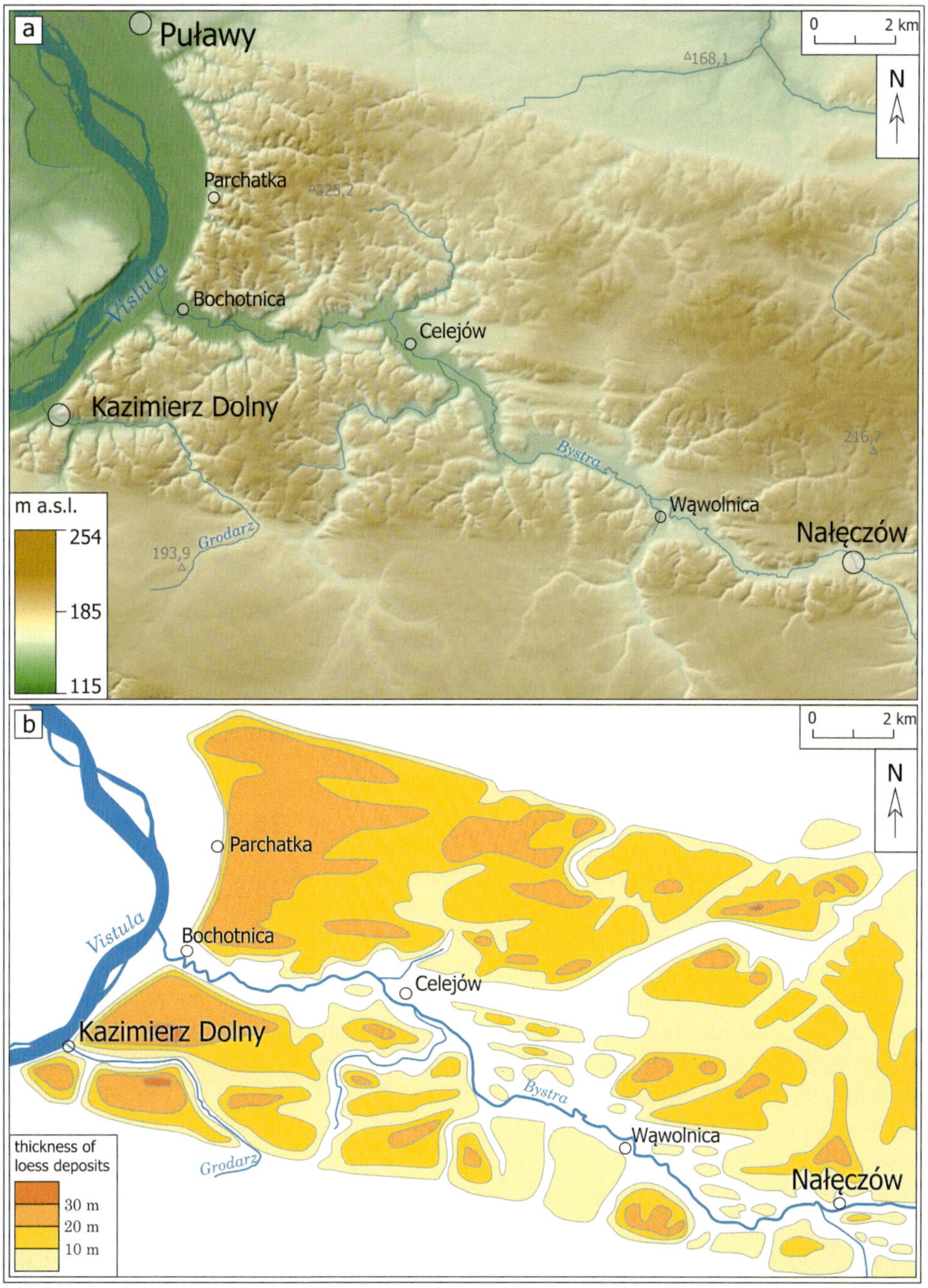

Fig. 26.2 Western part of the Nałęczów Plateau. **a** Morphology. **b** Thickness of loess deposits (after Harasimiuk and Henkiel 1976, modified)

slopes of the river valleys down to the foot of the slopes. In the vicinity of the Bystra valley, the loess cover typically has the form of small, elongated patches and caps covering kame hills and ridges from the Saalian glaciation (Harasimiuk and Henkiel 1976) (Fig. 26.2b).

The Holocene is represented by deposits in the floors of valleys of different orders. In dry valleys, these are mostly silty colluvial deposits, up to 2–3 m thick, originating from soil erosion. Colluvia of similar thickness, originating from erosion in the upper reaches, occur in the lower sections of the gullies. In the bottom series in the floors of river valleys, there are sandy-silty alluvial deposits and peats, filling the fossil river beds. In the top part, up to 3 m thick, alluvial deposits interfinger with different types of alluvial fan deposits originating from gully erosion (Zgłobicki and Rodzik 2007; Superson et al. 2016).

The mean annual temperature and mean annual precipitation volume decrease from W to E within the Nałęczów Plateau, from 7.8 to 7.4 °C and from 600 to 550 mm, respectively. The duration of the growing season (220–214 days) becomes shorter, while the average duration of snow cover increases (60–70 days), also from W to E (Kaszewski 2008). Air temperatures are noticeably lower, and the duration of snow cover is longer in gullies.

This phenomenon is manifested in the recurring name of "Zimny Dół" ("Cold Gully"). Nearly two-thirds of precipitation occur in the summer half-year; the mean number of days per year with high precipitation (≥10 mm) is considerable (15). Every 10 years on average, heavy rainfall with an intensity close to 50 mm/h occurs, and every 100 years, torrential rainfall bringing more than 100 mm/h takes place. Every few years, the spring snowmelt runoff exceeds 10 mm, while every dozen years, after a long, cold and snowy winter, it may exceed 25 mm (Janicki et al. 2010).

The Vistula River, with a mean discharge of 500 m^3/s in this area, is joined by two small tributaries from the Nałęczów Plateau. The Grodarz, a stream flowing into the Vistula in Kazimierz Dolny, has a discharge of 10–20 l/s, but it disappears in exceptionally dry periods. After heavy rainfall, however, due to orographic conditions of the catchment, the stream turns into a rapid river with a discharge that can reach 30 m^3/s or, during a hundred-year flood, even 100 m^3/s (Maruszczak et al. 1984). With a mean discharge of 1.2 m^3/s at the mouth, the Bystra flows into the Vistula in Bochotnica. It is a typical upland river, with a mean gradient of 2.8‰, ranging from 1.4 to 7.0‰ along its course (Michalczyk and Wilgat 2008). The 85% share of groundwater in the supply of the Bystra results in low variability of river runoff. With the annual precipitation of 600 mm, the total runoff is 125 mm. The specific runoff, amounting to 4.15 $l/s/km^2$, is typical of uplands in the Lublin Region (Michalczyk and Wilgat 2008).

The western part of Nałęczów Plateau is dissected by river valleys, the Bystra valley being the largest one. The mean depth of the valley is 40–50 m, but it reaches even 80 m near the mouth. The slopes of the Bystra valley are steep, usually > 15°, locally even > 30°, while the width of the valley floor varies from 200 to 600 m. The valleys of the Bystra's tributaries and the Grodarz valley are similarly controlled by tectonics.

The present-day relief of the top surface of the subregion is an outcome of deposition of loess cover and its subsequent modelling and dissection. During the deposition of loess, extensive shallow denudation troughs and closed depressions were formed. The relief in most of the subregion was impacted by erosion and denudation processes, particularly active at the end of the last glaciation (MIS 1), which led to the development of a system of dry valleys (Maruszczak 1961). Dry erosional–denudational valleys, in a complex with trough-shaped valleys and ablation troughs on slopes, usually form multi-branched systems up to a few kilometres long. The main forms are 10–20 m deep, while their sides are often asymmetrical. "Warm" slopes, with a southern aspect, are usually steeper and fragmented by ablation troughs (shallow hollows, whose floors have a very steep gradient). The upper reaches of dry valleys and their tributaries occupy trough-shaped valleys about 100 m long.

In the Holocene, the undulating loess relief was stabilized by woodland, first a boreal forest, then a *Tilio-Carpinetum* forest. Under the forest cover, relatively fertile, about 1.5-m-thick Luvisols developed. The Luvisols occur in the entire loess-covered area, but due to agricultural activity on the ridges and valley slopes, their profiles were reduced to a varying degree, which increased soil susceptibility to erosion. The soils in the floors of dry valleys and depressions were covered with colluvia. Initial soils predominate in the gullies, while alluvial soils predominate in the floors of river valleys (Klimowicz and Uziak 1994; Rodzik 2010). Among forest communities in the study area, *Tilio-Carpinetum* forests predominate. They were preserved mainly in areas where the relief does not permit tillage, i.e. within gully systems and on steep valley slopes. The tree stands are mainly composed of hornbeam (*Carpinus betulus*), with a smaller proportion of small-leaved lime (*Tilia cordata*), pedunculate oak (*Quercus robur*), Norway maple (*Acer platanoides*), and sycamore maple (*Acer pseudoplatanus*). The largest area, however, is occupied by synanthropic associations of fertile and warm habitats, i.e. cultivated crops, xerothermic grasslands, and scrublands (Fijałkowski 1972; Kucharczyk 1992).

26.3 Development of Gully Network

26.3.1 Phases of Gully Erosion

The thinning of the forests and cultivation of the land caused more intensive rainwater and snowmelt runoff, which—given the relief and climatic conditions of the western part of the Plateau—triggered gully erosion. The beginning of gully formation usually coincided with the beginning of the agricultural use of the land. This occurred as early as the Neolithic in some places (6–5 thousand years ago), but in the Bronze Age (4–3 thousand years ago) or in the Middle Ages in other places (Rodzik 2010; Dotterweich et al. 2012; Superson et al. 2014).

Rapid development of gullies occurred in the sixteenth and seventeenth centuries, in the age of manorial economy, when forests were cleared on a large scale to make space for cereal crops (mostly rye), for export, in a relatively cold and humid climate of the Little Ice Age. In the Age of Discovery, the demand for agricultural and forest crops grew substantially in Western Europe (Zientara et al. 1988). The town of Kazimierz Dolny was the port of shipment for the region: from here goods were shipped along the Vistula River to the seaport in Gdańsk (Gierszewski 1982). A major role in the development of gullies was then played by the network of dirt roads, both local ones, leading from settlements in the valleys to the fields on the plateau tops, as well as travel and trade routes of regional and even international significance. The combination of climate factors, i.e. frequent heavy rainfalls and spring thaws, and economic pressure resulted in the intensive gullying of the area. Particularly in the first half of the seventeenth century this led to the siltation of river valleys and destruction of settlement structures, among other consequences (Gardziel and Rodzik 2001; Zgłobicki and Baran-Zgłobicka 2011; Dotterweich et al. 2012; Superson et al. 2014). However, climatic, environmental, and, in particular, political factors (numerous wars from the mid-seventeenth century) led to economic decline and depopulation of villages, followed by partial reforestation of arable fields (Zientara et al. 1988).

The renewed development of the gullies, related to the transition from the three-field system to modern crop rotation using root crops (potatoes, sugar beet), occurred in the nineteenth century (Willaume 1964). At the end of the nineteenth century, farm holdings were restructured following the granting of freehold to peasants and partial parcellation of estates (Strzemski 1961). The fragmentation of land intensified gully erosion, even more so in the first half of the twentieth century due to the "hunger for land" resulting from wars, economic crisis, and overpopulation of villages. Trees and shrubs in gullies were cut to be used as fuel, while the slopes and flat floors of some gullies were taken over for pasturing as well as crop cultivation (Rodzik 2010; Superson et al. 2014).

Since the 1970s, the pressure on gullies has been decreasing. After the establishment of the Kazimierz Landscape Park in 1979, the introduction of orchards as well as currant, raspberry, chokeberry and gooseberry plantations was recommended. These crops are cultivated in turf and ensure a good protection of the soil in the fruiting years. However, in the first two or three years, they grow on fallow soil treated with herbicide, and such soil is highly susceptible to runoff and sheet erosion. Ornamental shrub plantations and cruciferous crops (broccoli and cabbage), popular in recent years, have a similar impact—they do not shield the soil surface until late in the summer (Fig. 26.3a). These changes in the crop structure, causing intensified runoff from the fields to the gullies, contributed to another phase of renewal of gully erosion (Fig. 26.3b) in the form of linear erosion, headcut erosion, and piping (Rodzik et al. 2009; Rodzik 2010). The floors and slopes of many existing gullies are still being dissected (Fig. 26.4a, b), sometimes with quite a high intensity (Kociuba et al. 2015), but new in-valley gullies do not form nowadays. Areas most susceptible to gully erosion have already been dissected, the dirt roads are reinforced or paved, while initial dissections that could initiate gullying within arable fields are quickly levelled by the landowners (Janicki et al. 2002).

26.3.2 Morphology of Gullies

The notion of gullies as applied to the Nałęczów Plateau traditionally includes all young, morphologically diverse erosional forms of Holocene age. They are distinguished by sharp outlines and steep slopes of their walls, usually from 30° to 50° and, in the case of sunken lanes (road gullies), reaching even 70°–80°. The depth of gullies usually reaches a few metres, locally a dozen metres or more. Two genetic types of gullies can be distinguished: the valley type ["doły" in Polish (plural)], and the road type, the so-called sunken lanes ["głębocznice" in Polish (plural)] (Figs. 26.5, 26.6).

26.3.2.1 Valley Gullies

Gullies of the first type are represented by multi-branched systems incised into the floors and slopes of dry valleys. Formed as a result of runoff from arable fields along rills and field boundaries, these gullies are usually "inserted"

Fig. 26.3 Effects of torrential runoff. **a** Evidence of runoff from a Thuja plantation—accumulation of material above the head of the gully and cutting the forehead. **b** Runoff and incision within the bottom of the gully (photograph **a** by J. Rodzik; **b** by G. Gajek)

Fig. 26.4 Erosion along gully edges. **a** Piping tunnel with hollows in the ceiling. **b** Enormous piping and evorsion pothole (photograph **a** by J. Rodzik; **b** by R. Kołodyńska-Gawrysiak)

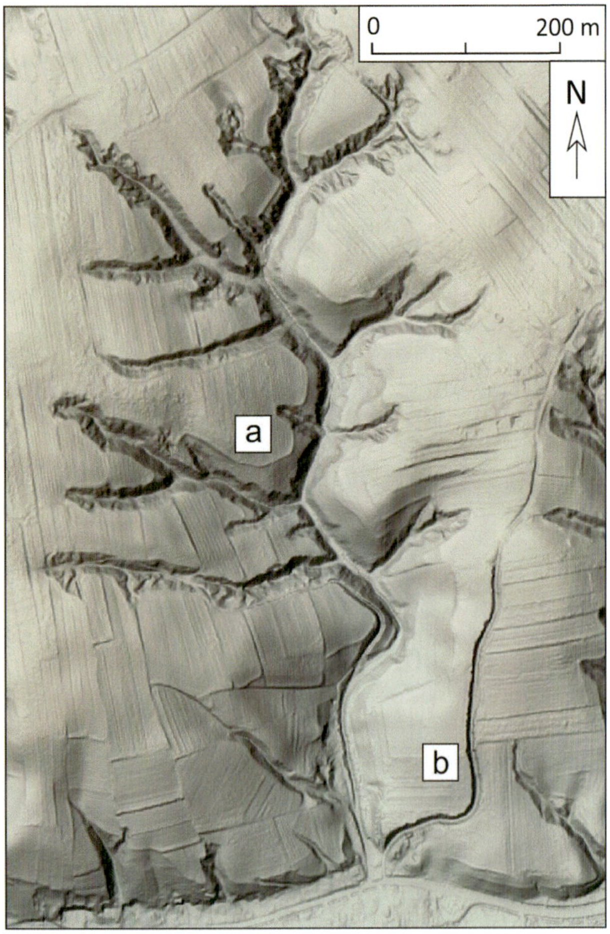

Fig. 26.5 Spatial pattern of main types of loess gullies—valley gullies (**a**) and a sunken lane (**b**)

into older and larger forms. Despite anthropogenic factors influencing their development, they are regarded as natural forms. They can be regarded as a response of various loess landforms to intensified anthropogenic pressure. While the size of individual gullies is not impressive and they are secondary forms in relation to dry valleys, these gully systems may extend over vast areas and are perceived as conspicuous landscape features. Depending on the field pattern and the layout of road network in the catchment, valley gullies form various systems: dendritic, comb-, feather- and fork-shaped (Rodzik and Gardziel 2004; Rodzik 2006). In complex systems of valley gullies, with a typical catchment area of about 0.5 km^2, one can usually distinguish the main form, about 1 km long, a few branches 200–500 m long, and numerous short forms (bank gullies) measuring 20–50 m and dissecting the edges of the larger gullies (Fig. 26.5). The main forms and branches are usually 5–15 m deep, while the bank gullies are several metres deep.

Main gullies and branches dissect not only the thin loess cover in the floors of dry valleys but also the underlying sand and till; locally they are incised into bedrock (gaize and/or opoka). Since they developed by headward erosion, the oldest part of the gully is the lower one, with a trapezoid section, a flat floor of depositional origin and a low gradient. In the middle reaches, erosion of the underlying rocks and deposits usually results in the formation of a V-shaped form (Fig. 26.7a). However, if till and clay with low permeability occur in the substrate, they inhibit vertical erosion

Fig. 26.6 Main genetic types of gullies of the Nałęczów Plateau. **a** The valley gully "Doły Podmularskie" near the town of Kazimierz Dolny, incised into an older erosion–denudation valley. **b** Road gully in the village of Wierzchoniów, incised into a convex slope of the Bystra river valley (photographs by J. Rodzik)

Fig. 26.7 Types of valley gullies. **a** V-shaped side gully of the "Zimny Dół" gully near the village of Celejów. **b** Wet, middle section of the gully in Kolonia Celejów. **c** Active side gully of the "Doły Podmularskie" gully near Kazimierz Dolny. **d** Badland within the slope of the "Zimny Dół" gully near Celejów (photographs by J. Rodzik)

(downcutting). This leads to the development of a gully with a flat, wet floor and a relatively high gradient while the gully sides are often shaped by soil creep and small landslides (Fig. 26.7b). On the other hand, the upper reaches of the main gully and its branches as well as bank gullies usually dissect a thicker loess cover, and they typically have steep slopes and a narrow floor with a varying gradient (Fig. 26.7c). They are shaped largely by piping (formation of wells, sinkholes and underground canals) initiated by tunnels of burrowing animals. Bank gullies can have no surface drainage (Fig. 26.4b) and are often suspended above higher-order landforms and separated from them by thresholds. They usually run parallel to one another (Fig. 26.7d) because they correspond to the layout of the fields in the catchment. Due to divisions of inheritance, this layout typically consists of relatively narrow plots perpendicular to the main gully or its branches (Zgłobicki et al. 2015a). The bank gullies, now separated by narrow ridges, form comb-shaped or feather-shaped badlands (Rodzik 2006; Rodzik et al. 2021).

Erosion in the upper reaches of a gully and dissection of its edges (Fig. 26.8a, b) is accompanied by deposition in the lower reaches, which leads to the raising of the gully floor and formation of a trapezoidal gully in cross-section (Fig. 26.8c). These processes occur mainly during the thaw. When the upper reaches become stable, rapid runoff after heavy rainfall can cause secondary incision of the depositional floor and rejuvenation of the gully (Fig. 26.8d). The regressive terrace, preserved to a varying degree, is the remnant of the former gully floor (Rodzik et al. 2009). During extreme heavy rainfall, old dissections can be reactivated and new ones can develop, while the intensity of erosion can reach 5–10 thousand t/km^2 of the gully catchment (Maruszczak et al. 1984).

26.3.2.2 Sunken Lanes

In the western part of the Nałęczów Plateau, a considerable concentration of road gullies [sunken lanes] can be observed. This special kind of gullies forms as a result of sheet erosion, linear erosion, and headcut erosion within intensively used dirt roads on slopes (Zgłobicki 1998; Gardziel and Rodzik 2001; Dotterweich et al. 2012; Rodzik et al. 2014). The movement of material by the hooves of pack animals, vehicle wheels and repair work also played a role, which is why sunken lanes are generally considered to be anthropogenic forms. They are characterized by an almost straight course and a box-shaped cross-section (Figs. 26.5, 26.6b). With a steep slope and considerable vehicular traffic, the average sinking rate of the road surface can reach several centimetres in a year (Nowocień 1996). This rate decreases as the sunken lane becomes wider, but after centuries of use, road gullies can be as much as 10 m

Fig. 26.8 Effects of gully erosion. **a** Effect of surface runoff from an arable field—a piping-evorsion pothole at the edge of the gully in the hamlet of Kolonia Stok. **b** Badlands forming along the edge of a gully in Mareczki near the village of Wąwolnica. **c** Accumulation in the bottom of the lower section of the gully in Kolonia Celejów leads to the formation of a trapezoidal gully cross-section. **d** Incision of the bottom of a trapezoidal gully in Kolonia Celejów and formation of a regression terrace (photographs **a**, **c**, **d** by J. Rodzik; **b** by W. Zgłobicki)

deep or more (usually a few metres), and their floors can be 3–4 m wide. Strong erosion of the gully floor sometimes led to the road having been relocated to elsewhere, resulting in the formation of a new sunken lane, while slope processes on the former gully sides have made an abandoned sunken lane similar to a typical gully. Gullies of this type often form complex systems in this region—fork-shaped in the case of gullies dissecting systems of dry valleys, or fan-shaped in the case of gullies dissecting slopes of the ridges (Rodzik and Gardziel 2004; Superson et al. 2014).

26.4 Environmental Consequences of Gully Erosion

The occurrence of a dense network of gullies entails various environmental consequences, including hydrological, geomorphological, and those related to land cover and land use. Deeply incised valley gullies often drain water from the neighbouring areas, which consequently become dry. At the same time, gullies force concentration of runoff and fast formation of water outflow (Fig. 26.3a), which accelerates the circulation of water in the landscape and increases the flood hazard, particularly during torrential rainfall and thaw. In general, dry gullies become channels of rapid episodic streams (Fig. 26.3b), while the discharge at their mouth can exceed 1 m^3/s (Rodzik et al. 2009). This leads to the occurrence of flash floods in the main valleys; these flash floods can pose a threat to buildings and infrastructure. In addition, the sediment transported from the gullies silts up the fields, roads, and properties located near the mouths of gullies. In the past, heavy torrential rainfalls led to the siltation of the market square and streets of Kazimierz Dolny, located at the mouth of the Grodarz valley, whose basin comprises a dense network of gullies (Maruszczak et al. 1984; Dotterweich et al. 2012).

Until recently, gullies were viewed exclusively as negative features of the agricultural landscape and it was attempted to restrict or control their development. In 1980s, there was a plan to build concrete steps across the floor of the loess gullies in the Grodarz catchment. However, the efficacy of such costly solutions raised serious doubts. There was even a plan to cover up and rehabilitate badlands formed by road gullies, so as to reclaim these areas for agriculture (Józefaciuk and Józefaciuk 1996). After the political and economic transformation and gradual restructuring of agriculture, such measures would not be economically viable.

In the agriculturally used loess areas of the western part of Nałęczów Plateau, gullies along with the adjacent areas, such as steep valley sides, constitute enclaves of forest communities (Kucharczyk 1992; Zgłobicki and Baran-Zgłobicka 2012). At present, these communities no longer form a continuous forest complex. Instead, they comprise relatively young tree stands consisting of hornbeam with lime, Norway maple, and oak. Nonetheless, they play an important role in the preservation of biodiversity in the agricultural landscape, with numerous plant species, as well as they act as sanctuaries and ecological corridors for animals. The impact of gullies on microclimate is a significant, though insufficiently studied aspect of their existence. Sheltered and shaded concave landforms, covered by forest communities, are characterized by lower temperatures, greater humidity and significantly lower wind speed in comparison with the adjacent areas.

At present, all valley gullies are covered by woodland because a different type of use is not cost-effective at all. The development of gullies, particularly their complex systems, thus leads to the exclusion of the adjacent areas from agricultural use. In multi-branched gully systems, forest communities occur not only in the gullies themselves, but often also in areas between the branch gullies. The succession of shrubs and trees results from difficult access to fields and less favourable conditions for crop cultivation, such as fragmentation of plots, presence of tall trees (lower availability of sunlight), and threat of erosion. Pioneer species are the first to colonize these areas; birch, aspen, and hornbeam predominate among trees. The abandonment of arable fields in the vicinity of gullies is often forced by the development of badlands and shifts of the field-forest boundary. The former boundary, elevated by colluvia in the past, now remains as a "false" bank of the gully (Rodzik et al. 2021). The occurrence of gullies influences the agrarian structure because not only it leads to the increase of the forest cover, but also results in a smaller proportion of large-area farm holdings and their lower economic productivity (Zgłobicki et al. 2015b).

26.5 Gullies as Touristic and Geotouristic Attractions

Loess gullies are a tourist attraction on the regional and national scale (Zgłobicki and Baran-Zgłobicka 2013). The occurrence of gullies is highlighted in guidebooks, and the local tourism organization operating in the Nałęczów Plateau is called "Kraina Lessowych Wąwozów" (=Land of Loess Gullies). Tourists visit gullies primarily because of their aesthetic appeal and the possibility of active tourism (hiking, biking) (Fig. 26.9a). Tourist paths and trails have been waymarked in some gullies. Road gullies enjoy the greatest popularity among tourists because of their good accessibility and aesthetic value. The Korzeniowy Dół sunken lane, where tourist traffic in summer reaches 200 people per hour, is an example of such a landmark in the western part of the Nałęczów Plateau (Fig. 26.9b).

Fig. 26.9 Tourist exploration of gullies, western part of the Nałęczów Plateau (photograph **a** by W. Zgłobicki; **b** by G. Gajek)

Gullies are important in geo-education, offering an opportunity to learn about topics such as geological structure (bedrock), active geomorphological processes, phases of gully erosion, and human influence on present-day erosion processes. Nevertheless, the development and use of gullies for geotourism remains very poor despite their high potential, even though it is relatively well-researched and recognized in comparison with other countries (Zgłobicki et al. 2015a, 2019; Warowna et al. 2016). The future establishment of the Małopolska Vistula Gap Geopark, whose concept has already been prepared, would be an opportunity for the development of geotourism. The geomorphological heritage of loess relief, of which gullies are an important element, is one of the key aspects of the Geopark (Warowna et al. 2016). The establishment of the Geopark would facilitate the preservation of gullies and their assets because uncontrolled economic and tourist activity leads to threats such as the misuse of gullies for off-road vehicular traffic and as waste dumping sites. Furthermore, paving of the floors of some road gullies is controversial. Whereas it improves their accessibility for vehicles and protects them against erosion, it has a negative impact on the image of these landforms and their scenic values.

26.6 Conclusions

The formation and development of gullies in the western part of the Nałęczów Plateau is a result of environmental response to the development of agriculture and destruction of the natural forest cover protecting the loess slopes against runoff and erosion. Erosion processes, particularly during torrential rainfall and thaw, have led to the development of a peculiar landscape whose dense network of different types of gullies is unique on the national and European scale. Their formation had some negative consequences for human activity, particularly for agriculture, local transport, and settlement development. On the other hand, gullies are now perceived as valuable natural landscape features and play an important role from the perspective of biodiversity, environmental education, and the development of tourism. Action should be taken to preserve the values of these unique landforms while limiting the negative consequences of gully development.

The gullies in the western part of the Nałęczów Plateau are landforms with a high potential dynamics of development, particularly during heavy rainfall and thaw. The total elimination of gully erosion is impossible, but it is feasible to reduce its extent, primarily by limiting runoff from the fields to the gullies. In the case of valley gullies, this can be achieved by changing the direction of cultivation of the adjacent fields (Rodzik et al. 2021). However, such a change is difficult to achieve because it would require the consolidation of land whose ownership structure is now highly fragmented. The removal of the most erosion-inducing crops, such as cruciferous vegetables, seems to be easier to implement. It would be advisable to leave a few metres of turf or shrubs as a buffer, which has actually been done in many locations.

In the case of road gullies, which usually do not have an additional catchment, attention should be focussed on protecting their floors against erosion and, if necessary, removing material resulting from landslides. In the case of gullies having an important transport function (access to fields and properties), paving or reinforcing the floors of selected gullies seems inevitable and in some cases this has already been done. This triggers conflicts between the local authorities, which respond to the needs of the local residents, and environmental organizations. It should be stressed, however, that strict conservation of sunken lanes, i.e. the exclusion of all transport functions and repair measures, would lead to the loss of their typical appearance and scenic values.

References

Dotterweich M, Rodzik J, Zgłobicki W, Schmitt A, Schmidtchen G, Bork H-R (2012) High resolution gully erosion and sedimentation processes, and land use changes since the Bronze Age and future trajectories in the Kazimierz Dolny area (Nałęczów Plateau, SE-Poland). Catena 95:50–62

Fijałkowski D (1972) Stosunki geobotaniczne Lubelszczyzny. Lubelskie Towarzystwo Naukowe, Ossolineum, Wrocław, p 285

Frankowski Z, Grabowski D (2006) Geologiczno-inżynieryjne i geomorfologiczne uwarunkowania erozji wąwozowej w lessach w rejonie Kazimierza Dolnego (wąwóz Opolska Droga). Przegl Geol 54:777–783

Gardziel Z, Rodzik J (2001) Drogi gruntowe jako stymulator przemian silnie urzeźbionego krajobrazu lessowego (w okolicy Kazimierza Dolnego). Probl Ekol Krajobr 10:305–311

Gawrysiak L, Harasimiuk M (2012) Spatial diversity of gully density of the Lublin Upland and Roztocze Hills (SE Poland). Annales UMCS B 67(1):27–43

Gierszewski S (1982) Wisła w dziejach Polski. Wydawnictwo Morskie, Gdańsk, p 278

Harasimiuk M, Henkiel A (1976) Wpływ budowy geologicznej i rzeźby podłoża na ukształtowanie pokrywy lessowej w zachodniej części Płaskowyżu Nałęczowskiego. Annales UMCS B 30(31):55–80

Janicki G, Rodzik J, Zgłobicki W (2002) Geomorphic effects of land use changes (a case of the Gutanów loess catchment, Poland). Geogr Čas 54:39–57

Janicki G, Kociuba W, Rodzik J, Zgłobicki W (2010) Ekstremalne procesy geomorfologiczne we wschodniej części Wyżyn Polskich—warunki występowania i oddziaływanie na rzeźbę. Prace Studia Geogr WGiSR UW 45:11–28

Józefaciuk C, Józefaciuk A (1996) Erozja wąwozowa i metody zagospodarowania wąwozów. Bibloteka Monitoringu Środowiska, Warszawa, p 150

Kaszewski BM (2008) Klimat. In: Uziak S, Turski R (eds) Środowisko przyrodnicze Lubelszczyzny. Lubelskie Towarzystwo Naukowe, Lublin, pp 75–111

Klimowicz Z, Uziak S (1994) An attempt at the reconstruction of the past and future soil cover in the loess area (on the basis of the Nałęczów Plateau). Annales UMCS B 49:229–242

Kociuba W, Janicki G, Rodzik J, Stępniewski K (2015) Comparison of volumetric and remote sensing methods (TLS) for assessing the development of a permanent forested loess gully. Nat Haz 79:139–158

Kucharczyk M (1992) Roślinność i flora Kazimierskiego Parku Krajobrazowego. In: Wilgat T (ed) System obszarów chronionych województwa lubelskiego. Wydawnictwo UMCS, Lublin, pp 76–81

Maruszczak H (1961) Le relief des terrains de loess sur le Plateau de Lublin. Annales UMCS B 15:93–122

Maruszczak H (1963) Warunki geologiczno-geomorfologiczne rozwoju erozji gleb w południowej części województwa lubelskiego. Wiad IMUZ III 4:19–44

Maruszczak H (1972) Wyżyny Lubelsko-Wołyńskie. In: Klimaszewski M (ed) Geomorfologia Polski, 1. PWN, Warszawa, pp 340–383

Maruszczak H (1973) Erozja wąwozowa we wschodniej części pasa wyżyn południowopolskich. Zesz Probl Post Nauk Roln 151:15–30

Maruszczak H (1995) Glacial cycles of loess accumulation in Poland during the last 400 ka and global rhythms of paleogeographical events. Annales UMCS B 50:127–156

Maruszczak H, Michalczyk Z, Rodzik J (1984) Warunki geomorfologiczne i hydrogeologiczne rozwoju denudacji w dorzeczu Grodarza na Wyżynie Lubelskiej. Annales UMCS B 39:117–145

Michalczyk Z, Wilgat T (2008) Wody. In: Uziak S, Turski R (eds) Środowisko przyrodnicze Lubelszczyzny. Lubelskie Towarzystwo Naukowe, Lublin, pp 113–209

Nowocień E (1996) Dynamika rozwoju wąwozów drogowych na obszarach lessowych. Pamiętnik Puławski 107:101–111

Pożaryski W, Maruszczak H, Lindner L (1994) Chronostratygrafia osadów plejstoceńskich i rozwój doliny Wisły środkowej ze szczególnym uwzględnieniem przełomu przez wyżyny południowopolskie. Prace PIG 147:1–58

Rodzik J (2006) Wąwozy—naturalne, czy kulturowe elementy krajobrazu? Probl Ekol Krajobr 18:377–382

Rodzik J, Furtak T, Zgłobicki W (2009) The impact of snowmelt and heavy rainfall runoff on erosion rates in a gully system, Lublin Upland, Poland. Earth Surf Proc Landf 34:1938–1950
Rodzik J, Krukowski M, Zagórski P (2014) The mechanism and stages of development of a field road gully in relation to changes in the surrounding land relief. Annales UMCS B 69(1):135–147
Rodzik J, Gardziel Z (2004) Układy krajobrazowe wąwozów kazimierskich. In: Kucharczyk M (ed) Współczesne problemy ochrony krajobrazu. Zarząd Zespołu Lubelskich Parków Krajobrazowych, Lublin, pp 85–92
Rodzik J, Ciupa T, Janicki G, Kociuba W, Tyc A, Zgłobicki W (2021) Współczesne przemiany rzeźby Wyżyn Polskich. In: Kostrzewski A, Krzemień K, Migoń P, Starkel L, Winowski M, Zwoliński Z (eds) Współczesne przemiany rzeźby Polski, wyd. II. Bogucki Wyd Nauk, Poznań, pp 333–414
Rodzik J (2010) Influence of land use on gully system development (case study: Kolonia Celejów loess catchment). In: Warowna J, Schmitt A (eds) Human impact on upland landscapes of the Lublin region. Kartpol, Lublin, pp 195–209
Rodzoś J, Gawrysiak L, Bochra A (2006) Rzeźba terenu a organizacja przestrzeni miejskiej Lublina. Annales UMCS B 60:35–45
Schmitt A, Rodzik J, Zgłobicki W, Russok C, Dotterweich M, Bork H-R (2006) Time and scale of gully erosion in the Jedliczny Dol gully system, south-east Poland. Catena 68:124–132
Strzemski M (1961) Przemiany środowiska geograficznego Polski, jako tła przyrodniczego rozwoju rolnictwa na ziemiach polskich (od połowy trzeciego tysiąclecia p.n.e. do naszych czasów). Kwart Hist Kult Mater 9:331–357
Superson J, Rodzik J, Reder J (2014) Natural and human influence on loess gully catchment evolution: A case study from Lublin Upland, E Poland. Geomorphology 212:28–40
Superson J, Rodzik J, Reder J, Zgłobicki W, Klimowicz Z, Franczak Ł (2016) Phases of alluvial fan development in a loess area, Lublin Upland, E Poland. Quat Intern 399:31–45
Warowna J, Zgłobicki W, Kołodyńska-Gawrysiak R, Gajek G, Gawrysiak L, Telecka M (2016) Geotourist values of loess geoheritage within the planned Geopark Małopolska Vistula River Gap, E Poland. Quat Intern 399:46–57
Willaume J (1964) Wieś lubelska przed uwłaszczeniem. Wydawnictwo Lubelskie, Lublin, p 101
Zgłobicki W, Baran-Zgłobicka B (2012) Impact of loess relief on land use mosaic in SE Poland. Catena 96:76–82
Zgłobicki W, Baran-Zgłobicka B (2013) Geomorphological heritage as a tourist attraction. A case study in Lubelskie Province, SE Poland. Geoheritage 5:137–149
Zgłobicki W, Rodzik J (2007) Heavy metals in the slope deposits of loess areas of the Lublin Upland (E Poland). Catena 71:84–95
Zgłobicki W, Baran-Zgłobicka B, Gawrysiak L, Telecka M (2015a) The impact of permanent gullies on present-day land use and agriculture in loess areas (E. Poland). Catena 126:28–36
Zgłobicki W, Gawrysiak L, Kołodyńska-Gawrysiak R (2015) Gully erosion as a natural hazard: the educational role of geotourism. Nat Haz 79(Suppl 1):159–181
Zgłobicki W, Poesen J, Cohen M, Del Monte M, García-Ruiz JM, Ionita I, Niacsu L, Machová Z, Martín-Duque JF, Nadal-Romero E, Pica A, Rey R, Solé-Benet A, Stankoviansky M, Stolz C, Torri D, Soms J, Vergari F (2019) The potential of permanent gullies in Europe as geomorphosites. Geoheritage 11:217–239
Zgłobicki W, Baran-Zgłobicka B (2011) Gullies as an indicator of human impact on loess landscape (Case study: North Western part of Lublin Upland, Poland). Z Geomorph NF 55, SI 1:119–137
Zgłobicki W (1998) Wąwozy drogowe północno-zachodniej części Płaskowyżu Nałęczowskiego. In: Dobrowolski R (ed) Główne kierunki badań geomorfologicznych w Polsce. Stan aktualny i perspektywy. IV Zjazd Geomorfologów Polskich, 3, Przewodnik wycieczkowy. Wydawnictwo UMCS, Lublin, pp 175–179
Zientara B, Mączak A, Ihnatowicz I, Landau Z (1988) Dzieje gospodarcze Polski do roku 1939. Wiedza Powszechna, Warszawa, p 552

Jan Rodzik is geomorphologist, Associate Professor at the Institute of Earth and Environmental Sciences, Maria Curie-Skłodowska University in Lublin. His main research interests are the impact of human activity on the transformation of relief in loess areas, especially on the development of gullies. In addition, he participates in research on the environmental conditions of former settlements and functioning of the polar environment (Spitsbergen). His main research area is the Lublin Upland in eastern Poland and neighbouring regions. He is a member of the Association of Polish Geomorphologists and the Polish Geographical Society, where in 2005–2012 he was the President of the Lublin Branch.

Grzegorz Gajek is geographer, Assistant Professor at the Institute of Earth and Environmental Sciences, Maria Curie-Skłodowska University in Lublin. His research concerns geoheritage, geotourism, and geoparks in Poland and Europe. He also conducted research in southern Spitsbergen (Arctic).

Wojciech Zgłobicki is Professor of Geography in the Institute of Earth of Environmental Sciences, Maria Curie-Sklodowska University in Lublin. He serves as Associate Dean for Student Affairs (from 2016) and Head of the Department of Geology, Soil Science and Geoinformation (from 2011). His research concerns gully and soil erosion, geochemistry of young sediments, geoheritage, and geotourism. He conducts research primarily in loess areas of southeastern Poland. He also participated in review studies covering Europe and other continents.

Chalk Karst in the Lublin Region

27

Radosław Dobrowolski

Abstract

The Lublin chalkland, located in eastern Poland, is one of only a few compact areas in the world where, on the basement of soft carbonate rocks of the Upper Cretaceous age, the so-called chalk karst has developed. This rare type of karst relief is characterized by specific underground drainage, lacking caves but with a rich assemblage of surface forms such as dolines, uvalas, karst valleys, poplavas and large karst depressions of polje type. The onset of the evolution of karst relief in this area, embracing the Lublin-Volhynia Uplands and their northern foreland, dates back to the beginnings of the Cenozoic, which is documented by the fillings of numerous paleodolines and/or paleouvalas. The most visible features in the modern morphology are small dolines, the density of which in the central part of chalkland, within the Chełm Hills, reaches 100 forms per 1 km^2. It is also one of the few areas in the world where the presence of a chalky substrate determined the development of karst lakes in periglacial conditions.

Keywords

Chalk karst · Landscape evolution · Fossil karst · Lakes

27.1 Introduction—What is Chalk Karst?

Chalk karst is a special type of karst phenomena, the nature of which is determined by lithological features of carbonate rocks represented by soft pelagic facies of the Late Cretaceous sea, mainly by chalk (Rodet 1991). Dissolution plays an important morphogenetic role in chalk karst areas, considerably more important than in non-karstic landscapes, but not always and not in all conditions is the dominant process. The low resistance of chalk to destructive factors results in the absence of underground drainage system typical for classical karst, and in consequence, in a very rare occurrence of typical underground karst forms. Caves have been found and described mainly in Normandy and southern England so far (Maurice et al. 2020; Nehme 2020). Significant morphological differences characterize also the surface forms of chalk karst (Maruszczak 1966). Taking into account that a significant part of karst forms is included in the surface drainage system, it can be concluded that karst phenomena that develop in chalk outcrops correspond to the concept of fluviokarst (sensu Roglic 1964).

Chalk karst terrains (chalklands) occur in Europe in several compact areas (Fig. 27.1), including Ireland (Drew 2015), SE England (Maurice 2009), Normandy in France (Rodet 1991), Belgium and Dutch borderland (Willems et al. 2007), eastern Poland (Maruszczak 1966), southern Belarus, north-western Ukraine (Lomaev 1979; Dobrowolski et al. 2000) and western Russia (Maksimovich 1969).

The Lublin chalkland is a traditional name for a vast area in eastern Poland where the Late Cretaceous soft porous carbonate rocks are commonly exposed on the land surface. It includes fragments of the Lublin Upland and the Volhynia Upland as well as their northern forelands—the Volhynia Polesie and Western Polesie Lowlands (Fig. 27.1). It is also the western periphery of the Lublin-Volhynia chalk karst, which is the largest compact area of chalk karst occurrence in Europe.

27.2 Geological and Hydrogeological Background

27.2.1 Lithology and Structure

Karst phenomena in the Lublin chalkland develop in the Late Cretaceous carbonate rocks (mainly Upper Maastrichtian, secondarily Campanian) filling a large,

R. Dobrowolski (✉)
Institute of Earth and Environmental Sciences,
Maria Curie-Skłodowska University, Al. Kraśnicka 2d, 20-718
Lublin, Poland
e-mail: radoslaw.dobrowolski@mail.umcs.pl

P. Migoń and K. Jancewicz (eds.), *Landscapes and Landforms of Poland*,
World Geomorphological Landscapes, https://doi.org/10.1007/978-3-031-45762-3_27

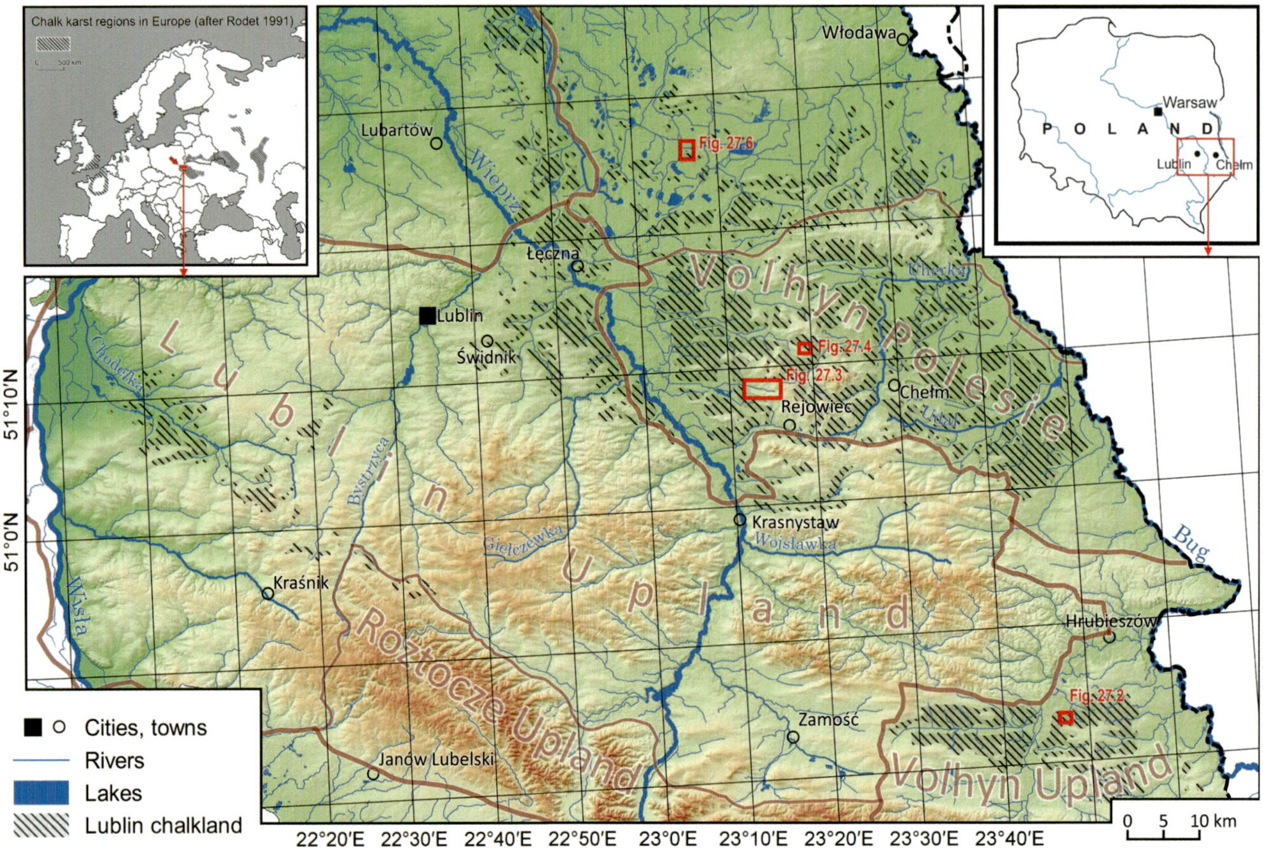

Fig. 27.1 Location of the Lublin chalkland in relation to chalk karst regions in Europe (after Rodet 1991) and macroregions of eastern Poland (after Solon et al. 2018)

asymmetric structure of the Lublin Trough (Hakenberg and Świdrowska 2001). Their thickness varies from 200 m in the eastern part to 1000 m along the trough axis. Lithologically, the Late Cretaceous massif is composed of alternating series of carbonate-siliceous rocks (gaizes, opokas) and carbonate rocks (mainly chalk and marly limestones). Boundaries of lithostratigraphic units are usually barely visible and difficult to recognize (Wyrwicka 1980). Considering the most favourable conditions for the development of karst phenomena, chalk is of key importance.

27.2.1.1 Chalk

The Lublin chalk is a white micritic carbonate rock characterized by bedding and earthy fracture. Its main component is calcium carbonate, the content of which in the Lublin chalkland often exceeds 95% and does not generally fall below 90% (Olszewski 1997). Due to the uniform chemical composition, its specific weight varies in a relatively narrow range, from 2.1 to 2.7 G/cm^3 (Rybicki and Rybicki 1973). Chalk is a very soft rock, susceptible to mechanical deformation. Its dry compressive strength varies from 22 kg/cm^3 in the near-surface zone to about 300 kg/cm^3 at a depth of several dozen metres. These indices decrease even five times when chalk is saturated with water. However, the analysis of total relative deformations in vertical profile indicates that with increasing depth this rock transforms from brittle into plastic, which significantly influences its fissuring degree. Weathered chalk, depending on the morphological situation, occurs in the form of fine breccia (brecciated chalk) or clayey residuum. Silty/clayey residuum is an effect of physical and chemical weathering of chalk. On the other hand, the brecciated chalk bedrock develops mainly by ice segregation associated with freezing and micro-cracking of moist, frost-susceptible chalk (Murton et al. 2006).

27.2.1.2 Structural Fissuring

The Late Cretaceous massif is strongly fissured. It is dissected by a dense net of fissures: (1) weathering fissures occurring in the top part to a depth of 2–5 m, and (2) tectonic cracks—faults and joints occurring as two orthogonal joint systems (NW–SE/NE–SW and N–S/W–E) (Dobrowolski 1995). Opening of fissures is vertically and horizontally varied depending on the present-day pattern of tectonic stress field in the rock massif. The nature of rock massif fissuring, orientation and vertical extent of the fissure system directly affect the course and rate of karstification. These features are closely related to the tectonics of the rock massif.

27.2.2 Hydrogeology

Late Cretaceous deposits are the main environment of groundwater occurrence in the study area. The fissure and fissure-pore waters are related to the system of tectonic and weathering fissures of the carbonate rock massif. The groundwater resources clearly vary depending on the local lithological diversity of rocks and the degree of massif fissuring. All lithological types of the Late Cretaceous massif are characterized by high total porosity, about 30–50% (Olszewski 1997), and low water permeability. The determined coefficients of permeability of rock monoliths range from 10^{-7} to 10^{-11} m/s (Rybicki and Rybicki 1973). Hydrogeological features of the massif are determined by the nature of its tectonics, especially the course and hydrogeological properties of faults constituting the main drainage zones (Woźnicka 2001). The highest values of transmissivity, reaching even 1000 m^2/h, are recorded near known active fault zones, especially of tensional nature (Krajewski and Motyka 1999). Thus, the direction of underground runoff following the local structural pattern of the area. The depth of strongly saturated zone in the Late Cretaceous massif does not exceed 100 m in chalk and 150 m in opokas (Krajewski and Motyka 1999). Below this boundary, due to the influence of geostatic pressure within the rock massif, a virtually impermeable horizon is formed.

The Quaternary aquifer, usually limited to river valleys and vast basin-like depressions, occurs locally as perched aquifer in interfluve areas. Perched waters act as a buffer in the recharge process of local groundwater circulation systems. Tills and clayey residua of chalk are the main confining layers.

27.3 Chalk Karst Relief

27.3.1 Surface Landforms

The present-day karst relief of the Lublin chalkland represents covered and reproduced karst types. Due to the lithologic properties of the chalk, forms typical of bare karst are absent. Arrays of covered and reproduced karst forms often co-occur in a mosaic arrangement, and their morphometric features do not show significant differences. Taking into account the absence of an underground drainage system typical of classic karst, and significant differences in relief, the discussed phenomena can be treated as incomplete karst (merokarst). The record of surface forms of the Lublin karst includes dolines, uvalas, karst valleys, poplavas (which are larger topographic basins than uvalas) and large karst depressions (polje type).

27.3.1.1 Dolines

Dolines (called *werteby* in the Lublin chalkland) are the most common, elementary forms of chalk karst (Fig. 27.2). These are small depressions, circular or oval in plan, with horizontal dimensions of several dozen, less often one hundred several dozen metres. Their depth usually does not exceed 5 m. The inclination of their slopes is relatively low, usually about 5°, rarely exceeding 10°. It depends on lithology (inclination is higher in marly limestones than in chalk), hypsometric situation (it is lower in less elevated areas than in the higher ones) and land use (it is the highest in forest areas). The most common dolines have poorly visible floors, in the upper part of the slopes gently passing into ridges separating the adjacent forms. Solution dolines are dominant, and collapse dolines are absent (Maruszczak 1966). The cryogenic-subsidence processes, related to the periglacial environment of Late Weichselian, were an important factor initiating the development of karst dolines (Dobrowolski 2006).

The dolines of the Lublin karst are young forms and were formed during the Late Weichselian and Holocene (Dobrowolski et al. 2010). Some of them are active in our times, as indicated by water absorption in their floors after intensive rainfalls and sudden thaws (Harasimiuk and Henkiel 1974). Most of them occur within the lowest morphological flattenings. In such locations, their density may exceed 50 forms/km^2 in the study area.

27.3.1.2 Uvalas

Uvalas are complex forms of surface karst formed by merging of dolines. They originate due to degradation of ridges separating elementary forms, being simultaneously filled by aggradation of organic mineral deposits in the floors of dolines. The resulting complex forms have uneven floors, and their irregular shapes reflect the size and spatial arrangement of primary dolines. Such uvalas should therefore be identified as forms of the compound doline type (sensu Jennings 1971). The longer axes of the uvalas reach 1 km; most often, however, their length ranges from 200 to 500 m. Like the dolines, they occupy low-lying topographic surfaces, mainly on the flattenings at the foot of slopes (Maruszczak 1966). The process of merging of dolines into uvalas, through the growth of peat in their floors, developed especially in the Atlantic phase of the Holocene.

27.3.1.3 Karst Valleys

Karst valleys are forms the status of which is similar or higher than that of uvalas. Their length can reach several kilometres (Fig. 27.3). They are formed by widening and then merging of adjacent dolines into one row. Hence, the groundplan of karst valleys has a very irregular shape, reflecting the number and size of the elementary forms they are composed of. Development process is similar as

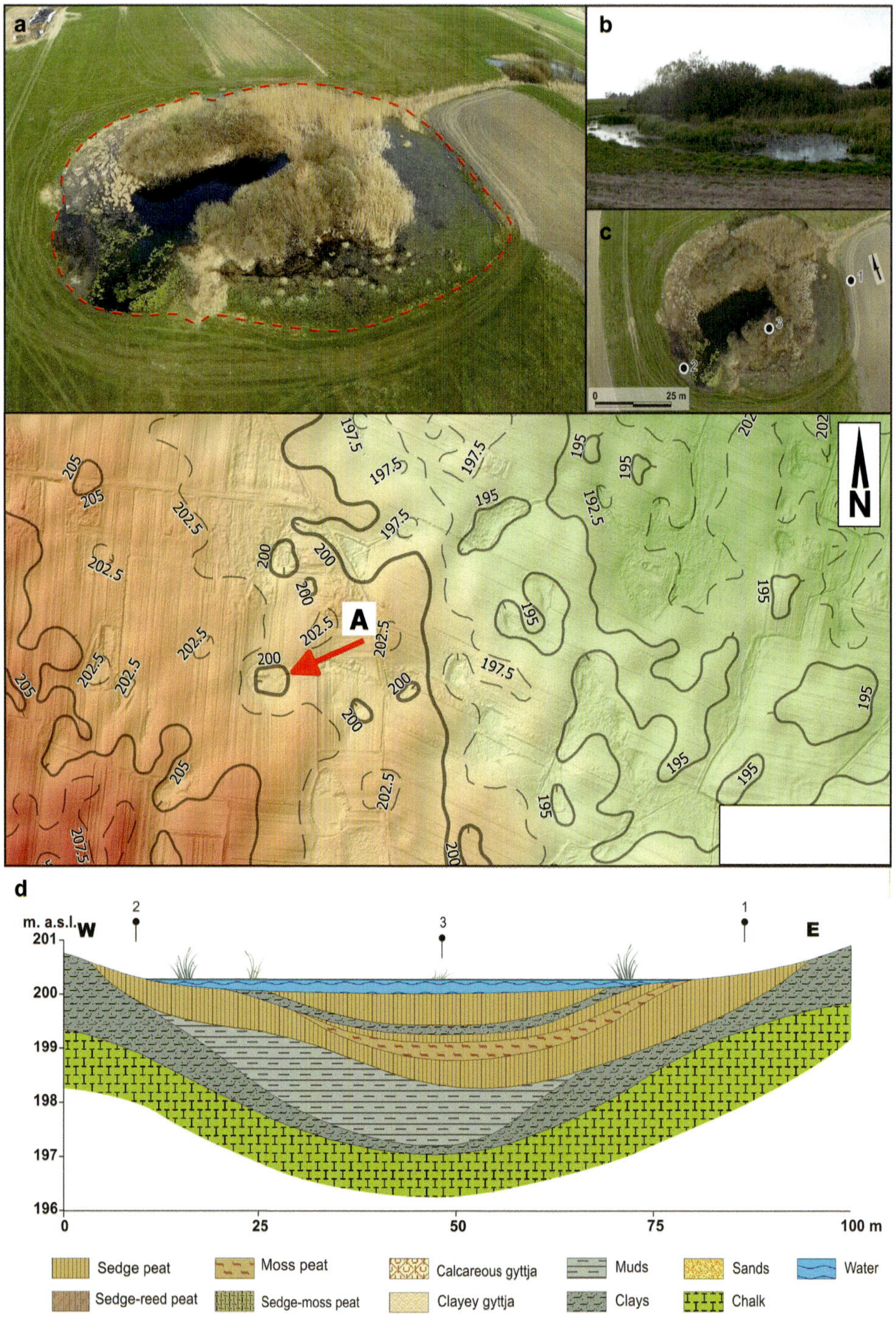

Fig. 27.2 Doline in Łysa Góra (after Łojek 2019; location as in Fig. 27.1). **a** Aerial photo. **b** Ground photo. **c** Location of drillings shown in geological cross-section. **d** Topographic situation in the vicinity of the site. **e** Geological cross-section through the doline (photographs by J. Łojek)

for the uvalas but in the final stage fluvial processes play the main role. That is why they are included in the present-day system of surface drainage. Their linear courses indicate direct relation with tectonic structures of the Late Cretaceous rock massif. They developed along the single meso-faults in the Meso-Cenozoic complex and open joints (Dobrowolski 1998). Geological cross-sections reveal complex nature of these forms; in the longitudinal profile, there is visible a dense net of fossil dolines, often reproduced in the Pleistocene cover, which are relatively deep compared with their horizontal dimensions. Both the thickness and the lithology of the deposits filling the individual doline segments are clearly diverse, which indicates different local conditions of sedimentation in adjacent elementary forms. According to Harasimiuk (1975), merging of dolines into karst valleys due to the aggradation of organic matter, as in the case of uvalas, most probably took place during the Atlantic phase of the Holocene.

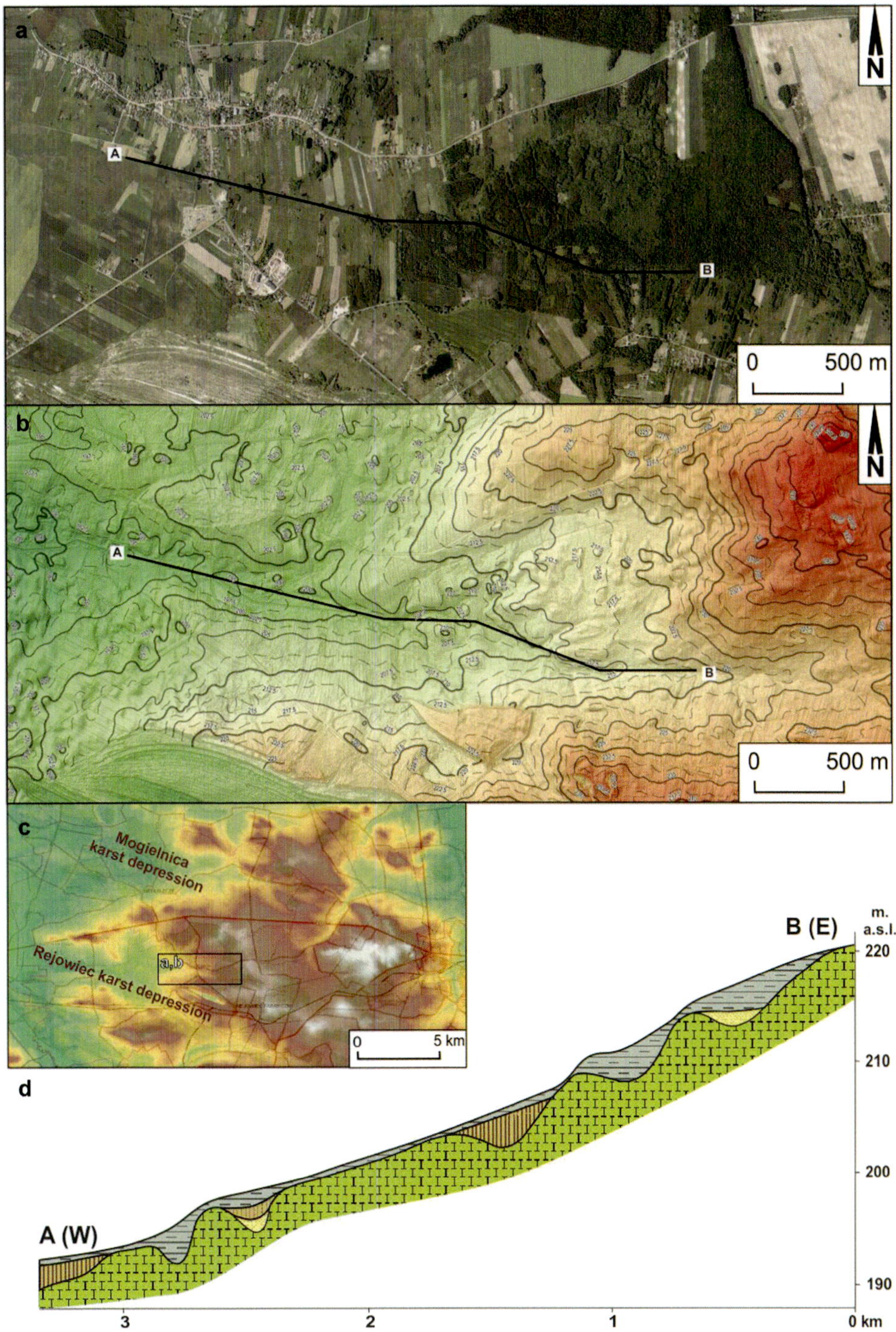

Fig. 27.3 Karst valley in Pawłów (after Harasimiuk 1975; as in Fig. 27.1). **a** Aerial photo with the location of geological cross-section (A–B) (source: geoportal.gov.pl). **b** Topographic situation in the vicinity of the site. **c** Situation of the site against the background of distribution of large karst depressions of the polje type in the central part of the Lublin chalkland. **d** Geological cross-section through the karst valley (description of symbols as in Fig. 27.2)

27.3.1.4 Poplavas

Poplavas are vast, basin-shaped depressions of oval or elliptic shape and flat floor (Fig. 27.4). These forms are larger and deeper than the uvalas, and usually much more regular in plan. Their horizontal dimensions range mainly from 1 to 3 km and the depths exceed 10 m. They develop as separate forms or by merging of smaller uvalas; in both cases, they are clearly controlled by tectonics. In the case of the Lublin chalk karst, it seems to be justified to distinguish between poplavas and uvalas. From the typological point of view, the poplavas should be treated as forms superior to the uvalas, and inferior to even larger karst depressions (poljes). However, while poljes are related to faults of regional importance, the poplavas developed rather along secondary faults (meso-faults) accompanying major dislocations. The results of morphometric analysis also provide reasons for their separate treatment (Dobrowolski 1998). Poplavas are usually branches of poljes; they tend to be oriented en échelon or perpendicularly to the poljes and are hydrographically related to them (open poplavas). However, they also

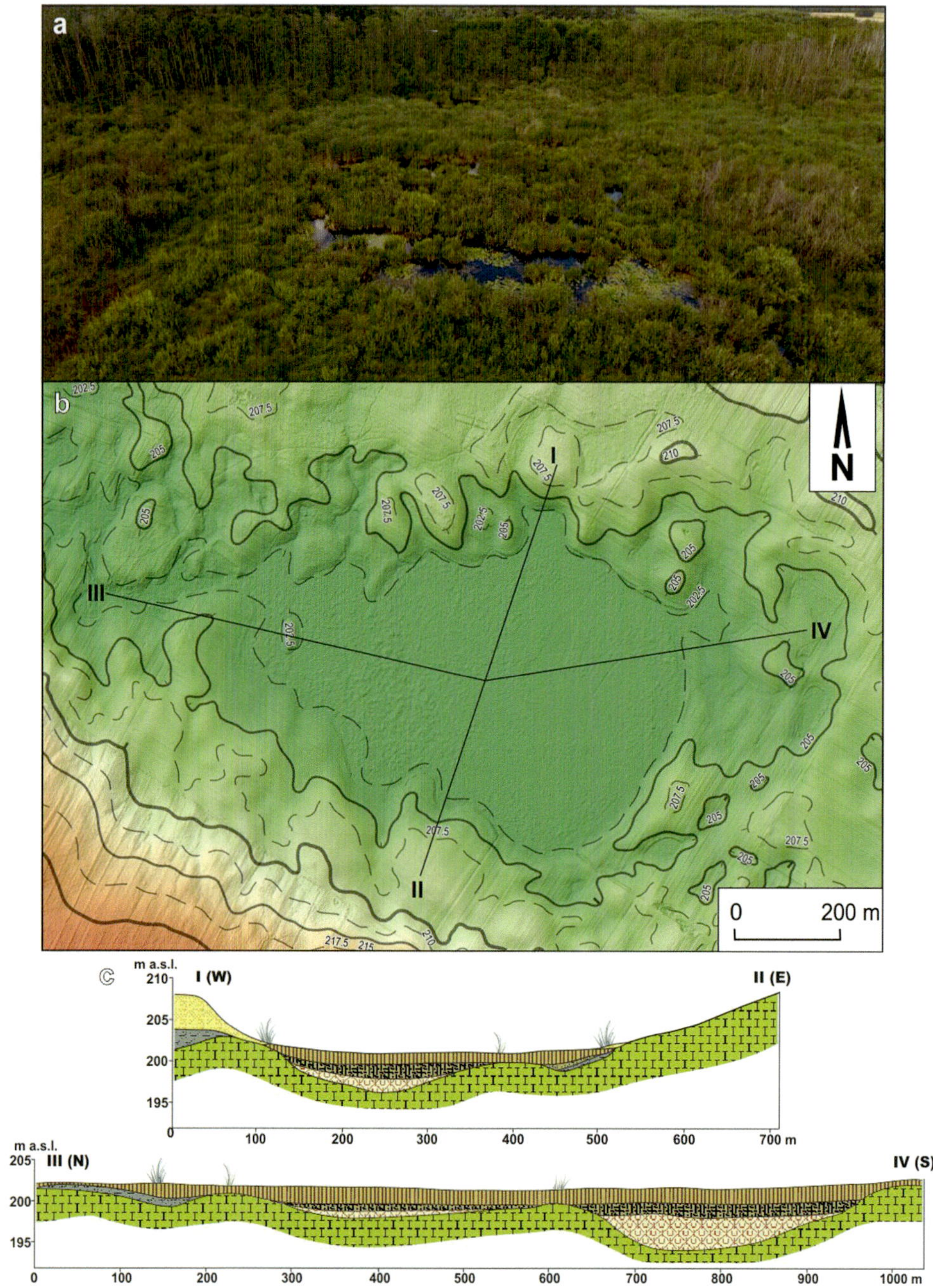

Fig. 27.4 Poplava in Nowosiółki (after Harasimiuk 1975; location as in Fig. 27.1). **a** Aerial photo (photograph by J. Łojek). **b** Topographic situation in the vicinity of the site with the location of geological cross-sections (I–II and III–IV). **c** Geological cross-sections through the poplava (description of symbols as in Fig. 27.2)

occur as separate forms in the vicinity of poljes, and show a similar orientation (closed poplavas).

27.3.1.5 Large Karst Depressions—Poljes

Large karst depressions in the Lublin chalkland are undoubtedly polygenic forms. However, according to Maruszczak (1966), karstification played an important role in their development. These forms are evidently determined by tectonics. It is because of their direct relation with faults that they were compared with the Dinaric poljes (Maruszczak 1966). At the same time, clear differences were found, especially with respect to hydrographic conditions. Originally probably partially without drainage, nowadays they are included in the surface drainage system of the area. Due to their size, they are treated as the chalk karst forms of the highest status. The lower, concave and strongly karstified parts of polje slopes are gently inclined and gradually turn into the flat surface of accumulation floor. In the Lublin chalkland, several such forms are recognized (Figs. 27.1, 27.3). They are surrounded by isolated remnant hills, clearly visible in the relief, rising up to 100 m above the floors of the poljes. It is supposed that their formation started in the Neogene, and an important stage of their development took place in the Middle and Late Pliocene (Maruszczak 1966).

27.3.2 Fossil Karst Forms

The term 'fossil karst' or 'paleokarst' is related to karst forms/features formed in the geologic past and representing different stages of karstification, which are now fully filled and thereby preserved beneath younger deposits. The set of paleokarst forms in the top part of the Late Cretaceous complex is accessible for direct observations, and contains pits, which are very different in respect of their origin, age and type. However, they have one common feature indicating a considerable role of karstification in morphogenesis, i.e. the occurrence of clayey karst cortex, which is the lining of each form. Despite a great morphological variability of paleokarst pits, and the occurrence of intermediate forms, the following five main morphotypes were distinguished in the study area (Fig. 27.5): (a) solution joints and solution interstratal fissures; (b) pockets; (c) pipes and kettles; (d) paleodolines; and (e) paleouvalas.

27.3.2.1 Karstified Fissures

Joints and interstratal fissures, i.e. fractures occurring in the top part of a carbonate rock massif, were enlarged by karst processes and filled with residual clay. The origin of these fractures is both tectonic (extension joints) and sedimentary (interstratal fissures). The opening of karst fissures is from several millimetres to several or rarely a dozen or so centimetres, and their vertical extent is from 0.5 to 10 m. Considering the initial nature of these forms (occurrence in a narrow zone near fissure and lack of allochthonous fills), they should be recognized as rather young, formed not earlier than in the latter part of the Late Weichselian. The development of karst processes was possible only after permafrost degradation and re-establishment of groundwater vertical circulation.

27.3.2.2 Pockets

Pockets are microstructures a dozen or so centimetres in diameter, and their depth rarely exceeds 0.5 m. These conical or funnel-shaped pits become narrower downwards and turn into karstified joints. All these forms are very similar but their fills are distinctly lithogenetically differentiated. Generally, two different types of fill can be distinguished. The first one consists of sandy and sandy-silty deposits of massive structure, locally with poorly defined stratification. Their textural features indicate that they were deposited quasi-syngenetically in relation to the development of forms. This deposition occurred in periglacial environment, when enlarged joints and weathering fissures were gravitationally filled. The second type of fill consists of glacial diamictic deposits of massive structure. Their lithology indicates cohesive, subaerial flows of supraglacial sediments (Dobrowolski 2006).

27.3.2.3 Pipes and Kettles

Pipes and kettles are vertical forms, round or elliptic in plan, 0.5–1.0 m in diameter, developed on open joints. Their depth varies from 1.0 m (kettles) to 10 m and more (pipes). These forms are usually filled with glaciofluvial sands and sands with gravels, the structure of which follows the pit shape. Massive structure occurs only in the top parts of fills, and grains are mat. Luminescence ages of the deposits indicate that these forms were filled during the Saalian glaciation (Dobrowolski and Fedorowicz 2007). The nature of main lithofacies points to rather fast filling of glaciofluvial erosional channels in subglacial environment (Dobrowolski 2006). The features of the uppermost deposits indicate that they were exposed to secondary aeolian reworking in periglacial conditions.

27.3.2.4 Paleodolines

Dolines are bowl-shaped or funnel-shaped pits 1.0–20.0 m deep, with diameters of between several and 150–200 m. Their occurrence is usually connected with the strongly fissured zones of the rock massif; sometimes they developed on tectonic wedges or in the proximity of faults.

Different ages and lithostratigraphic variability of the fills enable us to distinguish at least three generations of paleodolines. The oldest are dated for the Eocene; they developed in subaerial conditions of warm and humid climate. These forms are filled with the Early Oligocene series of quartz-glauconitic sands with silt interlayers. Horizontal or ripple lamination of these deposits indicates that they were sedimented from traction currents and suspension in small lakes (Dobrowolski 2006). Younger generation is represented by Middle Pliocene dolines filled with sandy-silty and sandy-clayey sediments. They were deposited from traction currents and suspension in lakes, which periodically dried out (Harasimiuk et al. 1981). Heavy mineral analysis indicates that sandy material was redeposited. It was delivered to the lakes from the neighbouring Sarmatian marine deposits (Harasimiuk 1975). The third, most numerous generation consists of dolines filled with Saalian glacigenic deposits. These deposits are usually massive diamictons of varied (rarely uniform) texture, i.e. clayey-silty, sandy-silty, sandy or gravely-sandy ones. Such depositional sequences should be related to single or repeated mass flows of supraglacial material, from cohesive to cohesionless ones.

Paleodolines, which developed through lateral amalgamation of adjacent small pockets, form a separate group. Consequently, their fills are typical of pockets: washed sandy-silty deposits of massive structure or with poorly defined stratification. The floors of dolines formed in this way are indistinct and uneven.

Fig. 27.5 Fossil landforms of the Lublin chalk karst. **a–g** Chełm chalk quarry: **a** Karstified fissures. **b, c** Pipes. **d–f** Kettles and pockets. **g, h** Paleodolines. **i–k** Rejowiec marly limestones quarry: **i** Paleouvala with Pliocene fills. **j, k** Examples of glaciotectonic transformation of Neogene paleouvala (arrows show the directions of truncations inside fills) (photographs by R. Dobrowolski)

27.3.2.5 Paleouvalas

Uvalas are complex paleokarst depressions. Such a depression consists of several adjacent dolines, which are incorporated into one morphological system. Uvalas are large; they are usually 5–25 m deep, and their length depends on the number of incorporated dolines so it varies from about one hundred to several hundred metres.

Paleouvalas underwent complex evolution. They are composed mostly of the Pliocene dolines, rarely Eocene ones. Lithofacial features of their fills and deposition environment resemble those of the two oldest doline generations described above. The Neogene deposits, the top part of which was truncated by erosion, are overlain by a glacial (Elsterian and/or Saalian) series of the Pleistocene. Both the Neogene and Quaternary deposits were strongly deformed by glaciotectonics.

27.4 Stages of Karst Evolution

The state of preservation of paleokarst forms and deposits representing successive development stages is fragmentary. However, based on them, it is possible to distinguish several main phases of activity of karst processes in the morphogenetic evolution of the Lublin chalkland. They are related to changes of climatic conditions and/or phases of tectonic activity in the area.

- Stage I—Paleogene (~65–23 Ma). It is related to the intensive uplift of the area during the Laramian phase of the Alpine cycle. Enhanced chemical weathering of carbonate rocks occurred in subaerial conditions in warm, humid climate. This stage is recorded at many sites by: (1) decalcified opokas, (2) regoliths, (3) paleodolines with the Early Oligocene fill. Vast karst paleodepressions of polje type, visible in the present-day morphology, were also formed in the Early Paleogene.
- Stage II—Early Neogene (~24–5 Ma). It should be related to the uplift of the Savian and Styrian phases of Alpine cycle. This stage is poorly documented. It is possible to deduce about the activity of karst processes mainly from the very varied relief of the sub-Sarmatian (Cretaceous and Oligocene) surface, with numerous paleodolines.
- Stage III—Late Neogene–Early Pleistocene—from the Middle Pliocene (~4–3.5 Ma) to the anaglacial phase of the Elsterian/Sanian_2 glaciation—MIS 12 (~480–470 ka). It is commonly considered the main development stage of karst processes in the study area, the evidence of which survived. It is well recorded at numerous sites of paleodolines and paleopipes, mainly filled with the Sarmatian deposits (Harasimiuk 1975). Significant dimensions of the identified dolines, exceeding 20 m in places, indicate that karst processes were efficient and long-lasting. Several phases of intensive development of karst forms occurred probably within this stage. They are related to the periods of climate moistening in the Middle and Late Pliocene and Eopleistocene, as well as to significant lowering of groundwater table and intensive erosion in river valleys as an effect of neotectonic uplift of the Attic, Rhodanian and Valachian phases.
- Stage IV—Pleistocene—with the alternating phases: glacial (mainly Elsterian/Sanian_2—MIS 12—480–440 ka and Saalian_I/Odranian MIS 8—300–250 ka), periglacial (Elsterian/Sanian_2—MIS 12, Saalian_II/Wartanian—MIS 6—190–115 ka and Weichselian/Vistulian—MIS 2–5d—115–11.7 ka) and interglacial (mainly Holsteinian/Mazovian—MIS 11—440–370 ka and Eemian—MIS 5e—130–115 ka). In cold periods, strong transformation of the pre-Pleistocene karst relief occurred and dolines and uvalas were glaciotectonically deformed (Dobrowolski and Terpiłowski 2006). Protokarst forms were formed in subglacial conditions, including the development of evorsion hollows, which were filled with glaciogenic deposits and changed by karst processes in the next stages (Dobrowolski 2006), and in periglacial conditions (karst transformation of primary cryoturbation structures—Dobrowolski 2006; Dobrowolski and Mroczek 2015). Warm interglacial periods favoured the development of karst processes sensu stricto.
- Stage V—Holocene—MIS 1 (from 11.7 ka)—the present-day development of karst processes occurring mainly in the warm and humid phases of the Holocene, especially in the Atlantic and Subatlantic phases (Harasimiuk 1975; Pazdur et al. 2001; Dobrowolski et al. 2010). It is mainly characterized by the evolution of dolines, which merge into uvalas or karst valleys (Dobrowolski 1998).

27.5 Lakes in the Lublin Chalkland–Karst versus Thermokarst Origin

The northern part of the Lublin chalkland (Volhynia Polesie and partially Western Polesie) is characterized by the occurrence of numerous lakes concentrated in a relatively small area, 200–300 km to the south of the zone of the maximum extent of the Weichselian ice sheet, which excludes their glacial origin. It is the only such area in Poland, and probably only one of the few in the world, where the presence of a chalky substrate determined the development of lakes in periglacial conditions. The origin of these lakes has been the subject of numerous discussions and polemics for many years. Generally, there are two separate concepts for the

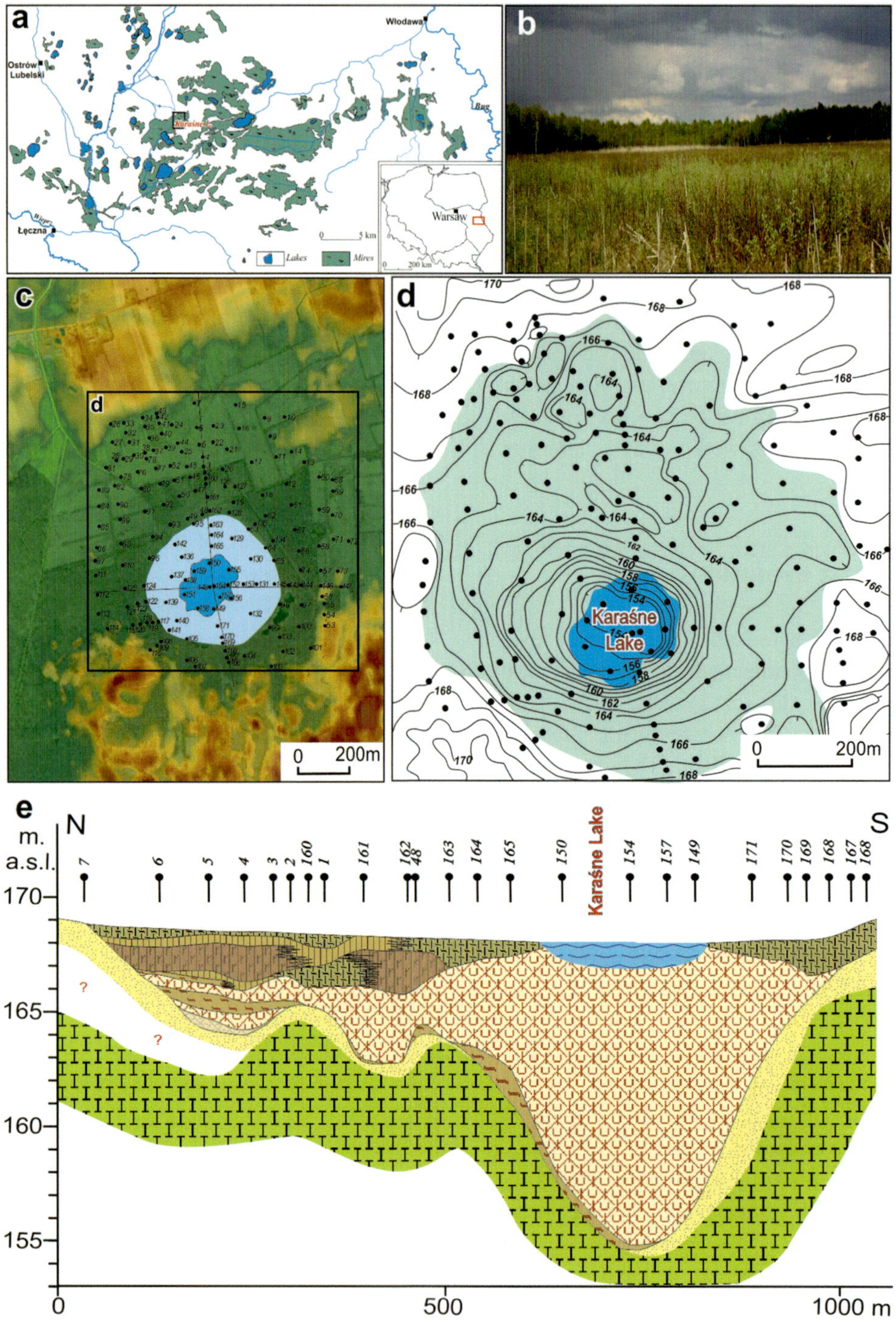

Fig. 27.6 Karaśne Lake (after Bałaga et al. 2002; location as in Fig. 27.1). **a** Map of lakes in the Lublin chalkland (Łęczna-Włodawa Lakland). **b** Ground photo (photograph by R. Dobrowolski). **c** Topographic situation in the vicinity of the lake with the location of geological drillings. **d** Paleomorphology of the lake with interpretation of primary (Late Weichselian) extent of the water body (blue area marks limit of the lacustrine deposits). **e** Geological cross-section through the lake (description of symbols as in Fig. 27.2)

development of these forms: karst (e.g. Wilgat 1954, 1994; Maruszczak 1966) and thermokarst (e.g. Wojtanowicz 1994). The arguments put forward by their followers, as a matter of fact often based on the same premises, suggest clear differences in morphogenetic mechanisms. However, there is no doubt about the main issues: (1) the sequence of events—the lakes were formed at the transition from the Late Weichselian to the Preboreal, and (2) spatial relationship between the lakes and the structure of the Late Cretaceous massif—location of the lake basins

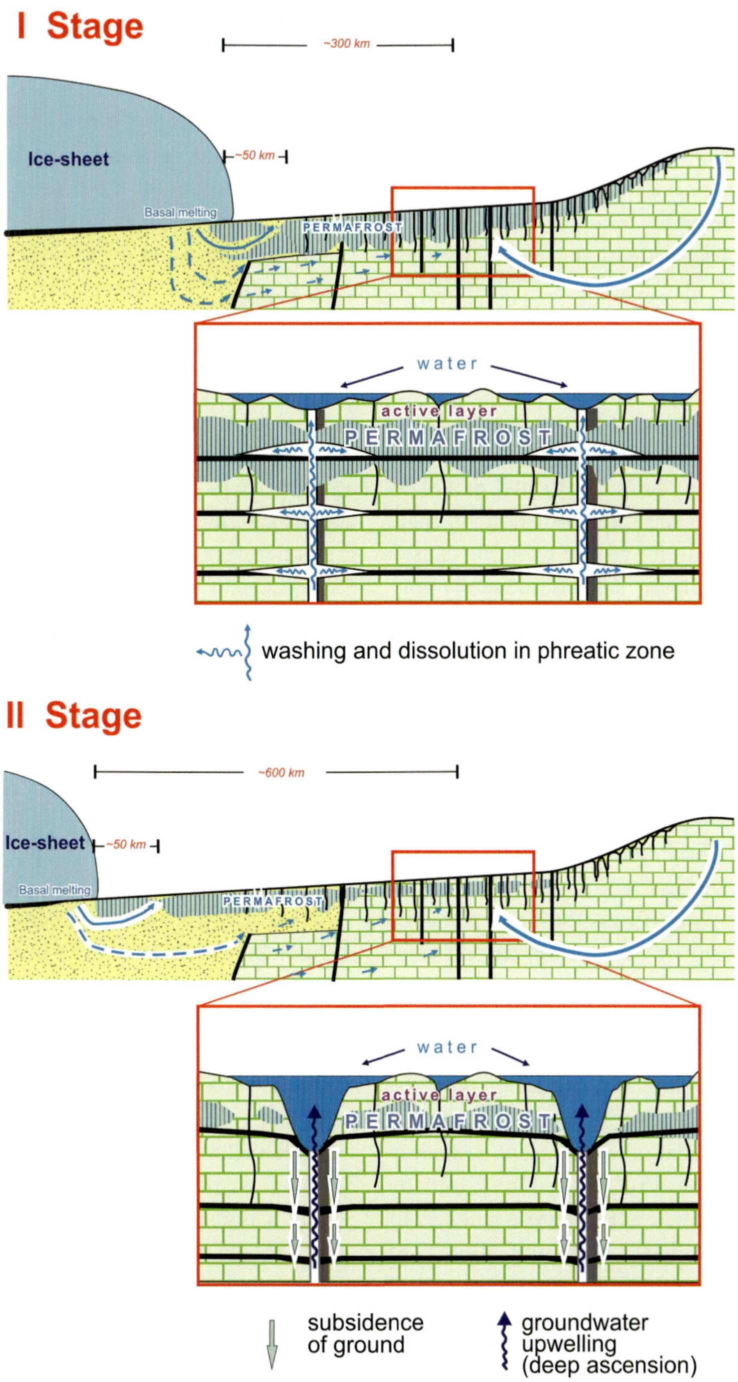

Fig. 27.7 Model of periglacial transformation of the Lublin chalkland (after Dobrowolski 2006). Stage I—Pleniglacial, with continuous permafrost in substratum. The supply of deeper aquifers was possible only (1) in the zone of basal melting of warm ice masses by the injections of subglacial water into permeable or/and strongly fissured substratum (shown on the model by blue arrows—continuous or dashed depending on the assumed flow intensity), (2) through taliks in discontinuous permafrost, and (3) by infiltration from the areas without permafrost (shown on the model by a thick blue arrow). The pressure of ice masses favoured sub-permafrost flow of water, as long as permafrost thickness and structural conditions permitted it (shown on the model by short blue arrows). Underground water migrated in the Cretaceous massif along fault zones, extension joints and interstratal fissures. They were widened then, both as a result of mechanical washing and chemical dissolution. The circulation of water in the near-surface zone was limited only to a thin active layer. Meltwater occurred on the surface in numerous but shallow depressions. **Stage II**—Late Glacial, with discontinuous permafrost in substratum. Milder climatic conditions in the kataglacial phase of glaciation contributed to the gradual degradation of permafrost, both in its upper part (increasing thickness of active layer) and in the bottom (geothermal energy flow). Permafrost degradation occurred easily along faults where the outflow of confined groundwater on the surface was accompanied by ground subsidence

was determined by the course of tectonic faults activated in the Pleistocene. Moreover, both concepts in their general assumptions are not mutually exclusive but require only corrections and supplements concerning the mechanisms of Late Weichselian morphogenesis in this area. In this context, clear identification of the mechanisms of Lake Basin formation in the Lublin chalkland becomes a key issue in considering the periglacial transformation of the chalk karst (Figs. 27.6, 27.7).

Climate changes at the end of the Weichselian and at the turn of the Holocene were very dynamic. During the period of ca. 4 ka (between 16 and 12 ka), the average annual air temperature in Central Europe increased by about 12–15 °C (Velichko et al. 2002). The climate changed especially rapidly in its eastern, more continental part. This had a significant influence on the transformation of chalk karst in the Lublin chalkland. Long-lasting warming caused gradual degradation of permafrost, and consequently fundamental reorganization of groundwater flow resulting in the complete unblocking of its vertical circulation. These phenomena occurred in stages (Fig. 27.7), which corresponded to climatic fluctuations, and selectively—depending on lithology (differences in the heat capacity of rocks) and tectonic conditions (differences in the fissuring of the Late Cretaceous massif). Complete permafrost degradation occurred probably not until the Preboreal but the rate of its decay was considerable; complete melting took place during only 1000–1500 years (*vide* Gerasimov and Velichko 1982). Such a rate of permafrost degradation caused catastrophic landscape changes, mainly large-scale ground subsidence.

In the context of the origin of lakes in the Lublin chalkland, the model of karst development assuming only the dissolution of chalk seems unlikely. The average rate of the present-day chemical denudation of the Lublin chalk is estimated by Świeca (2000) at 25–45 t km^{-2} a^{-1}, and by Maruszczak (2001) at 55 t km^{-2} a^{-1}. These values converted into the thickness of denuded layer give 10–18 and 22 mm ka^{-1}, respectively. Nevertheless, a significant part of the lakes have their basins fully or partially formed within the Late Cretaceous rocks. The model of thermokarst development, proposed by Wojtanowicz (1994), is not applicable to them due to the lack of right conditions for the development of segregated ground ice (=lack of thick mineral deposits). On the other hand, the development of intrusive ice in karstified bedding planes (=sill ice) and in fault fissures (=vein ice) is fully possible, and in extreme cases it results in the formation of pingo open systems, which may form on the consolidated substratum when the average annual air temperature is between –4 and –1 °C (Huijzer and Isarin 1997). Therefore, pingos may have formed in the study area during permafrost aggradation in the cold climatic phases of the Late Weichselian (Dobrowolski 2021). This process was probably synchronous with the glacioisostatic activation of some tectonic faults, which largely explains the position of lake basins on the slopes of the chalk hills (= horsts). Degradation of intrusive ice in the warm interphases of the Late Weichselian and in the Preboreal caused the unloading of considerable hydraulic gradients and resulted in efficient ascending groundwater supply accompanied by gradual subsidence of ground. The concentration of numerous depressions of various sizes and depths and similar age in a small area may have been the result of changing location and dynamics of groundwater upwelling within one dislocation zone. Similar cases were found and described in the present-day permafrost areas (Gurney 1998; Wu et al. 2005). Since the activation of deep ascension was preceded by the phase or phases of washing and dissolution in phreatic zone, the supply of ascending groundwater resulted in an increased carbonate deposition in the bottoms of the newly formed lakes (Dobrowolski 2006).

27.6 Conclusions

Several conclusions arise from the presented review of the Lublin chalk karst.

- The specificity of karst in the Lublin chalkland results obviously from the lithology of carbonate rocks, represented by chalk and its lithofacial varieties. Such carbonate soft rocks are very susceptible both to chemical dissolving and to mechanical deformation, especially as a result of frost activity. This influenced the morphogenesis in the subsequent stages of the development, especially during cold events. In consequence the Lublin chalkland is characterized by the co-occurrence of: (1) chalk karst complex, distinguished from the classic one by a different register of surface forms and limited occurrence or lack of underground forms, and (2) complex of gentle denudational landforms (subdued shapes of convex forms, slightly inclined slopes, levelled foothills).
- Groundwater flow in chalk is mainly fissure-/fracture-controlled and thus runs most of all along faults and joints. Thus, tectonics—distribution of faults and joints in the Meso-Cenozoic rock complex—has played an important role in the morphogenetic evolution of the area. The structural plan of the Late Cretaceous massif determined hydrogeological conditions and, consequently, the dynamics of karst processes, including the spatial arrangement of forms. Large karst depressions (polje type) and poplavas show direct dependence on faults, whereas smaller forms (dolines and uvalas) only show indirect relationships, mainly with the joint systems in the epikarst zone.

- Despite variable degree of preservation of karst forms and sediments, it is possible to reconstruct the main stages of chalk karst morphogenesis, from the Early Paleogene to the Holocene. Glacial processes related to the last ice sheet in this area—the Saalian/Odranian ice sheet—and periglacial processes related to the Weichselian cold period played a decisive role in the radical transformation of relief, determining its contemporary architecture. Their role was not limited to the destruction of old karst forms, but rather to their rebuilding and reorganization of groundwater circulation. They also played an important role in creating protokarstic forms (pipes, kettles, pockets).
- The most dynamic changes in the karst system of the Lublin chalkland took place during all glacial–interglacial transitions and were a consequence of long-term climate warming. Degradation of intrusive ice, being the primary infilling of interstratal fractures within the karstic bedrock, resulted in the discharge of considerable hydraulic pressures by the reactivation of groundwater circulation. The upwelling supply of groundwater induced intensive weathering and dissolution of carbonate bedrock and progressive ground subsidence and lake formation. The most readable karst forms in the contemporary morphology of this area are those from the Late Weichselian/ Holocene transition.

References

Bałaga K, Dobrowolski R, Rodzik J (2002) Lithostratigrafic record of the development of Lake Karaśne (Lublin Polesie, E Poland) in Late Glacial and Holocene. Limnol Rev 2:5–14

Dobrowolski R (1995) Drobne struktury tektoniczne w skałach górnokredowych wschodniej części Wyżyny Lubelskiej a dyslokacje podłoża platformy wschodnioeuropejskiej w kenozoiku. Ann Soc Geol Polon 65:79–91

Dobrowolski R (1998) Strukturalne uwarunkowania rozwoju współczesnej rzeźby krasowej na międzyrzeczu środkowego Wieprza i Bugu. UMCS Press, Lublin, p 88

Dobrowolski R (2006) Glacjalna i peryglacjalna transformacja rzeźby krasowej północnego przedpola wyżyn lubelsko-wołyńskich (Polska SE, Ukraina NW). UMCS Press, Lublin, p 184

Dobrowolski R (2021) "Ghost structures" in alkaline fen microrelief as a consequence of Late Glacial periglacial activity in chalklands—a case study from the Chełm Hills (East Poland). Ann Soc Geol Polon 91:347–362

Dobrowolski R, Bogucki A, Zaleski I (2000) Morfometryczne kryteria oceny związku powierzchniowych form krasowych z tektoniką na przykładzie podniesienia Lubomla (Ukraina NW). Przegl Geol 48:634–638

Dobrowolski R, Fedorowicz S (2007) Glacial and periglacial transformation of palaeokarst in the Lublin-Volhynia region (SE Poland, NW Ukraine) on the base of TL dating. Geochronometria 27:41–46

Dobrowolski R, Mroczek P (2015) Clay cortex in epikarst forms as an indicator of age and morphogenesis—case studies from Lublin-Volhynia chalkland (East Poland, West Ukraine). Geomorphology 247:66–75

Dobrowolski R, Terpiłowski S (2006) Influence of palaeokarst morphology on the formation of ice-pushed ridges: a case study from Rejowiec, eastern Poland. Boreas 35:213–221

Dobrowolski R, Ziółek M, Bałaga K, Melke J, Bogucki A (2010) Radiocarbon age and geochemistry of the infillings of small closed depressions from Western Polesie (Poland SE, Ukraine NW). Geochronometria 36:39–46

Drew DP (2015) Hydrogeology of lowland karst in Ireland. Q J Eng Geol Hydrogeol 41:61–72

Gerasimov IP, Velichko AA (eds) (1982) Paleogeography of Europe during the last one hundred thousand years: Atlas-monography, p 156. Nauka, Moscow (In Russian, with English summary)

Gurney SD (1998) Aspects of the genesis and geomorphology of pingos: perennial permafrost mouds. Prog Phys Geogr 22:307–324

Hakenberg M, Świdrowska J (2001) Cretaceous basin evolution in the Lublin area along the Teisseyre-Tornquist Zone (SE Poland). Ann Soc Geol Polon 71:1–20

Harasimiuk M (1975) Rozwój rzeźby Pagórów Chełmskich w trzeciorzędzie i czwartorzędzie. Prace Geogr IG PAN 115:94

Harasimiuk M, Henkiel A (1974) Zjawiska Krasowe We Wschodniej Polsce. Speleologia 8:37–49

Harasimiuk M, Jezierski W, Król T, Szwajgier W (1981) Tertiary and Quaternary fossil karst in the Cretaceous rocks of chalk type in eastern Poland. In: Proc. Europ. Regional Conf. Speleology, Sofia, Bulgaria, 22–28 Sept 1980, vol 2, pp 45–50

Huijzer AS, Isarin RFB (1997) The reconstruction of past climates using multi-proxy evidence; an example of the Weichselian Pleniglacial in northwest and central Europe. Q Sci Rev 16:513–533

Jennings JN (1971) Karst. MIT Press, London, p 252

Krajewski S, Motyka J (1999) Model sieci hydraulicznej w skałach węglanowych w Polsce. Biul PIG 388:115–138

Lomaev AA (1979) Karst geology of Volhyn-Podolia. Naukova Dumka, Kiev, p 132 (In Russian, with English summary)

Łojek J (2019) Warunki rozwoju krasowych form wertebowych w wyżynnej części międzyrzecza Wisły i Bugu w świetle analiz przestrzennych i rekonstrukcji paleośrodowiskowych. Maria Curie-Sklodowska University in Lublin, Faculty of Earth Sciences and Spatial Management, Lublin, PhD thesis, manuscript, p 185

Maksimovich GA (1969) Chalk karst. In: Fundamentals of karst studies, vol 2. Perm, pp 383–421 (In Russian, with English summary)

Maruszczak H (1966) Zjawiska krasowe w skałach górnokredowych międzyrzecza Wisły i Bugu. Przegl Geogr 38:339–370

Maruszczak H (2001) Rozwój rzeźby wschodniej części wyżyn metakarpackich w okresie posarmackim. Przegl Geogr 73:253–280

Maurice L (2009) Investigations of rapid groundwater flow and karst in the Chalk. Unpublished PhD Thesis, University College, London, p 453

Maurice L, Bunting S. Farrant A, Mathewson E (2020) BGS Karst Report Series: C4. Karst in the Chalk of the Chilterns and the Berkshire and Marlborough Downs. British Geological Survey Open Report, OR/20/044, p 48

Murton JB, Peterson R, Ozouf J-C (2006) Bedrock fracture by ice segregation in cold regions. Science 314:1127–1129

Nehme C (2020) Stratigraphical influence on chalk cave development in Upper Normandy, France: implications for chalk hydrogeology. Intern J Speleol 49:187–208

Olszewski A (1997) Charakterystyka przestrzeni porowej kredy piszącej rejonu Chełma. Przegl Geol 45:1282–1285

Pazdur A, Dobrowolski R, Mohanti M, Piotrowska N, Das S (2001) Radiocarbon time scale for deposition of the Holocene calcareous tufaceous sediments from Poland and India (Orissa). Geochronometria 21:85–96

Rodet J (1991) Les karsts de la craie—étude comparative. Thèse du Doctorat d'Etat. Université de Paris IV-Sorbonne, p 562

Roglic J (1964) Karst valleys in the Dinaric Karst. Erdkunde 18:113–116

Rybicki S, Rybicki J (1973) Własności inżyniersko-geologiczne utworów kredowych okolic Chełma Lubelskiego. Kwart Geol 17:301–309

Solon J, Borzyszkowski J, Bidłasik M, Richling A, Badora K, Balon J, Brzezińska-Wójcik T, Chabudziński Ł, Dobrowolski R, Grzegorczyk I, Jodłowski M, Kistowski M, Kot R, Krąż P, Lechnio J, Macias A, Majchrowska A, Malinowska E, Migoń P, Myga-Piątek U, Nita J, Papińska E, Rodzik J, Strzyż M, Terpiłowski S, Ziaja W (2018) Physico-geographical mesoregions of Poland: verification and adjustment of boundaries on the basis of contemporary spatial data. Geogr Polon 91:143–170

Świeca A (2000) Spatial variability of chemical denudation in the upland part of the Vistula and Bug interfluve. Kras Speleol 10:84–104

Velichko AA, Catto N, Drenova AN, Klimanov VA, Kremenetski KV, Nechaev VP (2002) Climate changes in East Europe and Siberia at the Late Glacial-Holocene transition. Quat Intern 91:75–99

Wilgat T (1954) Jeziora Łęczyńsko-Włodawskie. Ann UMCS B8:37–122

Wilgat T (1994) Spory wokół jezior Łęczyńsko-Włodawskich. In: Przewodnik Ogólnopolskiego Zjazdu PTG, Lublin, pp 122–129

Willems L, Rodet J, Fournier M, Laignel B, Dusar M, Lagrou D, Pouclet A, Massei N, Dussart-Baptista L, Compère P, Ek C (2007) Polyphase karst system in Cretaceous chalk and calcarenite of the Belgian-Dutch border. Z Geomorph NF 51:361–376

Wojtanowicz J (1994) O termokrasowej genezie Jezior Łęczyńsko-Włodawskich. Ann UMCS B49:1–18

Woźnicka M (2001) Przestrzenna zmienność warunków hydrogeologicznych północno-zachodniej części Wyżyny Lubelskiej na tle liniowych elementów strukturalnych. Przegl Geol 49:643–648

Wu Z, Barosh PJ, Hu D, Wu Z, Zhao X, Peisheng Y, Qisheng L, Zhou C (2005) Migrating pingos in the permafrost region of the Tibetan Plateau, China and their hazard along the Golmud-Lhasa railway. Eng Geol 79:267–287

Wyrwicka K (1980) Stratygrafia, facje i tektonika mastrychtu zachodniej części Wyżyny Lubelskiej. Kwart Geol 24:805–819

Radosław Dobrowolski is Professor of Earth Sciences, a physical geographer, geomorphologist, graduate of geographical studies at the Maria Curie-Skłodowska University in Lublin. Former dean of the Faculty of Earth Sciences and Spatial Management; since 1987 he has been working at the Department of Geomorphology and Palaeogeography (currently the head of this unit). An expert in structural geomorphology, paleoenvironmental reconstructions, geoarchaeology and environmental protection. Scientific interests focus on the morphogenesis and structural evolution of the Polish carbonate uplands, the European chalk karst, reconstruction of Late Glacial and Holocene natural environments of Central Europe, wetland evolution, environmental conditions of prehistoric settlement and the mediaeval Polish-Ruthenian borderland, as well as geodiversity and geoprotection.

Mużaków Rampart (Muskau Arch)—The Legacy of Glacial Processes and Mining in the UNESCO Global Geopark

28

Jacek Koźma and Piotr Migoń

Abstract

The Mużaków Rampart in south-western Poland is a unique glacitectonic structure of Middle Pleistocene (Elsterian) age that retained its morphological expression despite several hundred thousand years of denudation. The eastern arm of the resultant push moraine is located in Poland, whereas the remaining part is in Germany. Glacitectonic processes affected the substratum down to the depth of nearly 300 m, squeezing up the Miocene deposits, including the coal-bearing ones, and bringing them to the surface in the form of stacked thrust sheets. Subsequent weathering of outcrops of brown coal seams produced distinctive topographic hollows known as gieser. 150-year-long underground and open-cast coal and clay mining resulted in considerable anthropogenic transformation, including the origin of numerous lakes in and above former mining areas. Geoheritage and geodiversity values of the region are promoted in the transboundary Muskau Arch UNESCO Global Geopark, approved in 2011. The area includes the romantic landscape park in Bad Muskau/Łęknica, inscribed as a World Heritage property in 2004, whose design was considerably influenced by the landform pattern.

Keywords

Push moraine · Glacitectonism · Subsidence · Anthropic landforms · Geopark

J. Koźma (✉)
Stowarzyszenie "Geopark Łuk Mużakowa", Tadeusza Kościuszki 18, 68-208 Łęknica, Poland
e-mail: jkozma@wp.pl

P. Migoń
Institute of Geography and Regional Development, University of Wrocław, Pl. Uniwersytecki 1, 50-137 Wrocław, Poland
e-mail: piotr.migon@uwr.edu.pl

28.1 Introduction

The Mużaków Rampart, a relatively small area adjoining the valley of the Nysa Łużycka (Lusatian Neisse) River on the Polish-German border, clearly stands out in the lowland landscape of western Poland. It represents a distinctive relief, with a series of ridges of variable orientation, which nevertheless combine into a large-scale arcuate elevation, if seen in a ground plan. The arch-like outline becomes even more evident if the near-border area of Germany is considered, where the arch continues. Even though it is not particularly conspicuous from the ground perspective due to sheer spatial scale and modest relative height of c. 100 m, digital terrain models clearly show its exceptional size and distinctiveness (Fig. 28.1). It is also globally recognized for its geoheritage and geodiversity, having the status of a UNESCO Global Geopark, which is a transboundary German-Polish geopark, admitted to the Global Geopark Network in 2011.

In the broader geographical context, the Mużakow Rampart is part of a chain of end-moraine hills that stretch across central Europe, from the Netherlands in the west through Germany to Poland (Liedtke 1981). They indicate the maximum extent of the younger phase of the Saalian glaciation (Warthe), but cannot be simply considered as depositional end moraines from that period. The origin of the rampart as a large-scale landform is explained in general by dynamic glacial processes associated with an advancing Scandinavian Ice Sheet during one of earlier cold phases of the Middle Pleistocene. During the Warthe phase the ridge acted as a topographic barrier that the ice sheet was unable to cross and hence, sediments and landforms from different inland glaciations are superimposed. However, the contemporary topography also reflects subsequent re-modelling by non-glacial processes, mainly hillslope and fluvial. In addition, easy access to various mineral resources (brown coal, clay) within glacitectonic structures of the rampart allowed for an extensive

P. Migoń and K. Jancewicz (eds.), *Landscapes and Landforms of Poland*,
World Geomorphological Landscapes, https://doi.org/10.1007/978-3-031-45762-3_28

development of mining, which in turn resulted in a multitude of anthropogenic landform changes. Of particular interest are numerous lakes in abandoned open-cast pits, which together form an anthropogenic lakeland, with more than 340 water bodies. It differs significantly from other lakelands in the lowlands of Poland within the limits of the last inland glaciation, which are exclusively of natural origin. Thus, morphology of the area is a palimpsest of landforms of different ages and origins, with a significant contribution of human impact.

This chapter examines geomorphology of the Polish part of the Mużaków Rampart, presenting its geological foundations, diversity of landforms, and landscape evolution, considering both natural processes and anthropogenic impacts. It also presents the Muskau Arch UNESCO Global Geopark, outlining its history and geodiversity, with particular focus on the star attraction of the area—the geotouristic path established in the former brown coal mine 'Babina'. The area of interest covers about 150 km^2 and is entirely contained within the Lubuskie province, in its south-western corner. In Germany, the rampart continues to the west in the Federal Land of Saxony and then to the north, in Brandenburg, making a large horseshoe shape, which is nearly 30 km long from north to south and up to 25 km wide (Fig. 28.1).

28.2 Pre-Quaternary Geological Context

The Mużaków Rampart is located in an area where the solid basement rocks occur at depth of more than 220 m below the ground and have no direct influence on topography. The overlying Cenozoic deposits were laid down during a long interval from the Late Oligocene to the Pliocene and comprise several fine-grained clastic formations, deposited in alternating fluvial, lacustrine and coastal environments. These in turn consist of well-sorted quartz sands, silts and clay. Between various clastic series, there are brown coal seams of Early to Middle Miocene age. The thickest of them have from 8 to 15 m, with the average of 10 m. Some of these were continuous over large areas, whereas others occurred more locally, probably deposited in oxbow lakes and minor swamps. The youngest brown coal seam is dated for the terminal Middle Miocene, but sedimentation continued and the thick Poznań series of Late Miocene to Pliocene age sealed all coal-bearing strata fairly deep under the surface. As the area was generally far from active faults in the south, responsible for the uplift of the Sudetes mountain range along the present-day Polish-Czech border and the origin of tectonic grabens in their marginal parts, the Oligocene and Neogene strata were more or less horizontal and little deformed. Most of the Pliocene and early Quaternary was a period of predominant denudation, and the top part of the sedimentary succession must have been eroded away, but the overall low-lying topography at that time certainly limited the efficacy of erosion. However, this relatively simple, layer-cake geological structure was considerably disturbed during the Quaternary.

28.3 Quaternary Glacial History and the Origin of the Moraine Arch

28.3.1 Gross Relief Features

Before deciphering processes responsible for the present-day shape of the Mużaków Rampart and deformation of the Neogene deposits, it is useful to present the main topographic features of the area. The dominant landform is the rampart itself, which in the Polish part has north-to-south extension, turning slightly westward in its southernmost section, in the vicinity of the town of Łęknica. Its most distinctive section is between Łęknica and the village of Trzebiel, where the rampart is also the highest, reaching 182 m asl in the most elevated spot, next to the village of Nowe Czaple. The width of the rampart is up to 5 km, and more or less constant along its strike. In the south-west the rampart is cut by the water gap of the Nysa Łużycka River, whose floor is at 110–100 m asl, but the valley is a younger landscape element (see Sect. 28.3.4). The eastern limit of the arch is not clear-cut and the drop of elevation towards the adjacent rolling plain is gradual. The plain is incised by the Skroda River, which has cut a 10–20 m deep valley. Towards the west the topographic surface descends to the flat valley floor of the Nysa Łużycka River and is finally truncated by a distinct terrace riser, which is locally as high as 20 m. The elevation drop towards both the east and the west is thus 30–40 m (Fig. 28.2). To the north of Trzebiel, the rampart form is less evident, topography becomes more subdued, and the altitudes decrease to c 150 m asl. To the north of the village of Tuplice, the rampart essentially disappears, giving way to a plain punctuated by dunes.

The morphology of the rampart surface shows some diversity, even if anthropogenic landforms are disregarded. Following the principal SSW–NNE trend of the rampart, multiple second-order ridges occur parallel to one another (Fig. 28.3). They are typically 5–10 m high, locally up to 20 m, with spacing ranging from 200 m to 400–500 m from crest to crest. The intervening depressions are of two kinds. Some are broad and flat-floored, locally widening into less regular basins. They were naturally drained towards the Nysa Łużycka River in the west, but today many are occupied by artificial lakes that filled former open-cast mining areas. Others are straight, but very narrow, and are discussed in more detail in Sect. 28.3.3.

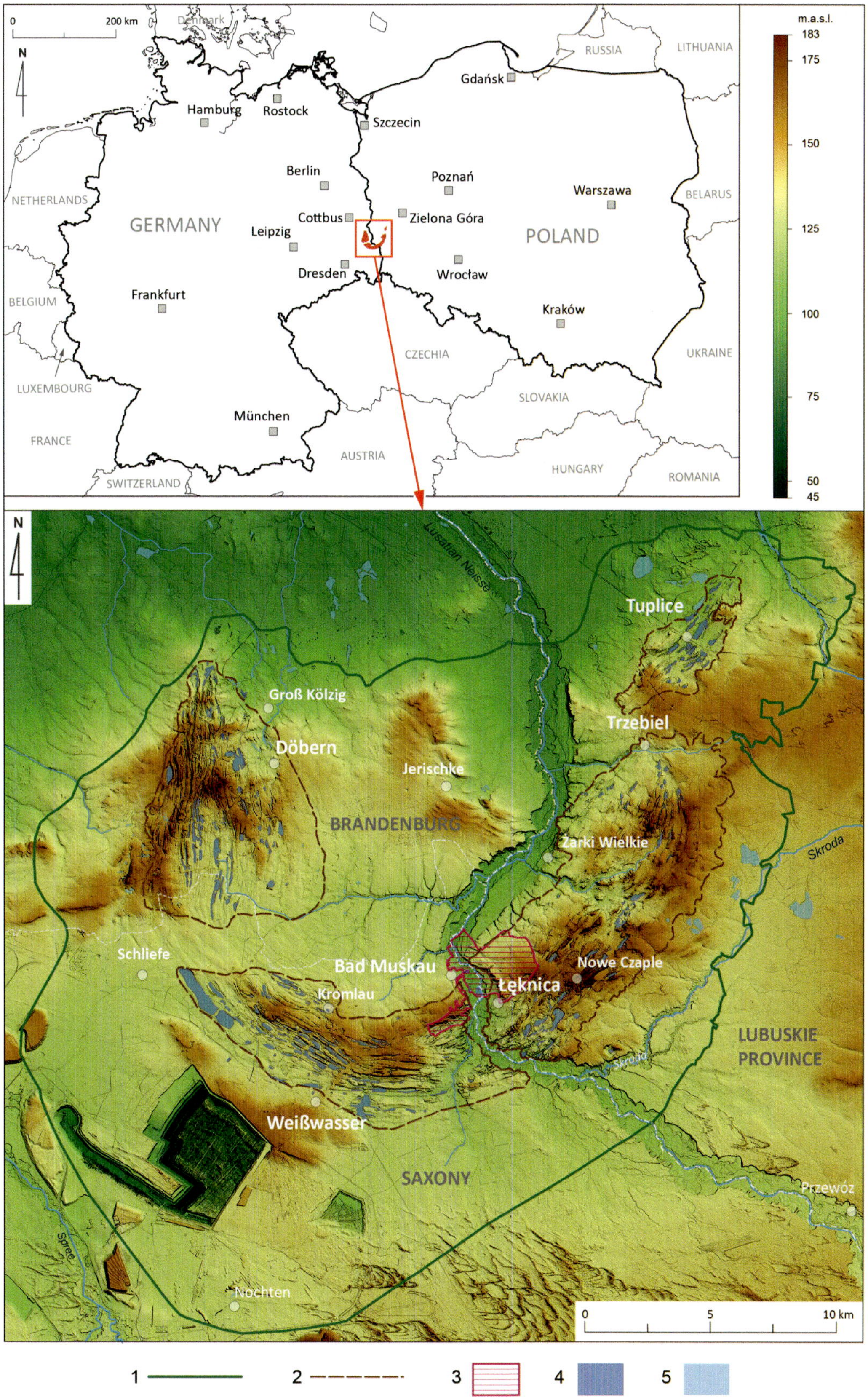

Fig. 28.1 Characteristic arcuate course of the push moraine of the Mużaków Rampart (Muskau Arch), hillshade model of relief built from LiDAR data, source of spatial data geoportal.pl, Lausitzer Braunkohle (Laubag). 1—boundary of the Polish-German UNESCO Global Geopark Muskau Arch, 2—boundary of a push moraine area, 3—Muskauer Park, 4—anthropogenic post-mining lakes, 5—natural lakes and fish ponds

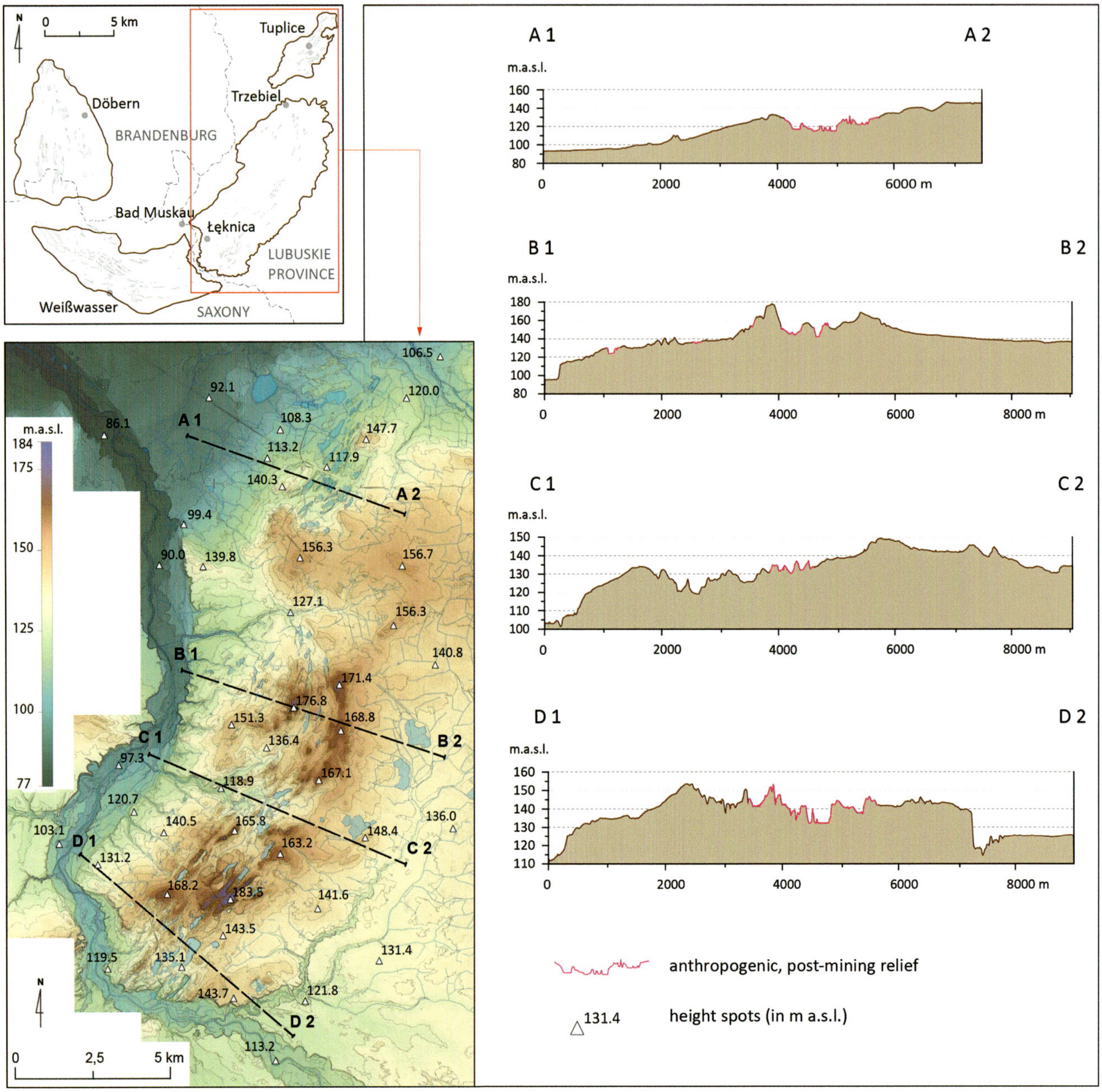

Fig. 28.2 Hypsometry and topographic profiles across the moraine of the Muskau Arch (Polish part)

28.3.2 Deciphering Glacial Processes

The arcuate shape of the Mużaków Rampart (Muskau Arch) was interpreted as resulting from glacial processes already in the beginning of the twentieth century (Keilhack 1907, 1909; Jentsch and Berg 1913). This large-scale topographic feature was but a part of a long line of terrain elevations extending from the Elbe river in the west to the Prosna river in the east, more than 400 km in total and reaching nearly 300 m asl in the highest place. Known as Fläming in the west, Lusatian Wall (Germ. *Lausitzer Grenzwall*) in the middle, and Silesian Ridge [Germ. *Schlesicher Landrücken*, Polish *Wał Trzebnicki* (=Trzebnica Rampart)] in the east, this belt of hilly terrain was an accepted southern marginal zone of the younger Saalian ice sheet, one that predated the Eemian Interglacial. Its initial interpretation as a simple end moraine did not stand the test of time, though, especially if confronted with the evidence from brown coal mines scattered along the hilly belt, especially common in the Mużaków Rampart and adjacent areas. The presence of brown coal seams cropping out at the surface, along with their variable dips and widespread contortions, indicates

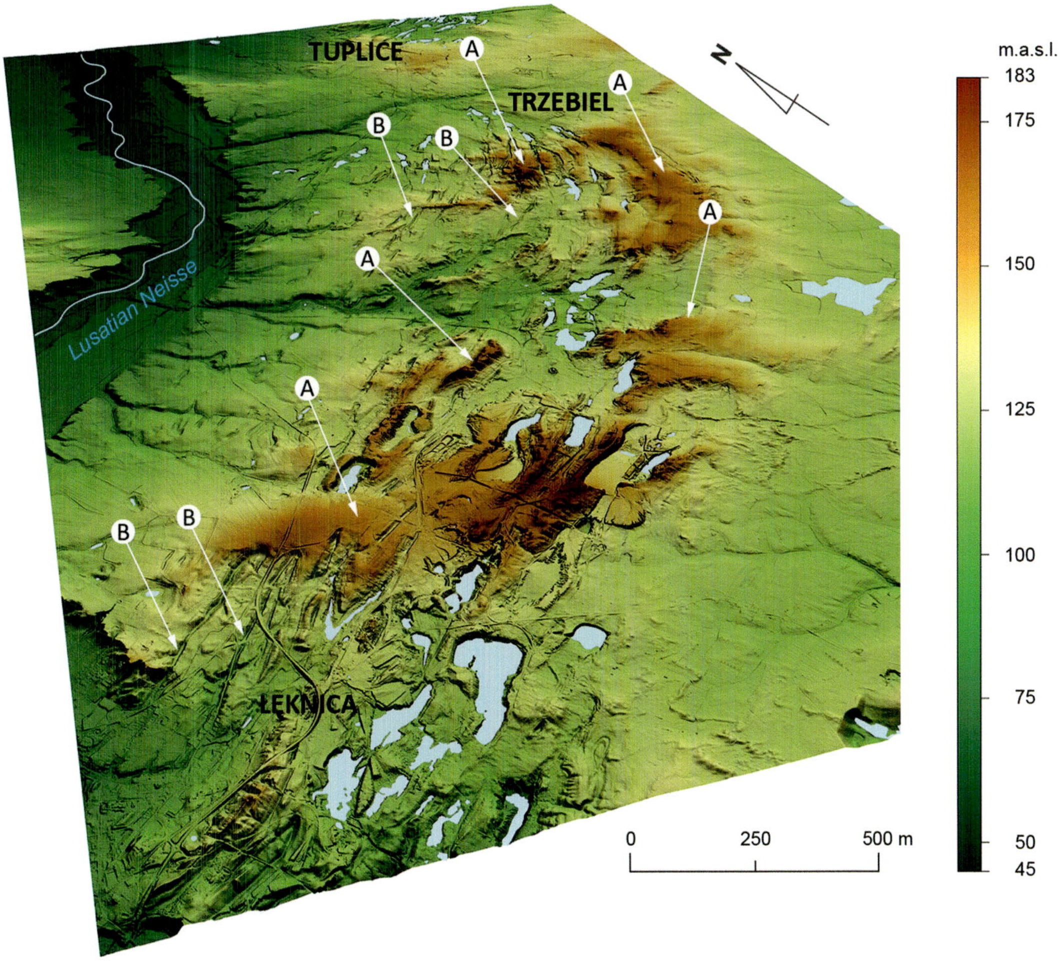

Fig. 28.3 Morphology of the Polish part of the Muskau Arch, visible on the gradient shaded digital terrain model (source of spatial data: geoportal.gov.pl). Elevations of accumulation moraines (**A**) are arranged diagonally across giesers—natural depressions arising from weathering of coal seams (**B**)

that significant deformation of the sedimentary succession had to take place to bring the Miocene brown coal to daylight. Moreover, the deformation had to affect large areas in a repetitive manner, to account for the occurrence of parallel lines of brown coal outcrops, continuous over distances of many kilometres. The only possible cause of this distortion of strata, reaching down to the early Miocene series, was glacitectonism. Thus, an end-moraine interpretation was replaced by one that emphasized deformation under the loading imposed by the advancing ice sheet and the Mużaków Rampart was re-assessed as a large-scale push moraine complex. However, it also became clear that the Warthe-phase ice sheet was not sufficiently thick and dynamic to cause deformation of the substratum reaching more than 150 m deep. Consequently, it is now accepted that the glacitectonically-produced elevations were already in existence during the Warthe ice-sheet advance rather than they are products of this advance (Kupetz 1997; Kupetz and Koźma 2015).

Thus, the history of the evolution of glacial landforms in the Mużaków Rampart region is multi-stage and complex. In the current understanding, it can be divided into several stages (Dyjor and Chlebowski 1973; Kupetz 1997; Kupetz and Koźma 2015):

(a) Multiple deformations during an ice-sheet advance in the Elsterian [equivalent to Marine Oxygene Isotope Stage (MOIS) 12], producing three main parallel arcs of push moraines. As the earliest of these was located most to the north, it was subsequently overridden by ice and largely obliterated. The subdued ridges near the village of Tuplice are interpreted as the remnant of this arc. Likewise, the younger arc located further south was later overridden by the ice and is faintly preserved, mostly on the German side of the Nysa Łużycka River valley (Fig. 28.4).
(b) The final push of the Elsterian ice sheet reached furthest south and produced the main body of the Mużaków

Rampart, as well as the topographic depression in its hinterland. Later on, the ice sheet continued advancing to the south, eventually reaching the marginal parts of the Sudetes mountain range.

(c) The older Saalian ice sheet (equivalent to MOIS 6) crossed the rampart, deeply eroding and truncating its most elevated parts, but not obliterating it completely.

(d) In the younger phase of the Saalian, the ice sheet only reached the Mużaków Rampart, without causing much change to the pre-existing landform. During recession, an end-moraine lobe was built within the hinterland basin, known as the Jerischke End Moraine. Its most distinctive section is in Germany, whereas in Poland it is marked by low WSW–ENE trending ridges to the east of the village of Żarki Wielkie, superimposed onto larger and higher elevations of SSW–NNE strike (Fig. 28.1).

(e) Further denudation by non-glacial processes gave the rampart complex its sub-recent appearance, including unblocking the flow of the Nysa Łużycka River to the north, then heavily modified by mining and industrial activities.

28.3.3 Primary and Secondary Effects of Glacitectonic Deformations

Extensive glacitectonism that occurred during the Elsterian glaciation had both direct and indirect geomorphic consequences. The former are related to the dynamics of the ice sheet and are manifested as landforms whose spatial pattern and other characteristics reflect glacial processes. The latter result from the specific course of denudation processes acting upon glacitectonic ridges and exploiting variable properties of sediments brought to the surface.

It is assumed that the principal style of glacitectonic deformation of the Mużaków Rampart is folding of unconsolidated layers of sediments under the load of many hundred metres thick, relatively fast-moving ice tongue, accompanied by large-scale thrusting in the frontal part (Kupetz 1997). The vertical extent of strata distortion was demonstrated to the depth of 236 m below the contemporary surface, which means an even greater depth at the time of deformation, given unknown amount of subsequent erosion. Squeezing-up of deformed sediments beyond the ice front resulted in the origin of stacked thrust sheets, each

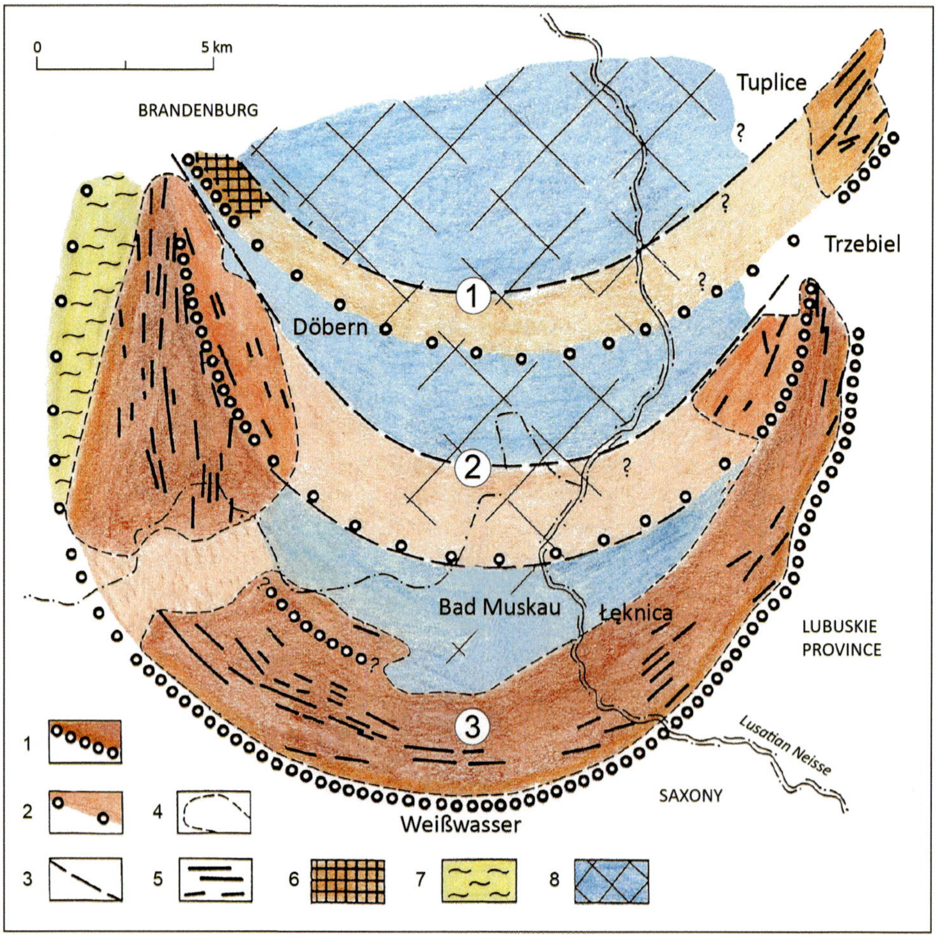

Fig. 28.4 Structural sketch map of the Muskau Arch (after Kupetz 1997). 1—morphologically visible outer border of the Muskau Arch (no. 3) and inner arcs (Kölzig-Tuplice Arch—no. 1, Döbern-Trzebiel Arch—no. 2), 2—inferred courses of distal borders of the arcs, where not evident in relief, 3—inferred proximal borders of the arcs, 4—part of the Muskau Arch evident in relief, 5—important glaciotectonic structures visible on aerial photographs, 6—large-scale glaciotectonic deformations recognized only in the subsurface, 7—part of the Muskau Arch that was eroded by a subglacial stream ("Bohsdorfer Rinne", Bohsdorf incision), 8—glacial depression in the hinterland (Bahrener Becken), continuation in Poland unknown

composed of clastic deposits and brown coal seams, which were dragged towards the surface. The parallel linear surface texture of those parts of the Mużaków Rampart, which were not transformed by human activities, is derived from the stacked arrangement of several glacitectonic thrust sheets. Other types of deformation have more local distribution and include diapiric structures, tight folds, and rootless rafts. Although the complete picture of a deformation style in a particular place can now be only obtained from dense boreholes (after wall exposures in former mines are no longer accessible due to land rehabilitation and vegetation succession), there are surface indicators, which assist in determining how the strata were deformed. Their presence is an indirect effect of glacitectonism.

As both clastic and coal deposits are involved in glacitectonic structures, the contemporary surface truncates both as well, exposing them to weathering and water-related processes. However, they do not react identically. Outcrops of brown coal are subject to more intense weathering, which leads to reduction in volume due to oxidation. As no comparable process occurs in sands or silts, long-term surface lowering typifies brown coal outcrops, resulting in the origin of specific terrain depressions (hollows), locally called *gieser* (or *gjeser*) (Fig. 28.5). It probably derives from an Upper Sorbian word *jezero*, meaning a lake or swamp that in turn indicates that these hollows are waterlogged and may host bodies of standing water (Potonié 1930).

Their typical depth is 3–5 m, but some may reach 20 m and be traced over several kilometres. The shapes and spatial patterns of gieser approximate the geometry of glacitectonic structures. Within thrust sheets gieser are 1–2 km long and keep constant width and strike, whereas those associated with diapiric structures have irregular courses and variable width, may not be continuous, and are shorter. There are also large depressions along the axis of anticlinal folds, which may be up to 2 km long and several hundred metres wide (Fig. 28.6). Thus, the ridge-and-trough topographic pattern of the Mużaków Rampart is to be seen as an indirect result of glacitectonism rather than its direct effect, even though it would be tempting to equate the ridges with the axis of individual thrust sheets.

28.3.4 Postglacial Re-Modelling—Activity of Hillslope, Fluvial and Aeolian Processes

The contemporary appearance of the Mużaków Rampart (anthropogenic transformation excluding) is the combined result of glacial processes during successive glaciations and non-glacial processes that occurred in interglacial periods, but particularly during the last cold stage of the Pleistocene and likely at the Pleistocene/Holocene transition (Fig. 28.7). The area was subject to various surface processes typical for periglacial environments and then, also associated with the decay of permafrost. Climate amelioration during the early Holocene resulted in the development of forest communities, which probably considerably reduced the efficacy of erosional and denudational processes, except for weathering responsible for the emergence of gieser.

Although periglacial transformation of relief was never subject to any detailed research in this area, its cumulative impact is best reflected in significant re-modelling of the original push moraine landform. Sandy, silty and clayey materials were subject to seasonal thawing, which favoured solifluction and surface wash, probably more effective on sandy substratum because of deeper extent of summer thawing. Dry valleys and gullies, present in various localities, especially dissecting the risers of fluvial terraces of the Nysa Łużycka River, indicate the role of

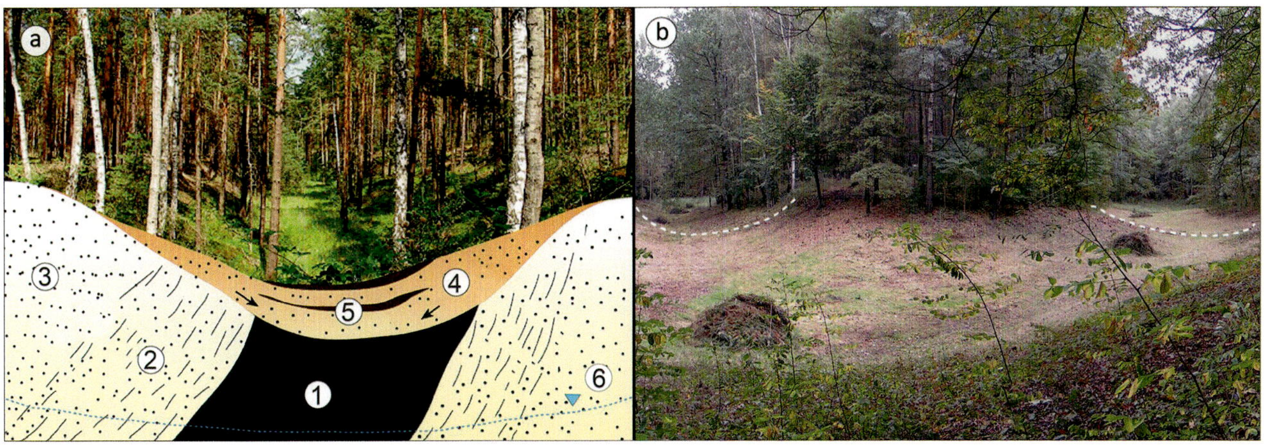

Fig. 28.5 Formation of gieser structure. **a** Schematic section (after Kupetz and Kupetz 2009). 1—inclined brown coal seams, 2—clay sands, silts, 3—Miocene fine-grained sands, 4—Holocene sands, 5—peat, peat soils, 6—groundwater table level. **b** Giesers in Kromlau Park (cross-sections highlighted by dashed lines), in the German part of the Muskau Arch

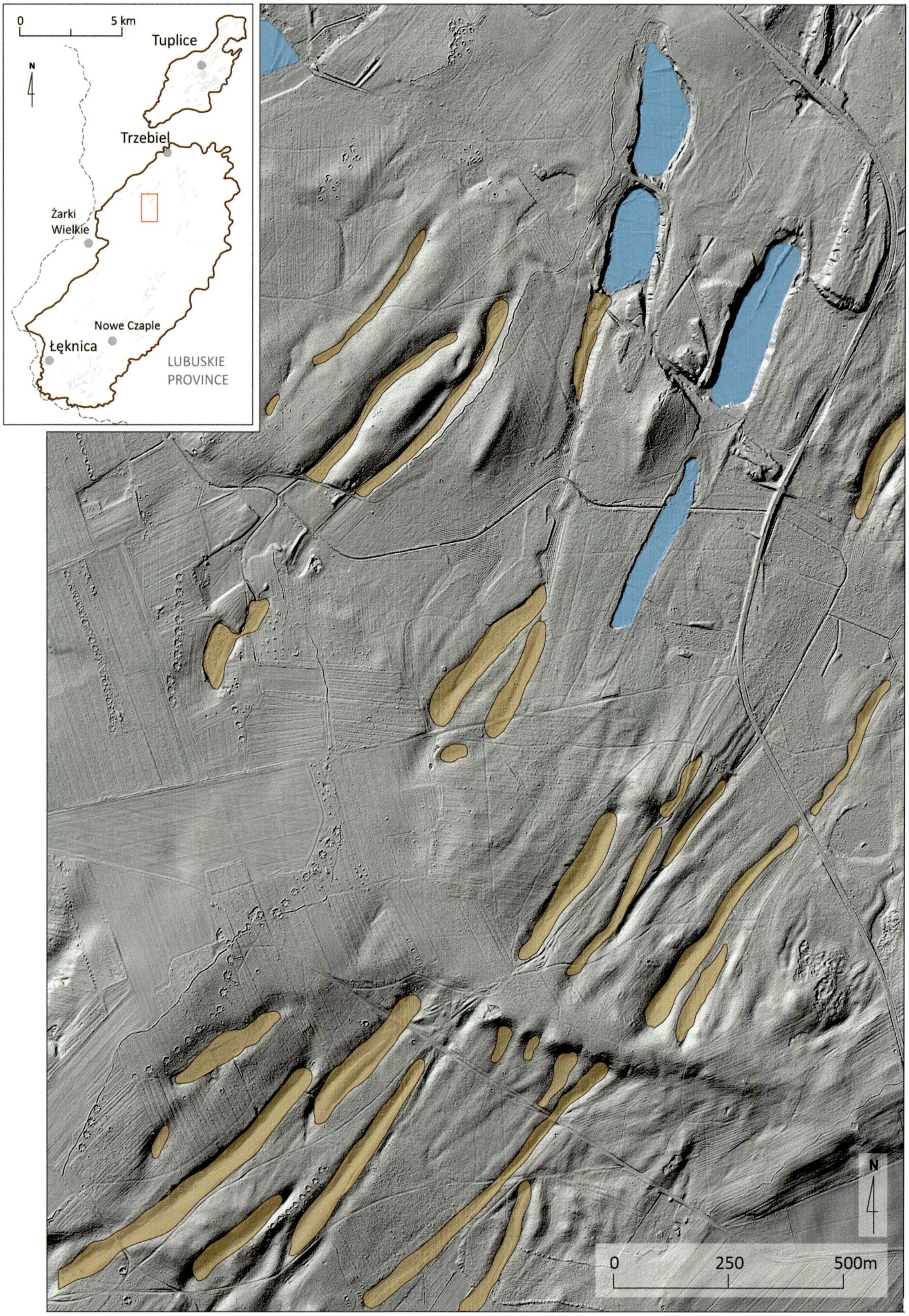

Fig. 28.6 Gieser depressions (light brown) and waterlogged opencast brown coal workings (light blue) follow the same direction, as visible on the high-resolution digital terrain model (source of spatial data: geoportal.gov.pl)

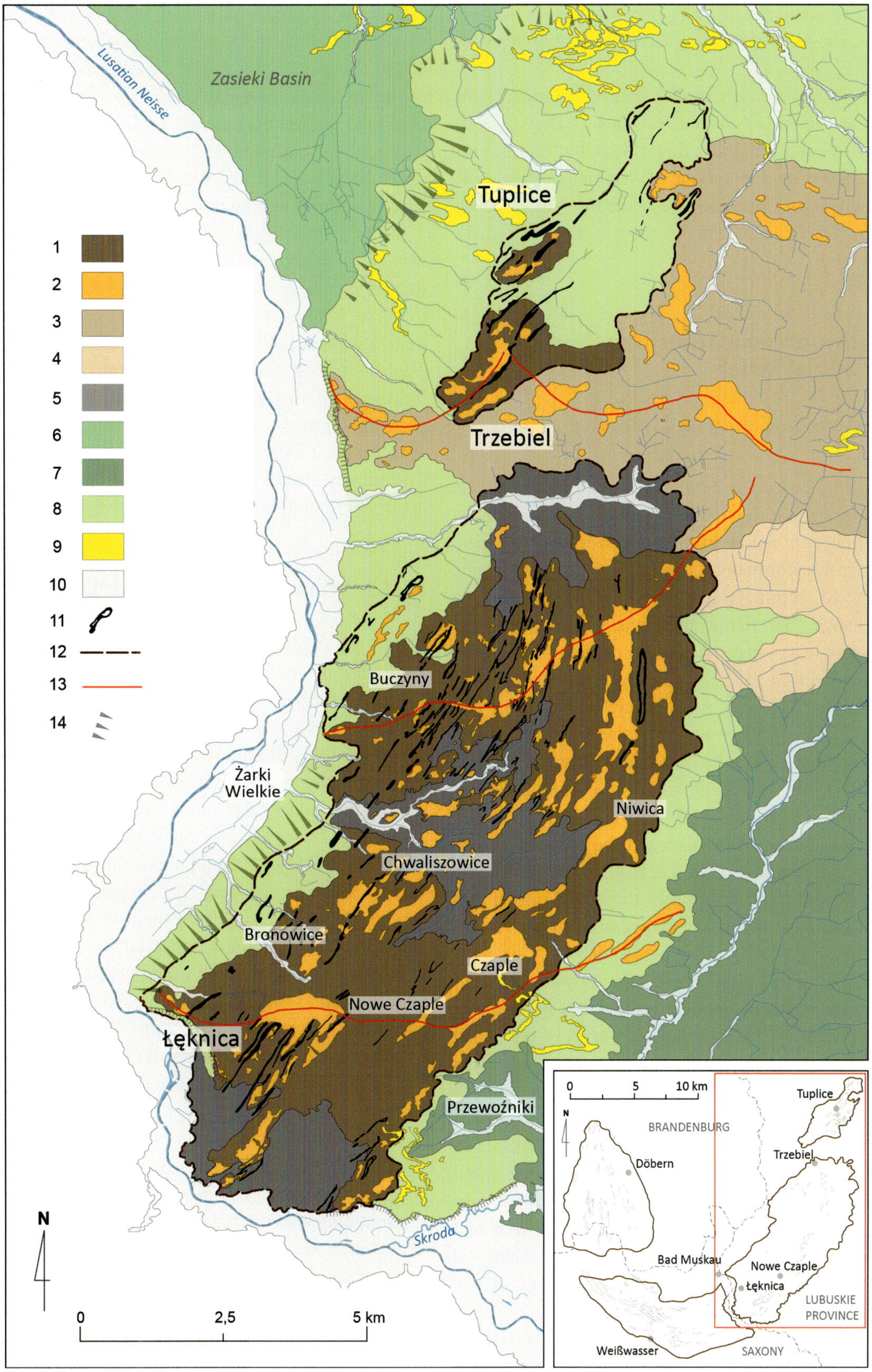

Fig. 28.7 Geomorphological sketch of the Polish part of the Muskau Arch. 1—push moraine upland, 2—moraine hills, (push moraine and accumulation moraine), 3—flat moraine upland, 4—denudation plain, 5—erosional depression, 6—alluvial fan of the Nysa Łużycka River, 7—accumulation terrace within an ice-marginal valley, 8—outwash plain, 9—dunes, 10—valley floors, 11—giesers, 12—boundary of the glaciotectonic structure, 13—probable ice-sheet extents during the Middle Pleistocene, 14—long slopes

seasonal and episodic runoff. They may reach 20 m deep. In a few places, the steps separating terraces were affected by landslides.

Dunes occur in the northernmost part of the area, rising above the flat surface of the Zasieki Basin. They vary in shapes, from regular parabolic dunes standing in isolation through conjoined parabolic dunes, asymmetric ones, with one arm much shorter than the other, to contorted ridges as long as 4 km. The height of the dunes varies too, from just a few to 15 m.

28.4 Anthropogenic Transformation of Glacial Relief

28.4.1 History of Mining and Industry

Although natural processes involved in re-modelling of the push moraine rampart undoubtedly played an important part, their effects are not as striking as those resulting from anthropogenic transformation of relief in the last 200 years or so (Fig. 28.8). Mining was the main activity leading to landform change, but minor modifications resulted also from processing of raw resources, development of transportation lines, agriculture, and landscaping. Consequently, the Mużaków Rampart abounds in anthropic landforms of different kinds and sizes, whereas the evolution of many others is indirectly conditioned by human activities.

Earlier, there may have been small-scale exploitation of local resources, such as bog iron ores, but alum clays and silts were certainly mined in 1825 and the onset of organized brown coal mining dates back to 1843, when the 'Julius' mine came into operation in the now German part of the arch structure (Schossig and Kulke 2006; Tischer 2020).

In the Polish section, the beginnings of mining activity are a decade younger. Initially, brown coal mining was carried out primarily underground, in vertical shafts and sloping galleries that followed the dip of brown coal seams involved in glacitectonic thrust sheets. These main galleries provided access to horizontal adits and chambers, located at different levels, where the actual coal retrieval took place. In this way, the depth of 50–60 m below the surface could have been reached. After works had been finished, the supporting timber walls were removed, allowing for roof and wall collapses into the voids. Subsequent technological developments regarding dewatering of large open-cast pits allowed for gradual replacement of underground mining by surface exploitation and since the beginning of the twentieth century most of mining occurred at the surface, resulting in considerable relief change. The largest mining area in the eastern section of the Mużaków Rampart, the 'Babina' mine, was located immediately to the east from the town of Łęknica (Fig. 28.8) and was in operation from 1921 to 1973. The year 1973 marks the end of brown coal mining in the area, but local exploitation of fire clays to supply a brickyard in Łęknica lasted until the early 1990s. Apart from brown coal and clays, other resources mined included glass sands (well-sorted quartz sands of Miocene age) and sand and gravel aggregates.

Most processing and use of raw resources took place locally, in brickyards, glass works and a power plant in the village of Nowe Czaple. To ensure efficient delivery to these industrial plants a dense network of transportation lines was built, including the main railway line from Tuplice to Bad Muskau and several branch lines serving individual mining fields. They were supplemented by road network. Ground levelling for expanding industrial facilities also occurred. Following closure of the mines and clay pits, the industry closed too and the former mining and industrial areas went into the stage of land rehabilitation.

28.4.2 Primary Anthropogenic Landforms

The most evident primary anthropogenic landforms are large terrain hollows indicating the location of former open-cast brown coal mines. Many are filled with standing water bodies, creating an anthropogenic lakeland. The largest in terms of area is the former 'C' pit in the 'Babina' mine, known as the lake Africa because of outline similarities, approximately 925 m × 450 m and covering more than 27 ha (Fig. 28.9a). Two large elongated hollows occur to the north-east of Nowe Czaple, measuring 1250 m × 350 m and 1100 m × 250 m, respectively. The shapes of open-cast pits vary and are more regular for the former brown coal mining areas, especially if these followed linear outcrops of coal seams within glacitectonic thrust sheets, whereas those left after clay and sand extraction are less so. The depth of pits varies too, from several metres to 15–20 m, if measured to the dry floor or the lake water level. However, the depth of lakes filling pits may be substantial, up to 24 m for the lake in the 'C' pit.

The second major category of primary anthropogenic landforms are heaps of waste material, deposited in the vicinity of the pits. The amount of waste rock was limited when mining was carried out mainly underground, but drastically increased along with the development of open-cast exploitation. Waste heaps occupy up to 10 ha and are up to 10 m high. However, some may not be evident in the contemporary landscape, since the material was dumped into disused pits, filling them partly or even totally.

Minor, but nevertheless conspicuous landforms, especially if seen on digital terrain models, are

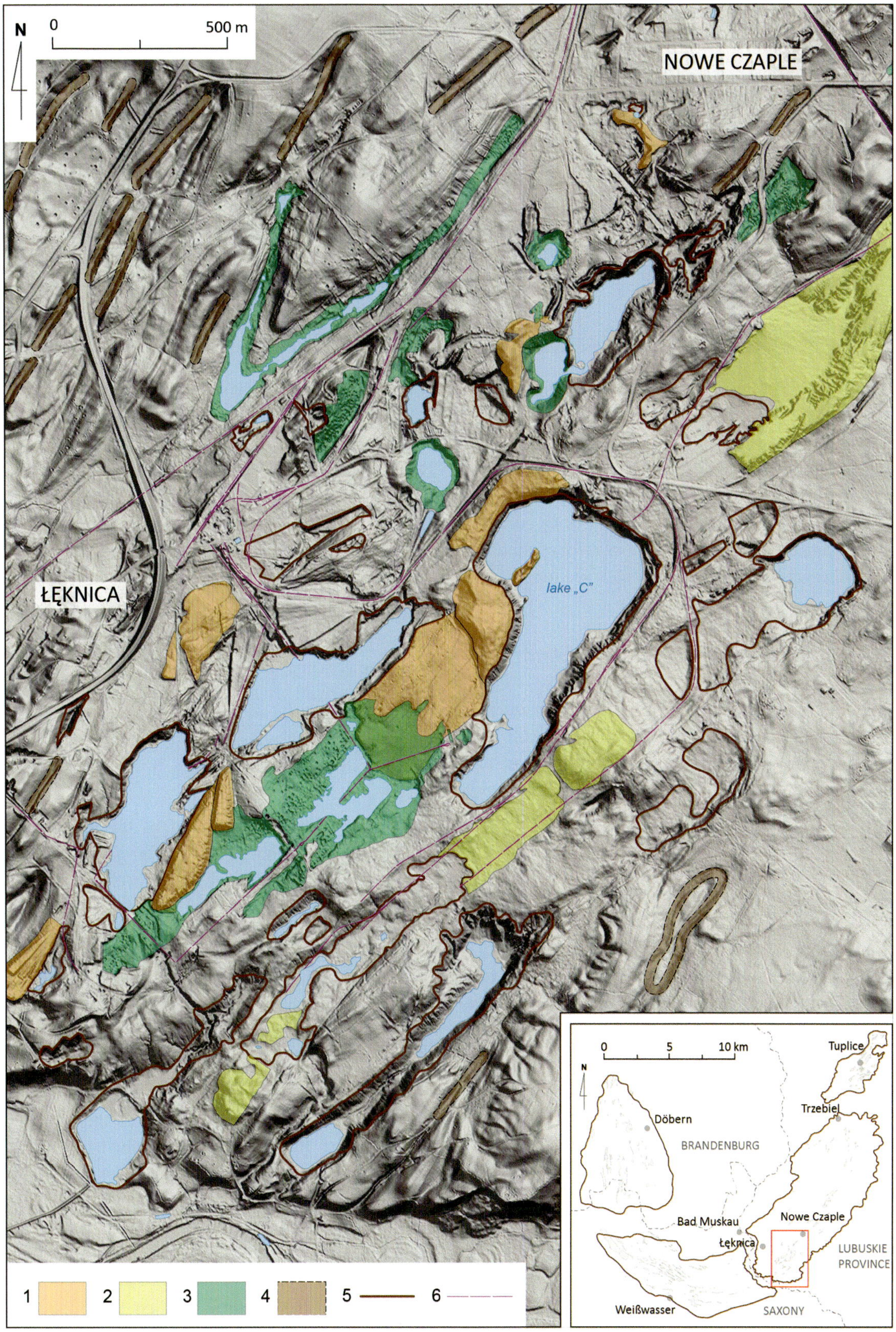

Fig. 28.8 Elements of post-mining landscape in the area of the former "Babina" mine, where a geotourist path was waymarked. 1—heaps and embankments of overburden rock, 2—recultivation embankments of opencast workings 3—subsidence basins over underground workings, 4—giesers, 5—morphological boundaries of opencast workings, 6—former transportation routes

Fig. 28.9 Post-mining antopogenic lakeland in the Polish part of the Muskau Arch. **a** Post-mining lake "C" (Africa), former open-cast brown coal mine. **b** Subsidence basin above the workings of the "Babina" underground mine (photographs by J. Koźma)

former transportation lines. These include railway and road embankments and, particularly, trenches, which are locally up to 10 m deep. They often follow straight lines connecting the former pits and industrial plants or railway stations, consisting of alternating elevated and trench sections.

28.4.3 Human-Induced Processes and Landforms

Mining activity contributed to landform change in indirect way too. Surface alterations are mainly related to the closure of underground chambers and galleries by wall and particularly roof collapses. These subsurface processes induce land subsidence above the former voids, producing topographic hollows whose dimensions may be substantial. Their shapes vary, reflecting the pattern of underground works. Hence, some are almost perfectly circular, whereas others have more irregular outlines or are distinctly elongated following the strike of a coal seam. Although the highest rates of ground subsidence were recorded in the first decade after exploitation was terminated, remnants of submerged forest in some of these hollows indicate that subsidence continues, although at a much lower rate.

Another process worth mentioning, which locally produces spectacular effects, is runoff erosion on non-vegetated parts of waste rock heaps and steep slopes of former open-cast pits (Fig. 28.10). The material exposed is usually highly erodible fine sand and silt, so that runoff produces erosional incisions, creating semi-anthropogenic badlands. Although they do not cover large areas, they add to geodiversity of the region and have geotourist and geo-educational potential. The most spectacular example is offered by a large waste heap near the village of Nowe Czaple, which is almost entirely turned into badland.

28.5 Muskau Arch Geopark

28.5.1 Towards the Transboundary Geopark

Unique geoscience-related values of the Mużaków Rampart, emerging from both its natural history as a complex glacitectonic feature exposing much older Neogene deposits, as well as mining legacy, were realized in both Germany and Poland in the 1990s, when geoheritage matters became an important part of Earth Science agenda and the geopark initiative was launched. Working in international cooperation, geologists from both countries began exploring geoheritage potential of the area, documenting geodiversity and strengthening efforts to explain this unusual landform assemblage. At the turn of the twenty-first century, the concept of the Muskau Arch Geopark was elaborated and presented (Rascher et al. 2000; Badura et al. 2003). An important component of the project was an inventory of geosites, carried out jointly for the German and Polish section of the arch structure. In the meantime, in 2001, the Polish part was declared a Landscape Park, which is a category of large-scale nature conservation. After several more years of preparatory work a National Geopark was formally established in the German part of the Mużaków Rampart in 2006, whereas in 2009 a certificate of a National Geopark was granted to the Polish part. An application for the status of a Global Geopark was submitted in 2007 and finally approved in 2011 as the Muskau Arch Geopark. Thus, a transboundary geopark was set up and whereas it was not the first Global Geopark in Germany, it has become the first Polish member of the Global (and European) Geopark Network. Since that time numerous initiatives to improve tourist infrastructure and

Fig. 28.10 Erosional forms within the post-mining landscape (anthropogenically-induced badlands). **a** Heavily dissected waste heap in the area of Nowe Czaple. **b** Gullied shore of the post-mining lake "C" (Africa) (photographs by J. Koźma)

enhance educational potential of the geopark have been launched and implemented in both countries, including waymarking of the 'Babina' Mine Trail, which has become the highlight of the Polish side of the Muskau Arch Geopark (see Sect. 28.5.3).

28.5.2 Geosites—Distribution and Diversity

The most recent inventories of geosites for the entire Muskau Arch Geopark include 10 thematic groups and nearly 800 localities, among which 124 occur within the territory of Poland (Koźma and Kupetz 2008; Heidenfelder et al. 2021; Kupetz et al. 2021). The following groups were distinguished: (1) glacitectonic structures, (2) glacial and periglacial landforms, (3) aeolian landforms, (4) fluvial landforms, (5) wetlands, including peat bogs, (6) springs, (7) sites of former exploitation of mineral resources, (8) post-mining lakes, (9) examples of use of erratic boulders and cobbles in building construction, (10) ex-situ geological objects in romantic parklands. Geosites are rather evenly distributed across the Polish part of the Mużaków Rampart, with minor clusters around Łęknica, where the 'Babina' mine was located, and the villages of Tuplice and Trzebiel.

Apart from sediment outcrops and landforms related to the former mining areas presented earlier, further examples of sites of geomorphological interest classified as geosites include:

- the granite Devil's Stone (*Diabelski Kamień*) near Trzebiel, which is the largest erratic boulder in the Polish section of the Geopark. It is 5.1 m long, 3.5 m wide and 2.5 m high, with unknown depth of the buried part. It is believed that minor anthropogenic alterations of the boulder's surface (regularly spaced holes and rills) testify to its ancient use as a cult site.
- a very regular parabolic dune at the outskirts of Tuplice, overlooking an artificial pond. The dune is representative of a larger cluster of aeolian landforms in the plain adjacent to the north-eastern tip of the Mużaków Rampart, easily accessible from the village. The arms are 250–350 m long and the width of the dune is 280 m. The height is 14 m.

28.5.3 Babina Mine Trail

The geotouristic trail within the former "Babina" brown coal mine near Łęknica was opened in 2012 and, after expansion in 2021, is approximately 10 km long, accessible for both hikers and bikers (Koźma et al. 2013). It uses a network of old transportation roads within the extensive mining area, with short paths to specific sites of interest. The main elements of geotourist infrastructure are 15 interpretation panels, erected at key localities more or less regularly spaced along the trail. The panels address various themes related to mining activity, geology of coal seams, origin and subsequent development (erosion, subsidence) of anthropogenic landforms, and hydrology, including hydrochemistry of springs and lakes. An observation tower was built next to the largest artificial lake, which makes it possible to see the full extent of the reservoir and widespread erosional re-modelling of its banks, otherwise difficult to appreciate. In addition, aerial photographs included in several panels show an impressive scale of anthropic transformation of the landscape.

28.5.4 The Muskau Park—World Heritage Property

Within the boundaries of the Muskau Arch UNESCO Global Geopark there is also a World Heritage site, namely the Muskau Park (Muskauer Park). It received its UNESCO status in 2004, based on criteria (i) and (iv) described in the Operational Guidelines for the Implementation of the World Heritage Convention. The property covers 560 ha on both sides of the Nysa Łużycka River and protects a unique example of a large-scale romantic landscape park, designed and created in 1815–1844, reflecting the vision of Prince Hermann von Pückler-Muskau how to enhance scenic values of the existing natural landscape (Stachańczyk 2020).

In the justification of the inscription it is emphasized that the idea was ahead of time and the park itself inspired many subsequent park and garden designers in Europe and North America (http://whc.unesco.org/en/list/1127; access date 2022-05-04). The core part of the estate is located on the German side of the river, with two palaces and auxiliary buildings. From there, the park spreads over the floodplain (Fig. 28.11a), with canals and ponds, likely in the place of former secondary river channels and oxbow lakes. Several bridges connect with the floodplain on the opposite side of the river, in Poland, which is much narrower and terminates against a 10-m-high riser, above which a higher terrace level occurs. Numerous paths criss-cross the terrace tread, whereas viewing points were arranged at its edge, including one next to the monument commemorating Prince von Pückler-Muskau that was built using erratic boulders from the surroundings (Fig. 28.11b). There the park extends over a much higher escarpment (>20 m) that links the terraced valley floor with the plateau and is locally incised by ravines and dry troughs. Thus, although the Muskau Park is recognized as a solely cultural property, it in fact demonstrates the close link between nature and culture, and shows how geomorphology may underpin the design of recreational grounds.

28.6 Conclusions

The Mużaków Rampart, even though small in terms of spatial extent, occupies a special position among geomorphological landscapes of the Polish lowlands. It represents a well-developed glacitectonic structure, which despite considerable age and denudation still clearly corresponds to the pattern of landforms. Being formed as a push moraine during the Elsterian glaciation, more than 400,000 years ago, it is a rare example of a landform inherited from this rather distant interval of the Quaternary. The exposure of Miocene brown coal seams to the atmosphere triggered alteration processes which resulted in the origin of unique medium-size landforms within the rampart—the gieser. Landforms resulting from mass movements, fluvial, and aeolian processes add to geodiversity of the region. The area is also known for large-scale anthropogenic transformation of relief, which is in principle the outcome of nearly 200 years of brown coal and clay mining history, both at the surface and underground. Mining-related landforms are both direct results of human activity (open-cast pits, waste rock heaps), as well as indirect consequences of land and bedrock disturbance. The latter include numerous subsidence hollows developed above former mining galleries, many currently filled with water. Thus, a unique anthropogenic lakeland developed in the Mużaków Rampart, with no comparable example elsewhere in Poland. Geoheritage and geodiversity values of the region were recognized more than 20 years ago and are currently promoted within the transboundary, German-Polish UNESCO Global Geopark Muskau Arch,

Fig. 28.11 The Nysa Łużycka River valley in the World Heritage Muskauer Park. **a** Wide valley floor. **b** Terrace riser, with the monument commemorating the Prince von Pückler-Muskau, made of erratic boulders (photographs by P. Migoń)

approved in 2011. In addition, the romantic landscape park in Bad Muskau/Łęknica was inscribed by UNESCO as a World Heritage property in 2004. It covers a fraction of the area described in this chapter and even though it is recognized as a solely cultural property, it demonstrates how the landform pattern may underpin world-class park and garden design.

References

Badura J, Gawlikowska E, Kasiński JR, Koźma J, Kupetz M, Piwocki M, Rascher J (2003) Geopark "Łuk Mużakowa"—proponowany transgraniczny obszar ochrony georóżnorodności. Przegl Geol 51:54–58

Dyjor S, Chlebowski Z (1973) Budowa geologiczna polskiej części Łuku Mużakowa. Acta Univ Wratisl Prace Geol-Miner 192(3):3–37

Heidenfelder W, Rascher J, Richter L, Koźma J, Löwen K, Rascher M, Kupetz M (2021) Das Geotopkataster des UNESCO Global Geoparks Muskauer Faltenbogen / Łuk Mużakowa, Teil 1: Revision 2020 Brandenburgische Geowiss. Beitr 28(1/2):57–76

Jentsch A, Berg G (1913) Die Geologie der Braunkohlenablagerungen im östlichen Deutschland. Abh Königl Preuß Geol Landesanst NF 72:141

Keilhack K (1907) Bericht über wissenschaftliche Ergebnisse: 7 Provinz Sachsen und des Grenzgebiet der Provinz Brandenburg und des Herzogtums Anhalt. Jhrb Preuß Geol Landesanst Bergak Berlin Für 1904 25:672–684

Keilhack K (1909) Begleitworte zur Karte der Endmoränen und Urstromtäler Norddeutschlands. Jahrb Königl Preuß Geol Landesanst Bergakad Berlin 30(1):507–510

Koźma J, Kupetz M (2008) The transboundary Geopark Muskau Arch (Geopark Łuk Mużakowa, Geopark Muskauer Faltenbogen). Przegl Geol 56:692–698

Koźma J, Sandak D, Bieniasz J (2013) Geotouristic "Babina" path as an example of sustainable development in the Muskau Arch Geopark. Rendiconti Online Soc Geol Italiana 28:93–96

Kupetz M (1997) Geologischer Bau und Genese der Stauchendmoräne Muskauer Faltenbogen. Brandenburgische Geowiss Beitr 4:1–20

Kupetz M, Sauer N, Koźma J, Brauer E, Jagiełło K, Löwen K, Nickel D (2021) Das Geotopkataster des UNESCO Global Geoparks Muskauer Faltenbogen/Łuk Mużakowa, Teil 2: Seine Anwendung in der Bildung für nachhaltige Entwicklung (BNE). Brandenburgische Geowiss Beitr 28:77–96

Kupetz M, Koźma J (2015) Europäischer and Globaler Geopark Muskauer Faltenbogen/Geopark Łuk Mużakowa—die weltweit am besten untersuchte Grundbruchmorane ("Stauchendmorane"). In: Rohling H-G (ed) GeoBerlin 2015. Dynamische Erde—von Alfred Wegener bis heute und in die Zukunft. Exkursionsfuhrer. Veröff dt Geol Gesell vol 255, p 113–135

Kupetz A, Kupetz M (2009) Der Muskauer Faltenbogen. Wanderungen in die Erdgeschichte (24). Verlag Dr. Friedrich Pfeil, München, p 224

Liedtke H (1981) Die nordischen Vereisungen in Mitteleuropa. Forsch Zur Deutschen Landeskunde 104:204

Potonié R (1930) Über den Muskauer Faltenbogen, seine Oberflächenformen und deren Abhängigkeit von der Beschaffenheit und der Tektonik der Braunkohle. Jhrb Preuß Geol Landesanst 51:392–416

Rascher J, Meier J, Kupetz M (2000) Der Geopark Muskauer Faltenbogen—Grundlagen, Stand, Perspektiven. Geowiss Mitt Thüringen Beih 10:75–85

Schossig W, Kulke M (2006) Braunkohlenbergbau auf dem Muskauer Faltenbogen. Beiträge zur Geschichte des Bergbaus in der Niederlausitz 6, Cottbus, p 200

Stachańczyk R (2020) Park Mużakowski/Muskauer Park. Przewodnik. Narodowy Instytut Dziedzictwa, Warszawa, p 128

Tischer F (2020) Der Braunkohlenbergbau im östlichen Teil des Muskauer Faltenbogens zwischen Muskau und Teuplitz. Weißwasser, p 240

Jacek Koźma holds a PhD in Earth Sciences. Until his retirement, he was a long-time employee of the Lower Silesian Branch of the Polish Geological Institute in Wrocław. His research interests include geological cartography of the Cenozoic formations, geoenvironmental cartography and mineral economy. He has also written extensively on issues related to geoparks and geological heritage in the context of geotourism. Since 2000, he has been conducting research and organisational activities related to the development of the UNESCO Global Geopark Muskau Arch. He is a former member of the UNESCO Global Geoparks Coordinating Committee.

Piotr Migoń is Professor of Geography at the University of Wrocław and former Director of the Institute of Geography and Regional Development. His main research interests are rock control in geomorphology, especially in sandstone and granite areas, mass movements in mountain terrain, geoheritage and geotourism. The Sudetes Mountains in Central Europe is his main research area, but he was also involved in projects in various other European countries (Czechia, Germany, Great Britain, Sweden, Portugal, Spain), in the United States, Mexico, Brazil, China, Jordan and Namibia. He used to be the Secretary of the International Association of Geomorphologists (1997–2001), its Vice-President (2009–2013) and member of Executive Committee (2013–2022); he was also the President of the Association of Polish Geomorphologists (2005–2008). He is the series editor of 'World Geomorphological Landscapes' series at Springer Nature and member of the editorial board of *Geomorphology* journal.

Polygenetic Lowland Relief Around the City of Łódź—Glacial Inheritance and Periglacial Transformation

29

Jan Goździk and Jacek Forysiak

Abstract

The Łódź region in central Poland straddles the Middle Polish Lowlands in the north and the Upland belt in the south. The most elevated area, above 200 m asl, occurs in the central part, whereas lower areas surround it in the west, north and east. The gross morphological framework of the area was shaped by glacial processes during the Saalian Glaciation (Middle Pleistocene), but was subsequently transformed by fluvial, lacustrine, and especially slope and aeolian processes under periglacial conditions of the late Pleistocene (Weichselian). The relief of the Łódź region is therefore polygenic. Two parts of this region are presented in more detail. These are the immediate vicinity of the city of Łódź with various glacial landforms reshaped by periglacial processes and the Szczerców Basin in the south, with vast sandy plains, moraine plateaus, broad valleys, and dune fields. Dry valleys and dune chains are typical inherited cold-climate landforms, whereas ice-wedge casts are typical periglacial ground deformation structures formed due to the presence of permafrost. The Holocene did not cause significant landform change, although human-induced modifications of the landscape are very legible.

Keywords

Glacial landforms · Periglacial · Fluvial landforms · Dunes · Human impact

J. Goździk · J. Forysiak (✉)
Department of Geology and Geomorphology, University of Łódź, Narutowicza 88, 90-139 Lodz, Poland
e-mail: jacek.forysiak@geo.uni.lodz.pl

29.1 Introduction

Very close to the geographical centre of Poland lies one of the biggest city in the country – Łódź. The surrounding area, which was long in research focus of geomorphologists based in this city, has been termed 'the Łódź region (Fig. 29.1). This notion was first used by Dylik (1961) and was more precisely defined by Klatkowa (1972), as an area of central Poland, whose core part is a peninsula-like northward extension of higher upland ground that rises more than 200 m asl, with the surrounding areas, where altitudes typically vary from 100 to 150 m asl. In the north, this surface descends to 70 m asl.

In geomorphological studies carried out in this region, much attention has been paid to morphogenetic processes that took place in the periglacial environment and their effects. Periglacial studies began in the late 1940s and are associated with a distinctive research school of Professor Jan Dylik, one of the most prominent figures in the history of Polish geomorphology. Initially, he had been conducting research in historical geography, working together with archaeologists, but then switched the field of study to explore geomorphology (Dylik 1948). During geomorphological fieldwork, he used experience gained from working with archaeologists, and thoroughly analysed sediment exposures. He noticed that in the area around the city of Łódź, which had been glaciated during the Saalian cold stage (MIS 6), landforms of glacial origin coexist with relief features, until then largely overlooked, that owed their origins to the periglacial environment of the Weichselian (last glacial, Vistulian) period.

In the Łódź region, several circumstances facilitated such studies. Particularly advantageous has been the existence of huge open-cast lignite mines of 'Bełchatów' and 'Adamów', and of numerous large gravel pits and clay pits accompanying brickworks, which allowed for the observation of exposed sediments over distances of many kilometres and to considerable depths exceeding 100 m. Near-surface

P. Migoń and K. Jancewicz (eds.), *Landscapes and Landforms of Poland*, World Geomorphological Landscapes, https://doi.org/10.1007/978-3-031-45762-3_29

sediments were also observed along long pipeline routes, including the Pilica–Łódź water pipeline, the gas pipeline running from northern Poland to the south, and in numerous pits associated with building industrial and residential infrastructure. This allowed for a more accurate reconstruction of processes that shaped the surface of the area.

The Łódź region lies within the Middle Polish Lowlands, with only its southern and south-eastern parts belonging to the Małopolska Upland (Fig. 29.1). Along the region's main north–south axis there are areas with elevations close to those found in the upland belt further south. These elevated areas are bounded to the west, north, and partly to the east, with a few exceptions, by areas with elevations typical of the lowlands. This difference in elevation, in the order of 100 m, favoured a variety of periglacial processes and impacted on variable intensity of some of them. The region has a nearly continuous cover of sediments from the Saalian glaciation, while pre-Quaternary rocks crop out at the surface in only a few places as small islets.

As a result of many years of research, especially in deep excavations of lignite open-cast pits, it has been possible to trace the evolution of the relief since the end of the penultimate glacial cycle (Saalian). Subsequently, it occurred exclusively in non-glacial conditions, as the maximum extent of the last Scandinavian ice sheet (Weichselian stage) was located further north (Fig. 29.1). Following the leadership of Jan Dylik, who noticed the major role of periglacial processes in transforming the glacial landscape, much attention has been paid to these processes by his collaborators and successors. Attempts have been made to reconstruct the course of periglacial processes, especially during the Weichselian. This made possible to estimate the scale of changes in the glacial landscape and to recognize new landforms, whose origin was attributed to periglacial processes.

29.2 Stages of Relief Evolution

The relief of the Łódź region is polygenic. In its evolution, various phases may be distinguished, during which dominant morphogenetic processes clearly changed following climate changes. During the Saalian glaciation, in its younger (Warthanian) stage glacial processes produced numerous glacial landforms, whereas glacigenic deposits formed a thick cover concealing older bedrock and sediments. This period is considered the first stage and the beginning of a cycle leading to the formation of contemporary polygenetic relief. Dylik (1973, p. 87) stated that '*the glacial cycle created the basis for later relief*'. Therefore, in this chapter, the main features of post-Saalian relief will be presented in more detail.

In the Late Saalian and in the following Eemian interglacial, non-glacial morphogenetic processes initiated the transformation of the post-glacial surface. However, only a few evident traces of these processes could have been recognized. The clearest record of Eemian sedimentation has been found in post-glacial closed depressions. The shallow ones were filled during this interglacial (Klatkowa 1972), whereas in deeper depressions the Eemian sediments constitute the floor of sedimentary complexes that continued to be deposited later, often up to the Holocene. Most of the sediments filling these depressions were deposited in a periglacial environment and will be discussed later. During the recession of Saalian ice sheet meltwaters formed new valleys, in particular ice-marginal valleys. These valleys were used for flow routing in the subsequent periods and the contemporary river system started to develop. However, it is difficult to distinguish fluvial forms from the short Eemian Interglacial, as the river network continued to evolve intensively in the following Weichselian period of much longer duration.

In the next glacial stage, the Weichselian, which lasted about 100,000 years, cold climate persisted and permafrost developed, with an active layer in the top, creating a periglacial environment that is of particular importance to morphological changes experienced in the Łódź region. The Weichselian comprises three principal periods (ages after Marks et al. 2016): the Early Glacial (110–71 ka), the Pleniglacial (71–16.8 ka) and Late Glacial (16.8–11.7 ka). The Pleniglacial is further divided into the Early (71–58 ka), Middle (58–32 ka) and Late Pleniglacial (32–16.8 ka). Throughout the Weichselian, significant changes in climatic conditions took place, exerting considerable influence on geomorphic process and resultant landforms. In the Early Glacial, climate was cooler than in the preceding Eemian Interglacial, but it remained mildly cool. However, in the Pleniglacial cooling continued and the Late Pleniglacial was typified by the coldest environment. During the Late Glacial (Late Weichselian), the climate gradually warmed up, although with brief returns of cold intervals, until the Holocene, when a temperate climate was established.

The evolution of permafrost in central Poland in the Weichselian has been reconstructed by Klatkowa (1996). At the beginning of the Early Glacial, isolated permafrost appeared, turning to discontinuous permafrost as the climate cooled. Only later in this period did there appear a continuous permafrost zone that lasted throughout the Pleniglacial. However, by the beginning of the Late Glacial, it began to disappear. The permafrost decayed during interstadials, but aggraded again following climate deterioration during stadials.

Climatic changes in the Weichselian resulted in significant changes in vegetation patterns. Based on the studies by Balwierz (2003), the Early Glacial (Early Weichselian) was dominated by dense covers of trees, shrubs and herbaceous

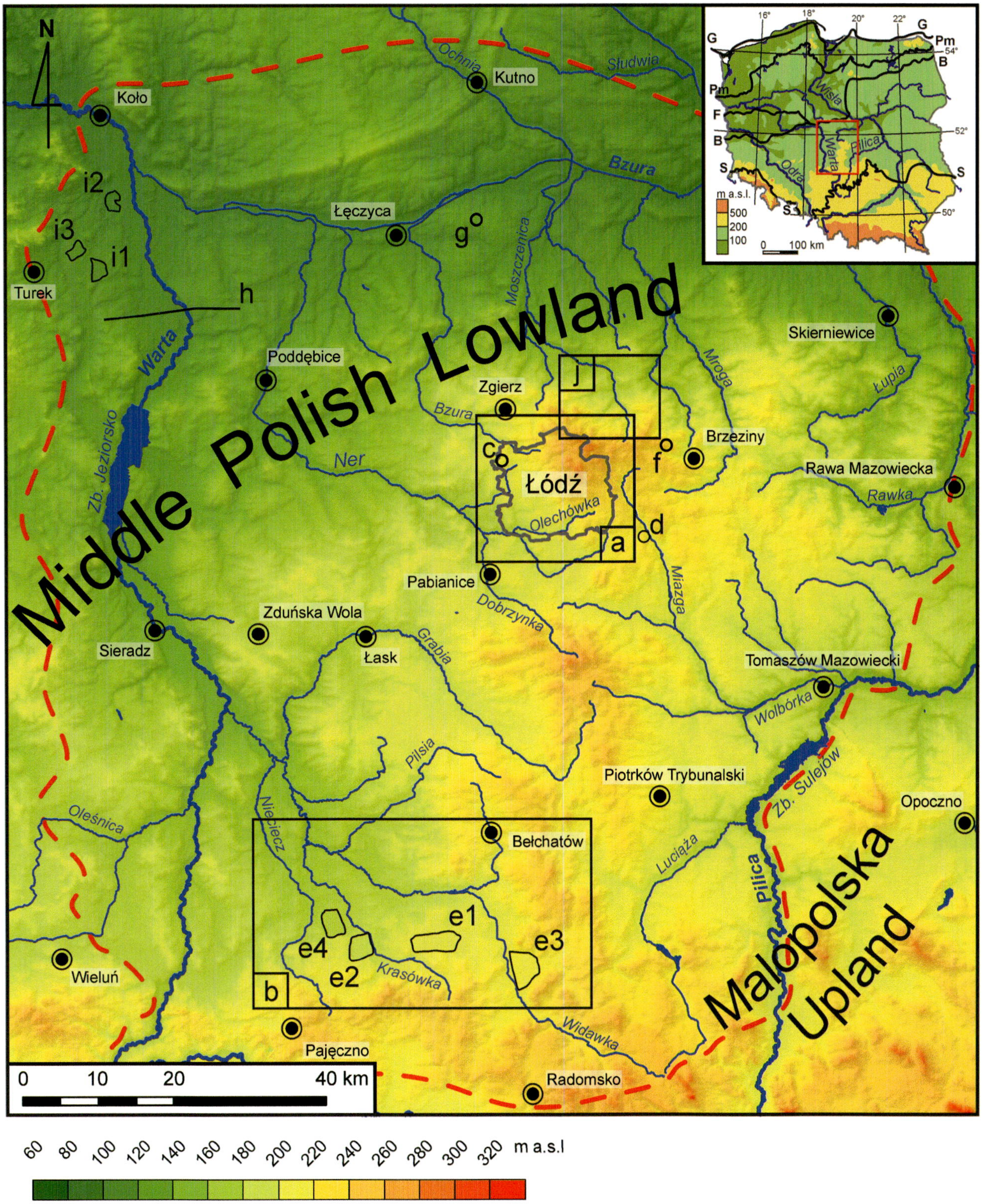

Fig. 29.1 Hypsometry of the Łódź region. Location of sites and areas shown in figures: a—Fig. 29.2, b—Fig. 29.3, c—Fig. 29.6a, d—Fig. 29.6c, e—Bełchatów open cast mine (e1—mine-field Bełchatów—Fig. 29.18, e2—mine field Szczerców, e3—mine heap Kamieńsk, e4—mine heap Pole Szczerców), f—Fig. 29.16, g—Fig. 29.7, h—Fig. 29.11, i—Adamów open cast mine (i1—mine-field Adamów, i2—mine field Koźmin—Fig. 29.12, i3—mine heap), j—Fig. 29.14. The red dashed line delimits the extent of Łódź region. The inset shows location of the region in Poland in relation to the extents of the ice sheets: S—Saalian Glaciation; Weichselian Glaciation: B—Leszno (Brandenburg) Phase; F—Poznań (Frankfurt) Phase, Pm—Pomeranian Phase, G—Gardno Phase

plants. Tree coverage declined in colder periods. The Early Pleniglacial was dominated by shrub tundra, and the Middle Pleniglacial—by tundra grassland and open plant communities restricted to moist habitats. In the coldest intervals of the Late Pleniglacial, cold desert prevailed (Goździk 2007). The presence of permafrost and vegetation cover in the Weichselian greatly influenced the course of morphogenetic processes that transformed the inherited glacial relief.

The principal features of the glacial landscape and its transformation as a result of periglacial morphogenetic processes will be exemplified mainly from two locations: the city of Łódź itself and its surrounding area (Fig. 29.2), and the Bełchatów area (Fig. 29.3). In the latter, it was required that the geological structure and relief of the area was correctly identified prior to the excavation of the lignite open-cast pit and building of the power plant. These two areas presented the best opportunities to observe the sediments that underlie the present-day landforms. The significant differences in relief between these areas allow for the most complete presentation of essential periglacial features of the Łódź region.

29.3 Basic Features of Glacial Morphology

The Saalian ice sheet entirely covered the Łódź area, and glacial and meltwater deposition created a nearly continuous sedimentary cover (e.g. Klatkowa 1972; Turkowska 2006; Rdzany 2009). Associated glacial landforms constituted canvas for subsequent morphogenetic processes. The

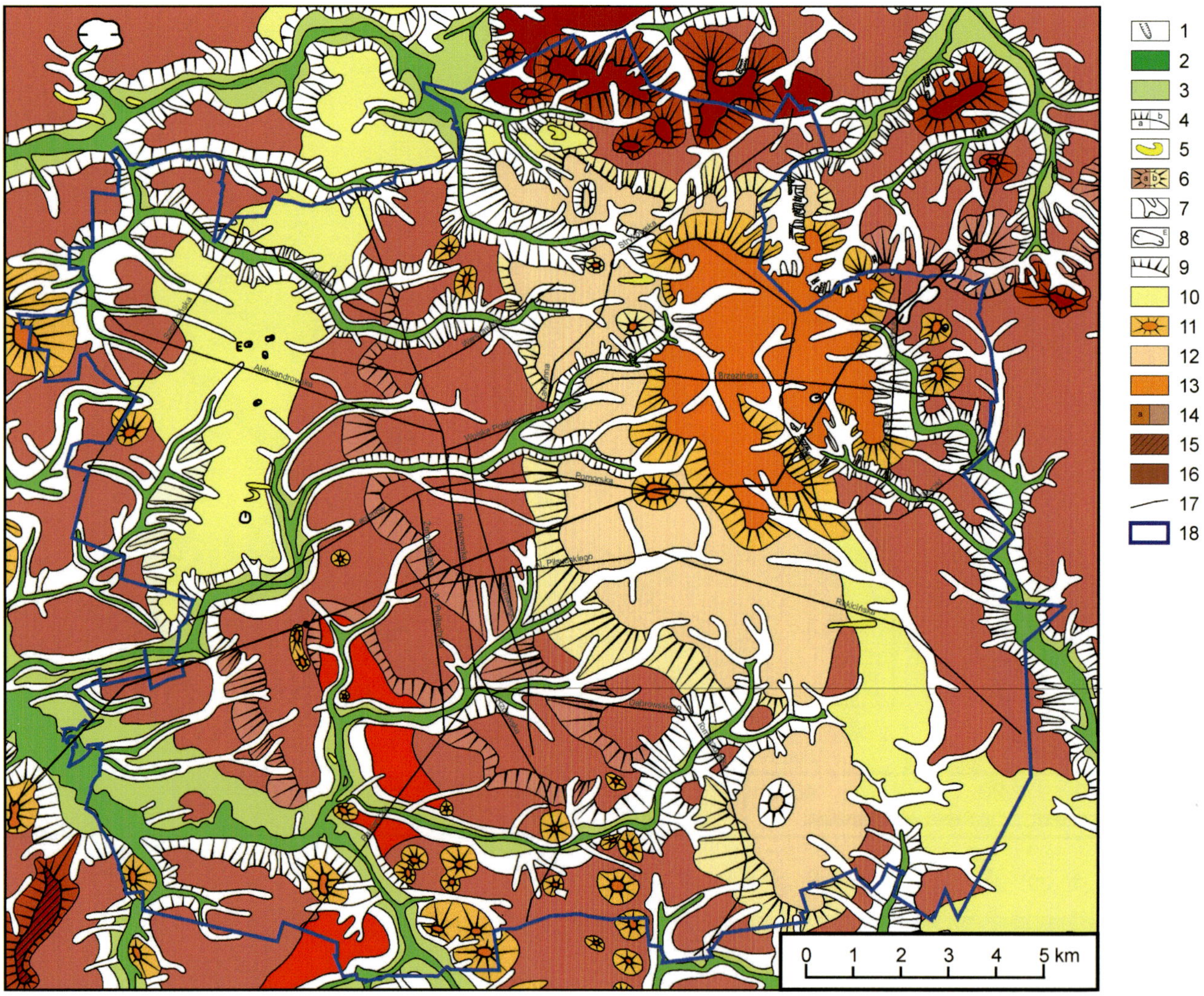

Fig. 29.2 Geomorphological map of the city of Łódź and its vicinity (after Goździk and Wieczorkowska 2011). 1—gully, 2—valley floor, 3—fluvial terrace, 4—valleys slopes: steep (a), gentle (b), 5—dune, 6—erosional remnant: of morainic upland (a), of erosional-denudational plain (level) (b), 7—denudational trough valley, 8—closed depression (E—with documented Eemian age of deposits), 9—long slopes, 10—sandur, 11—kame, 12—erosional-denudational plain (level), 13—kame plateau, 14—morainic upland, including melt-out areas (a), 15—glacitectonic rampart, 16—terminal moraine, 17—important streets, 18—city boundary

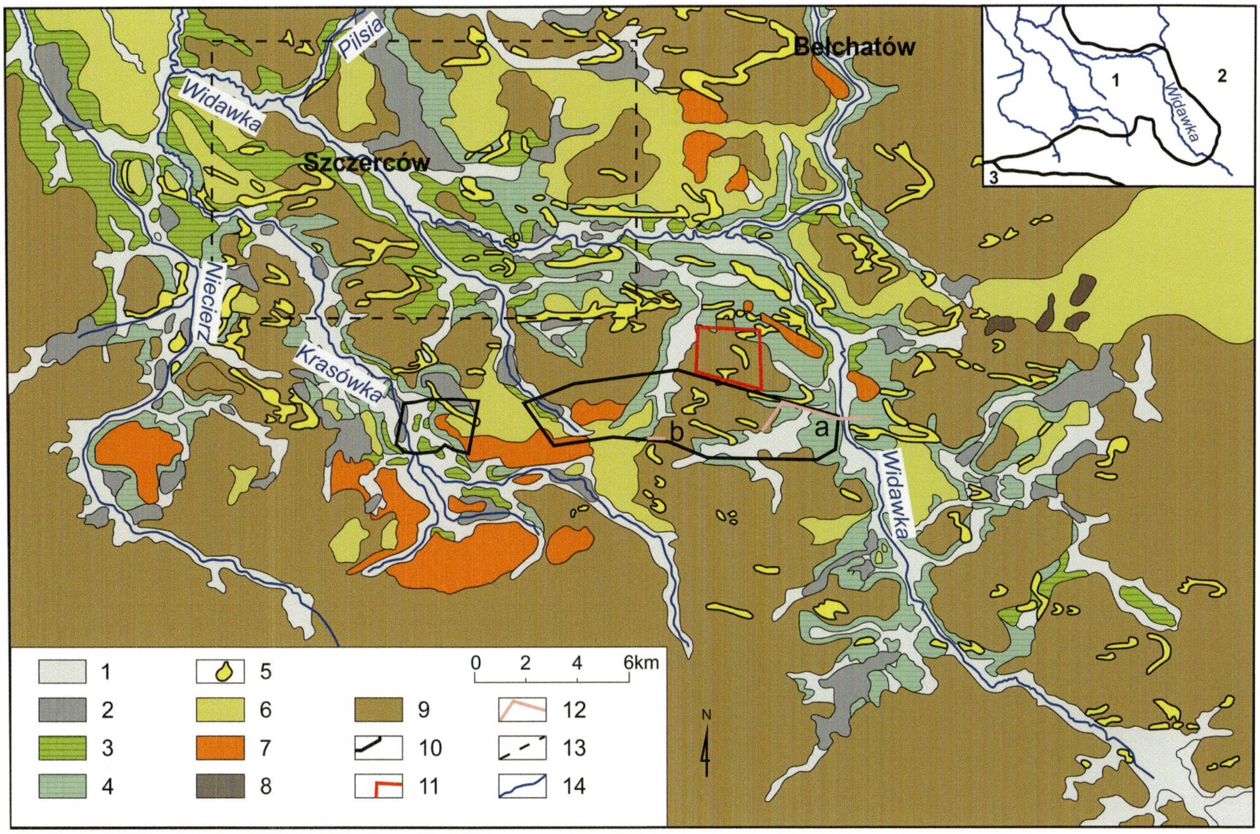

Fig. 29.3 Geological map of the Bełchatów area (after Baraniecka and Sarnacka 1971, modified). 1—sand and silt with organic material, 2—peat, 3—fluvial sand with silt of low terrace, 4—fluvial sand of high terrace, 5—aeolian sand in dunes, 6—glacifluvial sand and gravel, 7—sand and gravel in kames, 8—sand, gravel and till in end-moraines, 9—till, 10—limits of Bełchatów open cast mine, 11—boundaries of Bełchatów power plant, 12—location of cross-section (a—Fig. 29.9, b—Fig. 29.10), 13—extent of Fig. 29.4b, 14—rivers. Inset shows the limits of geomorphological regions after Gilewska (1986): 1—Szczerców Basin, 2—Bełchatów Upland, 3—Wieluń Upland

internal structure and origin of most of these forms can be recognized, even though they were later partly eroded or covered by younger sediments. The deposits belonging to this cover built the landforms of the post-glacial surface. Within morainic plateaus and uplands, but also within major valleys, fossil relief features from the Middle Pleistocene have also been reconstructed. However, landforms that dominate the contemporary relief were created during the Saalian glaciation.

In the peninsular upland, whose terminal part lies in the middle of the area (Fig. 29.1), large glacial landforms dominate. In the city of Łódź itself and its immediate surroundings, the depositional activity of the ice sheet created a specific morphological system, with an extensive kame plateau in the Stoki district of Łódź as the dominant major landform (Fig. 29.2; Goździk and Wieczorkowska 2011). Its culmination is at 265.8 m asl, one of the highest spots in the region. Its internal geological structure consists of a series of glacifluvial sands and gravels up to considerable 80 m thick, deposited in a single Saalian (Warthanian) depositional cycle. The plateau surface drops quite steeply to the west, to an erosional level created during the renewed transgression of the ice sheet (Rdzany 2009), followed by a series of steps extending as far as the Ner River valley (Figs. 29.2, 29.4a). Behind the plateau in the north there occurs a complex of morainic hills recognised as end-moraine landforms (Klatkowa 1972; Goździk and Wieczorkowska 2011), and characterised by large differences in elevation (Fig. 29.5), with characteristic topographic steps in their marginal zone. Their genesis was related to the glacitectonic influence of the Saalian ice sheet and its expansion over pre-existing glacial and glacifluvial deposits (Klatkowa 1972). Numerous glacitectonic structures have been found in the sediments of the kame plateau and adjacent glacial formations, supporting claims of the influence of an active ice sheet (Rdzany 2009). The image of the Saalian relief-forming stage is completed by sandur (outwash) surfaces, attesting to the last period of release of glacial meltwaters in this area.

South of Łódź, towards the town of Piotrków Trybunalski, runs the axial part of the morphological elevation, whose core consists of numerous kames bounded by

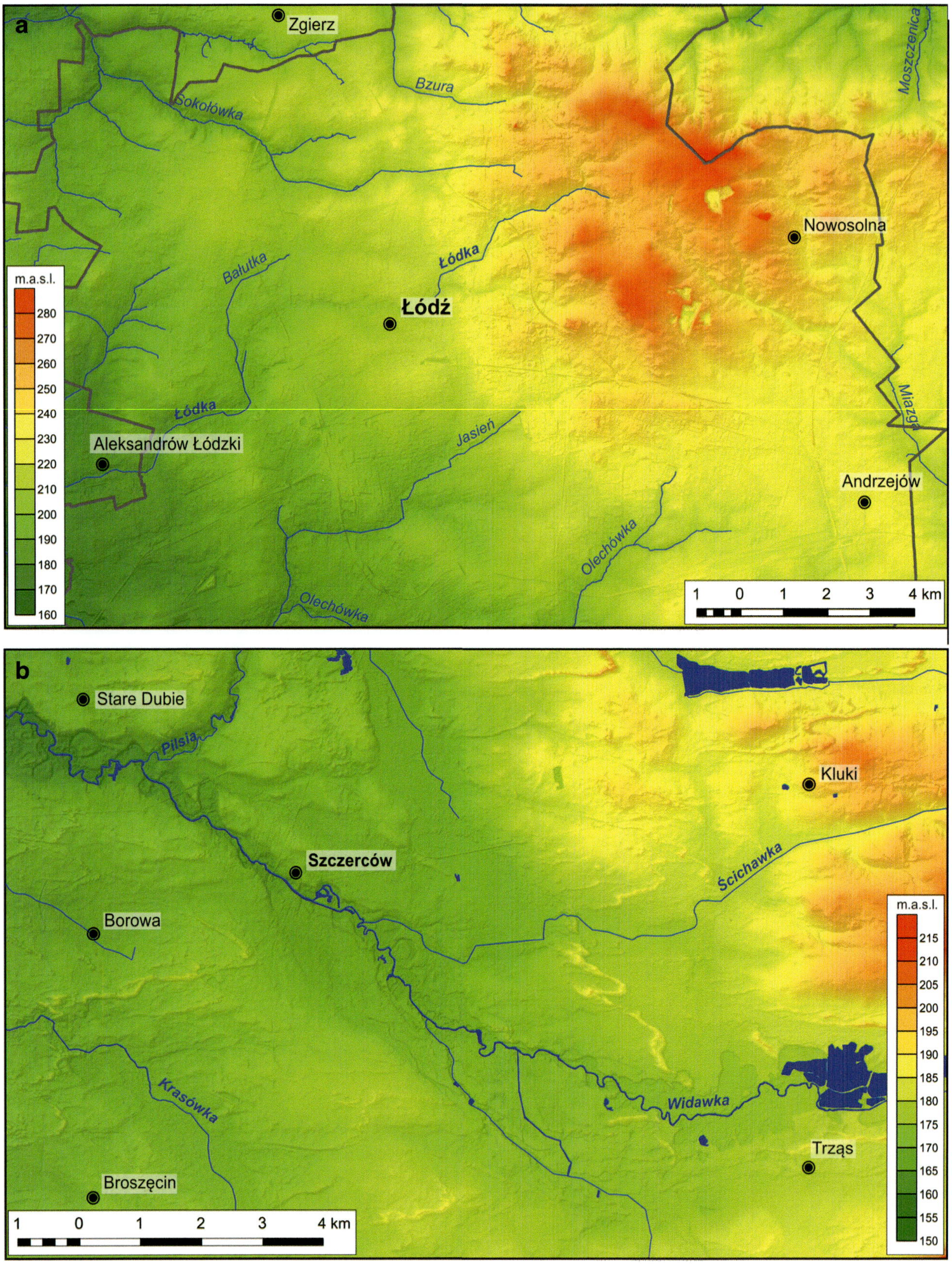

Fig. 29.4 Digital terrain models of representative areas within the Łódź region. **a** City of Łódź area. **b** Part of Szczerców Basin (note sinuous dune ridges on both sides of the Widawka River). Source: geoportal.gov.pl

Fig. 29.5 Morainic upland near Łódź. **a** Gently undulated water-divide areas with terminal moraine ridge visible in the distance. **b** Long slopes and wide part of a denudational valley (photographs by J. Forysiak)

moraine plateaus associated with the Saalian glaciation (Klatkowa 1972; Rdzany 2009). In the later stages of relief development, this area was intersected by river valleys descending to the east and west.

The southern part of the upland, the Bełchatów area, was selected as the second area for a detailed description. There, the morphological axis of the terrain runs first to the south-west and then westwards. The sequence of distinctive, elevated glacial forms encloses a vast, flat area to form the Szczerców Basin (Fig. 29.3). The surface of the basin slopes very slightly westwards. According to Gawlik (1970), the basin lies in a zone of lowered pre-Cenozoic bedrock, and during the recessions of successive ice sheets it played variable role. It acted as an area into which the glacier forefield was drained, and during subsequent transgressions it hosted an ice-dammed lake. The present-day geomorphology of the Szczerców Basin clearly shows the high-elevation glacigenic landforms around it, which formed during the Saalian glaciation, and a flat bottom zone. The kame knolls (Fig. 29.3), as well as hummocks considered to be end-moraines, are particularly well visible (Fig. 29.4b). The picture of the morainic plateaus and uplands is completed by relatively level terrains of clayey ground moraine with patches of glacial and glacifluvial sands, often with a thin cover of Late Weichselian aeolian sands. In the central part of the basin, post-glacial sediments lie several tens of metres below ground. Therefore, glacial bedrock of this part of the region has a completely different shape than in the Łódź part, which clearly influenced diverging pathways of periglacial processes during the Weichselian.

The period of deglaciation was associated with the formation of a river network. Proglacial waters have formed large valleys, the most important of them being the Pilica spillway leading north-east to the Warsaw Basin, the Warta-Widawka spillway in the western part of the region, which drained waters to the northwest, and the Warsaw–Berlin ice-marginal spillway that drained the northern part of the Łódź region (Fig. 29.1). The latter valley was a transit route for westward-flowing proglacial waters during the Saalian glaciation (Jewtuchowicz 1967). These major landforms are connected with the network of minor new river valleys that shaped the relief of the region in the subsequent developmental stages.

The wastage of the ice cover resulted in the appearance of numerous closed depressions, mainly in the areas of ground moraine and within sandur surfaces, similarly to areas within the limits of the last ice sheet. However, they are poorly marked in the modern-day relief of the Łódź region due to later transformations. Nonetheless, their occasional traces lend support to the hypothesis that they were numerous and varied in size.

29.4 Periglacial Transformation and the Origin of New Landforms

29.4.1 Frost Cracks, Ice and Sand Wedges, and Involutions

Climatic conditions have crucial influence on the characteristics of the periglacial environment and permafrost in particular. Periglacial structures associated with permafrost have been recognised in various deposits in the Łódź region, allowing for the recognition of climatic condition of the Weichselian. They indicate a significant variability of environmental conditions and regress of plant cover. During the Weichselian permafrost prevailed, which was either continuous, discontinuous or isolated, depending on changing climatic conditions. For example, continuous permafrost was present across the area throughout the entire Pleniglacial. Consequently, certain processes unique to permafrost areas could have occurred. Thermal contraction in winter created frost cracks, which converged into polygons. Water that subsequently filled the fissures froze in the following winter and this repetitive process led to the formation of ice wedges. During permafrost degradation, melting of these wedges created casts with secondary infilling, made of material from the sides and from above that initially surrounded the ice wedges. These structures were identified in the early days of periglacial research in the Łódź region (Pierzchałko-Dudkiewiczowa 1961; Dylik 1963). Later, more detailed analyses of structures of frost wedges were carried out. An exemplary ice wedge from the Łódź area (Fig. 29.6a) developed in the marginal part of a shallow depression within an outwash plain, where peat accumulated during the Eemian Interglacial. After ground ice melted, the wedge structure with secondary filling was formed (Goździk 1973). Ice wedges often developed in alluvial deposits, concurrently with sedimentation and rising surface elevation and hence, they are referred to as syngenetic (French and Goździk 1988; Fig. 29.6b).

In extremely cold conditions, when the cold desert prevailed, material was transported by wind in winter. This caused finer sand to be carried down into fissures, whereas coarse sand and gravel came to rest at the top (Fig. 29.6c). The repetition of this process produced sand wedges. Currently, these are being formed mainly in the ice-free parts of Antarctica (Péwé 1959). In the Łódź region, they formed only in specific intervals of the Upper (Late) Pleniglacial, when the cold desert prevailed (Goździk 1970, 1973; Fig. 29.6d, e).

The fissures are a good indicator of the downslope movement of sediments in the active layer. On slightly inclined surfaces, only the upper sections of the ice cracks/fissures, contained within the active layer, are inclined. On steeper slopes, striped patterns formed instead of polygonal fissures typical for horizontal surfaces. Within these complex structures, fissures were preserved if they were oriented roughly parallel to the slope, while others were blurred as the soil masses gradually slid downslope (Goździk 1973).

In the active layer within steeper slopes, material was transported by gelifluction. An example from the elevation of Góra Św. Małgorzaty near the town of Łęczyca (Figs. 29.1, 29.7) was presented by Dylik (1969). In the active layer, depositional structures were often deformed to create involutions (Fig. 29.7). There is a great variety of these structures, with dominating folds and frequent drop structures. Changes in relief recorded in the deformation of fissures on low-angle slopes and in involutions are small, but their occurrence may have influenced the development of certain features of the landscape, such as soil and water relations.

29.4.2 Lacustrine Morphogenesis

The Łódź region abounds in fossil closed depressions, even though these are difficult to discern in the contemporary landscape. Many of them have been studied over the past

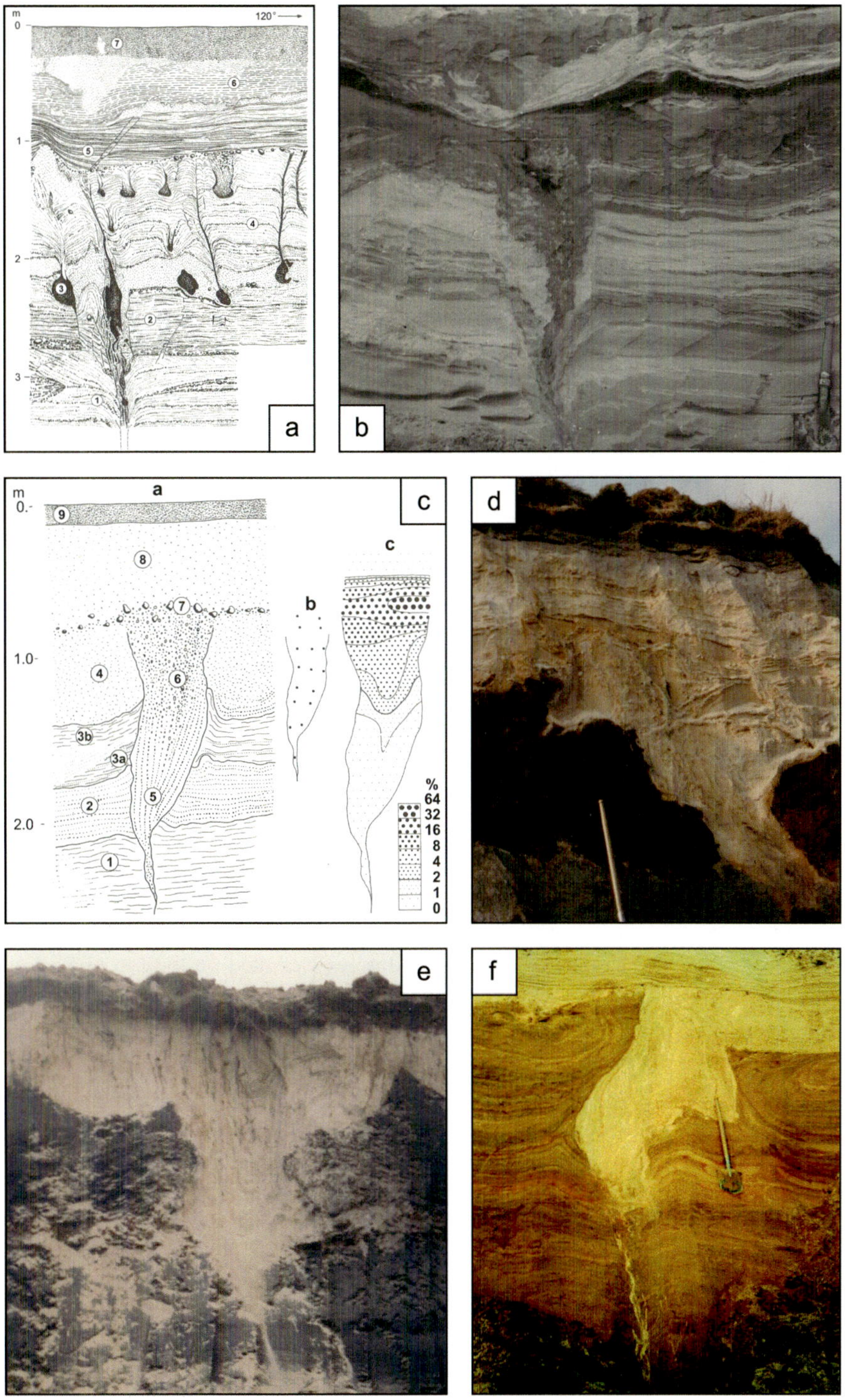

Fig. 29.6 Examples of periglacial structures. **a** Łódź-Teofilów. Drop structures in the buried active layer of permafrost and wedges with secondary fillings. 1—medium- and coarse sand with gravel, 2—medium sand, 3—peat with sand, 4—medium- and coarse sand, with pavement at the top, 5—laminated medium sand, 6—silt, 7—soil (after Goździk 1994). **b** Syngenetic wedge structure with secondary infilling in Middle Pleniglacial alluvial deposits exposed in the Bełchatów open-cast mine. **c** Sand wedge in the village Bedoń near Łódź. (a) lithology: 1—silt, 2—medium sand, 3a—fine sand with silt, 3b—silt, 4—fine- and medium sand, 5—medium sand, 6—sand with gravel, 7—pebbles, 8—medium- and fine sand, 9—sand with hummus, (b) location of samples, (c) share (%) of grains with a diameter greater than 2 mm (after Goździk 1994). **d** Sand-filled ice-wedge cast dissects a peat layer of Eemian age in the Bełchatów open-cast mine. **e** Sand-filled ice-wedge cast in the Bełchatów open-cast mine. **f** Complex wedge structure exposed in the Bełchatów open-cast mine. The upper part of wedge is interpreted as a primary sand wedge, the lower part as a composite-wedge cast, sand veins below (photographs by J. Goździk)

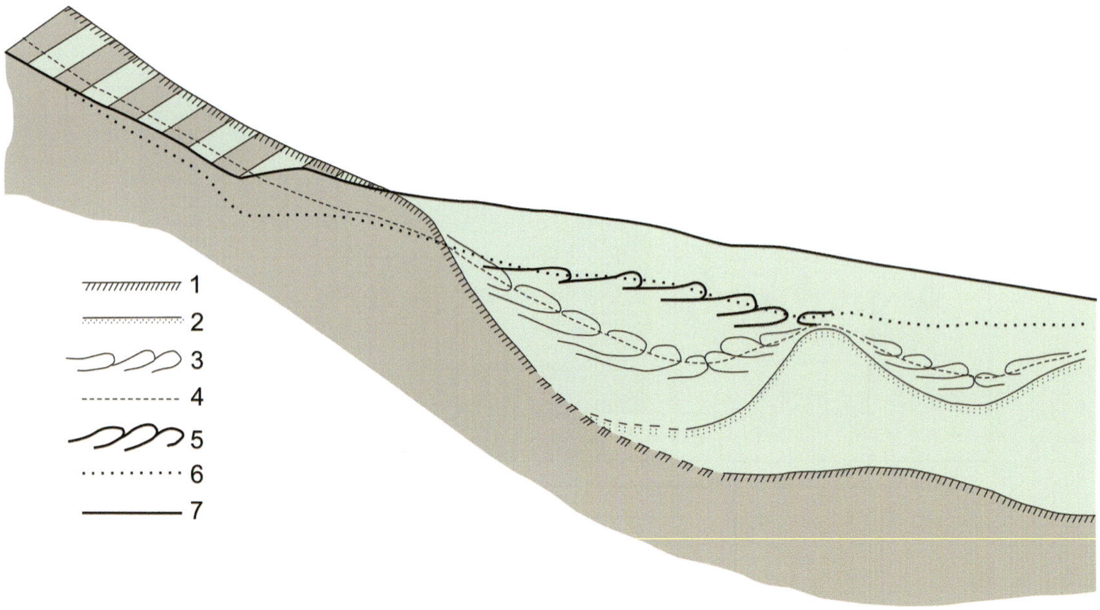

Fig. 29.7 Schematic diagram of periglacial slope development in Góra Św. Małgorzaty (after Dylik 1969). 1—primary glacigenic slope surface, 2—slope surface of Eemian and Early Weichselian fluvial sediments, 3—first generation of gelifluction lobes, 4—erosional surface truncating the first generation of gelifluction lobes, 5—second generation of gelifluction lobes developed in the top part of rhythmically-stratified slope deposits, 6—erosional surface truncating the second generation of gelifluction lobes, 7—present-day slope

several decades, with the best results achieved by works carried out in large excavations, and to a lesser degree if a dense network of drillings was available.

In the open-cast pit of the Bełchatów mine, closed depressions were visible on the top of the Saalian till. Some of these depressions occurred on flat surfaces of morainic plains, but more often they were present in the lower parts of slopes (Fig. 29.8). They were later incorporated into the valley network and many shallow depressions were filled with sediments as early as in the Eemian interglacial.

Deep depressions became filled with sediments over a long period. As a rule, sedimentation was continuous. Studies of these sediments allow one to trace the evolution of certain features of the natural environment. For example, palaeobotanical analyses have revealed changes in vegetation. One of the deepest closed depressions exposed in the mine is located on the southern slope of the pit, in the area of Kuców (Balwierz and Goździk 1999) (Fig. 29.9). The hollow was created on the top of Saalian till. A two-metre-thick layer of sand, with silt in the top part, originated during the period of deglaciation, rests directly on clay. At the beginning of the Eemian, a lake about 15 m deep existed. During this interglacial, clays accumulated in a forest environment, later turning into organic shales (about 2.5 m thick) as a result of compaction. These are overlain by Weichselian deposits accumulated in a periglacial

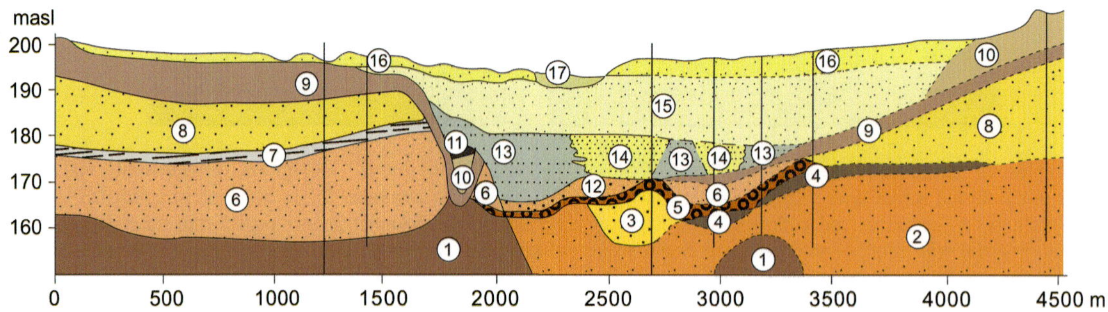

Fig. 29.8 Section through the Widawka River valley in the Bełchatów open-cast mine. Odranian stage (Saalian): 1—till, 2—medium and coarse sand, 3—sand with admixture of organic material, 4—till, 5—stone pavement, 6—medium and coarse sand, 7—varved clay, 8—medium sand, 9—till with medium sand layer, 10—medium and coarse sand and silt with clay; Eemian Interglacial: 11—peat with sand in the top; Weichselian: 12—medium and coarse sand and with gravel, 13—silt and sand, 14—medium and coarse sand, 15—medium and coarse sand (Pleniglacial), 16—medium sand with silt at the bottom, 17—fine, medium and coarse sand

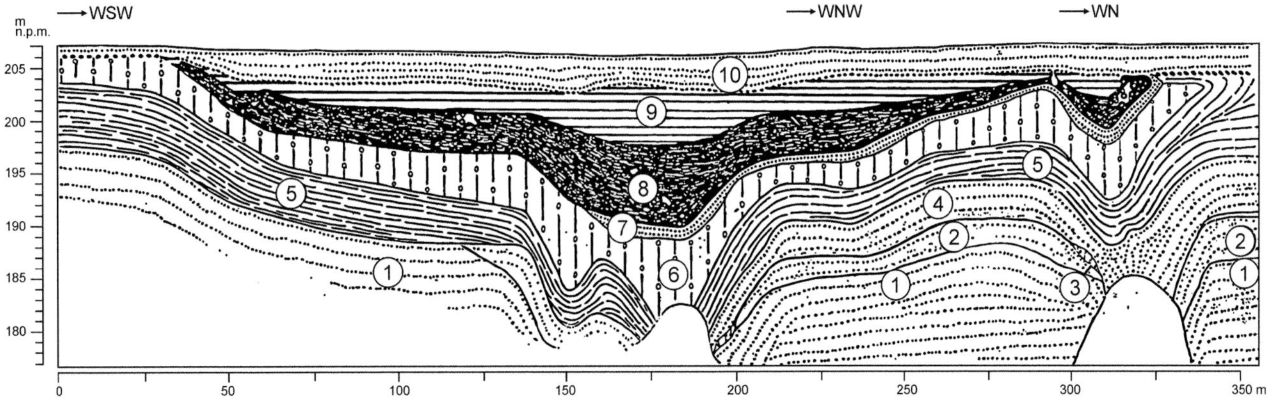

Fig. 29.9 Geological section across the depression with Eemian and Weichselian organic sediments at the Kuców 2 site, in the Bełchatów open-cast mine. Sanian stage (Elsterian): 1—medium and coarse sand with gravel; Wartanian stage (Saalian): 2—fine sand with admixture of medium sand, 3—till, 4—medium and coarse sand, 5—silt, 6—till, 7—medium and coarse sand with silty sand at the top; Eemian and Early Weichselian: 8—peat and organic silt; Pleniglacial: 9—silt with admixture of organic material decreasing upwards, 10—medium sand with silt laminae and gravel at the bottom (after Goździk 1986)

environment. In the Early Weichselian, sediments about 5 m thick accumulated and these are mainly organic silts and shales. These indicate that a dense plant cover with the participation of trees prevailed (Fig. 29.9; upper part of layer 8), and a clear change in sediment composition occurred when the deposition of a 5.5-m-thick layer of silt began in the Pleniglacial (layer 9). There was a significant admixture of organic material in the bottom part, but its share decreases upwards, comprising only a few percent in the top part. This reflects the onset of cooling and a change in vegetation to shrub tundra, and then to tundra grassland, with open plant communities limited to moist habitats. The dominant component of these silts is clastic material of aeolian origin. By the Upper Pleniglacial the depression was almost completely infilled and in its the coolest part the surface of the depression was only slightly concave. In the cold desert conditions, aeolian activity produced a layer of fine gravel and coarse sand linked with a stone pavement outside the depression. On the gravel layer, there is a sheet of sand of aeolian origin up to 3 m thick, with a few laminae of silt.

Among the documented fossil lake basins, a depression in the village of Józefów, identified in the early 1960s within moraine sediments of the Saalian glaciation, is of particular interest. It was filled with biogenic formations dated to the Eemian Interglacial and Early Weichselian. Above peat are mineral sediments considered to be of gelifluction origin (Dylik 1961), heavily deformed by segregated ice and thermokarst processes, associated with the development of pingos. The research currently conducted at this site confirms the presence of numerous periglacial structures and deformations resulting from the degradation of ground ice, but the verification of the earlier conclusions has not yet been completed.

29.4.3 Fluvial Morphogenesis

29.4.3.1 History of Fluvial Valleys

The post-glacial Saalian relief had a decisive influence on the layout of fluvial network in the Łódź region. This can be seen in the vicinity of Łódź itself (Fig. 29.2). The most elevated landform, the kame plateau, acts as a watershed. Its marginal zone and the less elevated surroundings constitute the source area of many rivers flowing westwards to the Warta spillway, to the north to the Warsaw–Berlin ice-marginal valley and to the south-east, to the Pilica valley. There are long dry valleys and troughs above the springs of these rivers. The valleys incised into the successive morphological steps around the kame plateau are fairly deep, especially in the edge zones of some of the steps, where they exceed 50 m deep (Turkowska 1988), as compared to the flat surfaces, where the depth is around 15–20 m (Fig. 29.4a).

The development of valleys and changes in morphogenetic processes and their effects can be best tracked in the Bełchatów mine excavations that exposed the sediments of the trunk valley of Widawka River and its several tributary valleys (Figs. 29.3, 29.8). In the area of Piaski, the central and western sections of the Widawka valley were excavated, whereas the eastern part was reconstructed from drillings. Information obtained from direct observations in excavations differs from the data obtained from even numerous drillings. The valley is quite deep, about 30 m, whereas the bottom of the contemporary valley lies barely about 10 m below the level of the surrounding plateau, proving higher intensity of erosional processes that formed the fossil valley. The erosional processes probably began in the Eemian Interglacial. However, it is worth paying attention to a mid-slope closed depression with an organic filling dated to the Eemian (Fig. 29.8). This dating means that incision of this depression and the lower part of the valley proceeded under periglacial conditions in the Early Weichselian. Pavement on the buried valley floor (Fig. 29.8) and medium and coarse sands with gravel inhibited further erosion.

Accumulation of an alluvial cover in this deep valley began in the Pleniglacial within the floodplain, sandy and silty sediments were rhythmically laminated, and higher up the slope, a cover of silts with inserts of organic material formed. Concurrently with sedimentation in the valleys, silty sedimentation occurred in large closed depressions. This is confirmed by radiocarbon dating and palaeobotanical analyses (Balwierz and Goździk 1999). Accumulation of a large amount of dust during this period suggests long-distance aeolian transport, probably from the more northerly areas that either had very poor vegetation or were semi-deserts. Morphoscopic analyses of grains from the sandy layers were performed (Fig. 29.10). The diagram shows that the proportion of rounded and mat (RM) to rounded and shiny (EL) grains in the bottom of the silty-sandy series is similar to that in the underlying till. However, the proportion of rounded and mat grains (RM) increases gradually upwards and shiny grains (EL) disappear almost completely towards the top, being then similar to proportions observed in the overlying sandy series.

Grain characteristics with this degree of rounding are comparable to those seen in deserts after long-distance transport. This way of aeolian transport affected the interaction between flowing water and sand dunes. The rhythmically-laminated silty-sandy sediments on the floodplain near the valley axis were incised by sand-transporting channels. In the top, the silty-sandy series and the sandy riverbed series were eroded. On the eroded surface, a more than 15-m-thick series of medium-grained sands with an admixture of coarse sands and fine gravel began to accumulate. Sedimentary structures indicate that the accumulation took place within a braided river. This series contains no organic material, which constitutes an important indicator that a cold desert predominated by that time. These sands have a high proportion of rounded mat grains (Goździk 1981). The development of these two series was accompanied by the development of syngenetic ice wedges (French and Goździk 1988). In the Late Weichselian, the sandy series in the upper axial part of the valley was eroded by the Widawka River, leaving a high terrace level. In this eroded part, organic sands, silts and peat began to accumulate in the Late Weichselian and the Holocene. At the transition between the Pleniglacial and Late Weichselian, aeolian cover sands accumulated on the surface of the terrace, on the valley slope and further up, on the plateau. The low volume of Late Weichselian and Holocene sediments is striking in comparison to those formed in the only slightly longer period spanning part of the Middle and Upper Pleniglacial.

It should be emphasised that the contemporary valleys of the Widawka River system in the Szczerców Basin are not very deep, being typically up to 5 and 10 m in the valley heads (Fig. 29.4b), whereas the fossil valleys are very deep, up to 25 m. In these valleys, the silty-sandy series reaches a similar thickness as in the Widawka valley, whereas the sand series is reduced. In the Struga Żłobnicka valley, Coope (in Kasse et al. 1998) analysed beetle assemblages in the silty-sandy sediments, which were dated using 14C yielding the age between 26,480 and 24,000 + 590 years BP.

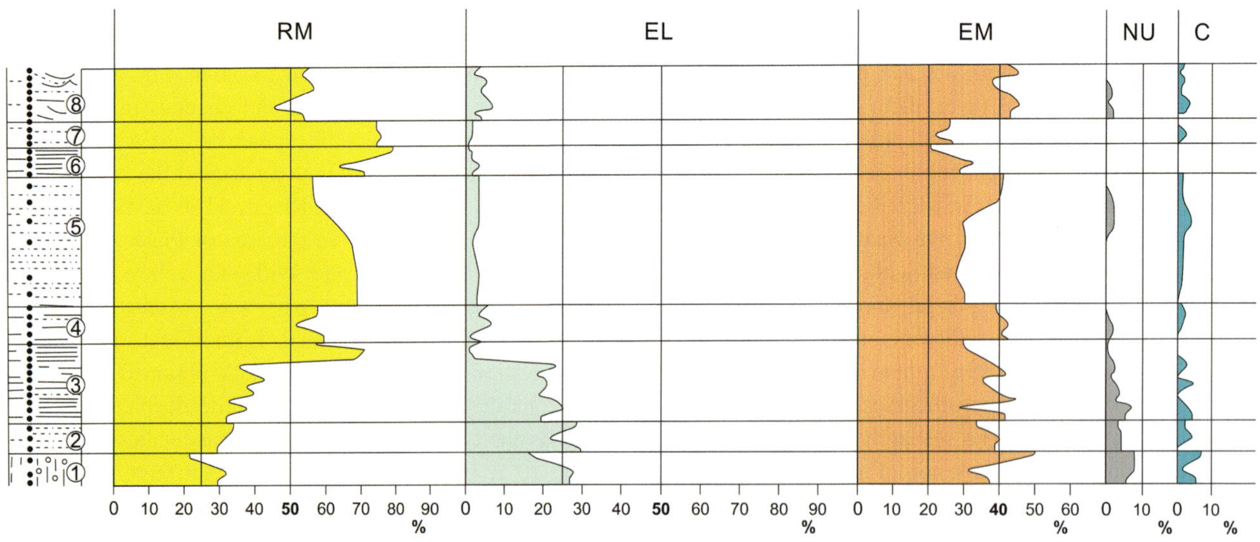

Fig. 29.10 Variability of the share of quartz grain types (morphoscopy, after Cailleux 1942) in a section exposed in the Bełchatów open-cast mine: RM—round mat, EL—shiny, EM—mat, NU—not abraded/fresh, C—crushed/broken (after Goździk 1980)

The results confirm that tundra dominated at the time and Coope concluded: '*The assemblage of fossil Coleoptera from Bełchatów is one of the most cold-adapted fauna so far recorded from Europe*'.

In the Warta valley, one of the main rivers in the region, the sedimentary history is similar to that in the Widawka valley, with the sediments having similar lithological features but being far more extensive. Cross-sections of the Warta River valley within the Uniejów Basin reveal that the valley was up to 10 km wide during the Weichselian (Figs. 29.1, 29.11). Sediments filling the valley have been recognised most comprehensively in two excavations in the open-cast mines of Adamów and Koźmin, where the total thickness of the fluvial series is up to 22 m (Fig. 29.12) (Petera 2002; Forysiak 2005). The lower segment of the alluvia is composed of sands and silts with numerous inserts of organic material (Fig. 29.13) and can be correlated with the silty-sandy series deposited in other valleys in the region. They contain numerous remnant structures from ice wedges and cryoturbations developed in permafrost conditions (Petera 2002). The upper, sandy segment deposited by a braided river is 10–15 m thick (Petera 2002; Forysiak 2005).

29.4.3.2 Periglacial Dry Valleys

In association with river valleys, there occur denudational troughs and dry valleys, lacking permanent streams and channel landforms. They are located above terrace levels and were formed by headward erosion and denudation rather than fluvial erosional processes. Dylik (1953) was the first to emphasize the role of periglacial processes in shaping these landforms, followed in more detail by Klatkowa (1965). Denudational troughs and valleys developed mainly in sandy and sandy-gravelly areas, which are easily permeable under contemporary conditions, but were frozen during cold periods of the Pleistocene and hence, the presence of permafrost made these sediments impermeable. Large volumes of thaw-water flowing down the inclined surfaces caused surface lowering, whereas erosion occurred if surface runoff was more concentrated. The efficacy of these processes increased in areas representing higher local relief such as the kame plateau in the north-eastern part of Łódź (Fig. 29.2), moraine hills and their northern foreland (Fig. 29.14). Some dry valleys in this area exceed 5 km in length (Fig. 29.2). Numerous denudational troughs and valleys were also formed on long slopes of larger inclination and within valley sides of deeply incised rivers (Klatkowa 1965; Turkowska 2006), being now suspended above terrace levels.

The lower reaches of dry valleys are filled with alluvial deposits, which connect with Early Pleniglacial alluvial deposits in the river valleys. However, these deposits wedge out in the middle reaches. This indicates that the lower reaches of denudational valleys were shaped during the Early Pleniglacial or at the beginning of Middle Pleniglacial. By contrast, the upper reaches were formed in periglacial conditions during younger period of Pleniglacial, due to headward erosion.

29.4.4 Aeolian Morphogenesis

29.4.4.1 Aeolian Sediments

The first person to draw attention to clear traces of intense aeolian activity in central Poland during the advance of the Weichselian ice sheet was André Cailleux, the author of an extensive study '*Les actions éoliennes périglaciares en Europe*' (Cailleux 1942). In the introduction to this study,

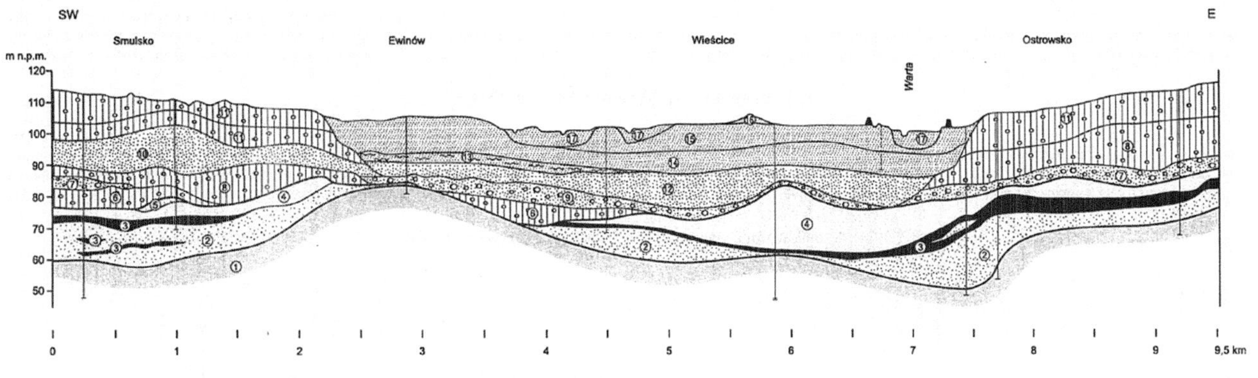

Fig. 29.11 Geological cross-section of the Warta River valley. Cretaceous: 1—gaize, marl, limestone; Neogene: 2—sand, 3—brown coal, 4—silt and clay, 5—silt; Quaternary: Elsterian: 6—till, 7—glacifluvial sand with gravel; Saalian: 8—till I, 9—glaciofluvial sand with gravel, 10—glacifluvial sand, 11—till II, 12—glacifluvial and fluvial sands; Weichselian: 13—fluvial silty sand, 14—fluvial sand with interbeddings of silt, 15—fluvial sand with organic mud admixtures, 16—aeolian sand; Holocene: 17—fluvial sand with mud (after Forysiak 2005)

Fig. 29.12 Late Quaternary deposits exposed in the Adamów open-cast mine. Fluvial sands of Weichselian age, Saalian till is visible in the foreground, on the lower excavation level (photograph by J. Petera-Zganiacz)

he described what drove him to undertake the study of periglacial aeolian processes. In the 1930s, while in Warsaw, he collected a handful of sand from alluvia in the Vistula River valley thinking that he had found an example of fluvial sand grains with shiny surfaces. To his amazement, the grains were dull and rounded, and thus being a clear result of aeolian reworking. He collected samples from Pleistocene sediments of northern, western and central Europe and discovered that there was a significant share of grains with aeolian imprint in the zone immediately adjacent to areas glaciated in the Weichselian cold stage. On land, this zone runs from the Netherlands and Belgium through Germany to Poland, now known as the European Sand Belt (Koster 1988; Zeeberg 1998). These grains occur most abundantly in central Poland and Cailleux (1942) believed that they had been wind-reworked during the transgression of the ice sheet.

Landforms associated with aeolian periglacial morphogenesis and clearly visible in the landscape are sand dunes. Central Poland is located in the middle of the European Sand Belt, where dunes occur alongside aeolian cover sands (Koster 1988; Zeeberg 1998). In the Łódź region, aeolian coversands occur in many places and occupy considerable areas, much larger than the dunes.

The aeolian coversands mantle various older landforms. The most extensive covers were formed on flat surfaces of morainic uplands, morainic plains, sandurs and wide high terraces, as well as on the tops of infills of closed depressions (Fig. 29.9). However, they were rarely included on the geological maps, which therefore underestimate their actual extent.

The opinion of Cailleux (1942) about the presence of a large number of rounded mat grains in the Pleistocene sediments of central Poland was confirmed by subsequent, more detailed analyses, but the share of these grains was found to vary greatly between sediments of different ages and origins. The share of RM grains is about 65% in the dune sediments of the Łódź region, but in the glacial substrate it drops to about 25% (Goździk 1981). The presence of wind-reworked grains in glacial sediments is not surprising, since in the Bełchatów region there are three alluvial and aeolian series analogous to the one described for the Weichselian, but originated during the previous glacial periods (Goździk 2007). The successive ice sheets eroded these alluvia, incorporating the grains. It is worth noting, however, that glacial sediments contain about 30% of rounded shiny grains (EL), while the dune sands contain only a few percent.

The share of each grain type in the dune sands is very similar to that found in the aeolian coversands, Upper Pleniglacial sandy fluvial series, and sand-wedge infills of the same age. These similarities and the difference in relation to the grains from their glacial substrate indicate that

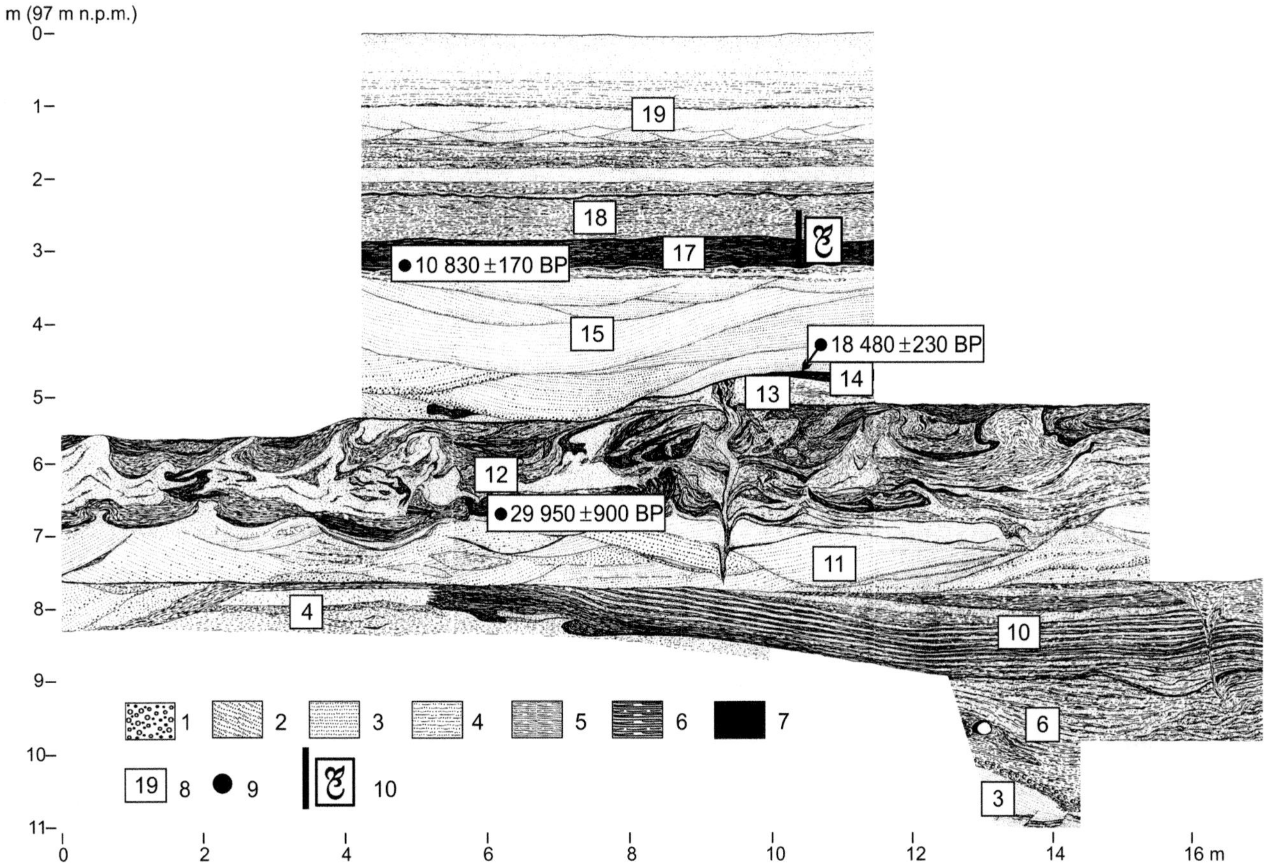

Fig. 29.13 Late Quaternary deposits exposed in the Koźmin open-cast mine: 1—gravels, 2—sand with gravels, 3—sand, 4—sand with silt, 5—silt, 6—organic silt, 7—peat, 8—number of the series, 9—location of radiocarbon datings, 10—location of the pollen profile (after Petera 2002)

a significant proportion of the material did not come from deflation or erosion of the local substrate, but was transported by wind under favourable conditions for long distances. Such conditions prevailed in the Middle and, above all, Late Pleniglacial as the climate was becoming colder. Gradually, tundra vegetation turned from shrub to grass, then to open plant community vegetation, and eventually disappeared. The sediments deposited during the maximum cooling contain no organic material, whereas they include features characteristic of cold deserts such as sand wedges and extensive stone pavements containing boulders with clear traces of aeolian abrasion (ventifacts), indicating a period favourable to aeolian activity. Sand was transported relatively easily by wind in lowland areas. In Central Europe, the dominant wind direction is—as it was in the Pleistocene—westerly, and the Central European lowlands, whose eastern part comprises the Central Polish lowlands, is likewise oriented east–west. It is through these lowlands that the European Sand Belt runs. Aeolian processes are difficult to accurately reconstruct, but some assumptions can be made based on the results of research in modern desert areas. They show that a high degree of rounding of quartz grains is only achieved after transport over hundreds of kilometres. It can therefore be assumed that most of the sandy material in the Pleniglacial sediments of the Łódź region came from long-distance transport.

29.4.4.2 Dune Fields

Spatial distribution of dunes in the region exhibits a certain regularity that suggests sand delivery by winds from the west. There are only a few dunes in or around the city of Łódź. To the north of the city, dunes are numerous in the Warsaw–Berlin ice-marginal spillway and on the lower morphological levels in the peripheral zone of the Łódź on the north-west and west side, but they do not go higher. On the upper levels, wind-transported sands do not form very extensive covers but are a component of relatively thin Late Pleniglacial alluvial series of the high terrace and/or fill sand wedges. This proves that the supply of aeolian sand to this area was limited. However, sand may also have been easily subject to further wind erosion and onward transport.

In the Bełchatów region, unlike in Łódź and its vicinity, there are numerous dunes, mainly in the Szczerców Basin. These are parabolic dunes, often with a remodelled north arm, linear dunes, as well as irregular mounds (Gawlik 1970; Figs. 29.3, 29.4b). The length of dune arms

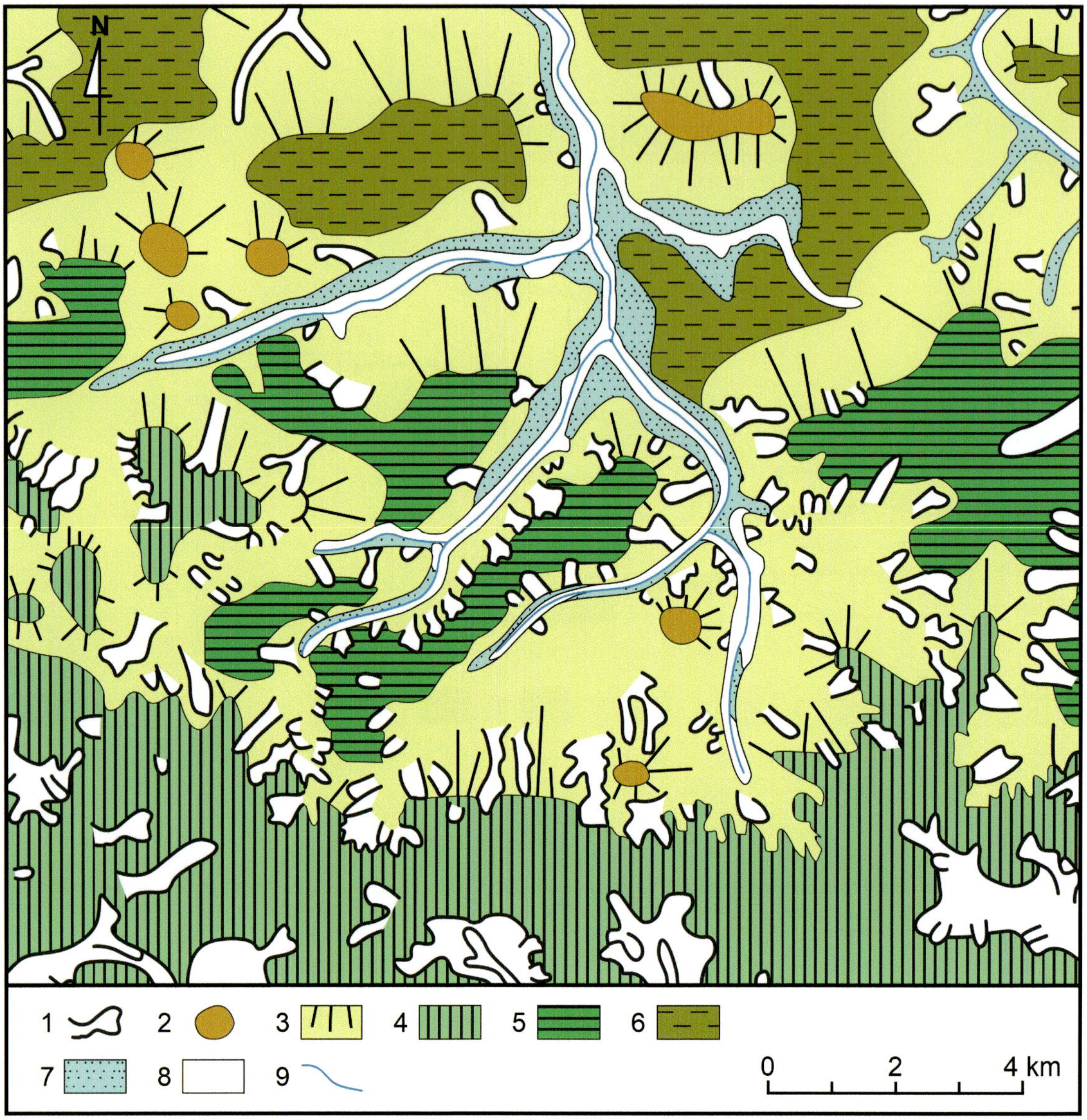

Fig. 29.14 Denudational troughs and valleys in the northern part of Łódź area, within the system of morphological levels after Klatkowa (1965). 1—denudational troughs and valleys, 2—isolated hills, 3—slopes, 4—moraine plateau level, 5—Smardzewski level, 6—Strykowski level, 7—fluvial terraces, 8—river valley floors, 9—rivers

varies, reaching up to 3.5 km, and the height of dunes is also variable, generally around 5 m in the western part of the Szczerców Basin, whereas towards the east it increases to 10 m or even 18 m. The orientation of the dunes and the asymmetry of their slopes clearly indicate the dominance of westerly winds (Gawlik 1970; Manikowska 1985). At the moment, dunes are usually covered by forest stands (Fig. 29.15a, b), but deforestation makes them looking similar to the original appearance during the Weichselian, under the conditions of a frozen desert.

Dunes and coversands are also found (though less often) in the Bełchatów Upland that surrounds the Szczerców Basin in the east and south-east. However, they disappear towards the south, near the border with the Małopolska Uplands, where the terrain rises. The aeolian covers have limited contribution to the relief, but they play an important role in shaping the landscape. For example, sands on the moraine surfaces determine soil type, and to some extent humidity, and hence, vegetation.

29.4.4.3 Aeolian-Fluvial Interactions

Sarnacka (in Baraniecka and Sarnacka 1971) emphasised that, among the deposits of Weichselian age, river sediments are the thickest and occur over large areas. She agreed with the prevailing opinion that sand was transported into the valleys by rivers eroding deposits in adjacent areas. Aggradation of valley bottoms '*facilitated the shifting of river beds and brought about the phenomenon*

Fig. 29.15 Dune landscapes. **a** Exposure in the longitudinal dune in Kuźnica Kaszewska (Szczerców Basin), located close to the Widawka River valley floor (photograph by J. Goździk). **b** Lee face of a dune in the eastern part of Szczerców Basin (photograph by J. Forysiak)

of braided rivers' (Baraniecka and Sarnacka 1971, p. 237). She indicated sections of three rivers at the western edge of the Szczerców Basin, where river beds were considerably shifted laterally to new locations. These are a SE–NW oriented section of the Widawka River and its similarly oriented tributaries, Krasówka and Nieciecz. More recent studies have indicated that heavy supply of sand to the valleys did not come from fluvial erosion of the glacial substrate, but was accomplished by long-distance wind transport (Goździk 1980).

The SE–NW-running sections of these three river valleys were oriented obliquely to westerly winds and crosswise to south-westerlies. This is important when considering aeolian–fluvial interactions. In modern desert areas, much attention is paid to this interaction, and Liu and Coulthard (2015) have distinguished its various types. Changes in the position of river valleys indicate that they occurred under the 'mostly aeolian dominant' interaction type, and such conditions probably existed here in the Late Pleniglacial In the section of the Widawka River valley oriented E–W (i.e., parallel to the prevailing wind), high terraces are extremely extensive along the wide valley (Figs. 29.3, 29.4b). It can be assumed that the channel was not pushed in one direction in the Late Pleniglacial, but wandered back and forth across the valley floor.

Processes involved in aeolian–fluvial interactions were largely terminated because of climate warming and accompanying development of plant cover in the Late Glacial, which resulted in the weakening of aeolian activity and the dominance of fluvial processes. However, the Late Weichselian is divided into three cold phases, separated by two warm intervals. In the Oldest Dryas modest amelioration of climate occurred, but it was still cool and tundra vegetation was present. Dunes were active and migrated, whereas low flat convex forms were built within aeolian coversands. In the warmer Bølling period, a park forest appeared in Central Poland and poor soils developed upon aeolian sandy covers (Manikowska 1985). However, high dunes were not overgrown and remained active, especially during the subsequent Older Dryas cool period. In the Allerød, warming enabled for expansion of mixed forest and stabilization of dunes began, along with the development of podzolic soil (Manikowska 1985). In the Younger Dryas cool period there was a short-term revival of aeolian processes, mainly within the dune fields.

29.5 Holocene Landform Transformation

The establishment of a continuous plant cover in the Holocene determined geomorphic stability and cessation of many geomorphological processes active before. A more pronounced activity was visible only in the river valleys, with channels adapting their patterns to changing conditions at the Weichselian-Holocene transition, transforming from braided rivers to meandering river systems (Turkowska 1988; Forysiak 2005). However, an increasing density of human occupation and the development of agricultural activities in the Neolithic, and then in the Bronze and Iron Ages, led to the activation of various morphodynamic processes (Twardy 2008). Initially, deflation and transformation of Late Weichselian dunes and aeolian covers or mechanical denudation on the cultivated slopes were localized, but starting from the Iron Age geomorphic transformation affected significant parts of the region, mainly the valleys of Warta, Pilica and Widawka rivers, and the Warsaw-Berlin ice-marginal valley (Twardy 2008). In the morainic uplands, especially in their central parts, as in the Łódź Upland, more intensive settlement and economic activity, with consequent relief transformation, appeared only in the Middle Ages (Dylik 1961; Twardy 2008). Human activities resulted in the development of younger aeolian forms, as well as in the incision of slopes inherited from the Weichselian period and deposition of sandy-silty fillings in denudational troughs and valleys, where gullies appeared, either as singular forms or clusters and networks (Twardy 2008; Figs. 29.16, 29.17). The Holocene is also the time of the development of wetlands in the region, especially in the low-lying areas, e.g., in the Warsaw-Berlin ice-marginal spillway and the Szczerców Basin (Fig. 29.3), where many deflation hollows and depressions enclosed by dunes in the Late Weichselian, as well as small fluvial forms in river valleys, have been subject to organic accumulation and peat formation.

A separate problem is human activity in modern times and direct transformations of the landscape as a result of the development of industrial and communication infrastructure, especially the exploitation of mineral resources. There are numerous pits left after exploitation of till, sand and gravel, both within the uplands, river valleys, and aeolian landforms. However, the greatest impact on the changes in hypsometry is exerted by the open-cast lignite mines (Figs. 29.12, 29.18) and the accompanying dumps. Currently, in the Bełchatów area one can see the largest height difference in Poland caused by human activity, reaching 400 m.

29.6 Conclusions

The relief of the Łódź region is polygenic. In the south-eastern and southern parts of the region, which lie within the Małopolska Upland, the activity of morphogenetic processes can be traced back to the Neogene. By contrast, within the bulk of the region, which lies within the Central Polish Lowlands, the present-day relief began to be shaped concurrently with the disappearance of the Saalian (Warthanian) ice sheet. Deglaciation left a nearly continuous cover of glacial and glacifluvial sediments that comprised various landforms. The topography created at that time was later transformed in the brief Eemian Interglacial and during the roughly one hundred thousand years long reign of periglacial conditions during the Weichselian cold stage. However, only a few traces of morphogenetic processes from the Eemian have remained. At that time, small glacial endorheic depressions were filled in and a river network began to form. By contrast, the effects of subsequent morphogenesis under periglacial conditions are much clearer. Larger endorheic depressions were filled with sediments and landforms characteristic for postglacial relief disappeared. Water filled the depressions, forming lakes, and the elimination of the depressions implied an important change in the landscape.

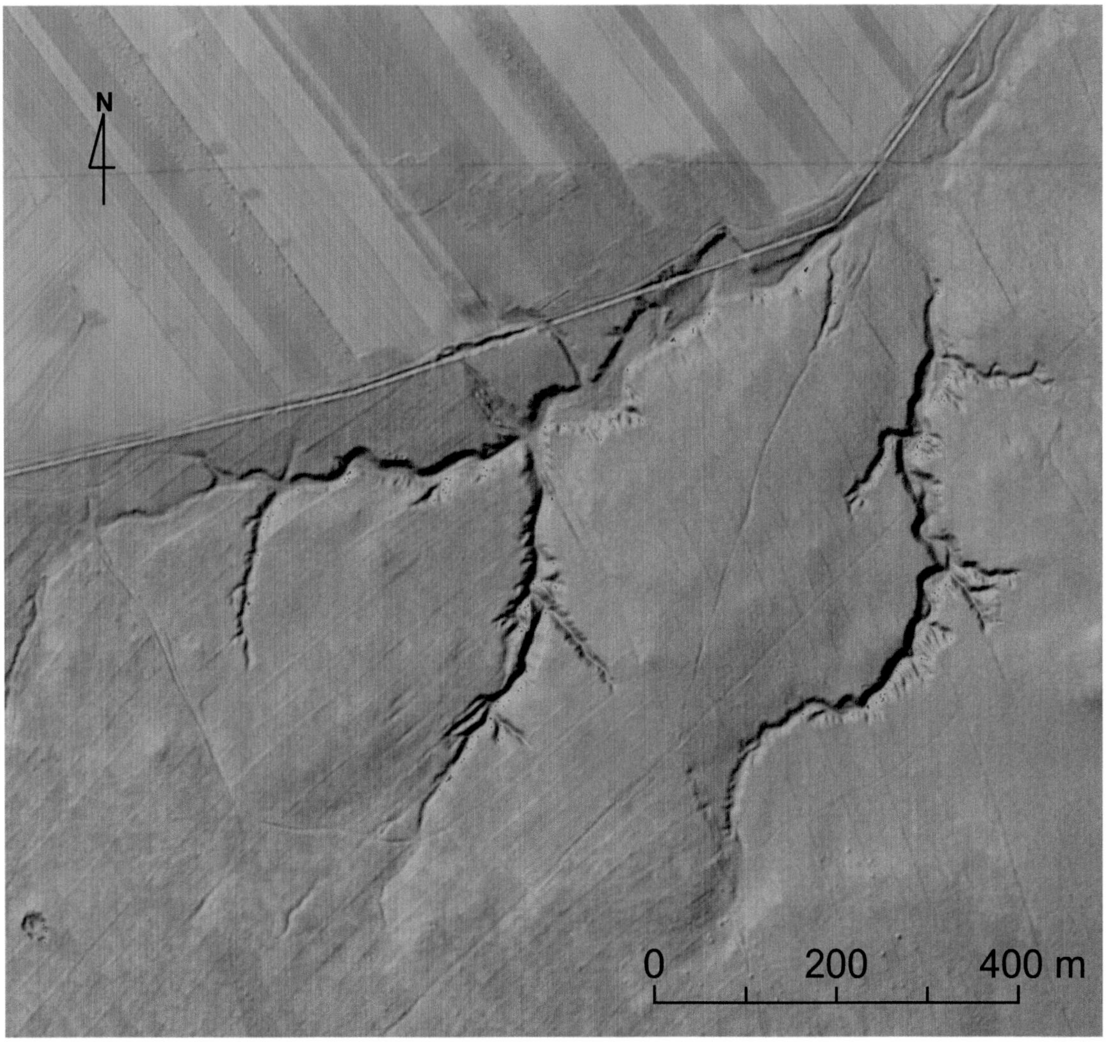

Fig. 29.16 System of late Holocene gullies in the nature reserve "Parowy Janinowskie" (ortophotomap superimposed on the digital terrain model). Source: geoportal.gov.pl

In the Early Weichselian deep valleys were formed and an efficient drainage system came into being. In the vicinity of the kame plateau of Łódź these valleys were incised to a depth of 50 m, whereas in the less elevated areas the depth of incision reached 15–20 m.

During the Middle Pleniglacial the depletion of vegetation led to the appearance of a grassy tundra, while aeolian dust brought in from distant northern regions, accumulated in river valleys and endorheic depressions. The intensity of aeolian processes depended very clearly on climatic conditions and related changes in the plant cover. In the coldest part of the Late Pleniglacial, when cold desert conditions prevailed, sand wedges commonly formed and sandy series devoid of organic material accumulated in the valleys. In this period the dunes were mobile, and it was only the change of climate and vegetation in the Late Weichselian that halted their activity.

The pathways of wind-driven sand movement were interrupted by rivers and hence aeolian–fluvial interactions took place. Sand transported by the wind was intercepted by rivers and subsequently deposited on the floodplain or in the channels. There, some sand was incorporated into the newly formed alluvia, and other was again subject to deflation and onward transport by wind.

Regional relief had considerable influence on the amount of sand transported by wind. In more elevated areas, for example, on and around the kame plateau of Łódź, a few dunes were formed and only aeolian coversands were deposited across small areas. By contrast, in the Szczerców Basin, numerous dunes formed and aeolian covers were laid

Fig. 29.17 Gully in the nature reserve "Parowy Janinowskie" (photograph by J. Twardy)

Fig. 29.18 Bełchatów open-cast coal mine represents the most significant relief transformation in the Polish Lowland (photograph by P. Czubla)

down across vast areas. The extensive shallow valleys and depressions between dunes on flat surfaces favoured the formation of peatlands.

Periglacial processes were decisive in relief transformation in the Łódź region. They obscured part of the glacial landforms and turned the region into an area of subdued glacial relief, known in Poland under the name of 'old glacial relief' and the Łódź region may be considered as a type area for this kind of geomorphological landscape of Poland.

Natural morphogenetic processes in the Holocene did not cause significant changes in the landscape of the Łódź region, except for localized areas, but the transformations induced by human activity are clearly visible in the shape of many medium-size and minor landforms.

References

Balwierz Z (2003) The Vistulian vegetation of central Poland (in Polish with English summary). Bot Guide Books 26:217–232

Balwierz Z, Goździk JS (1999) Excursion III. The Bełchatów opencast mine. Site 3. Eemian and Vistulian deposition in a closed depression—a long continuous record. Periglacial environments: past, present and future. In: Łódź Periglacial Symposium, pp 86–88

Baraniecka MD, Sarnacka Z (1971) Stratygrafia czwartorzędu i paleogeografia dorzecza Widawki. Biul Inst Geol 254:157–269

Cailleux A (1942) Les actions éoliennes périglaciares en Europe. Mem Soc Geol France 46:176

Dylik J (1948) Ukształtowanie powierzchni i podział na krainy podłódzkiego obszaru. Acta Geogr Univ Lodz 1:1–46

Dylik J (1953) O peryglacjalnym charakterze rzeźby środkowej Polski. Acta Geogr Univ Lodz 4:109

Dylik J (1961) The Łódź Region. In: Guide-Book of Excursion C, VIth Congress INQUA, Poland, pp 7–83

Dylik J (1963) Traces of thermokarst in the Pleistocene sediments of Poland. Bull Soc Sci Lettr De Łódź 14:1–16

Dylik J (1967) Główne elementy paleogeografii młodszego plejstocenu Polski Środkowej. In: Galon R, Dylik J (eds) Czwartorzęd Polski. PWN, Warszawa, pp 311–352

Dylik J (1969) Najstarszy interstadiał ostatniego pietra zimnego w Polsce (amersfoort). Kwart Geol 13:408–423

Dylik J (1973) Znaczenie badań wiecznej zmarzliny dla pełniejszego poznania peryglacjalnej morfogenezy w Polsce plejstoceńskiej. Czas Geogr 44:207–215

Forysiak J (2005) Rozwój doliny Warty między Burzeninem i Dobrowem po zlodowaceniu warty. Acta Geogr Lodz 99:116

French HM, Goździk J (1988) Pleistocene epigenetic and syngenetic frost fissures, Bełchatów, Poland. Can J Earth Sci 25:2017–2027

Gawlik H (1970) Geomorfologia Kotliny Szczercowskiej. Acta Geogr Lodz 26:104

Gilewska S (1986) Podział Polski na jednostki geomorfologiczne. Przegl Geogr 58:15–40

Goździk JS (1970) Geneza szczelin zmarzlinowych w regionie łódzkim w świetle analizy ich wypełnień. Acta Geogr Lodz 24:181–196

Goździk JS (1973) Origin and stratigraphical position of periglacial structures in Middle Poland. Acta Geogr Lodz 31:117

Goździk JS (1980) Zastosowanie morfoskopii i graniformametrii do badań osadów w Kopalni Węgla Brunatnego Bełchatów. Stud Reg 4(9):101–114

Goździk JS (1981) Less changements du processus eoliens dans la Pologne Centrale au cours du Vistulian (Würm). Rech Geogr Strasbourg 16(17):115–120

Goździk JS (1986) Structures de fentes à remplissage primaire sableux du Vistulien en Pologne et leur importance paléogéographique. Biul Perygl 31:70–105

Goździk JS (1994) Etudes des fentes de gel en Pologne Centrale. Biul Perygl 33:49–78

Goździk JS (2007) The Vistulian aeolian succession in central Poland. Sediment Geol 193:211–220

Goździk J, Wieczorkowska J (2011) Table VIII: Relief. In: The Łódź Atlas, Łódź

Jewtuchowicz S (1967) Geneza Pradoliny Warszawsko-Berlińskiej między Nerem a Moszczenicą. Prace Geogr IG PAN 62:102

Kasse C, Huijzer AS, Krzyszkowski D, Bohncke SJP, Coope GR (1998) Weichselian Late Pleniglacial and Late-glacial depositional environments, Coleoptera and periglacial climatic records from central Poland (Bełchatów). J Quat Sci 13:455–469

Klatkowa H (1965) Niecki i doliny denudacyjne w okolicach Łodzi. Acta Geogr Lodz 57:142

Klatkowa H (1972) Paleogeografia Wyżyny Łódzkiej i obszarów sąsiednich podczas zlodowacenia warciańskiego. Acta Geogr Lodz 28:220

Klatkowa H (1996) Symptoms of the permafrost presence in Middle Poland during the last 150 000 years. Biul Perygl 35:45–72

Koster EA (1988) Ancient and modern cold-climate aeolian sand deposition: a review. J Quat Sci 3:69–83

Liu B, Coulthard TJ (2015) Mapping the interactions between rivers and sand dunes: implications for fluvial and aeolian geomorphology. Geomorphology 231:246–257

Manikowska B (1985) O glebach kopalnych, stratygrafii I litologii wydm Polski Środkowej. Acta Geogr Lodz 52:137

Marks L, Gałązka D, Woronko B (2016) Climate, environment and stratigraphy of the Pleistocene last glacial stage in Poland. Quat Intern 420:259–271

Petera J (2002) Vistuliańskie osady dolinne w basenie uniejowskim i ich wymowa paleogeograficzna. Acta Geogr Lodz 83:174

Péwé TL (1959) Sand-wedge polygons (tesselations) in the McMurdo Sound region, Antarctica; a progress report. Am J Sci 257:545–552

Pierzchałko-Dudkiewiczowa Ł (1961) Nowostawy. Guide-Book of Excursion C. VIth Congress INQUA, Poland, pp 55–57

Rdzany Z (2009) Rekonstrukcja przebiegu zlodowacenia warty w regionie łódzkim. University of Lodz Press, Łódź, p 312

Turkowska K (1988) Rozwój dolin rzecznych na Wyżynie Łódzkiej w późnym czwartorzędzie. Acta Geogr Lodz 57:157

Turkowska K (2006) Geomorfologia regionu łódzkiego. University of Lodz Press, Łódź, p 206

Twardy J (2008) Transformacja rzeźby centralnej części Polski Środkowej w warunkach antropopresji. University of Lodz Press, Łódź, p 292

Zeeberg JJ (1998) The European sand belt XE "European Sand Belt" in eastern Europe—and comparison of Lateglacial dune orientation with GCM simulation results. Boreas 27:127–139

Jan Goździk is geographer and geomorphologist. Throughout the entire period of his research and teaching activity, he was associated with the Institute of Geography, currently the Faculty of Geographical Sciences, University of Łódź. His main research interests are periglacial environments, frost-related structures, and transformation of relief in the Pleistocene periglacial conditions. He also conducted research of aeolian forms and deposits. He participated in scientific internships and conducted research in various European countries, also in Canada and Siberia; presented the results during many international conferences organized by International

Geographical Union and INQUA. He participated in activities of IGU groups such as the Coordinating Committee for Periglacial Research, the Commission on Significance of Periglacial Phenomena, and the Commission on Frost Action Environment. He is author or co-author of about 150 papers, published in Polish, French and English.

Jacek Forysiak is geographer and geomorphologist, Associate Professor based at the Department of Geology and Geomorphology at the University of Łódź. His research deals with wetland deposits, fluvial geomorphology and stratigraphy of the Late Pleistocene and has been involved in projects in different geomorphological zones of the Central European Lowland. He is author or co-author of about 100 papers, member of scientific associations (e.g. Vice-President of the Association of Polish Geomorphologists from 2017) and committees of Polish Academy of Sciences (Committee of Geohraphical Sciences, Committee of Quaternary Research). He is the editor-in-chief of *Acta Geographica Lodziensia.*

30 Inland Dunes of the Kampinos National Park

Maciej Dłużewski and Barbara Woronko

Abstract

The Kampinos National Park (KNP) was founded in 1959 to protect valuable natural features within two inland dune belts interspersed with swamps. The KNP landscape is associated with glacier activity during the last Ice Age, when Scandinavian Ice Sheet has reached its maximum extent north of the present-day border of KNP, dammed the meltwater and proto-Vistula outflow, which led to the creation of a huge lake filled with alluvia. Then the proto-Vistula flowed along the ice margins to the west, forming up to 10 km wide sandy Warsaw–Berlin Proglacial Valley. Around 13 ka cal BP, when ice sheet retreated, the width of the proto-Vistula channel greatly decreased, exposing sandy terraces, which became the source of sediments for aeolian processes. The shape of the present-day dunes in the KNP indicates that they are related to the last phase of aeolian activity in Poland, to the Younger Dryas and locally to the Preboreal and historical times. Two belts of parabolic dunes, i.e. transverse dunes with arms directed toward the west, anchored by vegetation and high sand moisture, were formed, indicating westerly wind as the dominant dune-forming wind. The tallest dunes of the KNP reach up to 30 m height. They are V-, U- or even C-shaped simple transverse ridges or compound dunes, when they developed close to each other and their arms coalesced. Around 10 ka cal BP the Vistula shifted to its present position, while abandoned channels began to transform into two parallel belts of swamps. As a result two latitudinal dune belts separated by two swamp belts were formed.

Keywords

Parabolic dunes · Inland dunes · Warsaw basin · European sand belt · Warsaw-Berlin ice-marginal valley · Kampinos National Park

30.1 Introduction

The Kampinos National Park (KNP) is the second largest national park in Poland. It was established on 16 January 1959 to protect rich mosaic of habitats, from the very wet to the very dry, and particularly to protect one of the largest inland dune fields in Europe. As early as 1924, thanks to the efforts of Jadwiga and Roman Kobendza, the first nature reserves were created. Later on, in 1951, they set up a reserve breeding centre for elk, a symbol of the Park. The KNP is located in the Masovian Voivodship, in the south-western part of the Warsaw Basin (Figs. 30.1 and 30.2), between the left bank of the Vistula River and the Bzura River, near the western border of Warsaw, the capital of Poland. The area of the park is 38,544.33 ha and a buffer zone with an area of 37,756 ha stretches around the park. The most valuable parts of the park, which together constitute approximately 12% of its area (4636 ha), are under strict protection.

Stabilized inland dunes and marshy areas are the most characteristic elements of the KNP landscape. The dunes of the KNP are considered to be the best preserved complex of inland dunes in Europe (Kobendza and Kobendza 1958). This dune-dominated area is located in the middle part of the European Sand Belt (ESB), stretching from the Netherlands and Belgium in the west, across Germany, southern Denmark, Poland into Ukraine, Latvia and Russia in the

M. Dłużewski (✉)
Faculty of Geography and Regional Studies, University of Warsaw, Krakowskie Przedmieście 30, 00–927 Warsaw, Poland
e-mail: dluzewski@uw.edu.pl

B. Woronko
Faculty of Geology, University of Warsaw, Żwirki i Wigury 93, 02–089 Warsaw, Poland
e-mail: bworonko@uw.edu.pl

P. Migoń and K. Jancewicz (eds.), *Landscapes and Landforms of Poland*, World Geomorphological Landscapes, https://doi.org/10.1007/978-3-031-45762-3_30

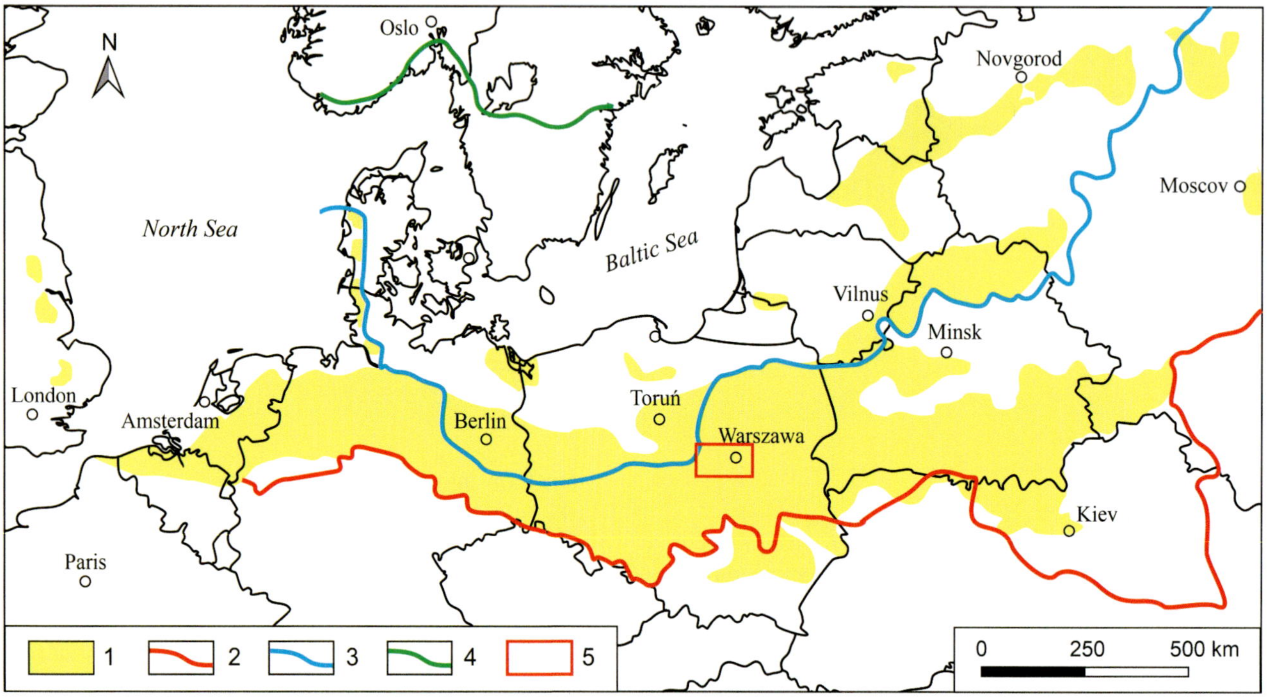

Fig. 30.1 Extent of the European Sand Belt (after Zeeberg 1998, simplified). 1—sand belt, 2—Saalian maximum ice margin, 3—Weichselian maximum ice margin, 4—Younger Dryas maximum ice margin, 5—location of KNP with its surroundings shown in details in Figs. 30.2 and 30.4

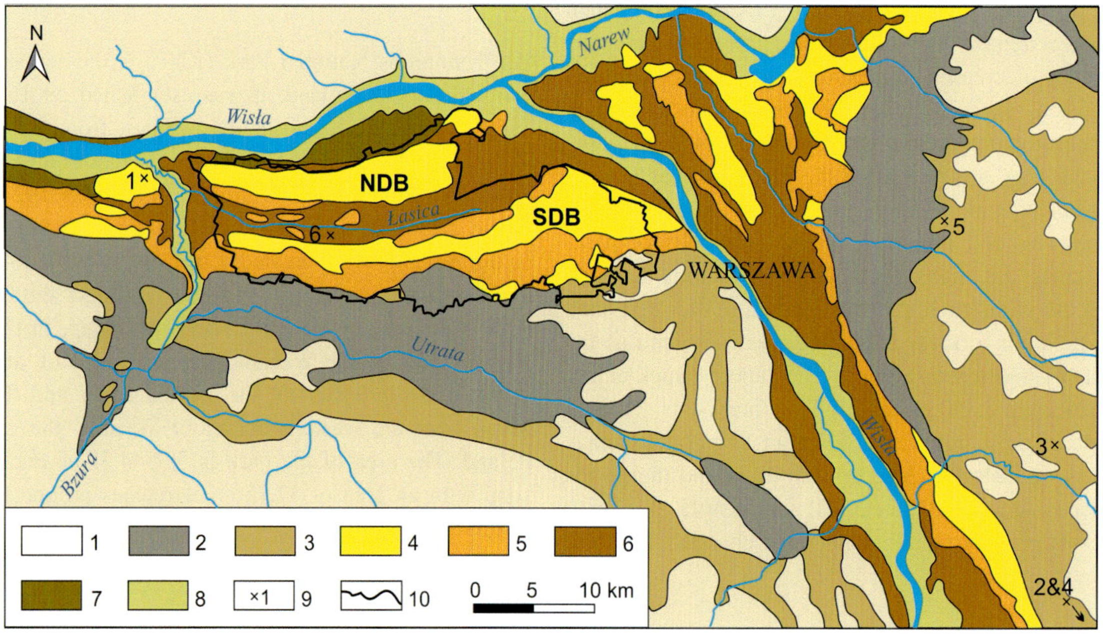

Fig. 30.2 Geomorphology of the Warsaw Basin (based on Nowak 1974; Baraniecka 1982; Baraniecka and Konecka-Betley 1987; Wiśniewski 1990 simplified). 1—postglacial plateau, 2—plains of ice-dammed deposits, 3—accumulation covers of upland river valleys and alluvial fans, Fluvial terraces: 4—Otwock terrace (with dunes), 5—Falenica terrace, 6—Praga terrace (sands and alluvial soils), 7—upper part of floodplain, 8—lower part of floodplain, 9—sites presented (1—Kamion, 2—Całowanie, 3—Wiązowna-Piekiełko, 4—Stężyca, 5—Cięciwa, 6—Narty), 10—border of the KNP, NDB—Northern Dune Belt, SDB—Southern Dune Belt

east (Koster 1988, 2005; Mycielska-Dowgiałło 1993; Zeeberg 1998; Kasse 1999) (Fig. 30.1). The ESB includes cold-climate inland aeolian sediments and landforms, inherited from the Pleistocene (Koster 1988). In Poland, dunes were widely developed in central and southern part of the country in lowland areas, mostly within large ice-marginal valleys (Germ. *Urstromtal*), including the Toruń-Eberswalde, Warsaw-Berlin, Głogów-Baruth and Wrocław-Magdeburg valleys. They also developed within basins filled with sandy alluvial deposits when Scandinavian Ice Sheet dammed outflow of the meltwater, as in the Toruń, Szczecin, Warsaw, Grudziądz, Grabów, and Sandomierz basins or within glaciofluvial outwash fans (Nowy Tomyśl and Kurpie) (Rotnicki 1970). The main dune-forming phases within the Polish lowland date back to the Oldest and/or Older Dryas, whereas the Younger Dryas and Preboreal were phases of dune transformation (Izmaiłow 1976; Nowaczyk 1986; Moska et al. 2022). Only in the areas located in northern Poland, as in the Szczecin and Toruń basins, the main dune-forming phase coincides with the Younger Dryas (Andrzejewski and Weckwerth 2010). All these dunes where formed by wind from the western sector during dry and cold climatic conditions, where aeolian sand transport has not been limited by high moisture content or vegetation in the source areas. Aeolian deposition was interrupted in the Bølling and Allerød interstadials as well as in the Younger Dryas-Preboreal period. It should be emphasized here, however, that the present-day shape of the dunes in the KNP, determined on the basis of DTM, results from the last phase of activity of aeolian processes, which took place in the Younger Dryas and locally in the Preboreal and historical times.

The KNP is a special natural and cultural object, also shaped by human activities (Kobendza and Kobendza 1958; Bomanowska et al. 2009). It is a fragment of the sixteenth century Kampinos Forest, formerly known as the Mazowiecka Forest, and KNP is identified with it (Andrzejewska et al. 2010; Owadowska et al. 2013). As the rivers Bug, Narew, Wkra and Bzura flowing into the Vistula River form here a hydrological node, in 1998, the European Council recognized the KNP as a European ecological site and a corridor of European importance. More than 73% of the Park's area is covered by forests, mainly pine. They cover the dunes, which currently results in their stabilization (Fig. 30.3). Due to its natural values and social importance, in 2000 the KNP with its buffer zone was recognized by UNESCO as a World Biosphere Reserve under the name of "Puszcza Kampinoska" (Owadowska et al. 2013). In 2004, the KNP became a part of the European Ecological Network Natura 2000 (PLC 140,001) (Okołów and Marczak 2008). The Park is a recreational and leisure base mainly for residents of Warsaw, who have 350 km of hiking trails, a 200-km-long bicycle trail and horse trails at their disposal.

30.2 Geology and Geomorphology

The KNP is located in the eastern part of the Warsaw-Berlin ice-marginal valley. In relation to the surrounding areas the park is a separate unit both in terms of geology and relief (Figs. 30.2 and 30.4). The most characteristic feature of the KNP is a latitudinal extent of landforms parallel to the Vistula River. Local relief in this area reaches up to 30 m. The highest point is 107.4 m asl, the lowest point is located at the mouth of the Bzura River to the Vistula River and is at 66.7 m asl.

The northern part of the Park includes the Vistula River valley with a floodplain terrace 1.5–5.0 km wide, which is locally shaped into two terrace levels (Baraniecka 1982) (Fig. 30.2). It rises about 3 m above the mean level of the Vistula River and is called a 'meadow terrace' (Owadowska et al. 2013). The surface of the floodplain terrace is criss-crossed by numerous oxbow lakes and paleochannels (Liwski et al. 1967).

To the south of the floodplain terrace the northern dune belt (NDB) occurs, forming a belt about 3.5 km wide (Figs. 30.2 and 30.4). The dunes were formed on the Otwock terrace, known in the northern part as the 'dune terrace' (Baraniecka 1982). Within the KNP, this terrace is located at an altitude of 73–80 m asl and is separated from the Vistula floodplain terrace by a 3 m high scarp, not clearly visible everywhere in the relief (Liwski et al. 1967). Elevation differences within the NDB reach 30 m. The development of the Otwock terrace was associated with the Last Glacial Maximum (LGM) and the subsequent ice-sheet recession. At that time, the front of the ice sheet was located about 50 km to the north-west of the KNP (Marks 2002, 2012). It blocked the valley and a large ice-dammed lake developed, filling the Warsaw Basin (Baraniecka 1982; Marks 2012). During the LGM, the Vistula River valley along with the tributary valleys were completely filled up with alluvial deposits. The accumulation took place even in upland areas. The end of the accumulation of sediments building the Otwock terrace dates back to 14 ka BP. During the period of ice-sheet recession, the Otwock terrace was subject to significant erosion. As a result, nowadays, the remnants of the Otwock terrace occur in the form of small 'islands' (Baraniecka 1982).

From the south, the NDB is bounded by the Northern Swamp Belt (NSB), from 4 to 5 km wide. Its bottom is located at 74 m asl (Fig. 30.4). The NSB extends in the form of a valley open in the east towards the Vistula River valley and towards the Bzura River valley in the west. Now, it covers the area of the valley of the Łasica River known for its narrow and elongated old oxbow lakes filled with peat (Liwski et al. 1967; Owadowska et al. 2013). The NSB is located within the Praga terrace of Allerød age (Fig. 30.4), which is predominantly covered with alluvial deposits (Baraniecka 1982).

Fig. 30.3 Present-day landscape of parabolic dunes covered with dense pine forest. **a** Google Earth image (November 2018) of fragment of the Northern Dune Belt. **b** Aeolian topography masked by vegetation. **c** Longitudinal cross section through parabolic dune with distinct gentle windward slope and steeper lee slope. **d** Outcrop of dune sand. Photos b, c, d taken within the Southern Dune Belt near Roztoka (central part of the KNP) (photographs by M. Dłużewski)

South of the NSB lies the Southern Dune Belt (SDB). Its width ranges from 1.5 to even 5 km and the length is about 30 km (Figs. 30.2 and 30.4). Most of the dunes in the SDB are developed on the surface of the Falenica terrace (Baraniecka 1982; Baraniecka and Konecka-Betley 1987; Fig. 30.2), although some enter the area of the Southern Swamp Belt (SSB). The terrace extends at an altitude of 76–73 m asl and the main phase of sedimentation coincides with the Oldest Dryas, with the termination in the Bølling (Baraniecka 1982).

From the south, the SDB is bounded by the Southern Swamp Belt which developed within deposits of the

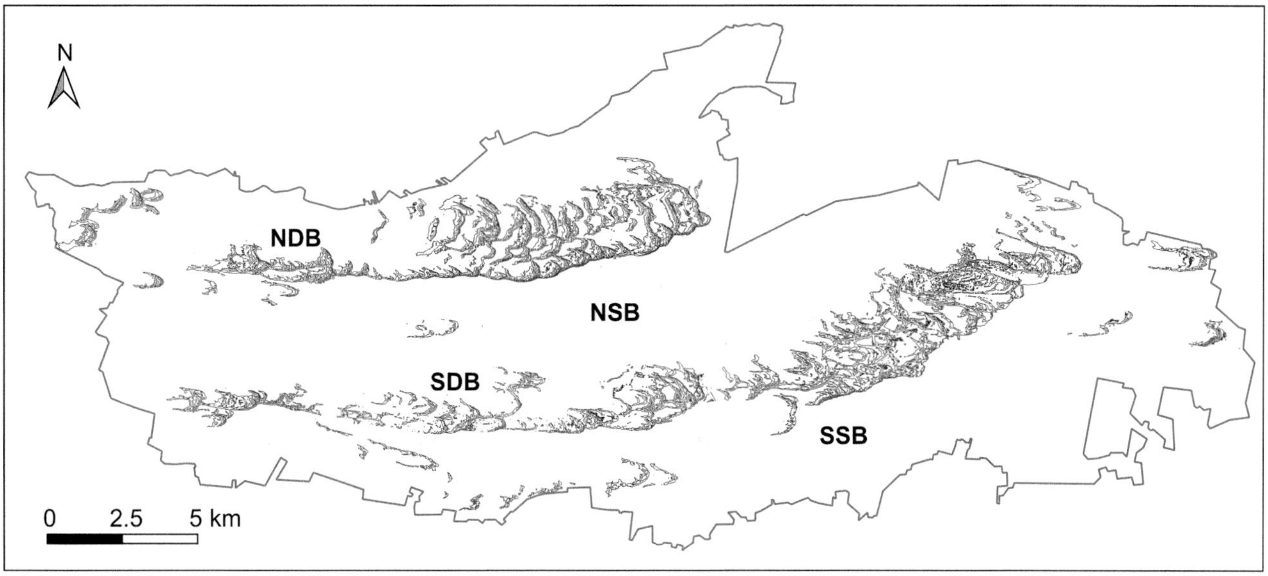

Fig. 30.4 The extent of the KNP with the Northern (NDB) and Southern (SDB) Dune Belts and Northern (SSB) and Southern (SSB) Swamp Belts. DTM based on Sentinel-2A data

Falenica terrace (Baraniecka 1982). The SSB is divided into three sub-belts separated by dunes (Liwski et al. 1967). The lowermost depressions are filled up with peat (Fig. 30.2). To the south there is an escarpment of the ice-marginal valley and the Błonie Plain, which in the southern part has the characteristics of an erosional terrace (Mycielska-Dowgiałło and Krzywobłocka-Laurow 1974).

30.3 Age of Dune-Forming Processes

A problem in the study of the stratigraphy of dune deposits in the KNP, and thus in the determination of periods of aeolian activity, is the long-term protection of the area, preventing us from obtaining sufficient data for the KNP area itself. Our knowledge of the KNP dune sediment stratigraphy is therefore based on sites often located outside the borders of the Park, in different parts of the Warsaw Basin (Fig. 30.2). These are the sites where soils separating aeolian dune sands were analyzed: (1) Kamion site near the town of Wyszogród, located on the left bank of the Bzura River, near the border with KNP, where the dunes are the western extension of the NDB, (2) Całowanie, Wiązowna-Piekiełko, Stężyca and Cięciwa sites located on the right bank of the Vistula River, and (3) Narty site located within NSB (Fig. 30.2), where a palynological analysis was performed. An important limitation of our knowledge is the lack of absolute dating of the dune sediments themselves. Therefore, the age of aeolian processes in the Warsaw Basin is determined on the basis of dating fossil soils, assuming that, regardless of the age of the terraces, dune-forming processes and then soil development took place in very short periods, almost simultaneously and difficult to separate (Konecka-Betley 1982, 2003).

The oldest soil in the Warsaw Basin was found in the stratigraphic profile of Wiązowna-Piekiełko, located in the right bank of the Vistula River and in Kamion. This soil was even older than the oldest stage of aeolian activity in this area. At the Wiązowna-Piekiełko site, the bottom of gley soil was dated at 16.3–15.7 ka cal BP and 16.3–15.5 ka cal BP, while for top of the soil profile dates of 16.0–14.8 ka cal BP and 15.6–14.4 ka cal BP were obtained (Konecka-Betley 1991). On the other hand, at the Kamion site, river sands are overlaid by a layer of sandy silt, about 0.8 m thick, with a 0.1 m layer of clayey silt with an admixture of organic matter at the top, which in fact is a preserved fossil brown soil. It was dated to 18.1–16.5 ka cal BP at the bottom and 17.1–15.4 ka cal BP at the top (Manikowska 1985; Dzieduszyńska and Forysiak 2019), indicating its formation coinciding with the first signal of improvement in climatic conditions, referred to as the Kamion Phase or the Epe interstadial in Denmark—a warm period before the Oldest Dryas (Kolstrup 1980; Konecka-Betley 1991; Manikowska 1991).

The first period of activity of aeolian processes in the Warsaw Basin occurred during the Oldest Dryas cooling (Kobendza and Kobendza 1958; Baraniecka 1982; Konecka-Betley 1991), dated to 17.2–14.2 ka cal BP (Dzieduszyńska and Forysiak 2019). At that time, silts and sands that build the river terraces started to deflate (Konecka-Betley 1991) and therefore sand that builds dunes from this period is very fine and fine-grained (Baraniecka 1982). This feature clearly distinguishes these sands from younger dune sands. Aeolian deposits from this period are recorded at the sites of Wiązowna-Piekiełko (Konecka-Betley 1991) and Kamion (Manikowska 1991). During the Bølling warming period, dated to 14.2–13.5 ka cal BP (Dzieduszyńska and Forysiak 2019), initial soil developed in the environment characterized by tundra vegetation with the presence of larger forest communities dominated by alder and pine (Manikowska 1982). However, as the aeolian sands were not fully inhabited by vegetation, the soil formed was not homogenous (Manikowska 1982), as for example at the Kamion site, where it is represented by a thin humus horizon with charcoal (Manikowska 1985, 1991). It has been dated to 15.1–13.6 ka cal BP in the Mariew site and to 16.1–14.8 ka cal BP and 15.7–14.8 ka cal BP in the Wiązowna-Piekiełko site (Konecka-Betley 1991). The soil separates the aeolian sediments of the Oldest and Older Dryas.

The second dune-forming phase took place during the Older Dryas (13.5–13.2 ka cal BP; Dzieduszyńska and Forysiak 2019) and is considered as the main phase of inland dune formation (Manikowska 1985, 1991; Dzieduszyńska and Forysiak 2019; Moska et al. 2022). The cooling that took place during this period triggered aeolian processes on a very large scale causing dunes development. In the Allerød interstadial (13.95–12.9 ka cal BP; Rasmussen et al. 2014), the interval which is believed to be the first period in the Late Vistulian, when the aeolian processes ceased, dunes were stabilized due to encroachment of forest vegetation (Konecka-Betley 1991). The soil horizons from this period found in the villages of Cięciwa and Całowanie are dated to 13.4–13.1 cal BP and 13.2–12.9 cal BP (Konecka-Betley 1991).

In the Younger Dryas (12.6–11.65 ka cal BP), a distinct cooling of the climate occurred (Dzieduszyńska and Forysiak 2019). This resulted in renewed intensification of aeolian processes in the Warsaw Basin (Baraniecka 1982; Manikowska 1982; Konecka-Betley 1991). Despite low wind energy most of existing dunes underwent final formation. The present-day dunes of the KNP mostly inherited their shape from that period. In the second half of the Younger Dryas, groundwater levels rose (Konecka-Betley 1991) and made the preservation of dune sands possible, but slight and local reshaping of dunes was also possible in Preboreal and historical times.

30.4 Dunes of the Kampinos National Park

30.4.1 Morphology

Parabolic dunes, that is, transverse dune with arms directed toward dominant wind direction, anchored by vegetation or/and high sand moisture, are the main type of dunes in the KNP. They were formed by westerly winds and occur as dune chains (Figs. 30.3a and 30.4). The NDB is divided into two chains (northern and southern) that differ in the morphology of the dunes. The southern chain is about two times longer but narrower than the northern one. The northern one includes more than 30 parabolic dunes, mostly U- or even C-shaped (Figs. 30.5a and 30.6a), with an average symmetry axis orientation equal to 286°, i.e., approximately parallel to the Vistula River valley, the orientation of which in the northern part of KNP is in the range of 240–270°. The windward slopes of mature dunes reach approximately 100 m long and the lee slopes (faced to the east)—approximately 40–50 m long. The length of horns approaches 350 m and the horn-to-horn distance varies between 700 and 1500 m. Wherever the interdune area was large enough, these dunes formed transverse ridges (Fig. 30.6a), but when they developed close to each other, their arms coalesced and compound dunes originated (Figs. 30.5a). They are larger but have similar heights, equal to 15–18 m. On large windward slopes of simple and compound dunes smaller parabolic dunes were formed (Fig. 30.5a). They have NW–SE axis showing that they were formed by secondary NW wind direction. The southern chain of the NDB is smaller in size and consists of V-shaped compound parabolic dunes which are on average two times higher (up to 27 m) than the dunes in the northern chain of the NDB (Fig. 30.6c). All their southern arms are coalescent and form a longitudinal ridge with avalanche slopes facing to the south, which suggests that this chain was also remodelled by NW winds. The distinct, almost straight line of slip face toe indicates that it was later modelled by water erosion in the area of the northern swamp belt.

The SDB is much longer than the NDB (Figs. 30.3b, c and 30.4). It is formed by more than forty simple or compound parabolic dunes. Contrary to the northern chain of the NDB, they are mostly V-shaped, but the height of mature dunes is equal to 15–19 m, same as of the dunes forming the northern chain of the NDB (Figs. 30.5b and 30.6b). Their windward slopes reach approximately 150 m long and the lee slopes (faced to the east)—approximately 50–70 m long. Horns length as well as the horn-to-horn distances vary around 1000 m on well-developed dunes. Their average symmetry axis orientation equals 278°. There are also smaller parabolic forms within their windward slopes (Fig. 30.5b).

The shape of a simple dune is usually expressed by the ratio of dune length to the horn-to-horn distance. In the case of mature dunes present in the NDB and SDB, it is 0.47 in the NDB (northern chain) and 1.31 in the SDB on average. It proves that dune morphologies in both dune belts strongly differ. This difference may suggests that in the period when the SDB was formed either the moisture of the

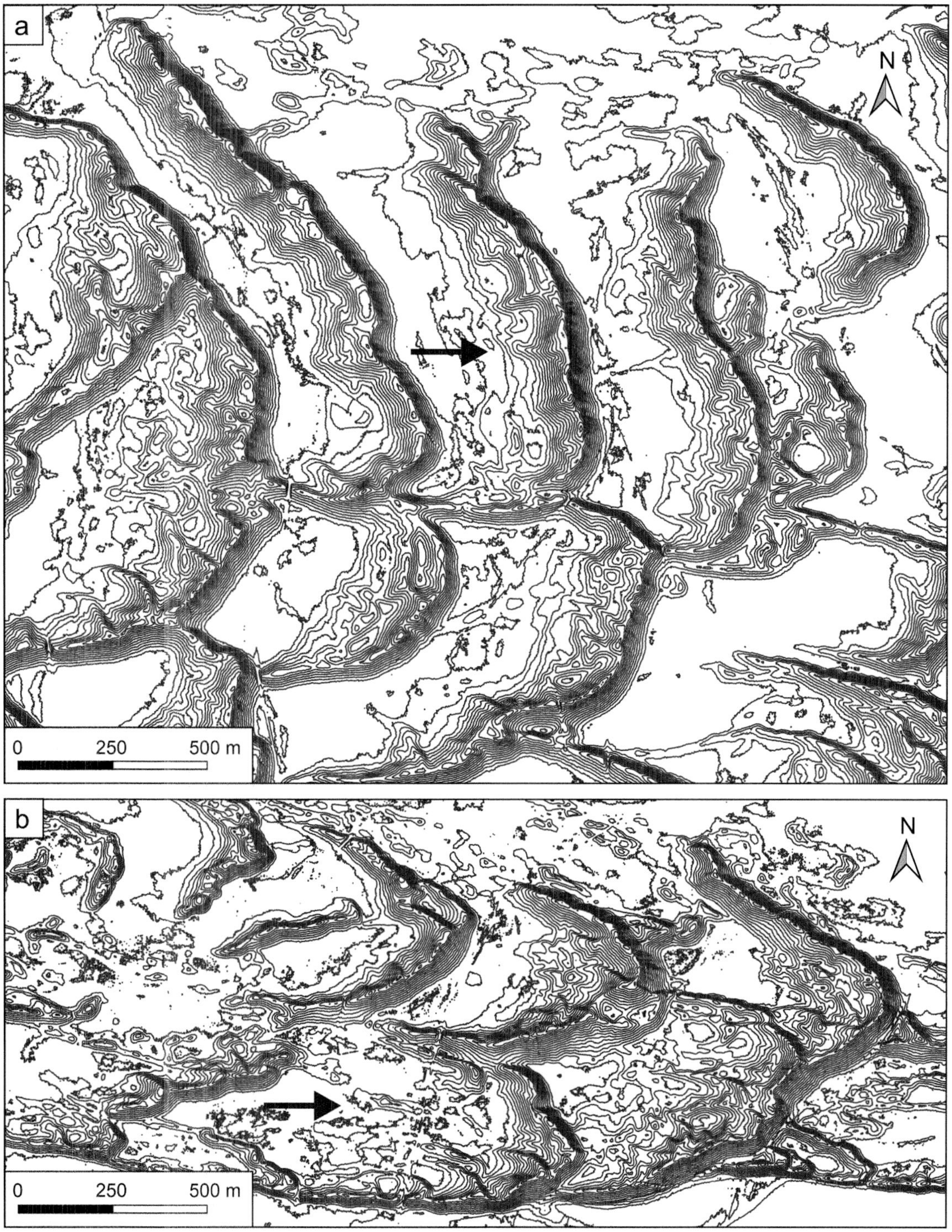

Fig. 30.5 Dune types within two dune belts in the Kampinos Forest. **a** U- and C-shaped parabolic dunes of the Northern Dune Belt. **b** V-shaped parabolic dunes of the Southern Dune Belt. Smaller V-shaped parabolic dunes formed on windward slopes are present in mature dunes of both dune belts. Arrow shows the dominant wind direction responsible for dune development. DTM based on Sentinel-2A data

underlying deposits was relatively higher that in the NDB and it caused that dune horns were efficiently established or that better conditions for plant growth occurred, favouring sand stabilization within the horns of parabolic dunes. Similar conditions probably occurred for smaller dunes of the southern chain of the NDB (Fig. 30.6c).

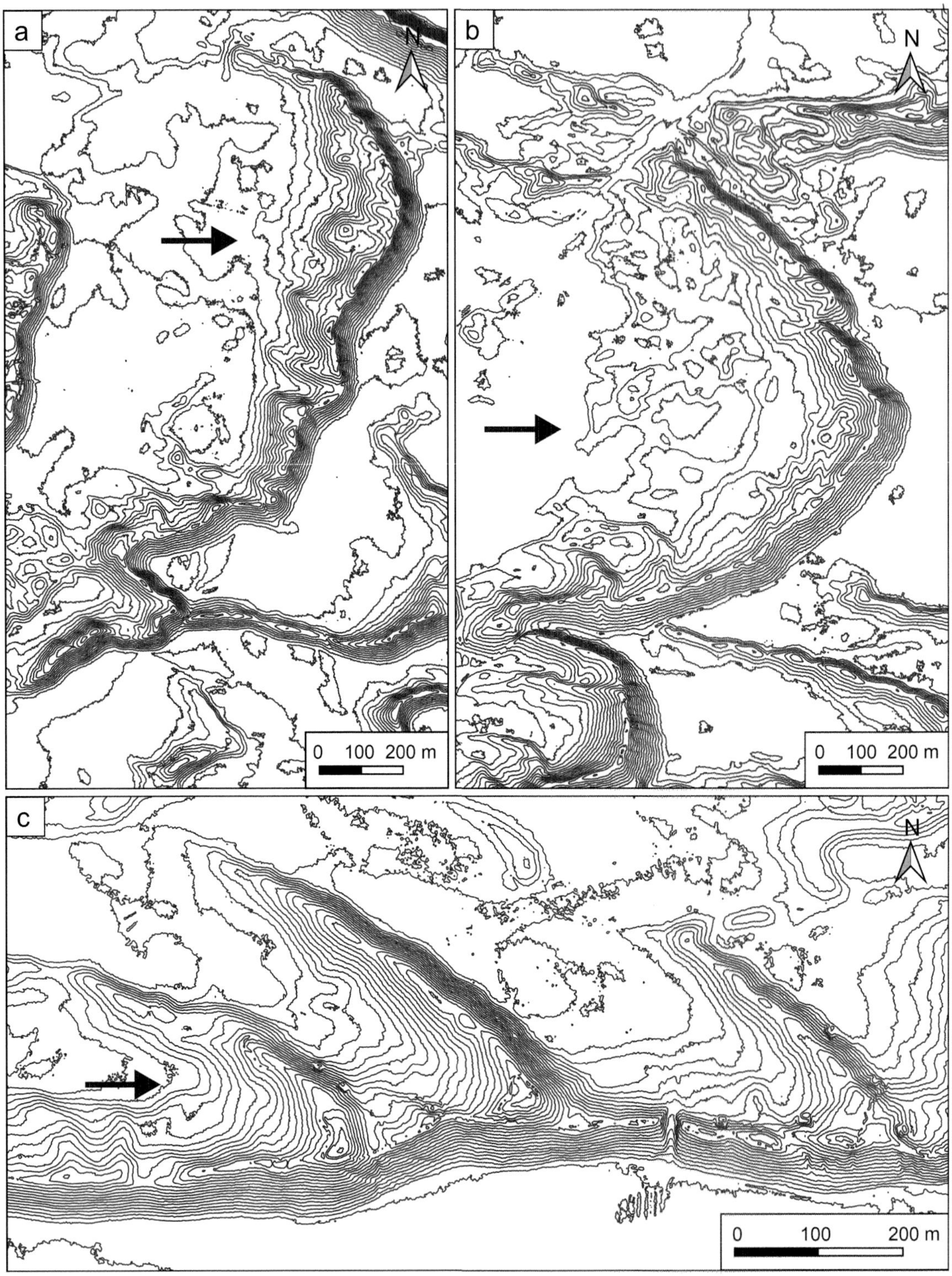

Fig. 30.6 Examples of different dunes occurring in the KNP. **a** Simple C-shaped parabolic dunes typical for the northern dune chain within the Northern Dune Belt (NDB). **b** Simple V-shaped parabolic dunes typical for the Southern Dune Belt. **c** Small compound V-shaped dunes forming longitudinal ridge typical for the southern chain of NDB. Arrow shows the dominant wind direction responsible for dune development. DTM based on Sentinel-2A data

30.4.2 Dune Sand Composition and Sand Grain Textures

Dunes of both the northern and southern dune belts are built of medium quartz sand with a small admixture (less than 2%) of feldspar (Fig. 30.7). The mean grain size is 0.35 and 0.27 mm for the NDB and the SDB, respectively. The smaller value for the SDB results from greater content of fine sand (up to 30%). Regardless the dune belt, the sand is moderately well-sorted (with sorting equal to 0.64–0.66 phi), but the grain size distribution in sand from the NDB is unimodal with symmetrical skewness ($S_k = -0.004$), whereas in the SDB it is bimodal and fine skewed ($S_k = 0.2$).

According to Reichelt's (1961) 4-grade scale of roundness, dune sand of both dune belts is dominated by well-rounded and rounded grains, with some admixture of subangular grains with slightly rounded edges (Fig. 30.7). Among the well-rounded grains, highly spherical and ellipsoidal grains are in majority. There are no angular grains (Mycielskia-Dowgiałło and Krzywobłocka-Laurow 1974).

Analysis of microtextures of sand grains from the size class of 0.5–0.8 mm, based on scanning electron microscope (SEM), showed that regardless of dune location, whether in the NDB or SDB, the aeolian grain surface microrelief is poorly developed (Figs. 30.8 and 30.9). Only a few percent of grains exhibit mechanically upturned plates, which are typical aeolian microtextures (Krinsley and Doornkamp 1973; Mahaney 2002; Szerakowska et al. 2018) (Figs. 30.8g-l and 30.9g-l) resulting from grain-to-grain collisions during saltation (Pye and Tsoar 2009). No grains were recorded, which at the same time would be very well-rounded and had the entire surface covered with well-developed mechanically upturned plates. Equidimensional and elongate depressions (20–250 μm) formed during grains collisions are common.

A high percentage of grains features smooth and glossy surfaces (Fig. 30.8e, f) resultant from chemical etching (Figs. 30.8b, e–g and 30.9d, f). These features are usually observed in microdepressions of different types. On convex parts of grains, crescentic gouges and V-shaped percussion cracks are developed (Fig. 30.8e, f). These mikrotextures result from grain-to-grain collisions in an aqueous environment (Krinsley and Doornkamp 1973; Lindé and Mycielska-Dowgiałło 1980; Mahaney 2002; Vos et al. 2014).

Fig. 30.7 Binocular photographs of dune quartz sand building the Northern Dune Belt (**a, b**) and the Southern Dune Belt (**c, d**). **a, c** Grain size class 0.5–0.8 mm. **b, d** Grain size class 0.3–0.5 mm

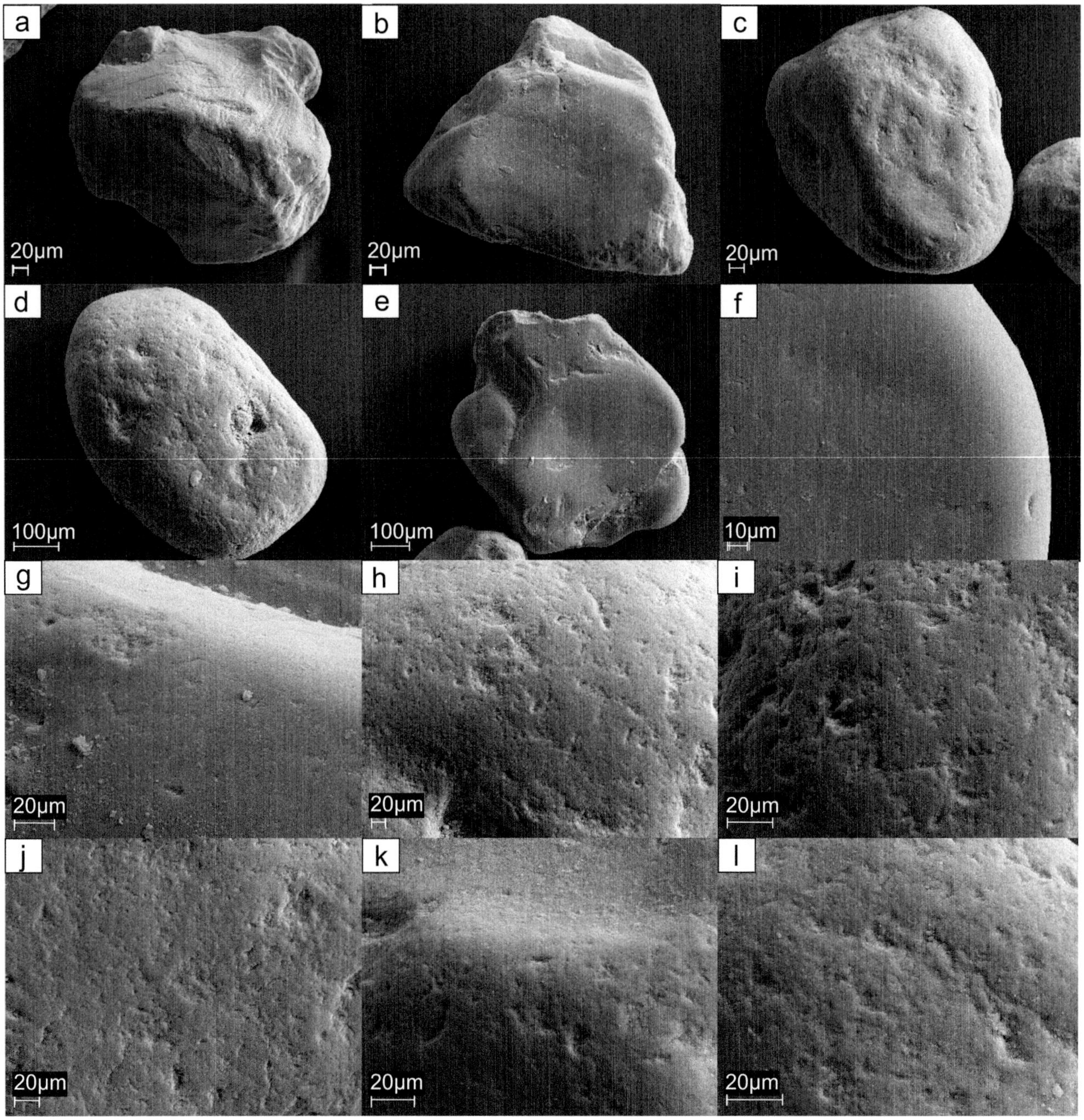

Fig. 30.8 SEM microphotographs of a sand-sized quartz grain surfaces typical for dune sand from the Northern Dune Belt. **a–c** Subangular grains with abrasion visible on the corners. **d** Well-rounded grain with low sphericity and poorly marked aeolian relief. **e** Subangular grain with a smooth surface type. **f** Fragment of quartz grain with a smooth surface type and crescentic gouges. **g-l** Fragments of grains with aeolian microtextures

30.4.3 Source of Dune Sand

The shape and roundness of quartz grains building the dunes in the KNP were inherited from deposits of aqueous environments, mainly fluvial, which were the source of younger dune sands. Mycielska-Dowgiałło and Krzywobłocka-Laurow (1974) showed that sand grains from fluvial deposits underlying the dunes are dominated by subrounded grains and the degree of roundness in dune sands is only slightly higher.

The textures and microtextures of dune sand grains, mainly their smooth and glossy surfaces, show that aeolian processes responsible for dune formation in the KNP were relatively short, and sand was transported along relatively

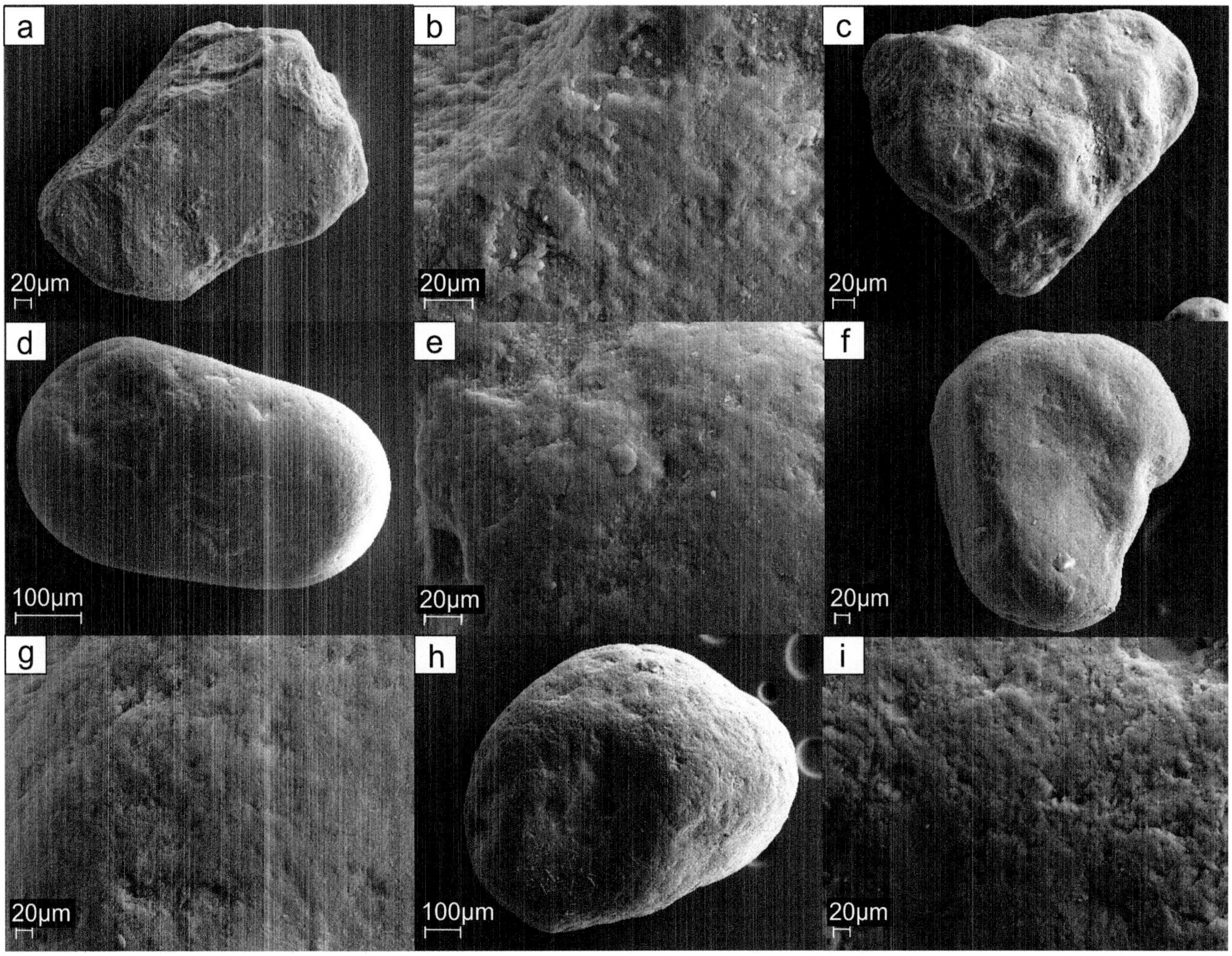

Fig. 30.9 SEM microphotograps of a sand-sized quartz grain surface typical for dune sand from the Southern Dune Belt. **a-c** Subangular grains with abrasion visible on the corners. **d** Well-rounded grain with low sphericity and poorly marked aeolian relief. **e** Fragment of grain with poorly marked aeolian relief on the corner. **f** Subangular grain with abrasion visible on the corners. **g** Fragment of a grain with well-developed mechanically upturned plates on the corner. **h** Well-rounded grain with mechanically upturned plates and microdepression. **i** Fragment of a grain with well-developed mechanically upturned plates on the corners

short distances. Grain transport under such conditions can only lead to slight remodelling of surface microtextures, but it will not cause the change of shape and roundness of grains entrained by wind from other environments.

Contrary to the dunes present in areas, which were not covered by an ice sheet during the Weichselian glaciation and whose sands exhibit high degree of aeolian abrasion (Woronko et al. 2015; Zieliński et al. 2015; Zieliński 2017; Kalińska-Nartiša et al. 2017; Moska et al. 2022), dunes of the KNP are built of sand that has undergone only slight aeolian reworking. Deposits of the Otwock, Falenica and Praga terraces were the source of the dune sand.

30.5 Degradation of Dunes in the Kampinos Forest

The oldest traces of human activity in the KNP date back to the first–second century, when iron ore smelting became common. Wood from the Kampinos Forest was most legitimately used to fire the stoves. Initially, people settled only at the outskirts of the forest. The first evidence of settlements located well inside the forest area dates back to the fifteenth century (Owadowska et al. 2013). Since then, the population has grown steadily, and thus more and more forest areas disappeared. In the nineteenth century, the lack of land for cultivation led to intense deforestation of even low-productive dune areas and drainage of swamps located in the area of today's KNP, that is, in the northern and southern swamp belt. The Łasica Canal and successive villages emerging in the drained areas are a record of the human activity in those days. The dune areas have been converted to pastures and sometimes even to arable land. As a result of the devastating agriculture even several hundred hectares were completely devoid of vegetation which favoured the reactivation of aeolian processes (Kobendza and Kobendza 1958). After the World War II, all these negative effects have grown even further and manifested themselves in ploughing of the dunes, massive felling of trees and converting forest areas into pastures. Kobendza and Kobendza (1958), who analysed the effects of human activity, described the landscape of the Kampinos Forest from those times as follows: '*... from the bird's eye view, it is strewn with white patches with flying sand that accompany almost every settlement*'. The soil cover was disturbed and destroyed and in the 1950s, the dunes were at various stages of destruction. On some of them this process had just started, while on others, deflation left only aeolian outliers. Between them vast deflation surfaces existed with some minor aeolian sand accumulation in the form of shadow structures formed behind plants or other obstacles. To counteract this process, new plantings were introduced on a large scale in the north-western part of the forest, which resulted in the dominance of a single-species pine stands within the dune areas (Owadowska et al. 2013).

Even though the KNP was established in 1959, it was not possible to prevent the destruction of this area and to preserve valuable natural values as, in the time in question, most of the area of the Kampinos Forest was in private hands. It was only in 1975 that the Council of Ministers, by the resolution No. 139/75 of July 18, 1975, decided to purchase private land in the territory of KNP through expropriation, and it was possible to buy the land from private owners.

30.6 Summary

The present-day landscape of KNP is connected with the retreat of ice sheet following the Last Glacial Maximum, the accumulation of alluvial deposits within the Warsaw Basin and then within the wide proto-Vistula valley, which during the Late Glacial and early Holocene cold-climate periods were transformed by aeolian processes. Thanks to these processes two dune belts separated by two swamp belts were formed. The present-day dune shape is inherited from the Younger Dryas. They are mostly V-, U- or even C-shaped parabolic dunes, that is, transverse dunes with arms directed toward the west. Locally their arms coalesced and these form compound dunes. The present-day dune shape shows that the dune-forming winds were blowing from the west. Recently, all dunes are stabilized by pine forest. The large system of well-developed dune chains indicates large supply of material during the period of dune formation and further evolution. However, the shape, roundness and texture of quartz grains confirm that the process of dune formation was short.

The KNP was established to protect the rich mosaic of habitats, from the very wet to the very dry. The dunes are considered to be the best preserved complex of inland dunes in Europe. Thanks to its considerable geodiversity they receive a great number of visitors each year.

References

Andrzejewska A, Ferchmin M, Kębłowska A, Otręba A (2010) Puszcza Kampinoska. Charakterystyka geobotaniczna Puszczy Kampinoskiej. In: Obidziński A (ed) Z Mazowsza na Polesie i Wileńszczyznę. Zróżnicowanie i ochrona szaty roślinnej pogranicza Europy Środkowej i Północno-Wschodniej. Polskie Towarzystwo Botaniczne, Warszawa, pp 41–56

Andrzejewski L, Weckwerth P (2010) Dunes of the Toruń Basin against palaeogeographical conditions of the late glacial and Holocene. Ecol Quest 12(Spec Iss):9–15

Baraniecka D (1982) Sytuacja geologiczna i rozmieszczenie wydm okolic Warszawy. Roczn Glebozn 33:33–57

Baraniecka MD, Konecka-Betley K (1987) Fluvial sediments of the Vistulian and Holocene in Warsaw Basin. In: Evolution of the Vistula river valley during the last 15000 years II. Geogr Stud, SI 4:151–171

Bomanowska A, Michalska-Hejduk D, Kirpluk I, Szczepko K, Olech B, Olszewski A, Marczak D (2009) Czynna ochrona terenów porolnych—propozycje, podejmowane działania, perspektywy. In: Michalska-Hejduk D, Bomanowska A (eds.) Rola Kampinoskiego Parku Narodowego w zachowaniu różnorodności biologicznej i krajobrazowej dawnych obszarów wiejskich. Kampinoski Park Narodowy, Izabelin, pp 148–153

Dzieduszyńska DA, Forysiak J (2019) Chronostratigraphy of the late Vistulian in central Poland and the correlation with Vistulian glacial phases. Stud Quat 36:137–145

Izmaiłow B (1976) The state of research in inland dunes in Poland. Zesz Nauk UJ, Prace Geogr 43:39–63

Kalińska-Nartiša E, Woronko B, Ning W (2017) Microtextural inheritance on quartz sand grains from Pleistocene periglacial environments of the Mazovian Lowland, Central Poland. Permafrost Perigl Proc 28:741–756

Kasse C (1999) Late pleniglacial and late glacial aeolian phases in The Netherlands. In: Schirmer W (ed) Dunes and fossil soils. GeoArchaeoRhein vol. 3, pp 61–82

Kobendza J, Kobendza R (1958) Rozwiewane wydmy Puszczy Kampinoskiej. In: Galon R (ed) Wydmy śródlądowe Polski. PWN, Warszawa, pp 95–171

Kolstrup E (1980) Climate and stratigraphy in northwestern Europe between 30,000 BP and 13,000 BP, with special reference to the Netherlands. Med Rijks Geol Dienst 32(15):181–238

Konecka-Betley K (1982) Gleby kopalne i reliktowe wydm okolic Warszawy. Roczn Glebozn 33:81–112

Konecka-Betley K (1991) Late Vistulian and Holocene fossil soils developed from aeolian and alluvial sediments of the Warsaw Basin. Z Geomorph NF, Suppl. 90:99–105

Konecka-Betley K (2003) Gleby. In: Andrzejewski R (ed) Kampinoski Park Narodowy. T. I. Przyroda Kampinoskiego Parku Narodowego. KPN, Izabelin, pp 97–133
Koster EA (1988) Ancient and modern cold-climate aeolian sand deposition: a review. J Quat Sci 3:69–83
Koster EA (ed) (2005) The physical geography of Western Europe. Oxford University Press, Oxford, p 438
Krinsley DH, Doornkamp JC (1973) Atlas of quartz sand surface textures. Cambridge Univ Press, Cambridge, p 91
Lindé K, Mycielska-Dowgiałło E (1980) Some experimentally produced microtextures on grain surfaces of quartz sand. Geogr Ann 62A:171–184
Liwski S, Maciak F, Karpinska J (1967) Torfowiska Puszczy Kampinoskiej. Rocz Glebozn 18(1):159–184
Mahaney WC (2002) Atlas of sand grain surface textures and applications. Oxford University Press, New York, p 256
Manikowska B (1982) Gleby kopalne w wydmach Polski Środkowej. Rocz Glebozn 33:119–133
Manikowska B (1985) O glebach kopalnych, stratygrafii i litologii wydm Polski środkowej. Acta Geogr Lodz 52:137
Manikowska B (1991) Vistulian and Holocene aeolian activity, pedostratygraphy and relief evolution in Central Poland. Z Geomorph NF, Suppl. 90:131–141
Marks L (2002) Last glacial maximum in Poland. Quat Sci Rev 21:103–110
Marks L (2012) Timing of the late Vistulian (Weichselian) glacial phases in Poland. Quat Sci Rev 44:81–88
Moska P, Sokołowski RJ, Jary Z, Zieliński P, Raczyk J, Szymak A, Krawczyk M, Skurzyński J, Poręba G, Łopuch M, Tudyka K (2022) Stratigraphy of the Late Glacial and Holocene aeolian series in different sedimentary zones related to the Last Glacial maximum in Poland. Quat Int 630:65–83
Mycielska-Dowgiałło E (1993) Estimates of late glacial and Holocene aeolian activity in Belgium, Poland and Sweden. Boreas 22:165–170
Mycielska-Dowgiałło E, Krzywobłocka-Laurow R (1974) Typy urzeźbienia powierzchni ziarn kwarcowych piasków Puszczy Kampinoskiej w oparciu o analizę w prześwietlonym i analizującym mikroskopie elektronowym. Roczn Pol Tow Geol 44:227–246
Nowaczyk B (1986) Wiek wydm, ich cechy granulometryczne i strukturalne a schemat cyrkulacji atmosferycznej w Polsce w późnym vistulianie i holocenie. Geografia 28, Wyd. Nauk UAM, Poznań, p 245
Nowak J (1974) Stratygrafia plejstocenu północnej części Kotliny Warszawskiej. Biul Inst Geol 268:91–164
Okołów G, Marczak D (2008) Kampinoski Park Narodowy. Natura 2000. Kampinoski Park Narodowy, Izabelin, p 2
Owadowska E, Andrzejewska A, Danyłow J, Kębłowska A, Marczak D, Olszewski A, Pepłowska-Marczak D (2013) Kampinos forest biosphere reserve. In: Kunz M, Nienartowicz A (eds) Biosphere reserves in Poland. UMK, Toruń, pp 144–164
Pye K, Tsoar H (2009) Aeolian sand and sand dunes. Springer, Berlin, p 458
Rasmussen SO, Bigler M et al (2014) A stratigraphic framework for abrupt climatic changes during the Last Glacial period based on three synchronized Greenland ice-core records: refining and extending the INTIMATE event stratigraphy. Quat Sci Rev 106:14–28
Reichelt G (1961) Über Schotterformen und Rundungsgradanalyse als Feldmethode. Petermanns Geogr Mitt 105:15–24
Rotnicki K (1970) Główne problemy wydm śródlądowych w Polsce w świetle badań wydmy w Węglewicach. Pr Komisji Geogr-Geol, PTPN, Poznań 11(2):146
Szerakowska J, Woronko B, Sulewska MJ, Oczeretko E (2018) Spectral method as a tool to examine microtextures of quartz sand-sized grains. Micron 110:36–45
Vos K, Vandenberghe N, Elsen J (2014) Surface textural analysis of quartz grains by scanning electron microscopy (SEM): from sample preparation to environmental interpretation. Earth Sci Rev 128:93–104
Wiśniewski E (1990) The lower Vistula valley. In: Evolution of the Vistula river valley during the last 15000 years III. Geography Study, SI, Vol. 5, pp 102–110
Woronko B, Zieliński P, Sokołowski RJ (2015) Climate evolution during the Pleniglacial and Late Glacial as recorded in quartz grain morphoscopy of fluvial to aeolian successions of the European Sand Belt. Geologos 21:89–103
Zeeberg J (1998) The European sand belt in eastern Europe—and comparison of Late Glacial dune orientation with GCM simulation results. Boreas 27:127–139
Zieliński P, Sokołowski RJ, Woronko B, Jankowski M, Fedorowicz S, Zaleski I, Molodkov A, Weckwerth P (2015) The depositional conditions of the fluvio-aeolian succession during the last climate minimum based on the examples from Poland and NW Ukraine. Quat Intern 386:30–41
Zieliński P (2017) Regional and local conditions of the late Vistulian aeolian deposition in the central part of the European sand belt. Wyd UMCS, Lublin, p 235

Maciej Dłużewski is Associate Professor of Geography at the University of Warsaw. His main research interests are aeolian geomorphology, especially dune morphology and dynamics in deserts and coastal areas of arid and temperate climate. His work also focuses on the impact of desertification on human life and regional development. Northern Sahara is his main research area, but he was also involved in projects related to past and present dune formation in Poland and in various deserts in Africa (Morocco, Tunisia, Libya, Egypt, Namibia), Middle East, Asia (China, Uzbekistan, Turkmenistan) and South America (Chile, Peru). He is an expert in World Heritage Nominations—Natural Sites and member of International Forum Sustainable Future of Life in the Desert. He used to promote geomorphological knowledge by organizing workshops in different landscape and climate zone around the World.

Barbara Woronko is a Professor of Geology at the University of Warsaw. Her main research interests include microtextures of sand- and silt-sized quartz grains, frost micro-weathering, glacial, periglacial and aeolian geomorphology, climate change and geoarcheology. Her main research area is the eastern Poland, but she has been also involved in projects conducted in other European countries (Czechia, Lithuania, Latvia, Estonia, Portugal, Belarus), Egypt, Sudan, Morocco, Oman and Lebanon. She is the Vice-President of the Association of Polish Geomorphologists (since 2021). She was a member of editorial board of the *Studia Quaternaria* journal.

31 The Valley of Anastomosing Narew River (NE Poland)—A Unique Fluvial Landscape of the European Lowlands

Tomasz Zieliński

Abstract

There are three basic types of rivers in the world. Apart from the frequent meandering and braided rivers, the anastomosing river type is one of the rarest. The valley of the anastomosing upper Narew River (NE Poland) is natural and therefore unique in Europe. The anastomosing river is a multi-channel river system of very low energy. The channels are relatively narrow and deep. The floodplain is built from sediments resistant to erosion and therefore the fluvial system is stable, with little change over hundreds of years. The floodplain is often submerged, and hence, it is an environment of dense aquatic vegetation. It is still unclear why this river reach obtained a specific anastomosing pattern. The article presents various hypotheses on this topic. The geological history of the Narew valley and its vicinity during the last million years is also presented. The river changed its character several times, in relation to climatic fluctuations occurring in the Pleistocene and Holocene. Other examples of Polish rivers with features of anastomosing pattern are also given, and a fluvial analogue of the Narew River in southern Africa has been found.

Keywords

Anastomosing river type · Fluvial geomorphology · Alluvial sediments · River metamorphosis · Polish Lowlands · Holocene

T. Zieliński (✉)
Institute of Geology, Adam Mickiewicz University, Krygowskiego 12, 61–680 Poznań, Poland
e-mail: tomasz.zielinski@amu.edu.pl

31.1 Introduction

River valleys are common forms in any landscape. They are formed by different types of rivers, whereas the river type itself depends on many environmental factors. These factors can be divided into two main groups. The first group includes internal factors: geomorphological and geological. In the mountains, where the tectonic movements were relatively recent (in terms of geological timescale), the longitudinal gradients of the valleys are high, and the hard, lithified rocks are on the surface. Old, mature landscapes are characterized by plains covered with loose deposits resulting from long-lasting weathering or delivered by glaciers or winds. The second group includes climate-controlled factors. The type and intensity of weathering, i.e. the grain size of the sediments available in the catchment, depend on the temperature. Amount of precipitation and its distribution over the year depend on the climatic zone. Well-developed vegetation (especially forests) can intercept a significant amount of rainfall supplied to the river. Thus, the climate determines the hydrological regime of the river, i.e. an average discharge, the frequency and size of floods, the existence of droughts. The discharge and flow velocity contribute to the hydrodynamic characteristics of the river. On the other hand, the hydrodynamics determines the grain size (and volume) of the transported and deposited sediments as well as the balance of accumulation-to-erosion of the fluvial environment. The final result of all these overlapping factors is the river channel pattern, i.e. the width and depth of the channel (channels) and the planform arrangement of channel network. Practically, the term *river channel pattern* denotes the *river type*.

For over a hundred years, river systems have been divided into two categories: braided and meandering. In the second half of the twentieth century, observations of rivers in North America, Australia and Africa prompted the researchers to distinguish another type of river—the anastomosing river system (Schumm 1972, 2005; Rust 1978).

P. Migoń and K. Jancewicz (eds.), *Landscapes and Landforms of Poland*,
World Geomorphological Landscapes, https://doi.org/10.1007/978-3-031-45762-3_31

While the braided river system represents the highest energetic environment, the anastomosing system corresponds to the lowest energetic environment. The meandering river in this classification occupies an intermediate position between the other two (Fig. 31.1). As a rule, anastomosing river valleys developed in the plains that were later colonized most rapidly and to the greatest extent by humans. Therefore, the courses of these rivers have been strongly changed anthropologically, especially in Western Europe. Today, relics of anastomosing river channels can only be found in the mouth of the Rhine in the Netherlands (Törnqvist 1993; Berendsen and Stouthamer 2000). For historical reasons (less intensive economic development), fragments of anastomosing river valleys are more common in Central Europe, mainly in Poland (the lowermost reaches of the Odra and Warta rivers and a part of the upper Narew). The Narew River valley has been preserved in an almost natural form (Fig. 31.2) and is therefore a geomorphological phenomenon on a European scale.

31.2 Geomorphological and Geological Setting

The Narew is a typical lowland river, the fifth longest Polish river, although with the source in Belarus (Fig. 31.3). Its length is 485 km, including 450 km in Poland. It drains a catchment area of approximately 75,000 km^2 located in the north-eastern part of the country. In Polish physico-geographical regionalization the Upper Narew River Valley divides two

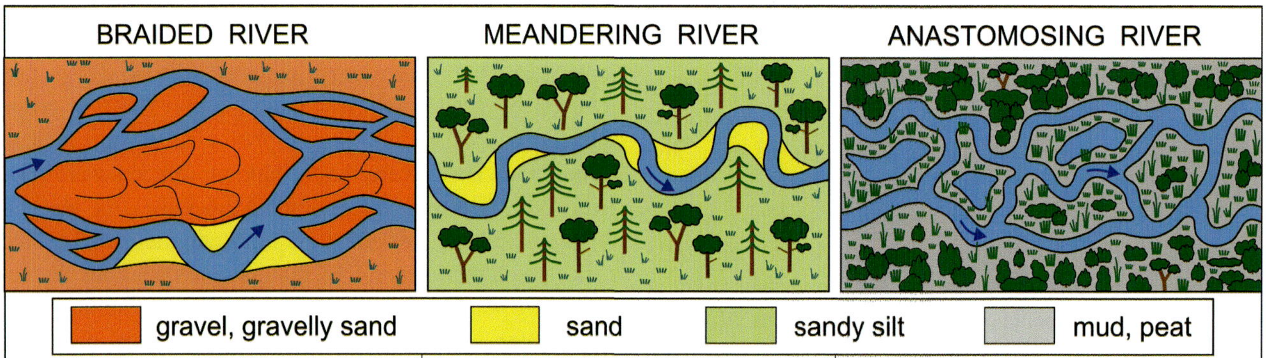

Braided river	Meandering river	Anastomosing river
	river pattern:	
multiple channels separated by bars	single channel	multiple channels separated by islands
	valley slope & flow energy:	
steep valley floor, high-energy flow	low-slope valley floor, moderate flow energy	very low slope, quiescent flow with stagnant phases
	climate & vegetation:	
mainly semi-dry (polar or tropical) climate, sparse vegetation	mainly temperate climate, well-developed vegetation (forest)	temperate or tropical climate, dense swamp thickets
	channel morphology:	
wide and shallow channels of low sinuosity	relatively narrow and deep channel, highly sinuous	very narrow and deep channels of low-to-high sinuosity
	channel stability:	
stability very low, frequent, abrupt channel shifting	moderate stability, slow but continuous channel migration	stability very high, permanent channel position
	deposition/erosion ratio:	
pronounced deposition (aggradation)	balanced deposition-to-erosion	distinct predomination of deposition over erosion
	sediment deposited:	
gravel, sandy gravel, coarse sand	sand in channel, sandy silt on floodplain	mud (or clayey sand) in channels, mud or peat on floodplain

Fig. 31.1 Most important features of three main river types: braided, meandering and anastomosing

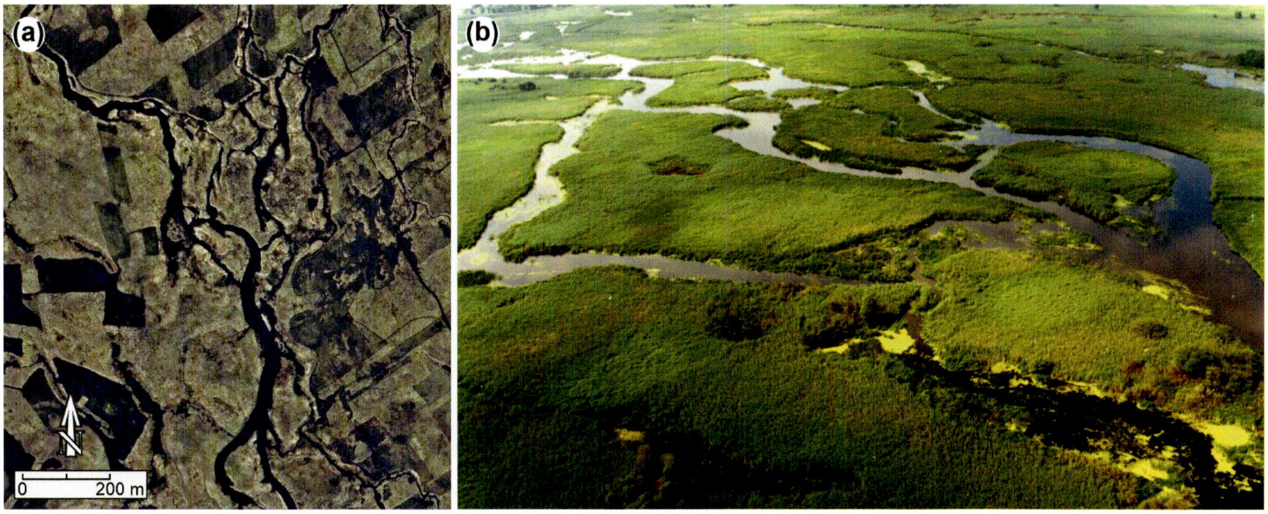

Fig. 31.2 Typical anastomosing pattern of the Narew River. The same section of the valley visible in the vertical satellite image (**a**) and in the oblique aerial photograph (**b**) (photograph a from Geoportal.pl; b by M. Doktor)

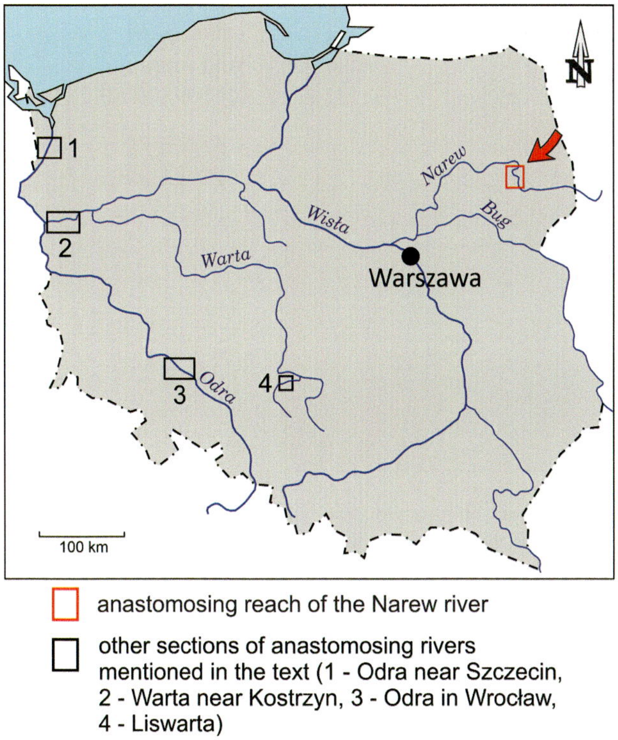

Fig. 31.3 Polish river network and location of the anastomosing reach of the Narew River

regions of the Northern Podlasie Plain: the Białystok Heights in the east and the Wysokie Mazowieckie Heights in the west. It is an area with humid continental climate, with average annual precipitation in the range of 550–560 mm. Due to the continental climate, the floods of the upper Narew are concentrated in spring and are associated with snow melting.

The described section of the valley is approximately 30 km long. The river flows from south to north, which is unusual in this region of Poland. The valley slope is small and amounts 0.00022 m m^{-1}, i.e. 0.22 ‰. In this section, the mean annual discharge is 15 m^3 s^{-1}, while flood discharge is approximately 75 m^3 s^{-1} (Mioduszewski et al. 2016). Upstream and downstream of the anastomosing section, the river is of the meandering type.

The valley is surrounded by areas with a thick (up to 200 m) cover of glacigenic Pleistocene deposits. It is mainly undulating till plain with an average height of 10–25 m above the valley floor (Fig. 31.4). In many places, mainly in topographic basins, the till is overlain with sands and glaciofluvial (outwash) gravels. This cover is not thick, usually a few metres. There are end-moraine hills within the till plain. These are elongated and latitudinally-oriented, up to several kilometres in length (Fig. 31.4). Their heights are 10–20 m above the surface of the till plain (Fig. 31.5). The Narew valley has a sinuous course, which depends on the morphology of the glacial-derived landscape. It consists of several widenings, so-called basins, within which the Narew channel belt is 2 km wide, and the network of anastomosing channels is best developed (Fig. 31.4). Usually 5 to 7 active channels are present. In narrowings, the number of channels is reduced to 2–4, and the river typically has a meandering pattern.

Due to significant natural (geomorphological and biological) values, this section of the valley is under protection, within the Narew National Park (Narwiański Park Narodowy). It covers the valley floor along the length of 35 km and has an area of 73.5 km^2.

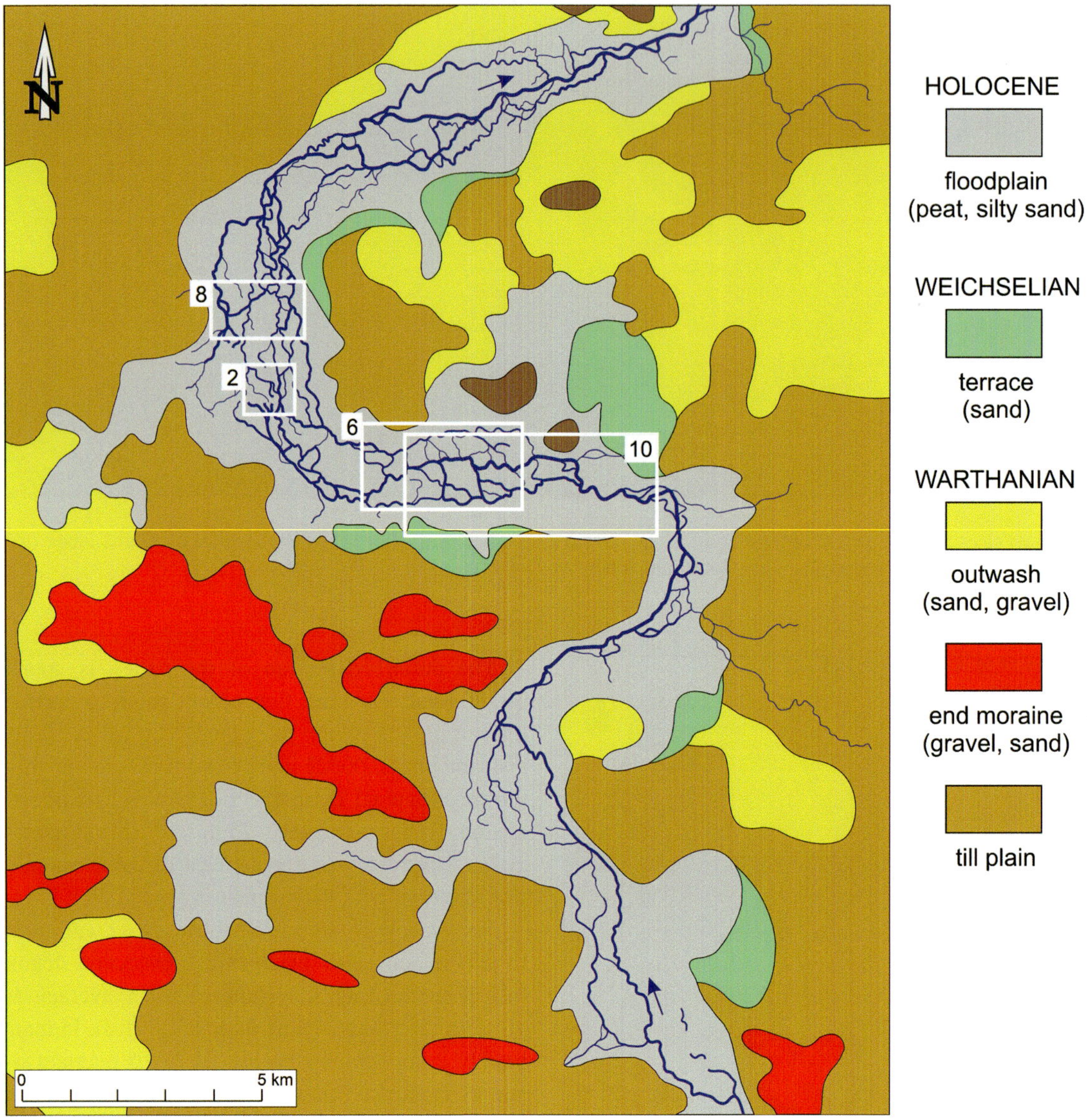

Fig. 31.4 Geological sketch of the Narew River valley within the Narew National Park and the surrounding area (based on Bałuk 1973, simplified). White rectangles show the location of vertical images (figure numbers relevant to this chapter are listed)

31.3 The Channels of Anastomosing Narew River

Two types of channels can be distinguished in the Narew fluvial system: main and secondary ones (Fig. 31.6). The main channels are 10–35 m wide and 4–5 m deep. They are therefore relatively narrow and deep. The flow velocity during average water stages is low, of the order of 10–30 cm s^{-1}. The banks of the channels are steep, often vertical. The bottom is sandy, often covered with underwater vegetation. Large-scale depositional forms—the bars—are rarely formed in the main channels. The length of elongated bars is comparable to the width of channels. Most often they occur in low-sinuosity river reaches, close to the banks or in the central parts of channels (Gradziński et al. 2003a; Kędzior et al. 2021). Less frequently, they are found in bends of high curvature. Their characteristic feature is that, unlike typical meandering river bars, which are accumulated at the inner banks of the meanders (so-called point bars), the bar forms in the Narew channel are located close to the outer banks of the bends. These are so-called concave-bank bars, which are considered indicator channel forms for low-energy anastomosing rivers.

There are many more secondary channels in the Narew fluvial system than the main ones. They are more sinuous (Fig. 31.7). The flows there are clearly slower, which means that the secondary channels are more easily occupied by aquatic vegetation. Low-velocity flows (up to 10 cm s^{-1})

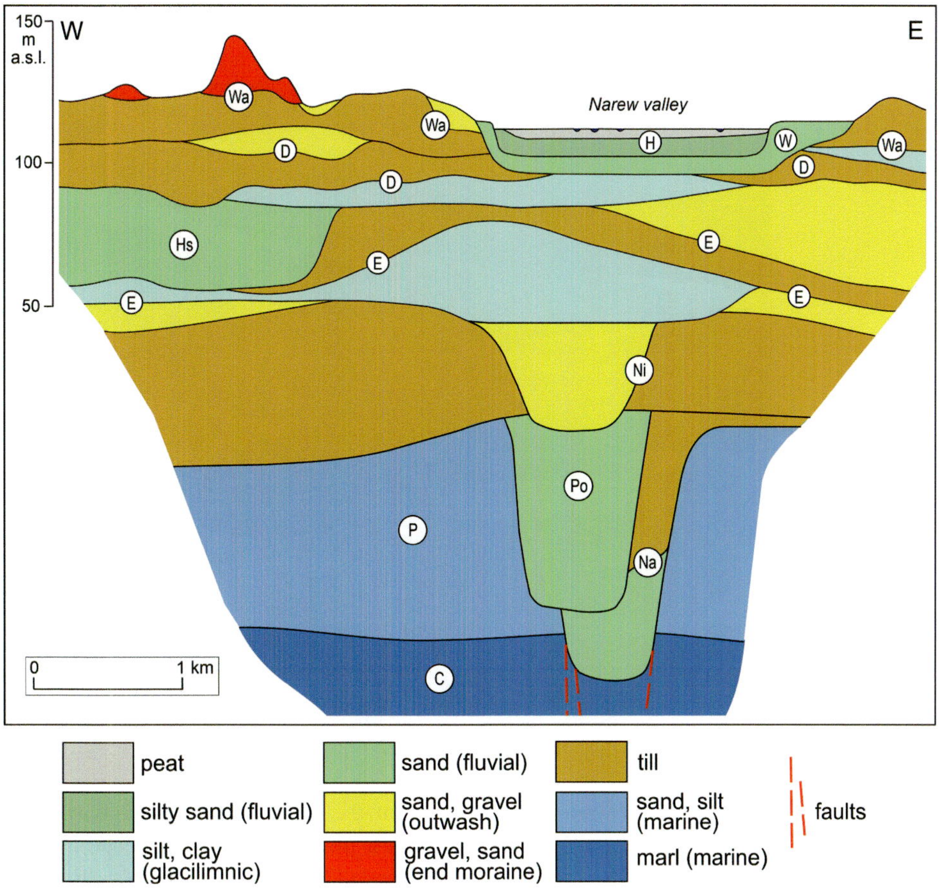

Fig. 31.5 Schematic geological cross-section through the Narew valley and adjacent area (based on Butrymowicz 2001, modified). Stratigraphic symbols: H—Holocene, W—Weichselian Glaciation, Wa—Warthanian Glaciation, D—Drenthian Glaciation, Hs—Holsteinian Interglacial, E—Elsterian Glaciation, Ni—Nidanian (Dorst) Glaciation, Po—Podlasian Glaciation, Na—Narewian Glaciation, P—Paleogene, C—Cretaceous

Fig. 31.6 River system of the Narew consists of the main channels (arrowed) and numerous secondary ones (photograph from Geoportal.pl)

Fig. 31.7 Secondary channels of the Narew fluvial system are often meandering (photograph from narwiański-park-narodowy.pl)

deposit the fine-grained sand, silty sand, and organic silt at the bottom. Many of the secondary channels are in the decay phase and the water often stagnates. The cause of channel abandoning is the accumulation of sand bodies at the convergence zone of the secondary channel with the main one. These are the so-called plug bars. Another reason for channel abandoning is the blockade by floating islands of accumulated aquatic plants—the plant jams. With time, the channels become completely cut off from the main flow arteries and begin to act as small basins of standing waters. Another type of basins are oxbow lakes, which accompany meandering reaches of the river (Fig. 31.8). They arise from the cutting of meanders' necks and re-directing the current to the newly eroded, shorter sections of the channel.

31.4 The Narew River Floodplain

As mentioned before, the floodplain is best developed in the broader reaches of the valley—in the basins. The floodplain surface is only slightly above the average water level in the river. Therefore, even the smallest floods can inundate it (Fig. 31.9). As a result, the entire floodplain is occupied by a dense cover of aquatic-plant vegetation (rush and reed communities). Deceased plants are a source of intense organic deposition. The upper part of the floodplain is made up of an organic-rich layer, 1–4 m thick. In the inter-channel islands, only peat occurs (sedge peat and reed peat) (Gradziński et al. 2003b; Kędzior et al. 2021). The wider areas of the floodplain, further away from the main channels, are made of peat sandwiched with organic silt and silty sand. The peat is underlain with fine-grained sands and silty sands, 5–10 m thick (Fig. 31.5). These are the oldest Holocene alluvial deposits, formed before the period of the development of dense vegetation. The surface of a floodplain is almost flat. There are no flood-derived forms in the immediate vicinity of the channel, neither natural levees nor crevasse splays. This is probably due to the fact that floods of the Narew River are not large, and the flows are characterized by a low concentration of sediment transported in suspension. The only few elevations within the floodplain are aeolian dunes (Fig. 31.10). They reach a height of several metres. Most often these are parabolic dunes, i.e. they have the planform of arches pointed towards the east. A characteristic feature is that the dunes are concentrated in the latitudinal sections of the valley. The orientation of the dunes and their location are due to the fact that from the end of the Ice Age (Pleistocene) to the early Holocene, in a cold sub-polar climate, intense winds blew from west to east, causing aeolian accumulation.

31.5 Searching for the Driver of the Narew Anastomosing Pattern

Unambiguous determination of the main reasons for the transition of rivers to anastomosing patterns is still a debatable issue in the literature of fluvial systems. The channel

Fig. 31.8 Fragment of the Narew valley: m—meanders, o—oxbow lakes (photograph from Geoportal.pl)

Fig. 31.9 After each flood, the floodplain remains submerged for a long time (photograph from narwiański-park-narodowy.pl)

pattern change is referred to as river metamorphosis, and it is useful to decipher which factors could have caused this relatively short reach of the river to adopt an anastomosing pattern. This issue can be reduced to the following problem: What mechanisms caused the origin of a dense network of numerous channels?

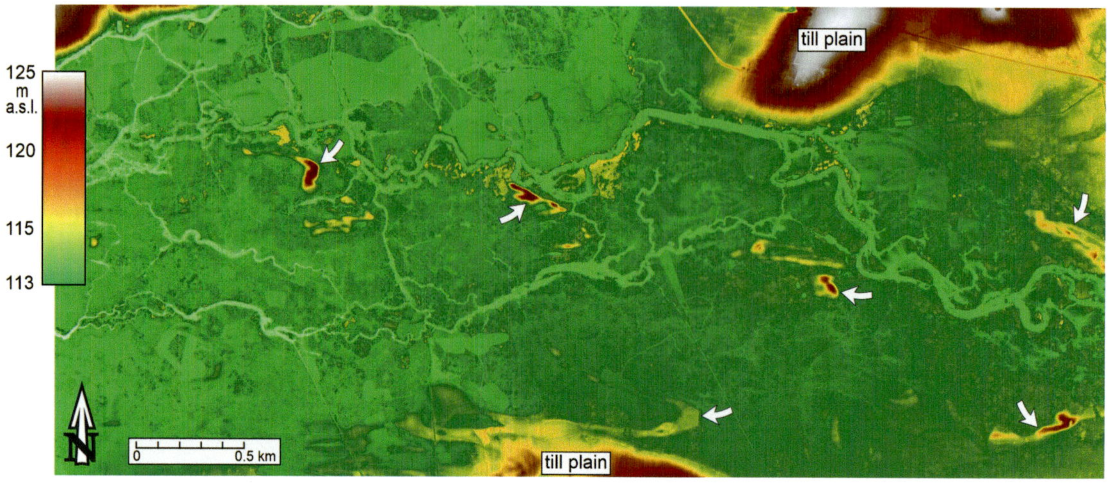

Fig. 31.10 Latitudinal reach of the Narew valley in DTM image. Dunes, most commonly of the parabolic type, are arrowed. Source: Geoportal.pl

The main geomorphological factor seems to be a low slope of the valley floor, i.e. high base level of the mentioned river reach. It mainly determines the low energy of the river environment, which is the basic feature of anastomosing rivers (Fig. 31.1; Teisseyre 1992). It can be hypothesized that the upper Narew River flows in an area where the vertical tectonic movements lead to the formation of a broad basin, from which the river runoff is hindered. The geological cross-section through the Quaternary deposits (Fig. 31.5) shows numerous fine-grained series of glacilimnic origin (i.e. clays and silts of ancient glacial lakes). This may support the hypothesis that for hundreds of thousands of years this region has been characterized by impeded drainage as a result of tectonic subsidence. The relationship between the basement tectonics and the anastomosing pattern of another Polish river—the Bug—has been proved by Harasimiuk et al. (1993). Another example of this relationship is the tectonically-controlled formation of the 'inland delta' of the Okavango River in South Africa (Smith et al. 1997). Moreover, a significant role in damming of the meridian-oriented Narew valley can also be played by the latitudinal rows of end-moraine hills (Fig. 31.4), through which the river had to incise.

Why the Narew River does not flow in a single channel, but in numerous ones? It is usually argued that the multi-channel network of anastomosing rivers is due to the fact that the waters cannot be contained in a single channel. The deposits of the valley floor are resistant to erosion, which means that the river cannot easily widen the channel. During a flood, when the entire floodplain is submerged, the flows cut new additional channels. The process of new channel incision in the valley floor is called avulsion. In the Narew River channels, bars are accumulated, reducing channel capacity. Aquatic vegetation plays the same role, especially plant jams obstructing the flow (Gradziński et al. 2003b). The peat that builds up a floodplain is an erosion-resistant material. It determines that the river does not erode laterally and does not increase the capacity of its channels. It was therefore concluded that the dense vegetation of reeds and the resulting thick layer of peat determined the anastomosing pattern of the river system (Gradziński et al. 2000, 2003b). Other researchers of anastomosing rivers located in Europe, North America, Africa, and Australia came to similar conclusions (Teisseyre 1992; Stanistreet et al. 1993; Nanson and Knighton 1996; Smith et al. 1997; Makaske 2001). It should also be remembered that the north-eastern part of Poland is a region with a continental climate. The winters are very cold and cause a thick layer of ice on the rivers. Spring floods, generated by melting snow, inundate the Narew valley when there is still ice in the channels. Such an ice blockage may also force incision of new channels into the floodplain. Moreover, one more (interesting!) factor favouring the formation of new channels cannot be omitted. The waterlogged valley floor is home to numerous animals, including beavers and elks. They move along their permanent paths. During a flood, these are the favoured lines of flow concentration and erosion of new channels on the floodplain (Gradziński et al. 2000).

It should be emphasized that the channel network of the Narew River is very stable. The channels have not changed their position for hundreds of years. This is due to the fact that the vertical growth of fluvial sediments, both in the channels and on the floodplain, occurs slowly but evenly and consistently. This style of sedimentation (vertical aggradation) is a typical feature of anastomosing river valleys (Fig. 31.1; Ghinassi et al. 2014). While the braided rivers shift their channels frequently and rapidly, and the sinuous channels of meandering rivers constantly migrate on the valley bottoms, the anastomosing river system may retain its unchanged planform for thousands of years.

31.6 Geological History of the Narew River Valley

Geological drillings show that there are 7 till horizons in the vicinity of the Narew valley (Butrymowicz 2001). They testify that the Scandinavian ice sheet appeared in this area seven times. In this regard, the Podlasie Plain is an excellent, even classic area for studying global climate changes reflected in glacial-interglacial alternations that have occurred in Europe in the last 1.5 million years.

The history of the pre-Narew valley began in the early Pleistocene, over a million years ago. At that time, during the oldest glaciation of the Polish area, the 40-m-deep valley of the pre-Narew river was cut in the marls of the Cretaceous substrate. Most likely, the origin of the fossil valley was related to tectonic processes, i.e. the existence of faults in the bedrock (Butrymowicz 2001) (Fig. 31.5). This oldest Pleistocene glaciation is called in the Polish stratigraphic division as the Narewian Glaciation. In eastern Poland, the first Scandinavian ice sheet (1.3–1.2 million years before present) reached the southernmost extent in the Central Europe. Large valleys, 40–60 m deep and 8 km wide, were formed here several times: in the Podlasian Interglacial (approx. 800–700 thousand years BP, i.e. Cromerian Interglacial in Western European stratigraphy), Nidanian Glaciation (700–650 ky BP, Glacial A/Dorst) and Holsteinian Interglacial (420–360 ky BP). Beside ancient valleys, glacial lakes were also formed, the waters of which were dammed up in depressions in front of the ice sheets during the Elsterian (500–430 ky BP), Drenthian (180–150 ky BP) and Warthanian (140–120 ky BP). The repeated concentration of river and lake waters confirms the hypothesis that the location and atypical orientation of the present-day Narew valley and the anastomosing channel pattern of this river have tectonic causes and are related to protracted crustal subsidence.

Although the initiation of the valley may be very old, the history of the present-day Narew River began with the retreat of the last ice sheet that reached this area. The receding front of the Warthanian ice sheet stagnated for a long time, with meltwaters flowing down from the ice sheet and depositing gravels and sands close to its margin. In this way, the end moraines were formed in the Northern Podlasie Plain (Fig. 31.4). When the front of the ice sheet began to retreat northwards again, the Narew valley was most likely used by a glacial river artery (proglacial river). This is evidenced by the covers of glacifluvial sands and gravels, which often accompany the valley (Figs. 31.4 and 31.5). After the warming of the Eemian Interglacial (130–115 ky BP), the Weichselian Glaciation began, but the ice sheet did not reach the Podlasie Plain. The ice-sheet margin was located 50–100 km north of the Narew valley. At that time, this region was characterized by polar climate, and with the disappearance of glaciation—the sub-polar one. Initially, vegetation practically did not exist, there was a 'polar desert'. Then, the area began to be conquered by sparse vegetation of the tundra (lichens, mosses, and small shrubs). Such a cold zone in the ice-sheet foreland is called the periglacial zone. Large volumes of water were released from the snow that melted in summer seasons. The poor vegetation cover did not intercept water, and the lower part of the ground was frozen and prevented water to soak into. Surface flows that flowed to the valley eroded the ground. Therefore, the Narew was a river with large discharges and frequent floods, transporting a lot of sediments. This periglacial river was the braided-type one (comp. Figure 31.1). The rule is that the rivers of cold climatic zones form braided fluvial systems. The overloaded flows have intensively deposited large amounts of sand, which builds the terrace of Weichselian age (Figs. 31.4 and 31.5).

About 10,000 years ago began a warm period that continues today—the Holocene. At the turn of the Holocene, still under periglacial conditions, aeolian accumulation took place. Strong winds blew the exposed glacifluvial sand covers (outwash) and accumulated sand on the damp valley floor. In this way, dunes were formed (Fig. 31.10), which testify to the last cold episode in the area. In the Holocene, the Narew catchment was covered with dense forests. Lush forest vegetation significantly influenced the hydrological regime of the river. The trees intercepted significant amounts of rainfall and protected the ground surface against erosion. Compared to the previous braided river, the Holocene Narew River was characterized by significantly lower discharges, less intense floods and a lower concentration of transported sediment. Therefore, the river metamorphosis to a meandering channel pattern took place (Gradziński et al. 2000). The meandering Narew River had lower energy and accumulated finer-grained sediments. They were only sands, often fine-grained (Fig. 31.5). Traces of this meandering phase of river development are still visible in the morphology of many sections of the contemporary Narew channels (Figs. 31.7 and 31.8). Finally, as a result of the overlapping of a number of factors that had been discussed earlier, the meandering river began to straighten its course and create numerous branches. In this way, an anastomosing river system was formed, nonetheless preserving traces of the previous meandering channel pattern. Climate change is believed to be the most important factor in river metamorphosis. At 5–7 ky BP, the warmest period of the Holocene took place, the so-called Holocene optimum. At that time, the greatest expansion of vegetation

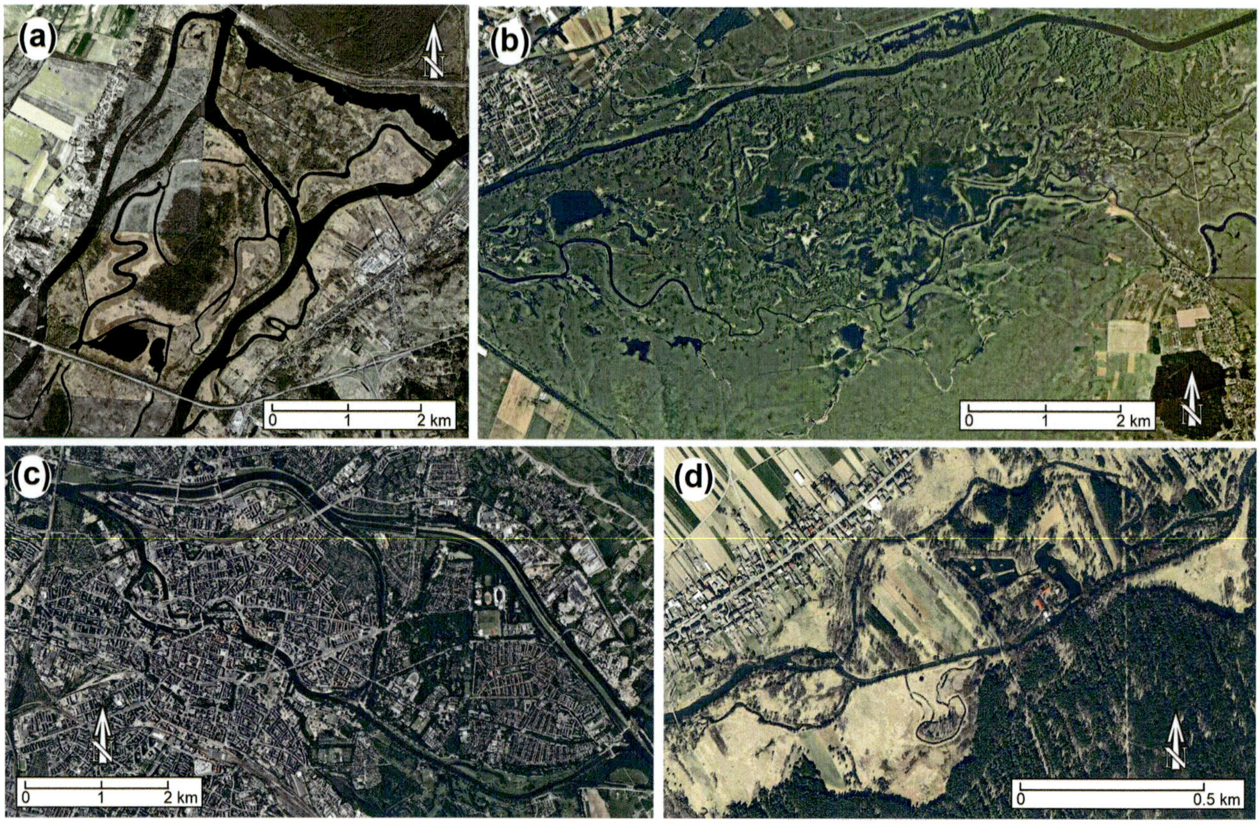

Fig. 31.11 Examples of anastomosing reaches of Polish rivers. **a** The lowermost reach of the Odra River near Szczecin. **b** The mouth of the Warta River to the Odra River. **c** The Odra River in Wrocław. **d** The Liswarta River (Kraków-Wieluń Upland) (photographs from Geoportal.pl)

in Europe took place. Thus, meandering-to-anastomosing metamorphosis of the Narew River occurred primarily as a result of the invasion of the dense vegetation cover at the bottom of its valley. This palaeoenvironmental statement has been proved by the dating of the lowest peat layers that build the Narew floodplain. For this purpose, the radiocarbon dating method was used. In this method, the age of the organic substance is calculated by examining the degree of radioactive decay of carbon isotope ^{14}C. The dates obtained were about 3000 years ago (Gradziński et al. 2003b). Thus, several thousand years of delay had to pass for the Narew River to change into a different river environment. In the 'clock' of geological history, a thousand years is just a minute.

31.7 Fluvial Analogues of the Narew River in Poland and in the World

Other contemporary examples of anastomosing rivers in Poland can be provided, although they are not as clear as in the Narew River valley. The reaches of anastomosing rivers exist in the lowest parts of the valleys, just before the river flows into the sea (as in the case of the Odra River, Fig. 31.11a) or to the main river (the Warta River case, Fig. 31.11b). In these cases, the main factor causing anastomosis is the proximity to the local base level, which has been rising in the recent geological past. The second largest Polish river—the Odra River—used to have an anastomosing channel pattern within the city of Wrocław, which is partly copied by the present-day network of regulated channels, although many original channels have disappeared due to urban expansion (Fig. 31.11c). The reason may be a tectonic fault, which occurs transversely in the bedrock under the valley and which could have acted as a factor reducing the slope of the Odra River. It is also interesting that the city was built within this multi-channel section of the valley, taking advantage of local geomorphic conditions. The branches of the Odra River facilitated access to river transport and the development of trade, and also played an important military role—buildings located on the islands were more easily defended. Anastomosing processes are visible not only on the main Polish rivers, but also on smaller ones: in Pomerania and central Poland (Fig. 31.11d) (Florek 1997; Słowik 2015). It may also be a smaller-scale, local phenomenon. Here, in turn, it is assumed that the metamorphosis of rivers to several anastomosing channels took place in historical times and was related to human activity. Mills built on rivers caused a local decrease in river gradient, which induced the river to anastomosing (Kaniecki and Brychcy 2010).

Where to look for analogues of the anastomosing Narew River outside Poland? Very low slope of the broad plain and the intense aquatic vegetation in the river and on the floodplain have made the Okavango River in Botswana a classic example of an anastomosing river system. Interestingly, the Okavango River does not debouch into any ocean or lake. In the tectonic depression, the river divides its flow into numerous channels, slows down and, in conditions of hot climate, finally disappears into the Kalahari savannah. In this way, the waterlogged 'inland delta' was formed—a unique area with characteristic, original flora and fauna. The Okavango National Park is recognized as one of the Seven Natural Wonders of Africa and is inscribed on the UNESCO World Heritage List. In this context, the valley of the anastomosing Narew River deserves to be known and protected as a unique fluvial landscape on European scale.

Finally, it is worth providing some geological information. Most of the coal basins (both hard and brown coal) are derived from environments resembling the Narew River valley. Wetlands of anastomosing river valleys, covered with lush vegetation (also woodland), were basins of intensive accumulation of layers of plant debris. After millions of years, thick layers of peat became the brown coal and then, after hundreds of millions of years, the hard coal.

31.8 Conclusions

The upper Narew valley is located in north-eastern Poland. It is surrounded by a low-relief landscape of primarily glacial origin. Although the river is relatively small (mean annual discharge = 15 $m^3 s^{-1}$), the width of its valley reaches several kilometres. The anastomosing Narew River is a multi-channel fluvial system (up to seven active channels per valley width), characterized by a very low energy of river flows (average flow velocity = 10–30 cm $s-^1$). The channels are narrow and deep, and their banks are often vertical. The secondary channels are numerous, often with stagnating water. The floodplain is very low in respect to the channel and is submerged with every flood. It is occupied by a dense cover of aquatic plants (rush and reed communities). Organic deposition resulted in the formation of a thick (up to 4 m) peat layer. The multi-channel river network was caused by the overlapping of two factors. Channel capacity is limited due to the accumulation of sand bars and plant jams that block the flow. On the other hand, the river cannot widen the channels because the thick layer of peat on the floodplain is resistant to erosion. Therefore, the flood flows cut new channels in the outer parts of the floodplain where the peat was thinner. This is how the avulsion occurred.

In the past centuries, many European rivers had an anastomosing pattern. Human activity has transformed these rivers into regulated canals. The upper Narew River, located in central-eastern Europe, has not changed in historical times. Therefore, it is an example of a natural anastomosing river, unique on a European scale. Due to significant geomorphological and biological values, the anastomosing section of the valley is protected by the Narew National Park.

References

Bałuk A (1973) Geological Map of Poland 1:200,000, sheet Łomża. Wydawnictwo Geologiczne, Warszawa

Berendsen HJA, Stouthamer E (2000) Late Weichselian and Holocene palaeogeography of the Rhine-Meuse delta, the Netherlands. Palaeogeogr Palaeoclimatol Palaeoecol 161:311–335

Butrymowicz N (2001) Detailed geological map of Poland 1:50,000, sheet Choroszcz. Polish Geological Institute, Warszawa

Florek W (1997) Climatic and anthropogenic impulses in the late Vistulian and Holocene development of the river channels and valleys of the Baltic Coastal Region and Pomerania. Landf Anal 1:41–50

Ghinassi M, Nemec W, Aldinucci M, Nehyba S, Özaksoy V, Filolini F (2014) Plan-form evolution of ancient meandering rivers reconstructed from longitudinal outcrop sections. Sedimentology 61:952–977

Gradziński R, Baryła J, Danowski W, Doktor M, Gmur D, Gradziński M, Kędzior A, Paszkowski M, Soja R, Zieliński T, Żurek S (2000) Anastomosing system of the upper Narew River, NE Poland. Ann Soc Geol Polon 70:219–229

Gradziński R, Baryła J, Danowski W, Doktor M, Gmur D, Gradziński M, Kędzior A, Paszkowski M, Soja R, Zieliński T, Żurek S (2003a) In-channel accretionary macroforms in the modern anastomosing system of the upper Narew River, NE Poland. Ann Soc Geol Polon 73:35–53

Gradziński R, Baryła J, Danowski W, Doktor M, Gmur D, Gradziński M, Kędzior A, Paszkowski M, Soja R, Zieliński T, Żurek S (2003b) Vegetation-controlled modern anastomosing system of the upper Narew River (NE Poland) and its sediments. Sediment Geol 157:253–276

Harasimiuk M, Rzechowski J, Szwajgier W (1993) Wpływ ruchów neotektonicznych na warunki rozwoju równi zalewowej i koryta Bugu w Obniżeniu Dubienki (Polesie Zachodnie). Ann Univ M Curie-Skłodowska (Lublin) B 48:123–132

Kaniecki A, Brychcy D (2010) Średniowieczne młyny wodne i ich wpływ na przemiany stosunków wodnych na przykładzie zlewni Obry Skwierzyńskiej. Bad Fizjogr A61:145–156

Kędzior A, Widera M, Zieliński T (2021) Ancient and modern anastomosing rivers: insights from sedimentological and geomorphological case studies of the Triassic, Neogene and Holocene of Poland. Geol Quart 65:54

Makaske B (2001) Anastomosing rivers: a review of their classification, origin and sedimentary products. Earth Sci Rev 53:149–196

Makaske B, Maathuis BHP, Padovani CR, Stolker Ch, Mosselman E, Jongman RHG (2012) Upstream and downstream controls of

recent avulsions on the Taquari megafan, Pantanal, SW Brazil. Earth Surf Proc Landf 37:1313–1326
Mioduszewski W, Napiórkowski J, Okruszko T (2016) Wody Narwiańskiego Parku Narodowego. In: Banaszuk P, Wołkowycki D (eds) Narwiański Park Narodowy. Krajobraz, Przyroda, Człowiek. Białystok—Kurowo, pp 51–64
Nanson GC, Knighton AD (1996) Anabranching rivers: their cause, character and classification. Earth Surf Proc Landf 21:217–239
Rust BR (1978) A classification of alluvial channel systems. In: AD Miall (ed) Fluvial Sedimentology. Canadian society of petroleum geologists Memoir, vol. 5, pp 187–198
Schumm SA (1972) Fluvial paleochannels. In: JK Rigby, WR Hamblin (eds) Recognition of ancient sedimentary environments. SEPM Special Publication, vol. 16, pp 98–107
Schumm SA (2005) River variability and complexity. Cambridge University Press, 223 p
Słowik M (2015) Is history of rivers important in restoration projects? The example of human impact on a lowland river valley (the Obra River, Poland). Geomorphology 251:50–63
Smith ND, McCarthy TS, Ellery WN, Merry CL, Rüther H (1997) Avulsion and anastomosis on the panhandle region of the Okavango Fan, Botswana. Geomorphology 20:49–65
Stanistreet IG, Cairncross B, McCarthy TS (1993) Low sinuosity and meandering bedload rivers of the Okavango Fan: channel confinement by vegetated levees without fine sediment. Sediment Geol 85:135–156
Teisseyre AK (1992) Rzeki anastomozujące—procesy i modele sedimentacji. Przegl Geol 4:241–248
Trnqvist TE (1993) Meandering and anastomosing fluvial systems in the Rhine-Meuse Delta (central Netherlands) controlled by sea-level rise and subsoil erodibility. J Sediment Petrol 63:683–693

Tomasz Zieliński is Professor of Geology at the Adam Mickiewicz University in Poznań. His research topics are sedimentology of clastic deposits, Quaternary geology, ancient and recent fluvial sedimentary environments. He is particularly interested in the conditions of formation of Pleistocene terminal moraines and sandurs. On the other hand, he studied fluvial successions and environments from the Mesozoic age, through the Neogene and Quaternary, to the present-day ones. He is, among others, the author of the book 'Sedimentology. Sediments of rivers and lakes', in which he reviewed the worldwide knowledge of sedimentology from the 1970s to the 2010s. Tomasz Zieliński is the editor-in-chief of the *Geologos* journal, which focuses on geological studies of central Europe and those areas of the world whose geology is still poorly explored.

32 Morasko—Meteorite Impact Craters in the Glacial Landscape

Witold Szczuciński and Wojciech Włodarski

Abstract

The Morasko meteorite impact craters are among approximately 200 documented impact structures on Earth. They were most likely formed in the mid-Holocene by the impact of an iron meteorite, which left at least seven impact craters and numerous meteorite pieces scattered over several km^2. The craters were formed on slopes of the push moraine, finally shaped around 18–19 ka BP, during the Poznań–Frankfurt retreat phase of the latest glaciation (Weichselian Glaciation). Thus, the modern landscape is of complex origin, with major glacial landforms overprinted with impact structures. The craters are circular, bowl-shaped depressions with various cross-sectional asymmetries resulting from the interplay between the impact process and the primary pre-impact topography of glacial and post-glacial landforms. The diameters of the craters range between c. 30 and 100 m. The processes associated with the impact and their effects have been recently subjected to numerous studies, including detailed morphological and geological investigations, numerical modelling of the impact, and the quest for indicators of impact's environmental effects. Although some issues are still unresolved, the Morasko meteorite impact is likely one of the best-studied small/moderate impacts globally.

Keywords

Impact structure · Meteorite crater · Glacial landforms · Push moraine

32.1 Introduction

The usually symmetrical depressions called craters belong to the most intriguing landforms. They dominate the landscape of many planets and moons with a solid surface, for instance, the Moon, Mercury, and Mars. Most of these landforms have resulted from the hypervelocity impacts of asteroids. However, their genesis was disputed for a long time, because on Earth, craters were usually associated with volcanic activity and, in recent decades, also with bomb explosions and ground-based nuclear tests (Melosh 1989). The rapid progress in studies on impact crater formation in the last few decades led to the conclusion that impact cratering may be the most important geological process shaping planetary surfaces (Osinski and Pierazzo 2013).

According to Osinski (2013), an impact crater is '*a generally circular crater formed by impact of an interplanetary body (projectile) on a planetary surface*'. In the case of old eroded craters, the term '*impact structure*' should be used. Impact craters may range from simple, small (metres to hundreds of metres), bowl-shaped depressions to large (up to hundreds of km) and complex multi-ringed impact basins. However, on Earth, many processes may form similar circular depressions (Branney 1995). For example, volcanic collapses may produce circular calderas, karstic subsidence produces sinkholes, melting of buried ice results in the origin of glacial kettle holes, and hydrothermal explosions also give rise to regular circular hollows, to name a few. Thus, identifying impact craters requires careful analyses of the local geological context and consideration of several indicators (French and Koeberl 2010). The latter include, for instance, evidence of high pressure and temperature experienced by rocks and minerals (shatter cones,

W. Szczuciński (✉)
Geohazards Research Unit, Institute of Geology, Adam Mickiewicz University, Krygowskiego 12, 61–680 Poznań, Poland
e-mail: witek@amu.edu.pl

W. Włodarski
Department of Palaeoenvironmental Research, Institute of Geology, Adam Mickiewicz University, Krygowskiego 12, 61–680 Poznań, Poland
e-mail: wojtekw@amu.edu.pl

P. Migoń and K. Jancewicz (eds.), *Landscapes and Landforms of Poland*, World Geomorphological Landscapes, https://doi.org/10.1007/978-3-031-45762-3_32

planar deformation features in minerals), specific crater geometry (e.g. the depth-to-diameter ratio in the order of 1:5 to 1:7 in case of new forms (Melosh 1989)), the occurrence of ejecta blanket (layer of material excavated from the crater during the impact and deposited in the form of symmetrical apron surrounding the crater) and the presence of the remnants of the projectile. Crater identification may be challenging in the case of small impact structures, where the amount of material subjected to high pressure and the temperature is small; thus, additional contextual evidence may be necessary.

In comparison with the surface of the Moon or Mercury, dotted with numerous craters, impact craters on Earth are much less common. The database 'Impact Earth' (Impact Earth 2022; Osinski et al. 2022) listed only approximately 200 confirmed impact craters and structures on Earth. The moderate number of craters is primarily due to the action of erosion, sedimentation, and tectonic processes combined. The identified impact structures vary in age and size (from tens of metres to almost 200 km). Most of the identified structures are relatively old and large; only 34 were formed in the Quaternary, and among them, 26 are less than 1 km in diameter. However, according to Hergarten and Kenkmann (2015), several hundreds of small to moderate size craters may still be discovered on Earth.

In Europe, about 50 impact craters and structures were documented (Impact Earth 2022). Most of them were identified in the areas built of the oldest continental crust and exposed for the longest time to the impacts. Some of them are relatively well-studied structures, for instance, Kaalijärv (Estonia), Ries (Germany), Mjølnir (Norway), and Boltysh (Ukraine). In Poland, a single confirmed impact site with well-preserved small craters is documented on Morasko Hill in Poznań, western Poland (Fig. 32.1).

Numerous finds of the Morasko meteorite fragments (the first one was found and identified as an iron meteorite in 1914) and topographic hollows now interpreted as impact craters (Pokrzywnicki 1964) have been the subject

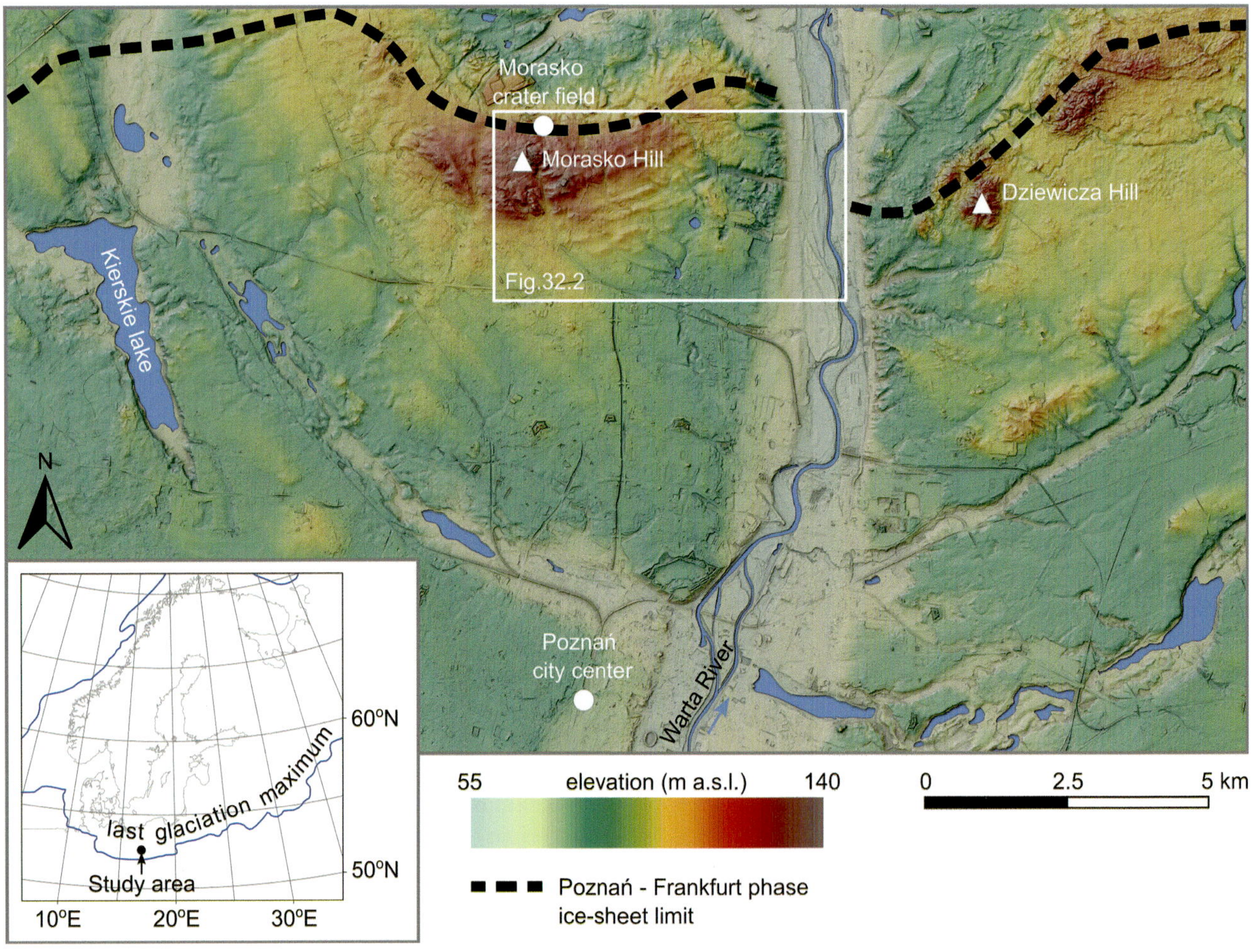

Fig. 32.1 Morphology of the area adjacent to the limit of the Poznań—Frankfurt phase of Weichselian ice-sheet near the city of Poznań. The inset map shows the position of the area in northern Europe and the limits of the Scandinavian ice sheet during the Last Glacial Maximum (Svendsen et al. 2004)

of many studies for over a century. A summary of the state of knowledge about the Morasko meteorite and the effects of its impact, as well as the controversy related to various interpretations, can be found in several review works, for example, by Pokrzywnicki (1964), Hurnik (1976), Stankowski (2001, 2008), Muszyński et al. (2012) and Szczuciński and Muszyński (2020). Among the major research topics were the composition and structure of the meteorite (e.g. Dominik 1976; Muszyński et al. 2012; Duczmal-Czernikiewicz and Michalska 2018), leading to the discovery of two new minerals, namely moraskoite (Karwowski et al. 2015) and czochralskiite (Karwowski et al. 2016). In addition, detailed studies were carried out on the morphology and structure of impact craters (e.g. Karczewski 1976; Stankowski 2001, 2008; Włodarski et al. 2017; Szokaluk et al. 2019) and the surrounding ejecta layer (Szokaluk et al. 2019), as well as on sediments filling the craters (Tobolski 1976; Stankowski 2008; Choiński et al. 2019). Moreover, numerical modelling of the impact provided likely physical boundary conditions of the event (Bronikowska et al. 2017), and the nearby lake sediments were studied to assess the environmental effects of the meteorite impact (Pleskot et al. 2018). Thus, despite many unanswered issues, the Morasko meteorite craters belong to the best-studied small impact craters worldwide. These studies provided new insights into the formation and identification of small impact craters in unconsolidated sediments applying, for instance, ejecta properties (Szokaluk et al. 2019) and charcoal characteristics (Losiak et al. 2022).

The present chapter aims to present an outline of the geomorphological context of the Morasko meteorite impact craters. They are interesting not only as a unique type of relief in Polish landscapes but also provide an example of an interplay of glacial landforms overprinted with impact features. They also offer a rare example of a relatively well-preserved small impact crater in unconsolidated sediments.

32.2 Regional Setting

The Morasko Hill (Góra Moraska) is one of the highest elevation in the Wielkopolska region (153.8 m asl; Fig. 32.1) and has a relatively complex history (Rachlewicz and Szczuciński 2022). It is a part of a series of moraine hills north of Poznań formed ca. 18,500 years BP (Kozarski, 1995), which mark the Poznań—Frankfurt retreat phase of the last glaciation, lasting probably 300–400 years. However, the origin of this elevation and the sediments forming it (Fig. 32.2) are much older.

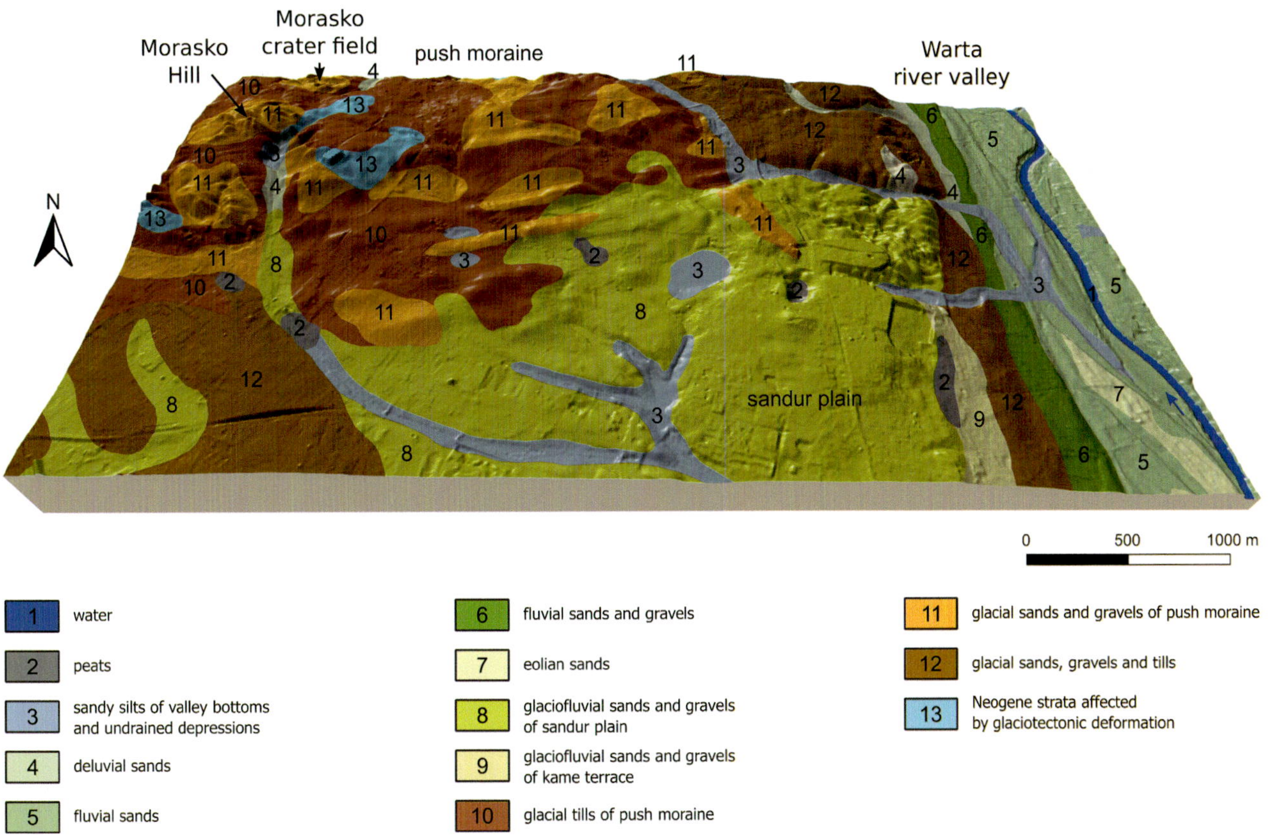

Fig. 32.2 Block diagram of an area marked in Fig. 32.1, with surface sediment distribution after Chmal (1990). The vertical exaggeration of the block diagram is 5x

The oldest deposits in the Morasko Hill area are clays and silts of the Neogene age, forming the Poznań Series (Chmal 1990). They were documented in the boreholes, at elevations of about 50 to 70 m asl in the vicinity of the hill. They are usually covered by a 20 to 50 m thick layer of variable glacial and glacifluvial deposits. However, within the Morasko Hill moraine, they reach an altitude of 130 m asl and locally outcrop at the surface (Fig. 32.2). This anomalously high position of these sediments is related to glacitectonic deformations (Stankowski 2001). The deformations probably took place already during the older (e.g., Saalian), more extensive glaciations (Chmal 1997; Stankowski 2001, 2008), which also left respective till layers (Widera and Chomiak 2019). Subsequent modifications during the latest Weichselian glaciation were relatively minor.

The Quaternary deposits in the region of Morasko Hill reach up to 100 m in thickness. They are mainly composed of various types of tills, glacifluvial sands and gravels forming, for instance, an outwash plain southward of the Morasko Hill moraine ridge (85–105 m asl). There are also hillslope deposits, river, aeolian, and peat deposits (Fig. 32.2). Most near-surface deposits were formed during the last glaciation, when particular ice streams within the ice sheet flowed southward (Szuman et al. 2021), reaching maximum limit during the Last Glacial Maximum around 20 ka BP, when the ice-sheet margin was ca. 70 km southwards, nearby the present-day town of Leszno. At that time, the ice-sheet thickness in the area of Poznań could reach even several hundred of metres (Kasprzak 2003). During the recession, also minor landforms were formed, for instance, small-scale ridges and undrained depressions probably related to the decay of ice-cored moraines (Ewertowski and Rzeszewski 2006).

After the ice sheet retreat, the area was still subjected to permafrost and periglacial processes, and it took several thousand years to have all the dead ice blocks and ground ice completely melt. Due to that, many hollows were formed, which now host lakes and peatlands (e.g., Stankowski 2008; Pleskot et al. 2018).

32.3 Discovery of Impact Craters

Pokrzywnicki (1964) was the first to suggest several circular depressions on the northern slopes of Morasko Hill (Figs. 32.3, 32.4 and 32.5), next to sites with iron meteorite (coarse-grained octahedrite) common findings, were of impact origin. However, this interpretation was only partially accepted by then for two main reasons. First, at that time there were no commonly accepted criteria for identifying impact craters (French and Koeberl 2010), and especially the criteria for small impact craters, formed under lower pressures and temperatures, were still under development. Second, these impact craters were formed in a glacial landscape of complex geology, the latter mainly due to glacitectonic processes. In such an area, it is common to find small hollows related to, for instance, melting of dead ice blocks. Moreover, although the sediments in the vicinity of the impact craters were found to be deformed in various ways, deformation was assigned mainly to glaciotectonic processes (Karczewski 1976). Thus, the early detailed studies of Karczewski (1976) and his students did not provide an unambiguous interpretation of the origin of the depressions.

For several decades, the main argument for the impact origin of at least some of the craters was the age of their sedimentary infill. As small lakes occupy the biggest craters, they also contain organic-rich deposits. Their ages post-date the formation of the craters, and they were established to be mid-Holocene (c. 5500–5000 years BP) based on pollen analysis (Tobolski 1976) and radiocarbon datings (Stankowski 2001, 2008). Thus, they are much younger than the deglaciation of the last ice sheet from the area (c. 18,000 years BP). The age of the impact was later partly supported by datings using luminescence methods (e.g., Stankowski et al. 2007).

More recent studies provided new lines of evidence supporting the impact origin of the craters. They included detailed studies on the morphology of the craters (Włodarski et al. 2017), geology (Szokaluk et al. 2019), dating (Szczuciński et al. 2016), and distinct characteristics of charcoal in the ejecta layer (Losiak et al. 2022). In addition, detailed geological mapping of the surroundings of the craters revealed the presence of ejecta deposits covering in some places the pre-impact palaeosoil (Fig. 32.5; Szokaluk et al. 2019). The latter's age appeared to overlap within ^{14}C uncertainty dating margins, with the oldest dates documented in the crater infill deposits, thus providing strong evidence of the mid-Holocene age of the impact craters (Szczuciński et al. 2016). These works, along with the recent documentation of charcoals typical for hot ejecta excavated from impact craters (Losiak et al. 2022) and other supporting evidence discussed in overview papers (e.g. Stankowski 2008; Muszyński et al. 2012; Szczuciński and Muszyński 2020) confirmed that at least the biggest hollows are impact craters.

The number of Morasko impact craters is still under debate. Various authors suggested different numbers, ranging from 7 to more than 11 (Pokrzywnicki 1964; Stankowski 2008; Włodarski et al. 2017; Walesiak 2021). However, only for the seven located close together (Figs. 32.3 and 32.4) exists some additional geological evidence, which in the majority is limited to the biggest crater (A, Fig. 32.5c). Numerical modelling of the impact (Bronikowska et al. 2017) suggested that the number of small impact craters could be higher, but they are probably not preserved in the modern landscape.

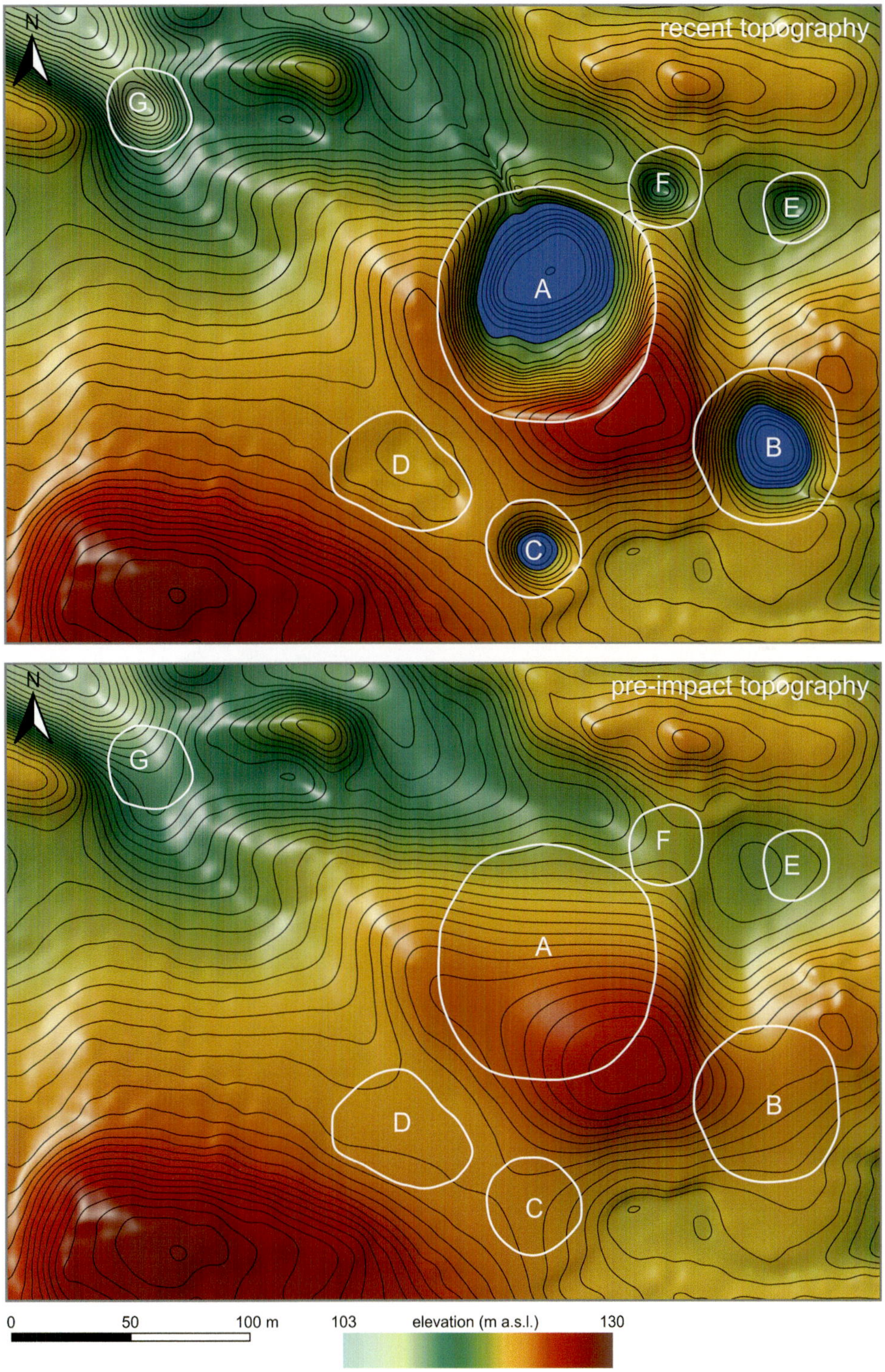

Fig. 32.3 DEM of recent topography of Morasko craters (upper panel) and reconstructed pre-impact topography (lower panel). The contour intervals are 5 m. White lines mark the rims of impact craters. The labels of craters follow those used in earlier studies (Pokrzywnicki, 1964; Karczewski, 1976; Stankowski, 2008). Data from Włodarski et al. (2017)

32.4 Morphology and Modifications of the Impact Craters

Although, at the first sight, the Morasko impact craters seem irregular in shape, they are in fact bowl-shaped, circular depressions (Figs. 32.3 and 32.4). The impact itself would likely generate symmetrical forms, whereas the present-day diversity of forms results from the interaction of impact processes with the pre-existing landscape and post-impact alterations. The numerical modelling study suggested that these craters could be formed by the impact of iron meteorites with diameters up to ca. 3 m and velocity in

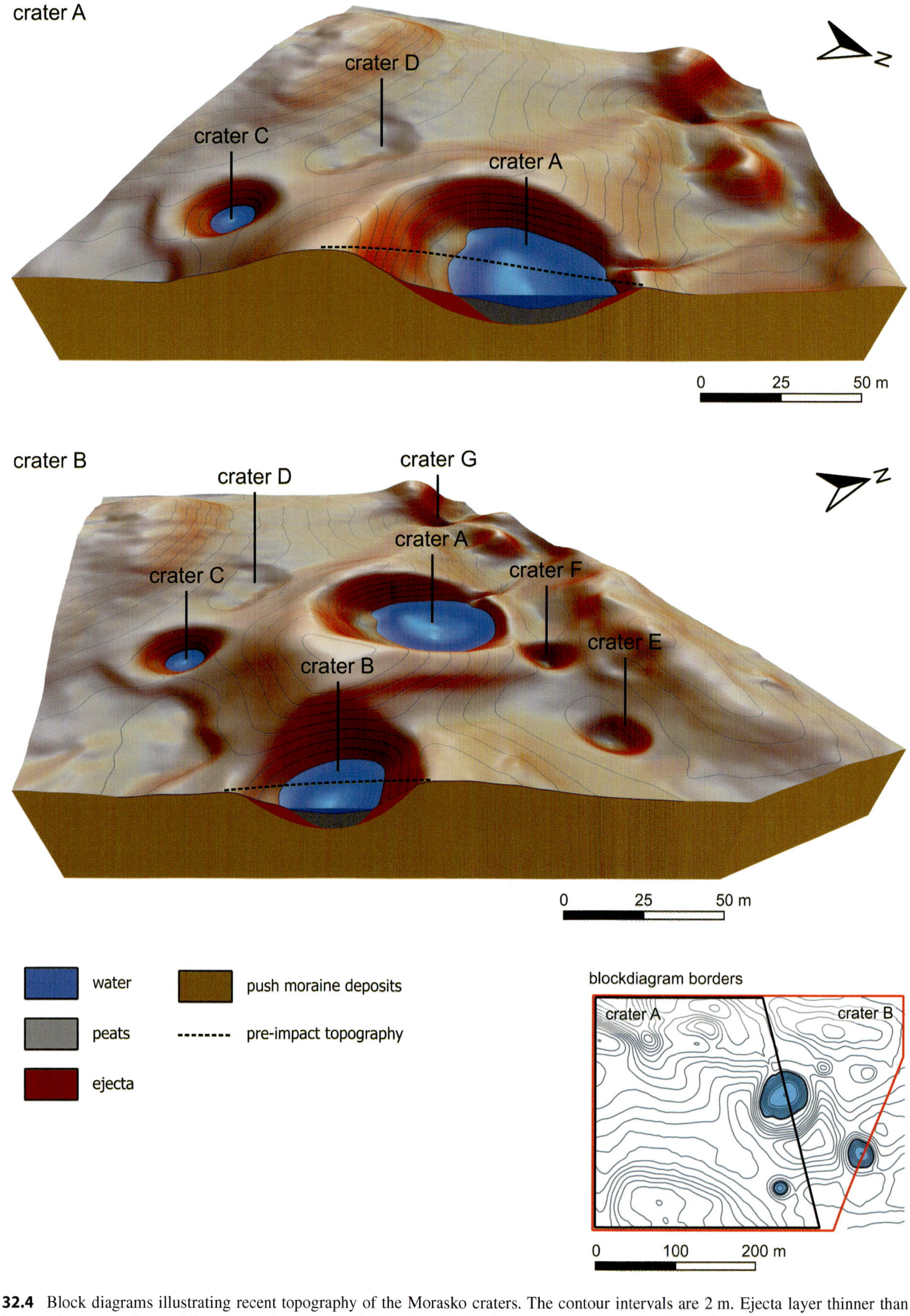

Fig. 32.4 Block diagrams illustrating recent topography of the Morasko craters. The contour intervals are 2 m. Ejecta layer thinner than 1 m, dominating the surroundings of the craters, is not marked. The ejecta in the impact craters is mainly due to its post-impact redeposition. Push moraine deposits are shown in a simplified way; in fact, they present a very complex nature, including also Neogene clays and Pleistocene till

order of 6 km s^{-1} (Bronikowska et al. 2017). The modelled spatial distribution of the craters, expected after pre-impact meteorite fragmentation in the atmosphere, is in accordance with the observed location of smaller craters distributed more or less symmetrically around the biggest one.

The best-studied seven craters (Figs. 32.3 and 32.4) are located on the northern slopes of the Morasko Hill. They are 30 to 96.5 m in diameter, and their average depths are 1.2 to 12.1 m. Ellipticity of the craters, defined as the ratio between the maximum and minimum crater diameters, ranges from 1.03 to 1.1 and is similar to that of other terrestrial impact structures, except for more elongated crater D. The unique geometric characteristics of the crater D may suggest more significant post-impact changes or possibly its formation due to the overlap of two smaller craters (Włodarski et al. 2017).

The detailed study by Włodarski et al. (2017) revealed that the maximum internal slope angles of the craters vary between 20 and 28°, except for the crater D, with more gentle slopes. This slope variability results from pre-existing topography and sediment types, impact processes, and post-impact modification. For instance, the observed elevation difference between the southern and northern rim of the biggest crater A results mainly from the fact that the crater was formed on a slope (Fig. 32.3). An example of post-impact modifications resulting from mass movements are small-scale lobe-shaped slope terraces in the lower parts of southern internal slopes near the crater-lake in crater A, which underline the present-day crater asymmetry (Figs. 32.3 and 32.4).

The slumping of slopes and accumulation at the floors of the craters led to the shallowing and widening of the craters. The biggest crater A is filled with a lake up to 2.6 m deep (Pokrzywnicki 1964) and with post-impact, mainly organic deposits, up to 4 m thick (Choiński et al. 2019). The morphometric analyses revealed that the present-day crater's diameters could be bigger than the original ones by up to almost 20% (Włodarski et al. 2017). Due to that, the present depth-to-diameter ratio is lower than 1:5 to 1:7, typical for fresh impact craters (Melosh 1989).

Geological studies (Szokaluk et al. 2019) revealed that the mass-moved materials include mainly ejecta produced due to the impact-induced fragmentation and excavation of the Pleistocene glacial tills and Neogene clays. The ejecta was deposited within the craters, where it was subjected to slumping and redeposition (Fig. 32.4), and around them. In the latter case, it forms up to 1 m thick sediment layer, underlain by older deposits and occasionally preserved palaeosoil (Fig. 32.5c).

Considering the age of the hollows and unconsolidated sediments in which they were formed, the preservation of the craters seems to be relatively good. It may be partly due

Fig. 32.5 The biggest crater (A) on the Morasko hill filled with a shallow lake (max depth 2.6 m). **a** View towards north-east. **b** The western rim of the crater, view towards the north. **c** Section from a trench dug next to the southern rim of the crater A, with remnants of organic matter-rich palaeosoil covered by ejecta layer (photographs by W. Szczuciński)

to their formation mainly in cohesive clays and tills and lack of significant anthropogenic alterations, as this area was likely continuously covered with oak-hornbeam forest during historical times (Macias and Dryjer 2010).

32.5 Present State and Visiting

The impact craters and valuable forest ecosystem next to them were subjected to protection in 1976 and are within the Nature Reserve 'Meteoryt Morasko'. The craters are easy to access; there is a parking and bus stop nearby and a well-marked path with guiding information panels. The impact craters are particularly well visible during late autumn to early spring, when foliage does not obscure the view.

Those interested in the meteorites themselves may see them in many museums. The closest to the impact site is the Muzeum Ziemi UAM (Museum of Earth at Adam Mickiewicz University) in Poznań, located within the building of the Faculty of Geographical and Geological Sciences of Adam Mickiewicz University (address: Bogumiła Krygowskiego 10, Poznań; https://muzeumziemi.amu.edu.pl/). The museum hosts a unique collection of meteorites, including the second-biggest specimen of the Morasko meteorite (261 kg). The biggest specimen (271.8 kg) is in a private collection.

32.6 Conclusions

The history of the discovery of small depressions on the northern slopes of the Morasko Hill is an interesting example of the development of a scientific hypothesis. These impact craters are examples of one of the best-studied small impact structures on Earth and provide many clues helpful for recognizing impact craters. There are certainly more craters on Earth (Hergarten and Kenkmann 2015); however, many are hard to identify due to post-formation changes. The case of Morasko may serve as an excellent example of how to identify them. Most of the information and publications related to Morasko meteorites and craters are listed on Wiki.meteoritica (2022) web page.

Acknowledgements Most of the research leading to this chapter was conducted within an international project '*Effects of meteorite impact in unconsolidated sediments—case of iron meteorite shower "Morasko", Poland*' funded by National Science Centre, Poland (2013/09/B/ST10/01666), the authors thank all the project participants for the collaborative work and fruitful discussions. The authors also thank Piotr Migoń, Kacper Jancewicz and two anonymous reviewers for useful comments on the chapter.

References

Branney MJ (1995) Downsag and extension at calderas: new perspectives on collapse geometries from ice-melt, mining, and volcanic subsidence. Bull Volcanol 57:303–318

Bronikowska M, Artemieva NA, Wunnemann K (2017) Reconstruction of the Morasko meteoroid impact—Insight from numerical modeling. Meteoritics Planet Sci 52:1704–1721

Chmal R (1990) Szczegółowa mapa geologiczna Polski 1:50000. Arkusz Poznań. PIG-PIB, Warszawa

Chmal R (1997) Objaśnienia do szczegółowej mapy geologicznej Polski w skali 1:50000. Arkusz Poznań (471). Polish Geological Institute, Warszawa

Choiński A, Muszyński A, Ptak M, Słowik M (2019) Reconstruction of the primary bottom of a unique crater lake in the "Meteoryt Morasko Reserve" (Poland). Bull Geogr, Phys Geogr Ser 17:5–16

Dominik B (1976) Mineralogical and chemical study of coarse octahedrite Morasko (Poland). Pr Miner PAN 47:61

Duczmal-Czernikiewicz A, Michalska D (2018) Mineralogy and microstructure of the Morasko meteorite crust. Planet Space Sci 164:44–53

Ewertowski M, Rzeszewski M (2006) Using DEM to recognize possible minor stays of Vistulian (Weichselian) ice-sheet margin in the Wielkopolska Lowland. Quaest Geogr 25A:7–21

French BM, Koeberl C (2010) The convincing identification of terrestrial meteorite impact structures: what works, what doesn't, and why. Earth Sci Rev 98:123–170

Hergarten S, Kenkmann T (2015) The number of impact craters on Earth: any room for further discoveries? Earth Planet Sci Lett 425:187–192

Hurnik H (ed) (1976) Meteorite Morasko and the region of its fall. Wydawnictwo Naukowe Uniwersytetu im. A. Mickiewicza, Poznań, Seria Astronomia 2, p 64

Impact Earth (2022) https://impact.uwo.ca/

Karczewski A (1976) Morphology and lithology of closed depression area located on the northern slope of Morasko Hill near Poznań. In: Hurnik H (ed) Meteorite Morasko and the Region of Its Fall. Seria Astronomia vol. 2. Adam Mickiewicz University Press, Poznań, pp 7–20

Karwowski Ł, Kusz J, Muszyński A, Kryza R, Sitarz M, Galuskin EV (2015) Moraskoite, $Na_2Mg(PO_4)F$, a new mineral from the Morasko IAB-MG iron meteorite (Poland). Miner Mag 79:387–398

Karwowski Ł, Kryza R, Muszyński A, Kusz J, Helios K, Drożdżewski P, Galuskin EV (2016) Czochralskiite, $Na_4Ca_3Mg(PO_4)_4$, a second new mineral from the Morasko IAB-MG iron meteorite (Poland). Europ J Miner 28:969–977

Kasprzak L (2003) Model sedymentacji lądolodu vistuliańskiego na Nizinie Wielkopolskiej. Wydawnictwo Naukowe UAM, Poznań, p 214

Kozarski S (1995) Deglacjacja północno-zachodniej Polski: warunki środowiska i transformacja geosystemu (ok. 20 ka—10 ka BP). Continuo, Wrocław, p 82

Losiak A, Belcher CM, Plado J, Joeleht A, Herd C, Kofman R, Szokaluk M, Szczuciński W, Muszyński A, Wild E, Baker S (2022) Small impact cratering processes produce distinctive charcoal assemblages. Geology 50:1276–1280

Macias A, Dryjer M (2010) Forest cover dynamics in the city of Poznań from 1830 to 2004. Quaest Geogr 29:47–57

Melosh HJ (1989) Impact cratering: a geologic process. Oxford University Press, New York, p 245

Muszyński A, Kryza R, Karwowski Ł, Pilski AS, Muszyńska J (eds) (2012) Morasko. The largest iron meteorite shower in Central Europe. Studia i Prace z Geografii i Geologii, 28. Bogucki Wydawnictwo Naukowe, Poznań, p 111

Osinski GR, Pierazzo E (eds) (2013) Impact cratering: processes and products. Wiley-Blackwell, Hoboken, p 316

Osinski GR, Grieve RAF, Ferrière L, Losiak A, Pickersgill AE, Cavosie AJ, Hibbard SM, Hill PJA, Bermudez JJ, Marion CL, Newman JD, Simpson SL (2022) Impact Earth: a review of the terrestrial impact record. Earth Sci Rev 232:104112

Osinski GR (2013) Processes and products of impact cratering: glossary and definitions. In: Osinski GR, Pierazzo E (eds) Impact cratering: processes and products. Wiley-Blackwell, Hoboken, pp 306–309

Pleskot K, Tjallingii R, Makohonienko M, Nowaczyk N, Szczuciński W (2018) Holocene paleohydrological reconstruction of Lake Strzeszyńskie (western Poland) and its implications for the central European climatic transition zone. J Paleolimnol 59:443–459

Pokrzywnicki J (1964) I. Meteoryty Polski. II. Katalog meteorytów w zbiorach polskich. Studia Geol Polon 15:9–140

Rachlewicz G, Szczuciński W (2022) Geologiczna historia Góry Moraskiej. In: Kaczmarek F (ed) Radojewo i Morasko. Kronika Miasta Poznania 2022/2, pp 338–352

Stankowski WTJ (2001) The geology and morphology of the natural reserve "Meteoryt Morasko." Planet Space Sci 49:749–753

Stankowski WTJ, Raukas A, Bluszcz A, Fedorowicz S (2007) Luminescence dating of the Morasko (Poland), Kaali, Ilumetsa and Tsõõrikmäe (Estonia) Meteorite Craters. Geochronometria 28:25–29

Stankowski WTJ (2008) Meteoryt Morasko—osobliwość obszaru Poznania. Wydawnictwo Naukowe UAM, Poznań, p 94

Svendsen JI, Alexanderson H, Astakhov VI, Demidov I, Dowdeswell JA, Funder S, Gataullin V, Henriksen M, Hjort C, Houmark-Nielsen M, Hubberten HW, Ingólfsson Ó, Jakobsson M, Kjær KH, Larsen E, Lokrantz H, Lunkka JP, Lyså A, Mangerud J, Matiouchkov A, Murray A, Möller P, Niessen F, Nikolskaya O, Polyak L, Saarnisto M, Siegert C, Siegert MJ, Spielhagen RF, Stein R (2004) Late quaternary ice sheet history of northern Eurasia. Quat Sci Rev 23:1229–1271

Szczuciński W, Szokaluk M, Bronikowska M, Jagodziński R, Muszyński A, Wünnemann K (2016) Identification and dating of small impact crater ejecta deposits, case of Morasko craters, Poland. In: 32nd IAS International meeting of sedimentology, 2016, Marrakech, Morocco, pp 296

Szczuciński W, Muszyński A (2020) Meteoryty, kratery uderzeniowe i inne ślady kosmicznej katastrofy w rejonie Moraska pod Poznaniem. Przegl Geol 68:637–644

Szokaluk M, Jagodziński R, Muszyński A, Szczuciński W (2019) Geology of the Morasko craters, Poznań, Poland—small impact craters in unconsolidated sediments. Meteorit Planet Sci 54:1478–1494

Szuman I, Kalita JZ, Ewertowski MW, Clark CD, Livingstone SJ (2021) Dynamics of the last Scandinavian Ice Sheet's southernmost sector revealed by the pattern of ice streams. Boreas 50:764–780

Tobolski K (1976) Palynological investigations of bottom sediments in closed depressions. In: Hurnik H (ed) Meteorite Morasko and the region of its fall. Seria Astronomia vol. 2. Adam Mickiewicz University Press, Poznań, pp 21–26

Walesiak T (2021) Kratery Morasko—obrona hipotezy ukośnego impaktu w kontekście dostępnej wiedzy i wyników badań. Acta Soc Metheorit Polon 12:108–128

Widera M, Chomiak L (2019) Various genetic types of end moraines in the city of Poznań and its close vicinity, central-western Poland. Civil Env Eng Rep 29:218–227

Wiki.meteoritica (2022) http://wiki.meteoritica.pl/index.php5/Kratery_Morasko

Włodarski W, Papis J, Szczuciński W (2017) Morphology of the Morasko crater field (western Poland): Influences of pre-impact topography, meteoroid impact processes, and post-impact alterations. Geomorphology 295:586–597

Witold Szczuciński works as a Professor in the Institute of Geology at the Adam Mickiewicz University in Poznań, where he leads the Geohazards Research Unit and Gamma Spectrometry Laboratory. His main research interests include contemporary sedimentary processes and sediments as archives of violent Earth's history. He is particularly interested in quantifying modern sedimentation processes using ^{210}Pb and ^{137}Cs dating techniques, and deciphering past environmental changes from sediments, especially those related to natural disasters (tsunamis, storms, floods, meteorite impacts, glacial surges). He conducts research on land and at sea, mainly in the polar regions (Svalbard, Greenland, Antarctic), East and South-East Asia, and Europe.

Wojciech Włodarski is Associate Professor of Earth Science at the Adam Mickiewicz University in Poznań and a specialist in structural geology and geomorphology. His research interest lies in glaciotectonics and neotectonics of glaciated terrains in the Polish lowlands during the Pleistocene. His major research focuses on the interplay between deformation and deposition processes based on detailed structural analysis of faults and folds and 3D GIS modelling of high-resolution topographic and stratigraphic data. He is an author and co-author of many scientific papers in high-impact peer-reviewed journals, including *Quaternary Science Reviews*, *Boreas*, *Geomorphology*, and *Estuaries and Coasts*.

Lednica Landscape Park—A Geoarchaeological Landscape in Central Wielkopolska

33

Lidia Żuk, Agnieszka Latocha-Wites and Andrzej Kowalczyk

Abstract

This chapter presents the Lednica Landscape Park as an example of distinct transformations of natural lowland relief that resulted from long-term, intense human activity. The study area represents a characteristic young-glacial landscape formed around 18,000 BP that is typical of central Wielkopolska and features moraine hills, plateaus and tunnel valleys. Human activity in this area was present from the recession of the ice sheet. However, its unique position is connected with the medieval period when it became a centre of the duchy built by the Piast dynasty that is considered the cradle of the Polish statehood. In particular, the establishment of a residential stronghold at the island of Ostrów Lednicki led to intensive transformations of the landscape around the Lednica Lake. Using results of interdisciplinary research that have been conducted in this area for decades, we present cases of direct landscape transformation related to medieval construction works and also indirect human impact on the topography and geomorphic processes. Moreover, several issues related to the protection of natural and cultural heritage of the Lednica Landscape Park are addressed.

Keywords

Young-glacial landscape · Human impact · Medieval period · Earthworks · Heritage protection

33.1 Introduction

The geoarchaeological landscape of central Wielkopolska in central-west Poland is presented on the example of the Lednica Landscape Park. It is an area unique in many respects on the national scale, and at the same time representative for the region, where anthropogenic transformations of topography time-wise related to the beginnings of Polish statehood have survived to this day. The natural environment represents a young-glacial landscape, which is the result of the presence of the last Scandinavian ice sheet in this area (Marks 2011). Afterwards, it was considerably modified by human activity. Direct and indirect human impact on the components of the natural environment has been recorded in this area already for prehistoric times, but the most intense and persisting transformations of the relief took place in the early Middle Ages (Kurnatowska and Kurnatowski 1993).

The entire cultural landscape has a significant symbolic meaning for the Polish state. In the tenth century CE, the central Wielkopolska, and especially the Gniezno Lakeland, was the centre of the duchy built by Mieszko I, which became the foundation of the Polish kingdom and Piast dynasty.[1] Due to the historical importance of this area, large-scale interdisciplinary research in the vicinity of the Lednica Lake has been carried out for decades. It has allowed for a good recognition of long-term natural and cultural changes in this area, and in particular, helped to determine the relationships between anthropogenic

L. Żuk (✉)
Faculty of Archaeology, Adam Mickiewicz University, Uniwersytetu Poznańskiego 7, 61–614 Poznań, Poland
e-mail: lidkazuk@amu.edu.pl

A. Latocha-Wites
Institute of Geography and Regional Development, University of Wrocław, pl. Uniwersytecki 1, 50–137 Wrocław, Poland
e-mail: agnieszka.latocha@uwr.edu.pl

A. Kowalczyk
Muzeum Pierwszych Piastów na Lednicy, Dziekanowice 32, 62–261 Lednogóra, Poland
e-mail: andrzej.kowalczyk@lednica.pl

[1] The Piast dynasty was the first historical ruling dynasty of Poland. The first documented Polish ruler was the duke Mieszko I (d. May 25, 992).

P. Migoń and K. Jancewicz (eds.), *Landscapes and Landforms of Poland*, World Geomorphological Landscapes, https://doi.org/10.1007/978-3-031-45762-3_33

activity and the transformations of the natural environment (Filbrandt-Czaja 1998), as reflected in the rich literature on this region (Kurnatowska and Wyrwa 2016).

The rich history of human presence in the area is documented not only by numerous archaeological finds, but also by a number of earthworks from various periods and with various functions, which have survived in good condition to this day. As a result, a unique anthropogenic landscape was created. In addition, a significant number of sites of biogenic sedimentation is an important asset of the Lednica Landscape Park and its surroundings. They constitute valuable geoarchives that allow for the detailed reconstruction of the evolution of relief, biotic conditions, palaeoclimate, palaeohydrology and the human impact, which was variable in time and space (Tobolski 1993).

The benchmark significance of the geoarchives and the exceptional character of the archaeological structures related to the residential and capital centre of the first Piast state at the turn of the tenth and eleventh centuries are of the highest historical and scientific value and of great importance for the cultural heritage of Poland (Kurnatowski 1993).

33.2 Setting and Location

The Lednica Landscape Park (LLP) is located in the Gniezno Lakeland mesoregion. The latter constitutes the eastern part of the Greater Poland Lakeland, which is a part of the larger unit of the South Baltic Lakelands (Fig. 33.1) (Richling et al. 2021). The whole area belongs to the Warta River basin.

The landscape park was established in 1988. It is located approximately 30 km east of the city of Poznań. The area of the park has an elongated, meridional shape, slightly more than 12 km long and approx. 6 km wide on average. It covers an area of 76.1 km^2. Only 10% of its area is covered by forests, mainly in the northern part of the LLP. This is the result of centuries-long deforestation, so that the agricultural landscape dominates nowadays, with vast arable fields and meadows dotted with small enclaves of trees and shrubs (Filbrandt-Czaja 1998). Together, they constitute nearly half of the park area. Water bodies of various sizes are also numerous and include lakes, ponds and lakelets. The central part of the LLP is occupied by the Lednica Lake, with an area of 3.4 km^2. It is located in a meridional glacial tunnel valley that is about 8.5 km long. The valley has a diversified outline with numerous troughs and shallows within the lake basin. The lake has an average depth of about 7.5 m and a catchment area of 38.4 km^2. There are five islands on the Lednica Lake, including the largest one—Ostrów Lednicki, with the early medieval stronghold and relicts of monumental architecture.

The process of landscape shaping of the LLP reaches back to the oldest farming communities and proceeded with different intensity in the subsequent centuries. Deforestation, which started in the Neolithic, was relatively small-scale for a long time. Intensive deforestation started only during the period of Roman influence (ca. first–mid-fifth century CE). Deforestation in the Lednica area was then significant and amounted to 22–29% (Kurnatowska and Kurnatowski 1991). From the end of the fourth century CE, a settlement regress was recorded in the Lednica Lake surroundings. It is only from the early Middle Ages that intensive transformations of the landscape can be noted, mainly related to the establishment of one of the most important residential and capital centres of the emerging state of the first Piasts in Ostrów Lednicki, in the seconnd half of the tenth century CE. These changes were related to significant deforestation in the Lednica region, with 63% of the area being affected in the mid-eleventh century CE (Kurnatowska and Kurnatowski 1991). Another characteristic feature of this period is the interference with topography. Although this concerned relatively small areas of the LLP and basically these changes affected landscape on a microscale, the strongholds constructed at this time caused permanent transformation of natural relief, mainly on the largest island on the Lednica Lake, and they are still prominent elements in the local landscape. In the subsequent centuries, the centre in Ostrów Lednicki was marginalized. However, it still played political and administrative functions, which is recorded in numerous mottes. The development of rural settlements from the late Middle Ages (Górecki 2001) launched a number of processes, which shaped the entire area of the LLP, giving it its present-day agricultural characteristic. In contrast to the local but pronounced and long-lasting impact on relief caused by the construction of medieval strongholds and mottes, later transformations of the landscape were more spatially extensive and related mainly to agricultural intensification along with settlement expansion. However, these transformations are less spectacular in the landscape and more difficult to capture as they are linked first of all to the long-term denudation and lowering of the morainic hills due to ploughing.

33.3 Geology and Main Features of Relief

The study area is characterized by a large thickness of Quaternary sediments, ranging from 25 m in the northern part to even 100 m in the southern part (Skoczylas 1989). Pleistocene sediments are represented mainly by different types of tills, in which four stratigraphic levels were distinguished: tills of the Sanian glaciations of the older and the younger stages (MIS 12), and glacial tills of the Odranian (MIS 6) and Weichselian (MIS 2) glaciations

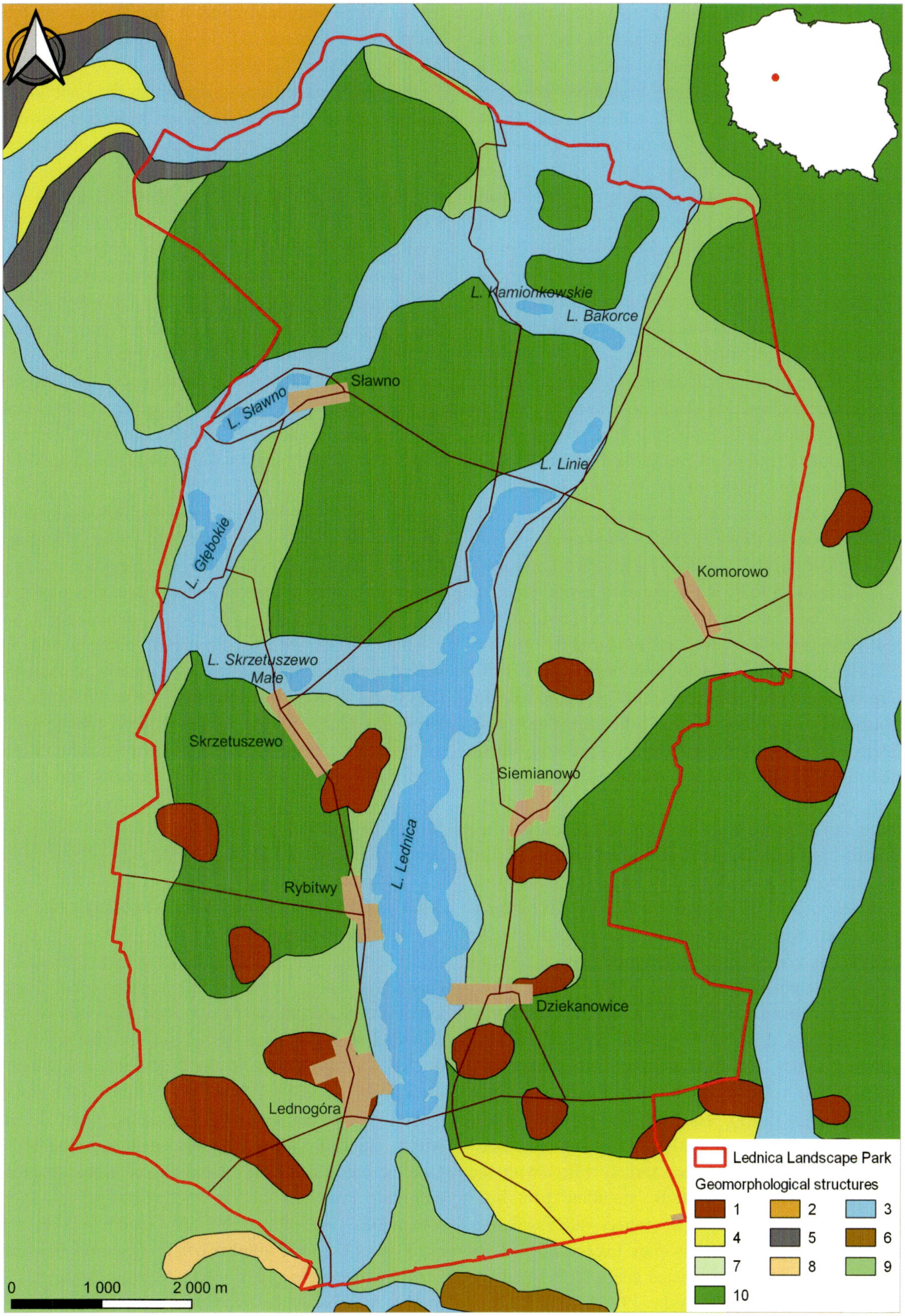

Fig. 33.1 Geomorphological map of the Lednica Landscape Park based on Krygowski (2007). Legend: 1—isolated morainic hills, 2—denudational hills, 3—flat-bottomed tunnel valleys, 4—outwash accumulation plains, 5—terraces, escarpments, 6—frontal moraine hills, 7—floodplains and basin floors, 8—eskers, 9—undulating moraine surface, 10—flat moraine surface. Dark blue colour indicates lakes (image by A. Latocha-Wites and L. Żuk)

(Skoczylas 1989; Marks 2011). Their total thickness is up to 60 m. Locally, there are layers of glacifluvial sand and gravel deposits of the Weichselian glaciation, as well as varve clays and silts. The Holocene formations are represented mostly by river sands, muds, and also peat, gyttjas and lacustrine sands and silts. The individual Quaternary sediments closely follow the topography. Most of the area is covered with tills of moraine plateaus, sands, gravels and boulders of terminal moraines, and glacifluvial sands and gravels within outwash plains, the range of which is limited to the south-east part of the LLP area. Glacifluvial deposits are also dispersed within the morainic plateaus (Przybylski 2008). Holocene peats and fluvial sands occur in the river valley floors (Skoczylas 1989).

The relief of the Lednica Landscape Park, as well as of the entire Greater Poland Lakeland, is related to the last Pleistocene glaciation and represents the young-glacial landscape. Its main features were formed around 18,000 BP within the marginal zone of the Poznań phase of the Weichselian ice sheet and later oscillations related to the ice-sheet recession (Kozarski 1962; Stankowski 1989). The effect of the first, the so-called Dzwonowo-Lednogóra oscillation (Kozarski 1962), is the zone of frontal moraine hills that define the southern edge of the Lednica Lake tunnel valley. From the south, they are bordered by flat surfaces of outwash accumulation. In the northern part of the LLP, there is a series of hills associated with the second, younger ice-sheet oscillation, the so-called Gniezno oscillation (Stankowski, 1989). By contrast, Bartkowski (1967) interpreted these landforms as the evidence of areal downwasting of the ice sheet. The newest research explains the landscape of the region as the terminal zone of an ice stream and this is the largest area in Poland with preserved subglacial landforms, as evidenced by lineations of the subglacial channel network (Przybylski 2008). The above mentioned two zones with several rows of small hills can be interpreted in this context as De Geer moraines as well as ribbed or Rogen moraines (Przybylski 2008).

The moraine plateaus, which are formed by the dominant glacial deposits in terms of surface area of the LLP, are undulating or flat, and their altitude ranges from 110 to 115 m asl. The relative heights of moraine hills are 10–15 m (Skoczylas 1989). The morainic plateau is dissected by flat-bottomed tunnel valleys of glacial or glacifluvial origin, elongated N–S, now largely filled with the waters of ribbon lakes or wetlands, including peat bogs (Krygowski 2007). The slopes of the tunnel valleys are varied, both gentle and steep; locally they exceed 30°, as for example in the case of basins located south of the Lednica Lake (Kowalewski and Wiśniewski 1994). The Lednica Lake is located in the largest tunnel valley in the LLP area. The valley results from subglacial erosion, as evidenced by both the complex outline of the lake and its longitudinal profile with numerous troughs and shallows and small islands (Stankowski 1989).

Interesting minor landforms within the Lednica Lake are shelves of lake terraces. They are located at different levels in relation to the water surface (as of the years 1985–1986): higher terraces are located 65–120 cm above the water level, lower terraces are 20–40 cm above the water level and underwater terraces are 30–80 cm below the water level. These terraces are abrasive and abrasive-accumulation levels, on which mineral deposits, mainly lacustrine sediments, were accumulated. The abrasive terraces are cut in glacifluvial and glacial deposits, and the underwater ones—in the lacustrine sands and gyttjas. They constitute geomorphological and sedimentological evidence of fluctuations in the water level of the Lednica Lake (Stankowski 1989). These fluctuations are also confirmed by lacustrine sediments found in the coastal zone of the Ostrów Lednicki island.

The varied topography and lithological diversity have also resulted in a variety of soil cover (Drzymała and Mocek 1994). Two main types are dominant: Luvisols and Phaeozems (black soils). Luvisols occur in elevated areas and their parent rock is sandy till, while black soils are concentrated in depressions, mainly along watercourses or within tunnel valleys, and are periodically excessively moist (Drzymała et al. 1993; Filbrandt-Czaja 1998). In addition, there are Histosols, Arenosols (called rankers according to the older systematics—they occur on alluvial fans and within the hillslope covers), and Podzols, the latter in the northern part of the park under pine forests (Drzymała and Mocek 1994).

33.4 Diversity of Geoarchaeological Landscapes

Nowadays, there are ca. 500 archaeological sites registered in the Lednica Landscape Park. The vast majority was recorded as surface scatters of pottery sherds which leave no other visible traces in the landscape, but which might indicate the presence of buried structures. Only four earthworks, visible as upstanding remains, were documented within the study area. Underwater surveys in the Lednica Lake revealed further 17 locations of archaeological sites, including structures such as wooden bridges and shore reinforcements (Fig. 33.2). The early medieval sites represent ca. 31% of the total number of all registered archaeological sites. Their large proportion and location in various parts of the LLP (in particular near Ostrów Lednicki) may indicate the scale of human transformation of the natural environment, mostly in medieval period.

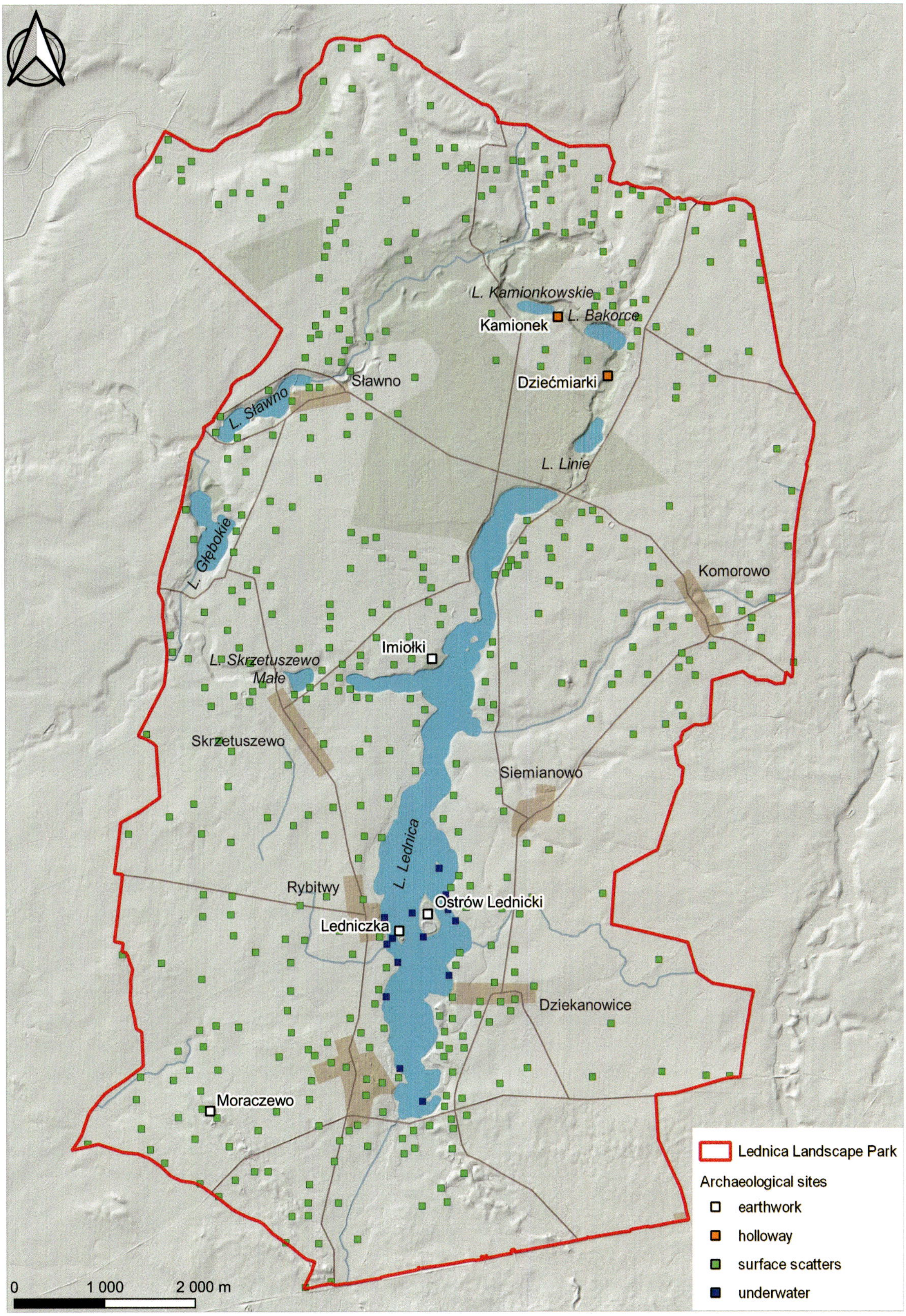

Fig. 33.2 Location of archaeological sites in the Lednica Landscape Park (background: shaded relief model, data source: geoportal.gov.pl, image by L. Żuk)

Fig. 33.3 Earthen banks of strongholds. **a**, **b** Moraczewo. **c**, **d** Ostrów Lednicki (photographs by A. Latocha-Wites)

33.4.1 Anthropogenic Landforms—Human Impact on Topography

Currently, the most visible forms of human transformation of the Lednica Landscape Park are early medieval strongholds. Originally, these were artificial embankments constructed with wood and earth (Fig. 33.3), usually surrounded by an outer ditch. Two strongholds were identified in the study area—in the village of Moraczewo and on the island of Ostrów Lednicki. Their massive banks are still preserved as outstanding earthworks and thus can be regarded as perfect case studies to illustrate the extent of human impact on the natural relief. The second category is a motte, an artificial mound on which residential buildings were constructed in the Middle Ages. Structures of this type were identified on the Ledniczka island and in the village of Imiołki. Although their impact on the natural relief is less evident than that of strongholds, their construction also caused a permanent transformation of the topography. Mounds still reach the height up to 7 m and dominate their immediate surroundings, which were also transformed as a result of construction of the motte (Sikora and Kittel 2017). This is especially evident in the case of the Ledniczka island. An interesting feature of the LLP is a group of objects that were earlier identified by archaeologists as strongholds or possible strongholds. Among those we can mention a structure in the village of Sławno, which in 1984 was registered in the sites and monuments record. However, contemporary analysis of the Airborne Laser Scanner (ALS) derivatives and field survey seem to indicate that this is a natural form. Other possible strongholds located on the peninsulas of the Głębokie Lake are also currently interpreted as natural features. In the following chapter four evident archaeological earthworks will be discussed as examples of various anthropogenic processes which shaped and modified the area of the Lednica Landscape Park (Table 33.1).

Table 33.1 General characteristic of Medieval earthwork sites in the Lednica Landscape Park

Location	Site type	Date	Landform	Area	Preserved structures	Protection
Moraczewo site 2, Łubowo commune	Stronghold	Early Medieval (9/10th to mid-10th century)	Isolated morainic hills	ca. 1.4 ha	Wooden and earthen banks up to 10 m high, ditch	Scheduled since 28.02.1972
Rybitwy (Ostrów Lednicki) site 1, Łubowo commune	Stronghold	Early Medieval (10th to 12th century)	Ostrów Lednicki island, Lake Lednica	1st phase ca. 0.36 ha; 2nd and 3rd phases ca. 2.3 ha	Wooden and earthen bank up to 7.4 m high	Scheduled since 26.03.1957 Historic Monument since 08.09.1994
	Palatium with a chapel	Early Medieval (2nd half of 10th to 12th century; three use phases)	Ostrów Lednicki island, south-west part of the stronghold	0.045 ha	Stone walls up to 2 m	
Rybitwy (Ledniczka) site 4, Łubowo commune	Motte	Medieval (14th to 15th century)	Ledniczka island, Lake Lednica	ca. 0.38 ha	Mound up to 7 m, moat and earthen bank	Scheduled since 26.03.1957
Imiołki site 22, Kiszkowo commune	Motte	Medieval (12th to 13th century)	Lake Lednica western shore	0.17 ha	Mound up to 7 m, ditch	Scheduled since 01.03.1972

Fig. 33.4 DEM of the medieval stronghold in Moraczewo: **1** bank-phase 1; **2** bank-phase 2; **3** ditch; **4** truncated moraine. Model composition: superposition of hillshade over hypsometric tint (data source: geoportal.gov.pl, image by L. Żuk)

33.4.1.1 Moraczewo—Early Medieval Stronghold

This is the oldest stronghold located in the Lednica Landscape Park, dated to the early-Piast dynasty period. It is located within isolated morainic hills which extend from north-west to south-east (Fig. 33.4). Its western part was partly cut into a side of a natural hill. From this point, the stronghold's bank is gradually lowering towards the north-east. Two phases of its development were distinguished, involving an older construction of wood and earthen embankment with a ditch (ca. 900–925 AD) and the later extension towards the east (ca. 925–950 AD). Around 950 AD the stronghold was burnt down and abandoned. This event coincides with the construction of the Piasts' residential stronghold at Ostrów Lednicki. In both phases, the remains of a considerable building, interpreted as a location for some ceremonial/ religious activities, were identified inside the stronghold (Strzyżewski et al. 2003). The stronghold is also regarded as a refugium for inhabitants of neighbouring settlements who could seek safety within the embankments in times of unrest and raids. Nowadays, only the earthen construction of the banks is preserved. However, after more than a millennium these structures still reach up to 10 m and dominate the local landscape. One can easily identify two embankments which indicate two phases of construction and the remains of the ditch. Moreover, ALS analysis reveals transformation of the nearby morainic hills into regular forms which have not yet been studied in details.

There can be little doubt that the construction of the bank required a considerable amount of oak wood to build boxes which were filled with sand. We can assume that both materials were obtained from the surrounding area. Some quantities of sand could have been extracted from the abutting hill which was reshaped as a result of the construction. ALS analysis also reveals destruction of postglacial hills in the vicinity of the stronghold. This may indicate the presence of sand extraction pits although there is no evidence that they were contemporary with the stronghold. Nonetheless, it can be assumed that the hills were used as a source of materials in different times and contexts.

One more feature requires special attention due to its unusual shape and close proximity to the stronghold. A regular form of a moraine with a truncated northern end clearly indicates human intervention. However, its origins and purpose remain unknown. Perhaps it was protecting the entrance to the stronghold though we cannot eliminate the possibility that those changes are of a later date and related to some other practices.

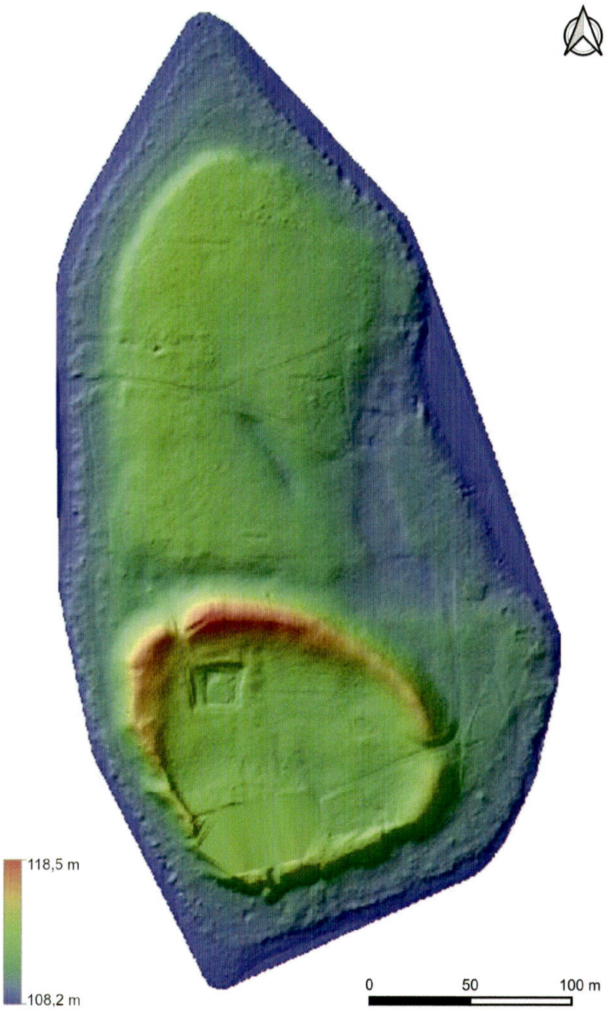

Fig. 33.5 DEM of the medieval stronghold at Ostrów Lednicki (phase 2). Stone structures of the palatium and the chapel in the southern part are protected by a roof which caused disturbance in surface modelling (see Fig. 33.8). Model composition: superposition of hillshade over hypsometric tint (data source: UAV laser scanning, image by L. Żuk)

33.4.1.2 Ostrów Lednicki—Early Medieval Stronghold

The site is located in the southern part of Ostrów Lednicki island (Fig. 33.5). Between 900 and 950 AD, a small fortress with a wood and earthen embankment was constructed in its south-western part. After 950 AD, an imposing stronghold was constructed and remained in use until the end of the twelfth century. It is regarded as one of the principal seats for the early Piast dynasty. Later its interior was turned into a cemetery for a settlement which continued to exist on the island till the end of thirteenth century.

The ruins of a chapel and palatium, thought to be the residence of the first rulers in the Piast dynasty, are located within the stronghold. The palatium was divided into two parts: a rectangular residential building and a chapel built on a plan of a Greek cross which was adjacent to the eastern part of the palatium. It is one of the contenders for the historical site of the personal baptism of the duke Mieszko I, his entire court and subordinated tribal elites.

The original morphology of the site was reconstructed on the basis of geological surveys which were conducted on the island for several decades. The shoreline prior to the construction of the stronghold was much more diversified, especially in the eastern and southern parts. An isolated site with biogenic accumulation existed in the middle of the island, while in the eastern part there was a bay separated from lake waters by a small peninsula. Another smaller bay was identified in the southern part of the island (Hildebrandt-Radke 2020).

Recently, archival archaeological and geological measurements were used to reconstruct the natural surface of Ostrów Lednicki (Kasprowicz 2016). The most distinguishable features of the original landform are two well-defined local elevations in the southern part of the island, separated by a depression which formed a bay. Equally evident is an area of higher ground in the northern part, more extensive than that in the south, with steep western and gentle eastern slopes. Its height was declining towards the lake's bay located in the middle-east part of the island. Between the northern and southern higher grounds there were two further elevated areas although less distinct than those described above. The natural relief of the south-east part of the island remains unknown due to a lack of measurements (Kasprowicz 2016).

The highest elevation located in the south-western part of the island was characterized by steep slopes underlain by clays and was surrounded by lake waters. In the 1st half of the tenth century a small stronghold was built on this spot, surrounded by a wooden and earthen embankment. Its construction caused the first transformation of the island, although its extent remains unknown (Górecki and Łastowiecki 2016; Wrzesiński and Kara 2016).

During the reign of Mieszko I, in the second half of the tenth century the embankment was dismantled, the area between it and the adjacent hill was levelled and an imposing residential complex was constructed. Apart from the aforementioned palatium and the chapel, the construction work involved the building of a church, two wooden bridges which connected the island with the mainland on the east and west sides, and a wooden breakwater on the south-west part of the stronghold (Górecki et al. 2020).

Archaeological excavations were conducted on different parts of the stronghold. This permitted fairly accurate estimations of the amount of building material which was used to construct the embankment and other structures. Several phases of bank reconstruction/rebuilding were identified due to different construction techniques. Nonetheless, the building material in each phase consisted of oak wood to build boxes which were filled mainly with fine-grained sand, sometimes mixed with clay and lake lime. As in Moraczewo, the bank construction required a considerable amount of wood, sand, gravel and clay. Archaeological and environmental analyses focused on microfossils demonstrate that loose materials used for bank construction were obtained from various locations. Sand, gravel and clay (in some cases also humus that formed cultural layers) were extracted both from the eastern and western shore of the lake (Makohonienko 1998). Current estimations indicate that approximately 80 ha of woodland would have been felled to construct the embankment 12 m high and 0.5 km long. Moreover, considerable amounts of oak were used to construct two bridges, 638 m in total length. Their construction and subsequent repair work required wood from approx. 4000–4800 mature oaks (approx. 10–15 ha of woodland) (Kola et al. 2016). Additional wood was used for the construction of shoreline reinforcements and buildings on the island, whereas the deforested areas were turned into arable land, contributing to the progressive agrarian erosion and denudation.

The cultural layers at Ostrów Lednicki have a thickness of 0.2–1.7 m (apart from wooden and earth embankments with a preserved height of up to 7 m). They are made of polygenetic mineral-organic deposits with numerous traces of human activity. They include mainly sands and—to a lesser extent—clay and gravel, as well as a high content of humus, fine plant detritus, wood and charcoal. Layers related to anthropogenic activity clearly differ from the underlying sediments of natural origin. Their accumulation in some places led to the levelling of the original relief (Schubert and Kowalewski 1998).

Anthropogenic transformations of the morphology, which were directly influenced by human activity, also include underwater cultural layers. These are sediments that were formed when the natural sedimentation process for a given water reservoir has been disturbed or was impacted by the direct influence of human activity. Such activities include, among others, the construction of bridges, dykes or the use of waters and coastal zones (Polcyn 1991). Such layers have been identified at Ostrów Lednicki next to the relics of wooden bridges, bridge abutments and former bank reinforcements (Kola and Wilke 1989; Polcyn 2003; Górecki 2010). The piles of the supporting structure of the west (Poznań) bridge, embedded in the bottom of the lake, favoured the accumulation of a cultural layer of considerable thickness (Kola and Wilke 1989). On the other hand, anthropogenic layers, 15–90 cm thick, related to the aquatic-terrestrial reinforcements along the western shore of the island, contributed to a change in its original morphology, which is clearly visible in the sediment structure (Górecki 2010). The construction layer with wooden structures has been dated to the 2nd half of the tenth century (Krąpiec 2003, after: Górecki 2010).

33.4.1.3 Ledniczka—Late Medieval Motte

A medieval motte (end of the fourteenth century to fifteenth century AD) is located on the island of Ledniczka (Fig. 33.6). In the southern part of the small island an artificial mound was constructed, separated from the northern part by a moat and an earthen bank. Archaeological and historical research indicate that in the fifteenth century it was a seat of the local noblemen, the Latalski family (Górecki et al. 1996).

The natural landform was reconstructed on the basis of archaeological excavations carried out in 1966 and 1989. This helped to establish that the top of the natural form was approx. 3 m above the lake at the highest point and was gently declining from south to north. It is formed by clay interlayered with lake lime.

Underwater survey in 2017 revealed the remains of wooden bridges which connected Ledniczka with the western bank of the mainland in the tenth and later in thirteenth/fourteenth centuries. Further archaeological survey on the island in 2021 revealed medieval stone and wooden constructions reinforcing the western shore and also the bridgehead. These structures may indicate an important, so far unknown role which Ledniczka played for the Lednica settlement complex in the tenth and later, in the fourteenth century (Pydyn et al. 2018).

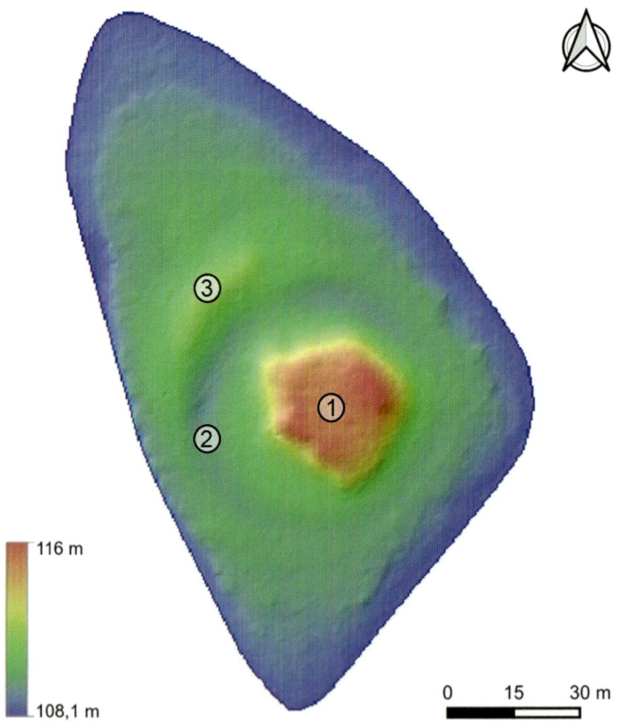

Fig. 33.6 DEM of the medieval motte at the Ledniczka island: **1** mound; **2** moat; **3** bank. Model composition: superposition of hillshade over hypsometric tint (data source: UAV laser scanning, image by L. Żuk)

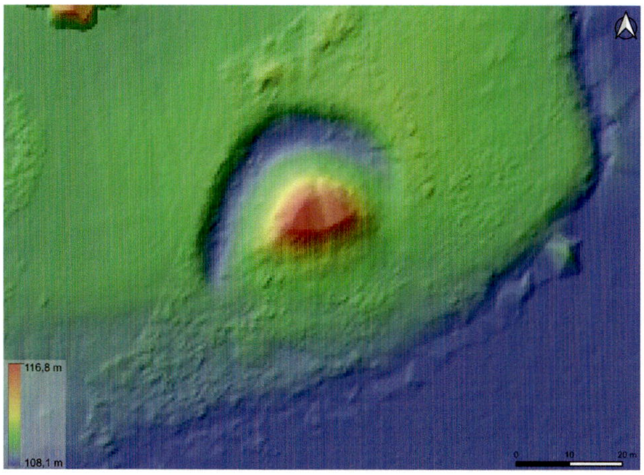

Fig. 33.7 DEM of the medieval motte in Imiołki. Model composition: superposition of hillshade over hypsometric tint (data source: geoportal.gov.pl, image by L. Żuk)

Like at Ostrów Lednicki, traces of earlier human activity were documented on the island. However, they were obliterated during the construction of the motte which was built on the highest point of the island and remains its most prominent feature, rising up to 7 m. Two buildings were documented during excavations in the centre of the mound. Their eastern and southern parts were surrounded by an additional earthen bank (1 m high) which was strengthened by a stone parapet to prevent earth slides (Górecki et al. 1996).

Due to the relatively small size of the mound and sparse settlement, we can assume that the material used for their construction (earth and stone) was extracted from the island. Typically, a ditch around the mound (now 1.8 m deep) would have also been a source of the earth and stone for constructing the mound itself. However, we cannot exclude the possibility that some construction material was also brought from the mainland, whose shore is only 100 m away from the island.

33.4.1.4 Imiołki—Medieval Motte

This medieval motte, dated to the twelfth–thirteenth century on the basis of surface scatters, is located on the western mainland shore of the Lednica Lake (Fig. 33.7). Little can be said about the natural landform and its transformation during construction of the motte due to insufficient survey. Nevertheless, it is one of the best examples of sites modified by human activity. It bears some similarity to the Ledniczka site (see Sect. 4.1.4), with the height of the mound of approx. 7 m. Unlike Ledniczka, the topmost part of the mound is much smaller and its flattening is less pronounced. This probably left much less space for buildings and one can only presume that some facilities were constructed there. It was also surrounded by a semicircular ditch which separated the structure from the mainland (Górecki et al. 1996). The ditch is still well preserved as an earthwork and we can assume that building material for the mound was also extracted from there.

33.4.2 Sources of Building Materials—Destructive Human Impact on Topography

The geomorphological cultural landscape of the Lednica Landscape Park is reflected not only in the anthropogenic earthworks, but also in the use of natural resources to build various constructions. One of the examples is the use of rock material to construct the early medieval monumental buildings in Ostrów Lednicki—the palatium and the church (Fig. 33.8). Erratic boulders constitute as much as 98% of the former and 82% of the latter (Skoczylas 1989). They were chosen in a thoughtful and selective manner. Although granites and gneisses are the most common among the erratic boulders in the study area, in both buildings quartzites, quartzitic sandstones and sandstones predominate, accounting for 58% (palatium) and 50% (church) of building material among several types of rocks distinguished in each construction (Skoczylas 1993).

An important building and cladding material of both buildings is also the travertine—in case of the church it constitutes nearly 11% of the building material, and for the palatium—nearly 2%. Presumably, there were many local deposits of travertine in the past, in the area of the present LLP or in its immediate vicinity. They may have been exploited or covered with lake sediments as a result of climate change and rising water level in the lake (Skoczylas 1989). The masonry mortars also used gyttja commonly found in the area, lacustrine chalk, calcium carbonate obtained locally from moraine clay, as well as secondary gypsum deposits (Skoczylas 1993). It is also interesting to note the use of bog iron as a building material for a church (it accounts for over 9% of the composition), which is rather rare (Skoczylas 1989). The local raw materials (erratic boulders) were the main building material, but gypsum, carbonate rocks and quartzitic sandstones were imported from a distance of 45 km to even over 100 km (Skoczylas 1993).

The exploitation of raw materials (boulders, sand, iron ore) must have led to the deformation of the original land surface, leading to the formation of concave forms. However, the current identification of these places is difficult or even impossible. It can only be assumed that the material was taken from the immediate vicinity of the erected structures due to the huge amounts needed, for example, for the construction of the embankments (boulders, stones, sand). Numerous contemporary sand or gravel

Fig. 33.8 Stone remains of the early medieval buildings at Ostrów Lednicki. **a** Palatium. **b** Church (photographs by A. Latocha-Wites)

pits—clearly visible on ALS derivatives—suggest that this material could have also been intensively exploited in the past. However, due to the nature of sandy sediments, the sites of their exploitation of medieval age cannot be expected to be preserved in the relief. The places where erratic boulders or pebbles were acquired are also not currently possible to capture in the field. On the other hand, one can see small water reservoirs south of the Lednica Lake, which have been created in the last 100 years as a result of exploitation of peat for heating purposes. After its completion, the flooded area underwent secondary ecological succession (Kowalewski and Wiśniewski 1994).

33.4.3 Indirect Human Impact on Topography and Geomorphic Processes

Apart from the clearly distinguished anthropogenic landforms and earthworks, human impact also indirectly contributed to the transformation of natural topography. It is mainly the result of enhanced denudation, erosion and accumulation processes initiated or intensified by various human activities.

An example of such processes is gully erosion associated with unpaved roads, which led to the formation of holloways (road gullies/sunken roads) and other forms associated with the long-term use of communication routes. In the area of the Lednica Landscape Park, two landforms deserve particular attention. These are the holloways in the northern part of LLP and a road escarpment in the vicinity of the stronghold in Moraczewo. Two spectacularly deep road gullies in the northern part of the LLP developed along sections of dirt roads descending to the valley floor, i.e., where the slopes increased. In terms of geomorphology, they are located at the junction of moraine plateaus and flat-bottomed tunnel valleys (Fig. 33.9). The increase in the road gradient along with the constant use, prevented the development of protective vegetation cover, which, in turn, favoured a greater intensity of gully erosion. The subsoil of both sunken roads is light clayey sand, which is conducive to water erosion and, at the same time, to soil creep observed on the sides of the gullies. Currently, it is not possible to determine the duration of use of these routes, but considering the significant depth of the gullies it can be assumed that we are dealing with a long-term use of both roads. The largest holloway was located in the village of Kamionek (Fig. 33.2). It is 130 m long and its sides are 1.5 to over 4 m high. The second sunken road was identified near the village of Dziećmiarki. It is 160 m long and 2–4 m deep, with a clear asymmetry of its slopes (Fig. 33.10).

In Moraczewo, a high road escarpment has formed in the place where an old road was incised into the slope of one of the moraine hills (Fig. 33.11). It reaches a maximum height of 4.3 m and stretches for a length of about 130 m. The relative height of the escarpment could additionally have increased as a result of other anthropogenic influences such as undercutting of its lower part as a result of ploughing on the slope below and building-up of its upper edge as a result of ploughing and/or washing material from a slope above it. Such processes are typical for agricultural terraces. The escarpment in Moraczewo can therefore be interpreted as a result of various human impacts and the landform can be classified as an escarpment of both a road and agricultural terrace edge, formed along the property or field boundary.

Increased anthropogenic erosion in croplands within the morainic hills was demonstrated on the basis of degraded soil profiles. Considerable truncation was observed in their upper parts, as evidenced by the appearance of glaciofluvial sands on the surface, which constituted their substrate (Kowalewski and Wiśniewski 1994) (Fig. 33.12).

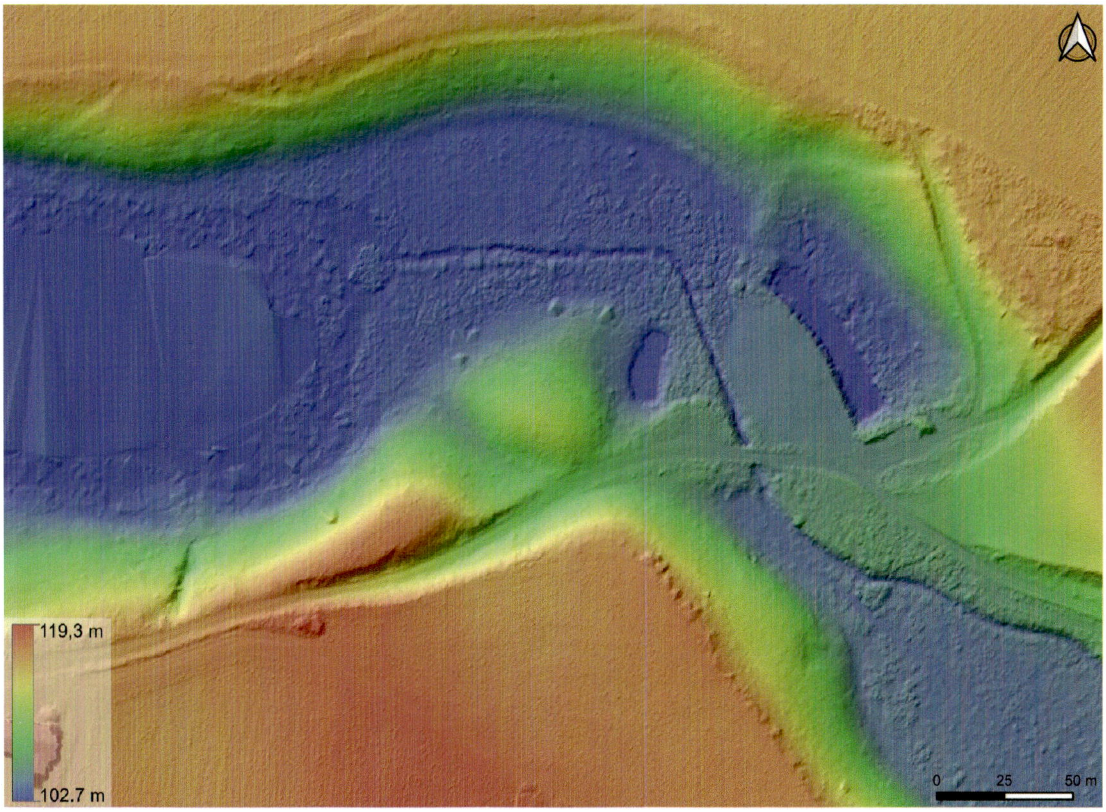

Fig. 33.9 DEM of a holloway near the village of Kamionek, located at the junction of a moraine plateau and a flat-bottomed tunnel valley. Model composition: superposition of hillshade over hypsometric tint (data source: geoportal.gov.pl, image by L. Żuk)

Fig. 33.10 Holloways/sunken roads. **a** Dziećmiarki. **b** Kamionek (photographs by A. Latocha-Wites)

Sediments that can be associated with anthropogenically intensified denudation have been identified in many places in the Lednica Landscape Park. The sediment thickness of the Lednica Lake reaches 19 m, and the oldest layers date back to the late glacial period (Tobolski and Litt 1994). Within the tunnel valleys, in lacustrine and biogenic sediments, sandy sediments were found in several places. They can be associated with deforestation and increased wash from the nearby agricultural slopes (Tobolski 1993; Kowalewski and Wiśniewski 1994). For the Lake Skrzetuszewo, it was determined that intensification of mechanical denudation occurred in the early Middle

Fig. 33.11 A road escarpment near Moraczewo (photograph by A. Kowalczyk)

Fig. 33.12 Selective soil erosion on arable lands near Moraczewo (photograph by A. Latocha-Wites)

Ages. It resulted in covering the fossil sediments of the littoral zone by slope materials. In turn, the limnic deposits of this lake record traces of even earlier denudation (Tobolski 1993). Lake ecosystems reacted even to small alternations of human economic activity—a short-term increase in the trophy of the Lake Skrzetuszewo was recorded in the Neolithic period, and another one in the Bronze and Iron Age, both related to the intensification of human-related development in the vicinity (Tobolski 1991).

There were also attempts to limit the intensity of natural processes, which included the erection of bank and abutment reinforcements of two bridges at Ostrów Lednicki (Kola and Wilke 1989; Górecki 2010). Although their basic functions were adequately defensive and functional, at the same time, indirectly, both the boxes and wooden piles in the shore and coastal zones (under water), as well as stone reinforcements, ensured the protection of the banks against abrasion and erosion (Fig. 33.13). Shore reinforcements of defensive importance were located along the south-west bank of Ostrów Lednicki. They can also be interpreted as a kind of breakwater and ice-floe breaker (Górecki 2010). The maximum width of the structure of bank reinforcements so far captured during archaeological surveys is 9 m. Apart from the wooden elements, their integral part were also granite pebbles placed between the piles and horizontal elements of the reinforcements. They formed the foundations of the structure and the barriers in the foreground for wooden reinforcements, both from the side of the land and the lake water. The diameters of the pebbles ranged from 10 to 40 cm (Górecki 2010).

Fig. 33.13 Wooden constructions at lake shores. **a** Remains of the piles of the wooden bridge, with the island of Ostrów Lednicki in the background. **b** Reconstruction of wooden reinforcements of the island banks (photographs by A. Latocha-Wites)

33.5 Geoarchaeological Heritage—Current Threats, Protection, Promotion and Education

In 1988 the Lednica Landscape Park was created to protect the landscape of the Lednica Lake with its natural and cultural heritage. In particular, the aims of the Lednica Landscape Park were defined as follows: (a) to protect and preserve the cultural landscape of the Lake Lednica in its current form, especially the landscape of the major water reservoir with its diversified shoreline and islands, and also woodland-and-farmland landscape with its diversified landforms in the northern part of the Park; (b) to protect and preserve valuable ecosystems with unique plant and animal species, in particular the ecosystem of the Lake Lednica, riparian forests, fenwoods and oak-hornbeam forests; (c) to protect and preserve cultural and historical heritage and their surroundings, in particular the remains of the Piasts' settlement centre.

Three different levels of heritage protection can be identified for archaeological sites within the Lednica Landscape Park. According to the Polish law, all archaeological sites are protected by definition regardless of their status. Moreover, earthworks in Moraczewo, Lednica and Ledniczka were registered as scheduled monuments in the 1950s and 1970s and are strictly protected. Due to the unique significance of the monumental complex at Ostrów Lednicki, it received in 1994 the status of a Historic Monument as one of the first such monuments in Poland. Currently this is the highest heritage status according to the national law. To protect its landscape, the lake shore near the island was also registered as a scheduled monument in 2006, which is a unique situation in Poland.

However, the Lednica Landscape Park has been undergoing several social and economic transformations in the last two decades. These processes were mainly caused by intensive modernization of rural areas. Thus, despite the precise definition of heritage status and protection forms for the cultural and natural heritage, it may lead to irreversible changes of the landscape. Three major threats were identified within the study area:

- Growing investment pressure, prompted by the attractiveness of areas around the Lake Lednica for housing;
- Modernization of farming practices with the introduction of heavy machinery and industrial farming;
- Intensive development of tourism.

Regarding earthwork sites, major problems related to their preservation and protection are connected with intensive development of tourism. A considerable growth of tourist visits is particularly noticeable for the Ostrów Lednicki island. For several decades, starting from the 1930s when the island was first opened for visitors and then in 1969, when the Museum of the First Piasts in Lednica was established, tourism had a minimal impact on the island and the monument complex (Wyrwa 2019). However, the tourist boom in the last decade (approx. 60,000 visitors per season) has caused a series of changes which affect the monumental complex. Several constructions imitating medieval settlement and also viewing platforms were built in different parts of the island to make it more attractive for tourists. These changes were introduced in cooperation with the heritage office and conformed with the existing heritage law. While their impact on archaeological structures is reduced to a minimum, they certainly affect the visual character of the island. However, the intensive development of tourism has also increased erosion processes in more exposed parts of the monument. For example, the formation of erosion furrows on steeper slopes (e.g. at passages that go through the banks of the stronghold) can be observed after rainfall. This called for actions to prevent further damage. Therefore, in 2010, tourists paths with hardened surfaces were made to direct movement through the archaeological structures. Erosion processes are closely monitored and their further development is prevented by ongoing maintenance works. Geomorphological analysis shows that limiting tourist movement to designated routes effectively protects the remaining area, including the majority of the embankments, against trampling, and thus against the destruction of the protective turf cover, the removal of which could trigger the processes of surface wash and linear erosion.

More complex issues arise from the necessity to find a balance between the protection of natural and cultural heritage of the Lednica Landscape Park, which is also a home to unique species of plants and animals. To some extent, this affects the possibility to carry out archaeological research which stems from the need to comply with the breeding season. Moreover, animal presence at archaeological structures may directly affect their preservation and condition. This can be observed on strongholds in particular. Within their banks, numerous burrows of various sizes and shapes were identified. Some of those structures, presumably belonging to badgers and foxes, can be of considerable size and their digging effectually throws out substantial amounts of building material of the banks. In case of small rodents (mice, bank voles) the scale of their activity can also lead to a noticeable reshaping of slope surfaces. Soil creep, surface wash and linear erosion are actively developing within the freshly excavated material, whereas accumulation below leads to further modification of slope surfaces. Flat surfaces inside the strongholds can be subject to intensive activity of moles, which also causes destruction of archaeological structures, disturbs stratigraphy of cultural layers, removes artifacts from their context and reshapes the surface of monuments. Similarly, beavers found the wooden constructions of medieval bridges, breakwaters and shore reinforcements particularly attractive for their dens. In the last few years there has been a noticeable increase in their activities on the shores of Ostrów Lednicki and Ledniczka islands where they construct dams using favourable conditions inside wooden structures of medieval constructions. Their activity can remove a protective layer of earth and sand, exposing wooden constructions to adverse weather phenomena which accelerate their destruction.

Undoubtedly, the richness, variety and well-being of fauna in the Lednica Landscape Park is a very positive phenomenon. However, heritage management still requires effective solutions to prevent the negative effects of animal presence in the archaeological structures. Currently, these activities are monitored and attempts are undertaken to estimate the effect of zoogenic activity, for example, the impact of beavers on wooden structures were documented during archaeological excavations which were carried out on Ledniczka in 2021.

Dissemination of knowledge about the archeological heritage of the Lednica Landscape Park, in particular about its earthwork structures, is varied and depends on numerous factors. For example, the stronghold in Moraczewo is located within a private property with restricted access for visitors. Ostrów Lednicki, Ledniczka and Imiołki have been maintained for more than 50 years by the Museum of the First Piasts at Lednica on behalf of the local government of Wielkopolska district. The museum's statutory aim is to preserve in the collective memory the knowledge of the significance of Ostrów Lednicki and other early medieval strongholds in the Wielkopolska region from the period of the formation of the Polish state. The Museum organises a number of activities (publications, exhibitions, educational activities, lectures or other cultural events) which aim to present the heritage of the first Piasts to wider public. These also include raising public awareness about the cultural landscape in the Lednica Landscape Park and the necessity to protect it. Tourists can visit Ostrów Lednicki (off the winter season) but Ledniczka and Imiołki are refuges for wildlife and are not accessible for visitors. Currently, interdisciplinary research is being carried out on the Ledniczka island and its surroundings. Knowledge about the motte on Ledniczka and also about the stronghold in Moraczewo is presented at educational and promotional events organized by the Museum for Ostrów Lednicki.

33.6 Conclusions

The Lednica Landscape Park has a unique geo-archaeological landscape, which is the result of transformation of the natural post-glacial relief by diverse and centuries-long human activities. Several phases of human impact (prehistoric, medieval and modern) were marked in a significant transformation of the topography, both directly (convex and concave anthropogenic landforms) and indirectly (the activation or modification of natural morphogenetic processes). On the one hand, the earthworks from the Middle Ages stand out in particular. They are associated with the foundations of the Polish statehood, which gives this area a symbolic value as a high-level national cultural heritage. On the other hand, numerous long-term interdisciplinary archaeological-historical-natural sciences research projects place this area at the country's forefront in terms of recognizing and reconstructing the complex human–environment interactions from prehistory to modern times. Additionally, the recent development of cognitive tourism and availability of the most important sites for sightseeing make this area also to play an important educational role.

References

Bartkowski T (1967) O formach strefy marginalnej na Nizinie Wielkopolskiej. Pr Kom Geogr-Geol PTPN 7(1):260

Drzymała S, Mocek A, Spychalski W (1993) Właściwości fizykochemiczne gleb zachodniej części Lednickiego Parku Krajobrazowego. In: Kurnatowska Z (ed) Przeszłość regionu Ostrowa Lednickiego i jego perspektywy. Prace Kom Archeol PTPN 12, pp 57–64

Drzymała S, Mocek A (1994) Przydatność rolnicza gleb Lednickiego Parku Krajobrazowego w aspekcie ochrony wód i krajobrazu. Stud Lednickie 3:309–326

Filbrandt-Czaja A (1998) Holoceńska historia roślinności północnej części Lednickiego Parku Krajobrazowego ze szczególnym uwzględnieniem czynnika antropogenicznego. Biblioteka Studiów Lednickich 3:9–41

Górecki J (2001) Gród na Ostrowie Lednickim na tle wybranych ośrodków grodowych pierwszej monarchii piastowskiej. Biblioteka Studiów Lednickich 7:29–41

Górecki J (2010) Brzegowe umocnienia obronne Ostrowa Lednickiego. Stud Lednickie 10:107–127

Górecki J, Łastowiecki M, Wrzesiński J (1996) Gródek Na Ledniczce. Stud Lednickie 4:197–246

Górecki J, Łastowiecki M (2016) Konstrukcje obronne Ostrowa Lednickiego. In: Kurnatowska Z, Wyrwa AM (eds) Ostrów Lednicki—Rezydencjonalno-stołeczny ośrodek pierwszych Piastów. Inst Archeologii i Etnologii PAN, Warszawa, pp 59–72

Górecki J, Tabaka A, Wyrwa A (2020) Ostrów Lednicki—ku perspektywie dalszych badań. Podsumowanie. In: Banaszak D, Kowalczyk A, Tabaka A, Wyrwa AM (eds) Ostrów Lednicki—Palatium Mieszka I i Bolesława Chrobrego. Chronologia i kontekst. Muzeum Pierwszych Piastów, Lednica, pp 351–363

Hildebrandt-Radke I (2020) Analiza sedymentologiczna oraz geochemiczna osadów. In: Banaszak D, Kowalczyk A, Tabaka A, Wyrwa AM (eds) Ostrów Lednicki—Palatium Mieszka I i Bolesława Chrobrego. Chronologia i kontekst. Muzeum Pierwszych Piastów, Lednica, pp 117–134

Kasprowicz T (2016) Rekonstrukcja stropu calca. In: Kurnatowska Z, Wyrwa A M (eds) Ostrów Lednicki—Rezydencjonalno-stołeczny ośrodek pierwszych Piastów. Inst Archeologii i Etnologii PAN, Warszawa, pp 55–58

Kola A, Radka K, Wilke G (2016) Mosty traktu „poznańskiego" i „gnieźnieńskiego" w świetle badań podwodnych. In: Kurnatowska Z, Wyrwa A M (eds) Ostrów Lednicki—Rezydencjonalno-stołeczny ośrodek pierwszych Piastów. Inst Archeologii i Etnologii PAN, Warszawa, pp 107–130

Kola A, Wilke G (1989) Sprawozdanie z archeologicznych badań podwodnych reliktów wczesnośredniowiecznego mostu „poznańskiego" (Rybitwy, stan 3A) w Jeziorze Lednickim w latach 1986–1987. Stud Lednickie 1:77–97

Kowalewski G, Wiśniewski R (1994) Zbiorniki akumulacji biogenicznej na południe od Jeziora Lednickiego. Stud Lednickie 3:277–308

Kozarski S (1962) Recesja ostatniego lądolodu z północnej części Wysoczyzny Gnieźnieńskiej a kształtowanie się pradoliny Noteci—Warty. Pr Komisji Geogr-Geol PTPN 2(3):153

Krąpiec M (2003) Wyniki analiz dendrochronologicznych prób drewna dębowego pochodzących z umocnień zachodniego brzegu wyspy; Rybitwy - Ostrów Lednicki, stan. 2, Kraków, maszynopis

Krygowski B (ed) (2007) Mapa geomorfologiczna Niziny Wielkopolsko-Kujawskiej, 1:300000. Instytut Paleogeografii i Geoekologii UAM, Poznań

Kurnatowska Z, Kurnatowski S (1991) Zasiedlenie regionu Lednicy w pradziejach i wczesnym średniowieczu w świetle dotychczasowych badań. In: Tobolski K (ed) Wstęp do paleoekologii Lednickiego Parku Krajobrazowego. Biblioteka Studiów Lednickich 1, pp 35–42

Kurnatowska Z, Kurnatowski S (1993) Przemiany zasiedlenia regionu Lednicy w starożytności i we wczesnym średniowieczu. In: Kurnatowska Z (ed) Przeszłość regionu Ostrowa Lednickiego i jego perspektywy. Poznańskie Towarzystwo Przyjaciół Nauk, Prace Komisji Archeologicznej vol 12, pp 23–26

Kurnatowska Z, Wyrwa AM (eds) (2016) Ostrów Lednicki—rezydencjonalno-stołeczny ośrodek pierwszych Piastów. Inst Archeologii i Etnologii PAN, Warszawa, p 349

Kurnatowski S (1993) Miejsce interdyscyplinarnych badań lednickich w badaniach regionalnych Wielkopolski. Uwagi metodyczne. In: Kurnatowska Z (ed) Przeszłość regionu Ostrowa Lednickiego i jego perspektywy. Pr Kom Archeol PTPN vol 12, pp 5–11

Makohonieko M (1998) Analiza palinologiczna konstrukcji wałowych wczesnośredniowiecznego grodu na Ostrowie Lednickim. In: Grygorowicz A, Tobolski K (eds) Podstawy rekonstrukcji wczesnodziejowego zespołu rezydencjonalno-obronnego i sakralnego na Ostrowie Lednickim. Homini, Bydgoszcz, pp 65–87

Marks L (2011) Quaternary Glaciations in Poland. In: Ehlers J, Gibbard PL, Hughes PD (eds) Quaternary Glaciations—Extent and Chronology. Developments in Quaternary Sciences, vol 15, pp 299–303

Polcyn M (1991) Podwodne warstwy kulturowe—problem definicji. Stud Lednickie 2:159–164

Polcyn M (2003) Archeobotaniczna interpretacja wczesnośredniowiecznych, podwodnych warstw kulturowych z Ostrowa Lednickiego i Giecza. Biblioteka Studiów Lednickich 9, Lednica, Poznań, p 78

Przybylski B (2008) Geomorphic traces of a Weichselian ice stream in the Wielkopolska Lowland, western Poland. Boreas 37:286–296

Pydyn A, Popek M, Dębicka D, Radka K (2018) Przeprawa mostowa na wyspę Ledniczka. Wstępne Wyniki Badań. Stud Lednickie 17:181–196

Richling A, Solon J, Macias A, Balon J, Borzyszkowski J, Kistowski M (eds) (2021) Regionalna geografia fizyczna Polski. Bogucki Wydawnictwo Naukowe, Poznań, p 608

Schubert T, Kowalewski G (1998) Wyniki badań geologicznych prowadzonych na Ostrowie Lednickim. Stud Lednickie 5:417–418

Sikora J, Kittel P (2017) Problem nieco zapomniany? Zaplecza rezydencji rycerskich typu motte. In Różański A (ed) Gemma Gemmarum. Studia dedykowane profesor Hannie Kóčce-Krenz. Wyd PTPN, Poznań, pp 561–589

Skoczylas J (1989) Budowa geologiczna i surowce mineralne regionu Jeziora Lednickiego. Stud Lednickie 1:209–224

Skoczylas J (1993) Pozyskiwanie surowców skalnych dla budowli lednickich. In: Kurnatowska Z (ed) Przeszłość regionu Ostrowa Lednickiego i jego perspektywy. Prace Kom Archeol PTPN, vol 12, pp 49–55

Stankowski W (1989) Morfogeneza Jeziora Lednickiego i jego obramowania (doniesienia wstępne). Stud Lednickie 1:225–231

Strzyżewski C, Łastowiecki M, Kara M (2003) Wczesnośredniowieczne grodzisko w Moraczewie pod Ostrowem Lednickim. Komunikat o wynikach badań archeologicznych w latach 1977–1981. Wielkopolskie Spraw Archeol 6:76–93

Tobolski K (ed) (1991) Wstęp do paleoekologii Lednickiego Parku Krajobrazowego. UAM, Poznań, p 157
Tobolski K, Litt T (1994) Voralleródzeitliche Seeablagerungen in Wielkopolska ("Grosspolen"). In: Lotter AF, Ammann B (eds) Festschrift Gerhard Lang. Diss Botanicae, vol 234, pp 487–496
Tobolski K (1993) Badania ekologicznej przeszłości Lednickiego Parku Krajobrazowego. In: Kurnatowska Z (ed) Przeszłość regionu Ostrowa Lednickiego i jego perspektywy. Prace Kom Archeol PTPN, vol 12, pp 13–22
Wrzesiński J, Kara M (2016) Stratygrafia, chronologia oraz wybrane zagadnienia socjotopografii najstarszego grodu na Ostrowie Lednickim. In: Kurnatowska Z, Wyrwa AM (eds) Ostrów Lednicki— Rezydencjonalno-stołeczny ośrodek pierwszych Piastów. Inst Archeologii i Etnologii PAN, Warszawa, pp 73–106
Wyrwa AM (2019) *Ad futuram rei* memoriam on the fiftieth anniversary of the Museum of the First Piasts at Lednica foundation The Historic Monument of the Nation's Culture. Stud Lednickie 18:13–78

Lidia Żuk is Associate Professor of archaeology at the Faculty of Archaeology, Adam Mickiewicz University in Poznań. She is particularly interested in the application of remote sensing methods (aerial photography, satellite imagery, airborne laser scanning) to landscape studies and heritage protection and management. She has been involved in several EU-funded projects, including ArcLand: ArchaeoLandscape Europe (2010–2014), RESEARCH: REmote SEnsing techniques for ARCHaeology (2018–2023) and TRIQUETRA: Toolbox for assessing and mitigating climate change risks and natural hazards threatening cultural heritage (2023–2025). Her main research area is Greater Poland, but she also participated in numerous workshops and fieldworks in Europe (UK, Germany, Serbia, Cyprus) and Africa (Sudan). She authored numerous articles and chapters in scientific monographs. She is a member of the European Association of Archaeologists and Honorary Secretary of the Aerial Archaeology Research Group (2007–2010 and 2020–2023).

Agnieszka Latocha-Wites is Associate Professor in the Institute of Geography and Regional Development at the University of Wrocław. She is also the coordinator of the Laboratory of Landscape Research. In her research she combines physical and human geography. Her main research interests are long-term human impact on landscape and geomorphic systems, persistence of anthropogenic signatures in depopulated areas, and transformations of cultural landscapes. Her main study area is Lower Silesia, mainly the Sudetes Mountains, but she also conducted her research in Ireland and in the Artic. She authored numerous articles, chapters and books, including travel books ('Ireland—my paths'; 'My Arctic Summer'). She is a member of the Executive Committee of the Polish Geographical Society (2020–2024) and member of editorial board of *Journal of Landscape Ecology*. She was a secretary of the Association of Polish Geomorphologists (2005–2011).

Andrzej Kowalczyk is a doctor of humanities in the field of archaeology, head of the department of the Early Piast Residence in Ostrów Lednicki and a qualified curator at the Museum of the First Piasts at Lednica. He is the author of more than 50 scientific and popular studies. His main research interests are medieval and modern archaeology. He is particularly interested in intercultural relations, as emerging from archaeological sources. He has conducted fieldwork at several archaeological sites, participated in underwater archaeological surveys, and is involved in interdisciplinary studies of the Lednica Landscape Park. He is the curator of several archaeological exhibitions. Since 2018, he has been the editorial secretary and managing editor of the journal *Studia Lednickie*.

The Middle Noteć Valley—The Landscape of an Ice-Marginal Valley and Its Surroundings

34

Leszek Kasprzak, Marek W. Ewertowski, Izabela Szuman, Jakub Kalita and Aleksandra M. Tomczyk

Abstract

Pradolinas (also called ice-marginal valleys, ice-marginal spillways, ice-marginal streamways, or *Urstromtäler*) are very wide (up to 25 km) and long (up to several hundred km) valleys oriented parallel to the former margin of an ice sheet. The best-known pradolinas in the European Lowland, related to Pleistocene Glaciations, are: Wrocław-Magdeburg-Bremen, Głogów-Baruth-Hamburg, Warsaw-Poznań-Berlin-Hamburg, and Toruń-Eberswalde-Hamburg ones, all running from east to west. The main cause of formation of pradolinas was the general terrain gradient in the European Lowland that directed drainage towards the north and west. Extraglacial rivers met waters generated by melting ice masses and jointly drained towards the west in the direction of the North Sea during subsequent deglaciation stages, because the northern direction was blocked by standstill ice margin and associated sediments. After the margin of the ice sheet retreated, fluvial system was successively extended towards the north. Pradolinas are a characteristic element of the contemporary lowland landscape and often used by modern-day rivers. This chapter focuses on the Toruń-Eberswalde Pradolina, which is characterised by prominent morphological expression, not only evident from digital elevation models and remote sensing data but also from the ground level. We describe the most characteristic elements of the morphology of a pradolina and its surroundings based on an example of the Middle Noteć River Valley (central Poland), which include a diversity of smaller-scale features such as terraces, erosional incisions into valley sides, dune fields, and anthropogenic modifications of the river system.

Keywords

Weichselian · Scandinavian ice sheet · MIS2 · Ice-marginal valley · Pradolina · Terraces

34.1 Introduction

Pradolinas (also called ice-marginal valleys, ice-marginal spillways, ice-marginal streamways, or *Urstromtäler*) are very wide (up to 25 km), and long (up to several hundred km) valleys oriented parallel to the former margin of an ice sheet (Galon 1961; Benn and Evans 2010). The best-known pradolinas related to Pleistocene Glaciations in the European Lowlands are: Wrocław-Magdeburg-Bremen, Głogów-Baruth-Hamburg, Warsaw-Poznań-Berlin-Hamburg, and Toruń-Eberswalde-Hamburg ones, characterised by an E–W orientation (Fig. 34.1). In this chapter, we used the term "*pradolina*" instead of "*ice-marginal valley*". "Pradolina" in Polish means an ancient (primaeval) valley, characterised by broad and flat bottom, formed on permafrost at some distance from the ice margin and collecting waters from glacial and non-glacial sources. "Ice-marginal valley" is a more general term associated with all large valleys routed along the ice margin (see Benn and Evans 2010).

The main cause of formation of pradolinas was the general terrain gradient in the European Lowland that dictated

L. Kasprzak · M. W. Ewertowski (✉) · I. Szuman · J. Kalita · A. M. Tomczyk
Faculty of Geographical and Geological Sciences, Adam Mickiewicz University, Krygowskiego 10, 61–680 Poznań, Poland
e-mail: evert@amu.edu.pl

L. Kasprzak
e-mail: l.kasp@amu.edu.pl

I. Szuman
e-mail: szuman@amu.edu.pl

J. Kalita
e-mail: jakub.kalita@uam.edu.pl

A. M. Tomczyk
e-mail: alto@uam.edu.pl

P. Migoń and K. Jancewicz (eds.), *Landscapes and Landforms of Poland*,
World Geomorphological Landscapes, https://doi.org/10.1007/978-3-031-45762-3_34

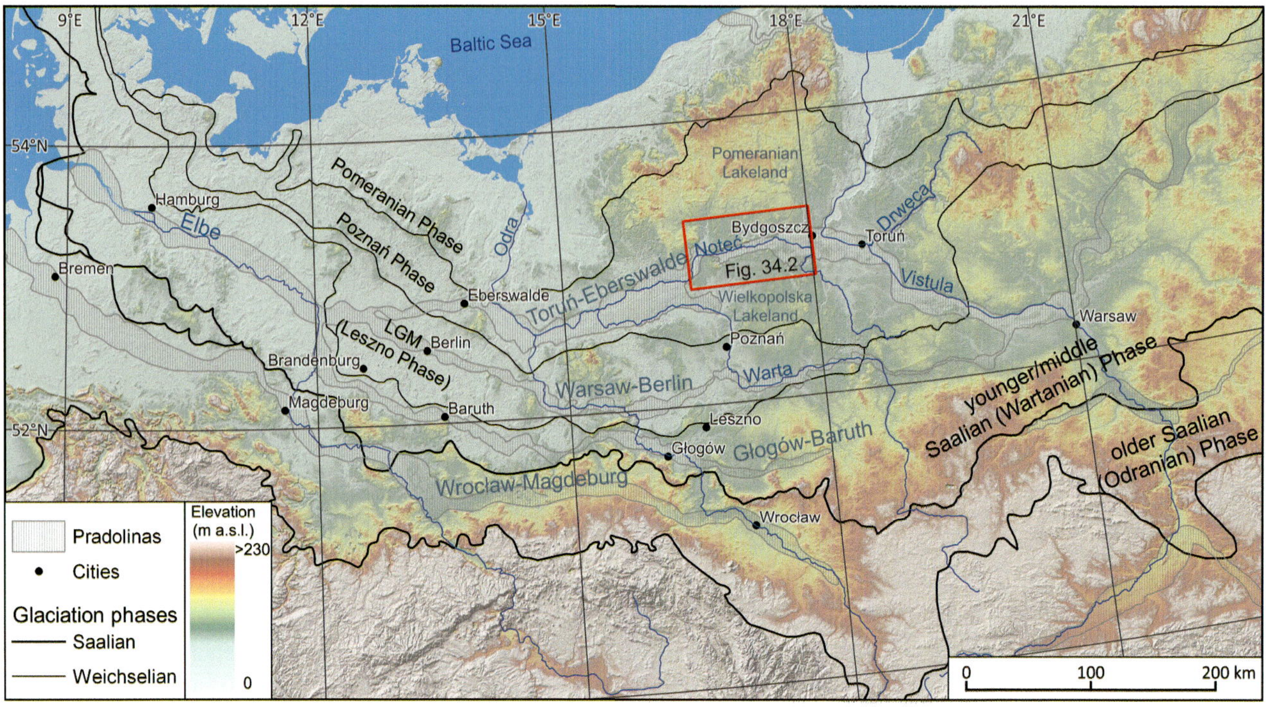

Fig. 34.1 Location of main pradolinas in the European Lowlands. Glaciations extents after Ehlers et al. (2011a), Ehlers et al. (2011b), Marks (2011), Böse et al. (2012)

drainage towards the north and west. During deglaciation stages following consecutive ice-sheet advances, extraglacial rivers met waters generated by the melting of ice masses and jointly drained towards the west in the direction of the North Sea, as the northern direction was blocked by the standstill ice margin and associated sediments (Galon 1961; Kozarski 1962a; Özmaral et al. 2022). After the margin of the ice sheet retreated, rivers recommenced their flows towards the north (Galon 1961). Pradolinas are a characteristic element still visible in the landscape and often used by modern-day rivers. Apart from the morphological expression, they also functioned as a particular depositional environment during deglaciation, characterised by large, but confined braided rivers collecting water from the Scandinavian Ice Sheet (SIS) and extraglacial sources, and affected by periglacial conditions developed on the foreland of SIS (Jahn 1975; Kozarski 1993, 1995b; Pisarska-Jamroży and Zieliński 2011; Weckwerth and Chabowski 2013; Pisarska-Jamroży 2015; Weckwerth and Pisarska-Jamroży 2015). This chapter focuses on the Middle Noteć Valley section of Toruń-Eberswalde Pradolina, which is characterised by a prominent morphological expression visible not only in digital terrain models (DTMs) and remote sensing data (Fig. 34.2) but also from the ground level.

The genesis and functioning of pradolinas in Poland generated some controversies related, among others, to the origin and preservation potential of these landforms, e.g., whether they were indeed formed during the last SIS advance or if they represent older structures rejuvenated by subsequent glaciations (cf. Galon 1961; Kozarski 1962a). Research conducted in the Toruń Basin in Central Poland indicated that river valleys functioned there during the whole Weichselian period, including time spans when the terrain has not been occupied by ice (Weckwerth 2013, 2018; Weckwerth and Chabowski 2013). The fluvial sedimentary environment varied from ephemeral sandy-bed meandering river systems through ephemeral and shallow perennial sand-bed braided rivers to the depositional environment of the deep perennial sand-bed braided rivers (Weckwerth 2018). The causes of these river system transformations were related to glacio-isostasy (which guided river incision) and climate cooling, which resulted in the development of continuous permafrost and increased the sediment supply into rivers (Weckwerth and Chabowski 2013; Weckwerth 2018). As a result, different styles of fluvial sedimentation reflecting warmer and cooler periods and SIS margin dynamics resulted in changes from meandering to braided-river patterns and differences in channel depths, widths and stream power (Weckwerth 2018). Moreover, it was usually indicated that during the Pomeranian Phase extraglacial rivers were important sources of water and sediments (Galon 1968; Wiśniewski 1990, 1992), a view that was recently challenged in several studies (e.g., Weckwerth 2013; Weckwerth and Chabowski 2013; Pisarska-Jamroży

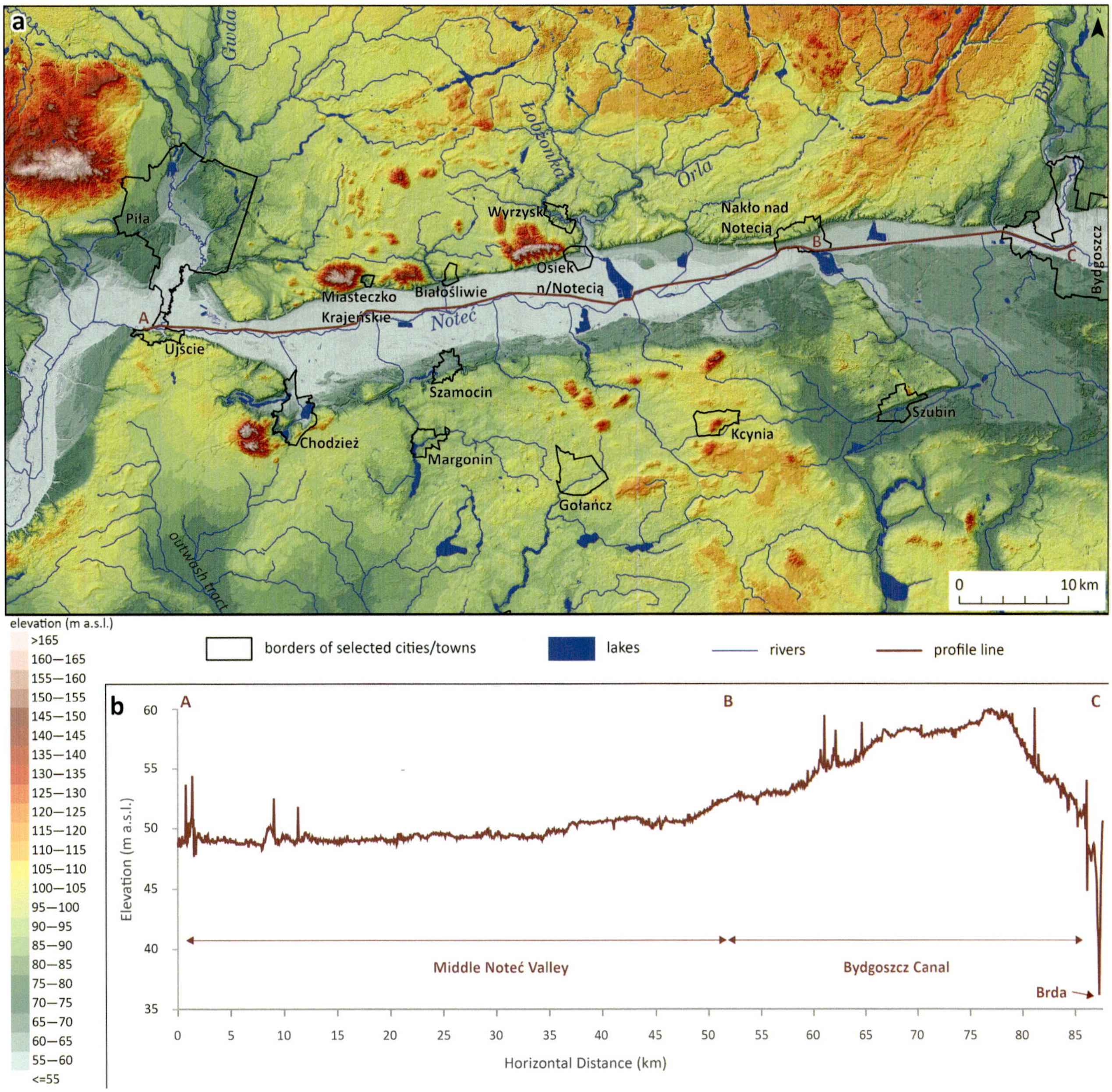

Fig. 34.2 Study area. **a** Location of the Bydgoszcz-Ujście section of the Toruń-Eberswalde Pradolina. **b** Longitudinal profile along the bottom of the Noteć Valley and Bydgoszcz Canal from the Brda River in the east to Ujście in the west (note that "peaks" in the profile are related to elevated roads and railroads crossing the valley)

2015) which demonstrated that the role of rivers coming from the south, in terms of water and sediment delivery, was marginal.

This chapter aims to describe the most characteristic elements of the morphology of a pradolina and its surroundings based on the central section (The Middle Noteć Valley—Fig. 34.2) of the Toruń-Eberswalde Valley, including a diversity of smaller-scale features such as terraces, erosional incisions into valley sides, dune fields, and anthropogenic modifications of the river system. The recent availability of high-resolution LiDAR-based DTMs for Poland, (e.g., Szuman et al. 2021b), combined with an increase of knowledge about the behaviour of the last SIS in Poland (e.g., Wysota et al. 2009; Szuman et al. 2018, 2021a; Weckwerth 2018; Roman 2019; Tylmann et al. 2019; Weckwerth et al. 2019; Sokołowski and Wysota 2020), and the functioning of ice sheets and glaciers in general (Benn and Evans 2010; Cuffey and Paterson 2010; Benn et al. 2019), have enabled us to update previously published morphological data (e.g., the elevation of terraces) and also to propose a new interpretation of landform development.

34.2 Geological Setting

The Toruń-Eberswalde Pradolina (TEP) extends over 400 km from its easternmost section in the Toruń Basin in the Wisła (Vistula) Valley to a bend of the modern-day Elbe River in the west, where it joins the Warsaw-Poznań-Berlin Pradolina (Fig. 34.1). Central section of TEP was glaciated during the Elsterian stage (MIS 12), Saalian complex stage (MIS 6–8), and Weichselian (MIS 2), and intensively eroded by river activity during warm interglacial periods, which is partly documented by a well-developed series of fluvial deposits in the western part of the Middle Noteć Valley (Kasprzak 1981). There were assumptions that the origin of the TEP was related to a standstill of the last SIS during the Pomeranian Phase (Galon 1961). However, it is also probable that older, pre-Quaternary structures controlled the functioning of the TEP and its redevelopment during successive glaciations (Gadomska 1957; Krygowski 1961; Kozarski 1962a; Włodek 1980; Weckwerth 2018). The latitudinal pre-Pleistocene Noteć furrow exists at around 0 m asl and is surrounded by elevated pre-Pleistocene surfaces (Kowalska 1960; Dzierżek 1997). The TEP is incised into till plains and fluvioglacial deposits from the Weichselian and Saalian glaciations, and filled with Weichselian fluvioglacial sands and gravels as well as Holocene fluvial sands. The primary stage of fluvial sedimentation was dated between 16 and 17 ka (Weckwerth 2010, 2013). The geology of the axial part of the TEP reveals up to 40 m deep palaeo-throughs/valleys filled with sediments as old as the Ferdynandów Interglacial (592–472 ka, MIS 15–13, Fig. 34.3) incised locally into sediments of Oligocene age (Włodek 1980), which indicates that at this position favourable conditions for the development of wide valleys existed long before the Weichselian period.

During the Pomeranian Phase of Weichselian glaciation, the depositional environment of the TEP was characterised by large braided river systems functioning under periglacial conditions, which is documented by the presence of sedimentary breccias generated by lateral thermo-erosion (Pisarska-Jamroży and Zieliński 2011; Pisarska-Jamroży 2015; Weckwerth and Pisarska-Jamroży 2015). Steep valley slopes, however, confined this braided river system. The estimated flow depths in the channels were between 1 and 3.2 m (Weckwerth 2013; Pisarska-Jamroży 2015). Palaeohydraulic calculations indicated the occurrences of sporadic large floods (megafloods) (Weckwerth 2013, 2018; Pisarska-Jamroży 2015).

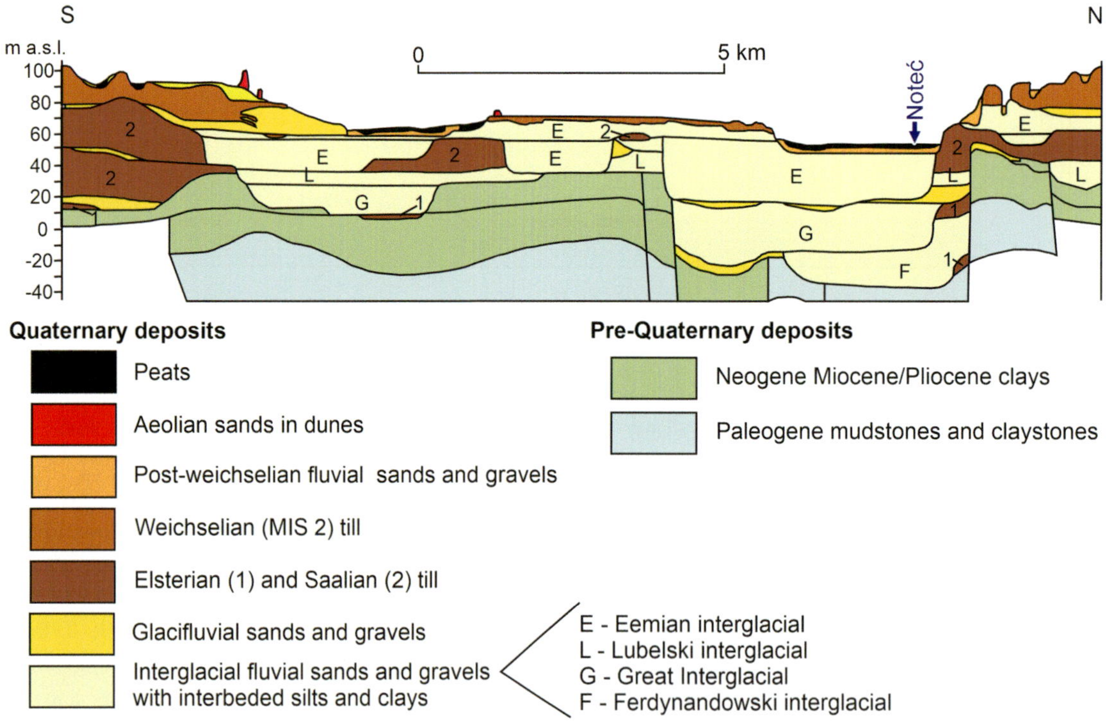

Fig. 34.3 Schematic geological profile of the Toruń-Eberswalde Pradolina based on Włodek (1998) and Kasprzak (2010). Several metres thick Weichselian till plain to the north and south is incised by up to 25 m deep and 10 km wide valley. Notice that river systems (recorded by fluvial sands and gravels) were active at this position during Great Interglacial, Saalian glaciation, and Eemian Interglacial, and even during the oldest interglacial period—Ferdynandowski Interglacial—fluvial activity is evident by the presence of through/valley incised into the pre-Quaternary deposits

34.3 Geomorphological Characteristics of the Central Section of the Toruń-Eberswalde Pradolina and Its Surroundings

The morphology of the Toruń-Eberswalde Pradolina reveals an interesting geomorphological pattern, i.e., the presence of alternating narrow (valleys or gates) and broad reaches (basins): Płock Basin, Toruń Basin, Middle Noteć Valley, Ujście Basin, Gorzów Basin, Kostrzyń Basin, Freienwalde Basin, and Eberswalde Valley (Kondracki 2000). In different sections, the width of the valley varies from 2 to 25 km. We focus on the section between the city of Bydgoszcz and the town of Ujście (Fig. 34.2), which is utilised by the contemporary Noteć River and the artificial Bydgoszcz Canal linking Odra and Wisła river systems. The terrain to the south of the TEP belongs to the region of Wielkopolska Lakeland (Krygowski 1961; Solon et al. 2018) and traditionally was attributed to, broadly speaking, the ice-marginal area of the Chodzież sub-phase of Vistulian glaciation (Kozarski 1962a). The landscape to the north of the TEP belongs to the Pomeranian Lakeland, and most moraines are associated with the so-called Noteć moraine chain (Galon 1961) or Wyrzysk oscillation (Kozarski 1962a). The TEP is a very large and wide landform, and as such, it is more evident in remote sensing data than from the ground (Fig. 34.4a, b). Nevertheless, it can also be observed from the ground-based level, e.g., from the viewpoints in the vicinity of towns of Ujście, Chodzież, or Szamocin (Fig. 34.4c).

34.3.1 Glacial Landforms

34.3.1.1 Areas to the North of the Toruń-Eberswalde Pradolina

The terrain to the north of the TEP consists of four morphological levels located at different elevations. The lowest level contains outwash fans related to several larger valleys (e.g., Brda, Gwda) and extends between 70 and 90 m asl. The character of fluvioglacial sedimentation changed through time in response to ice-mass dynamics (Mleczak and Pisarska-Jamroży 2021). Depending on the location, there are several outwash levels and most contain kettle-holes. Moreover, the uppermost level of the Gwda outwash fan has traces of probable megadunes, suggesting the occurrence of megafloods (Weckwerth et al. 2019). Recent studies also indicate that the Gwda outwash functioned over a relatively long period, with sedimentation starting as early as 20 ka (Mleczak and Pisarska-Jamroży 2021).

The second, most widespread level is located slightly higher than the outwash fans. It is characterised by an elevation of 90–100 m asl and genetically is a till plain. Most of the till plains are built of subglacial tills (1–3 m thick) covered by a relatively thin layer of sediment gravity-flow deposits (Kozarski and Kasprzak 1987; Kasprzak 2003), therefore representing a sequence from the advance to the recession of the ice sheet. Relatively flat till plains are located west of the town of Wyrzysk and near the town of Miasteczko Krajeńskie (Kasprzak 2010, 2017). Only recently, subtle lineations have been recognised here using detailed DTM (Szuman et al. 2021a; b). These lineations represent strongly elongated, streamlined bedforms (Fig. 34.5b) characterised by a width of 100–150 m and elevations of up to 1.5 m. The length of individual landforms is harder to measure as they are interrupted by small valleys and channels, but range roughly from 6 to 8 km. They are interpreted as mega-scale glacial lineations (MSGL) (Szuman et al. 2021b). The remaining part of the till plain (especially near the town of Nakło and Notecią) is more undulating, with local relief differences up to several tens of metres. Numerous small hillocks are visible, with relief between 2 and 5 m (Fig. 34.5c), and these represent classic hummocky topography sensu Evans (2009). The surface of the till plain is also interrupted by kettle-holes, former subglacial channels, as well as modern-day surface drainage and lakes. The most distinct drainage channels are probably inherited from the last glacial period and are used by the modern-day rivers: Orla/Łobżonka and Białośliwka (Fig. 34.2a). The channels of Orla and Łobżonka are narrow (200–400 m), steep, and cut deeply into the till plain, up to 25 m. These two channels join into a wider (about 1000 m) and deeper (up to 45 m) N–S oriented channel that enters the TEP near the town of Osiek and Notecią. The channel occupied by Białośliwka is narrow (100–200 m) and cut up to 20 m into the till plain. There are also numerous smaller channels (< 100 m wide, < 10 m deep), often oriented in a WNW–ESE direction. Altogether, this drainage system probably represents a network of subglacial-to-proglacial channels combined with smaller marginal channels developed along the stationary ice front (Szupryczyński 1966; Kasprzak 2017).

The third (upper middle) distinct morphological level is composed of very characteristic arcuate chains of hillocks located slightly north of the TEP. The first set of four chains of lower hillocks (5–10 m above the till plain level) is located just west of Wyrzysk (Fig. 34.5b). The second set consists of 7–8 arcuate chains of slightly higher hillocks (5–23 m above till plain level) located at the NW end of the streamlined bedforms (Fig. 34.5b). As they are built of subglacial tills, which in places have a significant thickness, Kasprzak (2003) suggested that they might represent older landforms overridden by ice during younger glaciations.

The highest morphological levels are formed by hills along the rim of the Noteć Valley, but there are also some isolated hills of similar height located more towards

Fig. 34.4 Middle Noteć Valley and representative outcrops of Pleistocene deposits. **a**, **b** Oblique visualisation of the valley using DTM: a view from the town of Nakło towards the west (**a**) and a view from the town of Ujście towards the east (**b**). **c** Ground-level view of the Toruń-Eberswalde Pradolina near the town of Szamocin. **d**, **e** Thick sequences of subglacial tills building moraines located at the northern rim of the Noteć Valley. **f** Till/sand breccia in Atanazyn (near Szamocin) developed as a result of lateral thermo-erosion of permafrost river banks. **g** Isolated kame hillock built of fine sands and silts. **h** Series of normal faults in kame hillock, which indicate deposition in the vicinity of dead-ice blocks

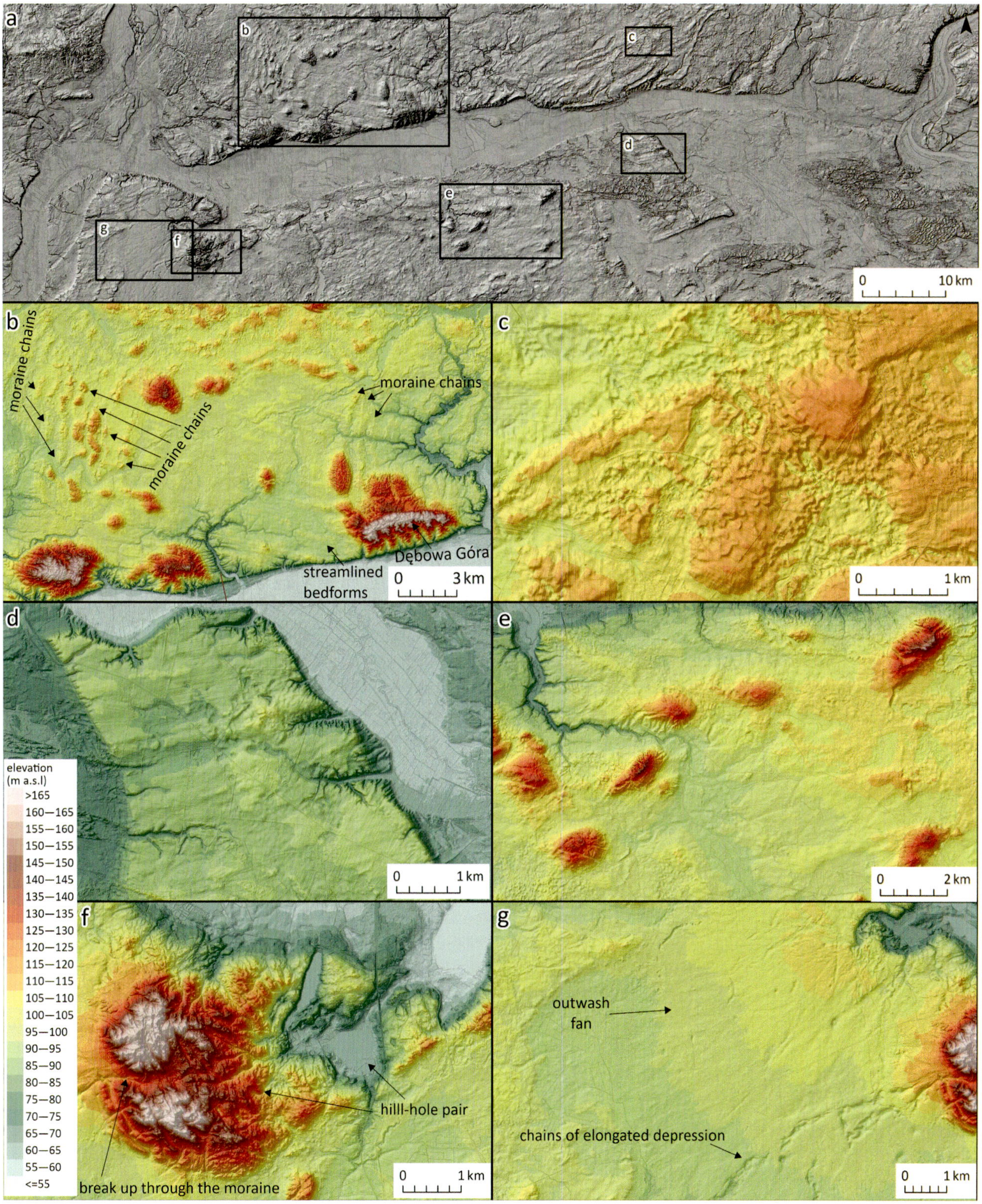

Fig. 34.5 Examples of glacial landforms along the northern and southern rim of the Noteć Valley. **a** Hillshade model with location of insets. **b** Till plain north of the Noteć Valley: streamlined bedforms in the central part are located between several chains of small moraine hillocks; the highest hills are glaciotectonic moraines. **c** Hummocky moraine to the north of the Noteć Valley. **d** Streamlined bedforms on an isolated remnant of a till plain. **e** Moraine hillocks to the south of the Noteć Valley. **f** Chodzież thrust-block moraine. **g** Large outwash fan developed at the mouth of subglacial channel indicating the probable occurrence of a glacial outburst flood; note the chains of elongated depressions which probably show the former presence of buried dead-ice blocks (see similar features in the foreland of modern Icelandic glaciers, e.g., Russell et al. 2006)

the north. The most impressive ones are Dębowa Góra (192.4 m) (Fig. 34.5b) and hills located near Miasteczko Krajeńskie (about 160 m) (Fig. 34.5b). There are also several other hills exceeding 150 m asl, which means that this morphological level occurs 50–90 m above the till plain level. Long axes of these terrain elevations are generally oriented latitudinally, and their length is several thousand metres. Analysis of the contact between the Noteć Valley and the hills indicates that river erosion did not change their orientation; however, their slopes are strongly dissected by numerous small valleys and gullies. Several outcrops along the rim of the Noteć Valley reveal the internal structure of these hills, which consists of very thick (up to 40 m) sequences of subglacial tills (Fig. 34.4d), often interweaving with sand and gravel deposits (Fig. 34.4e). According to Włodek (1980) and Ratajczak-Szczerba (2011b), the age of these tills is uncertain. Geological cores also informed about the existence of rafts of older sediments (Pliocene clay) within moraines and locally within the till plain (Szupryczyński 1966). At Dębowa Góra, Ratajczak-Szczerba (2011a) described a sequence of three deformed tills interweaving with sand. Based on structural analysis, the main axis of strain was inferred to be from the NE towards SW. Internal structures of other hills also indicate that the sediments were glaciotectonically deformed (Ratajczak-Szczerba 2011c, d), which is also confirmed by surface morphology (e.g., the occurrence of minor depressions and rafts within larger complexes, Fig. 34.5b). The age of these glaciotectonic features is uncertain. They might be related to the ice-margin standstill, which blocked extraglacial and proglacial water, therefore triggering the development of the Toruń-Eberswalde Pradolina as suggested by several authors (e.g., Kozarski 1995a; Kasprzak 2003). However, their large size and internal structures consisting of several layers of deformed tills and fluvioglacial sediments may also represent older features, only overridden and smoothened by the last ice sheet.

34.3.1.2 Areas to the South of Toruń-Eberswalde Pradolina

The relief south of TEP is slightly different than in the northern part, with distinct morphological levels more difficult to distinguish. On the north side, the Noteć Valley is bordered by till plains, whereas along the southern border, there are two morphological levels with outwash plains. The lower level is at an elevation of 67 to 70 m asl, only slightly higher than the Noteć River terraces, and is built of sands and gravels. Morphology of the terrain and structural measurements (Kozarski 1962a; Kasprzak 2003) indicate meltwater drainage toward the west. There are also some fluvioglacial depressions as low as 63 m asl, related to washout of the outwash plain by extraglacial waters draining from the south towards the TEP (Kasprzak 2010, 2017). The higher level of fluvioglacial plains (83–90 m asl) is built of a thin veneer of sands and gravels deposited over subglacial tills (Kasprzak 2010). These two levels of fluvioglacial deposits are separated by a more or less distinct erosional escarpment whose height is between several to several tens of metres.

The main level of the till plain in the south extends at an elevation range between 90 and 105 m asl and is characterised by a gentle drop in elevation from the east to both west and southwest. The till plain is relatively flat (compared to the northern side of the TEP) and has the transitional character from the near-continuous cover in the east (near the town of Kcynia) to isolated patches of tills surrounded by fluvioglacial deposits in the west (near the town of Szamocin). These isolated patches of the till plain form a belt parallel to the TEP. In some places, a subglacial till emerges from under the fluvioglacial deposits, which indicates that the flat morphology of till patches is only apparent, i.e., fluvioglacial deposits mask their low-lying fragments, suggesting that a large part of the till plains was remodelled by meltwater flows (Kasprzak 2017). Small, isolated hillocks often occupy central positions in the till plain patches, with relief of up to several metres, and are built of interweaving layers of sands and silts (Fig. 34.4g). Numerous minor normal faults (Fig. 34.4h) indicate deposition in contact with ice blocks; therefore, these features are interpreted as kames (Kasprzak 2010). The till plains are dissected by subglacial tunnel valleys, often containing lakes and modern-day river valleys, while their bottoms form a morphological level extending from 90 to 80 m asl.

As the till plain drops towards the southwest, four arcuate chains of small hillocks become visible located to the south and southwest from the town of Gołańcz (Fig. 34.2a). These hillocks are less conspicuous than their counterparts on the northern side of the TEP and are characterised by elevations between 96 and 100 m asl (6–8 m above the till plain level). They are interpreted as moraines related to the oscillations of the ice sheet (Kozarski 1962a). There is also an isolated remnant of the till plain at an elevation of 90 m asl a bit farther east, near the town of Szubin. This patch exhibits a smooth appearance and subtle lineations (Fig. 34.5d), similar to the lineated terrain west of Wyrzysk (north of TEP), but less conspicuous.

The uppermost morphological level is associated with the tops of the highest hillocks (e.g., near Kcynia), which are at elevations between 130 and 155 m asl (Fig. 34.5e). Their height above the till plain level (40–55 m) and strong dissection are comparable to the highest morphological level located to the north of the Noteć Valley.

One of the most spectacular glacial landscapes in the area is located near the town of Chodzież, where a Chodzież moraine complex (Fig. 34.5f) (with the highest point on the southern side of the TEP—Gontyniec at

191.6 m) and subglacial tunnel valleys with associated outwash fans occur (Fig. 34.5g). The Chodzież moraine complex is one of the best examples of a thrust-block moraine in central Poland (Kozarski 1959, 1995a). The hill-hole pair is clearly visible in the landscape (Fig. 34.5f), with local elevation differences exceeding 130 m, from the lake level at 61 m asl to the top of the moraine at 191.6 m asl. The Chodzież moraine is built of thrust blocks of tills, sands, gravels, and rafts of Neogene deposits (Kozarski 1959). Very characteristic is the presence of a depression filled with fluvioglacial sediments in the central part of the moraine, which has the morphological continuation in the form of a channel dissecting the moraine towards the west (Fig. 34.5f). At the mouth of this channel, a steep outwash fan has developed, which gradually transits into a gentler outwash surface. Numerous elongated depressions up to 10 m deep are visible (Fig. 34.5g). These depressions were interpreted by Kozarski (1977) as having developed because of naledi ice. However, considering that the channel breaks through the moraine, it is more likely that they represent a signature of large blocks of ice transported during a glacial outburst flood (Jökulhlaup). A similar landscape can be observed in the foreland of modern-day Skeiðarárjökull in Iceland, where numerous depressions developed within an outwash when the blocks of ice were transported during the flood and stranded as water transport capacity diminished and were subsequently buried under the glacifluvial sediment (see Russell et al. 2006). The subglacial tunnel valley (more than 1 km wide and at least 8 km long) immediately north of the Chodzież moraine complex fed a large outwash fan, which also bears traces of stranded ice blocks and obstacle marks (Fig. 34.5g). Both outwash levels (i.e., the one originating from the subglacial tunnel valley and that arising from the Chodzież moraine) join together, and the surface of fluvioglacial sediments dips gently towards the south, forming a narrowing outwash tract (Fig. 34.2a). Another subglacial tunnel valley (up to 1 km wide and 10 km long) is located south of the town of Margonin and also terminates with an impressive outwash fan (Fig. 34.2a).

34.3.2 Palaeoglaciological Interpretation

A traditional reconstruction of the Weichselian glaciation suggested that the vicinity of the Toruń-Eberswalde Pradolina was shaped by an oscillating ice-sheet margin, which left behind a series of moraine chains related to specific oscillations (Kozarski 1962a). The larger glaciotectonic structure of the Chodzież Moraine complex was attributed to the individual sub-phase of the glaciation (Kozarski 1959, 1986, 1995a). Kasprzak (2003, 2017) linked morphological differences between the landscape in the immediate vicinity of the TEP and the terrain located further north, within the range of the Pomeranian stage, to the different characteristics of the ice masses. The terrain north of the TEP was considered to be occupied by thicker ice masses, which were less dynamic and characterised by slower recession rates, whereas the immediate surroundings of the TEP and the terrain located south of it (i.e., Wielkopolska Lakeland) were associated with the activity of relatively thin "outlets" from the main SIS body (Kasprzak 2003).

The recent availability of detailed DTMs has enabled us to propose a more detailed interpretation of the character of ice masses (Szuman et al. 2021a, b). The reconstruction by Szuman et al. (2021a) indicates that during the LGM, ice masses advanced into the Wielkopolska from both NW (from the Odra Ice Stream) and NE (from the Wisła Ice Stream). In contrast, the terrain further north of the TEP was occupied by a less dynamic inter-stream zone. During the Poznań Phase and subsequent oscillations, these active ice streams that operated along the TEP, both from NE and NW, interacted zone near Chodzież. The last episode of glaciation is related to the activity of the Chodzież Ice Stream, which advanced from the Wisła Valley and left a clear morphological record in the form of mega-scale glacial lineations (Fig. 34.5b), closely spaced push moraines (Fig. 34.5b), thrust moraines and hill-hole pairs (Fig. 34.5f), disrupted drumlins, and hummocky moraines (Szuman et al. 2021a). Such morphological characteristics indicate the polythermal character of ice masses. Therefore, it is probable that meltwater was stored behind the frozen margin of the ice stream, and that this water ruptured moraines in the Chodzież area and created drainage just north of the moraine, forming a large outwash fan (Fig. 34.5g). During that time, the ice from the Odra Ice Stream was still present further west (see Fig. 34.6 in Szuman et al. 2021a), and the drainage was forced southward, developing an impressive outwash tract. Only after both ice streams had retreated was water drainage possible towards the west, and the formation of the Toruń-Eberswalde Pradolina began. As the thicker, less dynamic ice masses were still present on the line of the Pomeranian stage (Kasprzak 2003), and in addition, large outwash fans of Gwda and Brda Rivers blocked the drainage towards the north (Pisarska-Jamroży 2015; Mleczak and Pisarska-Jamroży 2021) water drainage was forced towards the west, utilising the lower-located bed of the former Chodzież Ice Stream.

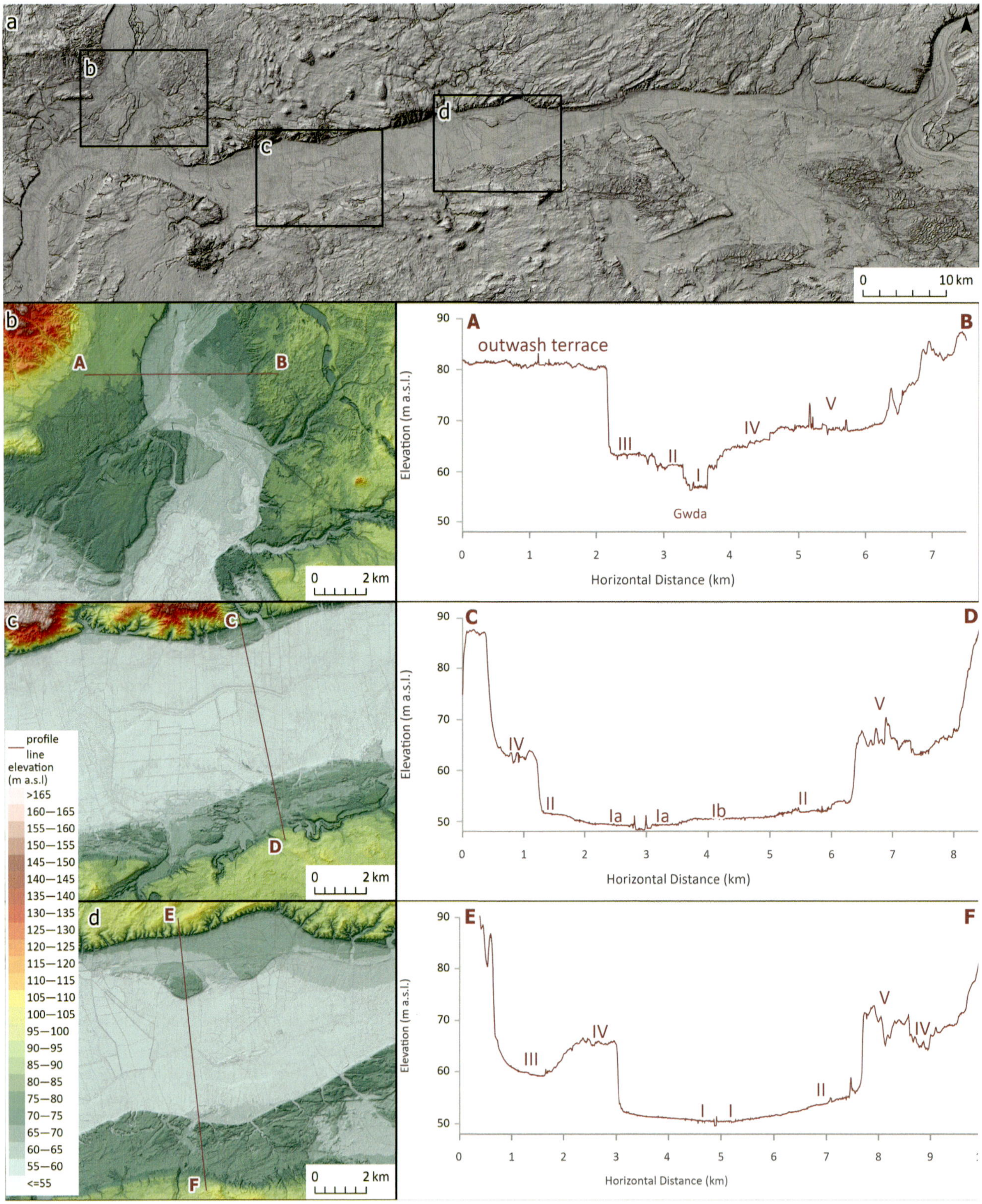

Fig. 34.6 Examples of modern-day morphology of the Toruń-Eberswald Pradolina. **a** Hillshade model with location of insets. **b** Lower section of the Gwda River, close to the Noteć Valley; note the presence of five levels of river terraces and the uppermost outwash-related terrace. **c** The Noteć Valley near Szamocin, the lowest (floodplain) terrace is relatively narrow and consists of two parts (0–1 m and 1–2 m above river level); terrace V is located high above the river (15–20 m) and covered with aeolian landforms. **d** The Noteć Valley east of Osiek, terrace III is located at 60 m asl and remnants of the terrace IV (about 65 m asl) are also visible

34.3.3 Morphology of Bydgoszcz-Ujście Section of the Toruń-Eberswalde Pradolina

34.3.3.1 General Features

The Middle Noteć River Valley is one of the narrowest sections of the Toruń-Eberswalde Pradolina. Its width varies from 2.5 km (between Ujście and Chodzież) to 11 km in the Ujście Basin (western part) and 10 km in the eastern part. Further to the east lies the Toruń Basin, with a width of up to 25 km. Elevation of the bottom of the TEP in the studied section, analysed in the direction from east to west, drops from 60 m asl near the western border of Bydgoszcz to 55 m asl (average slope: 0.37‰) near the junction of the Noteć River and Bydgoszcz Canal, and then, more gently, to 48.5 m asl near Ujście (average slope 0.11‰) (Fig. 34.2b). Valley sides are steep (45–75%), contrasting with the generally flat or gently inclined valley bottom (Fig. 34.6). The ratio of valley width to depth is 1000:3, suggesting activity of periglacial braided river systems and lateral erosion (Kozarski 1965), which is also confirmed in sediment assemblages found along the rim of the TEP, especially till/sand breccia facies (Fig. 34.4f) (Pisarska-Jamroży and Zieliński 2011; Weckwerth and Pisarska-Jamroży 2015). The largest relief differences occur on the northern side of the TEP, in places where terraces were eroded or have not been formed. Here, the moraines are elevated up to 90 m above the till plain level, which itself is located 30–40 m above the floor of the Noteć Valley. The most spectacular transition of the landscape is visible near Dębowa Góra (192.4 m) (Figs. 34.5b and 34.7c), where a change between moraines and the valley bottom located at 50 m asl (i.e., more than 140 m elevation difference) occurs along a horizontal distance of less than 2 km. Along the southern side, absolute elevation differences are smaller because a staircase of several terraces was formed. They have relatively gentle edges, and therefore, there is no abrupt change in elevation, but the elevation increases gradually from the bottom of the valley to the level of till plain (around 80–90 m asl), with more or less visible escarpments marking the edges of terraces (Fig. 34.6).

34.3.3.2 Terrace System

The studied section of the Toruń-Eberswalde Pradolina is characterised by the occurrence of a staircase of terraces (Kozarski and Szupryczyński 1958; Kozarski 1962a) (Fig. 34.6) that are also present along the tributaries (e.g., Gwda, Łobżonka/Orla, Brda Valleys) (Galon 1961, 1968). The number of reported terraces varies from 2 to 5 depending on the author and place of study; also varied is their reported elevation. Based on detailed DTMs, we attempt to provide consistent elevation ranges along the studied section of the TEP.

The lowest terrace (I) (i.e., floodplain) is slightly higher (0.5–2 m) than the modern-day river, corresponding with elevation ranges of the valley bottom of 48.5–55 m asl in the Noteć Valley and 49–61.5 m asl in the Gwda Valley. In the east, in the Brda Valley, which represents a river system younger than TEP (Weckwerth 2010), floodplains are located at lower absolute elevations, from 38.5 to 57 m asl. In the Noteć Valley, this accumulation floodplain extends on both sides of the modern-day river and is built of sands with a small admixture of gravels combined with clays and silts with the character of rhythmites; in places, the terrace is covered by Holocene peat (Kasprzak 2010). It is still possible to detect older levees and channels; however, agricultural activity has intensely transformed the original fluvial morphology. In some sections, the upper level of this terrace is visible (level Ib). It is located only slightly higher (1.5–2.5 m above the river level), but separated from the bottom of the valley by a gently inclined, clearly visible edge (Fig. 34.6c—levels Ia and Ib). In other sections, the floodplain more gently transits into the upper terraces (Fig. 34.6d).

The lower (II) terrace is located 3–5 m above the river level and visible only sporadically in the Noteć Valley (Fig. 34.6c, d). It has an elevation between 52 and 56 m asl. The lower terrace is built mainly by sands and gravels related to fluvial sedimentation, i.e., it has accumulation character and its origin was probably not associated with glacial meltwater but with post-glacial river activity. The internal structure consists of horizontally laminated sands with a mixture of gravels and, in places, silts (Kasprzak 2010). The corresponding level is better developed in the Brda and Gwda Valleys (Fig. 34.6b). The middle terrace (III) is located at 6–9 m above the river level (54 and 63 m asl). It builds the well visible bench east of Osiek and Notecią town (Fig. 34.6d). The upper terrace (IV) has an elevation of 10–12 m above the river level (60–65 m asl) and only small remnants of this level can be observed (Fig. 34.6d). Its surface is punctuated by kettle-holes and obstacle marks, indicating the presence of ice blocks during its development (Fig. 34.8c).

The uppermost terrace (V) is located 15–21 m above the river level (66 to 70 m asl) and can be traced along most of the southern side of the Middle Noteć Valley. It is built of very coarse sands with cross-laminations related to the activity of braided river systems. This level also contains numerous boulders and erosional traces associated with the degradation of the till plain by meltwater. There are also morphological traces of melt-out of ice blocks and channels, leading Kasprzak (2017) to suggest that this level is associated with outwash sedimentation. Similarly, Mleczak

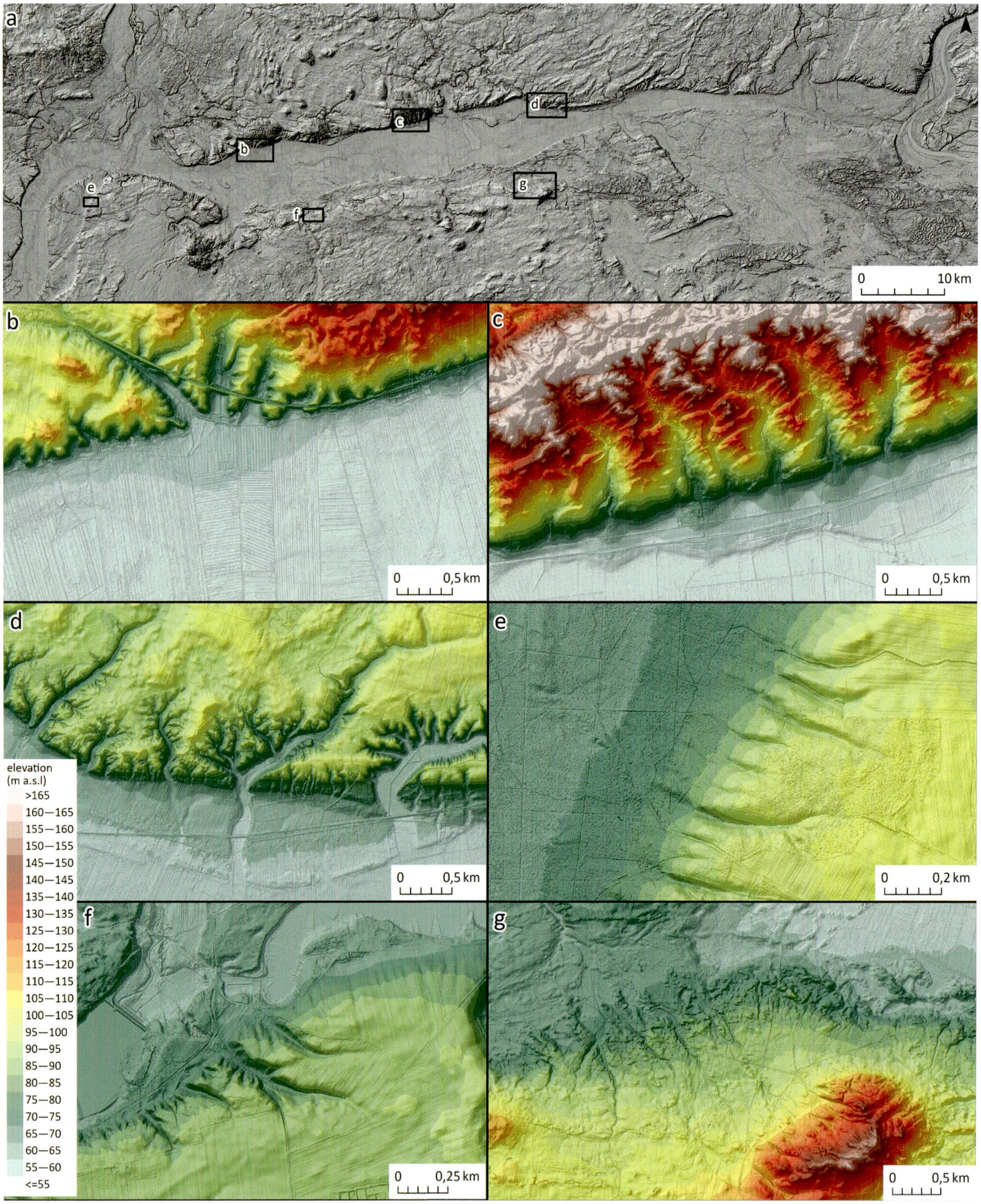

Fig. 34.7 Examples of gullies and ravines and alluvial fans which developed at their mouths. **a** Hillshade model with location of insets. **b** Flat-bottomed valley linking till plain and Toruń-Eberswalde Pradolina. **c** Series of ravines and small fans developed near Dębowa Góra. **d** Flat-bottomed valleys near Nakło and Notecią dissecting middle and upper terraces. **e**, **g** Examples of ravines developed along the southern rim of the Toruń-Eberswalde Pradolina; some of them are utilised by roads and tracks

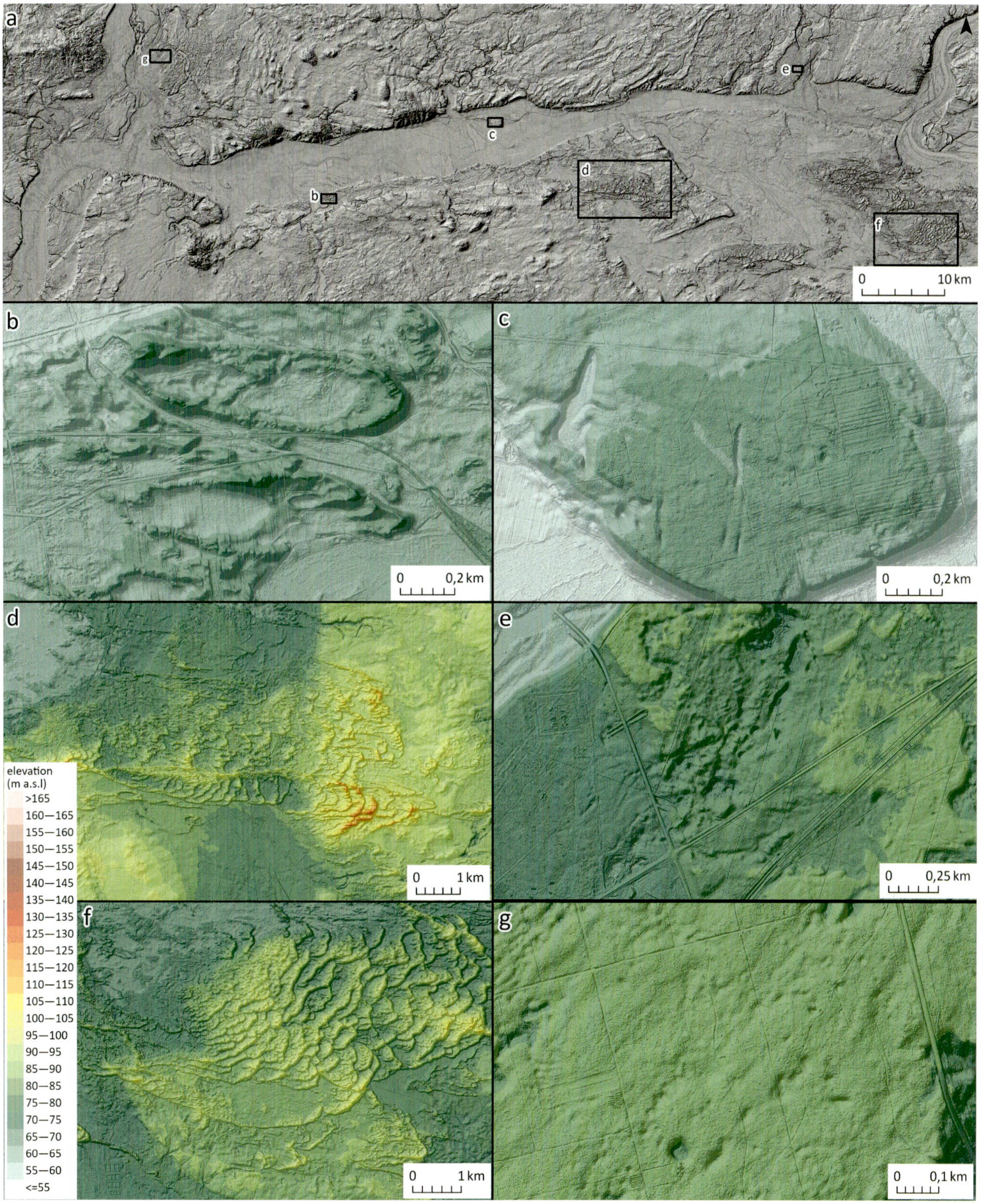

Fig. 34.8 Examples of aeolian landforms (b, d, f) and kettle-holes (c, e, g). **a** Hillshade model with location of insets. **b** Small dunes developed on the uppermost terrace in the Toruń-Eberswalde Pradolina. **c** Small depressions related probably to melting of dead ice on the remnant of terrace IV. **d** Large dune field which climbed onto one of the till plain remnants. **e, g** Chains of depression on the outwash-related terrace (e—Brda; g—Gwda). **f** Large dune field in the Toruń Basin

and Pisarska-Jamroży (2021) indicated that the upper terrace levels in the Gwda Valley are related to outwash sedimentation.

34.3.4 Minor Landscape Elements

The TEP developed due to the interplay between the ice sheet, meltwater, and extraglacial rivers. Apart from major landscape elements related to glaciation (Sect. 3.1) and the valley itself (Sect. 3.3), smaller geomorphic features have developed within the TEP and along its sides.

34.3.4.1 Gullies and Alluvial Fans

As the slopes in the landscape transformed by the last glaciation are an important element linking the till plains and outwash plains with the bottom of the river valleys, subglacial tunnels, and pradolinas, erosion and denudation landforms developing within the slopes provide an insight into the processes operating after deglaciation (Szpikowski et al. 2008). Lateral margins of the Toruń-Eberswalde Pradolina are dissected by numerous small erosional cuts, nivation hollows, and denudation valleys (Fig. 34.7) of variable age (Paluszkiewicz and Ratajczak-Szczerba 2013; Ratajczak-Szczerba and Paluszkiewicz 2015). Some are related to the Weichselian glacifluvial activity, while others were formed in late glacial and post-glacial periods. Jaworski (2018) also demonstrated that many denudation valleys started developing due to periglacial processes. Moreover, he indicated that some forms were rejuvenated during the late glacial period because of thermokarst erosion and permafrost degradation.

The morphology of larger forms (Fig. 34.7b) is characterised by steep valley slopes (20–45%) and flat bottoms up to 100 m wide. Some are dry, i.e., are no longer utilised by permanent drainage. Alluvial fans which formed at the mouth of these valleys are often incorporated into the Noteć River terraces and masked by river deposits, so they are now hard to distinguish based on morphology alone (Fig. 34.7b), which suggests that the most intensive period of denudation activity occurred before the formation of the Noteć River terraces. Another group dissects the middle and upper terraces (Fig. 34.7d), indicating active processes after the retreat of ice masses. Jaworski (2018) demonstrated that erosional-denudation flat-bottom valleys in the vicinity of Nakło nad Notecią were multi-stage landforms formed by mass movements and running water. He also suggested that they ceased developing in the Younger Dryas and Preboreal, which is indicated by their organic infill.

Highly characteristic are ravines dissecting the elevation of Dębowa Góra (Fig. 34.7c), which are relatively long (up to 2 km), and have steep sides (20–30%) being almost bottomless (i.e., there is an almost immediate transition between the opposite valley sides). A series of small fans developed at their mouths, which are deposited on top of the terraces, indicating active erosional processes after the formation of the TEP. Similar but slightly smaller features are also common in other sections of the valley rims (Fig. 34.7e–g). Many of these denudation valleys and erosion cuts are utilised as roads and tracks (Fig. 34.7b–c, e–f) as they offer a convenient way to cross the steep edge of the TEP.

34.3.4.2 Dunes and Dune Fields

Aeolian landforms include parabolic dunes, deflation hollows (blowouts), and plains of aeolian sands, located mostly to the south of the TEP but also within the Noteć Valley on most of the terrace levels, including the floodplains (Kozarski 1962b; Kasprzak 2010). Dunes are primarily parabolic, with the central parts extended in accordance with the dune-forming westerly winds. They can be found in isolation (Fig. 34.8b), but in many places, they form double, parallel ridges, where their continuity was breached due to sand movement. There are also dune fields (Fig. 34.8d, f), the most impressive of which is located in the Toruń Basin, south of the city of Bydgoszcz, and contains longitudinal, parabolic, and irregular dunes, currently covered by forest (Fig. 34.8f). Abundance of aeolian landforms on the left-bank side of the Toruń Basin is related to the availability of sandy glacifluvial and river deposits which facilitated dune-forming processes (Andrzejewski and Weckwerth 2010). Another large field of parabolic dunes is situated just south of the TEP, between Szubin and Kcynia, where they climb from the terraces of the TEP onto the more elevated till plain (Fig. 34.8d).

Individual dunes are mostly small, between 5 and 10 m high, up to 1000 m long and 200 m wide. Higher forms (up to 25 m high) are present in dune fields. On the western side of dunes, deflation hollows usually occur, locally filled with biogenic material (Kasprzak 2010). The most prominent depressions are up to several metres deep and 200–300 m in diameter. The size of the dune fields in the Toruń Basin, as well as the height of individual landforms (Fig. 34.8f), increase towards the east as the dune complexes join one another (Andrzejewski and Weckwerth 2010).

Dunes in the TEP and its vicinity were formed during the late stages of the last glacial period, in the Older and Younger Dryas, after ice sheet retreat, in cold and dry climatic conditions (Kozarski 1962b; Nowaczyk 1986). They represent classic inland dunes, formed with sand of outwash plains, and were constructed after the ice sheet retreated but before vegetation was re-established (Rotnicki 1970; Nowaczyk 1986, 2002). In the Toruń Basin, another dune-forming phase was also proposed for Preboreal period (Andrzejewski and Weckwerth 2010).

34.3.4.3 Kettle-Holes

Kettle-holes are abundant within outwash fans (Fig. 34.8e, g) (Chmal 2006), indicating that fluvioglacial deposition occurred at least partly during the ice retreat and outwash deposits buried parts of the glacier snout. They are also common in the upper terraces, indicating the previous position of blocks of ice transported by meltwater (Fig. 34.8c). However, they are uncommon within lower terraces in the studied section of the TEP, showing that these levels were formed in post-glacial conditions. Many of the kettle-holes are filled with organic sediments, and the dates from peat suggest that the infill started developing in the Younger Dryas, at ~11,000 ka (Wodziczko 1948; Przybylski 1961).

34.3.5 Landscape Evolution

While the modern-day landscape developed during and after the Weichselian glaciation (Fig. 34.9a), the pre-Quaternary relief probably impacted both SIS's behaviour and the formation of the TEP (Weckwerth 2010, 2018). The geological cross-section (Fig. 34.3) indicates that fluvial systems employed the pre-Quaternary Noteć furrow during several periods of the Pleistocene. The Middle Noteć Valley is located where large river systems existed during the Great Interglacial and Ferdynandowski Interglacial (Kowalska 1960; Dzierżek 1997; Weckwerth 2010). The evolution of drainage patterns was also related to the presence of salt structures and tectonic activity, probably rejuvenated by loading and unloading caused by the advancing and retreating ice masses (Piotrowski 1991; Liszkowski 1993; Weckwerth 2010; Lang et al. 2014). Part of older bedrock structures was uplifted due to post-glacial rebounding, which enhanced glacial erosion at the end of the Saalian glaciation (Marks 2005; Weckwerth 2018). As a result, Basins and valleys were formed during the following interglacial, and the general position of TEP was controlled by the existence of older valleys and basins (before LGM) (Makowska 1980; Weckwerth 2010). Former river valleys often guided the development of outwash sedimentation and probably facilitated ice advance from the Wisła Valley into the direction of Chodzież.

The present-day relief of the TEP results from the activity of water of glacial and fluvial origin, slope processes, and aeolian activity. The formation of the TEP started probably after the Chodzież sub-phase, i.e., after the retreat of the Chodzież Ice Stream towards NE and Oborniki and Pszczew Ice Streams towards NW (Szuman et al. 2021a). Large outwash fans developed to the north of TEP (Mleczak and Pisarska-Jamroży 2021), and water drainage (meltwater from the north and extraglacial rivers from the south) was directed utilising depression of the bed of the former Chodzież Ice Stream (Fig. 34.9b). Remnants of meltwater sedimentation are visible as outwash terraces (e.g., in Gwda Valley—Fig. 34.6b). In the southern part of the Toruń Basin, outwash levels form discontinued benches and are dated at 17 ka BP, i.e., earlier than the Pomeranian Phase (Weckwerth 2010). The first stage of TEP formation was probably related to the formation of a braided-river meltwater tract which is recorded as the terrace V, marking the transition from longitudinal outwash drainage to a more latitudinal course. The main stage of TEP development occurred during the Pomeranian Stage and was dated in the Toruń Basin at 16–17 ka BP (Weckwerth 2013). Most probably, the TEP evolved from west to east, linking with the Toruń Basin, which gathered water from the Wisła, Noteć, Drwęca, and Brda Valleys (Galon 1961; Kozarski 1962a; Wiśniewski 1990; Weckwerth 2010). Subsequent water flows at lower elevations are recorded in terraces III and IV (Fig. 34.6). Successive development of lower terraces (II) is an effect of post-glacial river activity (Fig. 34.9c). Floodplains (terrace I) most likely started forming in the Holocene, as suggested by organic infill from the bottom of the valley near Nakło nad Notecią (Jaworski 2018).

Glacier outburst flood probably impacted the formation of TEP in two ways. First, an outburst of subglacial meltwater was probably associated with the advance of the Chodzież Ice Stream and is recorded in the outwash fan near Chodzież (Fig. 34.5g). It is likely that after a surge-like advance and associated sudden drainage of meltwater, ice masses forming the Chodzież Ice Stream become inactive, therefore preserving the depression of the ice stream bed from being occupied by meltwater deposits at least for some time. Secondly, during the functioning of the TEP, high-energy waters, potentially associated with glacial outburst floods, were injected into TEP, periodically increasing erosion rates and forming deep gravel-bed braided channels (Pisarska-Jamroży 2015). In addition, periglacial conditions widespread in Central Poland after LGM (Kozarski 1995b; Kasprzak 2003; Ewertowski 2009; Ewertowski et al. 2017) also controlled the functioning of the TEP. Such conditions resulted in the domination of the lateral thermo-erosion, which widened the valley bottom and are also recorded in the formation of till-breccia, a characteristic element of sedimentary sequences in the TEP (Pisarska-Jamroży and Zieliński 2011; Weckwerth and Pisarska-Jamroży 2015).

After the SIS retreated north from the Pomeranian Stage, slopes of the TEP began degrading because of thermokarst processes, and the development of erosion and denudation forms started (Jaworski 2018) (Fig. 34.9d). Subsequent cooling during the Older and Younger Dryas resulted in the formation of dunes and dune fields (Kozarski 1962b; Nowaczyk 1986, 2002; Andrzejewski and Weckwerth

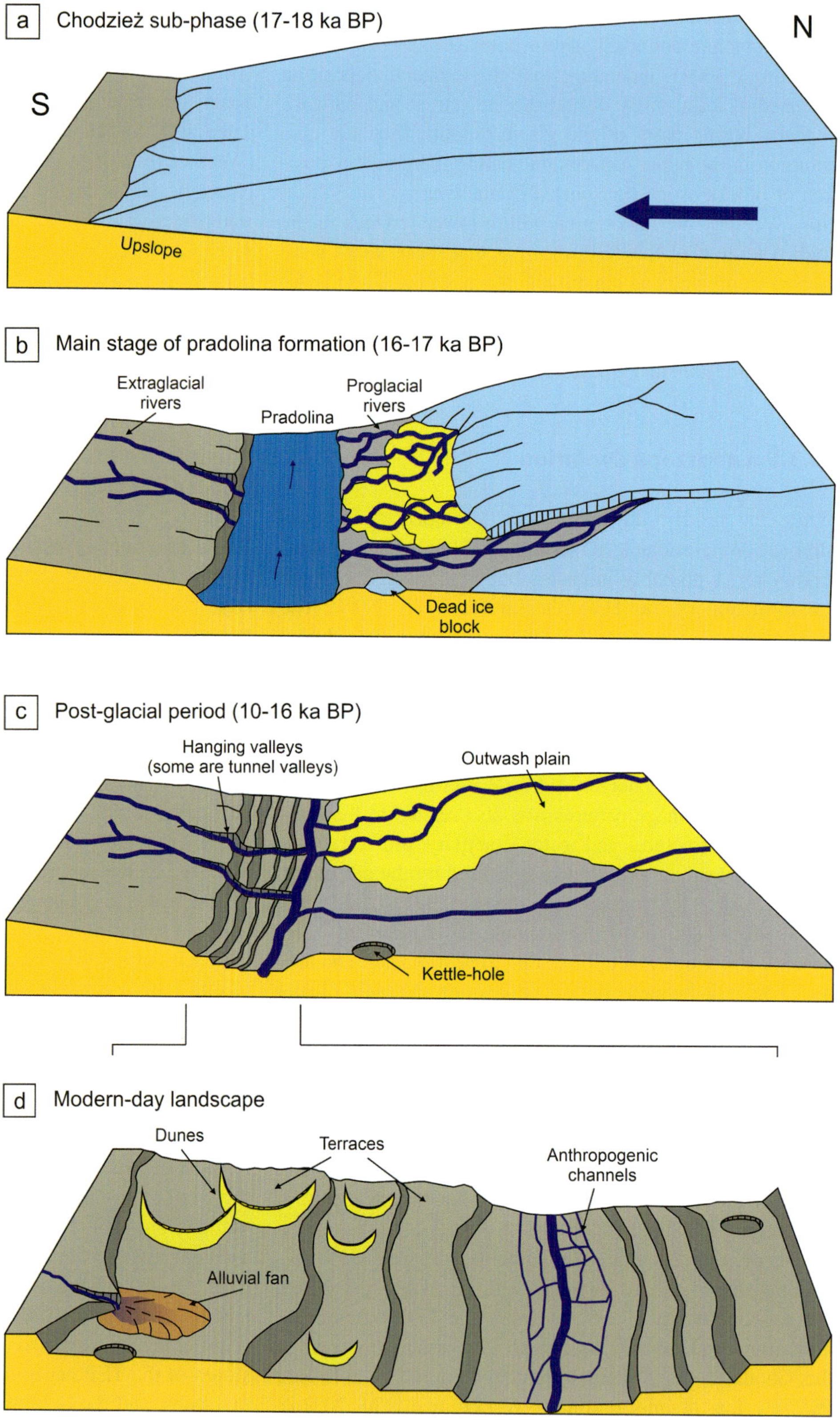

Fig. 34.9 Toruń-Eberswalde Pradolina as a multi-scale landscape feature: simplified model of development in the Nakło-Ujście section (Middle Noteć Valley)—see text for further explanations

2010). Finally, during the Holocene, V-shaped erosion valleys and small alluvial fans at their mounts developed (Jaworski 2018).

34.4 Anthropogenic Modifications of Relief of the Toruń-Eberswalde Pradolina

Human activities that strongly impacted the Noteć Valley started almost 1000 years ago. Relatively early deforestation caused an increase in erosion and surface runoff. Melioration activities began as early as in the thirteenth century; however, large-scale artificial transformations of water drainage started in the sixteenth century (Starkel 1999; Kaniecki 2007). As a result, the network of narrow channels and ponds constructed to regulate water circulation in the valley is quite dense (Fig. 34.10b). Some of the tributaries (e.g., Gwda River) still exhibit a natural meandering character (Fig. 34.10c); however, most of the course of the Noteć River channel was artificially straightened (Fig. 34.10d) and deepened, thereby enhancing drainage. The Bydgoszcz Canal (Fig. 34.10e) was built in 1777 between Bydgoszcz and Nakło nad Notecią, linking the Wisła and Noteć river systems. Despite severe anthropogenic modifications, traces of former channels are still visible in the landscape (Fig. 34.10d). Very impressive are the meandering channels preserved at the mouth of the Łobżonka River (Fig. 34.10f), where the modern-day river was channelised and is now used to charge a series of artificial fishponds.

Due to the lack of investment and maintenance, some of the melioration activities were abandoned after World War II, which resulted in re-naturalisation of large sections of the Middle Noteć Valley and the development of very diversified ecosystems. In consequence, the natural environment of the valley is protected in different forms, e.g., as Natura 2000 Special Protection Area "Middle Noteć Valley and Bydgoszcz Canal", Special Area of Conservation "Noteć Valley", and the landscape protected area "Noteć Valley". There are also numerous existing and proposed nature reserves and geosites (Ratajczak-Szczerba 2013).

34.5 Conclusions

In the present-day landscape, the Toruń-Eberswalde Pradolina constitutes a clearly visible morphological border between the Wielkopolska Lakeland and the Pomeranian Lakeland. The main element of the landscape is the till plain from the Weichselian glaciation dissected by the Middle Noteć River Valley, with glaciotectonic moraines located mainly along its northern rim, though also on the southern one, near Chodzież and Kcynia. Large outwash fans of Gwda and Brda join the TEP on the northern side, whereas in the south there are outwash plains related to meltwater activity, including impressive outwash fans and tracts near Chodzież probably developed due to glacier outburst flood.

The study area was covered by dynamic ice masses several times during the Pleistocene (Fig. 34.9a). The Weichselian glaciation largely rejuvenated an older glacial landscape, covering it with a layer of subglacial till widely observed on till plains located to the north and south of the Toruń-Eberswalde Pradolina (Fig. 34.3). Moraines situated along the southern side of the TEP were formed during the Chodzież sub-phase around 17.7 ka (Kozarski 1959, 1986, 1995a). The occurrence of streamlined bedforms (Fig. 34.5) on both the north and south rim of the Noteć Valley, combined with a clearly visible series of arcuate chains of moraines, indicate that the ice sheet advanced from the NE (from the Wisła Ice Stream) (Szuman et al. 2021a). Subsequent stages of valley development were related to local oscillations of the SIS, during which meltwater employed lowering of the terrain in the Ujście Basin (Kozarski 1959). Extensive outwash plains, developed before and during the Pomeranian Phase (Brda, Gwda, Drawa outwashes), delivered a large amount of water to the foreland of the ice sheet (Pisarska-Jamroży 2015; Mleczak and Pisarska-Jamroży 2021). As the outwash fans and ice masses blocked the extraglacial river flow towards the north, the joint drainage was redirected toward the west (Fig. 34.9b). The depositional environment consisted of large, braided rivers, which contributed to the degradation of the permafrost present in the banks, leading to lateral widening of the fluvial tract (Kozarski 1995b; Pisarska-Jamroży and Zieliński 2011; Pisarska-Jamroży 2015; Weckwerth and Pisarska-Jamroży 2015).

During the functioning of the Toruń-Eberswalde Pradolina, periods of lateral erosion and accumulation alternated with periods of incision, as reconstructed from a series of terraces (Galon 1961; Kozarski 1962a) (Fig. 34.9c). The incision was related to warmer interstadial periods of Bølling and Allerød, whereas lateral erosion and accumulation dominated during colder periods of the Pomeranian Phase, Oldest Dryas, Older Dryas, and Younger Dryas (Kozarski 1962a, 1986, 1988, 1995a). Galon (1961) also suggested that erosion rhythms were probably regular because vertical distances between terraces are very similar (about 4–5 m). Further development of the landscape included deepening of the valley which destroyed large parts of the higher terraces, activation of aeolian processes, organic sedimentation, and denudational processes on steep valley sides (Fig. 34.9d). Modern transformations of the landscape are related to human impact, comprising mainly changes in water circulation, accelerated denudation, and enhanced erosion processes.

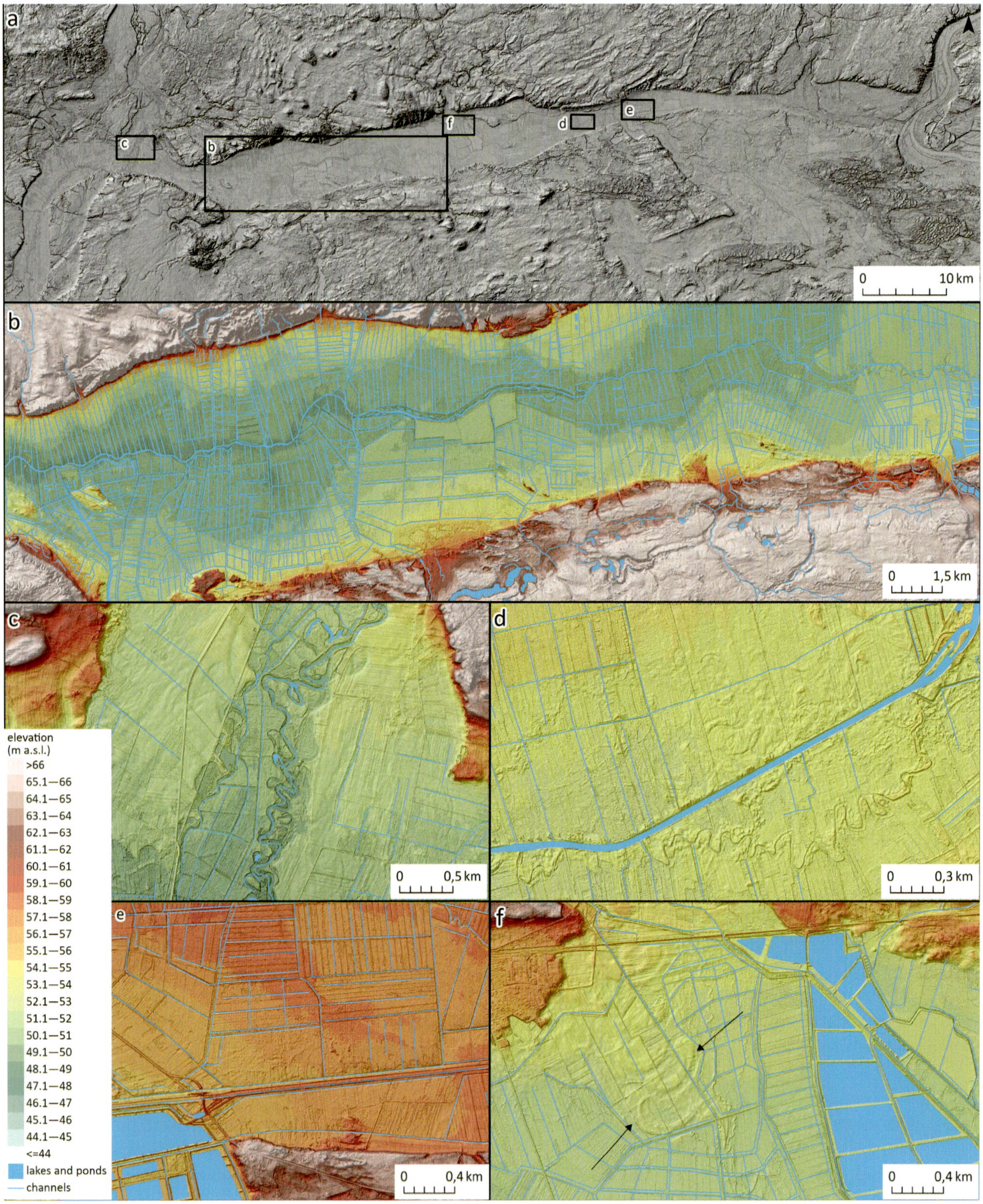

Fig. 34.10 Anthropogenic landscape modifications in the Toruń-Eberswalde Pradolina (note: elevation scale is different from those in Figs. 34.5, 34.6, 34.7 and 34.8). **a** Hillshade model with location of insets. **b** Dense network of artificial channels which were constructed to regulate water circulation in the valley. **c** The Gwda River still demonstrates meandering characteristics of a (semi)natural river. **d** The Noteć River was artificially straightened, but the course of an older, meandering channel is still visible to the south of the modern-day channel. **e** Artificial Bydgoszcz Canal, built in 1777, linked Wisła and Odra river systems. **f** Traces of former meandering river system (indicated by arrows) of the Łobżonka River (near Osiek n/Notecią)—the modern-day artificially straightened river is used to charge a large group of fish ponds

Acknowledgements The research was funded by the National Science Centre, Poland (Project number 2015/17/D/ST10/01975).

References

Andrzejewski L, Weckwerth P (2010) Dunes of the Toruń Basin against palaeogeographical conditions of the late glacial and Holocene. Ecol Quest 12:9–15

Benn DI, Evans DJA (2010) Glaciers and glaciation. Hodder Education, London, p 816

Benn DI, Fowler AC, Hewitt I, Sevestre H (2019) A general theory of glacier surges. J Glaciol 65(253):701–716

Böse M, Lüthgens C, Lee JR, Rose J (2012) Quaternary glaciations of northern Europe. Quat Sci Rev 44:1–25

Chmal R (2006) Objaśnienie do Szczegółowej Mapy Geologicznej Polski 1:50000, arkusz Śmiłowo (314), Państwowy Instytut Geologiczny, Warszawa

Cuffey K, Paterson W (2010) The physics of glaciers. Academic Press, p 704

Dzierżek J (1997) Geology of sub-quaternary basement and stratigraphy of quaternary sediments in the middle Notec river valley. Ann Soc Geol Polon 67:57–81

Ehlers J, Gibbard PL, Hughes PD (2011a) Quaternary glaciations—extent and chronology: a closer look. Developments in quaternary Sciences 15, Elsevier, Oxford, p 1108

Ehlers J, Grube A, Stephan H-J, Wansa S (2011b) Chapter 13—Pleistocene glaciations of North Germany—New Results. In: Ehlers J, Gibbard PL, Hughes PD (eds) Quaternary glaciations—extent and chronology: A closer look. Developments in Quaternary Sciences 15, Elsevier, Oxford, pp 149–162

Evans DJA (2009) Controlled moraines: origins, characteristics and palaeoglaciological implications. Quat Sci Rev 28:183–208

Ewertowski M (2009) Ice-wedge pseudomorphs and frost-cracking structures in Weichselian sediments, central-west Poland. Permafrost Perigl Proc 20:316–330

Ewertowski MW, Kijowski A, Szuman I, Tomczyk AM, Kasprzak L (2017) Low-altitude remote sensing and GIS-based analysis of cropmarks: classification of past thermal-contraction-crack polygons in central western Poland. Geomorphology 293:418–432

Gadomska S (1957) Utwory trzeciorzędowe i czwartorzedowe doliny Noteci w okolicy Krostkowa i Osieka n/Notecią koło Wyrzyska. Biul Państw Inst Geol 118:371–393

Galon R (1968) New facts and problems pertaining to the origin of the Noteć-Warta Pradolina and the valleys linked with it. Przegl Geogr 40:307–315

Galon R (1961) Morphology of the Noteć-Warta (or Toruń-Eberswalde) ice marginal streamway. Geogr Stud 29 Pol Acad Sci, Inst Geogr, Warsaw 129

Jahn A (1975) Problems of the periglacial zone. PWN, Warszawa, p 200

Jaworski T (2018) Późnoglacjalny i holoceński rozwój dolinek erozyjno-denudacyjnych na wybranych przykładach zboczy dolin i rynien w krajobrazie młodoglacjalnym Polski Północnej. Wyd Nauk UMK, Toruń, p 208

Kaniecki A (2007) Przemiany stosunków wodnych na obszarze Niziny Wielkopolskiej do końca XVIII wieku związane z antropopresją. In: Michalczyk Z (ed) Obieg wody w środowisku naturalnym i przekształconym. Wyd UMCS, Lublin, pp 304–318

Kasprzak L (1981) The facial differentiation of the tills of glacial series at Stare Kurowo, NW Poland. Quaest Geogr 1:73–90

Kasprzak L (2003) Model sedymentacji lądolodu vistuliańskiego na Nizinie Wielkopolskiej. Wyd Nauk UAM, Poznań, p 216

Kasprzak L (2010) Szczegółowa Mapa Geologiczna Polski 1:50000. Arkusz Szamocin (315). Państwowy Instytut Geologiczny—Państwowy Instytut Badawczy, Warszawa

Kasprzak L (2017) Objaśnienia do Szczegółowej Mapy Geologicznej Polski 1:50000. Akrusz Szamocin (315). Państwowy Instytut Geologiczny—Państwowy Instytut Badawczy, Warszawa

Kondracki J (2000) Geografia regionalna Polski. PWN, Warszawa, p 440

Kowalska A (1960) Paleomorfologia powierzchni podplejstoceńskiej niżowej części dorzecza Odry. Pr Geogr IG PAN 21:1–75

Kozarski S (1959) O genezie chodzieskiej moreny czołowej. Bad Fizj Pol Zach 5:45–72

Kozarski S (1962a) Recesja ostatniego lądolodu z północnej części Wysoczyzny Gnieźnieńskiej a kształtowanie się Pradoliny Noteci-Warty. Pr Kom Geogr-Geol 2, PTPN, Poznań, p 153

Kozarski S (1962b) Wydmy w pradolinie Noteci koło Czarnkowa. Bad Fizj Pol Zach 9:37–60

Kozarski S (1977) Ukierunkowane wytopiska na sandrach. Stud Geol Polon 52:235–248

Kozarski S (1986) Skale czasu a rytm zdarzeń geomorfologicznych vistulianu na Niżu Polskim. Czas Geogr 57:247–270

Kozarski S (1988) Time and dynamics of the last Scandinavian ice-sheet retreat from northwestern Poland. Geogr Polon 55:91–101

Kozarski S (1993) Late Plenivistulian deglaciation and the expansion of the periglacial zone in NW Poland. Geol Mijnb 72:143–157

Kozarski S (1995a) Deglacjacja północno-zachodniej Polski: warunki środowiska i transformacja geosystemu (~20 ka 10 ka BP). Continuo, Wroclaw, p 82

Kozarski S (1995b) The periglacial impact on the deglaciated area of northern Poland after 20 kyr BP. Biul Perygl 34:73–102

Kozarski S, Szupryczyński J (1958) Terasy pradoliny Noteci między Nakłem a Milczem. Przegl Geogr 4:671–684

Kozarski S, Kasprzak L (1987) Sedymentologiczne i geomorfologiczne podstawy analizy maksymalnego zasięgu ostatniego lądolodu w Wielkopolsce. Centralny Program Badań Podstawowych. Unpublished report from the project CPBP 03.13 Ewolucja środowiska geofizycznego Polski

Kozarski S (1965) Zagadnienie drogi odpływu wód pradolinnych z zachodniej części Pradoliny Noteci-Warty. Prace Kom Geogr-Geol 5, PPTN, Poznań, p 87

Krygowski B (1961) Geografia fizyczna Niziny Wielkopolskiej: geomorfologia. PWN, Poznań, p 204

Lang J, Hampel A, Brandes C, Winsemann J (2014) Response of salt structures to ice-sheet loading: implications for ice-marginal and subglacial processes. Quat Sci Rev 101:217–233

Liszkowski J (1993) The effects of Pleistocene ice-sheet loading-deloading cycles on the bedrock structure of Poland. Folia Quat 64:7–23

Makowska A (1980) Late Eemian with preglacial and glacial part of Vistulian glaciation in the lower vistula region. Quat Stud Poland 2:37–56

Marks L (2005) Pleistocene river systems in the southern peribaltic area as indication of interglacial sea level changes in the Baltic Basin. Quat Intern 130:43–48

Marks L (2011) Quaternary glaciations in Poland. In: Ehlers J, Gibbard PL, Hughes PD (eds) Quaternary glaciations—Extent and chronology: a closer look. Developments in Quaternary Sciences 15, Elsevier, Oxford, pp 299–303

Mleczak M, Pisarska-Jamroży M (2021) A record of deglaciation-related shifting of the proximal zone of a sandur—a case study from the Gwda sandur, NW Poland (MIS 2). J Palaeogeogr 10:12

Nowaczyk B (1986) Wiek wydm, ich cechy granulometryczne i strukturalne a schemat cyrkulacji atmosferycznej w Polsce w późnym vistulianie i holocenie. Geografia 28, Wyd Nauk UAM, Poznań, p 245

Nowaczyk B (2002) Litologiczny i morfologiczny zapis działalności wiatru w Polsce w ostatnich 30 tysiącach lat. Czas Geogr 73:275–311

Özmaral A, Abegunrin A, Keil H, Hepp DA, Schwenk T, Lantzsch H, Mörz T, Spiess V (2022) The Elbe Palaeovalley: evolution from an ice-marginal valley to a sedimentary trap (SE North Sea). Quat Sci Rev 282:107453

Paluszkiewicz R, Ratajczak-Szczerba M (2013) Charakterystyka geomorfologiczna późnoglacjalnych form denudacyjnych w obrębie południowej krawędzi Pradoliny Toruńsko-Eberswaldzkiej. In: Kostrzewski A, Rachlewicz G, Woszczyk M (eds) VI Seminarium Geneza, litologia i stratygrafia utworów czwartorzędowych. Wyd Nauk UAM, Poznań, pp 114–117

Piotrowski A (1991) The influence of sub-Quaternary basement on the development of Lower Odra Valley in Pleistocene and Holocene. Geol Quart 35:221–234

Pisarska-Jamroży M (2015) Factors controlling sedimentation in the Toruń-Eberswalde ice-marginal valley during the Pomeranian phase of the Weichselian glaciation: an overview. Geologos 21:1–29

Pisarska-Jamroży M, Zieliński T (2011) Genesis of a till/sand breccia (Pleistocene, Notec Valley near Atanazyn, central Poland). Sediment Geol 236:109–116

Przybylski T (1961) Późny glacjał w Pradolinie Toruńsko-Eberswaldzkiej. Bad Fizj Pol Zach 8:57–90

Ratajczak-Szczerba M (2011a) Budowa wewnętrzna moreny czołowej glacitektonicznie spiętrzonej kompleksu Dębowej Góry—stanowisko Krostkowko. Landf Anal 16:115–123

Ratajczak-Szczerba M (2011b) Charakterystyka geomorfologiczna obszaru położonego wzdłuż Doliny Środkowej Noteci. Landf Anal 16:99–106

Ratajczak-Szczerba M (2011c) Geologia, stratygrafia i interpretacja genetyczna glin lodowcowych wzniesienia czołowomorenowego w Wolsku Dolnym. Landf Anal 16:111–114

Ratajczak-Szczerba M (2011d) Stratygrafia, geneza i wiek osadów lodowcowych Wysoczyzny Krajeńskiej w stanowisku Dziembowo. Landf Anal 16:107–110

Ratajczak-Szczerba M (2013) Geo- i bioróżnorodność doliny środkowej Noteci i doliny dolnej Gwdy szansą rozwoju geoturystyki. Acta Geogr Silesiana 14:71–86

Ratajczak-Szczerba M, Paluszkiewicz R (2015) Analiza sedymentologiczna osadów denudacyjnych w niszy niwalnej w krawędzi Pradoliny Toruńsko-Eberswaldzkiej. Landf Anal 28:73–85

Roman M (2019) Ice-flow directions of the last Scandinavian ice sheet in central Poland. Quat Intern 501:4–20

Rotnicki K (1970) Główne problemy wydm śródlądowych w Polsce w świetle badań wydmy w Węglowicach. Pr Kom Geogr-Geol 11(2):146, PPTN, Poznań

Russell AJ, Roberts MJ, Fay H, Marren PM, Cassidy NJ, Tweed FS, Harris T (2006) Icelandic jökulhlaup impacts: implications for ice-sheet hydrology, sediment transfer and geomorphology. Geomorphology 75:33–64

Sokołowski RJ, Wysota W (2020) Differentiation of subglacial conditions on soft and hard bed settings and implications for ice sheet dynamics: a case study from north-central Poland. Int J Earth Sci 109:2699–2717

Solon J, Borzyszkowski J, Bidłasik M, Richling A, Badora K, Balon J, Brzezińska-Wójcik T, Chabudziński Ł, Dobrowolski R, Grzegorczyk I, Jodłowski M, Kistowski M, Kot R, Krąż P, Lechnio J, Macias A, Majchrowska A, Malinowska E, Migoń P, Myga-Piątek U, Nita J, Papińska E, Rodzik J, Strzyż M, Terpilowski S, Ziaja W (2018) Physico-geographical mesoregions of Poland: Verification and adjustment of boundaries on the basis of contemporary spatial data. Geogr Polon 91:143–170

Starkel L (ed) (1999) Geografia Polski: Środowisko przyrodnicze. PWN, Warszawa, p 669

Szpikowski J, Kostrzewski A, Mazurek M, Smolska E, Stach A (2008) Współczesne procesy kształtujące rzeźbę stoków. In: Starkel L, Kostrzewski A, Kotarba A, Krzemień K (eds) Współczesne przemiany rzeźby Polski. Instytut Geografii i Gospodarki Przestrzennej UJ, Kraków, pp 283–291

Szuman I, Kalita JZ, Ewertowski MW (2018) The impact of geothermal heat flux on the last Scandinavian ice sheet over W Poland and E Germany. Geogr Ann 100A:388–403

Szuman I, Kalita JZ, Ewertowski MW, Clark CD, Livingstone SJ (2021a) Dynamics of the last Scandinavian Ice sheet's southernmost sector revealed by the pattern of ice streams. Boreas 50:764–780

Szuman I, Kalita JZ, Ewertowski MW, Clark CD, Livingstone SJ, Kasprzak L (2021b) GIS dataset: geomorphological record of terrestrial-terminating ice streams, southern sector of the Baltic ice stream complex, last Scandinavian Ice Sheet, Poland. Earth Syst Sci Data 13:4635–4651

Szupryczyński J (1966) Objaśnienia do mapy geomorfologicznej 1:50000, arkusz Szamocin. Inst Geol, Warszawa

Tylmann K, Rinterknecht VR, Woźniak PP, Bourlès D, Schimmelpfennig I, Guillou V, Aster Team (2019) The local last glacial maximum of the southern Scandinavian ice sheet front: cosmogenic nuclide dating of erratics in northern Poland. Quat Sci Rev 219:36–46

Weckwerth P (2010) Evolution of the Toruń Basin in the late Weichselian. Landf Anal 14:57–84

Weckwerth P (2018) Fluvial responses to the Weichselian ice sheet advances and retreats: implications for understanding river paleohydrology and pattern changes in Central Poland. Int J Earth Sci 107:1407–1429

Weckwerth P, Chabowski M (2013) Heavy minerals as a tool to reconstruct river activity during the Weichselian glaciation (Toruń Basin, Poland). Geologos 19:25–46

Weckwerth P, Pisarska-Jamroży M (2015) Periglacial and fluvial factors controlling the sedimentation of pleistocene breccia in nw poland. Geogr Ann 97A:415–430

Weckwerth P, Wysota W, Piotrowski JA, Adamczyk A, Krawiec A, Dąbrowski M (2019) Late Weichselian glacier outburst floods in North-Eastern Poland: landform evidence and palaeohydraulic significance. Earth Sci Rev 194:216–233

Weckwerth P (2013) Ewolucja fluwialnych systemów depozycyjnych i jej uwarunkowania paleośrodowiskowe w Kotlinie Toruńskiej podczas zlodowacenia wisły. Wyd Nauk UMK, Toruń, p 206

Wiśniewski E (1990) Evolution of the vistula valley. Geogr Studies 5:146–153

Wiśniewski E (1992) Evolution of the lower vistula river valley in the upper Plenivistulian. Quaest Geogr 3:187–195

Włodek M (1980) Młodszy plejstocen w rejonie Nakła nad Notecią. Przegl Geol 28:453–456

Włodek M (1998) Szczegółowa Mapa Geologiczna Polski 1:50000. Arkusz Nakło n. Notecią (317). Państwowy Instytut Geologiczny, Warszawa

Wodziczko A (1948) Materiały do stratygrafii i analizy pyłkowej osadów w pradolinie Noteci. Bad Fizj Pol Zach 1:129–151

Wysota W, Molewski P, Sokołowski RJ (2009) Record of the Vistula ice lobe advances in the Late Weichselian glacial sequence in north-central Poland. Quat Intern 207:26–41

Leszek Kasprzak is Associate Professor at the Adam Mickiewicz University in Poznań, Poland, and the current head of the Department of Geomorphology. His main research interest includes Pleistocene and modern glacial processes, sediments and landforms, and geotourism. His fieldwork has been conducted in Poland, Sweden, and

Svalbard. He developed a regional sedimentation model of the Scandinavian Ice Sheet based on input data from geological field observations supplemented with quantitative investigations. He authored the Szamocin Sheet of Detailed Geological Map of Poland (1:50,000), which covers the central section of the Toruń-Eberswalde Pradolina. He was also one of the authors of the Geomorphological Map of Petuniabukta in Svalbard. Currently, he is focusing on modelling physical parameters of the last Scandinavian Ice Sheet and interactions between ice sheets and permafrost.

Marek Ewertowski is Associate Professor at the Faculty of Geographical and Geological Sciences at Adam Mickiewicz University in Poznań, Poland. He is interested in landscape response to deglaciation (i.e., paraglacial period). His approach combines geomorphology, glaciology, remote sensing (satellite, aerial, and UAV), and GIS to map and quantify landform changes following deglaciation at annual and decadal temporal scales. He is also involved in geomorphological mapping and reconstruction of Pleistocene glacial and periglacial features. Geographically, his works concentrate in Wielkopolska (western Poland), Svalbard, Iceland, Greenland, and New Zealand.

Izabela Szuman is Assistant Professor at the Faculty of Geographic and Geological Sciences, Adam Mickiewicz University in Poznań. Her research focuses on glacial geology and geomorphology, in relation to the last Scandinavian Ice Sheet. She was involved in projects concerning application of OSL and 10Be dating techniques in order to reconstruct the ice-sheet behaviour. Currently, she is studying glacial landforms in Siberia.

Jakub Kalita is Assistant Professor at the Faculty of Geographical and Geological Sciences, Adam Mickiewicz University in Poznań. His research focuses on the analysis of glacial landforms and geochronological databases in order to study and reconstruct ice-sheet behaviour. He also studied GNSS precise positioning techniques. Currently, his interests lie in applying numerical methods in geosciences.

Aleksandra Tomczyk is Assistant Professor at the Faculty of Geographical and Geological Sciences at Adam Mickiewicz University in Poznań. Her research focuses on understanding landscape dynamics in mountain areas at multiple temporal and spatial scales. Topics of her investigations include geomorphological mapping, quantification of geomorphological effects of glacier lake outburst flood, morphometry of alluvial and colluvial fans in Arctic, quantification of recreational human impacts, and assessment of local-scale geohazards in tropical mountains. She uses an approach that links geomorphological fieldwork with GIS and remote sensing (including unmanned aerial vehicles) to quantify landscape transformation and investigate its drivers. Geographically, her works concentrate in Svalbard, Iceland, Greenland, Peru, Colombia, the USA, and the Carpathian Mountains.

Glacial and Postglacial Landforms of the Drawsko Lakeland

35

Małgorzata Mazurek, Renata Paluszkiewicz
and Zbigniew Zwoliński

Abstract

The Drawsko Lakeland is situated in the morphogenetic young glacial zone of north-western Poland. In general terms, the landforms of this area have been shaped since the Pomeranian phase of Weichselian glaciation (MIS 2). The location of the Drawsko Lakeland in the zone of transitional temperate climate with a marked influence of the Baltic Sea determines the contemporary dynamics of geomorphological processes. There are over 250 lakes with an area of more than 1 ha in the Lakeland. Lake basins differ in size, shape, and genesis, and together with the areas of peatlands they determine the retention capacity of Pomerania region. The contemporary landscape of the Drawsko Lakeland is represented by a diverse complex of glacial landforms, including moraine hills and undulating moraine plateaus with crevasse-fill forms and kames, dissected by deep subglacial tunnel channels and valleys, and vast outwash plains with kettle hollows. Inherited glacial relief, characterised by high hypsometric diversity, is nowadays subjected to modifications as a result of the impact of fluvial (river channels, terrace systems), hillslope (channel heads, denudation hollows, erosional-denudational valleys), biogenic (peatlands) and anthropogenic processes (development of settlement, transport networks, mining activities, water management). Geodiversity of the lakeland is reflected in various types of geosites, which together with diverse vegetation determine unique and outstanding natural values of the region and underpin the diversity of nature conservation forms.

Keywords

Moraine plateau · Tunnel valley · Outwash plain · Glaciolacustrine plain · River valley · Erosional-denudational valley · Headwater alcove · Geodiversity

35.1 Introduction

The Drawsko Lakeland, located in north-western Poland, is the largest (2715 km^2, Solon et al. 2018) lake district in the Western Pomerania Lakeland macroregion (Fig. 35.1). Glacial and fluvioglacial erosion and accumulation that occurred during the Weichselian glaciation (MIS 2), followed by denudational and erosion processes in the Holocene (MIS 1), determined the shape of the contemporary relief of this area, referred to as a young glacial landscape. In the Drawsko Lakeland, glacial and fluvioglacial landscapes are hilly, undulating, and flat. They are superimposed by the Holocene valleys with floodplains and upper river terraces, as well as flat areas of biogenic accumulation. The relief is distinguished by high hypsometric diversity related to glacial, fluvioglacial, slope, fluvial, and biogenic landform complexes. With horizontally and vertically diverse sedimentary successions, such a variety of landforms creates a mosaic of soils, land cover, and land use. The topography, different lithologies of sediments, and high amounts of precipitation favour the development of peatlands (fens and bogs). In the Drawsko Lakeland, one can find acidophilic habitats of beech-oak forests with fertile lowland beech and mixed forests, while on outwash sands, there are suboceanic pine forest habitats. In waterlogged basins, there is a significant proportion of bog woodland habitats.

M. Mazurek (✉) · R. Paluszkiewicz · Z. Zwoliński
Institute of Geoecology and Geoinformation, Adam Mickiewicz University, Krygowskiego 10, 61–680 Poznań, Poland
e-mail: malgorzata.mazurek@amu.edu.pl

R. Paluszkiewicz
e-mail: reniach@amu.edu.pl

Z. Zwoliński
e-mail: zbzw@amu.edu.pl

P. Migoń and K. Jancewicz (eds.), *Landscapes and Landforms of Poland*,
World Geomorphological Landscapes, https://doi.org/10.1007/978-3-031-45762-3_35

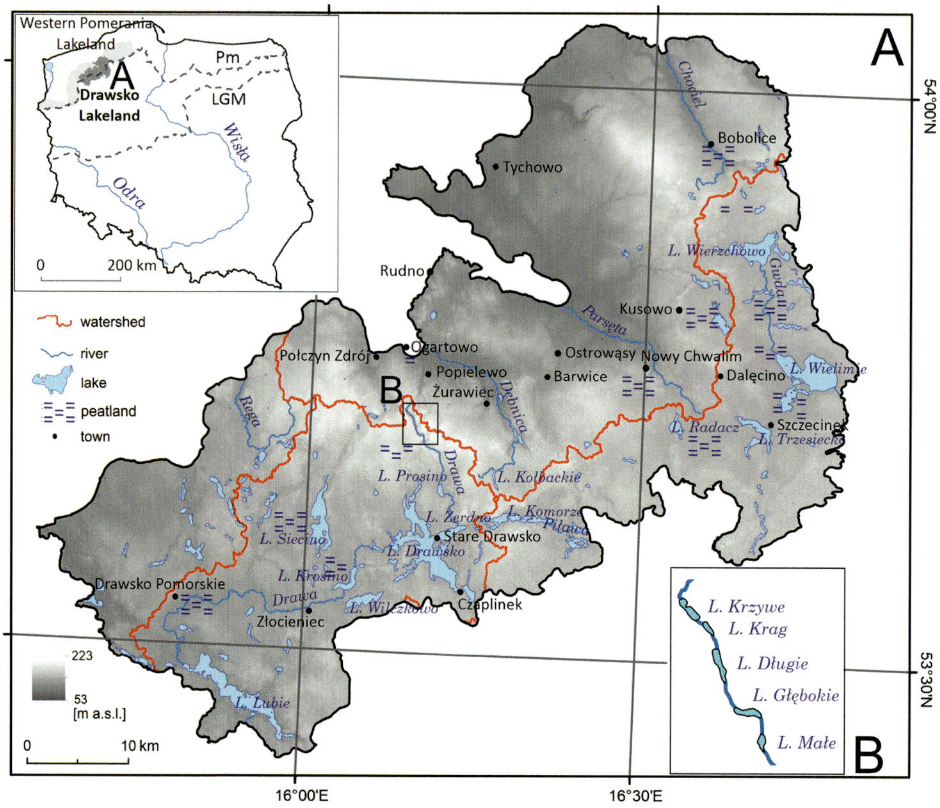

Fig. 35.1 General topographic features of the Drawsko Lakeland (based on Map of Hydrographic Division of Poland 2004) on the background of digital elevation model (DTED2, NGIA 2004). The upper inset map shows location of the Drawsko Lakeland (dark grey) within the Western Pomerania Lakeland (light grey) in Poland (after Solon et al. 2018). LGM—ice-sheet extent during the Last Glacial Maximum (MIS 2), Pm—ice-sheet extent during the Pomeranian phase of the Weichselian glaciation (MIS 2) according to Adamczyk et al. (2022). The lower inset map shows enlarged fragment of the Drawa Valley with five lakes separated by hill thresholds and kames (all maps elaborated by M. Stefaniak)

35.2 Geographical Setting

In the Drawsko Lakeland, altitudes exceed 200 m asl (Fig. 35.2). The highest elevations are Czarnkowie (223 m asl) and Wola Góra (219.2 m asl). The lowest position, at an altitude of approximately 53 m asl, is occupied by the valley bottom of the Parsęta River. Thus, elevation differences up to 173 m determine contemporary high-energy morphogenetic processes. The northern and eastern parts of the Drawsko Lakeland (55.9% of the total area) show 0–2° slopes, while the most significant slopes of 10–30° (0.7% of the area, Fig. 35.3) occur in the kettle and kame topography, in sections of tunnel and gap valleys, and along edges of moraine plateaus. Glacial tills prevail in terms of surface lithology. Other areas in the eastern, southern, and south-western parts are built of glacial sand and gravel, while in the central part of the Lakeland, there is a belt of sand, gravel, and clay of terminal moraines. Silts and clays are found in vast areas in the vicinity of the town of Złocieniec.

The zone of the highest frontal moraines of the Drawsko Lakeland constitutes a topographic watershed between the Odra Basin and coastal catchments. The rivers of the Drawsko Lakeland flowing from the southern slopes of the moraine hills are the Drawa and Gwda rivers with tributaries, while the other rivers drain the northern slopes. These are the Parsęta River with the Dębnica River, as well as the Rega River with tributaries (Fig. 35.1). The total length of the river network in the Drawsko Lakeland is 3241.3 km. The river network density, including both permanent and periodic watercourses, is 1.19 km^{-2}. The hydrological regime of the rivers is characterised by groundwater, rain, and snow supply, with floods in the spring months and the possibility of an increase in runoff in the summer months. The Lakeland is located in a zone with relatively high river unit outflow values (8–10 $dm^3\ s^{-1}\ km^{-2}$) in Poland, with a stable and high proportion of underground discharge, reaching 60–75% of the total (Stachý 1987).

An essential element of the young glacial landscape of the Drawsko Lakeland is lakes, whose basins belong to different morphogenetic types. Some are associated with tunnel valleys extended in NNW–SSE and ENE–WSW directions, whereas others occur within accumulation plains and take various shapes. They form separate lake systems or river–lake systems (Figs. 35.1 and 35.4). Lakes are often accompanied by biogenic plains. There are more than 250 lakes with an area of more than 1 ha. They differ in size, shape, and origin of the basin (Figs. 35.5 and 35.6; Table 35.1). The trough lakes are narrow, with notable depths, overdeepenings, and shallows. They have long, well-developed shorelines with peninsulas and numerous islands. Lakes within ground moraine and terminal moraine terrains occupy basins resulting from melting of dead-ice

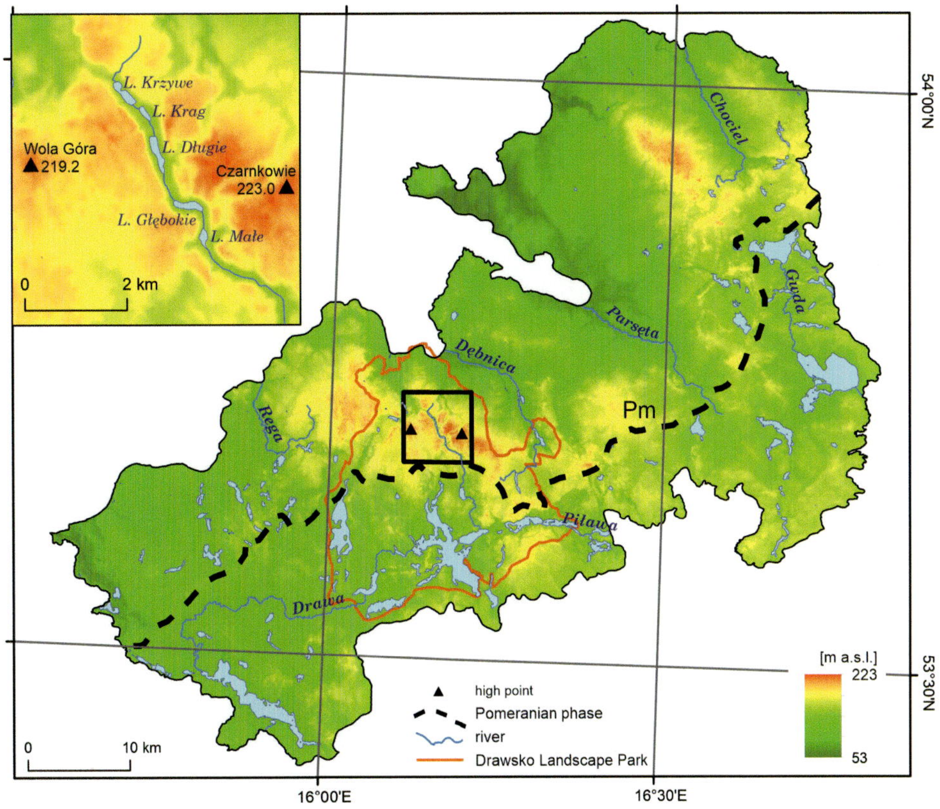

Fig. 35.2 Digital elevation model of the Drawsko Lakeland (DTED2, NGIA 2004). Pm—ice-sheet extent during the Pomeranian phase of the Weichselian glaciation (MIS 2). The borders of the Drawsko Landscape Park are marked by red line. The inset map shows the detailed DEM of the upper Drawa tunnel valley, i.e. the Valley of the Five Lakes reserve (black rectangle) (elaborated by M. Majewski)

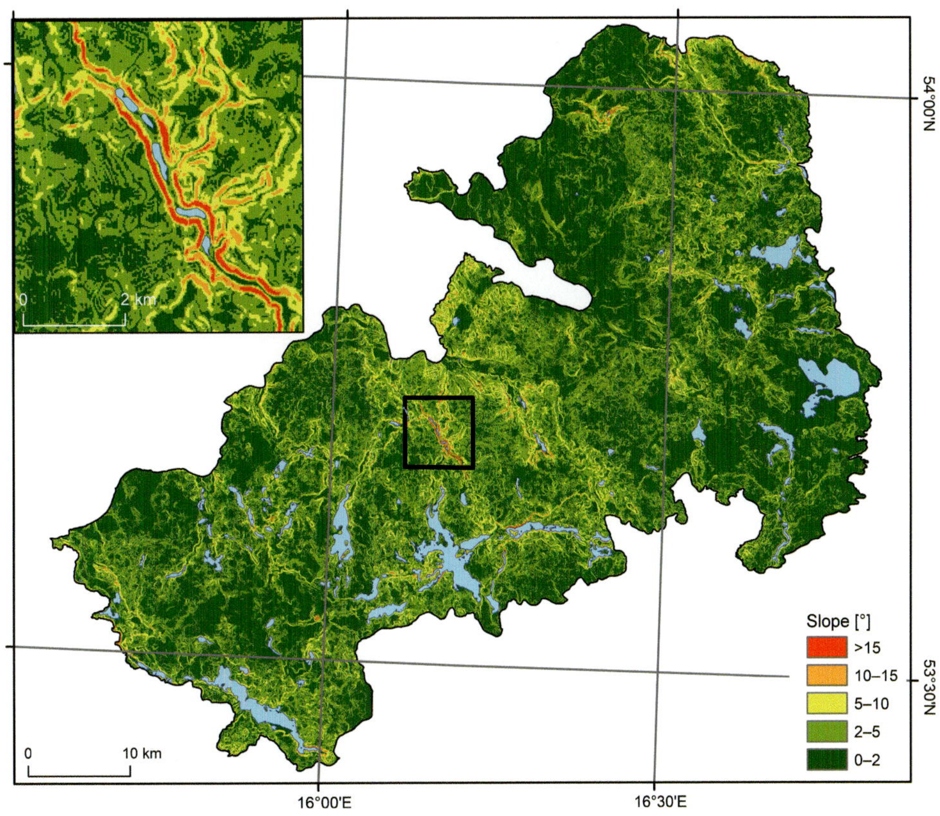

Fig. 35.3 Slope map of the Drawsko Lakeland. The inset map shows the detailed slope map of the upper Drawa tunnel valley, i.e. the Valley of the Five Lakes reserve (black rectangle) (elaborated by M. Majewski)

Fig. 35.4 Upper Drawa River connects the lakes of Głębokie and Długie situated in the winding tunnel valley. This section, called the Valley of the Five Lakes, is under protection as a nature reserve (photograph by M. Winowski)

Fig. 35.5 Kołbackie Lake located in the upper Dębnica tunnel valley is part of the Dębnica river–lake system (photograph by J. Borysiak)

blocks and have large surface areas, relatively shallow depths, and complex shorelines. The shoreline development index K ranges from 4.94 to 2.05 for both types of lakes. In addition, there are small oval kettle lakes in melt-out depressions.

An outstanding lake of the region is the Drawsko Lake, the largest and deepest lake in the Drawsko Lakeland (Fig. 35.6). In terms of depth, it is the second deepest in Poland, reaching 82.2 m (Choiński 2006). It has a significantly developed shoreline, with several long, narrow inlets (including Henrykowska, Rękicka, and Kluczewska) and several peninsulas. There are 13 small islands on the lake and one large island–Bielawa, with an area of 79.5 ha and permanent residents. In many parts, the lakeshores are steep, rising to approximately 40 m above the water level. The Drawa River flows through the Drawsko Lake.

The Drawsko Lakeland is located in a transitional temperate zone with a strong oceanic influence. The proximity of the Baltic Sea makes the winters warmer and summers cooler than in the central part of country. The mean annual temperature varies between 7 and 8 °C. Climatic diversity of the Drawsko Lakeland is

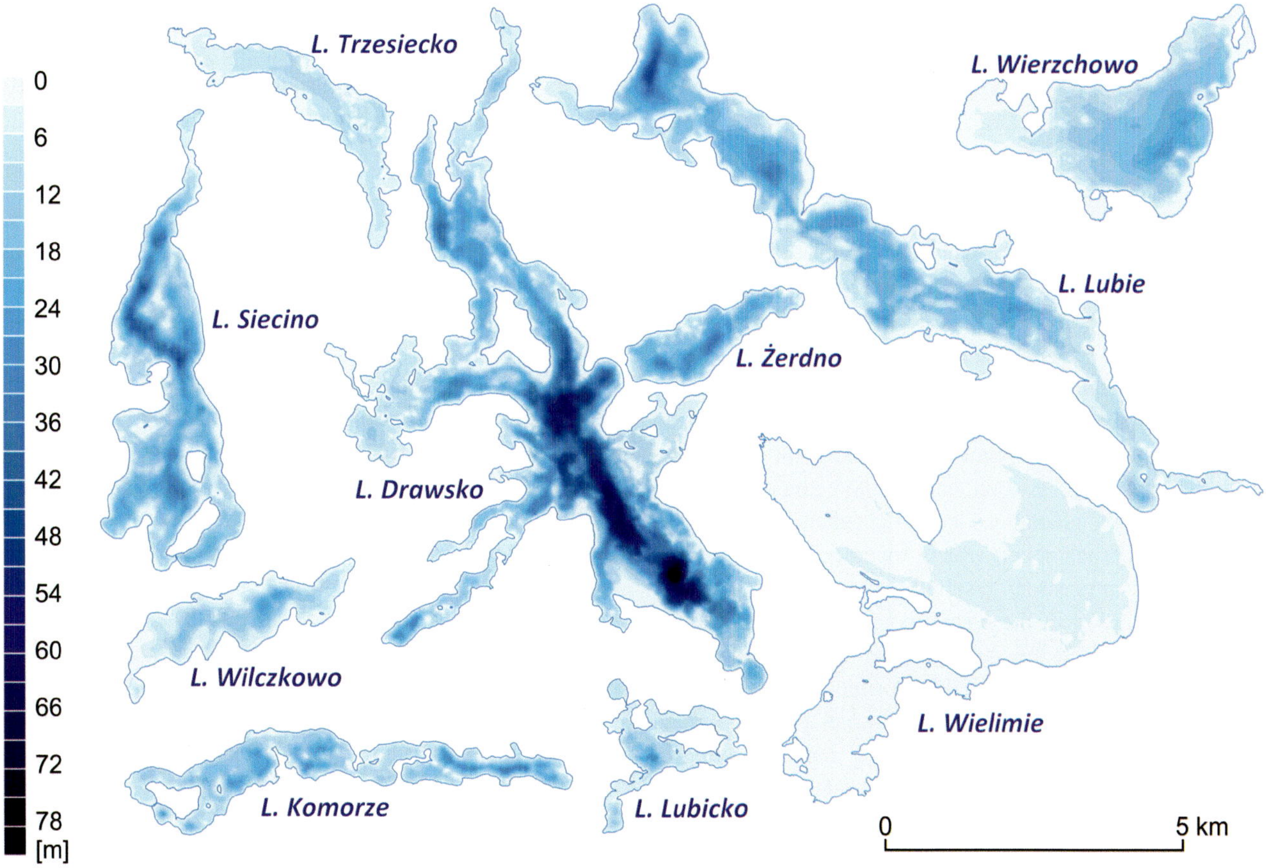

Fig. 35.6 Bathymetric plans of selected trough and moraine lakes within the Drawsko Lakeland (© EKO-MAP 2022)

Table 35.1 Morphometric parameters of selected lakes of the Drawsko Lakeland

Lake (origin)	Area	Water volume	Length of shoreline	Average depth	Maximum depth	Length	Average width	Island area
	(km^2)	(10^6 m^3)	(km)	(m)	(m)	(km)	(km)	(km^2)
Drawsko (trough lake, polygenetic)	17.97	331.4	74.2	17.2	82.2	10.1	1.78	0.849
Wielimie (kettle hole)	16.38	40.1	37.2	2.0	5.1	6.1	2.69	0.127
Lubie (trough lake)	14.87	169.9	46.5	11.8	46.0	13.9	1.07	0.234
Wierzchowo (moraine lake)	7.20	70.2	19.5	9.5	26.3	5.1	1.41	0.217
Siecino (trough lake)	7.25	104.4	25.6	13.9	44.1	7.2	1.01	0.148
Komorze (trough lake)	3.89	49.3	24.9	11.3	34.3	7.3	0.53	0.311
Wilczkowo (trough lake)	2.90	23.3	12.3	7.5	26.5	3.8	0.76	0.014
Trzesiecko (moraine lake)	2.80	16.0	15.9	5.2	12.3	5.3	0.53	0.121
Radacz (moraine lake)	1.5	8.2	6.3	2.8	10.5	3.5	0.43	no islands

Source Hydrographic Map of Poland 1:50,000

most pronounced in the total precipitation and its distribution, with relatively rare pouring rainfall and short-lasting snow cover. Exposure of the highest elevations to the north-western oceanic air masses and significant differences in relative altitudes bring annual precipitation of 700–800 mm, with two-thirds of it in the summer. The north-western slopes of moraine hills receive most rainfall each year. However, the annual precipitation totals are much lower on the south-eastern slopes, located in the rainfall shadow. In the spa town of Połczyn Zdrój the annual precipitation is 711 mm, whereas in the vicinity of the town of Szczecinek, it is 604 mm.

35.3 Pleistocene Inheritance in the Contemporary Landscape

The topography of the Drawsko Lakeland shows a strong relationship with the geological structure and the occurrence of main tectonic structures of north-western Poland (Dobracka and Piotrowski 2002). A characteristic element of the pre-Quaternary buried relief is deep erosive SE–NW-trending valleys, often replicated by younger tunnel valleys (Fig. 35.7, e.g. the upper Parsęta River, the Dębnica River). Pleistocene transgressions and recessions of the Scandinavian ice sheets, especially of the last one (i.e. the Weichselian glaciation, MIS 2), played a crucial role in transforming the relief that existed at the turn of the Quaternary. Glacial erosion and glaciotectonic processes during successive glaciations led to the enhancement of pre-Quaternary bedrock relief, which, together with the glacial accumulation processes, resulted in considerable elevation differentiation in the contemporary terrain. Currently, the elevation difference is twice that of the pre-Quaternary surface. Moreover, the Pleistocene was marked by the occurrence of glaciotectonic processes (Dobracka and Lewandowski 2002). On the other hand, accumulation processes during successive glacial and interglacial periods left sediment series of notable thickness, often exceeding 200 m in total (approximately 233 m near the village of Toporzyk; Dobracka 2017). In this area, five glacial complexes can be distinguished. These are tills, which together with glaciolacustrine and fluvioglacial sediments correspond to the following glaciations: the Nidanian (MIS 22), Sanian 1 (MIS 16), Sanian 2 (MIS 12), Odranian (MIS 6), and Weichselian (MIS 2) (Fig. 35.8).

The present-day relief of the Drawsko Lakeland results mainly from processes taking place during the transgression and frontal-areal deglaciation during the Pomeranian phase of the Weichselian glaciation (dated at ca 17–16 ka according to Marks 2012, 16.5 ka^{10}Be according to Tylmann et al. 2019). The Weichselian ice sheet, filling the Baltic Sea Basin, used tectonic predisposition of the bedrock and moved southwards in a series of ice streams, including those of the Odra, Parsęta, Słupia, and Vistula. In the Drawsko Lakeland area, the Parsęta ice stream occurred, and its advance direction was strictly conditioned by the Parsęta fossil valley and tectonics of the Permian-Mesozoic complex at depth (Karczewski 1989). The barrier to the advancing ice sheet, possibly moving in the form of a surge (Jania and Bukowska-Jania 1997), was the so-called lakeland ridge, a glaciotectonic ridge inherited from the Odranian glaciation, lying upon a bedrock elevation. In its core, there are glaciotectonically pushed Oligocene and Miocene deposits covered by Pleistocene deposits from the Odranian glaciation. Thus, the lakeland ridge played a

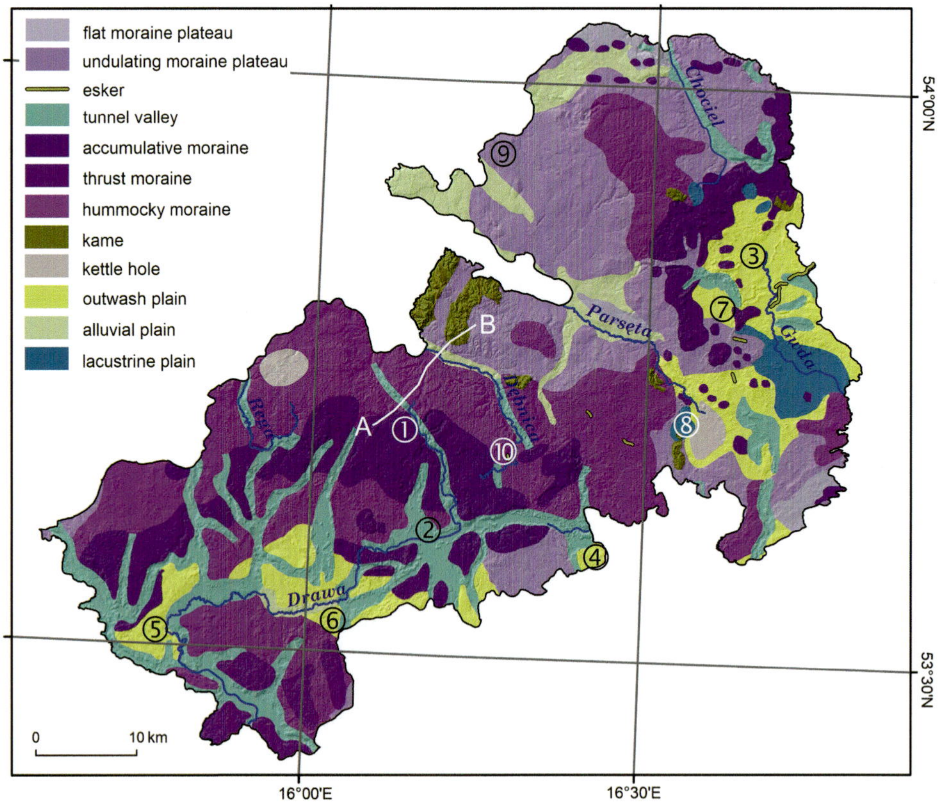

Fig. 35.7 Geomorphological map of the Drawsko Lakeland (after Starkel 1980, modified) on the background of hillshaded digital elevation model (DTED2, NGIA 2004; elaborated by M. Majewski). The geological cross-section A–B is shown on Fig. 35.8. Numbers 1–10 refer to geosites: 1—upper Drawa tunnel valley, 2—Drawsko Lake tunnel valley, 3—Gwda outwash plain, 4—Piława outwash plain, 5—Drawa outwash plain, 6—Złocieniec ice-dammed lake deposits, 7—Kusowo Baltic raised bog, 8—Nowy Chwalim peatland, 9—Trygław erratic boulder, 10—upper Dębnica River gap

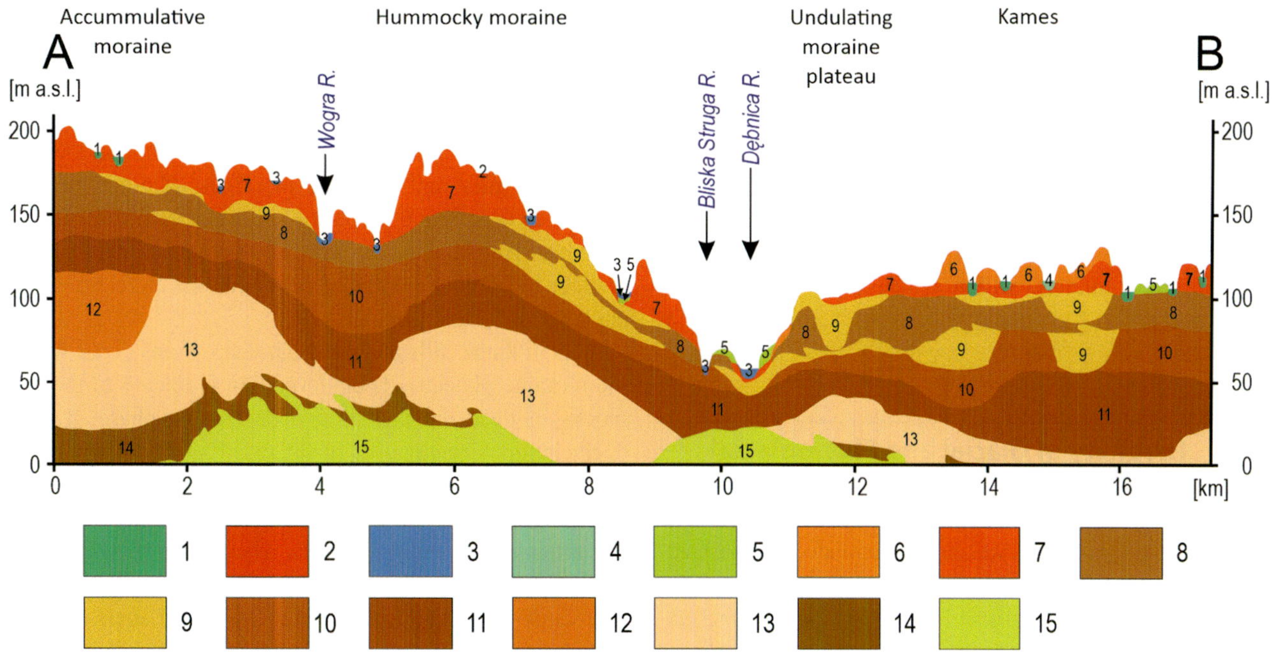

Fig. 35.8 Geological cross-section (after Dobracka 2009, modified). The line of cross section is marked on Fig. 35.7. Holocene (MIS 1): 1—biogenic deposits, 2—alluvia and sand of closed depressions, 3—fluvial sand and silt in valley floors and floodplains; Pleistocene, Weichselian glaciation (MIS 2): 4—sands and gravels, locally silts of melt-out basins, 5—fluvial and fluvioglacial sand and gravel of meltwater valley, 6—sand, gravel, silt and clay of kames, 7—end-moraine till (Pomeranian phase), 8—till (LGM and Poznań phase), 9—fluvioglacial and glaciolacustrine silt, sand, gravel and boulders; Wartanian upper stadial (MIS 6): 10—glacial till, fluvioglacial sand and gravel; Wartanian lower stadial (MIS 6): 11—glacial till and fluvioglacial sand, gravel and boulders; Lubavian interglacial (MIS 7): 12—sand with gravel, with organic debris and wood fragments and fluvial silt; Odranian glaciation (MIS 6): 13—glacial till and fluvioglacial and glaciolacustrine clay, silt, sand, gravel and boulders; Sanian 2 glaciation (MIS 12): 14—till; Miocene: 15—quartz-mica sandstones, silt, clay and brown coal

crucial role in controlling the extent of the ice sheet of the Pomeranian phase. The morphology of the marginal zone in the Drawsko Lakeland is complicated because there are two separate ice-sheet stagnant zones in the area, marked by two complexes of frontal moraines and/or moraine-kame hills. The first (outer, southern) zone is associated with the maximum extent of ice-sheet front in the Pomeranian phase, whereas the second (inner, northern) zone is associated with a more prolonged ice-sheet stagnation, marking the line of the highest elevations of the Pomeranian phase. After the transgression, multi-stage frontal and areal deglaciation occurred. It included the division of the ice sheet into an outer dead part to the south of the ridge, and an inner northern part, remaining in dynamic equilibrium between the alimentation and ablation of ice masses (Jania and Bukowska-Jania 1997). Supraglacial outflow prevailed during the transgression and stagnation of the ice-sheet front, leading to the accumulation of terminal moraines. As a result of frontal deglaciation, ice-moraine ridges were formed, transformed into ablation moraines after melting of ice cores. However, areal deglaciation led to the formation of dead-ice moraines, crevasse-fill forms, and undulating plateaus. In the foreland of most frontal moraine hills, vast outwash plains were built by proglacial waters flowing from ice tunnels. The proximal zone of the outwash plains also determined the extent of the front of the ice sheet.

Along with the recession of the Pomeranian phase ice sheet, the development of the river network on the northern slope of Pomerania began. In the Drawsko Lakeland, the development of valley network occurred through the gradual incorporation of depressions formed as a result of glacial, fluvioglacial, and periglacial processes into the fluvial outflow system. The end of the subaerial deglaciation meant the end of the most important stage in the formation of the current valley and drainage systems in Pomerania (Lewandowski and Nita 2008).

During the Late Glacial-Holocene transition, when permafrost still existed, intense slopewash, linear erosion, and solifluction were active under periglacial conditions. This led to the formation of numerous denudational and erosional-denudational valleys. Slope processes shaped the valley sides. Milder climatic conditions and gradual degradation of permafrost with simultaneous expansion of forest vegetation limited the role of surface runoff in relation to underground outflow. Research on biogenic sediments of the spring-fed fen in Bobolice (for location see Fig. 35.1) indicated final permafrost degradation and the consequent unblocking of vertical circulation of groundwater in the

Preboreal period (ca. 11,450 cal BP, Dobrowolski et al. 2019). In contrast, in a spring-fed fen in Ogartowo (for location see Fig. 35.1), located approximately 50 km to the west, there is evidence of the presence of permafrost in the substrate during the Preboreal period, that is, at approximately 10,900 cal BP (Dobrowolski et al. 2019). Late Glacial retouching mainly affected erosional scarp edges of moraine plateaus and slopes of moraine hills, kame stools, and subglacial tunnels, within which periglacial denudational troughs and erosional valleys with alluvial fans at their outlets were formed.

35.4 Diversity of Glacial and Fluvioglacial Landforms

Currently, the Drawsko Lakeland is dominated by the landscape formed by a diverse set of landforms of glacial origin, which include moraine hills and vast gently undulating glacial and fluvioglacial accumulation plains (Fig. 35.7). These landforms are dissected by deep tunnel valleys and ice-marginal valleys.

35.4.1 Moraine Plateaus

The Drawsko Lakeland is dominated by hummocky and undulating accumulative moraine plateaus that stretch to the north of the vast marginal zone (Figs. 35.7 and 35.9). The moraine plateau (70–110 m asl) descends with a distinct slope to the north, giving way to the bottoms of melt-out basins at 110–115 m asl (Barwice Basin) and 85–90 m asl (Połczyn Zdrój Basin) and to the next, lower level of the plateau (60–100 m asl, Fig. 35.10) (Karczewski 1985). The levels of moraine plateaus are diversified by hillocks of terminal moraines and dead-ice moraines. These elevations mark the consecutive stages of ice-sheet frontal recession. The moraine plateaus are built of ablation till with a large number of boulders, while at culminations, there are different sand and gravel deposits.

Sets of kame hills and melt-out depressions occur within the moraine plateaus. Their formation is related to the vast blocks of dead ice that melted during the phase of areal deglaciation. Crevasse-fill forms exist between these melt-out depressions, referred to by Karczewski (1989) as kame moraines. Kames are grouped into complexes that consist of numerous mounds and hills. They have an oval shape and a relatively flat top, and are built from sandy-gravelly and sandy-silty sediments. In the Drawsko Lakeland, there are also eskers that accumulated in the subglacial tunnels. These landforms take the shape of ridges and hill sequences, ranging from a few to several dozen metres in height, with steep slopes and winding ridgelines.

In the geological structure of frontal moraine hills and moraine plateaus in the Drawsko Lakeland, the proportion of erratic boulders found in large local clusters is significant. Various petrographic types of in situ erratic boulders make it possible to determine the area of their alimentation and indicate the direction of ice-sheet transgression. In the glacial sediments of the Pomeranian phase in NW Poland,

Fig. 35.9 Agricultural landscape of undulating morainic plateau near the towns of Barwice (upper panel) and Połczyn Zdrój (lower panel). Undulating moraine plateau is an extensive glacial surface with relative heights of 2–5 m and slopes of 2–5°, composed of subglacial tills (photographs by Z. Zwoliński)

Fig. 35.10 Landscape of a flat moraine plateau with a small kettle hole, which is a local accumulation base for denudational processes. Visible is strong waterlogging of slopewash sediments filling the bottom of the depression; covered with rush vegetation (photograph by J. Borysiak)

30–40% of Scandinavian erratic boulders are crystalline rocks, while 60–70% are sedimentary (Górska-Zabielska 2008). Their source areas are Småland and Blekinge regions in south-eastern part of Sweden and the Åland Islands (Górska-Zabielska 2013). The largest erratic boulder Trygław is located in the small town of Tychowo, on the northern border of the Drawsko Lakeland (Figs. 35.7—geosite no. 9, 35.11).

35.4.2 Tunnel Valleys

In Pomerania, a trellis pattern of tunnel and ice-marginal valleys is noticeable. Tunnel valleys run almost from NNW to SSE and from NW to SE and host trough lakes and fens. These are generally narrow (100–1000 m) and deep (30–80 m) incisions, with very steep slopes and diversified bottom morphology with numerous thresholds and

Fig. 35.11 Trygław (the name in honour of the Slavic god meaning "three heads") in Tychowo is the largest erratic boulder in Poland (photograph by P. Migoń)

overdeepenings, a few to a dozen kilometres long. The tunnel valleys are filled by lakes (e.g. Lubie Lake–the longest one, Siecino Lake) or are used by rivers (e.g. the upper Parsęta and the Chociel rivers) or river–lake system (the upper Drawa River, the Dębnica River) (Fig. 35.7).

Tunnel valleys were eroded by meltwater flowing almost perpendicular to the ice-sheet margin, and their course is often dictated by the presence of faults in the bedrock. During the multiple-stage retreat of the Pomeranian phase ice sheet, subglacial drainage ceased, subglacial tunnels were degraded and filled with dead ice. The occurrence of dead ice within the tunnel valleys protected these landforms from burial, whereas the final melting of ice caused exhumation of the valley and its appearance in the landscape. Consequently, the tunnel valleys are inconsistently incised into the moraine plateaus and outwash plains.

Among these erosional landforms, the upper Drawa tunnel valley, called the Valley of the Five Lakes from small lakes filling the deepest parts of the channel (Fig. 35.7—geosite no. 1) and separated by bottom thresholds and kames, stands out (Figs. 35.1, 35.2 and 35.4). The morphology of the tunnel valley indicates that its formation was the result of erosive activity of the extreme-pressure discharges of subglacial waters. These lakes are part of the Drawa River fluvial system. The tunnel valley cuts the zone of the moraine hills along approximately 8 km. The floor is relatively narrow (100–500 m wide), with steep slopes rising 30–70 m above it and dissected by erosional-denudational valleys (Fig. 35.3).

The polygenetic Drawsko Lake is a unique phenomenon that combines intersecting channels with meridional and latitudinal courses (Fig. 35.7—geosite no. 2, 35.12). The meridional tunnel valley is located within the wide zone of the Pomeranian phase limited by the outer, southern maximum extent of the Pomeranian phase and the inner, northern line of the highest elevations of the Pomeranian phase (Figs. 35.2 and 35.7). The maximum depths of the Drawsko tunnel valley are between 100 and 110 m (Fig. 35.6, Table 35.1). The present-day basin of Drawsko Lake is the result of subglacial erosion, probably during the earlier stages of Weichselian glaciation. This depression then widened and deepened due to glacial erosion during the transgression of the ice stream of the Pomeranian phase. It took several thousand years for the buried blocks of dead ice to melt and gradually this part of the lake was shaped. The course of the subglacial valley resembles the buried valley present in the pre-Quaternary relief (Dobracka and Lewandowski 2002; Lewandowski et al. 2008). During subsequent glacial-interglacial cycles, the valley experienced repeated phases of burial with fluvioglacial and glaciolacustrine deposits and dissection. In its central section, the tunnel valley of Drawsko Lake is crossed by the longest latitudinal ice-marginal valley of meltwater flowing parallel to the front of the subsequent ice sheets, with a total length of over 40 km. The latter is filled with lakes: Wilczkowo Lake—Henrykowska Inlet of Drawsko Lake—Żerdno Lake—Komorze Lake—Pile Lake (Marsz 1973, compare Figs. 35.1 and 35.12). In the latitudinal section of the marginal valley, north of Lake Komorze, 57 parallel curvilinear ridges in three swarms have been found by Adamczyk et al. (2022) at the research site of Łubowo. They suggest that these glacial curvilineations are erosional remnants of antecedent topography carved by subglacial channelized meltwater flows.

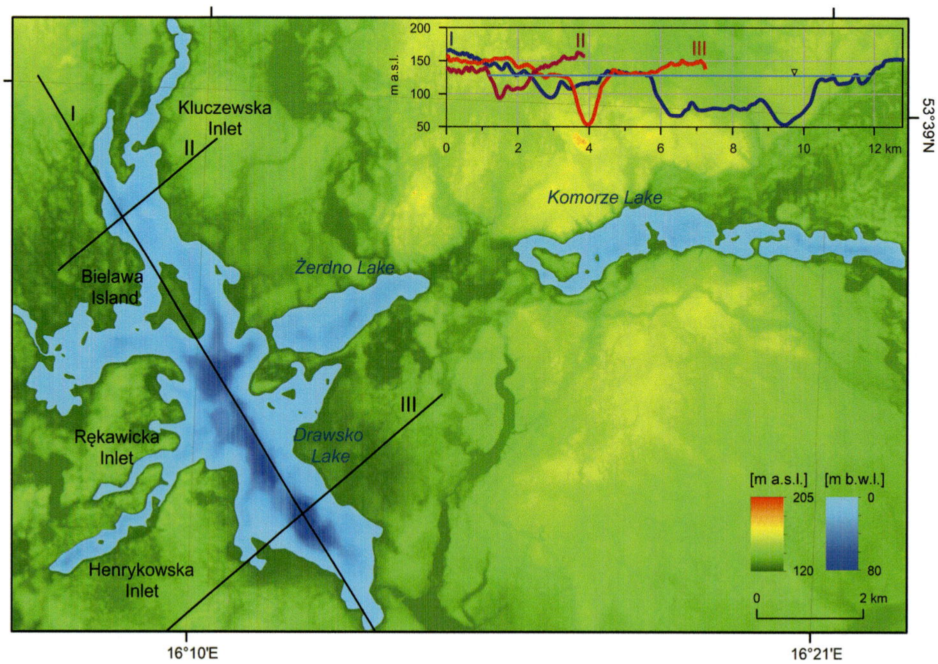

Fig. 35.12 Digital elevation model (LiDAR ISOK, Geoportal.pl) and bathymetric plan of the meridional Drawsko Lake intersected by the latitudinal ice-marginal valley of the Henrykowska Inlet—Żerdno Lake—Komorze Lake. The inset shows longitudinal (I) and transverse morphologic profiles (II, III) (elaborated by M. Majewski)

35.4.3 Outwash Plains

In the foreground of the marginal zone of the Pomeranian phase, there are multi-level outwash plains that were formed by proglacial waters (Karczewski 1989; Mojski 2005). Three large outwash plains occur in the Drawsko Lakeland: the Gwda, Piława, and Drawa outwash plains (Fig. 35.7—geosites no. 3–5), all with their proximal parts next to the Pomeranian frontal moraines (Maksiak and Mróz 1978). The outwash plains were accumulated in two stages: during the maximal (southern, outer line) and main (northern, inner line) extent of the Pomeranian phase formed during ice-front standstills and periods of significant ablation. Large amounts of meltwater flowing through the tunnels dissected the terminal moraine ridge, creating glacier gates. In their foreground, sandy-gravel material accumulated on the ice and between blocks of dead ice (Lewandowski et al. 2003, 2008; Popielski 2005, 2006).

The outflow of proglacial waters to the south in the eastern part of the Drawsko Lakeland formed the Gwda outwash plain (Fig. 35.13); in the central part, it accumulated the Piława outwash plain, and in the western part the Drawa outwash plain came into being. The proglacial waters continuing southward flowed into the Toruń-Eberswalde Ice-Marginal Valley (pol. *pradolina*), feeding a large drainage system that operated in the foreland of the Pomeranian marginal zone. Outwash plains developed at two primary levels. The upper (older) level begins at the terminal moraines and gently descends towards the south. Its surface is punctuated with melt-out forms of various sizes (pitted outwash plain), which indicates accumulation of outwash deposits in the presence of dead-ice blocks. An example of such a surface is the outwash plain north of the Komorze-Pile lake system (Szafraniec 2013). The lower (younger) outwash plain starts inside the arc of the frontal moraines, then dissects the marginal forms of the external arc and cuts into the proximal part of the upper outwash level to a depth of several metres. Sandy-gravel lithofacies dominate the internal structure of the outwash plains, and there is a clear relationship between the grain size of the sediments and the distance from the apex of the outwash. The proximal part has a large proportion of gravel, whereas the distal part is dominated by fine- and medium-grained sand. The maximum thickness of the Drawa outwash plain deposit is 23 m. The lowest thickness (4–5 m) occurs in the outwash plain at the edge of the Złocieniec glaciolacustrine plain.

Fig. 35.13 Fluvioglacial material extraction in the Gwda outwash plain near the village of Parsęcko. **a** Overview of the exposure. **b** Different facies of fluvioglacial deposits (photographs by Z. Zwoliński)

35.4.4 Glaciolacustrine Plains

In the foreground of the Pomeranian phase, in the vicinity of the settlements of Złocieniec and Wierzchowo, two separate proglacial glaciolacustrine basins existed. The Złocieniec glaciolacustrine plain covers an area of 26 km^2 (Fig. 35.7—geosite no. 6). The sediments accumulated in a basin with significant bottom denivelations, resulting in varying thicknesses. The glaciolacustrine series, represented by clays and varved silts up to a maximum thickness of 30 m, is covered with fluvioglacial sands of the Drawa outwash plain. In the vertical profile of the glaciolacustrine sediments, three primary lithofacial complexes are distinguishable, formed in specific sedimentary conditions (Paluszkiewicz 2004; Paluszkiewicz and Paluszkiewicz 2005). The initial phase of glaciolacustrine basin development is represented by fine-grained sand, which is the coarsest sediment in the basin. The main, longest stage is reflected in the sedimentary complex represented by varved clays, whose thickness ranges from 5 m in the eastern part to 15 m in the western part of the basin. The sediments of this complex were formed due to the suspension (parapelagic) falling in the annual rhythm of ice sheet ablation (Fig. 35.14).

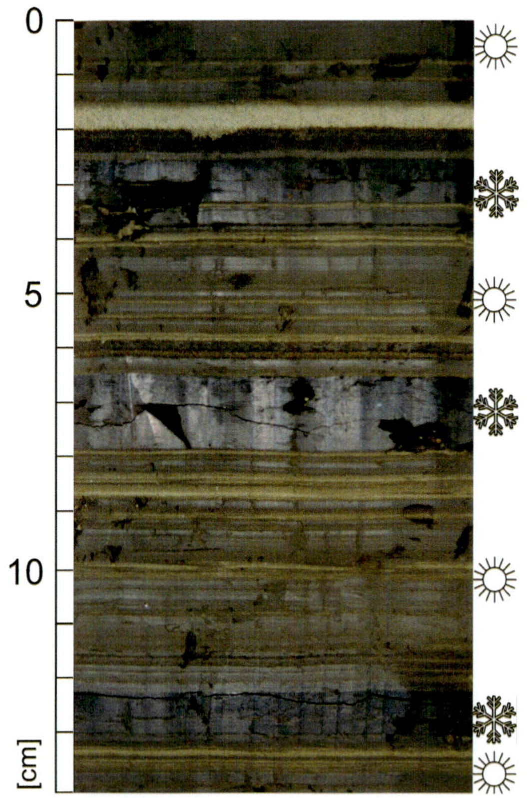

Fig. 35.14 Rhytmically-laminated varves in the sedimentological cross-section of the Złocieniec glaciolacustrine plain. Cyclothems of light and dark laminae correspond to the longest stage of lake development. Sun symbol—light summer laminae, snowflake symbol—dark winter laminae (photograph by R. Paluszkiewicz)

The final phase of development is represented by a massive series of coarse silty clay or fine silt at the top, disturbed by syn- and post-sedimentation processes. The number of varves preserved in the fossil depression could be the basis for estimating the duration of the glaciolacustrine basin functioning. Within the Złocieniec glaciolacustrine plain, more than 385 deposition cycles have been distinguished, referring to the annual sedimentation cycle, which can be called years (Paluszkiewicz 2004).

35.5 Holocene Landscape

The intensity of denudational processes and their morphogenetic role in the Drawsko Lakeland changed during the Holocene, depending on climatic conditions, related vegetation cover, water circulation rate, and human activity. The present-day relief was shaped mainly by slope denudational processes, fluvial erosion and accumulation, intense downcutting in the upper reaches of rivers, building-up of floodplains and bank erosion in the lower reaches of rivers, and both degradation and aggradation processes caused by human activity (Kostrzewski et al. 2021).

The vast moraine plateaus and outwash plains are dissected by river valleys that are generally oriented longitudinally and latitudinally, often following tunnel valleys or meltwater outflow routes (Sylwestrzak 1978). Hence, many valleys are polygenetic, and in their longitudinal profiles, one can observe alternating sections of fluvial, tunnel, and/or melt-out genesis (Piasecki 1982). These valleys dissect the moraine plateau to a depth of 40 m in the Dębnica River valley, up to 50 m in the Bliska Struga and Wogra River valleys, and up to 70 m in the upper Drawa River valley.

River valleys with a course similar to latitudinal are wider and usually include benches of upper terraces, often of Late Glacial age. River channels have a winding and/or meandering pattern, usually cut into their own alluvia of floodplains and less often directly into the Quaternary substrate (Fig. 35.15a). They have a significant gradient up to 23‰, and their bottoms often contain lag moraine gravels (pavements), forming an armoured bed (Fig. 35.15b). During flooding, overbank accumulation is locally observed. In the Drawsko Lakeland, the rhythm of fluvial processes is climatically conditioned, and in the last 3000–4000 years, it has been disturbed to varying degrees by human interference (Szpikowski 2010).

The revival of slope processes due to anthropogenic activity occurred in the Sub-Atlantic period and again in the early Middle Ages and the Little Ice Age (Szpikowski 2010). In the edge zones of moraine plateaus and on valley slopes, poorly branched erosional valleys developed (Fig. 35.15c), while in the upper sections of older denudational and erosional-denudational valleys, dells and

Fig. 35.15 Fluvial landforms in the Drawsko Lakeland. **a** Meandering river channel of the Lubiatówka River (tributary of the Dębnica River). The bottom and lower parts of the river valley are covered with sub-Atlantic oak-hornbeam forest (*Stellario holosteae-Carpinetum*) and the upper parts with beech forests (*Fagion sylvaticae*). **b** Armoured bed of the upper Dębnica River channel in the section of gap valley, covered with sub-Atlantic oak-hornbeam forest (*Stellario holosteae-Carpinetum*). **c** Ostrowąsy—one of the erosional-denudational valleys on the slopes of the moraine plateau covered with beech forest (*Fagion sylvaticae*). **d** Elliptic in plan, compact headwater alcove developed by seepage erosion next to the village of Rudno, located on the slope of a kame within beech forest (*Fagion sylvaticae*) (photographs a by J. Borysiak; b and d by M. Mazurek; c by R. Paluszkiewicz)

gullies developed, which were deepened and lengthened (Paluszkiewicz 2016).

Currently, the creation of new landforms of linear erosion, mainly of an episodic or ephemeral nature, has been observed on agriculturally used slopes. Their formation is favoured by morphological conditions, physical properties of the soils, poorly developed vegetation cover, and agrotechnical facilities and procedures (Kostrzewski et al. 1989). The development of denudational processes is limited by afforestation of high-energy relief zones, which results in the stabilisation of valley forms. Significant changes in the morphology of the valleys occur because of erosion processes triggered by extreme weather events.

Deepening of some denudational and erosional landforms has led to the dissection of aquifers and drainage of groundwater onto the surface, thus initiating permanent flows in formerly dry valleys. As a result of seepage erosion, headwater alcoves develop in the groundwater outflow zone, which constitutes the channel heads of the 1st order valleys (Mazurek 2010, 2011). Groundwater flows cause headward erosion (elongation of watercourses) and branching of an existing network through the evolution of tributary valleys. As a result of seepage erosion, headwater alcoves have developed, among which amphitheatric, poorly branched forms predominate (Fig. 35.15d). In spring hollows cut in loose sediments, mainly of glacial and fluvial origin, seepage erosion is manifested by undercutting of slopes and deepening of the bottom. In the footslope zones, emerging groundwater also leads to undercutting. These undercuts with inclinations of 38° to 46° are further shaped by gravity-driven processes. The slopes of headwater alcoves recede because of repeated episodes of seepage erosion and mass movements.

Research in the river catchments of the Drawsko Lakeland has made it possible to determine the domination of chemical denudation over mechanical one (Zwoliński 1989). In the upper Parsęta catchment, the rate of chemical denudation exceeds 100 t km^{-2} a^{-1}, whereas the rate of mechanical denudation does not exceed 10 t km^{-2} a^{-1} (Kostrzewski and Zwoliński 1992; Kostrzewski et al. 1994). The low outflow of suspended solids and bedload results from the limited contact of slopes with the river channels, which is manifested by low supply of mineral material from the catchment. By contrast, the high amount of dissolved substances included into fluvial transport is mainly through leaching of calcium carbonate-rich Quaternary sediments and groundwater transport. This process occurs most intensively at the foot of slopes and in topographic hollows with high groundwater level.

Among the modern morphogenetic processes shaping the Drawsko Lakeland, in addition to fluvial processes of bed and bank erosion and overbank sedimentation, one should also mention slopewash and linear wash (soil erosion), aeolian and niveo-aeolian processes, littoral processes within the lake basins, and biogenic accumulation in the form of fens and peat bogs (Figs. 35.1 and 35.16).

Fig. 35.16 Baltic-type raised bog of Kusowo Bagno—one of the best preserved active raised bogs in Poland. It was formerly a lake; a layer of gyttja is covered by a thin layer of fen peat, overlain by a thick (up to 12 m) layer of bog peat (Cedro and Sotek 2016). **a** Bog woodland (*Vaccinio uliginosi-Pinetum*), tourist path has been built in the southern part of the bog, while the northern part is under strict protection. **b** Phytocenosis of *Sphagno recurvi-Eriophoretum vaginati* (photographs by J. Borysiak)

Steep slopes of large trough lakes are modelled by wave activity. Mass movements and sheet/rill wash intensively modelled these slopes in both the Late Pleistocene and Holocene, and now they shape the littoral zones of the lakes. A typical process is the overgrowth of lakes, their transformation into fens, and creation of biogenic accumulation plains. Lake plains were formed on the outskirts of the lakes and are built by lake chalk, gyttja, sand, and silt. Calcareous tufa is formed due to leaching of calcium carbonate from glacial sediments and its precipitation in the zones of groundwater outflow, including spring-fed fens (Dobrowolski et al. 2019).

35.6 Anthropocene Landscape

Anthropogenic landforms of different origin and ages occur in the Drawsko Lakeland. Over the last 200 years, there has been a rapid increase in human interference in river systems. These activities included, among other things, regulation of river beds, drainage of wetlands (first in the river valleys and then in the moraine plateaus), and lowering of lake levels. A network of ditches and canals increased the natural river network in the drained areas, and as a result, the density of the local river network may have increased threefold, as in the Perznica River catchment, tributary of the Parsęta River (Szpikowski 2010). Regulatory and drainage works that changed the water-table level, length, slope, and longitudinal profile of river channels have also disturbed erosion–accumulation balance in the river valleys. Anthropogenic denudation, especially mechanical one, within the Drawsko Lakeland has been present since the Middle Ages and especially in the last 200 years (Szpikowski 2010).

There are numerous anthropogenic landforms in the Drawsko Lakeland such as road and rail embankments and cuttings, and levelled surfaces to allow buildings to be erected, sometimes made up of a thick series of anthropogenic ground. From the SW to the NE, a strip of ditches and embankments of an unfinished motorway from Berlin to Königsberg dating from the third and fourth decades of the twentieth century is currently used as a cycling path or is covered with forest. The same is true in the case of railroads closed down in recent years, whose tracks, often laid on embankments and in trenches, have been turned into cycling paths. In the Drawsko Lakeland, mining and mineral processing are developing locally, focused on exploiting natural sandy-gravel aggregates, clay and silt for construction ceramics, and lake marl, which is extracted periodically. In the village of Nowy Chwalim, the largest peatland in use (resource ca. 2.7 million m^3) in the

Drawsko Lakeland is found (Figs. 35.1 and 35.7—geosite no. 8, 35.17). The Drawsko Lakeland is partly occupied by military training grounds located south of the Lubie Lake. Covering an area of approximately 340 km^2, this is one of the largest areas used for training ground troops in Europe.

35.7 Geosites

The Drawsko Lakeland has considerable geodiversity and numerous existing and potential geosites. To protect the unique values of the abiotic nature of the Drawsko Lakeland, the Drawsko Landscape Park was created in its central part, covering 41.43 km^2 (Fig. 35.2). Trough lakes occupy more than 10% of the area. Numerous rivers have their beginnings in the moraine hills, the upper reaches of which look like streams more typical for the foothill belts of southern Poland. In addition, 18 nature reserves were established. Among them are the Baltic-type raised bog of Kusowo Bagno (Figs. 35.1 and 35.7—geosite no. 7, 35.16), Czarnówek Lake, Mire on Morzysław Mały Lake, Brown Soil, Prosino Lake, Green Mire, Toporzyk Mire, the Dębnica River Gap (Fig. 35.7—geosite no. 10), and Brzozowe Mire. In addition, there are three areas of protected landscape (Drawsko Lakeland, Szczecinek Lakes, and the surroundings of Kalisz Pomorski). Numerous tourist routes include nine hiking, fourteen cycling, and nine natural paths, and these are accompanied by other well-developed tourist facilities.

An exceptionally attractive nature reserve is the Valley of the Five Lakes, commonly known as the Połczyn Switzerland (Figs. 35.1, 35.3, 35.4 and 35.7—geosite no. 1), which occupies a tunnel valley with a winding course and steep slopes, as well as a string of depressions filled with waters of five lakes: Krzywe (Górne), Krąg (Okrągłe), Długie, Głębokie, and Małe, with a total area of 46.52 ha. Low peat fens occupy the lake shores. The Drawa River flows through the lakes and its source is located near the Krzywe Lake. Furthermore, the Drawa River flows through the lakes of Prosino, Żerdno, Drawsko, Krosino, and Lubie, creating one of the most interesting geotourist water routes.

River valleys are among the distinctive landscapes of the Drawsko Lakeland (Mazurek et al. 2010). The geodiversity of the Dębnica River catchment was assessed, considering elements such as landforms, lithology, relief energy, hydrographic elements, and insolation based on the digital elevation model (DEM) (Najwer et al. 2016). High (14.9% of the entire catchment area) and very high (8.9% of the area) geodiversity classes were mainly associated with typical glacial landforms of the marginal zone of the Pomeranian phase of the Weichselian glaciation and its denudation-fluvial transformation during the Holocene. This area has a highly varied topography with numerous culminations, erosional–denudational valleys dissecting the hillslopes, and headwater alcoves, which are the effects of groundwater seepage erosion (Mazurek and Paluszkiewicz 2013). The slopes of the moraine plateau are dissected by numerous erosional-denudational valleys. A distinctive geomorphological feature in the catchment of the Dębnica River is its narrow and deep gap through the accumulation moraine (Fig. 35.7—geosite no. 10). The gap was created as a result of the final formation of the outflow of waters towards the Baltic Sea, to the north.

Setting up a geopark of the "Post-Glacial Land of the Drawa and Dębnica" in the south-western part of the

Fig. 35.17 Peatland in the village of Nowy Chwalim is impacted by drainage and peat exploitation. As a result of peat extraction, fluvioglacial sands of the substrate were exposed (photograph by M. Mazurek)

Drawsko Lakeland is planned (Górska-Zabielska 2021). Geosites in the Drawsko Lakeland represent various landforms resulting from both accumulation and erosion, such as the terminal moraine in Popielewko, the moraine plateau of Trzy Garby near Połczyn Zdrój, the Bliska Struga Ravine in Zdroje, the upper Dębnica River Gap (Fig. 35.7—geosite no. 10), hydrographic objects (e.g. Drawsko Lake and Siecino Lake, the upper river parts of Drawa and Dębnica), artificial geological exposures (e.g. numerous sand and gravel pits), and sites of the occurrence of interesting rocks/sediments and minerals (e.g. erratic boulders, Petrographic Path in Żurawiec) (PGI-NRI 2022, Fig. 35.1).

Unusual geosites include clusters of erratic boulders found in many places in the Drawsko Lakeland and lapidaries (Górska-Zabielska 2021). Some erratic boulders are legally protected. The largest of these, Trygław in Tychowo (Fig. 35.7—geosite no. 9, 35.11), is a dark grey migmatite boulder with a very distinct streaky structure and large eye-shaped feldspars and garnets, as well as quartz, with visible glacial scratches. It is considered the largest in Poland. It has the circumference of approximately 50 m, is 16 m long, 11 m wide, and 7.8 m high (including 4 m under the ground level). Its volume is estimated for 700 m^3 and the weight for 2000 t. Scandinavian erratic boulders were used from the Middle Ages to build castles, towers and churches (Fig. 35.18), and they were later applied in residential and farm building foundations. The Medieval Castle of the Knights of the Order of St. John of Jerusalem in Stare Drawsko is an example of a building created using Scandinavian erratic boulders.

Fig. 35.18 Neogothic church in the village of Popielewo, built in 1876, is an example of the use of various types of erratic boulders obtained from the local moraine plateau for building purposes (photograph by R. Paluszkiewicz)

35.8 Conclusions

The presented characteristics of the Drawsko Lakeland document its great geomorphological and landscape diversity. This Lakeland stands out morphogenetically from the rest of Pomerania and the northern lowland part of Poland. The region is characterised by considerable geomorphic diversity, with the presence of large tunnel valleys cutting through the dominant Weichselian glacial plateaus. The area of the Lakeland is represented mainly by complexes of post-glacial (last glaciation), slope (Late Glacial), and fluvial (Holocene) landforms. The most characteristic sets of landforms are hypsometrically diversified moraine hills and ridges, reaching the highest elevations above 220 m asl, vast gently undulating areas of ground moraine plateaus, outwash plains, locally deeply incised subglacial tunnel valleys (up to 110 m), and river valleys usually with a meridional course and often with gaps. Late Glacial and Holocene retouching affected mainly erosional edges of moraine plateaus, slopes of moraine hills, kames and subglacial tunnel valleys remodelled by periglacial denudation, basins and multi-age erosional incisions with alluvial fans at their outlets. But the most spectacular feature of the Drawsko Lakeland is the high ratio of lakes (nearly 15%) to its total area. The high natural diversity, both geodiversity and biodiversity of the Drawsko Lakeland, results in the presence of valuable natural resources protected in the landscape park and nature reserves.

References

Adamczyk A, Wysota W, Piotrowski JA (2022) Inventory of glacial curvilineations (GCLs) at the southern periphery of the last Scandinavian Ice Sheet. Geomorphology 400:108094

Cedro A, Sotek Z (2016) Natural and anthropogenic transformations of a Baltic raised bog (Bagno Kusowo, North West Poland) in the light of dendrochronological analysis of *Pinus sylvestris L.* Forests 7(202):1–14

Choiński A (2006) Katalog jezior Polski. Wyd Nauk UAM, Poznań, p 599

Dobracka E, Lewandowski J (2002) Strefa marginalna fazy pomorskiej. In: Dobracki R, Lewandowski J, Zieliński T (eds) Plejstocen Pomorza Środkowego i strefa marginalna lobu Parsęty—IX Konferencja Stratygrafia plejstocenu Polski, PIG Oddział Pomorski, Szczecin i UŚ WNoZ, Sosnowiec, pp 109–117

Dobracka E, Piotrowski A (2002) Budowa geologiczna i rzeźba powierzchni podczwartorzędowej. In: Dobracki R, Lewandowski J, Zieliński T (eds) Plejstocen Pomorza Środkowego i strefa marginalna lobu Parsęty—IX Konferencja Stratygrafia plejstocenu Polski, PIG Oddział Pomorski, Szczecin, UŚ WNoZ, Sosnowiec, pp 85–92

Dobracka E (2009) Szczegółowa Mapa Geologiczna Polski w skali 1:50000, arkusz Połczyn-Zdrój (158), PIG, Warszawa
Dobracka E., 2017, Objaśnienia do Szczegółowej Mapy Geologicznej Polski w skali 1:50000, arkusz Połczyn-Zdrój (158), PIG, Warszawa, p 71
Dobrowolski R, Mazurek M, Osadowski Z, Alexandrowicz WP, Pidek IA, Pazdur A, Piotrowska N, Drzymulska D, Urban D (2019) Holocene environmental changes in northern Poland recorded in alkaline spring-fed fen deposits. A Multi-Proxy Approach. Quat Sci Rev 219:236–262
Górska-Zabielska M (2008) Obszary macierzyste skandynawskich eratyków przewodnich osadów ostatniego zlodowacenia północno-zachodniej Polski i północno-wschodnich Niemiec. Geologos 12:177–194
Górska-Zabielska M (2013) Lapidarium w Żurawcu na Pojezierzu Drawskim, Pomorze Środkowe. Przegl Geogr 85:435–454
Górska-Zabielska M (2021) In the footsteps of the ice sheet in the area of the planned geopark postglacial land of the Drawa and Dębnica rivers (the Drawskie Lakeland, Poland). Landf Anal 40:37–56
Jania J, Bukowska-Jania E (1997) Model procesów paleoglacjalnych fazy pomorskiej vistulianu w obrębie lobu Parsęty na podstawie badań obszarów aktualnie zlodowaconych. In: Bukowska-Jania E, Pulina M (eds) Studia nad środowiskiem geograficznym Bornego Sulinowa. PWN, Warszawa, pp 27–57
Karczewski A (1985) Spatial arrangement of morainic plateau levels of the northern sloping surface in Pomerania as a result of varying deglaciation. Quaest Geogr Spec Iss 1:99–107
Karczewski A (1989) Development and extent of the Pomeranian phase marginal zone within the Parseta Lobe during the Vistuian glaciation. Quaest Geogr Spec Iss 2:61–67
Kostrzewski A, Zwoliński Z (1992) Udział denudacji chemicznej i mechanicznej we współczesnym systemie geomorficznym górnej Parsęty (Pomorze Zachodnie). Prace Geogr IGPZ PAN 155:11–45
Kostrzewski A, Klimczak R, Stach A, Zwoliński Z (1989) Morphologic effects of heavy rainfall (24 may 1983 over relief features of the scarpland in the middle Parsęta valley, west Pomerania, Poland. Quaest Geogr Spec Iss 2:101–110
Kostrzewski A, Mazurek M, Zwoliński Z (1994) Dynamika transportu fluwialnego górnej Parsęty jako odbicie funkcjonowania systemu zlewni. Bogucki Wyd Nauk, Poznań, p 165
Kostrzewski A, Zwoliński Z, Andrzejewski L, Florek W, Mazurek M, Niewiarowski W, Podgórski Z, Rachlewicz G, Smolska E, Stach A, Szmańda J, Szpikowski J (2021) Współczesna ewolucja rzeźby młodoglacjalnej Niżu Polskiego. In: Kostrzewski A, Krzemień K, Migoń P, Starkel L, Winowski M, Zwoliński Z (eds) Współczesne przemiany rzeźby Polski (2nd ed), Bogucki Wyd Nauk Poznań, pp 483–571
Lewandowski J, Heliasz Z, Chybiorz R (2003) Objaśnienia do Szczegółowej Mapy Geologicznej Polski w skali 1:50000, Arkusz Łubowo (197), PIG, Warszawa, p 45
Lewandowski J, Heliasz Z, Chybiorz R (2008) Objaśnienia do Szczegółowej Mapy Geologicznej Polski w skali 1:50000, arkusz Czaplinek (196), PIG, Warszawa, p 50
Lewandowski J, Nita M (2008) Ewolucja systemu hydrograficznego i szaty roślinnej dorzecza górnej Piławy i górnej Drawy (Pomorze Środkowe). Przegl Geol 56:380–390
Maksiak S, Mróz W (1978) Czwartorzęd środkowej części Pojezierza Pomorskiego. Biul Inst Geol 300:97–152
Map of Hydrographic Division of Poland (2004) IMGW, Min. Środ., Warszawa
Marks L (2012) Timing of the Late Vistulian (Weichselian) glacial phases in Poland. Quat Sci Rev 44:81–88
Marsz A (1973) Niektóre zagadnienia geomorfologii bezpośredniego przedpola zasięgu stadiału pomorskiego na Pojezierzu Drawskim (na przykładzie obrzeżenia rynny marginalnej Drawsko–Pile). Bad Fizjogr Nad Pol Zach Ser A 26:97–143
Mazurek M (2011) Geomorphological processes in channel heads initiated by groundwater outflows (the Parsęta catchment, north-western Poland). Quaest Geogr 30(3):33–45
Mazurek M, Paluszkiewicz R (2013) Formation and development of a 1st-order valley network in postglacial areas (the Dębnica catchment). Landf Anal 22:75–87
Mazurek M, Paluszkiewicz R, Piotrowska I (2010) Walory turystyczne sieci dolinnej w dorzeczu Parsęty. Krajobraz a turystyka. Prace Kom Kraj Kultur 14:207–220
Mazurek M (2010) Hydrogeomorfologia obszarów źródliskowych (dorzecze Parsęty, Polska NW). Seria Geografia 92, Wyd Nauk UAM, Poznań, p 304
Mojski JE (2005) Ziemie polskie w czwartorzędzie. Zarys morfogenezy. PIG, Warszawa, p 404
Najwer A, Borysiak J, Gudowicz J, Mazurek M, Zwoliński Z (2016) Geodiversity and biodiversity of the postglacial landscape (Dębnica river catchment, Poland). Quaest Geogr 35(1):5–28
NGIA (2004) Performance specification digital terrain elevation data (DTED). 2000. MIL-PRF-89020B
Paluszkiewicz Re, Paluszkiewicz Ry (2005) Lithofacies differences in proglacial lake sediments: an example of the Pyrzyce and Złocieniec proglacial lakes (West Pomerania). Quaest Geogr 24:75–84
Paluszkiewicz R (2004) Warunki sedymentacji osadów rytmicznie warstwowanych w zastoisku złocienieckim na Pojezierzu Drawskim. PTPN Wydz Mat–Przyr Prace Kom Geogr-Geol 34, Poznań, p 101
Paluszkiewicz R (2016) Postglacjalna ewolucja dolinek erozyjno-denudacyjnych w wybranych strefach krawędziowych Pojezierza Zachodniopomorskiego. Studia i Prace z Geografii 55, Poznań, p 225
PGI-NRI [Polish Geological Institute] (2022) Geosites. Online: http://geoportal.pgi.gov.pl/portal/page/portal/geostanowiska/ (Accessed 15 Jan 2022)
Piasecki D (1982) Ewolucja dolin rzek Przymorza. Przegl Geogr 54:49–68
Popielski W (2005) Objaśnienia do Szczegółowej Mapy Geologicznej Polski w skali 1:50000, arkusz Barwice (159), PIG, Warszawa, p 38
Popielski W (2006) Objaśnienia do Szczegółowej Mapy Geologicznej Polski w skali 1:50000, arkusz Szczecinek (160), PIG, Warszawa, p 41
Solon J, Borzyszkowski J, Bidłasik M, Richling A, Badora K, Balon J, Brzezińska-Wójcik T, Chabudziński Ł, Dobrowolski R, Grzegorczyk I, Jodłowski M, Kistowski M, Kot R, Krąż P, Lechnio J, Macias A, Majchrowska A, Malinowska E, Migoń P, Myga-Piątek U, Nita J, Papińska E, Rodzik J, Strzyż M, Terpiłowski S, Ziaja W (2018) Physico-geographical mesoregions of Poland: verification and adjustment of boundaries on the basis of contemporary spatial data. Geogr Polon 91:143–170
Stachŷ J (ed) (1987) Polish hydrological atlas. Wyd Geol, Warszawa, p 79
Starkel L (1980) Przeglądowa mapa geomorfologiczna Polski w skali 1:500000, IGiPZ PAN, Warszawa
Sylwestrzak J (1978) Rozwój sieci dolinnej na Pomorzu pod koniec plejstocenu. Zakład Narodowy im. Ossolińskich, Gdańsk, p 161
Szafraniec JE (2013) Paleofood marks in sandur morphometry as the glacier surge result (NW Poland). Hydrol Res 44:264–280

Szpikowski J (2010) Antropogeniczne przekształcenia rzeźby zlewni Perznicy w neoholocenie (Pojezierze Drawskie, dorzecze Parsęty). Wyd Nauk UAM, Poznań, p 209

Tylmann K, Rinterknecht VR, Woźniak P, Bourles D, Schimmelpfenning I, Gullou V, Aster Team (2019) The local last glacial maximum of the southern Scandinavian ice sheet front: Cosmogenic nuclide dating of erratics in north Poland. Quat Sci Rev 219:36–46

Zwoliński Z (1989) Geomorficzne dostosowywanie się koryta Parsęty do aktualnego reżimu rzecznego. Dokument Geogr IGiPZ PAN 3–4:144

Małgorzata Mazurek is Associate Professor of Geography at Institute of Geoecology and Geoinformation, Faculty of Geographical and Geological Sciences, Adam Mickiewicz University in Poznan. Her main research in the field of hydrogeomorphology, environmental monitoring and geo-ecology, is focused on the evolution of valley and river network, geodiversity of groundwater outflows, chemical denudation in various morphoclimatic zones, Late Glacial and Holocene environmental changes in the light of the geochemistry of sediments of the alkaline spring-fed fens and urban geomorphology. The north-west of Poland, Pomerania, and Greater Poland are her main research areas, but she also conducts research in the polar zone, whereas field studies in the equatorial and subtropical zones provide further inspirations. She was the President of the Association of Polish Geomorphologists (2017–2021) and since 2019 is the Vice President of the Polish Geographical Society.

Renata Paluszkiewicz is Associate Professor of geography at the Institute of Geoecology and Geoinformation, Faculty of Geographical and Geological Sciences, Adam Mickiewicz University in Poznan. The fundamental trend of her research focuses on the young glacial area, especially in western Pomerania (NW Poland) concentrating on relief and sediments of proglacial meltwater valleys (Pomeranian Pradolina), ice-dammed lakes, the system of erosional-denudational valleys, denudation forms and related geomorphological processes. She also conducted observations of landscape evolution in other geographical regions, e.g. in the polar zone since 2001. She is a member of the Association of Polish Geomorphologists and the Poznań Society for the Advancement of Arts and Sciences. In 2017–2020, she was the head of the Cryosphere Research Unit at Adam Mickiewicz University.

Zbigniew Zwoliński is a geographer, geomorphologist, Professor at the Institute of Geoecology and Geoinformation, Faculty of Geographical and Geological Sciences, Adam Mickiewicz University in Poznań. The most important research issues investigated recently include studies of denudation and environmental changes in different morphoclimatic zones, development of the concept and implementation of geomorphological map of Poland in the scale of 100k and 500k, studies on morphology and sediments in urban areas, development of the concept of geodiversity and methods of its assessment, changes in land cover and land use in selected landscapes of Poland, uncertainty and sensitivity analysis in multi-criteria modelling, creation of geo-questionaries and geo-discussion applications for spatial planning purposes. He is author of over 300 publications, editor-in-chief of *Landform Analysis*, managing editor of *Quaestiones Geographicae*. Member of the Association of Polish Geomorphologists and its past President (2008–2011), Chairman of the Working Groups on Geodiversity and Global Denudation of the International Association of Geomorphologists. Awarded the Denys Brunsden Medal by the International Association of Geomorphologists.

Tunnel Valleys of the Tuchola Forest and Kashubian Lake District

36

Mirosław Błaszkiewicz and Karol Tylmann

Abstract

This chapter presents one of the most characteristic landforms of the glacial landscape of northern Poland—tunnel valleys. Three tunnel valleys located in the forefield of the maximum limit of the Scandinavian Ice Sheet during the Pomeranian Phase of the last glaciation (the outwash plain of the Tuchola Forest), and three tunnel valleys located behind this limit (the Kashubian Lake District) are analysed. All the landforms were created by subglacial erosional processes, mainly associated with the channelised flow of meltwaters flowing under hydrostatic pressure. One tunnel valley (the Wiecka tunnel channel) was formed in a single catastrophic outburst event involving bank-full subglacial discharge; another is a composite tunnel valley formed due to the coexistence of glacial and meltwater erosion over at least three last glaciations (the Żarnowiecka tunnel valley), and the remaining ones are complex systems of tunnel valleys formed due to steady-state meltwater erosion of the ice-sheet bed. All tunnel valleys were infilled with dead-ice blocks immediately after their formation, which protected reaches located within outwash plains against backfilling with sandur deposits. Outwash flow also significantly modified their morphology; hence, such landforms located within a sandur are called "palimpsest" tunnel valleys. Melting of buried dead-ice blocks in warm periods of the Late Glacial and at the beginning of the Holocene revealed the tunnel valleys within the glacial landscape, as well as the formation of lakes, which in most cases were incorporated into the existing fluvial systems. In this way, a period of intensive fluvial and lacustrine transformation of tunnel valleys began.

Keywords

Tunnel valleys · Tunnel channels · Subglacial drainage · Buried dead-ice blocks · Lake formation · Fluvial transformation · Tuchola Forest · Kashubian Lake District

36.1 Introduction

Tunnel valleys are among the most spectacular landforms to have originated during the advances and retreats of all Pleistocene ice sheets. Their morphology is very distinctive among glacial landforms. They are elongated incisions, usually hundreds of metres wide, and can be tens of kilometres long. Typically, tunnel valleys are deeply incised, locally by more than 100 m, into the surrounding moraine plateaus or outwash plains. The most characteristic feature of their morphology is a thalweg that undulates along the longitudinal profile, with numerous thresholds and overdeepenings, which clearly distinguishes them from river (fluvial) valleys. The largest tunnel valleys, more than 500 m deep, have been documented in Germany, and their genesis is related to the Elsterian glaciation (Stackebrandt 2009). However, they are not visible in the present-day landscape because during successive glacial/interglacial cycles they were completely filled with glacial, glaciolacustrine, lacustrine and fluvial sediments. These landforms are described as "buried tunnel valleys", in contrast to landforms that are visible in the contemporary relief, which are described as "open tunnel valleys" (Jørgensen and Sandersen 2006).

M. Błaszkiewicz (✉)
Department of Environmental Resources and Geohazards, Institute of Geography and Spatial Organization, Polish Academy of Sciences, Kopernika 19, 87-100 Toruń, Poland
e-mail: miroslaw.blaszkiewicz@geopan.torun.pl

K. Tylmann
Department of Geomorphology and Quaternary Geology, Faculty of Oceanography and Geography, University of Gdańsk, Bazynskiego 4, 80-309 Gdansk, Poland
e-mail: karol.tylmann@ug.edu.pl

P. Migoń and K. Jancewicz (eds.), *Landscapes and Landforms of Poland*,
World Geomorphological Landscapes, https://doi.org/10.1007/978-3-031-45762-3_36

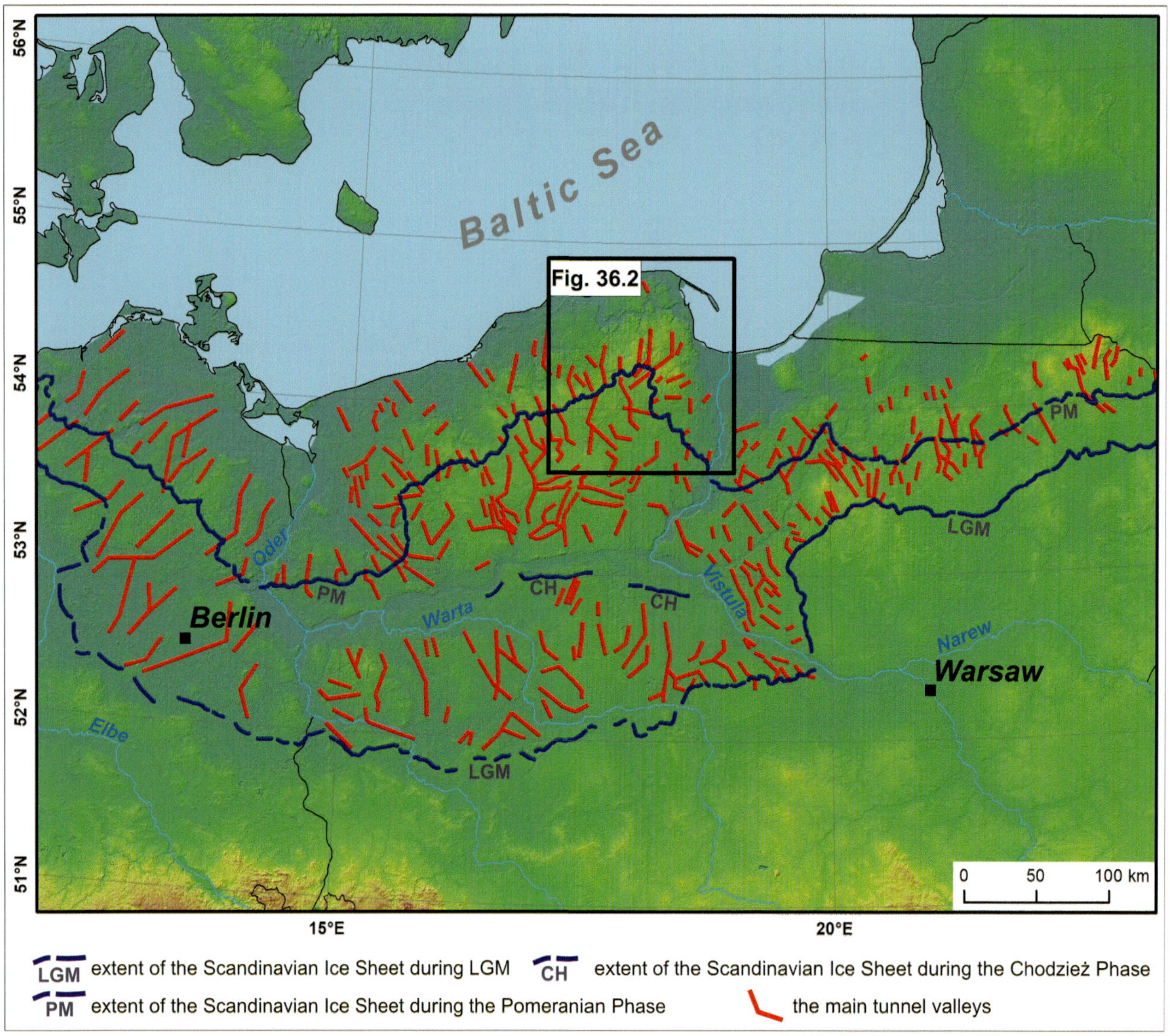

Fig. 36.1 Main tunnel valleys of the central European Lowland against the main ice-margin limits of the last Scandinavian Ice Sheet. Digital elevation model based on SRTM data with horizontal resolution of 80 m. LGM—Last Glacial Maximum

Elsterian buried tunnel valleys have also been reported in Poland, and one of the most widely known in the literature is a valley filled with annually laminated lake sediments from the Holstein interglacial reported at the Ossówka site in eastern Poland (Krupiński 1995; Nitychoruk et al. 2006, 2018). In the European Lowlands, many buried tunnel valleys from the Saalian glaciation filled with sediments of Eemian lakes have also been reported (Roman 2016). Tunnel valleys that formed during the last glaciation (Weichselian) are still very visible in the relief of the glacial landscape, despite protracted accumulation of lacustrine and fluvial deposits at their floors. The spatial distribution of these tunnel valleys (open valleys) is one of the most important indicators for mapping the maximum extent of the last Scandinavian Ice Sheet in the European Lowlands (Majdanowski 1947; Marks 2002) (Fig. 36.1).

This chapter presents six morphologically diverse tunnel valleys located within various geomorphological units (moraine plateaus and outwash plains) in the foreland of the Pomeranian Phase ice-sheet limit, as well as in the hinterland of this limit.

36.2 Origin of Tunnel Valleys

Since tunnel valleys were first recognized and linked with subglacial environments (Jentzsch 1884; Gottsche 1897), the processes of their formation and evolution have been the subject of many discussions and polemics. The most up-to-date summaries of current views on the genesis of tunnel valleys are included in, for example, the papers of Kehew et al. (2012), Van Der Vegt et al. (2014) or Livingstone and

Clark (2016). Currently, the most widely accepted idea of tunnel valley formation is that emphasizing the relationship with glaciofluvial erosion at the ice–bed interface caused by pressurised meltwater flowing under the glaciers and ice sheets (e.g., Ussing 1903; Majdanowski 1950; Galon 1965; Kozarski 1966; Piotrowski 1994). According to this concept, tunnel valleys are in fact glaciofluvial valleys, but formed under specific subglacial conditions in which pressurised meltwater can produce deep valleys with a very irregular and undulating thalweg. Thus, tunnel valleys often have an abrupt beginning and end, and may be sloped up-ice, which is also one of their most characteristic features. Because they are formed by subglacial pressurised meltwater, their spatial arrangements are usually related to equipotential hydrologic surfaces within ice sheets and subglacial hydrostatic gradients. Therefore, tunnel valleys are typically oriented parallel to the lines of ice flow and perpendicular to the ice margin (Bennett and Glasser 2009). Therefore, one of the most characteristic spatial arrangements of tunnel valleys is a radial network that formed under fan-shaped ice lobes.

Some authors previously interpreted the mechanism of tunnel valley formation as involving glacial erosion shaping the ice bed without the significant role of subglacial meltwater. These ideas assumed that the main morphogenetic factor for tunnel valleys was glacial erosion within narrow ice outlets (ice lobes) of ice sheets—the German term *Glaziellen* (Woldstedt 1952; Gripp 1964). Some authors linked the origin of tunnel valleys mainly to these narrow zones of ice flow, and they argued that glacial erosion had a significant role in the subglacial shaping of such valleys (Paterson 1981; Niewiarowski 1995). Then, as research accumulated, an idea emerged that subglacial meltwater erosion co-existed with glacial erosion and subglacial deformation of sediments during the formation of tunnel valleys (Boulton and Hindmarsh 1987).

Attempting a more detailed characterization and classification of tunnel valleys and their links with specific subglacial processes, Clayton et al. (1999) proposed to distinguish two basic types: tunnel channels and tunnel valleys. According to this classification, tunnel channels are formed by catastrophic, bank-full subglacial meltwater discharges. This is usually a one-time process that could occur very suddenly, and tunnel channels are formed relatively quickly. By contrast, tunnel valleys are a morphological expression of long-lasting and multiple instances of vertical and lateral erosion of subglacial rivers (steady-state formation), whose individual channels were smaller than the present tunnel valleys. In the case of tunnel valleys, glacial erosion and subglacial deformation of sediments are also possible during their formation (Clayton et al. 1999).

Another factor that played an important role in tunnel valley formation was the morphology and lithology of the terrain that was overridden by the ice sheet. Large, concave structures of preglacial relief (e.g., river valleys, lake basins, tectonic depressions) under the ice sheet sole became the paths of subglacial meltwater drainage, and could influence local directions of ice-stream flows (Molewski 1999; Kehew et al. 2012). Thus, in the case of some tunnel valleys, to some extent their re-production and transformation by subglacial erosion could occur.

36.3 Setting and Location

The tunnel valleys analysed in this chapter are located in north-central Poland, in an area entirely covered by the last Scandinavian Ice Sheet during the late Weichselian glaciation. The landforms lie within two morphogenetically distinct regions: the Tuchola Forest and the Kashubian Lake District. The border between these two adjacent regions is the line of the ice-sheet maximum extent during the Pomeranian Phase of the Weichselian glaciation within the western flank of the Vistula Ice Lobe (Fig. 36.2). Moraines of the Pomeranian Phase in northern Poland were recently dated on the basis of ^{10}Be cosmogenic dating of erratic boulders at ca. 16.5 ka (Rinterknecht et al. 2005; Tylmann et al. 2019). Most ice-marginal features correlated with the Pomeranian Phase in the western flank of the Vistula Ice Lobe area were destroyed by proglacial meltwater erosion, so the ice-sheet extent is usually delineated by deriving from the spatial distribution of the proximal edges of outwash plains, where in some cases ice-contact sedimentary scarps occur (Błaszkiewicz 2005).

The Tuchola Forest is one of the biggest complexes of outwash plains in the European Lowlands, and is correlated with proglacial meltwater outflow during ice-margin standstill and recession during the Pomeranian Phase of the Weichselian glaciation. It consists of two large outwash routes: the Brda sandur and the Wda sandur, within both of which several topographically diversified outwash levels linked to subsequent stages of ice-sheet recession have been distinguished (Galon 1953). The topography of the outwash plains' topography is diversified most significantly by systems of tunnel valleys that formed during the ice sheet recession from the ice-margin limit of the Chodzież Phase to the ice-margin limit of the Pomeranian Phase (Fig. 36.1). The floors of most of the tunnel valleys are currently occupied by lakes and fens, which typically cover lake sediments. During cooler periods of the Late Glacial, mainly during the Younger Dryas, the sand plains of the Tuchola Forest were subjected to aeolian processes, forming dunes (Kaiser et al. 2009).

The Kashubian Lake District is located within the western flank of the Vistula Ice Lobe area, behind the end moraines of the Pomeranian Phase of the Weichselian

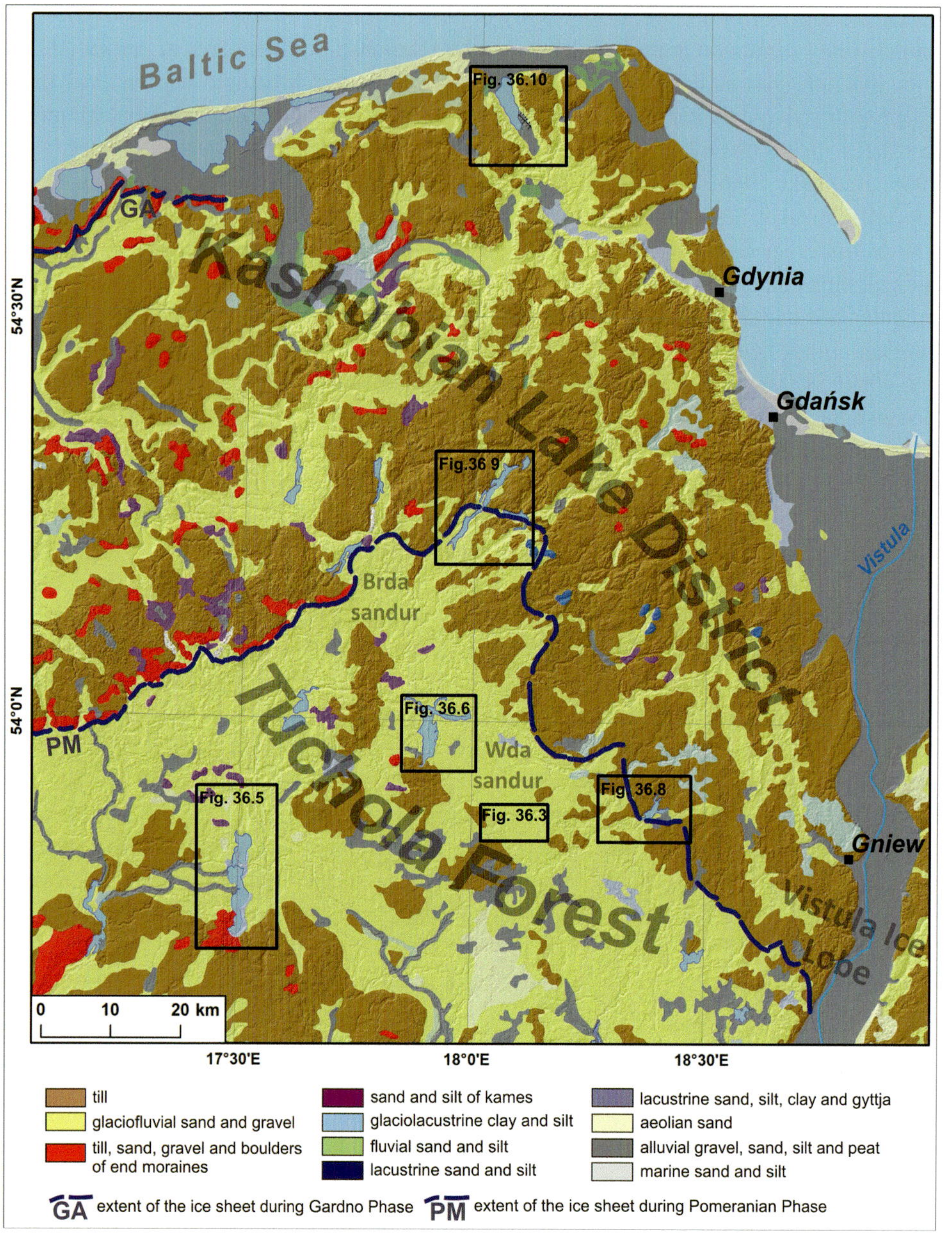

Fig. 36.2 Study area with the analysed tunnel valleys; geological map according to Marks et al. (2006)

glaciation (Fig. 36.2). It represents a morphogenetically diverse region consisting mainly of undulating and hummocky moraine plateaus with numerous kames, including an extensive ice-walled lake plain near the town of Gniew (Błaszkiewicz and Gruszka 2005). However, the most characteristic feature of the glacial landscape in this region is the occurrence of tunnel valleys that are arranged in systems incised into morainic surfaces. The western flank of the Vistula Ice Lobe area includes of a dense network of tunnel valleys oriented mainly NE–SW, which are probably part of a radial system of tunnel valleys that formed under the last ice sheet. For detailed presentation in this chapter three tunnel valleys from the Tuchola Forest and three tunnel valleys from the Kashubian Lake District were selected.

36.4 Examples of Tunnel Valleys

36.4.1 The Wiecka Tunnel Channel

36.4.1.1 Geomorphology

The Wiecka tunnel channel is incised into two morphologic levels of the Wda outwash plain. The higher one, where the valley originates, has elevations in the order of 140–142 m asl, whereas the lower one, where most of the valley is located, lies at 128–132 m asl (Błaszkiewicz 2005). The outwash plains surrounding the tunnel channel consist of several-metre-thick sands with a domination of well-sorted, fine- and medium-grained sand and admixture of coarse-grained sand. The outwash sequence was accumulated in shallow, sand-bed braided rivers. The Wiecka tunnel channel is a single landform extending over an area of ca. 8 km (Fig. 36.3a). The channel is relatively narrow with an average width of 350 m and a longitudinally undulating morphological profile. The northern fringe of the valley floor lies at 133 m asl, whereas the outlet of the channel is between 128 and 129 m asl (Fig. 36.3b). Within the largest overdeepenings of the channel floor (the Wiecko and Skrzynka lakes), the depth of the landform relative to the levels of the surrounding outwash plain (taking into account the depth of the lakes and the thickness of their bottom deposits) is up to 40 m. The slopes of the tunnel channel are typically steep, and in a few localities they are undulated by the occurrence of elongated ridges running along the directions of meltwater outflow on the outwash plain (Fig. 36.3a).

The Wiecka tunnel channel is dissected in its south-western part by the Wda River valley. The elevation of the floodplain within the valley corresponds to the elevation of the peat plain and the water table of the Wiecko and Skrzynka lakes within the tunnel channel (Fig. 36.4). A palaeo-delta has been recognized in the channel floor where the Wda River valley dissects the tunnel channel. The elevation of the palaeo-delta surface is the same as of the terrace

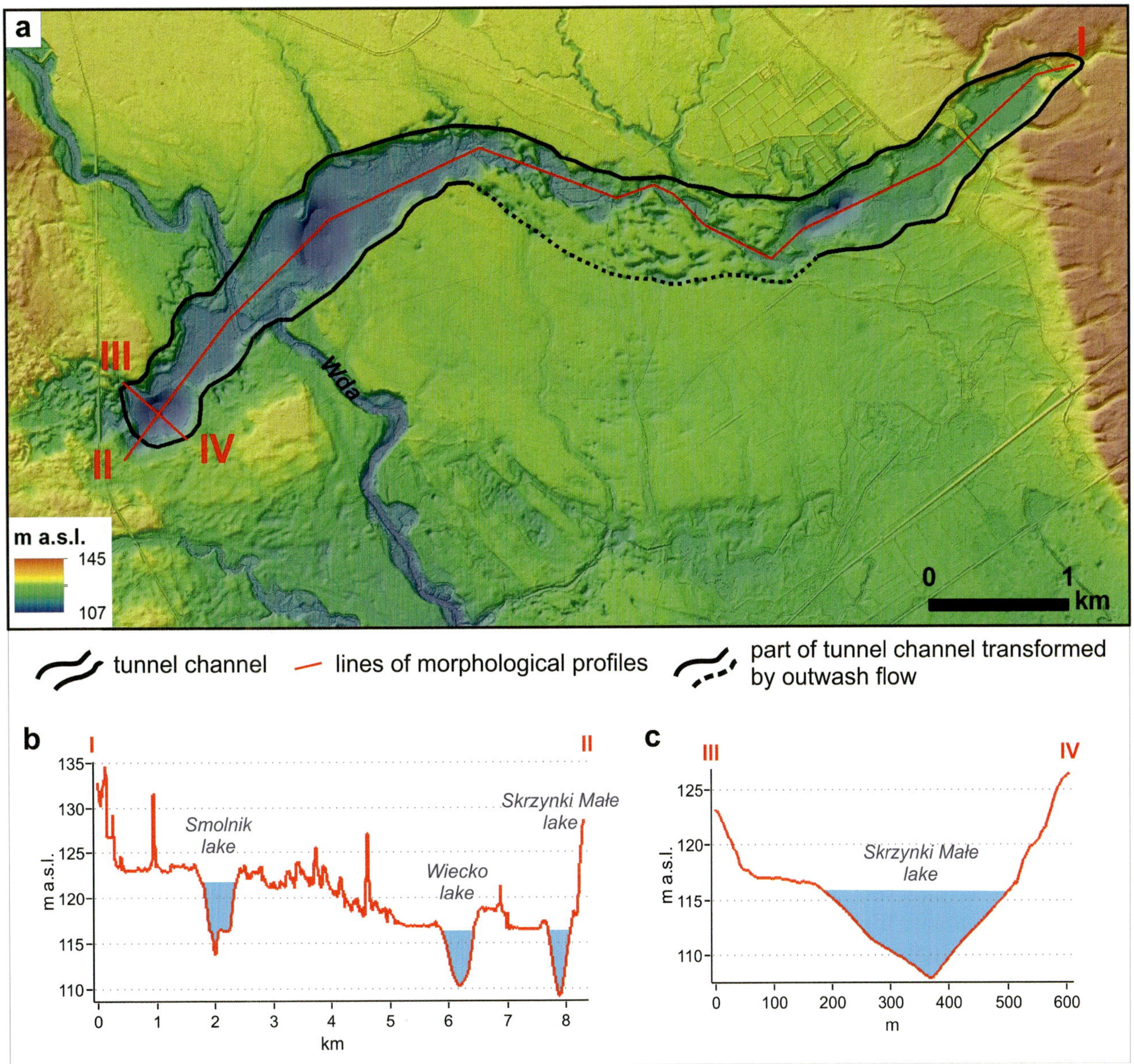

Fig. 36.3 The Wiecka tunnel channel. **a** Digital elevation model with the outline of tunnel channel. **b** Longitudinal profile. **c** Transverse profile. Digital terrain model was created using LiDAR elevation data with horizontal resolution of 1 m

Fig. 36.4 South-western part of the Wiecka tunnel channel with the lake basins of Wiecko (foreground) and Skrzynki Małe (background) (photograph by W. Stępień)

level within the valley. The inner structure of the palaeodelta reveals that below several metres of deltaic sand deposits a one-metre-thick layer of Allerød gyttja occurs (Błaszkiewicz 2005).

36.4.1.2 Origin and Evolution

The morphology of the valley (single landform without tributaries; relatively constant dimensions) and its geomorphological context indicate that it originated as a result of a catastrophic, bank-full discharge of pressurised subglacial meltwater. Therefore, it could be interpreted as a typical "tunnel channel" landform (according to Clayton et al. 1999) that formed relatively quickly as a result of a one-time process. After the subglacial meltwater stream was incised into the ice-sheet bed and after the initial formation of the tunnel channel, it was filled with dead-ice blocks that formed during deglaciation as a result of the collapse of the subglacial channel's roof, and/or as a result of supercooled meltwater freezing due to a rapid drop in pressure within the subglacial channel (Błaszkiewicz 2011). The occurrence of dead ice within the tunnel channel protected the landform from being infilled with sand deposits during subaerial meltwater flow and outwash plain formation after the ice-margin retreat from the maximum limit of the Pomeranian Phase ice sheet.

The dead-ice blocks were preserved under a thick cover of outwash sediments under permafrost for ca. 3000 years. The final melting of this ice at the beginning of the Allerød exhumed the tunnel channel, causing it to appear in the landscape. Within the deepest parts of the tunnel channel, lakes were formed that were then incorporated into the fluvial system of the Wda River (Błaszkiewicz 2005, 2011). However, the outwash flow over the undulating top of the dead-ice blocks buried within the tunnel channel resulted in the accumulation of fluvioglacial sediments of various thickness, which had a significant impact on the present-day morphology of the tunnel channel. The palimpsest nature of the Wiecka tunnel channel is highlighted by the occurrence of elongated landforms within the channel that clearly record fluvioglacial accumulation partly masking the original morphology of the channel after deglaciation (Kehew et al.1999).

36.4.2 The Charzykowska Tunnel Valley

36.4.2.1 Geomorphology

The Charzykowska tunnel valley lies in the west of the Brda outwash route, on the border with the Krajna moraine plateau (Galon 1953). The tunnel valley is mainly surrounded by two topographically distinct levels of an outwash plain, which were accumulated during the ice-margin standstill and retreat during the Pomeranian Phase. Around the southern fringe of the tunnel valley, near the outlet, hummocky moraines occur. In this zone, a series of hummocks and small hills occurs that have been interpreted as arcs of push end-moraines (Pasierbski 1975). More recent research involving a detailed geological mapping of the area (Urbański 2012) suggested that these hummocks and hills are kames superimposed on push end-moraines.

The Charzykowska tunnel valley is a 19-km-long valley with a relatively compact shape and N–S longitudinal orientation (Fig. 36.5a). Its width varies from 200 m at the northern fringe up to 2.6–2.8 km within the widenings, but the most frequent width of the valley is from 1.5 to 2.5 km. The depth of the incision is typically between 25 and 45 m. The longitudinal morphological profile of the valley floor reveals a highly undulating thalweg, with numerous depressions and thresholds. The valley floor lies at 125 m asl at the northern fringe, whereas at the outlet of the tunnel valley it is at 122 m asl, but the relative differences in elevations along the longitudinal profile are up to 34 m (Fig. 36.5b). The deepest parts of the valley are currently occupied by three main lakes: Charzykowskie, Karsińskie and Witoczno. They are currently a part of the Brda river fluvial system. Within the Charzykowska tunnel valley, river deltas and three lake terraces have been documented. The lake terraces are elevated 7–8 m, 3 m and 1 m above the present-day level of the lake.

36.4.2.2 Origin and Evolution

According to Pasierbski (1975, 1979) the Charzykowska tunnel valley is a polygenetic landform that formed in two main stages: (1) meltwater erosion and (2) glacial erosion. In the first stage, subglacial meltwater erosion during the recession of the ice sheet after the Chodzież Phase of the last glaciation caused the ice-sheet bed to be incised, forming a narrow tunnel channel. In the second stage, this initial channel was widened and deepened by glacial erosion of the ice lobe during a local re-advance of the ice sheet

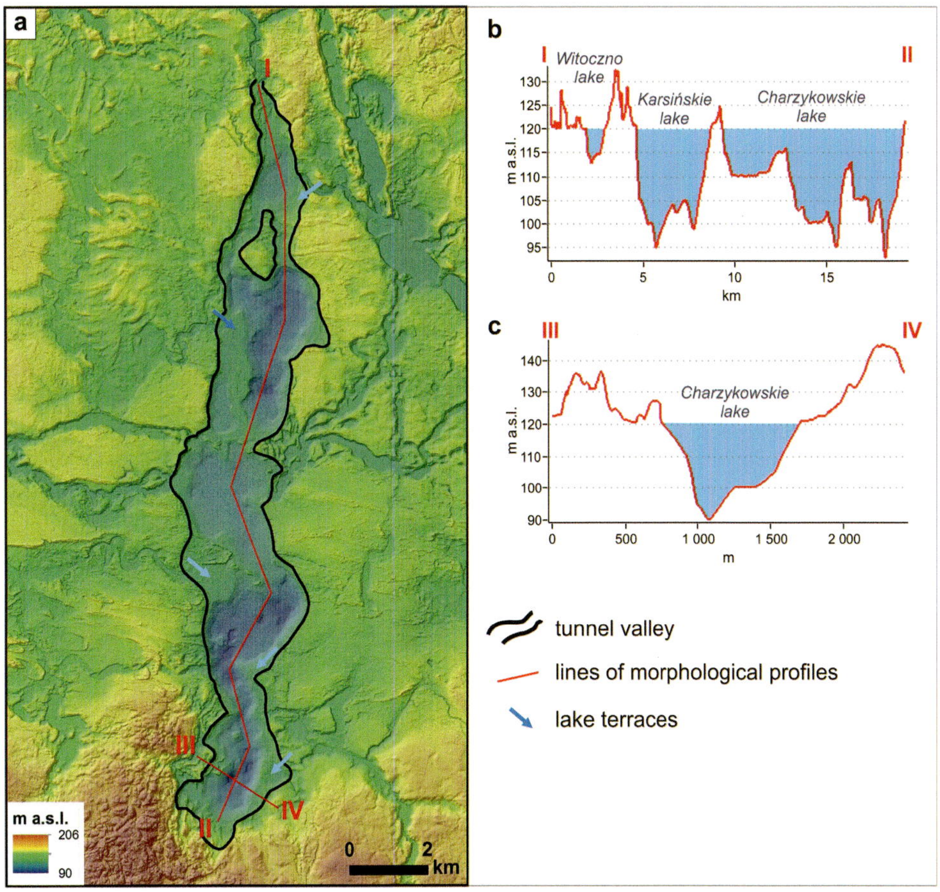

Fig. 36.5 The Charzykowska tunnel valley. **a** Digital elevation model with the outline of tunnel channel. **b** Longitudinal profile. **c** Transverse profile. Digital terrain model was created using LiDAR elevation data with horizontal resolution of 1 m

(Pasierbski 1975). Regardless of the main factor involved in the formation and shaping of the Charzykowska tunnel valley (meltwater erosion or glacial erosion), morphological relations characterising the whole complex tunnel-valley system indicate that at least two generations of tunnel channels occur here.

Geomorphologic relationships between the Charzykowska tunnel valley and the surrounding outwash plains show that during the outwash flow from the ice margin of the Pomeranian Phase, the tunnel valley was infilled with dead-ice blocks, and therefore preserved by being buried under fluvioglacial (sandur) sediments. This phase of preservation by dead-ice blocks is documented in the evolution of most tunnel valleys in northern Poland, including all the tunnel valleys presented in this chapter (Błaszkiewicz 2005, 2011). Palynological dating of the initial lake sediments deposited within the tunnel valley floor indicate that the final melting of the dead ice may have occurred in the Allerød (Nowaczyk 1994). An extensive lake was formed at this stage, as recorded by the highest lake terraces being located 7–8 m above the present level of the Charzykowskie lake. The lower lake terraces record subsequent water level fluctuations within the Charzykowskie lake caused by climate changes in the second part of the Late Glacial and the Holocene, and by the evolution of the local drainage network and incorporation of emerging lakes into the fluvial system of the Brda River. Lacustrine sedimentation present since the Late Glacial caused a more than 14-m-thick layer of lake sediments, mainly calcareous gyttja, to be deposited in the Charzykowskie lake basin (Tobolski 2010).

36.4.3 The Wdzydze-Gołuń Tunnel Valley

36.4.3.1 Geomorphology

The Wdzydze-Gołuń tunnel valley is located ca. 20 km south of the ice-margin limit of the Pomeranian Phase, within an extensive outwash plain that is part of the Wda outwash route (Fig. 36.2). Almost the entire tunnel valley is surrounded by the outwash plain, which lies on various levels, from ca. 143–148 m asl in the northern part to 140–143 m asl in the south. These outwash levels accumulated during the ice-margin recession from the Chodzież Phase ice-sheet limit and during the subsequent ice-margin standstill of the Pomeranian Phase. The southern fringe of the tunnel valley is surrounded by a wavy moraine plateau topped with kame hills (Petelski and Majewska 2008). The Wdzydze-Gołuń tunnel valley is part of a complex system of tunnel valleys dissecting the outwash plain levels in this part of the Wda outwash route. This tunnel valley is composed of two main incisions oriented perpendicular to each other: the Wdzydze tunnel valley and the Gołuń tunnel valley (Fig. 36.6a). This makes the spatial arrangement of the landform very peculiar, with a cross-like planar shape that differs greatly from that of the other tunnel valleys.

The tunnel valley occupied by the Wdzydze lake is 9 km long, and consists of two main channels oriented N–S. Its width varies from 100 m at the southern fringe (the valley outlet) up to 3 km within the wide, central part. The eastern channel of the valley is typically 1.0–1.5 km wide, whereas the western channel is slightly narrower, with a width usually ranging from 0.5 to 1.0 km. The depth of the Wdzydze tunnel valley is usually between 20 and 60 m, with a maximum depth of 75 m within the eastern channel. The longitudinal morphological profile of the valley floor reveals a very undulating thalweg with relative differences in elevation of up to 40 m within the eastern channel. The northern fringe of the valley floor lies at ca. 130 m asl, whereas the southern outlet of the Wdzydze tunnel valley is at ca. 150 m asl (Fig. 36.6b). The tunnel valley, occupied by the Gołuń lake and the Radolne lake, is 10 km long and oriented E–W. It is typically 500–700 m wide, being narrowest at its western fringe (ca. 220 m), whereas the widest segment occurs in the east (above 850 m). The depth of the valley is 15–25 m over most of its length, and the longitudinal morphological profile of the thalweg also undulates, but the height differences are not so great as within the thalweg of the Wdzydze tunnel valley (Fig. 36.6c). There are morphological ridges of a few metres high at the junctions between the two valleys, and the floor of the Gołuń tunnel valley is generally slightly higher than the northern part of the Wdzydze valley floor.

36.4.3.2 Origin and Evolution

The genesis of the Wdzydze-Gołuń tunnel valley, in terms of the origin of the landform itself and the formation of lakes and the evolution of the tunnel valley alike, has not yet been investigated in detail. The geomorphic context of this tunnel valley and its morphology suggest that the origin is probably related to long-lasting, complex subglacial meltwater erosion of the ice bed during the recession of the ice margin from the ice-sheet limit of the Chodzież Phase to the ice-sheet limit of the Pomeranian Phase.

The Wdzydze tunnel valley consists of two morphologically different, N–S oriented channels that may have formed as a result of quasi-parallel subglacial meltwater flows. The elongated hills that separate the channels are probably erosional remnants formerly dividing two main subglacial meltwater routes. They are superimposed by crevasse-like landforms, indicating that, within the ice sheet, crevasses may have opened within the zone of thinner ice between the two channels. The highly undulating thalweg of the channels, with deep depressions and a southern outlet higher than the northern fringe suggests that highly

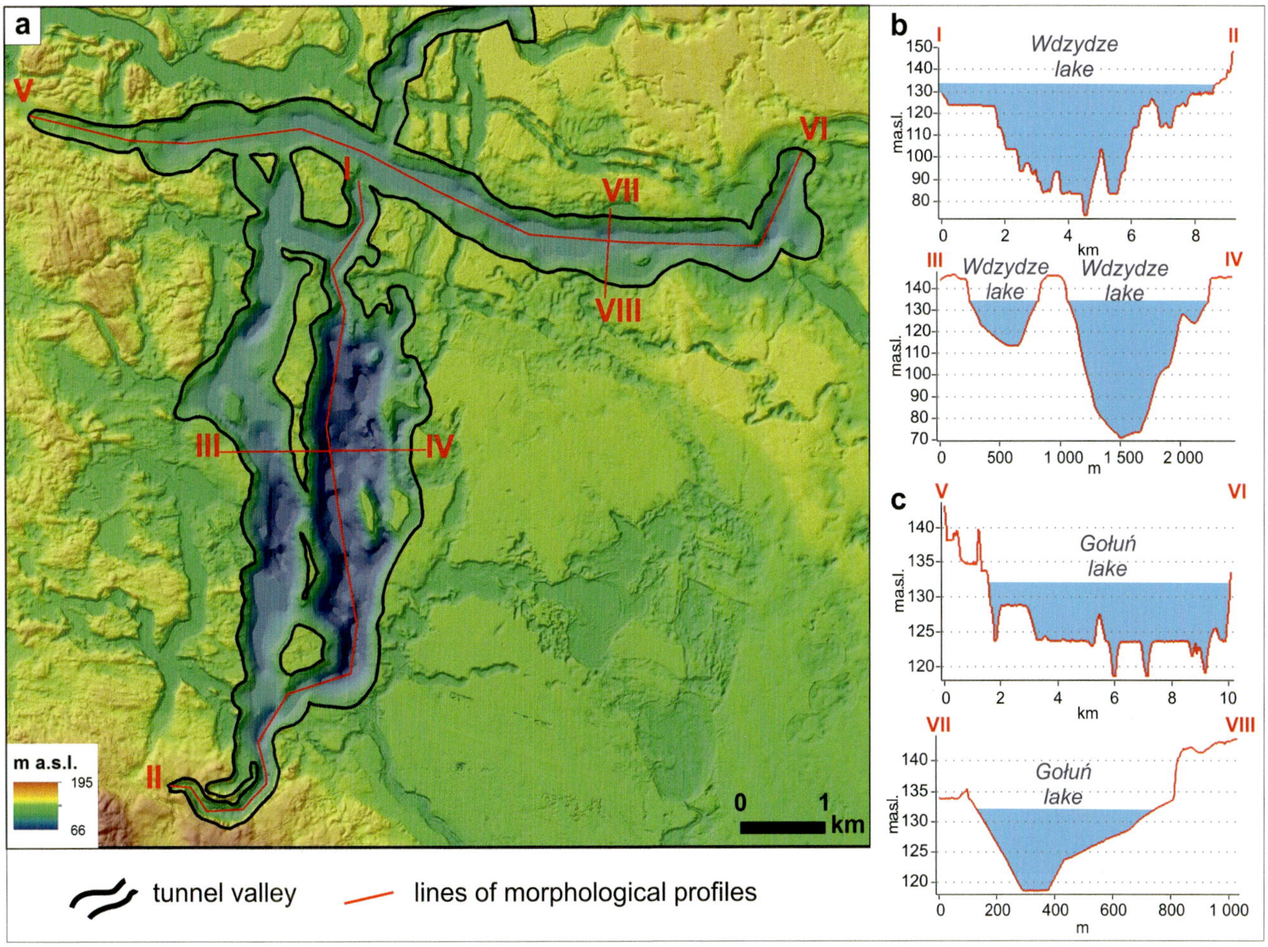

Fig. 36.6 The Wdzydze-Gołuń tunnel valley. **a** Digital elevation model with the outline of tunnel channel. **b** Longitudinal and transverse profiles of the Wdzydze tunnel valley. **c** Longitudinal and transverse profiles of the Gołuń tunnel valley. Digital terrain model was created using LiDAR elevation data with horizontal resolution of 1 m

pressurised subglacial meltwater flowing from the ice-sheet interior towards the ice margin was the main factor shaping the valley. In contrast, the Gołuń tunnel valley is oriented E–W and its thalweg is not so undulating, with much shallower depressions, suggesting that it is probably a kind of ice-marginal valley formed during later stages of deglaciation, when the ice margin was located around the northern fringes of the Wdzydze tunnel valley.

The location of the Wdzydze-Gołuń tunnel valley within the outwash plain area indicates that during the outwash accumulation in the Pomeranian Phase both landforms (Wdzydze and Gołuń valleys) were infilled with dead ice and thus preserved by burial. The ages of lakes located in other tunnel valleys of this region (Błaszkiewicz 2005) suggest that the dead-ice blocks that filled the Wdzydze-Gołuń tunnel valley finally melted in the Allerød, when the Wdzydze and Gołuń lakes initially formed. The contemporary characteristic outline of these lakes is the reason why their local names are "the Cross of the Wdzydze Lakes" or "the Kashubian Sea" (Fig. 36.7).

36.4.4 The Borzechowska Tunnel Valley

36.4.4.1 Geomorphology

The Borzechowska tunnel valley is located in the southern part of the Kashubian Lake District (Starogard Lake District) in the vicinity of the ice-sheet maximum limit of the Pomeranian Phase. In its northern and central parts the valley is incised into hummocky moraine plateaus, where a small drumlin field has been documented indicating active flow of the ice sheet during the Pomeranian Phase (Błaszkiewicz 2005). Along the south-eastern and southern edges of the tunnel valley there is an outwash plain that is connected with the Wda outwash route to the south. The Borzechowska tunnel valley, together with the tunnel valleys of Steklińskie and Sumińskie lakes, is connected southward to the network of tunnel valleys belonging to the present-day fluvial system of the Wda River (Fig. 36.2).

Fig. 36.7 Perspective view in the south direction of the Gołuń and Wdzydze lakes (photograph from LGD Stolem archive)

The tunnel valley is 11–14 km long with multiple channels oriented mostly NE–SW and ENE–WSW (Fig. 36.8a). Its width varies from ca. 130 m within the narrowing in the central part, up to 1.2 km within the widenings. The thalweg is highly undulating, as manifested by the longitudinal morphological profile of the valley floor with height differences up to 20–30 m (Fig. 36.8b). The depth of the incision is typically between 10 and 15 m in its western part, whereas in its eastern segment it is much deeper, up to 40 m (Fig. 36.8c). The deepest parts of the valley are currently occupied by three lakes: Borzechowskie Wielkie (max. depth 43 m), Steklińskie (max. depth 16 m) and Sumińskie (max. depth 7 m).

In the southern part of the tunnel valley and transverse to its course a second-order water divide between the catchments of the Wierzyca and Wda rivers is located. It is a rare morphological-hydrographic phenomena for the European Lowlands, with the divide located in the floor of the tunnel valley. Within the Borzechowskie lake basin a layer of lake-bottom sediments of up to 15 m thick, mainly comprising calcareous gyttja, has been documented. Lake bottom deposits of several metres thick overlain by a 1–3-m-thick peat layer also occur within the peatbogs adjacent to the lakes. In the zone where the Borzechowska tunnel valley contacts with the Wda River valley, the bottom lake deposits are intercalated with deltaic sediments, above which the meander belt was formed (Błaszkiewicz 2005).

36.4.4.2 Origin and Evolution

The Borzechowska tunnel valley and several other arteries connected to it to the north are a convergent system of subglacial meltwater outflow active during the maximum extent of the ice sheet in the Pomeranian Phase of the last glaciation. When the subglacial meltwater discharge had ceased, the already-formed tunnel valley was infilled with dead-ice blocks that preserved the landform despite subsequent fluvioglacial sedimentation. During the progressive recession of the ice margin from its maximum extent in the Pomeranian Phase, there was an outwash flow over the buried dead-ice blocks and the accumulation of sandur, which merged southward with the main outwash route of the Wda River.

The Borzechowska tunnel valley is the landform within which the asynchrony of dead-ice blocks melting and the formation of lake basins was documented. The final melting of dead ice and the formation of the Borzechowskie

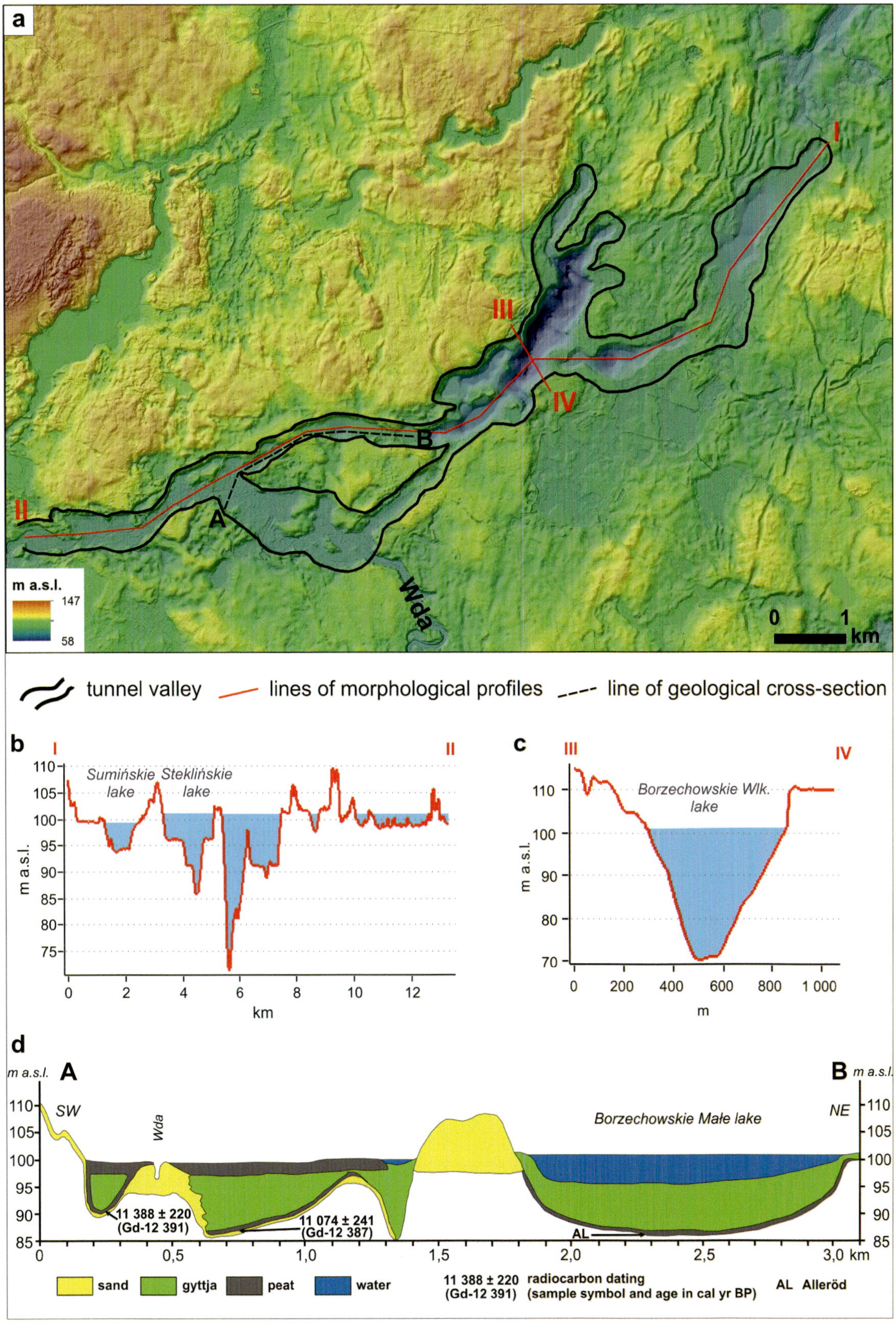

Fig. 36.8 The Borzechowska tunnel valley. **a** Digital elevation model with the outline of tunnel channel. **b** Longitudinal profile. **c** Transverse profile. **d** Geological cross-section A–B; radiocarbon dates calibrated with *IntCal20* calibration curve (after Reimer et al. 2020). Digital terrain model was created using LiDAR elevation data with horizontal resolution of 1 m

lakes (part of the tunnel valley located within the Wierzyca catchment) occurred in the Allerød, whereas in the southern part of the tunnel valley (drained by the Wda River) it occurred much later, in the second half of the Preboreal stage (Fig. 36.8d). This asynchrony was probably caused by a delay in thermokarst processes triggering complete melting of the dead ice in the southern part of the valley (Błaszkiewicz 2005). The Wda River incorporated previously-formed lakes into its fluvial system, so relatively quickly these basins were infilled with lake and deltaic deposits, over which a flood plain of the present-day Wda River developed.

36.4.5 The Raduńska Tunnel Valley

36.4.5.1 Geomorphology

The Raduńska tunnel valley belongs to a dense network of tunnel valleys incised into the moraine plateaus and outwash plains in the southern part of the Kashubian Lake District (Fig. 36.2). This network was formed mainly during the ice-margin standstill and recession of the Pomeranian Phase of the last glaciation as a result of intensive subglacial meltwater discharge relatively close to the ice margin, at the eastern flank of the Vistula Ice Lobe (Petelski and Moczulska 2006; Prussak et al. 2015). The Raduńska tunnel valley is located in very close proximity to the ice-sheet limit of the Pomeranian Phase. However, the northern segment of the valley lies to the north of this limit, within the extent of the Pomeranian Phase ice sheet, and its southern segment is located to the south of this limit, in front of the Pomeranian ice-sheet extent.

It is oriented NE–SW, incised deeply into a wavy moraine plateau and various levels of outwash plain, and occupied by two main lakes: Raduńskie Dolne and Raduńskie Górne (Fig. 36.9a). This tunnel valley has a relatively compact shape with one main, 16-km-long channel. It varies in width from 50 m at the valley outlet to 1.5 km in the central part. Despite the floor of the valley at its NE and SW fringes being at almost the same elevation (162 m asl), its thalweg is highly undulating and characterised by height differences up to 35 m along the longitudinal profile (Fig. 36.9b). The depth of the valley is typically in the

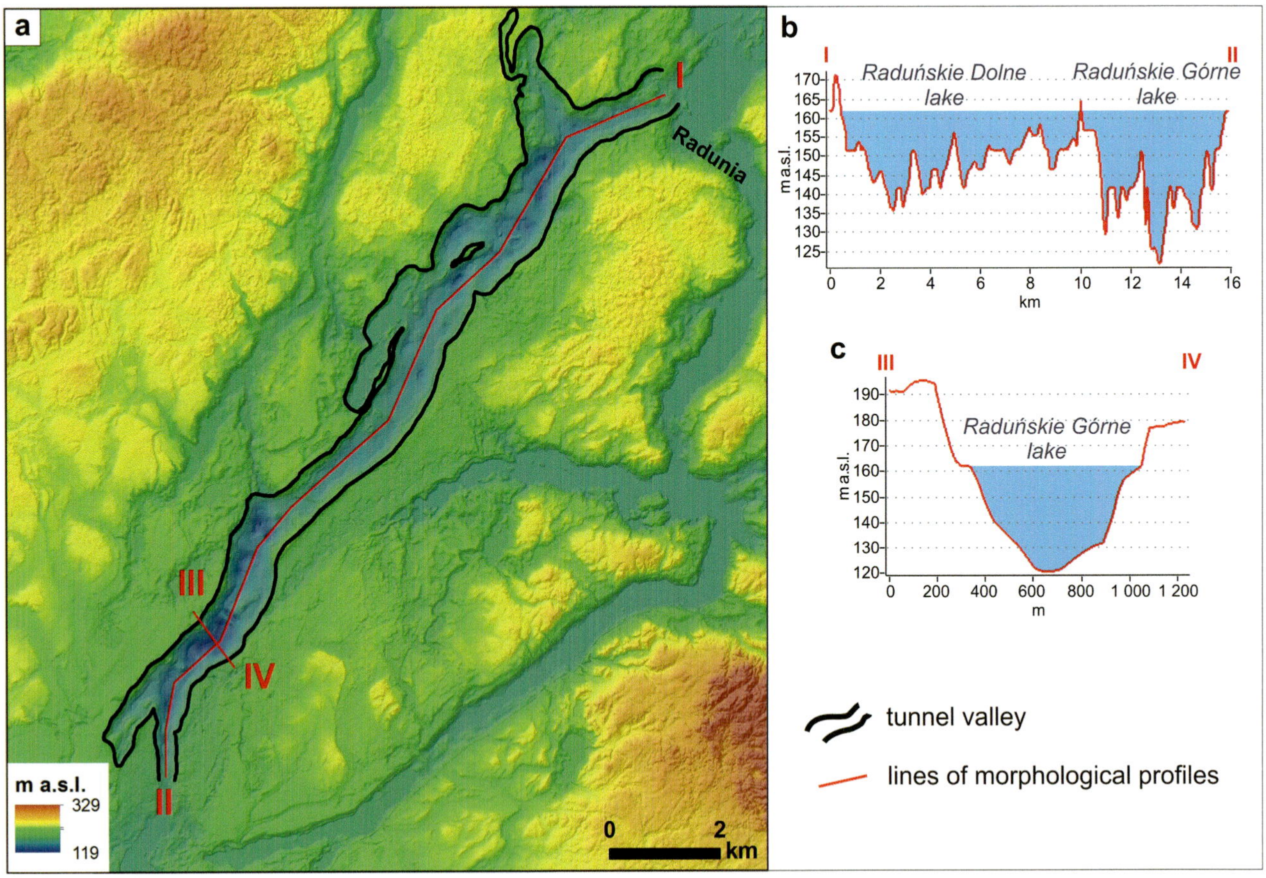

Fig. 36.9 The Raduńska tunnel valley. **a** Digital elevation model with the outline of tunnel channel. **b** Longitudinal profile. **c** Transverse profile. Digital terrain model was created using LiDAR elevation data with horizontal resolution of 1 m

order of 40–50 m, with maximum values of 70 m within an overdeepening in the south-western part (Fig. 36.9c). This makes the Raduńska tunnel valley one of the deepest incisions among the landforms described here.

36.4.5.2 Origin and Evolution

The Raduńska tunnel valley is located between the eastern flank of the Vistula Iced Lobe and the much smaller Bytów Ice Lobe to the west. These two ice lobes developed during the maximum extent of the ice sheet in the Pomeranian Phase and possibly partly also during the subsequent recession of the ice margin. The Raduńska tunnel valley was one of the main routes of subglacial meltwater drainage in the Pomeranian Phase in this region, and these meltwaters flowing southwards beyond the extent of the ice sheet formed the proximal parts of the Wda River outwash route. The southern segment of the valley must have formed before the main ice-margin standstill correlated with the Pomeranian Phase, probably during ice-sheet recession from the Chodzież Phase limit. However, the northern segment of the valley was probably formed during the ice-margin standstill in the Pomeranian Phase. The whole tunnel valley was infilled with dead ice after being shaped by subglacial meltwater, and was most likely also a route for subaerial outwash outflow during subsequent stages of deglaciation, as may be indicated by the lower outwash plain levels within the valley. According to Gołębiewski (1976), the final melting of dead-ice blocks and the beginning of the lake basin formation within the Raduńskie tunnel valley occurred in the Preboreal stage.

36.4.6 The Żarnowiecka Tunnel Valley

36.4.6.1 Geomorphology

The Żarnowiecka tunnel valley is the most northerly landform among those described in this chapter and lies in the north of the Kashubian Lake District, within the Żarnowiecka moraine plateau (Fig. 36.2). This tunnel valley dissects a moraine plateau, and only in the vicinity of its western edge small patches of outwash plains occur. The landform is open northwards, towards the Baltic Sea, where its peaty floor turns into a coastal plain. To the south, the tunnel valley ends with an ice-contact sedimentary scarp of the high outwash level.

The Żarnowiecka tunnel valley is not morphologically complicated, consisting of a single, deep, wide channel oriented NW–SE (Fig. 36.10a). The length of the tunnel valley is 10 km, and its width varies from 1.9 km at its outlet at

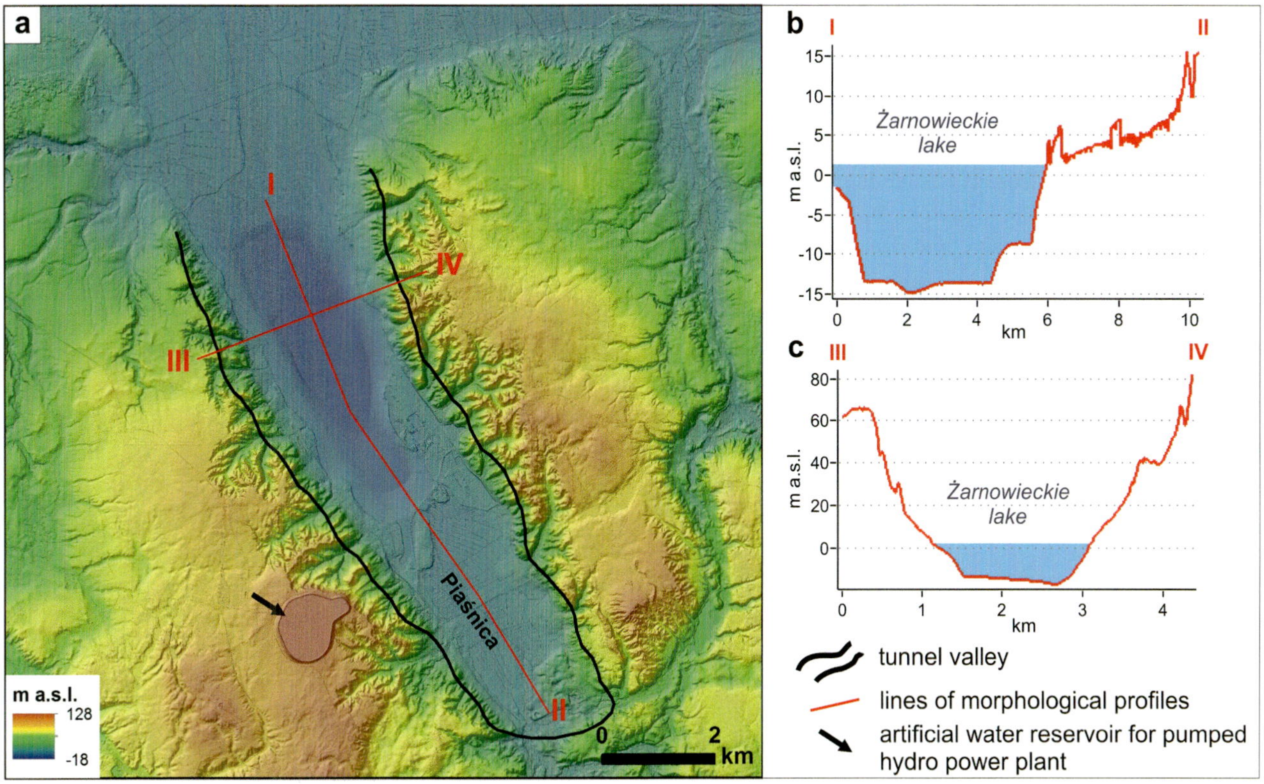

Fig. 36.10 The Żarnowiecka tunnel valley. **a** Digital elevation model with the outline of tunnel channel. **b** Longitudinal profile. **c** Transverse profile. Digital terrain model was created using LiDAR elevation data with horizontal resolution of 1 m

the south-eastern fringe to 3.7 km at the north-western end. Its undulating longitudinal morphological profile shows height differences along the thalweg up to 15 m. The north-eastern edge of the valley floor lies at ca. 0 m asl, whereas its south-eastern edge is ca. 15 m higher (Fig. 36.10b). The depth of the tunnel channel varies from 30 to 50 m in the south-east up to 80–100 m in the north-eastern part, where the Żarnowieckie lake is located (Fig. 36.10c).

Most of the tunnel valley floor is occupied by the Żarnowieckie lake and by fen peatbogs. A distinct morphological element of the tunnel valley are extensive terraces located 3–18 m above the current level of the Żarnowieckie lake, along both edges of the valley. These terraces have been interpreted as kame terraces (Pasierbski 1979) or, more recently, as lake terraces (Małka et al. 2021).

36.4.6.2 Origin and Evolution

Geological data indicate that along the axis of the Żarnowiecka tunnel valley there occurs a wide (up to 4 km) and deep (up to 322.9 m below sea level) buried valley incised into Cenozoic and Jurassic-Cretaceous sediments, reaching even Triassic beds (Małka et al. 2021). This buried depression extends southwards significantly beyond the tunnel valley that is visible in the contemporary landscape, and towards the north it continues towards the Baltic Sea (Kramarska et al. 2015). The floor of the buried valley is undulating, with height differences exceeding 200 m. The morphology of the valley and its relation to the early Pleistocene erosive bases suggest that it is a deeply incised, buried tunnel valley. Its origin was possibly related to cooperating processes of glacial erosion and subglacial meltwater erosion of the ice bed, probably during the Elsterian glaciation. During subsequent glacial–interglacial cycles the valley underwent repeated phases of burial by fluvioglacial and glaciolacustrine deposits, and dissection. The location of the Żarnowiecka tunnel valley that is currently visible within the landscape is directly related to the location of the buried tunnel valley that finally formed during the Saalian glaciation. The reproduction of this landform, and the final shaping of the open valley visible in the relief occurred during the ice-margin standstill and recession during the Gardno Phase of the last glaciation as a result of glacial erosion and subglacial meltwater erosion. The significant role of pressurised subglacial meltwater in the final shaping of the tunnel valley is indicated by the proximal zone of an outwash plain associated with its southern edge. During later stages of deglaciation, the tunnel valley was most likely also infilled with dead ice that spared it from burial. The dead ice covered with fluvioglacial sediments existed here under the periglacial conditions until the Allerød warming, when it finally melted and the Żarnowieckie lake was formed (Latałowa 2002).

36.5 Summary and Conclusions

Landforms presented in this chapter were formed during the general recession of the last Scandinavian Ice Sheet, within ice-marginal zones related to local and regional (e.g., the Pomeranian Phase) ice-sheet standstills, when the ice front was in the state of dynamic equilibrium. The main factor triggering the formation and evolution of tunnel valleys was erosion of the ice bed caused by the discharge of pressurised subglacial meltwater. This kind of erosion may occur rapidly as a result of high-magnitude discharge related to catastrophic subglacial drainage of meltwater stored within supra- or subglacial lakes (Fig. 36.11a). The rapid drainage of pressurised subglacial meltwater stored behind an ice margin frozen to its bed (the occurrence of subglacial permafrost) was also postulated as a possible interpretation of the "outburst" hypothesis for tunnel valley formation (e.g., Piotrowski 1994; Cutler et al. 2002; Hooke and Jennings 2006). This kind of one-time, catastrophic event may result in high-magnitude, bank-full discharge of meltwater within subglacial channels, producing tunnel channel landforms according to the terminology of Clayton et al. (1999). In our opinion, only one of the presented landforms may be interpreted as a typical tunnel channel, and it is the Wiecka tunnel channel. The other landforms are characterised by more complicated morphology and are parts of very complex systems of subglacial drainage, which suggests that they were formed by long-lasting and diversified discharge of subglacial meltwater. It could have been repeated catastrophic events (outbursts), and/or "steady-state" erosion of a soft-sediment bed in low-pressure subglacial channels (Fig. 36.11b). Therefore, they are interpreted as typical tunnel valleys according to the terminology of Clayton et al. (1999), not reflecting a single erosion event but rather being the result of multiple widenings and deepenings of the valley within subglacial channels. In this case, we cannot exclude the role of glacial erosion, which to some extent could have influenced the formation of tunnel valleys. The most complex history of origin and evolution typifies the Żarnowiecka tunnel valley. The present-day open valley was formed along a deeply incised buried valley that originated at least during the Elsterian glaciation. This landform was reproduced during subsequent glaciations, and it was probably finally shaped during the Gardno Phase of the last glaciation.

After being shaped within subglacial channels, the tunnel valleys presented in this chapter were most likely infilled with dead ice and/or ice that originated from supercooled subglacial meltwater (Fig. 36.11c, d). During deglaciation, this ice-fill was buried by morainic, fluvioglacial or glaciolacustrine sediments. In the case of tunnel valleys located within extensive outwash plains of the

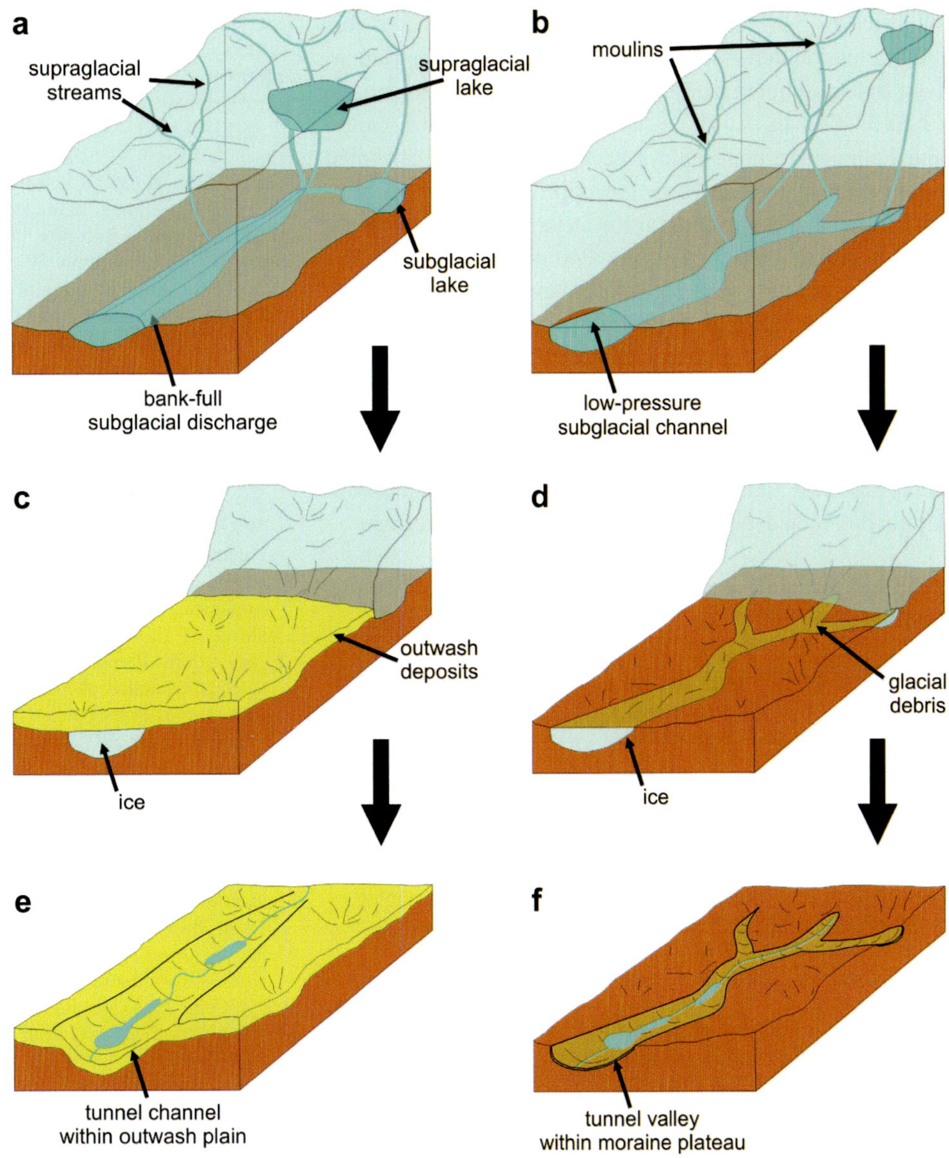

Fig. 36.11 Idealized model of the origin and evolution of tunnel valleys (channels). **a** Catastrophic, bank-full subglacial drainage of meltwater stored within supra- or subglacial lakes. **b** "Steady-state" erosion of ice-sheet bed in low-pressure subglacial channels. **c** Tunnel channel infilled with dead ice and/or ice that originated from supercooled subglacial meltwater, buried with thick layer of outwash sand and gravel. **d** Tunnel valley infilled with dead ice and/or ice that originated from supercooled subglacial meltwater and buried with a layer of glacial debris. **e** Tunnel channel visible within outwash plain after melting of buried ice. **f** Tunnel valley visible within moraine plateau after melting of buried ice

Tuchola Forest region, dead-ice fillings were additionally covered with layers of thick fluvioglacial sand and gravel (Fig. 36.11c). As a result of outwash flow over buried dead ice and its subsequent melting, the original morphology of these tunnel valleys was significantly modified after deglaciation, as indicated within the Wiecka tunnel channel by outwash ridges entering the valley. Therefore, we postulate that tunnel valleys located within extensive outwash plains are a kind of "palimpsest tunnel valleys" (Fig. 36.11e), in contrast to tunnel valleys located within moraine plateaus, whose morphology closely corresponds to the shape of subglacial channels carved into the ice bed (Fig. 36.11f).

Dead-ice fillings of tunnel valleys, buried by sediments, survived under periglacial conditions with continuous and patchy permafrost for several thousand years until the Allerød warming and, in the case of the Raduńska tunnel valley and part of the Borzechowska tunnel valley, until the beginning of Holocene. The final melting of the dead ice resulted in the appearance of the tunnel valleys within the landscape and initial formation of lakes in the deepest parts of the valleys. Most tunnel valleys in Poland, and all those presented in this chapter, were incorporated into the fluvial systems that formed during and after deglaciation, and they became important parts of polygenetic river valleys (Błaszkiewicz et al. 2015). The Charzykowska tunnel valley is part of the Brda River valley; the Wdzydze-Gołuń tunnel valley, the Wiecka tunnel valley and part of the Borzechowskie tunnel valley are parts of the Wda River valley; the Raduńska tunnel valley is included in the Radunia River valley; and the Żarnowiecka tunnel valley constitutes part of the Piaśnica River valley. Since incorporation into the river valleys, they have been subjected to fluvial transformation, and the degree of this transformation depends on the relation between the dimension and morphology of tunnel valleys and the magnitude of river discharges. The Charzykowska, Wdzydze-Gołuń and Raduńska tunnel valleys have been least transformed because they are located in the upper parts of respective river valleys. Despite the Żarnowiecka tunnel valley being located in the lower part of the Piaśnica River valley, the large disproportion between the dimension of tunnel valleys and the magnitude of the Piaśnica discharge has caused the fluvial transformation of this landform also to be low. The highest degree of fluvial transformation is typical for the Wiecka and Borzechowska tunnel valleys, which are located in the middle part of the Wda River valley. In the latter case, the tunnel valley floor was entirely filled with lake and deltaic deposits, and a meander belt was formed.

References

Bennett MR, Glasser NF (2009) Glacial geology: ice sheets and landforms, 2nd edn. Wiley-Blackwell, Oxford, p 385

Błaszkiewicz M (2005) Późnoglacjalna i wczesnoholoceńska ewolucja obniżeń jeziornych na Pojezierzu Kociewskim (wschodnia część Pomorza). Prace Geogr IGiPZ PAN 201, Warszawa, p 192

Błaszkiewicz M (2011) Timing of the final disappearance of permafrost in the Central European Lowland as reconstructed from the evolution of lakes in N Poland. Geol Quart 55:361–374

Błaszkiewicz M, Gruszka B (2005) Development and infill of Vistulian glacial Lake Gniew (N Poland): a sedimentological analysis. Geol Quart 49:449–462

Błaszkiewicz M, Piotrowski J, Brauer A, Gierszewski P, Kordowski J, Kramkowski M, Lamparski P, Lorenz S, Noryśkiewicz A, Ott F, Słowiński M, Tyszkowski S (2015) Climatic and morphological controls on diachronous postglacial lake and river valley evolution in the area of Last Glaciation, northern Poland. Quat Sci Rev 109:13–27

Boulton GS, Hindmarsh RCA (1987) Sediment deformation beneath glaciers: rheology and geological consequences. J Geophys Res 92:9059–9082

Clayton L, Attig JW, Mickelson DM (1999) Tunnel channels formed in Wisconsin during the last glaciation. In: Mickelson DM, Attig JW (eds) Glacial processes past and present: boulder, Colorado, vol 337. Geological Society of America, Special Papers, pp 69–82

Cutler PM, Colgan PM, Mickelson DM (2002) Sedimentologic evidence for outburst floods from the Laurentide Ice Sheet margin in Wisconsin, USA: implications for tunnel-channel formation. Quat Int 90:23–40

Galon R (1953) Morfologia doliny i sandru Brdy. Stud Soc Sci Torunensis C 1(6):1–53

Galon R (1965) Some new problems concerning subglacial channels. Geogr Pol 6:19–28

Gołębiewski R (1976) Osady denne jezior Raduńskich. GTN, Gdańsk, p 92

Gottsche C (1897) Die tiefsten Glacialablagerungen der Gegend von Hamburg. Vorläufige Mitteilung. Mitt Geogr Gesell Hamburg 13:131–140

Gripp K (1964) Erdgeschichte von Schleswig-Holstein. Wachholtz-Verlag, Neumünster, p 412

Hooke RL, Jennings CE (2006) On the formation of the tunnel valleys of the southern Laurentide ice sheet. Quat Sci Rev 25:1364–1372

Jentzsch A (1884) Über die Bildung der preussischen Seen. Z Deut Geol Ges Erdk 36:699–702

Jørgensen F, Sandersen PBE (2006) Buried and open tunnel valleys in Denmark—erosion beneath multiple ice sheets. Quat Sci Rev 25:1339–1363

Kaiser K, Hilgers A, Schlaak N, Jankowski M, Kühn P, Bussemer S, Przegiętka K (2009) Palaeopedological marker horizons in northern central Europe: characteristics of Lateglacial Usselo and Finow soils. Boreas 38:591–609

Kehew A, Nicks L, Straw W (1999) Palimpsest tunnel valleys: evidence for relative timing of advances in an interlobate area of the Laurentide ice sheet. Ann Glaciol 28:47–52

Kehew AE, Piotrowski JA, Jørgensen F (2012) Tunnel valleys: concepts and controversies—a review. Earth Sci Rev 113:33–58

Kozarski S (1966) The origin of subglacial channels in the North Polish and North German Plain. Bull Soc Amis Sci Lett Poznań B 20:21–36

Kramarska R, Kasiński JR, Czapowski G, Piwocki M, Słodkowska B (2015) Referencyjny profil osadów neogenu Pomorza Wschodniego w Łęczycach koło Lęborka (Północna Polska). Biul Państw Inst Geol 461:193–250

Krupiński KM (1995) Pollen stratigraphy and successions of vegetation during the Mazovian Interglacial. Acta Geogr Lodz 70:1–200

Latałowa M (2002) Ewolucja środowiska przyrodniczego w rynnie jeziora Żarnowieckiego na otaczających ją wysoczyznach w ciągu ostatnich około 11 tys. 14C lat BP. In: Geologia regionu gdańskiego. Przewodnik LXXIII Zjazdu Polskiego Towarzystwa Geologicznego. PIG, Gdańsk, pp 49–57

Livingstone SJ, Clark CD (2016) Morphological properties of tunnel valleys of the southern sector of the Laurentide Ice Sheet and implications for their formation. Earth Surf Dyn 4:567–589

Majdanowski S (1947) Rozmieszczenie, gęstość i kierunki rynien jeziornych na Niżu Polskim. Przegl Geogr 21:37–71

Majdanowski S (1950) Zagadnienie rynien jeziornych na Niżu europejskim. Bad Fizjogr Nad Polską Zach 2:35–122

Małka A, Jurys L, Maszloch E, Wirkus K, Pączek U (2021) Szczegółowa mapa geologiczna Polski 1:50,000, ark. Sławoszyno (5). Centr Arch Geol Państw Inst Geol, Warszawa

Marks L (2002) Last Glacial Maximum in Poland. Quat Sci Rev 21:103–110

Marks L, Ber A, Gogołek W, Piotrowska K (eds) (2006) Geological map of Poland 1:500,000. Ministerstwo Środowiska, PIG-PIB, Warszawa

Molewski P (1999) Rynna Gopła – problem jej genezy i roli w odpływie wód roztopowych podczas zlodowacenia vistuliańskiego. Stud Soc Sci Torunensis 10(6):180

Niewiarowski W (1995) Diagnostic features of subglacial channels of glacial and glacio-glaciofluvial origin, exampled by channels of the Chełmno-Dobrzyń and the eastern Gniezno Lakelands. Quaest Geogr Spec Iss 4:225–231

Nitychoruk J, Binka K, Ruppert H, Schneider J (2006) Holsteinian interglacial marine isotope stage 11? Quat Sci Rev 25:2678–2681

Nitychoruk J, Bińka K, Sienkiewicz E, Szymanek M, Chodyka M, Makos M, Ruppert H, Tudryn A (2018) A multiproxy record of the Younger Holsteinian Oscillation (YHO) in the Ossówka profile, eastern Poland. Boreas 47:855–868

Nowaczyk B (1994) Geomorfologia rynny glacjalnej jezior Małe Głuche i Duże Głuche; cechy fizyko- chemiczne kredy jeziornej: przyczyny zaniku jezior Małe Głuche i Duże Głuche. In: Nowaczyk B (ed) Geomorfologia i osady strefy litoralnej jezior, Instytut Badań Czwartorzędu UAM, Poznań, pp 67–70

Pasierbski M (1975) Uwagi o genezie niecki Jeziora Charzykowskiego. Acta Univ Nicolai Copernici Geogr 11:101–113

Pasierbski M (1979) Remarks on the genesis of subglacial channels in northern Poland. Eiszeit Gegenw 29:189–200

Paterson WSB (1981) The physics of glaciers. Pergamon Press, Oxford, p 380
Petelski K, Majewska A (2008) Szczegółowa mapa geologiczna Polski 1:50,000, ark. Karsin (127). Centr Arch Geol Państw Inst Geol, Warszawa
Petelski K, Moczulska G (2006) Szczegółowa mapa geologiczna Polski 1:50,000, ark. Egiertowo (53). Centr Arch Geol Państw Inst Geol, Warszawa
Piotrowski JA (1994) Tunnel-valley formation in northwest Germany—geology, mechanisms of formation and subglacial bed condition for the Bornhoved tunnel valley. Sediment Geol 89:107–141
Prussak W, Jurys L, Kaulbarsz D (2015) Szczegółowa mapa geologiczna Polski w skali 1:50,000, ark. Stężyca (52). Centr Arch Geol Państw Inst Geol, Warszawa
Reimer PJ, Austin WEN, Bard E, Bayliss A, Blackwell PG, Ramsey CB, Butzin M, Cheng H, Edwards RL, Friedrich M, Grootes PM, Guilderson TP, Hajdas I, Heaton TJ, Hogg AG, Hughen KA, Kromer B, Manning SW, Muscheler R, Palmer JG, Pearson Ch, van der Plicht J, Reimer RW, Richards DA, Scott EM, Southon JR, Turney CSM, Wacker L, Adolphi F, Büntgen U, Capano M, Fahrni SM, Fogtmann-Schulz A, Friedrich R, Köhler P, Kudsk S, Miyake F, Olsen J, Reinig F, Sakamoto M, Sookdeo A, Talamo S (2020) The IntCal20 northern hemisphere radiocarbon age calibration curve (0–55 cal kBP). Radiocarbon 62:725–757
Rinterknecht VR, Marks L, Piotrowski JA Raisbeck GM, Yiou F, Brook EJ, Clark PU (2005) Cosmogenic ^{10}Be ages on the Pomeranian Moraine, Poland. Boreas 4:186–191
Roman M (2016) Pojezierze eemskie: uwagi o genezie i zaniku jezior polodowcowych centralnej Polski. Acta Geogr Lodz 105:11–25
Stackebrandt W (2009) Subglacial channels of Northern Germany—a brief review. Z Deut Ges Geowiss 160:203–210
Tobolski K (2010) Preliminary information on the palynological research on sediments of Lake Charzykowskie (Zaborski Landscape Park). Stud Limn Telmat 4:29–34
Tylmann K, Rinterknecht V, Woźniak PP, Bourles D, Schimmelpfennig I, Guillou V, Aumaitre G, Keddadouche K (2019) Retreat of the southern front of the last Scandinavian Ice Sheet: dates and rates. In: INQUA congress, Dublin, Ireland. Abstract book for INQUA 2019, p 227
Urbański K (2012) Szczegółowa mapa geologiczna Polski 1:50,000, ark. Przechlewo (163). Centr Arch Geol Państw Inst Geol, Warszawa
Ussing NV (1903) Om Jyllands hedesletter og teorierne for deres Dannelse. Oversigt over Det Kongelige Danske Videnskabernes Selskabs Forhandlingar 2:1–152
Van Der Vegt P, Janszen A, Moscariello A (2014) Tunnel valleys: current knowledge and future perspectives. Geol Soc Lond Spec Publ 368:75–97
Woldstedt P (1952) Die Entstehung der Seen in den ehemals vergletscherten Gebieten. Eiszeit Gegenw 2:146–153

Mirosław Błaszkiewicz is Professor of Earth Sciences, geomorphologist and specialist in geology and paleogeography of the Quaternary. He conducts research in Poland and Germany, as well as Norway and Svalbard. He is the author of many monographs, scientific articles and sheets of the Detailed Geological Map of Poland. He was the PI of research projects of the National Science Center in Poland and the manager of the international ICLEA project on the Polish side. Since 1985 he has been an employee of the Institute of Geography and Spatial Organization of the Polish Academy of Sciences. In 2004–2021 he was the head of the Department of Environmental Resources and Geohazards of the Institute of Geography and Spatial Management, Polish Academy of Sciences, in Toruń.

Karol Tylmann is a glacial geomorphologist, specialist in palaeoglaciological reconstructions and palaeo-ice sheet dating with cosmogenic nuclides. He graduated from the Nicolaus Copernicus University in Toruń, Poland, when he completed his PhD in 2014. Currently he works as Assistant Professor at the Faculty of Oceanography and Geography, University of Gdańsk, Poland. He conducts research mainly in northern Poland, but also in Scandinavia and in the Georgian Caucasus. He is an author of several research articles as well as chapters in monographs. He was a Principal Investigator of the National Science Centre project "DatErr", and is involved in the project related to palaeo-ice streams record at the Baltic Sea floor.

Glacial Curvilineations of the Dobrzyń Lakeland—Spectacular Landscape Carved by Subglacial Meltwater Erosion

37

Wojciech Wysota, Aleksander Adamczyk, Jan A. Piotrowski and Piotr Weckwerth

Abstract

The Dobrzyń Lakeland (north-central Poland) is characterised by a remarkable subglacial landscape generated by the last Scandinavian Ice Sheet. This landscape exhibits consistently arranged groups of ridges and intervening troughs in a network of anabranching tunnel valleys terminating at the former ice margin. These ridges have been earlier described and interpreted as drumlins, although their geomorphic characteristics are at odds with landforms typically considered as drumlins. Recent studies revealed that these ridges represent a new type of glacial landscape referred to as "glacial curvilineations" (GCLs). GCLs occur in multiple and morphologically diverse swarms occupying tunnel valleys in the broad area referred to as the Zbójno GCL field. These swarms contain numerous elongated, winding and parallel ridges separated by troughs of exceptional regularity and high sinuosity. The GCLs are mostly composed of tills, interbedded with sand and gravel in places. The model of GCL formation invokes erosion by channelised meltwater flows beneath the ice sheet. It is proposed that longitudinal vortices within subglacial meltwater flows excavated troughs in the bed and left behind ridges as remnants of an older, antecedent landscape. The scale and coeval formation of the entire tunnel valley network suggests that this unique subglacial landscape results from a catastrophic drainage of meltwater stored in supraglacial or possibly subglacial lakes.

Keywords

Glacial curvilineations (GCLs) · Subglacial landscape · Last Scandinavian ice sheet · North-central Poland

37.1 Introduction

The glacial curvilineations landscape of the Dobrzyń Lakeland is a unique subglacial landsystem generated under the Vistula ice lobe at the southern margin of the Scandinavian Ice Sheet (SIS) in north-central Poland (Niewiarowski et al. 1995; Wysota et al. 2009; Wysota and Molewski 2011; Fig. 37.1). It includes numerous sets of enigmatic landforms consisting of elongated, winding and parallel ridges occurring in tunnel valleys[1] that form a complex and extensive geomorphic network (Lesemann et al. 2010, 2014). These landforms, previously described by many researchers, were interpreted as drumlins, although neither the shapes of the individual hills nor the geomorphology of the entire assemblage of the hills correspond to what is typically considered as drumlins and drumlin fields, respectively. These features in the Dobrzyń Lakeland were known in the literature as "the Zbójno drumlin field".

W. Wysota (✉) · P. Weckwerth
Faculty of Earth Sciences and Geospatial Management, Nicolaus Copernicus University in Toruń, Lwowska 1, 87–100 Toruń, Poland
e-mail: wysota@umk.pl

P. Weckwerth
e-mail: pweck@umk.pl

A. Adamczyk
State Water Holding Polish Waters, Regional Water Management Authority in Bydgoszcz, Mickiewicza 15, 85–071 Bydgoszcz, Poland
e-mail: ale.adamczyk@gmail.com

J. A. Piotrowski
Department of Geoscience, Aarhus University, Høegh-Guldbergs Gade 2, 8000 Aarhus C, Denmark
e-mail: jan.piotrowski@geo.au.dk

[1] In the chapter, the term "tunnel valley" is used in the broadest non-genetic sense referring to all valleys eroded by subglacial meltwater including "tunnel channels" in the sense of Clayton et al. (1999).

P. Migoń and K. Jancewicz (eds.), *Landscapes and Landforms of Poland*,
World Geomorphological Landscapes, https://doi.org/10.1007/978-3-031-45762-3_37

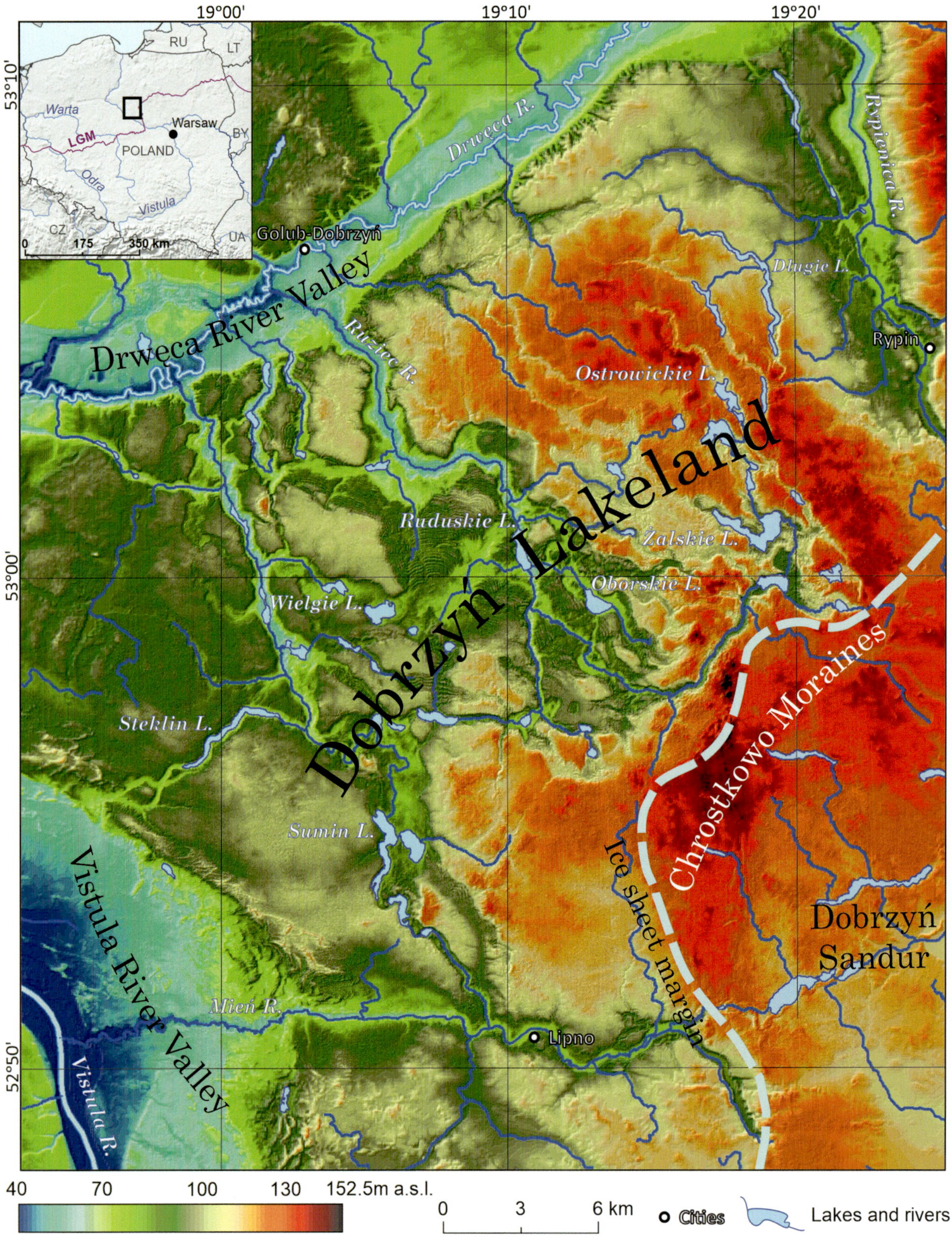

Fig. 37.1 Digital terrain model of the Dobrzyń Lakeland, with the Drwęca River valley to the north, Vistula River valley to the west and an ice-marginal zone (broken line) marking a former standstill of the Scandinavian Ice Sheet (SIS). Anabranching tunnel valley network with glacial curvilineations (GCLs) crossing the area from NW to SE forms a broad area of complicated morphology referred to as the Zbójno GCL field. Inset shows location of the study area in relation to maximum extent of SIS during the Late Weichselian glaciation (LGM)

Almost 100 years ago, a Polish geologist Wiktor Nechay was the first to write about this unique subglacial landscape. Nechay (1927) mapped regular assemblages of parallel landforms (Fig. 37.2), which he described as "drumlin hills with elliptical shape, more or less elongated, with a rounded crest and uniformly gentle slopes, separated by trough depressions in which lakes occur" (Nechay 1927, p. 97, *in Polish*). He suggested that these unusual features were formed by glacial erosional-depositional processes.

More detailed studies of this subglacial landscape appeared in the following decades (e.g. Jewtuchowicz 1956; Niewiarowski 1957; Liberacki 1961; Lamparski 1972; Baranowski 1977, 1979; Wysota 1993, 1994; Olszewski 1997; Chutkowski and Olszewski 2008; Głębicki and Marks 2009). Both the morphology and internal structure of these hills—referred to as drumlins in accordance with Nechay's earlier interpretation—were studied. Numerous hypotheses of their formation were put forward: an erosional and depositional model (Jewtuchowicz 1956), an erosional and deformational–depositional model (Olszewski 1997; Chutkowski and Olszewski 2008), a model invoking ablation and sediment squeezing around dead-ice blocks (Lamparski 1972; Głębicki and Marks 2009) or a subglacial meltwater erosion model (Niewiarowski 1957; Niewiarowski et al. 1995; Liberacki 1961). Additionally, some hypotheses have invoked multi-process mechanisms combining various subglacial processes as mentioned above (Wysota 1993, 1994).

More recent studies (Lesemann et al. 2010, 2014) showed that these hills represent a new landform type referred to as glacial curvilineations (GCLs) and interpreted them as having been eroded from older landscapes and sediments by channelised subglacial meltwater flows. Thanks to the availability of high-resolution digital terrain models (DTMs), other occurrences of GCLs have been documented in Poland and in the neighbouring areas of Germany and Denmark covered by the SIS (Weckwerth et al. 2019; Adamczyk et al. 2022). Recently, GCL landscapes were also recognised along the southern sector of the Laurentide Ice Sheet in the northern states of the USA (Clark and Livingstone 2018).

The main aim of the chapter is to illustrate the geomorphology of the GCL landscape in the north-western part of the Dobrzyń Lakeland. The morphology, morphometry and geological composition of the GCLs are presented. Furthermore, the origin of the GCL landforms and the entire GCL landscape is addressed.

37.2 Regional Setting

The Dobrzyń Lakeland is located near the confluence of the Drwęca and Vistula (Weichsel) River valleys in north-central Poland (Fig. 37.1). It is a low-relief moraine plateau mostly consisting of flat, partly undulating till plains hosting ice-marginal and dead-ice features formed during the recession of the SIS around 17–18 ka ago (Niewiarowski et al. 1995; Wysota et al. 2009; Wysota and Molewski 2011). The moraine plateau surface rises gently from ca. 90–100 m asl near the Vistula River valley in the west to ca. 130–140 m asl close to the Rypienica River in the east.

The moraine plateau consists of Pleistocene sediments, usually 30–80 m in thickness, overlying Miocene clay and sand formations (Dzierżek 2007, 2008; Wysota 2007, 2009, 2017; Dzierżek and Szymanek 2014, 2015; Wysota and Sokołowski, 2017, 2018). The Pleistocene sediment succession is mostly represented by three to five till layers often separated by glaciofluvial and glaciolacustrine deposits, and less frequently by fluvial successions. The surface of the moraine plateau is typically composed of Late Weichselian till that is up to 10 m thick (Wysota 2007, 2009; Dzierżek 2008, 2009; Dzierżek and Szymanek 2014, 2015; Wysota and Sokołowski 2017, 2018).

The moraine plateau is dissected by prominent tunnel valleys, which include regular groups (swarms) of GCLs (Fig. 37.2) consisting of long, parallel, winding ridges separated by troughs extending over several kilometres (Lesemann et al. 2010, 2014; Adamczyk 2023). The tunnel valleys start near the Drwęca River valley and extend to the ice-marginal sedimentary complex in the south-east that marks a temporary standstill of the Vistula ice lobe during ice-sheet decay (Niewiarowski et al. 1995).

GCL landforms occur within tunnel valleys of variable and complex morphologies (Figs. 37.1 and 37.3) comprising simple tunnel valleys (STVs) and compound tunnel valleys (CTVs) (Adamczyk et al. 2022). The CTVs are large "trunk" valleys of up to 20 km long and 7 km wide that are incised as much as 40 m into the moraine plateau surface and exhibit steep margins and undulating floors, sometimes with thresholds and overdeepenings containing lakes (Lesemann et al. 2010; Adamczyk 2023). Much smaller than these are the STVs, which are up to 10 km long and 30 m deep, mostly narrow (up to 1 km) winding valleys with characteristic box-like cross profiles termed "incipient valleys" by Lesemann et al. (2010). They are usually tributaries to the trunk valleys but also appear as separate tunnel valleys, often parallel to each other, on the surface of moraine plateau. In the central part of the GCL area, simple and compound tunnel valleys merge and create broad (up to 7 km wide), elongated (up to 8 km long) depressions called "glacial basins" (Olszewski 1997) that extend towards an ice-marginal zone in the south-east (Figs. 37.1 and 37.3). Within these depressions, the more continuous and longest GCL swarms of exceptional regularity occur (Lesemann et al. 2010; Adamczyk et al. 2016, 2022). Regionally, simple and compound tunnel valleys form an integrated anabranching tunnel valley network (tunnel valley complex)

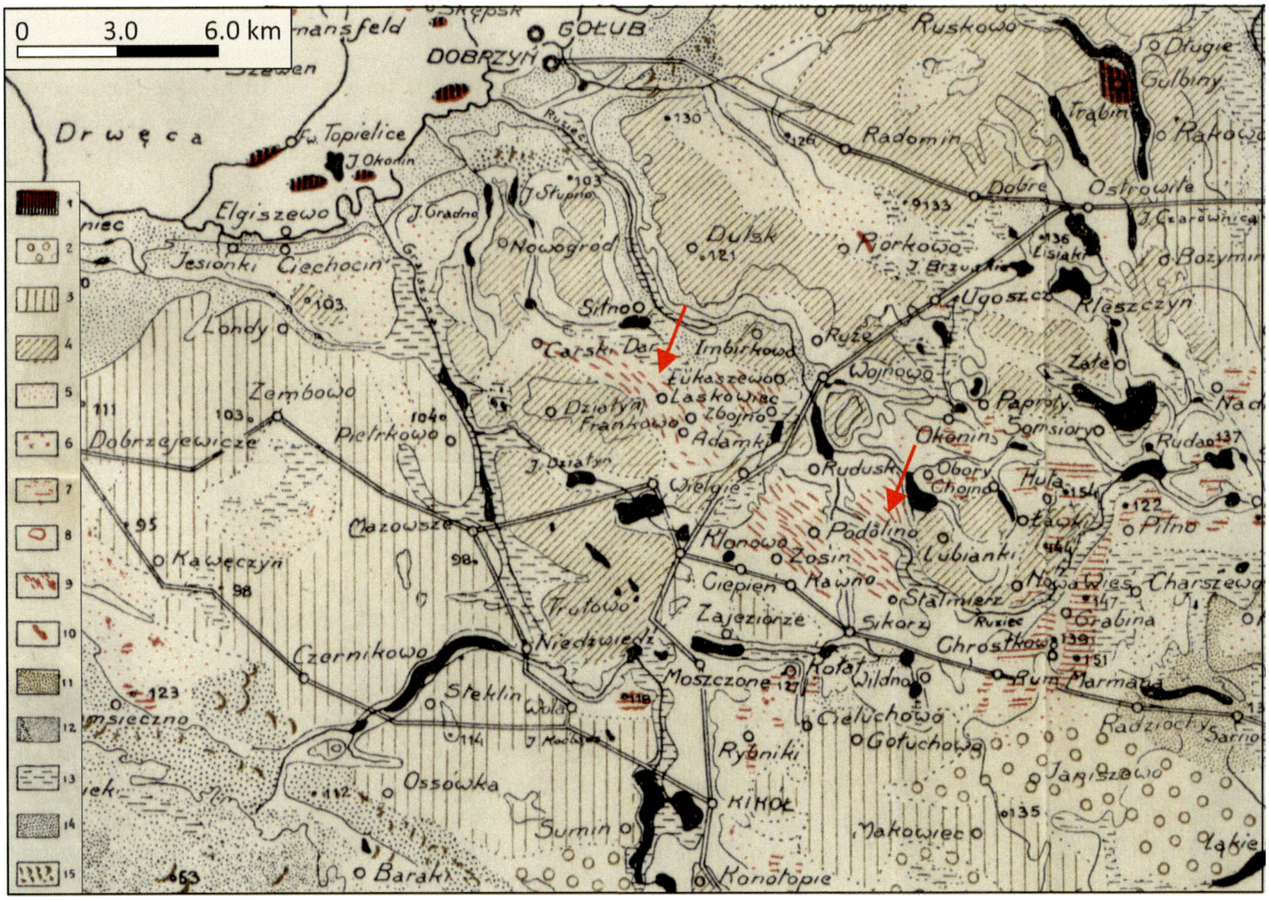

Fig. 37.2 Major glacial features of the NW part of the Dobrzyń Lakeland on the first geological map of this area (Nechay 1927). Note groups of landforms (red arrows) that were considered as drumlins. Explanations: 1—Lignite formation and grey clays (Miocene), 2—Ground moraine with sand and boulders, 3—Ground moraine with sand admixture, 4—Massive ground moraine, 5—Sands and gravels, 6—Concentration of cobbles and boulders, 7—End moraines, 8—Denuded end moraines, 9—Drumlins, 10—Eskers, 11—Sandur, 12—Valley sands, 13—Peat and swamps, 14—Aeolian sands, 15—Dunes

carved into the remnants of moraine plateaus of various sizes.

The tunnel valleys typically have down-ice rising longitudinal profiles with a distinct topographic gradient rising from about 50 m asl at the heads of the valleys in the north-west to about 150 m asl at the top of the ice-marginal sedimentary complex in the south-east. This gives a rise of around 100 m over a distance of 20 km along the ice-flow line (Lesemann et al. 2010, 2014).

The tunnel valleys converge into an extensive ice-marginal sedimentary complex extending from the town of Lipno in the south-west to the town of Rypin in the north-east (Nechay 1927; Pasierbski 1984; Niewiarowski et al. 1995; Dzierżek 2007, 2008, 2009; Dybowski 2009; Lesemann et al. 2010, 2014; Dzierżek and Szymanek 2014, 2015). This complex consists of a broad, apron-like accumulation zone composed of coalescing outwash fans formed at the mouths of the tunnel valleys. The most impressive topographic features in this zone occur east of the village of Chrostkowo and are known as the Chrostkowo Moraines (Nechay 1927). They form an arcuate sedimentary belt including highly elevated (up to 150 m asl) oval and elongated hills with numerous kettle holes (Figs. 37.2 and 37.3). The sediments in the ice-marginal zone consist mainly of stratified gravel, sand and silt, occasionally interbedded with boulderly gravel and diamictons, with a total thickness of more than 20 m (Dybowski 2009; Dzierżek and Szymanek 2014; Lesemann et al. 2010, 2014).

An extensive outwash area called the "Dobrzyń sandur" extends to the east and south of the ice-marginal zone (Niewiarowski et al. 1995; Wysota 1999). The Dobrzyń sandur consists of unconfined outwash fans in the proximal part and a vast outwash plain in the more distal area. The sandur surface is shaped by numerous features of different sizes and origins including proglacial spillways, mega-scale bedforms (megadunes), obstacle marks, palimpsest tunnel valleys, and kettle holes (Dzierżek and Szymanek 2014; Sobiech 2019).

37.3 Morphology and Morphometry of Glacial Curvilineations

The broad area including multiple GCL swarms (sets) within both simple and compound tunnel valleys was described—in reference to previous studies on the Dobrzyń Lakeland—as the "Zbójno GCL field" (Lesemann et al. 2010, 2014; Adamczyk et al. 2016, 2022; Adamczyk 2023). In this field, the digital terrain model revealed 21 GCL swarms (Fig. 37.3) occupying 83.6 km^2 and representing 23% of the entire field area of nearly 358 km^2 (Adamczyk et al. 2016, 2022; Adamczyk 2023).

The GCL sets are characterised by various morphologies (Lesemann et al. 2010; Adamczyk et al. 2016, 2022; Adamczyk 2023). Within compound valleys, the GCLs form quasi-continuous swarms over the entire valley width, though in places this pattern is disrupted by post-glacial fluvial incisions (Figs. 37.3 and 37.4). In simple valleys GCLs constitute less continuous swarms (Figs. 37.3 and 37.5). The swarms with the largest average areas—and the greatest width—are located in the central part of the Zbójno GCL field. The GCL sets on its periphery are smaller and contain fewer individual forms per swarm. The average area of a swarm is 4 km^2, the largest one (no. 9 in Fig. 37.3) has

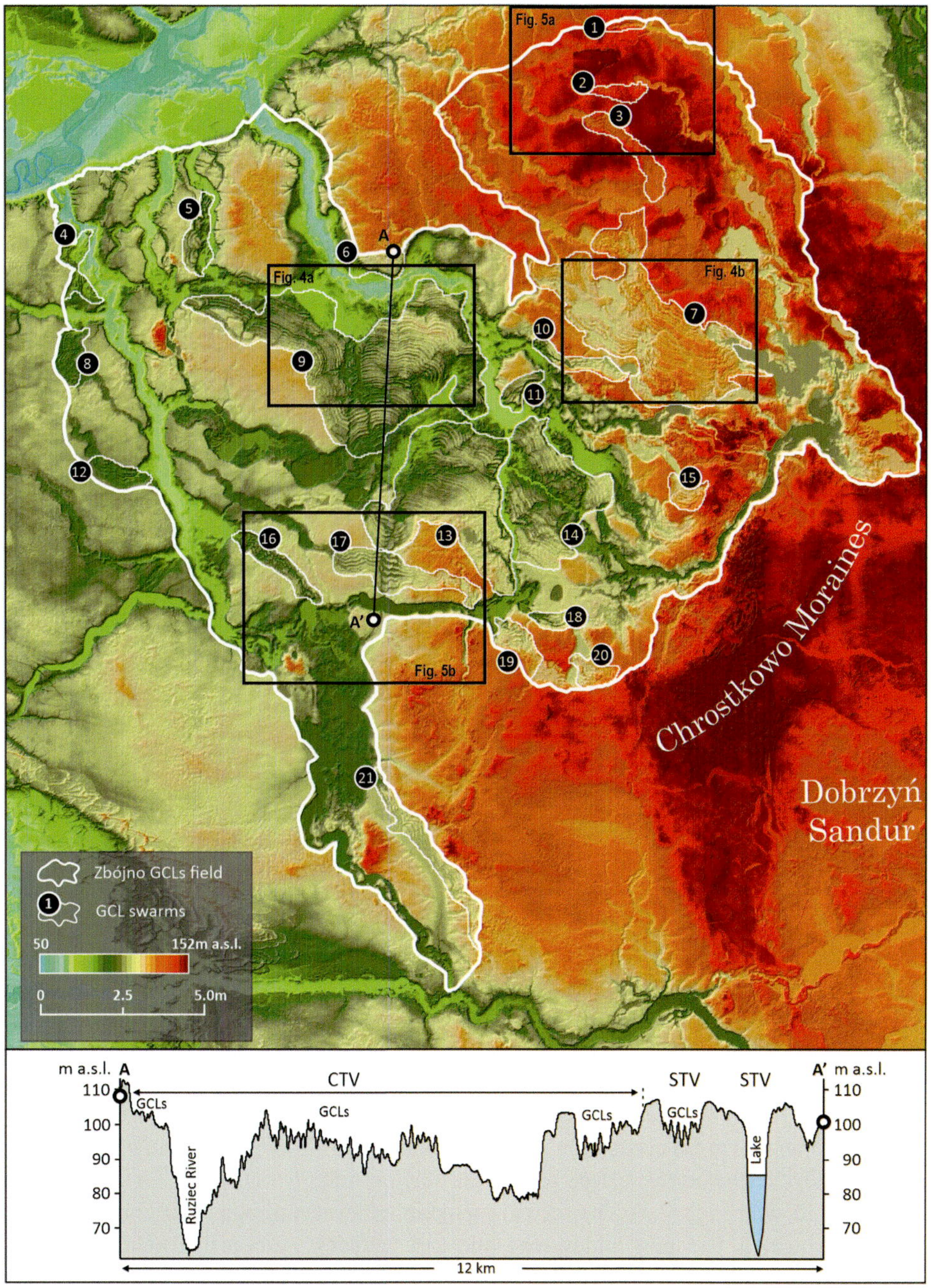

Fig. 37.3 Topographic relationships within the Zbójno GCL field. Above: distribution of GCL swarms. Below: topographic profile across the central part of the field showing GCL sets in simple tunnel valleys (STVs) and compound tunnel valleys (CTVs)

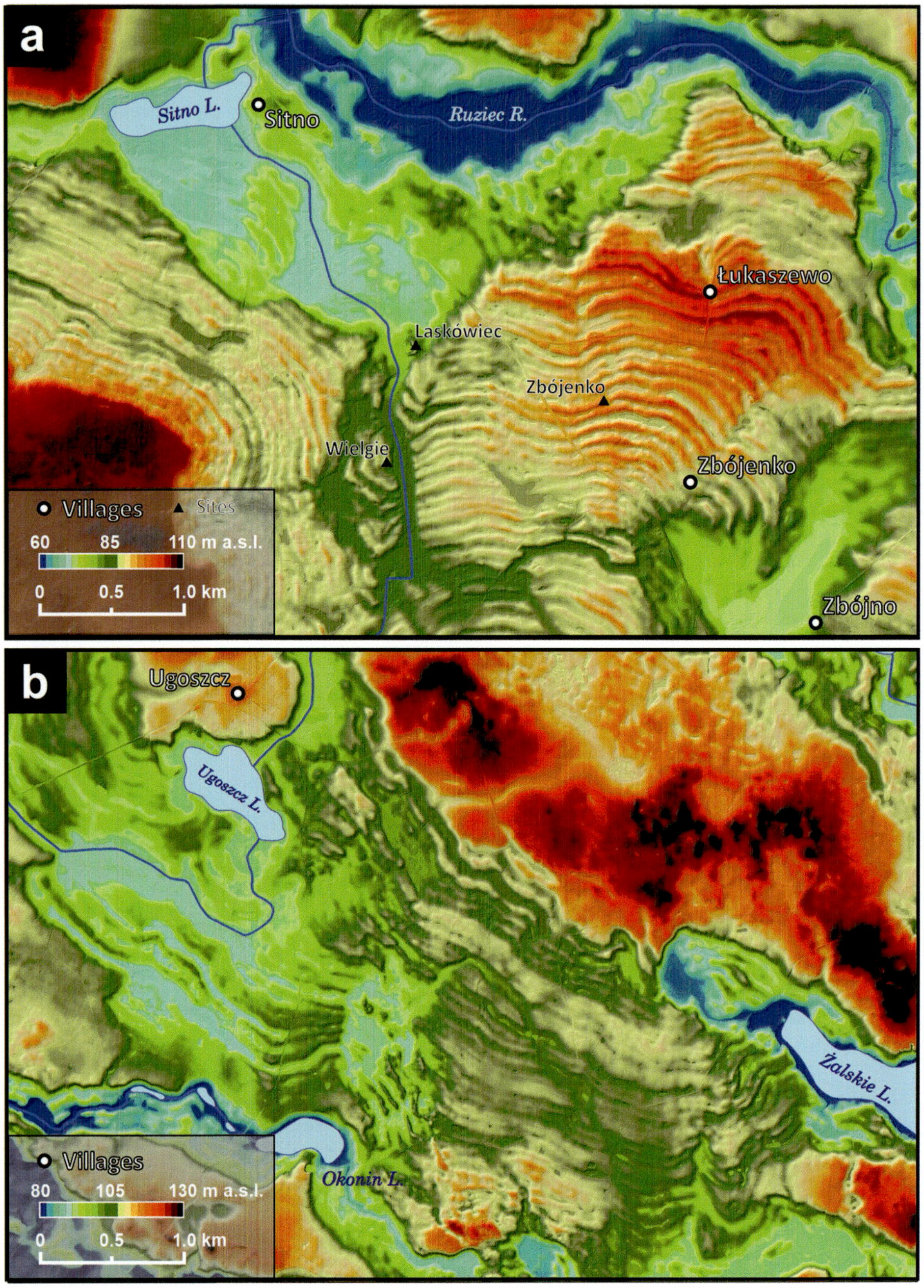

Fig. 37.4 Patterns of glacial curvilineations in compound tunnel valleys. **a** Central part of the Zbójno GCL field (swarm 9). **b** Eastern part (swarm 7). Black triangles indicate locations of sites discussed in the text and shown in Figs. 37.8c,d and 37.9

an area of 21.2 km^2, and the smallest (no. 1 in Fig. 37.3) covers 0.4 km^2. The average length of a swarm is 3.3 km, while the average width is 1.4 km. Swarm 9 is the longest (10 km) and widest (5.5 km) of them all.

The GCL swarms in the Zbójno field occur at four topographic levels (Adamczyk 2023): 75–90, 90–100, 100–115 and 115–126 m asl (Fig. 37.3). GCL sets located between 75 and 90 m asl occupy the proximal parts of compound valleys in the north-western part of the Zbójno field (swarms 4, 5, 8 in Fig. 37.3). The second topographic level (90–100 m asl) is the most widespread one and includes GCLs grouped in a broad depression in the central part of the field (swarms 6, 9, 11, 13, 14, 16, 17 and 18; Figs. 37.3, 37.4a and 37.5b). The third topographic level, 100–115 m asl, contains the easternmost and southernmost GCL sets (swarms 7, 10, 15, 19, 20 and 21; Figs. 37.3 and 37.4b). The highest topographic level (115–126 m asl) hosts GCL swarms 1, 2 and 3, which occupy the bottoms of narrow simple valleys in the north-eastern part of the Zbójno field (Figs. 37.3 and 37.5a).

The GCL sets contain parallel ridges separated by troughs that continue for more than 1 km, although a large proportion of the forms, despite being fragmented, shows continuity over the distances of up to 10 km (Lesemann

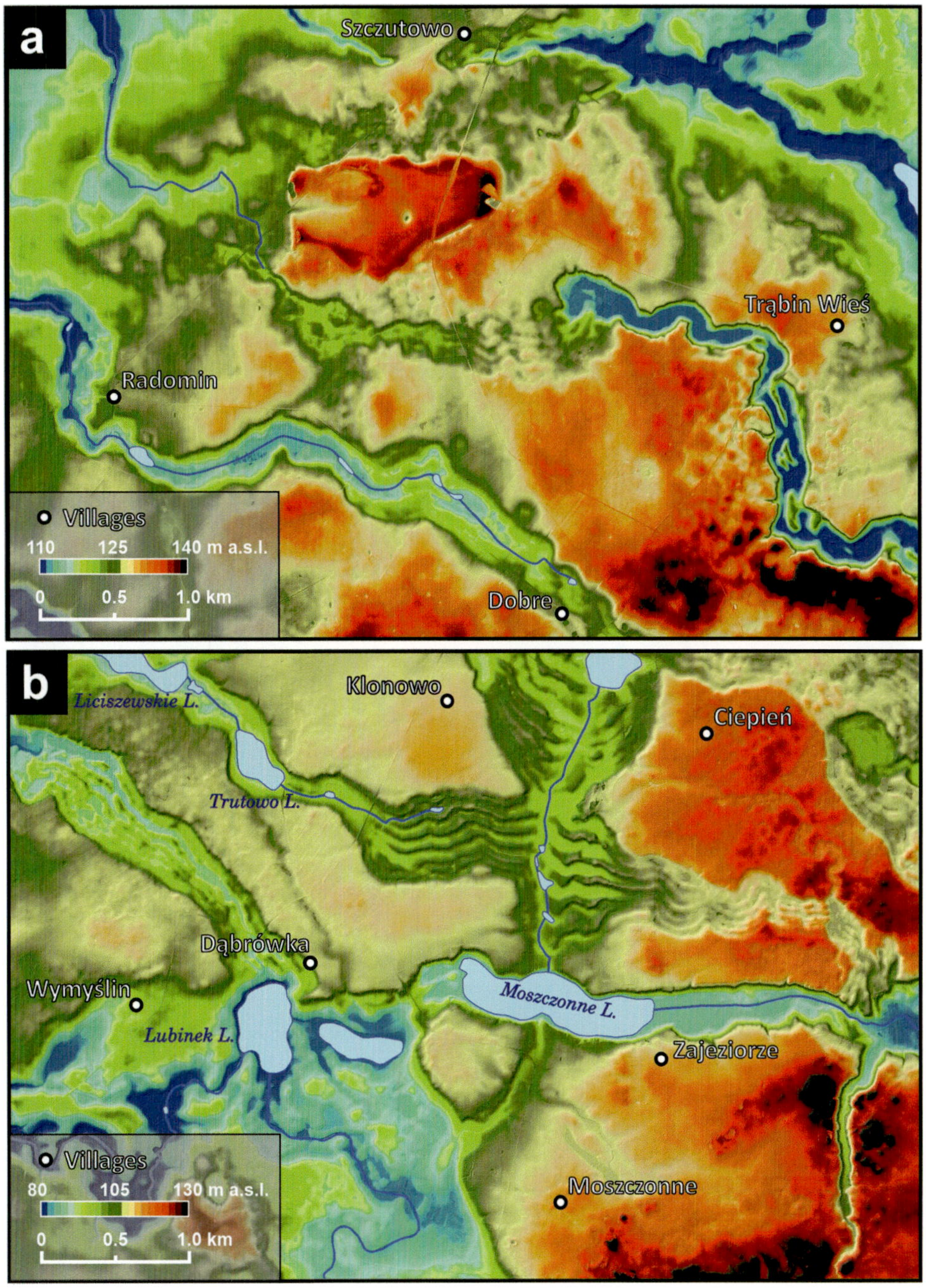

Fig. 37.5 Patterns of glacial curvilineations in simple tunnel valleys. **a** North-eastern part of the Zbójno GCL field (swarms 1, 2 and 3). **b** South-western part (swarms 16 and 17)

et al. 2010; Adamczyk 2023). Most individual ridges are from 1 to 3 km long, while the longest forms reach almost 9 km. The longest GCLs are found in the central part of the Zbójno field and in the middle parts of the individual swarms (Fig. 37.6a). The height of GCL ridges varies and typically reaches up to 5 m above the surrounding trough floors, but in a few groups (e.g. swarm 9) their height exceeds 10 m. In places, the highest GCL ridges occur 1–2 m below the neighbouring till plateau (Lesemann et al. 2014; Adamczyk 2023; Adamczyk et al. 2022). The GCL troughs have irregular longitudinal profiles with numerous thresholds and overdeepenings that contain closed depressions occupied by small lakes and peat bogs (Fig. 37.7). The widths of individual forms vary between 50 and 120 m, with a maximum of over 200 m.

The characteristic feature of the GCLs is their parallelism, both with the neighbouring ridges and troughs, and the margins of tunnel valleys or dissected remnants of the moraine plateau (Lesemann et al. 2010; Adamczyk 2023; Adamczyk et al. 2022). The regularity of the GCL distribution is highest within the largest groups in the central part of the Zbójno field (Figs. 37.3, 37.4 and 37.7a), but it is also visible in narrow and elongated swarms (Figs. 37.3 and 37.5). The number of individual forms in the GCL sets

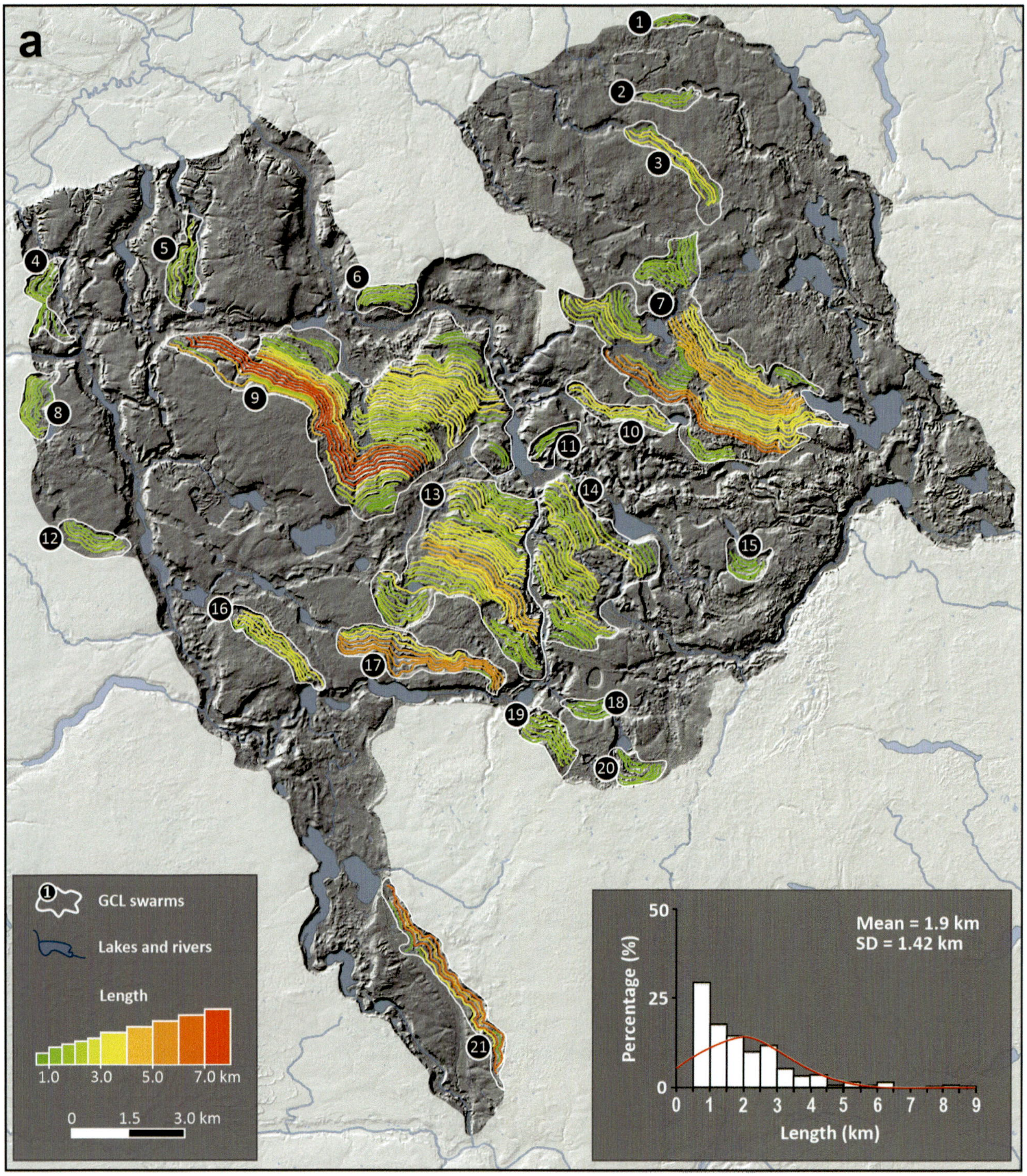

Fig. 37.6 Morphometric parameters of troughs separating glacial curvilineation ridges in the Zbójno field. **a** Lengths of GCLs. **b** Horizontal sinuosity indices (after Adamczyk 2023). Numbers in circles indicate GCL swarms

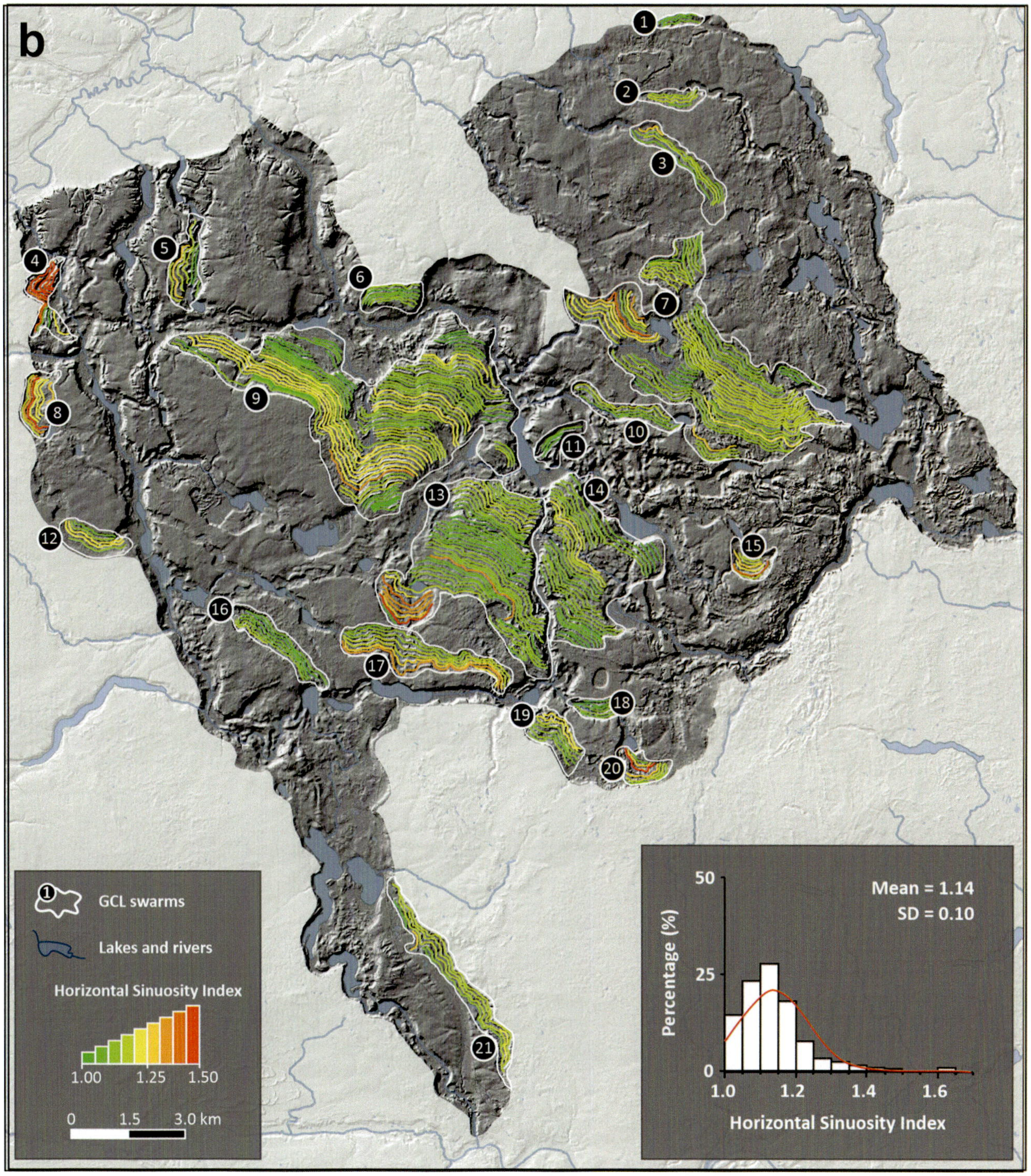

Fig. 37.6 (continued)

Fig. 37.7 Typical landscape of glacial curvilineations in the central part of the Zbójno field. **a** Aerial view of GCLs north of Zbójno (north-west view). Note multiple parallel GCL ridges with undulating crestlines (black arrows) and the intervening troughs (blue arrows) including depressions occupied by small lakes and peat bogs. Red arrow indicates the Zbójenko exposure discussed in the text and shown in Figs. 37.8c and 37.9a. **b, c** GCL ridges and troughs near the village of Łukaszewo (photographs a by T. Karasiewicz; b and c by W. Wysota)

varies greatly and ranges from just a few in narrow swarms to dozens of forms in large swarms (Adamczyk 2023; Adamczyk et al. 2022). In the largest GCL swarms (no. 7, 9, 13 and 14 in Fig. 37.3), the number of individual ridges and intervening troughs ranges from 20 to over 60. The total number of individual GCLs in the whole Zbójno field is 255.

Another characteristic feature of the GCLs is their horizontal sinuosity (Lesemann et al. 2010; Adamczyk 2023), which clearly corresponds to the morphology of the tunnel valleys and the moraine plateau margins (Fig. 37.6b). The vast majority of the individual GCLs have a horizontal sinuosity index[2] up to 1.2 (Adamczyk 2023). The indices are above 1.2 mostly for groups of short forms in the marginal parts of swarms, especially in the western and southern parts of the field. It is peculiar that slightly more winding forms occur more frequently in simple tunnel valleys than in compound valleys.

37.4 Geological Composition of Glacial Curvilineations

Already, Nechay (1927) noted that some GCL ridges in the Zbójno area consist of till underlain by sands and gravels locally overlying a lower till. He also pointed out that some ridges are composed nearly entirely of till. In the same area, Jewtuchowicz (1956) identified two layers of till separated by up to several metres thick glaciofluvial sand and gravel (Fig. 37.8a). He documented the discontinuous nature of the upper till layer and the orientation of elongated stones in it usually normal to the morphological axis of the ridges. Similar features of the geological composition were found in a few ridges within the central and southern parts of the Zbójno area by Lamparski (1972), Olszewski (1997) (see Fig. 37.8b–d), and more recently by Lesemann et al. (2010, 2014).

Detailed geological investigations in the entire Zbójno field (Dzierżek 2007, 2008; Wysota 2007, 2009; Dzierżek and Szymanek 2014, 2015; Wysota and Sokołowski 2017, 2018) have shown that the GCLs are mostly composed of till(s) with thicknesses of more than 10 m. Regionally, the internal structure of the ridges reveals two, in places even three tills interbedded with discontinuous layers of sand and gravel with cobbles, mostly stratified and partially disturbed, with thicknesses from 0.5 m to several metres. The tills are typically brown or dark brown sandy or clay-rich massive diamictons. The geological composition of the GCL ridges and the lithological-petrographic properties of tills that build them correspond with the sediment characteristics of the surrounding moraine plateau (Dzierżek 2007; Wysota 2007; Wysota and Sokołowski 2017).

The numerous depressions occurring within the troughs separating the GCLs are usually filled with colluvial and organic sediments (Dzierżek 2007, 2008; Wysota 2007, 2009; Dzierżek and Szymanek 2014, 2015; Wysota and Sokołowski 2017, 2018). Recently, in one such closed depression north of Zbójno, Karasiewicz et al. (2019) documented organic sediments (peat and gyttja) with a thickness of almost 6 m that formed during the Late Glacial and Holocene.

The internal structures and sedimentological features of the GCL deposits were studied in detail in exposures at Zbójenko, Wielgie and Laskówiec located in the central part of the Zbójno field (Olszewski 1997; Lesemann et al. 2010, 2014). The Zbójenko exposure (see Figs. 37.4a and 37.7a for location) is characterised by a range of sediments including three subglacial till layers with glaciofluvial interbeds (Figs. 37.8c and 37.9a). The tills—the lowest one being less than 1 m thick, the middle one being 0.5–1 m thick, and the upper one being up to 3 m thick—have diverse compositions, from clayey to sandy diamicton (Olszewski 1997; Lesemann et al. 2014). The till beds are typically massive or weakly stratified and contain discontinuous, thin intercalations of sorted sand. The orientation of the elongated stones within these till layers shows a generally high fabric clustering, both parallel and transverse to the ridge axis (Lesemann et al. 2014). The intervening glaciofluvial deposits consist of sand and gravel with cobbles, usually horizontally and cross-bedded, only slightly disturbed in places.

The Wielgie site (see Fig. 37.4a for location) is an example of GCLs almost entirely built of till (Lesemann et al. 2014). The exposure reveals two distinct layers of subglacial tills (upper and lower) of up to 4.5 m in total thickness (Fig. 37.9b). Irregular bodies of glaciofluvial sediments, mostly deformed sand and gravel, occur at the base of the lower till. The till fabric commonly shows poor clustering with orientation of elongated clasts approximately normal to the ridge axis (Lesemann et al. 2014). Particular attention is drawn to the fact that the upper till layer is truncated by the ground surface of the ridge (Fig. 37.9b).

The Laskówiec exposure reveals a ca. 10-m-thick succession of waterlaid deposits and two subglacial tills within the ridge core (Olszewski 1997; Wysota 2007) (Figs. 37.8d and 37.9c). At the base of the section, a 2-m-thick layer of mostly massive or horizontally bedded sand and gravel occurs (Fig. 37.9c). It is overlain by a 1.5-m-thick till, which is a massive, dark brown, clay-rich diamicton. Further up, there is a ca. 4.5-m-thick layer of parallel-bedded, ripple-laminated sand with dispersed plant detritus. This sand layer is covered by a 1.3-m-thick, light-brown, massive, sand-rich till. The top of the succession exposes horizontally and cross-bedded sand and gravel with silty intercalations. The regional distribution of the sediments and the OSL ages obtained from the sand layers indicate that a significant part of the sedimentary succession, including both till layers, derives from the Saalian (penultimate) glaciation (Wysota 2007, 2009).

[2]The horizontal sinuosity index of GCL troughs can be treated as equivalent to the meandering river sinuosity measurement, i.e. the ratio of the inter-ridge trough length to the smallest distance between the beginning and the end of an inter-ridge trough (Adamczyk et al. 2017; Adamczyk 2023).

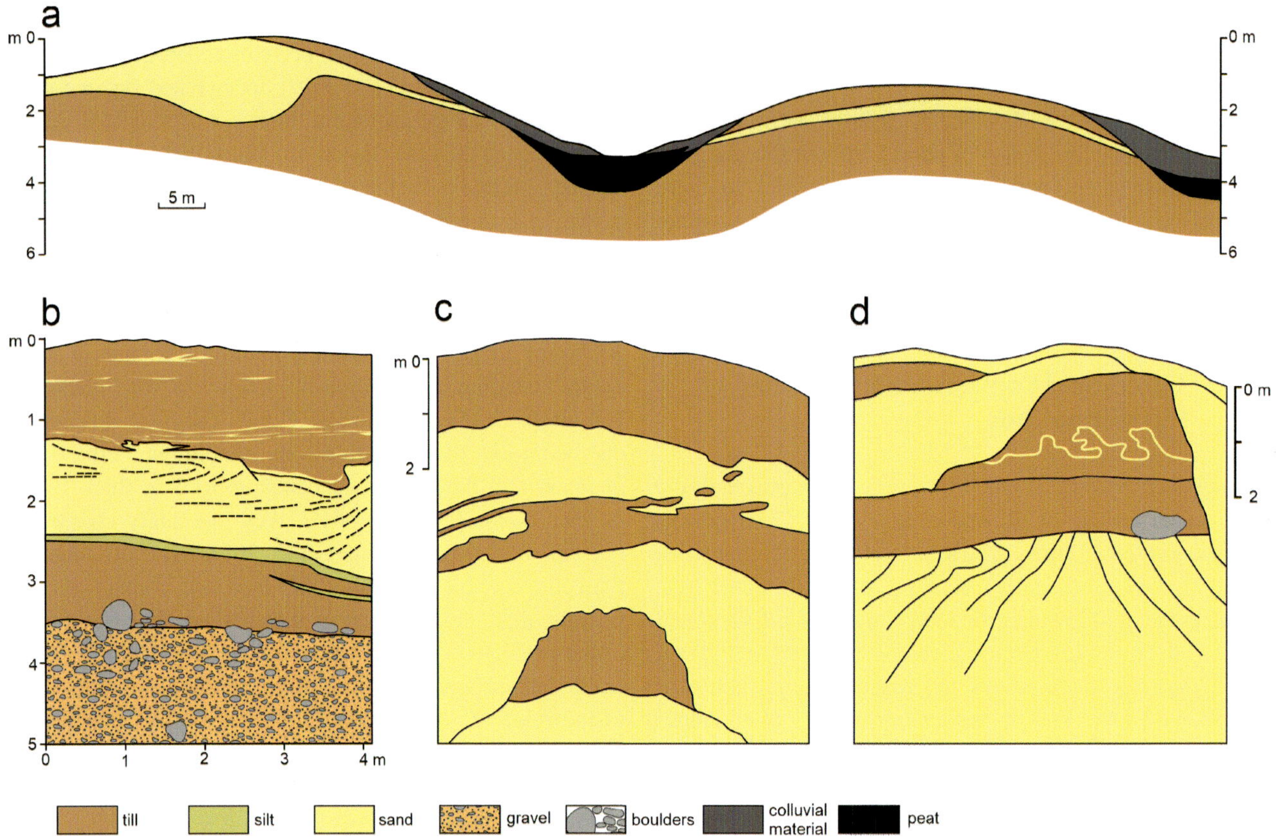

Fig. 37.8 Geological composition of GCLs based on previous studies. **a** Geological cross-section through GCLs near Zbójno (Jewtuchowicz 1956, modified). **b** Internal composition of a GCL near Nowa Wieś (Lamparski 1972, modified). **c, d** Simplified internal composition of GCLs at Zbójenko and Laskówiec (see Figs. 37.4a and 37.7a for location) (after Olszewski 1997, modified)

37.5 Origin of the Landscape of Glacial Curvilineations

For decades, the ridges in the Zbójno area have been considered as drumlins. The following main hypotheses and geological processes accounting for their formation have circulated. Nechay (1927), the first investigator of these forms, assumed three stages of their formation, specifically (1) creation of initial subglacial hills (proto-drumlins); (2) deposition of sorted sediments (sand and gravel) on the previously smoothed landforms and (3) deposition of till covering the ridges during deglaciation. He supposed that individual groups of ridges were formed under narrow "intra-ice streams" functioning along subglacial depressions during the ice-sheet advance. Jewtuchowicz (1956) considered subglacial erosion and subsequent deposition of till on a varied substrate to be the predominant processes of GCL formation.

Explanations similar to those of Nechay (1927) and Jewtuchowicz (1956) were provided by Olszewski (1997). He assumed that the ridges in the Zbójno field were formed subglacially as result of multi-cyclic and complex erosional and deformational–depositional processes during three successive ice-sheet advances. In reference with Nechay's interpretation, Olszewski (1997) considered that these landforms were finally shaped by extremely dynamic "basal ice streams" during the ice-sheet re-advance at the Late Weichselian stage.

Lamparski (1972) also interpreted the landforms in the Zbójno field as drumlins but excluded their formation as a result of processes under active ice. Rather, he considered them as a product of supraglacial melting and sediment deposition in stagnant ice accompanied by sediment squeezing around dead-ice blocks.

A completely different hypothesis was suggested by Niewiarowski (1957) and Liberacki (1961), although they too referred to the ridges as drumlins. Niewiarowski (1957) considered that the elongated ridges at the floors of the narrow tunnel valleys in the south-western part of the Zbójno area were erosional features formed by subglacial meltwater flows. He defined these landforms as drumlinoids. Similarly, Liberacki (1961) assumed that the regular ridges near Zbójno may have been eroded by subglacial meltwaters.

Fig. 37.9 Examples of the internal composition and sediment architecture of GCLs. **a** Interbedded layers of till and glaciofluvial deposits at the Zbójenko site. **b** Succession of two tills at the Wielgie exposure. Truncation of the upper till suggests that deposits pre-date the landforming event and thus the GCL ridge has an erosional origin (Lesemann et al. 2014). **c** Succession of waterlaid deposits and two till layers at Laskówiec exposure; location of sites in Figs. 37.4a and 37.7a (photographs a and c by W. Wysota; b by J. Lesemann)

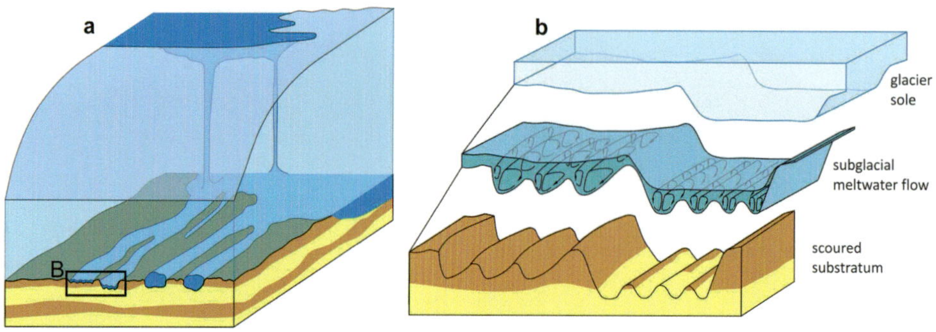

Fig. 37.10 Model of GCLs landscape formation based on Lesemann et al. (2010), modified. **a** Supraglacial and possibly subglacial water storage and subsequent drainage of meltwater along the ice/bed interface. **b** Scouring of curvilineations by longitudinal vortices within subglacial meltwater flows and progressive flow channelisation at late stages of tunnel valley erosion

The above hypotheses, albeit attractive, unfortunately did not explain why the morphology of these landforms did not resemble classical drumlins at all, nor their characteristic occurrence within a complex tunnel valley network. The hypothesis of Lesemann et al. (2010, 2014) explains the development of this spectacular landscape, considering the origin of the whole subglacial landsystem in relation to the ice-marginal sedimentary complex. They showed that these ridges are not drumlins, but represent a new type of glacial landscape, for which the term "glacial curvilineations" (GCLs) was coined.

Based on the geological and geomorphic evidence, Lesemann et al. (2010, 2014) suggested that GCLs were likely formed by channelised meltwater flows beneath the ice. One major argument was that the GCLs occur in tunnel valleys, which form an anabranching network in a moraine plateau. At the mouths of these tunnel valleys, there are thick accumulations of meltwater deposits including interbeds of very coarse material with well-rounded boulders (of up to 2 m in diameter), documenting flows of high erosional capacity. Moreover, Lesemann et al. (2010, 2014) pointed to the lack of correspondence between the morphology of GCLs and their internal composition, suggesting that the deposits pre-date the landforming event. This is also supported by the stratigraphical evidence of deposits older than the Weichselian glaciation in the GCLs (Wysota 2007, 2009) and the orientation of the elongated stones in the tills being unrelated to the orientation of the ridge axes.

The model proposed by Lesemann et al. (2010, 2014) for the formation of the GCL landscape assumes abrupt drainage of sub- or supraglacial meltwater reservoirs and erosion by pressurised water flows in a network of subglacial channels (Fig. 37.10a). It was hypothesised that the GCL ridges developed as a result of the action of longitudinal vortices within subglacial meltwater flows. Pairs of counter-rotating vortices excavated troughs in the pre-existing substratum and generated GCL ridges as remnants of an antecedent landscape (Fig. 37.10b). The formation of the GCLs continued with the incision of the moraine plateau surface and the development of tunnel valleys, from incipient to trunk valleys. There was a progressive incision resulting in the formation of several topographic levels with GCLs within wide tunnel valleys in response to fluctuating subglacial meltwater flows. The erosional event was followed by emplacement of melt-out till and occasionally fine-grained kame deposits, draping the GCLs morphology in places (Lesemann et al. 2010, 2014).

37.6 Conclusions

Lesemann et al. (2010, 2014) considered subglacial meltwater erosion to be the best unifying model for the formation of the Dobrzyń Lakeland—one that was capable of creating an erosional landscape (tunnel valleys and GCLs) and a depositional landscape (ice-marginal sedimentary complex). Because GCLs and tunnel valleys have a common genesis, it w2014as suggested that the whole tunnel valley network formed coevally from a single or several meltwater discharge events. Based on the tunnel valley morphology, Lesemann et al. (2010, 2014) suggested rather broad (a hundred to thousands of metres wide) and shallow (metres to tens of metres deep) flows with discharges capable of removing material from these valleys and re-depositing it in front of the ice sheet.

The scale and contemporaneous development of an anabranching tunnel valley network suggests that the subglacial erosional relief of the Dobrzyń Lakeland is the result of an abrupt release of a great amount of meltwater from supraglacial and possibly subglacial lakes, likely during a single, catastrophic event. With reference to the recently documented glacial megaflood events in NE Poland (Weckwerth et al. 2019; Weckwerth and Wysota 2023, this volume) and also to the modelling of catastrophic subglacial drainage in Canada (Beaney and Hicks 2000), it is suggested that the discharge through the tunnel valley network in the Dobrzyń Lakeland could have been in the order of a million cubic metres per second.

References

Adamczyk A (2023) Geomorfologia i geneza krętych form glacjalnych (glacial curvilineations - GCLs) na Niżu Europejskim. Manuscript of PhD thesis, Nicolaus Copernicus University, p 138

Adamczyk A, Wysota W, Karasiewicz MT, Piotrowski JA, Lesemann JE (2016) Site Zbójenko—geomorphology and postglacial evolution of glacial curvilineations (GCL) landscape in the Dobrzyń Plateau, north-central Poland. In: Sokołowski RJ, Moskalewicz D (eds) Quaternary geology of North-Central Poland: from the Baltic coast to the LGM limit. Wydział Oceanografii i Geografii Uniwersytetu Gdańskiego, Gdynia, pp 99–122

Adamczyk A, Wysota W, Sobiech M (2017) A morphometric analysis of an exceptionally long and complex tunnel valley: a case study of the Byszewo Landform, NW Poland. Z Geomorph 61(Suppl 2):27–44

Adamczyk A, Wysota W, Piotrowski JA (2022) Inventory of glacial curvilineations (GCLs) at the southern periphery of the Scandinavian Ice Sheet. Geomorphology 400:108094

Baranowski S (1977) Regularity of drumlins distribution and the origin of their formation. Studia Geol Polon 52:53–68

Baranowski S (1979) The origin of drumlins as an ice-rock interface problem. J Glaciol 23:435–436

Beaney CL, Hicks FE (2000) Hydraulic modelling of subglacial tunnel channels, south-east Alberta, Canada. Hydrol Proc 14:1434–1446

Chutkowski K, Olszewski A (2008) The problem of the genesis and relief of glacial basins: the case of the Ugoszcz-Żałe area, the Dobrzyń Moraine Plateau. Landf Anal 6:3–16

Clark CD, Livingstone SJ (2018) Glacial curvilineations found along the southern sector of the Laurentide Ice sheet and a hypothesis of formation involving subglacial slope failure in tunnel valleys and subglacial lakes. Earth Surf Proc Landf 43:1518–1528

Clayton L, Attig JW, Mickelson DM (1999) Tunnel channels formed in Wisconsin during the last glaciations. In: Mickelson DM, Attig JW (eds) Glacial processes past and present. Geol Soc Amer Spec Pap 337:69–82

Dybowski M (2009) Zapis dynamiki lądolodu subfazy dobrzyńskiej ostatniego zlodowacenia w świetle budowy wewnętrznej wzgórz chrostkowskich. Manuscript of the Master thesis, Nicolaus Copernicus University, Toruń, p 66

Dzierżek J (2007) Objaśnienia do szczegółowej mapy geologicznej Polski 1:50 000, arkusz Lipno. Państwowy Instytut Geologiczny, Warszawa, p 58

Dzierżek J (2008) Szczegółowa mapa geologiczna Polski 1:50 000, arkusz Lipno. Państwowy Instytut Geologiczny, Warszawa

Dzierżek J (2009) Paleogeografia wybranych obszarów Polski w czasie ostatniego zlodowacenia. Acta Geogr Lodz 95:112

Dzierżek J, Szymanek M (2014) Objaśnienia do szczegółowej mapy geologicznej Polski w skali 1:50 000, arkusz Skępe. Państwowy Instytut Geologiczny-PIB, Warszawa, p 54

Dzierżek J, Szymanek M (2015) Szczegółowa mapa geologiczna Polski w skali 1:50 000, arkusz Skępe. Państwowy Instytut Geologiczny-PIB, Warszawa

Głębicki P, Marks L (2009) The origin of drumlins at Zbójno, Central Poland. Geologija, Vilnius 51(3–4):67–68

Jewtuchowicz S (1956) Struktura drumlinów w okolicach Zbójna. Acta Geogr Univ Lodz 7:74

Karasiewicz TM, Hulisz P, Noryśkiewicz AM, Stachowicz-Rybka R (2019) Post-glacial environmental history in NE Poland based on sedimentary records from the Dobrzyń Lakeland. Quat Intern 501A:193–207

Lamparski Z (1972) Geneza form drumlinowych okolic Zbójna (Pojezierze Dobrzyńskie). Acta Geol Polon 22:139–158

Lesemann JE, Piotrowski JA, Wysota W (2010) 'Glacial curvilineations' produced by longitudinal vortices in subglacial meltwater flows. Geomorphology 120:153–161

Lesemann JE, Piotrowski JA, Wysota W (2014) Genesis of the 'glacial curvilineation' landscape by meltwater processes under the former Scandinavian Ice Sheet, Poland. Sedim Geol 312:1–18

Liberacki M (1961) Drumlins near Zbójno. In: Guide-Book of Excursion, Part I, North Poland, VI th INQUA Congress, Warszawa, pp 115–117

Nechay W (1927) Utwory polodowcowe Ziemi Dobrzyńskiej. Spraw Państw Inst Geol 4(1–2):61–144

Niewiarowski W (1957) Wybrane zagadnienia z badań geomorfologicznych na arkuszu Mazowsze. Dokument Geogr IG PAN 3:24–37

Niewiarowski W, Olszewski A, Wysota W (1995) Role of subglacial features in glacial morphogenesis of the Kujawy-Dobrzyń subphase area in the southern and eastern part of the Chełmno-Dobrzyń Lakeland. Quat Stud Poland 13:65–76

Olszewski A (1997) Drumlins of the northwestern Dobrzyń Moraine Plateau: location, structure and morphogenesis. Quat Stud Poland 14:71–83

Pasierbski M (1984) Struktura moren czołowych jako jeden ze wskaźników sposobu deglacjacji obszaru ostatniego zlodowacenia w Polsce. Wydawnictwo UMK, Toruń, p 170

Sobiech M (2019) Geomorfologia i formowanie glacimarginalnych stożków sandrowych w obszarze ostatniego zlodowacenia w Polsce. Manuscript of PhD thesis, Nicolaus Copernicus University, Toruń, p 169

Weckwerth P, Wysota W, Piotrowski JA, Adamczyk A, Krawiec A, Dąbrowski M (2019) Late Weichselian glacier outburst floods in North-Eastern Poland: landform evidence and palaeohydraulic significance. Earth Sci Rev 194:216–233

Weckwerth P, Wysota W (2023) The unique landscape originated by cataclysmic glacial floods at the Weichselian glaciation decline in north-eastern Poland. In: Migoń P, Jancewicz K (eds) Landscapes and landforms of Poland. Springer, Cham, pp 000–000

Wysota W (1993) Geneza drumlinów w środkowo-wschodniej części Pojezierza Chełmińsko-Dobrzyńskiego. Przegl Geogr 65:332–362

Wysota W (1994) Morphology, internal composition and origin of drumlins in the southeastern part of the Chełmno-Dobrzyń Lakeland, North Poland. Sedim Geol 91:345–364

Wysota W (1995) Structure and mechanisms of the formation of drumlins in glacial channels: a case study of the mid-eastern part of the Chełmno-Dobrzyń Lakeland (North Poland). Quat Stud Poland 13:109–124

Wysota W (1999) Ice sheet maximum of the Vistulian Glaciation in the mid-eastern part of the Chełmno-Dobrzyń Lakeland, northern Poland. Geol Quart 43:189–202

Wysota W (2007) Objaśnienia do Szczegółowej mapy geologicznej Polski w skali 1:50 000 arkusz Golub-Dobrzyń (323). Państwowy Instytut Geologiczny, Warszawa, p 43

Wysota W (2009) Szczegółowa mapa geologiczna Polski w skali 1:50 000 ark. Golub-Dobrzyń (323). Państwowy Instytut Geologiczny, Warszawa

Wysota W, Molewski P (2011) Chronologia i zasięgi nasunięć lądolodu na obszarze lobu Wisły podczas stadiału głównego ostatniego zlodowacenia. Przegl Geol 59:214–225

Wysota W, Molewski P, Sokołowski RJ (2009) Record of the Vistula ice lobe advances in the Late Weichselian glacial sequence in north-central Poland. Quat Intern 207:26–41

Wysota W, Sokołowski RJ (2017) Objaśnienia do Szczegółowej mapy geologicznej Polski w skali 1:50 000, ark. Rypin. Państwowy Instytut Geologiczny-PIB, Warszawa, p 47

Wysota W, Sokołowski RJ (2018) Szczegółowa mapa geologiczna Polski w skali 1:50 000, ark. Rypin. Państwowy Instytut Geologiczny-PIB, Warszawa

Wojciech Wysota is Professor of Quaternary Geology at Nicolaus Copernicus University (NCU) in Toruń. His research focuses mainly on morphology and origin of glacial landscapes in northern Poland, in particular landforms and sediments of the subglacial setting. Other research interests concern sedimentary environments and stratigraphy of Late Pleistocene successions in Poland. He has been Member of the Committee for Quaternary Research of the Polish Academy of Sciences (2011–2015) and Vice-President of the Peribaltic Working Group (TERPRO Commission) of the International Union for Quaternary Research (2012–2016). He has held the position of the Dean of the Faculty of Earth Sciences, NCU (2012–2016) and NCU Vice-Rector for International Cooperation (2016–2020). Since 2020, he has been NCU's Vice-Rector for Research.

Aleksander Adamczyk graduated in geography at Nicolaus Copernicus University in Toruń. Since 2018, he has been associated with the State Water Holding, Polish Waters Regional Water Management Authority in Bydgoszcz. His main research interests are glacial geomorphology, especially subglacial water erosional forms. He is the author and co-author of articles about glacial curvilineations (GCLs) and glacial lake-outburst floods in north-eastern Poland and also Member of the Polish Geographical Society and the Association of Polish Geomorphologists.

Jan A. Piotrowski is Professor of Quaternary Geology at the Aarhus University, Denmark, and at the Nicolaus Copernicus University in Toruń, Poland. His research mainly concerns glacial processes, sediments and landforms with the focus on subglacial environments of past continental ice sheets in Europe and North America, utilising combined field studies, laboratory experiments and numerical modelling. Other areas of interest include hydrogeology and global environmental changes during the Quaternary. He was Vice-President of the International Union for Quaternary Research (2003–2007) and Vice-President and President of INQUA Commission on Glaciation (1995–2003). Since 2000, he has been editor-in-chief of *Boreas* journal and member of multiple international committees, boards and panels for research administration and evaluation.

Piotr Weckwerth is Associate Professor at Faculty of Earth Sciences and Spatial Management, Nicolaus Copernicus University in Toruń. His main research interests are fluvial and glacigenic processes, sediments and landform origin, Quaternary paleogeography and geology, sedimentology of clastic sediments, palaeohydraulic reconstructions, geological and hydrogeological modelling and palaeoenvironmental investigations in archaeological studies. His main research areas are fluvial systems and glacier marginal zones and forefields of contemporary and formerly glaciated areas in Central Europe, Spitsbergen and Iceland. He is the current President of the Association of Polish Geomorphologists (2021–2024).

Suwałki Lakeland—Best Example of Lowland Glacial Geomorphology in Poland

38

Ewa Smolska, Katarzyna Pochocka-Szwarc and Piotr Szwarczewski

Abstract

The landscape of Suwałki Lakeland in NE Poland was created during the last, Weichselian glaciation. After the last ice sheet melted, a wealth of glacial landforms remained. These are both terminal moraines, reflecting the position of the ice-sheet margin, and landforms from areal deglaciation. Long and deep tunnel valleys are common. The deepest reaches of these tunnel valleys are occupied by lakes, including the Hańcza Lake, which is the deepest lake in Poland and in the Central European lowlands. Tunnel valleys are related to the activity of subglacial meltwaters and within the Masurian ice lobe they are generally oriented from the NW to the SE, while those within the Lithuanian ice lobe run almost meridionally or from the NE to the SW. The depressions of complex origin are the Szeszupa and Wigry basins. The mosaic of small and large landforms with significant elevations creates an extremely varied relief, unique among other Polish lakelands.

Keywords

Suwałki lakeland · Last glacial relief · Moraine plateau · Tunnel valley · Outwash plain · Depression

E. Smolska (✉) · P. Szwarczewski
Faculty of Faculty of Geography and Regional Studies, University of Warsaw, Krakowskie Przedmieście 30, 00–927 Warszawa, Poland
e-mail: e.smolska@uw.edu.pl

P. Szwarczewski
e-mail: pfszwarc@uw.edu.pl

K. Pochocka-Szwarc
Polish Geological Institute-National Research Institute, Rakowiecka 4, 00–975 Warszawa, Poland
e-mail: katarzyna.pochocka-szwarc@pgi.gov.pl

38.1 Introduction

The Suwałki Lakeland is located in NE Poland, next to the Polish-Lithuanian border (Fig. 38.1). It is a region characterized by a young glacial landscape, morphologically the most diversified in Poland. It includes numerous glacial hills, deep tunnel valleys and depressions of various sizes and origin. The slopes of hills and tunnel valleys are locally fairly steep (over 20°) and local relief is significant, up to 30–50 m. In such a varied landscape, there are picturesquely situated lakes, which occupy tunnel valleys and kettle holes. The postglacial history in the late Pleistocene and later—in the Holocene—involved erosion and denudation processes, and from the Bronze Age onwards, also human activity, mainly related to agriculture. However, geomorphic effects of the latter are comparatively minor and anthropic changes of the postglacial relief are limited.

In order to protect the values of the postglacial landscape, the Suwałki Landscape Park was established in 1976. There are four protected areas of outstanding natural, geological and landscape values within the park. In the south-eastern part of the region, one of the largest parks in Poland, the Wigry National Park (150.9 km^2), established in 1989, is located. In 1991, the protected landscape area "Dolina Rospudy" was set up with an area of 237 km^2.

The area has long been of interest to researchers dealing with geography and geology. The first geomorphological observations in the central part of the Jeleniewo Hills were carried out by Pietkiewicz (1928), who was the first to provide the description of the Turtul esker. He also marked the ice-sheet limits during the last glaciation (Weichselian). Geomorphological issues were studied by Kondracki and Pietkiewicz (1967), whereas the geological map was elaborated by Ber (1967). Morphogenesis (origin and evolution) of the Suwałki Lakeland was presented by Ber (2000), whereas the synthesis of geological structure and geomorphology of the Suwałki region is available on the Suwałki sheet of the geological map of Poland at the scale of

P. Migoń and K. Jancewicz (eds.), *Landscapes and Landforms of Poland*,
World Geomorphological Landscapes, https://doi.org/10.1007/978-3-031-45762-3_38

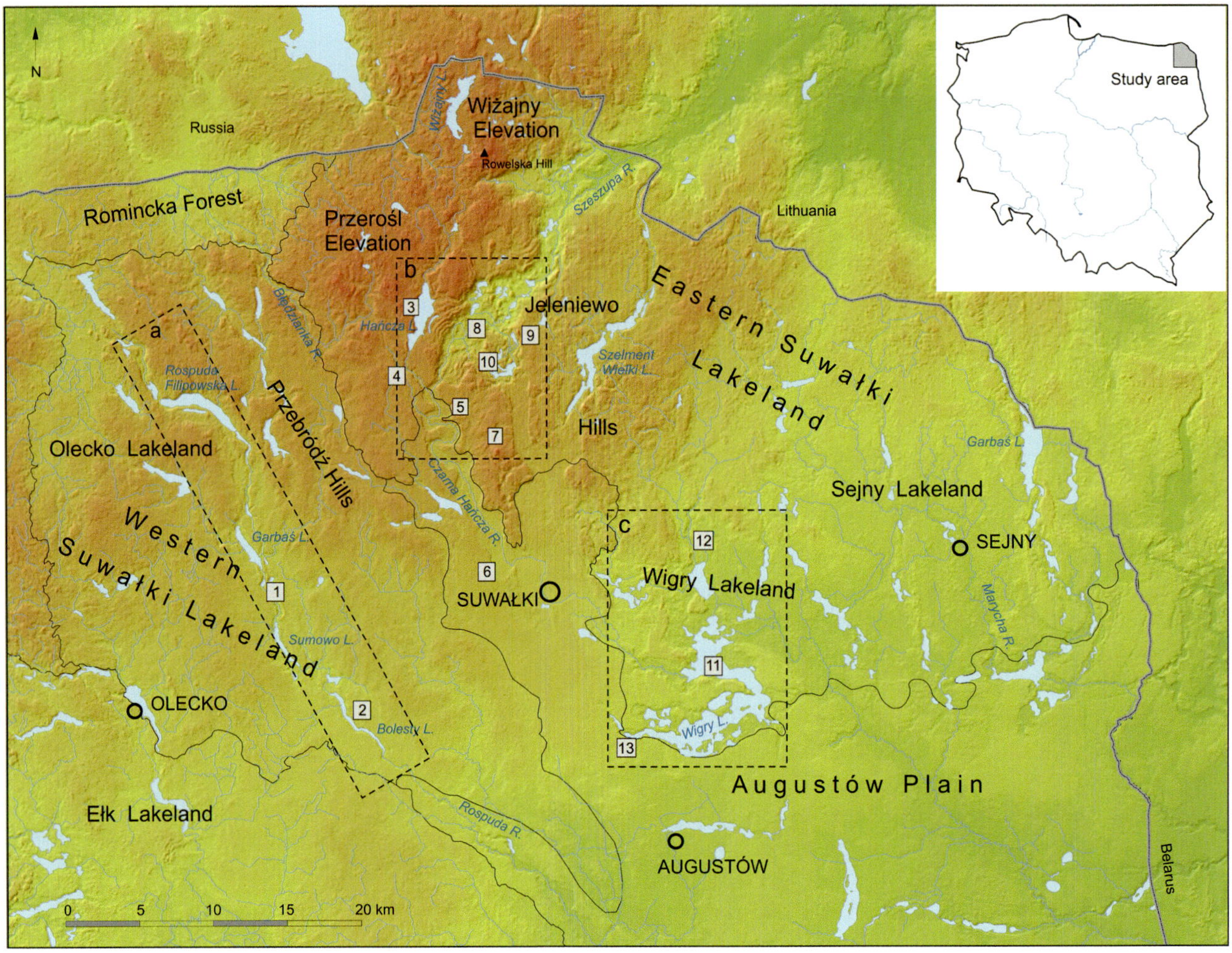

Fig. 38.1 Location and general relief of the Suwałki Lakeland. Letters and numbers indicate specific localities discussed in the text: a—Rospuda tunnel valley (Fig. 38.4); b—Szeszupa Depression (Fig. 38.6); c—Wigry Basin; 1—Bakałarzewo esker, 2—Bolesty Lake, 3—Hańcza Lake, 4—Czarna Hańcza and Kozikówka tunnel valleys, and Bachanowo boulder field, 5—Large paleomeander of the Czarna Hańcza River, 6—Large palaeomeander of the Czarna Hańcza River near Potasznia, 7—Mt. Krzemieniucha, 8—Szeszupa Depression, 9—Cisowa Hill, 10—Zamkowa Hill, 11—Wigry Lake, 12—Wiatrołuża Valley and Pietronajć Lake, 13—Czarne Lake and Ukleja Bay of the Wigry Lake

1:200,000 (Pochocka-Szwarc 2020a; Pochocka-Szwarc and Jóźwik 2020). Various palaeoenvironmental topics relevant to the southern part of Suwałki Lakeland and the Augustów Plain are addressed in the monograph edited by Weckwerth et al. (2020a). Studies presenting geology and geomorphology of this area also include geological-tourist maps of the Suwałki Landscape Park (Pochocka-Szwarc et al. 2014) and the Wigry National Park (Rychel et al. 2010) developed and published by the Polish Geological Institute-National Research Institute.

This chapter outlines geological setting of the Suwałki Lakeland, provides an overview of its Quaternary history and offers detailed presentation of some of the most interesting landforms and geomorphological issues of this region.

38.2 Geographical Location and Geological Setting

The area belongs to the Lithuanian Lakeland (Fig. 38.1). Among its various regions, the Western Suwałki and Eastern Suwałki Lakelands (delimited by the Błędzianka and Czarna Hańcza rivers) and the northern part of the Augustów Plain (Kondracki 2009; Solon et al. 2018) are located within the borders of Poland. This area is therefore also called the Suwałki and Augustów Lakeland.

The division of the Suwałki Lakeland into the western and eastern parts refers to the main directions of ice-sheet movement during the last glaciation and the presence of the Masurian and Lithuanian lobes, which had a significant impact on the geomorphology of the region (Ber 2000)

(Fig. 38.2). Within these main lobes, during the recessive Pomeranian stage (Pm3), smaller secondary lobes emerged, influencing the formation of glacial relief, especially the outflow routes of subglacial waters (Fig. 38.3). The varied relief of individual parts of the Suwałki Lakeland is reflected in local names such as extensive "humps"—Wiżajny and Przerośl elevations, clusters of hills separated by depressions—Jeleniewo Hills and Przebródź Hills, and clusters with concentrations of lakes—Wigry Lakeland and Sejny Lakeland (Fig. 38.1).

The region is crossed by the water divide between the Biebrza, Nemunas and Pregoła catchments. The main river of the western part of the lake district is Rospuda River, which is a tributary of the Biebrza River (Vistula River basin). The Czarna Hańcza River, like Rospuda flows towards southeast, while the Szeszupa River drains towards northeast, both rivers belonging to the Nemunas River catchment area. The course of many rivers follows the orientation of long and deep subglacial tunnel valleys. These rivers connect the lakes which occupy their floors.

The Suwałki Lakeland is situated within the geological structure of the East European platform in the border zone between the Peribaltic Depression and the Masuria-Suwałki elevation (Ryka 1998; Żelaźniewicz et al. 2011). The basement is built of crystalline rocks of Palaeo and Mesoproterozic age and includes granites and gneisses, diorites, norites and anorthosites. According to Szewczyk and Nawrocki (2011), the presence of an Anorthosite Massif, which is characterized by very low heat flux values, creates conditions for the preservation of relict permafrost, documented at a depth of over 300 m in the Szypliszki area (the Udryń borehole). In no place in the region, basement rocks crop out at the surface. They are covered by sedimentary rocks, with a total thickness from 857 m in the vicinity of Krzemianka to 1064 m in the central part of the Szeszupa Basin (Wiszniewska and Petecki 2014). The sedimentary cover is represented by Palaeozoic (Cambrian, Ordovician, Silurian, Permian), Mesozoic and Cenozoic formations.

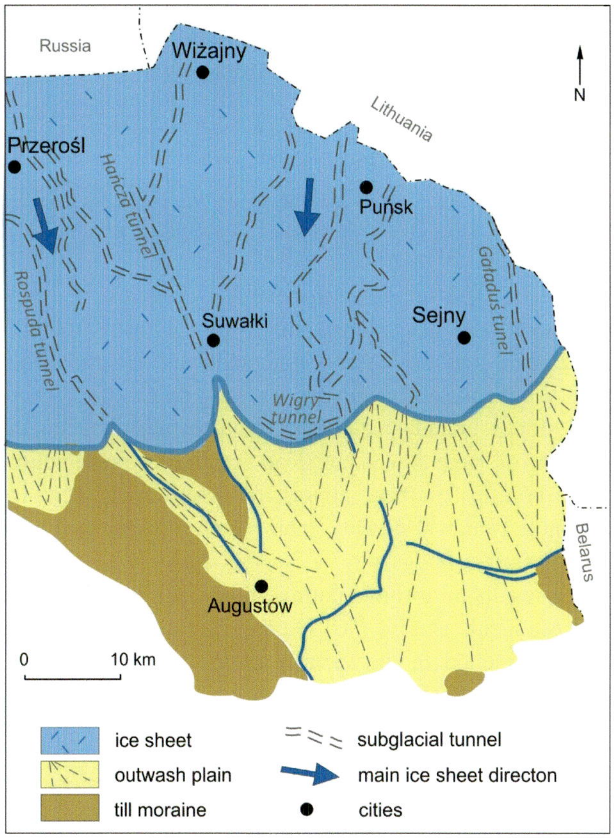

Fig. 38.2 Extent of the Pomeranian phase of the Weichselian glaciation with subglacial tunnel valleys and outwash plains (based on Pochocka-Szwarc et al. 2014; modified)

Pleistocene sediments reach the biggest thickness in the area of Hańcza Lake and in the vicinity of the village of Wiżajny (260–280 m asl) (Ber 2000; Pochocka-Szwarc 2020b). They are represented by tills, fluvioglacial sands and gravels, clays and silts from ice-dammed lakes that were deposited during successive glaciations (Lindner et al. 2013): during the Early Pleistocene—the Nidian (MIS 22); Middle Pleistocene—the Sanian 1 and Sanian 2 (Elsterian glaciation—MIS 16, MIS 12), Odranian and Wartanian (Saalian glaciation MIS 10–6) and Late Pleistocene—Weichselian glaciation (MIS 5d–2).

The contemporary landscape was created during the Weichselian glaciation, mainly during its youngest phase on the territory of Poland—the Pomeranian phase and its recessional-oscillational stages, further shaped by melting of dead-ice blocks at the end of the Pleistocene and the beginning of the Holocene (Ber 2000). Erosion and denudation processes in the Holocene only slightly changed the relief.

38.3 Main Landforms and Their Origin

The Wiżajny elevation in the NW part of the Eastern Suwałki Lakeland (Fig. 38.1) reaches the highest altitude within the region. It is a glacial plateau (250–298 m asl) with the Rowelska Hill as the highest spot. The plateau is built mainly of till and its surface is punctuated by numerous kettle holes. Some of the highest hills are terminal moraines, such as the Krzemieniucha Hill (289 m asl). Low-lying areas occur in the Wigry and Sejny Lakelands (130–150 m asl), with the lowest point located in the southern part of the Sejny Lakeland, in the Marycha River valley, at 122 m asl.

The extent of Masurian and Lithuanian lobes during the Pomeranian phase is best recognizable in the southern part of the Suwałki Lakeland (Fig. 38.3), where it is determined by terminal moraines and the main places of

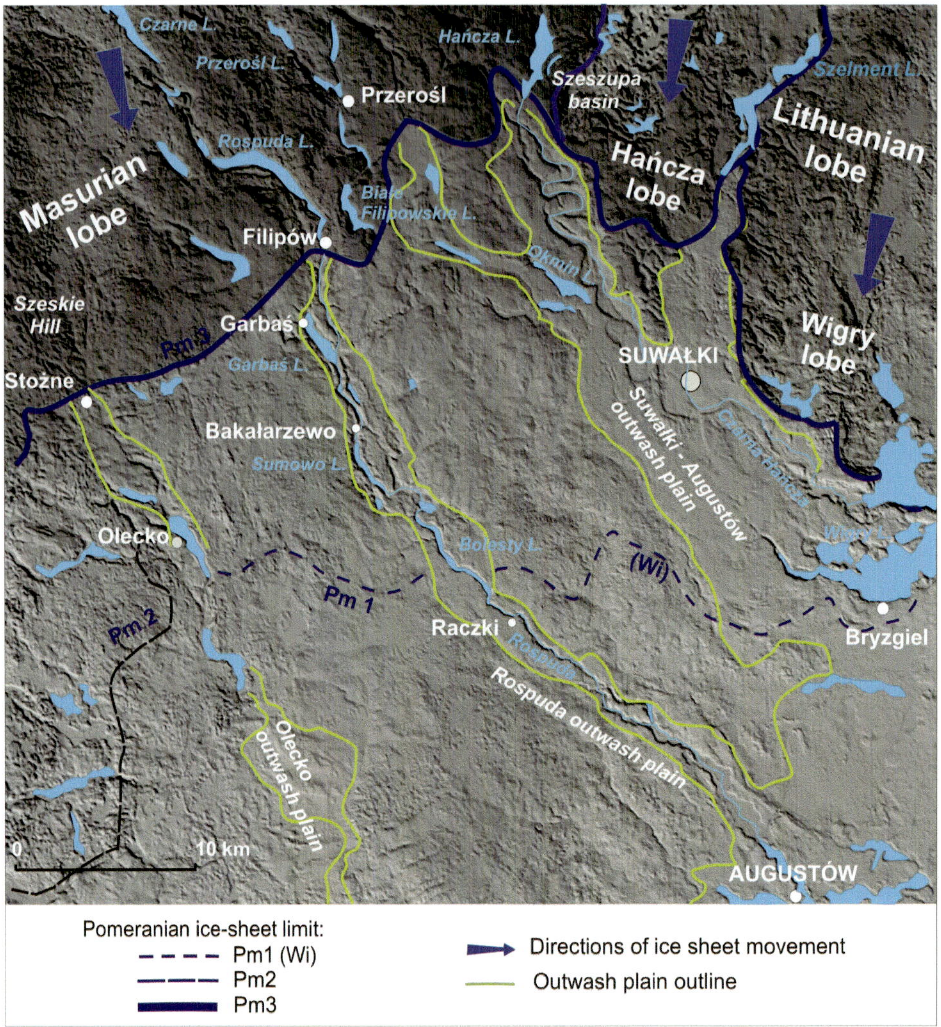

Fig. 38.3 Extent of ice-sheet lobes and outwash plains during the Pomeranian phase (after Pochocka-Szwarc and Krzyszkowski 2015; modified)

outflow of subglacial waters. A characteristic feature of the Suwałki Lakeland are long and deep tunnel valleys. Their course corresponds to the principal orientation of ice lobes (Figs. 38.2 and 38.3). Tunnel valleys that formed within the Masurian lobe are oriented NW–SE, while those associated with the Lithuanian lobe are oriented NNE–SSW (Ber 2000). These different orientations are perfectly visible in the arrangement of lakes and eskers accompanying the tunnel valleys (Fig. 38.1). Numerous lakes occur within the tunnel valleys, the largest of them in terms of area being the Wigry Lake (21.75 km^2), Szelment Wielki Lake (3.56 km^2), Hańcza Lake (3.11 km^2) and Wiżajny Lake (2.6 km^2). The deepest lake is the Hańcza Lake (105.6 m), which is also the deepest in Poland. The depths of other lakes are also impressive: Wigry Lake—74.2 m, Garbaś Lake—48 m, Szelment Wielki Lake—45 m and Rospuda Filipowska Lake—38.9 m.

Other characteristic elements of the landscape are extensive concave forms such as the Szeszupa Depression or the Wigry Basin, with a lake located in the central part (Fig. 38.1). The main morphological features of these depressions are steep slopes that separate them from the surrounding surfaces of morainic plateaus. Height differences are significant, typically reaching up to 40 m. These steep slopes include ice-contact features and indicate the presence of large blocks of dead ice during deglaciation. In the bottoms of these depressions there occur dead-ice moraines, kames and lakes, mainly kettle lakes, lacustrine terraces and peat bogs.

The results of absolute dating of selected boulders using the cosmogenic isotope ^{36}Cl indicate that at 19.7 ka BP the ice sheet melted in the vicinity of Wigry Lake. Towards the north, in the Szeszupa Depression and in the Czarna Hańcza tunnel valley, ice disappeared later, from 17.9 to 14.4 ka BP (Dzierżek and Zreda 2007; Dzierżek 2009). The ages of bottom sediments of selected lakes indicate that lacustrine accumulation in these lakes started mostly between 14 and 12.5 ka, less often later (Stasiak 1971; Kupryjanowicz 2007; Kinder et al. 2019; Lauterbach et al. 2010; Gałka and Apolinarska 2014; Makri et al. 2021).

Many lakes contain laminated sediments, which are particularly valuable in geochronology and palaeonvironmental studies (Tylmann et al. 2013).

The late glacial and Holocene history is related primarily to the development of river valleys, which mostly use glacial tunnel valleys (e.g. Rospuda, Czarna Hańcza in the section from Lake Hańcza to Turtul) or parts of outwash plains (e.g. Czarna Hańcza in the middle section). The second type of river valleys consists of melt-out depressions connected by short water gap sections, such as the upper Szeszupa, Błędzianka or Marycha valleys. In the Neo-Holocene, gullies developed on slopes of considerable elevations. Soil and gully erosion are mostly anthropogenically conditioned, reflecting acceleration of erosional processes after deforestation and occupation of slopes for agricultural use.

38.4 Specific Landforms

38.4.1 The Rospuda Tunnel Valley

The hydrographic axis of the Western Suwałki Lakeland is the Rospuda tunnel valley, with a total length of almost 30 km, oriented from NW to SE (Fig. 38.1). This long subglacial meltwater outflow route was created within the Masurian lobe (Figs. 38.2 and 38.3). During the recession of this lobe, new sections of the subglacial tunnel were revealed successively to the north (Fig. 38.4). Within them, water emerged from the glacier gate accumulated sand and gravel deposits of the proximal part of an outwash fan (Pochocka-Szwarc and Krzyszkowski 2015). The width of this outwash form ranged from 0.5 to 4–5 km. The thickness of sand and gravel deposits is up to 17 m.

Along the tunnel valley eskers occur (Fig. 38.4). These are winding ridges alternating with hillocks with a total length of some 7 km near the village of Bakałarzewo, from the Lake Garbaś to the southern end of the Lake Sumowo. Nowadays, the Rospuda River flows within the former tunnel valley. The valley has steep slopes, 10–15 m high, but in esker sections the slopes are twice as high. There are nine lakes located in the bottom of this valley, including the Czarne Lake (27.2 m max depth), Rospuda Filipowska Lake (38.9 m max depth), Garbaś Lake (48.0 m max depth) and Bolesty Lake (16.2 m max depth), all being flow-through lakes. The reach of the Rospuda River in the bottom of the tunnel valley ends in the Rospuda Lake in the vicinity of the town of Augustów (Augustów Plain).

The Rospuda is a meandering river. Within the narrow sections of the valley, between the lakes, the channel is similar to upland rivers. Locally, in the sections where peat bogs occur, higher values of sinuosity index are recorded. The slopes of the tunnel valley, especially in the parts where lakes occur, are incised by gullies (Fig. 38.5).

38.4.2 The Hańcza Lake Basin

The Hańcza Lake, 4.5 km long and up to 1.2 km wide, is located in the Eastern Suwałki Lakeland, within a tunnel valley of a meridional course, located between the Masurian lobe and the Lithuanian lobe (Fig. 38.1). The deepest part of the valley is filled by the Hańcza Lake (Figs. 38.6 and 38.7a). The greatest depths (more than 80 m) occur in the central part of the lake. The previously mentioned maximum depth of 108.5 m has not been confirmed and the measured depth is 105.6 m (Popielarczyk et al. 2015). The average depth of the lake is 38.7 m.

The unique character of the Hańcza Lake basin lies not only in its considerable depth, but also the presence of very steep slopes, both above and below the water level, and boulder beaches. The slopes on the west side make the lake shore almost inaccessible from the land side. The shore breaks off with a steep cliff, which is an underwater continuation of the tunnel valley slope above the water level, reaching a total height of 140 m. This is the largest local difference in elevation. Underwater "walls" or escarpments are mainly built of tills. Underwater photos of these units show numerous boulders and glacitectonic disturbance zones (Pochocka-Szwarc et al. 2012, Pochocka-Szwarc 2020b). Locally, there are shallow parts of the lake with boulders and they can reach up to 10 m from the shoreline (Fig. 38.7b). The boulders layer is up to 2 m thick and therefore only a little of rush vegetation occurs along the lake perimeter. In terms of trophic state, the Hańcza Lake is an oligotrophic lake. Its waters are fed by underwater springs, and to a lesser extent by surface runoff and by waters of the Czarna Hańcza River that flows through it.

The greatest depths at the bottom of the lake most likely have an eversional origin associated with intense erosion of the bedrock. As shown by modelling exercise, the velocities of subglacial waters were about 30 times higher than the velocities of flows in contemporary shallow aquifers (Piotrowski 1997). As a result of such dynamic processes, a subglacial tunnel valley of Hańcza Lake was formed, with a significant depth (evorsion hollow) of about 100 m. The overflow of a part of subglacial waters from the Szeszupa Depression could have contributed to the formation of such a large depression (Weckwerth et al. 2020a).

The total thickness of bottom sediments in the Hańcza Lake is about 5 m. They were deposited from the Allerød to the end of the Holocene. The oldest, initial phase of the lake formation took place about at 12.9 ka (Lauterbach et al. 2010).

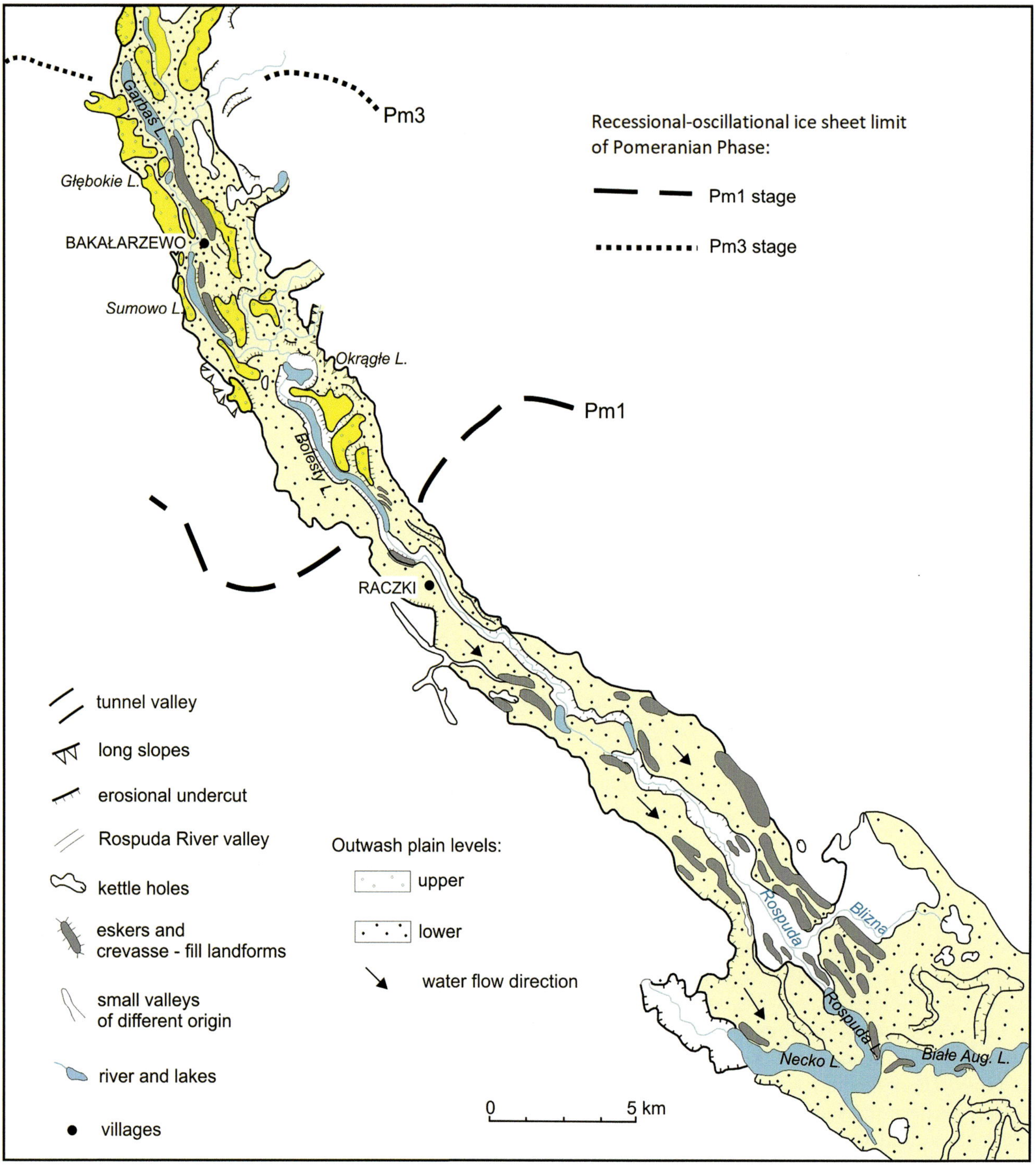

Fig. 38.4 Geomorphological sketch of the Rospuda tunnel valley (after Ber 2007; simplified)

38.4.3 The Czarna Hańcza River Valley

The source of the Czarna Hańcza River is located to the north from the Hańcza Lake. The river, 142 km long in total, flows through two large and deep lakes: Hańcza and Wigry (Fig. 38.1). The Czarna Hańcza River, downstream of the Lake Hańcza, is locally almost like a mountainous river, with a significant slope of 3.2–15 permille, the average slope between lakes Hańcza and Wigry being 2.2 permille. To the south of the Hańcza Lake, the river flows in the bottom of a narrow tunnel valley, incised up to 20 m (Fig. 38.6). This valley connects with another one of

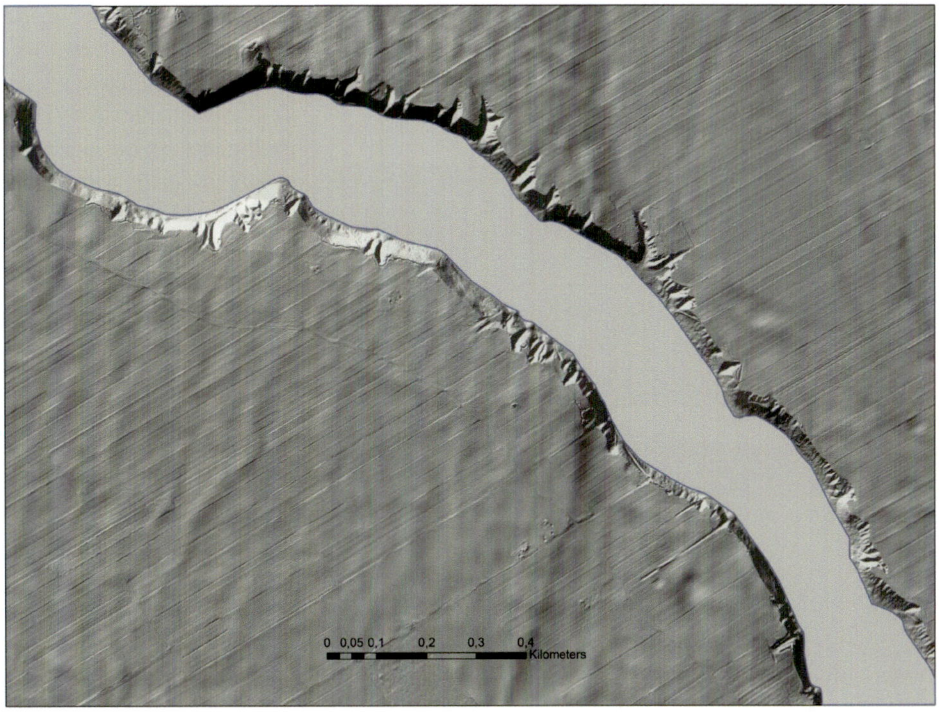

Fig. 38.5 One of the lakes within the Rospuda tunnel valley—Bolesty Lake with numerous gullies on the slopes

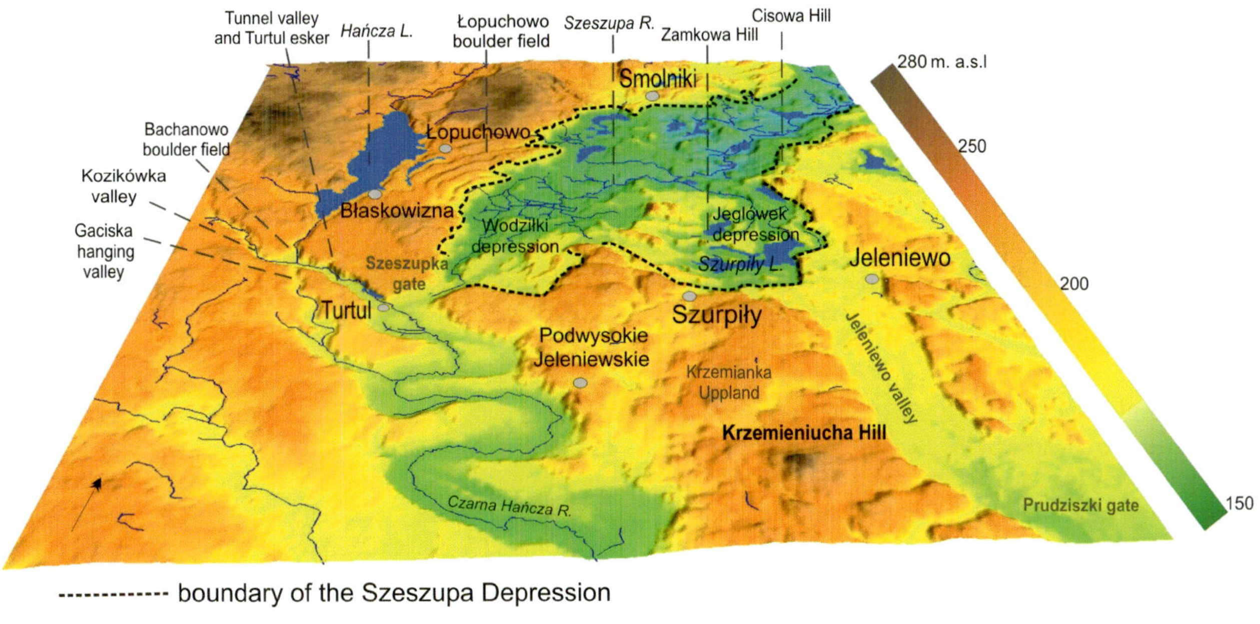

Fig. 38.6 Szeszupa Depression and the surrounding plateaus

NW–SE orientation, now used by the small Kozikówka River, a tributary of the Czarna Hańcza River. Below the confluence of these two rivers, the tunnel valley more than doubles its width, to some 150 m, and its depth increases to 30 m. Within the bottom of the Kozikówka (tunnel) valley and in the Czarna Hańcza valley, the Turtul esker is located. It consists of 13 hills, up to 12 m high, built of sand and gravel packages about 15 m thick (Pietkiewicz 1928; Ber 2000). Some of these hills are now islands in the middle of an artificial Turtul pond (Fig. 38.8a). On the western side of the Lake Hańcza tunnel valley, another esker existed (Bachanowo esker), but was completely exploited for aggregates (Fig. 38.6).

How this complex set of landforms was probably created is shown on Fig. 38.9. In the Turtul area, an ice gate was formed (Fig. 38.9a), through which meltwater emerged

Fig. 38.7 Hańcza Lake. **a** General view in northerly direction. **b** Boulder-gravelly coastal shallows (photograph a by M. Żarski; b by K. Pochocka-Szwarc)

causing accumulation of the oldest part of the outwash fan. The waters flowed from two different directions: (1) through the subglacial tunnel from the NNE and these accumulated the sediments of the latter Bachanowo esker; and (2) those flowing through the tunnel from the NW, creating the Turtul esker (Lisicki 1993; Pochocka-Szwarc et al. 2014). After the collapse of the Bachanowo tunnel roof, the meltwaters were directed to the south (Fig. 38.9b). They incised into the previously formed outwash fan, creating a meltwater valley known as the Gaciska Valley (Fig. 38.8b). After melting of the ice that protected the Turtul tunnel, the outflow direction changed again and the Gaciska Valley was no longer used by meltwaters (Fig. 38.9c). Its bottom became a hanging one and is now located about 15 m above the bottom of the tunnel valley with the Turtul esker (Fig. 38.8b) (Pochocka-Szwarc et al. 2014).

The considerable size of the outwash fan suggests that its accumulation was the joint effect of meltwater coming from the Hańcza Lake tunnel valley and from the western part of the contemporary Szeszupa Depression. Next to the confluence of Czarna Hańcza and Kozikówka valleys, an area with an exceptionally large number of erratic boulders occurs, called the Bachanowo boulder field, protected as a geological reserve and a geosite (Fig. 38.8c). There are about 10,000 boulders with the circumferences of 0.5–3 m. A few of them have a circumference of 6–8.7 m. Currently, it is considered that they may come from the washed-off sediments of the outwash fan. Dating the boulders using the cosmogenic isotope ^{36}Cl yielded dates 14.4 ± 1.0 ka and 14.7 ± 2.0 ka, confirming the association with the outwash (Dzierżek and Zreda 2007).

The post-glacial history of the Czarna Hańcza valley includes incision into outwash deposits and the formation of erosional river terraces. They were formed at the end of the Pleistocene (in the late glacial phase) and their development may have been associated with the formation of the valley of Nemunas River, tributary of which is the Czarna Hańcza River.

The Czarna Hańcza River has characteristic large-radius palaeomeanders (Figs. 38.6 and 38.10). In the area of the last glaciation limit, such a channel pattern is very rare. Large discharges that formed this channel were supplied by the Szeszupa basin, from which glacial meltwater flowed towards the Czarna Hańcza valley through the "Szeszupka gate" (Fig. 38.6). In the vicinity of the village of Potasznia (Figs. 38.1 and 38.6), such palaeochannels of Czarna Hańcza with a radius of some 500 m functioned up to 9530 ± 180 years BP (Monderewicz and Smolska 1999). Later, some large meanders were cut off, whereas others are still used today but by a river with smaller meanders.

38.4.4 The Szeszupa Depression

To the east of the Hańcza Lake, a vast depression, drained by the Szeszupa River heading to the northeast, is located (Fig. 38.1). Its floor lies at an altitude of 140–150 m asl, while the surrounding moraine plateau reaches 220–230 m asl. Terminal moraine hills of the former Szeszupa lobe rise above the plateau level by 25–50 m, with the highest Krzemieniucha Hill (289 m). Slopes of the Szeszupa Depression show significant steepness, in many places exceeding 20° or even reaching up to 30°. Numerous kame hills occur in the bottom of the basin, the highest of which are located in the northern part. Other landforms include dead-ice moraines, kame terraces, glaciolacustrine plains and numerous kettle holes of various sizes. The largest kettle holes are filled by lakes.

The southern part of the Szeszupa Depression has the floor located slightly higher than its other parts (some 150–160 m asl) and includes smaller, inset depressions such as the Wodziłki and Jeglówek ones, the latter hosting the Szurpiły Lake (Fig. 38.6). The most characteristic feature of the Szeszupa Depression is the occurrence of complexes of elongated ramparts. The morphological axes of these forms are arranged festoon-like (i.e. amphitheatrically) and they

Fig. 38.8 Landforms associated with the tunnel valley of Czarna Hańcza. **a** General view of the tunnel valley with esker hills in the Turtul pond. **b** Gaciska hanging valley. **c** Bachanowo boulder field—a nature reserve and a geosite (photographs a and c by M. Żarski; b by K. Pochocka-Szwarc)

run to the edge of the depression (Fig. 38.6). They are separated by narrow depressions up to 10 m deep, filled with mineral and organic sediments (Micun 2014). They occur among others in the Wodziłki Depression, in the vicinity of the village of Łopuchowo (between the basin and the Hańcza Lake), as well as in the NE edge of the depression, to the east of the village of Smolniki. Formerly, these landforms were considered as terminal or push moraines (Ber 2000). However, according to recent re-assessment by Weckwerth et al. (2019), Wysota et al. (2020), they have subglacial origin and are described as glacial curvilineations. Being among the most peculiar glacigenic landforms, they are discussed separately in this volume (Weckwerth and Wysota 2023; see chapter 39).

Locally, at the edge of the depression, dead-ice moraines occur, reaching a relative height of up to 30 m in relation to the surrounding morainic plateau. In the north, around the Jaczno Lake, relative relief approaches 90 m. Typically, the peripheral slopes of the depression descend in several steps with the height of some 20–40 m each. One of dead-ice moraines is the Zamkowa Hill, located in the Jegłowek Depression, rising over 40 m above the water level of the surrounding lakes (Fig. 38.6) (Ber 1967; Pochocka-Szwarc et al. 2014). This hill has been anthropogenically transformed. As archaeological research has shown, a Yatving settlement complex existed on the hill and in its vicinity from the mid-9th to the twelfth century and is considered the most powerful Yatvingian fortress in the Prussian territory (Sawicka 2010; Engel 2020).

A distinctive regional landmark is the Cisowa Hill, located at the eastern edge of the Szeszupa Depression (Fig. 38.6). Due to the conical shape of its upper part, it is very often called "the Suwałki Fujiyama" (Fig. 38.11). This hill has also other names, such as Góra Gulbieniska or Góra Sypana ("poured"). The latter name refers to the Kopane ("dugged") Lake located at its foot and suggests its anthropogenic origin (it is a legend probably from the Yatvings times—850–1350 AD). In reality, the Cisowa Hill is built of sand and gravel deposits. Since its internal structure is unknown, it was considered as a terminal moraine or a kame (Ber 1967). The elongated and meridional shape of the lower part of this hill, parallel to the direction of the eastern edge of the depression, indicates that deposition may have occurred in an ice crevasse. The top part of Cisowa Hill could have been changed anthropogenically (Pochocka-Szwarc et al. 2014).

The origin of the Szeszupa Depression is not fully researched and understood and interpretations have changed through time. It was considered to be the final depression of the Szeszupa lobe (Ber 2000), but recently it has been explained as an extensive subglacial tunnel valley (Weckwerth et al. 2020b). There is no doubt, however, that the last evolutionary stage involved melting of numerous blocks of dead ice, as indicated by ubiquitous kame forms. Disappearance of dead ice was concurrent with the establishment of the recent drainage network, with the upper Szeszupa River flowing towards the NE. The Szeszupa Valley thus is a young fluvial system that connects a number of smaller melt-out depressions via short water gap sections. Within the melt-out depressions, the gradient is very low and meanders occur, while in the gap sections the river is winding and the gradient is steeper, 4–9 permille.

The slopes of the Szeszupa Depression are cut by gullies and locally the density of these erosional landforms

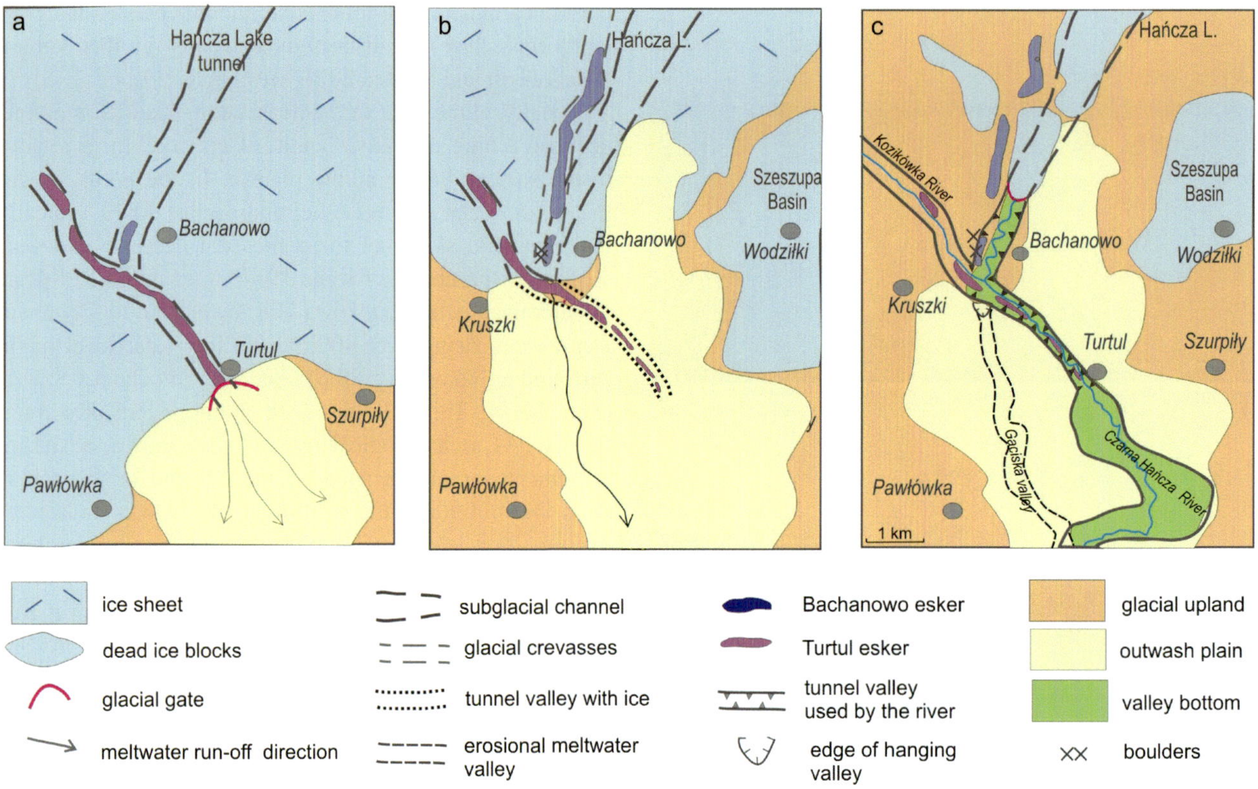

Fig. 38.9 Evolution of the tunnel valley of Czarna Hańcza and Kozikówka (after Pochocka-Szwarc et. al. 2014)

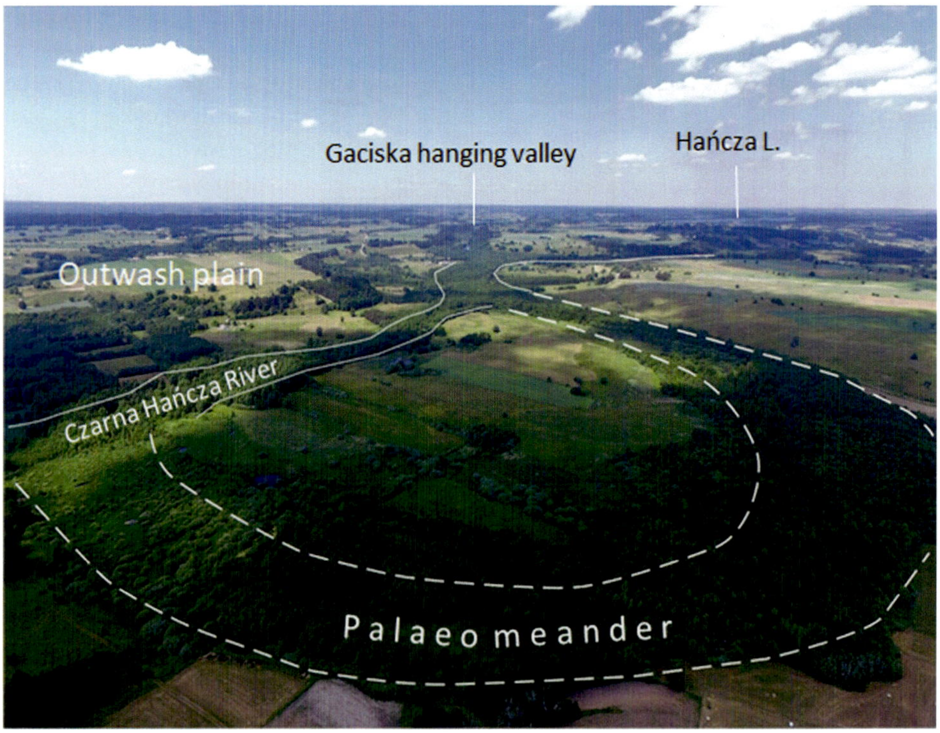

Fig. 38.10 Large palaeomeander of Czarna Hańcza River in the foreground (dashed lines) and the contemporary valley with the river (continuous line) in the distance (photograph by M. Żarski)

Fig. 38.11 View of the Szeszupa Depression and the Cisowa Hill from the viewpoint in the village of Smolniki, with several lakes visible in the foreground (photograph by E. Smolska)

is considerable, reaching 1 km/km^2. Most gullies have a V-shaped cross-profile, with their steep slopes usually forested or covered by grass. Some of the larger gullies (over 1 km long) have a flat bottom and smaller tributary side forms. There are short but distinct fans at the mouth of the gullies. The 3500–3200 BP ages of buried soil or peat covered by the alluvial-deluvial fans indicate the beginning of accumulation. The next stages of gully erosion are clearly correlated with phases of settlement expansion in the area, developed by the Balts (700 BC–600 AD), Yatvings (850–1350 AD) and finally, after the war with the Teutonic Order in 1410 AD, by Polish-Lithuanian settlers. The last stage of significant erosional activity in the gullies coincides with the Little Ice Age (Smolska 2007).

A characteristic feature of the young glacial landscape, apart from lakes, is the occurrence of numerous small closed depressions. At present they are becoming less and less visible in the landscape as they have been incorporated into farmlands. Formerly these waterlogged depressions, with shrub vegetation, were not used by farmers. Currently, only after heavy rainfall they fill with water and eroded soil from the surrounding slopes. These forms are filled with gyttja and peat deposits at the bottom and covered with colluvial deposits up to 2 m thick. The top layer (up to 1 m) is the result of soil transfer due to water and tillage erosion (Smolska and Szwarczewski 2008).

38.4.5 The Wigry Lake Basin

The Wigry Lakeland covers the area of the Wigry Basin, a sand and gravel outwash plain in the vicinity of this basin, and the morainic plateau located to the north and northeast (Fig. 38.1). The landscape of this area, located within the former Lithuanian ice-sheet lobe, evolved during several stages of the Pomeranian phase of the Weichselian glaciation (Fig. 38.3). Landform diversity includes tunnel valleys and eskers from the period of ice-sheet advance, as well as a number of dead-ice moraines, kames, kame terraces from the dead-ice melting stage, and a glaciolacustrine plain. Similarly to the Szeszupa Depression, arch-shaped hills separated by elongated depressions occur in the marginal zone of the Wigry Basin. The heights of these hills correspond to the surrounding moraine plateau. Some of these hills were formerly considered as drumlins. Their similarity to the glacial curvilineations recognized in the Szeszupa Depression suggests their subglacial origin (Weckwerth et al. 2020b). They are well visible in the SE and SW part of the Wigry Basin.

The Lake Wigry has a complex tunnel valley and melt-out origin (Ber 2000; Rutkowski and Krzysztofiak 2009). Its name comes from the Lithuanian language (vingrys—meaning "winding") and perfectly describes the outline of the lake basin (Fig. 38.12). The lake has a very varied shoreline

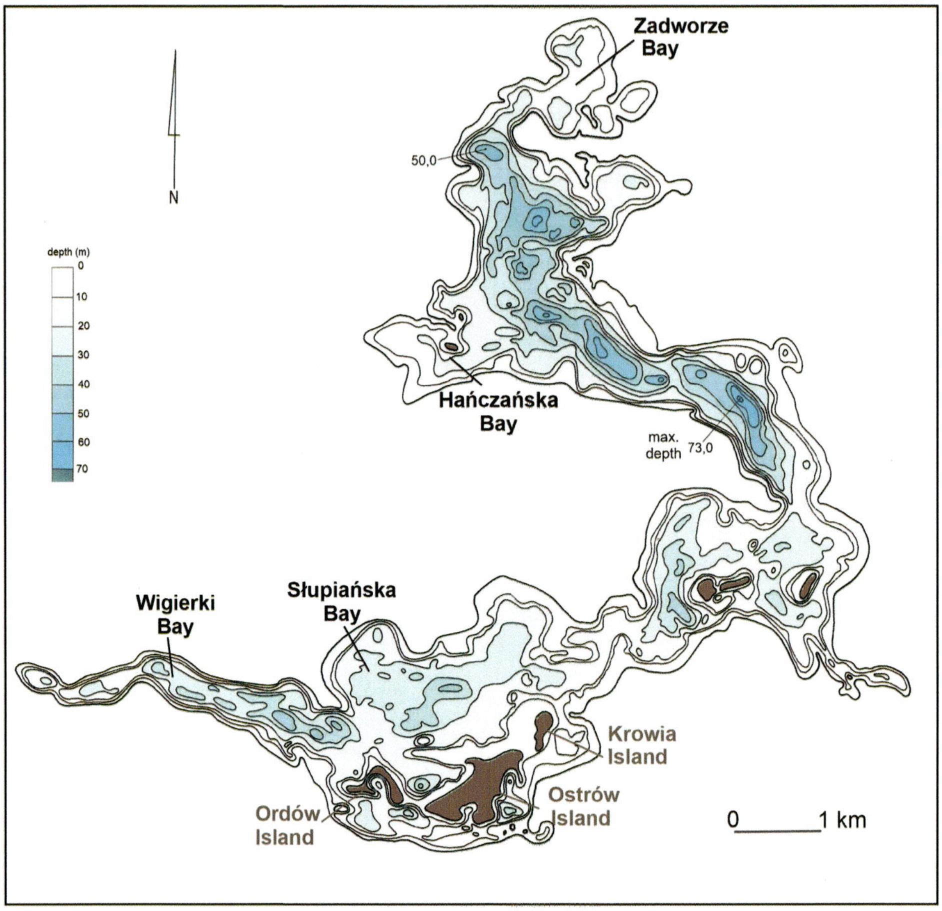

Fig. 38.12 Bathymetry of Lake Wigry (based on Jańczak 1999)

with numerous peninsulas, whereas islands are dead-ice moraine hills. More dead-ice moraine elevations are not visible on the surface, completely underwater (Figs. 38.12 and 38.13). The bottom of the lake is also very diverse. The lake consists of several smaller basins called "plos" (Fig. 38.12) and its bottom bathymetry includes elongated zones of considerable depth (more than 40 m) and shallows. The Wigry Lake ranks 10th in Poland in terms of surface area (21.7 km^2) and 5th in terms of maximum depth (74.2 m), whereas the average depth is relatively small—15 m.

During the accumulation of outwash deposits, the Wigry Basin was filled with dead ice. After the ice melted, a large lake was formed in this area, much larger than its contemporary successor. Its past extent is evidenced by the distribution of silty and silty-clayey lacustrine deposits (30 km^2). The formation of Lake Wigry started in Allerød period. The bottom sediments of the Wigry Lake were investigated by Rutkowski et al. (2009). The largest area is covered by carbonate-organic deposits (carbonate gytja) characteristic for profundal zone. Pure carbonate sediments (lacustrine chalk) predominate in the littoral zone. In the shore zones, clastic sediments occur. The results of high-resolution seismic surveys made possible to identify underwater landslides in the bottom of the lake. They occur in the tunnel valleys parts of the lake (e.g. in the Wigierka Bay and in the Szyja ploso), where the slopes are the steepest.

38.4.6 The Pietronajć Lake in the Wiatrołuża Valley—A Possible Pingo

The Wiatrołuża River flows into the Wigry Lake from the north (Fig. 38.1). This river uses the bottom of a tunnel valley that is incised by some 20–30 m and includes the minor eskers (Fig. 38.14a). In the edge zone of the valley, within its bottom, the strikingly circular Pietronajć Lake is located (Fig. 38.14a, c). The lake is 140 m wide in diameter and surrounded by a sandy rampart reaching up to 19 m high (Fig. 38.14d). Rutkowski et al. (1998) explained the lake as a relic of a pingo form that existed here in the period when the area was covered by tundra. Usually, pingo forms occur in groups. The detailed DTM analyses proved the probable existence of further similar forms, located not only in the Wiatrołuża Valley, but also within the Czarne Lake basin and in the Ukleja Bay (Fig. 38.14b). These sites need further detailed research.

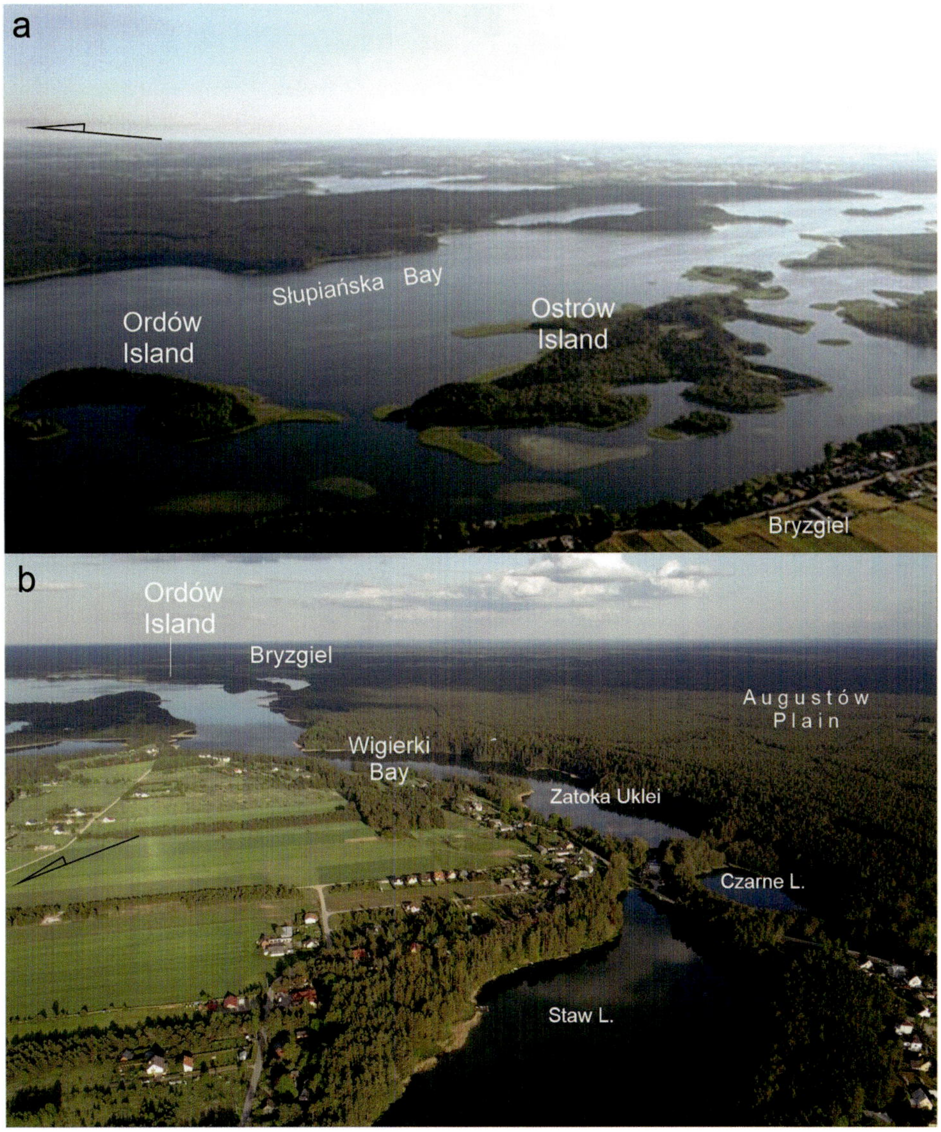

Fig. 38.13 Landscapes of the Wigry Basin. **a** View of Lake Wigry from the village of Bryzgiel towards the north. **b** South-western part of the lake basin, with minor adjacent lakes and the outwash surface of the Augustów Plain on the right (south). Source: http://www.sot.suwalszczyzna.eu/kat/o-suwalszczyznie/suwalszczyzna-z-lotu-ptaka

38.5 Conclusions

The main feature of the Suwałki Lakeland is the very diversified geomorphology created during the last glaciation. The landscape has been only slightly changed by human economic activity. Being relatively natural and apparently more diverse than other lakelands in central and northern Poland in terms of both landform complexity and local relief, the lakeland offers considerable opportunities for geotourism. There are extensive hills with a varied surface morphology and depressions of a complex origin. Long tunnel valleys with steep slopes are very characteristic and there is a considerable number of glacial hills of various origin such as terminal moraines, dead-ice moraines, eskers and kames that constitute local terrain culminations. The latter provides great views from its top parts. A large number of glacial lakes at different elevations typifies the area. Many are small and hidden amidst the hilly landscape, whereas the large ones are located within the tunnel valleys and have considerable area and depth.

The unique landscape of the Suwałki Lakeland is protected. The Wigry National Park is located in the eastern part of the region and the Lake Wigry is also part of the planned geopark "Augustowski Channel—Augustowskie Sandurs" (Krzywicki and Pochocka-Szwarc 2012). Due to the beauty of the Rospuda valley with the accompanying esker hills, a protected landscape area "Rospuda Valley" was created. Diversified morphologically and in terms of origin, the Szeszupa Basin and its surroundings is protected as the Suwałki Landscape Park. The geological reserves

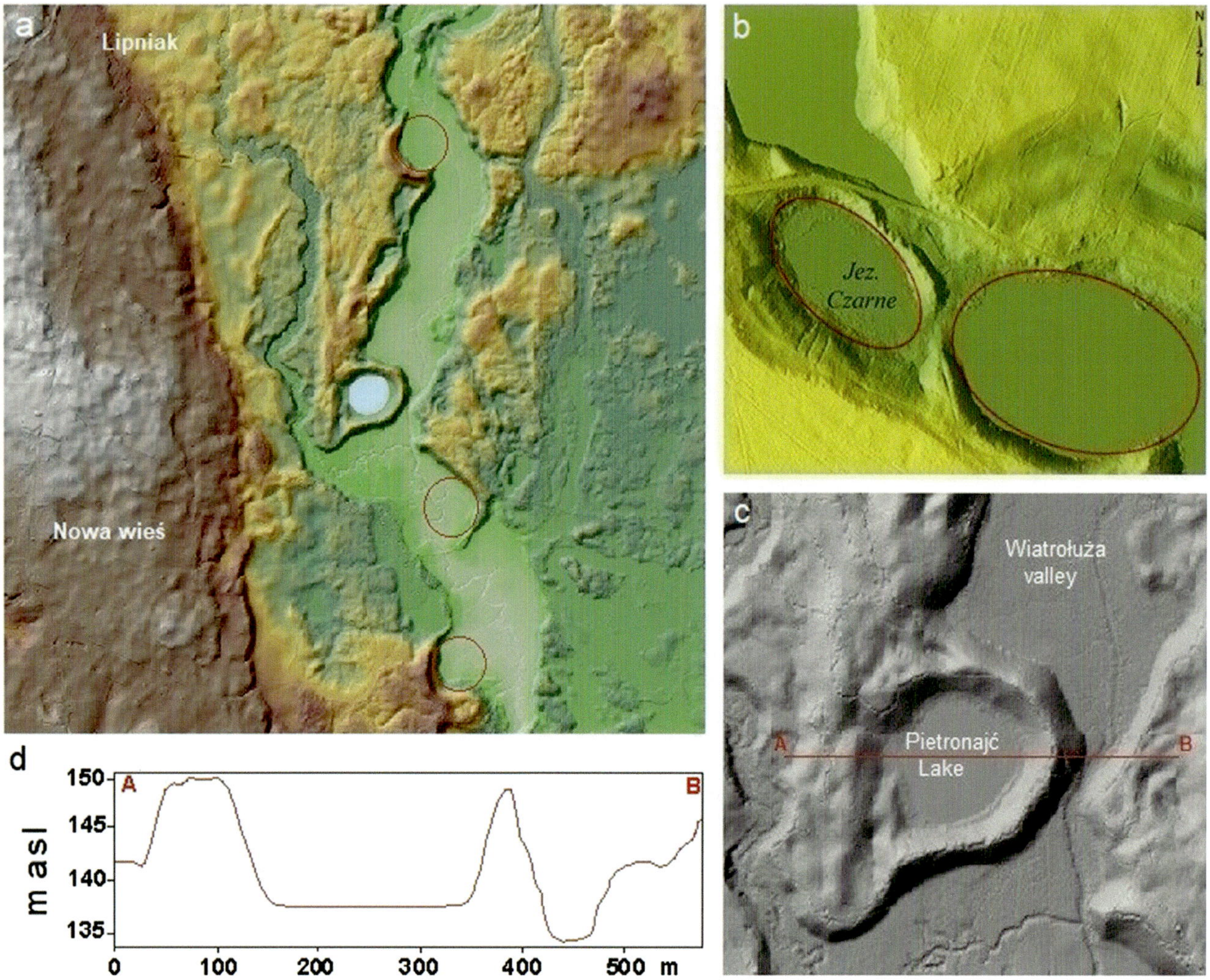

Fig. 38.14 Possible pingo remnants in the vicinity of Wigry Lake basin. **a** Location of Pietronajć Lake and other post-pingo depressions (red ovals) in the Wiatrołuża River Valley. **b** Oval outlines of the Ukleja Bay of the Wigry Lake and Czarne Lake—possible post-pingo remnants. **c** Pietronajć Lake and the surrounding embankments on a high-resolution DTM. **d** Topographical profile across the Pietronajć Lake and the Wiatrołuża Valley (profile line on c) (after Rutkowski et al. 1998; Zaborowski 2017)

of the Bachanowo and Łopuchowo boulder fields represent landscapes similar to those known from Scandinavia. The deepest lake in Poland, Hańcza Lake, is also the nature reserve. Some of the places described in this chapter are considered as geosites and have been included into the register compiled by the Polish Geological Survey (Centralny Rejestr Geostanowisk Polski—see https://cbdgportal.pgi.gov.pl/geostanowiska/). In addition, individual higher hills are viewpoints that show the varied topography of the area with lakes and swamps, surrounded by steep slopes with gullies. The natural values of the Suwałki Lakeland have been included in the project Geopark Yotvings—Polish-Lithuanian cross-border area (Graniczny et al. 2008; Sygar and Zgłobicki 2022).

References

Ber A (1967) Szczegółowa mapa geologiczna Polski 1:50 000 ark. Jeleniewo (72). Inst Geol, Warszawa

Ber A (2000) Plejstocen Polski północno-wschodniej w nawiązaniu do głębszego podłoża i obszarów sąsiednich. Prace Państw Inst Geol – PIB 170, p 89

Ber A (2007) Objaśnienia do Szczegółowej Mapy Geologicznej 1:50.000 ark. Augustów. Państwowy Instytut Geologiczny-PIB. Warszawa

Centralny Rejestr Geostanowisk Polski. Available at: https://cbdgportal.pgi.gov.pl/geostanowiska/. Access date: 2023-06-09

Dzierżek J (2009) Paleogeografia wybranych obszarów Polski w czasie ostatniego zlodowacenia. Acta Geogr Lodz 95:112

Dzierżek J, Zreda J (2007) Timing and style of deglaciation of northeastern Poland from cosmogenic ^{36}Cl dating of glacial and glaciofluvial deposits. Geol Quart 51:203–216

Engel M (2020) Jaćwieskie ośrodki grodowe. Państw. Muzeum Archeologiczne – Wydz Archeologii UW, Warszawa, p 624

Gałka P, Apolinarska K (2014) Climate change, vegetation development and level fluctuations in Lake Purwin (NE Poland) during the last 8600 cal. yr BP based on a high resolution plant macrofossil record and stable isotope data. Quart Intern 328–329:213–225

Graniczny M, Kowalski Z, Czarnogórska M, Krzeczyńska M, Pupienis D, Satkunas J (2008) Projected Geopark Yotvings—Polish-Lithuanian cross border area. Przegl Geol 56(8/1):611–613

Jańczak J (ed) (1999) Atlas jezior Polski, vol 3. Bogucki Wydawn Nauk, Poznań, p 240

Kinder M, Tylmann W, Bubak I, Fiłoc M, Gasiorowski M, Kupryjanowicz M, Mayr C, Sauer L, Voellering U, Zolitschka B (2019) Holocene history of human impacts inferred from annually laminated sediments in Lake Szurpiły, northeast Poland. J Paleolimnol 61:419–435

Kondracki J (2009) Geografia regionalna Polski. PWN, Warszawa, p 441

Kondracki J, Pietkiewicz S (1967) Czwartorzęd północno-wschodniej Polski. In: Dylik J, Galon R (eds) Czwartorzęd Polski. PWN, Warszawa, pp 206–258

Krzeczyńska M, Pochocka-Szwarc K, Rychel J (2013) Wigierski Park Narodowy. Wycieczka tematyczna. In: Zrozumieć Ziemię, PIG-PIB. Available at: http://geoportal.pgi.gov.pl/portal/page/portal/zrozumiec_ziemie/wycieczki/wigierski_1. Access date: 2023-06-09

Krzywicki T, Pochocka-Szwarc K (2012) Projekt geoparku „Kanał Augustowski – Augustowskie sandry". In: Kształtowanie przestrzeni turystycznej transgranicznego produktu turystycznego „Kanał Augustowski – wodny cud natury". Materiały konferencyjne, Augustów, 18 maja 2012, pp 8–9

Kupryjanowicz M (2007) Postglacjal development of vegetation in the vicinity of the Wigry Lake. Geochronometria 27:53–66

Lauterbach S, Brauer A, Andersen N, Danielopol DL, Dulski P, Huls M, Milecka K, Namiotko T, Plessen B, von Grafenstein U (2010) Multi-proxy evidence for early to mid-Holocene environmental and climatic changes in northeastern Poland. Boreas 40:57–72

Lindner L, Marks L, Nita M (2013) Climatostratigraphy of interglacials in Poland: Middle and Upper Pleistocene lower boundaries from a Polish perspective. Quart Intern 292:113–123

Lisicki S (1993) Deglacjacja Pojezierza Suwalskiego w okresie schyłku plejstocenu In: Przewodnik 45 Zjazdu Polskiego Towarzystwa Geologicznego na Ziemi Suwalsko-Augustowskiej, 9–11 września 1993 r. Wydawn Geol, Warszawa, pp 81–86

Makri S, Lami A, Tu L, Tylmann W, Vogel H, Grosjean M (2021) Holocene phototrophic community and anoxia dynamics in meromictic Lake Jaczno (NE Poland) using high-resolution hyperspectral imaging and HPLC data. Biogeosci 18:1839–1856

Micun K (2014) Rola zagłębień bezodpływowych jako lokalnych zbiorników sedymentacyjnych w krajobrazie młodoglacjalnym Suwalskiego Parku Krajobrazowego. Inż Ekol 40:196–207

Monderewicz W, Smolska E (1999) Analiza morfologiczno-litologiczna paleomeandru w Potaszni. In: Dynamika procesów stokowych i fluwialnych w rzeźbie młodo glacjalnej w świetle wybranych cech sedymentologicznych osadów, Jeleniewo k/Suwałk 13-17 września 1999. Fac Geogr Regional Studies, University of Warsaw, pp 51–52

Pietkiewicz S (1928) Pojezierze Suwalszczyzny Zachodniej (zarys morfologii lodowcowej). Przegl Geogr 8:168–222

Piotrowski JA (1997) Subglacial groundwater flow during the last glaciation in northwestern Germany. Sediment Geol 111:217–224

Pochocka-Szwarc K (2020a) Objaśnienia do arkusza Suwałki Mapy geologicznej Polski 1:200 000. Państwowy Instytut Geologiczny – PIB, Warszawa

Pochocka-Szwarc K (2020b) Morphological and geological outlines of Lake Hańcza. In: Weckwerth P, Kalińska E, Wysota W (eds) Glacial megaflood landforms and sediments in North-Eastern Poland. UAM, Toruń, pp 41–48.

Pochocka-Szwarc K, Ber A, Szpygiel P, Szpygiel J (2012) Jezioro Hańcza – wstępne wyniki geologicznej interpretacji obrazów sonarowych. Przegl Geol 61:187–194

Pochocka-Szwarc K, Jóźwik K (2020) Mapa geologiczna Polski w skali 1:200 000, mapa A. arkusz Suwałki. Państw Inst Geol – PIB, Warszawa

Pochocka-Szwarc K, Krzyszkowski D (2015) The outwash plain of the Rospuda river valley—a record of depositional environments. Studia Quat 32:63–78

Pochocka-Szwarc K, Przasnyska J, Rychel J, Tekielska A (2014) Mapa geologiczno-turystyczna Suwalskiego Parku Krajobrazowego 1:25 000. Państw Inst Geol – PIB, Warszawa

Popielarczyk D, Templin T, Łopata M (2015) Using the geodetic and hydroacustic measurements to investigate the bathymetric and morfometric parameters of Lake Hańcza (Poland). Open Geosci 7:854–869

Rutkowski J, Krzysztofiak L (ed) (2009) Jezioro Wigry. Historia jeziora w świetle badań geologicznych i peleoekologicznych. Stowarzyszenie „Człowiek i Przyroda", Suwałki, p 288

Rutkowski J, Król K, Lemberger M (1998) The pingo remnant in Suwałki Region (NE Poland). Quat Stud Poland 15:50–60

Rutkowski J, Prosiewicz D, Aleksander-Kwaterczak U, Krzysztofiak L (2009) Osady jeziora Wigry. In: Rutkowski J, Krzysztofiak L (eds) Jezioro Wigry. Historia jeziora w świetle badań geologicznych i peleoekologicznych. Stowarzyszenie „Człowiek i Przyroda", Suwałki, pp 83–109

Rychel J, Pochocka-Szwarc K, Ber A (2010) Wigierski Park Narodowy. Mapa geologiczno-turystyczna 1:30,000. Państw Inst Geol – PIB, Warszawa

Ryka W (1998) Geologic position of the Suwałki Anorthosite Massif. In: Geologia Suwalskiego Masywu Anortozytowego, północno-wschodnia Polska. Prace Państw Inst Geol 161:19–26

Sawicka L (2010) Szurpiły, St. 8 („Mosiężysko"), Woj. Podlaskie. Badania w latach 2008–2010. Światowit (8)49/Fasc B:263–268

Smolska E (2007) Development of gullies and sediment fans in last-glacial areas on the example of the Suwałki Lakeland (NE Poland). Catena 71:122–131

Smolska E, Szwarczewski P (2008) Zróżnicowanie wiekowe pokryw stokowych jako efekt niesynchronicznego zasiedlenia Pojezierza Suwalskiego i Sejneńskiego. Botanical Guidebooks 30:157–168

Solon J, Borzyszkowski J, Bidłasik M, Richling A, Badora K, Balon J, Brzezińska-Wójcik T, Chabudziński Ł, Dobrowolski R, Grzegorczyk I, Jodłowski M, Kistowski M, Kot R, Krąż P, Lechnio J, Macias A, Majchrowska A, Malinowska E, Migoń P, Myga-Piątek U, Nita J, Papińska E, Rodzik J, Strzyż M, Terpiłowski S, Ziaja W (2018) Physicogeographical mesoregions of Poland: verification and adjustment of boundaries on the basis of contemporary spatial data. Geogr Polon 91:143–170

Stasiak J (1971) Geneza basenów sedymentacyjnych na obszarach sandrowych. Zesz Probl Post Nauk Roln 107:103–112

Sygar S, Zgłobicki W (2022) Geoheritage resources in Polish Landscape Parks as a basis for developing a network of geoparks. Land 11:2277

Szewczyk J, Nawrocki J (2011) Deep-seated relict permafrost in northeastern Poland. Boreas 40:385–388

Tylmann W, Zolitschka B, Enters D, Ohlendorf C (2013) Laminated lake sediments in northeast Poland: distribution, preconditions for formation and potential for paleoenvironmental investigation. J Paleolimnol 50:487–503

Weckwerth P, Wysota W (2023) Unique landscape originated by cataclysmic glacial floods at the *Weichselian glaciation decline* in north-eastern Poland. In: Migoń P, Jancewicz K (eds) Landscapes and landforms of poland. Springer, Cham, pp 000–000
Weckwerth P, Wysota W, Piotrowski JA, Adamczyk A, Krawiec A, Dąbrowski M (2019) Late Weichselian glacier outburst floods in North-Eastern Poland: landform evidence and palaeohydraulic significance. Earth Sci Rev 194:216–233
Weckwerth P, Kalińska E, Wysota W (eds) (2020a) Glacial megaflood landforms and sediments in North-Eastern Poland. UAM, Toruń, p 107
Weckwerth P, Wysota W, Piotrowski JA, Krawiec A (2020b) Pleistocene glacial megaflood landform system in NE Poland. In: Weckwerth P, Kalińska E, Wysota W (eds) Glacial megaflood landforms and sediments in North-Eastern Poland. UAM, Toruń, pp 49–59
Wiszniwska J, Petecki Z (2014) Mezoproterozoiczne złoże rud tytanomagnetytu w Suwalskim Masywie Anortozytowym i jego środowisko geologiczne. Górn Odkr 2–3:44–59
Wysota W, Weckwerth P, Adamczyk A, Piotrowski J. A, Sokólska A (2020) Morphology and origin of glacial curvilineations (GCLs) east of Hańcza Lake, Suwałki Lakeland. In: Weckwerth P, Kalińska E, Wysota W (eds) Glacial Megaflood Landforms and Sediments in North-Eastern Poland. UAM, Toruń, pp 61–69
Zaborowski Z (2017) Czytanie rzeźby polodowcowej nad Wigrami cz. IV - "Takie pingo...". Kwartalnik Wigry. Wigierski Park Narodowy, Krzywe-Suwałki:6–9
Żelaźniewicz A, Aleksandrowski P, Buła Z, Karnkowski PH, Konon A, Oszczypko N, Ślączka A, Żaba J, Żytko K (2011) Regionalizacja tektoniczna Polski. Kom Nauk Geol PAN, Wrocław, p 64

Ewa Smolska PhD, is geomorphologist and sedimentologist, works at the Faculty of Geography and Regional Studies, University of Warsaw, as Associate Professor. Her main research interests are soil and gully erosion, contemporary and past dynamics of fluvial processes, and geoarchaeology. She conducted her research mainly in north-eastern Poland, where she deals with, among others, monitoring of water, soil erosion and fluvial transport. Other research areas include Iceland and Tunisia. In the years 2002–2006, she was the head of the Sedimentology Laboratory and since 2006 she has been the head of the Geomorphology Department.

Katarzyna Pochocka-Szwarc PhD, is a graduate of the Faculty of Geology of the University of Warsaw and has been working at the Polish Geological Institute-National Research Institute for many years. She is an experienced geologist—cartographer. Her specialty is Quaternary geology. Recently, she also deals with the geology of sediments of Cretaceous, Paleogene and Neogene age. She is an author of several detailed geological maps of Poland at the scale of 1:50,000 from the Mazury and Suwałki regions and from West Polesie, as well as an author of overview geological maps at larger scales (1:200,000, 1:500,000). She is an author and co-author of scientific articles in the field of Quaternary geology as well as of geotourist maps for the area of north-eastern Poland. She conducts popularizing activities in the field of geology, organizing conferences, geological exhibitions and giving papers and lectures.

Piotr Szwarczewski PhD, is geomorphologist, Associate Professor at the Faculty of Geography and Regional Studies in the University of Warsaw. His main research interests are human impact and climate change record in the sediments, palaeogeography of the Quaternary period, archival topographical maps analysis, and geochemistry. The Masovian Lowland, Suwałki Lakeland and Nida River basin are his main research areas in Poland, but he was also involved in research projects in Iceland, Lithuania, Slovakia, Italy, and in the North African countries.

Unique Landscape Originated by Cataclysmic Glacial Floods at the Weichselian Glaciation Decline in North-Eastern Poland

39

Piotr Weckwerth and Wojciech Wysota

Abstract

Rapid climate changes during the decline of the Weichselian glaciation caused a sudden increase in meltwater activity and significantly transformed land relief in north-eastern Poland. The sudden release of meltwater from glacial lakes was responsible for the development of the Megaflood Landform System that includes subglacial and proglacial subsystems as a unique landscape pattern in north-eastern Poland. These subsystems, respectively, constitute a part of subglacial and marginal landsystems. The main pathways of subglacial floodwater flow are represented by exceptionally large tunnel valleys with subsequent erosional features attributed to catastrophic drainage of subglacial lakes located likely in south-western Lithuania. Features indicative of subglacial outburst floods are glacial curvilineations (GCLs), which appear as large ridges separated by inter-ridge troughs. The proglacial landscape is represented by main (first-order) landforms, including two floodwater flow pathways in the proglacial area: the Eastern and Western Spillways. The secondary components of this subsystem comprise a suite of landforms including outburst fans, scabland-like topography and mega-scale superimposed bedforms developed in the area of outwash plain. The Megaflood Landform System recognised in north-eastern Poland is the youngest geomorphic landscape in the European Lowland, and originated from cataclysmic glacial lake-outburst floods that were one of the major factors controlling the formation of the European valley system.

P. Weckwerth (✉) · W. Wysota
Faculty of Earth Sciences and Spatial Management, Nicolaus Copernicus University in Toruń, Lwowska 1, 87-100 Toruń, Poland
e-mail: pweck@umk.pl

W. Wysota
e-mail: wysota@umk.pl

Keywords

Glacial lake-outburst floods · Tunnel valleys · Spillways · Megadunes · Obstacle marks · Scabland-like landforms

39.1 Introduction

Catastrophic geological processes shaping the Earth's landscape have repeatedly been initiated by rapid climate changes of global nature. Such changes occurred during the Weichselian glaciation, causing a sudden increase in ice-sheet dynamics and meltwater activity in both subglacial and proglacial settings, which significantly transformed the relief of north-eastern Poland (Weckwerth et al. 2019b). These landscape transformations took place during the decline of the Weichselian, in the Pomeranian Phase, dated to 16–17 ka, and were accompanied by the collection of considerable quantities of meltwaters in subglacial and/or supraglacial lakes in the southern periphery of the Scandinavian Ice Sheet (SIS). Sudden releases of meltwater from these lakes generally happened shortly after an ice sheet reached its maximum extent (see Marren 2005) and were responsible for the development of extensive outwash plains in north-eastern Poland, where the Augustów Plain (the Suwałki-Augustów sandur) and the Suwałki Lakeland are located (Fig. 39.1).

In these regions, sediments and landscape features indicate that glacial lake-outburst floods were exceptional in their evolution as a response to climate changes at the end of the Weichselian (Zieliński 1993; Weckwerth et al. 2019b). These processes significantly affected the transformation of landscape pattern as a result of meltwater erosion and accumulation in the SIS forefield and in the subglacial environment. The transformation rate was related to the capacity and type of subglacial drainage, and the ability of meltwaters to collect and flow in sub, supra and proglacial settings.

P. Migoń and K. Jancewicz (eds.), *Landscapes and Landforms of Poland*, World Geomorphological Landscapes, https://doi.org/10.1007/978-3-031-45762-3_39

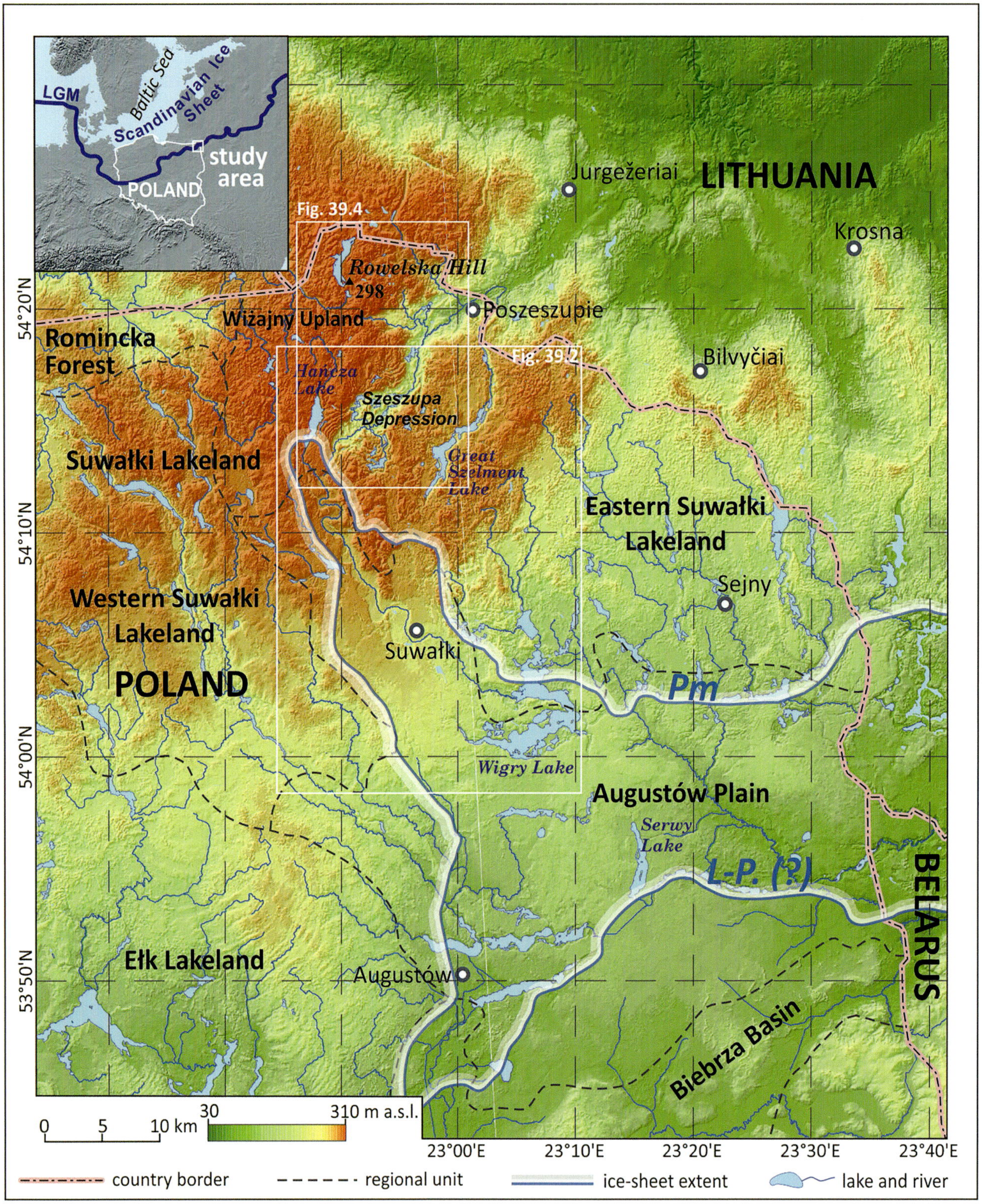

Fig. 39.1 Morphology and regional units of north-eastern Poland (after Solon et al. 2018) and the presumable extent of the Scandinavian ice sheet during or shortly after the Pomeranian Phase (*Pm*) of the Weichselian glaciation (modified from Ber 1967; Lisicki 1993; Dzierżek and Zreda 2007), when the glacial lake-outburst floods occurred. Presumable ice-sheet extent during the Leszno-Pomeranian Stadial (*L-P*) after Krzywicki (1997)

Geomorphological evidence of catastrophic glacial floods has been discovered in many formerly glaciated areas of North America, in Siberia and Central Asia, and in Northern Europe. Considering a few examples of glacial lake-outburst floods in Central Europe, which are older than the Eemian interglacial (Gibbard 2007; Jakobsson et al. 2007; Meinsen et al. 2011; Winsemann et al. 2016, 2018; Rdzany and Frydrych 2018; Lang et al. 2019), only the landscape in north-eastern Poland represents a complete pattern of subglacial and proglacial features related to a glacial outburst flood from the Weichselian glaciation (Weckwerth et al. 2019b). Looking from a broader perspective, exploration of this unique system is of interest in terms of understanding the former ice-sheet dynamics and investigating controls on glacial landscape and river valley network evolution in the European Lowland. This chapter aims to show the complexity of the glacial megaflood landform system in NE Poland and to give broader insights into the dynamics of the SIS and landscape evolution in the European Lowland.

39.2 Regional Setting

The Suwałki Lakeland and Augustów Plain represent landscapes shaped by ice-sheet and meltwater activity during the Weichselian glaciation and transformed postglacially as a result of fluvial, aeolian, periglacial and denudation processes (Smolska 2007a, b, 2008). The northern part of the Augustów Plain hosts the Hańcza River valley and divides the Suwałki Lakeland into the Western Suwałki Lakeland (WSL) and the Eastern Suwałki Lakeland (ESL) (Figs. 39.1 and 39.2). The

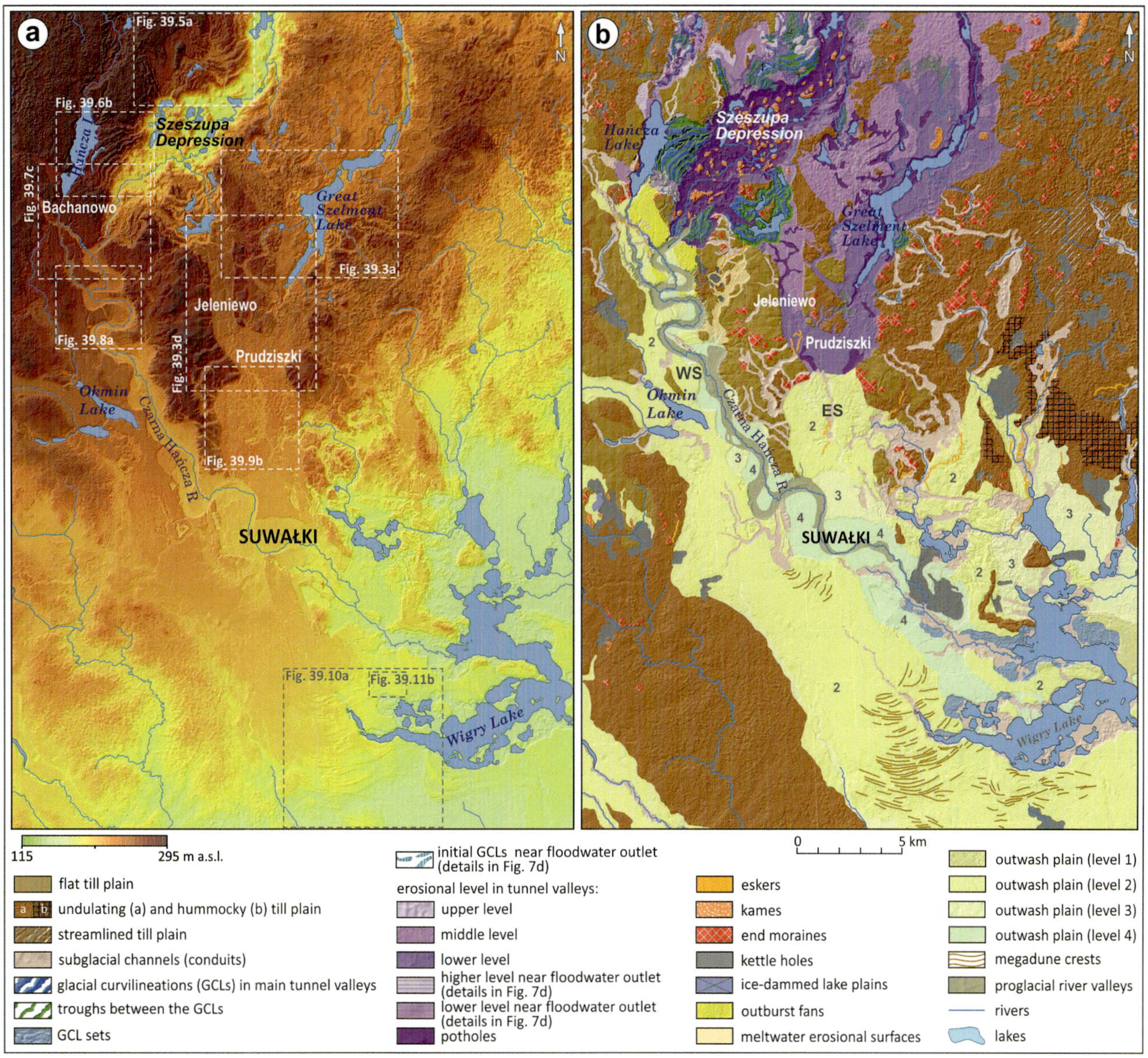

Fig. 39.2 Morphology (**a**) and Landform origin (**b**) in the Megaflood Landform System (after Weckwerth et al. 2019a, b, modified; WS—Western Spillway, ES—Eastern Spillway)

ground surface of these regions is mostly composed of sediments from the Middle Weichselian (the Świecie Stadial) and Late Weichselian (Leszno-Pomeranian or Main Stadial) stadials (Ber 2000; Krzywicki 2002). The age of glacigenic deposits of the Świecie Stadial was assessed as 67–50 ka BP (Krzywicki 2002), while the age of the maximum ice-sheet extent during the Leszno-Pomeranian Stadial was estimated to be about 20–19 ka based on cosmogenic nuclide exposure dating (Rinterknecht et al. 2005, 2006; Dzierżek and Zreda 2007; Marks 2012). It seems that the onset of deglaciation of north-eastern Poland did not take place before 18 ka (Dzierżek and Zreda 2007; Tylmann et al. 2019). During the SIS advance in the Leszno-Pomeranian Stadial, the ice-sheet front was subdivided into three lobes, namely the Rospuda Lobe in the west, the Wigry Lobe in the middle and the Sejny Lobe in the east (Ber 2000). Subsequent deglaciation was gradual (Ber 1966, 1982, 2000; Lisicki 1994) and recorded by end moraines of the Wigry subphase, followed by three recessional-oscillational stages. The ice standstill during the Pomeranian phase resulted in the complex, characteristic glacial landscape of the Suwałki Lakeland (Ber 1974, 1982, 2000). Its surface rises northwards and reaches its culmination in the Wiżajny Upland, in the north-eastern part of the WSL, where the Rowelska Hill (289 m asl) marks the highest point. Large height differences occur in the Suwałki Lakeland, up to 140 m between the Wiżajny Upland and the Szeszupa Depression (Fig. 39.1).

The main element of the Suwałki Lakeland landscape is a till plain (moraine plateau) that consists mostly of glacial deposits (tills). Its relief is diversified by end moraines, thrust and squeezed-up moraines with specific festoon-like relief, eskers and crevasse features (Kondracki and Pietkiewicz 1967; Ber 1974, 1982, 2000). Numerous concentric dead-ice moraines and kames were also formed. Moreover, large tunnel valleys dissect the moraine plateau, and these are the Hańcza Lake Tunnel Valley, the Szeszupa Tunnel Valley, the Jeleniewo Tunnel Valley and the Great Szelment Lake Tunnel Valley.

In the forefields of the former ice-sheet margin, outwash fans and outwash plains were formed during the Late Weichselian (Bogacki 1976, 1980; Zieliński 1989, 1993; Weckwerth et al. 2019b). These constitute the Augustów Plain, which is a vast, usually flat or slightly undulating plain formed through the activity of proglacial meltwater (Kondracki and Pietkiewicz 1967; Ber, 1974; Bogacki 1976, 1980; Zieliński 1989, 1993). Its surface descends gently from ca. 190–170 m asl near the town of Suwałki to ca. 130–140 m asl near the town of Augustów. A large number of tunnel valleys associated with the maximum ice-sheet limit of the late Weichselian glaciation intersect the outwash plain (Ber 1974, 1982, 2000; Krzywicki 2002).

It used to be assumed (Liszkowski 1993; Ber 2000) that the contemporary landscape of the Suwałki Lakeland and the Augustów Plain have developed mainly due to glaciotectonic deformations dependent on the geological structure and neotectonics of the basement. However, the most recent studies clearly show that catastrophic glacial lake-outburst floods significantly contributed to the current landscape in these regions (Weckwerth et al. 2019b). This is manifested by the occurrence of unique glacial flood-related landforms such as megadunes (giant and downstream-propagating bedforms), obstacle marks and linear clusters of kettle holes.

39.3 Lowland Landform System of Glacial Floods

The subglacial and proglacial erosional and depositional effects of glacial lake-outburst floods are well-represented in north-eastern Poland as a unique landscape pattern (Weckwerth et al. 2019b). This pattern forms a coherent system—a Megaflood Landform System—that includes subglacial and proglacial subsystems, both generated by catastrophic glacial floods (Evans 2003).

39.3.1 Subglacial Landscape

39.3.1.1 The Main Pathways of Subglacial Floodwater Flow

Although the interpretation of field evidence of subglacial floods is subject to controversy (e.g. Evans et al. 2006), a suite of erosional landforms in NE Poland is directly related to catastrophic subglacial flooding that affected the ice-sheet bed and produced erosional features.

The largest area in the subglacial megaflood subsystem is occupied by three erosional topographic levels that occur at 235–255 m asl (upper level), 190–220 m asl (middle level) and 170–210 m asl (lower level) and are incised into undulating till plains (Figs. 39.2b and 39.3). These levels partly constitute the bottom of three exceptionally large (up to 7.6 km wide) tunnel valleys: (1) the Great Szelment Lake Tunnel Valley, (2) the Jeleniewo Tunnel Valley and (3) the Szeszupa Tunnel Valley. These valleys evolved as main subglacial meltwater flow pathways during the glacial lake-outburst floods and have secondary erosional features attributed to these catastrophic drainages. These features were identified on the basis of their morphology, size and spatial pattern. Depositional landforms within these large tunnel valleys evolved during deglaciation, as climate warming caused the ice sheet to melt and decay.

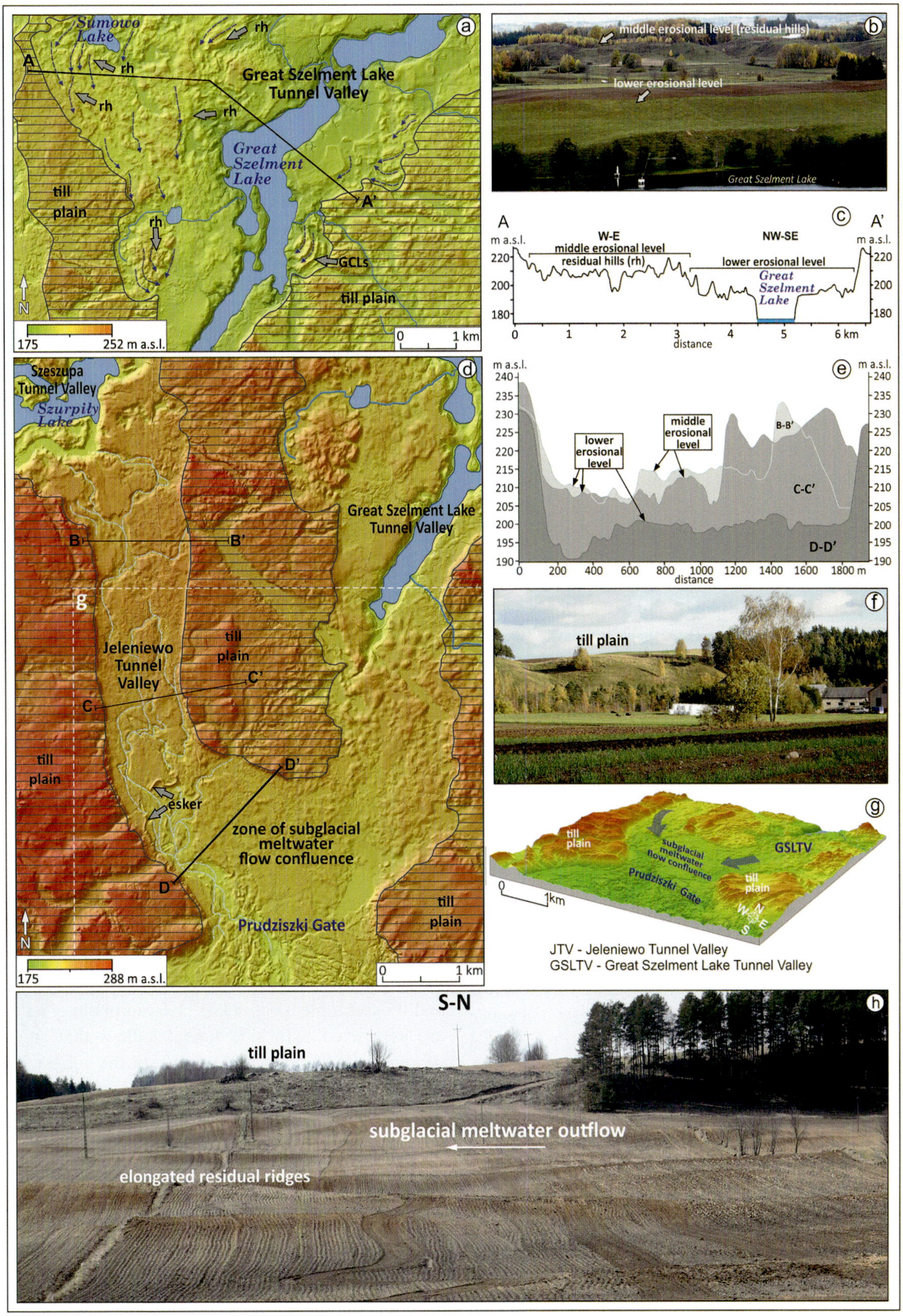

Fig. 39.3 Great Szelment and Jeleniewo Tunnel Valleys. **a** Morphology of southern part of the Great Szelment Lake Tunnel Valley (GCLs—glacial curvilineations, rh—residual hills). **b** Lower and middle erosional levels located to the west of the Great Szelment Lake. **c** Transverse profile showing erosional levels and transitional character of the slope between the flat tunnel valley bottom and the undulating till plain, manifested by the occurrence of erosional remnants. **d** Morphology of the Jeleniewo Tunnel Valley (JTV). **e** Transverse profiles showing two erosional levels in the JTV (location on panel d). **f** Distinct eastern slope of the JTV near the Prudziszki Gate. **g** 3-D model showing morphology of the Prudziszki Gate as an outlet of floodwater from both the Great Szelment Lake and Jeleniewo Tunnel Valleys. **h** Elongated residual ridges in the bottom of the JTV (photographs by P. Weckwerth)

(a) *The Great Szelment Lake Tunnel Valley*

The Great Szelment Lake Tunnel Valley is deepened locally by the Great Szelment Lake subglacial channel (conduit) and includes erosive levels that are up to 9.3 km wide and incised 6–24 m into the till plain, which surrounds them to the east (Fig. 39.2). The erosive levels have surfaces shaped by turbulent subglacial meltwater outflow, with southward (downstream) increasing height (e.g. Shaw 2002; Burke et al. 2012; Kehew et al. 2012; Lesemann et al. 2014; Hermanowski et al. 2019). These surfaces are diversified by eskers, potholes and scours, and elongated residual ridges (Figs. 39.2 and 39.3a, b).

The eastern slope of the Great Szelment Lake Tunnel Valley is distinct, especially in its southern reach, while the western slope has a transitional character between a flat tunnel valley bottom and undulating till plain (Fig. 39.3a–c). This is manifested by the occurrence of erosional remnants (residual hills; pseudo-eskers sensu Smed 1962; Galon 1965) between parallel small channels located at 210–218 m asl westward of the Great Szelment Lake and near to the Sumowo Lake (Fig. 39.3a). These channels are poorly defined, suggesting transition from sheet flow to channelised outflow of subglacial floodwaters along the western flank of the Great Szelment Lake Tunnel Valley (see Johnson 1999; Kehew et al. 2012). These features also imply an initial phase of glacial curvilineations (GCLs) development, similar to those recognised in central Poland, indicative for subglacial multi-phase meltwater releases (Lesemann et al. 2010; Wysota et al. 2023). Well-developed GCLs in the Great Szelment Lake Tunnel Valley were identified in the south-eastern vicinity of this lake (Fig. 39.3a, c).

(b) *The Jeleniewo Tunnel Valley*

The second pathway of subglacial meltwater flow, the Jeleniewo Tunnel Valley, has a length of 6.5 km (Figs. 39.2 and 39.3d). The valley runs north to south, from the southern vicinity of the Szurpiły Lake, where its width is 1600 m. The northern and central parts of the Jeleniewo Tunnel Valley are oriented north–south and are 1.3–1.6 km wide. The southern part turns south-eastwards and connects to the Szelment Lake Tunnel Valley, where it forms an almost 2-km wide outlet valley that ends at the Prudziszki Gate (Fig. 39.3d–g).

The Jeleniewo Tunnel Valley is bounded by steep slopes, ranging from 20 m high in the north to 35 m high in the south (Fig. 39.3d, e). Its bottom is characterised by a slightly undulated surface with elongated residual ridges. It drops southward from 216 to 196 m asl and is associated with side valleys (subglacial conduits) that cut into the moraine plateau (Fig. 39.3d–h). Moreover, the bottom of the Jeleniewo Tunnel Valley is incised by narrow (50–150 m) secondary channels up to 10 m deep forming an anastomosing pattern and stretching from the area near the Szurpiły Lake and the Szeszupa Tunnel Valley to the Prudziszki Gate (Fig. 39.3d).

We interpret the Jeleniewo Tunnel Valley as a broad subglacial tunnel responsible for the discharge of large amounts of meltwaters that exited the ice sheet through the ice-marginal outlet at the Prudziszki Gate. These meltwaters deeply incised the glacier bed, and high meltwater pressure probably caused the top of a distal section of the subglacial tunnel to break and collapse. As a result, an ice-walled open channel (an ice-canyon) was formed most likely around the Prudziszki Gate, which was an outlet of floodwaters from both the Great Szelment Lake and Jeleniewo Tunnel Valleys (Fig. 39.3d, g) (see Russell and Knudsen 1999; Russell et al. 2001; Weckwerth et al. 2019b; Wysota et al. 2020a).

(c) *The Szeszupa Tunnel Valley*

The next subglacial meltwater flow pathway identified in the study area is the Szeszupa Tunnel Valley, whose origin is primarily subglacial (Fig. 39.2b). This valley is also named the Szeszupa Depression, with the bottom that consists of landforms related mostly to dead-ice topography, kames and eskers (see Ber and Maksiak 1969; Rychel and Pochocka-Szwarc 2020). The bottom of the Szeszupa Tunnel Valley lies 40–90 m below the surrounding till plains and erosional levels and is characterised by an undulated surface, the height of which increases southward (Figs. 39.4, 39.5 and 39.6). This bottom is also punctuated by many potholes which exist in three widenings (northern, middle and southern basins; Figs. 39.2 and 39.4). Specific subglacial features in the Szeszupa Tunnel Valley are GCLs sets, located along its western flank. These are described in detail in the next subsection.

The northern basin in the Szeszupa Tunnel Valley is located between the villages of Rutka-Tartak, Rowele and Poszeszupie (Fig. 39.4). Its morphology and extent are associated with side tunnel valleys that suggest the development of connecting nodes in the subglacial drainage network. In this network, the Szeszupa Tunnel Valley was the main and the deepest subglacial meltwater flow pathway (Fig. 39.2). The middle basin occurs between the village of Ługiele and the Pobondzie Lake and represents the zone of wide erosional levels characterised by segments dominated by GCls which pass into the main, flat-bottomed and meandering channel (Fig. 39.5a–c). The remnants of higher and middle erosional levels, as well as the large-scale GCLs, split here into smaller-scale GCLs (Fig. 39.5a, b).

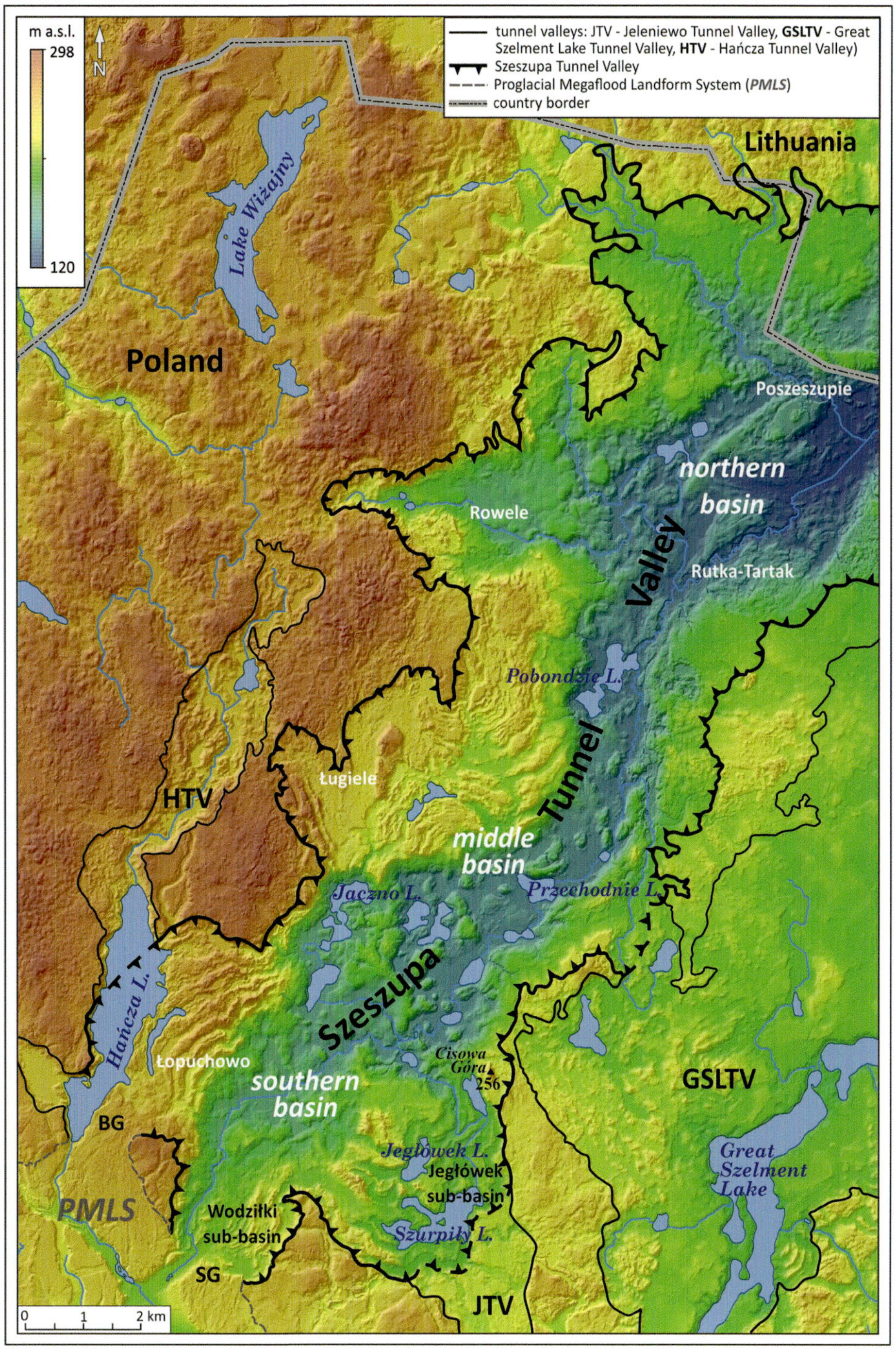

Fig. 39.4 Morphology and the main units of the Szeszupa Tunnel Valley (BG—Bachanowo Gate, SG—Szeszupka Gate)

The southern basin lies between the lakes of Szurpiły and Hańcza (Figs. 39.4 and 39.5d), and includes the Wodziłki and Jegłówek sub-basins (Ber and Maksiak 1969), which are adjacent to potholes and scours in the tunnel valley bottom (Fig. 39.4). The morphology of both Wodziłki and Jegłówek sub-basins also stands out due to glacial curvilineations (GCLs) which compose clusters of parallel ridges characterised by different lengths, widths and heights (Figs. 39.2b and 39.4). The GCLs in the Wodziłki sub-basin (10 individual features) are 400–890 m long and 30–75 m wide, lying between 196 and 208 m asl and dropping in elevation towards the Szeszupa Tunnel Valley axis (Figs. 39.4 and 39.5e). Between the ridges run parallel troughs up to 14 m deep. The GCLs in the Jegłówek sub-basin are hardly visible in the landscape, representing erosional remnants of the till plain. These remnant ridges are up to 300 m wide and 1.5 km long, and are separated by parallel channels up to 38 m deep that are partly infilled by water (e.g. Szurpiły and Jegłówek lakes) and peats and gyttja.

Fig. 39.5 Szeszupa Tunnel Valley. **a** Morphology of erosional levels in the middle basin of the Szeszupa Tunnel Valley. **b** Geomorphological map of the middle part of the valley and location of selected features presented in panel c (see Fig. 39.2 for explanation of colours). **c** Landscape in the middle part of the valley (view from Mt. Cisowa Góra towards the north). **d** Southern basin in the valley (view from Łopuchowo towards the east). **e** GCLs in the Wodziłki sub-basin (photographs by P. Weckwerth)

The origin of the Szeszupa Tunnel Valley is similar to the Great Szelment Lake Tunnel Valley. Both are attributed to erosion by a series of subglacial flood events from the drainage of sub or supraglacial lakes (see Jørgensen and Sandersen 2006; Lesemann et al. 2010; Kehew et al. 2012). Repeated subglacial meltwater outbursts through subglacial valleys are inferred to have formed: (1) the large width of both the Szeszupa and the Great Szelment Lake tunnel valleys with different topographic erosional levels; (2) initial and/or well-developed glacial curvilineations of different size lying within these levels; and (3) subsequent subglacial meltwater channels (sensu Glasser and Bennett 2004). All these features form a large subglacial landform subsystem, in which the Szeszupa Tunnel Valley must have functioned as an outburst flood subglacial flow pathway for the longest time because its bottom is incised into erosional levels that compose the bottoms of the Jeleniewo and Great Szelment tunnel valleys (Figs. 39.2 and 39.4). A model for the subglacial megaflood subsystem formation most likely involves drainage of subglacial lakes that probably existed in southwestern Lithuania.

39.3.1.2 Glacial Curvilineations as Indicative Features for Subglacial Outburst Floods

Glacial curvilineations (GCLs) represent ridges of exceptionally large width (up to 400 m) and length (up to 3 km), separated by inter-ridge troughs, and interpreted as erosional remnants of antecedent substratum carved by flows of subglacial meltwaters (Lesemann et al. 2010, 2014; Wysota et al. 2020b). These features occur in the subglacial megaflood subsystem as two clusters on the erosional level located along the western flank of the Szeszupa Tunnel Valley (Fig. 39.4).

The first GCLs set lies near the village of Łopuchowo, eastwards of the Hańcza Lake Tunnel Valley (Figs. 39.4 and 39.7). Six arcuate and parallel ridges are 100–400 m wide and 1.5–3 km long each, with uneven crest lines undulating by up to 20 m (Sokólska 2014; Wysota et al. 2020b). Most of these ridges occur at an altitude of 210–260 m asl. The largest ridges have the height similar to the neighbouring till plain (Fig. 39.6a–c). The longer axes of parallel ridges are sinuous, and their southern ends connect with the Bachanowo outwash fan or drop towards the Szeszupa Tunnel Valley (Fig. 39.6b, c).

The second GCLs cluster is located near the hamlet of Lizdejki and also comprises arched and parallel ridges of up to 240 m wide and 1.4 km long (Fig. 39.5a, b). These ridges have undulated top surfaces that drop from 233 to 213 m asl towards the Szeszupa Tunnel Valley. In the area between Lizdejki, Smolniki and Bondziszki, GCLs join with the remnants of higher or middle erosional levels, and the latter may pass into GCLs of smaller scale (Figs. 39.5a, b and 39.6d, e).

The origin of parallel and arcuate ridges located to the west of the Szeszupa Tunnel Valley has been interpreted as end moraines and drumlins (Pietkiewicz 1928; Kondracki and Pietkiewicz 1967) or as glaciotectonic ridges (squeezed-up or thrust moraines) formed as a result of intensive lateral deformations by narrow ice sheet lobes advancing into pre-existing elongated depressions during the Late Weichselian (Ber 1968, 1974, 1982, 1986, 1989, 2000). Considering the geomorphological characteristics, the subglacial meltwater origin of GCLs in the western vicinity of the Szeszupa Tunnel Valley is supported by: (1) parallel, non-intersecting arrangement of ridges and troughs; (2) large lateral sinuosity (similar to tunnel valleys or eskers); (3) irregular longitudinal profiles (overdeepenings) of inter-ridge troughs, typical of subglacial conduits; and (4) the existence of boulder pavements on some crests and slopes (Fig. 39.6a) (Wysota et al. 2020b). These pavements were formed as a result of selective outwash of fine fraction and thus comprise only the coarsest elements, which together with GCLs ridges may be indicative of massive meltwater erosion during subglacial flooding. High altitude of both GCLs sets, which is almost at the level of the neighbouring till plains, tends to exclude the possibility that these forms were created by subglacial slope failures in the tunnel valley or near the shore of a subglacial lake (Clark and Livingstone 2018).

39.3.2 Proglacial Landscape

High-energy discharges of glacial lake-outburst floods caused rapid changes (probably during just a few weeks, considering icelandic glacial floods; e.g. Björnsson 1992) in the proglacial landscape associated with the decline of the Weichselian glaciation in NE Poland. These events were among the major factors controlling landform patterns and sediment distribution in the proglacial megaflood subsystem. This subsystem is represented by the main (first-order) landforms including the floodwater flow pathways. These are the Eastern and Western Spillways, the morphology of which enable the floodwater discharge to be assessed in the range of 276×10^3 to 861×10^3 m^3 s^{-1} (Weckwerth et al. 2019b). These spillways pass into a broad outwash plain (level 2) located to the south of Suwałki (Fig. 39.2).

The secondary components of the subglacial megaflood subsystem (second-order landforms) are represented by features developed within the main floodwater routes. Their formation was dependent on time and geomorphic processes controlled by downstream distance from floodwater outlets (Weckwerth et al. 2019b, 2021). As a result, the second-order landforms compose a continuum associated with changes in the proglacial bedform pattern resulting from a progressive decrease in flow energy along the glacial flood pathways (e.g. Zieliński 1989, 1993; Baker 2002; Carrivick 2007; Carling 2013). Nevertheless, landform continuum may be resolved into three zones, where the dominant geomorphic processes were: (1) erosion with subsequent deposition at the front of floodwater outlets; (2) transportation and channel-bed deposition along spillways; and (3) unconfined (or semi-confined) outwash plain development and aggradation of mega-scale, superimposed bedforms.

39.3.2.1 Floodwater Outlets and Outburst Terraces in the Western Spillway

The Western Spillway (WS) has a length of 12.1 km and its width varies from 2 km in the proximal reach to 4.5 km in the distal part near Suwałki (Fig. 39.2). Meltwaters from glacial lake-outburst floods exited the ice sheet

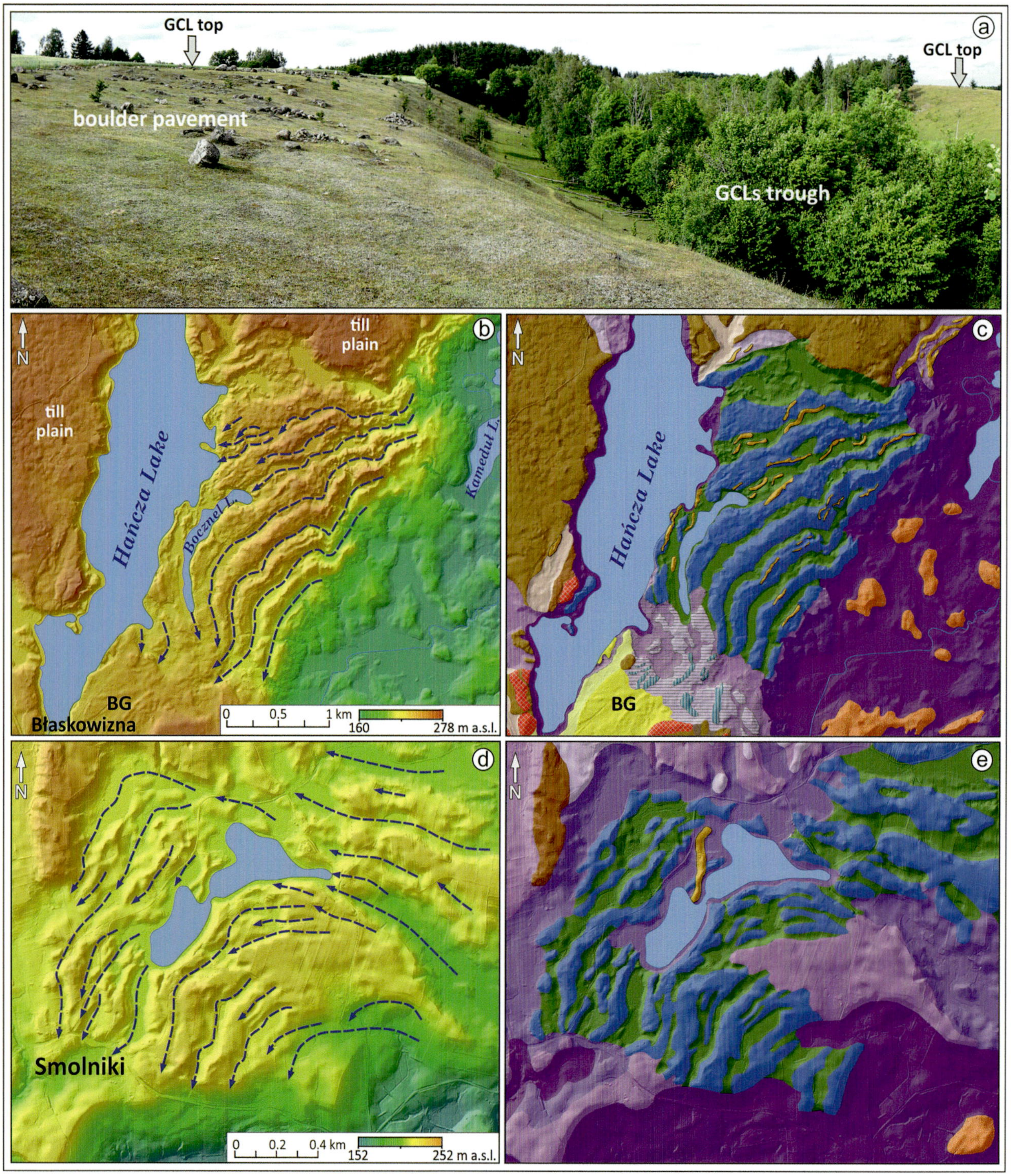

Fig. 39.6 Glacial curvilineations (GCLs). **a** Two parallel ridges of mega-scale GCLs near Łopuchowo with boulder pavement occupying their crests and slopes (photograph by P. Weckwerth). **b**, **c** Morphology and landform origin in the area of GCls cluster near Łopuchowo (see Fig. 39.2 for explanation of colours) (BG—Bachanowo Gate). **d**, **e** Complex system of GCLs of different scale in the area between Lizdejki, Smolniki and Bondziszki (blue dashed lines with arrows indicate the direction of subglacial meltwater outflow)

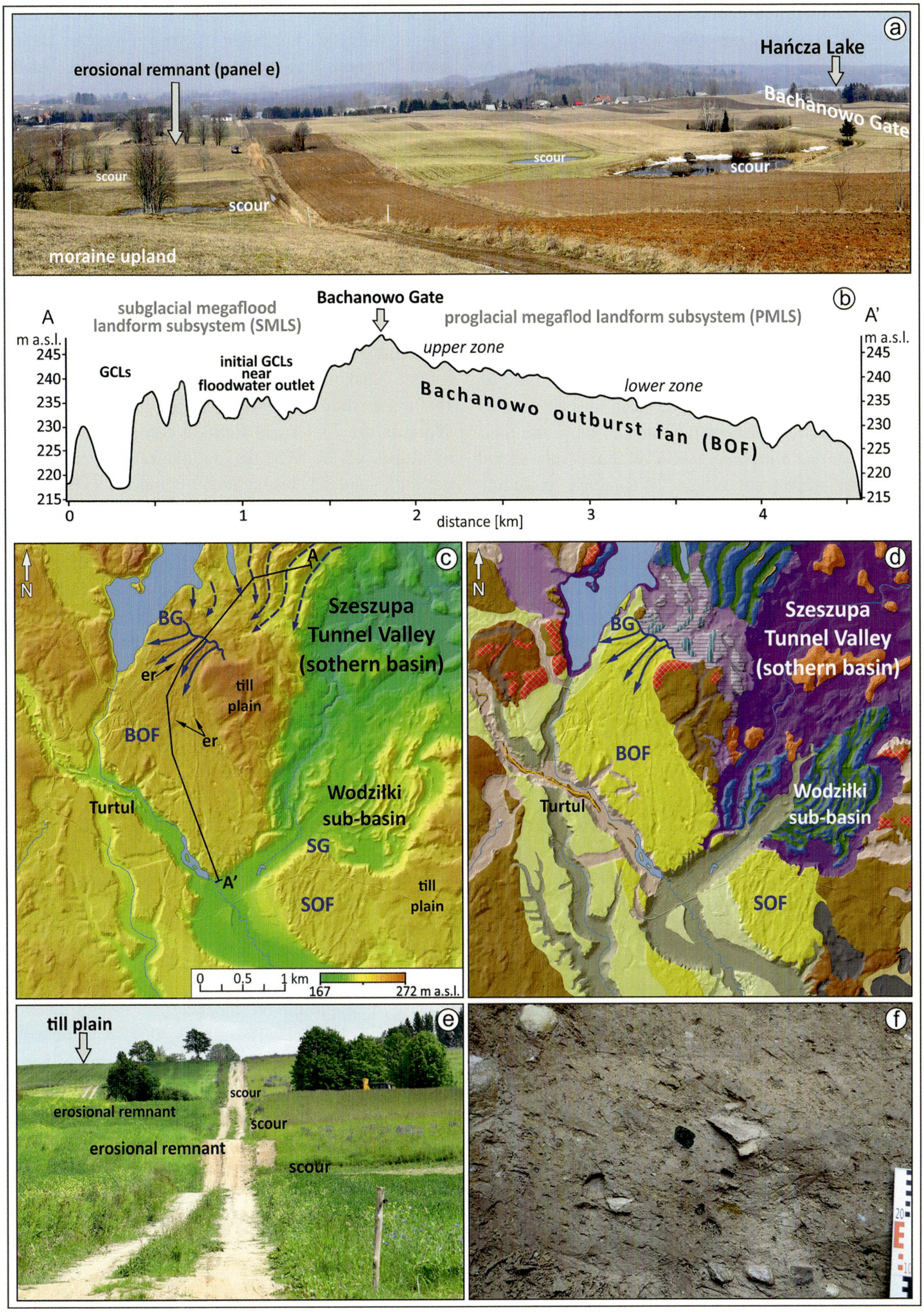

Fig. 39.7 Morphology of flood outburst fans. **a** Landscape of the proximal part of the Bachanowo outburst fan (BOF) with erosional remnants separated by scours near the Bachanowo Gate. **b** Longitudinal profile showing morphology of the BOF and GCLs of different scale developed under subglacial condition near floodwater outlet. **c, d** Morphology and landform origin in the area of outburst fans (see Fig. 39.2 for explanation of colours) (BOF—Bachanowo outburst fan, SOF—Szeszupka outburst fan, BG—Bachanowo Gate, SG—Szeszupka Gate, er—elongated residual ridges). **e** Scours in upper zone of the BOF separated by elongated erosional remnants built of glacial till similar to those in the surrounding till plain. **f** Glacial till which builds elongated ridges (erosional remnants) located in the proximal part of the BOF (photographs by P. Weckwerth)

through two ice-marginal outlets: (1) the Bachanowo Gate and (2) the Szeszupka Gate (Weckwerth et al. 2019b). These gates are associated with outwash (outburst) fans developed at the mouth of the Szeszupa Tunnel Valley, and have southwards-sloping surfaces, from 245 to 215 m asl (Fig. 39.7a–d). Two outburst fans continue southwards as an outwash plain forming the highest (main) outburst terrace (outwash plain level 2) and middle and lower terraces inset into it (outwash plain levels 3 and 4, respectively; Fig. 39.2b) (Ber 1967; Pochocka-Szwarc et al. 2017; Weckwerth et al. 2019b).

(a) *Bachanowo and Szeszupka Outburst Fans*

The Bachanowo and Szeszupka outburst fans are incised into the till plain and thus confined by slopes which are up to 18 m high (Figs. 39.2 and 39.7c, d). Both fans are located close to the mouth of the Szeszupa Tunnel Valley, and their distal parts form one morphological level that is intersected by a system of subsequent subglacial conduits, including the Turtul subglacial channel and a well-developed esker located in its bottom (Fig. 39.7c, d).

The Bachanowo Outburst Fan (BOF) has a surface with an average gradient of 0.0079 that drops downstream from 248 m asl (fan apex) to 224 m asl (near the village of Turtul; Fig. 39.7b–d). On the basis of morphological characteristics and the shape of longitudinal profile, the BOF can be divided into two zones (Fig. 39.7b). The upper zone is up to 1.6 km long and up to 1.4 km wide, with a concave longitudinal profile, while the lower zone is up to 1.12 km wide and 1.5 km long, with a uniform surface slope. Moreover, the BOF surface is characterised by secondary downstream elongated ridges (erosional remnants) and channels of braiding pattern (Fig. 39.7a, c–e). The width of ridges ranges between 23 and 61 m and their height does not exceed 3 m. The Szeszupka Outburst Fan (SOF) has a width up to 1.1 km, and its length (1.2 km) was reduced due to meltwater erosion forming lower morphological levels in the Western Spillway (Fig. 39.7b, c). The SOF surface has a gradient of 0.005, is incised up to 16 m into the till plain and has elongated ridges up to 2 m high separated by channels of braided pattern, similar to those in the BOF.

The surfaces of both BOF and SOF are covered in places by boulder pavements, while elongated erosional remnants in the upper zone of the BOF are built of diamicton (till) similar to that in the surrounding till plain (Fig. 39.7e, f). The middle and distal parts of outburst fans are composed of gravels with boulder horizons, indicative of pulses of floodwater outflow. The thickness of flood-related sediments does not exceed 4.5 m in the channels and flat parts of both fans. Considering these primary morphological and geological characteristics, the fans that originated from a glacier lake-outburst flood at the mouth of the Szeszupa Tunnel Valley are attributed to erosion by a series of floodwater outbursts (e.g. riverbed scouring) and repeated deposition of coarse sediments between and behind erosional remnants and in the distal parts of outburst fans (Weckwerth et al. 2019a).

(b) *Outburst Flood Terraces*

The bottom of the Western Spillway comprises three main topographic levels which correspond to the outwash plain levels 2–4 developed due to glacial floods (Fig. 39.2). These levels form three terraces T1–T3 in the WS (Figs. 39.2 and 39.8). The higher terrace (T1) drops from 226 to 195 m asl, the middle one (T2) lies between 212 and 185 m asl while the lower one (T3) lies at 200–177 m asl. Whereas the T1 terrace forms a continuous level confined by slopes of the WS, the middle and lower outburst terraces (T2 and T3, respectively) compose levels within five huge meander bends (valley bends) in the upper reach of the spillway (Figs. 39.2 and 39.8a–f). These bends are characterised by amplitudes ranging between 1.4–1.9 km, and the wavelength of 0.6–1.3 km. The bend axis has a length up to 900 m and corresponds to the geometry of the largest meandering rivers in the world (Stølum 1998; Constantine et al. 2010) and large incised meanders in the Volga and Dniepr river valleys formed by the catastrophic drainage of ice-dammed lakes (Sidorchuk et al. 2011; Panin et al. 2014, 2020). The lowermost topographic level, lying below the T3 terrace, constitutes the bottom of a proglacial river valley (Figs. 39.2 and 39.8b). This bottom is being currently transformed by migration of the Czarna Hańcza river meanders, which are however far less ample than the enclosing valley bends (Fig. 39.8a, b). Thus, the meanders of the Czarna Hańcza River are manifestly underfit in relation to the extent of valley bends.

The origin of outburst flood terraces T1–T3 was most likely related to the transition from a wide flood terrace (T1) to a single-thread meandering system of floodwater outflows (terraces T2 and T3) due to vertical incision caused by a decrease in both sediment supply and floodwater discharge (Sklar and Dietrich 2001; Turowski et al. 2008; Panin et al. 2020). Similar features, e.g. outburst flood terraces and large meanders, evolved due to strong erosion during glacial lake-outburst floods, were identified in the Dniepr and Volga river valleys (Panin et al. 2014, 2020). Moreover, the discharge variability in the WS resulted in the development of chute cut-offs, most likely due to headward incision and swale enlargement aligned to the main scroll bar direction (Fig. 39.8a–c, e). Their different evolutionary stages were recognised near the hamlet of Zarzecze Jeleniewskie, where the largest chute is 14 m

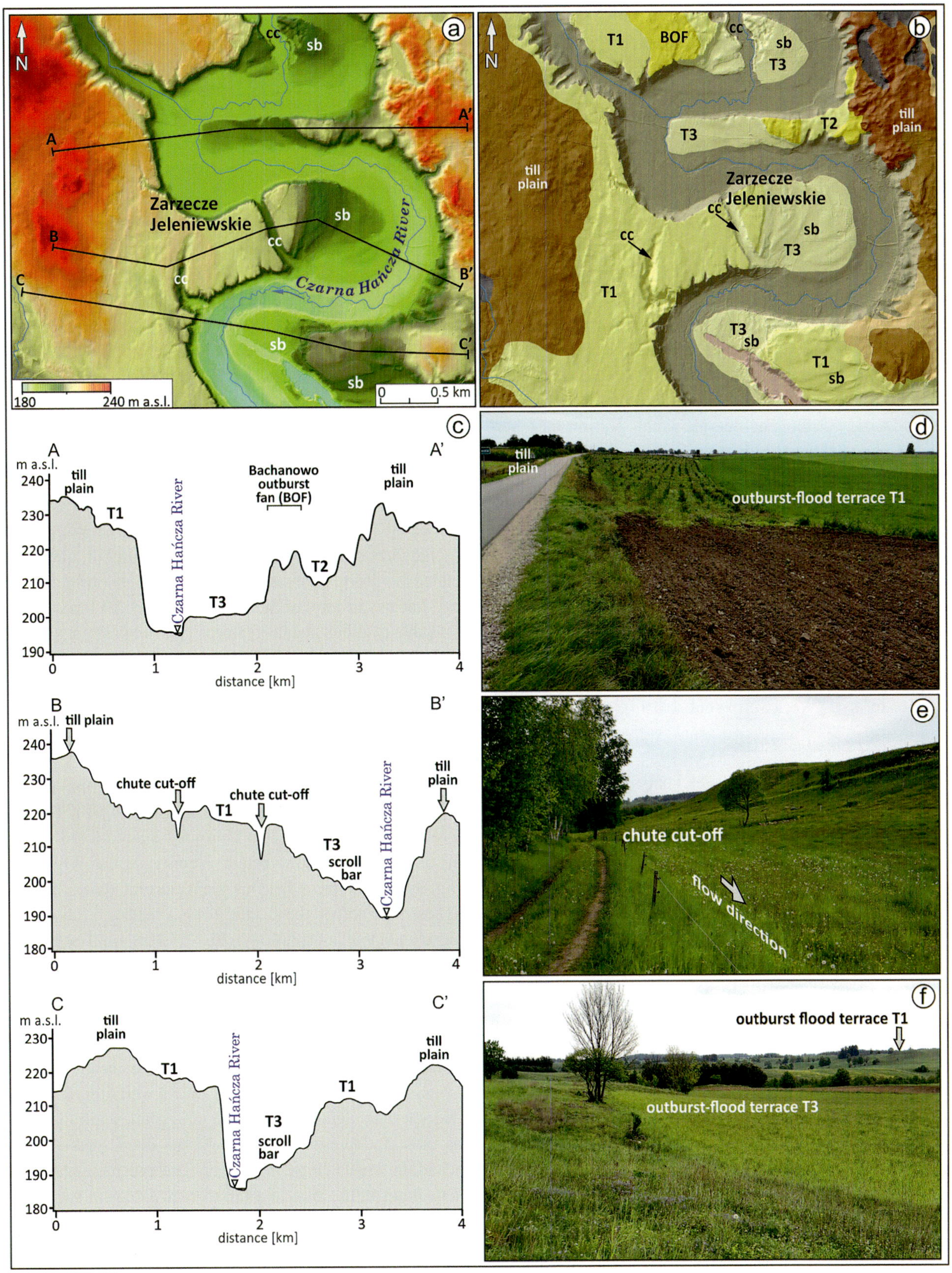

Fig. 39.8 Morphological features of the Western Spillway. **a**, **b** Landscape morphology and landform origin of upper reach of the spillway (cc—chute cut-off, sb—scroll bar; see Fig. 39.2 for explanation of colours; direction of floodwater outflow from north to south). **c** Cross-profiles showing morphology of outburst terraces T1–T3 (outwash plain levels 2–4 in Fig. 39.2). **d** Transitional contact between the outburst terrace T1 and the till plain. **e** Largest chute cut-off near Zarzecze Jeleniewskie. **f** Lowermost outburst terrace T3 with scroll bar near Zarzecze Jeleniewskie (photographs by P. Weckwerth)

deep, 110 m wide and its bottom drops downstream from 204 to 197 m asl over a distance of 600 m (Fig. 39.7c, e).

39.3.2.2 Eastern Spillway and Its Scabland-Like Landscape

(a) *Spillway Morphology*

Meltwaters from glacial lake-outburst floods also exited the ice sheet further to the east, shaping the morphology of the ice-proximal part of the Eastern Spillway (ES) (Figs. 39.2 and 39.9). The Eastern Spillway is 6.1 km long and up to 3.9 km wide. This flooding route starts at the Prudziszki Gate and ends in the northern vicinity of Suwałki (Fig. 39.9a). The bottom of the ES drops from 195 to 185 m asl, providing a gradient of 0.00286.

The morphology of the ES varies with the distance from the Prudziszki Gate. The ice-proximal portion is characterised by scabland-like topography (butte-and-basin topography, see Baker 1973, 2002), while in the more distal locations obstacle marks and dry channels predominate (Weckwerth et al. 2019b, 2021). Considering the height of the erosional terrace and marks on the side slopes of the ES, the floodwater depth must have been up to 22 m. Moreover, morphological insights including hydraulic radius and gradient enable for the calculation of flow velocity, which may have ranged between 4.6 and 14.3 m $s-^{1}$, as well as floodwater discharge in the range of 276×10^3 to 861×10^3 m^3 $s-^{1}$ (Weckwerth et al. 2019b).

(b) *Scabland-Like Landscape*

In the front of the Prudziszki Gate a butte-and-basin (scabland-like) landscape developed with different types of features, including (1) parallel and longitudinal furrows, (2) transverse furrows, (3) single and grouped potholes and (4) obstacle marks (Weckwerth et al. 2019b, 2021) (Fig. 39.9a–e). These features were created under conditions of a highly turbulent and supercritical flow regime (see Baker 1973; Baker and Nummedal 1978; Komar 1987; Benito 1997; Alho et al. 2010; Panin et al. 2020) in the proximal portion of the ES when unit stream power (amount of energy the water is exerting on the bottom of the river) and bed shear stress (the force of moving water against the bed of the channel) were 5.892 W m^{-2} and 473 N $m-^{2}$, respectively (Weckwerth et al. 2019b). Such settings were sufficient to form scabland-like topography when floodwater activity affected a permanently frozen ground (Weckwerth et al. 2020c) and a low-resistance riverbed composed of soft glacigenic sediments (see Alho et al. 2010; Carling et al. 2017).

The dominant morphological features of the butte-and-basin topography are transverse furrows, with the major axes oriented perpendicular to the palaeoflow direction (Fig. 39.9c–e). These features are between 12.7 and 346 m long (mean 89 m), 9–63 m wide (mean 26 m) and 0.3–12 m deep (mean 3.6 m; Weckwerth et al. 2021). The transverse furrows occur as isolated depressions or comprise longitudinal clusters (trains) oriented parallel to the direction of floodwater palaeoflow (Fig. 39.9c, d). These clusters represent short cyclic steps related to processes of upstream migrating troughs and scour infill in hydraulic jump zones (Fig. 39.9d, e) (Weckwerth et al. 2021).

39.3.2.3 The Broad Outwash Plain with Features Indicating Outburst Flood

The Western and Eastern spillways merge around Suwałki, forming a broad depositional surface (Fig. 39.2b). This surface is located to the south of Suwałki and represents the semi-confined or unconfined, sandy-gravelly outwash plain level 2, which is up to 11.5 km wide (Weckwerth et al. 2019b). The outwash plain level 2 is surrounded to the east and west by undulating till plains and intersected by tunnel valleys associated with the maximum ice-sheet limit of the Late Weichselian glaciation (Ber 1974, 1982, 2000; Krzywicki 2002). The distal part of the outwash plain level 2, located to the west and south-west of the Wigry Lake, is characterised by a comprehensive suite of megaflood-related landforms consisting of megadunes, obstacle marks and linear clusters of kettle holes (Fig. 39.10a). Finally, remnants of older glacial features and outwash fans are also seen (Ber 2000; Sobiech 2019).

The depositional landforms, most diagnostic of glacial outburst, are the megadunes, which are concentrated in four fields occurring downstream of the confluence of the Western and Eastern spillways (Fig. 39.2). An area south of Suwałki is particularly significant in terms of giant megadune recognition (Weckwerth et al. 2019b). The best-developed bedforms of this kind are situated 6 km south of Suwałki, in the south-western vicinity of the Wigry Lake (Fig. 39.10). Here, a field of giant current dunes hosts the highest number of these landforms creating a clearly visible element of morphology (Fig. 39.10a, b). The edges of giant two-dimensional (2-D), asymmetrical megadunes have the shape of parallel arcs and orientation indicating a NW–to–SE megaflood palaeocurrent (Fig. 39.10a). The maximum recorded megadune height is 8.5 m, and the maximum length is 420 m, while the mean values are 3.1 m and 163 m, respectively (Weckwerth et al. 2019b). Characteristic features of these megadunes are symmetrical crestal depressions up to 1.5 m deep, present along megadune crests (Fig. 39.10a, c). The morphometry of the megadunes in the area of a broad outwash plain suggests an outburst flood discharge of up to 2.2×10^3 m^3 s^{-1}, which places it among the largest glacial megafloods anywhere in the world (see Baker 2002; Herget 2005; Meinsen et al. 2011; Carling 2013).

Megadunes located in the vicinity of the Wigry Lake originated from depositional processes related to the reduction in flow energy and competence within a broad outwash

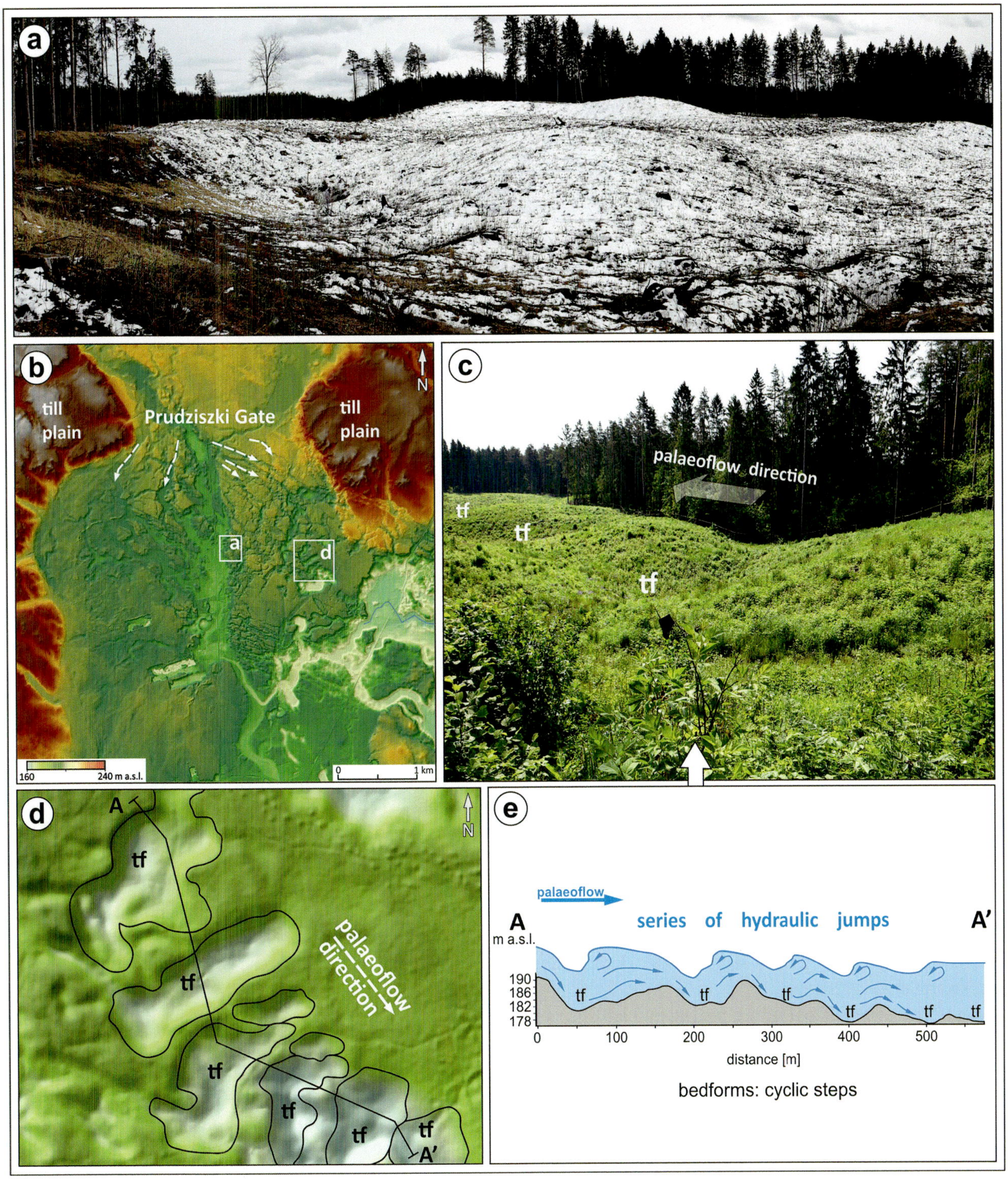

Fig. 39.9 Palaeohydrological evidence of meltwater floods. **a** Grouped potholes in the scabland-like topography to the south from the Prudziszki Gate (direction of floodwater flow according to the viewing direction). **b** Morphology of Eastern Spillway proximal reach with butte-and-basin (scabland-like) landscape (white arrows indicate local direction of floodwater flow). **c, d** Longitudinal cluster (train) of transverse furrows (tf) oriented parallel to the direction of floodwater palaeoflow. **e** Cyclic steps developed due to upstream migrating troughs and scour infill in hydraulic jump zones (after Weckwerth et al. 2021, modified) (photographs by P. Weckwerth)

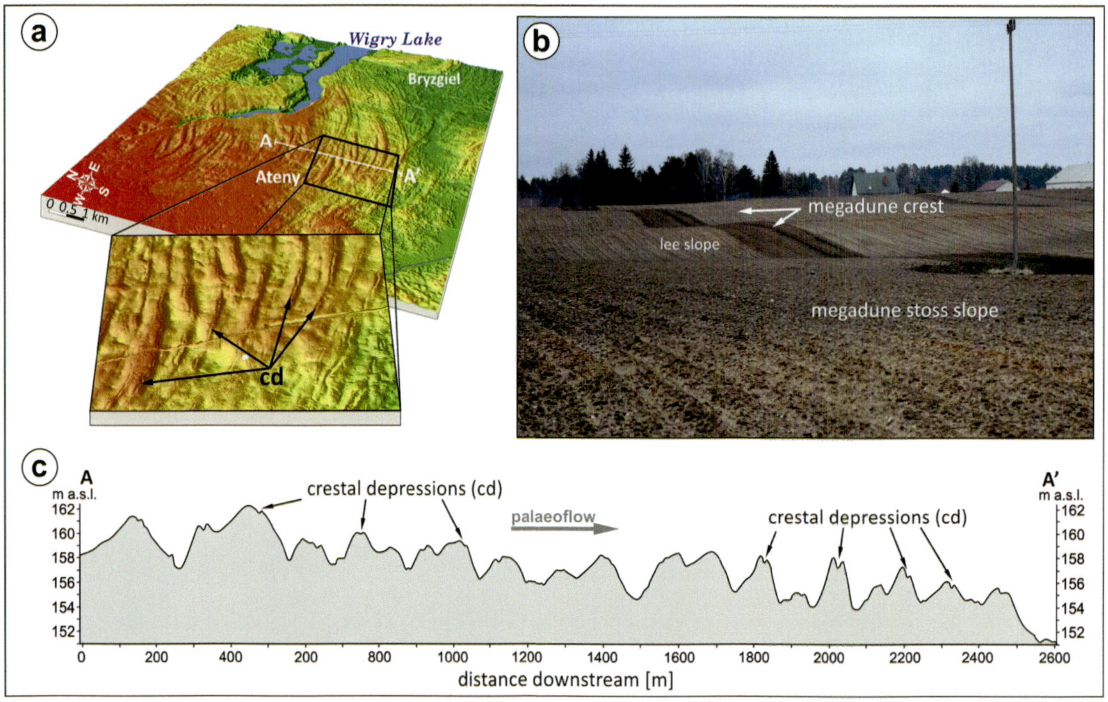

Fig. 39.10 Megadunes in the proglacial megaflood landscape. **a** Morphology of megadunes in the south-western vicinity of the Wigry Lake. **b** Megadune field near Bryzgiel (southern vicinity of the Wigry Lake) (photograph by P. Weckwerth). **c** Longitudinal profile showing morphology of megadunes near Ateny (south-western vicinity of the Wigry Lake)

plain. The width of this plain rapidly increases and thus creates an accommodation space (unconfined outwash plain) for aggradation of mega-scale, superimposed bedforms (Weckwerth et al. 2020b). Their evolution was recognised through sedimentological and geophysical investigations, and refers to foreset deposition on the lee-side of the megadunes and deposition from superimposed bedforms on a megadune stoss slope. The crestal depressions evolved as a result of the development of both north- and south-dipping normal faults oriented parallel to the megadune crest, representing syn- and meta-sedimentary deformations (Weckwerth et al. 2020a, b).

Landforms independent of the proglacial bedform continuum are obstacle marks and linear clusters of kettle holes that developed as a result of erosional, depositional and deformational processes during the waning stages of glacial floods. Obstacle marks were recognised in the ES and eastward of the Wigry Lake (Weckwerth et al. 2019b). These unique outwash landscape features coexist with kettle holes and comprise a proximal scour hollow, lateral scours and shadow ridges (Fig. 39.11a). As recognised in the contemporary glaciated areas affected by glacial floods (e.g. Russell 1993; Fay 2002; Russell and Knudsen 2002), the obstacle marks in NE Poland were formed by flow around stranded ice blocks inside floodwater flow pathways. These ice blocks were previously detached from the ice margin by pressurised meltwater that was associated with the rapid outburst of floodwaters (see Russell et al. 2010). Scour marks recognised in the nearby Suwałki area developed under waning glacial flood conditions when the flow depth was between 3 and 10 m and the mean velocity was around 9 ms^{-1} (Weckwerth et al. 2019b).

The obstacle marks in the area of Suwałki coexist with kettle holes that also form linear clusters in the northern and eastern vicinity of the Wigry Lake (Fig. 39.11b). These clusters comprise up to 22 enclosed depressions which are slightly elongated and U- and V-shaped, and have a mean length of 29 m and width of 32.5 m. Kettle-hole chains are oriented N–S and NW–SE, indicating floodwater palaeoflow direction similar to that inferred from the orientation of megadune crests. Flow-parallel kettle clusters have formed due to the melting of ice blocks that had previously been transported and then grounded in the final stage of the flood. According to the kettle-hole dimensions, the flow depth necessary for the transport of giant ice blocks was 21 m (Weckwerth et al. 2019b). The linear pattern of kettle holes, indicative of glacial outburst floods, directly refers to the flow competence. Ice blocks that formed a flow-parallel chain were transported around large, already grounded blocks and into their lee-sides and grounded here as a result of lower velocities (Fay 2002).

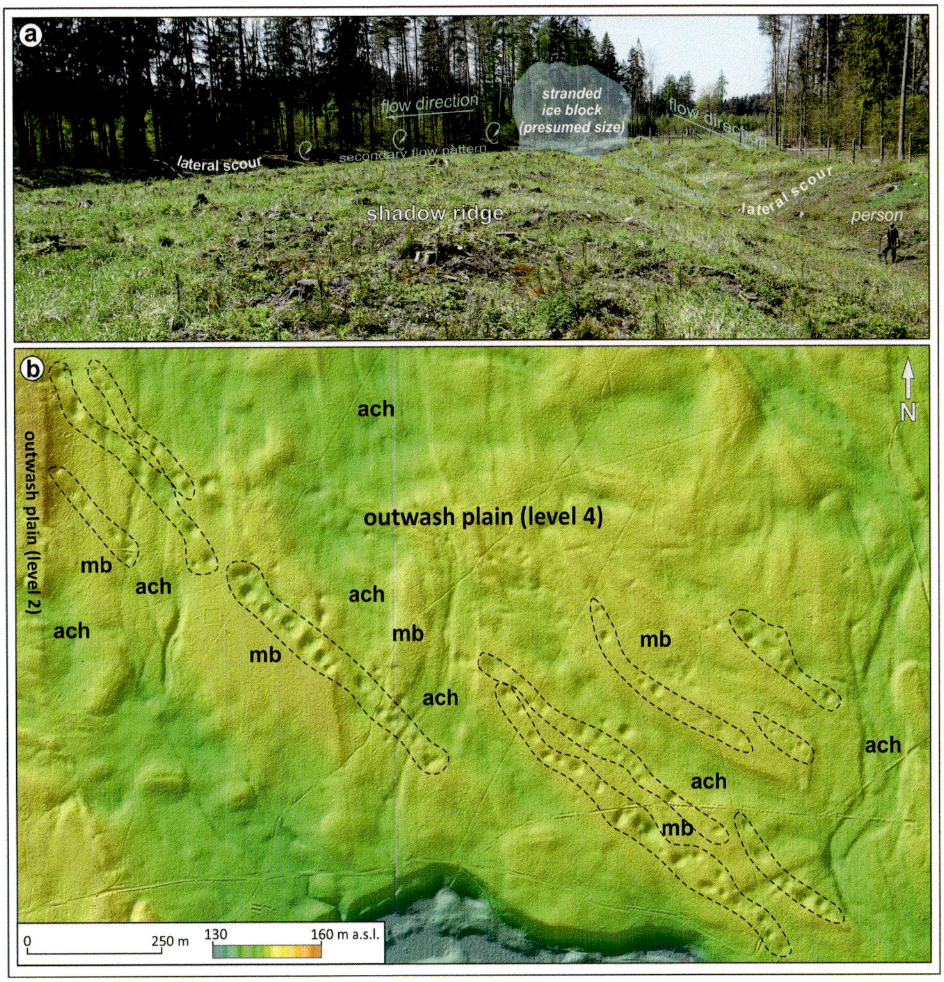

Fig. 39.11 Evidence of megaflood flow on the outwash plain. **a** Obstacle mark near the Wigry Lake with two well-developed lateral scours and a shadow ridge (photos by P. Weckwerth). **b** Kettle-hole chains (black dashed lines) on the surface of outwash plain (level 4), with many abandoned channels of braided pattern (ach) dividing mid-channel bars (mb)

39.4 Sources of Glacial Megafloods and Their Contribution to the Valley System Development

The recognition of floodwater sources at the southern periphery of the SIS during the Weichselian is all the more important given that nowhere in the European Lowland have flood-related landforms of this age been identified. Because landforms associated with lakes dammed by the ice front do not exist in the Megaflood Landform System in NE Poland, the presumed sources of floodwater are sub and/or supraglacial lakes, similar to those in contemporary glaciated areas such as Iceland, Greenland and Antarctica (e.g. Maizels 1989; Russell et al. 2001, 2010; Zwally et al. 2002, 2006; Carrivick et al. 2004; Shepherd et al. 2009; Sergienko 2013; Andrews et al. 2014; Stevens et al. 2015). While the location of supraglacial lakes is problematic, the subglacial source of floodwaters was inferred from the morphology and course of the tunnel valleys. The possible location of subglacial lakes is suggested by the course of the Great Szelment Lake and Szeszupa tunnel valleys, which join near the north-eastern border of Poland and continue north-eastwards as a single, main tunnel valley (Fig. 39.2). In south-western Lithuania, this main tunnel valley meets the up-ice sloping surface and large depressions that are enclosed to the south and located between the towns of Jurgežeriai, Bilvyčiai and Krasna. These depressions may relate to the location of subglacial lakes that stored the meltwaters that were repeatedly funnelled into the main tunnel valley between Jurgežeriai and Poszeszupie (Figs. 39.1 and 39.4).

Preliminary identification of the likely source of the glacial outburst flood inferred from the orientation of mega-dune crests near the Serwy Lake (Fig. 39.1) indicates that the floodwaters must also have come from the east (Weckwerth et al. 2019b). The east-to-west catastrophic meltwater outflow responsible for the development of these bedforms was most likely related to the outwash plains development in southern Lithuania (Blažauskas et al. 2007; Guobyte and Satkūnas 2011) or to the drainage of the Skidel ice-dammed lake, which existed in the forefield of the maximum extent of the last SIS in Belarus (Marks 2002,

2012; Pavlovskaya 2004; Baltrūnas et al. 2007; Marks and Pavlovskaya 2007).

The Suwałki-Augustów outwash plain, where the subglacial megaflood subsystem is located, was formed during the retreat of the Scandinavian ice sheet in the Late Weichselian (Ber 1967, 1968, 1982; Marks 2002, 2012; Dzierżek and Zreda 2007; Dzierżek 2009). The meltwater flowed from the north to the south shortly after 16 ka ago (Pomeranian Phase) and probably also earlier (around 18–19 ka), during the Poznań Phase (Voznyachuk and Valchik 1978; Marks 2002, 2012; Pavlovskaya 2004; Rinterknecht et al. 2005; Blažauskas et al. 2007; Dzierżek and Zreda 2007; Marks and Pavlovskaya 2007; Dzierżek 2009; Guobyte and Satkūnas 2011; Tylmann et al. 2019, 2022). These floodwaters discharged in both the Vilnius–Warsaw–Poznań–Berlin and Toruń–Eberswalde ice-marginal spillways and contributed to the final shaping of the river valley system of the European Lowland (Marks 2012; Weckwerth 2013; Molewski 2014; Weckwerth et al. 2019b). Although the glacial flood-related landforms from the Weichselian glaciation are a unique type of landscape developed due to cataclysmic fluvial processes, they may also be considered as large meltwater inputs to the ocean that possibly influenced global climate changes (e.g. Busschers et al. 2007; Toucanne et al. 2015).

39.5 Conclusions

The abrupt termination of the Weichselian glaciation, which was related to the sudden and catastrophic loss of glacier mass, caused a peak discharge of glacial meltwater in northeastern Poland. These events occurred shortly after the ice sheet reached here its maximum extent. As a result, the pattern of landscape elements represents subglacial and proglacial landforms of both erosional and depositional origin that reflect an exceptionally large magnitude of the repeated glacial lake-outburst floods in NE Poland. These landforms include first- and second-order features, which compose an internally coherent Megaflood Landform System.

A significant part of the Megaflood Landform System was formed in a subglacial setting as a result of floodwater erosion that caused the development of giant tunnel valleys and associated glacial curvilineations. These landforms are indicative of subglacial flooding events and were also helpful in the recognition of probable location of subglacial lakes in south-western Lithuania and ice-marginal zones in southern Lithuania as the most likely flood sources.

The landscape developed in proglacial settings reveals elements indicative of floodwater erosion in the area of floodwater outlets (scabland-like landscape and outburst fans), proglacial confined flooding routes (spillways) and broad outwash plains with fields of megadunes. One can also argue that the Megaflood Landform System is the youngest geomorphic landscape of the European Lowlands to have originated from cataclysmic glacial outburst floods, but the understanding of factors controlling its evolution and of the impact of outburst flood on the formation of the European valley system should become subject of geomorphological and sedimentological investigations in the near future.

Acknowledgements This study was carried out as part of a scientific project financed by the National Science Centre, Poland (project No. 2018/31/B/ST10/00976).

References

Alho P, Baker VR, Smith LN (2010) Paleohydraulic reconstruction of the largest Glacial Lake Missoula draining(s). Quat Sci Rev 29:3067–3078

Andrews L, Catania G, Hoffman MJ, Gulley JD, Lüthi MP, Ryser C, Hawley RL, Neumann TA (2014) Direct observations of evolving subglacial drainage beneath the Greenland Ice Sheet. Nature 514:80–83

Baker VR (1973) Paleohydrology and sedimentology of the Lake Missoula flooding in eastern Washington. Geol Soc Am Spec Pap 144:1–79

Baker VR (2002) High-energy megafloods: planetary settings and sedimentary dynamics. In: Martini IP, Baker VR, Garzon G (eds) Flood and megaflood deposits: recent and ancient examples, vol 32. IAS Special Publications, pp 3–15

Baker VR, Nummedal D (1978) The channeled Scabland. Planetary Geology Program, Washington, DC, National Atmospheric and Space Administration, Office of Space Science, p 186

Baltrūnas V, Švedas K, Pukelytė V (2007) Palaeogeography of South Lithuania during the last ice age. Sediment Geol 193:221–231

Benito G (1997) Energy expenditure and geomorphic work of the cataclysmic Missoula flooding in the Columbia River Gorge, USA. Earth Surf Proc Land 22:457–472

Ber A (1966) Strefa marginalna fazy pomorskiej zlodowacenia północnopolskiego na Pojezierzu Suwalskim. Kwart Geol 10:1132–1133

Ber A (1967) Szczegółowa mapa geologiczna Polski 1:50,000, arkusz Jeleniewo. Institute of Geology, Warszawa

Ber A (1968) Objaśnienia do Szczegółowej mapy geologicznej Polski 1:50,000, arkusz Jeleniewo. Institute of Geology, Warszawa, p 56

Ber A (1974) Czwartorzęd Pojezierza Suwalskiego. Biul Inst Geol 269:23–106

Ber A (1982) Marginal zones and deglaciation during the North-Polish Glaciation in the Suwałki-Augustów Lakeland. Biul Inst Geol 343:71–89

Ber A (1986) Szczegółowa mapa geologiczna Polski 1:50,000, arkusz Suwałki. Państwowego Instytutu Geologicznego, Warszawa

Ber A (1989) Stratigraphy of the Quaternary of the Suwałki Lakeland and its substrate based on recent data. Kwart Geol 33:463–478

Ber A (2000) Plejstocen Polski północno-wschodniej w nawiązaniu do głębszego podłoża i obszarów sąsiednich. Prace PIG 170:5–84

Ber A, Maksiak S (1969) Formy marginalne i formy martwego lodu w Zagłębieniu Szeszupy na Pojezierzu Suwalskim. Biul Inst Geol 220:347–356

Björnsson H (1992) Jokulhlaups in Iceland: prediction, characteristics and simulation. Ann Glaciol 16:95–106

Blažauskas N, Jurgaitis A, Šinkūnas P (2007) Patterns of Late Pleistocene proglacial fluvial sedimentation in the SE Lithuanian Plain. Sediment Geol 193:193–201

Bogacki M (1976) Współczesne sandry na przedpolu Skeidararjökull (Islandia) i plejstoceńskie sandry w Polsce północno-wschodniej. Rozprawy Uniw Warszawskiego 99:165

Bogacki M (1980) Types of outwash forms in North-East Poland. Geogr Polon 43:25–34

Burke MJ, Brennand TA, Perkins AJ (2012) Evolution of the subglacial hydrologic system beneath the rapidly decaying Cordilleran Ice Sheet caused by ice-dammed lake drainage: implications for meltwater-induced ice acceleration. Quat Sci Rev 50:125–140

Busschers FS, Kasse C, van Balen RT, Vandenberghe J, Cohen KM, Weerts HJT, Wallinga J, Bunnik FPM (2007) Late Pleistocene evolution of the Rhine-Meuse system in the southern North-Sea Basin: imprints of climate change, sea-level oscillation and glacio-isostacy. Quat Sci Rev 26:3216–3248

Carling PA (2013) Freshwater megaflood sedimentation: what can we learn about generic processes? Earth Sci Rev 125:87–113

Carling PA, Perillo M, Best J, Garcia MH (2017) The bubble bursts for cavitation in natural rivers: laboratory experiments reveal minor role in bedrock erosion. Earth Surf Proc Land 42:1308–1316

Carrivick JL (2007) Hydrodynamics and geomorphic work of jökulhlaups (glacial outburst floods) from Kverkfjöll volcano, Iceland. Hydrol Proc 740:725–739

Carrivick JL, Russell AJ, Tweed FS (2004) Geomorphological evidence for jökulhlaups from Kverkfjöll volcano, Iceland. Geomorphology 63:81–102

Clark CD, Livingstone SJ (2018) Glacial curvilineations found along the southern sector of the Laurentide Ice sheet and a hypothesis of formation involving subglacial slope failure in tunnel valleys and subglacial lakes. Earth Surf Proc Land 43:1518–1528

Constantine JA, McLane SR, Dunne T (2010) A mechanism of chute cutoff along large meandering rivers with uniform floodplain topography. Geol Soc Am Bull 122:855–869

Dzierżek J (2009) Paleogeografia wybranych obszarów Polski w czasie ostatniego zlodowacenia. Acta Geogr Lodz 95:1–112

Dzierżek J, Zreda M (2007) Timing and style of deglaciation of North Eastern Poland from cosmogenic 36Cl dating of glacial and glaciofluvial deposits. Geol Quart 51:203–216

Evans DJA (ed) (2003) Glacial landsystems. Arnold, London, p 544

Evans DJA, Rea BR, Hiemstra JF, Ó Cofaigh C (2006) A critical assessment of subglacial mega-floods: a case study of glacial sediments and landforms in south-central Alberta, Canada. Quat Sci Rev 25:1638–1667

Fay H (2002) The formation of ice-block obstacle marks during the November 1996 glacier outburst flood (jökulhlaup), Skeiðarársandur, southern Iceland. In: Martini IP, Baker VR, Garzon G (eds) Flood and megaflood deposits: recent and ancient, vol 32. International Association of Sedimentologists, Special Publications, pp 85–97

Galon R (1965) Some new problems concerning sub-glacial channels. Geogr Pol 6:19–28

Gibbard PL (2007) Palaeogeography: Europe cut adrift. Nature 448:259–260

Glasser NF, Bennett MR (2004) Glacial erosional landforms: origins and significance for palaeoglaciology. Prog Phys Geogr 28:43–75

Guobute R, Satkūnas J (2011) c. In: Ehlers J, Gibbard, PL, Hughes PD (eds) Quaternary glaciations: extent and chronology—a closer look. Elsevier, Amsterdam, pp 231–246

Herget J (2005) Reconstruction of ice-dammed lake outburst floods in the Altai mountains, Siberia. Geol Soc Am Spec Pap 386:118

Hermanowski P, Piotrowski JP, Szuman I (2019) An erosional origin for drumlins of NW Poland. Earth Surf Proc Landf 44:2030–2050

Jakobsson M, Björck S, Alm G, Andrén T, Lindeberg G, Svensson N-O (2007) Reconstructing the Younger Dryas ice dammed lake in the Baltic Basin: bathymetry, area and volume. Glob Planet Change 57:355–370

Johnson MD (1999) Spooner Hill, northwest Wisconsin: high relief hills carved by sub-glacial meltwater of the Superior Lobe. In: Mickelson DM, Attig JW (eds) Glacial processes past and present, vol 337. Geological Society of America, Special Papers, pp 83–92

Jørgensen F, Sandersen PBE (2006) Buried and open tunnel valleys in Denmark—erosion beneath multiple ice sheets. Quat Sci Rev 25:1339–1363

Kehew AE, Piotrowski JA, Jørgensen F (2012) Tunnel valleys: concepts and controversies—a review. Earth Sci Rev 113:33–58

Komar PD (1987) Selective grain entrainment by a current from a bed of mixed sizes: a reanalysis. J Sediment Petrol 57:203–211

Kondracki J, Pietkiewicz S (1967) Czwartorzęd północno-wschodniej Polski. In: Dylik J, Galon R (eds) Czwartorzęd Polski. PWN, Warszawa, pp 206–258

Krzywicki T (1997) Mapa rozprzestrzenienia młodoplejstoceńskich glin zwałowych w południowej części Puszczy Augustowskiej jako podstawa wyznaczenia zasięgu ostatniego zlodowacenia w północno-wschodniej Polsce. Centr Arch Geol Państw Inst Geol, Warszawa

Krzywicki T (2002) The maximum ice sheet limit of the Vistulian glaciation in north-eastern Poland and neighbouring areas. Geol Quart 46:165–188

Lang J, Alho P, Kasvi E, Goseberg N, Winsemann J (2019) Impact of Middle Pleistocene (Saalian) glacial lake-outburst floods on the meltwater-drainage pathways in northern central Europe: insights from 2D numerical flood simulation. Quat Sci Rev 209:82–99

Lesemann J-E, Piotrowski JA, Wysota W (2010) "glacial curvilineations": new glacial landforms produced by longitudinal vortices in subglacial meltwater flows. Geomorphology 120:153–161

Lesemann J-E, Piotrowski JA, Wysota W (2014) Genesis of the 'glacial curvilineation' landscape by meltwater processes under the former Scandinavian Ice Sheet, Poland. Sediment Geol 312:1–18

Lisicki S (1993) Deglacjacja Pojezierza Suwalskiego wokresie schyłku Plejstocenu. In: Juskowiak O (ed) Przewodnik 64 Zjazdu Polskiego Towarzystwa Geologicznego. Państw Inst Geol, Warszawa, pp 81–86

Lisicki S (1994) Objaśnienia do Szczegółowej mapy geologicznej Polski 1:50,000. Arkusze: Sejny (110) i Veisiejai (111). Państw Inst Geol, Warszawa, p 48

Liszkowski S (1993) The effects of Pleistocene ice-sheet loading-deloading cycles on the bedrock structure of Poland. Folia Quat 64:7–23

Maizels JK (1989) Sedimentology and paleohydrology of Holocene flood deposits in front of a jökulhlaup glacier, south Iceland. In: Beven K, Carling P (eds) Floods: hydrological, sedimentological and geomorphological implications. Wiley, New York, pp 239–252

Marks L (2002) Last Glacial Maximum in Poland. Quat Sci Rev 21:103–110

Marks L (2012) Timing of the Late Vistulian (Weichselian) glacial phases in Poland. Quat Sci Rev 44:81–88

Marks L, Pavlovskaya IE (2007) Development of meltwater outflow during Last Glacial Maximum in the Middle Neman valley region, central Europe. Quat Int 167–168:269

Marren PM (2005) Magnitude and frequency in proglacial rivers: a geomorphological and sedimentological perspective. Earth Sci Rev 70:203–251

Meinsen J, Winsemann J, Weitkamp A, Landmeyer N, Lenz A, Dölling A (2011) Middle Pleistocene (Saalian) lake outburst floods in the Münsterland Embayment (NW Germany): impacts and magnitudes. Quat Sci Rev 30:2597–2625

Molewski P (2014) Paleograficzne uwarunkowania odpływu wód z zastoiska warszawskiego doliną Bachorzy i pradoliną warszawsko-berlińską w czasie stadiału głównego zlodowacenia Wisły. Land Anal 25:105–114

Panin AV, Adamiec G, Arslanov KA, Bronnikova MA, Filippov VV, Sheremetskaya ED, Zaretskaya NE, Zazovskaya EP (2014) Absolute chronology of fluvial events in the Upper Dnieper river

system and its palaeogeographic implications. Geochronometria 41:278–293
Panin AV, Astakhov VI, Lotsari E, Komatsu G, Lang J, Winsemann J (2020) Middle and Late Quaternary glacial lake-outburst floods, drainage diversions and reorganization of fluvial systems in northwestern Eurasia. Earth Sci Rev 201:103069
Pavlovskaya IE (2004) Late Pleistocene evolution of hydrographical network recorded at geosites in the middle Neman area (Western Belarus). Polish Geol Inst Spec Pap 13:167–174
Pietkiewicz S (1928) Pojezierze Suwalszczyzny Zachodniej (Zarys morfologii lodowcowej). Przegl Geogr 8:168–222
Pochocka-Szwarc K, Lisicki S, Włodek M, Ber A (2017) Szczegółowa mapa geologiczna Polski 1:50,000, ark. Jeleniewo (reambulacja), Państwowy Instytut Geologiczny – Państwowy Instytut Badawczy, Warszawa
Rdzany Z, Frydrych M (2018) Record of glacial outburst floods in marginal zones and forelands of Scandinavian glaciations in Poland. Acta Univ Lodz Folia Geogr Phys 17:33–40
Rinterknecht VR, Marks L, Piotrowski JA, Raisbeck GM, Yiou F, Brooke EJ, Clark PU (2005) Cosmogenic ^{10}Be ages on the Pomeranian Moraine, Poland. Boreas 34:186–191
Rinterknecht VR, Clark PU, Raisbeck GM, Yiou F, Bitinas A, Brook EJ, Marks L, Zelčs V, Lunkka J-P, Pavlovskaya IE, Piotrowski JA, Raukas A (2006) The last deglaciation of the southeastern sector of the Scandinavian Ice Sheet. Science 311:1449–1452
Russell AJ (1993) Obstacle marks produced by flow around stranded ice blocks during a glacier outburst flood (jökulhlaup) in west Greenland. Sedimentology 40:1091–1111
Russell AJ, Knudsen Ó (1999) Controls on the sedimentology of the November 1996 jökulhlaup deposits, Skeiðarársandur, Iceland. In: Smith ND, Rogers J (eds) Fluvial sedimentology VI, vol 28. International Association of Sedimentologists, Special Publications, pp 315–329
Russell AJ, Knudsen Ó (2002) The effects of glacier-outburst flood flow dynamics in ice-contact deposits: November 1996 jökulhlaup, Skeiðarársandur, Iceland. In: Martini IP, Baker VR, Garzon G (eds) Flood and megaflood deposits: recent and ancient, vol 32. International Association of Sedimentologists, Special Publications, pp 67–83
Russell AJ, Knight PG, van Dijk TAGP (2001) Glacier surging as a control on the development of proglacial fluvial landforms and deposits, Skeiðarárjökull, Iceland. Glob Planet Change 28:163–174
Russell AJ, Tweed F, Roberts M, Harris T, Gudmundsson M, Knudsen O, Marren P (2010) An unusual jökulhlaup resulting from subglacial volcanism, Sólheimajökull, Iceland. Quat Sci Rev 29:1363–1381
Rychel J, Pochocka-Szwarc K (2020) Geomorphological features and relict permafrost in the Szeszupa depression and the Cisowa Hill area. In: Weckwerth P, Wysota W, Kalińska E (eds) Glacial megaflood landforms and sediments in North-Eastern Poland. Wyd Nauk UMK, Toruń, pp 29–39
Sergienko OV (2013) Glaciological twins: basally controlled subglacial and supraglacial lakes. J Glaciol 59(213):3–8
Shaw J (2002) The meltwater hypothesis for subglacial bedforms. Quat Int 90:5–22
Shepherd A, Hubbard A, Nienow P, McMillan M, Joughin I (2009) Greenland ice sheet motion coupled with daily melting in late summer. Geophys Res Lett 36:L01501
Sidorchuk AYu, Panin AV, Borisova OK (2011) Surface runoff to the Black Sea from the East European Plain during Last Glacial Maximum–Late Glacial time. In: Buynevich IV, Yanko-Hombach V, Gilbert AS, Martin R (eds) Geology and geoarchaeology of the Black Sea region: beyond the flood hypothesis, vol 473. Geological Society of America, Special Papers, pp 1–25
Sklar LS, Dietrich WE (2001) Sediment and rock strength controls on river incision into bedrock. Geology 29:1087–1090
Smed P (1962) Studier over den fynske øgruppes glaciale landskabsformer. Medd Dansk Geol Foren 15:74
Smolska E (2007a) Development of gullies and sediment fans in last-glacial areas on the example of the Suwałki Lakeland (NE Poland). CATENA 71:122–131
Smolska E (2007b) Extreme rainfalls and their impact on slopes—evaluation based on soil erosion measurements (as exemplified by the Suwałki Lakeland, Poland). Geogr Pol 80:151–163
Smolska E (2008) Channel response to flood flows on example of the Szeszupa River in the last-glacial area (NE Poland). Quaest Geogr 27A(2):63–72
Sobiech M (2019) Geomorfologia i formowanie glacimarginalnych stożków sandrowych w świetle analiz GIS. PhD thesis manuscript, Nicolaus Copernicus University in Toruń, p 350
Sokólska A (2014) Morfologia i geneza wałowych form polodowcowych na obszarze Suwalskiego Parku Krajobrazowego. MSc manuscript, Nicolaus Copernicus University in Torun, p 48
Solon J, Borzyszkowski J, Bidłasik M, Richling A, Badora K, Balon J, Brzezińska-Wójcik T, Chabudziński Ł, Dobrowolski R, Grzegorczyk I, Jodłowski M, Kistowski M, Kot R, Krąż P, Lechnio J, Macias A, Majchrowska A, Malinowska E, Migoń P, Myga-Piątek U, Nita J, Papińska E, Rodzik J, Strzyż M, Terpiłowski S, Ziaja W (2018) Physicogeographical mesoregions of Poland: Verification and adjustment of boundaries on the basis of contemporary spatial data. Geographia Pol 91:143–170
Stevens LA, Behn MD, McGuire JJ, Das SB, Joughin I, Herring T, Shean DE, King MA (2015) Greenland supraglacial lake drainages triggered by hydrologically induced basal slip. Nature 522:73–76
Stølum HH (1998) Planform geometry and dynamics of meandering rivers. Geol Soc Am Bull 110:1485–1498
Toucanne S, Soulet G, Freslon N, Jacinto RS, Dennielou B, Zaragosi S, Eynaud F, Bourillet J-F, Bayon G (2015) Millennial-scale fluctuations of the European Ice Sheet at the end of the last glacial, and their potential impact on global climate. Quat Sci Rev 123:113–133
Turowski JM, Hovius N, Hsieh M-L, Lague D, Chen M-C (2008) Distribution of erosion across bedrock channels. Earth Surf Proc Land 33:353–363
Tylmann K, Rinterknecht VR, Woźniak PP, Bourles D, Schimmelpfennig I, Guillou V, Aster Team (2019) The local Last Glacial Maximum of the southern Scandinavian Ice Sheet front: cosmogenic nuclide dating of erratics in northern Poland. Quat Sci Rev 219:36–46
Tylmann K, Rinterknecht VR, Woźniak PP, Guillou V,ASTER Team (2022) Asynchronous dynamics of the last Scandinavian Ice Sheet along the Pomeranian ice-marginal belt: A new scenario inferred from surface exposure 10 Be dating. Quatern Sci Rev 294:107755
Voznyachuk LN, Valchik MA (1978) Morphologiya, stroiyeniye i istoriya razvitiya dolinee Nemana v neopleistotsene. Nauka i Tekhnika, Minsk, p 211
Weckwerth P (2013) Ewolucja fluwialnych systemów dyspozycyjnych i jej uwarunkowania paleośrodowiskowe w Kotlinie Toruńskiej podczas zlodowacenia Wisły. Wyd. Nauk UMK, Toruń, p 206
Weckwerth P, Greń K, Sobota I (2019a) Controls on downstream variation in surficial sediment size of an outwash braidplain developed under high Arctic conditions updates (Kaffioyra, Svalbard). Sediment Geol 387:75–86
Weckwerth P, Wysota W, Piotrowski JA, Adamczyk A, Krawiec A, Dąbrowski M (2019b) Late Weichselian glacier outburst floods in North-Eastern Poland: Landform evidence and palaeohydraulic significance. Earth Sci Rev 194:216–233
Weckwerth P, Piotrowski JA, Wysota W, Krawiec A, Adamczyk A, Kalińska E, Chabowski M (2020a) Mature and immature megadunes: insights from morphology and sedimentary successions of mega-scale bedforms south of Wigry Lake (NE Poland). In: Weckwerth P, Wysota W, Kalińska E (eds) Glacial megaflood landforms and sediments in North-Eastern Poland. Wyd Nauk UMK, Toruń, pp 99–107
Weckwerth P, Wysota W, Chabowski M, Krawiec A, Kalińska E, Adamczyk A (2020b) Morphology, geological structure and development of two-dimensional megadunes formed on an outwash

plain in NE Poland. In: Weckwerth P, Wysota W, Kalińska E (eds) Glacial megaflood landforms and sediments in North-Eastern Poland. Wyd Nauk UMK, Toruń, pp 89–97

Weckwerth P, Wysota W, Krawiec A, Piotrowski JA, Adamczyk A, Chabowski M, Kalińska E (2020c) Megaflood track morphology and sediments: a case study from the proximal part of the Eastern Spillway, north of Suwałki (NE Poland). In: Weckwerth P, Wysota W, Kalińska E (eds) Glacial megaflood landforms and sediments in North-Eastern Poland. Wyd Nauk UMK, Toruń, pp 81–88

Weckwerth P, Kalińska E, Wysota W, Krawiec A, Adamczyk A, Chabowski M (2021) What does transverse furrow train in scabland-like topography originate from? The unique records of upper-flow-regime bedforms of a glacial lake-outburst flood in NE Poland. Quat Int 617:40–58

Winsemann J, Alho P, Laamanen L, Goseberg N, Lang J, Klostermann J (2016) Flow dynamics, sedimentation and erosion of glacial lake outburst floods along the Middle Pleistocene Scandinavian Ice Sheet (northern central Europe). Boreas 45:260–283

Winsemann J, Lang J, Polom U, Loewer M, Igel J, Pollok L, Brandes C (2018) Ice-marginal forced regressive deltas in glacial lake basins: geomorphology, facies variability and large-scale depositional architecture. Boreas 47:973–1002

Wysota W, Weckwerth P, Adamczyk A (2020a) Morphology and geological features of Jeleniewo subglacial valley, north of Suwałki. In: Weckwerth P, Wysota W, Kalińska E (eds) Glacial megaflood landforms and sediments in North-Eastern Poland. Wyd Nauk UMK, Toruń, pp 71–79

Wysota W, Weckwerth P, Adamczyk A, Piotrowski JA, Sokólska A (2020b) Morphology and origin of glacial curvilineations (GCLs) east of Hańcza Lake, Suwałki Lakeland. In: Weckwerth P, Wysota W, Kalińska E (eds) Glacial megaflood landforms and sediments in North-Eastern Poland. Wyd Nauk UMK, Toruń, pp 61–69

Wysota W, Adamczyk A, Piotrowski JA, Weckwerth P (2023) Glacial curvilineations of the Dobrzyń Lakeland: a spectacular landscape carved by subglacial meltwater erosion. In: Migoń P, Jancewicz K (eds) Landscapes and landforms of Poland. Springer, Switzerland, pp 000–000

Zieliński T (1989) Lithofacies and palaeoenvironmental characteristics of Suwałki outwash (Pleistocene, Northwest Poland). Ann Soc Geol Pol 59:249–270

Zieliński T (1993) Sandry Polski północno-wschodniej – osady i warunki sedymentacji. Uniw Śląski, Katowice, p 96

Zwally HJ, Abdalati W, Herring T, Larson K, Saba J, Steffen K (2002) Surface melt-induced acceleration of Greenland Ice Sheet flow. Science 297:218–222

Zwally HJ, Giovinetto MB, Li J, Cornejo HG, Beckley MA, Brenner AC, Saba JL, Yi D (2006) Mass changes of the Greenland and Antarctic Ice Sheets and shelves and contribution to sea level rise: 1992–2002. J Glaciol 51:509–527

Piotr Weckwerth is Associate Professor at Faculty of Earth Sciences and Spatial Management, Nicolaus Copernicus University in Toruń. His main research interests are fluvial and glacigenic processes, sediments and landform origin, Quaternary paleogeography and geology, sedimentology of clastic sediments, palaeohydraulic reconstructions, geological and hydrogeological modelling and palaeoenvironmental investigations in archaeological studies. His main research areas are fluvial systems and glacier marginal zones and forefields of contemporary and formerly glaciated areas in Central Europe, Spitsbergen and Iceland. He is the current President of the Association of Polish Geomorphologists (2021–2024) and Member of the Committee for Quaternary Research of the Polish Academy of Sciences (2017-2023).

Wojciech Wysota is Professor of Quaternary Geology at Nicolaus Copernicus University (NCU) in Toruń. His research focuses mainly on morphology and origin of glacial landscapes in northern Poland, in particular landforms and sediments of the subglacial setting. Other research interests concern sedimentary environments and stratigraphy of Late Pleistocene successions in Poland. He has been Member of the Committee for Quaternary Research of the Polish Academy of Sciences (2011–2015) and Vice-President of the Peribaltic Working Group (TERPRO Commission) of the International Union for Quaternary Research (2012–2016). He has held the position of the Dean of the Faculty of Earth Sciences, NCU (2012–2016) and NCU Vice-Rector for International Cooperation (2016–2020). Since 2020, he has been NCU's vice-rector for research.

Wolin Island–Outstanding Geodiversity on the Polish Coast

40

Zbigniew Zwoliński, Andrzej Kostrzewski, Marcin Winowski and Małgorzata Mazurek

Abstract

Wolin, the only large island in Poland, is a unique place on the map of geodiversity of Poland. Quaternary glacial and fluvioglacial sediments, mainly from the last phase of the Scandinavian glaciation, have been deposited on the Paleozoic and Mesozoic bedrock. The Wolin End Moraine, which dominates the landscape, is accompanied by a ground moraine, a coastal dune plain and three spits. A spectacular landscape has formed on the seaward side of the island by the highest cliffs in Poland, whereas the lagoon, the retrogressive delta and numerous chalk rafts are further characteristic landmarks. The island has many geosites that enrich the regional geodiversity and preserve geoheritage of the area.

Keywords

Wolin End Moraine · Cliff · Sandbar · Sand dune · Retrogressive delta

40.1 Introduction

The island of Wolin is the largest in Poland and together with the adjacent Uznam (Usedom) island, largely in Germany, it forms a distinctive mesoregion in the northern part of the Szczecin Lowland in north-western Poland (Fig. 40.1). Both islands constitute the southern shore of the Pomeranian Bay of the Baltic Sea. On the eastern and southern sides, the island of Wolin is separated from the mainland by the Dziwna River (Strait) and the Szczecin Lagoon respectively, while on the western side, the island is separated from the island of Uznam by the Świna River (Strait). In fact, both are river channels draining water from the Odra River and the Szczecin Lagoon to the Baltic Sea. The central part of the island is protected as the Wolin National Park (109.4 km^2), whereas the eastern near-shore terrains are included within the Odra Mouth and Szczecin Lagoon Natura 2000 area.

The surface area of the island is 265 km^2 and the length of the shoreline facing the Baltic Sea is nearly 33 km, including 11.5 km of high coast developed as a cliff coast. Following Marsz (1966), nine distinctive microregions can be distinguished within the island. These are: 1—Świna Gate, 2—Wolin Ridge (Wolin End Moraine), 3—Lubin-Wapnica Hummocks, 4—Wolin Lakeland, 5—Dargobądz Plain, 6—Mokrzyca Hills, 7—Kodrąb Depression, 8—Dziwna Sandbar, 9—Rów Peninsula (Fig. 40.2). Although this subdivision was presented as physico-geographical, it is in fact strongly based on the recognition of geomorphological diversity, as reflected by names of particular units (see also Fig. 40.3). The core of the island is made of various sediments and landforms of Pleistocene age, ranging from moraine hills and plains to fluvioglacial landforms. During the Holocene three striking peninsulas in the west, north-east and south-east have formed. The youngest elements of relief include cliff shores facing both the open Baltic Sea and the Szczecin Lagoon, wetlands, and anthropogenic landforms. However, the most spectacular geomorphic feature, even at the global scale, is the Retrogressive Świna Delta.

Z. Zwoliński (✉) · A. Kostrzewski · M. Winowski · M. Mazurek
Institute of Geoecology and Geoinformation, Adam Mickiewicz University, ul. Krygowskiego 10, 61-680 Poznań, Poland
e-mail: zbzw@amu.edu.pl

A. Kostrzewski
e-mail: anko@amu.edu.pl

M. Winowski
e-mail: marcin.winowski@amu.edu.pl

M. Mazurek
e-mail: malgorzata.mazurek@amu.edu.pl

P. Migoń and K. Jancewicz (eds.), *Landscapes and Landforms of Poland*,
World Geomorphological Landscapes, https://doi.org/10.1007/978-3-031-45762-3_40

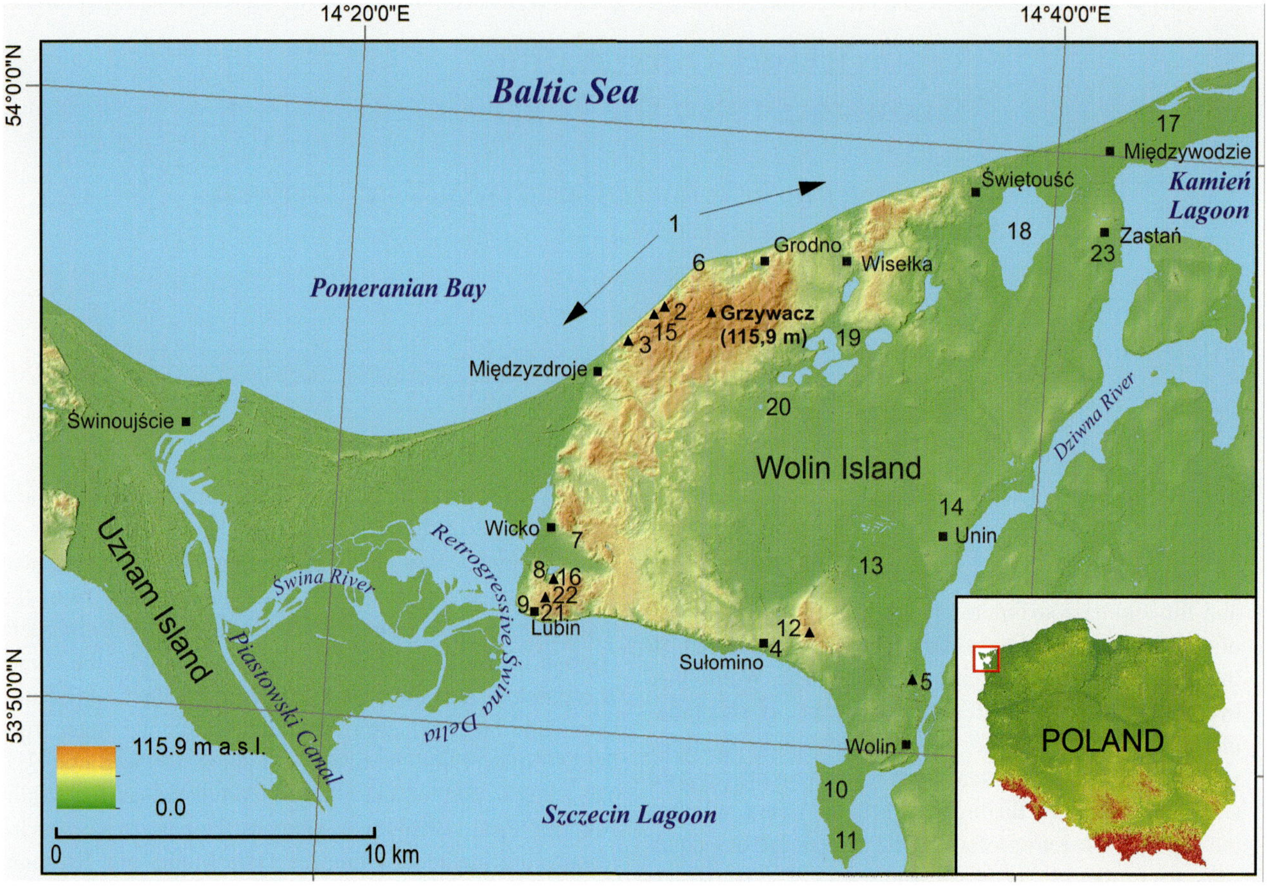

Fig. 40.1 Digital elevation model of the Wolin Island (*source* ALS, ISOK, Geoportal-2, GUGiK). TV—Trzciągowska Valley. Triangles mark significant elevations and squares mark towns and villages. Numbers from 1 to 23 indicate the location of the geosites listed in Table 40.2

Geological research of the island began in the early twentieth century (Keilhack 1912, 1914; Schulte 1924; Hartnack 1926) and was continued in the second half of the century (Matkowska et al. 1977; Ruszała et al. 1979). Since the 1970s, geomorphological research is focused on the morphodynamics of cliffed coasts (Subotowicz 1982; Kostrzewski and Zwoliński 1986, 1988, 1994, 1995, 1998; Furmańczyk 1994; Winowski 2008; Dudzińska-Nowak 2015; Kostrzewski et al. 2015, 2017; Winowski et al. 2024).

The Wolin island, apart from being a popular holiday destination, is a very attractive place from scientific and educational point view due to its geodiversity and biodiversity, which underpin landscape values related to geomorphology, climate and vegetation patterns. The cliffed coast within the boundaries of the Wolin National Park lacks any counter-erosion measures, ensuring natural pace of cliff evolution. Consequently, unique natural values of the highest cliffs in Poland can be appreciated by the visitors and are considered as the main assets of the island.

40.2 Geological Setting

The geological structure of the island and the progress in its recognition were summarized by Matkowska et al. (1977), Ruszała et al. (1979), Będkowski et al. (2009) and, most recently, by Schiewe (2021). The deep basement is made of Lower Permian rocks, mainly of sedimentary origin, such as conglomerates, coarse-grained sandstones, dolomites and anhydrites, but there are also occurrences of volcanic rocks (rhyolites). The top of Lower Permian series, which is around 1250 m thick, occurs at the depth of ca. 3 km. The stratigraphic succession above it includes Upper Permian limestones, dolomites and anhydrites, Triassic and Jurassic sandstones, mudstones, shales, limestones and marls, with intercalations of gypsum, siderites and lignites. The thickness of Triassic and Jurassic sediments reaches 1100 m and 1000 m on average, respectively. The top of the Jurassic strata at the highest points is at 16–60 m below the sea level. The thickness of Cretaceous sediments, which include shales, mudstones, glauconitic sandstones, marls, marly

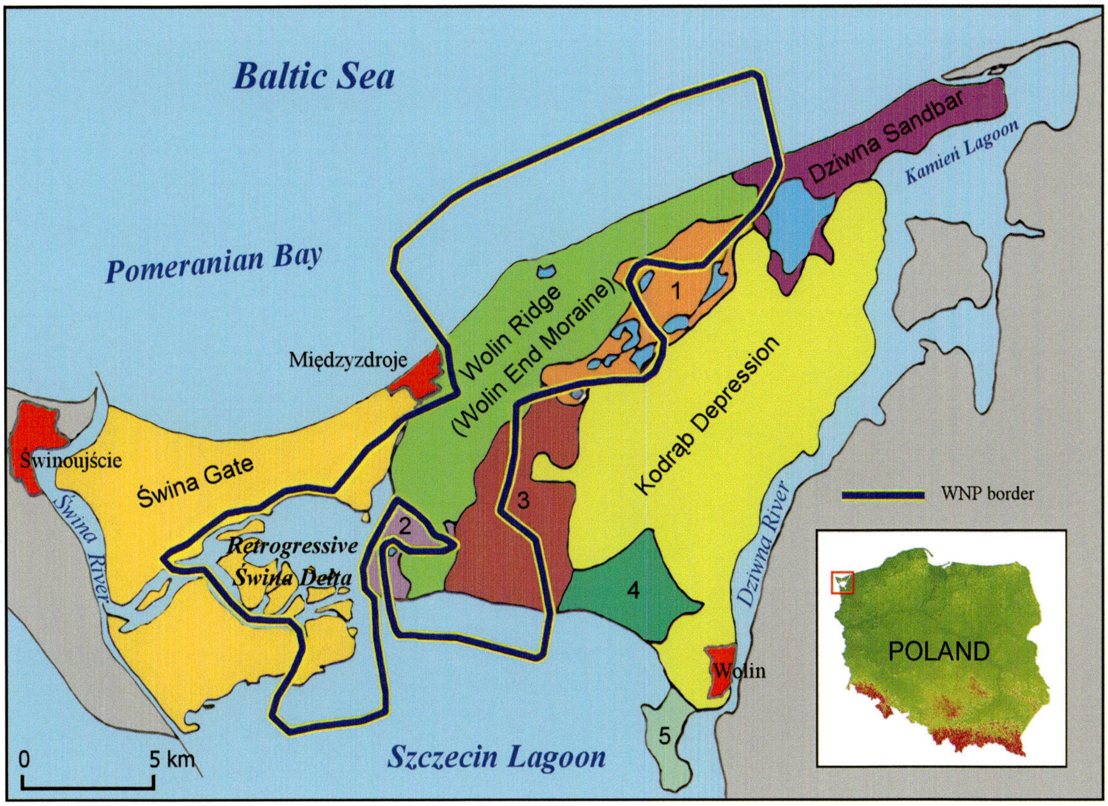

Fig. 40.2 Physical-geographical regionalization of the Wolin Island (after Marsz 1966): 1—Wolin Lakeland, 2—Lubin-Wapnica Hummocks, 3—Dargobądź Plain, 4—Mokrzyca Hills, 5—Rów Peninsula. Thick line shows the boundary of the Wolin National Park

limestones and chalk, is 100 m at maximum. The north-eastern part of the island lies within the anticline structure of Kamień Pomorski, which strikes NW–SE and represents a western branch of the Pomeranian Anticlinorium. The remaining part of the island belongs to the Szczecin Basin and includes synclinal structures. These large-scale deformations originated during the Laramian tectonic phase (Cretaceous/Palaeogene transition), whereas some faults were reactivated later, during Alpine movements, causing vertical displacements of basement blocks.

Quaternary deposits occur as a continuous cover overlying directly the Jurassic and Cretaceous rocks, as well as a small patch of Neogene sediments in the southern tip of the island. Their thickness reaches 165 m within the Wolin End Moraine rampart. Glacial and fluvioglacial sediments left during three continental glaciations, by the Sanian (MIS 16 and 12), Wartanian (MIS 6) and Weichselian (MIS 2) ice sheets, were recognized on the island. Tills from the Sanian glaciation form one horizon, strongly dissected by erosion, from a few to 20 m thick, although in local depressions it may reach the thickness of 50 m. Fluvial sands and gravels from the subsequent interglacial (Holsteinian) are very widespread, being from just several to 34 m in thickness. Wartanian tills, up to 100 m thick, build the basal parts of the highest elevations on the island, within the Wolin End Moraine. They can be subdivided into three horizons, separated by fluvioglacial sands and gravels, locally also by sands and silts deposited in ice-dammed lakes. Most common, however, are glacigenic deposits from the last glaciation (Weichselian), present all over the island. They include continuous blankets of till, 1–20 m thick, and patches of fluvioglacial sands and gravels. Rafts of Cretaceous rocks and older tills locally occur within the Weichselian tills. Tills are locally overlain by sands and gravels of kames and eskers, as well as sands of ice-dammed lakes in the low-lying areas.

Holocene deposits occur as fillings of internally drained terrain depressions, in the near-shore parts of lake basins, and in the floors of subglacial troughs and minor valleys. Their thickness does not exceed 20 m. However, typically they reach a few metres and are represented by fluvial gravels, sands and silts, sand barrier sediments, gravelly sandy beach materials of marine origin, aeolian cover sands, dune sands, humic sands, deltaic sediments of fluvial and marine origin, lacustrine sands, silts and marls, and biogenic deposits (peat, gyttja). Locally, in built-up areas and along transportation routes, anthropogenic embankments and grounds are important. Ongoing deepening of navigation channels and excavation of bottom sediments produces a specific type of deposit, i.e. dredge spoil, piled along the banks of the Świna Strait and two canals: Mieliński and Piastowski.

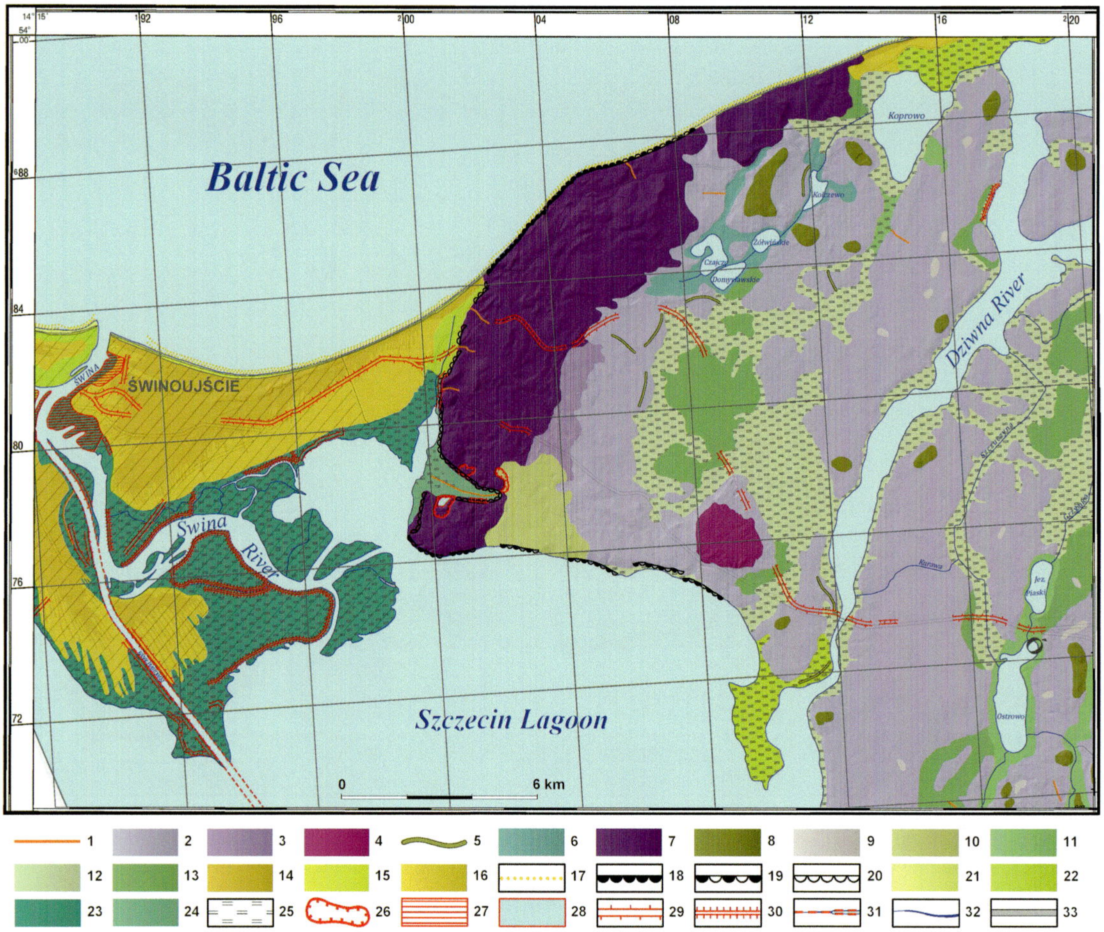

Fig. 40.3 Digital geomorphological map of the Wolin Island. 1—denudational-erosional valley (gully), 2—flat moraine plateau, 3—undulating moraine plateau, 4—overridden landform, 5—esker, 6—tunnel valley, 7—accumulative moraine, 8—kame, 9—kettle hole, 10—outwash fan, 11—ice-marginal valley, 12—floodplain, 13—river terraces, 14—group of stabilized dunes, 15—coastal dune, 16—dune plain, 17—sand bar, 18—active cliff, 19—stabilized cliff, 20—dead cliff, 21—beach, 22—spit, 23—regressive delta, 24—coastal plain, 25—biogenic plain, 26—excavation and opencast pit, 27—anthropogenic flat, 28—anthropogenic water reservoir, 29—embankment, 30—flood embankment, 31—canal, 32—watercourse, 33—national road (after Karczewski 1968; © Główny Geodeta Kraju)

40.3 Geomorphological Landscapes

In terms of elevation, the Wolin Island may be subdivided into three units (Fig. 40.1). The highest one occurs in the north-western part of the island, associated with the Wolin End Moraine (Wolin Ridge). Mean altitudes are around 40 m asl, varying between 0 and 115.9 m asl. The second unit covers the level and undulating ground moraine terrain and minor outwash fans in the eastern part of the island. Altitudes vary within the 0–30 m asl range, with the mean of only 4.5 m asl. Least elevated are three minor peninsular areas and the Świna Gate. Mean altitudes are as low as 0.3 m asl (Rów Peninsula), rising to 2.7 m asl in the Świna Gate.

40.3.1 Wolin End Moraine (Wolin Ridge) and Lubin-Wapnica Hummocks

The most diverse terrain on the island is the Wolin End Moraine (Wolin Ridge; Figs. 40.1, 40.2 and 40.3). Characteristic geomorphic elements include long slopes and numerous local elevations and hollows, some with slopes of considerable steepness, within the 10–30° range (Żynda 1962), as evident from the geomorphons map (Fig. 40.4). The origin of the terminal moraine ridge is fairly complex, with a few different stratigraphic levels hidden within its corpus (Fig. 40.5). Borówka et al. (1982, 1986, 2000) offered evidence of glacitectonic structures affecting till and sand-and-gravel beds within the Wolin

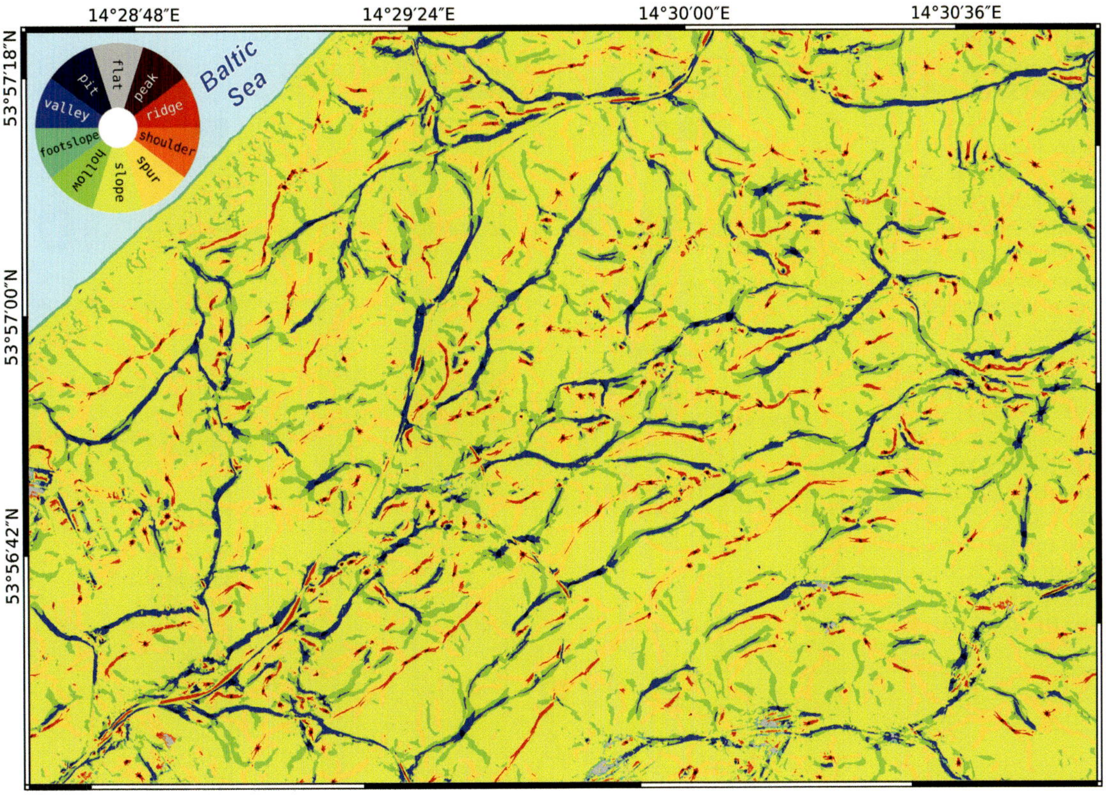

Fig. 40.4 Geomorphons map of a fragment of the Wolin End Moraine to show varied relief of the area with alternating hills and valleys (map courtesy of J. Jasiewicz)

End Moraine. However, geologists tend to interpret the Wolin Ridge as a large kame plateau (Krygowska and Krygowski 1965; Matkowska et al. 1977; Ruszała et al. 1979; Będkowski et al. 2009; Schiewe 2021). Numerous discussions regarding the origin of this large landform, carried out during field meetings, have not ended in a conclusive way. Whatever the origin, it is nevertheless clear that the Wolin End Moraine is the most attractive morphological part of the island. It rises up to 115.9 m asl (Mt. Grzywacz). Marine undercutting of the northern slopes of the moraine ridge has led to the emergence of an active cliffed coast facing the Baltic Sea. Cliff sections reveal the internal structure of the moraine, which consists of the following horizons (Fig. 40.5). The bottom part is made of grey tills, overlain by a thin bed of brown till, which is in turn covered by a thick series of fluvioglacial sands that also plays the role of an aquifer. The top of the succession is made of aeolian cover sands and blowout dusts (Borówka et al. 1982, 1986). The latter deposits are unique and according to Prusinkiewicz (1971), they are stratified accumulations built of alternating aeolian sediments, from a few millimetres to 15–20 cm thick, and organic beds composed of beech leaves (Fig. 40.6) supplied by the *Cephalanthero rubrae-Fagetum* forest communities that grow on the cliff top (Tylkowski et al. 2023). Undercut cliffs on the Baltic side are up to 90 m high, whereas those facing the Szczecin Lagoon do not exceed 50 m.

40.3.2 Abrasive Coastal Landforms

40.3.2.1 Cliffed Coast of the Baltic Sea

Active cliffs occur in the coastal belt between the seaside resort of Międzyzdroje in the west and the village of Międzywodzie in the east. They have a total length of 11.5 km. Two longest sections are located in the west (Międzyzdroje–Grodno; 5.5 km) and east (Wisełka–Międzywodzie; 4.6 km), separated by a dune coastline and inactive cliffs, of the total length of 3.2 km. The cliffed coast of the Wolin island represents the highest cliffs in Poland, reaching 93.4 m asl at Gosań Hill (Fig. 40.7). The origin and lithology of sediments exposed in the cliff faces show that this type of coastline belongs to the category of post-glacial cliffs (Kostrzewski and Zwoliński 1988). Considering lithology, three main variants of cliffs may be recognized. Fluvioglacial and fluvial deposits

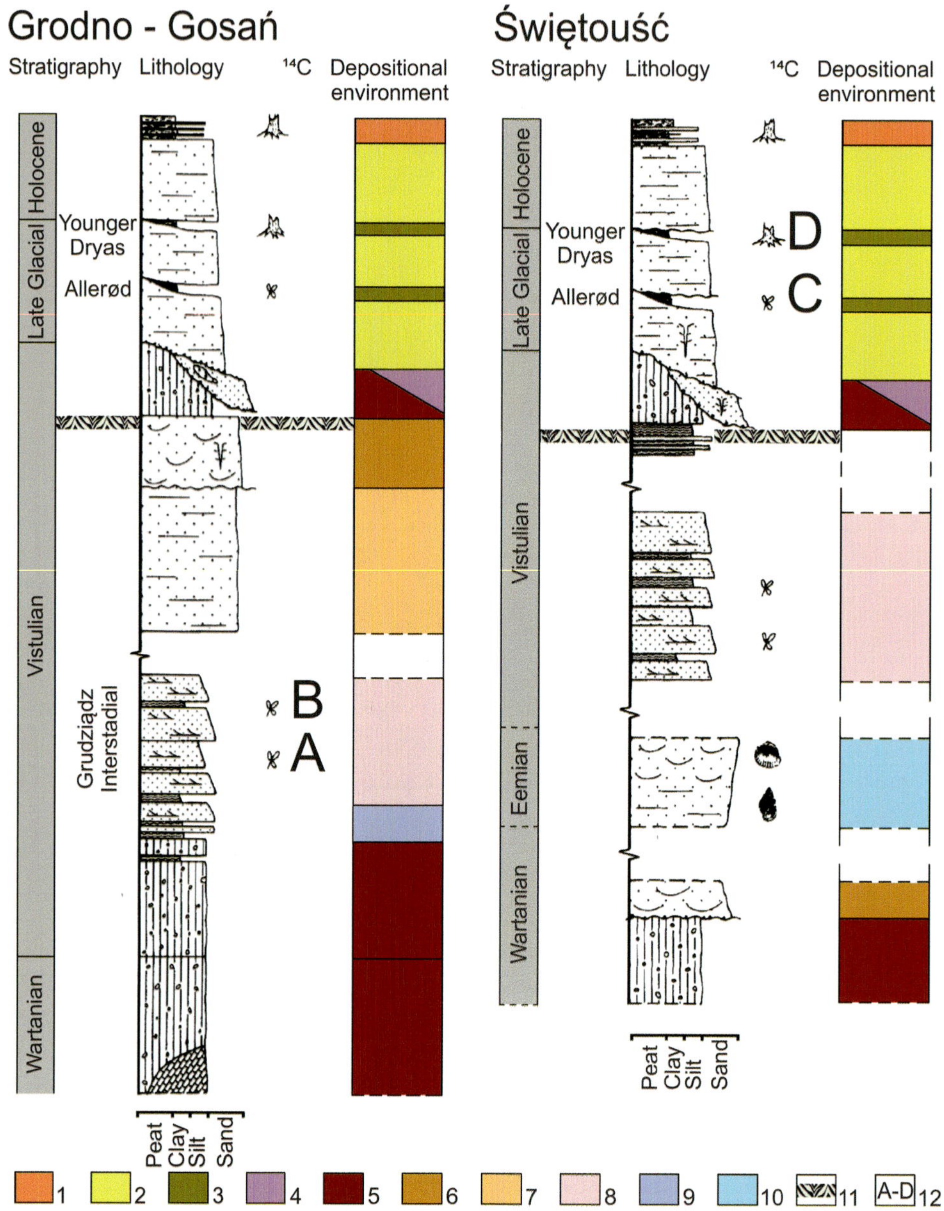

Fig. 40.5 Generalized lithostratigraphic log from the Wolin End Moraine in Grodno-Gosań and Świętouść (after Borówka et al. 2000). Depositional environments: 1—soil/aeolian, 2—aeolian, 3—soil/bog, 4—slope, 5—glacigenic, 6—fluvioglacial, 7—fluvial (?), 8—lacustrine/fluvial/marine (?), 9—glaciolacustrine, 10—marine, 11—top of the glaciotectonically disturbed deposits, 12—^{14}C samples: A—>35,400 BP (Gd-11278), >30,700 BP (Gd-0027), >29,300 BP (Gd-11376), >26,100 BP (Gd-9948), >25,900 BP (Gd-9933), B—31,200 ±2000 BP (Gd-10622), 30,100 ±600 BP (Gd-12063), 27,100 − 2400/+3400 BP (Gd-9662), 25,900 ±360 BP (Gd-12057), C—11,590 ±270 BP (Gd-631), D—1880 ±65 BP (Gd-1062)

build sandy cliffs, inclined 30–40° on average. Tills are the main constituent of loamy cliffs, which are the steepest ones, up to 80°, locally even vertical. Mixed sandy-loamy cliffs are built of tills in the lower part and fluvioglacial sands and gravels in the upper part (Winowski et al. 2024). Accordingly, they are steeper in the lower sections and become less inclined towards the cliff top.

Cliff faces are subjected to contemporary remodelling by a range of processes, resulting in a rich inventory of medium-size and minor landforms, reflecting variable lithology of the cliffs. To account for this diversity, a special key to geomorphic and morphodynamic mapping was proposed (Kostrzewski and Zwoliński 1986, 1988). Mapping allowed one to distinguish several morphodynamic types of cliffs (Winowski et al. 2024). Loamy cliffs are mainly shaped by earth flows and detachments of large till packages, whereas sandy cliffs are modelled by landslides and particle falls, contributing to the build-up of sandy cones. Within mixed cliffs two sub-types are recognized, moulded by landslides and flows or landslides and falls, in both cases acting in combination (Fig. 40.8).

Cliff erosion is primarily conditioned by the dynamics of the sea, including factors such as the frequency of storm surges, the significant wave height, and the direction of wave approach to the shore (Winowski 2019). Abrasion processes show distinctive seasonality within a year (Kostrzewski and Zwoliński 1988; Kostrzewski et al. 2015) and are mainly controlled by the frequency of storms. The dynamics of the sea is at its highest in autumn and winter (beginning of November to mid-February), with storm

Fig. 40.6 Unique deposits of blowout dusts (soil type after Prusinkiewicz (1971) and called 'embankment by the cliff' (in Polish)) are layered accumulations built of alternating fine mineral aeolian sediments with a thickness of several millimetres to 15–20 cm and organic deposits composed of fallen beech leaves supplied by the *Cephalanthero rubrae-Fagetum* forest communities that grew on the top of the cliff. **a** Multi-year layers of blowout dusts just below the cliff top and lying on beds of the aeolian cover sands. **b** Contemporary aeolian cover from summer 2023, which will be covered with leaves in autumn 2023 (photograph **a** by M. Mazurek; **b** by P. Czyryca)

surges accounting for 7.7% of that period. In spring (mid-February–mid-June) the activity of the sea decreases, with storms present over 4.6% of this time interval. Finally, summer and autumn time (mid-June to the end of October) is typified by the lowest sea dynamics, with storms experienced during a mere 4% of the length of the period. Wave energy degrades cliffs at different rates. Sandy cliffs are degraded more quickly, while loamy cliffs characterized by increased resistance to degradation processes are degraded more slowly.

Another important factor contributing to cliff morphodynamics and its degradation is rainfall and its distribution. Heavy rain triggers overland flow, earth flows and landslides. In the Wolin Island, June and July are the rainiest months, hence most conducive to water erosion. A significant part of summer rainfall occurs as relatively brief, but intense downpours, related to the passage of cold fronts associated with thunderstorms. By contrast, water erosion is least efficient in winter (January–February), reflecting low rainfall and low temperatures, often below 0 °C. Winowski (2014) showed that landslides within cliffs are generated by bi-weekly rainfall in excess of 90 mm, including two consecutive days with the total rainfall above 40 mm. In turn, earth flows require the moisture content of tills above 16% to occur.

A factor not to neglect in evaluating cliff morphodynamics is groundwater circulation pattern and the presence of springs and seepage. The structure and hydrogeology of the sedimentary package exposed in the cliffs make them particularly prone to sliding and slumping. Seepage typically occurs along the contact of till below and fluvioglacial sands above. Increasing delivery of water to the

Fig. 40.7 Views of the cliffed coast of the Wolin Island from the Baltic Sea. **a** View from the drone towards the east. **b** View of the cliff built of fluvioglacial and aeolian sediments. **c** View of the cliff built of grey glacial till (lower part) and brown glacial till (upper part) (photograph **a** by M. Winowski; **b** by Z. Zwoliński; **c** by M. Mazurek)

Fig. 40.8 Dominant processes shaping the cliffs of the Wolin Island are landslides and slumps. **a** Steep cliff faces with landslide material removed from the beach. **b** Landslide formed after prolonged rainfall (photograph **a** by M. Winowski; **b** by M. Mazurek)

cliff-building deposits results in the reduction of strength and increasing likelihood to generate shearing surfaces, determining thereby the morphodynamic type of a cliff (Fig. 40.9). Sliding processes associated with groundwater outflow occur mainly within flow- and slide-dominated cliffs (Winowski 2008; Winowski et al. 2024). They are responsible for the most pronounced changes in cliff morphology, with the amount of top scarp retreat during individual events up to 10 m. The occurrence of springs within the cliffs results in the enhancement of seepage erosion, producing spring hollows. However, these are rather minor and short-lived landforms, as conditioned by the geology

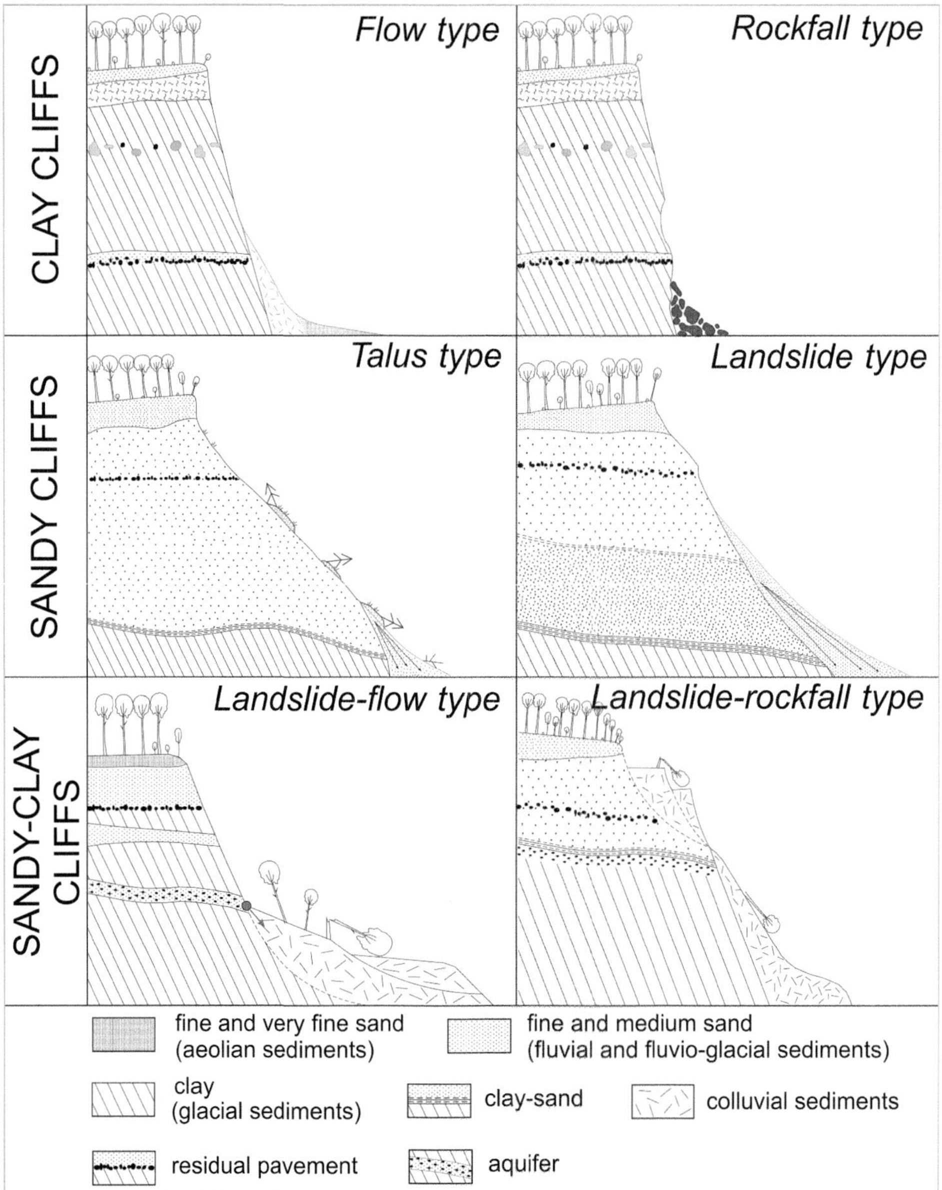

Fig. 40.9 Morphodynamic types of the cliff coast of the Wolin Island as typical example for the Southern Baltic (after Winowski et al. 2024)

of cliff deposits, location and limited efficacy of groundwater outflows, as well as obliteration by abrasion and mass movements.

The combined contribution of all these factors and processes results in systematic cliff retreat inland. Long-term measurements carried out since 1984 indicate generally low rates of retreat of the cliff top (0.22 m a^{-1}), decreasing over time (Kostrzewski et al. 2015). Cliff sections dominated by loamy deposits are eroded least efficiently (0.12 m a^{-1}). This type of substratum shows relatively high resistance against degradation processes, delaying the response of cliff top to the base undercutting. Many repetitive storm events have to occur to allow the retrogressive erosion to reach the cliff top. On the other hand, much higher retreat rates have been recorded on sandy cliffs (0.32 m a^{-1}), which are least cohesive and subject to rapid degradation. During storm surges, they are quickly undercut, which initiates dry granular flows and landslides. Material displacement occurs throughout the entire length of the cliff, significantly reducing the response time of the top. Increasing rates of retreat (0.28 m a^{-1}) also typify sandy-loamy cliffs.

The analysis of data derived from aerial laser scanning (ALS) for the 2012–2020 period allowed us to quantify the magnitude of cliff erosion (Winowski et al. 2022). The loss from a 9.6 km long cliff amounted to 510,000 m^3 of sediment (6.6 m^3 m^{-1} a^{-1} on average), although with significant spatial differences (Fig. 40.10). Sandy sections with thick

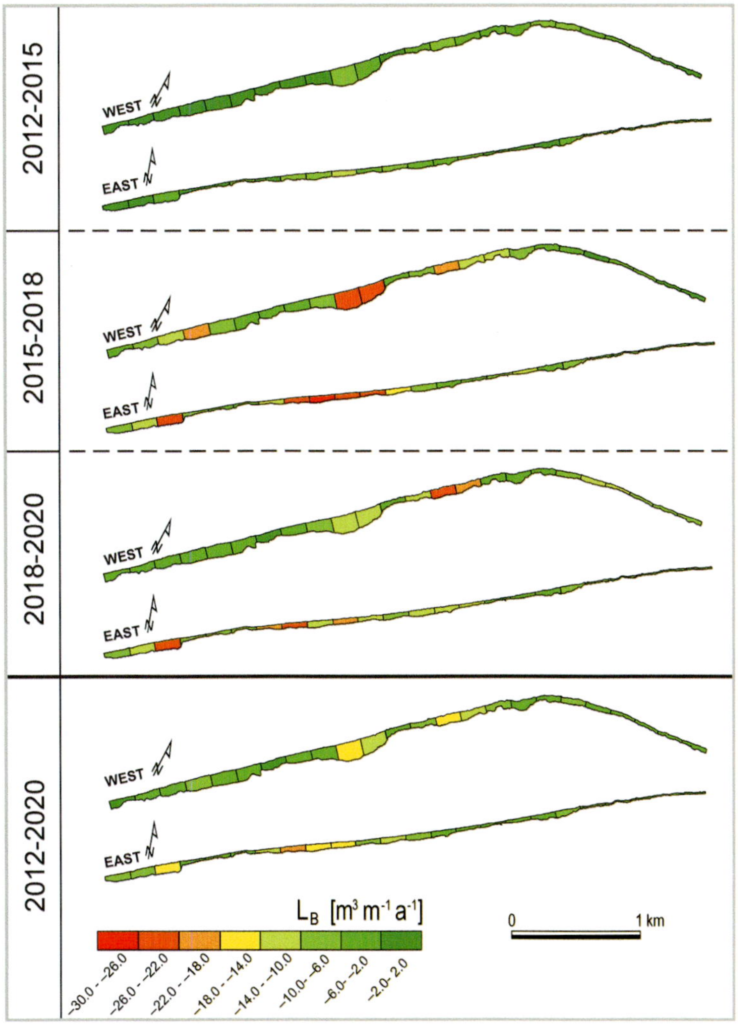

Fig. 40.10 Spatio-temporal differentiation of the length-normalized sediment budget (L_B) along the seaward side cliff coast divided into 200-m sectors in the western (Międzyzdroje–Grodno) and eastern (Wisełka–Międzywodzie) parts of the Wolin Island during the 2012–2020 period, subdivided into three intervals (after Winowski et al. 2022)

colluvial covers were eroded most (average 18 m^3 m^{-1} a^{-1}), whereas stable vegetated cliffs and resistant loamy cliffs suffered least (average 1.3 m^3 m^{-1} a^{-1}). A comparative analysis with other coastal sections of the Baltic Sea revealed that cliff erosion at the Wolin Island is several times faster than in Schleswig-Holstein in Germany (Averes et al. 2021), but several times slower than that of cliffs in the Ustka Bight in the central section of the Polish Baltic coast (Frydel et al. 2017).

Diverse rates of cliff retreat in the Wolin Island reflect complex morphodynamic regime of cliffed coasts, which varies both in time and space due to differences in lithology, morphometric characteristics and aspect. The measured rates of cliff top retreat are rarely representative for the entire cliff face. For example, to trigger mass movements within loamy cliffs, a larger number of storm episodes are required than within sandy cliffs. Complete remodelling of cliff faces during individual events occurs mainly within low cliffs. By contrast, high cliffs become entirely reshaped only during storm surges of extreme height (Kostrzewski and Zwoliński 1998).

40.3.2.2 Cliffed Coast of the Szczecin Lagoon

The southern part of the Wolin End Moraine is known as the Lubin-Wapnica Hummocks (Figs. 40.1, 40.2, 40.3 and 40.11). In terms of general geological structure and relief, this area does not differ much from the more northerly part of the terminal moraine. However, a distinctive feature is the occurrence of numerous rafts of Cretaceous sedimentary rocks incorporated into sandy and loamy moraine deposits (Fig. 40.12). Nearly a dozen of localities of Cretaceous glacial rafts within the whole island were described by Alexandrowicz (1966). One of them was entirely quarried away in the twentieth century, leaving a hollow currently occupied by the Turkusowe Lake (Turquoise Lake). The lake covers the surface area of 6.74 ha, which is 21.2 m deep and its bottom is 18.6 m below the sea level. The name alludes to the colour of lake waters, caused by the dispersal of sunlight reflected from the white Cretaceous chalk in the lake bottom and calcareous lake sediments. Karstic landforms—small dolines and vertical pipes—were recognized in the chalk rafts present within the south-facing valley sides near the villages of Wicko and Wapnica (Zwoliński 1975).

Fig. 40.11 Less active cliff on the southern side of the island facing the Szczecin Lagoon, developed on the backside of the Lubin-Wapnica Hummocks (photograph by M. Winowski)

Fig. 40.12 Raft of Cretaceous sedimentary rocks incorporated into sandy and loamy moraine deposits in the Baltic Sea cliff near Grodno (photograph by M. Mazurek)

The western section of the southern coastline of the Wolin island is made of an old, largely inactive cliff shore nearly 7 km long. It is included into three different geomorphological units. The westernmost part belongs to the Lubin-Wapnica Hummocks (Fig. 40.11). These cliffs are the highest, up to 50 m, and up to 35–40° steep if the substratum is sandy. However, in more cohesive loamy deposits, scarps may reach the inclination of 75°. The continuity of the cliff top is interrupted by deep erosional-denudational valleys, with the depth of incision up to 30 m. The height of the cliff decreases eastward and the moraine hills in the hinterland become replaced by the outwash surface of the Wolin sub-phase (Fig. 40.3), with no clear morphological boundary between the two. The landscape of this geomorphological unit is also characterized by numerous erosion-denudation incisions. Still further to the east, the outwash surface gives way to the rolling moraine upland and then, to the flat moraine. The inland side of the cliff is poorly diversified morphologically, whereas the marginal zone is incised by shallow minor erosional-denudational valleys. The cliff itself is rarely more than 10 m high. In the east, the cliff disappears and a depositional coastal flat is present.

In the boundary zone between the Wolin End Moraine and the Lubin-Wapnica Hummocks, an interesting landform of west–east extension occurs, whose origin has not yet been clarified. It is known under the name of Trzciągowska Valley (Figs. 40.2 and 40.3). It is nearly 2.5 km long and up to 1.5 km wide in the distal part, next to the shoreline of the Szczecin Lagoon. The valley sides are locally up to 75 m high. The valley floor lies almost at the contemporary sea level, between 0 and 5 m asl, and is made of mineral-organic deposits. Three different views regarding the origin of the valley form exist, none being supported by unequivocal evidence. It is interpreted as: (a) an old sea embayment; (b) glacial erosional feature; (c) least probably, a landform of tectonic origin.

40.3.3 Depositional Coasts of the Wolin Island

40.3.3.1 Świna Gate and Retrogressive Świna Delta

The microregion of the Świna Gate (also called Przytór Peninsula), geomorphologically equal to the Świna Gate Spit (Przytór Spit), is one of a few characteristic depositional coasts in the Polish coastal zone (Łabuz and Olechnowicz 2004; Figs. 40.2, 40.3 and 40.13). Consequent to marine, aeolian and fluvial accumulation processes occurring over the past 6000 years since the Littorina Transgression (Borówka et al. 2005), two sandy spits originated, growing towards the east (anchored on the Uznam Island) and to the west (from the Wolin Island). In time, they joined together cutting off the Szczecin Lagoon from the Baltic Sea (Musielak and Osadczuk 1995; Osadczuk 2004).

The Wolin spit extends in east-west direction, between the towns of Międzyzdroje and Świnoujście in the north, and the Wicko Małe Lake and the Świna River in the south (Fig. 40.1). It is 12 km long and from 2.5 km wide in the east to about 6.5 km in the west. The spit borders the Pomeranian Bay in the north, whereas in the south, it gives way to the picturesque islets of the Retrogressive Świna Delta. Quasi-parallel dune ridges, shaped by marine and aeolian depositional processes in the Holocene, typify the morphology of the spit (Fig. 40.13). The surface is relatively flat. The highest elevations, up to 27 m asl, occur in the northern part of the spit, within the highest dune ridge parallel to the shoreline of the Pomeranian Bay. The average elevation of the spit area does not exceed 3 m asl.

Research conducted by Keilhack (1912, 1914) allowed one to conclude that dune ridges of various ages developed during the formation of the Świna Gate. Based on studies of soil maturity, three dune generations were distinguished, coeval with three evolutionary phases of the origin of the Świna Gate. These are so-called brown, yellow and white dunes (Prusinkiewicz and Noryśkiewicz 1966) (Fig. 40.13). The best developed soils occur within the brown dunes, which are the oldest, whereas a yellow to white illuvial level is characteristic for the youngest white dunes. More recent studies led to the recognition of one more dune generation within the white dunes, named as 'white dunes I' (Osadczuk 2004). The model of morphological evolution of the Świna Gate, proposed by German researchers more than 100 years ago, was subsequently confirmed by the OSL dating (Reimann et al. 2011). The dunes were shaped around 6.62 ± 0.06 ka BP, soon after the main phase of the Littorina Transgression. Over time, the accumulation rates were decreasing, from 4.2 ± 1.9 to 2.6 ± 0.6 m a^{-1} for brown dunes, and reached a minimum of 1.3 ± 0.4 m a^{-1} for yellow dunes during the Subatlantic Transgression at ca. 1200 year BP. Subsequent acceleration of spit expansion, with accretion rates of 2.0 ± 0.2 m a^{-1}, coincided with the Little Ice Age and occurred in 1550–1850. The current rate of accretion in white dunes is ca 1.8 ± 0.2 m a^{-1}, which results in one of the widest beaches in Poland.

Nowadays, the growth of dunes does not occur in the Świna Gate, having been replaced by degradation resulting from natural erosion and anthropogenic interference. The largest erosional losses are noted within dune ridges facing the Baltic Sea. During storm surges of moderate magnitude, they amount to 5–10 m, but may reach 12 m during truly extreme events. In terms of volume, average 7.4–8.9 m^3 of beach sand and average 2.1–4.0 m^3 of dune sand is being removed to the littoral zone during storm surges in the years 2000–2020 (Łabuz 2022).

The area of the Świna Gate is of strategic importance for the Polish economy. The Świna River is the main entrance to the commercial harbour in the city of Szczecin. On the eastern side of the Świna mouth, the Świnoujście Liquefied Natural Gas (LNG) Terminal has been built, which is crucial for the energy supply system and energy security of Poland. Plans for the future include building of a deep-water container harbour, allowing for handling the largest ships entering the Baltic Sea. The development of industrial plants necessitates the expansion of transportation routes, such as the current building of a

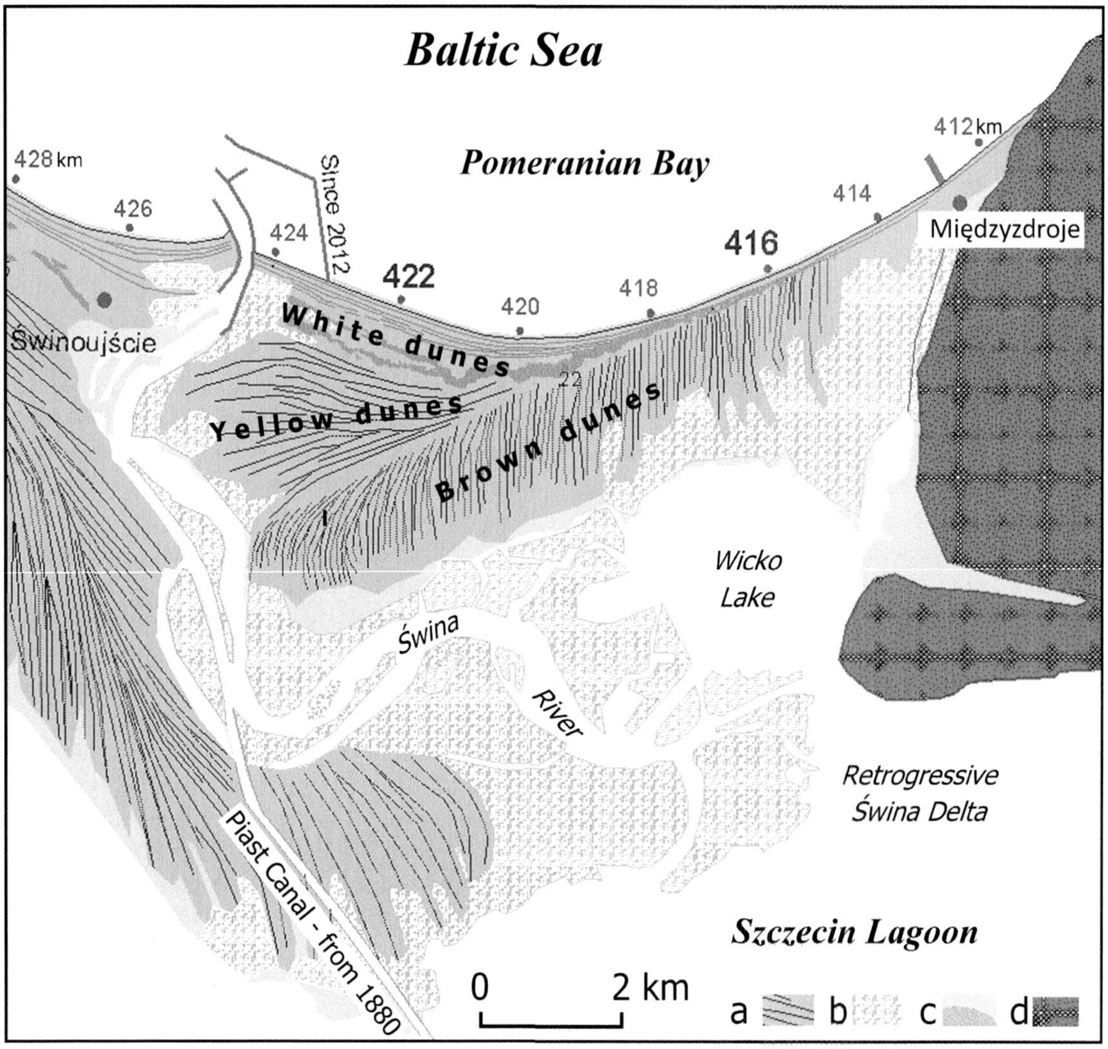

Fig. 40.13 Morphology of the Świna Gate Sandbar (after Łabuz 2022, modified). **a** White (0–0.45 ka), yellow (0.59–1.18 ka) and brown (1.65–5.39 ka) sandbar dune ridges. **b** Wetlands. **c** Lowlands and lakes. **d** Moraine. The OSL age of the dunes was given in ka before 2008 (after Reimann et al. 2011). Piast Canal—a water channel with a length of 8 km and a depth of 12.5 m dug in the south-eastern part of the island of Uznam in the years 1874–1880 in order to improve navigation between the harbour of Szczecin and the Baltic Sea. The beginning of the accumulation of fine-sea sediments in the form of the Retrogressive Świna Delta correlates with the period after the Littorina Transgression

motorway, including underwater road tunnel connecting the town of Świnoujście with the rest of the country (Fig. 40.14). Thus, this unique dune area is currently being intensively degraded by road and rail infrastructure and the expansion of the Świnoujście harbour on the eastern side of the Świna River.

To the south, the Świna Gate borders the Retrogressive Świna Delta (Fig. 40.15). This specific delta is caused by marine sedimentation during extreme storm events and covers an area of more than 30 km^2, with 44 wetland and marshy islands with numerous peninsulas, fens and tidal waters between the islets. The islands are covered with reeds (which grow up to 4 m tall), meadows and other low vegetation. The largest island is Wielki (Great) Krzek (3.2 km^2, Fig. 40.1).

40.3.3.2 Dziwna Sandbar

The Dziwna Sandbar (also called the Międzywodzie Peninsula) is a spit that separates the mouth of the Dziwna River from the Baltic Sea (Fig. 40.16a). The contemporary mouth and open sea connection are artificial, whereas the original channel was partially silted up, with the remaining reach making the Martwa Dziwna (Dead Dziwna) Lake. The sandbar is built of dune ridges and hummocks on the sea-facing side, derived from beach sands. The opposite inland side is occupied by waterlogged meadows and bogs.

40.3.3.3 Rów Peninsula

The Rów Peninsula is the youngest morphological element of the entire island, dated to the late Holocene (Fig. 40.16b).

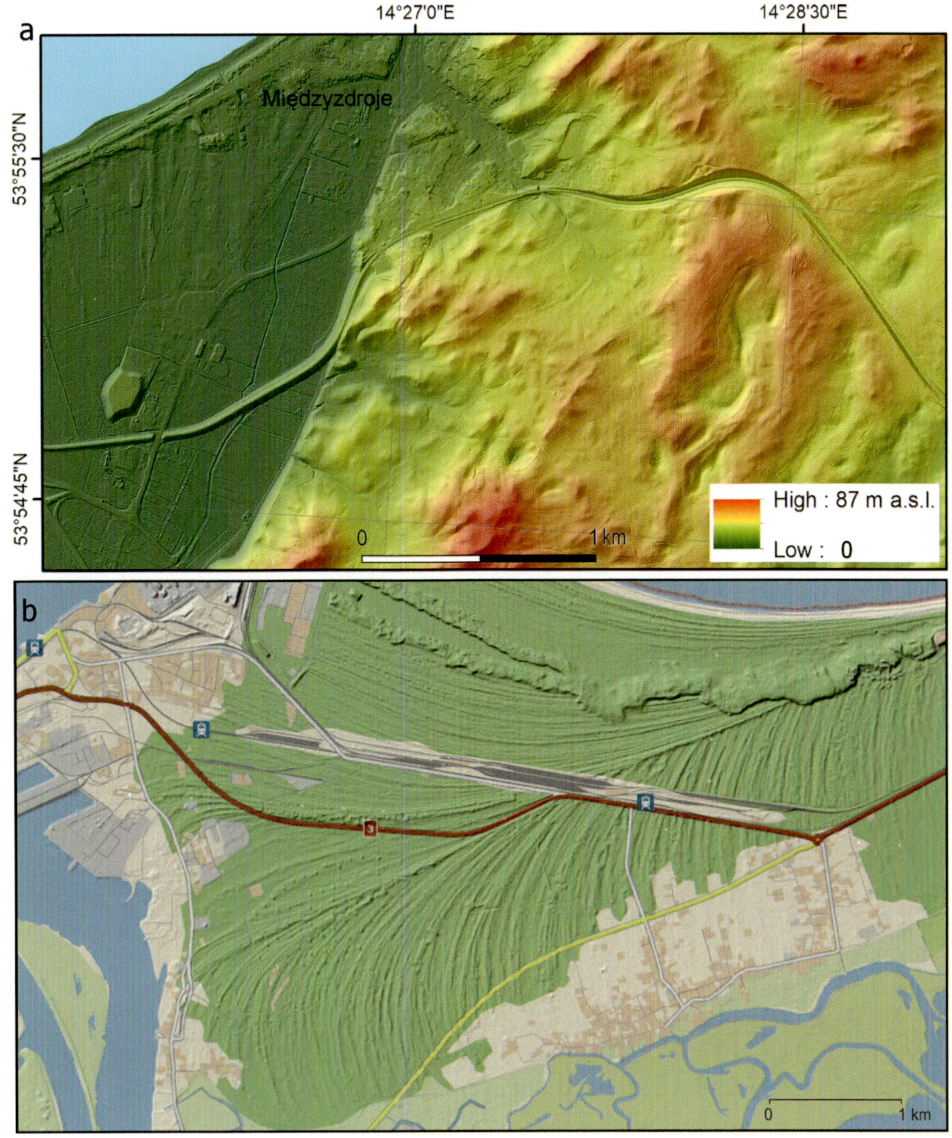

Fig. 40.14 Geodiversity loss within the Holocene dune field in the Świna Gate Spit and within the Pleistocene Wolin End Moraine, caused by the development of transportation infrastructure. **a** Railway embankments, cuts and street network in the town of Międzyzdroje). **b** Road, rail and harbour infrastructure in the eastern part of the town of Świnoujście (source: Geoportal-2, GUGiK)

It is being currently enlarged through the deposition of fluvial silts supplied by the Odra River to the Szczecin Lagoon. The peninsula is rooted in the most southward-projecting part of the ground moraine plateau in the eastern part of the island, where it reaches 21 m asl. This proximal part is a geoarchaeological site known as the Wzgórze Wisielców hill (Hangman's Hill), with 93 earth burial mounds from the ninth to tenth centuries. The remaining part of the peninsula is a wetland, with habitats of water birds.

40.3.4 Wolin Lakeland

The Wolin Lakeland is as scenic and attractive as is the Wolin End Moraine. It involves two subglacial troughs of SW–NE strike, located between the high-elevation moraine ridge in the north-west and a low-lying ground moraine plateau in the east (Figs. 40.1, 40.2, 40.3 and 40.17). Within the troughs numerous overdeepenings occur, filled by the strings of lakes. As many as ten lakes have their bottom parts below the sea level (Table 40.1). The Recze Lake is the deepest (12 m), including in respect to the sea level (bottom altitude – 9.4 m). The Koprowo Lake, the northeastern most one, is the largest within the lakeland, occupying 450 ha. Its origin is complex. Besides the role of glacial erosion, it used to be a sea embayment, subsequently plugged by the Dziwna Sandbar. Some of the lakes are connected by the Lewińska Struga stream (12.8 km long), which debouches into the Kamień Lagoon. Despite low discharges (0.01–0.4 $m^3\ s^{-1}$ in different cross sections), chemical denudation is significant, 76 $m^3\ km^{-2}\ a^{-1}$ on average.

Fig. 40.15 Retrogressive Świna Delta. **a** Panoramic view from the east (photograph by M. Winowski). **b** Orthophotomap (source: Geoportal-2, GUGiK)

40.3.5 Kodrąb Depression, Mokrzyca Hills and Dargobądz Plain

The Kodrąb Depression constitutes the central section of the eastern part of the Wolin Island. It is represented by a ground moraine, flat or gently undulating, which in terms of surface area is the largest geomorphological unit within the island (Fig. 40.3). It borders the Koprowo Lake basin and the Dziwna Sandbar in the north, the troughs of the Wolin Lakeland in the west, the Mokrzyca Peatbogs in the south, and the subglacial trough of Dziwna River in the east (Fig. 40.18). Altitudes vary from 2 to 20 m asl and local relief rarely reaches 10 m. Surface morphology of the ground moraine includes kames and eskers, as well as melt-out depressions. Both kames and eskers form oval hummocks 400–1500 m long, reaching maximum elevation of 20 m asl. They are most distinct in the marginal zones of the moraine terrain, both in the west (next to villages of Kołczewo and Warnowo) and the east (settlements of Sierosław, Unin and Wolin), and are directly superimposed on the basal till of ground moraine. Melt-out hollows left by decaying blocks of dead ice occur as shallow glaciolacustrine basins filled with sandy-silty sediments. Nowadays, they are biogenic plains, with ongoing growth of the peat layer.

The Mokrzyca Hills in the south-eastern part of the island (Figs. 40.2 and 40.3) are remnant landforms inherited from the penultimate inland glaciation and built of grey tills (Wartanian). These landforms were reshaped during the last glaciation. Maximum elevation reaches 67 m asl.

The Dargobądz Plain lies on the central-south tip of the island and is largely an outwash plain surface (Fig. 40.3). The sandur was deposited during the ultimate phase of ice-sheet withdrawal from the area of contemporary Poland. The southern terminus of the outwash plain is mildly undercut by the waters of the Szczecin Lagoon and a low sandy cliff has formed.

40.4 Geosites

Geosites in the Wolin Island, recognized by the State Geological Survey, include eight thematic categories: natural geological outcrops, artificial geological outcrops,

Fig. 40.16 Coastal depositional landforms. **a** Dziwna Sandbar in the north-eastern part of the island. **b** Rów Peninsula in the southernmost tip of the island (photographs by M. Winowski)

mineralogical and/or petrological localities, mining legacy, accumulation landforms, hydrological objects, geoarchaeological sites, and viewing points (PIG-PIB 2022). All geosites are summarized in Table 40.2, and although their distribution on the island is not uniform, they represent well the geodiversity and geoheritage of this island.

There are still many places on the island that deserve to be called geosites and should be covered by geoconservation. These include, e.g. the Retrogressive Świna Delta, brown, yellow and white dunes of the Świna Spit, Lake Grodno, and spring alcoves on the southern shore of the island.

40.5 Final Remarks

The Wolin Island has many landscape charms and fascinating geotouristic values that attract earth scientists and nature lovers. The main attraction is the Wolin National Park, which encompasses a variety of geoecosystems, including primarily cliffs, dunes, peat bogs, wetlands, and forests. It is also the habitat of many rare and protected plant and animal species. The post-glacial relief (terminal and ground moraines, outwash plain, biogenic accumulation plains, post-glacial lakes) determines its significant attractiveness and gives the island a unique landscape character. Impressive and picturesque cliffs, which are the highest in Poland (over 90 m) offer magnificent panoramic views of the Baltic Sea and are popular objects for landscape photography. On the Wolin Island, a large area is occupied by Holocene dunes of various ages, which, unfortunately, are being degraded by the development of transport network and harbour infrastructure. The existing geosites will not protect the entire island from human impact, but together with the area of the national park they are able to protect the geodiversity and biodiversity of these spectacular landscapes for future generations. Hiking and cycling trails that allow visitors to explore the remarkable natural, historical, and cultural attractions serve this purpose. The entire island is proposed as the future Geopark Wolin.

Fig. 40.17 Wolin Lakeland. **a** Orthophotomap (source: Geoportal-2, GUGiK). **b** Panoramic view (photograph by M. Winowski)

Table 40.1 Selected features of Wolin lakes

Lake name	Area (ha)	Perimeter (km)	Mean depth (m)	Maximum depth (m)	Crypto-depression (m bsl)	Volume (10^3 m^3)	Origin
Czajcze	77.6	5.55	2.9	6.0	4.5	2316.7	Trough lake
Domysłowskie	47.3	3.08	1.9	3.1	1.7	1018.0	Trough lake
Gardno	2.4	1.58	2.6	7.3	–	62.4	Moraine lake
Kołczewo	46.3	3.68	2.2	4.0	3.4	1081.0	Trough lake
Koprowo	483.0	9.80	1.6	3.1	3.1	7792.9	Seaside lake
Rabiąż (Waronowo Wschodnie)	18.2	1.78	1.5	2.8	1.0	280.1	Trough lake
Recze	4.5	0.77	–	12.0	10.1		Kettle hole[a]
Turkusowe	5.6	1.11	9.2	21.2	18.5	513.7	Post-mining lake
Warnowo Zachodnie	12.7	1.33	1.3	2.1	0.4	160.7	Trough lake
Wisełka	19.2	1.94	3.0	6.7	1.0	593.3	Trough lake
Żółwinskie	43.9	2.80	1.8	3.0	1.3	741.3	Trough lake

[a]Meteorite origin is also considered

Fig. 40.18 Dziwna River, locally known as the Dziwna Strait, uses the subglacial trough separating the island of Wolin from the mainland (photograph by M. Winowski)

Table 40.2 List of geosites located on the Wolin island (according to PIG-PIB 2022)

No.	Geosite name	Description
1	Międzyzdroje-Wisełka cliff	The western section of the cliffed coastline
2	Gosań hill	The highest cliff in the Polish sector of the south Baltic coast (93.4 m asl)
3	Kawcza Góra hill	Terrain elevation within the Wolin End Moraine, 61 m asl
4	Sułomino and cliff facing the Szczecin Lagoon	The southern coastline of the island. The cliff is built of glacigenic sands and gravels from the Pomeranian Phase of the Weichselian glaciation
5	Srebrne Wzgórze hill	Former esker, largely quarried away for sand and gravel. It is a 2-km-long ridge of running north-south, built of deposits laid down by subglacial waters, in a crevasse or in glacial tunnel
6	Głazy Piastowskie (Piast Boulders)	Erratic boulders representing various rock types, including granites, pegmatites, porphyries, gneisses and hornfelses, washed away from the retreating cliff face; monument of abiotic nature
7	Chalk quarry in Wicko	Anthropogenic outcrop of Turonian sedimentary rocks: clayey marls, marly chalk and chalk proper, with remnants of inoceramus shells and small siliceous concretions; karstic pipes and dolines
8	Turkusowe Lake (Turquoise Lake)	Artificial lake as a result of rainwater filling the former mining pit within the Cretaceous glacial raft; 21.5 m deep
9	Cement factory in Lubin	Ruins of an old cement plant, in which chalk, limestone and marl from the Trzciągowska Valley were used
10	Rów Peninsula	Biogenic accumulation area, with Holocene peat bogs on fluvio-marine deposits

(continued)

Table 40.2 (continued)

No.	Geosite name	Description
11	Wzgórze Wisielców hill (Hangman's Hill)	Near the town of Wolin, the southern tip of an esker, 21 m high, on the left bank of the Dziwna River; burial mounds from the ninth–tenth centuries
12	Mokrzyca hills	Moraine hills due to superimposition of glacigenic sediments from the main stadial of the Weichselian glaciation on the pre-existing elevation from an older glaciation; push moraine
13	Mokrzyca Peatbog	Peat accumulation area of Holocene age (fen), anthropogenically modified, with a dense network of drainage ditches
14	Unin hills	Kame hummock, built of meltwater-derived sediments superimposed on ground moraine till; with layered structure involving sand, silt and gravel
15	Biała Góra hill	Elevation within the Wolin End Moraine, 68 m asl
16	Piaskowa Góra hill	The highest spot within the steep southern bank of the Turkusowe Lake; in the vertical profile one can distinguish, from the bottom up, pale grey chalk with inoceramus, with chalk with flints, grey till and sands with erratics
17	Martwa Dziwna Lake	An old mouth of the Dziwna river into the Baltic Sea, naturally silted; in the ninth century a new connection was excavated, whereas an old mouth was filled with rubble and a lake was formed
18	Koprowo Lake	Nearshore lake, cut off from the Pomeranian Bay by the Dziwa Sandbar; the shallowest lake in the Wolin Lakeland
19	Lakes of the Wolin Lakeland	Natural lakes, with the bottoms below the sea level, within two glacial troughs (Wisełka Lake trough and Kołczewo Lake trough), which connect with the melt-out depression occupied by Czajcze, Domysławskie and Żółwińskie Lake; in addition: Zatorek, Rabiąż and Recze lakes
20	Łuniewo nature reserve	Łuniewo Lake occupies a basin within ground moraine, originated due to dead ice melting; filled with sandy-silty sediments and peat bogs
21	Lubin hillfort	A hill built of till, 50 m asl, in the southernmost part of the Wolin End Moraine, used for religious and defensive reasons in medieval times
22	Zielonka hill	End-moraine hill, 81 m asl, within the Lubin-Wapnica Hummocks, with panoramic view of 44 islets of the Retrogressive Świna Delta
23	Oil well in Zastań	Site of hydrocarbon exploitation. Oil occurs within the main dolomite; part of the oil-bearing area is located below the Kamień Lagoon

References

Alexandrowicz Z (1966) Utwory kredowe w krach glacjalnych na wyspie Wolin i w okolicy Kamienia Pomorskiego. Pr Geol PAN 35:7–103

Averes T, Hofstede JLA, Hinrichsen A, Reimers HC, Winter C (2021) Cliff retreat contribution to the littoral sediment budget along the Baltic Sea coastline of Schleswig-Holstein, Germany. J Mar Sci Eng 9(8):870

Będkowski Z, Siwy-Będkowska K, Wąsowicz A, Król J, Pasieczna A, Kwecko P, Bojakowska I, Tomassi-Morawiec H (2009) Objaśnienia do Mapy Geośrodowiskowej Polski 1:50,000, ark. Międzywodzie (75) i Wolin (114). PIG, Warszawa, p 67

Borówka RK, Gonera P, Kostrzewski A, Zwoliński Z (1982) Origin, age and paleogeographic significance of cover sands in the Wolin end moraine area, North-West Poland. Quaest Geogr 8:19–36

Borówka RK, Gonera P, Kostrzewski A, Nowaczyk B, Zwoliński Z (1986) Stratigraphy of eolian deposits in Wolin Island and the surrounding area, North-West Poland. Boreas 15:301–309

Borówka RK, Goslar T, Pazdur A (2000) Age of glacitectonic structures on the Wolin Island in the light lithostratigraphic data and radiocarbon dating. Geochronometria 18:35–40

Borówka RK, Osadczuk A, Witkowski A, Wawrzyniak-Wydrowska B, Duda T (2005) Late Glacial and Holocene depositional history in the eastern part of the Szczecin Lagoon (Great Lagoon) basin—NW Poland. Quat Int 130:87–96

Dudzińska-Nowak J (2015) Metody ochrony zachodniego wybrzeża Polski i ich wpływ na zmiany brzegu w latach 1938–2011. Wyd Nauk Uniw Szczecińskiego, Szczecin, p 174

Frydel J, Mil L, Szafarin T, Koszka-Maroń D, Przyłucka M (2017) Zmienność czasowa i zróżnicowanie przestrzenne wielkości i tempa erozji klifu Zatoki Usteckiej w rejonie Orzechowa. Land Anal 34:3–14

Furmańczyk K (1994) Współczesny rozwój strefy brzegowej morza bezpływowego w świetle badań teledetekcyjnych południowych wybrzeży Bałtyku. Wyd Nauk Uniw Szczecińskiego, Szczecin, p 140

Hartnack W (1926) Die Küste Hinterpommerns unter besonderer Berücksichtigung der Morphologie. Jb Geogr Ges Greifswald 43/44 Beih 2:1–324

Karczewski A (1968) Wpływ recesji lobu Odry na powstanie i rozwój sieci dolinnej Pojezierza Myśliborskiego i Niziny Szczecińskiej. Prace Kom Geogr-Geol PTPN 8(3):1–106

Keilhack K (1912) Die Verlandung der Swinepforte. Jb Konig Preuss Geol Land 31:209–244

Keilhack K (1914) Erläuterungen zur Geologischen Karte von Preussen und benachbarten Bundesstaaten. Blatt Misdroy, Lebbin, Caseburg, Swinemünde, Berlin

Kostrzewski A, Zwoliński Z (1986) Operation and morphologic effects of present-day morphogenetics processes modelling the cliffed coast of Wolin Island, N.W. Poland. In: Gardiner V (ed) International geomorphology 1986. Wiley, Chichester, pp 1231–1252

Kostrzewski A, Zwoliński Z (1988) Morphodynamics of the cliffed coast, Wolin Island. Geogr Pol 55:69–81

Kostrzewski A, Zwoliński Z (1994) Bałtyckie wybrzeże klifowe wyspy Wolin – stan aktualny, tendencje rozwoju. Klify 1:81–97

Kostrzewski A, Zwoliński Z (1995) Present-day morphodynamics of the cliff coasts of Wolin Island. J Coast Res SI 22:293–303

Kostrzewski A, Zwoliński Z (1998) Wpływ sztormów na rzeźbę wybrzeża klifowego wyspy Wolin. In: Pękala K (ed) Główne kierunki badań geomorfologicznych w Polsce. Stan aktualny i perspektywy. IV Zjazd Geomorf Pol, Lublin, pp 129–132

Kostrzewski A, Zb Z, Winowski M, Tylkowski J, Samołyk M (2015) Cliff top recession rate and cliff hazards for the sea coast of Wolin Island (Southern Baltic). Baltica 28(2):109–120

Kostrzewski A, Zwoliński Z, Winowski M, Tylkowski J (2017) Zróżnicowanie przestrzenne i zmienność czasowa morfodynamiki wybrzeża klifowego wyspy Wolin w latach 1984–2016. In: Kostrzewski A, Winowski M (eds) Geoekosystem Wybrzeży Morskich 3. Monitoring funkcjonowania i przemian wybrzeży morskich w warunkach zmian klimatu i narastającej antropopresji, Stacja Monit Środ Przyr, pp 133–142

Krygowska L, Krygowski B (1965) Kilka spostrzeżeń dotyczących struktury klifu w Grodnie na Wolinie. Bad Fizjograf Pol Zach 15:167–170

Łabuz TA (2022) Storm surges versus shore erosion: 21 years (2000–2020) of observations on the Świna Gate Sandbar (southern Baltic coast). Quaest Geogr 41(3):5–31

Łabuz TA, Olechnowicz P (2004) Rekonstrukcja rozwoju rzeźby akumulacyjnego brzegu wydmowego na podstawie struktur sedymentacyjnych – studium przypadku z Mierzei Bramy Świny. In: Błaszkiewicz M, Gierszewski P (eds) Rekonstrukcja i prognoza zmian środowiska przyrodniczego w badaniach geograficznych. Pr Geogr IGiPZ PAN 200:237–248

Marsz A (1966) Próba regionalizacji fizycznogeograficznej Wyspy Wolin. Bad Fizjograf Pol Zach 17(A):58–108

Matkowska Z, Ruszała M. and Wdowiak M (1977) Objaśnienia do Szczegółowej Mapy Geologicznej Polski, ark. Świnoujście i Międzyzdroje, 1:50,000. PIG, Warszawa, p 56

Musielak S, Osadczuk K (1995) Evolution of the Świna Gate. J Coast Res SI 22:305–308

Osadczuk K (2004) Geneza i rozwój wałów piaszczystych Bramy Świny w świetle badań morfometrycznych i sedymentologicznych. Wyd Nauk Uniw Szczecińskiego, Szczecin, p 211

PIG-PIB (Polish Geological Institute—National Research Institute) (2022) Portal CBDG—geostanowiska. Online: geoportal.pgi.gov.pl/portal/page/portal/geostanowiska/. Accessed 6 Aug 2022

Prusinkiewicz Z (1971) Naspy przyklifowe – nowy typ gleb morskiego pobrzeza. Zesz Nauk UMK Geogr 26(8):133–157

Prusinkiewicz Z, Noryśkiewicz B (1966) Problem of age of podsols on brown dunes of Bay Bars of River Swina in the light of a palynological analysis and dating by radiocarbon 14C. Zesz Nauk UMK Geogr 5:75–88

Reimann T, Tsukamoto S, Harff J, Osadczuk K, Frechen M (2011) Reconstruction of Holocene coastal foredune progradation using luminescence dating—an example from the Swina barrier (southern Baltic Sea, NW Poland). Geomorphology 132:1–16

Ruszała M, Dobracka E, Piotrowski A (1979) Objaśnienia do Szczegółowej Mapy Geologicznej Polski 1:50,000, ark. Wolin (114) i ark. Międzywodzie (75). PIG, Warszawa, p 89

Schiewe M (2021) Objaśnienia do szczegółowej mapy geologicznej Polski 1:50,000, ark. Międzyzdroje (113). PIG, Warszawa, p 46

Schulte L (1924) Blatt Wollin. Erläuterungen zur geologischen Karte von Preussen Ländern, v. 255(29–8), Berlin

Subotowicz W (1982) Litodynamika brzegów klifowych wybrzeża Polski. Ossolineum, Wrocław, p 154

Tylkowski J, Paluszkiewicz R, Winowski M, Czyryca P, Kostrzewski A, Mazurek M, Rachlewicz R (2023) Effects of geomorphological processes and phytoclimate conditions change on forest vegetation in the Pomeranian Bay coastal zone (Wolin National Park, West Pomerania). Quaest Geogr 42:141–160

Winowski M (2008) Geomorfologiczne skutki tajania pokrywy śniegu na wybrzeżu klifowym wyspy Wolin. Land Anal 9:222–225

Winowski M (2014) Aktywność procesów osuwiskowych na wybrzeżu klifowym wyspy Wolin w warunkach oddziaływania zdarzeń hydrometeorologicznych o wysokim potencjale morfogenetycznym (Zatoka Pomorska—Bałtyk Południowy). Land Anal 28:87–102

Winowski M (2019) Uwarunkowania hydrologiczne abrazji brzegów morskich wyspy Wolin w okresie 2010–2019 (Bałtyk Południowy). In: Michniewicz J (ed) Varia. Prace z zakresu geografii, vol 3. Bogucki Wyd Nauk, Poznań, pp 156–168

Winowski M, Kostrzewski A, Zwoliński Z (2024) Morphodynamic types of Southern Baltic postglacial cliff coasts. Geomorphology (in preparation)

Winowski M, Tylkowski J, Hojan M (2022) Assessment of moraine cliff spatio-temporal erosion on Wolin Island using ALS data analysis. Remote Sens 14:3115

Zwoliński Z (1975) Nowe obserwacje nad krą glacjalną turońskiej kredy marglistej w Wicku (Wolin). Stanowisko: Wicko. In: IX Ogólnopol Zjazd Stud Kół NaukGeogr, Poznań-Świnoujście, pp 66–67

Żynda S (1962) Wyniki wstępnych badań nad moreną czołową wyspy Wolin. Bad Fizjogr Pol Zach 9:159–168

Zbigniew Zwoliński is a geographer, geomorphologist, Professor at the Institute of Geoecology and Geoinformation, Faculty of Geographical and Geological Sciences, Adam Mickiewicz University in Poznań. The most important research issues investigated recently include studies of denudation and environmental changes in different morphoclimatic zones, development of the concept and implementation of geomorphological map of Poland in the scale of 100k and 500k, studies on morphology and sediments in urban areas, development of the concept of geodiversity and methods of its assessment, changes in land cover and land use in selected landscapes of Poland, uncertainty and sensitivity analysis in multi-criteria modelling, creation of geo-questionaries and geo-discussion applications for spatial planning purposes. He is author of over 300 publications, editor-in-chief of *Landform Analysis*, managing editor of *Quaestiones Geographicae*. Member of the Association of Polish Geomorphologists and its past President (2008–2011), Chairman of the Working Groups on Geodiversity and Global Denudation of the International Association of Geomorphologists. Awarded the Denys Brunsden Medal by the International Association of Geomorphologists.

Andrzej Kostrzewski geographer, geomorphologist, Professor at the Institute of Geoecology and Geoinformation, Faculty of Geographical and Geological Sciences, Adam Mickiewicz University in Poznań. Currently, he is the Head of the Center for Integrated Monitoring of the Natural Environment. His main research topics include the study of denudation and environmental changes under climate change and human impact in different morphoclimatic zones, especially within the lowland areas of the Wolin Island and the Pomeranian Lakeland in northern Poland, where he established field stations. Other research issues are the functioning of the natural environment in extreme conditions in desert and polar areas. Author of over 350 publications, editor-in-chief of *Quaestiones Geographicae*. Member of the Association of Polish Geomorphologists, Polish Geographical Society, and scientific committees of the Polish Academy of Sciences. He was Chairman of the Working Group on Small Catchment of the IAG. Past President of the Association of Polish Geomorphologists.

Marcin Winowski is Assistant Professor of Geography at the Institute of Geoecology and Geoinformation, Faculty of Geographical and Geological Sciences, Adam Mickiewicz University in Poznań. His main research interests are focused on cliffed coast morphodynamics, especially in post-glacial landscapes, in relation to climate change. He also investigates water erosion processes in central and northern Poland. In his research, he uses remote sensing techniques, such as laser scanning and drones. He used to be the secretary of the Small Catchment Working Group at the International Association of Geomorphologists (2008–2013), as well as of the Commission of Contemporary Geomorphological Processes within the Association of Polish Geomorphologists (since 2014). He was also a member of the Executive Committee of the Association of Polish Geomorphologists (2014–2021).

Małgorzata Mazurek is Associate Professor of Geography at the Institute of Geoecology and Geoinformation, Faculty of Geographical and Geological Sciences, Adam Mickiewicz University in Poznań. Her main research in the field of hydrogeomorphology, environmental monitoring and geoecology, is focused on the evolution of valley and river network, geodiversity of groundwater outflows, chemical denudation in various morphoclimatic zones, Late-Glacial, and Holocene environmental changes in the light of the geochemistry of sediments of the alkaline spring-fed fens and urban geomorphology. The north-west of Poland, Pomerania and Greater Poland are her main research areas, but she also conducts research in the polar zone, whereas field studies in the equatorial and subtropical zones provide further inspirations. She was the President of the Association of Polish Geomorphologists (2017–2021) and since 2019 is the Vice-President of the Polish Geographical Society.

Sandy Coast Landforms of the Łeba Barrier, Słowiński National Park

41

Joanna Rotnicka and Krzysztof R. Borówka

Abstract

The Słowiński National Park, located on the southern Baltic Sea coast, is known primarily for active dune fields, located in the central part of the sandy Łeba Barrier. The barrier itself constitutes a part of a former barrier-lagoon coast, which began to form about 6000 cal yr BP as a result of sea level changes. From the beginning, its surface was modelled by the wind, which led to the formation of dunes of different types. The most common dunes are incipient and established foredunes, which stretch as sandy ridges parallel to the coastline and are up to 15 m high. Smaller and ephemeral aeolian landforms developed on the beach include shadow dunes and/or nebkhas, and sand patches which may evolve into chains of small barchans up to 1.5 m height. Inland, different transverse dunes occur such as non-active parabolic dune anchored by vegetation and active simple or compound crescentic dunes with smaller crescentic ridges superimposed on their windward slopes and crestal areas. They formed due to unimodal wind regimes dominated by westerly winds and moved or still move eastward. Between the dunes depressions or hollows generated by aeolian deflation occur. Recently, due to the encroachment of vegetation, sand sources have been reduced, the height of migrating dunes gradually decreases and in consequence they transform into large dome dune.

J. Rotnicka (✉)
Institute of Geology, Adam Mickiewicz University in Poznań, Bogusława Krygowskiego 12, 61-680 Poznań, Poland
e-mail: joanna.rotnicka@amu.edu.pl

K. R. Borówka
Institute of Marine and Environmental Sciences, University of Szczecin, Mickiewicza 18, 70-380 Szczecin, Poland
e-mail: ryszard.borowka@usz.edu.pl

Keywords

Active dune fields · Foredune · Aeolian sand transport · Słowiński National Park · Łeba Barrier · Baltic Sea coast

41.1 Introduction

The Słowiński National Park was established in 1967 to protect one of the most attractive stretches of the southern Baltic Sea coast, and above all to protect one of the largest active dune fields in Europe. The area of the park is 327.44 km^2, of which 34.12% are the coastal waters. The park encompasses a barrier coast situated between the village of Rowy and the town of Łeba. The sandy Łeba Barrier, approximately 33 km long, extends between the mouth of Łupawa River in the west and the mouth of Łeba River in the east (Rotnicki 2008). It separates the Łebsko and the Gardno lakes (the third and eighth largest lake in Poland, respectively) and the adjacent wetland plains from the Baltic Sea (Fig. 41.1a). The barrier is known for the presence of a system of large dunes which were active in the past and in part still are. The seaward side features beaches of a varying width, which adjoin foredune ridges. The upper part of the beach and the foredune ridges are colonized by psammophilous vegetation, mostly marram grass (*Ammophila arenaria).* Further inland, large, mostly inactive dunes occur, as well as deflation hollows between them, locally waterlogged. These areas are usually occupied by the coastal pine forest (*Empetro nigri-pinetum*) and marshy pine forest (*Vaccinio uliginosi-Pinetum sylvestris*) associations. The largest active dunes are completely vegetation-free, whereas less active forms are vegetated by psammophytes only.

Today, there are only two fields of active dunes on the Łeba Barrier—Czołpino and Łeba dune fields, both located in the central part of the barrier (Fig. 41.1b, c). They are

P. Migoń and K. Jancewicz (eds.), *Landscapes and Landforms of Poland*,
World Geomorphological Landscapes, https://doi.org/10.1007/978-3-031-45762-3_41

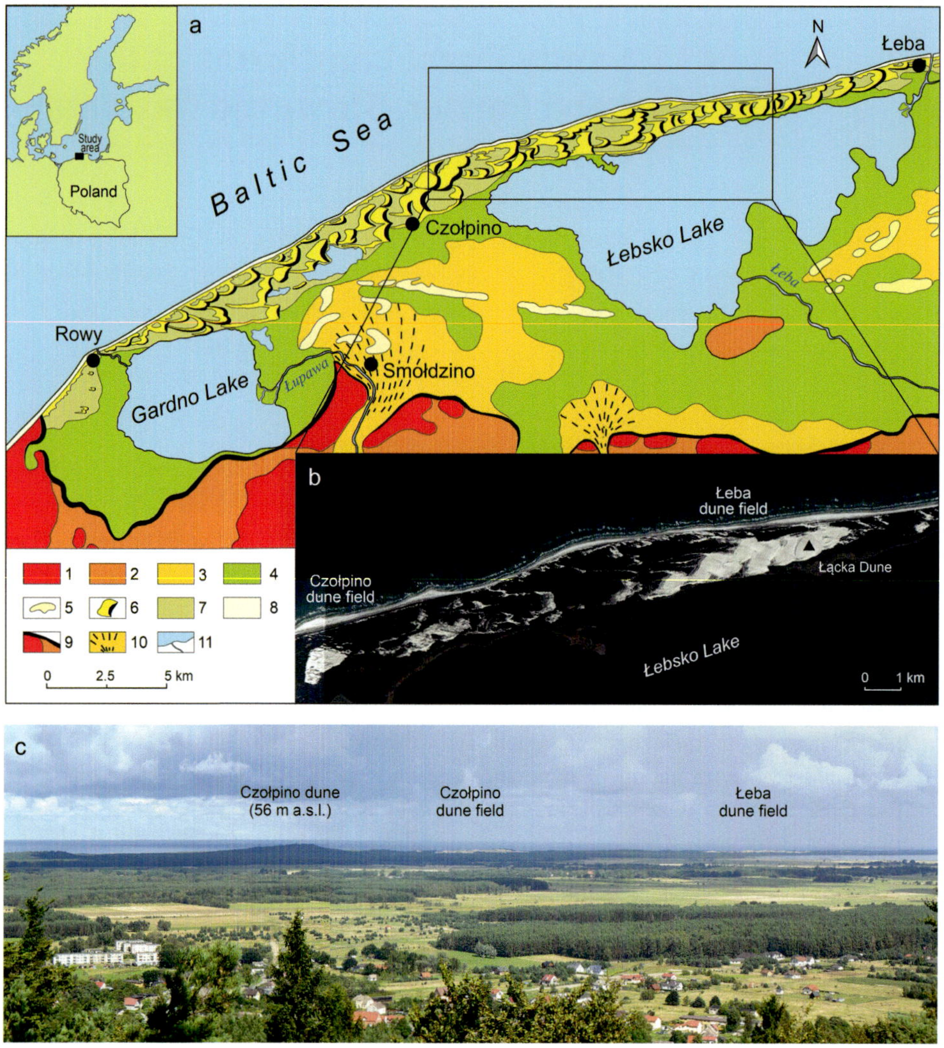

Fig. 41.1 Słowiński National Park and its surroundings. **a** Geomorphological map (modified from Miszalski 1973; Rotnicki 1995): 1—morainic plateau (Pomeranian Phase), 2—end-moraine (Gardno Phase), 3—sandy plain (Late Vistulian), 4—peat plain (Late Glacial and Holocene), 5—inland dunes (Late Glacial and Holocene), 6—coastal dunes (Holocene), 7—barrier sandy plain (Holocene), 8—beach, 9—erosional scarp, 10—alluvial fan, 11—lakes and rivers. **b** Active dunes on the Łeba Barrier (source: Google Earth). **c** Landscape of the Łebsko-Gardno lowland, with dunes on the Łeba Barrier in the distance (photograph by J. Rotnicka, taken from Rowokół hill)

dominated by active transverse crescentic dunes with episodically formed smaller crescentic ridges superimposed on their windward slopes and crestal areas. In such cases, according to the dune classification by Lancaster (1992), they are termed compound dune or draa (Borówka 1980b, 2001) (Fig. 41.1b). Outside the active dune fields parabolic dunes are most common. They consist of two trailing arms pointing upwind and anchored by vegetation, a deflation basin between the arms, and a depositional lobe at the downwind end. In the Łeba Barrier, there are also two types of dune ridges. Most numerous are foredunes extending along the northern limit of the barrier and occasionally forming systems of parallel ridges separated by elongated swales. The other type of elongated dune ridges is remnants of arms of parabolic dunes (Miszalski 1973).

41.2 Setting

41.2.1 Formation of the Łeba Barrier

The Łeba Barrier began to form when, following the main Littorina Sea transgression stage, the Baltic Sea level was low. According to Rotnicki (2009), the low sea level prevailed within 6100–3900 cal yr BP. It is to that period that the oldest peat layers and fossils soils overlaying the beach-dune deposits in the central part of the barrier have been dated (Borówka 1990). Initially, this depositional system was located a few kilometres north of the present shore and moved southward as the sea level was rising. It shifted to its present position between 4500 and 3000 cal yr BP (Rotnicki et al. 2009), but a tendency for slow erosion

Fig. 41.2 Recent changes of the Łeba Barrier coastline. **a** Coastal erosion and accretion between 1958 and 1984 (modified from Borówka and Rotnicki 1995). **b** Radiocarbon age of tree trunks and peat outcrops on the Łeba Barrier beach (modified from Borówka and Rotnicki 1995). **c, d** Peat outcrops and subfossil tree trunks on the Łeba Barrier beach (photographs by J. Rotnicka)

of the northern shore of the barrier (Fig. 41.2a) and southward expansion as the adjacent coastal lakes are being overgrown and infilled by dune sand (Borówka and Rotnicki 1995) are still visible. The southward migration of the barrier has left its trace in, *inter alia*, peat outcrops at the contemporary beaches along with subfossil tree trunks, frequently preserved in their life position (Fig. 41.2c, d). The peats and tree trunks have been dated to approximately 4730–1060 yr PB (Fig. 41.2b), that is 5500–1000 cal yr BP (Borówka and Rotnicki 1995; Krąpiec and Florek 2005).

The landscapes of the Łeba Barrier are also characterized by two large lakes bordering it to the south (Fig. 41.1a). Despite similar coastal position, they are of different origins. While the Łebsko Lake and smaller lakes are former lagoons, the Gardno Lake was originally a freshwater lake formed within a morainic upland, which, along with the rising sea level, was fed several times with salt water (Lutyńska 2008).

41.2.2 Climate Characteristics

The Łeba Barrier and its surroundings are influenced by maritime climate. Based on 20-year data set (2000–2019) from Łeba meteorological station, the highest mean air temperature was recorded in June and July and the lowest—in January (Fig. 41.3a). The seasonal mean temperatures equalled to 10.0 °C in spring, 16.9 °C in summer, 6.9 °C in autumn, and 0.8 °C in winter. The annual mean precipitation amounted to 684 mm, the highest values being recorded in July and August and the lowest ones in April (Fig. 41.3b). The seasonal mean precipitation was 115 mm in spring, 251 mm in summer, 196 mm in autumn, and 122 mm in winter. The mean monthly air humidity varies throughout the year from 70 to 90%.

The Łeba Barrier area is affected by frequent strong winds, usually from the western sector (Fig. 41.3c). Only in spring, there is a fairly large share of easterly winds, whose speed, however, rarely exceeds 8 m/s. The highest wind speeds, up to 30 m/s and more, are typically recorded in the autumn–winter season. South-westerly winds prevail, but it is also the period of rare but strong winds from the northern sector, which generate high storm surges during which water level rises up to 1.6 m above the mean sea level (Wiśniewski and Wolski 2009). These storms are responsible for the most significant coastal erosion.

According to the Fryberger and Deans' method (1979), which is widely used to assess wind energy, potential sand transport and types of dune formed under given wind conditions, the Łeba Barrier coast is characterized by the

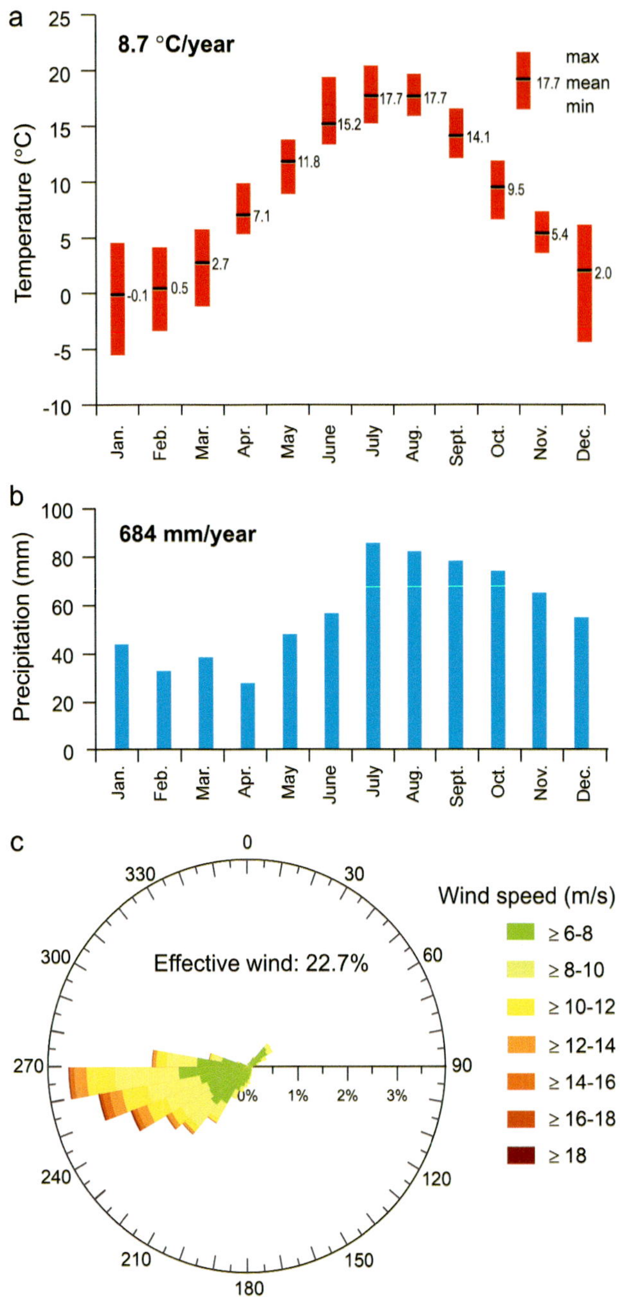

Fig. 41.3 Climate data for the Łeba Barrier. **a** Mean monthly temperature. **b** Precipitation. **c** Distribution of effective winds ($v \geq 6$ m/s). Based on data from 2000 to 2019, Łeba Meteorological Station, Institute of Meteorology and Water Management

annual sand drift potential (DP) of 444 VU (vector unit) and resultant drift potential (RDP) of 385 VU. The resultant drift direction (RDD) equals to 74°. The wind directional variability (RDP/DP) of 0.87 indicates that the wind regime is wide unidirectional. According to Fryberger and Deans (1979) classification of wind energy based on DP and RDP/DP values, the coast belongs to the group of high-energy wind environments.

41.3 Beaches

The beach of the Łeba Barrier is tideless, wave-dominated, low-gradient, and tens of metres wide, locally very wide up to 100 m or narrow, less than 20 m. The position of the shoreline, and thus the width and height of the beach, is determined by the sediment budget in the littoral zone. If it is negative, the beach is eroded, and if positive, it builds up (Psuty 1988; Furmańczyk 1994). On the tideless coast of the Baltic Sea, this budget results from the interaction between wind, waves, and wind-wave-induced nearshore circulation (longshore and rip currents). Sediment transport along the coast generated by the predominant westerly/south-westerly winds, which cause alongshore, eastward-dominated, sediment flux, is the main component of the sediment budget (Rotnicka 2013a; Zhang et al. 2013).

On a significant part of the shoreface of the Łeba Barrier, there are usually two or three sandbars parallel to the shore (Fig. 41.4a). They play an important role in the evolution of the beach as during storm periods they promote breaking of incident waves further away from the shoreline and thus they reduce the wave energy and forces otherwise exerted directly on the shore. During calm weather and mild waves, the sandbars migrate towards the beach and eventually the shallowest of them welds to the beach face (Fig. 41.4a). This is one of the important processes responsible for beach widening (Greenwood and Davidson-Arnott 1979).

41.3.1 Aeolian Transport on Beach

Beach morphology and beach surface are significantly modified by aeolian processes. However, depending on the year, effective winds, i.e. winds generating sand transport ($v \geq 6$ m/s), constitute between 15 and 34% of all winds (Rotnicka 2013a; Rotnicka and Dłużewski 2019), but the time when aeolian transport occurs on the beach is still reduced due to wave run-up and sometimes snow cover, both factors responsible for limiting the sand source.

In autumn and winter, which are the season of strongest winds, the sand is moist for most of the time and temporarily even frozen. Sources of moisture include rainfall, wave run-up and capillary rise. Therefore, sand moisture is the factor that has the greatest impact on the intensity and effects of aeolian processes on the beach. However, field measurements of sand transport rate, i.e. the mass of sand transported along a stretch of a defined width in unit time, showed that moisture in the range of 1–5% (in some cases up to 10%) does not suppress aeolian transport (Rotnicka 2011, 2013b). Moreover, during alongshore winds, when the fetch is unlimited and the conditions of saturated flux

Fig. 41.4 Main types of aeolian landforms on the Łeba Barrier coast. **a** Incipient foredune and the first sandbar welding to the beach, with the second and third sandbar visible underwater. **b** Two generations of foredunes (1—incipient, 2—established foredunes). **c** Well-developed nebkhas and shadow dunes on the beach (photographs by J. Rotnicka)

(relatively constant sand concentration in the wind boundary layer) are achieved, the sand transport rate over the moist beach surface is 20–45% higher than above the dry surface (Rotnicka 2013b) (Fig. 41.5). In some sections of the Łeba Barrier coast, beach surface is covered with pebbles which form lags with a density of up to 70%. In such settings, the thickness of the layer in which the sand grains saltate exceeds 1 m and the sand transport intensity is equal or even slightly greater than above bare sand surface (Rotnicka and Dłużewski 2022).

41.3.2 Beach Aeolian Mesoforms

The most common forms occurring on moist sandy surfaces are regularly spaced sand patches (every 15–17 m; Rotnicka 2013a) (Fig. 41.6). They form during alongshore and oblique onshore winds and following the diagram of sand patch evolution by Kocurek et al. (1992), all stages of their development can be observed on the Łeba Barrier beach. In the first stage, they are a few centimetres thick (Fig. 41.6a). They grow fast within hours and in the second stage they show asymmetry, in which the stoss slope is steeper than the lee slope (Fig. 41.6b). As the thickness of the sand patch increases, this asymmetry reverses to the one typical for transverse dunes. Under favourable conditions, meaning a more abundant source of dry sand, they evolve into small barchans or crescentic ridges, 1–1.5 m high (Fig. 41.6c). They occur in chains, with crest-to-crest wavelengths equal to 14–17 m (Tomczak 2021). While sand patches in various stages of development are often found on sandy beaches, well-developed barchans are unique bedforms specific for the Łeba Barrier. Their formation in this area is favoured by low beach sand moisture related to the lack of tides in the Baltic Sea and domination of alongshore wind. They develop mostly in spring and summer, when the thickness of surface layer of dry sand is usually greatest. Their lifetime is in the order of days as they are prone to erosion due to change of active wind direction, flooding by waves during onshore winds, or increase of sand moisture during rainfall.

41.4 Foredunes

The beach is backed by WSW–ENE oriented foredune ridges. On the Łeba Barrier coasts, where the beach is broad and thus, the source of sand is abundant, successive

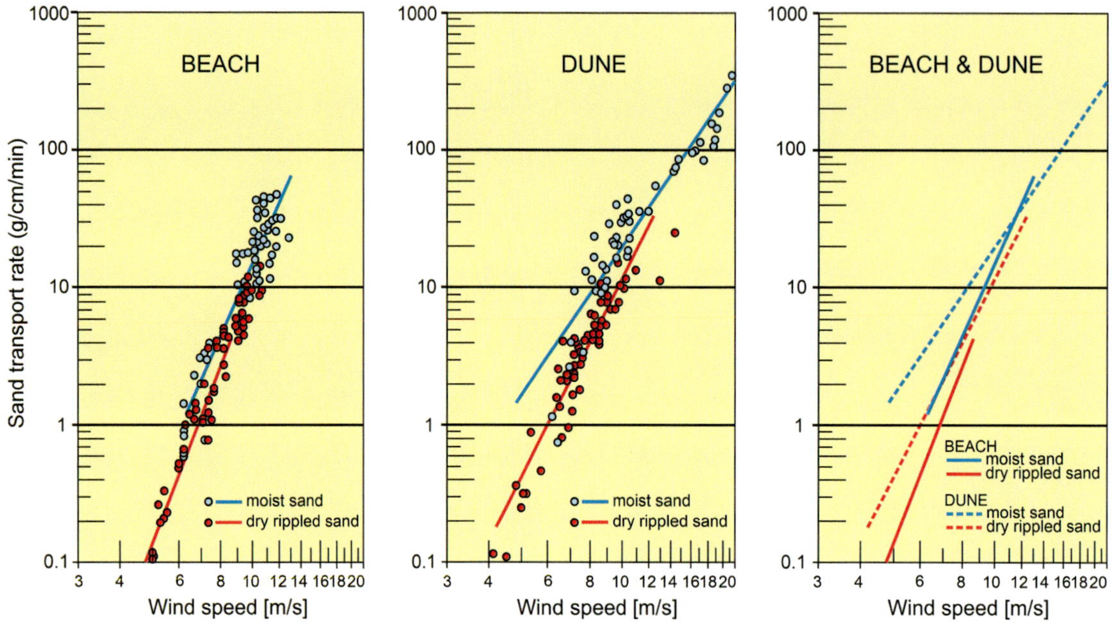

Fig. 41.5 The relationship between wind speed and sand transport rate on moist and dry sand surfaces on the beach and windward slope of transverse dunes of the Łeba dune field (modified from Borówka 1980b, Rotnicka 2013a)

foredune ridges, parallel to the coastline develop and up to 4–5 generations can be recognized (Fig. 41.4b). If beaches are narrow, the foredunes decay and/or are destroyed.

41.4.1 Foredune Development

Foredunes generally develop in the upper part of a wide beach under prevailing onshore winds, when sand transported landwards is trapped by vegetation. As the coastline orientation of the Łeba Barrier changes from 50°–230° in its western part to 90°–270° in the eastern part (Fig. 41.1a), the prevailing winds contributing to foredune development are both alongshore and oblique onshore winds (compare Fig. 41.3c) (Rotnicka 2013a). The process of foredune development proceeds in two stages. In the first stage, shadow dunes and/or nebkhas form on the uppermost backshore vegetated by sparse marram grass (mainly *Amophila arenaria*) (Fig. 41.4c). Over time, they laterally coalesce and grow in height, and eventually evolve into a continuous dune ridge forming a new foredune (incipient foredune, Hesp 1983) (Fig. 41.4b). Its further growth results in trapping of most of the sand supplied from the beach, and this cuts off the supply of sand to the older foredune located landwards, which therefore becomes stabilized by vegetation (established foredune, Hesp 1983) (Fig. 41.4b). Thus, the height and size of an established foredune depend on how long it acted as an incipient foredune. Recent incipient foredune ridges on the Łeba Barrier achieve the height of 12 m, whereas the established ones are up to 15 m. The width of the ridges varies between 30 and 50 m.

The size of incipient foredunes, their shapes, and rates of development depend on many factors. On the Łeba Barrier coast, the most important are wind regime and the availability of sand, which in turn depends on beach width and height. In the case of onshore winds, which are responsible for the supply of sand to the incipient foredune, the amount of sand actually delivered to the dune depends on the fetch distance, i.e. a downwind distance along which the aeolian transport intensity increases and eventually, at some critical point, achieves maximum/full saturation under given wind conditions. However, on the beach, the available fetch distance is imposed by the width of the beach and the angle of wind approach to the coast. In the case of a narrow beach, it is shorter than the critical fetch, and therefore the amount of sand transported towards the foredune is smaller than in the case of a wide beach and long fetch distance (Davidson-Arnott and Dawson 2001). Therefore, under given wind regime, changes of coastline orientation may result in different growth rates of adjacent incipient foredunes.

On the Łeba Barrier, the most frequent effective winds are from the direction of 240°–270°. Beach profiles of the Łeba Barrier, investigated in two places of similar beach width but differing in the orientation of the shoreline by 20°, clearly show the difference in the growth rate of incipient foredunes between parts of the coast dominated by alongshore versus onshore winds (Dłużewski et al. 2023) (Fig. 41.7).

Fig. 41.6 Mesoscale aeolian landforms developing on a beach. **a** Beginning of sand patch formation on moist sandy surfaces. **b** The second stage of formation of sand patches with steeper stoss- and gentle lee-slopes. **c** Small barchans regularly spaced every 13–17 m. The arrow shows wind direction (photographs by J. Rotnicka)

Due to topographic forcing and steering of the airflow in the beach-foredune system, the foredune shape itself also impacts the amount of sand delivered from the beach to the dune ridge (e.g. Arens et al. 1995; Walker et al. 2006). The higher the foredune and the steeper its seaward slope, the more pronounced is this effect. On the Łeba Barrier, if the foredune is high enough, oblique onshore winds, whose angle of approach is up to 25° in relation to the coastline, are steered to alongshore winds (Tomczak 2021) and they do not feed the foredunes with sand but cause alongshore sand drift. Thus, the differences in the foredune growth rates in two parts of the coast described above are further enhanced, as much greater percentage of onshore winds in the eastern part of the Łeba Barrier is steered to alongshore than in the western part.

The foredune ridge also causes that the velocity of onshore winds declines towards the foredunes, reaching the lowest values at the foot and in the lower part of the seaward slope of the foredune, which results in reduced intensity of sand transport, sand deposition, and the development of an aeolian ramp (Fig. 41.8). The more extensive and higher it is, the easier is the transfer of sand from the beach to the foredunes (Christiansen and Davidson-Arnott 2004). If the aeolian ramp is anchored high on the seaward slope of the foredune

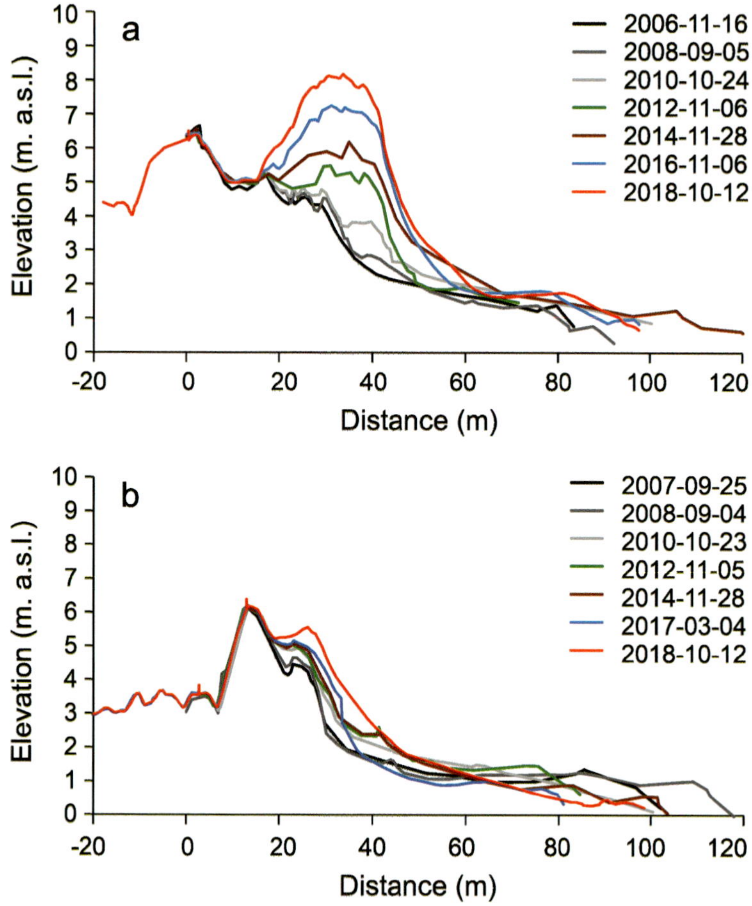

Fig. 41.7 Differences in growth rates of incipient foredunes in two sections of the Łeba Barrier differing in the coastline orientation by 20°. **a** Significant foredune growth under effective wind regime dominated by oblique onshore and alongshore winds, coastline orientation of 50°–230° (at the 205.5 km of the Polish coastline). **b** Slight foredune growth under effective wind regime dominated by alongshore winds, which feed the foredune with less sand, coastline orientation of 70°–250° (at the 201.2 km of the Polish coastline) (after Dłużewski et al. 2023)

(Fig. 41.8a), it reduces the original inclination of this slope from as much as 25° (and more) to less than 14° (Tomczak 2021). Such a slope deflects the near-surface air flow to alongshore to a much lesser extent and additionally favours building up of the aeolian ramp—usually by formation of small transverse dunes with crests at an acute angle to the axis of a foredune ridge, which migrate up the aeolian ramp (Fig. 41.8b, c). The aeolian ramp also develops at cliffed establish foredunes, leading to dune reactivation (Fig. 41.8d).

41.4.2 Decay of the Foredune

The beach width fluctuates along the Łeba Barrier and in places where it becomes narrow, the foredunes are no longer supplied with sand and during storms they are destroyed by high-energy waves. Recently, in some part of the barrier, the coastline and the foot of the foredune have retreated by almost 100 m during less than 10 years (Fig. 41.9). The most devastating storms are generated by wind from the northern sector, during which the sea level rises most significantly. These storms result in foredune scarping and formation of dune cliffs as well as overwashing and foredune breaching (Fig. 41.9c).

Along the Łeba Barrier, there are zones marked by nearly continuous development of foredunes, as well as zones where the beach and foredunes suffer intensive destruction during storms and neither the beach nor the foredunes are quickly rebuilt during inter-storm periods. Zones of erosion and foredune development alternate along the coast and are regularly spaced every 1.2–1.8 km (Rotnicka 2013a; Michałowska and Głowienka 2022). Some of these zones move eastwards, consistently with the resultant direction of littoral sediment flux.

The other process responsible for the destruction of the foredunes is deflation and formation of blowouts, which are a very common feature of this part of the Polish Baltic coast. They develop within established foredunes, beginning from the foredune crest and extending towards the landward slope. Newly formed blowouts are strongly elongated (through blowout according to Hesp and Hyde 1996), with a long axis at acute angle to the axis of foredune ridge and parallel to the dominant wind direction (Fig. 41.10). They reach the depth up to 6 m and length up to 65 m. In places of active cliffs, whose edges move landwards and reach the blowouts, the latter open to the sea and wind flow approaching from both seaward and alongshore directions is directed into the blowout. Further, airflow steering and

Fig. 41.8 Aeolian ramps. **a** Well-developed ramp, highly anchored on the seaward slope of the incipient foredune. **b, c** Build-up of the ramp by small transverse dune formed by oblique onshore winds. **d** Ramp developed at the foot of a cliffed, establish foredune (photographs by J. Rotnicka)

forcing occur within the blowouts. It results in intensive landward transport of sand and formation of depositional lobes migrating landward.

41.5 Stabilized and Active Dune Fields

Between the mouths of the Łeba and Łupawa rivers, Miszalski (1973) identified 78 large dunes with relative altitudes usually higher than several metres, the maximum height (the Czołpino Dune) being about 56 m asl. These forms are represented by transverse dunes, including parabolic dune anchored by vegetation and simple or compound crescentic dunes. They are formed in wind regime characterized by a narrow range of wind directions (western sector in the case of Łeba Barrier) and their crests are oriented transverse to the resultant direction of sand transport. The dunes are separated by depressions or hollows generated by deflation. Landforms arising from deflation are also present on the surface of partly established compound crescentic dunes. The largest deflation hollows can be waterlogged or they even feature ephemeral lakes appearing during wet periods and particularly during heavy storm surges which raise the groundwater level on the barrier (Borówka 1990).

The most active field of migrating dunes (Łeba dune field) occurs in the central part of the barrier (Fig. 41.1b, c). At present, the field has an area about 3 km^2, extending over about 4.5 km, and consists of eight large transverse crescentic dunes. They can be termed compound crescentic dunes or draa (Borówka 2001) when smaller crescentic ridges develop on their windward slopes and crestal areas. Their absolute altitudes vary, peaking at about 27–32 m asl, and they are about 250–650 m long (Sławik et al. 2014; Ludwig et al. 2017) (Fig. 41.11a, b). These dunes move eastward, in the resultant drift direction, at a mean speed of about 10–12 m/yr (Ptak and Rudowski 2014; Ludwig et al. 2017). Their windward slopes change during the year, the changes being induced, in addition to the wind regime, by varying sand moisture content. Small crescentic ridges develop mostly during dry seasons at distances necessary for a full saturation of the airflow. They are usually 0.3–2.5 m high (relative altitude) and 10–60 m long (Borówka 2001). The smallest forms are ripples, which are the most common aeolian microforms emerging on all wind-exposed sandy surfaces. The length and height of ripples depend primarily on the average grain size of wind-transported sand (Wilson 1972; Borówka 1980b). As the mean grain size increases, the relative height and length of ripples increase too, from 0.1 to 1.6 mm and 1.5 to 18 cm, respectively (Borówka 2001). They emerge exclusively when sand surface is dry and rapidly disappear during wet weather. When sand surface is moist, sand patches form on the windward slopes of transverse crescentic dunes (Fig. 41.11c).

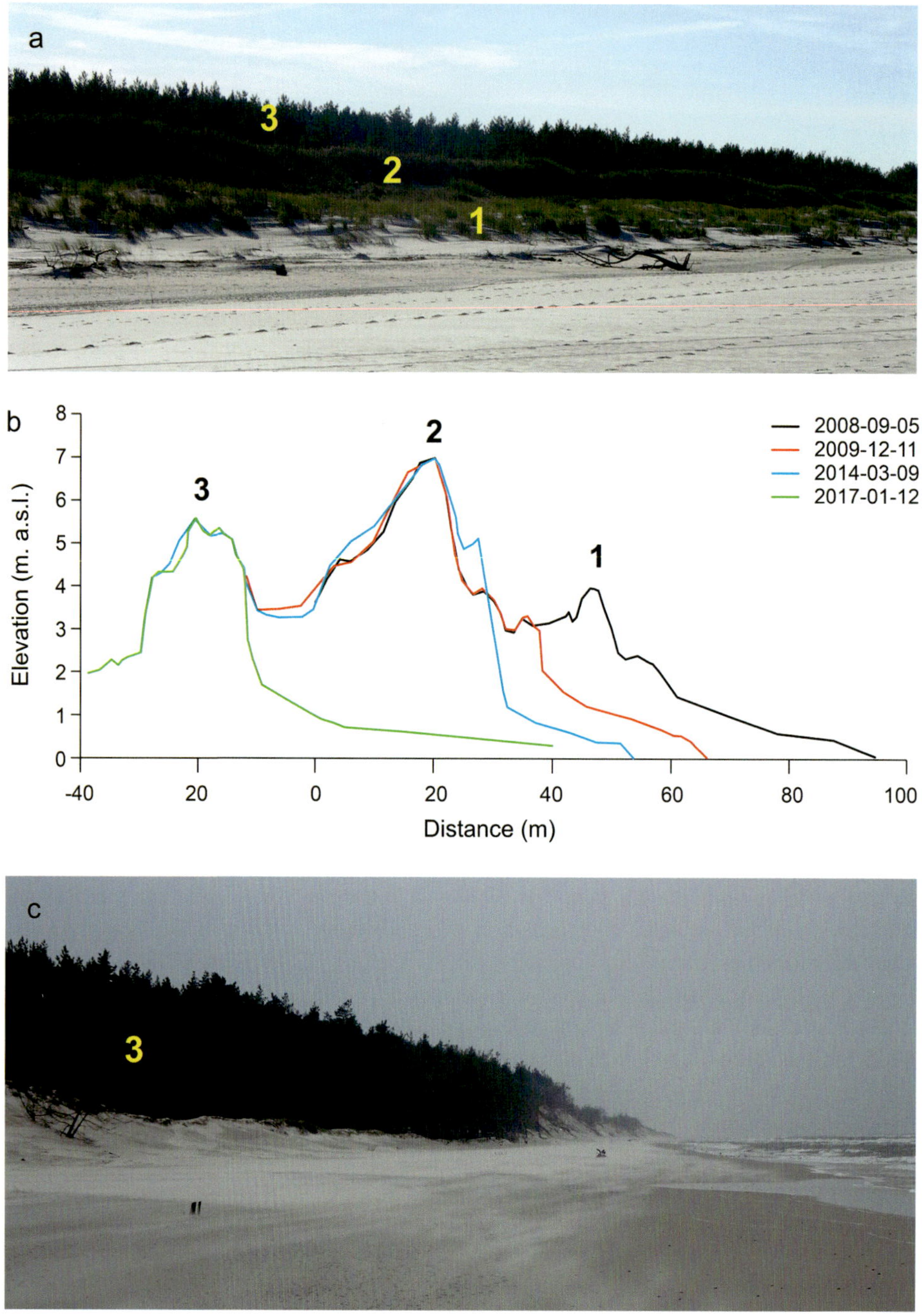

Fig. 41.9 Coastline retreat and foredune destruction by severe storms, the middle part of the Łeba Barrier (at the 203.5 km of the Polish coastline). **a** Photo showing three foredune generations that existed in 2008. **b** Coastal retreat due to storms that took place in October 2009, December 2013, and January 2017 (maximum sea-level rise recorded was higher by 1.1 m, 1.11 m and 1.43 m, respectively, than the mean sea level). **c** Photo showing the lack of two youngest foredunes and scarping of the oldest one (February 2017) (photographs by J. Rotnicka)

The non-vegetated active dunes are areas of the most intensive aeolian sand transport. As on the beach, at the same wind speeds, the thickness of saltation layer is smaller above dry than above moist sand surfaces. The cloud of grains saltating above a dry surface with ripples shows: (1) a high concentration (40–45%) of grains in the near-surface layer (up to 1 cm), and (2) a relatively small thickness of sand flux, about 13 cm at the wind speed of 8 m/s. On the other hand, if grains saltate on a smooth moist surface, as little as 30% of the total mass is transported in the near-surface layer, and the sand flux is thicker, about 20 cm at an 8 m/s wind speed (Borówka 1980b). It also results in higher sand transport rate over the moist windward slopes of transverse crescentic dune than above dry surfaces (Fig. 41.5). Thus, during autumn and winter, when the sand is moist for most of the time, the amount of sand transported by winds (particularly westerly winds) is higher than during spring and summer. In addition, important for the aeolian transport intensity is the kinematic viscosity and density of the air, both factors being temperature dependent. Hence, the aeolian transport intensity is somewhat higher at lower air temperatures (autumn–winter), as observed in the active dune field (Borówka 1980a, b).

Fig. 41.10 Blowouts developed in established foredunes. **a** Trough blowouts with long axis nearly parallel to the coastline. **b** Retreating cliffs often reach the blowouts, making them open to the sea (photographs by J. Rotnicka)

Measurements of aeolian sand transport intensity along the longitudinal profile of a windward side of the transverse crescentic dune showed the near-surface airflow to become fully saturated with sand faster if the sand surface is dry. As a consequence, high air temperature and low sand moisture enhance sand accumulation on the upper part of the windward side. Both the height of the dune surface between the crest and the brink and the length of the avalanche slope increase at the expense of the brink migration to the east (Fig. 41.12a). As already mentioned, faster saturation of near-surface airflow is also conducive to the development of small crescentic ridges on the windward side of dunes. Such conditions prevail primarily in spring and summer. On the other hand, in autumn and winter, when the dune sand is usually moist, the airflow does not attain full saturation, deflation being the dominant process. In the upper part of the windward side, it leads to levelling off of their longitudinal profile. The length of the avalanche slope decreases which leads to faster brink migration (Borówka 1980b) (Fig. 41.12b).

The annual changes of longitudinal profiles of transverse crescentic dunes, associated with variations in sand accumulation and alterations of deflation areas, are reflected in the internal structure of the dunes (Borówka 1980a). The uppermost part of the dune longitudinal cross section, above the average height of their autumn–winter profile, shows the presence of a system of sedimentary structures characterized by a very high lateral and vertical variability in lamination types, as well as in directions and angle of dip of the laminae (Fig. 41.13). This cross-stratification was produced by migration of small crescentic ridges on the dune's windward slope, rapidly responding to short-term changes in wind direction. In the middle part of the dune cross section, there is only one cross-strata set whose thickness more or less coincides with the average height of the autumn–winter avalanche slope. The laminae in this set dip generally eastward. The dune bed and the deflation depressions between the dunes usually feature a few cross-lamination sets, several tens of centimetres thick. Individual layers are separated by flat or slightly undulating erosional

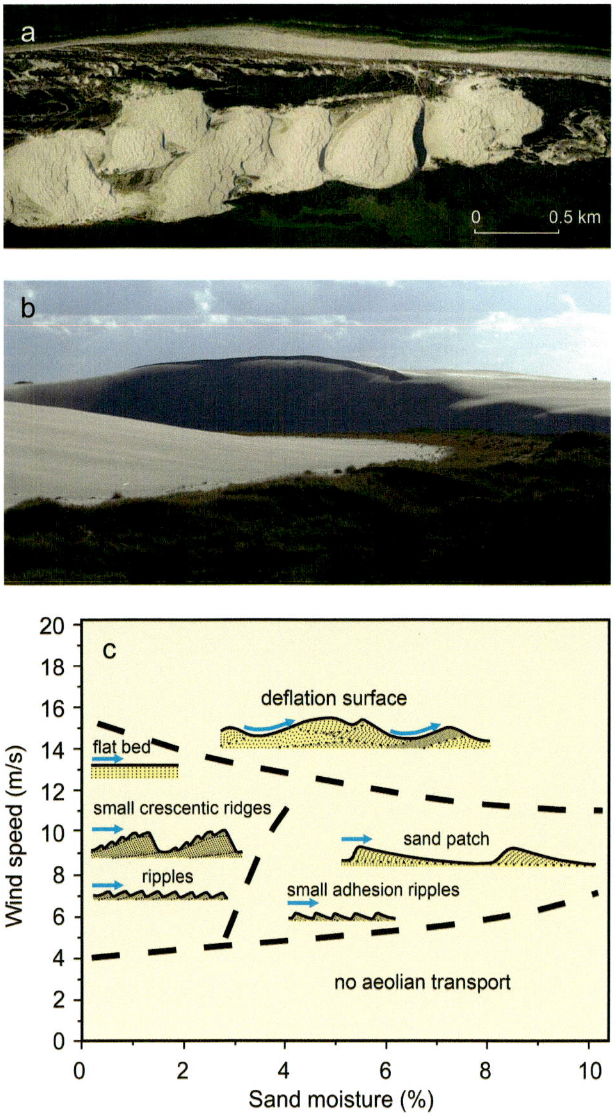

Fig. 41.11 Aeolian geomorphology of the Łeba dune field. **a** Transverse crescentic compound dunes (draa) with small crescentic ridges, 0.3–2.5 m high and 10–60 m long, developed on their windward sides (source: samper.pl/prezydencja/galeria). **b** Example of a high singular crescentic dune with distinct slip face (photograph by J. Rotnicka). **c** Ephemeral micro- and meso-scale aeolian bedforms developed on the windward side in relation to differences in wind speed and sand moisture (modified from Borówka 1990)

bounding surfaces (Fig. 41.13). These surfaces emerged as a result of deflation on the lowermost part of the windward side of the migrating dunes. With time, however, the vertical extent of deflation changes, being induced by various factors, particularly the long-term variations in the Łeba Barrier groundwater level affected by the changing sea level (Borówka 1990). Such a model of the transverse crescentic dunes structure on the Łeba Barrier has been recently confirmed by georadar surveys (Ludwig et al. 2017).

Much smaller (0.5 km^2, about 1 km long) field of migrating dunes (Czołpino dune field) occurs in the western part of the barrier (Fig. 41.1b, c). The maximum height of the dunes, which are nowadays partly covered with vegetation, reaches 37 m asl. The area is dominated by: (i) transverse active crescentic dunes, (ii) deflation hollows where fossil soil horizons and tree trunks are excavated, and (iii) small hills which are remnants of older dune eroded by wind and stabilized now by vegetation. The active dunes move eastward at a speed of 4 m/yr on average (Michałowska 2007). Due to the spread of vegetation, the sand source and aeolian processes in this area have become limited, which results in the decrease of height of active dunes. In the eastern limit of the dune field, the height of the slip face of the highest active dune decreased from 40 m in 1964 to 27 m in 2004 (Michałowska 2007), and in 2020 its central part nearly disappeared.

41.6 Phases of Intensification of Aeolian Activity on Łeba Barrier in the Holocene

The active dune fields, as well as many other sites on the Łeba Barrier, feature outcrops of fossil soils (Fig. 41.14a, b) (Tobolski 1975; Borówka 1975, 1990; Borówka and Tobolski 1979). Analysis of lithostratigraphic position and geomorphology of these soils and phytogenic deposits, in association with their radiocarbon ages, allowed one to distinguish four distinct stages of intensification of aeolian activity on the Łeba Barrier (Borówka 1990).

The oldest aeolian series is known from the environs of the village of Czołpino and from several other sites in the central part of the Łeba Barrier. At Czołpino, the series is overlain by a fossil soil containing numerous pieces of charcoal dated to 3340 ± 130 yr BP (3900–3270 cal yr BP). This deposit revealed the presence of numerous fragments of pottery and a well-preserved vessel from the Bronze Age V (1000–700 yr BC). The structure of sands making up the oldest aeolian series does not let itself to unequivocal identification of type of dunes developing at that time (Borówka 1990). It seems, however, that they were low foredune ridges or ridge-like parabolic dune arm remnants (Fig. 41.14c).

The second stage was identified at numerous sites (Tobolski 1975, 1980; Borówka 1990). Radiocarbon dating made it possible to determine the duration of the stage at 3300–2500 cal yr BP. It cannot be ruled out, however, that the stage was shorter, particularly when the age of peat layers with aeolian intercalations is taken into account (Fig. 41.14c). On the other hand, there are areas where sands attributed to this aeolian series are underlain by a fossil soil archaeologically dated at a period older than

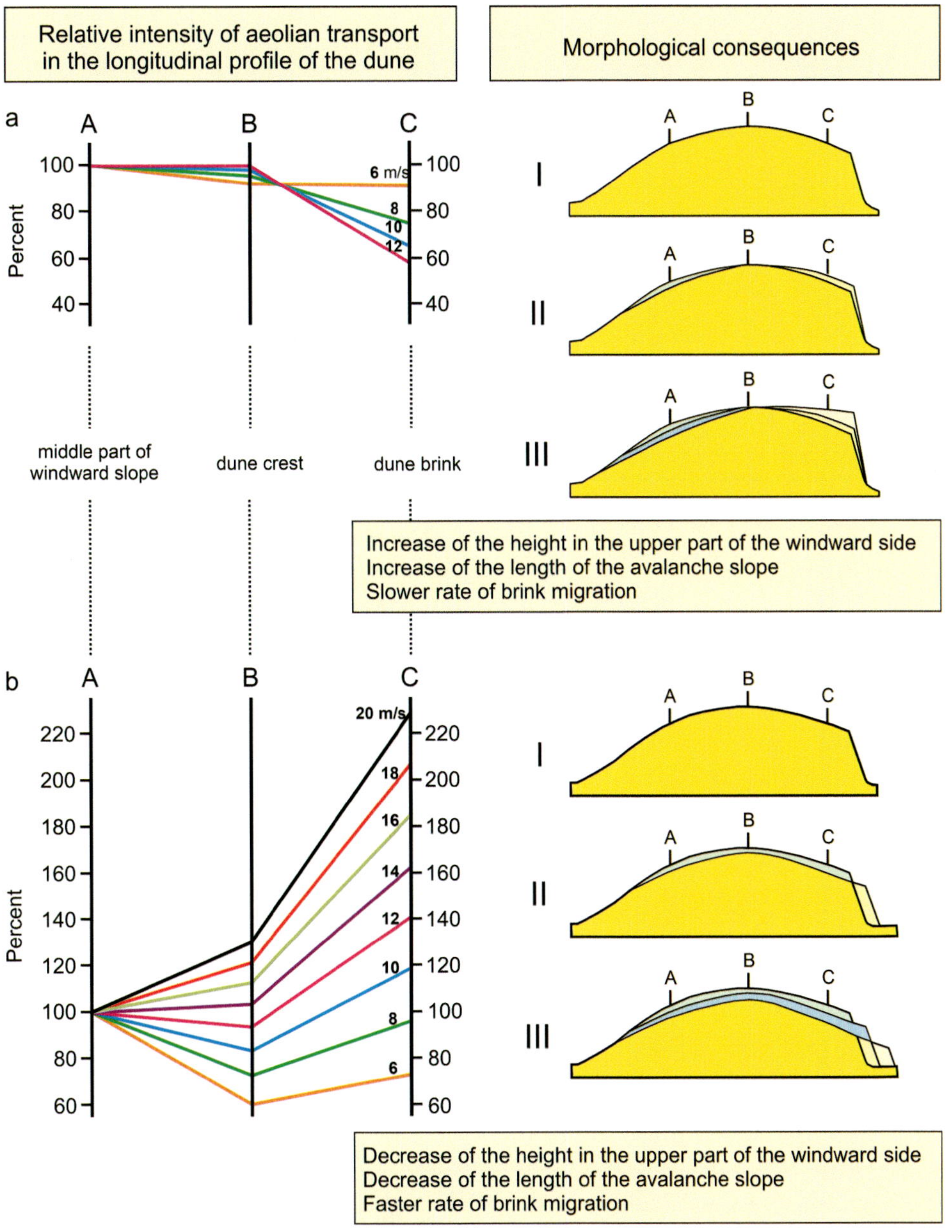

Fig. 41.12 Changes in dune morphology due to changes in the intensity of aeolian sand transport in particular zones of the windward side (A, B, C) at various wind speed and in condition of dry rippled (**a**) and moist surfaces (**b**) (modified from Borówka 1980b, 1990)

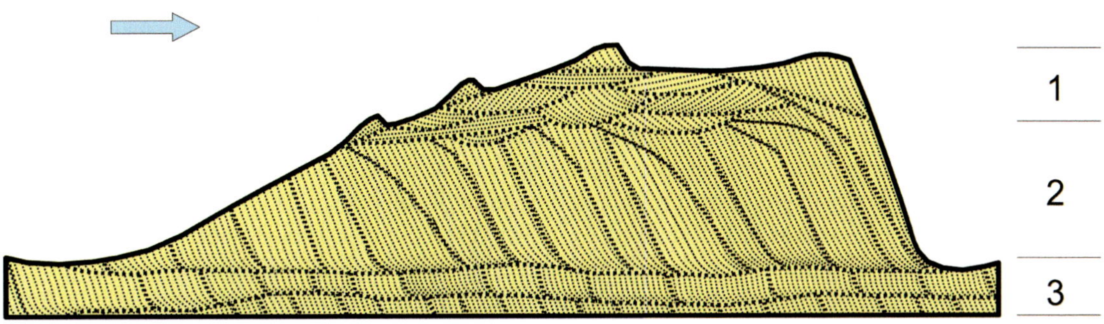

Fig. 41.13 Internal structure of the draa-type dune located in the Łeba dune fields (modified from Borówka 1980b, 1990). 1—coset of cross-stratification produced by migration of small crescentic ridges, 2—set of cross-strata produced by migration of transverse crescentic dunes, 3—coset of cross-strata separated by flat or slightly undulating erosional bounding surfaces present in the dune bed and deflation depressions

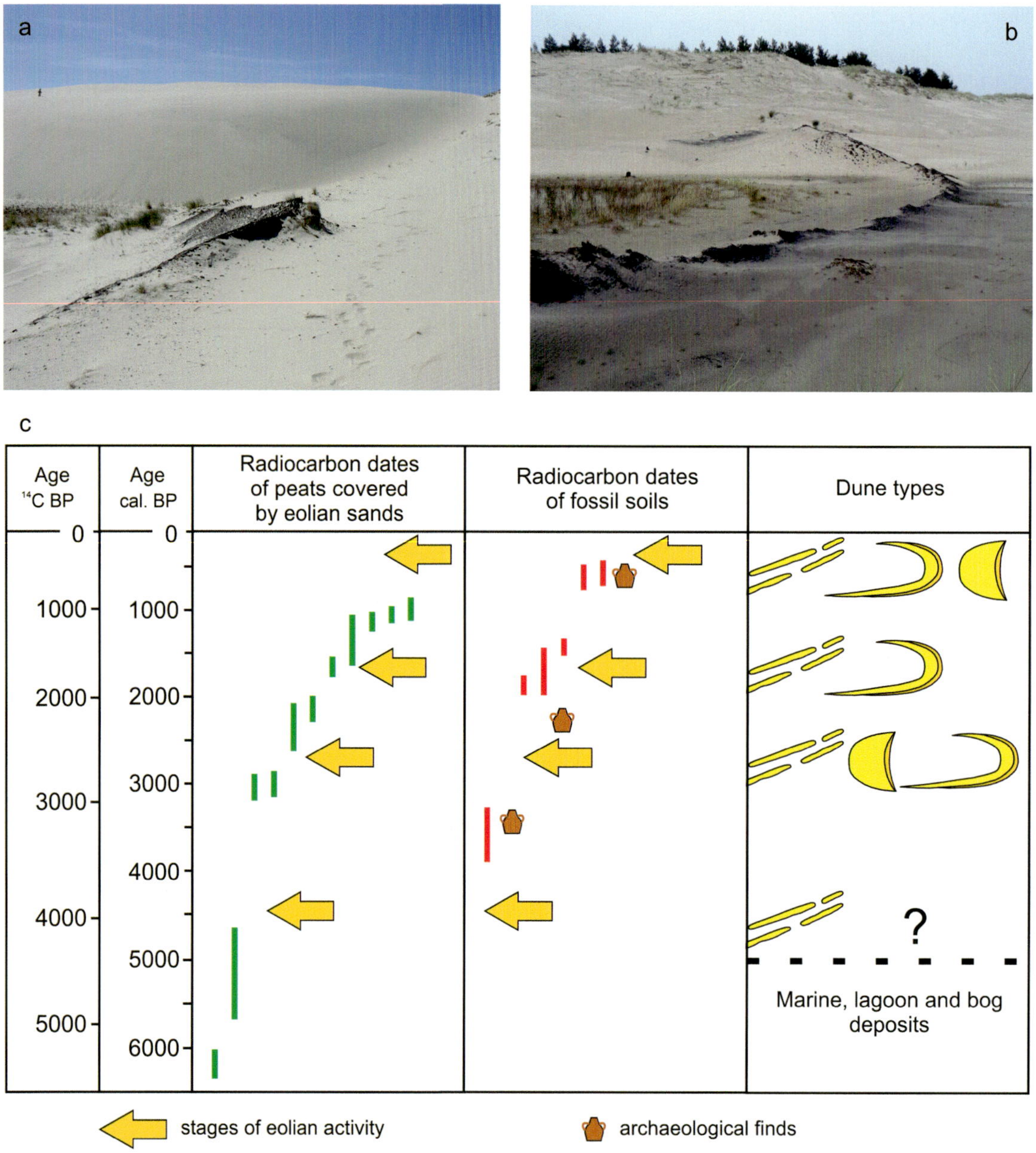

Fig. 41.14 Holocene evolution of the Łeba dunes field. **a, b** Outcrops of fossil soils in the central part of the barrier (photographs by J. Rotnicka). **c** Stages of aeolian activity on the Łeba Barrier (modified from Borówka 1990)

3000 years (Borówka and Tobolski 1979). The series are, however, covered by a well-developed fossil soil dated to about 2000 years. This younger soil is known from numerous sites at the Łeba Barrier and shows a very well-developed profile typical of the podzolic soil which, according to Tobolski (1975, 1980), was forming underneath the oak forest canopy. The structure of aeolian sands covered by this soil shows that, in the second phase of intensification of aeolian activity, the central part of the Łeba Barrier hosted transverse dunes, particularly simple or compound crescentic dunes.

The third stage of aeolian processes is most distinctly seen near Czołpino, although its traces are also found in the central part of the Łeba Barrier. East of the lighthouse-bearing Czołpino Dune, the stage is represented by a large parabolic dune more than 20 m high (Borówka 1975), covered by a fossil soil dated to 1540 ± 50 yr BP (1530–1330 cal yr BP). The soil was formed underneath a canopy of beech tree and pine forests (Tobolski 1975).

Between 1500 and 500 years BP (from the fifth until the fourteenth century), aeolian activity distinctly abated, to intensify again at the turn of the fifteenth century (Borówka

1990) due to human activity. The dune-forming processes have been since observed to proceed at a stable, fairly intensive pace.

Charcoal and numerous archaeological artefacts in both the fossil soils and peat deposits indicate that the periodically recurring intensification of aeolian activity on the Łeba Barrier is closely associated with human activities which, with some intermissions, have occurred there for at least the last three millennia (Borówka 1990).

In recent years, there has been a tendency to gradually decrease the height of migrating dunes on the Łeba Barrier and their transformation into large dome dunes. It is caused by limited sand supply caused by encroachment of vegetation (Michałowska et al. 2016), and probably the loss of dune sand as a result of backfilling the northern part of the Łebsko Lake.

41.7 Landforms of the Słowiński National Park as a Tourist Attraction

Wide beaches, active and stabilized dune fields, and lakes are the main attractions of the Słowiński National Park and many of the places are easily accessible via a network of tourist trails, with a total length of about 170 km. Some of them are also cyclist-friendly.

The largest active dune field (3 km^2), the Łeba dune field, is about 8.5 km west from the town of Łeba. The tourist trail leads from the town to the foot of the active dune field, and to the top of the highest active dune in the NP, called Łącka dune or simply Łączka, about 30 m in height. The smaller active dune field (0.5 km^2), Czołpino dune field, is located in the central part of the NP, less than 2 km to the north-east of Czołpino, where the museum of Słowiński NP is located. The tourist trail leads through a pine forest, crosses the dune field, and ends at the sea cost.

The area north of Czołpino is known for the highest dune in the NP, the Czołpino Dune (56 m asl). It can be reached by a tourist trail also starting at the museum of Słowiński NP. On top of this dune, there is 25 m high Czołpino lighthouse which allows the visitors to see the Łeba Barrier, the Czołpino dune field and Łebsko Lake to the west, and the belt of end-moraines to the south. Its highest elevation, called Rowokół (114.8 m asl), is located close to the village of Smołdzino. The lookout tower on top of it allows tourists to admire significant part of the Słowiński NP.

The undeniable attraction of the Słowiński National Park are its wide beaches that stretch along 33-km-long barrier coast. They are up to 100 m wide, built of white (quartz) sand, and backed by foredune ridges. However, along some parts of the coast, where the beach is narrow, the foredunes are destroyed by storm waves and dune cliffs can be observed. In the central part of the barrier (at the 208–209 km of the Polish coastline), "fossil forest" composed of well-preserved tree trunks in life position (Fig. 41.2c, d) is exposed on the beach. However, due to wave action, its appearance changes from year to year.

References

Arens SM, van Kaam-Peters HME, van Boxel JH (1995) Air flow over foredunes and implications for sand transport. Earth Surf Proc Landf 20:315–332

Borówka RK (1975) Problem of the morphology of fossil dune forms on the Łeba Bar. Quaest Geogr 2:39–51

Borówka RK (1980a) Present-day dune processes and dune morphology on the Łeba Barrier, Polish coast of the Baltic. Geogr Ann 62A:75–82

Borówka RK (1980b) Współczesne procesy transportu i sedymentacji piasków eolicznych oraz ich uwarunkowania i skutki na obszarze wydm nadmorskich. PWN, Warszawa-Poznań, p 126

Borówka RK (1990) The Holocene development and present morphology of the Łeba Dunes, Baltic coast of Poland. In: Nordstrom KF, Psuty N, Carter B (eds) Coastal dunes: form and process. Wiley, Chichester, pp 289–313

Borówka RK (2001) Struktura wewnętrzna wydm Łebskich jako efekt zmienności warunków meteorologicznych. In: Rotnicki K (ed) Przemiany środowiska geograficznego nizin nadmorskich południowego Bałtyku w vistulianie i holocenie. Bogucki Wyd. Nauk., Poznań, pp 89–93

Borówka RK, Rotnicki K (1995) Shoreline changes of the Łeba Barrier in modern times. In: Rotnicki K (ed) Polish coast: past, present and future. J Coast Res 2(Spec Iss):271–274

Borówka RK, Tobolski K (1979) Nowe stanowiska archeologiczne na Mierzei Łebskiej i ich znaczenie dla paleogeografii tego obszaru. Bad Fizj Pol Zach 32A:21–29

Christiansen MB, Davidson-Arnott RGD (2004) Rates of landward sand transport over the foredune at Skallingen, Denmark and the role of dune ramps. Danish J Geogr 104:31–43

Davidson-Arnott RGD, Dawson JC (2001) Moisture and fetch effects on rates of aeolian sediment transport, Skallingen, Denmark. In: Proceedings of the Canadian coastal conference, CCSEA, pp 309–321

Dłużewski M, Rotnicka J, Hesp PA, Tomczak JO, Dubis L (2023) Impact of coastline orientation on the dynamics of foredune growth (Łeba Barrier, south Baltic Sea coast, Poland). Miscellanea Geogr – Reg Stud Dev 27(3):1–10

Furmańczyk K (1994) Współczesny rozwój strefy brzegowej morza bezpływowego w świetle badań teledetekcyjnych południowych wybrzeży Bałtyku. Uniwersytet Szczeciński, Rozprawy i Studia, p 161

Fryberger SG (1979) Dune forms and wind regime. In: McKee ED (ed) A study of global sand seas. Government Printing Office, Washington, US, pp 137–169

Greenwood B, Davidson-Arnott RGD (1979) Sedimentation and equilibrium in wave-formed bars: a review and case study. Can J Earth Sci 16:312–332

Hesp PA (1983) Morphodynamics of incipient foredunes in New South Wales, Australia. In: Brookfield M, Ahlbrandt T (eds) Eolian processes and sediments. Elsevier, Amsterdam, pp 325–342

Hesp PA, Hyde R (1996) Flow dynamics and geomorphology of a trough blowout. Sedimentology 43:505–525

Krąpiec M, Florek W (2005) Subfosylne odziomki i pnie drzew na plażach w rejonie Rowów. In: Florek W (ed) Geologia i geomorfologia Pobrzeża i południowego Bałtyku, vol 6, pp 145–154

Kocurek G, Townsley M, Yeh E, Havholm K, Sweet ML (1992) Dune and dune field development on Padre Island, Texas, with implications for interdune deposition and water-table-controlled accumulation. J Sedim Petrol 62:622–635
Lancaster N (1992) Relations between dune generations in the Gran Desierto, Mexico. Sedimentology 39:631–644
Lutyńska M (2008) Fazy rozwoju jeziora Gardno na podstawie analizy okrzemkowej i geochemicznej. In: Rotnicki K, Jasiewicz J, Woszczyk M (eds) Holoceńskie przemiany wybrzeży i wód południowego Bałtyku – przyczyny, uwarunkowania i skutki. Wydawnictwo Tekst, Poznań-Bydgoszcz, pp 35–42
Ludwig J, Lindhorst S, Betzler C, Bierstedt SE, Borówka RK (2017) Sedimentary rhythms in coastal dunes as a record of intra-annual changes in wind climate (Łeba, Poland). Aeolian Res 27:67–77
Michałowska (2007) Spatial and temporal analysis of environmental changes in the Słowiński National Park between 1964 and 2004. Archiwum Fotogrametrii, Kartografii i Teledetekcji 17b:473–483
Michałowska K, Głowienka E, Hejmanowska B (2016) Temporal satellite images in the process of automatic efficient detection of changes of the Baltic Sea coastal zone. IOP Conf Ser Earth Environ Sci 44:042019
Michałowska K, Głowienka E (2022) Multi-temporal analysis of changes of the southern part of the Baltic Sea coast using aerial remote sensing data. Rem Sens 14:1212
Miszalski J (1973) Współczesne procesy eoliczne na Pobrzeżu Słowińskim; studium fotointerpretacyjne. Instytut Geografii, Polska Akademia Nauk, Warszawa, p 150
Psuty NP (1988) Sediment budget and dune/beach interaction. J Coast Res 3(Spec Iss):1–4
Ptak A, Rudowski S (2014) Piaskowe giganty. Geodeta 9:22–23
Rotnicka J (2011) Impact of beach surface type on the rate of sand transport by wind. J Coast Res 64(Spec Iss):2058–2062
Rotnicka J (2013a) Transport eoliczny na plaży morza bezpływowego: natężenie, uwarunkowania i wpływ na tworzenie wydm przednich (na przykładzie bariery łebskiej). Bogucki Wyd Nauk, Poznań, p 159
Rotnicka J (2013b) Aeolian vertical mass flux profiles above dry and moist sandy beach surfaces. Geomorphology 187:27–37
Rotnicka J, Dłużewski M (2019) A method to derive long-term coastal wind data from distant weather station to improve aeolian sand transport rate prediction. Aeolian Res 38:24–38
Rotnicka J, Dłużewski M (2022) Vertical profiles of aeolian mass flux above different sand surfaces and sand surfaces covered with pebbles. Catena 212:1–17
Rotnicki K (1995) An outline of geomorphology and main problems of the Upper Quaternary of the Gardno-Łeba Coastal Plain. In: Rotnicki K (ed) Polish coast: past, present and future. J Coast Res 22(Spec Iss):213–219
Rotnicki K (2008) Przemiany budowy geologicznej rzeźby terenu Słowińskiego Parku Narodowego i jego otuliny. In: Florek W (ed) Słowiński Park Narodowy. 40 lat ochrony unikatowej przyrody i kultury. Słowiński Park Narodowy, Smołdzino, pp 145–183
Rotnicki K (2009) Identyfikacja, wiek i przyczyny holoceńskich ingresji i regresji Bałtyku na polskim wybrzeżu środkowym. Słowiński Park Narodowy, Smołdzino, p 100
Rotnicki K, Alexandrowicz SW, Pazdur A, Goslar T, Borówka RK (2009) Stages of the formation of the Łeba barrier-lagoon system on the basis of the geological cross-section near Rąbka (Southern Baltic Coast, Poland). Studia Quat 26:3–24
Sławik Ł, Ptak A, Wróblewski R, Rudowski S (2014) Ruchome wydmy Słowińskiego Parku Narodowego – studium przypadku monitoringu teledetekcyjnego. In: Rudowski R, Sitkiewicz P, Wróblewski R (eds) II Sympozjum Morskiej Geomorfologii. Poziom morza, linia brzegowa. Instytut Morski, Gdańsk, pp 16–20
Tobolski K (1975) Studium palinologiczne gleb kopalnych Mierzei Łebskiej w Słowińskim Parku Narodowym. Prace Komisji Biologicznej 41, PTPN, Poznań, p 76
Tobolski K (1980) The fossil soils of the coastal dunes on the Łeba Bar and their palaeogeographical interpretation. Quaest Geogr 6:83–97
Tomczak J (2021) Uwarunkowania rozwoju barchanów w systemie plaża-wydmy przednie morza bezpływowego (na przykładzie Mierzei Łebskiej). BSc dissertation, University of Warsaw, p 75
Walker IJ, Hesp PA, Davidson-Arnott RGD, Ollerhead J (2006) Topographic steering of alongshore airflow over a vegetated foredune: Greenwich Dunes, Prince Edward Island, Canada. J Coast Res 22:1278–1291
Wilson IG (1972) Aeolian bedforms—their development and origin. Sedimentology 19:173–210
Wiśniewski B, Wolski T (2009) Katalog wezbrań i obniżeń sztormowych poziomów morza oraz ekstremalne poziomy wód na polskim wybrzeżu. Wyd Nauk Akademii Morskiej, Szczecin, p 156
Zhang W, Deng J, Harff J, Schneider R, Dudzińska-Nowak J (2013) A coupled modeling scheme for longshore sediment transport of wave-dominated coasts—a case study from the southern Baltic Sea. Coast Eng 72:39–55

Joanna Rotnicka is Associate Professor of Geology at the Adam Mickiewicz University in Poznań. Her research interests are sedimentology and geomorphology. Initially, she was involved in research of Cretaceous shallow marine rock successions exposed in the Sudetes Mountains, south Poland. Since 2002, she has focused on aeolian and coastal geomorphology with particular emphasis on the mechanisms of aeolian sand transport and coastal dune formation. Her main research area is South Baltic Sea coast, but she was also involved in projects on the Atlantic coast of Brazil, Morocco, and Iceland, and in the Sahara desert. She used to be the vice-Director of the Institute of Geology at Adam Mickiewicz University in Poznań (2016–2019), as well as the Secretary (2003–2017) and President (2017–2022) of the Poznań Branch of the Polish Geological Society. She is also a member of editorial board of *Geological Quarterly* journal.

Ryszard Krzysztof Borówka is Professor of Earth Sciences at the University of Szczecin and a former dean of the Faculty of Natural Sciences (2002–2008) and the Faculty of Earth Sciences (2008–2016). Currently, he works at the Institute of Marine and Environmental Sciences. His main research interests relate to paleogeography and chronostratigraphy of the late Quaternary, as well as sedimentology and geochemistry of various sedimentation environments, in particular aeolian, marine, lagoon, lacustrine, and swampy. His main research area is the Polish Lowland, but he was also involved in projects in various other regions of Poland (Łódź Upland, Holy Cross Mountains, Tatra and Podhale, Eastern Sudetes), different European (Germany, Denmark, Estonia, Moldova) and Asian countries (Mongolia, China–South China Sea and Hainan Island), and in the Eastern Pacific—Clarion-Clipperton Fracture Zone. He is a member of editorial colleges of several international journals (*Geochronometria, Oceanological and Hydrobiological Studies, Bulletin of Geography—Physical Geography Series, Acta Geographica Lodziensia*).

42 Vistula River Delta-Plain—A Region of Fluvial, Coastal, and Land Reclamation Impact on Landscape Development

Damian Moskalewicz and Piotr Paweł Woźniak

Abstract

Vistula River delta-plain, which includes the Żuławy area and the western part of Vistula Sandy Barrier, represents a unique, relatively flat landscape, shaped by fluvial, coastal, and anthropogenic interactions. The decay of the last ice sheet and the development of the Baltic Sea triggered the evolution of the Vistula River mouth that lasted for over 15 thousand years. In the past 700 years, human activity intensely transformed the natural character of the region. Characteristic landforms that distinctly contribute to the landscape include river channels with associated alluvial forms and artificial dykes, anthropogenic constructions like terps or fortifications, the relict Druzno lake, and sandy barriers insulating low-lying region from the Baltic Sea. The region is likely to evolve in the near future due to relentless human impact and ongoing climate change that put the Vistula River delta-plain in danger of coastal flooding.

Keywords

River network evolution · Relic delta lake · River mouths · Sandy barriers · Natural hazards · Human impact · Urban landscape

42.1 Introduction

The Vistula (Wisła) River delta-plain, located in northern Poland (Fig. 42.1), is an example of the landscape formed by multiple geological processes in the transitional zone between terrigenous and marine environments. The lower part of the Vistula River valley developed after the decay of the last Fennoscandian Ice Sheet. However, the river delta with the contemporary complex channel network was shaped over the last several thousand years in the Holocene (Cyberski 1995; Mojski 1995). The natural development of the coastal area related to the post-glacial sea-level rise created limited accommodation space between the front of the Vistula delta and the Vistula Sandy Barrier (Uścinowicz et al. 2021; Fig. 42.1). Consequently, the low-lying area was covered by fine-grained alluvial sediments, likely to be dominated by dense vegetation growth. Land reclamation and interferences with hydrological events led to the subsequent evolution of Vistula River mouths on the landward side of the Vistula Sandy Barrier and expansion of built-up areas, expressed in the contemporary landscape. Complex relationships between fluvial, coastal, and land reclamation processes that shaped the landscape make this region very interesting for a geomorphologist despite the relatively flat landscape.

42.2 Delimitation and Hypsometric Diversity

The Vistula River delta-plain covers two neighbouring geomorphological and historical regions: the Żuławy Alluvial Plain (pol. *Żuławy Wiślane*) and the western part of the Vistula Sandy Barrier that developed simultaneously with the river mouths (Fig. 42.1). The source of the term "żuławy" most probably comes from an old Polish word "żuł", which means "organic mud", or from an old Prussian word "solov", which means "island" (Jujka 1992). Both words are well linked to the genesis of the region. The

D. Moskalewicz (✉) · P. P. Woźniak
Department of Geomorphology and Quaternary Geology, University of Gdańsk, Bażyńskiego 4, 80–309 Gdańsk, Poland
e-mail: damian.moskalewicz@ug.edu.pl

P. P. Woźniak
e-mail: piotr.wozniak@ug.edu.pl

P. Migoń and K. Jancewicz (eds.), *Landscapes and Landforms of Poland*, World Geomorphological Landscapes, https://doi.org/10.1007/978-3-031-45762-3_42

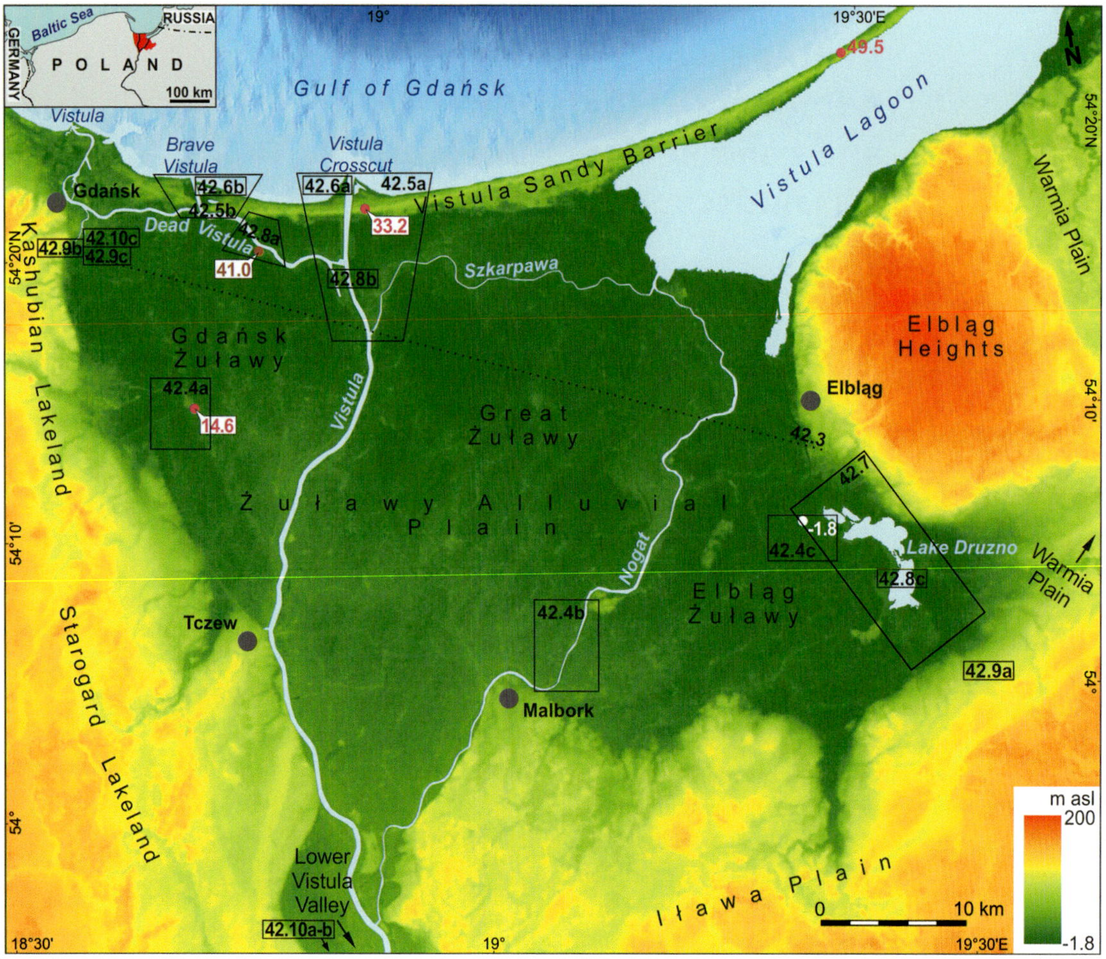

Fig. 42.1 Location and sub-division of the Vistula River delta-plain region. Points of the highest (pink) and the lowest (white) altitude of natural origin within the delta as well as the highest artificial elevation (brown) are marked. Rectangles with numbers mark the location of places and areas shown in Figs. 42.4, 42.5, 42.6, 42.7, 42.8, 42.9 and 42.10 (not to scale). Dotted line marks the location of geological cross-section (Fig. 42.3). DEM source: SRTM and ASTER GDEM (resolution of 30 m), European Environmental Agency

Żuławy Alluvial Plain (area of 1740 km^2) is divided by the Vistula and Nogat rivers into three sub-regions: Gdańsk Żuławy, Great Żuławy (also called Malbork Żuławy), and Elbląg Żuławy. The region expresses a classic delta shape and is surrounded by moraine plateaus of the Kashubian Lakeland and Starogard Lakeland in the west, the Iława Plain in the south, and the moraine plateau of the Elbląg Heights, the Warmia Plain, and the Vistula Lagoon in the east and northeast. The Vistula Sandy Barrier is a narrow belt extending along the northernmost part of the Vistula River delta-plain, at the contact with the Gulf of Gdańsk.

The Żuławy Alluvial Plain is a flat area with altitudes ranging from values close to the current sea level in the northern and east-northern parts up to 10 m above sea level (asl) at the southern corner (Fig. 42.1). An extensive, northern, and eastern part of the region is below the current sea level (Fig. 42.2), including the lowest point in Poland (1.8 m below the sea level, at the village of Raczki Elbląskie, Fig. 42.1). The highest point of natural origin, reaching 14.6 m asl, is located on one of the small morainic remnants at the village of Grabiny-Zameczek (Figs. 42.1 and 42.2). However, an industrial heap on the outskirts of Gdańsk reaches 41 m asl (Fig. 42.1). The Vistula Sandy Barrier shows a landscape composed of dune ridges of up to 30 m high. In the part adjacent to the Żuławy Alluvial Plain, the highest dune has an altitude of 33.2 m asl, while the highest dune in the whole Vistula Sandy Barrier reaches 49.5 m asl (Fig. 42.1). Due to that, an altitude contrast is visible in the landscape of the northern part of the Vistula River delta-plain. Such an effect is even better seen on contact between the alluvial plain with the surrounding morainic plateaus, because the latter reach up to 100–150 m asl, and even close to 200 m asl in the case of the Elbląg Heights (Fig. 42.1).

Fig. 42.2 Palaeogeographic development and main geomorphic features of the Vistula River delta-plain region. Compilation based on Cyberski (1995), Mojski (1995), Uścinowicz (2003), Plit (2010), Jegliński (2013)

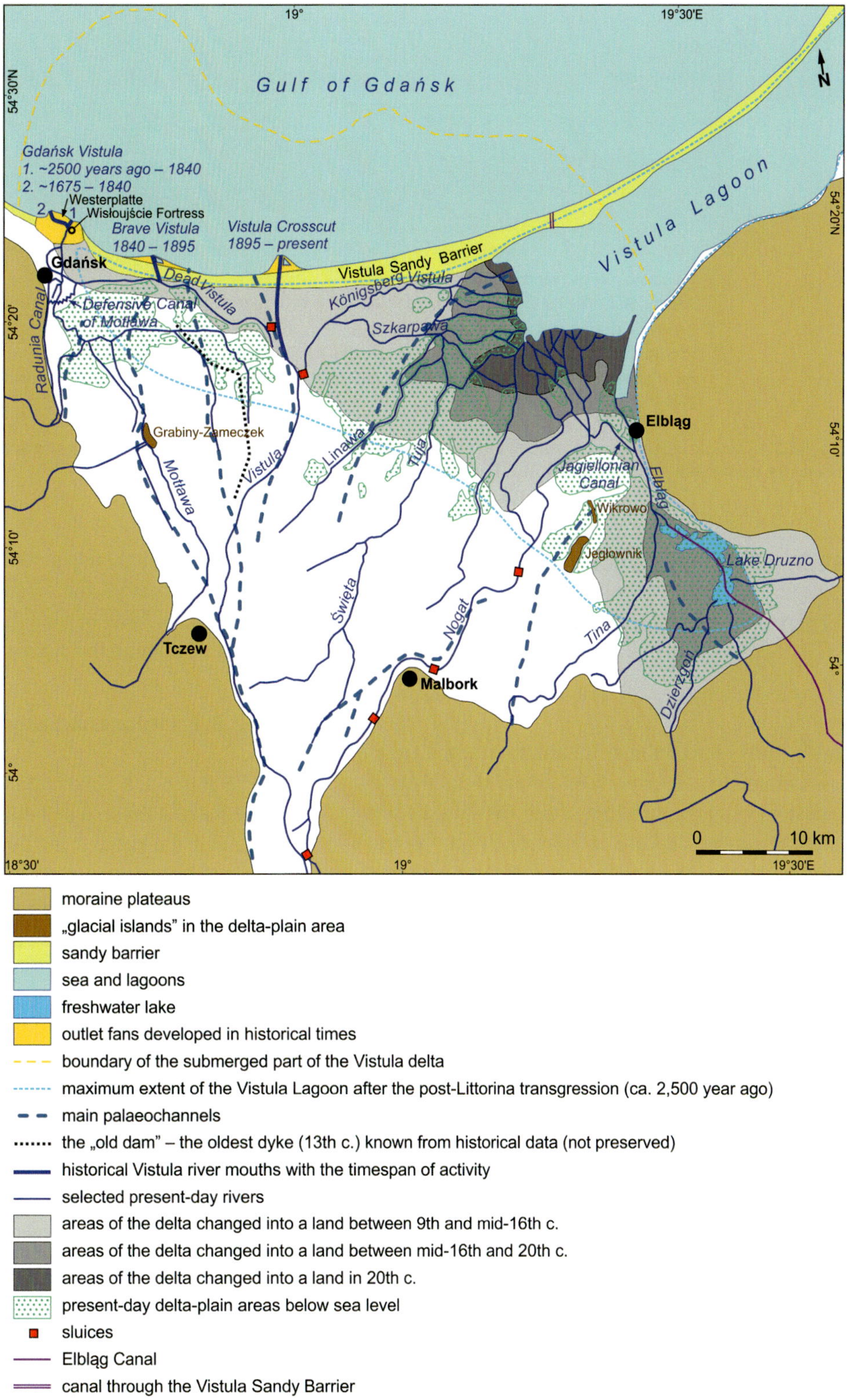

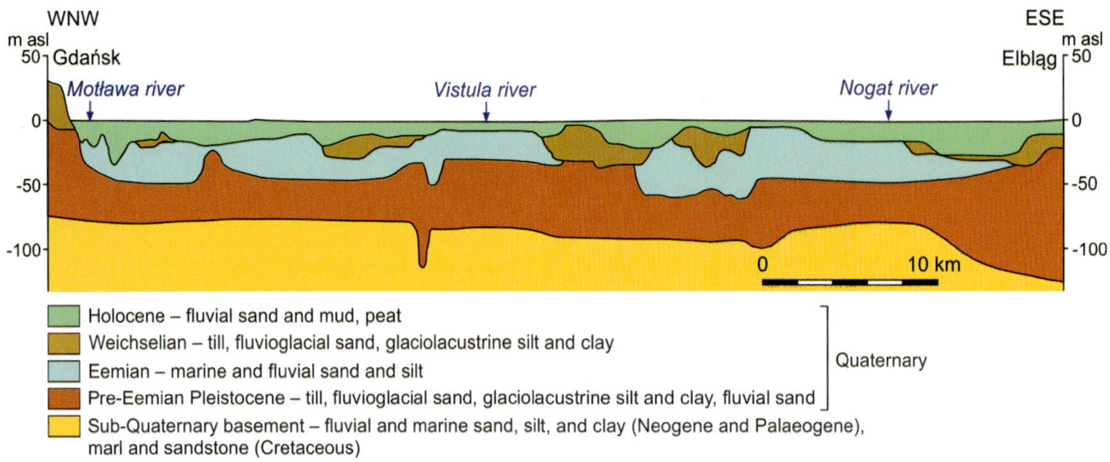

Fig. 42.3 Simplified geological cross-section along the northern part of the Vistula River delta-plain according to (Mojski 1995), with amendments; for the location of cross-section, see Fig. 42.1

42.3 Development of the Delta Plain

42.3.1 Pre-Holocene Stages

The substratum of the Pleistocene in the Żuławy region is not uniform, usually consisting of Palaeogene clastic deposits—mainly glauconite or quartz-rich sands or muds—or, less commonly, of Upper Cretaceous rocks—mainly sands, sandstones, marls, and limestones (Makowska 1998; Asch 2005). Neogene deposits are rare and occur as small sandy patches. In the Pleistocene, the region experienced several ice sheet advances interrupted by interglacial stages. In the Eemian, the last interglacial (126,000–110,000 years ago), marine transgression intruded almost into the whole area. Marine processes led to the erosion of older deposits and the accretion of new sediments. At several locations, Pleistocene deposits contain blocks of Neogene, Palaeogene, and Cretaceous series and lie directly on the Upper Cretaceous rocks (Sylwestrzak 1976). All these facts suggest that the topographic basin might be of erosional origin. The substratum of the delta sediments comprises till and glaciofluvial sands of the Last Glaciation. However, in the northern part, they occur only locally and deposits of the Eemian Sea dominate (Fig. 42.3). Buried channels of the Vistula River are incised to about 25 m below sea level (bsl) in the southern part and to about 35 m bsl in the northern part of the delta (Mojski 1988).

The last ice sheet advanced into the area ca. 25,000 years ago (Tylmann et al. 2019) and covered it until about 16,000 years ago (Tylmann et al. 2022). During the deglaciation, small ice-dammed lakes were formed in front of the retreating ice sheet. The Pleistocene fluvial discharge of Vistula catchment area was directed mainly to the west. The progressive melting of the ice sheet led to the formation of the Baltic Ice Lake, which covered the entire southern part of the contemporary Baltic Sea. The maximum water level of the lake was higher than the present sea level by about 1.0 – 1.5 m (Uścinowicz 2003).

42.3.2 Holocene Stages

The Baltic Ice Lake still existed in the early Holocene and the Vistula River, which has changed its course to the north, formed the delta front far north into the Gulf of Gdańsk (Fig. 42.2). At that time, the Baltic Sea basin experienced a glacioisostatic rebound which interfered with the post-glacial sea-level rise, leading to rapid changes in the regional sea level (rises and falls) and dynamic evolution of the coastal zone (Uścinowicz 2003). The northernmost position (20 km to the north from the current mouth) was reached 10,200 years ago (the beginning of the pre-boreal stage), when the Baltic Ice Lake was drained and changed into the Yoldia Sea (named after marine bivalve molluscs), with a water level 50–52 m below the current one. After that, rapid sea-level rise occurred and the northernmost part of the delta was submerged (Fig. 42.2). At that time, the inland area still had the character of a lakeland. Lakes evolved, e.g. the pre-Druzno lake changed into peatland about 9,000 years ago, at the beginning of the Boreal stage, due to the decrease in the groundwater level (Tylmann et al. 2007).

During the Littorina transgression (8500–5000 years ago, the end of the Boreal stage and the entire Atlantic stage), the sea level rose from about 28 to 2.5 m below the current one (Uścinowicz 2003). As a consequence, marine waters entered the northern parts of the delta. The maximum extent of the sea only slightly exceeded the modern coastline (from several hundred metres to about 2 km). Coastal abrasion and consequent sediment transport along the shores of the Gulf of Gdańsk supplied sediments necessary to form new landforms, including the Vistula Sandy Barrier and accompanying dune ridges. The formation of the oldest dune fields began around 7000–6000 years ago and they evolved in the following millennia. The development of the highest dunes was associated with cooling and aridification of climate during the Sub-boreal stage, around 4200 years ago (Uścinowicz et al. 2021).

The significant sea-level rise and the formation of the Vistula Sandy Barrier separating the delta plain from the Gulf of Gdańsk caused rapid growth of the internal delta in the southern part of the region (Jegliński 2016). As an effect of the sea-level rise, related to the post-Littorina transgression, an extensive water body (Vistula Lagoon) was created between the northern and northeastern parts of the delta and the Vistula Sandy Barrier. In the meantime, the former Lake Druzno re-emerged as the lagoon bay. About 2500 years ago, at the beginning of the Sub-Atlantic stage, the lagoon reached its maximum extent (Fig. 42.2). About 3000–2500 years ago, the Vistula River formed a new mouth north of the city of Gdańsk. It showed varied intensity of growth over time until 1840 CE (Jegliński 2016).

Before humans began to exert a significant influence on the development of the delta, the rivers flew quite freely. In the western part, they were heading north to the Gulf of Gdańsk (Fig. 42.2). Their alluvia filled a narrowing lagoon limited from the north by the Vistula Sandy Barrier. Consequently, the Vistula River turned west, eventually breaking through the barrier to the Gulf of Gdańsk. However, the location of its mouth was frequently changing, forming outlet fans much further to the north than today and building up the contemporary submerged part of the delta (Fig. 42.2). In the eastern part of the region (i.e. the contemporary Great Żuławy and Elbląg Żuławy), delta rivers headed northeast, to the much larger Vistula Lagoon (Fig. 42.2). The catchment area of most rivers was limited to the delta plain, where their springs were located. During floods, they often changed their courses (avulsion occurred). Even the Nogat River (Fig. 42.2) was flowing this way and became an important branch of the Vistula much later, in medieval times. Modest flow and limited sediment load resulted in slow expanse of this part of the delta. Therefore, the Lake Druzno was directly connected to the Vistula until about 2500 years ago. After this time, the Elbląg River became its link with the Vistula Lagoon (Tylmann et al 2007).

42.3.3 Anthropic Transformation of the Delta

According to historical records (Plit 2010), the first human interference with natural processes took place in the thirteenth century, when a 20-km-long dyke (the so-called "old dam") was built in the Gdańsk Żuławy (Fig. 42.2). Intense transformations started at the beginning of the fourteenth century, when the Teutonic Order captured the towns of Gdańsk and Tczew. Under their supervision, until the end of the fourteenth century, dykes along most rivers of the Gdańsk Żuławy were completed; only dykes along the Nogat River were finished later, in the fifteenth century (Majewski 1969). The takeover of most of the waters of the Vistula by the Nogat led in the fifteenth and sixteenth centuries to long-term disputes between the towns of Gdańsk and Elbląg ("war for water"), accompanied by investments transforming the hydrographic system (Cyberski 1995). First, at the end of the fifteenth century, the town of Elbląg dug the Jagiellonian Canal connecting the mouth of the Nogat River with the Elbląg River (Fig. 42.2) to strengthen the role of Elbląg as a commercial port. Then, in the mid-sixteenth century, by order of the Polish king, the division of water was regulated between Vistula and Nogat arms by digging a new canal.

The development of agriculture started in medieval times, both in the delta and in the surrounding uplands, leading to the transformation of forests and meadows into cultivated fields (mainly to grow cereals). This intensified denudation of these areas and, consequently, caused faster filling of the delta with sediments, observed mainly in its eastern part (Cyberski 1995). The Teutonic Order colonized the Żuławy promoting German settlement. Concentrated villages were established, and some of them are still present in the rural landscape of the Gdańsk Żuławy and the Great Żuławy (e.g. Figure 42.4a). This process was related to the polderization of wetlands and deforestation of the whole delta plain. Further transformation of the Żuławy Alluvial Plain was associated with the arrival of the Mennonites, Dutch members of an orthodox branch of the Protestant church, who settled after 1531 CE and remained until the beginning of nineteenth century, when, with the new Prussian policy, they lost their privileges and consequently, left the region (Klugiewicz 1995). Although they were a minority among the inhabitants of the established villages, due to their skills and economic privileges, they had a leading role in managing land reclamation and agricultural economy. Environmental transformation was manifested in the development of a characteristic cultural landscape, with dispersed rural settlements. Artificial residential mounds known as *terps* and a regular network of ditches were constructed (Fig. 42.4). The dispersed settlements (Fig. 42.4b, c) contrasted with concentrated villages established earlier, i.e. during the Teutonic colonization (see, for example, Suchy Dąb village in Fig. 42.4a).

The development of the Vistula River delta-plain and hydro-technical modernization in the nineteenth century resulted in further significant changes. Bucket wheels driven by windmills were replaced by much more efficient

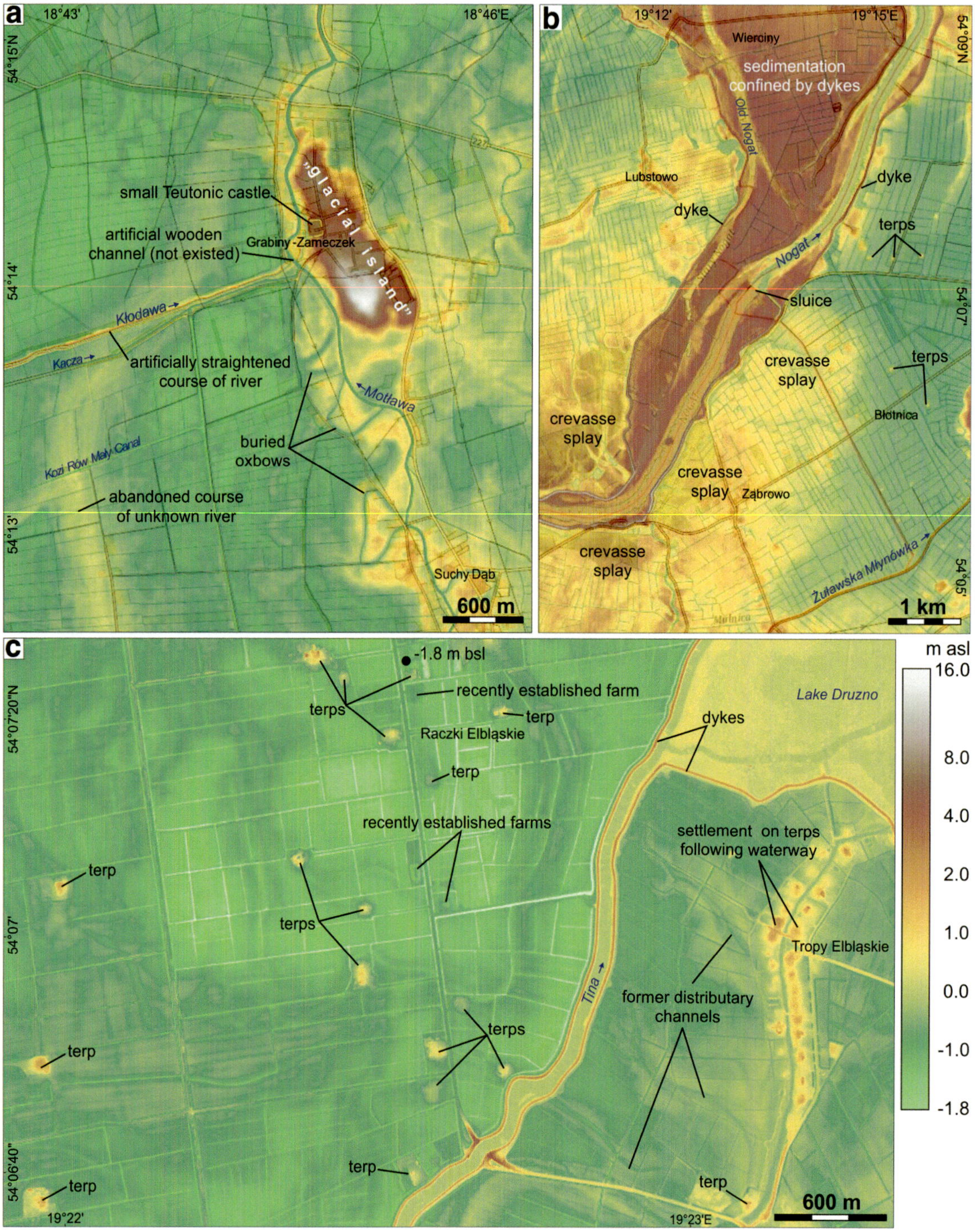

Fig. 42.4 Features of the geomorphological landscape of the Vistula River delta-plain region illustrated with DTM of 1 m × 1 m resolution (source maps: geoportal.gov.pl). The rectangular pattern of drainage ditches is most common. Irregular courses point on main channels and drainage ditches that replaced the former streams. **a** Examples from the Gdańsk Żuławy area: the abandoned course of an unknown deltaic river crosses artificially straightened course of the Kłodawa River, which then joins the Motława River (formerly led over it in an artificial wooden channel). It delivered water from the neighbouring upland to the watermill of a small Teutonic castle located on a "glacial island". Additionally, buried oxbows of the Motława River are visible. **b** Examples from the Great Żuławy area: area of alluvial sedimentation confined by dykes with a remnants of older course of the Nogat River and sluice; the area outside the dykes of the Nogat is at lower absolute elevation; crevasse splays mark sediment deposition after breaking levees or dykes with floodwaters; dispersed small artificial mounds (called terps) mark the location of single farms of Mennonite colonization. **c** Examples from the Elbląg Żuławy area: an area where the lowest point (1.8 m bsl) at the village of Raczki Elbląskie is located; the Mennonite settlement on terps shows dispersed pattern, while recently established farms are not constructed on them; in places, where the Mennonite settlement followed a waterway, terps are very close to one another; due to draining and subsequent peat compaction, most of the area is below the level of the Tina River channel and the Lake Druzno, both surrounded by dykes; note well-recognizable marks of former distributary channels of Tina River, i.e. from period when the Lake Druzno was more extensive (see shaded area surrounding the lake in Fig. 42.2)

Fig. 42.5 Pseudo-3D geovisualization of selected Vistula River mouths. Source: Google Earth Pro; due to the oblique view not to scale. **a** The Vistula Crosscut—an artificial mouth made in 1895; the Gdańsk Vistula has been cut by a dyke with sluice and renamed to the Dead Vistula. **b** Brave Vistula—a natural river mouth formed in 1840, when the Gdańsk Vistula was ice-jammed and the river itself drained a path through mid- to late-Holocene dune ridges of the Vistula Sandy Barrier. It remained active until the opening of the Vistula Crosscut

pumps with a steam engine, which allowed to merge small polders into larger structures (Klugiewicz 1995). In the mid of nineteenth century (1844–1860), the Elbląg Canal was constructed to connect the lakeland area in the south with the harbour in Elbląg via Lake Druzno and the Elbląg River. It was used for transport of timber and other goods (Fig. 42.2). In order to prevent floods as effectively as possible in 1895, the waters of the Vistula were directed by an artificial outlet to the Gulf of Gdańsk (Figs. 42.2 and 42.5a). In the following years, natural arms of Nogat, Szkarpawa, and Gdańsk Vistula, called the Dead Vistula since that time, were isolated with sluices. Three locks were built later to raise the water level to maintain the navigability of the Nogat River (Fig. 42.2). In 1945 the retreating German troops blew up selected dykes and flooded the region. In the following years, new colonization began and further modernization of hydro-technical structures occurred. Currently, the main human intervention is the construction of a canal through the Vistula Sandy Barrier, finished in 2022 (Fig. 42.2). The final impact of this structure on the natural environment is still unknown. Pre-construction reports suggested that a trough will interrupt the morphodynamics of the coastal zone. However, the influence of natural processes on a transport route will be limited, mainly to the impact of severe storm surges on lithodynamics (Kaczmarek et al. 2008).

Fig. 42.6 Landscapes of Vistula River mouths. **a** The distal part of the contemporary Vistula River mouth (Vistula Crosscut). Dynamic interactions between fluvial and marine processes have led to the formation and migration of sandy shoals, densely populated by birds and seals. **b** A view on the Bird Paradise Lagoon, the Messynian Sandy Barrier, and the Gulf of Gdańsk (Brave Vistula). At the left side of the picture, inactive washovers and dune field are visible. Active, sandy washover is visible in the middle part of the picture. It progrades into the lagoon during severe storm events (photographs by P.P. Woźniak)

42.3.4 Historical Vistula River Mouths

About 3000–2500 years ago, the Vistula River started forming the delta of Gdańsk Vistula exactly in the area of the historical city of Gdańsk (Fig. 42.2). Sediment supply was sufficient to form a wide sandy barrier, steeply prograding towards the Baltic Sea. Early phases of the mouth development are inferred from geological cross-sections and OSL ages of dunes located on the sandy barrier (Jegliński 2013; Jegliński and Uścinowicz 2015). In the fourteenth century, the shoreline advanced over the current position of the Wisłoujście Fortress (Fig. 42.2). Until the turn of the nineteenth century, river outflow was straightforward, directed to the northeast. The development of the underwater part of the delta resulted in the formation of shoals close to the mouth, similar to the modern Vistula crosscut (Figs. 42.1 and 42.6a). This process led to formation of the island of Westerplatte and caused bi-directional outflow of the Vistula to the Baltic Sea. Due to west-directed littoral transport along the Vistula Sandy Barrier and sustained high sediment supply, in the middle of the nineteenth century, the eastern arm of the mouth faded out, the island of Westerplatte changed into the peninsula, and the Gdańsk Vistula turned westward before reaching the Baltic Sea (Fig. 42.2).

High sediment supply, wave action, and related evolution of the barrier limited clear and unhindered outflow of river waters. In consequence, during winters, the terminal parts of the river channel were covered by ice and the flow of the Vistula was impeded. This led to multiple floods across the Vistula Delta region. In 1840, the Gdańsk Vistula River was jammed and river waters raised to levels that threatened local communities to be flooded again, but at that time, the river itself cut a path through the Vistula Sandy Barrier, forming a new mouth called the Brave Vistula (Figs. 42.2 and 42.5b, Basiński 1995). This event urged hydro-technical engineers to close the Gdańsk Vistula branch and direct all drainages through the new river mouth. The coastal area next to the Brave Vistula mouth experienced very dynamic evolution in the subsequent 70 years, during which a new Vistula delta front developed. Due to the positive sediment budget at this site, the Messynian Sandy Barrier was formed (Figs. 42.5b and 42.6b). Unfortunately, hydro-technical efforts did not resolve the flooding problem in the Vistula delta-plain region. Thus, as mentioned before, at the end of the nineteenth century, local authorities decided to form a new, wide, and straight river mouth for the Vistula. Construction works continued until 1895, when a new gate for drainage was opened (Cyberski 1995), called the Vistula Crosscut (Figs. 42.2 and 42.4a). The length of the channel was shortened to around 5 km from around 14 km (Brave Vistula) and 25 km even before (Gdańsk Vistula). Since that time, the direct impact of floods on the region has weakened. A new delta front started to form, prograding into the Baltic Sea.

42.4 Characteristic Landscapes

42.4.1 "Glacial Islands" in Deltaic Landscape

Fluvial erosion followed by long-term fluvial and biogenic sedimentation has caused burial of the pre-Holocene relief in the delta area. However, in three locations, the remnants of the Pleistocene surface are visible. They form small hills,

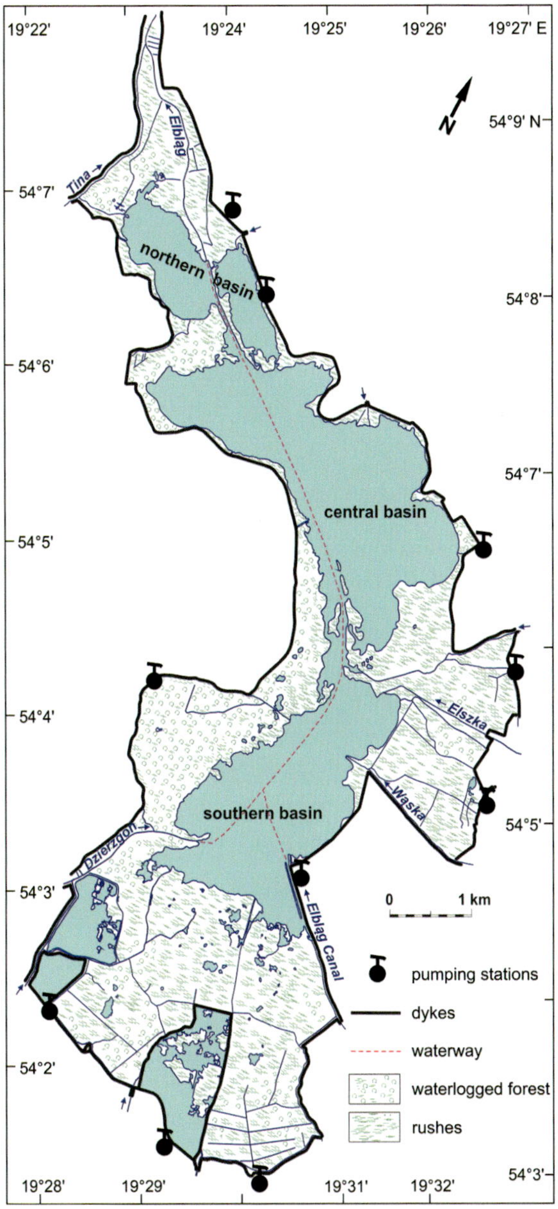

Fig. 42.7 Map of the contemporary Lake Druzno (after Borowiak and Borowiak 2002, with amendments). Sediment supply by streams coming from the surrounding moraine plateaus (Figs. 42.1 and 42.2) and autochthonic biogenic sedimentation caused that area of the lake was reduced and divided into nearly separated basins. The lake is very shallow, usually with a depth not exceeding 1 m, and for this reason, re-deposition of recent sediments is common. Therefore, the waterway was deepened to 3.5 m and requires dredging every few years

called "glacial islands", protruding through younger delta deposits. On their surface, both glaciofluvial sands and till were recognized (Makowska 1998). They reveal high elongation and are oriented in accordance with the general course of rivers in the vicinity (Figs. 42.2 and 42.4a). These properties indicate that the "glacial islands" may be erosional remnants, most likely isolated by fluvial action (Figs. 42.3 and 42.4). On the other hand, sub-Quaternary geological structures and surface features (Makowska 1998) suggest an involvement of neotectonic processes in their formation. One of these islands is located in the western part of the delta, in the Gdańsk Żuławy, whereas two others are in the eastern part, in the Great Żuławy area (Fig. 42.2). On each of the "glacial islands", there is a village, giving it its name. The Grabiny-Zameczek "island" reaches a height of 14.6 m asl and represents the highest point within the Vistula delta plain (Fig. 42.1). Its length is 1.7 km, the maximum width is 0.6 km, and the area is 0.6 km^2. The Jegłownik "Island" is lower (10 m asl), but the largest (4 km × 1.3 km, 3.5 km^2), while the Wikrowo "island" is the smallest (1.6 km × 0.4 km, 0.5 km^2) and the lowest (5 m asl).

All "glacial islands" are surrounded by areas at or even below sea level, so that their relative heights coincide with the absolute ones. They are fairly flat, with gentle slopes, and moderately distinctive in the landscape. However, their absolute heights were sufficient to be a convenient place to establish a settlement since the Neolithic. It should be assumed that the "glacial islands" of Jegłownik and Wikrowo were real islands in the Neolithic period, emerging from the waters of Vistula Lagoon that was much larger at that time (Fig. 42.2). In medieval times, they were surrounded by wetlands and rivers. The Teutonic Order used the convenient and defence location of the "glacial island" of Grabiny-Zameczek to establish a small castle with an accompanying water mill (Fig. 42.4a).

42.4.2 Alluvial Landforms

River channels and associated floodplains forming the delta through the past thousands of years are the main fluvial landforms of the Żuławy region. The terminal part of the Vistula River channel is shallow (up to 4 m deep in thalweg zone), around 300–400 m wide and of relatively low sinuosity, but with the tendency to avulsion. The meandering pattern of smaller rivers is better pronounced due to their higher sinuosity. Traces of palaeomeanders are still visible close to the smaller rivers (Fig. 42.4a), but also next to the major rivers such as the Vistula, where oxbow lakes occur on the floodplain and along the past course of the Nogat River downstream of the town of Malbork (Old Nogat, Fig. 42.4b). Crevasse splays composed of fine-grained deposits are visible near the channels, where floodwaters breached the natural levees or, since the fifteenth century, the dykes (Fig. 42.4b).

Another type of landforms characteristic of this region are sub-deltas formed in the eastern part of the delta plain. The Szkarpawa and Nogat rivers form adjacent distributary channel systems (Fig. 42.2). Dykes were constructed along these rivers, and the area between them has been turned into

Fig. 42.8 Characteristic landscapes of the Vistula River delta-plain region. **a** Flat area cut by a channel network, commonly revealing rectangular pattern. In the background are visible the Dead Vistula, industrial phosphogypsum heap and the heights of the Kashubian Lakeland. **b** Dyke along the Vistula Crosscut. The flat area on the right is located at a similar level or even lower (during high water) as the water level in the river. **c** Lake Druzno—a relict delta lake; slopes of the Elbląg Heights and the outlet cone of Wąska river (the area in trees and rushes in the left part of the picture) in the background (photographs by P.P. Woźniak)

polders. Modern outlet cones formed by small rivers near the Lake Druzno are less than 1 km wide and long (Figs. 42.7 and 42.8c). Traces of older, larger fans and their distributary channels, which developed towards the formerly much larger Lake Druzno, are still well recognizable on DTMs (Fig. 42.4c). Such a manifestation of preserved fluvial landforms, composed mainly of clastic deposits, is the result of drainage that lasted for hundreds of years and led to the compaction of adjacent peats and muds rich in organic matter in areas, which are now located at much lower elevations than these fluvial landforms (Fig. 42.4a, c).

42.4.3 Anthropogenic Landforms

A noticeable element of the landscape of the Vistula delta-plain is the dense hydrographic network (about 10 km per km^2), with the dominant role (85%) of anthropogenic water courses. Drainage ditches form a regular, rectangular network (Figs. 42.3 and 42.6a). Irregular systems are rare and occur where the drainage ditch follows the course of a former natural stream (Fig. 42.4a, c). In the fourteenth century, the Teutonic Order dug the Radunia Canal along the western edge of the Żuławy Alluvial Plain (Fig. 42.2), leading water to mills in the city of Gdańsk. At the southeastern edge of the region, the Elbląg Canal heads up to the moraine plateau. Its trail crosses the Lake Druzno (Fig. 42.7), influencing limnological properties of this reservoir (see Sect. 4.4). It includes several slipways that allow for gradual overcoming the altitude difference between the lake and the moraine upland further south, being now an attractive hydro-technical waterway structure, from the tourist point of view (Fig. 42.9a).

A distinctive features of the Żuławy Alluvial Plain landscape are dykes. Their total length is nearly 1000 km (Cebulak 1976). They follow all rivers and canals in the Gdańsk and Elbląg Żuławy area. The exception is the central part (the area between the Vistula, Nogat, and Szkarpawa rivers), where only the Tuja River is lined with dykes (Cebulak 1976). They have been built and maintained since medieval times, being repeatedly rebuilt and raised after floods. Today, they reach a height of 5 m and locally

Fig. 42.9 Examples of significant anthropogenic changes in the Vistula River delta-plain region during historical times. **a** Całuny slipway of the Elbląg Canal (1844–1860) in the rural landscape at the contact with the Iława Plain. **b** Earth bastion (note people on it for scale) in the city of Gdańsk as part of a defence system from the seventeenth century. **c** Defence moats branching from the Motława River in Gdańsk (photographs by P.P. Woźniak)

even 10 m in the case of the Vistula River. Therefore, they are distinctive landforms (Fig. 42.8b). In some places, the difference in height between the dykes and the surrounding area reaches several metres (Fig. 42.4b).

The construction of artificial levees (dykes) allowed for the creation of polders, which are flat areas, often located below sea level, drained by pumping stations (originally Dutch *polre*; "*pol*" meaning piece of land elevated from surrounding area). They constitute as much as 2/3 of the Żuławy area (Klugiewicz 1995). The areas of individual polders vary. In the eastern part of the delta plain, they are much smaller than in the central part, where many of them were merged in the nineteenth and twentieth centuries. Most often, polders were used as cultivated fields rather than pastures due to the fertility of alluvial soils. Elements complementing the geomorphological landscape of the Żuławy Alluvial Plain are terps (Fig. 42.4b, c)—artificial residential mounds with the height of 0.5 m to approx. 2 m, made of clay, straw, and animal excrement (Fac-Beneda 2010). They were constructed by the Dutch (Mennonite) settlers. Their size is small, adjusted to the size of a single farm. The settlement network is dispersed. Only if the village was founded along the river that served as the main transport route, terps occur as rows of mounds (Fig. 42.4c).

The most spectacular system of anthropic landforms in the region is the system of fortifications in the city of Gdańsk at the northwestern edge of the Żuławy region, built in the seventeenth century as one of the largest defence systems in Europe at that time. It consists of various defence structures, among which high earth bastions dominate the local landscape (Fig. 42.9b). Only two of them have been preserved up to day, with the higher one reaching 21 m asl. Bastions in the eastern part were demolished at the end of the nineteenth century, but their outline is still visible in the city's topography. Earth structures are accompanied by the Defence Channel of Motława River, with a zigzag course adapted to the layout of bastions (Figs. 42.2 and 42.9c). The modification of the course of this river and hydro-technical solutions allowed one to control its flow to protect against backwater or to flood the foreground of the fortress during the war. The highest contemporary landform in the urban landscape of Gdańsk is an industrial heap of phosphogypsum which reaches 41 m asl (Fig. 42.8a).

42.4.4 Lake Druzno

The Lake Druzno, located in the eastern part of the Vistula River delta-plain, is an example of a relict delta lake with strong anthropogenic influence (Fig. 42.1). The lake is a reservoir fed by freshwater from the catchment area, with periodic input of brackish water from the Vistula Lagoon via the Elbląg River. The lake receives freshwater not only from streams, but also from the Elbląg Canal and numerous canals discharging water from polders through pumping stations. The natural outflow of water from the lake is only through the Elbląg River (Fig. 42.7). The Lake Druzno experiences significant seasonal changes in water level. Maximum fluctuations are about 165 cm, with mean annual amplitudes of about 95 cm (Borowiak and Borowiak 2002). The highest water level is usually observed in summer (July or late September), whereas the lowest one occurs in late winter (February) and late spring (late May–early June) (Fac-Beneda 2013). As a result, the lake area ranges from 12.6 to 29.8 km^2 (Borowiak and Borowiak 2002). These fluctuations are mainly related to variations in the water level of the Vistula Lagoon, changes in stream inflow, and canal draining management. The Lake Druzno is a unique ecosystem, abundant in varied flora and fauna, including disappearing species, and is protected as a nature reserve.

The lake is very shallow, with a mean depth of 1.2 m. Most of its area forms a vast littoral zone with extremely lush aquatic vegetation. In addition, a specific type of forest grows on the waterlogged terrain along the southern and western shores of the lake—the alder carr (Fig. 42.7c). Allochthonic sediment supply by streams, autochthonic biogenic sedimentation, and expanding phytogenetic barriers led to the division of the lake into three distinct morphological basins (Fig. 42.7). The basin of Lake Druzno is separated from the surrounding terrain by dykes (Fig. 42.4c). Due to drainage and subsequent peat compaction, most of the area surrounding the lake, especially in the west, lies below the sea level (up to 1.8 m bsl) and thus below the water level in the lake (0.1 m asl). This requires discharging water from polders through 10 pumping stations and 19 drainage locks. Ground compaction also causes dyke subsidence. The height of some of them has decreased from 1.8 m to 1.2 m during the last two hundred years, but in recent years, they have been restored to their original state (Fac-Beneda 2013).

Although the streams flowing into the lake are small, their fans easily expand due to a flat morphology of the lake basin (Figs. 42.7 and 42.8c). Deposition of siliciclastic sediments is dominant. They are composed mostly of silt, with addition of sand. Admixture of sand is more common closer to the stream outlets (Czarnecka et al. 2005). High organic matter content is expected in conditions of significant primary production, resulting from the vast littoral zone densely filled with aquatic vegetation. However, due to the shallowness of sedimentary basins combined with large open water surface, wind-induced waves cause good oxygen conditions leading to the reduction of organic components in the sedimentary record (Czarnecka et al. 2005). Bottom sediments are contaminated by intrusions of brackish water from the Vistula Lagoon, via the Elbląg River (Tylmann et al. 2007). They are also influenced by wind-induced waves and bioturbation, leading to sediment mixing. In addition, the waterway of the Elbląg Canal, which was deepened to 3.5 m, requires dredging every few years. During such works, the extracted mud is deposited in the shallows, contributing to the increasing thickness of lake sediments. Considering the average rate of sedimentation (5 mm/year) and the average depth of the lake (1.2 m), the Lake Druzno would transform into peatbogs in about 250 years (Gołębiewski and Tylmann 2002). This period, however, is likely to be significantly longer due to the expected compaction of sediments, ongoing inflow of water from the catchment, and possible sea-level rise.

42.4.5 Sandy Barrier Landforms

The Żuławy region is separated from the Baltic Sea by the 1–2.5 km wide Vistula Sandy Barrier, whose landscape is dominated by several generations of dune ridges covered mainly by pine forest (Mojski 1995) (Figs. 42.5 and 42.6). Current interactions between coastal, marine, and fluvial environments are well exemplified at the Messynian Sandy Barrier—a narrow sandy landform, which is the youngest part of the Vistula Sandy Barrier, located in its western part (Fig. 42.5a). The Messynian Sandy Barrier is approximately

2 km long and about 0.35 km wide. Its beach is 42 m wide and has 3.4° slope on average. Coastal dune ridges are elongated and up to 5 m high. In the western part of the barrier, foredunes were transformed into a sea-protection dyke. At the backside of the dune ridges, a 100–200 m wide washover terrace faces the Bird Paradise Lagoon (pol. *jezioro Ptasi Raj*) (Figs. 42.5b and 42.6b). Besides well-developed dunes and deflation hollows (blow-outs), landforms of the Messynian Sandy Barrier were formed due to storm activity. According to the classification by Morton and Sallenger Jr (2003), four types of landforms related to storm events can be distinguished: storm scarps, channel incisions, perched fans, and washover terraces. All of them occur within the Messynian Sandy Barrier.

Storm scarps are common erosional landforms at the edge of foredunes and the upper beach. They are continuous, linear features, tens to hundreds of metres long and 0.2–3 m high. The typical characteristic of this landform is a highly inclined slope, reaching 60°. Lower scarps occur in locations where the beach is wide, and dunes are poorly developed next to pine forest in the eastern part of the barrier. The highest storm scarps occur in the central part of the barrier, where dune ridges are well-developed.

Channel incisions are erosional features that cut through the dune ridges and form sediment transport pathways perpendicularly to the coast, from the foreshore and the beach to the back-barrier lagoon or coastal plain. They form during extreme storm surges, if the sea level is high enough to reach and breach the top of the dune ridge and marine inflow may pass through the ridge. In the eastern part of the Messynian Sandy Barrier, marine waters flow into the lagoon through an active channel, subdivided into two passes, commonly called gates. At several locations, dune ridges were partially breached by stormwater, leaving dune remnants—small hillocks, being relict landforms after dune ridges. Their height is about half of the surrounding dune ridges, and the perimeter of individual forms is 3–10 m.

Perched fans are small lobate or elongate washover features oriented perpendicularly to the coastline, which occur only in the inner part of the barrier. They form during storm surges when waves approach the top of the dune ridges and partially transport sediment in landward direction. Aeolian processes usually rework the surface of perched fans after deposition. The inner structure of perched fans shows planar stratification, with an angle depending on the shape of an older perched fan or dune ridge surface.

Washover deposits that reach the lagoon form a vast belt of flat, commonly lobate, sandy depositional forms, called washover terraces. Most of them were created in the early stages of barrier development, when morphological conditions favoured frequent marine inflows. In the western part of the barrier, washovers completely stopped to expand after the construction of the dyke. However, in the eastern part of the barrier, an active washover terrace is still increasing its dimensions and expanding into the lagoon. The inner structure of these landforms usually show planar, low-angle cross-stratification, or horizontal stratification. In the channel zone, cross-stratification or ripple lamination may also appear due to deeper, directed inflow. The upper surface of active washovers indicates a remnant of aeolian reworking of previously deposited sands.

42.5 Natural Hazards

The Vistula River delta-plain region was strongly influenced by anthropogenic interference, leading to the origin of a complex, semi-artificial hydrographic system. However, despite efforts put into the construction of dykes, the area is not entirely secure from floods (Fig. 42.10a, b). In historical times, the greatest damage was caused by floods at the end of winter or in early spring when ice jams developed along the river system. Historical data indicate that the number of dyke breaks during floods increased towards recent times, from about 20 in the fifteenth and sixteenth centuries, through 26 in the seventeenth century to 40 in the eighteenth century (Cyberski 1995). The nineteenth century, with 28 floods, stood out with a slightly smaller number, but in the spring of 1829, the most tragic flood in the history of Gdańsk and Gdańsk Żuławy occurred (commemorated with floodmarks, e.g. Figure 42.10c), and in winter of 1840, the Gdańsk Vistula created a new mouth (Figs. 42.2 and 42.6b). As outlined before, the flood influence decreased in the northern part of the delta plain after the construction of an artificial Vistula Sandy Barrier crosscut at the end of the nineteenth century (Fig. 42.5a). A significant threat, still important today, is the water level rise at the mouths (the backwater), caused by stormy conditions, associated with winds blowing from the north.

The region of the Żuławy Alluvial Plain is an extremely low-lying area. Only 25% of its surface is located over the 5 m above the sea level, while almost 30% of terrain lies at the sea level or below it (Fig. 42.2). Thus, this region is endangered by current climate change trend and possible sea-level rise impact that includes an increase in the probability of marine flooding. Kirezci et al. (2020) determined global hotspots which may undergo marine inundation. The southeastern Baltic coasts, including the Vistula River delta-plain, were classified as an endangered area, which up to the year 2100 may face extreme sea levels of + 1.5–2.5 m within a projected 100-year return period of episodic flooding and inundation of 1–10 km^2 per km of the shoreline. For comparison, older storm surge risk models based on instrumental data for this region indicated that within the same 100-year return period, storm surges might reach the level of about + 1.7 m (Wolski et al. 2014).

Fig. 42.10 Floods in the Vistula River delta-plain region. **a**, **b** Contemporary flood (summer of 2007) induced by heavy and long-lasting rains—a view from the western edge of the Vistula valley in the upstream direction, a few km south of the Vistula River delta-plain region. Floodwaters cover the area between the steep western slope of the valley on the right (**a**) and dyke on the left (**b**). **c** Floodmark of the high water of the Motława river in Gdańsk on April 11, 1829, during an early spring flood. The flood was caused by ice jam on the Vistula River, but intensified due to long-lasting rains and backwater inflows induced by a storm surge in the Gulf of Gdańsk. On this day, the water reached 3.36 m asl in Gdańsk, being the most devastating flood in the history of the Gdańsk Żuławy. The water level of the Motława river visible in the picture is close to 0 m asl; on the left, automatic storm-gate preventing backwater inflows is visible (photographs by P.P. Woźniak)

During the past hundreds of years, coastal communities suffered from marine flooding, and the Vistula Sandy Barrier was disconnected multiple times due to the storm surges (Moskalewicz et al. 2020). Considering the upper sea-level limit for a 100-year return period of extreme surge (2.5 m) according to Kirezci et al. (2020), the analysis of Paprotny and Terefenko (2017) suggests that in the Vistula delta-plain region, 50–100% of the area is under the risk of inundation and the threat may affect 40–100% of the population. The sedimentary record of storm surges in the western part of the Messynian Sandy Barrier shows that the most intense storm activity occurred over a hundred years ago (Moskalewicz et al. 2020). On the other hand, satellite images from the past 30 years show the stepwise retreat of the coastline and the formation of washover features along the barrier (Figs. 42.5b and 42.6b).

In 2009, a substantial storm surge raised sea levels notably. A higher water level was also observed in the downstream part of the Vistula River and several localities experienced flooding due to dyke damage (Cieśliński and Chromniak 2010). Marine inundation threat is not the only concern for the communities living in the Żuławy Alluvial Plain. Low terrain and the presence of a large river make this region vulnerable to other kinds of floods as well. Up to the beginning of the twentieth century, the region was repeatedly flooded due to heavy rains, ice jams, or even intentional dyke breaching during warfare (Prabucka 2009).

42.6 Conclusions

The Vistula River delta-plain region represents a landscape that seems extremely flat, especially in contrast with the neighbouring uplands. The topography of the basin occupied by the delta was primarily shaped by Pleistocene glacial erosion as well as fluvial erosion and accumulation.

The region also experienced Eemian and Holocene marine transgressions. In the Holocene, deltaic rivers deposited a huge amount of alluvial sediments, and associated peat formation occurred, which led to the formation of flat landscape. A significant part of the region is below the current sea level. The only natural area with distinct elevations is the Vistula Sandy Barrier, where dune fields had been developing during the mid- and late-Holocene. However, about 700 years of human contribution to landscape evolution, enforced by the attempts of flooding control and the growing need for settlement and agriculture, transformed the Vistula delta into a diverse landscape with considerable anthropic signature. A dense drainage network composed of ditches, canals, and rivers surrounded by dykes, but also polders and artificial residential mounds, are a hydro-technical heritage that has modified the natural morphology of the flat, endless landscape. Even more interesting seems to be the landscape of the southern and eastern parts of the city of Gdańsk, with the Dead Vistula channel, earth bastions, and canals. These landforms are complemented by a relict delta lake, river outlets of different origins, and sandy barrier landforms. The Vistula River delta-plain region represents one of the youngest landscapes in Poland, which will be subject to further change due to the predicted sea-level rise. At least 50% of its area is at risk of marine floods. Additionally, climate instability may increase the threat caused by river floods due to heavy rains and ice jams. These challenges mean that despite long-lasting efforts to control the environment of the delta, natural factors will remain important for the future landscape evolution.

References

Asch K (2005) The 1:5.000.000 international geological map of Europe and adjacent areas. BRG, Hannover

Basiński T (1995) Origin of the "Wisła Śmiała" (Brave Vistula) mouth in the Vistula Delta. J Coastal Res 22:160–163

Borowiak D, Borowiak M (2002) Hydrochemical distinctive characteristics of Lake Druzno. Limnol Rev 2:45–56

Cebulak K (1976) System wodno-melioracyjny. In: Augustowski B (ed) Żuławy Wiślane. GTN, Gdańsk, pp 319–350

Cieśliński R, Chromniak Ł (2010) Hydrologiczne i hydrochemiczne efekty sztormu na polskim wybrzeżu i w delcie Wisły w dniach 14–15 października 2009 roku. In: Ciupa T, Suligowski R (eds) Woda w badaniach geograficznych. Gdańsk, pp 81–90

Cyberski J (1995) Hydrography of Żuławy Wiślane (Vistula Delta) and its changes over the historical period. J Coastal Res 22:152–159

Czarnecka K, Tylmann W, Woźniak PP (2005) Recent sediments of Lake Druzno (south basin). Limnol Rev 5:45–51

Fac-Beneda J (2010) Hydrograficzne ślady bytności Mennonitów we wschodniej części delty Wisły. In: Kaniecki A, Baczyńska A (eds) Zmiany stosunków wodnych w czasach historycznych, woda—środowisko—zmiany. Studia i Prace z Geografii i Geologii, vol 9. Poznań, pp 73–83

Fac-Beneda J (2013) The hydrological characteristics of Lake Druzno. In: Nitecki C (ed) Lake Druzno. Natural monograph. Mantis, Olsztyn, pp 15–31

Gołębiewski R, Tylmann W (2002) Współczesne osady denne jeziora Druzno. In: Drwal J (ed) Wody delty Wisły. Część wschodnia. GTN, Gdańsk, pp 167–180

Jegliński W (2013) Development of the Gulf of Gdansk coast in the area of the Dead Vistula mouth. Przegl Geol 61:587–595

Jegliński W, Uścinowicz S (2015) Location and geological setting of Wisłoujście Fortress. In: Dąbal J, Krawczyk K, Widerski T (eds) Wisłoujście fortress, Gdańsk. Archeological and architectural investigations, 2013–2014. Uniwersytet Gdański, Muzeum Historyczne Miasta Gdańska, pp 1–10

Jegliński W (2016) Origin and development of Vistula River mouth in Gdańsk. In: Abstract volume & trip guidebook. The 13th Colloquium on Baltic Marine Geology, PGI-NRI, Warsaw, pp 113–116

Jujka P (1992) Żuławy Wiślane, mapa turystyczna 1:100.000. Demart, Warszawa

Kaczmarek LM, Biegowski J, Gaca K, Gąsiorowski D, Kaźmierski J, Ostrowski R, Perfumowicz T, Pruszak Z, Schönhofer J, Skaja M, Szmytkiewicz M, Szmytkiewicz P (2008) Analiza procesów hydro- i litodynamicznych w rejonie planowanego przekopu przez Mierzeję Wiślaną i predykcja wpływu przekopu na brzeg morski wraz z oceną intensywności zapiaszczania (zamulania) toru wodnego na odcinku od przekopu do portu w Elblągu. Raport końcowy z realizacji projektu badawczego rozwojowego, IBW PAN, Gdańsk (unpubl)

Kirezci E, Young IR, Ranasinghe R, Muis S, Nicholls RJ, Lincke D, Hinkel J (2020) Projections of global-scale extreme sea levels and resulting episodic coastal flooding over the 21st century. Sci Rep 10:11629

Klugiewicz J (1995) Polderyzacja terenów depresyjnych. Towarzystwo Wolnej Wszechnicy Polskiej, Bydgoszcz, p 452

Majewski J (1969) Rozwój hydrograficzny delty Wisły w okresie historycznym. Przegl Geofiz 14:3–42

Makowska A (1998) Szczegółowa mapa geologiczna Polski, arkusz Elbląg Południe. Państwowy Instytut Geologiczny, Warszawa

Mojski JE (1988) Development of the vistula river delta and evolution of the Baltic Sea, an attempt to chronological correlation. Geol Surv Finland Spec Pap 6:39–51

Mojski JE (1995) Geology and evolution of the vistula delta and vistula bar. J Coast Res 22:141–149

Morton RA, Sallenger AH Jr (2003) Morphological impacts of extreme storms on sandy beaches and barriers. J Coast Res 19:560–573

Moskalewicz D, Szczuciński W, Mroczek P, Vaikutiene G (2020) Sedimentary record of historical extreme storm surges on the Gulf of Gdańsk coast, Baltic Sea. Mar Geol 420:106084

Paprotny D, Terefenko P (2017) New estimates of potential impacts of sea level rise and coastal floods in Poland. Nat Hazards 85:1249–1277

Plit J (2010) Naturalne i antropogeniczne przemiany krajobrazów delty Wisły. Pr Komisji Krajobrazu Kulturowego 13:13–28

Prabucka M (2009) Znaki Wielkich Wód na Żuławach Wiślanych. UMK, Toruń, p 163

Sylwestrzak J (1976) Rozwój paleogeograficzny. In: Augustowski B (ed) Żuławy Wiślane. GTN, Gdańsk, pp 133–174

Tylmann W, Gołębiewski R, Woźniak PP, Czarnecka K (2007) Heavy metals in sediments as evidence for recent pollution and quasi-estuarine processes: an example from Lake Druzno, Poland. Environ Geol 53:35–46

Tylmann K, Rinterknecht VR, Woźniak PP, Bourlès D, Schimmelpfennig I, Guillou V, ASTER Team (2019) The local last glacial maximum of the southern Scandinavian Ice Sheet front: Cosmogenic nuclide dating of erratics in northern Poland. Quat Sci Rev 219:36–46

Tylmann K, Rinterknecht VR, Woźniak PP, Guillou V, ASTER Team (2022) Asynchronous dynamics of the last Scandinavian ice sheet along the Pomeranian phase ice-marginal belt: a new scenario inferred from surface exposure ^{10}Be dating. Quat Sci Rev 294:107755

Uścinowicz S (2003) The Southern Baltic relative sea level changes, glacio-isostatic rebound and shoreline displacement. Pol Geol Inst Spec Pap 10:1–79

Uścinowicz S, Adamiec G, Bluszcz A, Jegliński W, Szpiganowicz G (2021) Holocene development of the Vistula Spit (Baltic Sea coast) based on multidisciplinary investigations. Holocene 31:658–671

Wolski T, Wiśniewski B, Giza A, Kowalewska-Kalkowska H, Boman H, Grabbi-Kaiv S, Hammarklint T, Holfort J, Lydeikaite Z (2014) Extreme sea levels at selected stations on the Baltic Sea coast. Oceanologia 56:259–290

Damian Moskalewicz is Assistant Professor and Postdoctoral Researcher in the Department of Geomorphology and Quaternary Geology, University of Gdańsk. His main research interests cover the reconstruction of Quaternary environments from sedimentary records, mineralogy and petrography of Quaternary deposits, and geological processes and geomorphology of the coastal zone. He conducts research in northern Poland, the southern Baltic coastal zone, the Baltic Sea bottom, and Fennoscandia. He is a member of the International Union for Quaternary Research, Polish Association of Geomorphologists, and Gdańsk Scientific Society.

Piotr Paweł Woźniak is Associate Professor in the Department of Geomorphology and Quaternary Geology, University of Gdańsk. His main research interests are glacial geomorphology, petrography of Quaternary deposits, and reconstruction of Quaternary environments from sedimentary records. Additionally, he supports archaeologists by archaeopetrographical analyses. The northern Poland is his main research area, but he was also involved in projects in the Baltic States and Fennoscandia. He is the president of the Gdańsk Branch of the Polish Geographical Society (since 2005) and was the secretary of the Earth Sciences Division of the Gdańsk Scientific Society (2004–2022). He is also a member of the Committee for Quaternary Research of the Polish Academy of Sciences and the Polish Association of Geomorphologists. He is a member of the Topical Advisory Panel of *Quaternary* journal.

Hel Peninsula—A Distinctive Spit on the Polish Baltic Coast

43

Radosław Wróblewski, Maria Rucińska, Patryk Sitkiewicz and Stanisław Rudowski

Abstract

The Hel Peninsula is one of the most geomorphologically interesting landforms on the Polish coast. It is a sandy barrier that divides the Gulf of Gdańsk from Puck Bay and Puck Lagoon. Its head is near the town of Władysławowo and continues southeast into the centre of the Gulf of Gdańsk. It is unique due to its length (35 km long), the volume of sediments, and the depth of the basin (60 m bsl) that it is prograding into. The Hel Peninsula is a classic continuation spit with the erosional neck and the accumulative hook. The neck is narrow, from 150 up to 300 m, low-lying, up to a few metres above sea level, relatively flat, with unvaried relief. The hook is from 1 to 3 km wide and covered by well-developed coastal dunes (up to 22 m asl). The shape of older dune formations indicates previous stages of hook development.

Keywords

Sandy barrier · Spit · Washover fans · Baltic coast

R. Wróblewski (✉) · M. Rucińska
Department of Geophysics, University of Gdańsk, al. Marszałka Piłsudskiego 46, 81-378 Gdynia, Poland
e-mail: radoslaw.wroblewski@ug.edu.pl

M. Rucińska
e-mail: maria.rucinska@ug.edu.pl

P. Sitkiewicz
Department of Geomorphology and Quaternary Geology, University of Gdańsk, ul. Bażyńskiego 4, 80-309 Gdańsk, Poland
e-mail: patryk.sitkiewicz@ug.edu.pl

S. Rudowski
Maritime Institute, Gdynia Maritime University, Długi Targ 41/42, 80-830 Gdańsk, Poland
e-mail: srudowski@im.umg.edu.pl

43.1 Introduction

Poland is bordered to the north by the Baltic Sea, which is one of the world's youngest seas. The earliest stage of the Baltic occurred after the Scandinavian ice sheet retreated, about 12,000–10,000 years ago. The sea has often and significantly changed its character during the Holocene. Periodically, it was even isolated from the global ocean due to the isostatic rebound (post-glacial upward movement of land once overloaded by the Pleistocene ice sheets) that limited seawater exchange between the Baltic and the North Seas. For example, such conditions have occurred during the stage of Ancylus Lake, about 9000–8000 years ago The Baltic became an ocean-linked sea again as a result of sea-level rise about 7000 years ago when the Danish straits were created. During the next stages of the Baltic Sea development, especially the one known as the Littorina Sea (7–5 ka BP), transgressive formation of the Baltic southern shoreline in a shape similar to the present has occurred.

The modern Baltic is a semi-closed tideless sea. In the northern part of the sea, rocky shores are dominant by contrast to the southern shore built by glacial and Holocene sediments such as sands, gravels, and till. Three main types of coast can be distinguished on the Baltic shore in Poland. Dune coasts are the most common. These are divided into sections separated by cliffs that undercut the seaward sides of moraine plateaus and hills. The last and rare type of the Polish coast is an alluvial plain. The most intensive development of the Southern Baltic coast occurs during storm surges due to wind-generated direct wave impact on the shore and the nearshore seabed. As a result, beach, dune, and cliff abrasion can be frequently observed. On most sections of the Polish shore, coastal erosion prevails over processes of marine and aeolian accumulation.

The Polish seashore of the Southern Baltic has a smooth shape (Fig. 43.1). It extends from the Szczecin Lagoon with the islands of Wolin and Uznam in the southwest to the northernmost cape of Poland, located within the Swarzewo

P. Migoń and K. Jancewicz (eds.), *Landscapes and Landforms of Poland*,
World Geomorphological Landscapes, https://doi.org/10.1007/978-3-031-45762-3_43

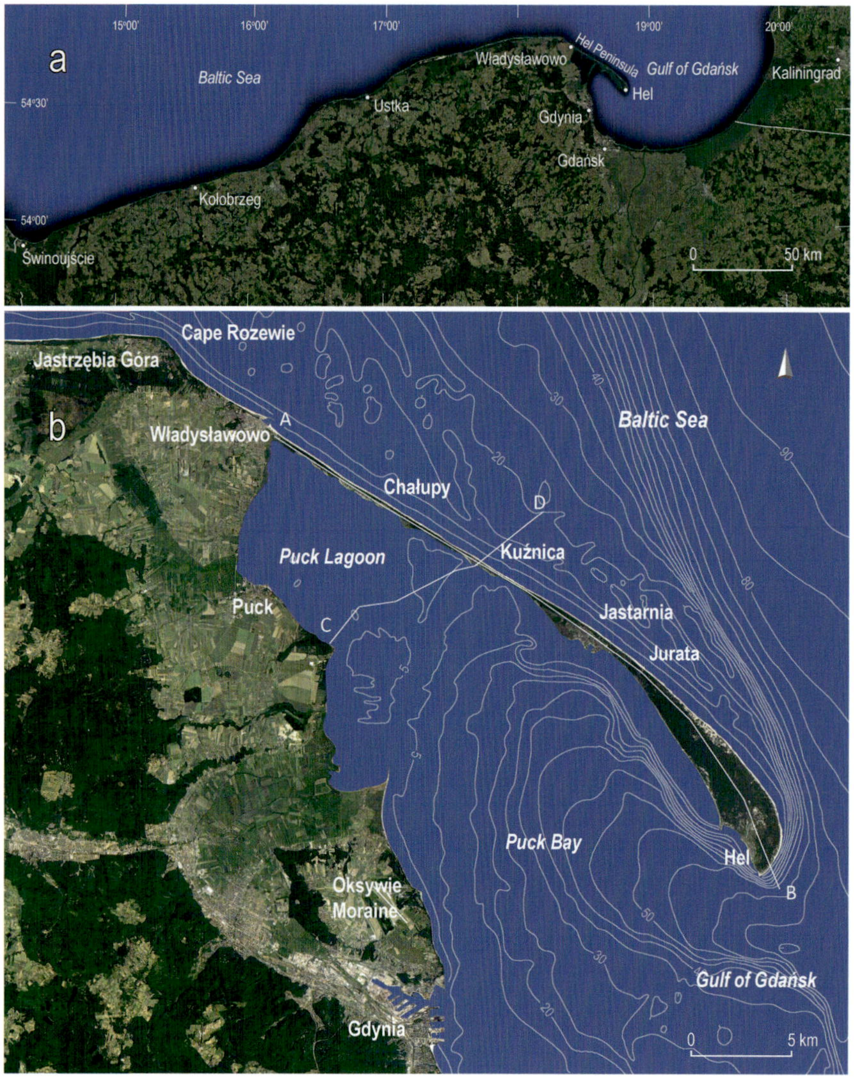

Fig. 43.1 Location of the Hel Peninsula. **a** Within the Polish coast. **b** Hel Peninsula closing Puck Bay from the northeast; locations of cross-sections A–B (Fig. 43.3) and C–D (Fig. 43.4) are shown (sources: geoportal.gov.pl, google com)

Moraine upland near the town of Jastrzębia Góra in the northeast. Further to the east, a semi-circular bay—the Gulf of Gdańsk—occurs which extends to the western endpoints of the Kaliningrad Oblast, which is part of the Russian Federation. However, looking at the map of Poland, following the borders of the country and especially the shoreline, it is impossible not to note one more detail—a few-hundred-metres-narrow and a 35-km-long strip of the land projecting into the Gulf of Gdańsk and ending in a 3-km-wide hook. It is the Hel Peninsula (Fig. 43.2)—one of the most exciting geomorphological features of the Polish coast.

The Hel Peninsula separates the Gulf of Gdansk and Puck Bay (with Puck Lagoon, Fig. 43.1b). It is located to the east of the town of Władysławowo, extending towards southeast. The Hel Peninsula is a unique sandy barrier due to its length, the volume of sediments, and the depth of the basin (60 m bsl) that it is prograding into. People have tried to discover and understand this landform for over 100 years. Along with new data obtained, more accurate information about its geological structure is provided and new hypotheses of the genesis of the Hel Peninsula are created.

Moreover, it is a remarkable part of the Southern Baltic coast in respect of the settlement history dating back to the seventeenth century, when the local population struggled with natural processes forming this piece of land. This is documented by historical records containing many references to sea storms during which seawater overflow through the sandy barrier into Puck Bay occurred. The places of expected seawater overflows were marked on historical maps. Sometimes, it was misinterpreted that the Hel Peninsula was an island or an island chain in historical times, but this is not true (Tomczak 2004).

The aim of the chapter is to present the current knowledge of the genesis, geological structure, and modern geomorphological processes that occur on the Hel Peninsula.

Fig. 43.2 Hel Peninsula from the air. **a** The neck—view from the village of Chałupy towards the end of the spit. **b** The hook—view of the tip of the peninsula, with the town of Hel on the left (photographs by Tomasz Golec—Helski Dron)

43.2 Geological Structure of the Hel Peninsula

Jurassic rocks are observed below the depth of 200 m and Cretaceous rocks are found below the depth of 100 m in the Hel Peninsula region (Fig. 43.3) (Tomczak 1995a, b, 2005). Eocene and Miocene sediments are located above them in the neck of the Hel Peninsula. The top surface of the Miocene deposits is located at 50 m deep, and Pleistocene sediments occur above. In the hook of the Hel Peninsula, Pleistocene sediments (mainly till, clay, silt, sand, and gravel) overlie Cretaceous rocks. Holocene sediments are mainly sands and sands with silt (Fig. 43.4). The sequences of Holocene sediments are about 10 m thick in the neck and about 100 m thick in the hook.

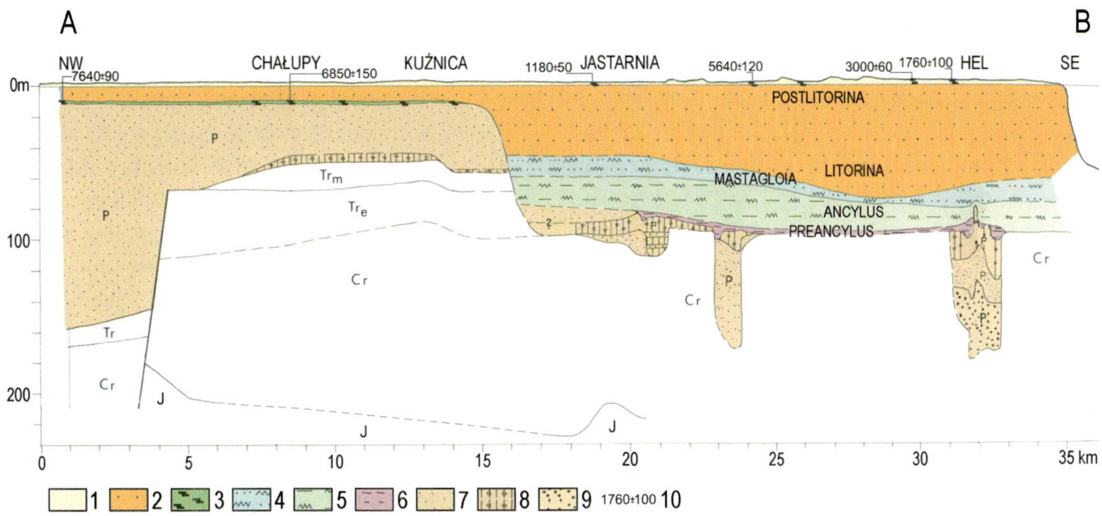

Fig. 43.3 Geological cross-section along the Hel Peninsula (after Tomczak 1995b); for location see Fig. 43.1. 1—sandy sediments, partly aeolian; 2—sandy sediments of Littorina Sea and post-Littorina Sea; 3—Holocene terrestrial sediments; 4—sandy and silty sediments of the Mastogloia Sea; 5—silt and clay of Ancylus Lake; 6—clay of pre-Ancylus Lake; 7—Pleistocene sandy sediments; 8—Pleistocene till; 9—Pleistocene coarse grain sediments, 10—C14 BP ages; P—Pleistocene; Tr_m—Miocene; Tr_e—Eocene; Cr—Cretaceous; J—Jurassic

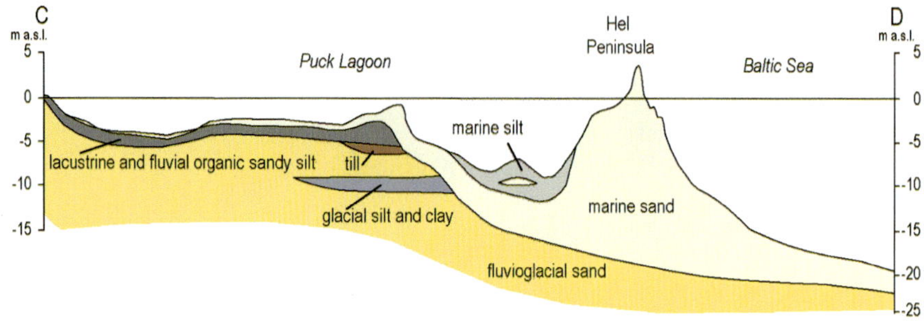

Fig. 43.4 Geological cross-section through Puck Lagoon and across the Hel Peninsula (after Kramarska et al. 1995); for location, see Fig. 43.1

Holocene sediments of the Hel Peninsula surface are relatively uniform. These are mainly fine and medium sands. Sand with gravel as well as peat with silt is observed only in a few places. The unification of sediment characteristics is the result of a complex system of sediment transport that occurs along and transversely to the shore. It is also the result of frequent changes in marine dynamic conditions. However, the nature of sedimentary structures and their sequences differ very clearly, because these sediments are deposited in different, often very diverse conditions: calm wave action, high energetic storm surges, and aeolian processes (Wróblewski 2009).

43.3 Geomorphology of the Hel Peninsula

The Hel Peninsula is a sandy barrier mainly formed under the conditions of wave action and sea currents of the non-tidal sea. The geological structure and the development of the Hel Peninsula are the effects of along-shore and cross-shore transport of littoral sediments. As a result, the Hel Peninsula is simultaneously subjected to longitudinal growth in the southeast direction and migration towards the southwest, towards Puck Bay (Bączyk 1963; Rosa 1963; Tomczak 2005).

The Hel Peninsula is a spit-type sandy barrier. It has two clearly visible main features: erosive neck and accumulative hook (in the sense of Davis 1896). The northwestern part of the Hel Peninsula is the neck. It is narrow, from 150 up to 300 m, low-lying, and relatively flat. Landforms of this section are divided into zones parallel to the shoreline. The proximal seaward zone is a sandy beach, 30–70 m wide. The next zone is formed by coastal dunes parallel to the beach and the shoreline. The interior of this section of the Hel Peninsula is covered by dune fields, developed on the surface of washover fans (Wróblewski 2001, 2009). In some of the lowest-lying points (1.0–1.5 m asl), peat bogs are observed. The interior is tens to hundreds of metres wide. It reaches narrow and poorly developed beaches located on the Puck Lagoon side shore (the name Puck Lagoon applies to the shallowest, western part of Puck Bay; Fig. 43.1). In many places, small dunes occur, but these are also much less developed than sandy barriers located on the open-sea side.

The shapes of the shoreline on both sides of the Hel Peninsula are very different. The open-sea side shoreline is regular and almost straight. The shoreline of Puck Lagoon is very varied with many small bays and capes. The shore of the Hel Peninsula neck on the side of Puck Bay is erosive. It is the transit section of littoral sediments. Sediment transport occurs parallel and cross-shore contemporaneously.

The storm surge washovers of the barrier and the development of washover fans are important for the evolution of the neck of the Hel Peninsula. The washover fans are the base of the marine and aeolian accumulation (Figs. 43.5 and 43.6).

The movement of the barrier in the direction of Puck Bay is observed in the neck of Hel Peninsula. In the condition of a relatively stable sea level, the cross-shore profile is retained, but the barrier is migrating (Wróblewski 2009).

The distal part of the Hel Peninsula is a hook. Its shape is similar to a drop or a club. The hook is about 1 km wide in the town of Jurata and widens in the southeastern direction, up to 3 km in the vicinity of the town of Hel. The relief of the hook is varied. Modern foredunes are observed near the beach, while older foredunes are located in the interior of the barrier (Tomczak 2005). The shape of older foredunes mimics the previous shorelines. These are indicators of the gradual growth of the hook in the southeast direction. In the distal parts of the hook, the shore is subjected to the action of refracted waves. These are responsible for the curvature of the shoreline, the shape of the beach, and the position of coastal dunes.

The highest dunes (up to 22 m asl) are located on the seashore, near the beach between the towns of Jastarnia and Hel, as well as along some parts of the Puck Bay shore. On both sides of the Hel Peninsula hook (Puck Bay side and open-sea side), beaches and coastal dunes are observed. However, much wider beaches and larger sand dunes occur on the open-sea side of the hook. The shore is in an accumulation state that prevails over erosion processes. In the nearshore of the hook, longshore transport from the northwest is observed. Marine accumulation occurs there, especially at the tip of the Hel Peninsula (Rucińska-Zjadacz and Rudowski 2015). The barrier is growing in the direction of the central part of the Gulf of Gdańsk (Rucińska-Zjadacz et al. 2018).

Relief of the Hel Peninsula has been changed by humans in many places as a result of coastal protection. The changes

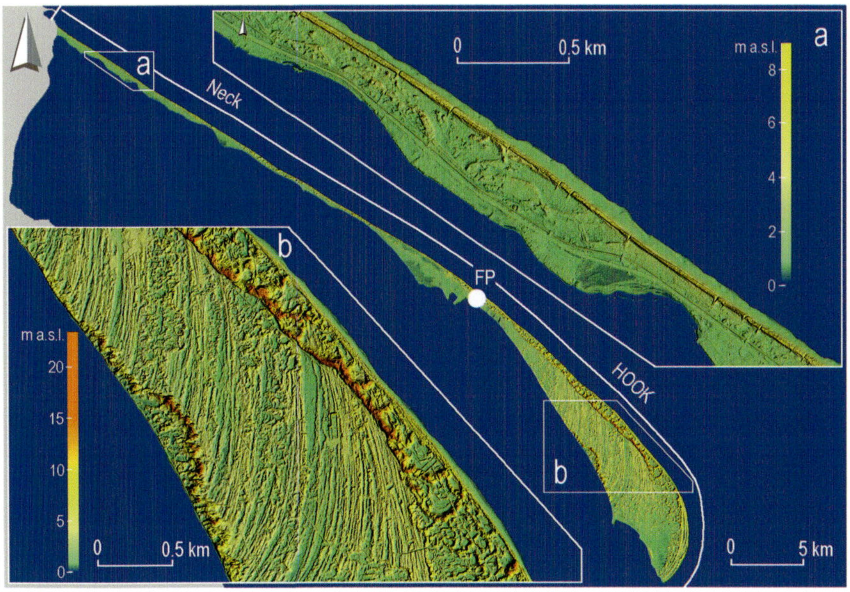

Fig. 43.5 Relief of the Hel Peninsula based on the LiDAR data (source: geoportal.gov.pl). **a** The neck—narrow section with one foredune ridge along the seaward beach and small aeolian forms in the interior. **b** Central part of the hook—clearly visible foredunes in the shape of elongated hills mark the location of older shorelines; FP—fulcrum point

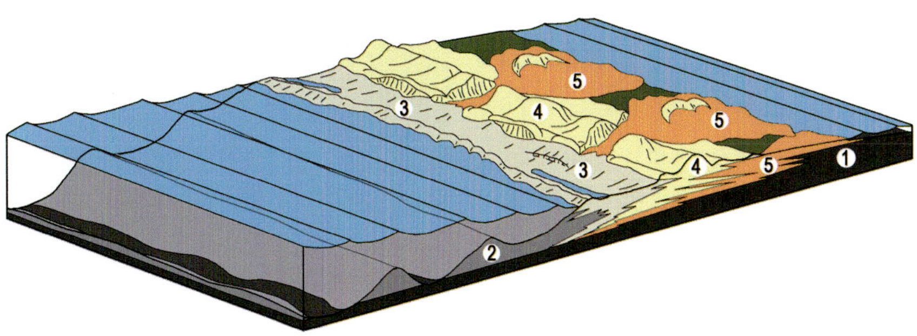

Fig. 43.6 Diagram of the structure of the sandy barrier in the advanced stage of development. 1—lagoon facies, 2—nearshore facies, 3—beach facies, 4—aeolian facies, 5—washover fan facies (after Wróblewski 2009, modified)

mainly affect the shore in the neck. Sand replenishment was conducted many times within the beach and the dunes. Groynes were redeveloped between Władysławowo and Kuźnica. Gabion seawalls were fitted into the beach and dunes relief in the Kuźnica region. Rock armour was built in the waterline of Puck Bay. The neck is therefore no longer a free landform, developing by natural processes connected to the storm surges, especially washover the sandy barrier.

The development of the Hel Peninsula is strongly related to the evolution of the Baltic Sea. According to Rosa (1963), the beginning of the Hel Peninsula formation took place during the Ancylus Lake stage. Two littoral drifts met in the place when a pre-Hel Peninsula was formed. One of them (the stronger one) transported sediments from the Cape of Rozewie in the west towards the east, whereas the second one moved from the Oksywie Moraine towards the north. As a result of the accumulation of deposits of these two streams of marine debris, submarine sandy shallow was made. It became the base and a core element for the further formation of the spit. The present thickness of the deposits in the hook, from the town of Jurata to the tip of Hel, is about 100 m (Fig. 43.3) (Tomczak 2004). In the neck, the thickness is much less, from a few up to tens of metres. These deposits are located above the top surface of Pleistocene fluvioglacial sands (Kramarska et al. 1995).

During the transgression that occurred in the stage of the Littorina Sea (7–5 ka BP), Baltic seashores and a shallow sea bottom were intensively eroded (Tomczak 1995a). As a consequence, a high volume of sediments was activated and became a dynamic layer for coastal processes. These sediments were transported by a strong littoral current that occurred from west to east direction, parallel to a young and still forming shore. The main zone of accumulation of these sediments was located in the extension of the shoreline where its smooth line was broken by the shape of the Gulf of Gdańsk. As a result of the deposition processes, a sand spit was formed. It was a pre-Hel barrier in the latitudinal shape directed towards the centre of the Gulf of Gdańsk (Rosa 1963; Mojski 2002). Permanent sea-level rise increased coastal processes, conduced to washover fan formation, and intensified the development of aeolian landforms. These conditions led to gradual elongation and migration of the expanding spit towards the southeast. During the last phase of the Littorina transgression (about 5.5 ka BP), the Baltic Sea flooded the land where Puck Lagoon is located at present (Kramarska et al. 1995).

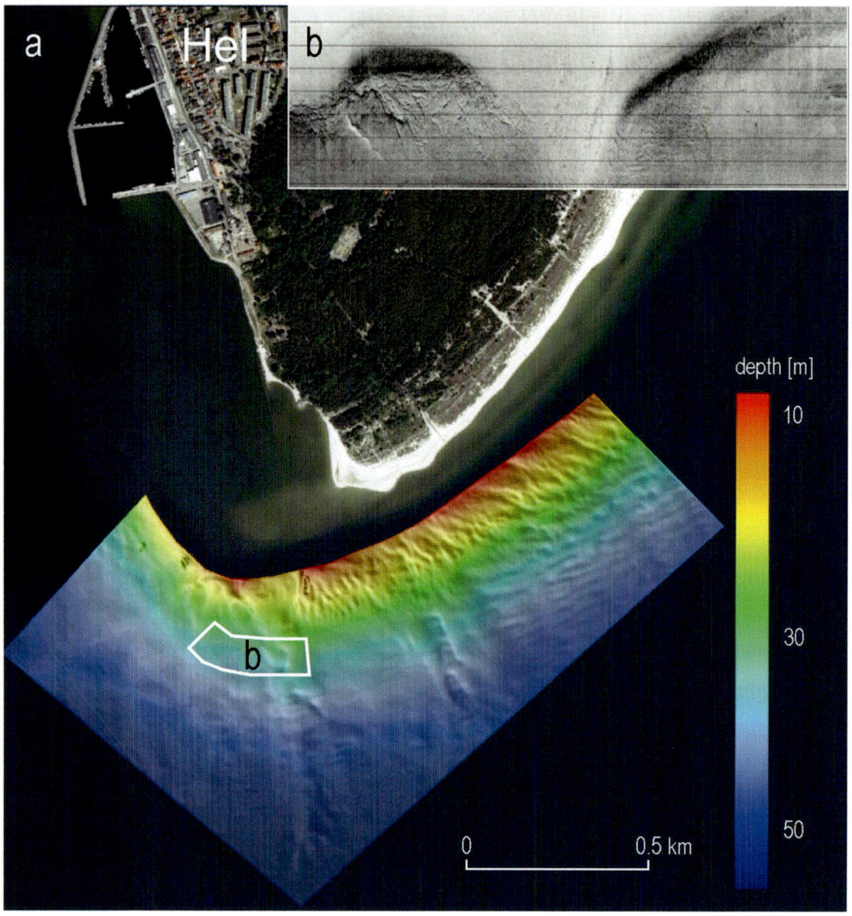

Fig. 43.7 Relief of the submarine slope of the tip of the Hel Peninsula (**a**) and landslide forms (crown, scarp, head) in the side-scan sonar images (**b**) (after Rudowski et al. 2016, modified)

At the stage of the post-Littorina Sea, after relative sea-level stabilization, the Hel Peninsula was an elongated spit that reached the Jurata point about 5.64 ka BP (Tomczak 2005). The neck was narrow and low-lying above sea level. One chain of small coastal dunes was located there. The neck was frequently ripped by seawater during storm surges and washover fans were formed above the sediments of the Puck Lagoon (Wróblewski 2001, 2009; Uścinowicz 2003; Rucińska-Zjadacz et al. 2009). The Hel Peninsula expanded very fast, but the growth rate decreased with the increasing depth of the basin (the Gulf of Gdańsk), into which the spit prograded (Tomczak 1995a; Uścinowicz 2003).

The Hel Peninsula is still developing. It is especially clearly visible at the tip of the hook. The tip of Hel accumulates and migrates towards the central part of the Gulf of Gdańsk (Rucińska-Zjadacz and Rudowski 2015). The submarine base of the tip is gradually expanding (Rucińska-Zjadacz et al. 2018), whereas wide beaches and coastal dunes are developing in the inland part of the hook. Accumulation of the base of the tip of the Hel Peninsula is mainly related to the processes that occur on its relatively steep submarine slopes. The submarine slope of the tip of Hel Peninsula (Fig. 43.7) is a unique landform, if compared to other sandy barriers, due to the steepness (5–8° on average and up to 20° in some places), height (about 50 m over nearby sea bottom), and because of slope movements observed there such as submarine landslides (Rudowski et al. 2016; Rucińska-Zjadacz and Wróblewski 2018).

43.4 Future Development of the Hel Peninsula

In the geological time scale, barriers are ephemeral, but very dynamic landforms. In the process of spit-type barrier formation, which is the Hel Peninsula, the most important factor is the supply volume of littoral drift as well as the possibility of transport and accumulation under the conditions involving the action of waves and nearshore currents in relation to short- and long-term changes in sea level. The development of sandy barriers takes place both during sea transgressions and regressions. Favourable for their development are frequent, but not rapid changes in sea level. When the sea level gradually rises, barriers tend to move inland (Wróblewski 2009).

Along with the projected sea-level rise, changes in atmospheric circulation over the southern Baltic Sea, and intensification of storm surges (Miętus et al. 2004), further development of the Hel Peninsula will likely follow the scheme of the development of this type of landforms. Intensification of coastal processes is possible, especially in the neck. During extremely strong storm surges, the neck may be temporarily ripped and seawater may invade into Puck Lagoon. Based on the results of the analysis of present-day changes of the Hel Peninsula as well as its geological evolution, it can be concluded that the landform moves towards its hinterland (Puck Bay direction) and extends towards the centre of the Gulf of Gdańsk.

References

Bączyk J (1963) Geneza Półwyspu Helskiego na tle rozwoju Zatoki Gdańskiej. Dokumentacja Geograficzna IG PAN 6:1–176

Davis WM (1896) The outline of Cape Cod. Proc Am Acad Arts Sci 31:303–332

Kramarska R, Uścinowicz S, Zachowicz J (1995) Origin and evolution of the Puck Lagoon. Polish Coast—past, present and future. J Coast Res Spec Iss 22:187–191

Miętus M, Filipiak J, Owczarek M (2004) Klimat wybrzeża południowego Bałtyku. Stan obecny i perspektywy zmian. In: Cyberski J (ed) Środowisko polskiej strefy południowego Bałtyku. Gdańskie Towarzystwo Naukowe, Gdańsk, pp 11–45

Mojski JE (2002) Czwartorzęd regionu gdańskiego. Przegl Geol 50:705–708

Rosa B (1963) O rozwoju morfologicznym wybrzeża Polski w świetle dawnych form brzegowych. Stud Soc Sci Torunensis V:1–172

Rucińska-Zjadacz M, Rudowski S (2015) Morpholithodynamic conditions of the tip of the Hel Peninsula, Baltic Sea. Oceanol Hydrobiol Stud 44:181–192

Rucińska-Zjadacz M, Wróblewski R (2018) The complex geomorphology of a barrier spit prograding into deep water, Hel Peninsula, Poland. Geo-Mar Lett 38:513–525

Rucińska-Zjadacz M, Rudowski S, Wróblewski R (2009) Bariery piaszczyste w Zatoce Puckiej In: Kostrzewski A, Paluszkiewicz R (eds) Geneza, litologia i stratygrafia utworów czwartorzędowych V, Seria Geografia 88. Wydawnictwo Naukowe UAM, pp 493–504

Rudowski S, Rucińska-Zjadacz M, Wróblewski R, Sitkiewicz P (2016) Submarine landslides on the slope of a sandy barrier: A case study of the tip of the Hel Peninsula in the Southern Baltic. Geol Quart 60:407–416

Rucińska-Zjadacz M, Rudowski S, Wróblewski R (2018) Seismic architecture of the tip of the Hel Peninsula, Poland. Baltica 31:63–72

Rudowski S, Rucińska-Zjadacz M, Wróblewski R, Sitkiewicz P (2016) Submarine landslides on the slope of a sandy barrier: a case study of the tip of the Hel Peninsula in the Southern Baltic. Geol Quart 60:407–416

Tomczak A (1995a) Budowa geologiczna strefy brzegowej. Półwysep Helski i Mierzeja Wiślana. In: Mojski JE (ed) Objaśnienia do: Atlas Geologiczny Południowego Bałtyku 1:500,000. Sopot - Warszawa, PIG, pp 48–51

Tomczak A (1995b) Budowa geologiczna strefy brzegowej. Półwysep Helski i mierzeja Wiślana. In: Mojski JE (ed) Atlas Geologiczny Południowego Bałtyku 1:500,000, PIG, plate XXXIV

Tomczak A (2004) Wybrane zagadnienia z przeszłości geologicznej i przyszłości Półwyspu Helskiego. In: Mat. Konf. Nauk. "Stan i zagrożenie Półwyspu Helskiego". Gdańskie Towarzystwo Naukowe - Komitet Badań Morza PAN - Instytut Oceanografii, Uniwersytet Gdański, pp 6–8

Tomczak A (2005) Wybrane zagadnienia z przeszłości geologiczne i przyszłości Półwyspu Helskiego. In: Cyberski J (ed) Stan i zagrożenie Półwyspu Helskiego. Gdańskie Towarzystwo Naukowe, Gdańsk, pp 13–58

Uścinowicz S (2003) Relative sea level changes, glacio-isostatic rebound and shoreline displacement in the Southern Baltic. Pol Geol Inst Spec Pap 10:1–79

Wróblewski R (2001) Facje stożków przelewowych w rozwoju barier piaszczystych. In: Florek W (ed) Geologia i geomorfologia Pobrzeża i południowego Bałtyku 4. Akademia Pomorska, Słupsk, pp 183–192

Wróblewski R (2009) Lithodynamical facies of sandy barrier—Hel Peninsula as an example. Oceanol Hydrobiol Stud 38(Suppl 1):147–161

Radosław Wróblewski is Assistant Professor at the Department of Geophysics, Faculty of Oceanography and Geography, University of Gdańsk. His main research interest is geomorphology and seabed geology, also in relation to coastal dynamics. He focuses on the practical application of bathymetric, backscattering, sonar, and seismoacoustic data in seabed research. His main research area is the Baltic Sea.

Maria Rucińska is Assistant Professor at the Department of Geophysics, Faculty of Oceanography and Geography of the University of Gdańsk. In her scientific work, she focuses on marine geomorphology, with particular emphasis on the development of large accumulation forms such as sandy barriers and sandy ridges. Her research interests also include hydrography and the application of non-invasive methods of seabed research in marine geology.

Patryk Sitkiewicz is Assistant Professor at the Department of Geomorphology and Quaternary Geology, Faculty of Oceanography and Geography, University of Gdańsk. His main research interest is coastal geomorphology, especially the impact of human activity and nearshore seabed processes on shore changes. He focuses on practical aspects of geological research and its utility in Integrated Coastal Zone Management. He develops new GIS tools to automate the process of beach observation using remote sensing data. His main research area is the dune coast of the Southern Baltic.

Stanisław Rudowski is Professor at the Maritime Institute, Gdynia Maritime University. His research is focused on marine geology, Quaternary geology, and marine sedimentology of the Baltic and polar regions. From 2016 to 2020, he was the Director for Scientific Affairs at the Maritime Institute in Gdańsk. He was a Manager of the fifth winter expedition to Polish Polar Station in Hornsund, Svalbard (1982–1983). He is a member of the Association of Polish Geomorphologists.

Index

P. Migoń and K. Jancewicz (eds.), *Landscapes and Landforms of Poland*,
World Geomorphological Landscapes, https://doi.org/10.1007/978-3-031-45762-3

Q

R

S

GPSR Compliance

The European Union's (EU) General Product Safety Regulation (GPSR) is a set of rules that requires consumer products to be safe and our obligations to ensure this.

If you have any concerns about our products, you can contact us on ProductSafety@springernature.com

In case Publisher is established outside the EU, the EU authorized representative is:

Springer Nature Customer Service Center GmbH
Europaplatz 3
69115 Heidelberg, Germany

Batch number: 10756636

Printed by Printforce, the Netherlands